LIST OF REFERENCE TABLES (*continued*)

Vibrational Spectroscopy

Electron Absorption Spectroscopy

Second Edition

Organic Structural Spectroscopy

Joseph B. Lambert
Northwestern University
Trinity University

Scott Gronert
Virginia Commonwealth University

Herbert F. Shurvell
Queen's University

David A. Lightner
University of Nevada, Reno

Prentice Hall
Boston Columbus Indianapolis New York San Francisco Upper Saddle River
Amsterdam Cape Town Dubai London Madrid Milan Munich Paris Montréal
Toronto Delhi Mexico City São Paulo Sydney Hong Kong Seoul Singapore Taipei Tokyo

Editor in Chief: Adam Jaworski
Marketing Manager: Erin Gardner
Assistant Editor: Carol DuPont
Managing Editor, Chemistry and Geosciences: Gina M. Cheselka
Project Manager, Science: Maureen Pancza
Art Project Managers: Connie Long and Ronda Whitson
Design Director: Jayne Conte
Cover Designer: Suzanne Behnke
Senior Manufacturing and Operations Manager: Nick Sklitsis
Operations Specialist: Maura Zaldivar
Composition/Full Service: Nesbitt Graphics, Inc.

Cover Art: Greg Gambino and Mark Landis

Credits and acknowledgments borrowed from other sources and reproduced, with permission, in this textbook appear on the appropriate page within the text.

The spectra in this book were reproduced from *The Aldrich Library of FT-IR Spectra,* 2nd ed., and *The Aldrich Library of 13C and 1H FT-NMR Spectra,* courtesy of the Aldrich Chemical Company.

Printed in the United States
10 9 8 7 6 5 4 3 2 1

Library of Congress Cataloging-in-Publication Data

Organic structural spectroscopy / Joseph B. Lambert . . . [et al.]. — 2nd ed.
p. cm.
Includes bibliographical references.
ISBN-13: 978-0-321-59256-9 (alk. paper)
ISBN-10: 0-321-59256-5 (alk. paper)
1. Organic compounds—Analysis. 2. Spectrum analysis. I. Lambert, Joseph B.
QD272.S6O74 2011
543'.5—dc22

2010031837

ISBN 10: 0-321-59256-5
ISBN 13: 978-0-321-59256-9

Prentice Hall
is an imprint of

www.pearsonhighered.com

BRIEF CONTENTS

CONTENTS

PART III Vibrational Spectroscopy 325

PREFACE

In the dozen years since the publication of the first edition of *Organic Structural Spectroscopy*, no subject in this field has made more advances than mass spectrometry. To reflect these changes, Part II has been entirely rewritten and its position moved forward in the second edition. During the period between the editions, the field has progressed quickly as new technologies have been developed and become widely available. This new section still emphasizes the fundamentals of mass spectrometry including electron impact ionization. In addition, newer methods and instrumentation, such as modern fragmentation approaches and biopolymer analysis, are described in detail, and a chapter on the quantitative aspects of mass spectrometry has been added. Finally, the majority of problems in these chapters have been revised. In summary, the new mass spectrometric material contains

- Added descriptions of new fragmentation techniques, including electron capture dissociation (ECD), electron transfer dissociation (ETD), and infrared multi-photon dissociation (IRMPD)
- Added discussion of sample preparations, including gel separation approaches
- A new section on proteomics applications
- A new chapter on quantification, including isotopic-labeling strategies

The section on nuclear magnetic resonance (Part I) also has been significantly revised to include:

- Expanded treatment of chemical shift basics
- Coverage of solvent effects
- Expanded coverage of the influence of rate processes on NMR spectra, the nuclear Overhauser effect, and INEPT-related pulse sequences
- A new section on shaped pulses
- New material on the pulse sequences Heteronuclear Single Quantum Correlation (HSQC), Heteronuclear Multiple Bond Correlation (HMBC), and Parallel Acquisition Nuclear magnetic resonance All-in-one Combination of Experimental Applications (PANACEA), which enable faster determination of structures
- Numerous new problems, always drawn directly from the literature

To allow for these additions and explanations, the sections on infrared and on ultraviolet-visible spectroscopy (Parts III and IV) have been condensed, while retaining all fundamental information and the extensive tables. This edition represents the state of the art of spectroscopic structure determination today.

Joseph B. Lambert
jlambert@northwestern.edu

Scott Gronert
sgronert@vcu.edu

Herbert F. Shurvell
shurvell@chem.queensu.ca

David A. Lightner
lightner@unr.edu

Organic Structural Spectroscopy

CHAPTER 1

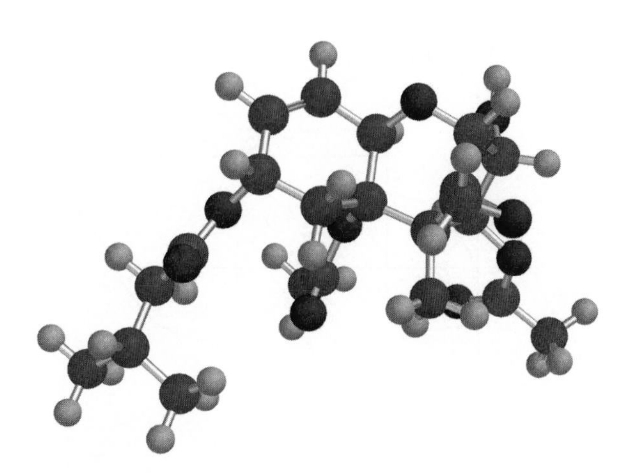

Introduction to Structural Spectroscopy

1-1 THE SPECTROSCOPIC APPROACH TO STRUCTURE DETERMINATION

"How can I determine the molecular structure of this unknown material?" This question is asked by the synthetic chemist after completing any chemical reaction. It is also foremost in the mind of the natural product chemist extracting molecules from plants in hopes of developing new medicinal materials. And the forensic chemist isolating drugs or toxins from a suspect or victim. And the environmental chemist examining the effects of materials contaminating soil, bodies of water, or the atmosphere. And the archaeological chemist tracing dietary information from food residues in pottery. And the biological chemist examining metabolic processes in the body. And the geochemist studying molecules extracted from geological strata as potential energy sources. The quest for structural information on solids, liquids, or gases, on crystalline, powdered, or glassy materials, on mixtures or pure compounds is a continuing challenge to chemists of every type.

Diffraction methods offer the ideal solution to many structural problems by providing data that often may lead to a complete structure. Equipment for neutron or electron diffraction, however, is not widely available, and these methods apply only to relatively small molecules. Although X-ray diffraction is common, the technique is restricted to materials in the solid state, normally crystals, for optimal results. Crystallographic methods cannot be applied to mixtures. None of these diffraction methods can provide the quick structural information that the synthetic chemist needs to continue to the next step or the analytical chemist needs to report to the lawyer, athlete, or geologist about the structures of specific molecules. Because crystallography is relatively time consuming, it cannot produce in a timely fashion the sheer quantity of results often needed. Various forms of spectroscopy can provide a wide array of structural information in the most rapid fashion possible, for all phases of matter, and on mixtures as well as on pure compounds. The equipment is routinely available, ranging from relatively inexpensive (thousands of dollars) to expensive (millions of dollars). The process of structural elucidation by spectroscopic methods is deductive. One or more spectroscopic experiments are carried out, and structural conclusions are reached by analyzing the resulting data. Sometimes a complete molecular structure is required; sometimes partial structures suffice. This text examines the four most common and useful methods, with emphasis on nuclear magnetic resonance (NMR) spectroscopy and mass spectrometry (MS). Each provides its own special kind of data that apply to molecular structure.

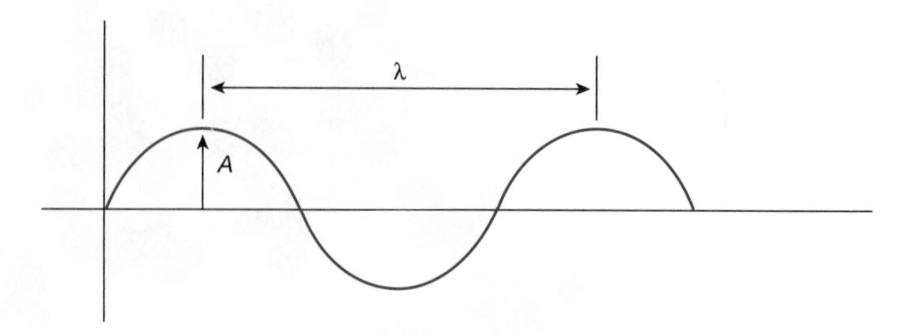

FIGURE 1-1 One and one-half cycles of electromagnetic radiation.

1-2 THE ELECTROMAGNETIC SPECTRUM

Most spectroscopic methods use electromagnetic waves to create the information on which structural deductions are made. Propagation of energy through space is characterized by both electrical and magnetic properties, and so the phenomenon is referred to as *electromagnetic radiation*, or, loosely, *light*. Such radiation has wave properties, in that its magnitude fluctuates sinusoidally over time (Figure 1-1) as it moves through space at the speed of light (velocity $c = 2.998 \times 10^8$ m s^{-1} in vacuum). The length of one full cycle is called the wavelength (λ), which corresponds, for example, to the distance from crest to crest or trough to trough in the diagram. The number of full-cycle fluctuations that occur over some period of time (usually one second) is called the frequency (υ). Wavelength and frequency are inversely related, and the speed of light is the constant of proportionality ($\upsilon = c/\lambda$). Thus the longer the wavelength, the shorter the frequency. More energetic waves fluctuate more rapidly in space, so that energy (ΔE) and frequency are directly related, with Planck's constant ($h = 6.624 \times 10^{-34}$ J • s) as the constant of proportionality ($\Delta E = h\upsilon$). The maximum excursion from the zero point is the amplitude of the wave (A in Figure 1-1) and may be thought of as its intensity.

Light has been given various names according to its wavelength (Figure 1-2). Cosmic, γ, and X rays have the shortest wavelengths (highest energies), and radiofrequency waves have the longest wavelengths (lowest energies). Because there is a sequence of related phenomena that differ in wavelength, the whole series is called the *electromagnetic spectrum*. As drawn in Figure 1-2, wavelength increases from left to right, whereas frequency and energy increase from right to left.

NMR, vibrational, and electronic spectroscopies involve absorption of electromagnetic energy, respectively from the radiofrequency, infrared (IR), and ultraviolet/visible (UV–Vis) regions of the electromagnetic spectrum. Each *absorption spectroscopy* promotes normal or ground state molecules into higher energy or excited states. For NMR, only the spin state of the nucleus is changed. Infrared absorption gives rise to well-defined molecular vibrations, and UV–Vis absorption results in excited electronic states. The energy absorbed (ΔE) is roughly 100 kcal mol^{-1} for electronic excitation, 10 kcal mol^{-1} for vibrational excitation, and 10^{-6} kcal mol^{-1} for nuclear spin excitation. Each of these spectroscopies has its preferred units for wavelength, frequency, and energy, developed historically. Mass spectrometry involves measurement of mass numbers rather than energy absorption (hence it is a spectro*metry* rather than a spectro*scopy*) and is not associated with a region of the electromagnetic spectrum.

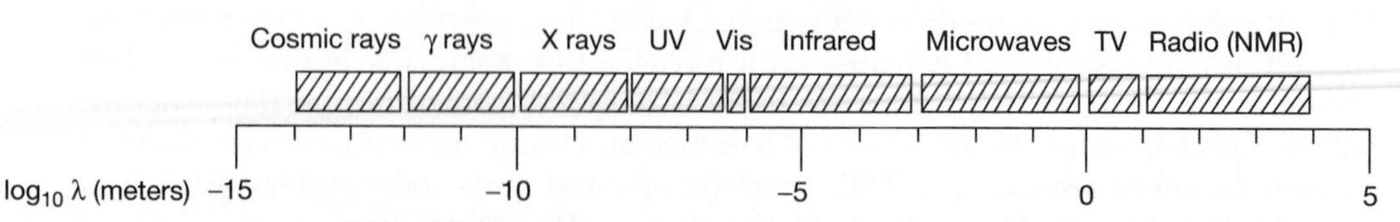

FIGURE 1-2 Names given to various regions of the electromagnetic spectrum; the plot is linear in the logarithm of wavelength (λ).

1-3 MOLECULAR WEIGHT AND MOLECULAR FORMULA

Determining the molecular weight and molecular formula is an important step in structural analysis. The traditional, nonspectroscopic procedures for obtaining such information include lowering the freezing or melting point, elevating the boiling point, and measuring vapor pressure changes. Although equipment for these classic physical chemistry methods is neither expensive nor complex, it is rarely available to the synthetic chemist. Today, molecular weight is obtained more reliably and easily by mass spectrometry (Part II).

High resolution mass spectrometry also can provide the molecular formula directly, that is, the identity and number of each atom in the molecule—for example, $C_6H_{10}O$ for cyclohexanone. More traditionally, chemists used quantitative elemental analysis to derive the *empirical formula* (the lowest integer ratio of elements present), which is related to the molecular formula by an integral factor. Thus the empirical formula of ethane is CH_3, but its molecular formula is $(CH_3)_2$ or C_2H_6; the empirical formula of cyclohexane is an uninformative CH_2, but its molecular formula is $(CH_2)_6$ or C_6H_{12}.

Almost all journals that publish synthetic chemistry research require accurate elemental analyses of new compounds as proof of purity. The data are obtained generally by submitting samples to a commercial laboratory for analysis, although many research laboratories have their own equipment. These analyses provide the percentages of carbon, hydrogen, nitrogen, and other elements (oxygen usually is calculated by difference from 100%). These elemental percentages are converted to the empirical formula by the following procedure. Division of an atomic percentage X_i by the atomic weight M_i of the ith atom gives the relative number of atomic equivalents ($E_i = X_i/M_i$) for each element in the compound. The numbers for all atoms (E_i) are converted to the simplest ratio through division by the smallest value of E: E_i/E_{min}. These numbers are then multiplied by the lowest integer n that yields whole numbers for all the ratios (nE_i/E_{min}). The whole numbers are the atomic ratios in the empirical formula. The procedure is illustrated as follows.

Atoms	**Carbon**	**Hydrogen**	**Oxygen**
Elemental analysis (X_i)	48.57	8.19	43.24
Relative atoms ($E_i = X_i/M_i$)	$\frac{48.57}{12.01} = 4.04$	$\frac{8.19}{1.008} = 8.13$	$\frac{43.23}{16.00} = 2.70$
Simplest ratio (E/E_{min})	$\frac{4.04}{2.70} = 1.50$	$\frac{8.13}{2.70} = 3.01$	$\frac{2.70}{2.70} = 1.00$
Lowest integer ratio (nE_i/E_{min})	$2 \times 1.50 = 3$	$2 \times 3.01 \sim 6$	$2 \times 1.00 = 2$
Empirical formula	C_3	H_6	O_2

As noted, the result of such calculations is the empirical rather than the molecular formula. Moreover, the procedure is poor at distinguishing small differences in carbon and hydrogen, it is subject to experimental error in the percentages (particularly that of oxygen), and it is very sensitive to the purity of the sample. For these reasons, high resolution mass spectrometry is preferred, although it is subject to its own set of limitations, as described in Part II. The conservative approach is to obtain both mass spectral and elemental analytical data.

The molecular formula itself provides important structural information beyond the simple facts of elemental identity and number. Introduction of a ring or a double bond into an alkane structure reduces the number of hydrogen atoms by two. Thus the formula for the homologous series of alkanes is C_nH_{2n+2}, whereas that for alkenes or monocyclic alkanes is C_nH_{2n}. Detailed examination of the molecular formula can provide the number of sites of unsaturation. Each ring is one such site, each double bond is one site, and each triple bond is two sites.

The total number of unsaturations is termed the *unsaturation number* (or index of hydrogen deficiency) (U), as given by eq. 1-1,

$$U = C + 1 - \frac{1}{2}(X - N) \tag{1-1}$$

in which C is the number of tetravalent elements (carbon, silicon, etc.), X is the number of monovalent elements (hydrogen and the halogens), and N is the number of trivalent elements (nitrogen, phosphorus, etc.). The expression is an elaboration of the formula for the alkane series, for which the unsaturation number is zero. Divalent elements such as oxygen and sulfur extend chains but do not alter the number of unsaturations, so they do not appear in the formula. Trivalent elements add a bond, and univalent elements subtract a bond, as indicated in the last term. For example, the unsaturation number for cyclohexanone ($C_6H_{10}O$) is $[6 + 1 - {}^1\!/_2(10)] = 2$, one for the ring and one for the carbonyl group.

1-4 STRUCTURAL ISOMERS AND STEREOISOMERS

In the most detailed sense, the structure of a molecule includes all its bond lengths, valence angles, and torsional angles, which in turn define atom connectivity, stereochemistry, and conformation. This type of information normally is available only from crystallography or some forms of multidimensional NMR spectroscopy.

In the absence of quantitative knowledge of bond lengths and angles, the spectroscopic method is to deduce the structure by identifying components of the molecule, by showing that specific atoms are connected to each other, and by determining the relationship between atoms in space. The spectroscopic objective is to obtain enough such information to overdetermine the structure.

The first step usually is to determine the molecular formula, because it defines a family of molecules or *isomers* to be considered. This family is composed of molecules that differ in the connectivity of atoms (*structural isomers*) and molecules that differ only in the spatial relationships of atoms (*stereoisomers*). For example, molecules **1-1**, **1-2**, and

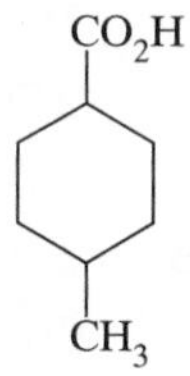

CO_2H CH_3 **1-1**

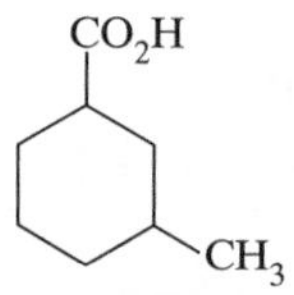

CO_2H CH_3 **1-2**

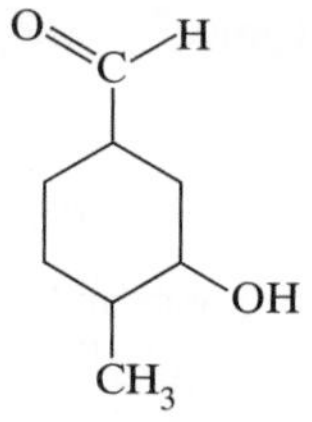

O C H OH CH_3 **1-3**

1-3 all have the molecular formula $C_8H_{14}O_2$, but they clearly have different structures. Molecules **1-1** and **1-2** differ only in the locational relationship of the carboxyl and methyl groups and are said to be *positional isomers*. Molecule **1-3** differs more radically because it possesses alcohol and aldehyde functionalities instead of the carboxylic acid, so it is a *functional isomer* of the first two molecules. To distinguish among this set of molecules, spectroscopy must (1) show how the various carbon atoms are connected and (2) distinguish between organic functional groups.

A closer look at molecule **1-1** reveals that the methyl and carboxyl groups can be located on the same side or on opposite sides of the six-membered ring, as indicated by the perspective drawings **1-4** and **1-5**. These two forms are different molecules, as bond cleavage is necessary to change one into the other. They have the same connectivity or constitution, so they are not structural isomers. Because they differ in the spatial arrangement of atoms, they are stereoisomers. Another type of stereoisomer

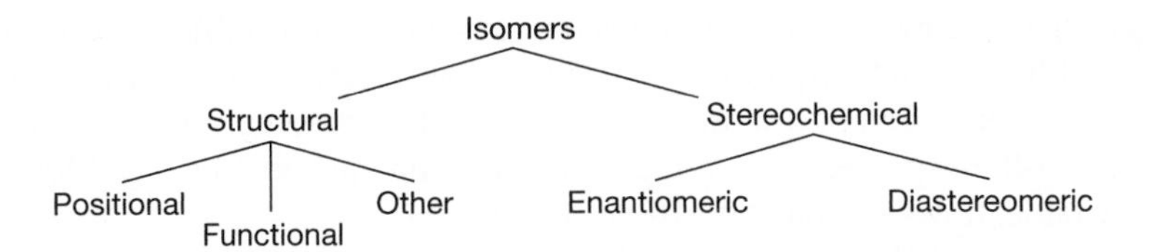

FIGURE 1-3 The isomer tree.

includes pairs of molecules that are nonsuperimposable mirror images, such as **1-6** and **1-7** (two forms of lactic acid). Such molecules are called *enantiomers* and are said

1-4 **1-5** **1-6** **1-7**

to be chiral. Stereoisomers that are not enantiomers, such as **1-4** and **1-5**, are called *diastereoisomers.*

When the six-membered ring of **1-5** is depicted more three dimensionally, it assumes a chair shape (**1-8**). By a series of rotations about single bonds, this molecule can switch all axial positions into equatorial positions, and vice versa, to produce the molecule **1-9.** These

1-8 **1-9**

molecules are stereoisomers, but because they can interconvert by rotations about single bonds they are called *conformational isomers* or *conformers*, which constitute a special category of diastereoisomers or enantiomers. All forms of spectroscopy can contribute to assigning stereochemical properties to molecules. Figure 1-3 illustrates the relationships among various classes of isomers.

The remainder of this book describes the spectroscopic methods used to obtain this structural information: molecular formula, identity of functional groups, atom connectivities, and arrangements of atoms in space. The successful result leads to the qualitative structure of the molecule, including the spatial arrangements of the atoms. The following section outlines what structural information each spectroscopic method provides.

1-5 CONTRIBUTIONS OF DIFFERENT FORMS OF SPECTROSCOPY

1-5a Nuclear Magnetic Resonance Spectroscopy

Nuclear magnetic resonance spectroscopy provides information about the types, numbers, connectivities, and interactions of particular atoms. Thus, it ideally can give the entire structure, including numbers of certain atoms, identity of functional groups, relationships between atoms, and spatial arrangements. For example, it can show that ethanol, CH_3CH_2OH, has two types of carbons in the ratio 1:1 and three types of hydrogens in the ratio 3:2:1, and that the methyl and methylene groups are bonded together rather than separated by oxygen. For this simple molecule, the entire structure is deduced. Moreover, NMR spectroscopy can provide information on the hydrogen bonding properties of the hydroxy group and consequently on molecular aggregation. The NMR experiment applies only to nuclei that have the quantum mechanical property of spin, through excitation of transitions between different quantized spin states. The most commonly examined nuclei are hydrogen (often referred to as the proton, ^{1}H) and carbon-13 (^{13}C). Most elements in the periodic table, however, have NMR-active isotopes.

Examination of the spin properties of hydrogens, carbons, and other nuclei in organic, biological, organometallic, and inorganic molecules characterizes each nucleus according to a physical parameter called the chemical shift (Chapter 3). Partial or even complete structures may be derived from analysis of chemical shifts alone. Interactions between NMR-active nuclei, as measured by the coupling constant (Chapter 4) and the relaxation time (Chapter 5), provide information about connectivities between atoms. Because coupling and relaxation depend on the distance between nuclei within a molecule, stereochemical information also is obtainable. Furthermore, because chemical reactions move nuclei from one position to another within a molecule or even to a new molecule, NMR is used to follow the kinetic course of many kinds of reactions.

The NMR experiment may be applied to molecules in any state of matter, but routine applications are carried out on liquids. For analysis of protons, the NMR experiment may be applied to microgram quantities and, for carbons, to milligram quantities. The samples may be mixtures. Thus NMR constitutes a very general approach to structural elucidation, as described in Part I of this volume.

1-5b Mass Spectrometry

The major function of mass spectrometry in organic chemistry is to provide the molecular formula of an unknown molecule. As this is the single important structural characteristic generally unavailable through NMR spectroscopy, the two methods are complementary. MS, however, can provide a vast array of other information, including functional group identity and connectivity. Both nominal (nearest integer) and exact (five or more significant figures) molecular weights may be obtained. In addition to intact ionic versions of the molecule, fragment ions often are generated, and their structures can prove helpful in deducing the complete molecular structure. The MS experiment involves generating ions in the gas phase and measuring their mass-to-charge ratio and relative abundances. Because isotopes differ in mass, they can be recognized by MS, thus allowing compounds to be analyzed according to their isotopic makeup.

Mass spectra can be taken on samples in any state of matter, including fragile, thermally labile solids. Mixtures are examined by MS with instruments that combine mass analysis with separations based on gas chromatography (GC–MS), liquid chromatography (LC–MS), or a second stage of mass spectrometry (MS–MS). The sensitivity of MS is unmatched by other common spectroscopic techniques. Work is carried out routinely at the submicrogram level ($<10^{-6}$ g), and quantities as small as femtomoles (10^{-12} mol) are detectable.

Chemical reactions invariably occur during the MS experiment. It is possible to follow their kinetics, to measure their thermodynamic parameters, and to prepare new compounds in the mass spectrometer. Part II describes the use of MS to obtain molecular weights, to analyze fragment ions, and to follow gas phase reactions.

1-5c Vibrational Spectroscopy

Infrared and Raman spectra provide a wealth of information about the structure of a molecule, particularly its symmetry and the identity of functional groups present. The spectra result from vibrations that occur naturally but predictably within molecules. The high sensitivity of IR spectroscopy and the ease of sample preparation contribute to its widespread use. Grating infrared spectrometers are inexpensive, compact, and so simple to use that an undergraduate student can obtain a spectrum after just a few minutes of instruction. Although Raman spectrometers are more complex and expensive, they can provide important information for structural analysis.

The IR and Raman spectra of any compound are unique and can be used as fingerprints for purposes of identification. Most important, vibrations of many functional groups such as carbonyl consistently give rise to features within well-defined ranges in the spectra, regardless of the overall structure of the molecule containing the group. Consequently, vibrational spectroscopy is particularly useful in the identification of functionalities, as described in Part III.

1-5d Electronic Spectroscopy

Electronic absorption spectroscopy provides information about conjugation within a molecule—for example, between two double bonds, between a double bond and a carbonyl group, or within an aromatic group. In addition, some unusual unconjugated functional groups (nitroso, thiocarbonyl) are uniquely characterized. The experiment measures the energy and probability of promoting a molecule from its ground electronic state to an electronically excited state. Excitation involves moving an electron from an occupied molecular orbital to a higher, unoccupied orbital. Since an organic molecule typically has numerous occupied and unoccupied molecular orbitals, many different electronic excitations are possible. The associated transition energies normally are found in the UV and vis regions of the electromagnetic spectrum (Section 1-2). Transitions of most unconjugated functionalities generally fall outside this range and are invisible to this technique without special equipment.

UV–vis spectroscopy is used qualitatively and quantitatively to detect and characterize conjugated functional groups, according to the position and intensity of the absorption band, as described in Part IV. The sensitivity of the UV–vis experiment is quite high, with detectability up to 10^{-9} M. Analysis of UV–vis absorption properties can characterize interactions between neighboring groups, such as that occurring in conjugated ketones and polyenes, and it can be used to study chemical or photochemical reactions that alter the functionality.

Problems

1-1 Specify whether the following pairs of molecules are structural isomers or stereoisomers.

(a) CH_3 CH_2 **(b)** CH_3NH CH_3 H_2N CH_2CH_3

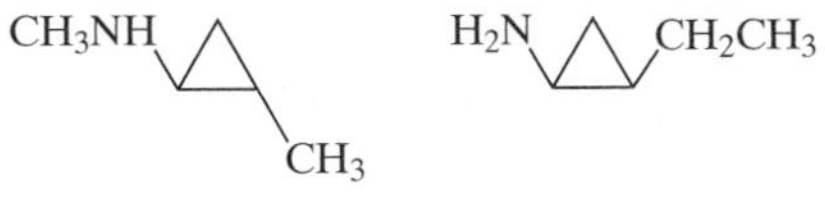

(c) H_3C CH_3 H_3C H_3C

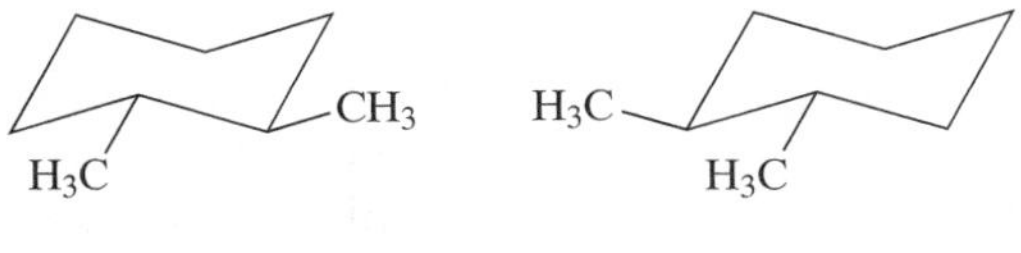

(d) $C_6H_5CH_2OCCH_3$ (O) $C_6H_5OCCH_2CH_3$ (O) **(e)** H OH HO H

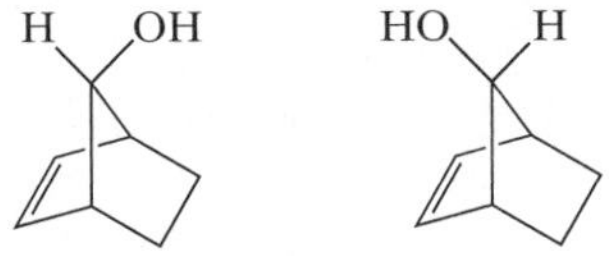

1-2 Carry out the following operations for each of the molecules in Problem **1-1.**

(a) Write down the molecular formula and calculate elemental percentages for C, H, N, and O (when present).

(b) Determine the unsaturation number.

(c) Specify the functional groups present.

1-3 For the molecules in Problems **1-1a**, **1-1d**, and **1-1e**, determine the number and relative proportions of different types of carbon atoms. Do the same with the hydrogens. Atoms are said to be different if they cannot be interconverted by a symmetry operation such as reflection through a mirror plane or rotation about an axis.

1-4 Determine the empirical formula of the molecules that gave the following elemental analyses.

(a) C, 59.89; H, 8.12; O, 31.99

(b) C, 66.02; H, 10.34; N, 10.96; O, 12.68

PART I

Nuclear Magnetic Resonance Spectroscopy

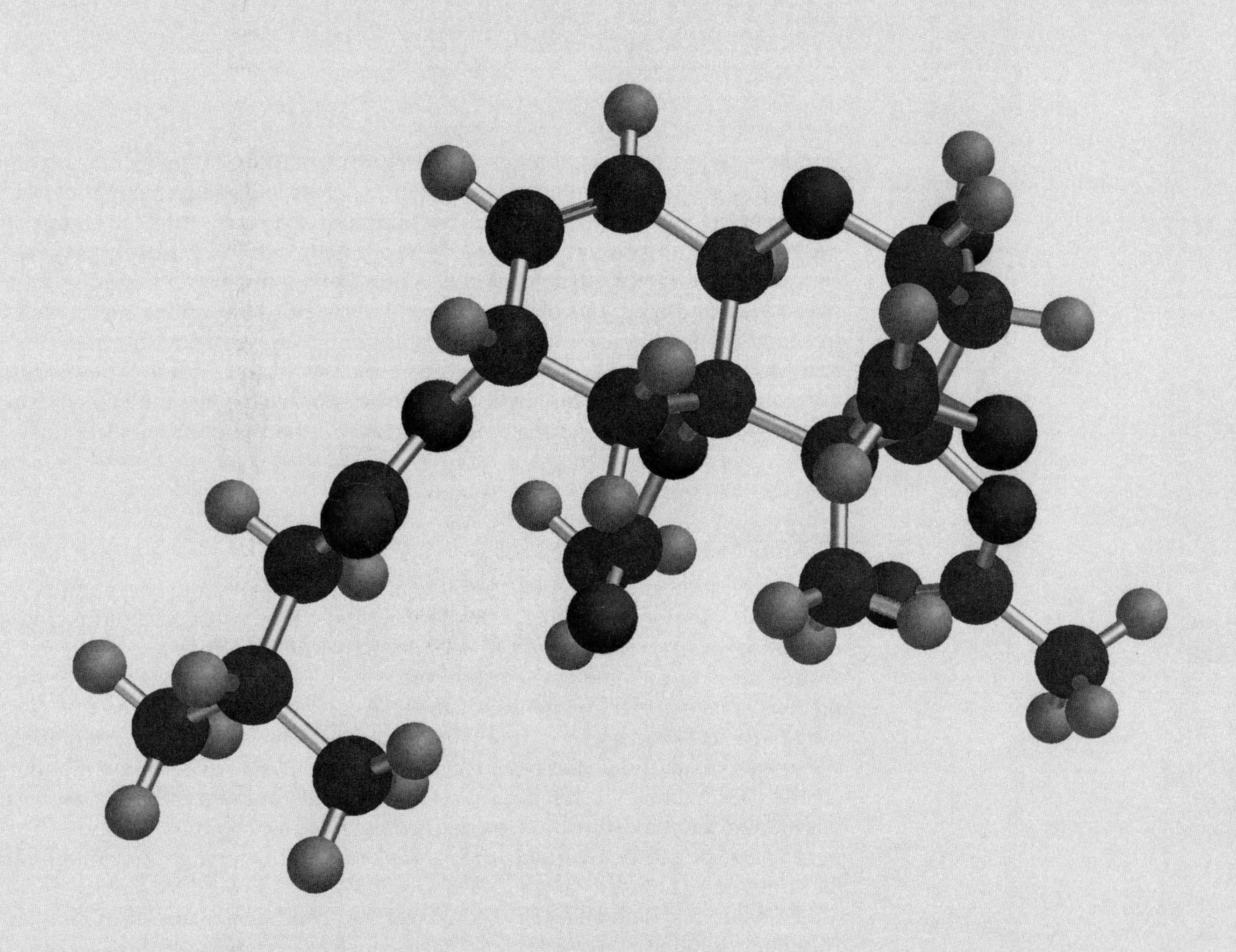

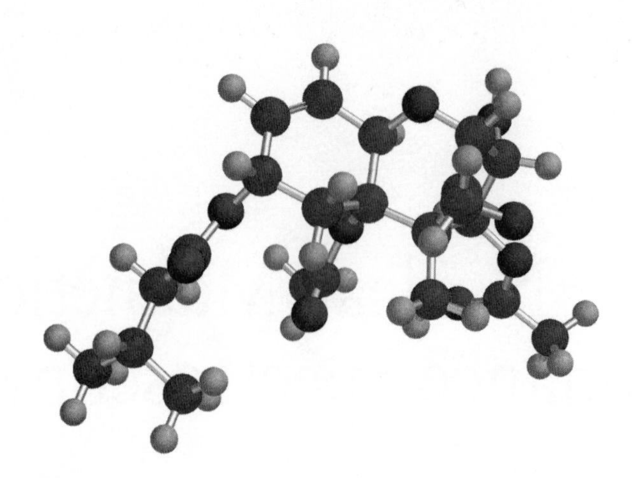

Introduction

Structure determination of almost any organic or biological molecule, as well as that of many inorganic molecules, begins with nuclear magnetic resonance (NMR) spectroscopy. During its existence of more than half a century, NMR spectroscopy has undergone several internal revolutions, repeatedly redefining itself as an increasingly complex and effective structural tool. Aside from X-ray crystallography, which can uncover the complete molecular structure of some pure crystalline materials, NMR spectroscopy is the chemist's most direct and general tool for identifying the structure of both pure compounds and mixtures as either solids or liquids. The process often involves performing several NMR experiments to deduce the molecular structure from the magnetic properties of the atomic nuclei and the surrounding electrons.

2-1 MAGNETIC PROPERTIES OF NUCLEI

The simplest atom, hydrogen, is found in almost all organic compounds and is composed of a single proton and a single electron. The hydrogen atom is denoted ^{1}H, in which the superscript signifies the sum of the atom's protons and neutrons, that is, the atomic mass of the element. For the purpose of NMR, the key aspect of the hydrogen nucleus is its angular momentum properties, which resemble those of a classical spinning particle. Because the spinning hydrogen nucleus is positively charged, it generates a magnetic field and possesses a *magnetic moment* μ, just as a charge moving in a circle creates a magnetic field (Figure 2-1). The magnetic moment μ is a vector, because it has both magnitude and direction, as defined by its axis of spin in the figure. The NMR experiment exploits the magnetic properties of nuclei to provide information on molecular structure.

The spin properties of protons and neutrons in the nuclei of heavier elements combine to define the overall spin of the nucleus. When both the atomic number (the

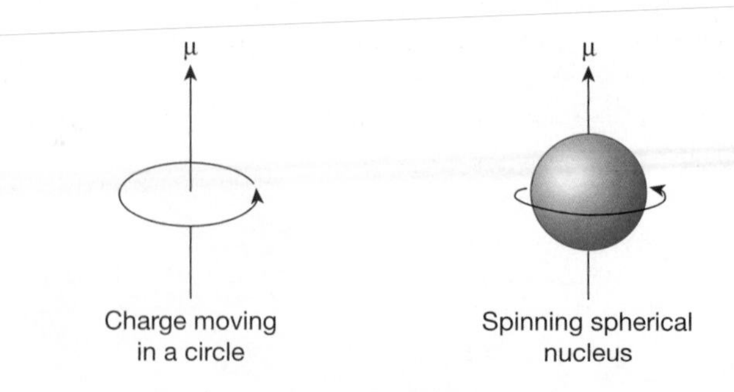

FIGURE 2-1 Analogy between a charge moving in a circle and a spinning nucleus.

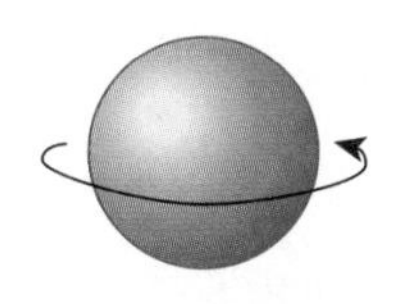

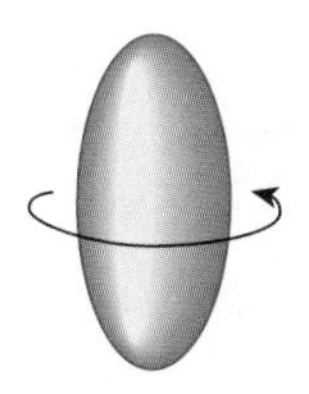

FIGURE 2-2 Three classes of nuclei.

number of protons) and the atomic mass (the sum of the protons and neutrons) are even, the nucleus has no magnetic properties, as signified by a zero value of its *spin quantum number I* (Figure 2-2). Such nuclei are not considered to be spinning. Common nonmagnetic (nonspinning) nuclei are carbon (^{12}C) and oxygen (^{16}O), which therefore are invisible to the NMR experiment. When either the atomic number or the atomic mass is odd, or when both are odd, the nucleus has magnetic properties that correspond to spin. For spinning nuclei, the spin quantum number can take on only certain values, which is to say that it is quantized. Those nuclei with a spherical shape have a spin I of $\frac{1}{2}$, and those with a nonspherical, or quadrupolar, shape have a spin of 1 or more (in increments of $\frac{1}{2}$). Common nuclei with a spin of $\frac{1}{2}$ include ^{1}H, ^{13}C, ^{15}N, ^{19}F, ^{29}Si, and ^{31}P. Thus, many of the most common elements found in organic molecules (H, C, N, P) have at least one isotope with $I = \frac{1}{2}$ (although oxygen does not). The class of nuclei with $I = \frac{1}{2}$ is the most easily examined by the NMR experiment. Quadrupolar nuclei ($I > \frac{1}{2}$) include ^{2}H, ^{11}B, ^{14}N, ^{17}O, ^{33}S, and ^{35}Cl.

The magnitude of the magnetic moment produced by a spinning nucleus varies from atom to atom in accordance with the equation $\mu = \gamma \hbar I$. The quantity $\hbar$ is Planck's constant h divided by 2π, and γ is a characteristic of the nucleus called the *gyromagnetic* or *magnetogyric ratio*. The larger the gyromagnetic ratio, the larger is the magnetic moment of the nucleus. Nuclei that have the same number of protons, but different numbers of neutrons, are called *isotopes* (^{1}H/^{2}H, ^{14}N/^{15}N). The term *nuclide* is generally applied to any atomic nucleus.

To study nuclear magnetic properties, the experimentalist subjects nuclei to a strong laboratory magnetic field B_0 with units of tesla (T). In the absence of this laboratory field, nuclear magnets of the same isotope have the same magnetic energy. When the B_0 field is imposed along a direction designated as the z axis, the energies of the nuclei in a sample are affected. There is a slight tendency for magnetic moments to move along the general direction of B_0 $(+z)$ over the opposite direction $(-z)$ (this motion will be more fully defined presently). Nuclei with a spin of $\frac{1}{2}$ assume only these two modes of motion. The splitting of spins into specific groups has been called the *Zeeman effect*.

The interaction is illustrated in Figure 2-3. At the left is a magnetic moment with a $+z$ component, and at the right is one with a $-z$ component. The nuclear magnets are not actually lined up parallel to the $+z$ or $-z$ direction. Rather, the force of B_0 causes the magnetic moment to move in a circular fashion about the $+z$ direction in the first case and about the $-z$ direction in the second, a motion called *precession*. In terms of vector analysis, the B_0 field in the z direction operates on the x component of μ to create a force in the y direction (Figure 2-3, inset in the middle). The force is the cross, or vector, product between the magnetic moment and the magnetic field, that is, $\mathbf{F} = \boldsymbol{\mu} \times \mathbf{B}$. The nuclear moment then begins to move toward the y direction. Because the force of B_0 on μ is always perpendicular to both B_0 and μ (according to the definition of a cross product), the motion of μ describes a circular orbit around the $+z$ or the $-z$ direction, in complete analogy to the forces present in a spinning top or gyroscope.

As the process of quantization allows only two directions of precession for a spin-$\frac{1}{2}$ nucleus (Figure 2-3), two assemblages or *spin states* are created, designated as $I_z = +\frac{1}{2}$ for precession with the field $(+z)$ and $I_z = -\frac{1}{2}$ for precession against the field $(-z)$. The assignment of signs (+ or −) is entirely arbitrary. The $I_z = +\frac{1}{2}$ state has a slightly

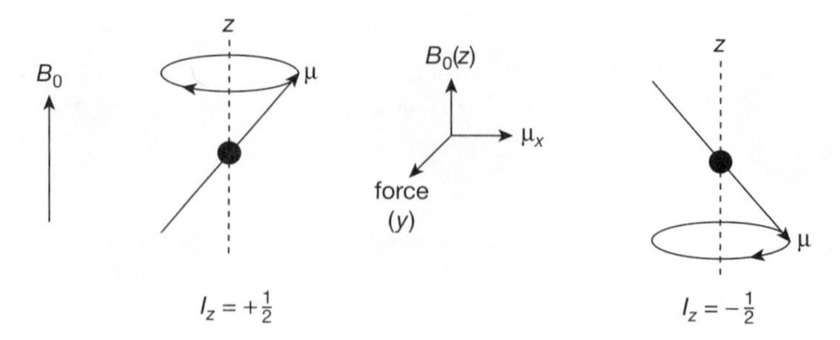

FIGURE 2-3 Interaction between a spinning nucleus and an external magnetic field B_0.

lower energy. In the absence of B_0, the precessional motions are absent, and all nuclei have the same magnetic energy.

The higher proportion of nuclei with $+z$ precession in the presence of B_0 is defined by Boltzmann's law, eq. 2-1,

$$\frac{n\left(+\frac{1}{2}\right)}{n\left(-\frac{1}{2}\right)} = \exp\left(\frac{\Delta E}{kT}\right) \tag{2-1}$$

in which n is the population of a spin state, k is Boltzmann's constant, T is the absolute temperature in kelvins (K), and ΔE is the energy difference between the spin states. Figure 2-4a depicts the energies of the two states and the difference ΔE between them.

The precessional motion of the magnetic moment around B_0 occurs with angular frequency ω_0, called the *Larmor frequency*, whose units are radians per second (rad s^{-1}). As B_0 increases, so does the angular frequency, that is, $\omega_0 \propto B_0$. The constant of proportionality between ω_0 and B_0 is the gyromagnetic ratio γ, so that $\omega_0 = \gamma B_0$. The natural precessional frequency can be expressed as linear frequency in Planck's relationship $\Delta E = h\nu_0$ or as angular frequency $\Delta E = \hbar\omega_0$ ($\omega_0 = 2\pi\nu_0$). In this way, the energy difference between the spin states is related to the Larmor frequency by eq. 2-2.

$$\Delta E = h\nu_0 = \hbar\omega_0 = \gamma\hbar B_0 \tag{2-2}$$

Thus, as the B_0 field increases, the difference in energy ΔE between the two spin states increases, as illustrated in Figure 2-4b.

The foregoing equations indicate that the natural precession frequency of a spinning nucleus ($\omega_0 = \gamma B_0$) depends only on the nuclear properties contained in the gyromagnetic ratio γ and on the laboratory-determined value of the magnetic field B_0. For a proton in a magnetic field B_0 of 7.05 T, the frequency of precession is 300 megahertz (MHz), and the difference in energy between the spin states is only 0.0286 cal mol^{-1} (0.120 J mol^{-1}). This value is extremely small in comparison with the energy differences

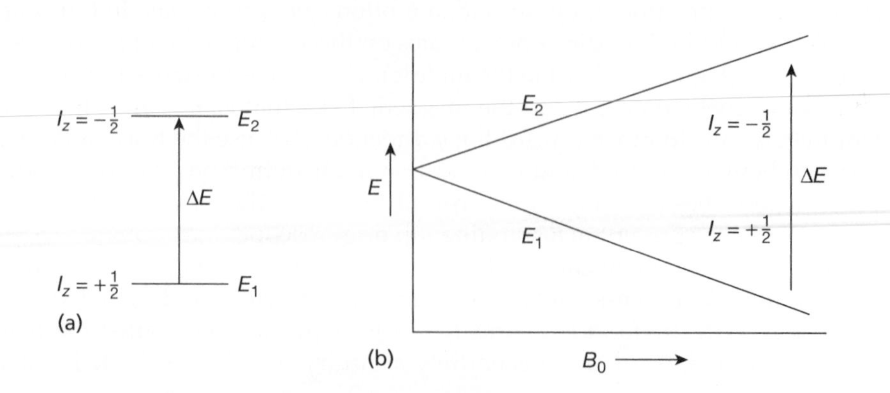

FIGURE 2-4 The energy difference (ΔE) between spin states as a function of the external magnetic field B_0.

between vibrational or electronic states. At a higher field, such as 14.1 T, the frequency increases proportionately, to 600 MHz in this case.

In the NMR experiment, the two states illustrated in Figure 2-4 are made to interconvert by applying a second magnetic field B_1, in the radiofrequency (rf) range. When the frequency of the B_1 field is the same as the Larmor frequency of the nucleus, energy can flow by absorption or emission between this newly applied field and the nuclei. Absorption of energy occurs as $+\frac{1}{2}$ nuclei become $-\frac{1}{2}$ nuclei, and emission occurs as $-\frac{1}{2}$ nuclei become $+\frac{1}{2}$ nuclei. Since there is an excess of $+\frac{1}{2}$ nuclei at the beginning of the experiment, absorption outweighs emission, to give a net absorption of energy. The process is called *resonance*, and the absorption may be detected electronically and displayed as a plot of frequency vs. amount of energy absorbed. Because the resonance frequency ν_0 is highly dependent on the structural environment of the nucleus, NMR has become the structural tool of choice for chemists. Figure 2-5 illustrates the NMR spectrum for the protons in benzene. Absorption is represented by a peak directed upward from the baseline.

Because gyromagnetic ratios vary among elements and even among isotopes of a single element, resonance frequencies also vary ($\omega_0 = \gamma B_0$). There is essentially no overlap in the resonance frequencies of different isotopes. At the field strength (11.755 T) at which protons resonate at 500 MHz, ^{13}C nuclei resonate at 125.8 MHz, ^{15}N nuclei at 50.7 MHz, and so on.

The magnitude of the gyromagnetic ratio γ also has an important influence on the intensity of the resonance. The difference in energy, $\Delta E = \gamma \hbar B_0$ (eq. 2-2), between the two spin states is directly proportional not only to B_0, as illustrated in Figure 2-4b, but also to γ. From Boltzmann's law (eq. 2-1), when ΔE is larger, there is a greater population difference between the two states. A greater excess of $I_z = +\frac{1}{2}$ spins (the E_1 state) means that more nuclei are available to flip to the E_2 state, so the resonance intensity is larger. The proton has one of the largest gyromagnetic ratios, so its spin states are relatively far apart, and the value of ΔE is especially large. The proton signal, consequently, is very strong. Many other important nuclei, such as ^{13}C and ^{15}N, have much smaller gyromagnetic ratios and hence have smaller differences between the energies of the two spin states (Figure 2-6). Thus, their signals are much less intense.

When spins have values greater than $\frac{1}{2}$, more than two spin states are allowed. For nuclei with $I = 1$, such as ^{2}H and ^{14}N, the magnetic moments may precess about three directions relative to B_0: parallel ($I_z = +1$), perpendicular (0), and opposite (-1). In general, there are $(2I + 1)$ spin states—for example, six for $I = 5/2$ (^{17}O has this spin). The values of I_z extend from $+I$ to $-I$ in increments of 1 $[+I, (+I - 1), (+I - 2) \ldots -I]$. Hence, the energy-state picture is more complex for quadrupolar than for spherical nuclei.

In summary, the NMR experiment consists of immersing magnetic nuclei in a strong field B_0 to distinguish them according to their values of I_z ($+\frac{1}{2}$ and $-\frac{1}{2}$ for spin-$\frac{1}{2}$

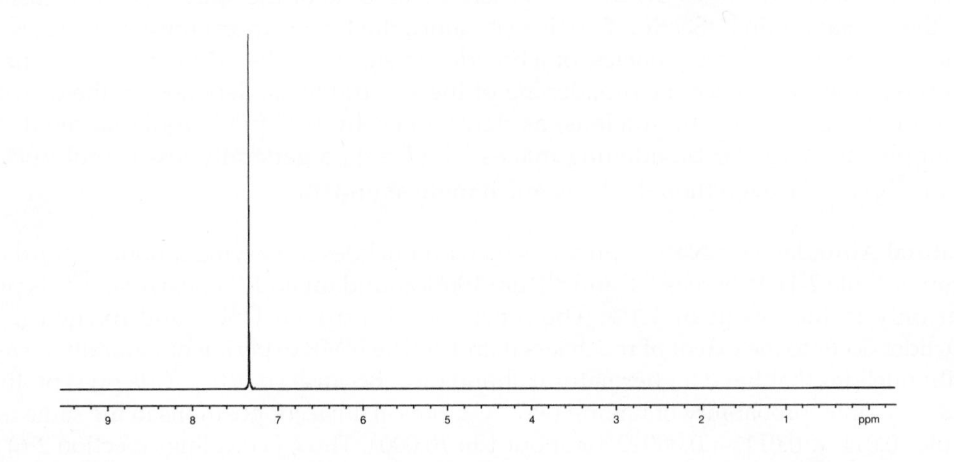

FIGURE 2-5 The 300 MHz ^{1}H spectrum of benzene.

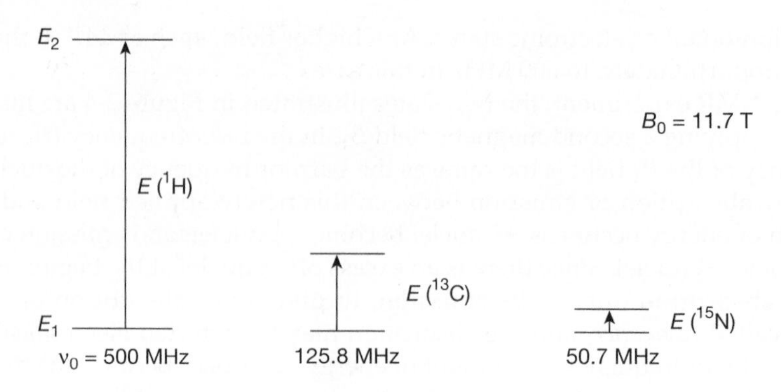

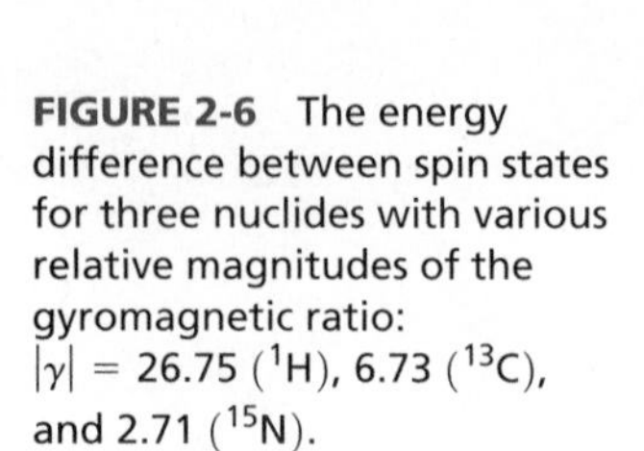
FIGURE 2-6 The energy difference between spin states for three nuclides with various relative magnitudes of the gyromagnetic ratio: $|\gamma| = 26.75$ (^{1}H), 6.73 (^{13}C), and 2.71 (^{15}N).

nuclei), followed by the application of a B_1 field whose frequency corresponds to the Larmor frequency ($\omega_0 = \gamma B_0$). This application of energy results in a net absorption, as the excess $+\frac{1}{2}$ nuclei are converted to $-\frac{1}{2}$ nuclei. The resonance frequency varies from nuclide to nuclide according to the value of the gyromagnetic ratio γ. The energy difference between the I_z spin states, $\Delta E = h\nu$, which determines the intensity of the absorption, depends on the value of B_0 (Figure 2-4) and on the gyromagnetic ratio of the nucleus ($\Delta E = \gamma\hbar B_0$) (Figure 2-6).

2-2 COMMONLY STUDIED NUCLIDES

Which nuclei (in this context, "nuclides") are useful in chemical problems? The answer depends on one's area of specialty. Certainly, for the organic chemist, the most common elements are carbon, hydrogen, oxygen, and nitrogen. The biological chemist would add phosphorus to the list. The organometallic or inorganic chemist would focus on whichever elements are of potential use, possibly boron, silicon, selenium, tin, mercury, platinum, or some of the low-intensity nuclei, such as iron and potassium. The success of the experiment depends on several factors, which are listed in Table 2-1 for a variety of nuclides and described in the following sections.

Spin. The overall spin of the nucleus (second column in Table 2-1) is determined by the spin properties of the protons and neutrons, as discussed in Section 2-1. By and large, spin-$\frac{1}{2}$ nuclei exhibit more favorable NMR properties than do quadrupolar nuclei ($I > \frac{1}{2}$). Nuclei with odd mass numbers have half-integral spins ($\frac{1}{2}$, $\frac{3}{2}$, etc.), whereas those with even mass and odd charge have integral spins (1, 2, etc.). Quadrupolar nuclei have a unique mechanism for relaxation that can result in extremely short relaxation times (Section 2-4), as discussed in Section 5-1. The relationship between lifetime (Δt) and energy (ΔE) is given by the Heisenberg uncertainty principle, which states that their product is constant: $\Delta E\ \Delta t \sim \hbar$. When the lifetime of the spin state, as measured by the relaxation time (Section 2-4), is very short, the larger uncertainty in energies implies a larger band of frequencies, or a broadened signal, in the NMR spectrum. The relaxation time, and hence the broadening of the spectral lines, depends on the distribution of charge within the nucleus, as determined by the quadrupole moment. For example, quadrupolar broadening makes ^{14}N ($I = 1$) a generally less useful nucleus than ^{15}N ($I = \frac{1}{2}$), even though ^{14}N is much more abundant.

Natural Abundance. Nature provides us with nuclides in varying amounts (third column in Table 2-1). Whereas ^{19}F and ^{31}P are 100% abundant and ^{1}H nearly so, ^{13}C is present only to the extent of 1.1%. The most useful nitrogen (^{15}N) and oxygen (^{17}O) nuclides occur to the extent of much less than 1%. The NMR experiment naturally is easier with nuclides that have higher natural abundance. Because so little ^{13}C is present, there is a very small probability of having two ^{13}C atoms at adjacent positions in the same molecule ($0.011 \times 0.011 = 0.00012$, or about 1 in 10,000). Thus, J couplings (Section 2-6) are

TABLE 2-1 NMR Properties of Common Nuclei[a]

Nuclide	Spin	Natural Abundance (N_a) (%)	Natural Sensitivity (N_s) (for equal numbers of nuclei) (vs. ^{1}H)	Receptivity (vs. ^{13}C)	NMR Frequency (at 7.05 T)	Reference Substance
Proton	$\frac{1}{2}$	99.985	1.00	5680	300.00	$Si(CH_3)_4$ or DSS[b]
Deuterium	1	0.015	0.00965	0.0082	46.05	$Si(CD_3)_4$
Lithium-7	$\frac{3}{2}$	92.58	0.293	1540	116.59	LiCl
Boron-10	3	19.58	0.0199	22.1	32.23	$BF_3 \cdot Et_2O$
Boron-11	$\frac{3}{2}$	80.42	0.165	754	96.25	$BF_3 \cdot Et_2O$
Carbon-13	$\frac{1}{2}$	1.108	0.0159	1.00	75.45	$Si(CH_3)_4$ or DSS[b]
Nitrogen-14	1	99.63	0.00101	5.69	21.68	NH_3 (*l*) or CH_3NO_2
Nitrogen-15	$\frac{1}{2}$	0.37	0.00104	0.0219	30.41	NH_3 (*l*) or CH_3NO_2
Oxygen-17	$\frac{5}{2}$	0.037	0.0291	0.0611	40.67	D_2O
Fluorine-19	$\frac{1}{2}$	100	0.833	4730	282.28	CCl_3F
Sodium-23	$\frac{3}{2}$	100	0.0925	525	79.36	NaCl (*aq*)
Aluminum-27	$\frac{5}{2}$	100	0.0206	117	78.17	$Al(NO_3)_3$
Silicon-29	$\frac{1}{2}$	4.70	0.00784	2.09	59.60	$Si(CH_3)_4$
Phosphorus-31	$\frac{1}{2}$	100	0.0663	377	121.44	85% H_3PO_4
Sulfur-33	$\frac{3}{2}$	0.76	0.00226	0.0973	23.03	CS_2 or $(NH_4)_2SO_4$
Chlorine-35	$\frac{3}{2}$	75.53	0.0047	20.2	29.39	NaCl (*aq*)
Chlorine-37	$\frac{3}{2}$	24.47	0.00274	3.8	24.47	NaCl (*aq*)
Potassium-39	$\frac{3}{2}$	93.1	0.000509	2.69	14.00	KCl (*aq*)
Calcium-43	$\frac{7}{2}$	0.145	0.00640	0.0527	20.19	$CaCl_2$ (*aq*)
Iron-57	$\frac{1}{2}$	2.19	0.0000337	0.0042	9.71	$Fe(CO)_5$
Cobalt-59	$\frac{7}{2}$	100	0.277	1570	71.18	$K_3Co(CN)_6$
Copper-63	$\frac{3}{2}$	69.09	0.0931	365	79.55	$Cu(CH_3CN)_4{}^+BF_4{}^-$
Germanium-73	$\frac{9}{2}$	7.76	0.00140	0.617	10.46	$Ge(CH_3)_4$
Selenium-77	$\frac{1}{2}$	7.58	0.00693	2.98	57.21	$Se(CH_3)_2$
Rhodium-103	$\frac{1}{2}$	100	0.0000312	0.177	9.56	$Rh(acac)_3$
Cadmium-113	$\frac{1}{2}$	12.26	0.011	7.6	66.58	$Cd(CH_3)_2$
Tin-119	$\frac{1}{2}$	8.58	0.0517	25.2	111.87	$Sn(CH_3)_4$
Tellurium-125	$\frac{1}{2}$	7.0	0.0315	12.5	94.65	$Te(CH_3)$
Platinum-195	$\frac{1}{2}$	33.8	0.00994	19.1	64.49	Na_2PtCl_6 (*aq*)
Mercury-199	$\frac{1}{2}$	16.84	0.00567	5.42	53.73	$Hg(CH_3)_2$
Thallium-205	$\frac{1}{2}$	70.50	0.192	769	173.05	$Tl(NO_2)_3$ (*aq*)
Lead-207	$\frac{1}{2}$	22.6	0.00920	11.8	62.76	$Pb(CH_3)_4$

[a]For a more complete list, see J. B. Lambert and F. G. Riddell, *The Multinuclear Approach to NMR Spectroscopy*, Dordrecht, The Netherlands: D. Reidel, 1983.
[b]DSS stands for 2,2-dimethyl-2-silapentane-5-sulfonic acid [also called 3-(trimethylsilyl)-1-propanesulfonic acid], which, as its sodium salt, is used as the reference standard in aqueous media.

not easily observed between two ^{13}C nuclei in ^{13}C spectra, although procedures to measure them have been developed.

Natural Sensitivity. Nuclides have differing sensitivities to the NMR experiment (fourth column in Table 2-1), as determined by the gyromagnetic ratio γ and the energy difference $\Delta E\,(=\gamma \hbar B_0)$ between the spin states (Figure 2-6). The larger the energy difference, the more nuclei are present in the lower spin state (see eq. 2-1), and hence the more are available to absorb energy. With its large γ, the proton is one of the most sensitive nuclei, whereas ^{13}C and ^{15}N, unfortunately, are rather weak (Figure 2-6). Tritium (^3H) is useful to the biochemist as a radioactive label. It has $I = \frac{1}{2}$ and is highly sensitive. Since it has zero natural abundance, it must be introduced synthetically. As a hydrogen label, deuterium also is useful, but it has very low natural sensitivity and must be introduced synthetically. Nuclei that are of interest to the inorganic chemist vary from poorly sensitive, such as iron and potassium, to highly sensitive, such as lithium and cobalt. Thus, it is important to be familiar with the natural sensitivity of a nucleus before designing an NMR experiment.

Receptivity. The intensity of the signal for a spin-$\frac{1}{2}$ nucleus is determined by both the natural abundance (in the absence of synthetic labeling) and the natural sensitivity of the nuclide. The mathematical product of these two factors is a good measure of how amenable a specific nucleus is to the NMR experiment. Because chemists are quite familiar with the ^{13}C experiment, the product of natural abundance and natural sensitivity for a nucleus is divided by the product for ^{13}C to give the factor known as receptivity (fifth column in Table 2-1). Thus, the receptivity of ^{13}C is, by definition, 1.00. The ^{15}N experiment is seen to be about 50 times less sensitive than that for ^{13}C, since the receptivity of ^{15}N is 0.0219.

In addition to these factors, Table 2-1 also contains the NMR resonance frequency at 7.05 T (the sixth column). The last column contains the reference substance for each nuclide, for which in most cases $\delta = 0$ (Section 2-3).

2-3 THE CHEMICAL SHIFT

The remaining sections in this chapter discuss the various factors that determine the content of NMR spectra. Uppermost is the location of the resonance in the spectrum, the so-called resonance frequency ν_0, which depends on the molecular environment as well as on γ and B_0. This dependence of ν_0 on structure is the ultimate reason for the importance of NMR spectroscopy in organic chemistry.

The electron cloud that surrounds the nucleus also has charge, motion, and, hence, a magnetic moment. The magnetic field generated by the electrons alters the B_0 field in the microenvironment around the nucleus. The actual field present at a given nucleus thus depends on the nature of the surrounding electrons. This electronic modulation of the B_0 field is termed *shielding* and is represented quantitatively by the Greek letter σ. The actual field at the nucleus becomes B_{local} and may be expressed as $B_0(1 - \sigma)$, in which the electronic shielding σ is positive for protons. This variation of the resonance frequency with shielding has been termed the *chemical shift*.

By substituting B_{local} for B_0 in eq. 2-2, the expression for the resonance frequency in terms of shielding becomes eq. 2-3.

$$\nu_0 = \frac{\gamma B_0(1 - \sigma)}{2\pi} \tag{2-3}$$

Decreased shielding thus results in a higher resonance frequency ν_0 at constant B_0, since σ enters the equation after a negative sign. For example, the presence of an electron-withdrawing group in a molecule reduces the electron density around a proton, so that there is less shielding and, consequently, a higher resonance frequency than in the case of a molecule that lacks the electron-withdrawing group. Hence, protons in fluoromethane (CH_3F) resonate at a higher frequency than those in methane (CH_4), because the fluorine atom withdraws electrons from around the hydrogen nuclei.

Figure 2-7 shows the separate NMR spectra of the protons and the carbons of methyl acetate ($CH_3CO_2CH_3$). Although 98.9% of naturally occurring carbon is the nonmagnetic ^{12}C, the carbon NMR experiment is carried out on the 1.1% of ^{13}C, which has an I of $\frac{1}{2}$. Because of differential electronic shielding, the ^1H spectrum contains separate resonances for the two types of protons ($O{-}CH_3$ and $C{-}CH_3$), and the ^{13}C spectrum

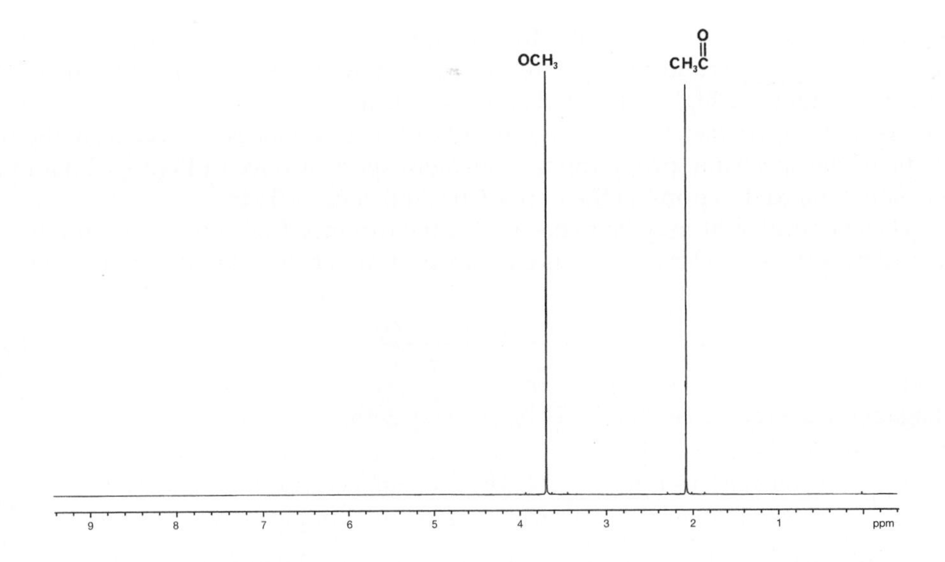

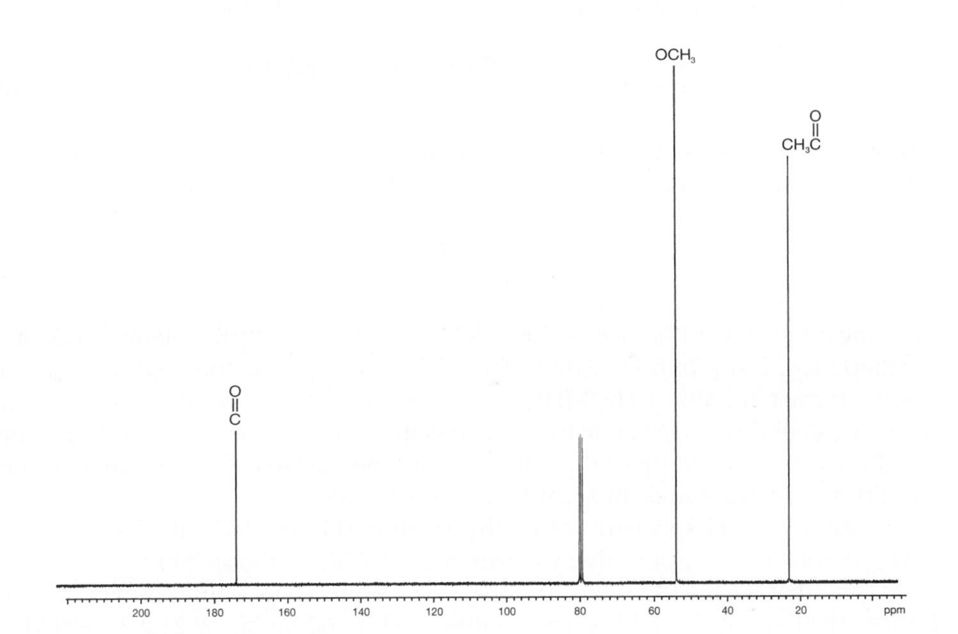

FIGURE 2-7 The 300 MHz ^{1}H spectrum (top) and the 75.45 MHz ^{13}C spectrum (bottom) of methyl acetate in $CDCl_3$. The ^{13}C resonance at δ 77 is from solvent. The ^{13}C spectrum has been decoupled from the protons.

contains separate resonances for the three types of carbons (O—CH_3, C—CH_3, and carbonyl) (Figure 2-8).

The proton resonances may be assigned on the basis of the electron-withdrawing abilities, or electronegativities, of the neighboring atoms. The ester oxygen is more electron withdrawing than the carbonyl group, so the O—CH_3 resonance occurs at a higher frequency than (and to the left of) the C—CH_3 resonance. By convention, frequency in the spectrum increases from right to left, for consistency with other forms of spectroscopy. Therefore, shielding increases from left to right, because of the negative sign before σ in eq. 2-3.

The system of units depicted in Figure 2-7 and used throughout this book has been developed to overcome the fact that chemical information often is found in small differences between large numbers. An intuitive system might have been absolute frequency—for example, in hertz (H_z). At the common field of 7.05 T, for instance, all protons resonate in the vicinity of 300 MHz. A scale involving numbers like 300.000064, however, is cumbersome. Moreover, frequencies would vary from one B_0 field to another (eq. 2-3). For every element or isotope, a reference material has been chosen and assigned a relative frequency of zero. For both protons and carbons, the substance is tetramethylsilane [$(CH_3)_4Si$, usually called TMS], which is soluble in most organic solvents, is unreactive,

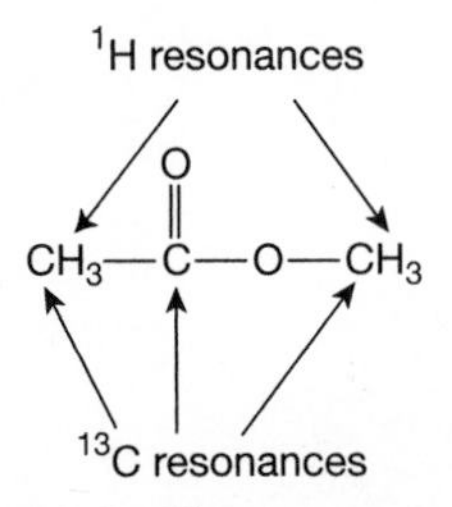

FIGURE 2-8 The resonances expected for methyl acetate.

has a strong signal, and is volatile. In addition, the low electronegativity of silicon means that the protons and carbons are surrounded by a relatively high density of electrons. Hence, they are highly shielded and resonate at very low frequency. Shielding by silicon is so strong, in fact, that the proton and carbon resonances of TMS are at the right extreme of the spectrum, providing a convenient spectral zero. In Figures 2-5 and 2-7, the position marked "0 ppm" is the hypothetical position of TMS.

The chemical shift may be expressed as the distance from the chemical reference standard by writing eq. 2-3 twice—once for an arbitrary observed nucleus i as in eq. 2-4a,

$$\nu_i = \frac{\gamma B_0(1 - \sigma_i)}{2\pi} \tag{2-4a}$$

and again for the reference, that is, TMS, as in eq. 2-4b.

$$\nu_r = \frac{\gamma B_0(1 - \sigma_r)}{2\pi} \tag{2-4b}$$

The distance between the resonances in the NMR frequency unit (hertz, which is equivalent to cycles per second) then is given by the formula in eq. 2-5.

$$\Delta\nu = \nu_i - \nu_r = \frac{\gamma B_0(\sigma_r - \sigma_i)}{2\pi} = \frac{\gamma B_0 \Delta\sigma}{2\pi} \tag{2-5}$$

This expression for the frequency differences still depends on the magnetic field B_0. To have a common unit at all B_0 fields, the chemical shift of nucleus i is defined by eq. 2-6,

$$\delta = \frac{\Delta\nu}{\nu_r} = \frac{\sigma_r - \sigma_i}{1 - \sigma_r} \sim \sigma_r - \sigma_i \tag{2-6}$$

in which the frequency difference in hertz (eq. 2-5) is divided by the reference frequency in megahertz (eq. 2-4b). In this fashion, the constants including the field B_0 cancel out. The δ scale is thus in units of Hz/MHz, or parts per million (ppm) of the field. Because the reference shielding is chosen to be much less than 1.0 $[(1 - \sigma_r) \sim 1]$, δ corresponds to the differences in shielding of the reference and the nucleus. An increase in σ_i therefore results in a decrease in δ_i, in accordance with eq. 2-6.

As seen in the ^{1}H spectrum of methyl acetate (Figure 2-7), the δ value for the C—CH_3 protons is 2.07 ppm (always written as "δ 2.07" without "ppm," which is understood) and that for the O—CH_3 protons is 3.67 ppm. These values remain the same in spectra taken at any B_0 field, such as either 1.41 T (60 MHz) or 21.2 T (900 MHz), which together represent the extremes of spectrometers currently in use. Chemical shifts in hertz, however, vary from field to field. Thus, a resonance that is 90 Hz from TMS at 60 MHz is 450 Hz from TMS at 300 MHz, but always has a δ value of 1.50 ppm ($\delta = 90/60 = 450/300 = 1.50$). Note that a resonance to the right of TMS has a *negative* value of δ. Also, since TMS is insoluble in water, other internal standards are used for this solvent, including the sodium salts of 3-(trimethylsilyl)-1-propanesulfonic acid $[(CH_3)_3SiCH_2CH_2CO_2Na]$ and 3-(trimethylsilyl)propionic acid [or 2,2-dimethyl-2-silapentane-5-sulfonic acid, DSS, $[(CH_3)_3Si(CH_2)_3SO_3Na]$.

In the first generation of commercial spectrometers, the range of chemical shifts, such as those in the scale at the bottom of Figures 2-5 and 2-7, was generated by varying the B_0 field while holding the B_1 field, and hence the resonance frequency, constant. As eq. 2-3 indicates, an increase in shielding (σ) requires B_0 to be raised in order to keep ν_0 constant. Since nuclei with higher shielding resonate at the right side of the spectrum, the B_0 field in this experiment increases from left to right. Consequently, the right end came to be known as the high field, or upfield, end, and the left end as the low field, or downfield, end. This method was termed *continuous-wave* (CW) *field sweep*. Although the method is rarely used today, its vestigial terms such as "upfield" and "downfield" remain in the NMR vocabulary.

Modern spectrometers vary the B_1 frequency while the B_0 field is kept constant. An increase in shielding (σ) reduces the right side of eq. 2-3, so that ν_0 must decrease in

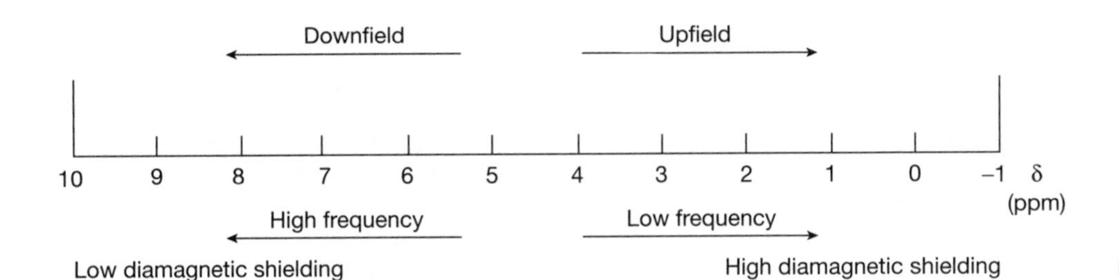

FIGURE 2-9 Spectral conventions.

order to maintain a constant B_0. Thus, the right end of the spectrum, as noted before, corresponds to lower frequencies for more shielded nuclei. The general result is that frequency increases from right to left, whereas field increases from left to right. Figure 2-9 summarizes the terminology. The right end of the spectrum still is often referred to as the high field or upfield end, in deference to the old field-sweep experiment, although it is more appropriate to call it the low-frequency or more shielded end.

The chemical shift is considered further in Chapter 3.

2-4 EXCITATION AND RELAXATION

To understand the NMR experiment more fully, it is useful to consider Figure 2-3 again—this time in terms of a collection of nuclei (Figure 2-10). At equilibrium, the $I_z = +\frac{1}{2}$ nuclei precess around the $+z$ axis, and the $-\frac{1}{2}$ nuclei precess around the $-z$ axis. Only 20 spins are shown on the surface of the double cone in the figure, and the excess of $+\frac{1}{2}$ over $-\frac{1}{2}$ nuclei is exaggerated (12 to 8). The actual ratio of populations of the two states is given by the Boltzmann equation (eq. 2-1). Inserting the numbers for $B_0 = 7.04$ T yields the result that, for every million spins, there are only about 50 more with $+\frac{1}{2}$ than $-\frac{1}{2}$ spin. If the magnetic moments are added vectorially, there is a net vector in the $+z$ direction, because of the excess of $+\frac{1}{2}$ over $-\frac{1}{2}$ spins. The sum of all the individual spins is called the *magnetization* (**M**) (**boldface** letters in this context connote a vectorial parameter). The boldface arrow pointing along the $+z$ direction in Figure 2-10 represents the resultant **M**. Because the spins are distributed randomly (or incoherently) around the z axis, there is no net x or y magnetization; that is, $M_x = M_y = 0$, and, hence, $\mathbf{M} = M_z$.

Figure 2-10 also shows the vector that represents the B_1 field placed along the x axis. When the B_1 frequency matches the Larmor frequency of the nuclei, some $+\frac{1}{2}$ spins turn over and become $-\frac{1}{2}$ spins, so that M_z decreases slightly. The vector B_1 exerts a force on **M** whose result is perpendicular to both vectors (inset at lower right of the figure;

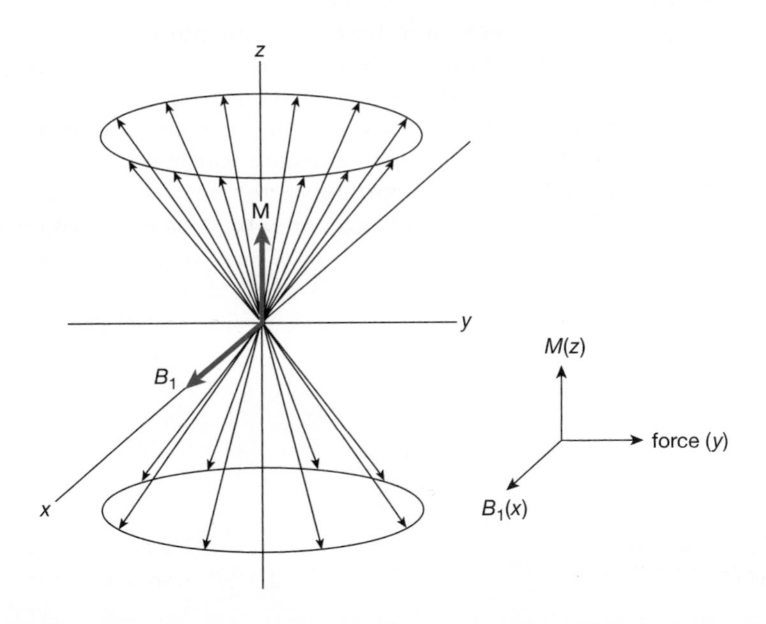

FIGURE 2-10 Spin-$\frac{1}{2}$ nuclei at equilibrium, prior to application of the B_1 field.

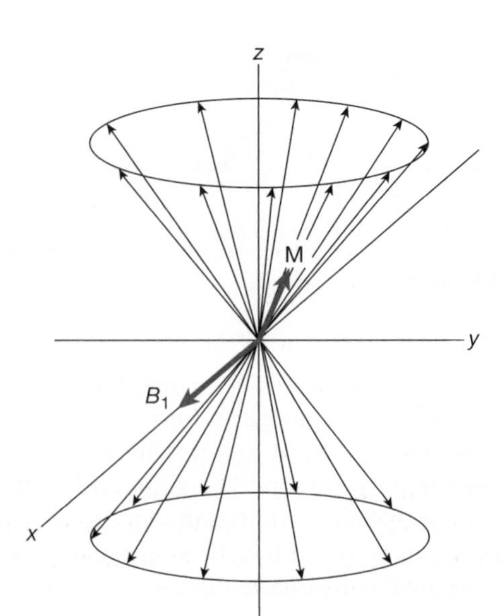

FIGURE 2-11 Spin-$\frac{1}{2}$ nuclei immediately after application of the B_1 field.

the force arises from the cross product $\mathbf{F} = \mathbf{M} \times \mathbf{B}$). If B_1 is turned on very briefly, the magnetization vector **M** is tipped only slightly off the z axis, moving toward the y axis, which represents the mutually perpendicular direction. Figure 2-11 illustrates the result.

The 20 spins of Figure 2-10 include 12 nuclei of spin $+\frac{1}{2}$ and 8 of spin $-\frac{1}{2}$. These same nuclei, as shown in Figure 2-11, comprise 11 of spin $+\frac{1}{2}$ and 9 of spin $-\frac{1}{2}$ after application of the B_1 field. Thus, only one nucleus has changed its spin. The decrease in M_z is exaggerated in the figure, but the tipping of the magnetization vector off the axis is clearly apparent. The positions on the circles, or *phases*, of the 20 nuclei are no longer random, because the tipping requires bunching of the spins. Thus, the phases of the spins now have some *coherence*, and there are x and y components of the magnetization. The xy component of the magnetization is the signal detected electronically as the resonance. It is important to appreciate that the so-called absorption of energy as $+\frac{1}{2}$ nuclei become $-\frac{1}{2}$ nuclei is not measured directly.

The B_1 field in Figures 2-10 and 2-11 oscillates back and forth along the x axis. As Figure 2-12 illustrates from a view looking down the z axis, B_1 may be considered either (1) to oscillate linearly along the x axis at so many times per second (with frequency ν) or (2) to move circularly in the xy plane with angular frequency $\omega\,(= 2\pi\nu)$ in radians per second. The two representations are vectorially equivalent. Resonance occurs when the frequency and phase of B_1 match that of the nuclei precessing at the Larmor frequency.

Figure 2-11 represents a snapshot in time, with the motion of both the B_1 vector and the precessing nuclei frozen. In real time, each nuclear vector is precessing around the z axis, so that the magnetization **M** also is precessing around that axis. Another way to look at the frozen frame is to consider that the x and y axes are rotating at the frequency of the B_1 field. In terms of Figure 2-12, the axes are following the circular motion. Consequently, B_1 appears to be frozen in position along the x axis, instead of oscillating in the fashion shown in that figure. This *rotating coordinate system* is used throughout the

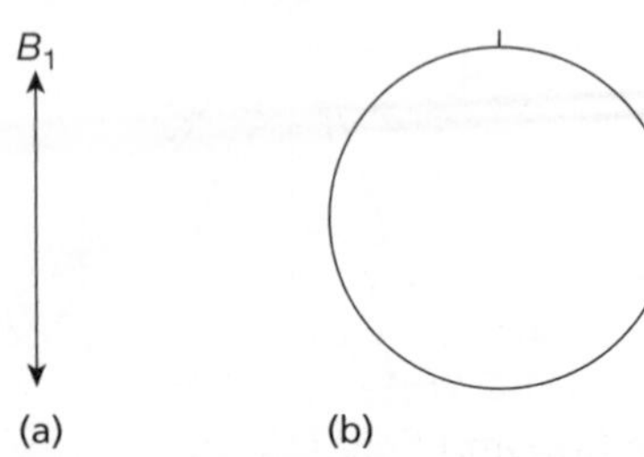

FIGURE 2-12 Analogy between (a) linearly and (b) circularly oscillating fields.

book to simplify magnetization diagrams. In the rotating frame, the individual nuclei and the magnetization **M** no longer precess around the z axis but are frozen for as long as they all have the same Larmor frequency matched by the frequency (B_1) of the rotating x and y axes.

Application of B_1 at the resonance frequency results in both energy absorption ($+\frac{1}{2}$ nuclei become $-\frac{1}{2}$) and emission ($-\frac{1}{2}$ nuclei become $+\frac{1}{2}$). Because initially there are more $+\frac{1}{2}$ than $-\frac{1}{2}$ nuclei, the net effect is absorption. As B_1 irradiation continues, however, the excess of $+\frac{1}{2}$ nuclei disappears, so that the rates of absorption and emission eventually become equal. Under these conditions, the sample is said to be approaching *saturation*, so the experiment should come to an end. The situation is ameliorated, however, by natural mechanisms whereby nuclear spins return to equilibrium from saturation. Any process that returns the z magnetization to its equilibrium condition with the excess of $+\frac{1}{2}$ spins is called *spin–lattice*, or *longitudinal, relaxation* and is usually a first-order process with time constant T_1. For a complete return to equilibrium, relaxation also is necessary to destroy magnetization created in the xy plane. Any process that returns the x and y magnetizations to their equilibrium condition of zero is called *spin–spin*, or *transverse, relaxation* and is usually a first-order process with time constant T_2.

Spin–lattice relaxation (T_1) derives from the existence of local oscillating magnetic fields in the sample that correspond to the resonance frequency. The primary source of these random fields is other magnetic nuclei that are in motion. As a molecule tumbles in solution in the B_0 field, each nuclear magnet generates a field caused by its motion. If this field is at the Larmor frequency, excess spin energy of neighboring spins can pass to this motional energy as $-\frac{1}{2}$ nuclei become $+\frac{1}{2}$ nuclei. The resonating spins are relaxed back to their initial state, and the absorption experiment can be repeated to increase intensities through signal averaging.

For effective spin–lattice relaxation, the tumbling magnetic nuclei must be spatially close to the resonating nucleus. For ^{13}C, attached protons (C—H) provide effective spin–lattice relaxation. A carbonyl carbon or a carbon attached to four other carbons relaxes very slowly and is more easily saturated because the attached atoms are nonmagnetic (motion of the nonmagnetic nuclei ^{12}C and ^{16}O provides no relaxation). Protons are relaxed by their nearest neighbor protons. Thus, CH_2 and CH_3 groups are relaxed by geminal protons (HCH), but CH protons must rely on vicinal (CH—CH) or more distant protons.

Spin–lattice relaxation also is responsible for generating the initial excess of $+\frac{1}{2}$ nuclei when the sample is first placed in the probe. In the absence of the B_0 field, all spins have the same magnetic energy. When the sample is immersed in the B_0 field, magnetization begins to build up as spins flip from the effect of interactions with surrounding magnetic nuclei in motion, eventually creating the equilibrium ratio with an excess of $+\frac{1}{2}$ over $-\frac{1}{2}$ spins.

For x and y magnetization to decay toward zero (spin–spin, or T_2, relaxation), the phases of the nuclear spins must become randomized (see Figures 2-10 and 2-11). The mechanism that gives the phenomenon its name involves the interaction of two nuclei with opposite spin. The process whereby one spin goes from $+\frac{1}{2}$ to $-\frac{1}{2}$ while the other goes from $-\frac{1}{2}$ to $+\frac{1}{2}$ involves no net change in z magnetization and hence no spin–lattice relaxation. The switch in spins results in dephasing, because the new spin state has a different phase from the old one. In terms of Figure 2-11, a spin vector disappears from the surface of the upper cone and reappears on the surface of the lower cone (and vice versa) at a new phase position. As this process continues, the phases become randomized around the z axis, and xy magnetization disappears. This process of two nuclei simultaneously exchanging spins is sometimes called the flip-flop mechanism.

Spin–spin relaxation also arises when the B_0 field is not perfectly homogeneous. Again in terms of Figure 2-11, if the spin vectors are not located in exactly identical B_0 fields, they differ slightly in Larmor frequencies and hence precess around the z axis at different rates. As the spins move faster or more slowly relative to each other, eventually their relative phases become randomized. When nuclei resonate over a range of Larmor frequencies, the line width of the signal naturally increases. The spectral line width at half height and the spin–spin relaxation are related by the expression $w_{1/2} = 1/\pi T_2$. Both mechanisms (flip-flop and field inhomogeneity) can contribute to T_2.

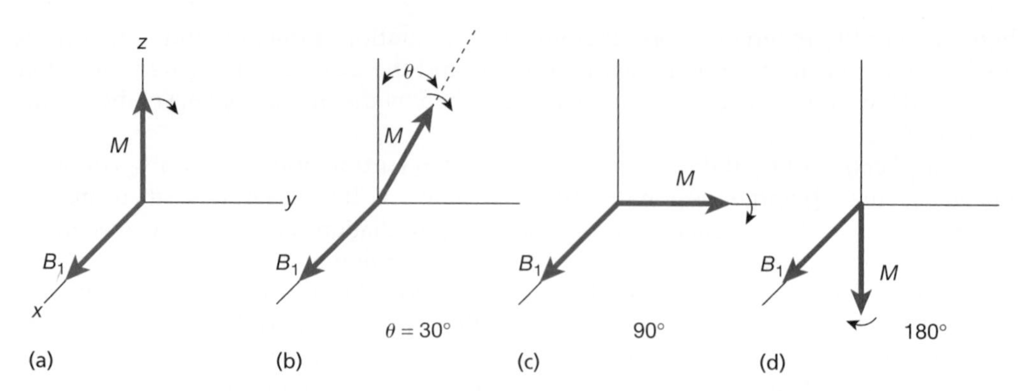

FIGURE 2-13 Net magnetization M as a function of time, following application of the B_1 field.

The subject of relaxation is discussed further in Section 5-1.

2-5 PULSED EXPERIMENTS

In the pulsed NMR experiment, the sample is irradiated close to the resonance frequency with an intense B_1 field for a very short time. For the duration of the pulse, the B_1 vector on the x axis (in the rotating coordinate system) exerts a force (see inset in Figure 2-10) on the **M** vector, which is on the z axis, pushing the magnetization toward the y axis. Figures 2-13a and 2-13b respectively simplify Figures 2-10 and 2-11 by eliminating the individual spins. Only the net magnetization vector **M** is shown in Figure 2-13.

As long as the strong B_1 field remains on, the magnetization continues to rotate, or precess, around B_1 on the x axis. The strength of the B_1 field is such that, when it is on, it forces precession to occur preferentially around its direction (x) rather than around the natural direction (z) of the weaker B_0 field. Consequently, the primary field present at the nuclei is B_1, so the expression for the precession frequency becomes $\omega = \gamma B_1$. More precisely, this equation holds at the resonance frequency $\omega_0 = \gamma B_0$. Farther and farther from the resonance frequency, the effect of B_1 wanes, and precession around B_0 returns. A full mathematical treatment requires inclusion of terms in both B_0 and B_1, but, qualitatively, our interest focuses on the events at or near the resonance frequency.

The angle θ of rotation of the magnetization increases as long as B_1 is present (Figure 2-13). A short pulse might leave the magnetization at a 30° angle relative to the z axis (Figure 2-13b). A pulse three times as long (90°) aligns the magnetization along the y axis (Figure 2-13c). A pulse of double this duration (180°) brings the magnetization along the $-z$ direction (Figure 2-13d), meaning that there is an excess of $-\frac{1}{2}$ spins, or a population inversion, a truly unnatural situation. The exact angle θ caused by a pulse thus is determined by its duration t_{p}. The angle θ therefore is ωt_{p}, in which ω is the precessional frequency in the B_1 field. Since $\omega = \gamma B_1$, it follows that $\theta = \gamma B_1 t_p$.

If B_1 irradiation is halted when the magnetization reaches the y axis (a 90° pulse), and if the magnetization along the y direction is detected over time at the resonance frequency, then the magnetization would be seen to decay (Figure 2-14). Alignment of the magnetization along the y axis is a nonequilibrium situation. After the pulse, xy magnetization decays by spin–spin relaxation (T_2). At the same time, z magnetization reappears by spin–lattice relaxation (T_1). The reduction in y magnetization with time shown in the figure is called the ***Free Induction Decay*** (FID) and is a first-order process with time constant T_2.

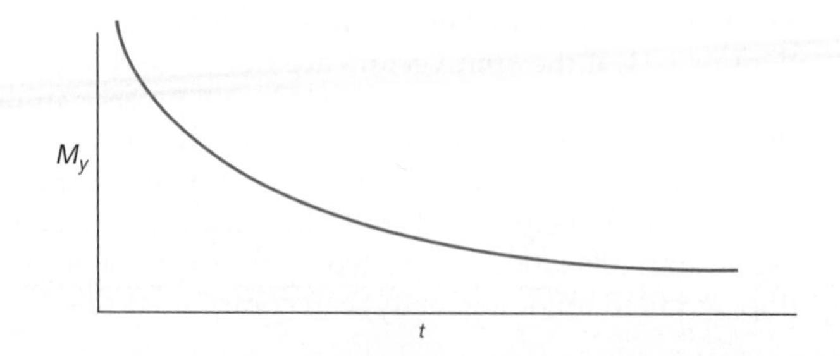

FIGURE 2-14 Time dependence of the magnetization M following a 90° pulse.

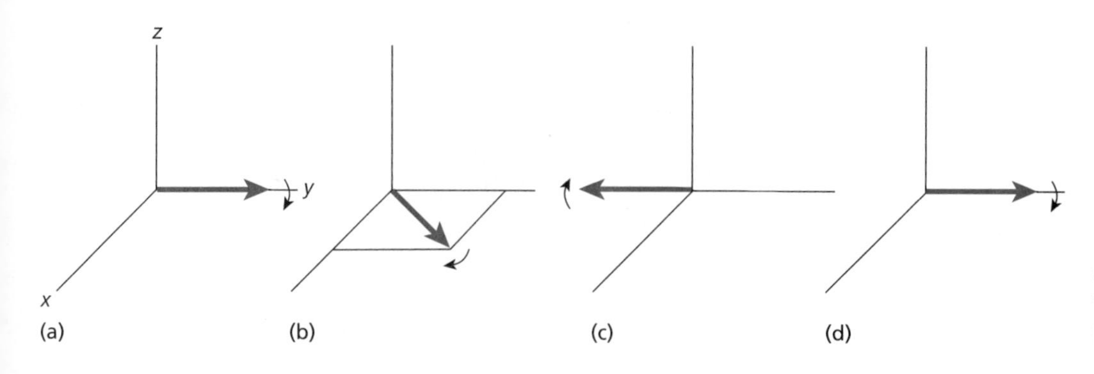

FIGURE 2-15 Induced magnetization along the y axis as a function of time after a 90° pulse.

The illustration in Figure 2-14 is unrealistic, because it involves only a single type of nucleus for which the resonance frequency γB_0 corresponds to the frequency of rotation of the x and y axes. Most samples have quite a few different types of protons or carbons, so that several resonance frequencies are involved, but the rotating frame can have only a single, or reference, frequency. What happens when there are nuclei with resonance frequencies different from the reference frequency? First, imagine again the case of a single resonance whose Larmor frequency corresponds to the reference frequency. At the time the 90° B_1 pulse is turned off, the spins are lined up along the y axis (Figure 2-15a). Precession then returns to the z axis at the Larmor frequency $\omega_0 = \gamma B_0$. In the rotating coordinate system, the x and y axes are rotating at γB_0 around the z axis. Nuclei with resonance frequency $\omega_0 = \gamma B_0$ appear not to precess about the z axis because their frequency of rotation matches that of the rotating frame and they remain lined up along the z axis (Figure 2-15a). In the case of a second nucleus with a different Larmor frequency ($\omega \neq \omega_0 = \gamma B_0$), the nuclear magnets move off the y axis within the xy plane (Figure 2-15b). Only nuclei precessing at the reference frequency γB_0 appear to be stationary in the rotating coordinate system. As time progresses, the magnetization continues to rotate within the xy plane, reaching the $-y$ axis (Figure 2-15c) and eventually returning to the $+y$ axis (Figure 2-15d). The detected y magnetization during this cycle first decreases, then falls to zero as it passes the $y = 0$ point, next moves to a negative value in the $-y$ region (Figure 2-15c), and finally returns to positive values (Figure 2-15d). The magnitude of the magnetization thus varies periodically like a cosine function. When it is again along the $+y$ axis (Figure 2-15d), the magnetization is slightly smaller than at the beginning of the cycle, because of spin–spin relaxation (T_2). Moreover, it has moved out of the xy plane (not shown in the figure) as z magnetization begins to return through spin–lattice relaxation (T_1). The magnetization varies as a cosine function with time, continually passing through a sequence of events illustrated by Figure 2-15.

Figure 2-16a shows what the FID looks like for the protons of acetone, when the reference frequency ω_0 is not the same as the Larmor frequency ω of acetone. The horizontal distance between adjacent maxima is the reciprocal of the difference between the Larmor frequency and the B_1 frequency $[(\Delta\omega)^{-1} = (\omega - \omega_0)^{-1}]$. The intensities of the maxima decrease as y magnetization is lost through spin–spin relaxation. Because the line width of the spectrum is determined by T_2, the FID contains all the necessary information to display a spectrum: frequency and line width, as well as overall intensity.

Now consider the case of two nuclei with different resonance frequencies, each different from the reference frequency. Their decay patterns are superimposed, reinforcing and interfering to create a complex FID, as in Figure 2-16b for the protons of methyl acetate [$CH_3(C{=}O)OCH_3$]. By the time there are four frequencies, as in the carbons of 3-hydroxybutyric acid [$CH_3CH(OH)CH_2CO_2H$] shown in Figure 2-16c, it is impossible to unravel the frequencies visually. The mathematical process called Fourier analysis matches the FID with a series of sinusoidal curves and exponential functions to obtain from them the frequencies, line widths, and intensities of each component. The FID is a plot in time (see Figures 2-14 and 2-16), so that experiment is said to occur in the *time domain.* The experimentalist, however, wants a plot of frequencies, so the spectrum must be transformed to a *frequency domain,* as shown in Figures 2-5 and 2-7. The Fourier transformation (FT) from time to frequency domain is carried out rapidly by computer, and the experimentalist need not examine the FID.

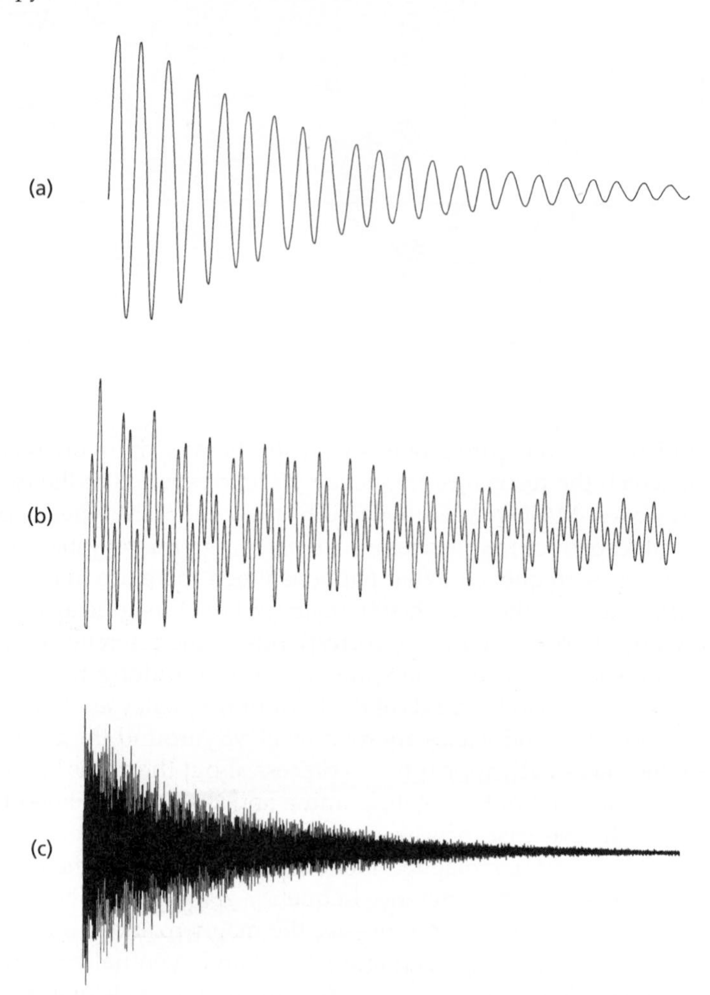

FIGURE 2-16 Free induction decay for the ^{1}H spectra of (a) acetone and (b) methyl acetate. (c) Free induction decay for the ^{13}C spectrum of 3-hydroxybutyric acid. All samples are without solvent.

2-6 THE COUPLING CONSTANT

The form of a resonance can be altered by the presence of a distinct, neighboring, magnetic nucleus. In 1-chloro-4-nitrobenzene (**2-1**), for example, there are two types of protons,

NO_2 H_A H_X Cl

2-1

labeled A and X, which are, respectively, ortho to nitro and ortho to chloro. For the time being, we will ignore any effects from the identical A and X protons across the ring. Each proton has a spin of $\frac{1}{2}$ and therefore can exist in two I_z spin states, $+\frac{1}{2}$ and $-\frac{1}{2}$, which differ in population only in parts per million. Almost exactly half the A protons have $+\frac{1}{2}$ X neighbors and half have $-\frac{1}{2}$ X neighbors. The magnetic environments provided by these two types of X protons are not identical, so the A resonance is split into two peaks (Figures 2-17a and 2-18). By the same token, the X nucleus exists in two distinct magnetic

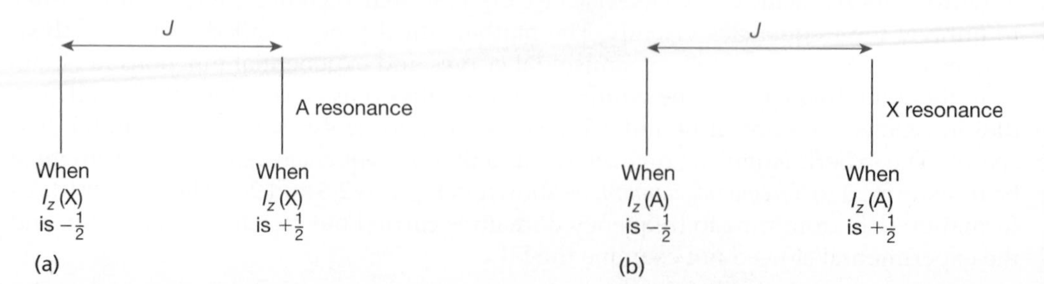

FIGURE 2-17 The four peaks of a first-order, two-spin system (AX).

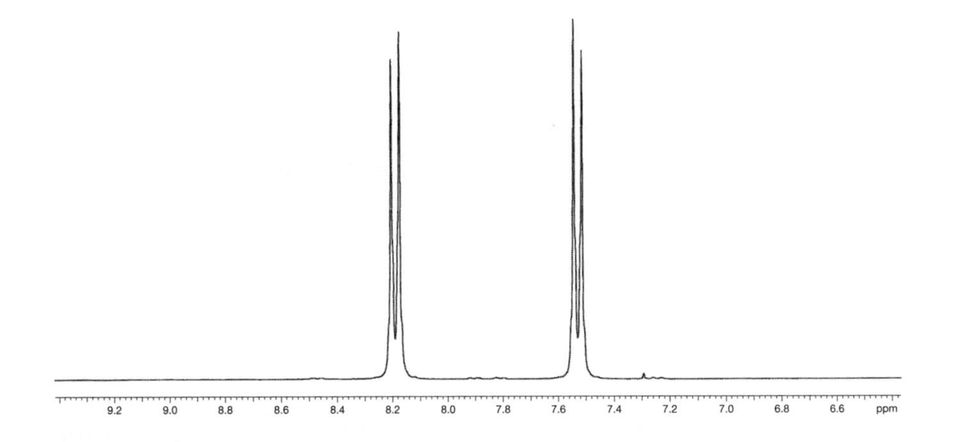

FIGURE 2-18 The 300 MHz ^{1}H spectrum of 1-chloro-4-nitrobenzene in $CDCl_3$.

environments, because the A proton has two spin states. The X resonance also is split into two peaks for the same reason (Figures 2-17b and 2-18). Quadrupolar nuclei, such as the nuclei of the chlorine and nitrogen atoms in molecule **2-1**, often act as if they are nonmagnetic and may be ignored in this context. This phenomenon is considered in greater detail in Section 5-1.

The influence of neighboring spins on the multiplicity of peaks is called *spin–spin splitting, indirect coupling*, or *J coupling*. The distance between the two peaks for the resonance of one nucleus split by another is a measure of how strongly the nuclear spins influence each other and is called the *coupling constant J*, measured in hertz. In 1-chloro-4-nitrobenzene (**2-1**), the coupling between A and X is 10.0 Hz, a relatively large value of J for two protons. In general, when there are only two nuclei in the coupled system, the resulting spectrum is referred to as AX. Notice that the splitting in both the A and the X portions of the spectrum is the same (Figure 2-18), since J is a measure of the interaction between the nuclei and must be identical for both nuclei. Moreover, J is independent of B_0, because the magnitude of the interaction depends only on nuclear properties and not on external quantities such as the field. Thus, in 1-chloro-4-nitrobenzene (**2-1**), J is 10.0 Hz when measured either at 7.05 T (300 MHz, Figure 2-18) or at 14.1 T (600 MHz).

For two nuclei to couple, there must be a process whereby information about spin is transferred between them. The most common mechanism involves the interaction of electrons along the bonding path between the nuclei. (See Figure 2-19 for an abbreviated coupling pathway over two bonds.) Electrons, like protons, act like spinning particles and have a magnetic moment. The X proton (H_X) influences, or polarizes, the spins of its surrounding electrons, making the electron spins favor one I_z state very slightly. Thus, a proton of spin $+\frac{1}{2}$ polarizes the electron to $-\frac{1}{2}$. The electron in turn polarizes the other electron of the C—H bond, and so on, finally reaching the resonating A proton (H_A). This mechanism is discussed further in Section 4-3. Because J normally represents an interaction through bonds, it is a useful parameter for drawing conclusions about molecular bonding, such as bond order and stereochemistry.

Additional splitting occurs when a resonating nucleus is close to more than one nucleus. For example, 1,1,2-trichloroethane (**2-2**) has two types of protons, which we

$$ClCH^A_2—CH^XCl_2$$

2-2

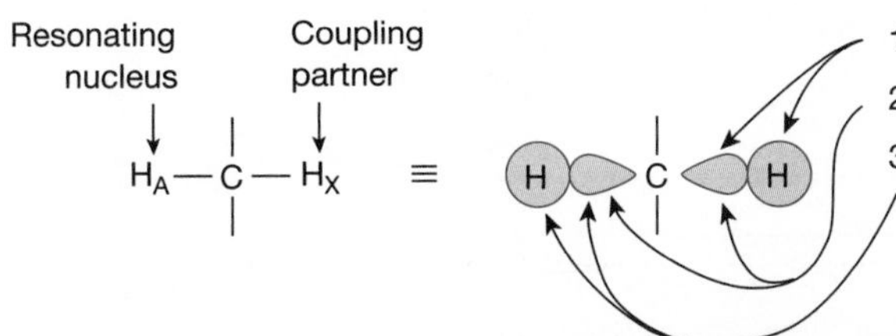

FIGURE 2-19 The mechanism of indirect spin–spin coupling.

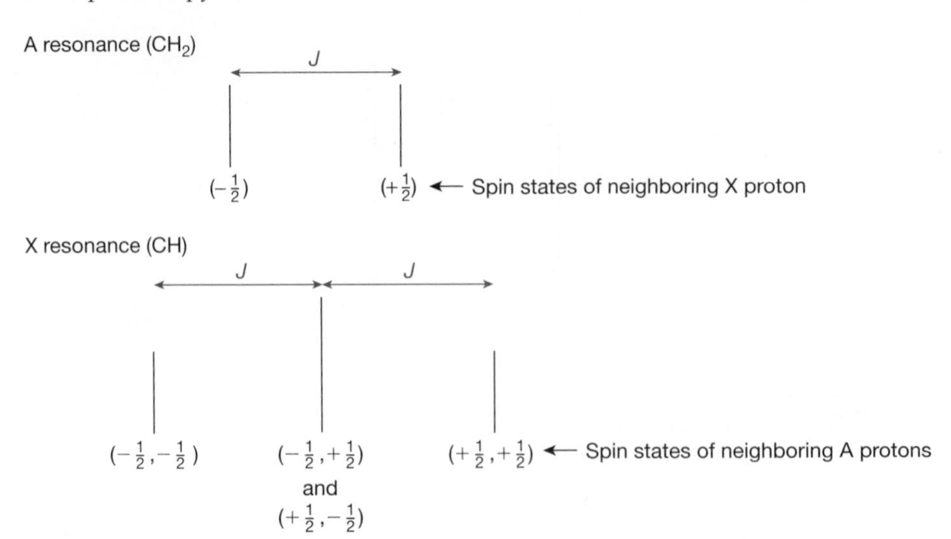

FIGURE 2-20 The five peaks of a first-order, three-spin system (A_2X).

label H_A (CH_2) and H_X (CH). The A protons are subject to two different environments, from the $+\frac{1}{2}$ and $-\frac{1}{2}$ spin states of H_X, and therefore are split into a 1:1 doublet, analogous to the situations shown in Figures 2-17 and 2-18. The X proton, however, is subject to *three* different magnetic environments, because the spins of H_A must be considered collectively: both may be $+\frac{1}{2}$ (+ +), both may be $-\frac{1}{2}$ (− −), and one may be $+\frac{1}{2}$ while the other is $-\frac{1}{2}$. The two equivalent such possibilities are (+ −) and (− +). The three different A environments—(+ +), (+ −)/(− +), (− −)—therefore result in three X peaks in the ratio 1:2:1 (Figure 2-20). Thus, the spectrum of **2-2** contains a doublet and a triplet and is referred to as A_2X (or AX_2 if the labels are switched) (Figure 2-21). The value of *J* is found in three different spacings in the spectrum, between any two adjacent peaks of a multiplet (Figures 2-20 and 2-21).

As the number of neighboring spins increases, so does the complexity of the spectrum. The identical ethyl groups in diethyl ether form an A_2X_3 spectrum (Figure 2-22). The methyl protons are split into a 1:2:1 triplet by the neighboring methylene protons, as in the X resonance of Figure 2-20. Because the methylene protons are split by three methyl protons, there are four peaks in the methylene resonance. The neighboring methyl protons can have all positive spins (+++), two spins positive and one negative (in three ways: ++−, +−+, and −++), one spin positive and two negative (also in three ways: +−−, −+−, −−+), or all negative spins (−−−). The result is a 1:3:3:1 quartet (Figure 2-23). The triplet–quartet pattern seen in Figure 2-22 is a reliable and general indicator for the presence of an ethyl group.

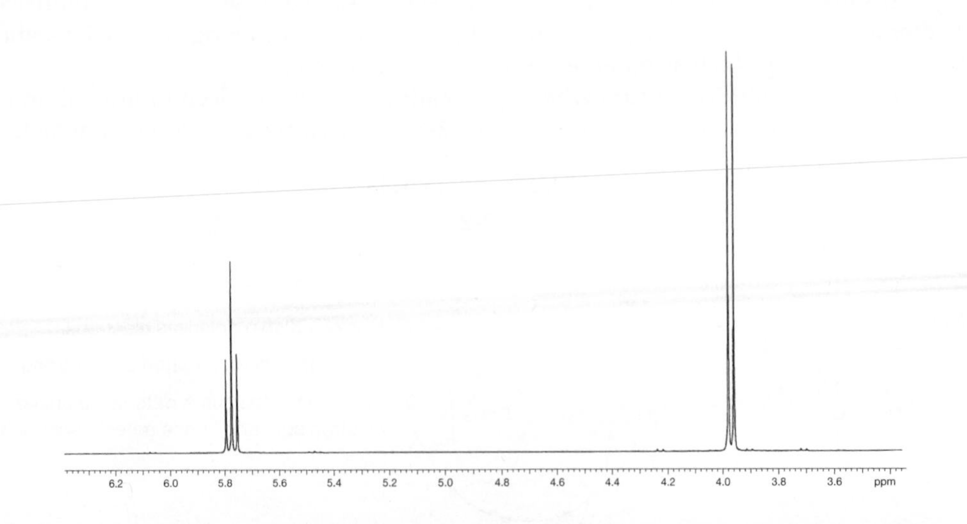

FIGURE 2-21 The 300 MHz ^{1}H spectrum of 1,1,2-trichloroethane in $CDCl_3$.

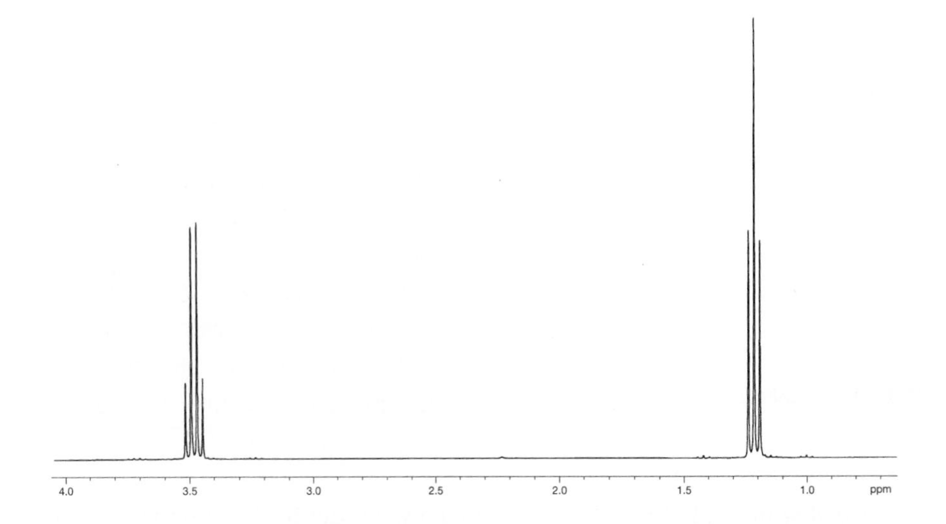

FIGURE 2-22 The 300 MHz ^{1}H spectrum of diethyl ether in $CDCl_3$.

The splitting patterns of larger spin systems may be deduced in a similar fashion. If a nucleus is coupled to n equivalent nuclei with $I = \frac{1}{2}$, there are $n + 1$ peaks, unless second-order effects discussed in Chapter 4 are present. The intensity ratios, to a first-order approximation, correspond to the coefficients in the binomial expansion and may be obtained from Pascal's triangle (Figure 2-24), since arrangements of the two I_z states are statistically independent events. Pascal's triangle is constructed by summing two horizontally adjacent integers and placing the result one row lower and between the two integers. Zeros are imagined outside the triangle. The first row (1) gives the resonance multiplicity (an unsplit singlet) when there is no neighboring spin, the second row (1:1) gives the multiplicity when there is one neighboring spin, and so on. We have already seen that two neighboring spins give a 1:2:1 triplet and three give a 1:3:3:1 quartet. Four neighboring spins are present for the CH proton in the arrangement $—CH_2—CHX—CH_2—$ (X is nonmagnetic), and the CH resonance is a 1:4:6:4:1 quintet (AX_4). The CH resonance from an isopropyl group, $—CH(CH_3)_2$, is a 1:6:15:20:15:6:1 septet (AX_6). Several common spin systems are given in Table 2-2.

Except in cases of second-order spectra (Sections 4-1 and 4-9), coupling between protons that have the same chemical shift does not lead to splitting in the spectrum. It is for this reason that the spectrum of benzene in Figure 2-1 is a singlet, even though each proton is coupled to its two ortho neighbors. For the same reason, protons within a methyl group normally do not cause splitting of the methyl resonance. Examples of unsplit spectra (singlets) include those of acetone, cyclopropane, and dichloromethane.

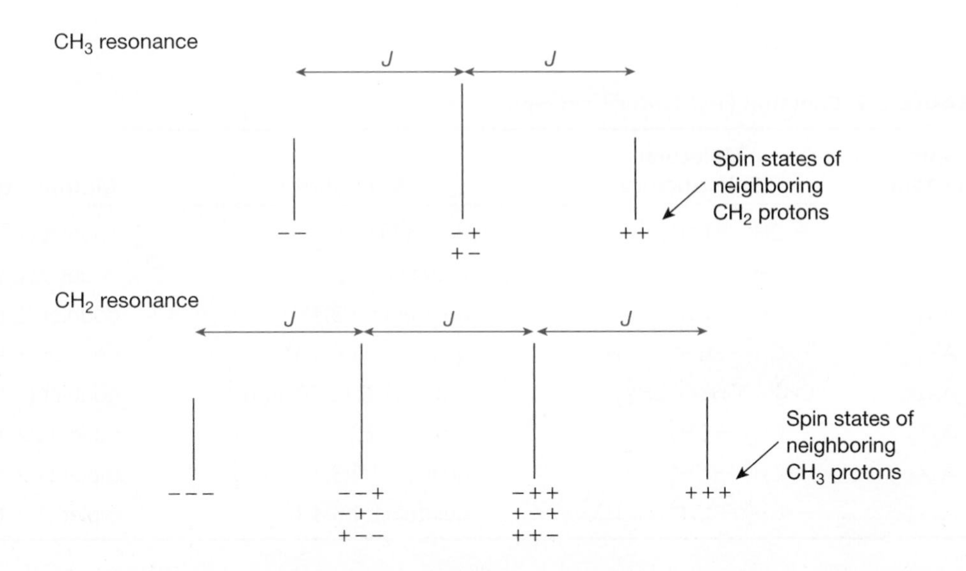

FIGURE 2-23 The seven peaks of the first-order A_2X_3 spectrum.

```
                          1
                        1   1
                      1   2   1
                    1   3   3   1
                  1   4   6   4   1
                1   5  10  10   5   1
              1   6  15  20  15   6   1
            1   7  21  35  35  21   7   1
          1   8  28  56  70  56  28   8   1
        1   9  36  84 126 126  84  36   9   1
      1  10  45 120 210 252 210 120  45  10   1
```

FIGURE 2-24 Pascal's triangle.

The absence of splitting between coupled nuclei with identical resonance frequencies is quantum mechanical in nature and requires a lengthy theoretical explanation.

Almost all the coupling examples given so far have been between vicinal protons, over three bonds (H—C—C—H). Coupling over four or more bonds usually is small or unobservable. It is possible for geminal protons ($—CH_2—$) to split each other, provided that each proton of the methylene group has a different chemical shift. Such geminal splittings are observed when the carbon is part of a ring with unsymmetrical substitution on the upper and lower faces, when there is a single chiral center in the molecule, or when an alkene lacks an axis of symmetry ($XYC{=}CH_2$). These couplings are discussed further in Chapter 4.

Coupling can occur between ^1H and ^{13}C, as well as between two protons. Because ^{13}C is in such low natural abundance (about 1.1%), these couplings are not important in analyzing ^1H spectra. In 99 of 100 cases, protons are attached to nonmagnetic ^{12}C atoms. Small satellite peaks from the 1.1% of ^{13}C sometimes can be seen in ^1H spectra. In the ^{13}C spectrum, carbon nuclei are coupled to nearby protons. The largest couplings occur with protons that are directly attached to the carbon. Thus, the ^{13}C resonance of a methyl carbon is split into a quartet, that of a methylene carbon (CH_2) into a triplet, and that of a methine carbon (CH) into a doublet. A quaternary carbon is not split by one bond coupling. Figure 2-25 (top) shows the ^{13}C spectrum of 3-hydroxybutyric acid [$CH_3CH(OH)CH_2CO_2H$], which contains a carbon resonance with each type of multiplicity. From right to left are seen a quartet (CH_3), a triplet (CH_2), a doublet (CH), and a singlet (CO_2H). Hence, the splitting pattern in the ^{13}C spectrum is an excellent indicator of each of these types of groupings within a molecule.

Instrumental procedures, called *decoupling*, are available by which spin–spin splittings may be removed. These methods, discussed in Section 5-3, involve irradiating one

TABLE 2-2 Common First-Order Spin–Spin Splitting Patterns

Spin System	Molecular Substructure	A Multiplicity	X Multiplicity
AX	$—CH^A—CH^X—$	doublet (1:1)	doublet (1:1)
AX_2	$—CH^A—CH_2^X—$	triplet (1:2:1)	doublet (1:1)
AX_3	$—CH^A—CH_3^X$	quartet (1:3:3:1)	doublet (1:1)
AX_4	$—CH_2^X—CH^A—CH_2^X—$	quintet (1:4:6:4:1)	doublet (1:1)
AX_6	$CH_3^X—CH^A—CH_3^X$	septet (1:6:15:20:15:6:1)	doublet (1:1)
A_2X_2	$—CH_2^A—CH_2^X—$	triplet (1:2:1)	triplet (1:2:1)
A_2X_3	$—CH_2^A—CH_3^X$	quartet (1:3:3:1)	triplet (1:2:1)
A_2X_4	$—CH_2^X—CH_2^A—CH_2^X—$	quintet (1:4:6:4:1)	triplet (1:2:1)

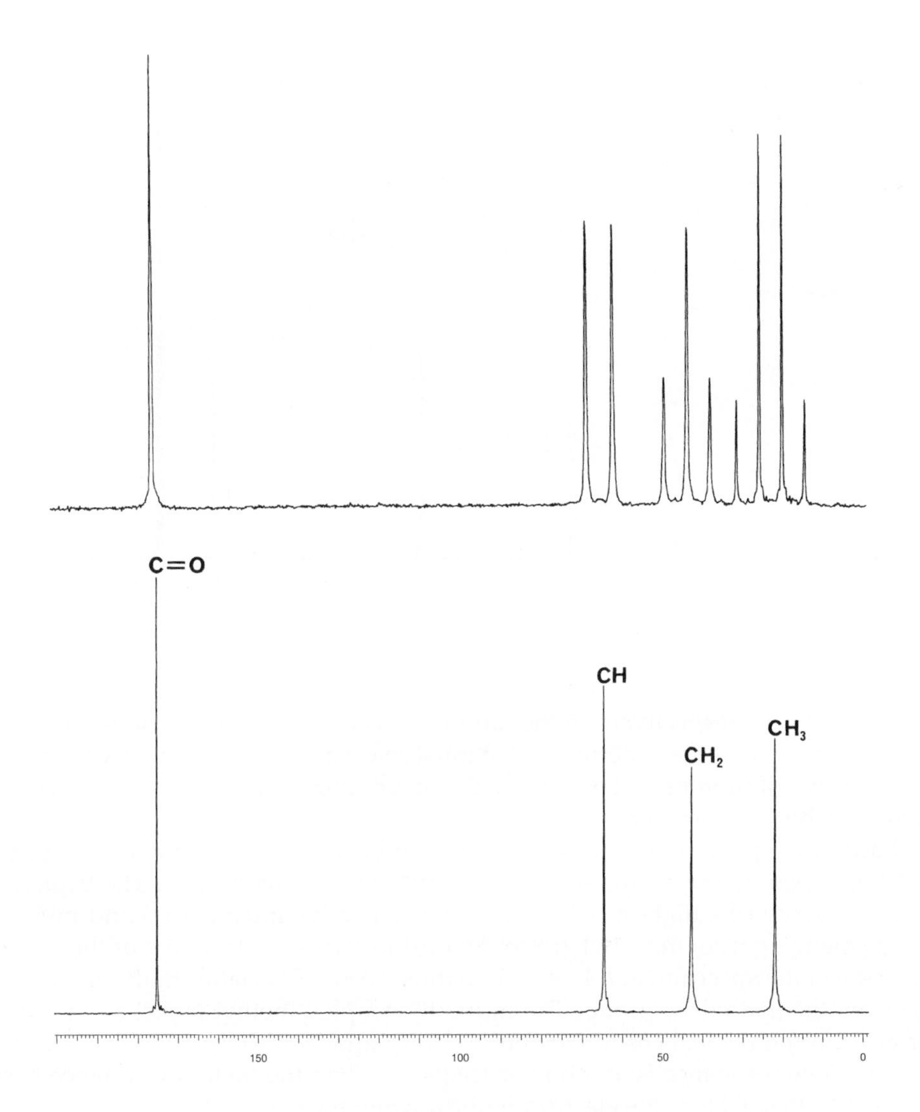

FIGURE 2-25 Top: The 22.6 MHz ^{13}C spectrum of 3-hydroxybutyric acid, $CH_3CH(OH)CH_2CO_2H$, without solvent. Bottom: The ^{13}C spectrum of the same compound with proton decoupling.

nucleus with an additional field (B_2) while observing another nucleus resonating in the B_1 field. Because a second field is being applied to the sample, the experiment is called *double resonance*. This procedure was used to obtain the ^{13}C spectrum of methyl acetate in Figure 2-7 (bottom) and the spectrum of 3-hydroxybutyric acid at the bottom of Figure 2-25. It is commonly employed to obtain a very simple ^{13}C spectrum, in which each carbon gives a singlet. Measurement of both decoupled and coupled ^{13}C spectra then produces in the first case a simple picture of the number and types of carbons and in the second case the number of protons to which they are attached (Figure 2-25). Coupling is treated in more detail in Chapter 4.

2-7 QUANTIFICATION AND COMPLEX SPLITTING

The signal detected when nuclei resonate is directly proportional to the number of spins present. Thus, the protons of a methyl group (CH_3) produce three times the signal of a methine proton (CH). This difference in intensity can be measured through electronic integration and exploited to elucidate molecular structure. Figure 2-26 illustrates electronic integration for the ^1H spectrum of ethyl *trans*-crotonate [$CH_3CH{=}CH(CO)OCH_2CH_3$)]. The vertical displacement of the continuous line above or through each resonance provides a measure of the area under the peaks. The vertical displacements show that the doublet at δ 5.84, the quartet at δ 4.19, and the

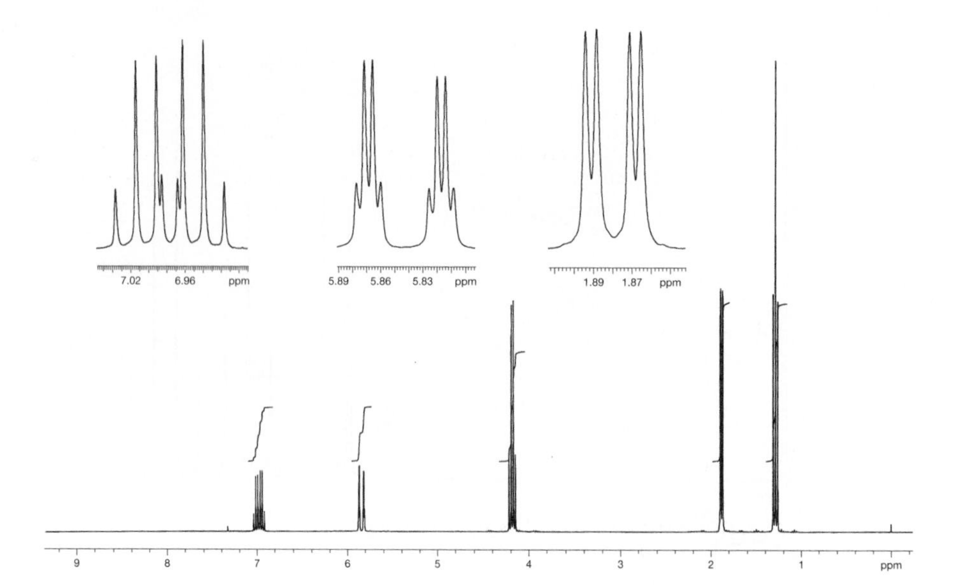
7.02
6.96
ppm
5.89
5.86
5.83
ppm
1.89
1.87
ppm
9
8
7
6
5
4
3
2
1
ppm

FIGURE 2-26 The 300 MHz ^{1}H spectrum of ethyl *trans*-crotonate in $CDCl_3$. The upper multiplets are expansions of three resonances.

triplet at δ 1.28 are, respectively, in the ratio 1:2:3. The integration provides only relative intensity data, so that the experimentalist must select a resonance with a known or suspected number of protons and normalize the other integrals to it. Integrals are provided digitally by the spectrometer.

Each of the peaks in the ^{1}H spectrum of ethyl crotonate illustrated in Figure 2-26 may be assigned by examining the integral and the splitting pattern. The triplet at the lowest frequency (the highest field, δ 1.28) has a relative integral of 3 and must come from the methyl part of the ethyl group. Its J value corresponds to that of the quartet in the middle of the spectrum at δ 4.19, whose integral is 2. This latter multiplet then must come from the methylene group. The mutually coupled methyl triplet and methylene quartet form the resonances for the ethyl group attached to oxygen ($—OCH_2CH_3$). The methylene resonance is at a higher frequency than the methyl resonance because CH_2 is closer than CH_3 to the electron-withdrawing oxygen.

The remaining resonances in the spectrum come from protons coupled to more than one type of proton. Coupling patterns, then, are more complex, as seen in the three spectral expansions in Figure 2-26. The highest-frequency (lowest-field) resonance (δ 6.98) has an intensity of unity and comes from one of the two alkenic ($—CH=$) protons. This resonance is split into a doublet ($J = 16$ Hz) by the other alkenic proton, and then each member of the doublet is further split into a quartet ($J = 7$ Hz) by coupling to the methyl group on carbon with a crossover of the inner two peaks. Stick diagrams (often called a tree) are useful in analyzing complex multiplets, as in Figure 2-27 for the resonance at δ 6.98.

The resonance of unit integral at δ 5.84 is from the other alkenic proton and is split into a doublet ($J = 16$ Hz) by the proton at δ 6.98. There is a small coupling

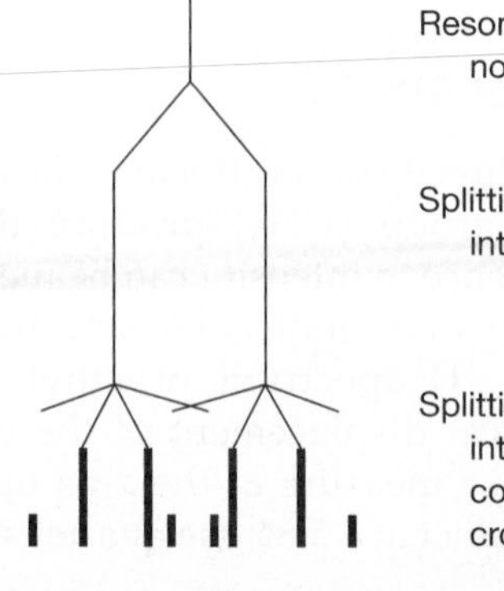
Resonance at δ 6.98 if there were no coupling
Splitting by the other CH═ (δ 5.84) into a 1/1 doublet (J = 16 Hz)
Splitting by each member of the doublet into a 1/3/3/1 quartet (J = 7 Hz) by coupling to CH3 (δ 1.88) (note the crossover of the two middle peaks)

FIGURE 2-27 Overlapping peaks for the resonance at δ 6.98, which arise when nuclei are unequally coupled to more than one other set of spins.

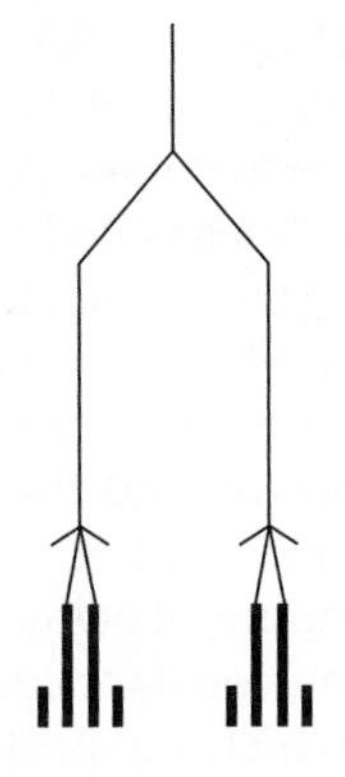
Resonance at δ 5.84 if there were no coupling
Splitting by the other CH═ (δ 6.98) into a 1/1 doublet (J = 16 Hz) (note same splitting as for δ 6.98 resonance)
Splitting by each member of the doublet (J = 1 Hz) into a 1/3/3/1 quartet by coupling to CH3 (δ 1.88)

FIGURE 2-28 Overlapping peaks for the δ 5.84 resonance (not to scale).

(1 Hz) over four bonds to the methyl group, giving rise to a quartet (Figure 2-28). The significance of these differences in the magnitude of couplings is discussed in Chapter 4. The resonance at δ 6.98 can be recognized as originating from the proton closer to the methyl. Thus J is larger because it is over three, rather than four, bonds.

The resonance at δ 1.88 has an integral of 3 and hence comes from the remaining methyl group, attached to the double bond. Because it is split by both of the alkenic protons, but with unequal couplings (7 and 1 Hz), four peaks result (Figure 2-29). This grouping is called a doublet of doublets; the term *quartet* normally is reserved for 1:3:3:1 multiplets. The two unequal couplings in the resonance at δ 1.88 correspond precisely to the quartet splittings found, respectively, in the two alkenic resonances. The final assignments are as follows:

$$\begin{array}{l} \delta \quad 1.9 \quad 7.0 \quad 5.8 \qquad 4.2 \quad 1.3 \\ \mathrm{CH_3CH{=}CHCO_2CH_2CH_3} \end{array}$$

Integration also may be used as a measure of the relative amounts of the components of a mixture. In this case, after normalizing for the number of protons in a grouping, the proportions of the components may be calculated from the relative integrals of protons in different molecules. An internal standard with a known concentration may be included. Comparisons of other resonances with those of the standard then can provide a measure of the absolute concentration.

2-8 DYNAMIC EFFECTS

According to the principles outlined in the previous sections, the ^{1}H of methanol (CH_3OH) should contain a doublet of integral 3 for the CH_3 group (coupled to OH) and a quartet of integral 1 for the OH group (coupled to CH_3). Under conditions of high

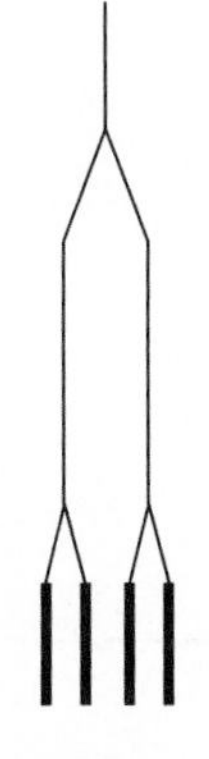
Resonance at δ 1.88 if there were no coupling
Splitting by CH═ with J = 7 Hz (δ 6.98)
Splitting by CH═ with J = 1 Hz (δ 5.84)

FIGURE 2-29 Overlapping peaks for the δ 1.88 resonance (not to scale).

purity or low temperature, such a spectrum is observed (Figure 2-30, bottom). The presence of a small amount of acidic or basic impurity, however, can catalyze the intermolecular exchange of the hydroxyl proton. When this proton becomes detached from the molecule by any mechanism, information about its spin states is no longer available to the rest of the molecule. For coupling to be observed, the rate of exchange must be considerably slower than the magnitude of the coupling, in hertz. Thus, a proton could exchange a few times per second and still retain coupling. If the rate of exchange is faster than J, no coupling is observed between the hydroxyl proton and the methyl protons. Hence, at high temperatures (Figure 2-30, top), the ^{1}H spectrum of methanol contains only two singlets. If the temperature is lowered or the amount of acidic or basic catalyst is decreased, the exchange rate slows down. The coupling constant continues to be absent until the exchange rate reaches a critical value at which the proton resides sufficiently long on oxygen to permit the methyl group to detect the spin states. As can be seen from the figure, the transition from *fast exchange* (upper) to *slow exchange* (lower) can be accomplished for methanol over a temperature range of 80°C. Under most spectral conditions, minor amounts of acid or base impurities are present, so hydroxyl protons do not usually exhibit couplings to other nuclei. The integral is still unity for the OH group, because the amount of catalyst is small. Sometimes the exchange rate is intermediate, between fast and slow exchange, and broadened peaks are observed. Amino protons (NH or NH_2) exhibit similar behavior.

A process that averages coupling constants also can average chemical shifts. A mixture of acetic acid and benzoic acid can contain only one ^{1}H resonance for the CO_2H groups from both molecules. The carboxyl protons exchange between molecules so rapidly that the spectrum exhibits only the average of the two. Moreover, if the solvent is water, exchangeable protons such as OH in carboxylic acids or alcohols do not give

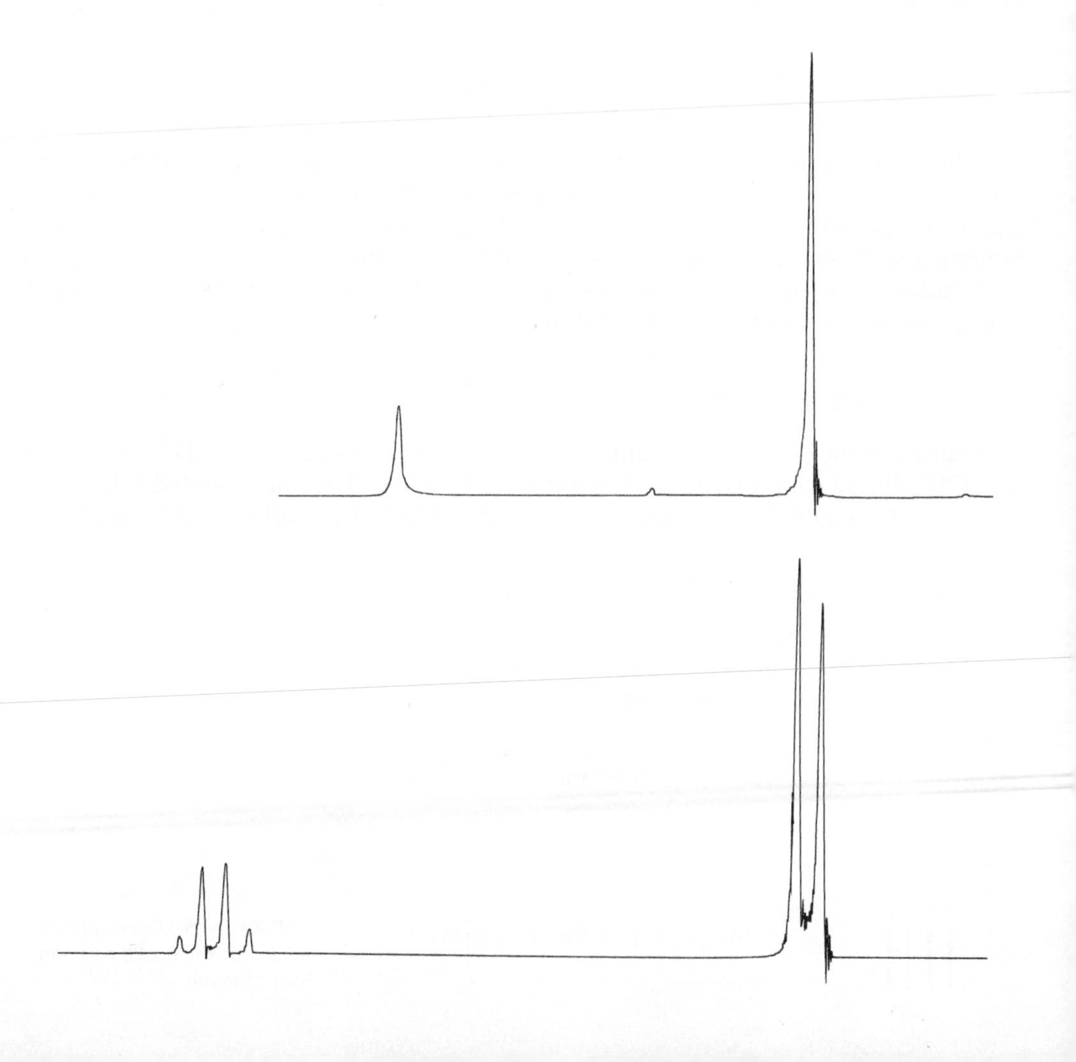

FIGURE 2-30 The 60 MHz ^{1}H spectrum of CH_3OH at +50°C (top) and at −30°C (bottom).

separate resonances. Thus, the ^{1}H spectrum of acetic acid (CH_3CO_2H) in water contains two, not three, peaks: the water and carboxyl protons appear as a single resonance whose chemical shift falls at the weighted average of those of the pure materials. If the rate of exchange between $—CO_2H$ and water could be slowed sufficiently, separate resonances would be observed.

Intramolecular (unimolecular) reactions also can influence the appearance of the NMR spectrum if the rate is comparable to that of chemical shifts. The molecule cyclohexane, for example, contains distinct axial and equatorial protons, yet the spectrum exhibits only one sharp singlet at room temperature. There is no splitting, because all protons have the same average chemical shift. Flipping of the ring interconverts the axial and equatorial positions. When the rate of this process is greater (in s^{-1}) than the chemical-shift difference between the axial and equatorial protons (in hertz, which, of course, is s^{-1}), the NMR experiment does not distinguish the two types of protons, and only one peak is observed. Again, this situation is called *fast exchange*. At lower temperatures, however, the process of ring flipping is much slower. At –100°C, the NMR experiment can distinguish the two types of protons, so two resonances are observed (*slow exchange*). At intermediate temperatures, broadened peaks are observed that reflect the transition from fast to slow exchange. Figure 2-31 illustrates the spectral changes as a function of temperature for cyclohexane, in which all protons but one have been replaced by deuterium to remove vicinal proton–proton couplings and simplify the spectrum (**2-3**).

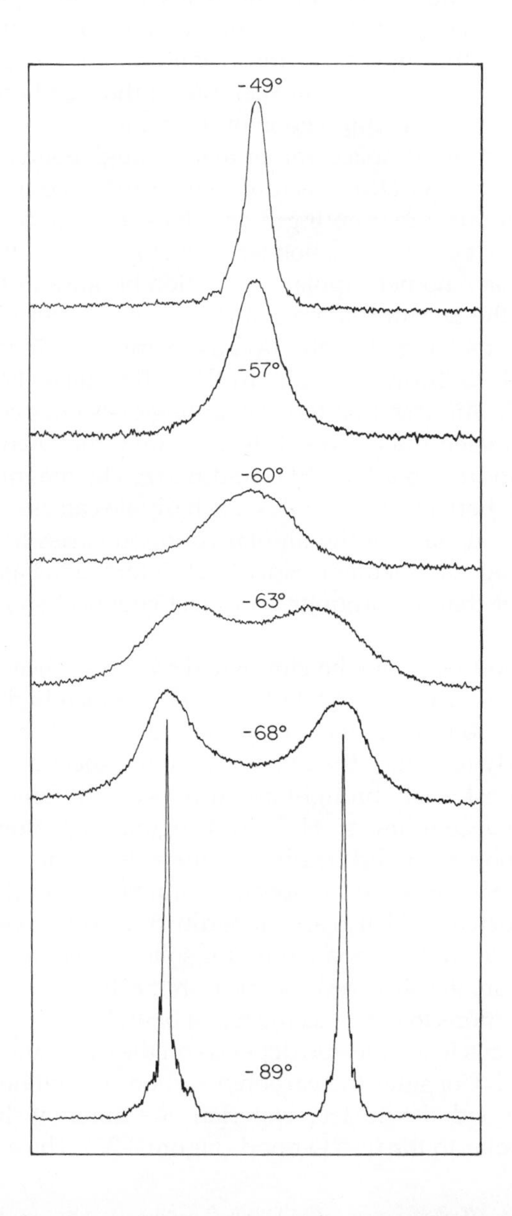

FIGURE 2-31 The 60 MHz ^{1}H spectrum of cyclohexane-d_{11} as a function of temperature. (Reproduced with permission from F. A. Bovey, F. P. Hood III, E. W. Anderson, and R. L. Kornegay, *J. Chem. Phys.*, **41**, 2042 [1964].)

2-3

These processes that bring about averaging of spectral features occur reversibly, whether by acid-catalyzed intermolecular exchange or by unimolecular reorganization. NMR is one of the few methods for examining the effects of reaction rates when a system is at equilibrium. Most other kinetic methods require that one substance be transformed irreversibly into another. The dynamic effects of the averaging of chemical shifts or coupling constants provide a nearly unique window into processes that occur on the order of a few times per second. This subject is examined further in Section 5-2.

2-9 SPECTRA OF SOLIDS

All of the examples and spectra illustrated thus far have been for liquid samples. Certainly, it would prove useful to be able to take NMR spectra of solids, so why have we avoided discussing solid samples? Under conditions normally used for liquids, the spectra of solids are broad and unresolved, providing only modest amounts of information. There are two primary reasons for the poor resolution of solid samples. In addition to the indirect spin–spin (J) interaction that occurs between nuclei through bonds, nuclear magnets also can couple through a direct interaction of their nuclear dipoles. This *dipole–dipole*, or *direct*, *coupling* occurs through space, rather than through bonds. The resulting coupling is designated with the letter D, and is much larger than J coupling.

In solution, dipoles are continuously reorienting themselves through molecular tumbling. Just as two bar magnets have no net interaction when averaged over all mutual orientations, two nuclear magnets have no net dipolar interaction because of the randomizing effect of tumbling. Thus, the D coupling normally averages to zero in solution. The indirect J coupling does not average to zero, because tumbling cannot average out an interaction that takes place through bonds. In the solid phase, however, nuclear dipoles are held rigidly in position, so that the D coupling does not average to zero. The dominant interaction between nuclei in solids is, in fact, the D coupling, which is on the order of several hundred to a few thousand hertz. The magnitude of the interaction depends on the angle between the dipoles. Each dipole can assume any relative angle in the solid, so the actual value of the dipolar coupling varies from zero to the maximum value in the optimal arrangement. Since such interactions assume a range of values and are much larger than J couplings and most chemical shifts, very broad signals are produced.

As with the J coupling, the D coupling may be eliminated by the application of a strong B_2 field. Power levels required for the removal of D must be much higher than those for J decoupling, since D is two to three orders of magnitude larger than J. High-powered decoupling is used routinely to reduce the line width of the spectra of solids. In practice, dipolar decoupling is most easily brought about between different spin-$\frac{1}{2}$ nuclei, as when ^{13}C is observed with decoupling of ^{1}H. The analogous all-proton experiment is more difficult, since both observed and irradiated nuclei have the same frequency range. Thus, the acquisition of solid-state ^{13}C spectra is much simpler than that of ^{1}H spectra. Quadrupolar nuclei such as ^{27}Al also are more difficult to observe.

The second factor that contributes to line broadening for solids is *chemical shift* or *chemical shielding anisotropy*. In solution, the observed chemical shift is the average of the shielding of a nucleus over all orientations in space, as the result of molecular tumbling. In a solid, shielding of a specific nucleus in a molecule depends on the orientation of the molecule with respect to the B_0 field. Consider the carbonyl carbon of acetone. When the B_0 field is parallel to the C=O bond, the nucleus experiences a different shielding from when the B_0 field is perpendicular to the C=O bond (Figure 2-32). The ability of

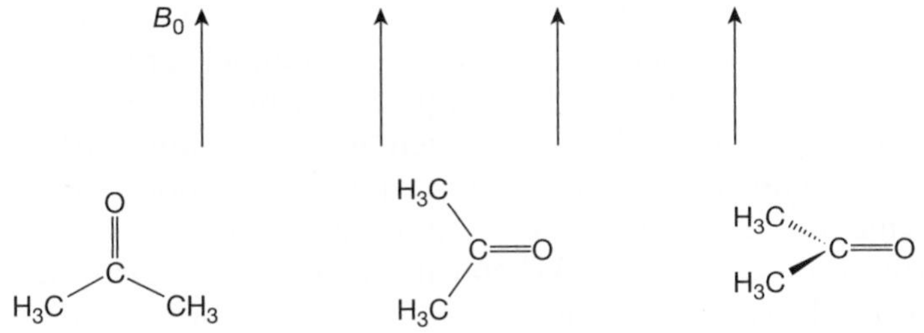

FIGURE 2-32 Anisotropy of shielding in the solid state.

electrons to circulate and give rise to shielding varies according to the arrangement of bonds in space. Differences between the abilities of electrons to circulate in the three arrangements shown in the figure, as well as in all intermediate arrangements, generate a range of shieldings and hence a range of resonance frequencies.

Double irradiation does not average shielding anisotropy, since the effect is entirely geometrical. The problem is removed largely by spinning the sample to mimic the process of tumbling. The effects of spinning are optimized when the axis of spin is set at an angle of 54°44′ to the direction of the B_0 field. This is the angle between the edge of a cube and the adjacent solid diagonal. Spinning of a cube along this diagonal averages each Cartesian direction by interconverting the x, y, and z axes, just as tumbling in solution does. When the sample is spun at that angle to the field, the various arrangements of Figure 2-32 average, and the shieldings are reduced to the isotropic chemical shift. The technique therefore has been called *Magic-Angle Spinning* (MAS). Because shielding anisotropies are generally a few hundred to several thousand hertz, the rate of spinning must exceed this range in order to average all orientations. Typical minimum spinning rates are 2–5 kHz, but rates up to 50 kHz are possible.

The combination of strong irradiation to eliminate dipolar couplings and MAS to eliminate shielding anisotropy results in ^{13}C spectra of solids that are almost as high in resolution as those of liquids. Spectra of protons or of quadrupolar nuclei in a solid can be obtained, but require more complex experiments. Figure 2-33 shows the ^{13}C spectrum of polycrystalline β-quinolmethanol clathrate. The broad, almost featureless spectrum at the top (Figure 2-33a) is typical of solids. Strong double irradiation (Figure 2-33b) eliminates dipolar couplings and brings out some features. MAS in addition to decoupling (Figure 2-33c) produces a high-resolution spectrum.

Relaxation times are extremely long for solids because the motion of nuclei necessary for spin–lattice relaxation is slow or absent. Carbon-13 spectra of solids could take a very long time to record because the nuclei must be allowed to relax for several minutes between pulses. The problem is solved by taking advantage of the more favorable properties of the protons that are coupled to the carbons. The same double-irradiation process that eliminates J and D couplings is used to transfer some of the proton's higher magnetization and faster relaxation to the carbon atoms. The process is called *Cross Polarization* (CP) and is standard for most solid state spectra of ^{13}C. After the protons are moved onto the y axis by a 90° pulse, a continuous y field is applied to keep the magnetization

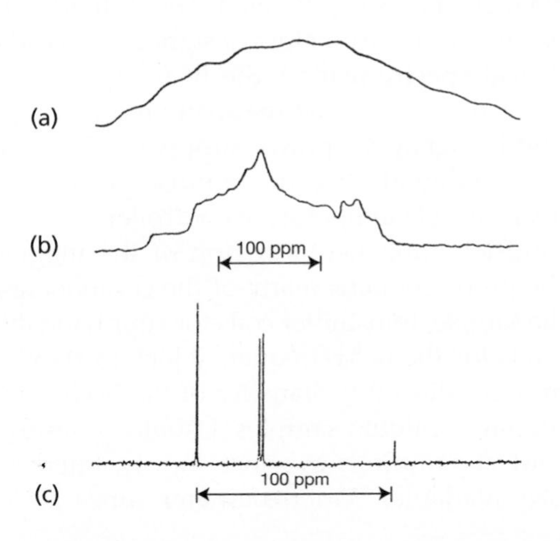

FIGURE 2-33 The ^{13}C spectrum of polycrystalline β-quinolmethanol clathrate (a) without dipolar decoupling, (b) with decoupling, and (c) with both decoupling and magic-angle spinning. (Reproduced with permission from T. Terao, *JEOL News*, **19**, 12 [1983].)

processing about that axis, a process called *spin locking*. The frequency of this field ($\gamma_H B_H$) is controlled by the operator. When the ^{13}C channel is turned on, its frequency ($\gamma_C B_C$) can be set to equal the ^1H frequency (the *Hartmann–Hahn condition*, $\gamma_H B_H = \gamma_C B_C$). Both protons and carbons then are precessing at the same frequency and hence have the same net magnetization, which, for carbon, is increased over that used in the normal pulse experiment. Carbon resonances thus have enhanced intensity and faster (proton-like) relaxation. When carbon achieves maximum intensity, B_C is turned off (ending the *contact time*) and carbon magnetization is acquired, while B_H is retained for dipolar decoupling.

The higher resolution and sensitivity of the experiment with CP/MAS opened vast new areas to NMR. Inorganic and organic materials that do not dissolve could be subjected to NMR analysis. Synthetic polymers and coal were two of the first materials to be examined. Biological and geological materials, such as wood, humic acids, and biomembranes, became general subjects for NMR study. Problems unique to the solid state—for example, structural and conformational differences between solids and liquids—also could be examined.

2-10 EXPERIMENTAL METHODS

2-10a The Spectrometer and the Sample

Although a wide variety of NMR instrumentation is available, certain components are common to all, including a magnet to supply the B_0 field, devices to generate the B_1 pulse and to detect the resulting NMR signal, a probe for positioning the sample in the magnet, hardware for stabilizing the B_0 field and optimizing the signal, and computers for controlling much of the operation and processing the signals.

Early NMR machines relied on electromagnets. Although a generation of chemists used them, the magnets had low sensitivity and poor stability. Currently most research-grade instruments use a superconducting magnet and operate at fields of 3.5–21.1 T (150–900 MHz for protons). These magnets provide high sensitivity and stability. In addition, the very high fields produced by superconducting magnets result in better separation of resonances, because chemical shifts and hence chemical shift differences increase with field strength. The superconducting magnet resembles a solid cylinder with a central axial bore hole. The direction of the B_0 field (z) is aligned with the axis of the cylinder, which is composed of a double Dewar jacket. The outer jacket is filled with liquid nitrogen and the inner with liquid helium, so that the magnet coils may be kept at approximately 4.2 K. The central bore, with a diameter of 53–89 mm, is maintained at room temperature (or controlled during variable temperature experiments). The sample tube for liquids is aligned along the z axis of the superconducting cylinder, whereas solid samples are rotated rapidly at the magic angle. In *cryoprobes*, the receiver coils and preamplifiers are kept at temperatures as low as 27 K to minimize noise and increase spectral sensitivity.

Separate from the magnet is a console that contains, among other components, the transmitter of the various applied magnetic fields (B_1, B_2, etc.). The B_1 field of the transmitter in the pulse experiment is 1–40 mT. In spectrometers designed to record the resonances of several nuclides (multinuclear spectrometers), the B_1 field must be tunable over a range of frequencies. The console also contains a broadband receiver, a preamplifier to enhance the inherently weak NMR signal, the power supply for the shim coils to adjust the B_0 magnetic field, an analog-to-digital converter, a pulsed field generator to produce shaped pulses, and possibly a variable temperature controller.

The sample is placed in the most homogeneous region of the magnetic field by means of an adjustable probe. The probe contains many of the components already alluded to, including the holder for the sample, transmitter coils for supplying the applied magnetic fields, receiver coils, and coils for the field/frequency lock to stabilize the B_0 field. The size of the probe determines the allowable diameter of the NMR tube, which commonly is 5 mm but can exceed 10 mm for liquid samples. Usually a sample volume of 500–650 μl is used for liquids, but, if sample is in short supply, microtubes that require a much smaller volume are available. Wider-diameter tubes are used for low-sensitivity nuclei and for samples in low concentration due to poor solubility or availability. In practice, ^{1}H spectra may be obtained on less than 1 μg of dissolved material, although the result depends on the molecular weight, the B_0 magnetic field, and other factors. For ^{13}C spectra, milligram samples are feasible. The current limits for ^{1}H and ^{13}C spectra have been extended, respectively, down to nanogram and microgram ranges.

For liquids, the sample must have good solubility in a solvent, which must have no resonances in the regions of interest. For protons, $CDCl_3$, D_2O, DMSO-d_5, and acetone-d_6 are traditional NMR solvents. If the spectrum is to be recorded above or below room temperature, the solvent chosen must not boil or freeze during the experiment. Only NMR tubes of high quality should be used. The tube should not be either over- or underfilled. Overfilling results in reduced sensitivity, whereas underfilling adversely affects the field homogeneity over the sample. The sample should be free of solid particles, which may be removed by filtration through a small plug of glass wool or a tissue inserted into the tip of a pipette. Sample degassing is required for the highest resolution and for measurements of spin–lattice relaxation times.

2-10b Optimizing the Signal

The NMR experiment is plagued by the dual problems of sensitivity and resolution. Peak separations of <0.5 Hz may need to be resolved, so the B_0 field must be uniform to a very high degree (for a separation of 0.5 Hz at 500 MHz, field homogeneity must be better than 1 part in 10^9). Corrections to field inhomogeneity may be made for small gradients in B_0 by the use of *shim coils*. For example, the field along the z direction might be slightly higher at one point than at another. Such a gradient may be compensated for by applying a small current through a coil built into the probe. Shim coils are available for correcting gradients in all three Cartesian coordinates as well as higher-order gradients (z^2, x^3, and so on) and combination gradients [xz, $(x^2 - y^2)$, and so on].

Signal optimization requires proper placement of the sample tube in the probe. A sample that is too far or not far enough does not expose all the material to the receiver coils. Inserting a sample too far also can damage components located just below the optimal tube location. In some instruments, the sample is spun along the axis of its cylinder at a rate of 20–40 Hz by an air flow to improve homogeneity within the tube. Spinning improves resolution because a nucleus at a particular location in the tube experiences a field that is averaged over a circular path. In the superconducting magnet, the axis of the tube is in the z direction (in electromagnets it is spun along the y direction). Spinning does not average gradients along the axis of the cylinder, so shimming is required primarily for z gradients for a superconducting magnet or y gradients for an electromagnet. Today, many high-quality magnets no longer require sample spinning.

All magnets are subject to field drift, which can be minimized by electronically locking the field to the resonance of a substance contained in the sample. Because deuterated solvents are used quite commonly for this purpose, an *internal lock* normally is at the deuterium frequency for both ^{1}H and ^{13}C spectra. In some instruments, the field is locked to a sample contained in a separate tube permanently located elsewhere in the probe. This *external lock* is usually found only in spectrometers designed for a highly specific use, such as taking spectra only of solids. The deuterium lock signal in essence is a parallel NMR system with both transmitter and receiver at the ^{2}H frequency. The lock is established by varying the frequency of the lock transmitter until it matches that of the deuterated solvent. The stability of the lock signal is sensitive to the presence of suspended particles in the sample, which may be removed by filtration.

Since there is an excess of only some 50 spin-$\frac{1}{2}$ nuclei per million, the NMR experiment is inherently insensitive. The sensitivity of a given experiment depends on the natural abundance and the natural sensitivity of the observed nucleus (related to the magnetic moment and gyromagnetic ratio). The experimentalist has no control over these factors. The sample size may be increased by using a larger-diameter container, and the field strength may be increased (sensitivity increases with the $\frac{3}{2}$ power of B_0). Collapse of peaks through decoupling effectively enhances sensitivity because, for example, several lines in a multiplet become concentrated at a single frequency, as shown in Figure 2-25 (the baseline clearly has less noise in the lower, decoupled spectrum).

Decoupling also can bring about sensitivity enhancement through the nuclear Overhauser effect (see Section 5-4). Complex manipulation of pulses can raise sensitivity, as is discussed in Section 5-6. Use of a cryoprobe provides significant enhancement of sensitivity through the advantages of maintaining electronics at very low temperatures.

Routine improvement of signal to noise is achieved through multiple scanning or signal averaging. In this procedure, the digitized NMR spectrum is stored in computer memory. The spectrum is recorded multiple times and is stored in the same locations. Any signal present is reinforced, but noise tends to cancel out. If n such scans are carried out and added digitally, the theory of random processes states that the signal amplitude is proportional to n but the noise is proportional to $\sqrt{n}$. The signal to noise ratio (S/N) therefore increases by $n/\sqrt{n}$ or $\sqrt{n}$. Thus 64 scans added together theoretically enhance S/N by a factor of about 8. Multiple scanning is routine for most nuclei and obligatory for many, including ^{13}C and ^{15}N. For ^{1}H spectra, the number of scans normally is a multiple of 4, since this is the length of the CYCLOPS phase cycle used to minimize imperfections from quadrature signal detection (Section 5-8). Usually 4 to 128 scans suffice to obtain a good ^{1}H signal. Block sizes of 32 or 64 are employed commonly for ^{13}C and other nuclei with low receptivity.

2-10c Spectral Parameters

In the pulsed experiment, spectral resolution is controlled directly by the amount of time taken to acquire the signal. To distinguish two signals separated by $\Delta\nu$ (in hertz), acquisition of data must continue for at least $1/\Delta\nu$ seconds. For example, a desired resolution of 0.5 Hz in a ^{13}C spectrum requires an *acquisition time* (t_a) of 1/0.5 or 2.0 s. Sampling for a longer time would improve resolution; for example, acquisition for 4.0 s could yield a resolution of 0.25 Hz. Thus longer acquisition times are necessary to produce narrow lines. The tail of the FID contains mostly noise mixed with signals from narrow lines. If sensitivity is the primary concern, either shorter acquisition times should be used or the later portions of the FID may be reduced artificially by a weighting function. In this way noise is reduced, but at the possible expense of lower resolution. Figure 2-34 illustrates this problem. Increased weighting of the data from longer times (from the bottom to the top of the figure) results in lower resolution and broader peaks.

After acquisition is complete, a *delay time* is necessary to allow the nuclei to relax before they can be examined again for the purpose of signal averaging. For optimal results, the delay time is on the order of three to five times the spin–lattice relaxation time. For a typical ^{13}C relaxation time of 10 s, a total cycle thus might take up to 50 s. Such cycle times are excessive and can be reduced by using a pulse angle θ that is less than 90° (see Figure 2-13, but usually referred to as α). When spins move through a smaller angle, longitudinal relaxation can occur more rapidly and the delay time can be shortened. The choice of α depends on T_1, which often is unknown. For ^{1}H experiments, α values

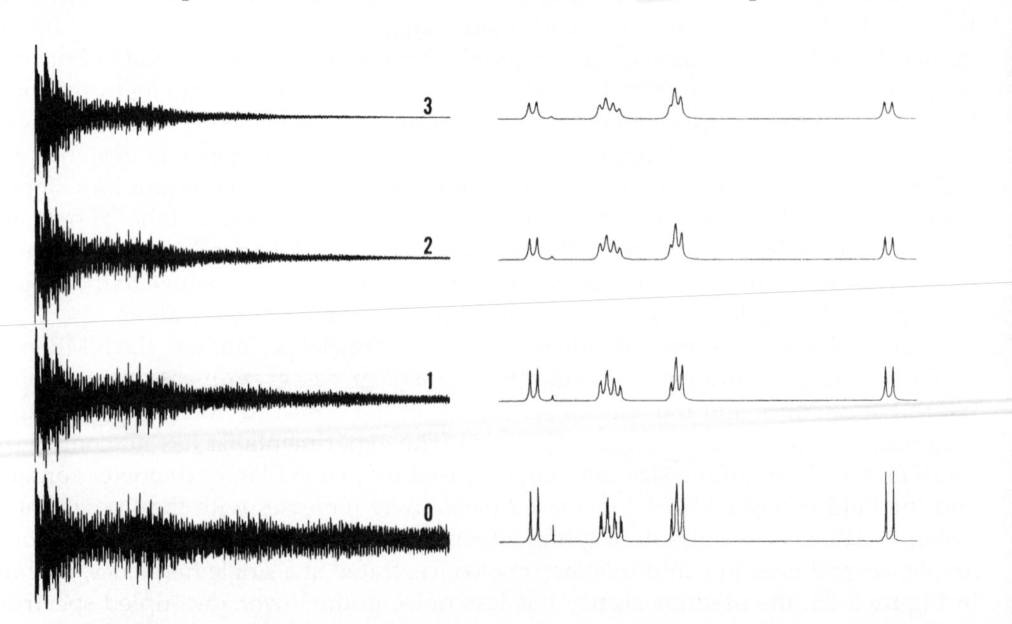

FIGURE 2-34 The effect of multiplying a free induction decay by a line broadening factor of 3, 2, 1, and 0 Hz. (Reprinted with permission from J. W. Cooper and R. D. Johnson, *FT NMR Techniques for Organic Chemists*, IBM Instruments Inc., 1986.)

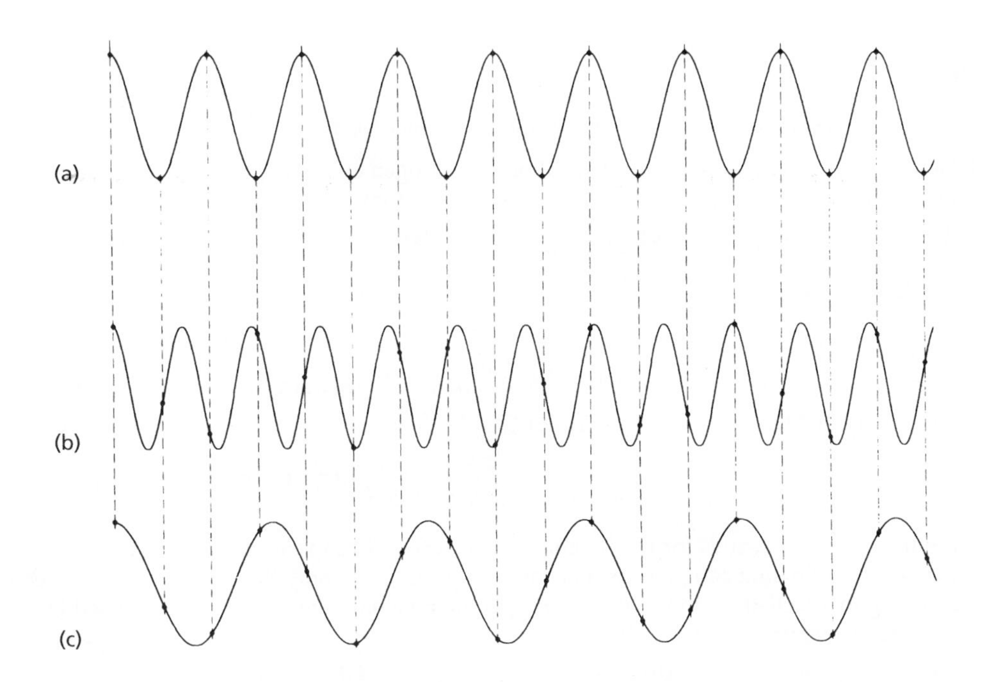

FIGURE 2-35 (a) Sampling a sine wave exactly two times per cycle. (b) Sampling a sine wave less than twice per cycle. (c) The lower-frequency sine wave that contains the same points as in (b) and is sampled more than twice per cycle. The frequency of (b) is not detected but appears as an aliased peak at the frequency of (c).

of 60–90° are common. For ^{13}C, values of 40–45° are used for protonated carbons. To ensure observation of quaternary carbons, values as low as 30° may be used. Proper balance between α and t_a (to optimize resolution) can result in essentially no delay time, and the next pulse cycle can be initiated immediately after acquisition.

The detected range of frequencies (the *spectral width*) is determined by how often the detector samples the value of the FID: the *sampling rate*. The FID is made up of a collection of sinusoidal signals (see Figure 2-16). A single, specific signal must be sampled at least twice within one sinusoidal cycle to determine its frequency. For a collection of signals up to a frequency of N, the FID thus must be sampled at a rate of $2N$ Hz. For example, for a ^{13}C spectral width of 20,000 Hz (200 ppm at 100 MHz), the signal must be sampled 40,000 times per second. Figure 2-35 illustrates this situation. The top signal (a) is sampled exactly twice per cycle (each dot is a sampling). The higher-frequency signal in (b) is sampled at the same rate, but not often enough to determine its frequency. In fact, the lower-frequency signal in (c) gives exactly the same collection of points as in (b). A real signal from the points in (b) is indistinguishable from an *aliased* or *foldover signal* with the frequency in (c). For example, if the ^{13}C spectrum is sampled only 20,000 times per second, for a spectral width of 10,000 Hz (100 ppm at 75 MHz), a signal with a frequency of 150 ppm (15,000 Hz) appears as a distorted, aliased peak in the 0–100 ppm region. The position of the aliased peak depends on the mode of detection (Section 5-8). Simply increasing the spectral width (sw) changes the position of a putative aliased signal and allows its identification.

If a signal is sampled 20,000 times per second, the detector spends 50 μs on each point. The reciprocal of the sampling rate is called the *dwell time*, which signifies the amount of time between sampling. Reducing the dwell time means that more data points are collected in the same period. If the acquisition time is 4.0 s (for a resolution of 0.25 Hz) and the sampling rate is 20,000 times per second (for a spectral width of 10,000 Hz), the computer must store 80,000 data points. Today's computers are sufficiently powerful that spectral width and resolution are not limited by computer capabilities. The number of data points n_p and the sweep width sw are related to the acquisition time t_a by eq. 2-7.

$$t_a = \frac{n_p}{2(\mathrm{sw})} \tag{2-7}$$

For a ^1H experiment with $n_p = 32\text{k}$ (which means 32,768) and $(\mathrm{sw}) = 4000$ Hz (10 ppm at 400 MHz), t_a becomes 4.1 s, allowing a resolution of about 0.25 Hz ($1/t_a$). For a ^{13}C experiment with $n_p = 32\text{k}$ and $(\mathrm{sw}) = 22{,}000$ Hz (220 ppm at 100 MHz), t_a becomes 0.74 s, allowing a resolution of about 1.3 Hz.

Problems

In the following problems, assume fast rotation around all single bonds.

2-1 Determine the number of chemically different hydrogen atoms and their relative proportions in the following molecules. Do the same for carbon atoms.

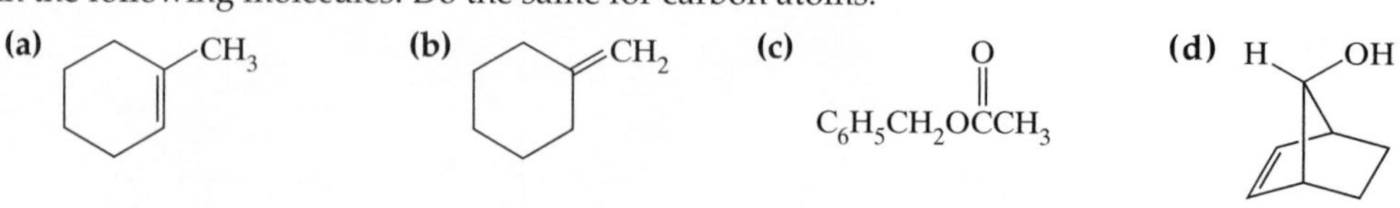

2-2 What is the expected multiplicity for each proton resonance in the following molecules?

(a) $ClCH_2CH_2CH_2Cl$ **(b)** $BrCH(CH_3)_2$ **(c)** $C_6H_5OC(=O)CH_2CH_3$ **(d)** H O Cl / H H

2-3 Predict the multiplicities for the ^1H and the ^{13}C resonances in the absence of decoupling for each of the following compounds. For the ^{13}C spectra, give only the multiplicities caused by coupling to attached protons. For the ^1H spectra, give only the multiplicities caused by coupling to vicinal protons (HCCH).

(a) $CH_3CH_2CH_2OC(=O)CH_3$ **(b)** C_6H_5—CH_2CH_2Br **(c)** (six-membered ring containing O)

(d) $N(CH_2CH_3)_3$ **(e)** CH_3(H)C=C(H)CO_2CH_3

2-4 For each of the following 300 MHz ^1H spectra, carry out the following operations. **(i)** From the elemental formula, calculate the unsaturation number U (eq. 1-1). **(ii)** Calculate the relative integrals for each group of protons. Then convert the integrals to absolute numbers by selecting one group to be of a known integral. **(iii)** Assign a structure to each compound. Be sure that your structure agrees with the spectrum in all aspects: number of different proton groups, integrals, and splitting patterns.

(a) C_4H_9Br

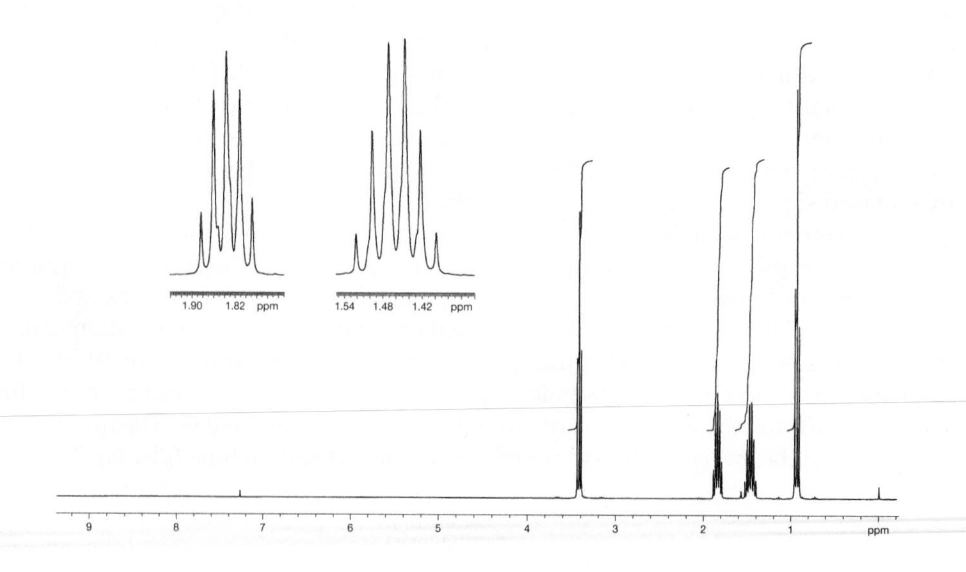

(b) $C_7H_{16}O_3$ (The resonance at δ 1.2 is a 1:2:1 triplet, and that at δ 3.6 is a 1:3:3:1 quartet.)

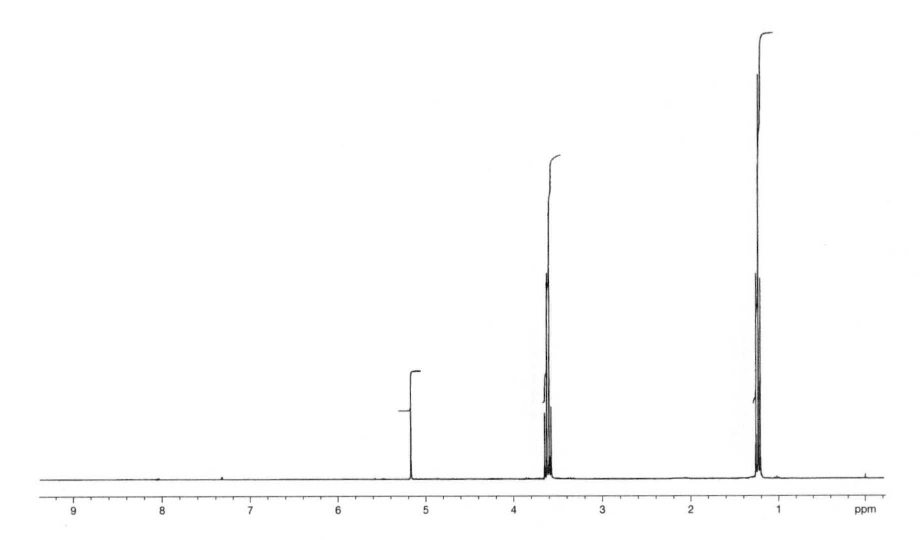

(c) $C_5H_8O_2$ (Ignore stereochemistry at this stage.)

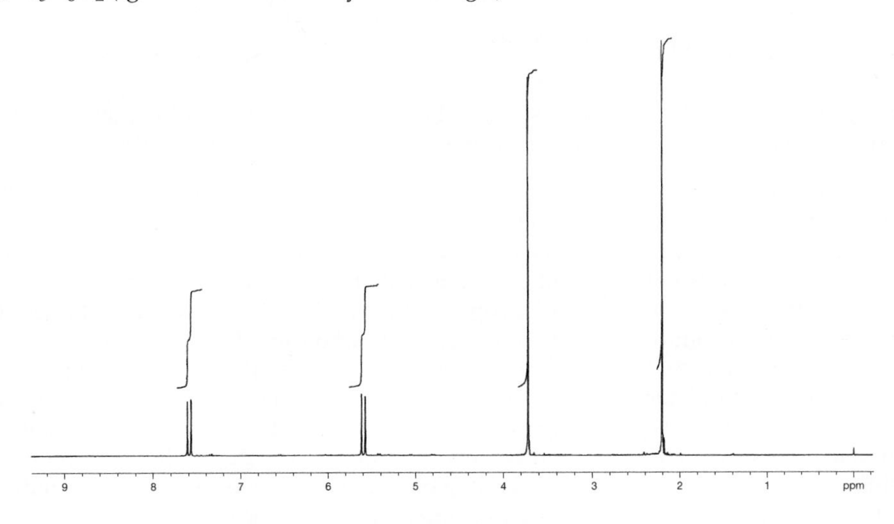

(d) $C_9H_{11}O_2N$ (*Hint*: The highest-frequency resonances (δ 6.6–7.8) come from a *para*-disubstituted phenyl ring. They are doublets. The resonance at δ 4.3 is a quartet, and that at δ 1.4 is a triplet.)

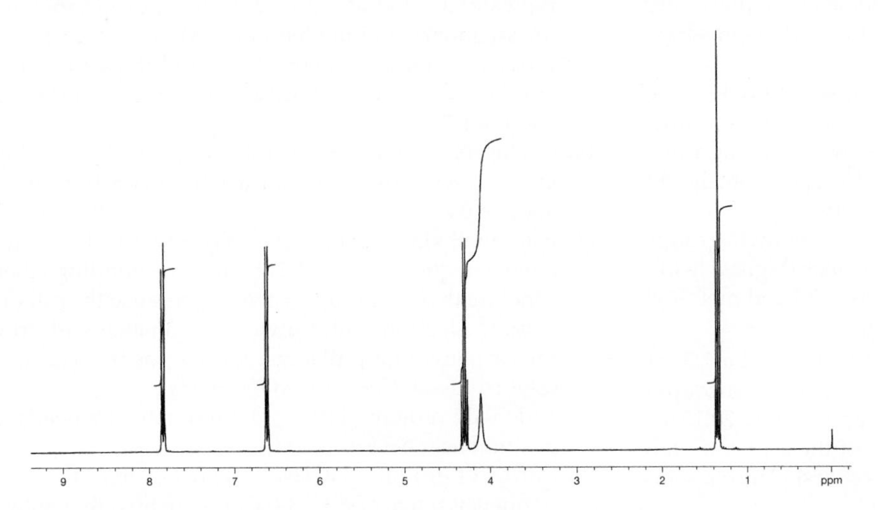

(e) C_5H_9ON (The resonances at δ 1.2 and 2.6 are triplets.)

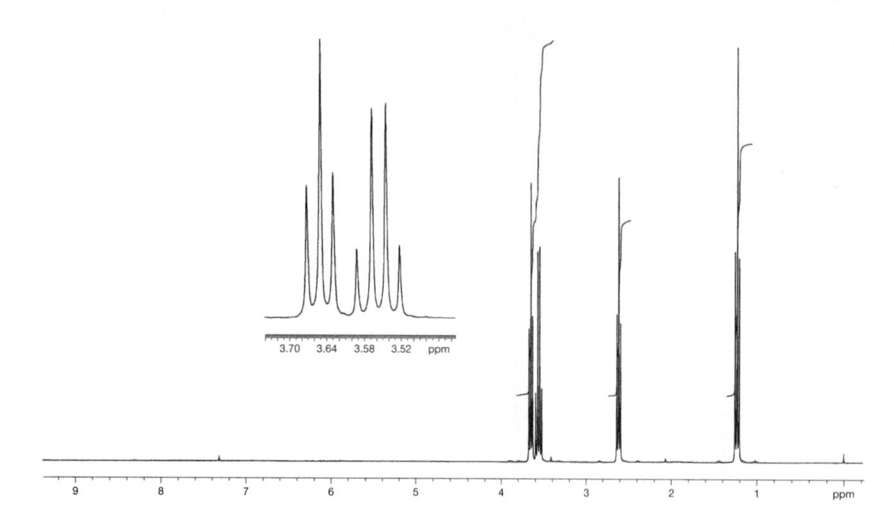

Tips on Solving NMR Problems

Spoiler alert: Some of these tips may be used directly in solving the foregoing problems and therefore should be studied only after you have made every effort to complete the problems. Many of these tips anticipate material that is covered in Chapters 3 and 4.

1. The elemental formula of a compound is obtained only rarely from elemental analysis, as described in Chapter 1. Usually it is obtained from high-resolution mass spectrometric experiments (Part II). Often the information is not available and must be inferred from the NMR spectra. The ^{13}C spectrum, for example, provides a reasonable count of all carbon atoms. As carbon functionalities are deduced from the NMR spectra, they should be written down for later assembly of the entire molecule from its constituents. Evidence for heteroatoms is obtained from mass spectrometry and from analysis of chemical shifts.
2. If the elemental formula is provided, the first step is calculation of the unsaturation number from eq. 1-1. As unsaturations are deduced during analysis of the NMR spectra, they should be enumerated and compared with the unsaturation number until all are accounted for.
3. Upon completion of the problem, double check that every chemical shift, coupling constant, and integral is in agreement with the concluded structure.
4. The first overview of the ^1H spectrum should determine whether there are aromatics, alkenes, and saturated functionalities (with and without electron-withdrawing functionalities). A similar overview of the ^{13}C spectrum should indicate whether there are carbonyl groups.
5. For a given type of substitution X, methyl groups ($CH_3—X$) are found at the lowest frequency (highest field), followed by methylene groups ($—CH_2—X$) and methinyl groups ($>CH—X$), both in ^1H and ^{13}C spectra.
6. Ethyl groups are indicated by a 1:2:1 triplet and a 1:3:3:1 quartet with a common coupling constant. Isopropyl groups are indicated by a 1:1 doublet and a 1:6:15:20:15:6:1 septet (sometimes the smallest peaks of the septet are too small to be observed, so the resonance at first glance resembles a quintet).
7. Electron-withdrawing groups such as NO_2, Cl, Br, OH, NH_2, C=O, and C=C shift saturated ^1H and ^{13}C resonances according to their respective electronegativities. The effect is attenuated rapidly with distance, so that in 1-bromopropane ($BrCH_2CH_2CH_3$) the effect is largest for the first hydrogen or carbon and decreases thereafter.
8. Unsaturated [NO_2, (C=O)R, CN] or lone-pair–bearing (R_2N:, RO:, Cl:) substituents attached to double bonds and aromatic rings exert resonance effects that can shift alkenic and aromatic resonances to either higher or lower frequency, depending on whether the effect is electron withdrawal or electron donation by resonance.
9. Primary ethers ($CH_2—OR$) and alcohols ($CH_2—OH$) are found typically at around δ 3.7, with secondary cases (CH—O) at higher frequency and methyl cases ($CH_3—O$) at lower frequency. Primary amines ($CH_2—N$) are found typically at around δ 2.7, with the appropriate variations for secondary and methyl cases. Methyls next to carbonyl groups are found at about δ 2.1, and those next to double bonds with only hydrocarbon substituents are at about δ 1.7.
10. Cyclopropane methylene protons ($C—CH_2—C$) are found at the lowest frequency of any hydrocarbon, usually at about δ 0.2.
11. Ester methylene groups [(C=O)—O—CH_2] are found at higher frequency (ca. δ 4.2) than corresponding ether or alcohol methylene groups (ca. δ 3.7) because the ester oxygen is more electron withdrawing. Resonance withdrawal by the carbonyl group places a formal positive charge on the ester oxygen [$(C—O^-)=O^+—CH_2$].
12. Aldehyde protons [H(C=O)] resonate at a nearly unique position near δ 9.8.
13. Carboxyl protons [HO(C=O)] resonate at the very high frequency range of δ 12–14. Consequently, they may be out

of the normally observed range. If suspected, the spectral width should be adjusted accordingly.

14. Protons on nitrogen or oxygen in amines and alcohols (OH, NH, NH_2) are usually observed in chloroform solution as broad, unsplit peaks in the range δ 1–4, because of exchange phenomena.
15. Exchangeable protons may be identified by adding a drop of D_2O to a chloroform solution. Two layers are formed, with the small aqueous layer on top. When the tube is shaken, the OH and NH protons exchange with D_2O and become OD and ND, which are not observed in the spectrum. Thus the spectrum is recorded before and after addition of the drop of D_2O, and resonances from exchangeable protons disappear. The aqueous layer at the top should be out of the receiver coils and not detected.
16. The $n + 1$ rule for identifying the multiplicity of neighboring proton functionalities has limited utility. Currently, carbon connectivities are more likely to be determined by the two-dimensional COSY experiment (Chapter 6) and carbon substitution patterns (CH_3 vs. CH_2 vs. CH vs. C) by the DEPT experiment (Chapter 5).

Bibliography

2.1 L. M. Jackman and S. Sternhell, *Applications of Nuclear Magnetic Resonance Spectroscopy in Organic Chemistry*, 2nd ed., Oxford, UK: Pergamon Press, 1969.

2.2 C. Brevard and P. Granger, *Handbook of High Resolution Multinuclear NMR*, New York: John Wiley & Sons, 1981.

2.3 R. K. Harris, *Nuclear Magnetic Resonance Spectroscopy*, London: Pitman Publishing, Ltd., 1983.

2.4 C. A. Fyfe, *Solid State NMR for Chemists*, Guelph, Ontario: C. F. C. Press, 1983.

2.5 A. E. Derome, *Modern NMR Techniques*, Oxford, UK: Pergamon Press, 1987.

2.6 R. Freeman, *A Handbook of Nuclear Magnetic Resonance*, New York: Longman Scientific & Technical, 1988.

2.7 F. A. Bovey, L. W. Jelinski, and P. A. Mirau, *Nuclear Magnetic Resonance Spectroscopy*, 2nd ed., San Diego: Academic Press, 1988.

2.8 W. S. Brey (ed.), *Pulse Methods in 1D and 2D Liquid-Phase NMR*, New York: Academic Press, 1988.

2.9 T. C. Farrar, *Pulse Nuclear Magnetic Resonance Spectroscopy*, 2nd ed., Chicago: Farragut Press, 1989.

2.10 L. D. Field and S. Sternhell (eds.), *Analytical NMR*, Chichester, UK: John Wiley & Sons, Ltd., 1989.

2.11 H. Duddeck and W. Dietrich, *Structure Elucidation by Modern NMR*, Darmstadt, Germany: Steinkopf Verlag, 1989.

2.12 R. J. Abraham, J. Fisher, and P. Loftus, *Introduction to NMR Spectroscopy*, New York: John Wiley & Sons, 1992.

2.13 E. Breitmaier, *Structure Elucidation by NMR in Organic Chemistry*, Chichester, UK: John Wiley & Sons Ltd., 1993.

2.14 J. K. M. Sanders and B. K. Hunter, *Modern NMR Spectroscopy*, 2nd ed., Oxford, UK: Oxford University Press, 1993.

2.15 H. Günther, *NMR Spectroscopy*, 2nd ed., Chichester, UK: John Wiley & Sons, Ltd., 1995.

2.16 T. D. W. Claridge, *High-Resolution NMR Techniques in Organic Chemistry*, Amsterdam: Pergamon, 1999.

2.17 S. Braun, H.-O. Kalinowski, and S. Berger, *100 and More Basic NMR Experiments*, New York: Wiley-VCH, 1999.

2.18 E. D. Becker, *High Resolution NMR*, 3rd ed., New York: Academic Press, 2000.

2.19 D. D. Laws, H.-M. L. Bitter, and A. Jerschow, *Angew. Chem., Int. Ed. Engl.*, **41**, 3096–3129 (2002).

2.20 J. B. Lambert and E. P. Mazzola, *Nuclear Magnetic Resonance Spectroscopy: An Approach to Principles, Applications, and Experimental Methods*, Upper Saddle River, NJ: Pearson Prentice Hall, 2004.

2.21 Y.-C. Ning and R. R. Ernst, *Structural Identification of Organic Compounds with Spectroscopic Techniques*, Weinheim, Germany: Wiley-VCH, 2005.

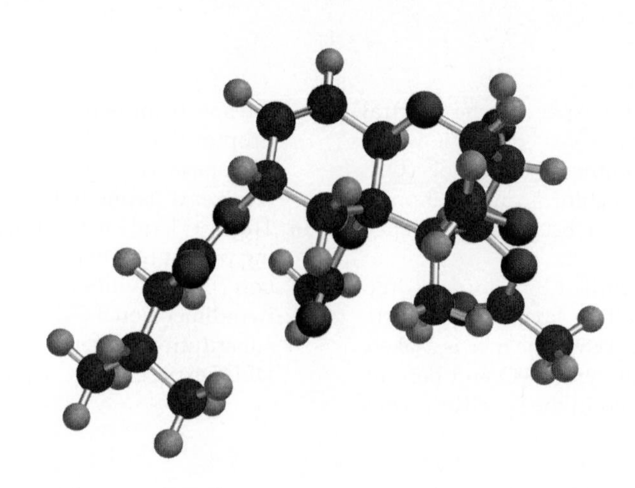

The Chemical Shift

3-1 FACTORS THAT INFLUENCE PROTON SHIFTS

Interpreting the location of a resonance in the ^{1}H NMR spectrum in terms of structure requires understanding several contributing factors. Chemical shifts vary according to structure because different nuclei experience different shielding by magnetic fields produced by their surrounding electrons and more distant electrons. The magnetic field at a nucleus is altered by the electrons, from B_0 to a value $B_0(1 - \sigma)$, in which the quantity σ is called the *shielding* (Figure 3-1a and b). When the magnetic field ($-B_0\sigma$) induced by the electrons opposes the static B_0 field, the effect is said to be *diamagnetic* (σ^d).

Local Fields. Shielding by the electrons that surround the resonating nuclei is said to arise from *local fields*, which may be assessed by considering electron density. For the proton, the electronic effects of physical organic chemistry (electronegativity and conjugation) conveniently describe the role of structure vis-à-vis electron density. In this way, both the atom to which the proton is attached and more distant atoms can modulate the electron density at the proton and hence alter the shielding effect.

The effects of electronegativity, usually called polar or inductive effects (not entirely synonymous terms but so treated herein), are manifested in the following fashion. An attached or nearby electron-withdrawing atom or group such as —OH or —CN decreases the electron density, and hence the diamagnetic shielding, and moves the resonance of the attached proton toward the left of the chart (to a higher frequency, or downfield; Figure 2-9). By contrast, an electron-donating atom or group increases the diamagnetic shielding and moves the resonance toward the right of the chart (to a lower frequency, or upfield). Although the effects of shielding on chemical shift are more properly described in terms of frequency, we include field terminology parenthetically, since it still enjoys wide, though inappropriate, usage.

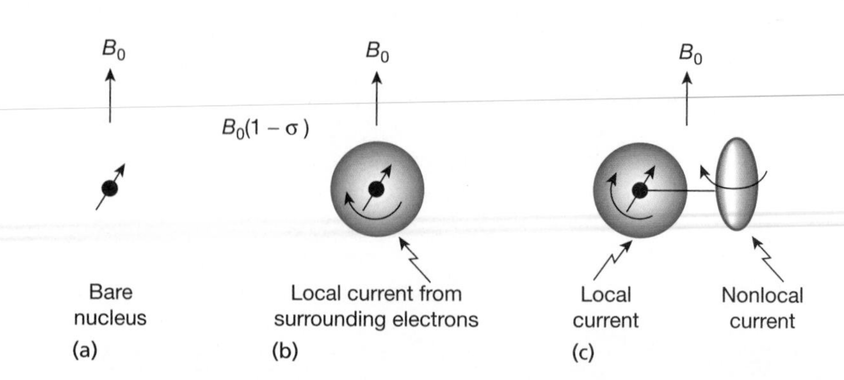

FIGURE 3-1 Shielding by local (b) and nonlocal (c) currents.

The progressive replacement of hydrogen with chlorine on methane moves the chemical shift to higher frequency (downfield) because of the ability of chlorine to remove electron density from the remaining protons: δ 0.23 for CH_4, 3.05 for CH_3Cl, 5.30 for CH_2Cl_2, and 7.27 for $CHCl_3$. The trend for a series of methyl resonances often can be explained in the same fashion by the polar effect. The chemical shifts for the series CH_3X for which X is F, HO, H_2N, H, Me_3Si, and Li, respectively, are δ 4.26, 3.38, 2.47, 0.23, 0.00 (TMS, the standard), and −0.4 (this last value is considerably dependent on the solvent; the minus sign indicates a lower frequency than TMS). This trend follows the electronegativity of the atom attached to CH_3.

Electron density is influenced by conjugation (resonance or mesomerism), as well as by polar effects, as seen in unsaturated molecules such as alkenes and aromatics. The donation of electrons through resonance by a methoxy group increases the electron density at the β position of a vinyl ether (**3-1**) and at the para position of anisole

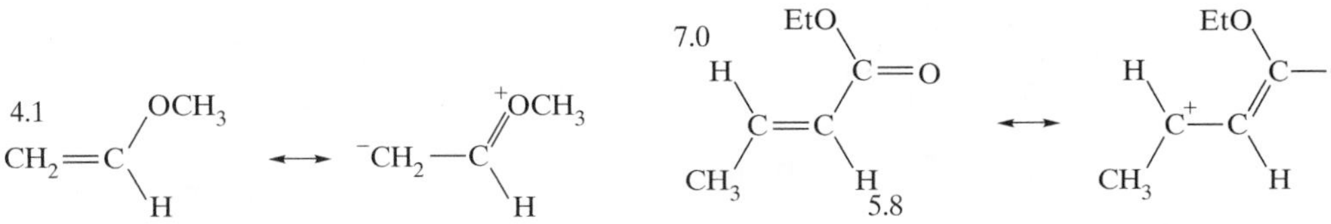

4.1 $CH_2{=}C(OCH_3)H$ ⟷ $^{-}CH_2{-}C(=\overset{+}{O}CH_3)H$

3-1

7.0 H(CH$_3$)C=C(H 5.8)C(=O)EtO ⟷ H(CH$_3$)$\overset{+}{C}$—C(H)=C(EtO)—O$^-$

3-2

($C_6H_5OCH_3$). Thus, the chemical shift of the β protons in **3-1** is at about δ 4.1, in comparison with δ 5.28 in ethene. The resonance frequency decreases, as is expected with the increased shielding from electron donation. The electron-withdrawing polar effect of CH_3O is overpowered by the resonance effect. Groups such as nitro, cyano, and acyl withdraw electrons by both resonance and induction, so they can bring about significant shifts to higher frequency (downfield). Ethyl *trans*-crotonate (**3-2**; Section 2-7) illustrates this effect. The electron-withdrawing group shifts the β proton strongly to higher frequency (δ 7.0) than in ethene. Although the α proton is not subjected to this strong resonance effect, it is close enough to the electron-withdrawing carboethoxy group to be shifted slightly to higher frequency (δ 5.8) by the polar effect.

Hybridization of the carbon to which a proton is attached also influences electron density. As the proportion of s character increases from sp^3 to sp^2 to sp orbitals, bonding electrons move closer to carbon and away from the protons, which then become deshielded. For this reason, ethane resonates at δ 0.86, but ethene resonates at δ 5.28. Ethyne (acetylene) is an exception in this regard, as we shall see presently. Hybridization contributes to chemical shifts in strained molecules, such as cyclobutane (δ 1.98) and cubane (δ 4.00), for which hybridization is intermediate between sp^3 and sp^2.

Nonlocal Fields. Induction, resonance, and hybridization modulate the density of the electrons immediately surrounding a proton, as the result of local electron currents around the nucleus (Figure 3-1b). In the absence of changes in electron density, purely magnetic effects of substituents also can have a major influence on proton shielding, but only when the groups have a nonspherical shape. Figure 3-1c illustrates the combined effects of local fields and nonlocal fields. The group giving rise to the nonlocal field could be, for example, methyl, phenyl, or carbonyl, and the resonating nucleus need not be attached directly to that group. To see why a spherical or isotropic ("same in all directions") group contributes no nonlocal effect, consider a proton attached to such a group, for instance, chlorine. The local effect arises from the electrons that surround the resonating proton. The electrons in the substituent, which are not around the proton, also precess in the applied field (Figure 3-1c). They induce a magnetic field in the molecule that opposes B_0 and that can have a nonzero value at the position of the proton. The nonlocal, induced field is represented by magnetic lines of force in Figure 3-2. If the bond from the spherical substituent to the resonating proton is parallel to the direction of B_0, as in Figure 3-2a, the lines of force from the induced field oppose

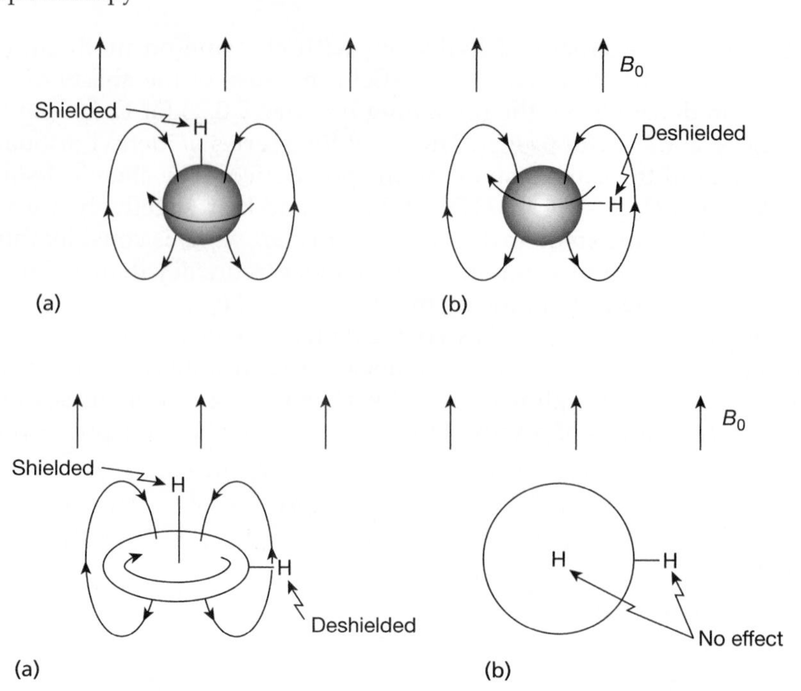

FIGURE 3-2 Shielding by a spherical (isotropic) group.

FIGURE 3-3 Shielding by an oblate ellipsoid, the model for aromatic rings.

B_0 at the proton, thereby shielding it. If the bond from the substituent to the proton is perpendicular to B_0, as in Figure 3-2b, the induced lines of force reinforce those of B_0, deshielding the proton. Because the group is isotropic, the two arrangements are equally probable. As the molecule tumbles in solution, the effects of the induced field cancel out. Other orientations cancel each other in a similar fashion. Thus, an isotropic substituent has no effect over and above what it provides to local currents from induction or resonance.

Most substituents, however, are not spherical. The flat shape of an aromatic ring, for example, resembles an oblate ellipsoid, and the elongated shape of single or triple bonds resembles a prolate ellipsoid. For a proton situated at the edge of an oblate ellipsoid, such as a benzene ring, there again are two extremes (Figure 3-3). When the flat portion is perpendicular to the static field (Figure 3-3a), a proton at the edge is deshielded, since the induced lines of force reinforce the B_0 field. For the same geometry, a proton situated over the middle of the ellipsoid is shielded, as the induced lines of force oppose B_0. For this geometry, the induced field is large because aromatic electrons circulate easily above and below the ring. When the ring is parallel to B_0 (Figure 3-3b), however, induced currents would have to move from one ring face to the other. In this orientation, little current or field is induced from this geometry. The cancelation seen for a spherical substituent as the molecule tumbles in solution does not occur for aromatic rings. A group that has appreciably different currents induced by B_0 from different orientations in space is said to have *diamagnetic anisotropy*. Because an oblate ellipsoid has the larger effect for the geometry shown in Figure 3-3a, a proton at the edge of an aromatic ring is deshielded and one at or over the center is shielded. It is for this reason that benzene resonates at such a high frequency (low field, δ 7.27), compared with the frequency of alkenes (e.g., ethene, at δ 5.28).

Just as the local effect can result in either shielding (from electron donation) or deshielding (from electron withdrawal), the nonlocal effect also can have either sign, depending on whether the nonlocal field enhances or diminishes the static magnetic field. Figure 3-4 illustrates how the diamagnetic effect of benzene is shielding (+) above and below the ring, but deshielding (−) around the edge. This effect was modeled quantitatively by McConnell as the influence of a magnetic dipole on the point in space at which a proton resides. He derived the formula in eq. 3-1

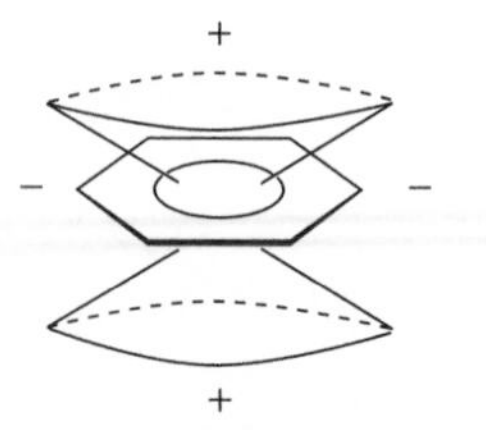

FIGURE 3-4 Shielding geometry for a benzene ring.

$$\sigma_A(r, \theta) = \frac{(\chi_L - \chi_T)(3\cos^2\theta - 1)}{3r^3} \qquad \textbf{(3-1)}$$

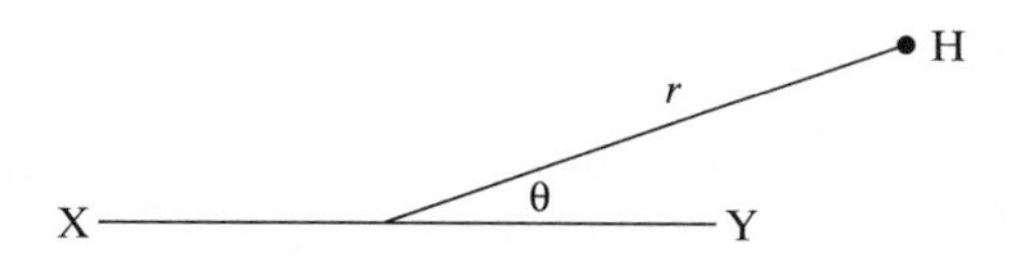

for the shielding σ_A by an anisotropic group (represented by the dipole X–Y) of a hydrogen atom at an arbitrary point in space with polar coordinates (r, θ). In this equation, r is the distance from the midpoint of X–Y to that point, θ is the angle between the line connecting the atoms X and Y and the line from the midpoint of X–Y to the point (r, θ), and χ_L and χ_T are the diamagnetic susceptibilities of the group along its longitude and its transverse, as, for example, in Figure 3-3a and b, respectively. The effect changes sign at the null point at which the angle θ is 54°44′, the so-called magic angle at which the expression $(3\cos^2\theta - 1)$ goes to zero.

Although the protons of benzene reside in the deshielded portion of the cone, molecules have been constructed to explore the full range of the effect. The methylene protons of methano[10]annulene (**3-3**) are constrained to positions above the aromatic 10π electron

−0.5

3-3

9.3 −3.0

3-4

system and consequently are shielded to a position (δ −0.5) at an even lower frequency (higher field) than TMS. [18]Annulene (**3-4**) has one set of protons around the edge of the aromatic ring that resonates at a deshielded position of δ 9.3 and a second set located toward the center of the ring that resonates at a shielded position of δ −3.0.

The presence of $(4n + 2)$ π electrons is a requirement for the existence of a diamagnetic circulation of electrons. Pople showed that an external magnetic field can induce an opposite, or paramagnetic, circulation in a $4n$ π electron system. Under such circumstances, the conclusions drawn from the configuration in Figure 3-3a are reversed; that is, outer protons are shielded and inner protons deshielded. The spectrum of [16]annulene is consistent with this interpretation (inner protons at δ 10.3, outer at 5.2). The most dramatic example is the [12]annulene **3-5**. The bromine atom was included to prevent conformational interconversion of the inner and outer protons. The indicated inner proton of **3-5** with $4n$ electrons resonates at δ 16.4, compared with δ −3.0 for the inner protons of [18]annulene (**3-4**) with $4n + 2$ electrons.

16.4 Br

3-5

The two arrangements of a prolate ellipsoid, used as a model for a chemical bond, may be considered in a similar fashion (Figure 3-5). In this case, it is not always clear which

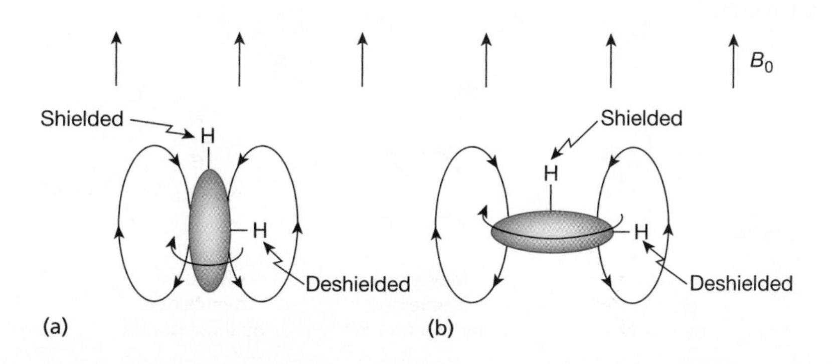

FIGURE 3-5 Shielding by a prolate ellipsoid, the model for a chemical bond.

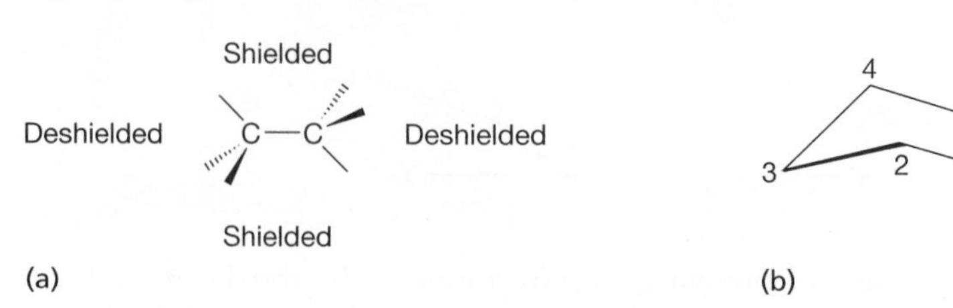

FIGURE 3-6 (a) Shielding region around a carbon–carbon single bond. (b) Geometry of a cyclohexane ring.

arrangement has a stronger induced current (or higher diamagnetic susceptibility χ). The π electrons of acetylene (ethyne) provide one clear-cut example. When the axis of the molecule is parallel to the B_0 field (as in Figure 3-5a), the π electrons are particularly susceptible to circulation around the cylinder. The alternative arrangement in Figure 3-5b is ineffective for acetylene and therefore does not provide a canceling effect. The acetylenic proton is attached to the end of the array of electrons and hence is shielded. For this reason, the acetylene resonance (δ 2.88) falls between those of ethene (δ 0.86) and ethene (δ 5.28). The effects of hybridization thus are superseded by those of C≡C diamagnetic anisotropy. In terms of the McConnell equation (eq. 3-1), the longitudinal susceptibility is much greater than the transverse susceptibility, that is, $\chi_L > \chi_T$. Thus, shielding is positive at the end of the bond ($\theta = 0°$), resulting in a shift to lower frequency (higher field).

Circulation of charge around a carbon–carbon single bond is less strong than that of a triple bond, and the larger current occurs when the axis of the bond is perpendicular to B_0 (as in Figure 3-5b), that is, $\chi_T > \chi_L$. Thus, a proton at the side of a single bond is more shielded than one along its end (Figure 3-6a). The axial and equatorial protons of a rigid cyclohexane ring exemplify these arrangements (Figure 3-6b). The two protons illustrated are equivalently positioned with respect to the 1,2 and 6,1 bonds, which thus produce no differential effect. The 1-axial proton, however, is in the shielding region of the more distant 2,3 and 5,6 bonds (darkened), whereas the 1-equatorial proton is in their deshielding region. In general, axial protons are more shielded and resonate at lower frequency (higher field) than do equatorial protons, typically by about 0.5 ppm.

The higher-frequency position for methine (CH) compared with methylene (CH_2) protons and for methylene compared with methyl (CH_3) protons [$(CH_3)_2C\mathbf{H}X$ to $CH_3C\mathbf{H}_2X$ to $C\mathbf{H}_3X$] for a single X group may be attributed to the anisotropy of the additional C—C bonds (Figure 3-7), although changes in hybridization also may contribute.

The highly shielded position of cyclopropane resonances (δ 0.22, vs. 1.43 in cyclohexane) may be attributed either to an aromatic-like ring current or to the anisotropy of the C—C bond that is opposite to a CH_2 group in the three-membered ring. The effect is much larger than the indicated 1.2 ppm (1.4 − 0.2), because the sp^2 cyclopropane carbon orbital to hydrogen (compared with the sp^3 orbital in cyclohexane) deshields the proton. A cyclopropane ring also can shield more distant hydrogens. In spiro[2.5]-octane (**3-6**), the indicated equatorial proton resonates 1.2 ppm to lower frequency (higher field) than does the axial proton. Since H_{ax} normally is about 0.5 ppm to lower frequency than H_{eq}, the differential effect is 1.7 ppm. In **3-6**, H_{eq} is perched over the shielding region of the cyclopropane ring, so it undergoes a very strong shift to lower frequency (upfield).

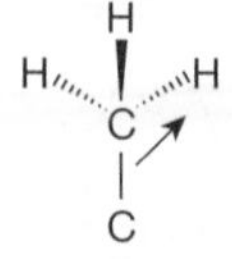

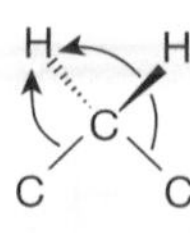

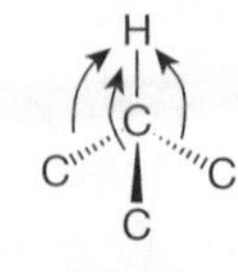

FIGURE 3-7 Shielding properties of methyl, methylene, and methine groups.

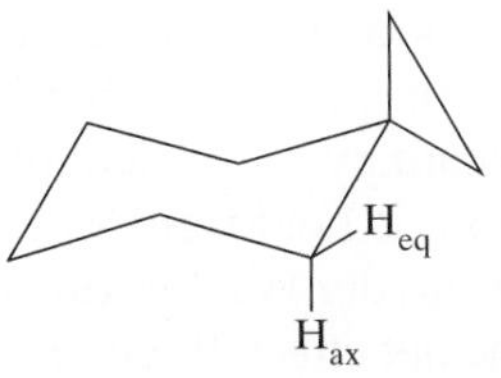

3-6 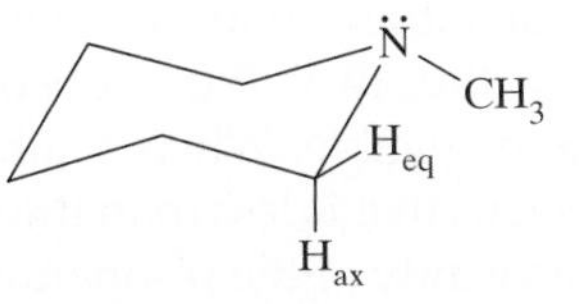 **3-7**

Most common single bonds (C—O, C—N) have shielding properties that parallel those of the C—C bond. There appears to be a sign reversal, however, for the C—S bond. In all these heteroatomic cases, the geometry is more complex than that for the C—C bond. In some instances, a lone electron pair can have a special effect. In *N*-methylpiperidine (**3-7**), the axial lone pair shields the vicinal H_{ax} by an $n \rightarrow \sigma^*$ interaction without any effect on H_{eq}. As a result, $\Delta\delta_{ae}$ increases to about 1.0 ppm or more.

The anisotropy of double bonds is more difficult to assess, because they have three nonequivalent axes. Thus, the McConnell equation, with only two axes, does not apply. Protons situated over double bonds are, in general, more shielded than those in the plane (Figure 3-8), both for alkenes (Figure 3-8a) and for carbonyl groups (Figure 3-8b). The position of the methylene protons in norbornene (**3-8**) may be explained in this fashion,

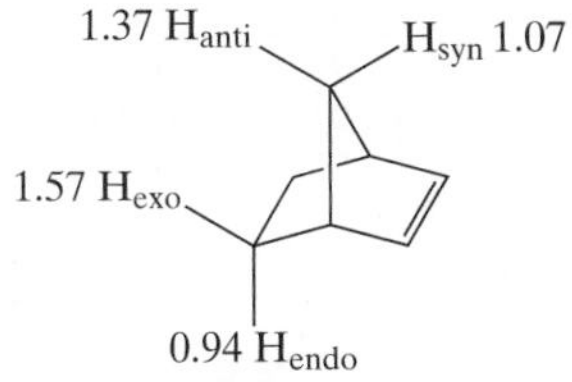

1.37 H_{anti} H_{syn} 1.07
1.57 H_{exo}
0.94 H_{endo}

3-8

since the syn and endo protons, respectively, are shielded with respect to the anti and exo protons. The highly deshielded position of aldehydes (δ ca. 9.8) is attributed to a combination of a strong polar effect and the diamagnetic anisotropy of the carbonyl group.

The nonspherical array of lone pairs of electrons may exhibit diamagnetic anisotropy, although, alternatively, the effect may be considered a perturbation of local currents. A proton that is hydrogen bonded to a lone pair invariably is deshielded. Thus, the hydroxyl proton in ethanol as a dilute solute in a non–hydrogen-bonding solvent such as CCl_4 resonates at δ 0.7, but, in pure ethanol with extensive hydrogen bonding, it resonates at δ 5.3. Carboxylic protons (CO_2H) resonate at extremely high frequency (low field, δ 11–14), because every proton is hydrogen bonded within a dimer or higher aggregate. Lone-pair anisotropy also has been invoked to explain trends in ethyl groups (CH_3CH_2X). The resonance position of a CH_2 group attached to X is well explained by the polar effect [for X = F (δ 4.36), Cl (3.47), Br (3.37), I (3.16)], but the trend for the more distant methyl group is opposite (in the same order, δ 1.24, 1.33, 1.65, and 1.86). As the size of X increases, the lone pair moves closer to the methyl group and deshields it more strongly.

In summary, functional group effects on proton chemical shifts may be explained largely by these two general effects. (1) Electron withdrawal or donation by induction (including hybridization) or by resonance alters the electron density and hence the local field around the resonating proton. Higher electron density shields the proton and moves its resonance position to lower frequency (downfield). (2) Diamagnetic anisotropy of nonspherical substituents is largely responsible for the proton resonance positions of aromatics, acetylenes, aldehydes, cyclopropanes, cyclohexanes, alkenes, and, possibly, hydrogen-bonded species.

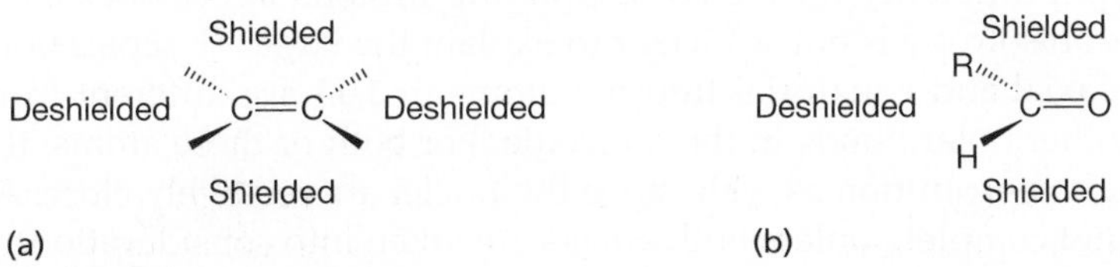

Shielded
Deshielded C=C Deshielded
Shielded
(a)

Shielded
Deshielded R C=O H
Shielded
(b)

FIGURE 3-8 Shielding properties of (a) the C=C double bond and (b) the C=O double bond.

Two other, less general diamagnetic effects also influence chemical shifts. As discussed, a substituent may deshield a neighboring proton—that is, decrease σ^d—by polar electron withdrawal. If the substituent is distant by a sufficient number of bonds, the effect becomes negligible. When a substituent atom is held rigidly at a distance from the resonating nucleus that is less than the sum of the van der Waals radii, the atom repels electrons from the vicinity of the resonating atom. The net effect, therefore, is a decrease in σ^d. The nucleus is deshielded, and its resonance is shifted downfield. The phenomenon arises from the mutual repulsion of induced dipoles (by van der Waals, London, or dispersion forces). The magnitude of the van der Waals effect falls off very rapidly with increasing internuclear distance and depends critically on the size and polarizability of the nuclei. A proton is not deshielded by another proton until they are within about 2.5 Å of each other. A bromine atom can deshield a proton from a much greater distance, and a fluorine atom is intermediate.

The 0.2-ppm downfield shift of the *tert*-butyl protons in *ortho*-di-*tert*-butylbenzene with respect to the position of the *tert*-butyl resonances in the meta and para isomers has been attributed to the van der Waals effect. A very dramatic example is seen in the partial cage compound **3-9**, in which the chemical shifts of H_b and H_c are δ 3.55 and

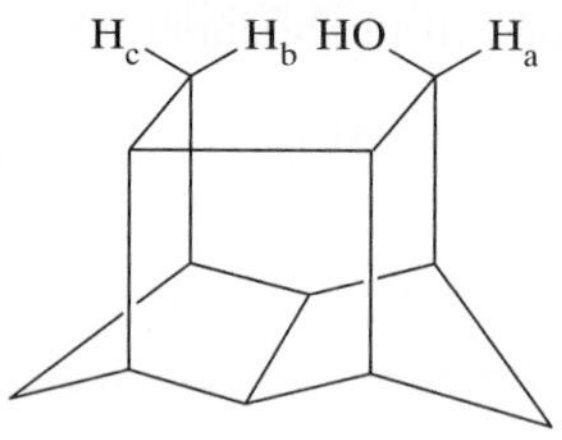

3-9

0.88, respectively. By comparison, the methylene protons of cyclohexane resonate at about δ 1.4. The oxygen atom therefore deshields H_b by more than 2 ppm. Interestingly, the electron density displaced from H_b is in part shifted to H_c, which is shielded by about 0.5 ppm.

A polar bond in a molecule generates an electric field that can have an appreciable value at the position of a nearby resonating nucleus. This electric field distorts the electronic structure around the nucleus and causes deshielding by diminishing σ^d. Unlike the inductive effect, the electric-field effect can be derived from a polar group that is many bonds removed from the resonating nucleus. For a significant effect, the polar bond must be reasonably close to the nucleus, but need not be in van der Waals contact.

Many examples of electric-field shielding have come from ^{19}F spectra, since the larger shifts of the fluorine nuclei magnify the effect. The ^{19}F resonance of 1-chloro-2-fluorobenzene (**3-10**) is more than 20 ppm to higher frequency than a simple polar effect

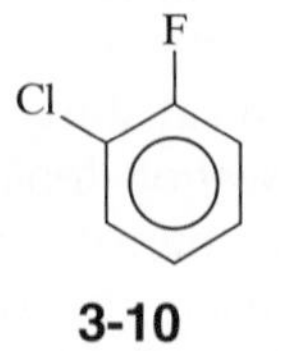

3-10 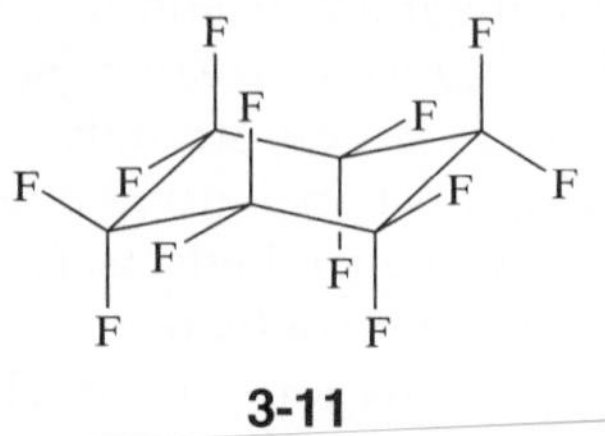 **3-11**

could explain. The interpretation that this large additional shift is due to the electric field of the C—Cl bond has been substantiated by calculations. The 18-ppm chemical shift between the axial and equatorial fluorines in perfluorocyclohexane (**3-11**) compares with a 0.5-ppm difference for the corresponding protons in cyclohexane. The effect of diamagnetic anisotropy is not sufficient to explain the large ^{19}F separation. The electric fields at the axial and equatorial fluorine atoms in **3-11** are different from the electric field at the other polar bonds in the molecule. For both of these atoms, there may be a van der Waals contribution as well, since the nuclei are relatively close. An analysis of ^{19}F shifts is not complete unless both effects are taken into consideration.

3-2 PROTON CHEMICAL SHIFTS AND STRUCTURE

The assignment of structure on the basis of NMR spectra requires knowledge of the relationship between chemical shifts and functional groups. Normally, both proton and carbon spectra are recorded and analyzed. This section considers the relationship between proton resonances and structure. Figure 3-9 summarizes the resonance ranges for common proton functionalities.

3-2a Saturated Aliphatics

Alkanes. Cyclopropane has the lowest-frequency position (δ 0.22) of any simple hydrocarbon, because of a ring current or the anisotropy of the carbon–carbon bonds. Unsubstituted methane has essentially the same chemical shift (δ 0.23). The progressive addition of saturated carbon–carbon bonds to methane results in a shift to higher frequency (downfield), as in the series consisting of ethane (CH_3CH_3, δ 0.86), propane ($CH_3\mathbf{CH_2}CH_3$, δ 1.33), and isobutane [$(CH_3)_3\mathbf{CH}$, δ 1.56] (Figure 3-7). The resonance position of cyclobutane is at an unusually high frequency (δ 1.98), because of the lower s character of the carbon orbitals. Cyclic structures other than cyclopropane and cyclobutane have resonance positions similar to those of open-chain systems, for example, δ 1.43 for cyclohexane. In complex natural products such as steroids or alkaloids, a large number of structurally similar alkane protons lead to overlapping resonances in the region δ 0.8–2.0, the analysis of which requires the highest possible field.

Functionalized Alkanes. The presence of a functional group alters the resonance position of neighboring protons according to the polar effect of the group and its diamagnetic anisotropy. Ethane (δ 0.86) is a useful point of reference for methyl groups. The replacement of one methyl group in ethane with hydroxyl yields methanol (CH_3OH), whose resonance position is δ 3.38. The electron-withdrawing effect of the oxygen atom is the primary cause of the large shift to higher frequency (lower field). Just as in unfunctionalized alkanes, methylene groups ($CH_3\mathbf{CH_2}OH$, δ 3.56) and methine groups [$(CH_3)_2\mathbf{CH}OH$, δ 3.85] are found at progressively higher frequencies. In general, methylene and methine protons resonate to higher frequency by about 0.3 and 0.6 ppm, respectively, than do analogous methyl groups (CH_3X vs. $—CH_2X$ vs. $>CHX$). There is considerable variation from one case to another, depending on the remainder of the structure, so that resonances for a given functionality can range over about 1 ppm. Ether resonances are similar to those for alcohols (CH_3OCH_3, δ 3.24). Ester alkoxy groups, however, usually resonate at even

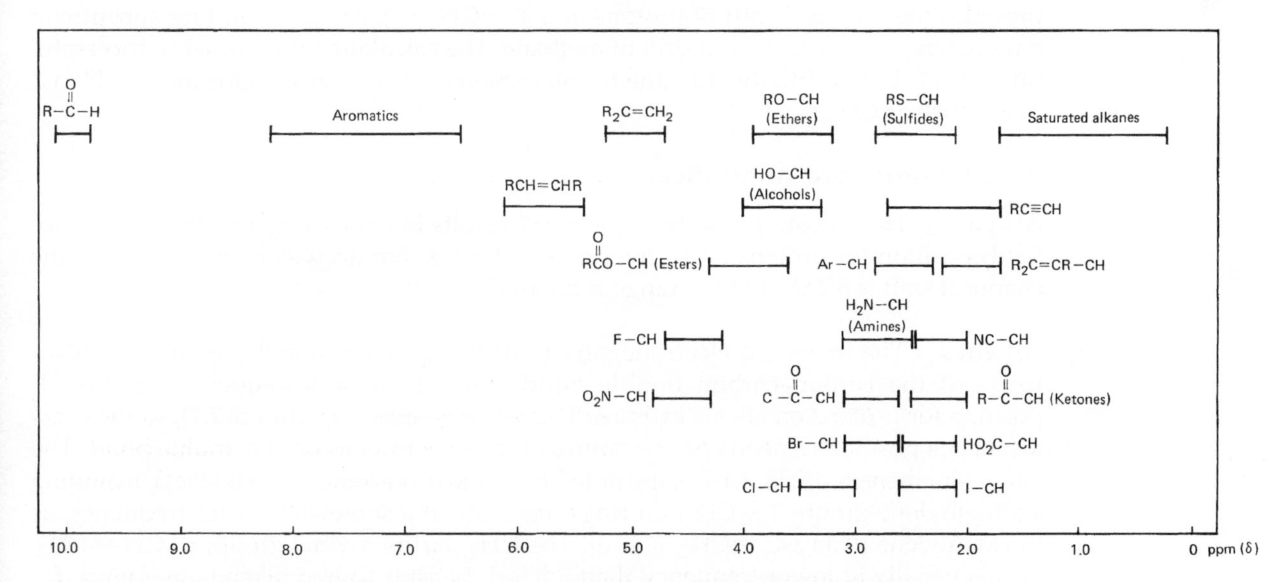

FIGURE 3-9 Proton chemical shift ranges for common structural units. The symbol CH represents methyl, methylene, or methine, and R represents a saturated alkyl group. The range for $—CO_2H$ and other strongly hydrogen-bonded protons is off scale to the left. The indicated ranges are for common examples; actual ranges can be larger.

higher frequency [$CH_3O(CO)CH_3$, δ 3.67)], because the attached oxygen is more electron withdrawing as the result of ester resonance (**3-12**).

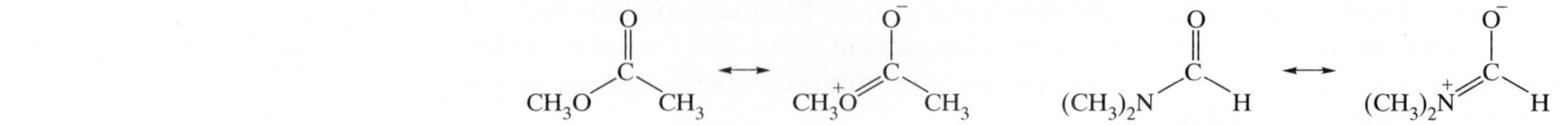

3-12 **3-13**

Since nitrogen is not so electron withdrawing as oxygen, amines resonate at a somewhat lower frequency (higher field) than ethers: δ 2.42 for methylamine (CH_3NH_2 in aqueous solution). Introducing a positive charge through quaternization induces increased electron withdrawal and causes a shift to still higher frequency, as with $(CH_3)_4N^+$ (δ 3.33), compared with $(CH_3)_3N$ (δ 2.22). An intermediate charge, as is produced in amides through resonance, results in an intermediate shift, as with *N,N*-dimethylformamide (**3-13**, δ 2.88).

The lower electronegativity of sulfur means that sulfides are at a lower frequency (higher field): δ 2.12 for dimethyl sulfide (CH_3SCH_3). Halogens move resonances to higher frequency according to the electronegativity of the atom: δ 2.15 for CH_3I, 2.69 for CH_3Br, 3.06 for CH_3Cl, and 4.27 for CH_3F. In all these cases, the shifts probably also are affected by the anisotropy of the C—X bond, but this factor is hard to assess and is somewhat diminished by free rotation in open-chain systems. Other electron-withdrawing substituents also cause shifts to higher frequencies, for example, cyano in acetonitrile (CH_3CN, δ 2.00) and nitro in nitromethane (CH_3NO_2, δ 4.33). Electron-donating atoms, such as silicon in TMS (δ 0.00), cause shifts to lower frequencies (higher field).

Methyl groups attached to carbon–carbon double and triple bonds are found usually in the region δ 1.7–2.5, as for the allyl protons in isobutylene [$(CH_3)_2C{=}CH_2$, δ 1.70], the propargylic protons in methylacetylene ($CH_3C{\equiv}CH$, δ 1.80), and the benzylic protons in toluene ($C_6H_5CH_3$, δ 2.31). Methyl groups on carbon–oxygen double bonds are found in the region δ 2.0–2.7, as for acetone [$CH_3(CO)CH_3$, δ 2.07], acetic acid (CH_3CO_2H, δ 2.10), acetaldehyde (CH_3CHO, δ 2.20), and acetyl chloride (CH_3COCl, δ 2.67). These functionalities exhibit an appreciable range determined by further substitution.

Empirical correlations between structure and proton chemical shift have been developed for common structural units (Section 3-2e). The earliest, called Shoolery's rule (eq. 3-2),

$$\delta = 0.23 + \Delta_X + \Delta_Y \qquad \textbf{(3-2)}$$

provides the chemical shift of protons in a $Y{-}CH_2{-}X$ group by adding substituent parameters Δ_i to the chemical shift of methane. The calculation is reasonably successful for CH_2XY, but additivity fails due to interactions between groups for many CHXYZ cases, such as $CHCl_3$.

3-2b Unsaturated Aliphatics

Alkynes. The anisotropy of the triple bond results in a relatively low frequency (upfield) position for protons on sp-hybridized carbons. For acetylene (ethyne) itself, the chemical shift is δ 2.88, and the range is about δ 1.8–2.9.

Alkenes. The increased electronegativity of the sp^2 carbon and the modest anisotropy of the carbon–carbon double bond result in a high-frequency (low-field) position for protons on alkene carbons. The range is quite large (δ 4.5–7.7), as the exact resonance position depends on the nature of the substituents on the double bond. The value for ethene is δ 5.28. 1,1-Disubstituted hydrocarbon alkenes (vinylidenes), including exomethylene groups ($=CH_2$) on rings, resonate at a somewhat lower frequency, as in isobutylene [$(CH_3)_2C{=}CH_2$, δ 4.73]. The CH_2 part of a vinyl group, $-CH{=}CH_2$, also is usually at lower frequency than δ 5.0. 1,2-Disubstituted alkenes, as found, for example, in endocyclic ring double bonds ($-CH{=}CH-$) and trisubstituted double bonds, generally resonate at even higher frequency than δ 5.0, as in *trans*-2-butene ($CH_3CH{=}CHCH_3$, δ 5.46). Angle strain on the double bond moves the resonance

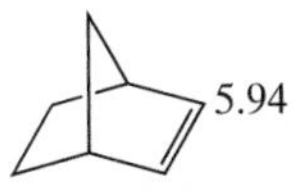

5.94 5.78 6.42

3-14 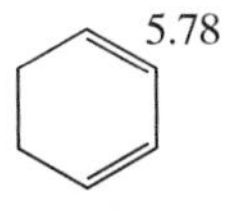 **3-15**  **3-16**

position to even higher frequency, as with norbornene (**3-14**, δ 5.94). Conjugation usually moves the resonance position to higher frequency, as in 1,3-cyclohexadiene (**3-15**, δ 5.78). The double bonds in 1,3-cyclopentadiene (**3-16**) are both strained and conjugated, so the chemical shift is at still higher frequency, δ 6.42. The phenyl ring of styrene ($C_6H_5CH{=}CH_2$) withdraws electrons from the double bond by the polar effect, so the position of the nearer CH (α) proton is moved to higher frequency, δ 6.66. The more distant CH_2 (β) protons are nonequivalent and resonate at δ 5.15 and 5.63. The anisotropy of the aromatic ring is largely responsible for the difference between the β protons. The closer cis proton is shifted to higher frequency. The nonaromatic portion of styrene, $-CH{=}CH_2$, is a *vinyl* group, and the term should be restricted to that structure. The term *alkenic*, not vinylic, should be used generically for protons on double bonds.

Carbonyl groups are strongly electron withdrawing by both induction and resonance. Thus, the β protons on double bonds conjugated with a carbonyl group have very high frequency (downfield) resonances, for example, δ 6.83 in the α,β-unsaturated ester *trans*-$CH_3CH_2O_2CCH{=}CHCO_2CH_2CH_3$. Compounds **3-17** and **3-18** illustrate the

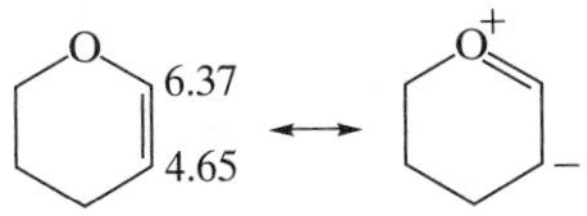

6.37 4.65 5.93 6.88

3-17 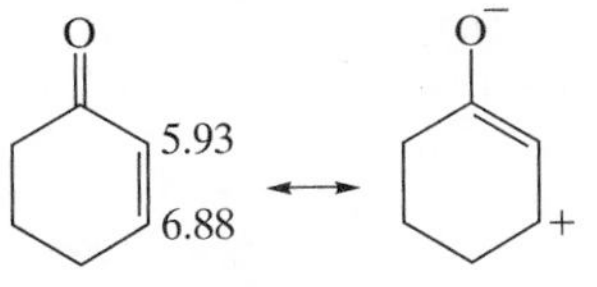 **3-18**

effects of conjugation on alkene chemical shifts. Whereas the alkenic protons of cyclohexene resonate at a normal δ 5.59, the oxygen atom in the unsaturated ether **3-17** donates electrons to the β position by resonance and moves the β proton to lower frequency (δ 4.65). The oxygen withdraws electrons inductively from the α position, whose proton resonance moves to higher frequency (δ 6.37). In contrast, the carbonyl group in the unsaturated ketone **3-18** withdraws electrons from the β position by resonance and so moves the β proton to higher frequency (δ 6.88). In this case, the polar effect of the carbonyl group causes a small shift to higher frequency for the α proton (δ 5.93).

These effects were quantified originally in the empirical approach of Tobey and of Pascual, Meier, and Simon, who used the formula of eq. 3-3

$$\delta = 5.28 + Z_{gem} + Z_{cis} + Z_{trans} \quad (3\text{-}3)$$

to calculate the chemical shift of a proton on a double bond (Section 3-2e). Substituent constants Z_i for groups geminal, cis, or trans to the proton under consideration are added to the chemical shift of ethene. The Z_i values may be found in references 3.5 and 3.6 and are exemplified by the program listed as reference 3.19. Although the parameters Z_i incorporate inductive and resonance effects, steric effects can cause deviations from observed positions.

Aldehydes. The aldehydic proton is shifted to very high frequency (low field) by induction and diamagnetic anisotropy of the carbonyl group. For acetaldehyde (CH_3CHO), the value is δ 9.80, and the range is relatively small, generally $\delta\,10 \pm 0.3$.

3-2c Aromatics

Diamagnetic anisotropy of the benzene ring augments the already deshielding influence of the sp^2 carbon atoms to yield a very high frequency (low field) position for benzene, δ 7.27. Polar and resonance effects of substituents are similar to those in alkenes.

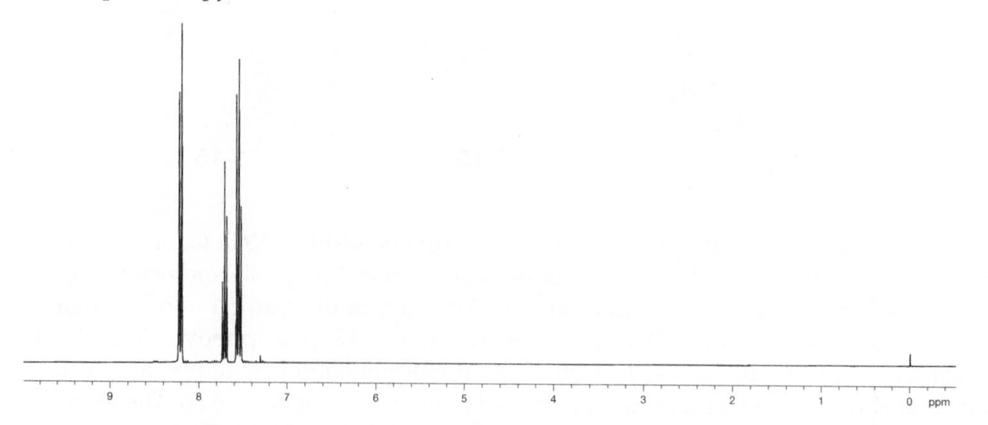

FIGURE 3-10 The 300 MHz ^{1}H spectrum of nitrobenzene in $CDCl_3$.

For toluene ($C_6H_5CH_3$), the electronic effect of the methyl group is small, and all five aromatic protons resonate at about δ 7.2. A narrow range is typical for aromatic rings with saturated hydrocarbon substituents (arenes). Conjugating substituents, however, result in a large spread in the aromatic resonances and in spectral multiplicity from spin–spin splitting. For nitrobenzene (Figure 3-10), the polar effect of the nitro group (**3-19**) moves all resonances to higher frequency with respect to benzene or toluene, but

3-19 **3-20**

3-21 **3-22**

the ortho and para protons are shifted further by electron withdrawal through resonance. By contrast, the methoxy group in anisole (**3-20**) donates electrons by resonance, so the ortho and para positions are at lower frequency than that for benzene. The α protons in heterocycles generally are shifted to high frequency, as in pyridine (**3-21**) and pyrrole (**3-22**), largely because of the polar effect of the heteroatom.

Aromatic proton chemical shifts also may be treated empirically, provided that no two substituents are ortho to each other (producing steric effects). The shift of a particular proton is obtained by adding substituent parameters to the shift of benzene, as shown in eq. 3-4

$$\delta = 7.27 + \Sigma S_i \quad \textbf{(3-4)}$$

and in Table 3-1 (which was compiled from many sources by Jackman and Sternhell). For example, the spectrum of 4-chlorobenzaldehyde (**3-23**) contains two doublets, centered at δ 7.75 and 7.50. The calculated position for H_a is 7.79 (7.27 + 0.58 − 0.06) and for H_b is 7.50 (7.27 + 0.02 + 0.21). Similarly, the observed resonances for 4-methoxybenzoic acid (**3-24**) are at δ 8.08 and 6.98, and the calculated positions are 7.98 for H_a and 6.98 for H_b. Spectral assignments of aromatics therefore often can be made with confidence. For multiply substituted aromatic rings, the identity of the substituents, but not their relative

3-23 **3-24**

TABLE 3-1 Substituent Parameters for Aromatic Proton Shifts

Substituent	S_{ortho}	S_{meta}	S_{para}
CH_3	−0.17	−0.09	−0.18
CH_2CH_3	−0.15	−0.06	−0.18
NO_2	0.95	0.17	0.33
Cl	0.02	−0.06	−0.04
Br	0.22	−0.13	−0.03
I	0.40	−0.26	−0.03
CHO	0.58	0.21	0.27
OH	−0.50	−0.14	−0.40
NH_2	−0.75	−0.24	−0.63
CN	0.27	0.11	0.30
CO_2H	0.80	0.14	0.20
CO_2CH_3	0.74	0.07	0.20
$COCH_3$	0.64	0.09	0.30
OCH_3	−0.43	−0.09	−0.37
$OCOCH_3$	−0.21	−0.02	−0.13
$N(CH_3)_2$	−0.60	−0.10	−0.62
SCH_3	0.37	0.20	0.10

positions, may be known. Calculation of chemical shifts for all possible substitution possibilities and comparison with the observed positions then can produce a structural assignment, as exemplified in reference 3.20.

3-2d Protons on Oxygen and Nitrogen

The NMR properties of protons attached to highly electronegative atoms such as oxygen or nitrogen are influenced strongly by the acidity, basicity, and hydrogen-bonding properties of the medium. For hydroxyl protons, minute amounts of acidic or basic impurities can bring about rapid exchange, as illustrated in Figure 2-30. Such protons then are averaged with other exchangeable protons, either in the same molecule or in other molecules, including the solvent. Only a single resonance is observed for all the exchangeable protons at a weighted-average position. In addition, no coupling is observed with other protons in the molecule. The resonance may vary from being quite sharp to having a characteristically slightly broadened shape, depending on the exchange rate. A convenient experimental procedure to identify hydroxyl resonances in organic solvents such as $CDCl_3$ is to add a couple of drops of D_2O to the NMR tube. Shaking the tube briefly results in exchange of the OH protons with deuterium, which is in molar excess. The aqueous layer separates out, usually to the top in halogenated solvents, and is located above the receiver coil. Consequently, OH resonances in the original spectrum may be identified by their absence in the two-phase case (converted largely to OD). In highly purified basic solvents, such as dimethyl sulfoxide, exchange is slow, and coupling between OH and adjacent protons can be observed.

At infinite dilution in CCl_4 (no hydrogen bonding), the OH resonance of alcohols may be found at about δ 0.5. Under more normal conditions of 5% to 20% solutions, hydrogen bonding results in resonances in the δ 2–4 range. More acidic phenols (ArOH) have resonances at higher frequency (lower field), δ 4–8. If the phenolic hydroxyl can hydrogen bond fully with an ortho group, the position moves to δ 10 or higher. Most carboxylic acids (RCO_2H) exist as hydrogen-bonded dimers or oligomers, even in dilute solution. Because essentially every OH proton is hydrogen bonded, the acid protons resonate at the very high frequency range of δ 11–14 (δ 11.37 for acetic acid, CH_3CO_2H). Other highly hydrogen-bonded protons also may be found in this range, such as sulfonic acids (RSO_3H) or the OH proton of enolic acetylacetone. Because of the variable position and appearance of hydroxy resonances, including that of water, one must be very careful of spectral assignments for the group.

Protons on nitrogen have similar properties, but the slightly lower electronegativity of nitrogen results in lower frequency (higher field) shifts than those of analogous OH protons: δ 0.5–3.5 for aliphatic amines, δ 3–5 for aromatic amines (anilines); δ 4–8 for amides, pyrroles, and indoles; and δ 6–8.5 for ammonium salts. The most common nuclide of nitrogen is ^{14}N, which is quadrupolar and possesses unity spin (Sections 2-2 and 5-1). The three resulting spin states could conceivably split the resonance of attached protons into a 1:1:1 triplet, but such a pattern is seen only in highly symmetrical cases, such as NH_4^+ or NMe_4^+. Otherwise, the rapid relaxation of quadrupolar nuclei averages the spin states. The resonance of a proton on nitrogen thus can vary from a triplet to a sharp singlet, depending on the relaxation rate, but the most common result is a broadened resonance representing incomplete averaging. In some cases, the broadening can render NH resonances almost invisible. In addition, amino protons can exchange rapidly with solvent or other exchangeable protons to achieve an averaged position.

3-2e Programs for Empirical Calculations

Table 3-1 contains data from which empirical calculations of aromatic chemical shifts can be made, and similar data are available for alkane and alkene structures. Corrections, however, must be applied in order to avoid nonadditivity caused primarily by steric effects. Thus, three groups on a saturated carbon atom, two large groups cis to each other on a double bond, or any two ortho aromatic groups can cause deviations from standard parameters. Empirical calculations are possible for any structural entity, so that the eclipsing strain in cyclobutanes, the variety of steric interactions in cyclopentanones, or the variations in angle strain in norbornanes may be taken into account.

Commercial software for carrying out these calculations, listed in the references at the end of this chapter and based on hundreds of thousands of chemical shifts in a database, is widely available. The procedure is begun by drawing the structure of the compound under study. The program then searches the database for molecules with protons whose structural environment resembles that of the compound under study. From the available data, the program calculates and displays the expected proton spectrum. Such information is extremely valuable, because the amount of empirical data available from the program vastly exceeds either the amount resident in the minds of most experimentalists or even in all published compilations.

The approach, however, is subject to four limitations. (1) The specific skeleton or functional groups may not exist in the database. (2) The database may not include sufficient information to assess steric effects that can lead to nonadditivity within an available series. (3) Solvent effects (Section 3-3) are not fully taken into consideration. (4) Coupling constants are calculated from simple relationships, such as the Karplus equation (Section 4-6). Usually, the program provides a list of the compounds used to calculate chemical shifts, so that the experimentalist can judge their relevancy. Sometimes, the compound under study in fact proves to be in the database, so that the real spectrum is reproduced. If not, the experimentalist always should review the structures of the compounds used for the calculations and decide whether they are sufficiently similar to trust the calculations. Despite considerable progress in this field, these calculations still do not accurately (all lines within <0.1 ppm of the observed positions) reproduce most proton spectra. Semiempirical and ab initio calculations of spectra do not suffer from most of these drawbacks. Density functional theory (DFT) in particular has been used with some success by Bagno (reference 3.21).

3-3 MEDIUM AND ISOTOPE EFFECTS

Medium Effects. The observed shielding of a particular nucleus consists of intramolecular components σ_{intra} (already discussed with regard to protons in Section 3-1) and intermolecular components σ_{inter} (eq. 3-5).

$$\sigma = \sigma_{\text{intra}} + \sigma_{\text{inter}} \tag{3-5}$$

Buckingham, Schaefer, and Schneider pointed out five sources of intermolecular shielding (eq. 3-6).

$$\sigma_{\text{inter}} = \sigma_B + \sigma_W + \sigma_E + \sigma_A + \sigma_S \quad \textbf{(3-6)}$$

We shall consider each contribution in turn and then give several illustrations of the effects of the medium.

The solvent has a bulk diamagnetic susceptibility that is dependent on the shape of the sample container. Thus, the solvent in a spherical container shields the solute to an extent that is slightly different from the shielding afforded by the solvent in a cylindrical container. The differences are expressed by eq. 3-7,

$$\sigma_B = \left(\frac{4}{3}\pi - \alpha\right)\chi_V \quad \textbf{(3-7)}$$

in which α is a geometric parameter and χ_V is the volume susceptibility of the solvent. For a sphere, $\alpha = 4\pi/3$, so there is no effect. The effect does not disappear for a cylinder ($\alpha = 2\pi/3$), for which σ_B is $2\pi\chi_V/3$. Normally, the solute and the standard (TMS) are present in the same solution. Under these circumstances, they experience parallel bulk effects, and no correction for σ_B on the relative shift is necessary. Since internal standards are common, the effect of bulk susceptibility is largely ignored. A correction would be necessary only if chemical shifts had to be compared for data obtained without an internal standard from containers of different shape—for example, the normal cylinder and a spherical microtube—or for solvents with different volume susceptibilities.

Close approach of the solute and the solvent can distort the shape of the electron cloud around a proton and deshield it, even when both components are nonpolar. Such a phenomenon (σ_W in eq. 3-6) is analogous to the van der Waals effect on the chemical shift. The magnitude is rarely more than 0.3 ppm. If chemical shifts are measured from the resonance of an internal standard, the contribution from σ_W should affect the solute and the standard similarly. Chemical shifts so measured should be largely independent of σ_W.

A polar solute, or even a nonpolar molecule with polar groups, induces an electric field in the surrounding dielectric medium. This reaction field, proportional to $(\epsilon - 1)/(\epsilon + 1)$ (ϵ is the dielectric constant), can influence the shielding of protons elsewhere in the molecule. Generally, the effect (σ_E) is largest for protons close to the polar group. The sign can be either positive or negative because of an angular dependence, but more often it is negative (indicating deshielding). This effect is not compensated for by the use of an internal standard. Even within the solute molecule, the effect can be quite variable for different protons. For polar molecules in solvents of high dielectric constant, σ_E can range up to 1 ppm. The effect may be minimized by the use of solvents with small dielectric constants.

An anisotropic solvent will not orient itself completely randomly with respect to the solute. Thus, even a nonpolar molecule such as methane will be exposed preferentially to the shielding face of benzene or to the deshielding side of acetonitrile. In general, aromatic (dishlike) solvents induce shifts (σ_A) to lower frequency (upfield) and rodlike solvents (acetylenes, nitriles, and CS_2) induce shifts to higher frequency. In the absence of any special solute–solvent interaction σ_S (charge transfer, dipole–dipole, or hydrogen bond), the internal standard and the solute exhibit similar anisotropic shifts, so that σ_A is compensated for in the δ value. As often as not, however, there is a special interaction between the solvent and a polar group in the solute molecule. As a result, the anisotropic solvent has a different effect on different protons, and solute resonances can undergo real shifts up to 0.5 ppm with respect to the internal standard. In some cases, only certain protons near a functional group are affected. The solvent effect then can be used to bring about differential shifts within the spectrum of a molecule. This technique is useful in spectral analysis to alter, for example, a case of accidental overlap. Because dishlike and rodlike solvents cause shifts in opposite directions, the investigator has some control over the relative movements of the resonances.

The chemical shift alterations caused by aromatic molecules have been termed *Aromatic Solvent-Induced Shifts*, with the acronym ASIS. The chemical shift in the aromatic solvent is compared with the resonance position in $CDCl_3$ via eq. 3-8.

$$\Delta_{C_6H_6}^{CDCl_3} = \delta_{CDCl_3} - \delta_{C_6H_6} \tag{3-8}$$

Because the shift is usually to lower frequency (upfield), Δ normally is positive. Although anisotropic shifts are frequently strongest close to a polar group, they may be differentiated from electric-field effects by the dependence of the latter on the dielectric constant of the solvent.

Specific interactions between solvent and solute, such as hydrogen bonding, can cause quite large effects (σ_S). It is not known whether the ASIS is caused by a time-averaged cluster of solvent molecules about a polar functional group or by a 1:1 solute–solvent charge transfer complex. In the latter case, the ASIS is more legitimately classified as σ_S rather than σ_A.

In an early study of solvent shifts, Buckingham, Schaefer, and Schneider examined the solute methane. For this molecule, σ_E and σ_S are zero and σ_B may be calculated or allowed to cancel by use of an internal standard. Thus, only σ_W and σ_A should affect the solute chemical shift. The authors plotted the difference between the chemical shift of methane in a given solvent and that in the gas phase vs. the heat of vaporization of the solvent at the boiling point. The latter quantity was taken as a measure of the van der Waals interaction. More than a dozen solvents, including neopentane, cyclopentane, hexane, cyclohexane, the 2-butenes, ethyl ether, acetone, $SiCl_4$, and $SnCl_4$, fell on a straight line with a small negative slope. The shifts to higher frequency (downfield) from gaseous methane ranged from 0.13 (neopentane) to 0.32 ppm ($SnCl_4$), largely as a function of atomic polarizability. The linear relationship with ΔH_v indicates that these shifts are due solely to σ_W. Well above this line are the dish-shaped aromatic molecules: benzene, toluene, and chlorobenzene, as well as nitromethane and nitroethane, whose nitro group serves as an anisotropic oblate ellipsoid. Below the line are the rodlike molecules acetonitrile, methylacetylene, dimethylacetylene, butadiyne, and carbon disulfide. Deviations from the Δv-vs.-ΔH_v line are due to true anisotropic shifts (σ_A), since the nonpolar, isotropic methane molecule has no direct σ_S interactions with the solvent. The largest upward displacement from the line was nitrobenzene (0.72 ppm), and the largest downward displacement was NC—C═C—CN (0.53 ppm).

Molecule **3-25** provides an interesting example of the electric-field effect. Table 3-2 contains the resonance positions of the methyl groups on C-8, C-10, and C-13 in cyclohexane ($\epsilon = 2.02$) and in dichloromethane ($\epsilon = 9.1$). These solvents were chosen for a low σ_A effect; σ_W and σ_B are discounted by the use of an internal standard. The methyl groups close to the ether linkage (C-8, C-13) are shifted about 0.07 ppm to higher

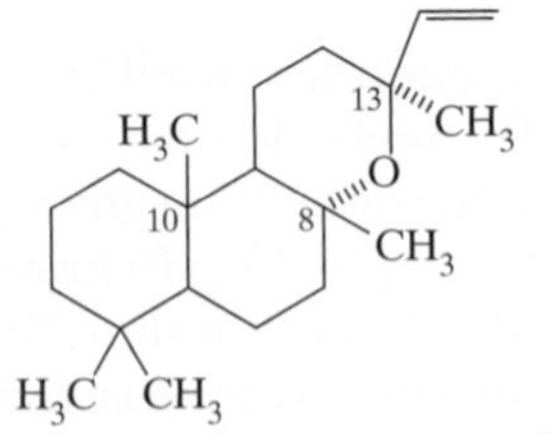

3-25

TABLE 3-2 Methyl Chemical Shifts (δ) as a Function of Solvent (Laszlo)

Solvent	ϵ	v_{10}	v_8	v_{13}
Cyclohexene	2.02	0.855	1.01	1.12
CH_2Cl_2	9.1	0.852	1.07	1.20

frequency by an electric-field effect. There is little or no shift of the C-10 methyl resonance, since the reaction field diminishes rapidly with distance.

A large ASIS appears on the relative chemical shifts of the nonequivalent methyl groups in *N,N*-dimethylformamide (**3-26**). The distance between the two methyl peaks

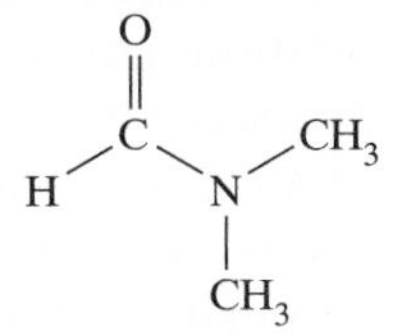

3-26

increases by up to 1.7 ppm upon replacement of $CHCl_3$ by benzene. Interestingly, the lower-frequency peak is responsible for almost all the shift. This observation was explained in terms of a short-lived 1:1 solvent–solute complex in which one of the methyl groups is situated over the benzene ring and the other is directed away from the ring. The model of a 1:1 complex between cyclohexanone and benzene has been used to justify the shift of the 2,6-axial protons (located above the benzene ring) to lower frequency and the negligible shift of the 2,6-equatorial protons (near the region of no effect at $\theta = 55°44'$). These shifts also have been explained in terms of a time-averaged solvent cluster, which produces the same effect without recourse to a short-lived 1:1 complex.

Isotope Effects. Isotopic changes within a molecule can alter the chemical shifts of neighboring nuclei. The methyl group in toluene is 0.015 ± 0.002 ppm higher frequency than the corresponding protons in toluene-α-d_1. The protons in cyclohexane are 0.057 ppm higher frequency than the single proton in cyclohexane-d_{11}. The effect is larger for other nuclei, such as ^{19}F, but falls off very rapidly with distance. The isotope shift is caused by differences in zero point vibrational energies between the H and D systems. Differential isotope effects have been exploited in a study of the ring reversal of 1,4-dioxane (eq. 3-9).

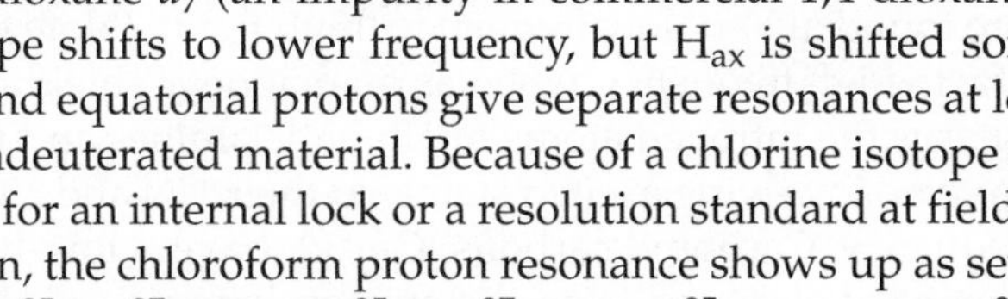

(3-9)

In undeuterated dioxane, H_{eq} and H_{ax} coincidentally have the same chemical shift (at the field studied), so they cannot be differentiated at low temperatures (Sections 2-8 and 5-2). In 1,4-dioxane-d_7 (an impurity in commercial 1,4-dioxane-d_8), both H_{ax} and H_{eq} exhibit isotope shifts to lower frequency, but H_{ax} is shifted somewhat more. As a result, the axial and equatorial protons give separate resonances at low temperatures, in contrast to the undeuterated material. Because of a chlorine isotope effect, chloroform is a poor substance for an internal lock or a resolution standard at fields above about 9.4 T. At high resolution, the chloroform proton resonance shows up as several closely spaced peaks, due to $CH(^{35}Cl)(^{37}Cl)_2$, $CH(^{35}Cl)_2(^{37}Cl)$, $CH(^{35}Cl)_3$, and $CH(^{37}Cl)_3$.

3-4 FACTORS THAT INFLUENCE CARBON SHIFTS

Carbon is the defining element in organic compounds, but its major nuclide (^{12}C) has a spin of zero. The advent of pulsed Fourier transform methods in the late 1960s made the examination of the low-abundance nuclide ^{13}C (1.11%) a practical spectroscopic technique. The low probability of having two adjacent ^{13}C nuclei in a single molecule [$(0.0111)^2 = 0.0001$, or 0.01%] removes complications from carbon–carbon couplings. When carbon–hydrogen couplings are removed by decoupling techniques, the spectrum is essentially free of all spin–spin coupling, and one singlet arises for each distinct type of carbon. Analysis therefore is simpler for ^{13}C than for ^{1}H spectra. Integration, however, is less reliable, because carbons have a much larger range of relaxation times than do protons and because the decoupling field perturbs intensities (Chapter 5). Interpretation of ^{13}C chemical shifts nonetheless is straightforward and often more useful than that of proton shifts.

Diamagnetic shielding (σ^d), which is responsible for proton chemical shifts, is caused by circulation of the electron cloud about the nucleus, as depicted in Figure 3-1b. Hindrance to free electron circulation creates an additional mechanism called *paramagnetic shielding* (σ^p). Although s electrons circulate freely, 2p electrons have angular momentum that can hinder free circulation. Because σ^p serves to reduce σ^d, the two mechanisms have opposite signs. Protons are surrounded solely by s electrons (lacking angular momentum) and consequently exhibit only diamagnetic shielding. Carbon nuclei (and almost all other nuclides as well) additionally are surrounded by p electrons and exhibit both forms of shielding. In this context, the word *paramagnetic* should not be confused with its common usage to describe molecules with unpaired electrons. Some authors have referred to shielding in closed shell molecules for nuclides with p electrons as the *second-order paramagnetic effect*, yielding first place to phenomena associated with unpaired electrons.

The paramagnetic component can be quite large. Whereas the chemical-shift range for protons is only a few parts per million, paramagnetic shifts can extend over a range of hundreds or even thousands of parts per million for other nuclei. Qualitatively, angular momentum can arise from excited electronic states and from π bonding. The effects are larger when electron density about the nucleus increases. These three considerations were gathered by Ramsay, Karplus, and Pople into the simple empirical relationship of eq. 3-10.

$$\sigma^p \propto -\frac{1}{\Delta E} < r^{-3} > \Sigma Q_{ij} \tag{3-10}$$

The quantity ΔE is the average energy of excitation of certain electronic transitions, such as the n $\rightarrow \pi^*$ transition for many ^{13}C and ^{15}N nuclei. The radial term $<r^{-3}>$ includes the average distance r from the nucleus of the 2p electrons (for second row elements like carbon). This term serves as a measure of electron density. Finally, ΣQ_{ij} represents π bonding to carbon. The negative sign in the equation indicates that paramagnetic shielding is in the opposite direction from σ^d.

Structural changes can affect all three components of the equation. The quantity ΔE in eq. 3-10 represents the weighted-average energy difference between the ground and certain excited states. Because of symmetry considerations, the $\pi \rightarrow \pi^*$ transition often is excluded. Low-lying excited states (with small ΔE) make the largest contribution, since ΔE appears in the denominator. Saturated molecules, such as alkanes, typically have no low-lying excited states (and hence possess a large ΔE), so that σ^p is small and alkane carbon resonances are found at very low frequency (high field). (Note that paramagnetic shielding causes shifts to high frequency, whereas diamagnetic shielding causes shifts to low frequency.) Similarly, the nitrogen atoms in aliphatic amines and the oxygen atoms in aliphatic ethers have no low-lying excited states, so their ^{15}N and ^{17}O resonances also are found at low frequency. Carbonyl carbons, C$=$O, have a low-lying excited state involving the movement of electrons from the oxygen lone pair to the antibonding π orbital that generates a paramagnetic current. This n $\rightarrow \pi^*$ transition causes the large shift to high frequency that characterizes carbonyl groups—up to 220 ppm from the zero of TMS. Even larger shifts to high frequency, viz., δ 335, have been observed for carbocations, R_3C^+.

The radial term in eq. 3-10 is responsible for effects related to electron density that parallel polar effects on proton chemical shifts. Paramagnetic shielding is larger when the p electrons are closer to the nucleus. Thus, substituents that donate or withdraw electrons influence the paramagnetic shift. Electron donation increases repulsion between electrons, which can be relieved by an increase in r. The paramagnetic shielding then decreases, causing a shift to lower frequency (upfield). Similarly, electron withdrawal permits electrons to move closer to the nucleus, increasing the paramagnetic shielding and causing a shift to higher frequency. Hence, placing a series of electron-withdrawing atoms on carbon results in progressively higher frequency shifts, as in the series CH_3Cl (δ 25), CH_2Cl_2 (δ 54), $CHCl_3$ (δ 78), and CCl_4 (δ 97). The situation is qualitatively similar to that for protons, but the numbers are much larger because the shift is from the paramagnetic term. Substituent effects in both cases, however, generally follow the electronegativity of groups attached to carbon.

Electronegativity is a measure of the ability of a nucleus to attract electrons. A highly electronegative element, such as oxygen, attracts p electrons more than does carbon and reduces the value of r. Thus, ^{17}O shifts are correspondingly larger than ^{13}C shifts. A plot of the ^{17}O shifts of aliphatic ethers versus the ^{13}C shifts of the analogous alkanes is linear with a slope of about 3. The linearity shows that oxygen and carbon chemical shifts are sensitive to the same structural factors, and the slope indicates that oxygen is more sensitive to these factors, because its 2p electrons are closer to the nucleus.

The third factor in eq. 3-10, ΣQ_{ij}, is related to charge densities and bond orders and can be considered a measure of multiple bonding. The greater the degree of multiple bonding, the greater is the shift to high frequency (low field). This term provides a rationale for the series ethane (δ 6), ethene (δ 123), and the central sp-hybridized carbon of allene (δ 214). Arene shifts are similar to those of alkenes (benzene, δ 129). The effects of diamagnetic anisotropy on carbon chemical shifts are similar in magnitude to the effects on protons, but are small in relation to the range of carbon shifts. The chemical shifts of alkynes do not follow this pattern, but are at an intermediate position (δ 72 for acetylene), because their linear structure has zero angular momentum about the C≡C axis.

Interpretation of the chemical shifts of most elements other than hydrogen is accomplished by analyzing the three factors in eq. 3-10: accessibility of certain excited states, distance of the p electrons, and multiple bonding. For carbon, the shifts of alkanes, alkenes, arenes, alkynes, and carbonyl groups and the effects of electron-donating or electron-withdrawing groups may be interpreted in this fashion. There are exceptions, the most prominent being the effect of heavy atoms. The series CH_3Br (δ 10), CH_2Br_2 (δ 22), $CHBr_3$ (δ 12), and CBr_4 (δ −29) defies any explanation based on electronegativity, unlike the analogous series given before for chlorine. The same series with iodine is monotonic to lower frequency (δ −290 for CI_4), that is, opposite to the chlorine series. This so-called heavy-atom effect has been attributed to a new source of angular momentum from *spin–orbit coupling*. These anomalous shifts to low frequency (high field) can be expected when nuclei other than hydrogen have a heavy-atom substituent.

3-5 CARBON CHEMICAL SHIFTS AND STRUCTURE

Figure 3-11 illustrates general ranges for ^{13}C chemical shifts.

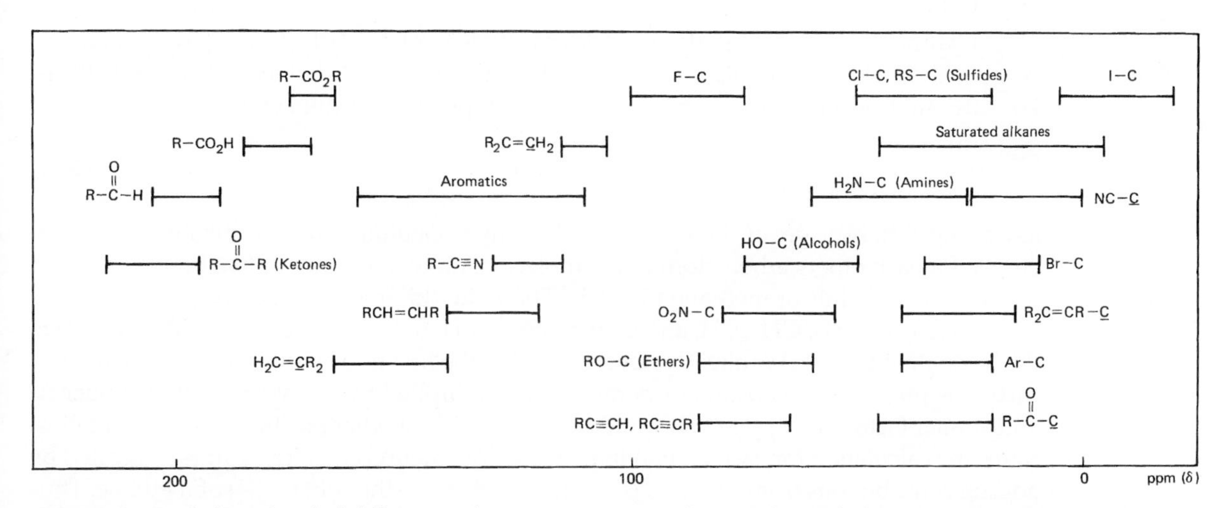

FIGURE 3-11 Carbon chemical shift ranges for common structural units. The symbol C represents methyl, methylene, methine, or quaternary carbon; R represents a saturated alkyl group. The indicated ranges are for common examples; actual ranges can be larger.

3-5a Saturated Aliphatics

Acyclic Alkanes. The absence of low-lying excited states and of π bonding minimizes paramagnetic shielding and places alkane chemical shifts at very low frequencies (high fields). Methane itself resonates at δ −2.5. The series ethane (CH_3CH_3), propane ($CH_3\mathbf{CH_2}CH_3$), and isobutane [$(CH_3)_3\mathbf{CH}$] follows a steady trend to higher frequency (δ 5.7, 16.1, 25.2), similar to the trend for proton shifts in the series methyl, methylene, and methine. Replacement of H by CH_3 adds about 9 ppm to the chemical shift of the attached carbon. The effect is similar for replacement by saturated CH_2, CH, or C. Because the added methyl group is attached directly to the resonating carbon, the shift has been termed the *α effect* (**3-27**). The effect is not restricted to the replacement of H by carbon.

α effect

—C—H → —C—X

3-27

β effect

—C—Y—H → —C—Y—X

3-28

Any group X that replaces hydrogen on a resonating carbon atom causes a relatively constant shift that depends primarily on the electronegativity of X.

The replacement of hydrogen by CH_3 (or by CH_2, CH, or C) at a β position (**3-28**) also causes a constant shift of about +9 ppm. Thus, the central carbon in pentane ($CH_3CH_2\mathbf{CH_2}CH_2CH_3$) is shifted by the α effects of the two methylene groups and by the *β effects* of the two methyl groups to δ 34.7. The replacement of a γ hydrogen (**3-29**)

γ effect

—C—Z—Y—H → —C—Z—Y—X

3-29

by CH_3 (or by CH_2, CH, or C) causes a shift of about −2.5 ppm (to low frequency, or upfield). Unlike the α and β effects, this *γ effect* has an important stereochemical component. Because of the α, β, and γ effects, the alkane chemical shift range is relatively large. Methyl resonances in alkanes are typically found at δ 5–20, depending on the number of β substituents; methylene resonances are at δ 15–35, and methine resonances at δ 25–45.

Carbon-13 chemical shifts lend themselves conveniently to empirical analysis, because these shifts are easily measured and tend to have well-defined substituent effects. For saturated, acyclic hydrocarbons, Grant developed the formula of eq. 3-11

$$\delta = -2.5 + \Sigma A_i n_i \qquad \textbf{(3-11)}$$

as an empirical measure of chemical shifts. For any resonating carbon, a substituent parameter A_i for each other carbon atom in the molecule, up to a distance of five bonds, is added to the chemical shift of methane (δ −2.5). There are different substituent parameters for carbons (any of CH_3, CH_2, CH, and C) that are α (9.1), β (9.4), γ (−2.5), δ (0.3), or ϵ (0.1) to the resonating carbon. We already have alluded to the first three figures. If more than one α carbon is present, the substituent parameter is multiplied by the appropriate number n_i, and similar factors are applied for multiple substitution at other positions. Figure 3-12 illustrates the calculation for each carbon in pentane. The methyl chemical shift is calculated by adding contributions from single α, β, γ, and δ carbons to the shift (−2.5) of methane. Thus the shift of the 2 carbon is calculated by adding contributions of two α carbons, one β carbon, and one γ carbon. Usually, the observed shifts are calculated to within 0.3 ppm, providing a reliable means for spectral assignment.

$\overset{\alpha}{\mathbf{C}}H_3\overset{\beta}{C}H_2\overset{\gamma}{C}H_2\overset{\delta}{C}H_2CH_3$ $\delta = -2.5 + 9.1 + 9.4 - 2.5 + 0.3 = 13.8$ (obs. 13.9)

$\overset{\alpha}{C}H_3\overset{\alpha}{\mathbf{C}}H_2\overset{\beta}{C}H_2\overset{\gamma}{C}H_2CH_3$ $\delta = -2.5 + (9.1 \times 2) + 9.4 - 2.5 = 22.6$ (obs. 22.8)

$\overset{\beta}{C}H_3\overset{\alpha}{C}H_2\mathbf{C}H_2\overset{\alpha}{C}H_2\overset{\beta}{C}H_3$ $\delta = -2.5 + (9.1 \times 2) + (9.4 \times 2) = 34.5$ (obs. 34.7)

FIGURE 3-12 Calculation of the ^{13}C chemical shifts of pentane.

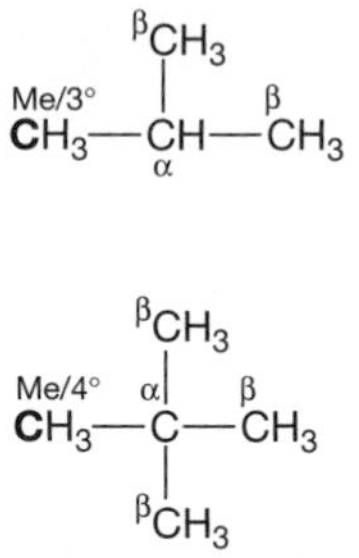

$^{\beta}CH_3$, Me/3°, $\mathbf{C}H_3—\overset{\alpha}{C}H—\overset{\beta}{C}H_3$ $\delta = -2.5 + 9.1 + (9.4 \times 2) - 1.1 = 24.3$ (obs. 24.3)

$^{\beta}CH_3$, Me/4°, $\mathbf{C}H_3—\overset{\alpha}{C}(^{\beta}CH_3)—\overset{\beta}{C}H_3$ $\delta = -2.5 + 9.1 + (9.4 \times 3) - 3.4 = 31.4$ (obs. 31.7)

FIGURE 3-13 Calculation of the ^{13}C chemical shifts of the indicated carbon in 2-methylpropane (isobutane) and in 2,2-dimethylpropane (neopentane).

There are complications, however. Corrections must be applied if branching is present, because eq. 3-11 applies rigorously only to straight chains. The resonance position of a methyl group is corrected for the presence of an adjacent tertiary (CH) carbon by adding −1.1 and for an adjacent quaternary carbon by adding −3.4. Methylene carbons have corrections of −2.5 and −7.2, respectively, for adjacent tertiary and quaternary carbons. Methine carbons have respective corrections of −3.7, −9.5, and −1.5 for adjacent secondary, tertiary, and quaternary carbons. Finally, quaternary carbons have corrections of −1.5 and −8.4 for adjacent primary and secondary carbons. Corrections for adjacent tertiary and quaternary carbons undoubtedly are significant, but are not known accurately. For example, the methyl group in isobutane (first calculation in Figure 3-13) is adjacent to a tertiary carbon. The methyl chemical shift is calculated by adding the contributions of one α carbon, two β carbons, and the correction of −1.1, since the methyl group is adjacent to a tertiary center. In the second calculation in Figure 3-13 (for neopentane), the methyl group is adjacent to a quaternary center.

It is noteworthy that the γ effect of a carbon substituent is negative (−2.5). A γ carbon can be either gauche or anti (Figure 3-14) to the resonating carbon, and the proportion of conformers can vary from molecule to molecule. The value −2.5 is a weighted average for open-chain conformers and does not serve accurately for all situations. For a pure *γ-anti effect*, the shift is about +1, and for a pure *γ-gauche effect*, it is about −6. The average value of −2.5 measured by Grant clearly indicates a mix of the two conformations. Hydrocarbons with unusually large deviations from the average mix may give poor results with eq. 3-11. The α and β effects are determined by fixed geometries and have no stereochemistry component.

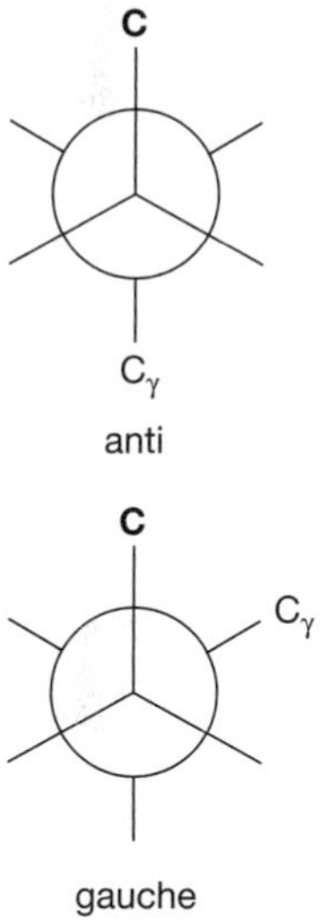

FIGURE 3-14 The anti and gauche geometries in a butane fragment.

Cyclic Alkanes. With a resonance position of δ −2.6, cyclopropane has the lowest-frequency resonance of hydrocarbons. Cyclobutane resonates at δ 23.3, and the remaining cycloalkanes generally resonate within 2 ppm of cyclohexane, at δ 27.7. The fixed stereochemistry represented by cyclohexane requires an entirely new set of empirical parameters that depend on the axial or equatorial nature of any substituents, as well as on the distance from the resonating carbon. Table 3-3 lists Grant's parameters for methyl substitution that

TABLE 3-3 Substituent Parameters for Methyl Substitution on Cyclohexane

Stereochemistry	α	β	γ	δ
Equatorial	5.6	8.9	0.0	−0.3
Axial	1.1	5.2	−5.4	−0.1

are added to the value for cyclohexane (δ 27.7). The substituent parameter for a γ-axial methyl is large and negative (-5.4), reflecting the pure gauche stereochemistry between the perturbing and resonating carbons. A γ-equatorial group represents a γ-anti effect and has little perturbation in this case. Corrections again are needed for branching. For two α methyls that are geminal (both on the resonating carbon), the correction is -3.4; for two β methyls that are geminal, it is -1.2. Thus, the calculated resonance position for C2 of 1,1,3-trimethylcyclohexane (**3-30**) is $27.7 + 5.2 + (8.9 \times 2) - 1.2 = 49.5$ (observed

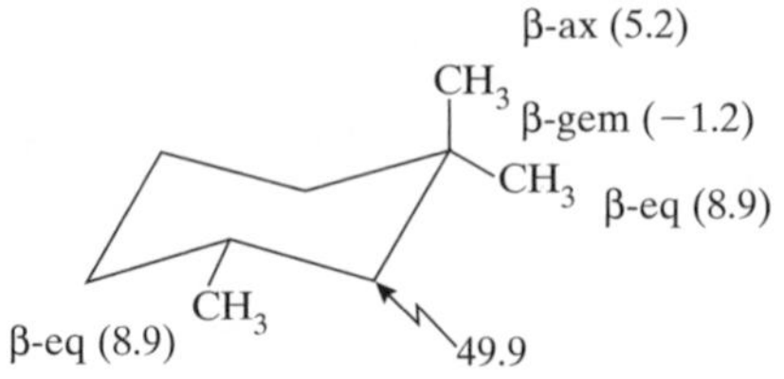

β-ax (5.2)
CH_3 β-gem (−1.2)
CH_3 β-eq (8.9)
β-eq (8.9) CH_3
49.9

3-30

value, 49.9). A pair of vicinal, diequatorial methyls that are respectively α and β to the resonating carbon require a correction of -2.3, and similar axial–equatorial vicinal methyls require a correction of -3.1.

Functionalized Alkanes. The replacement of a hydrogen atom on carbon with a heteroatom or an unsaturated group usually results in shifts to higher frequency (downfield) because of polar effects on the radial term. The effect parallels the same structural change on ^{1}H chemical shifts but arises from a different mechanism. Strongly electron-withdrawing groups have large positive α effects. In the halogen series CH_3X, the methyl chemical shifts are δ 75.4 for X = fluorine, 25.1 for chlorine, 10.2 for bromine, and -20.6 for iodine. Multiple substitution results in larger effects—δ 77.7 for $CHCl_3$. Recall that the α effect of heavy atoms such as iodine or bromine is influenced by a spin–orbit mechanism and hence may not follow the simple order of electronegativity. The general range for the α halogen effect in hydrocarbons extends from the values given above for the simple CH_3X systems to about a 25-ppm higher-frequency (downfield) shift for CH_2X and CHX systems, since the α and β effects of the unspecified hydrocarbon pieces contribute to the shift to higher frequency.

Methanol (CH_3OH) resonates at δ 49.2, and the range for hydroxy-substituted carbons is δ 49–75. Dimethyl ether [$(CH_3)_2O$] resonates at δ 59.5, and the range for alkoxy-substituted carbons is δ 59–80. The ether range is translated a few parts per million to higher frequency (downfield) from alcohols, because each ether must have one additional β effect with respect to the analogous alcohol.

The lower electronegativity of nitrogen moves the amine range to a somewhat lower frequency (upfield). Methylamine in aqueous solution resonates at δ 28.3, the range for amines extending some 30 ppm to higher frequency. The amine range is larger than the alcohol range because nitrogen can carry up to three substituents, with the possibility of more α and β effects. Dimethyl sulfide resonates at δ 19.5, acetonitrile at δ 1.8, and nitromethane at δ 62.6, with the respective ranges for thioalkoxy, cyano, and nitro substitution extending some 25 ppm to higher frequency. The anomalous low-frequency position for cyano substitution is related to the cylindrical shape of the group and its reduced angular momentum.

An attached double bond has only a small effect on a methyl group. The position for the methyls of *trans*-2-butene (*trans*-$CH_3CH{=}CHCH_3$) is δ 17.3, and that for the methyl of toluene ($C_6H_5CH_3$) is δ 21.3. The range for carbons on double bonds is about δ 15–40. Methyls on carbonyl groups are at a slightly higher frequency (lower field): δ 30.8 for acetone and δ 31.2 for acetaldehyde, with a range of about δ 30–45.

Introducing heteroatoms or unsaturation into alkane chains requires completely new sets of empirical parameters that depend on the substituent, on its distance from the resonating carbon (α, β, γ), and on whether the substituent is terminal (**3-31**) or internal (**3-32**) (Table 3-4). These numbers represent the effect on a resonating carbon of replacing a

TABLE 3-4 Carbon Substituent Parameters for Functional Groups

	Terminal X (3-31)			Internal X (3-32)		
X	α	β	γ	α	β	γ
F	68	9	−4	63	6	−4
Cl	31	11	−4	32	10	−4
Br	20	11	−3	25	10	−3
I	−6	11	−1	4	12	−1
OH	48	10	−5	41	8	−5
OR	58	8	−4	51	5	−4
OAc	51	6	−3	45	5	−3
NH_2	29	11	−5	24	10	−5
NR_2	42	6	−3			−3
CN	4	3	−3	1	3	−3
NO_2	63	4		57	4	
$CH{=}CH_2$	20	6	−0.5			−0.5
C_6H_5	23	9	−2	17	7	−2
C≡CH	4.5	5.5	−3.5			−3.5
(C=O)R	30	1	−2	24	1	−2
(C=O)OH	21	3	−2	16	2	−2
(C=O)OR	20	3	−2	17	2	−2
$(C{=}O)NH_2$	22		−0.5	2.5		−0.5

From F. W. Wehrli, A. P. Marchand, and S. Wehrli, *Interpretation of Carbon-13 NMR Spectra*, 2nd ed., Chichester, UK: John Wiley & Sons, Ltd., 1988.

Terminal

3-31

Internal

3-32

hydrogen atom at the respective position with a group X. With the exception of cyano, acetyleno, and the heavy atom iodine, the α effects are determined largely by the electronegativity of the substituent. It is interesting that the β effects are all positive and generally of similar magnitude (6 to 11 ppm) and that the γ effects are all negative and generally of similar magnitude (−2 to −5 ppm). Although the details are not entirely understood, it is clear that simple polar considerations do not dominate the β and γ effects.

To use the substituent parameters given in Table 3-4, one adds the appropriate values to the chemical shift of the carbon in the unsubstituted hydrocarbon analogue, rounding off to the nearest part per million. As seen in Figure 3-15, the chemical shift of the 1 carbon of 1,3-dichloropropane may be calculated from the value (16) for the methyl carbon of propane and from the figures in Table 3-4. The chemical shift of the β carbon of cyclopentanol similarly may be calculated from the value (27) for cyclopentane.

$\overset{\alpha}{Cl}—CH_2CH_2CH_2—\overset{\gamma}{Cl}$ $\delta = 16 + 31 - 4 = 43$ (obs. 42)

βOH

$\delta = 27 + 8 = 35$ (obs. 34)

FIGURE 3-15 Calculation of the ^{13}C chemical shifts of the indicated carbon in 1,3-dichloropropane and in cyclopentanol.

3-5b Unsaturated Compounds

The effects of diamagnetic anisotropy on a carbon and a proton have similar magnitudes, but the much larger paramagnetic shielding renders the phenomenon relatively unimportant for carbon. Thus, benzene (δ 128.4) and the alkenic carbon of cyclohexene (δ 127.3) have almost identical carbon resonance positions, in contrast to the situation with their protons. The full range of alkene and aromatic resonances is about δ 100–170.

Alkenes. Alkenic carbons that bear no substituents ($=CH_2$) resonate at low frequency (high field), such as isobutylene [$(CH_3)_2C=CH_2$] at δ 107.7, and have a range of about δ 104–115 for hydrocarbons. Alkenic carbons that have one substituent ($=CHR$), like those in *trans*-2-butene (δ 123.3), resonate in the range δ 120–140. Finally, disubstituted alkenic carbons ($=CRR'$), like that in isobutylene δ (146.4), resonate at the highest frequency δ (140–165). Polar substituents on double bonds, particularly those in conjugation with the bond, can alter the resonance position appreciably. Unsaturated ketones, such as **3-33** and **3-34**, have lower-frequency α resonances and higher-frequency β resonances. The effect is

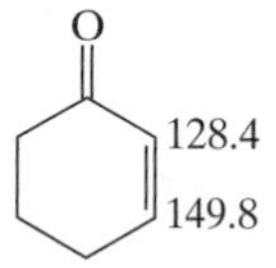

O
128.4
149.8
3-33

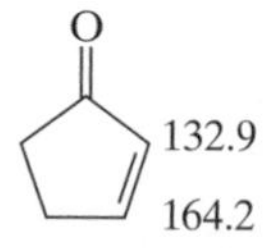

O
132.9
164.2
3-34

reduced in acyclic molecules. Electron donation, as in enol ethers, reverses the effect: $CH_2=CHOCH_3$ [$\delta(\alpha)$ 153.2, $\delta(\beta)$ 84.2]. Electron donation or withdrawal alters the radial term through resonance (mesomerism).

Alkene chemical shifts may be estimated from substituent parameters added to the shift for ethene (δ 123.3). For α, β and γ carbons on the same end of the double bond as the resonating carbon, respective increments of 10.6, 7.2, and -1.5 are added. For α', β', and γ' carbons on the opposite end of the double bond from the resonating carbon, respective increments of -7.9, -1.8, and -1.5 are added. An increment of -1.1 is added if any two substituents are cis to each other. Thus, the chemical shift of the unsubstituted alkene carbon in 1-butene ($CH_3CH_2CH=\mathbf{CH_2}$) is calculated to be $123.3 - 7.9 - 1.8 = 113.6$ (observed value, δ 113.3), and that of the substituted carbon ($CH_3CH_2\mathbf{CH}=CH_2$) is $123.3 + 10.6 + 7.2 = 141.1$ (observed value, δ 140.2).

Alkynes and Nitriles. An alkyne carbon that carries a hydrogen substituent ($\equiv CH$) generally resonates in the narrow range δ 67–70. An alkyne carbon that carries a carbon substituent ($\equiv CR$) resonates at a slightly higher frequency (δ 74–85), because of α and β effects from the R group. Effects of conjugating, polar substituents expand the total range to δ 20–90. Nitriles resonate in the range δ 117–130 (acetonitrile is at δ 116.9). The $n \rightarrow \pi^*$ transition pushes the range to high frequency.

Aromatics. Alkyl substitution, as in toluene (**3-35**), has its major (α) effect on the ipso carbon. Because this carbon has no attached proton, its relaxation time is much longer than those of the other carbons, and its intensity is usually lower. Conjugating substituents like nitro (**3-36**) have strong perturbations on the aromatic resonance positions, as the result of a combination of traditional α, β, γ effects and changes in electron density through delocalization (**3-19, 3-20**). A similar interplay of effects is seen in the resonance positions of pyridine (**3-37**) and pyrrole (**3-38**).

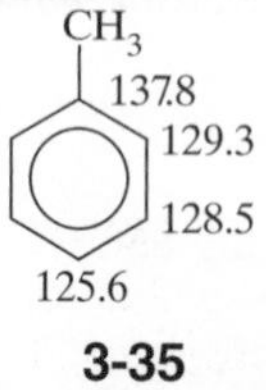

CH_3
137.8
129.3
128.5
125.6
3-35

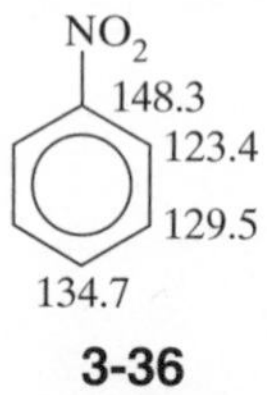

NO_2
148.3
123.4
129.5
134.7
3-36

TABLE 3-5 Carbon Substituent Parameters for Aromatic Systems

X	Ipso	Ortho	Meta	Para
CH_3	8.9	0.7	−0.1	−2.9
CH_2OH	13.3	−0.8	−0.6	−0.4
$CH{=}CH_2$	9.5	−2.0	0.2	−0.5
CN	−19.0	1.4	−1.5	1.4
CO_2CH_3	1.3	−0.5	−0.5	3.5
CHO	9.0	1.2	1.2	6.0
$CO{-}CH_3$	7.9	−0.3	−0.3	2.9
F	35.1	−14.1	1.6	−4.4
Cl	6.4	0.2	1.0	−2.0
Br	−5.4	3.3	2.2	−1.0
I	−32.0	10.2	2.9	1.0
NH_2	19.2	−12.4	1.3	−9.5
OH	26.9	−12.6	1.8	−7.9
OCH_3	30.2	−15.5	0.0	−8.9
SCH_3	10.2	−1.8	0.4	−3.6
NO_2	19.6	−5.3	0.8	6.0

From J. B. Stothers, *Carbon-13 NMR Spectroscopy*, New York: Academic Press, 1973.

135.9
123.9
150.2
N

3-37

108.0
118.4
N
H

3-38

Aromatic resonances may be calculated empirically by adding increments to the benzene chemical shift (δ 128.4) for each substituent that is ipso, ortho, meta, or para to the resonating carbon (Table 3-5).

3-5c Carbonyl Groups

Carbonyl groups have no direct representation in proton NMR spectra, so carbon NMR provides unique information for their analysis. The entire carbonyl chemical shift range, δ 160–220, is well removed to high frequency from that of almost all other functional groups, on account of the effect of the $n \rightarrow \pi^*$ transition on the magnitude of the paramagnetic shift. Like aromatic ipso carbons and nitriles, carbonyl carbons other than those in aldehydes carry no attached protons and hence relax more slowly and tend to have low intensities.

Aldehydes resonate toward the middle of the carbonyl range, at about δ 190–205, with acetaldehyde (CH_3CHO) at δ 199.6. Unsaturated aldehydes, in which the carbonyl group is conjugated with a double bond or phenyl ring, are shifted to lower frequency (upfield): benzaldehyde (C_6H_5CHO) at δ 192.4 and $CH_2{=}CHCHO$ at δ 192.2. The α, β, and γ effects of substituents on ketones add to the carbonyl chemical shift and hence are found at the high-frequency end of the carbonyl range. Their overall range is δ 195–220, with acetone at δ 205.1 and cyclohexanone at δ 208.8. Again, adjacent unsaturation shifts the resonances to lower frequency.

Carboxylic derivatives fall into the range δ 155–185. The resonances for the series carboxylate (CO_2^-), carboxyl (CO_2H), and ester (CO_2R) often are well defined, as for sodium acetate (δ 181.5), acetic acid (δ 177.3), and methyl acetate (δ 170.7). The range for esters is about δ 165–175, and that for acids is δ 170–185. Acid chlorides are at slightly lower frequency (higher field): δ 160–170, with δ 168.6 for acetyl chloride [$CH_3(CO)Cl$]. Anhydrides have a similar range: δ 165–175, with δ 167.7 for acetic anhydride [$CH_3(CO)O(CO)CH_3$]. Lactones overlap the ester range, with the six-membered lactone at δ 176.5. Amides have a similar range: δ 160–175, with δ 172.7 for acetamide [$CH_3(CO)NH_2$]. Oximes have a larger range, from δ 145–165. The central carbon of allenes ($R_2C{=}\mathbf{C}{=}CR_2$) falls into the ketonic range, δ 200–215, but the outer carbons have a much lower frequency range, δ 75–95.

3-5d Programs for Empirical Calculation

The facility and accuracy of empirical calculations for carbon have been exploited through commercial computer programs for general predictions of ^{13}C chemical shifts (see the references at the end of the chapter). As with ^1H calculations (Section 3-2e), the results are only as good as the data set used in their creation. The programs assume that the effects of multiple substitution are additive, unless specific corrections have been incorporated. Unconsidered or nonadditive phenomena, such as conformational and other steric effects, can cause unexpected deviations between observed and calculated chemical shifts.

3-6 TABLES OF CHEMICAL SHIFTS

Representative chemical shifts are given in Tables 3-6 through 3-10, drawn from references at the end of this chapter.

TABLE 3-6 Methyl and Methylene Groups

	δ (^1H)		δ (^{13}C)			δ (^1H)		δ (^{13}C)	
	CH_2	CH_3	CH_2	CH_3		CH_2	CH_3	CH_2	CH_3
CH_3Li		−0.4		−13.2	$(CH_3)_2NCHO$		2.88		36.0
CH_3CH_3		0.86		5.7			2.97		30.9
$(CH_3)_3CH$		0.89		25.2	CH_3Cl		3.06		25.1
$(CH_3)_4C$		0.94		31.7	$(CH_3)_2O$		3.24		59.5
$(CH_3)_3COH$		1.22		29.4	$(CH_3)_4N^+$		3.33		55.6
$CH_3CH{=}CH_2$		1.72		18.7	CH_3OH		3.38		49.2
$CH_3C{\equiv}CH$		1.80		−1.9	$CH_3CO_2CH_3$		3.67		51.0
$(CH_3)_3P{=}O$		1.93		18.6	$CH_3OC_6H_5$		3.73		54.8
CH_3CN		2.00		0.3	CH_3F		4.27		75.4
$CH_3CO_2CH_3$		2.01		18.7	CH_3NO_2		4.33		57.3
$CH_3(CO)CH_3$		2.07		30.2	$(CH_3CH_2)_2S$	2.49	1.25	26.5	15.8
CH_3CO_2H		2.10		18.6	$CH_3CH_2NH_2$	2.74	1.10	36.9	19.0
$(CH_3)_2S$		2.12		19.5	$CH_3CH_2C_6H_5$	2.92	1.18	29.3	16.8
CH_3I		2.15		−20.6	CH_3CH_2I	3.16	1.86	0.2	23.1
CH_3CHO		2.20		31.2	CH_3CH_2Br	3.37	1.65	28.3	20.3
$(CH_3)_3N$		2.22		47.3	CH_3CH_2Cl	3.47	1.33	39.9	18.7
$CH_3C_6H_5$		2.31		21.3	$(CH_3CH_2)_2O$	3.48	1.20	67.4	17.1
CH_3NH_2		2.42		30.4	CH_3CH_2OH	3.56	1.24	57.3	15.9
$CH_3(SO)CH_3$		2.50		40.1	CH_3CH_2F	4.36	1.24	79.3	14.6
$CH_3(CO)Cl$		2.67		32.7	$CH_3CH_2NO_2$	4.37	1.58	70.4	10.6
CH_3Br		2.69		10.2	$BrCH_2CH_2Br$	3.63		32.4	
$(CH_3)_4P^+$		2.74		11.3	$HOCH_2CH_2OH$	3.72		63.4	
$CH_3(SO_2)CH_3$		2.84		42.6	$ClCH_2CH_2Cl$	3.73		51.7	

TABLE 3-7 Saturated Ring Systems

		^1H	^{13}C			^1H	^{13}C
Cyclopropane		0.22	−2.6	Oxane (tetrahydropyran)	(α)	3.52	68.0
Cyclobutane		1.98	23.3		(β)	1.51	26.6
Cyclopentane		1.51	26.5		(γ)		23.6
Cyclohexane		1.43	27.7	Pyrrolidine	(α)	2.75	47.4
Cycloheptane		1.53	29.4		(β)	1.59	25.8
Cyclopentanone	(α)	2.06	37.0	Piperidine	(α)	2.74	47.5
	(β)	2.02	22.3		(β)	1.50	27.2
Cyclohexanone	(α)	2.22	40.7		(γ)	1.50	25.5
	(β)	1.8	26.8	Thiirane		2.27	18.9
	(γ)	1.8	24.1	Tetrahydrothiophene	(α)	2.82	31.7
Oxirane		2.54	40.5		(β)	1.93	31.2
Tetrahydrofuran	(α)	3.75	69.1	Sulfolane	(α)	3.00	51.1
	(β)	1.85	26.2		(β)	2.23	22.7
				1,4-Dioxane		3.70	66.5

TABLE 3-8 Alkenes

		^1H	^{13}C
$CH_2{=}CHCN$	(α)	{5.5−6.4}	107.7
	(β)		137.8
$CH_2{=}CHC_6H_5$	(α)	6.66	112.3
	(β)	5.15, 5.63	135.8
$CH_2{=}CHBr$	(α)	6.4	115.6
	(β)	5.7−6.1	122.1
$CH_2{=}CHCO_2H$	(α)	6.5	128.0
	(β)	5.9−6.5	131.9
$CH_2{=}CH(CO)CH_3$	(α)	{5.8−6.4}	138.5
	(β)		129.3
$CH_2{=}CHO(CO)CH_3$	(α)	7.28	141.7
	(β)	4.56, 4.88	96.4
$CH_2{=}CHOCH_2CH_3$	(α)	6.45	152.9
	(β)	3.6−4.3	84.6
$CH_3{}^4CH{=}{}^3CCH_3{=}{}^2CH{=}{}^1CH_2$	(1)	5.02	
	(2)	6.40	
	(4)	5.70	
$(CH_3)_2C{=}CHCO_2CH_3$	(α)	—	114.8
	(β)	5.62	155.9
Cyclopentene		5.60	130.6
Cyclohexene		5.59	127.2
1,3-Cyclopentadiene		6.42	132.2, 132.8
1,3-Cyclohexadiene		5.78	124.6, 126.1
2-Cyclopentenone	(α)	6.10	132.9
	(β)	7.71	164.2
2-Cyclohexenone	(α)	5.93	128.4
	(β)	6.88	149.8
exo-Methylenecyclohexane	($=CH_2$)	4.55	106.5
	($C=$)	—	149.7
Allene	($=CH_2$)	4.67	74.0
	($=C=$)	—	213.0

TABLE 3-9 Aromatics

	^{1}H			^{13}C			
	o	*m*	*p*	*i*	*o*	*m*	*p*
$C_6H_5CH_3$	7.16	7.16	7.16	137.8	129.3	128.5	125.6
$C_6H_5CH{=}CH_2$	7.24	7.24	7.24	138.2	126.7	128.9	128.2
$C_6H_5SCH_3$	7.23	7.23	7.23	138.7	126.7	128.9	124.9
C_6H_5F	6.97	7.25	7.05	163.8	114.6	130.3	124.3
C_6H_5Cl	7.29	7.21	7.23	135.1	128.9	129.7	126.7
C_6H_5Br	7.49	7.14	7.24	123.3	132.0	130.9	127.7
C_6H_5OH	6.77	7.13	6.87	155.6	116.1	130.5	120.8
$C_6H_5OCH_3$	6.84	7.18	6.90	158.9	113.2	128.7	119.8
$C_6H_5O(CO)CH_3$	7.06	7.25	7.25	151.7	122.3	130.0	126.4
$C_6H_5(CO)CH_3$	7.91	7.45	7.45	136.6	128.4	128.4	131.6
$C_6H_5CO_2H$	8.07	7.41	7.47	130.6	130.0	128.5	133.6
$C_6H_5(CO)Cl$	8.10	7.43	7.57	134.5	131.3	129.9	136.1
C_6H_5CN	7.54	7.38	7.57	109.7	130.1	127.2	130.1
$C_6H_5NH_2$	6.52	7.03	6.63	147.9	116.3	130.0	119.2
$C_6H_5NO_2$	8.22	7.48	7.61	148.3	123.4	129.5	134.7

	^{1}H			^{13}C		
	α	β	Other	α	β	Other
Naphthalene	7.81	7.46	—	128.3	126.1	—
Anthracene	7.91	7.39	8.31	130.3	125.7	132.8
Furan	7.40	6.30	—	142.8	109.8	—
Thiophene	7.19	7.04	—	125.6	127.4	—
Pyrrole	6.68	6.05	—	118.4	108.0	—
Pyridine	8.50	7.06	7.46	150.2	123.9	135.9

TABLE 3-10 Carbonyl Compounds

	^{1}H(CH_3)	^{1}H(other)	^{13}C(C=O)
$H(CO)OCH_3$	3.79	8.05 (HCO)	160.9
$CH_3(CO)Cl$	2.67	—	168.6
$CH_3(CO)OCH_2CH_3$	2.02 (CH_3CO)	4.11 (CH_2),1.24 (CH_3C)	169.5
$CH_3(CO)N(CH_3)_2$	2.10 (CH_3CO)	2.98 (CH_3N)	169.6
CH_3CO_2H	2.10	1.37 (HO)	177.3
$CH_3CO_2^-Na^+$	—	—	181.5
$CH_3(CO)C_6H_5$	2.62	—	196.0
$CH_3(CO)CH{=}CH_2$	2.32	5.8–6.4 ($CH{=}CH_2$)	197.2
$H(CO)CH_3$	2.20	9.80 (HCO)	199.6
$CH_3(CO)CH_3$	2.07	—	205.1
2-Cyclohexenone	—	5.93, 6.88 ($CH_\alpha{=}CH_\beta$)	197.1
2-Cyclopentenone	—	6.10, 7.71 ($CH_\alpha{=}CH_\beta$)	208.1
Cyclohexanone	—	1.7–2.5	208.8
Cyclopentanone	—	1.9–2.3	218.1

Problems

3-1 A trisubstituted benzene possessing one bromine and two methoxy substituents exhibits three aromatic resonances, at δ 6.40, 6.46, and 7.41. What is the substitution pattern?

3-2 Octahedral cobalt complexes, CoL_6, have three filled t_{2g} and two empty e_g molecular orbitals. The ^{59}Co chemical shifts of several such complexes are linear, with the wavelength of the first absorption (longest wavelength) in the UV/visible spectrum. Explain the linearity in terms of Ramsey's equation for shielding.

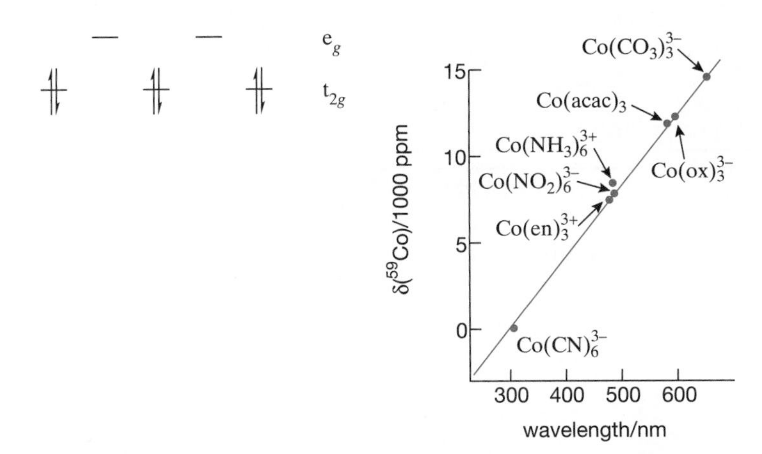

3-3 Calculate the expected ^{13}C resonance positions for all the carbon atoms in the following molecules. Ignore the δ and ε effects.

(a) $CH_3CH_2CH(CH_3)CH(CH_3)_2$ **(b)** $ICH_2CH_2CH_2Br$
(c) $CH_3CH_2CH(NO_2)CH_3$ **(d)** $(CH_3)_3CCN$

3-4 Of the α and β protons of naphthalene (see structure at right), which should resonate at a higher frequency? Why? Compare both resonance positions with that of benzene.

H_α H_β

3-5 The —OH proton resonance is found at δ 5.80 for phenol in dilute $CDCl_3$ and at δ 10.67 for 2-nitrophenol in dilute $CDCl_3$. Explain.

3-6 Derive the structures of the compounds that have the ^1H (300 MHz) and ^{13}C (75 MHz) spectra shown in parts (a)–(g). The 1:1:1 triplet at δ 78 in the ^{13}C spectra is from the solvent $CDCl_3$ [used in all cases except (f)].

(a) $C_4H_6O_2$

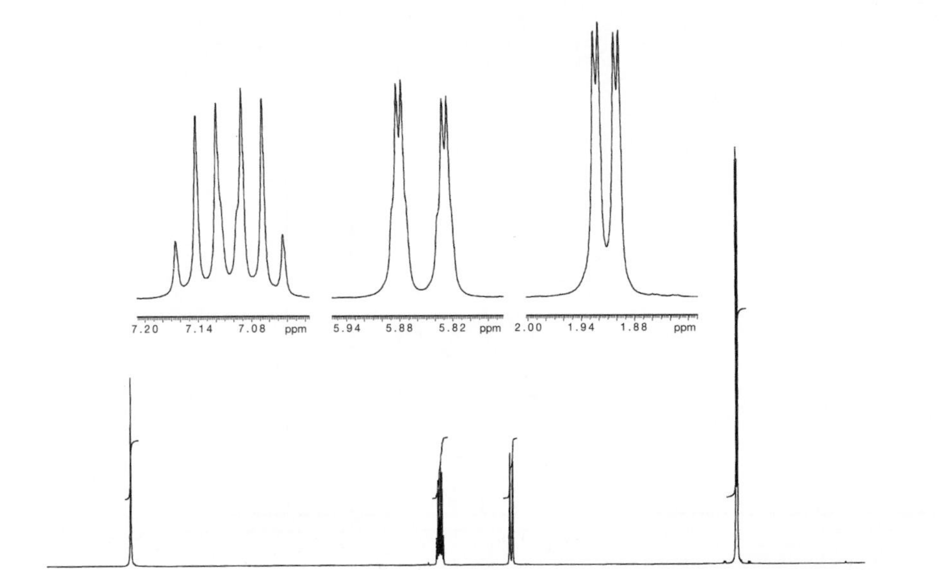

(continued)

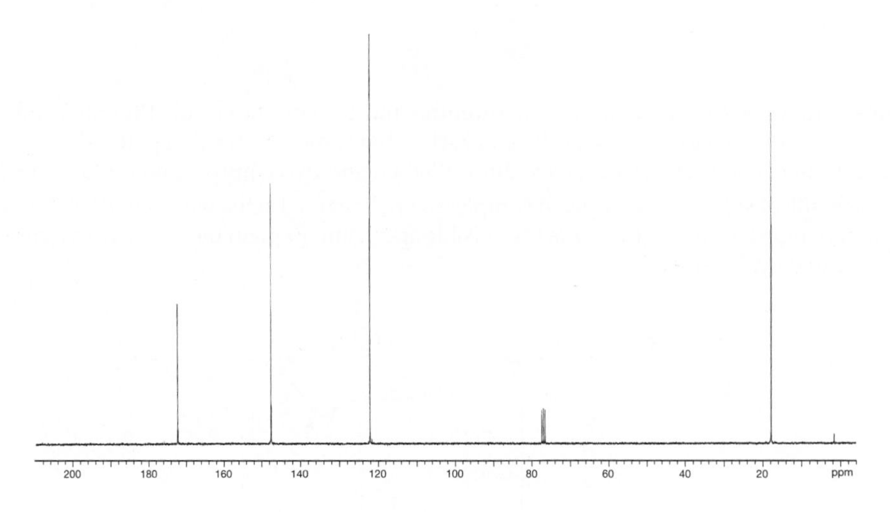
200
180
160
140
120
100
80
60
40
20
ppm

(b) $C_4H_8Cl_2$

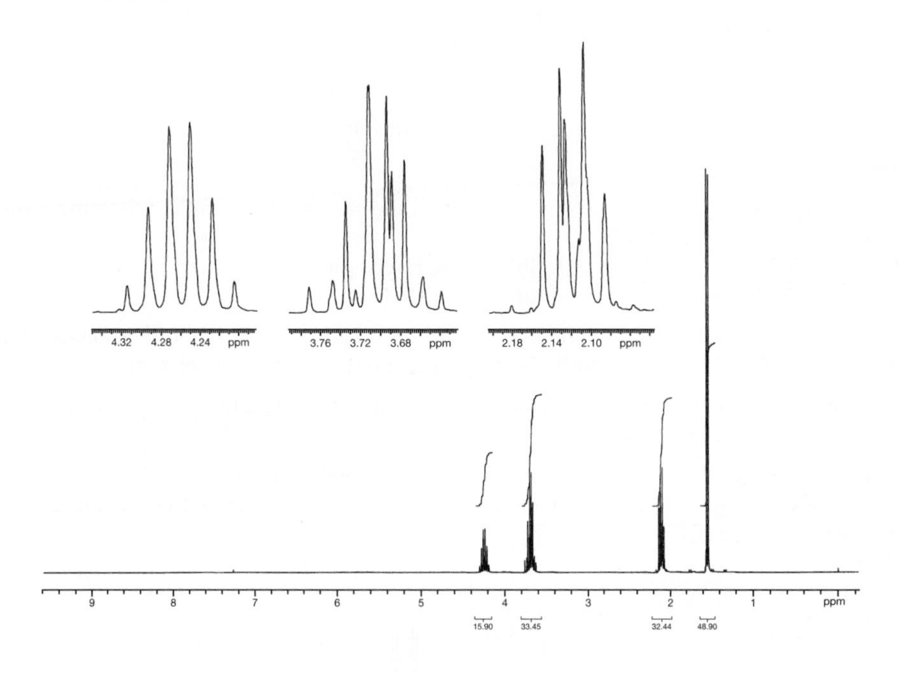
4.32
4.28
4.24
ppm
3.76
3.72
3.68
ppm
2.18
2.14
2.10
ppm
9
8
7
6
5
4
3
2
1
ppm
15.90
33.45
32.44
48.90

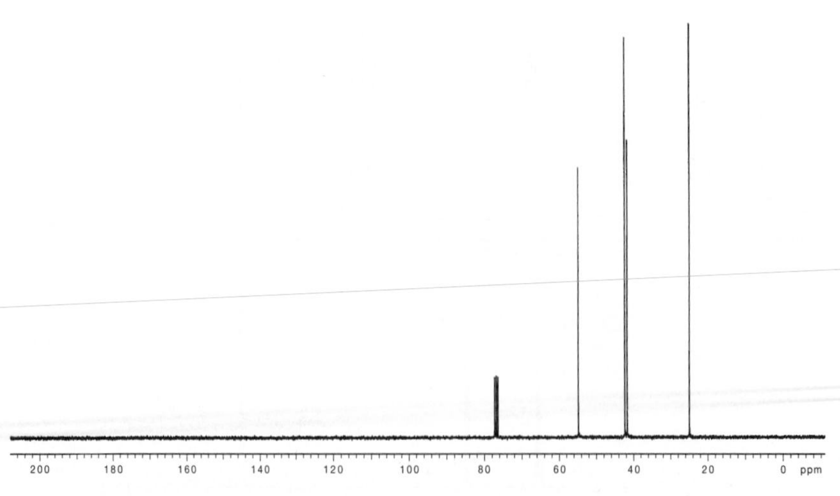
200
180
160
140
120
100
80
60
40
20
0 ppm

(c) C_5H_9OCl

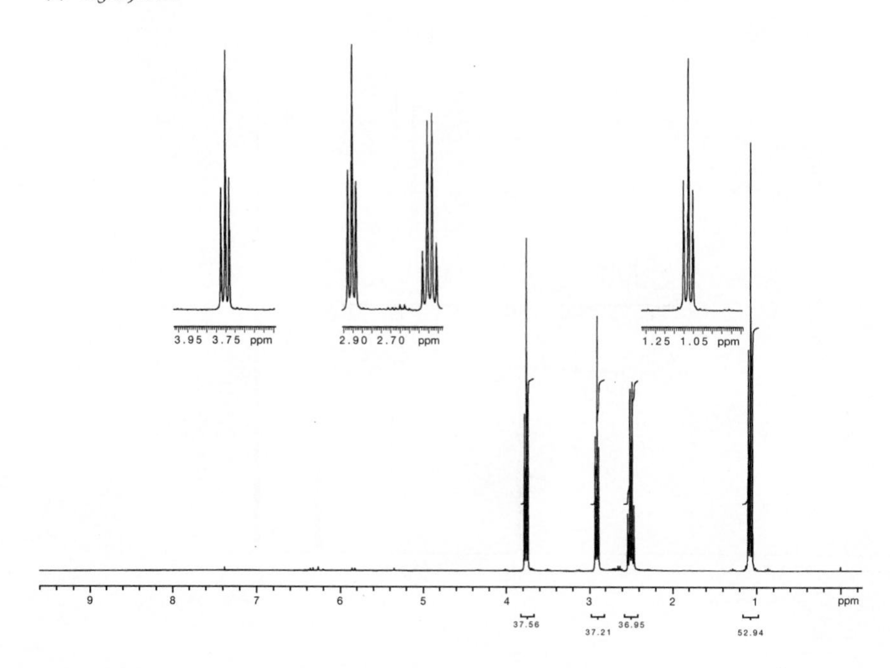

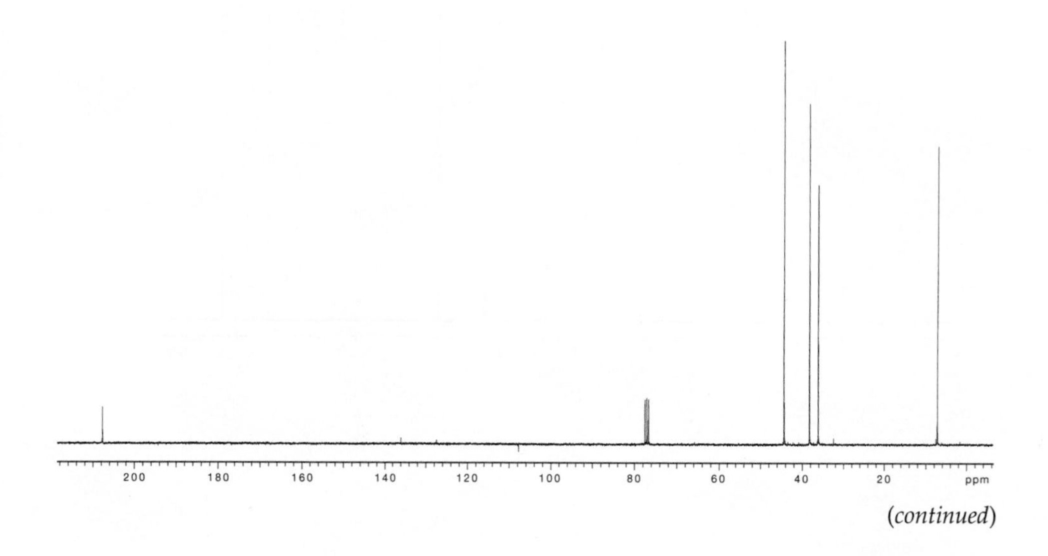

(*continued*)

(d) $C_4H_8O_2$

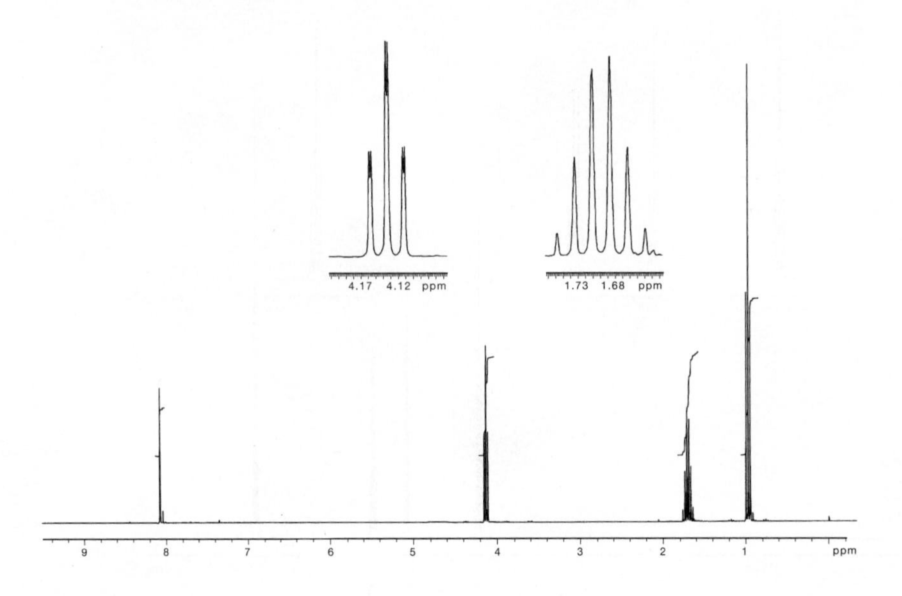
4.17 4.12 ppm
1.73 1.68 ppm
9 8 7 6 5 4 3 2 1 ppm

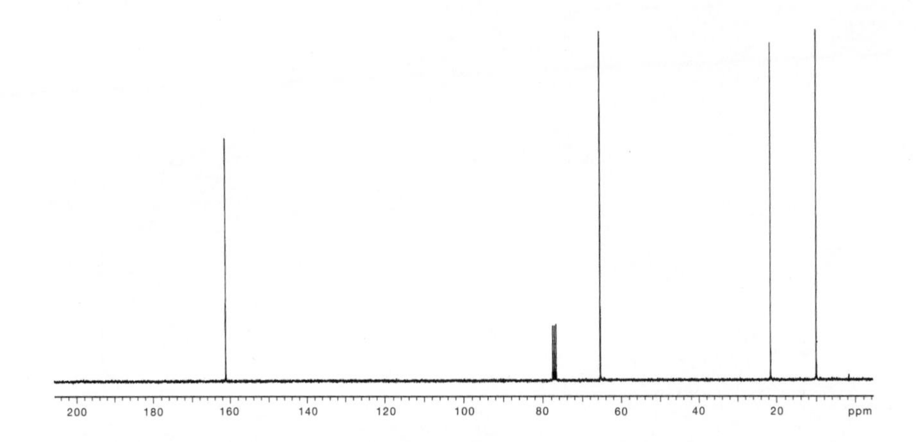
200 180 160 140 120 100 80 60 40 20 ppm

(e) $C_{10}H_{12}O_3$ (The peak at δ 6.87 disappears after adding D_2O and shaking.)

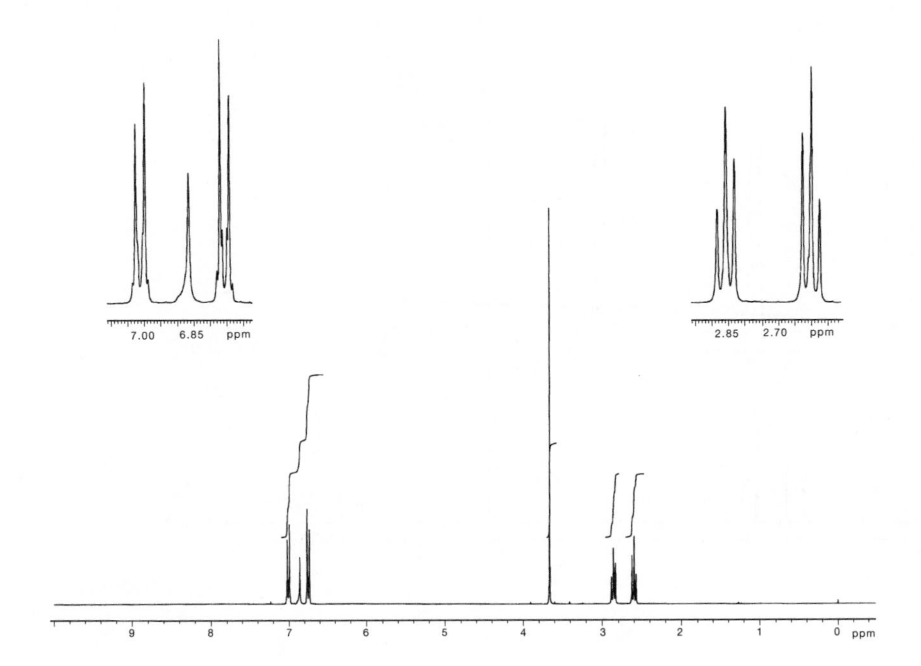

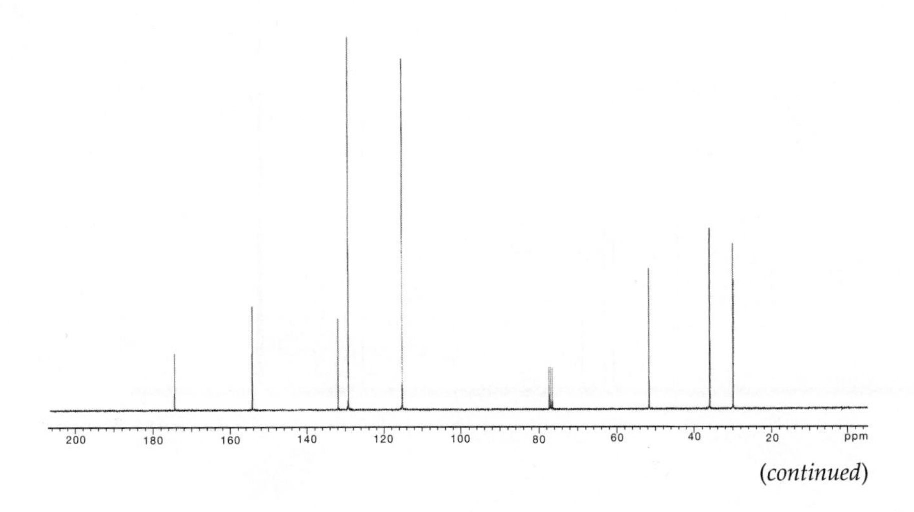

(continued)

(f) $C_8H_7BrO_3$ (A ^1H resonance of unit integral at δ 12 is not shown; the ^1H signals at δ 2.05 and the ^{13}C signals at δ 30 are from the solvent acetone-d_5).

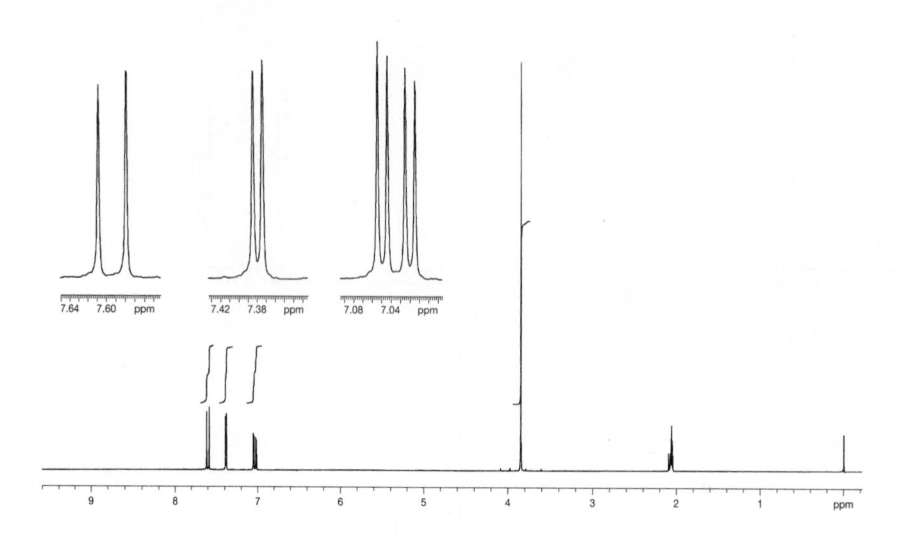

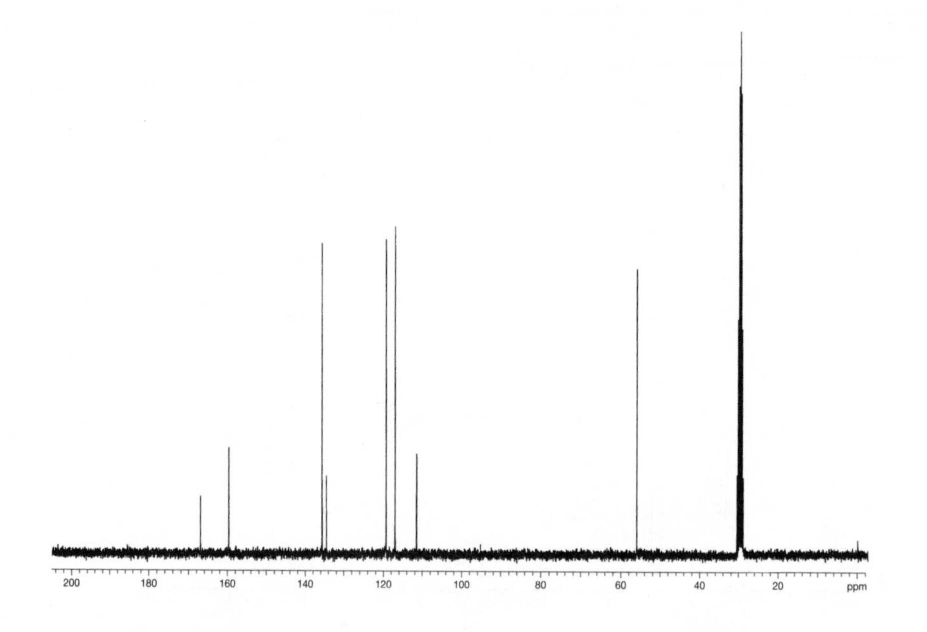

(g) C_6H_7NO

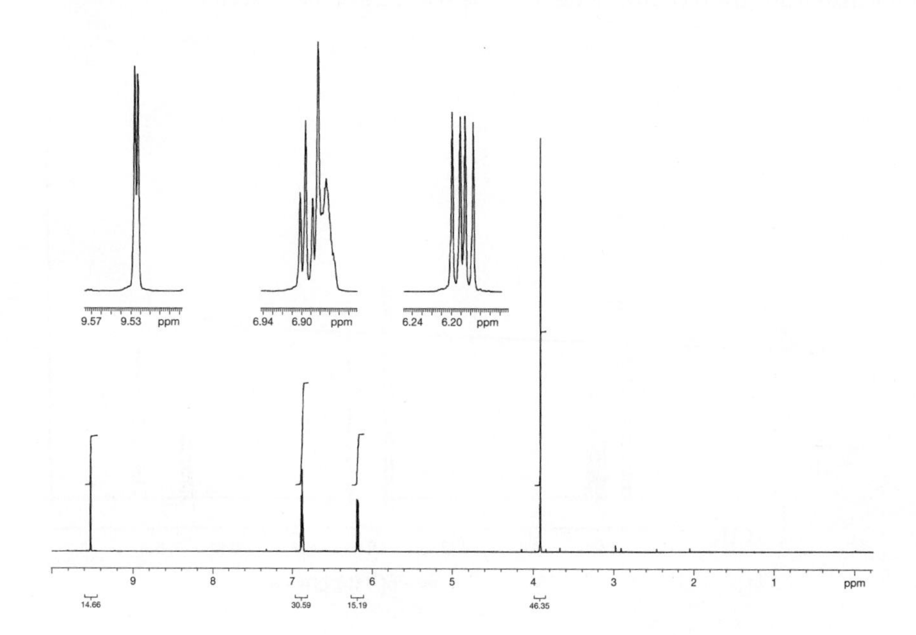
9.57
9.53
ppm
6.94
6.90
ppm
6.24
6.20
ppm
9
8
7
6
5
4
3
2
1
ppm
14.66
30.59
15.19
46.35

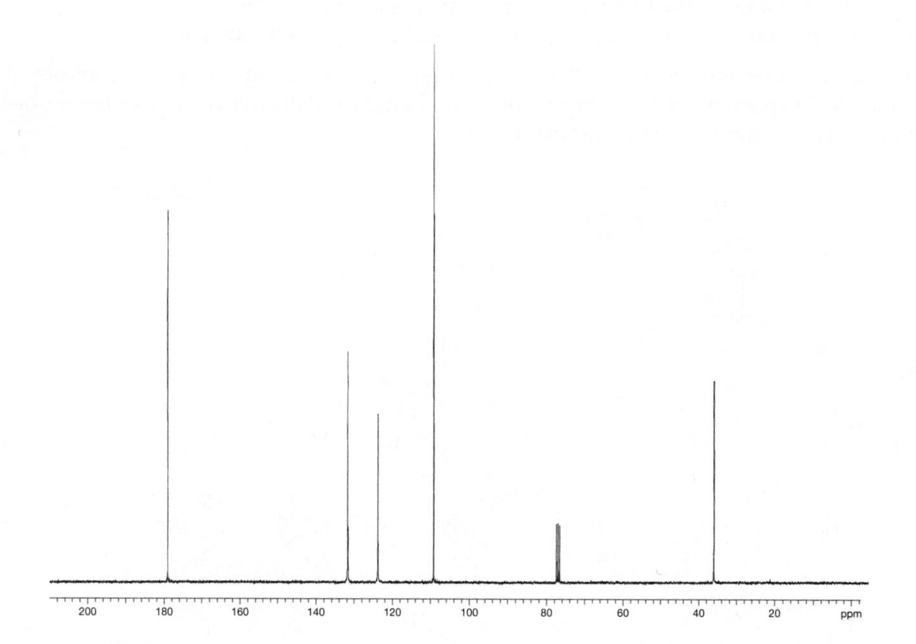
200
180
160
140
120
100
80
60
40
20
ppm

3-7 The aromatic and alkenic portions of the 500 MHz ^{1}H spectrum of the potassium channel activator bimakalin are given below, along with the structure of the molecule.

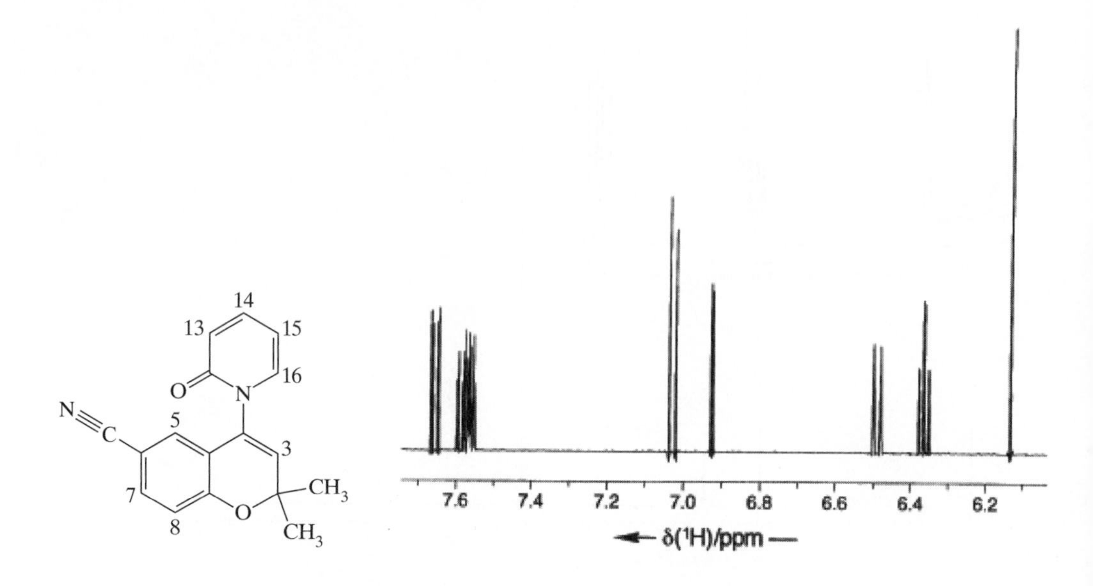

Reproduced with permission from K. G. R. Pachler, *Magn. Reson. Chem.*, **36**, 437 (1998).

(a) Assign these eight resonances to the appropriate protons strictly by the coupling patterns. Discuss your assignments.

(b) Explain the position of the three resonances clustered around δ 7.6.

(c) Why is the proton with chemical shift δ 6.12 found at so low a frequency?

3-8 The bark resin of *Commiphora mukul*, called the guggul tree, is thought to lower cholesterol and fat intake. The primary active components were found to be the five molecules below, belonging to a class of compounds called cembrenes.

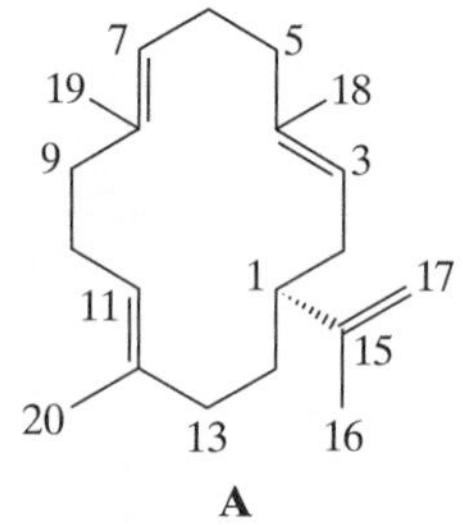

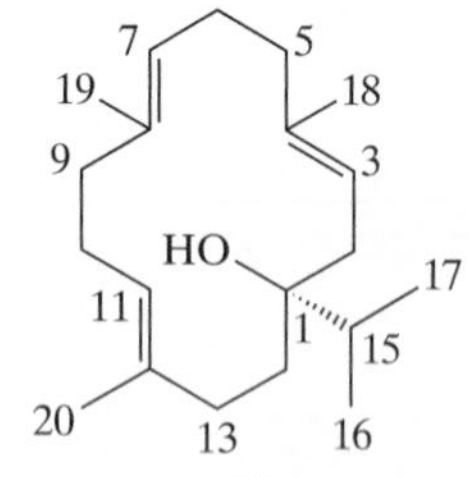

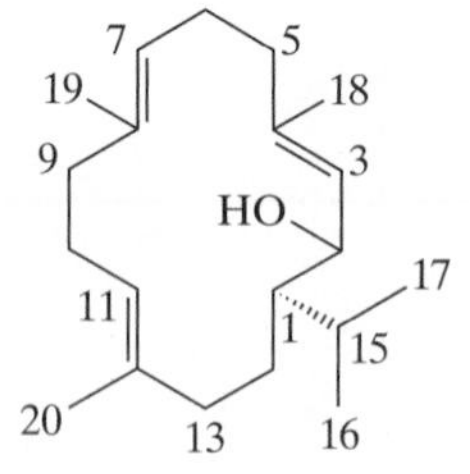

Adapted from S. Bai and M. Jain, *Magn. Reson. Chem.*, **46**, 792–793 (2008).

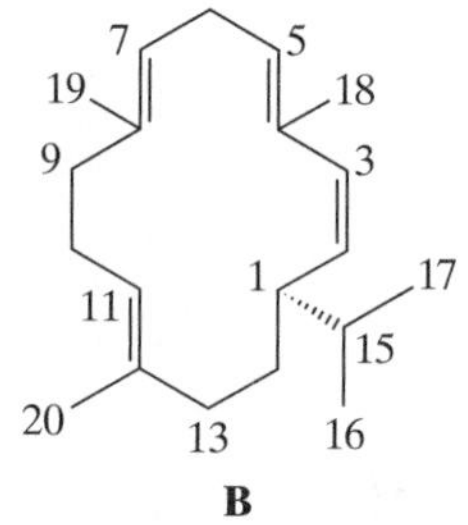

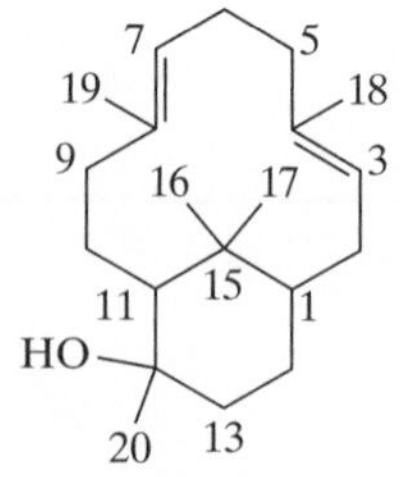

From the following table of the already assigned ^{1}H chemical shifts for these compounds, assign structures A–E to NMR spectra. In the table, for each proton the chemical shift is given in units of δ, then the multiplicity, and finally the values of any coupling constants in Hz. Also try this exercise without any prior knowledge of the peak assignments.

Proton	Cpd1	Cpd2	Cpd3	Cpd4	Cpd5
1	2.06 m	1.63 m	n/a	1.28 m	1.45 m
2	2.02 m	5.19 dd 15.5, 9.8	2.22 m	4.58 d 8.7	2.72 m
	1.98 m		2.18 m		
3	5.22 t 7.6	6.09 d 15.5	5.27 t 7.4	5.33 d 8.7	5.69 d 12.5
4	n/a	n/a	n/a	n/a	n/a
5	2.14 m	5.55 t 7.6	2.22 m	2.17 m	2.09 m
	2.19 m			2.20 m	2.23 m
6	2.19 m	2.44 dd 3.4, 1.5	2.12 m	2.12 m	2.05 m
	2.29 m	3.07 br s	2.31 m	2.20 m	2.42 m
7	5.00 t 6.2	5.13 d 10.6	4.91 t 6.8	5.04 t 5.8	4.91 d 10.5
8	n/a	n/a	n/a	n/a	n/a
9	2.08 m	2.06 m	1.95 m	1.94 m	2.11 m
		2.23 m	2.14 m	2.13 m	2.26 m
10	2.14 m	2.03 m	2.10 m	2.07 m	1.33 m
		2.32 m		2.18 m	1.49 m
11	5.08 t 6.2	4.90 d 6.8	5.01 t 7.4	4.94 dd 7.9, 7.2	2.27 m
12	n/a	n/a	n/a	n/a	n/a
13	1.96 m	2.02 m	1.94 m	2.05 m	1.86 m
14	1.69 s	1.69 m	1.65 m	1.69 m	2.00 m
15	n/a	1.51 m	1.72 m	1.79 m	n/a
16	1.68 s	0.88 d 6.8	0.96 d 6.8	0.99 d 6.8	0.81 s
17	4.68 s	0.84 d 6.8	0.94 d 6.8	0.96 d 6.8	0.74 s
	4.74 s				
18	1.59 s	1.82 s	1.56 s	1.60 s	1.56 s
19	1.62 s	1.62 s	1.59 s	1.59 s	1.53 s
20	1.57 s	1.54 s	1.61 s	1.56 s	1.28 s

Further Tips on Solving NMR Problems

See the Spoiler Alert in Chapter 2, *Tips on Solving NMR Problems.*

1. Empirical correlations provide a useful guideline for peak assignments but should not be construed as completely reliable. Problems of nonadditivity can be resolved only through the introduction of interaction factors, which in general have been ignored in calculations of proton chemical shifts. Most reliable for protons probably are the calculations of aromatic chemical shifts (see Problem 3.1), provided that no two substituents are ortho to each other. Empirical correlations for ^{13}C chemical shifts have been more successful (Problem 3.3), and many examples of interaction factors have been introduced. With commercial programs for the empirical calculation of spectra, the results can never be more accurate than the models on which they were based. Consequently, any such calculation should always include a direct assessment of the model molecules on which the calculation was based.

2. A doublet of quartets (dq) may be observed when a CH_n group is coupled to both a methyl and a methine group, as in CH_3**CH**—CH or CH_3**CH**=CH. The resulting pattern, however, can vary from four to eight peaks, depending on the relative values of $J(CH—CH_3)$ and $J(CH—CH)$. If the latter (doublet) coupling is zero, the resulting pattern is a quartet. Such an event could happen if the CH—CH dihedral angle is close to 90° (Karplus relationship, Section 4-6). If the doublet coupling is rather larger than the quartet coupling, as could occur for CH_3**CH**=CH with a trans arrangement between the alkenic protons (again, Section 4-6), the maximum of eight peaks would be observed. When the two couplings are comparable, anything from five to eight peaks can occur, depending on the precise relative values. For the example in Problem 3-6a, four of the peaks overlap in the middle to reduce the pattern to six peaks. This sort of situation is nicely analyzed by the use of stick diagrams or trees as described in Section 2-6 and is treated thoroughly in reference 4.3.

3. The proton integrals and the carbon peak intensities in Problem 3-6a illustrate the reliability of the former and the unreliability of the latter. Although each carbon peak represents a single carbon, the intensities vary by more than a factor of 2. The carbonyl carbon of the CO_2H group has no attached proton, relaxes more slowly, and has the smallest intensity. The intensities of the carbons bearing protons vary with the number of attached and neighboring protons and with spectral parameters such as the delay time t_d (Section 2-10c).

4. Carbon-13 spectra measured in $CDCl_3$ usually include a peak from the solvent at about δ 78. This resonance is a 1:1:1 triplet, because the carbon is coupled to the deuterium, which exists in three equally populated spin states (+1, 0, and −1).

5. A pair of 1:2:1 triplets often is a diagnostic for the isolated bismethylene group, $—CH_2CH_2—$. Deviations from first-order conditions can distort the intensities of the inner and outer peaks of both triplets (Section 4-2).

6. The H(C=O) group (formyl) usually is associated with aldehydes, which exhibit a sharp singlet at about δ 9.8. Formates, however, contain the same grouping: $H(C{=}O)OCH_3$ (methyl formate). A formate formyl proton is found at lower frequency (higher field) than the aldehyde proton, typically around δ 8.0 (Problem 3-6d).

7. The relative locations of alkyl groups in oxygen-containing molecules often are not clear from the splitting patterns (Problem 3-6e). The distinctions to be drawn are between ether/ester ($CH_3O—$) functionalities and carbonyl [$CH_3(C{=}O)—$] functionalities. These may be distinguished by the observation of resonances at δ ca. 2.1 for carbonyl functionalities, at δ ca. 3.2 for methyl ethers, and at δ ca. 3.7 for methyl esters. Higher frequencies are observed with more substitution on carbon, e.g., $RCH_2O—$ or $RCH_2(C{=}O)—$.

8. To distinguish ketones and aldehydes from esters, acids, amides, and their ilk, it is useful to remember that δ 190 provides something of a borderline. Ketones and aldehydes usually are at higher frequency than 190, and the carboxylic acid family usually is at lower frequency than 190. Problems 3-6a, c, d, e, and f provide examples.

9. Disubstituted aromatics (Problem 3-6e) can occur with ortho, meta, or para patterns:

These may be distinguished by the knowledge that J(ortho) > J(meta) > J(para). This order follows the number of bonds (three, four, and five, respectively), but the more complicated actual situation is discussed in Sections 4-5, 4-6, and 4-7. The para coupling, over five bonds, is almost always negligible. In the para-substituted case, meta couplings (HCCCH) are small and between equivalent protons, so the spectrum is dominated by the large ortho couplings (HCCH). The spectrum usually reduces to a simple two-spin AX spectrum of four peaks (Section 2-6 and Figure 2-18), sometimes with complications from magnetic nonequivalence (Section 4-2). If the para X and Y substituents are identical, the spectrum, of course, becomes a singlet. The meta-substituted case often exhibits a singlet or a closely spaced triplet for the proton between the substituents (the smaller splitting is from coupling to two meta protons). The spectrum for the remaining three protons is either an AMX spectrum (two doublets and a triplet) or an A_2X spectrum (a doublet and a triplet), depending on whether the X and Y substituents are identical. The doublets from the outer protons can be further split by meta couplings. The ortho-substituted case is the most complicated, with an AGRX spectrum when X and Y are different and an AA′XX′ spectrum when X and Y are the same. The AA′XX′ spectrum is an example of second-order effects, discussed in Section 4-2 and exemplified by 1,2-dichlorobenzene in Figure 4-3.

10. Trisubstituted aromatics (Problem 3-6f) can occur with 1,2,3, 1,2,4, or 1,3,5 patterns:

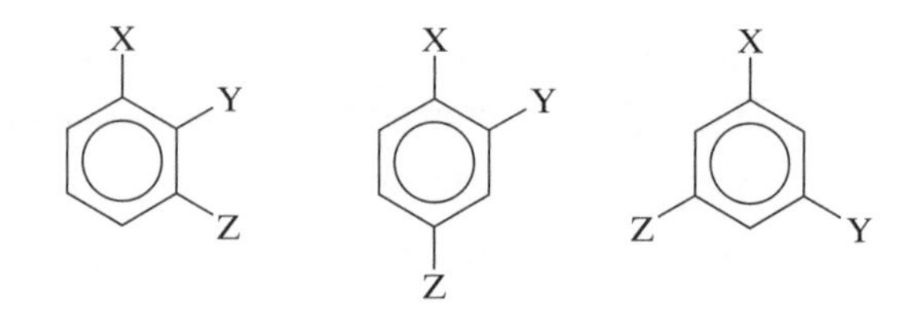

There is great variability in the appearances of these spectra, and we will consider only the cases when all three substituents are different. The cases when all three substituents are the same or when only two are the same can be worked out easily. For 1,3,5 substitution, the spectrum contains three singlets or closely spaced doublets (from meta couplings). For 1,2,3 substitution, the AMX spectrum contains two widely spaced doublets (or doublets split into closely spaced doublets from the meta coupling) for the protons adjacent to X and Z. The middle proton (para to Y) is usually a 1:2:1 triplet from two equal meta couplings. The appearance of the 1,2,4 spectrum is highly diagnostic for this substitution pattern. The peak between Y and Z is a singlet or a closely spaced doublet from the meta coupling. The proton next to Z is a widely spaced doublet from the ortho coupling, usually split by the meta coupling into closely spaced doublets (hence, dd). The proton next to X is a simple doublet from the ortho coupling, without any further coupling. The spectrum in Problem 2-6f shows this characteristic widely spaced doublet, narrowly spaced doublet, and doublet of doublets. These three substitution patterns all offer several structural isomers. The relative locations of the X, Y, and Z groups may not be assigned with coupling constant analysis, but empirical chemical shift correlations sometimes can accomplish this task (Problem 3-1).

11. Problem 3-6e illustrates the difficulties of getting a handle on OH and NH protons, which can give broad resonances at variable positions. The ^{14}N atom (Problem 3-6g) also broadens the resonance of the α protons on the ring. Thus a broadened resonance should not automatically be assumed to be from OH or NH functionalities, which must be assigned by the D_2O test in $CDCl_3$.

12. Although $N—CH_3$ protons usually resonate at δ ca. 2.4, this is the case only for amines. Changing the charge on or the hybridization of the nitrogen alters the chemical shift. Thus methyl groups in ammonium salts $[(R_3N—CH_3)^+]$ resonate at δ ca. 3.3, methyl groups in amides resonate at δ ca. 2.9, and methyl groups on pyrroles at δ ca. 3.7, as in Problem 3-6g. The ^{13}C chemical shifts of $N—CH_3$ groups exhibit similar variations.

Bibliography

Chemical Shift Conventions

3.1 R. K. Harris, E. D. Becker, S. M. Cabral de Menezes, P. Granger, R. E. Hoffman, and K. W. Zilm, *Magn. Reson. Chem.*, **46**, 582–596 (2008).

Diamagnetic Anisotropy

3.2 R. C. Haddon, *Fortsch. Chem. Forsch.*, **16**, 105 (1971).
3.3 C. W. Haigh and R. B. Mallion, *Progr. NMR Spectrosc.*, **13**, 303 (1979).

Van der Waals Shielding

3.4 F. H. A. Rummens, *NMR Basic Princ. Progr.*, **101** (1975).

Empirical Correlations

See also references in Chapter 2.
3.5 C. Pascual, J. Meier, and W. Simon, *Helv. Chim. Acta*, **49**, 164 (1966).
3.6 S. W. Tobey, *J. Org. Chem.*, **34**, 1281 (1969).
3.7 G. J. Martin and M. L. Martin, *Progr. NMR Spectrosc.*, **8**, 163 (1972).
3.8 D. J. Craik, *Ann. Rep. NMR Spectrosc.*, **15**, 1 (1983).
3.9 E. C. Friedrich and K. G. Runkle, *J. Chem. Educ.*, **61**, 830 (1984).
3.10 D. W. Brown, *J. Chem. Educ.*, **62**, 209–212 (1985).
3.11 P. S. Beauchamp and R. Marquez, *J. Chem. Educ.*, **74**, 1483–1485 (1997).
3.12 L. J. Tilley, S. J. Prevoir, and D. A. Forsyth, *J. Chem. Educ.*, **79**, 593–600 (2002).
3.13 H. J. Shine and P. Rangappa, *Magn. Reson. Chem.*, **45**, 971–979 (2007).

Programs for Empirical Calculations

3.14 ACD NMR Predictor: https://www.acdlabs.com/solutions/clue.html
3.15 ChemNMR: http://www.cambridgesoft.com/software/details/?ds=7
3.16 ChemWindow: http://www.bio-rad.com/prd/en/US/adirect/biorad?ts=1&cmd= BRCatgProductDetail&vertical=INF&catID=203371
3.17 gNMR: http://www.adeptscience.co.uk/products/lab/gnmr/
3.18 NMRPen: http://chempensoftware.com/nmrpen.htm
3.19 Alkenes: http://webcampus.stthomas.edu/chem/nmr/Calculator/olefin/index.html
3.20 Aromatics: http://www.stolaf.edu/depts/chemistry/courses/toolkits/380/js/nmr1/

DFT Spectral Calculations

3.21 A. Bagno, *Chem. European J.*, **7**, 1652–1661 (2001).

Solvent Chemical Shifts

3.22 H. E. Gottlieb, V. Kotlyar, and A. Nudelman, *J. Org. Chem.*, **62**, 7512–7515 (1997).

Solvent Effects

3.23 P. Laszlo, *Progr. NMR Spectrosc.*, **3**, 231 (1967).
3.24 J. Ronayne and D. H. Williams, *Ann. Rev. NMR Spectrosc.*, **2**, 83 (1969).
3.25 J. Homer, *Appl. Spectrosc. Rev.*, **9**, 1 (1975).

Isotope Effects

3.26 H. Batiz-Hernandez and R. A. Bernheim, *Progr. NMR Spectrosc.*, **3**, 63 (1967).
3.27 S. Berger, *NMR Basic Princ. Progr.*, **22**, 1 (1990).

Carbon-13 Spectra

3.28 J. B. Stothers, *Carbon-13 NMR Spectroscopy*, New York: Academic Press, 1973.
3.29 G. C. Levy, R. L. Lichter, and G. L. Nelson, *Carbon-13 Nuclear Magnetic Resonance Spectroscopy*, 2nd ed., New York: Wiley-Interscience, 1980.
3.30 W. Bremser, B. Franke, and H. Wagner, *Chemical Shift Ranges in ^{13}C NMR*, Weinheim, Germany: Verlag Chemie, 1982.
3.31 E. Breitmaier and W. Völter, *Carbon-13 NMR Spectroscopy*, 3rd ed., Weinheim, Germany: VCH, 1987.
3.32 H.-D. Kalinowski, S. Berger, and S. Braun, *Carbon-13 Spectroscopy*, Chichester, UK: John Wiley & Sons, Ltd., 1988.
3.33 F. W. Wehrli, A. P. Marchand, and S. Wehrli, *Interpretation of Carbon-13 NMR Spectra*, 2nd ed., Chichester, UK: John Wiley & Sons, Ltd., 1988.
3.34 K. Pihlaja and E. Kleinpeter, *Carbon-13 NMR Chemical Shifts in Structural and Stereochemical Analysis*, New York: John Wiley & Sons, 1994.

Other Nuclei (General)

3.35 R. K. Harris and B. E. Mann, *NMR and the Periodic Table*, London: Academic Press, 1978.
3.36 J. B. Lambert and F. G. Riddell (eds.), *The Multinuclear Approach to NMR Spectroscopy*, Dordrecht, The Netherlands: D. Reidel, 1983.

3.37 P. Laszlo (ed.), *NMR of Newly Accessible Nuclei*, New York: Academic Press, 1983.

3.38 J. Mason (ed.), *Multinuclear NMR*, New York: Plenum Press, 1987; B. E. Mann, *Ann. Rev. NMR Spectrosc.*, **23**, 141–207 (1991).

Other Nuclei (Specific)

3.39 *Deuterium*. H. H. Mantsch, H. Saitô, and I. C. P. Smith, *Progr. NMR Spectrosc.*, **11**, 211 (1977).

3.40 *Tritium*. J. A. Elvidge, in *Isotopes: Essential Chemistry and Applications*, J. A. Elvidge and J. R. Jones (eds.), London: The Chemical Society, 1980, p. 152.

3.41 *Boron-10,11*. W. L. Smith, *J. Educ. Chem.*, **54**, 469 (1977). B. Wrackmeyer, *Ann. Rep. NMR Spectrosc.*, **20**, 61 (1981).

3.42 *Nitrogen-14,15*. M. Witanowski and G. A. Webb (eds.), *Nitrogen NMR*, London: Plenum Press, 1973. G. C. Levy and Lichter, *Nitrogen-15 Nuclear Magnetic Resonance Spectroscopy*, New York: John Wiley & Sons, 1979. W. von Philipsborn and R. Müller, *Angew. Chem., Int. Ed. Engl.*, **25**, 383 (1986). M. Witanowski, L. Stefaniak, and G. A. Webb, *Ann. Rep. NMR Spectrosc.*, **25**, 1 (1993).

3.43 *Oxygen-17*. *Oxygen-17 NMR Spectroscopy in Organic Chemistry*, D. W. Boykin (ed.), Boca Raton, FL: CRC Press, Inc., 1990.

3.44 *Fluorine-19*. R. G. Mooney, *An Introduction to ^{19}F NMR Spectroscopy*, London: Heyden and Son, 1970. C. H. Dungan and J. R. Van Wazer, *Compilation of Reported Fluorine-19 Chemical Shifts*, New York: Wiley-Interscience, 1970. V. Wray, *Ann. Rep. NMR Spectrosc.*, **14**, 1 (1983).

3.45 *Sodium-23*. P. Laszlo, *Angew. Chem., Int. Ed. Engl.*, **17**, 254 (1978).

3.46 *Magnesium-25 and calcium-47*. S. Forsén and B. Lindman, *Ann. Rep. NMR Spectrosc.*, **11A**, 183 (1981).

3.47 *Aluminum-27*. J. W. Akitt, *Ann. Rep. NMR Spectrosc.*, **21**, 1 (1988).

3.48 *Silicon-29*. H. C. Marsmann, *NMR Basic Princ. Progr.*, **17**, 65 (1981). E. A. Williams, *Ann. Rep. NMR Spectrosc.*, **15**, 235 (1983). E. A. Williams, in *The Chemistry of Silicon Compounds*, S. Patai and Z. Rappoport (eds.), Chichester, UK: John Wiley & Sons, Ltd., 1989, Chapter 8; 2nd ed., 1998, Chapter 6.

3.49 *Phosphorus-31*. D. G. Gorenstein, *Phosphrus-31: Principles and Applications*, New York: Academic Press, 1984. *Phosphorus-31 NMR Spectroscopy in Stereochemical Analysis*, J. G. Verkade and L. D. Quin, (eds.), Deerfield Beach, FL: VCH, 1987.

3.50 *Sulfur-33*. G. Barbarella, *Progr. NMR Spectrosc.*, **25**, 317 (1993).

3.51 *Halogens*. B. Lindman and S. Forsén, *NMR Basic Princ. Progr.*, **12**, 1 (1976).

3.52 *Germanium-73*. Y. Takeuchi and T. Takayama, *Ann. Rep. NMR Spectrosc.*, **54**, 155–200 (2005).

3.53 *Selenium-77 and Tellurium-123,125*. M. Mauvin, in *Proceedings of the 4th International Conference on the Organic Chemistry of Selenium and Tellurium*, 1983, p. 406.

3.54 *Cadmium-113*. P. D. Ellis, *Science*, **221**, 1141 (1983).

3.55 *Tin-117,119*. B. Wrackmeyer, *Ann. Rep. NMR Spectrosc.*, **16**, 285 (1986).

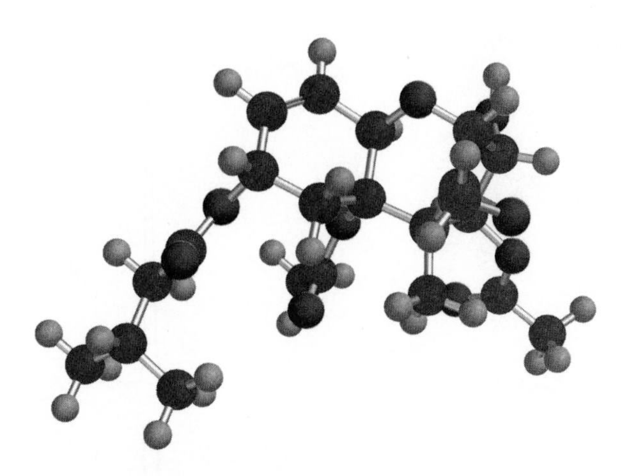

CHAPTER 4

The Coupling Constant

4-1 FIRST-ORDER SPECTRA

Most spectra illustrated up to this point are said to be *first order*. For a spectrum to be first order, the frequency difference ($\Delta\nu$) between the chemical shifts of any given pair of nuclei must be much larger than the value of the coupling constant J between them, approximately $\Delta\nu/J > 10$. In addition, an important symmetry condition discussed in the next section must hold. First-order spectra exhibit a number of useful and simple characteristics:

- Spin–spin multiplets are centered on the resonance frequency.
- Spacings between adjacent components of a spin–spin multiplet are equal to the coupling constant J.
- Multiplicities that result from coupling exactly reflect the $n + 1$ rule for $I = \frac{1}{2}$ nuclei ($2nI + 1$, in general). Thus, two equivalent, neighboring protons split the resonating nucleus into three peaks.
- The intensities of spin–spin multiplets correspond to the coefficients of the binomial expansion given by Pascal's triangle for spin-$\frac{1}{2}$ nuclei (see Figure 2-24).
- Nuclei with the same chemical shift do not split each other, even when the coupling constant between them is nonzero.

When the chemical shift difference is less than about 10 times J ($\Delta\nu/J \leq 10$), *second-order* effects appear in the spectrum, including deviations in intensities from the binomial pattern and other exceptions from the preceding characteristics. By the Pople notation, nuclei that have a first-order relationship are represented by letters that are far apart in the alphabet (AX), and those that have a second-order relationship are represented by adjacent letters (AB). Figure 4-1 illustrates the progression for two spins from AB to nearly AX. When $\Delta\nu/J$ is 0.4, the spectrum is practically a singlet. Intensity distortions enhance peak heights toward the center of the multiplet. A second-order multiplet typically leans toward the resonances of its coupling partner. The peak intensities within a multiplet are not equal even when $\Delta\nu/J = 15$. Thus the commonly quoted first-order criterion of $\Delta\nu/J > 10$ is not always adequate. With the wide availability of proton frequencies of 300 MHz and higher, first-order spectra have become common but are by no means universal.

4-2 CHEMICAL AND MAGNETIC EQUIVALENCE

In addition to meeting the requirement that compares chemical shift differences with coupling constants ($\Delta\nu/J$), first-order spectra must pass a symmetry test. *Any two chemically equivalent nuclei must have the same coupling constant to any other nucleus.* Nuclear pairs that fail this test are said to be *magnetically nonequivalent*, and their spectral

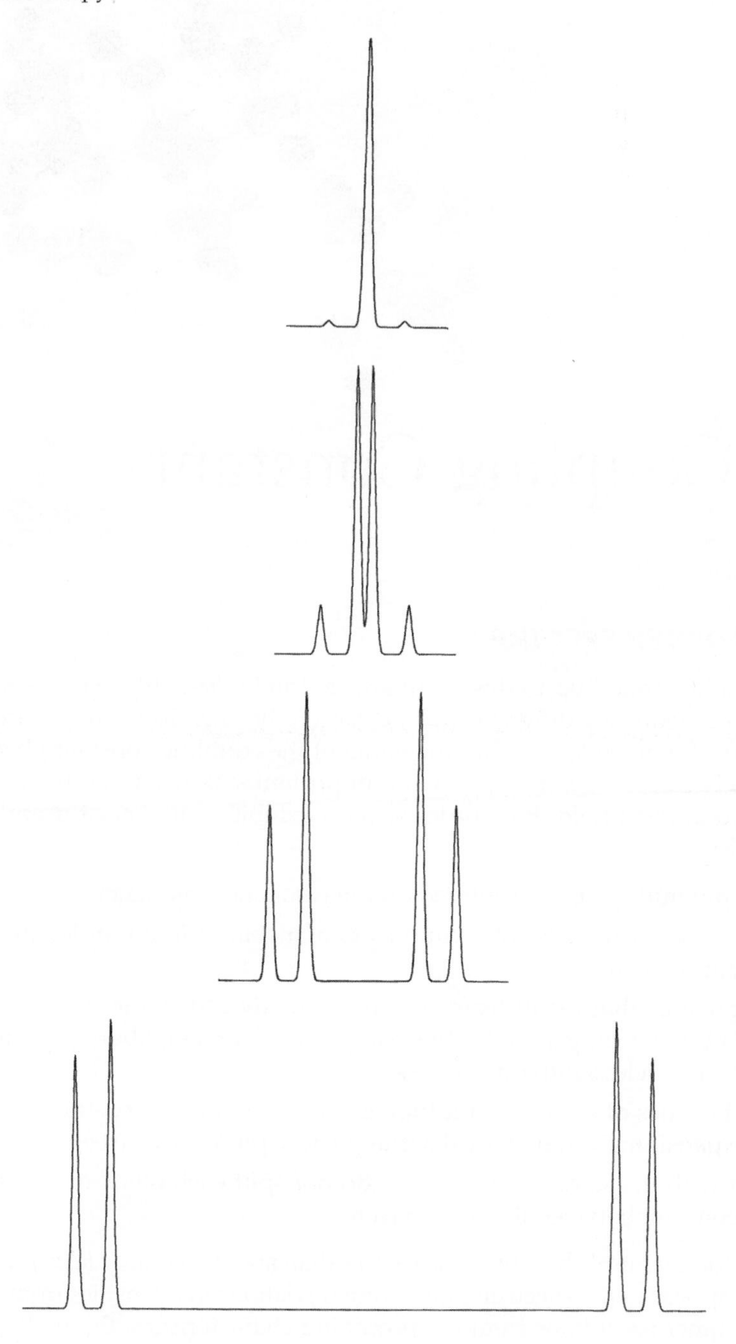

FIGURE 4-1 The two-spin spectrum with $\Delta\nu/J$ values of 0.4 (top), 1.0, 4.0, and 15.0.

appearance is second order. To apply this test, it is useful to understand the role of symmetry in the NMR spectrum.

Nuclei are *chemically equivalent* if they can be interchanged by a symmetry operation of the molecule. Thus the two protons in 1,1-difluoroethene (**4-1**) or in difluoromethane

4-1 **4-2** **4-3**

(**4-2**) may be interchanged by a 180° rotation. Nuclei that are interchangeable by rotational symmetry are said to be *homotopic*. Rotation about carbon–carbon single bonds is so rapid that the chemist rarely considers the fact that the three methyl protons in CH_3CH_2Br are not, in fact, equivalent by symmetry (compare nuclei A and X in **4-3**). Rapid C—C rotation, however, results in an average environment in which they are equivalent. Dynamic effects are considered more thoroughly in Section 5-2.

Nuclei related by a plane of symmetry are called *enantiotopic*, provided there is no rotational axis of symmetry. For example, the protons in bromochloromethane (**4-4a**) are

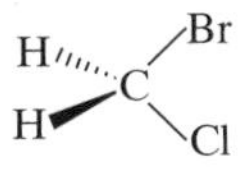

4-4a 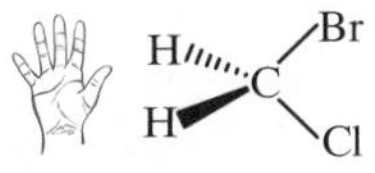 **4-4b**

chemically equivalent and enantiotopic because they are related by the plane of symmetry containing C, Br, and Cl. If the molecule is placed in a chiral environment, this statement no longer holds true. Such an environment may be created by using a solvent composed of an optically active material or by placing the molecule in the active site of an enzyme, as represented for **4-4b**, in which bromochloromethane has a small hand placed to one side. The protons are no longer equivalent because the hand is a chiral object. Because the plane of symmetry is lost in a chiral environment, the nuclei are not enantiotopic. They have become chemically nonequivalent (no symmetry operation can interchange them). Enantiotopic nuclei may be expected to become chemically nonequivalent and give distinct resonances in an optically active solvent. In a biological context, enantiotopic protons may be rendered nonequivalent by an enzyme and may exhibit distinct chemical properties, such as acidity.

The term enantiotopic was coined because replacement of one proton of the pair by another atom or group, such as deuterium, produces the enantiomer (nonsuperimposable mirror image, **4-4c**) of the molecule that results when the other proton is replaced by

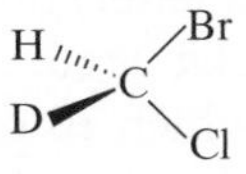

4-4c 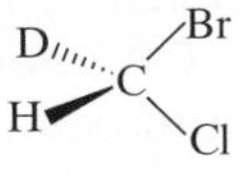 **4-4d**

the same group (**4-4d**). A pair of homotopic nuclei treated in this fashion produce identical molecules (superimposable mirror images). Enantiotopic or homotopic protons need not be on the same carbon atom. Thus the alkenic protons in cyclopropene (**4-5**) are homotopic, but those in 3-methylcyclopropene (**4-6**) are enantiotopic. Chemically equivalent

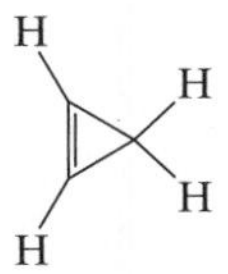

4-5 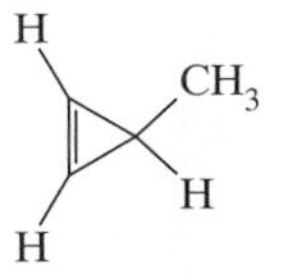 **4-6**

nuclei (either homotopic or enantiotopic) are represented by the same letter in the spectral shorthand of Pople. Cyclopropene (**4-5**) is A_2X_2, as is difluoromethane (**4-2**), since the two fluorine atoms have spins of $\frac{1}{2}$. The ring protons of 3-methylcyclopropene (**4-6**) constitute an AX_2 group.

To be *magnetically equivalent*, two nuclei must be chemically equivalent and have the same coupling constant to any other nucleus. The latter test is more stringent than that for chemical equivalence, because it is necessary to go beyond considering just the overall symmetry of the molecule. The first two molecules discussed in this chapter provide contrasting results. In difluoromethane (**4-2**) each of the two hydrogens has the same coupling to a specific fluorine atom because both hydrogens have the same spatial relationship to that fluorine. Consequently, the protons are magnetically equivalent.

By the same token, the two fluorine atoms also are magnetically equivalent by reference to coupling to either proton, and the spin system is labeled A_2X_2.

In 1,1-difluoroethene (**4-1**), however, the two protons do not have the same spatial relationship with respect to a given fluorine. Therefore, they have different couplings [J(HCCF)], one a J_{cis} and the other a J_{trans} (**4-7**), and are said to be magnetically

J_{cis} H_A F_X C=C $H_{A'}$ $F_{X'}$ J_{trans}

4-7

nonequivalent. The group of spins is represented by the Pople notation AA'XX', so that the two distinct couplings may be denoted by J_{AX} and $J_{AX'}$. In contrast, an A_2X_2 system such as difluoromethane (**4-2**) or cyclopropene (**4-5**) has only one coupling, J_{AX}. In an AA'XX' system, J_{AX} and $J_{A'X'}$ are the same, as are $J_{AX'}$ and $J_{A'X}$. Any spin system that contains nuclei that are chemically equivalent but magnetically nonequivalent is, by definition, second order. Moreover, raising the magnetic field cannot alter basic structural relationships between nuclei, so that the spectrum remains second order at the highest accessible fields. Nuclei that do not have the same chemical shift (*anisochronous*) also are magnetically nonequivalent because they resonate at different resonance frequencies, a so-called *chemical shift criterion*. *Isochronous* nuclei that are magnetically nonequivalent by having unequal couplings to another nucleus are said to fail the *coupling constant criterion*.

The AA'XX' notation may be parsed as follows. The chemical shifts of the A and X nuclei are very far from each other (opposite ends of the alphabet). The A and A' nuclei are chemically equivalent (denoted by the same letter) but magnetically nonequivalent (indicated by the prime), as are the X and X' nuclei. Figure 4-2 illustrates the proton AA' part of the spectrum of 1,1-difluoroethene, in which 10 peaks are visible. This appearance is quite different from the simple 1:2:1 triplet expected in the first-order case (e.g., the methyl triplet in Figure 2-22). The multiplicity of peaks in Figure 4-2 even permits measurement of $J_{AA'}$, the coupling between the equivalent protons. Such a measurement is impossible in first-order systems. The presence of splittings between equivalent

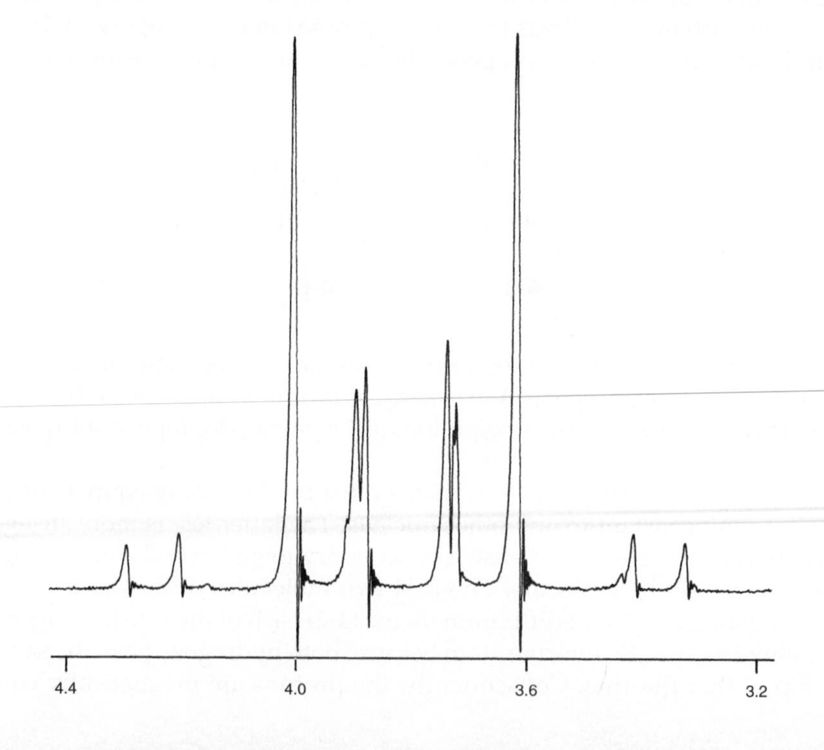

FIGURE 4-2 The 90 MHz ^{1}H spectrum of 1,1-difluoroethene in $CDCl_3$.

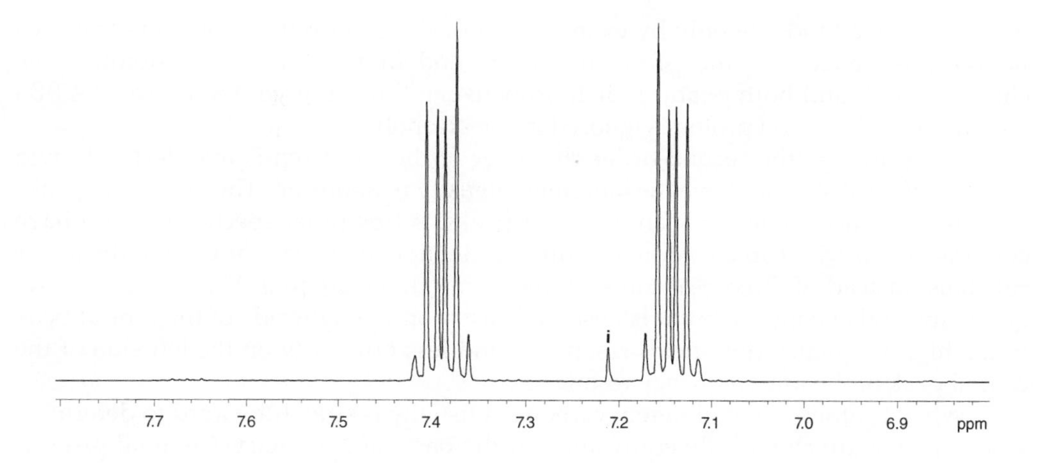

FIGURE 4-3 The 300 MHz ^{1}H spectrum of 1,2-dichlorobenzene in $CDCl_3$. An impurity is signified by the letter *i*.

nuclei in second-order spectra emphasizes that such couplings do exist but are not manifested in first-order spectra.

Magnetic nonequivalence is not uncommon for isochronous nuclei. The spin systems for both para- and ortho-disubstituted benzene rings formally are AA'XX' (or AA'BB' if the chemical shifts are close). Figure 4-3 illustrates the proton spectrum of 1,2-dichlorobenzene (**4-8**, in which H_A and $H_{A'}$ are seen to have different couplings to H_X),

4-8 **4-9**

which is AA'XX' and relatively complex. Constraints of a ring frequently convey magnetic nonequivalence, as, for example, in propiolactone (**4-9**, in which H_A and $H_{A'}$ also have different couplings to H_X). Even open-chain systems such as 2-chloroethanol ($ClCH_2CH_2OH$, Figure 4-4) contain magnetically nonequivalent spin systems, although

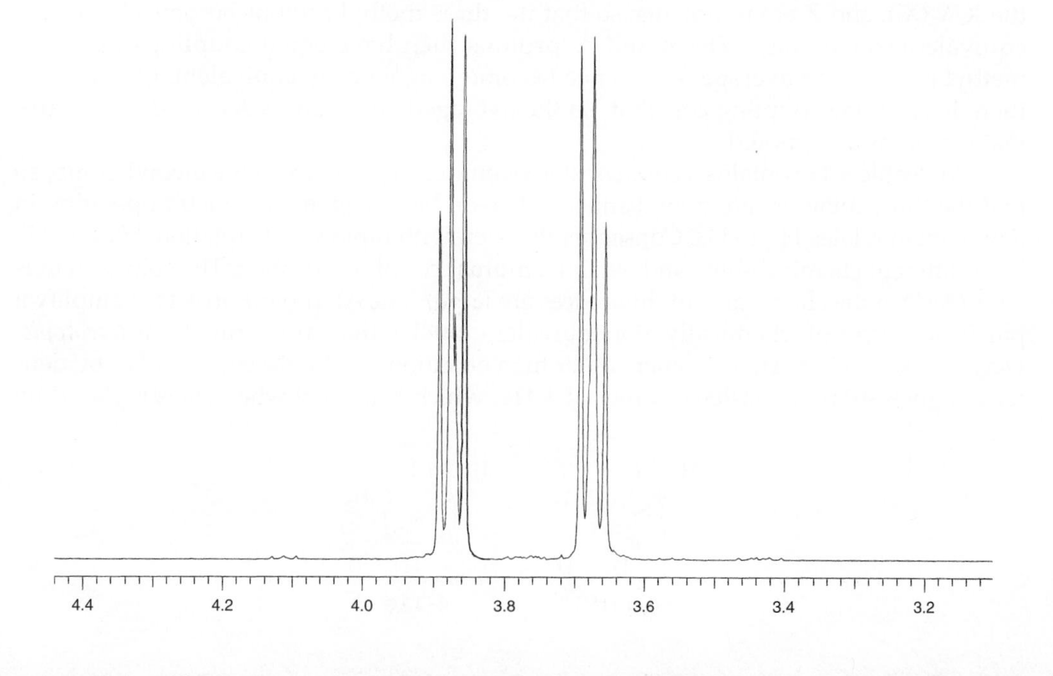

FIGURE 4-4 The 300 MHz ^{1}H spectrum of 2-chloroethanol (methylene resonances only) in $CDCl_3$.

they are understandable only by examination of the contributing rotamers (see two paragraphs hence and the problems at the end of the chapter). Propiolactone, chloroethanol, and both *o*- and *p*-dichlorobenzene thus all give AA'XX' (or AA'BB') spectra (if the hydroxyl proton is ignored in the alcohol).

In Figure 4-4, the second-order character of the spectrum is manifested in two ways. First, peaks do not have the binomial intensity relationship. Thus the inner peaks of each resonance are larger than the outer peaks. A first-order spectrum would have comprised two 1:2:1 triplets. More careful examination, however, shows that the $n + 1$ rule fails. Instead of three peaks in each resonance, there are four. The fourth peak requires spectral expansion, but it is observed easily on the right side of the central peak of the high-frequency (low-field) resonance and less obviously on the left side of the central peak of the low-frequency resonance.

When protons are on different carbons, it usually is straightforward to determine whether they are chemically equivalent on the basis of symmetry. Geminal protons (those on the same carbon, CH_2) can be more subtle. Consider the protons of ethylbenzene ($C_6H_5CH_2CH_3$) and of its β-bromo-β-chloro derivative ($C_6H_5CH_2CHClBr$). Rotation about the saturated C—C bond generates three rotamers for each molecule, which may be represented by the Newman projections shown in **4-10** and **4-11**. For **4-10**,

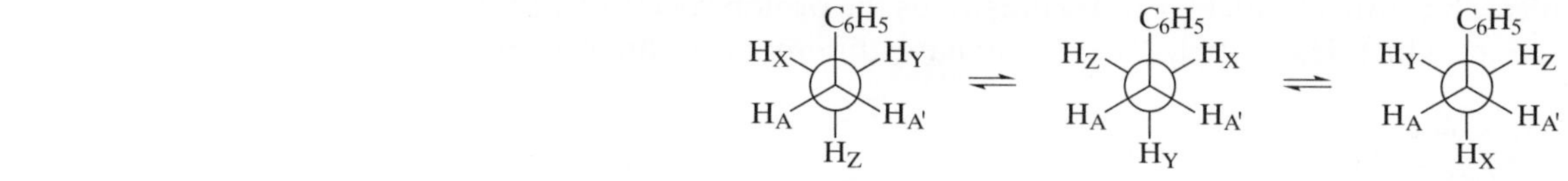

4-10a ⇌ **4-10b** ⇌ **4-10c**

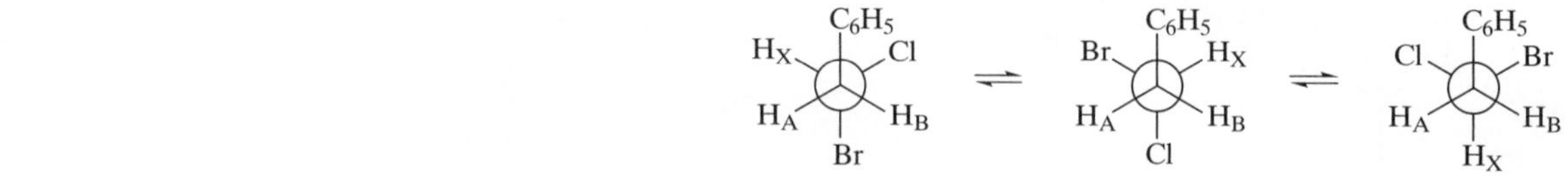

4-11a ⇌ **4-11b** ⇌ **4-11c**

the three rotamers are identical. In the rotamer **4-10a**, H_A and $H_{A'}$ are chemically equivalent and enantiotopic by reason of the plane of symmetry. When methyl rotation is slow, H_A and $H_{A'}$ are magnetically nonequivalent, because each would couple unequally with either H_X or H_Y. The plane of symmetry requires that H_Y should be labeled $H_{X'}$, but we retain the distinct lettering to illustrate the effect of methyl rotation. Thus the frozen structure **4-10a** would exhibit an AA'XX'Z spectrum. Rapid methyl rotation averages the X, Y (X'), and Z environments, so that the three methyl protons become chemically equivalent, on average. The A and A' protons then have equal couplings to all the methyl protons, on average, and hence become magnetically equivalent. On average, there is only one coupling constant, so the averaged spectrum is A_2X_3, when the aromatic protons are ignored.

Molecule **4-11** contains a chiral or stereogenic center in place of the methyl group, so that the three rotamers are now distinct (**4-11a–c**). Moreover, no symmetry operation in any of them relates H_A to H_B. Consequently, even with rapid C—C rotation, H_A and H_B have different chemical shifts and exhibit a mutual coupling constant. The spin system is ABX (AMX if the chemical shift differences are large). The AB protons in **4-11** exemplify a particular type of chemically nonequivalent nuclei that are termed *diastereotopic*. Diastereoisomers are stereoisomers other than enantiomers. Replacement of H_A by deuterium gives **4-11d**, a diastereoisomer of **4-11e**, which is formed when H_B is replaced by

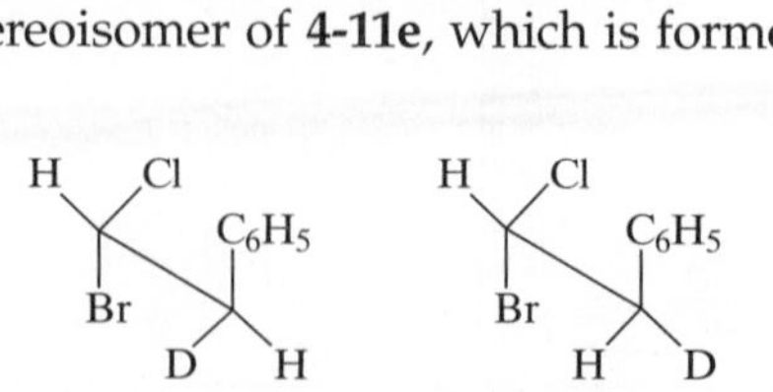

4-11d **4-11e**

deuterium. The deuterated derivative has two stereogenic centers. In general, the protons of a saturated methylene group are diastereotopic when there is a stereogenic center elsewhere in the molecule because no symmetry operation relates the two protons. The protons in **4-4b** are diastereotopic because the hand provides the stereogenic center. Accidental degeneracy can occur when the chemical shift difference is small or unobservable, so that diastereotopic protons can appear to be equivalent in the spectrum.

Methyl groups in an isopropyl group can be diastereotopic when stereogenic centers are present in the molecule, as in α-thujene (**4-12**). The proton resonance then appears as a

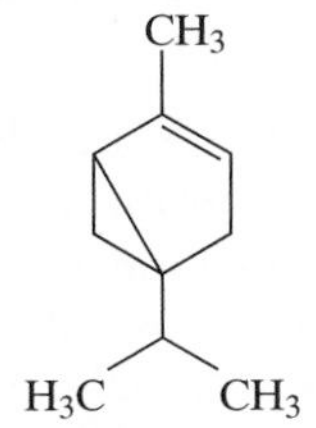

CH_3

H_3C CH_3

4-12

H_A

$O—C—CH_3$

$CH_3—CH$

H_B

$O—CH_2—CH_3$

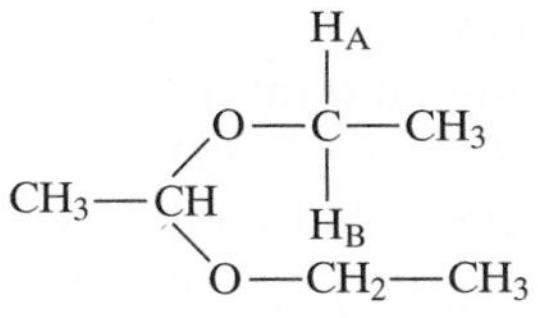

4-13

pair of doublets (coupled to the methine proton), and the carbon resonance appears as two singlets (proton decoupled). A stereogenic center is not necessary for methylene protons to be diastereotopic. The diethyl acetal of acetaldehyde (**4-13**) contains diastereotopic protons because the symmetry axis of the molecule is not a symmetry axis for the CH_2 protons. This situation may be understood by examining the rotamers or by replacing H_A with deuterium. This latter operation creates two stereogenic centers, $—OCHD(CH_3)$ and $—OCH(CH_3)O—$, and the resulting molecule is a diastereoisomer of the molecule in which H_B is replaced with deuterium. The methylene protons in *cis*-1,2-dichlorocyclopropane (**4-14a**) are diastereotopic because **4-14b** and **4-14c** are diastereoisomers. The axial

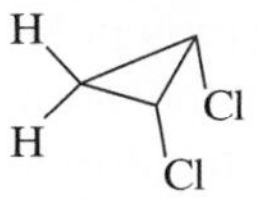

H H Cl Cl

4-14a

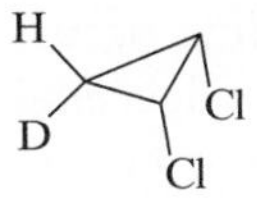

H D Cl Cl

4-14b

D H Cl Cl

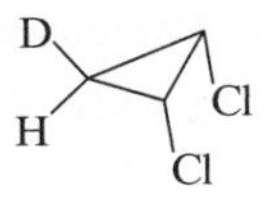

4-14c

and equatorial protons on a single carbon in ring-frozen cyclohexane are diastereotopic, because cyclohexane-axial-*d* and cyclohexane-equatorial-*d* are diastereoisomers. When ring flipping is fast on the NMR time scale, the geminal protons become equivalent on average. Thus the diastereotopic nature of protons can depend on the rate of molecular interconversions.

4-3 SIGNS AND MECHANISMS

Spin–spin coupling arises because information about nuclear spin is transferred from nucleus to nucleus via the electrons. Exactly how does this process occur? Several mechanisms have been considered, but the most important is the *Fermi contact interaction*. Both nuclei and electrons are magnetic dipoles, whose mutual interactions normally are described by the point–dipole approximation, as used, for example, by McConnell in his analysis of diamagnetic anisotropy (eq. 3-1). Fermi found that this approximation breaks down when dipoles are very close (comparable to the radius of a proton). Under these circumstances, when the nucleus and electron in essence are in contact, their interaction is described by a new mechanism, the Fermi contact term. The energy of the interaction is proportional to the gyromagnetic ratios of the nucleus and of the electron, the scalar (dot) product of their spins (**I** for a nucleus, **S** for an electron), and the probability that the electron is at the nucleus (the square of the electronic wave function evaluated with zero distance from the nucleus): $E_{FC} \propto -\gamma_n\gamma_e\mathbf{I}\cdot\mathbf{S}\psi^2(0)$. Because the nuclear and electronic gyromagnetic ratios have opposite signs ($\gamma_{H,C} > 0$, $\gamma_e < 0$), the stabler arrangement is when the nucleus and the electron are antiparallel (spins paired).

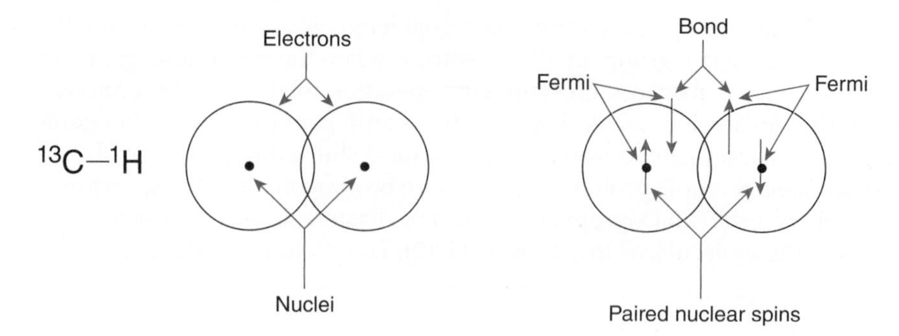

FIGURE 4-5 Diagram of the Fermi contact mechanism for the indirect coupling of two spins.

According to this model, an electron in a bond X—Y, in which both X and Y are magnetic, spends a finite amount of time at the same point in space as nucleus X. If nucleus X has a spin of $I_z = +\frac{1}{2}$, then by the Fermi contact mechanism the opposite $(-\frac{1}{2})$ spin is favored for the electron. In this way, the nuclear spin polarizes the electron spin (gives one spin state a higher population). The electron in turn shares an orbital in the X—Y bond with another electron, which, by the Pauli exclusion principle, must have a spin of $+\frac{1}{2}$ when the spin of the first electron is $-\frac{1}{2}$. This second $(+\frac{1}{2})$ electron by the Fermi mechanism polarizes the Y nucleus to prefer a spin of $-\frac{1}{2}$. With two Fermi interactions, the energy of the coupling interaction then is proportional to $\gamma_A\gamma_X\gamma_e^2(\mathbf{I}_A \cdot \mathbf{S})\psi_A^2(0)(\mathbf{I}_X \cdot \mathbf{S})\psi_X^2(0)$. All the constants move into the proportionality constant, designated J, which itself is proportional to the product of the two nuclear gyromagnetic ratios, $\gamma_A\gamma_X$. Thus whenever X has a spin of $+\frac{1}{2}$, a spin of $-\frac{1}{2}$ is slightly favored for Y, as shown in Figure 4-5 for a ^{13}C—^{1}H coupling (an upward-pointing arrow represents a $+\frac{1}{2}$ spin, and a downward-pointing arrow a $-\frac{1}{2}$ spin). Since the bonding electrons are used to pass the spin information, the contact term is not averaged to zero by molecular tumbling.

When one spin slightly polarizes another spin oppositely, as in the above model for coupling across X—Y, the coupling constant J between the spins is said by convention to have a positive sign. A negative coupling occurs when spins polarize each other in the same (parallel) direction. Qualitative models analogous to that shown in Figure 4-5 indicate that coupling over two bonds, as in H—C—H, is negative, while coupling over three bonds, as in H—C—C—H, is positive. There are numerous exceptions to this qualitative model, but it is useful in understanding that J has sign as well as magnitude.

High resolution NMR spectra normally are not dependent on the absolute sign of coupling constants. Simultaneous reversal of the sign of every coupling constant in a spin system results in an identical spectrum. Many spectra, however, depend on the *relative signs* of component couplings. For example, the general ABX spectrum is determined in part by three couplings, J_{AB}, J_{AX}, and J_{BX} (Section 4-8). Different spectra can be obtained when J_{AX} and J_{BX} have the same sign (both positive or both negative) from when they have opposite signs (one positive and the other negative), even when the magnitudes are the same.

The usual convention for referring to a coupling constant is to denote the number of bonds between the coupled nuclei by a superscript on the left of the letter J and any other descriptive material by a subscript on the right or parenthetically. A two-bond *(geminal)* coupling between protons, then, is $^2J_{HCH}$ or ^2J(HCH), and a three-bond *(vicinal)* coupling between a proton and a carbon is $^3J_{HCCC}$ or ^3J(HCCC). Beyond three bonds, couplings between protons are said to be *long range.*

4-4 COUPLINGS OVER ONE BOND

The one-bond coupling between carbon-13 and protons is readily measured from the ^{13}C spectrum when the decoupler is turned off. Although usually unobserved because of decoupling, this coupling can provide useful information and illustrates several important principles. Because a p orbital has a node at the nucleus, only electrons in s orbitals can contribute to the Fermi coupling mechanism (s orbitals have a maximum in electron density at the nucleus). For protons, all electrons reside in the 1s orbital, but, for other nuclei, only that proportion of the orbital that has s character can contribute to coupling. When a proton is attached to an sp^3 carbon atom (25% s character), ${}^1J({}^{13}C-{}^1H)$ is about half as large as that for a proton attached to an sp carbon atom (50%). The alkenic CH (sp^2, 33%) coupling is intermediate. The values of ${}^1J$ for methane (sp^3), ethene (sp^2), benzene (sp^2), and ethyne (sp) are 125, 157, 159, and 249 Hz, respectively. These numbers define a linear relationship between the percentage of s character of the carbon orbital and the one-bond coupling (eq. 4-1).

$$\%s(C-H) = 0.2J({}^{13}C-{}^1H) \qquad \textbf{(4-1)}$$

The zero intercept of this equation indicates that there is no coupling when the s character is zero, in agreement with the Fermi contact model.

The one-bond CH coupling ranges from about 100 to 320 Hz, and much of this variation may be interpreted in terms of the *J*–s relationship of eq. 4-1. The coupling constant in cyclopropane (162 Hz) demonstrates that the carbon orbital to hydrogen is approximately sp^2 hybridized. Intermediate values in hydrocarbons may be interpreted in terms of fractional hybridization. The coupling of 144 Hz for the indicated CH bond in tricyclopentane (**4-15**) corresponds to 29% s character ($sp^{2.4}$), 160 Hz in cubane (**4-16**)

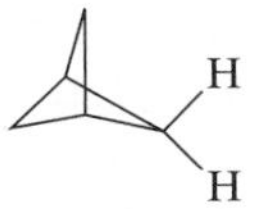

4-15 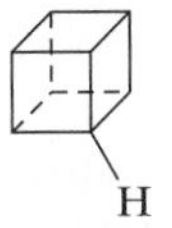 **4-16** 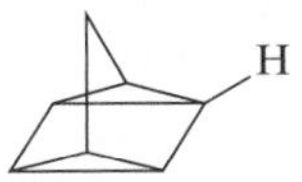 **4-17**

to 32% s character (sp^2), and 179 Hz in quadricyclane (**4-17**) to 36% s character ($sp^{1.8}$). Although the *J*–s relationship works well for hydrocarbons, there is some question as to its applicability to polar molecules. Variations in the effective nuclear charge, in addition to hybridization effects, may alter the coupling constants.

Just as the resonance frequency of a nucleus is proportional to its gyromagnetic ratio γ, the coupling constant between two nuclei, as noted above, is proportional to the product of both gyromagnetic ratios, $J(X-Y) \propto \gamma_X\gamma_Y$. Nuclei with very small gyromagnetic ratios, such as ${}^{15}N$, tend to have correspondingly small couplings. Furthermore, in this case, $\gamma({}^{15}N)$ has a negative sign, whereas $\gamma({}^{13}C)$ and $\gamma({}^1H)$ are positive. As a result, one-bond couplings between nitrogen-15 and hydrogen have a negative sign. This sign does not represent an exception to the Fermi model described above (Figure 4-5) but reflects the negative sign of the gyromagnetic ratio.

One-bond couplings have been studied for numerous other nuclei, as described in the references at the end of the chapter. When neither nucleus is ${}^1H$, the coupling constant depends on the product of the s characters of the orbitals from both nuclei that form the bond. This increased number of unknowns makes it more difficult, for example, to interpret ${}^1J({}^{13}C-{}^{13}C)$ or ${}^1J({}^{13}C-{}^{19}F)$ in terms of hybridization. Moreover, nuclei with lone pairs, as in ${}^1J({}^{31}P-{}^{31}P)$ or ${}^1J({}^{13}C-{}^{15}N)$, appear to have further complexities. The one-bond coupling between two carbon atoms is readily measured by the technique known as INADEQUATE (Section 5-7) and is extremely useful in mapping carbon connectivities in complex molecules. Further examples of one-bond couplings are given in Table 4-1 (page 104).

4-5 GEMINAL COUPLINGS

The geminal coupling between two protons (H–C–H) may be measured directly from the spectrum when the coupled nuclei are chemically nonequivalent, thus constituting, for example, the AB or AM part of an ABX, AMX, ABX_3, and so on, spectrum. If the relationship is first order (AM), the coupling may be measured by inspection. In second-order cases (AB), the spectrum must be simulated computationally, unless the two spins are isolated (a *two-spin system*) (Section 4-8). When nuclei are chemically equivalent but magnetically nonequivalent, as in the AA' part of an AA'XX' spectrum, their coupling constant is accessible. In such cases, computational methods normally are necessary to measure $J_{AA'}$.

Splittings are not observed between coupled nuclei when they are magnetically equivalent, but the coupling constant may be measured by replacing one of the nuclei with deuterium. For example, in dichloromethane-*d* ($CHDCl_2$) the geminal H–C–D coupling is seen as the spacing between the components of the 1:1:1 triplet (deuterium has a spin of 1). Since coupling constants are proportional to the product of the gyromagnetic ratios of the coupled nuclei, *J*(HCH) may be calculated from *J*(HCD), as shown in eq. 4-2.

$$J(\mathrm{HH}) = \frac{\gamma_H}{\gamma_D} J(\mathrm{HD}) = 6.51 J(\mathrm{HD}) \tag{4-2}$$

Geminal couplings depend strongly on the angle formed by the three atoms, H–C–H, as seen in the cyclic hydrocarbon series [cyclohexane (–12.6 Hz), cyclopentane (–10.5 Hz), cyclobutane (–9 Hz), cyclopropane (–4.3 Hz)], or by comparison of acyclic alkanes (methane, −12.4 Hz) with acyclic alkenes (ethene, +2.3 Hz). Note that the sign of the coupling constant is important. Although most geminal couplings are negative, many of those for sp^2 carbons are positive.

The typical range for alkanes is −5 to −20 Hz, and for alkenes is +3 to −3 Hz (or 0–3 Hz if the sign is ignored). Electron withdrawal by induction tends to make the coupling constant more positive, as pointed out by Pople and Bothner-By. For alkanes the negative coupling thus decreases in absolute value (becoming less negative), from −12.4 Hz for methane to −10.8 Hz for CH_3OH, −9.2 Hz for CH_3I, and −5.5 Hz for CH_2Br_2. Electron donation makes the coupling more negative (higher absolute value), as in −14.1 Hz for TMS (Me_4Si). Analogous substitution on sp^2 carbon changes the coupling profoundly, as in the effect of electron withdrawal in the structure $H_2C{=}X$, for example, +2.3 Hz when X is CH_2 (ethene), +17 Hz when X is N-(*tert*-butyl) (an imine), and +40 Hz when X is O (formaldehyde). These (positive) couplings are becoming more positive, paralleling the change to less negative for negative couplings.

These effects of withdrawal or donation of electrons through the σ bonds (induction) can be augmented or diminished by π effects such as hyperconjugation. Pople and Bothner-By found that lone pairs of electrons can donate electrons and make *J* more positive, whereas the π orbitals of double or triple bonds can withdraw electrons and make *J* less positive (or more negative). The above-mentioned large increase in the geminal coupling of imines or formaldehyde compared with ethene results from reinforcement of the effects of σ withdrawal and π donation, as illustrated in structure **4-18**. The effect of π

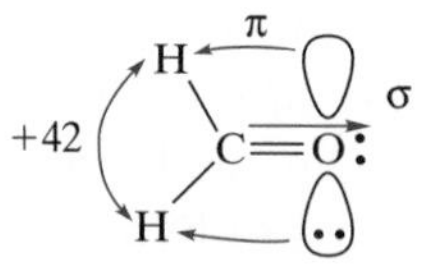

+42 π σ H C=O H

4-18

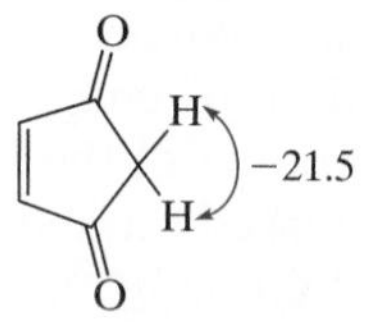

O H −21.5 H O

4-19

withdrawal occurs for carbonyl, nitrile, and aromatic groups, as in the values for acetone (−14.9 Hz), acetonitrile (–16.9 Hz), and dicyanomethane (–20.4 Hz). The π effect is somewhat reduced by free rotation in open-chain systems, so that particularly large effects are created by constraints of rings, as seen in **4-19** and **4-20**, as well as in the α protons of

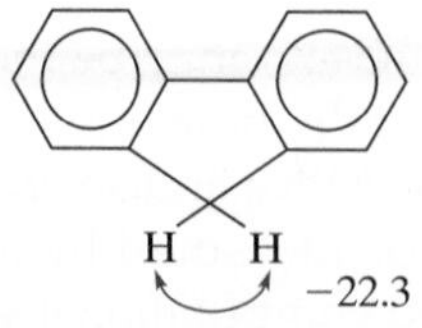

H H −22.3

4-20

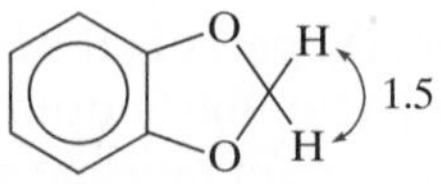

O H 1.5 O H

4-21

cyclopentanones and cyclohexanones. Structure **4-21** illustrates an example of how π donation by lone pairs makes J more positive. This effect also explains the difference in the geminal couplings of three-membered rings: cyclopropane [$(CH_2)_3$, −4.3 Hz] and oxirane [$(CH_2)_2O$, +5.5 Hz]. Although the difference in the absolute value of the couplings is only 1.2 Hz, the difference when signs are taken into consideration is almost 9 Hz.

Geminal couplings between protons and other nuclei also have been studied. The H–C–^{13}C coupling responds to substituents in much the same way as does the H–C–H coupling. The HCC couplings are smaller, however, because of the smaller gyromagnetic ratio of ^{13}C. Typical values in alkanes are −4.8 Hz in **H**—CH_2—**CH_3** and +1.2 Hz in **H**—CCl_2—**$CHCl_2$**. Couplings from hydrogen to sp^2 carbon (**H**—CH_2—**C**=) are typically −4 to −7 Hz, as in acetone [5.9 Hz, **H**—CH_2—**C**(=O)CH_3]. When the intermediate carbon is sp^2 [**H**—(C=X)—**C**], the coupling becomes larger and positive, as in aldehydes [+26.7 Hz for acetaldehyde, **H**—(C=O)—**CH_3**], but 5–10 Hz in alkenes [(**H**—(C=**C**R_2)—].

Unlike the proton–proton case, the proton–carbon geminal coupling pathway can include a double bond (**H**—C=**C**), and factors not considered by Pople and Bothner-By become important. For sp^2 carbons, these couplings often are small (−2.4 Hz for ethene, **H**—CH=**C**H_2). With proper substitution, however, stereochemical differences may be observed, as in *cis*-dichloroethene (**4-22**, 16.0 Hz) and *trans*-dichloroethene (**4-23**, 0.8 Hz).

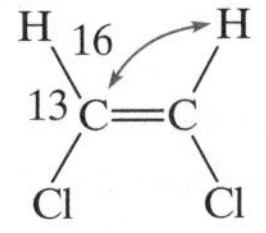

4-22 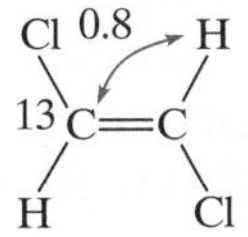 **4-23**

Such differences between alkene stereoisomers are common and may be exploited to prove stereochemistries. The geminal couplings in aromatics (**H**—C=**C**) are 4–8 Hz. For sp carbons, the coupling becomes quite large, 49.3 Hz in ethyne (**H**—C≡**C**H) and 61.0 Hz in **H**—C≡**C**—OPh.

The two-bond coupling between hydrogen and nitrogen-15 strongly depends on the presence and orientation of the nitrogen lone pair. The H–C–^{15}N coupling in imines is larger and negative when the proton is cis to the lone pair but smaller and positive for a proton trans to the lone pair, as in **4-24**. Thus ^2J(HCN) is a useful structural diagnostic

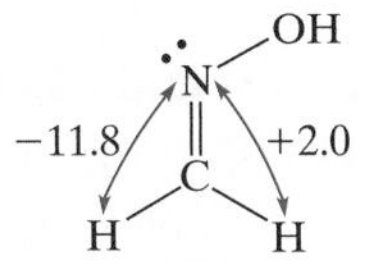

4-24 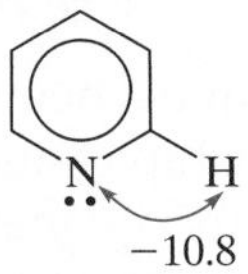 **4-25**

for syn–anti isomerism in imines, oximes, and related compounds. In saturated amines with rapid bond rotation, however, values typically are quite small and negative (−1.0 Hz for methylamine, CH_3NH_2). The cis relationship between the nitrogen lone pair and hydrogen also is found in heterocycles such as pyridine (**4-25**), in which the coupling constant is −10.8 Hz.

Two-bond couplings between ^{15}N and ^{13}C follow a similar pattern and also can be used for structural and stereochemical assignments. The carbon on the same side as the lone pair in imines again has a larger, negative coupling (−11.6 Hz in **4-26**). The isomer

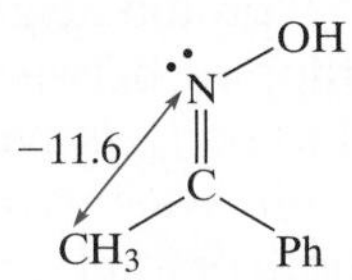

4-26 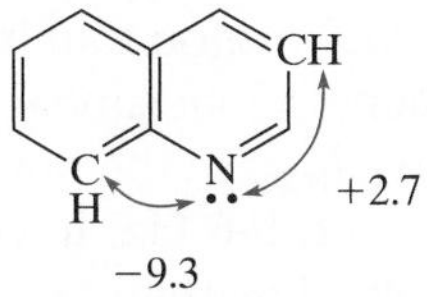 **4-27**

of **4-26**, in which the methyl is syn to hydroxyl (anti to the lone pair), has a ^2J(CCN) of only 1.0 Hz. The two indicated carbons in quinoline (**4-27**) have couplings, respectively, of −9.3 and +2.7 Hz, as one is syn and the other anti to the nitrogen lone pair.

Couplings between ^{31}P and hydrogen also have been exploited stereochemically. The maximum positive value of ^2J(HCP) is observed when the H—C bond and the phosphorus lone pair are eclipsed (syn), and the maximum negative value when they are orthogonal or anti. The situation is similar to that for couplings between hydrogen and ^{15}N, but signs are reversed as a result of the opposite signs of the gyromagnetic ratios of ^{15}N and ^{31}P. The heterocycle **4-28** exhibits a coupling of +25 Hz between ^{31}P and

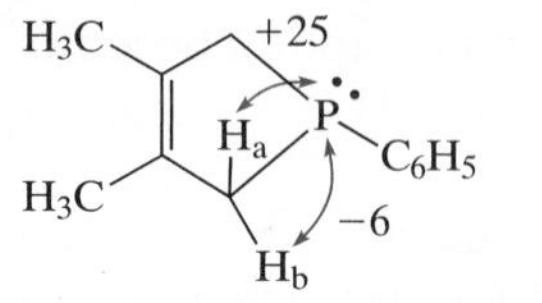

4-28

H_a (syn) and of −6 Hz between ^{31}P and H_b (anti). The coupling also is structurally dependent, as it is larger for P(III) than for P(V): 27 Hz for $(CH_3)_3P$: and 13.4 Hz for $(CH_3)_3P{=}O$.

Geminal H–C–F couplings are usually close to +50 Hz for an sp^3 carbon (47.5 Hz for CH_3**CH_2F**) and +80 Hz for an sp^2 carbon (84.7 Hz for $CH_2{=}$**CHF**). Geminal F–C–F couplings are quite large (+150–250 Hz) for saturated carbons (240 Hz for 1,1-difluorocyclohexane), but less than 100 Hz for unsaturated carbons (35.6 Hz for $CH_2{=}CF_2$).

4-6 VICINAL COUPLINGS

Coupling constants between protons over three bonds have provided the most important early stereochemical application of NMR spectroscopy. In 1961, Karplus derived a mathematical relationship between ^3J(HCCH) and the H—C—C—H dihedral angle ϕ. The simple form of eq. 4-3,

$$^3J = \begin{cases} A\cos^2\phi + C & (\phi = 0°–90°) \\ A'\cos^2\phi + C' & (\phi = 90°–180°) \end{cases} \tag{4-3}$$

illustrated in Figure 4-6, offers chemists a general and easily applied qualitative tool. The cosine-squared relationship results from strong coupling when orbitals are parallel. They can overlap at the synperiplanar (ϕ = 0°–30°) or antiperiplanar (ϕ = 150°–180°) geometries. When orbitals are staggered or orthogonal (ϕ = 60°–120°), coupling is weak.

The additive constants C and C' usually are neglected, as they are thought to be less than 0.3 Hz. When the constants A and A' can be evaluated, quantitation is possible. The inequality of A and A' ($A < A'$) means that J is different at the syn maximum on the left and the anti maximum on the right of Figure 4-6. Unfortunately, these multiplicative constants vary from system to system in the range 8–14 Hz (larger for alkenes). Because of this variation, quantitative applications cannot be transferred easily from one structure to another.

The Karplus equation provides useful qualitative interpretations in a number of very fundamental systems. In chair cyclohexanes (Figure 4-7), J_{aa} is large (8–13 Hz) because ϕ_{aa} is close to 180°, whereas J_{ee} (0–5 Hz) and J_{ae} (1–6 Hz) are small because ϕ_{ee} and ϕ_{ae} are close to 60°. An axial proton that has an axial proton neighbor can easily be identified by its large J_{aa}. When cyclohexane rings are flipping between two chair forms, J_{aa} is averaged with J_{ee} to give a J_{trans} in the range 4–9 Hz, and J_{ae} is averaged with J_{ea} to give a smaller J_{cis}, still in the range 1–6 Hz. In complex spin systems, axial proton resonances sometimes can be recognized by their larger linewidth or total spread than those of equatorial protons, because J_{aa} increases the width or spread more than does J_{ae}.

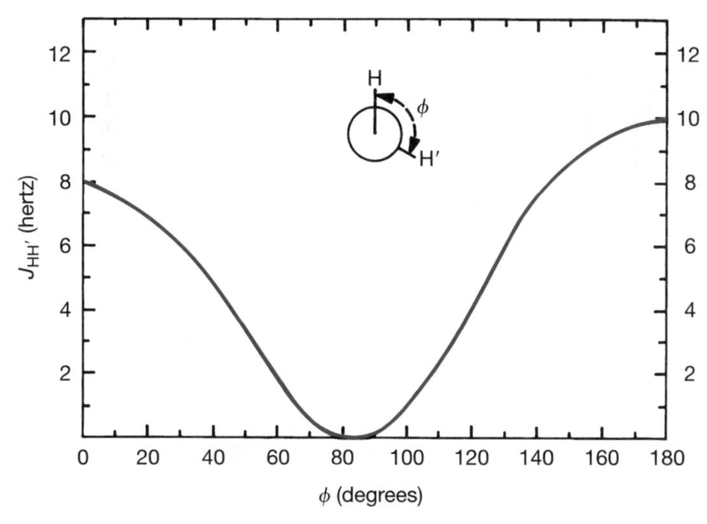

FIGURE 4-6 The vicinal H—C—C—H coupling constant as a function of the dihedral angle ϕ.

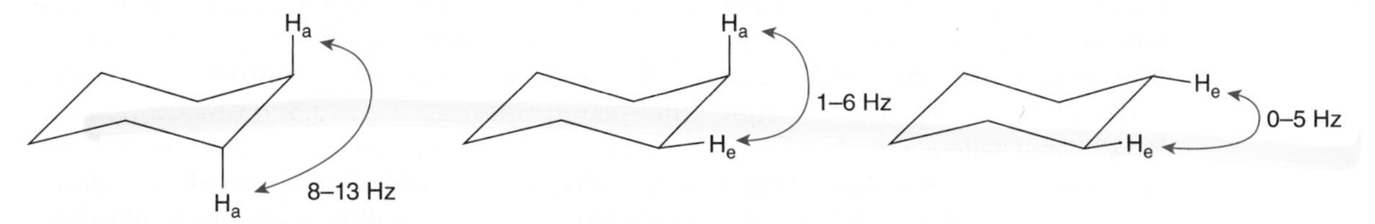

FIGURE 4-7 Coupling magnitudes between vicinal protons in chair six-membered rings.

In three-membered rings (**4-29**), J_{cis} ($\phi = 0°$) is always larger than J_{trans} ($\phi = 120°$), as a glance at the Karplus plot in Figure 4-6 will indicate. For the parent cyclopropane,

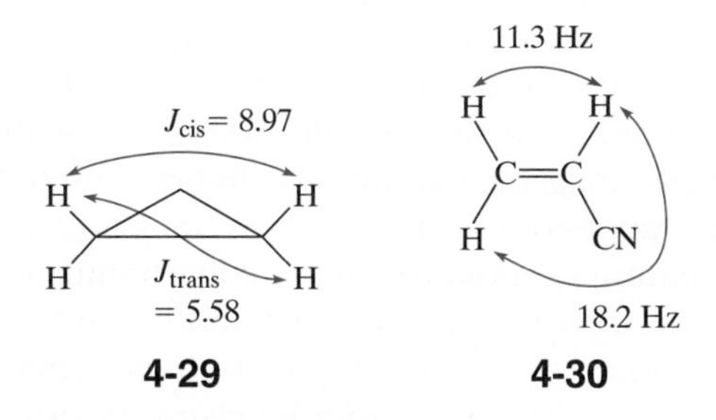

$J_{cis} = 8.97$ Hz and $J_{trans} = 5.58$ Hz. In four-membered rings, the cis coupling usually is larger than the trans coupling, but the two quantities can be close enough to be ambiguous. In five-membered rings, either J_{cis} or J_{trans} can be the larger, because the dihedral angles are toward the center of the Karplus curve.

In alkenes, J_{trans} ($\phi = 180°$) is always larger than J_{cis} ($\phi = 0°$), for example, 18.2 and 11.3 Hz in acrylonitrile (**4-30**). The spectrum of this compound is illustrated in Figure 4-8. The vinyl resonance is composed of three groupings, which comprise an AMX, ABX, or ABC spectrum, depending on the field strength. The three couplings (J_{AM}, J_{AX}, J_{MX}) may be assigned by inspection with the knowledge that $^3J_{trans} > {^3J_{cis}} > {^2J_{gem}}$. The reader should carry out the measurements for **4-30** from Figure 4-8. The 12 peaks are located at δ 5.647, 5.684, 5.708, 5.744, 6.090, 6.093, 6.128, 6.132, 6.211, 6.215, 6.273, and 6.278. Finally, ortho protons in aromatic rings have a dihedral angle of 0°, so $^3J_{ortho}$ is generally quite large (6–10 Hz) and can be distinguished from the smaller J_{meta} and J_{para}.

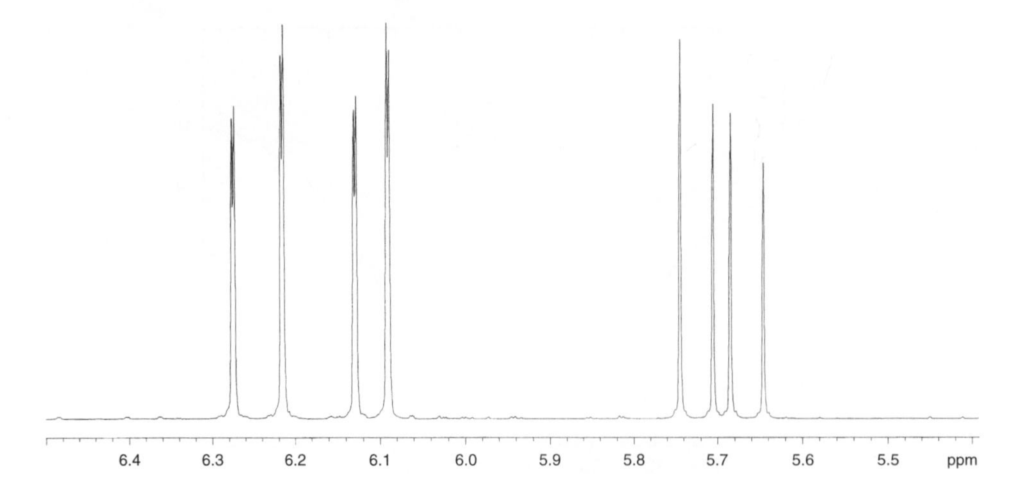

FIGURE 4-8 The 300 MHz ^{1}H spectrum of acrylonitrile ($CH_2{=}CHCN$) in $CDCl_3$.

Despite the potentially general application of the Karplus equation to dihedral angle problems, there are quantitative limitations. The vicinal H–C–C–H coupling constant depends on the C—C bond length or bond order, the H—C—C valence angle, the electronegativity of substituents on the carbon atoms, and the orientation of these substituents, in addition to the H—C—C—H dihedral angles. All of these other factors contribute to the multiplicative factors A and A'. A properly controlled calibration series of molecules must be rigid (monoconformational) and have unvarying bond lengths and valence angles. Several approaches have been developed to take the only remaining factor, substituent electronegativity, into account. One approach is to derive the mathematical dependence of ^3J on electronegativity. Another is empirical allowance by the use of chemical shifts that depend on electronegativity in a similar fashion as ^3J. A third approach is to eliminate the problem through the use of the ratio (the R value) of two ^3J coupling constants that respond to the same or related dihedral angles and that have the same multiplicative dependence on substituent electronegativity, which divides out in R. These more sophisticated versions of the Karplus method have been used quite successfully to obtain reliable quantitative results.

The existence of factors other than the dihedral angle results in ranges of vicinal coupling constants at constant ϕ even in structurally analogous systems. Saturated hydrocarbon chains (**H**—C—C—**H**) exhibit vicinal couplings in the range 3–9 Hz, depending on substituent electronegativity and rotamer mixes (J = 3.06 Hz for $Cl_2CHCHCl_2$ and 8.90 Hz for CH_3CH_2Li). Higher substituent electronegativity always lowers the vicinal coupling constant. In small rings, the variation is almost entirely the result of substituent electronegativity, with cis ranges of 7–13 Hz and trans ranges of 4–10 Hz in cyclopropanes. Coupling constants in oxiranes (epoxides) are smaller because of the effect of the electronegative oxygen atom. Couplings across a double bond (**H**—C=C—**H**) depend strongly on the valence angles, as well as on the electronegativity of the other two substituents. In cycloalkenes, the value varies from 1.3 Hz in cyclopropene to 8.8 Hz in cyclohexene, all with ϕ = 0°. In acyclic alkenes, J_{trans} has a range of 10–24 Hz and J_{cis} of 2–19 Hz. Because the ranges overlap, the distinction between cis and trans isomers is fully reliable only when both isomers are in hand. When bonds are intermediate between single and double bonds, ^3J is proportional to the overall bond order, as in $^3J_{12}$ = 8.6 Hz and $^3J_{23}$ = 6.0 Hz in naphthalene.

The ortho coupling in benzene derivatives varies over the relatively small range of 6.7–8.5 Hz, depending on the resonance and polar effects of the substituents. The presence of heteroatoms in the ring expands the range at the lower end down to 2 Hz, because of the effects of electronegativity (pyridines) and of smaller rings (furans, pyrroles).

When one carbon is sp^3 and one is sp^2 [H—C—C(=X)—H] the range is 5–8 Hz for freely rotating acyclic hydrocarbons (X = CR_2) and 1–5 Hz for aldehydes (X = O). The value varies in hydrocarbon rings from −0.8 Hz in cyclobutene to +3.1 Hz in cyclohexene and +5.7 Hz in cycloheptene. For the central bond in dienes (**H**—C(=X)—C(=Y)—**H**), the range is 10–12 Hz for transoid systems (X, Y = CR_2).

When constrained to rings, the pathway is cisoid and the coupling is 1.9 Hz in cyclopentadiene and 5.1 Hz in 1,3-cyclohexadiene. In α, β-unsaturated aldehydes (X = O, Y = CR_2), the coupling is about 8 Hz if transoid and 3 Hz if cisoid.

The H–C–X–H (X = O, N, S, Si, etc.), H–C–C–C, H–C–C–F, H–C–N–F, and C–C–C–C couplings also follow Karplus-like relationships. The ^3J(H–C–O–P) couplings are useful in determining backbone conformations of nucleotides. The ^3J(C–C–C–C) couplings have a range of values (3–15 Hz) that is larger than the two bond case [the range for ^2J(C–C–C) is 1–10 Hz]. The F–C–C–F and H–C–C–P couplings appear not to follow the Karplus pattern.

4-7 LONG-RANGE COUPLINGS

Coupling between protons over more than three bonds is said to be *long range*. Sometimes coupling between ^{13}C and protons over more than one bond also is called long range, but the term is inappropriate for ^2J(CCH) and ^3J(CCCH). Long-range coupling constants between protons normally are less than 1 Hz and frequently are unobservably small. In at least two structural circumstances, however, such couplings commonly become significant.

σ–π Overlap. Interactions of C—H(σ) bonds with π electrons of double and triple bonds and aromatic rings along the coupling pathway often increase the magnitude of the coupling constant. One such case is the four bond *allylic coupling*, **HC—C=CH**, with a range of about +1 to −3 Hz and typical values close to −1 Hz. Larger values are observed when the saturated C—H_a bond is parallel to the π orbitals (**4-31**). This σ–π

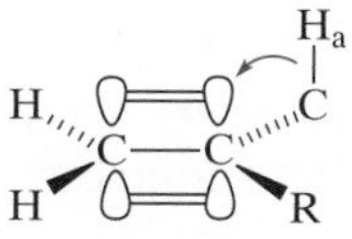

H
H
H_a
C
R
4-31

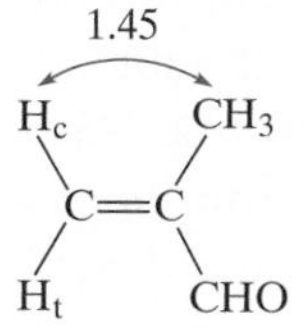

1.45
H_c
CH_3
H_t
CHO
4-32

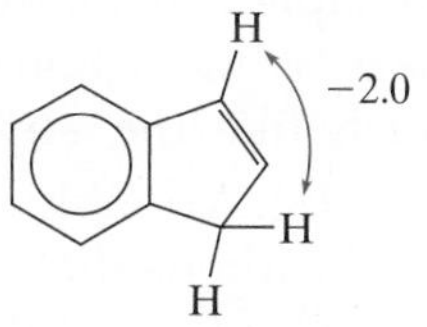

H
−2.0
H
H
4-33

overlap enables coupling to be transmitted more effectively. When the C—H_a bond is orthogonal to the π orbitals, there is no σ–π contribution and couplings are small (< 1 Hz). In acyclic systems, the dihedral angle is averaged over both favorable and unfavorable arrangements, so an average ^4J is found, as in 2-methylacryloin (**4-32**, $|^4J|$ = 1.45 Hz). Ring constraints can freeze bonds into the favorable arrangement, as in indene (**4-33**, ^4J = −2.0 Hz).

The five-bond doubly allylic coupling (also called *homoallylic*), **HC—C=C—CH**, depends on the orientation of two C—H bonds with respect to the π orbitals. For acyclic systems such as the 2-butenes, ^5J typically is 2 Hz, with a range of 0–3 Hz. When both protons are well aligned, the coupling can be quite large, as in the planar 1,4-cyclohexadiene (**4-34**), for which the cis coupling is 9.63 Hz and the trans coupling is 8.04 Hz.

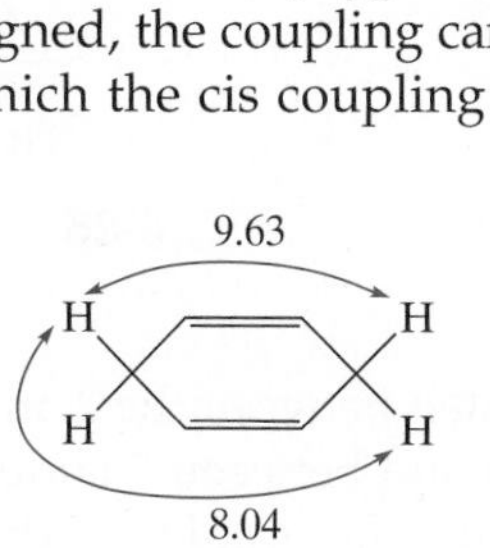

9.63
H H
H H
8.04
4-34

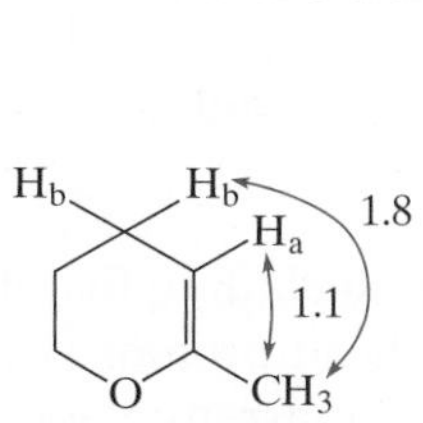

H_b H_b
H_a
1.8
1.1
O
CH_3
4-35

These couplings were measured by appropriate deuterium labeling. It is not unusual for the doubly allylic coupling to be larger than the allylic, as in **4-35** [^4J(CH_3—H_a) = 1.1 Hz, ^5J = 1.8 Hz].

Coupling constants are particularly large in alkynic and allenic systems, in which σ–π overlap can be very effective. In allene itself (CH_2=C=CH_2), ^4J is −7 Hz. In

1,1-dimethylallene, ^5J decreases to 3 Hz. Allene stereochemistry locks in a favorable arrangement for $\sigma-\pi$ overlap, as illustrated by the pair of arrows from the CH_2 group into the π orbitals of **4-36**. In both propyne (methylacetylene, ^4J = 2.9 Hz) and 2-butyne

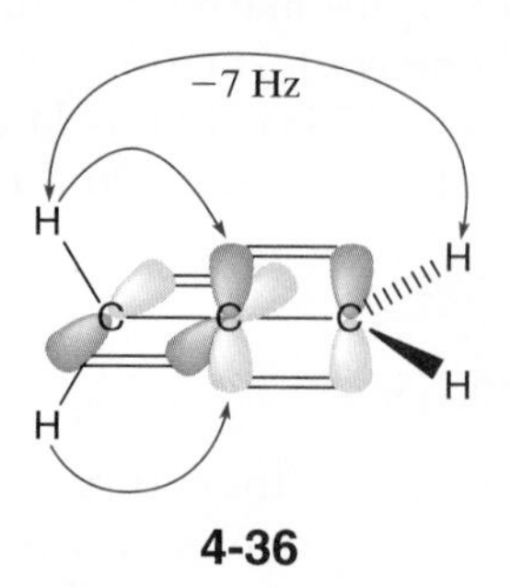

4-36

(dimethylacetylene, ^5J = 2.7 Hz), the long-range coupling is enhanced because the triple bond imposes no steric limitations on $\sigma-\pi$ overlap. Appreciable long-range couplings have been observed over up to seven bonds in polyalkynes.

Conjugated double bonds provide a more complicated situation. In butadiene, there are two four-bond (–0.86, –0.83 Hz) and three five-bond (+0.60, +1.30, +0.69) couplings. In aromatic rings, the meta coupling is a ^4J (range 1–3 Hz) and the para coupling is a ^5J (range 0–1 Hz). In benzene itself, $^3J_{\text{ortho}}$ is 7.54 Hz, $^4J_{\text{meta}}$ is 1.37 Hz, and $^5J_{\text{para}}$ is 0.69 Hz. None of these couplings in butadiene and benzene involves $\sigma-\pi$ overlap, as they all occur within a plane. Protons on saturated carbon atoms attached to an aromatic ring ($CH_3—C_6H_5$) couple with all three types of protons on the ring. These *benzylic couplings* depend on the $\sigma-\pi$ interaction between the substituent C—H bonds and the aromatic π electrons, much like the allylic coupling ($^4J_{\text{ortho}}$ = 0.6–0.9 Hz, $^5J_{\text{meta}}$ = 0.3–0.4 Hz, $^6J_{\text{para}}$ = 0.5–0.6 Hz). A doubly benzylic coupling can take place between protons on different saturated carbons that are attached directly to the benzene ring, as in xylenes ($CH_3—C_6H_4—CH_3$, $^5J_{\text{ortho}}$ = 0.3–0.5 Hz).

Zigzag Pathways. In the second major category of long-range coupling, enhanced values often are observed between protons that are related by a planar W or zigzag pathway. This geometry is seen, for example, in the 1,3-diequatorial arrangement between protons in six-membered rings (**4-37**, ^4J = 1.7 Hz). The norbornane framework (**4-38**) contains

4-37 **4-38**

several W arrangements, including that illustrated between the 2 and 6 exo protons, but also between the bridgehead protons (1 and 4) and between 3-endo and 7-anti protons.

In the planar, zigzag arrangement, there is favorable overlap between parallel C—H and C—C bonds, analogous to the optimal vicinal coupling at ϕ = 180°. The zigzag pathway is entirely within the σ framework but is important for many π systems, including aromatic meta couplings (hence the enhanced ^4J = 1.37 Hz in benzene, **4-39**). Five-bond zigzag pathways similarly can give rise to enhanced long-range couplings, such as the ^5J = +1.3 Hz in 1,3-butadiene (**4-40**) and the ^5J = 0.9 Hz coupling between the indicated protons in quinoline (**4-41**). Zigzag pathways over up to six bonds have been found to exhibit couplings.

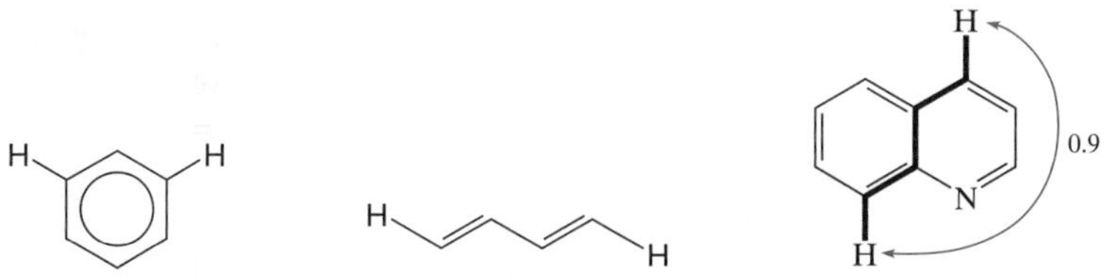

4-39 4-40 4-41

Through-space Coupling. Although coupling information always is passed via electron-mediated pathways, in some cases part of the through-bond pathway may be skipped, as in allylic (**4-31**) and benzylic couplings with $\sigma-\pi$ overlap. Two nuclei that are within van der Waals contact in space over any number of bonds can interchange spin information if at least one of the nuclei possesses lone pair electrons. These so-called through-space couplings are found most commonly, but not exclusively, in H–F and F–F pairs. The six-bond CH_3–F coupling is negligible in **4-42** (H–F distance 2.84 Å)

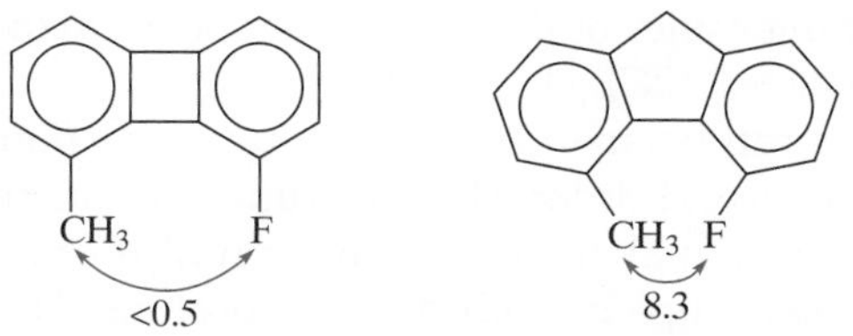

CH_3 F <0.5
4-42

CH_3 F 8.3
4-43

but is 8.3 Hz in **4-43** (1.44 Å) (the sum of the H and F van der Waals radii is 2.55 Å). In the latter case, coupling information is probably passed from the proton through the lone pair electrons to the fluorine nucleus. Such a mechanism very likely is important in the geminal F–C–F coupling, which is unusually large. Values of ^2J(FCF) are larger for sp^3 CF_2 (ca. 200 Hz) than for sp^2 CF_2 (ca. 50 Hz), as the smaller tetrahedral angle brings the fluorine atoms closer together.

4-8 SPECTRAL ANALYSIS

We have not said much about how coupling constants are extracted from spectra. Measurement is straightforward when the spectrum is first order, as chemical shifts correspond to the midpoint of a resonance multiplet. The midpoint falls between the components of a doublet from coupling to one other spin, is coincident with the middle peak of a triplet from coupling to two other spins, and so on. The coupling constant corresponds to the distance between adjacent peaks in the resonance multiplet. These ideal characteristics may fail in second-order spectra. Because most nuclei other than the proton have very large chemical shift ranges and because these nuclei often are in low natural abundance and hence do not show coupling to each other, second-order analysis is a consideration primarily for proton spectra alone. For protons, spectra measured above 500 MHz are usually first order from the $\Delta\nu/J$ criterion. Magnetic nonequivalence (Section 4-2), however, is independent of field and produces second-order spectra such as AA'XX' even with the most expensive superconducting magnet.

The AX spectrum consists of two doublets, all components with equal intensities (the spectra in Figure 2-18 and at the bottom of Figure 4-1 are very close to first order). The doublet spacing is J_{AX}, and the midpoints of the doublets are ν_A and ν_X. The second-order, two-spin (AB) system also contains four lines, but the inner peaks are always more intense than the outer peaks (see Figures 4-1 and 4-9). The coupling constant (J_{AB}) still is obtained directly and accurately from the doublet spacings, but no specific peak position or simple average corresponds to the chemical shifts. The A chemical shift (ν_A) occurs at the weighted-average position of the two A peaks

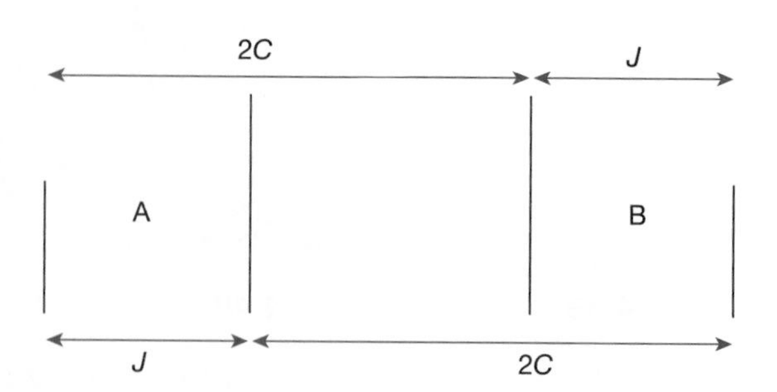

FIGURE 4-9 Notation for spacings in a second-order, two-spin system (AB).

(and similarly for ν_B). The chemical shift difference $\Delta\nu_{AB}$ ($\nu_B - \nu_A$) is most easily calculated from eq. 4-4,

$$\Delta\nu_{AB} = (4C^2 - J^2)^{\frac{1}{2}} \tag{4-4}$$

in which 2C is the spacing between alternate peaks (see Figure 4-9). The values of ν_A and ν_B then are determined readily by adding $\frac{1}{2}\nu_{AB}$ to and subtracting it from the midpoint of the quartet. The ratio of intensities of the larger inner peaks to the smaller outer peaks is given by the expression $(1 + J/2C)/(1 - J/2C)$.

Three-spin systems can be analyzed readily by inspection only in the first-order cases AX_2 and AMX. The second-order AB_2 spectrum can contain up to nine peaks: four from spin flips of the A proton alone, four from spin flips of the B protons alone, and one from simultaneous spin flips of both A and B protons. The ninth peak is called a *combination line* and ordinarily is forbidden and of low intensity. Although these patterns may be analyzed by inspection, recourse normally is made to computer programs. The other second-order, three-spin systems (ABB', AXX', ABX, and ABC) and almost all second-order systems of four spins (AA'BB', AA'XX', ABXY, etc.) or larger are seldom able to be analyzed by inspection, and computer methods are employed.

Most spectrometers today provide the software for spectral calculation for up to seven spins. The first step is a trial and error procedure of approximating the chemical shifts and coupling constants in order to match the observed spectrum through computer simulation. Chemical shifts are varied until the widths and locations of the observed and calculated multiplets agree approximately. Then the coupling constants or their sums and differences are varied systematically until a reasonable match is obtained. This method is relatively successful for three and four spins but is difficult or impossible for larger systems.

Refinements of direct calculations or of this trial and error procedure utilize iterative computer programs. The program of Castellano and Bothner-By (LAOCN-5) iterates on peak positions but requires assignments of peaks to specific spin flips. The program of Stephenson and Binsch (DAVINS) operates directly on unassigned peak positions.

4-9 SECOND-ORDER SPECTRA

Second-order spectra are characterized by peak spacings that do not correspond to coupling constants, by nonbinomial intensities, by chemical shifts that are not at resonance midpoints, or by resonance multiplicities that do not follow the $n + 1$ rule (see Figures 4-1, 4-2, and 4-3). Even when the spectrum has the appearance of being first order, it may not be. Lines can coincide in such a way that the spectrum assumes a simpler appearance than seems consistent with the actual spectral parameters (a situation termed *deceptive simplicity*). For example, in the ABX spectrum, the X nucleus is coupled to two nuclei (A and B) that are *closely coupled* ($\Delta\nu_{AB}/J \ll 10$). Under these circumstances, the A and B spin states are fully mixed, and X responds as if the nuclei were equivalent. Thus the ABX spectrum resembles an A_2X spectrum, as if $J_{AX} = J_{BX}$. Figure 4-10 illustrates this situation. When $\Delta\nu_{AB} = 3.0$ Hz (Figure 4-10a), the calculated example looks like a first-order A_2X spectrum with one coupling constant, even though $J_{AX} \neq J_{BX}$. When $\Delta\nu_{AB} = 8.0$ Hz (Figure 4-10b), a typical ABX spectrum is obtained. Deceptive simplicity sometimes, but

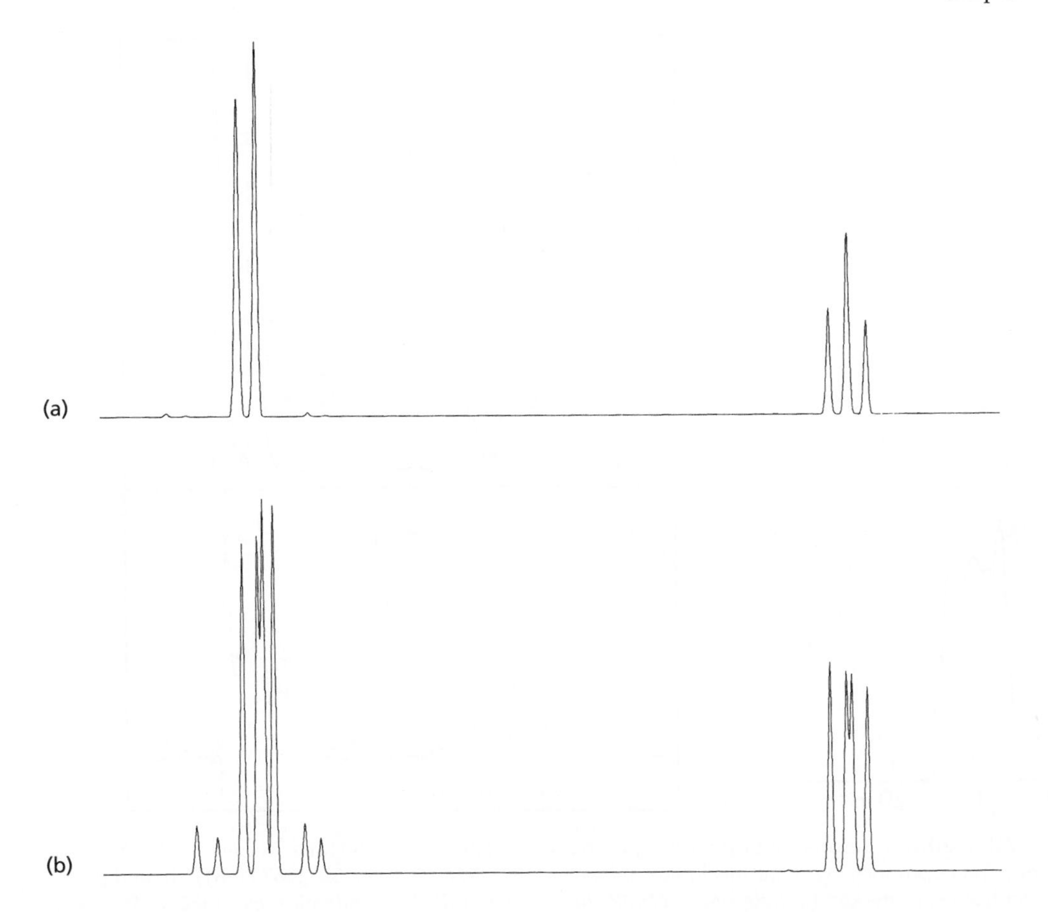

FIGURE 4-10 (a) A deceptively simple ABX spectrum: $\nu_A = 0.0$ Hz, $\nu_B = 3.0$ Hz, $\nu_X = 130.0$ Hz; $J_{AB} = 15.0$ Hz, $J_{AX} = 5.0$ Hz, $J_{BX} = 3.0$ Hz. (b) The same parameters, except $\nu_B = 8.0$ Hz. The larger value of $\Delta\nu_{AB}$ removes the deceptive simplicity and produces a typical ABX spectrum.

not always, can be removed by use of a higher field. When the spectrum is deceptively simple, only sums or averages of coupling constants may be measured. Actual coupling constants are impossible to obtain.

The AA'XX' spectrum often is observed as a deceptively simple pair of triplets, resembling A_2X_2. In this case, it is the A and A' nuclei that are closely coupled ($\Delta\nu_{AA'} = 0$ Hz and $J_{AA'}$ is large). Such deceptive simplicity is not eliminated by raising the field because A and A' are chemically equivalent. The chemist should beware of the pair of triplets that falsely suggests magnetic equivalence (A_2X_2) and equal couplings ($J_{AX} = J_{AX'}$), when the molecular structure suggests AA'XX'. Sometimes the couplings between A and X may be observed by lowering the field to turn the AA'XX' spectrum into AA'BB' with a larger number of peaks that may permit a complete analysis.

A particularly subtle example of second-order complexity occurs in the ABX spectrum (or, more generally, $A_xB_yX_z$) when A and B are very closely coupled, J_{AX} is large, and J_{BX} is zero. With no coupling to B, the X spectrum should be a simple doublet from coupling to A. Since A and B are closely coupled, however, the spin states of A and B are mixed, and the X spectrum is perturbed by the B spins (the phenomenon has been termed *virtual coupling,* which is something of a misnomer, since B is not coupled to X). As an example in a slightly larger but analogous spin system, the CH and CH_2 protons of β-methylglutaric acid (**4-44**) are closely coupled. Although the CH_3 group is coupled

$$\mathrm{HO_2C{-}CH_2{-}\underset{\displaystyle CH_3}{\overset{\displaystyle H}{C}}{-}CH_2{-}CO_2H}$$

4-44

only to the CH proton, its resonance is much more complicated than a simple doublet (Figure 4-11a). The CH and CH_2 protons are closely coupled, so their spin states

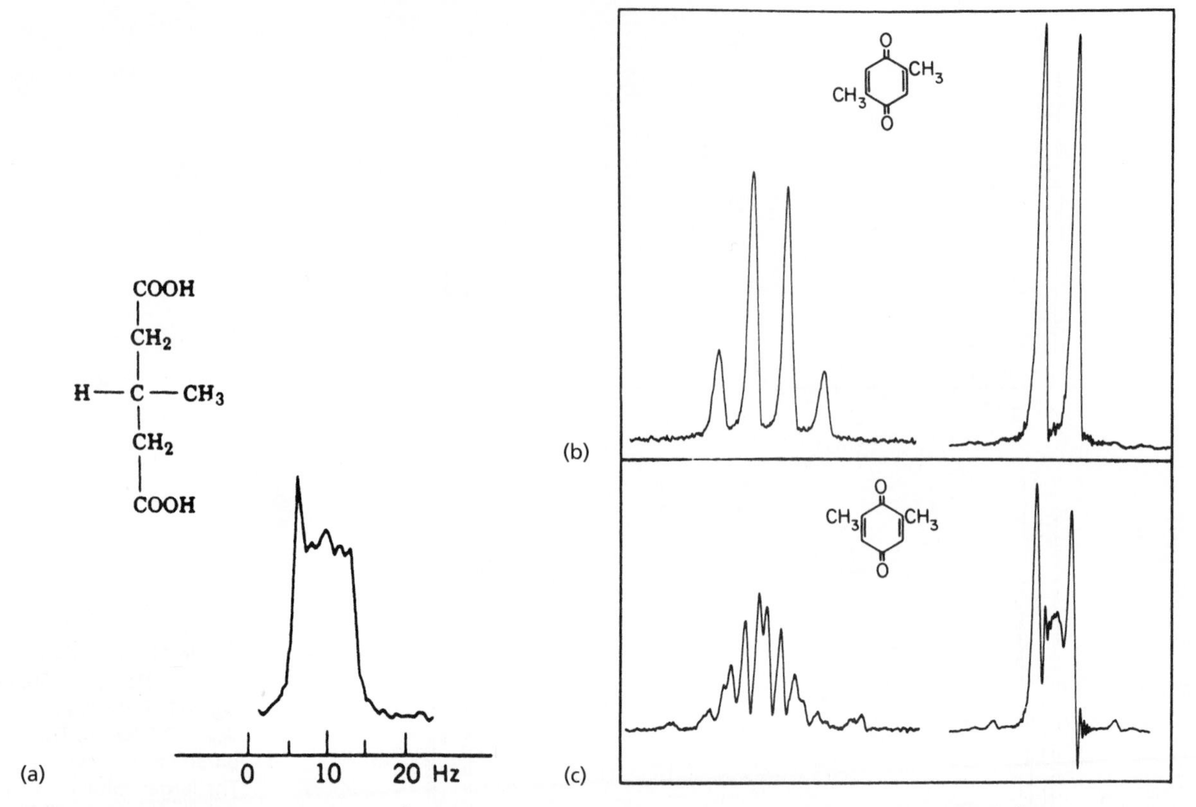

FIGURE 4-11 (a) The 60 MHz methyl ^{1}H resonance of β-methylglutaric acid. (b) and (c) The 60 MHz ^{1}H spectra of the 2,5- and 2,6-dimethylbenzoquinones. ([a] Reproduced with permission from F. A. L. Anet, *Can. J. Chem.*, **39**, 2267 [1961]; [b] and [c] reproduced with permission from E. D. Becker, *High Resolution NMR,* 2nd ed., Orlando, FL: Academic Press, 1980, p. 166.)

are mixed. The CH_3 group interacts with a mixture of CH and CH_2 spin states, even though $J = 0$ between CH_3 and CH_2. This problem is eliminated at a higher field, at which the CH and CH_2 resonances are well separated. The methyl group, unmixed with the CH_2 spin states, then couples cleanly with CH.

The dimethylbenzoquinones provide a further example of virtual coupling. The proton spectrum of the 2,5-dimethyl isomer (**4-45**, Figure 4-11b) contains a first-order

4-45 **4-46**

methyl doublet and an alkene quartet. The spectrum of the 2,6 isomer (**4-46**, Figure 4-11c) is much more complicated. The alkenic protons in both molecules are equivalent (AA'). In **4-45** they are coupled only to the methyl protons ($J_{AA'} = 0$ Hz), but in **4-46** they are closely coupled to each other because of the zigzag pathway. The multiplicity of the methyl resonance is perturbed not only by the adjacent alkenic proton but also by the proton on the opposite side of the ring. In other words, **4-45** is $(AX_3)_2$ but **4-46** is $AA'X_3X'_3$. This effect is not altered at higher field because A and A' are chemically equivalent.

Sometimes proton spectra are second order even at 500 MHz or higher (aside from the AA' case). In addition, some institutions still have access only to iron core, 60 MHz spectrometers, which produce largely second-order proton spectra. These spectra may be clarified somewhat by the use of *paramagnetic shift reagents.* These molecules contain unpaired spins and form Lewis acid–base complexes with dissolved substrates. The

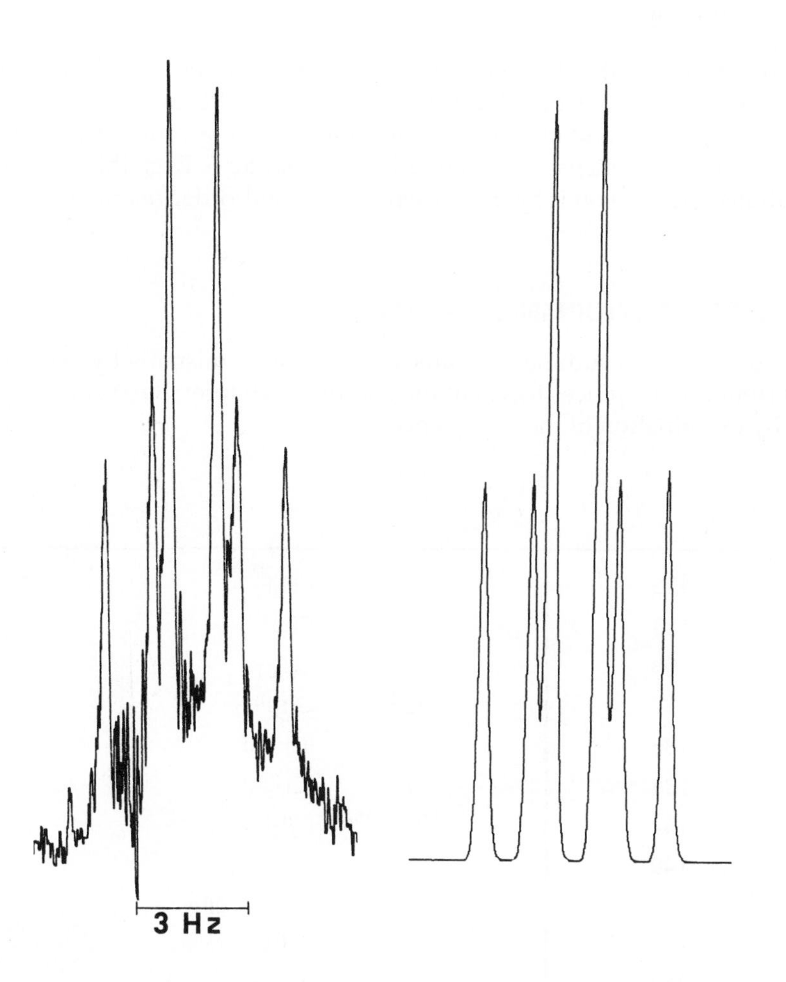

FIGURE 4-12 The 90 MHz ^1H spectrum of cyclopropene, showing the observed (left) and calculated (right) high-frequency ^{13}C satellite of the alkenic protons. (Reproduced with permission from J. B. Lambert, A. P. Jovanovich, and W. L. Oliver, Jr., *J. Phys. Chem.* **74**, 2221 [1970]. Copyright 1970 American Chemical Society.)

unpaired spin exerts a strong paramagnetic shielding effect (hence to higher frequency or downfield) on nuclei close to it. The effect drops off rapidly with distance, so that those nuclei in the substrate that are closest to the site of acid–base binding are affected more. Consequently, the shift to higher frequency varies through the substrate and hence leads to greater separation of peaks. Two common shift reagents contain lanthanides: tris(dipivalomethanato)europium(III)·2(pyridine) [called Eu(dpm)$_3$ without pyridine] and 1,1,1,2,2,3,3-heptafluoro-7,7-dimethyloctanedionatoeuropium(III) [or Eu(fod)$_3$]. Shift reagents are available with numerous rare earths as well as other elements. Almost all organic functional groups that are Lewis bases have been found to respond to these reagents. When the shift reagent is chiral, it can complex with enantiomers and generate separate resonances from which enantiomeric ratios may be obtained.

Spectral analysis can sometimes be facilitated by taking advantage of dilute spins present in the molecule. Earlier in this chapter, cyclopropene (**4-5**) was mentioned as an example of an A_2X_2 spectrum, and in Section 4-6 the vicinal coupling between the protons on the double bond (J_{AA}) was quoted as being 1.3 Hz. How was such a coupling constant between two chemically equivalent protons measured? Its small value prohibits the use of deuterium, as J_{HD} would be only 0.2 Hz for cyclopropene. For 1.1% of the molecules, the double-bond spin system is H—^{12}C=^{13}C—H. The proton on ^{12}C resonates at almost the same position as the molecules with no ^{13}C. The large one-bond ^{13}C–^{1}H coupling produces multiplets, called *satellites*, on either side of the centerband and separated from it by about $\frac{1}{2}J$(CH). Small isotope effects can shift the center of the satellites. The separation of each satellite from the centerband serves as an effective chemical shift difference, so that the H-H coupling between H—^{12}C and H—^{13}C is present in the satellite. Figure 4-12 shows the satellite spectrum of the alkenic protons of cyclopropene. The satellite is a doublet of triplets, since the alkenic proton on ^{13}C is coupled to the other alkenic proton and to the two methylene protons. In this way, ^3J(HC=CH) can be measured between the normally equivalent hydrogens. Other dilute spins produce satellite spectra that are commonly observed in proton spectra, including ^{15}N, ^{29}Si, ^{77}Se, ^{111}Cd, ^{113}Cd, ^{117}Sn, ^{119}Sn, ^{125}Te, ^{195}Pt, and ^{199}Hg.

The most general and effective method for analyzing complex proton spectra involves the use of two dimensions, as described in Chapter 6. Even this method, however, has limitations imposed on it by the presence of second-order relationships.

4-10 TABLES OF COUPLING CONSTANTS

Tables 4-1 through 4-5 summarize values of coupling constants by class of structure, extracted from the references found at the end of the chapter. Further examples may be obtained by examination of these references.

TABLE 4-1 One-Bond Couplings

Coupling	Compound	J
$^{13}C—^{1}H$	CH_3CH_3	125
	$(CH_3)_4Si$	118
	CH_3Li	98
	$(CH_3)_3N$	132
	CH_3CN	136
	$(CH_3)_2S$	138
	CH_3OH	142
	CH_3F	149
	CH_3Cl	150
	CH_2Cl_2	177
	$CHCl_3$	208
	Cyclohexane	125
	Cyclobutane	136
	Cyclopropane	162
	Tetrahydrofuran (α, β)	145, 133
	Norbornane (C1)	142
	Bicyclo[1.1.1]pentane (C1)	164
	Cyclohexene (C1)	157
	Cyclopropene (C1)	226
	Benzene	159
	1,3-Cyclopentadiene (C2)	170
	$\mathbf{CH_2}{=}CHBr$	197
	Acetaldehyde (CHO)	172
	Pyridine (α, β, γ)	177, 157, 160
	Allene	168
	$CH_3C{\equiv}\mathbf{CH}$	248
	$(CH_3)_2\mathbf{C^+H}$	164
	$HC{\equiv}N$	269
	Formaldehyde	222
	Formamide	191
$^{13}C—^{19}F$	CH_2F_2	235
	CF_3I	345
	C_6F_6	362
$^{13}C—^{31}P$	CH_3PH_2	9.3
	$(CH_3)_3P$	−13.6
	$(CH_3)_4P^+\ I^-$	56
$^{13}C—^{15}N$	CH_3NH_2	−4.5
	$C_6H_5NH_2$	−11.4
	$CH_3(CO)NH_2$	−14.8
	$CH_3C{\equiv}N$	−17.5
	Pyridine	+0.62
	$CH_3HC{=}N—OH$ (E, Z)	−4.0, −2.3
$^{15}N—^{1}H$	CH_3NH_2	−64.5
	$CH_3(CO)NH_2$	−89
	Pyridinium	−90.5
	$HC{\equiv}N^+H$	−134
	$(C_6H_5)_2C{=}NH$	−51.2
$^{15}N—^{15}N$	Azoxybenzene	12.5
	Phenylhydrazine	6.7
$^{15}N—^{31}P$	$C_6H_5NHP(CH_3)_2$	53.0
	$C_6H_5NH(PO)(CH_3)_2$	−0.5
	$[(CH_3)_2N]_3P{=}O$	−26.9
$^{13}C—^{13}C$	CH_3CH_3	35
	$CH_3(CO)CH_3$	40
	CH_3CO_2H	57
	$CH_2{=}CH_2$	68
	$CH{\equiv}CH$	171
$^{31}P—^{1}H$	$C_6H_5(C_6H_5CH_2)(PO)H$	474
$^{31}P—^{31}P$	$(CH_3)_2P—P(CH_3)_2$	−179.7
	$(CH_3)_2(PS)(PS)(CH_3)_2$	18.7

TABLE 4-2 Geminal Proton–Proton (H—C—H) Couplings (Hz)

Compound	J (Hz)	Compound	J (Hz)
CH_4	−12.4	Oxirane	+5.5
$(CH_3)_4Si$	−14.1	$CH_2{=}CH_2$	+2.3
$C_6H_5CH_3$	−14.4	$CH_2{=}O$	+40.22
$CH_3(CO)CH_3$	−14.9	$CH_2{=}NOH$	9.95
CH_3CN	−16.9	$CH_2{=}CHF$	−3.2
$CH_2(CN)_2$	−20.4	$CH_2{=}CHNO_2$	−2.0
CH_3OH	−10.8	$CH_2{=}CHOCH_3$	−2.0
CH_3Cl	−10.8	$CH_2{=}CHBr$	−1.8
CH_3Br	−10.2	$CH_2{=}CHCl$	−1.4
CH_3F	−9.6	$CH_2{=}CHCH_3$	2.08
CH_3I	−9.2	$CH_2{=}CHCO_2H$	1.7
CH_2Cl_2	−7.5	$CH_2{=}CHC_6H_5$	1.08
Cyclohexane	−12.6	$CH_2{=}CHCN$	0.91
Cyclopropane	−4.3	$CH_2{=}CHLi$	7.1
Aziridine	+1.5	$CH_2{=}C{=}C(CH_3)_2$	−9.0

TABLE 4-3 Vicinal Proton–Proton (H—C—C—H) Couplings (Hz)

Compound	J (Hz)	Compound	J (Hz)
CH_3CH_3	8.0	$CH_2{=}CH_2$ (cis, trans)	11.5, 19.0
$CH_3CH_2C_6H_5$	7.62	$CH_2{=}CHLi$ (cis, trans)	19.3, 23.9
CH_3CH_2CN	7.60	$CH_2{=}CHCN$ (cis, trans)	11.75, 17.92
CH_3CH_2Cl	7.23	$CH_2{=}CHC_6H_5$ (cis, trans)	11.48, 18.59
$(CH_3CH_2)_3N$	7.13	$CH_2{=}CHCO_2H$ (cis, trans)	10.2, 17.2
CH_3CH_2OAc	7.12	$CH_2{=}CHCH_3$ (cis, trans)	10.02, 16.81
$(CH_3CH_2)_2O$	6.97	$CH_2{=}CHCl$ (cis, trans)	7.4, 14.8
CH_3CH_2Li	8.90	$CH_2{=}CHOCH_3$ (cis, trans)	7.0, 14.1
$(CH_3)_2CHCl$	6.4	$ClHC{=}CHCl$ (cis, trans)	5.2, 12.2
$ClCH_2CH_2Cl$ (neat)	5.9	Cyclopropene (1–2)	1.3
$Cl_2CHCHCl_2$ (neat)	3.06	Cyclobutene (1–2)	2.85
Cyclopropane (cis, trans)	8.97, 5.58	Cyclopentene (1–2)	5.3
Oxirane (cis, trans)	4.45, 3.10	Cyclohexene (1–2)	8.8
Aziridine (cis, trans)	6.0, 3.1	Benzene	7.54
Cyclobutane (cis, trans)	10.4, 4.9	C_6H_5Li (2–3)	6.73
Cyclopentane (cis, trans)	7.9, 6.3	$C_6H_5CH_3$ (2–3)	7.64
Tetrahydrofuran ($\alpha-\beta$: cis, trans)	7.94, 6.14	$C_6H_5CO_2CH_3$ (2–3)	7.86
Cyclopentene (3–4: cis, trans)	9.36, 5.72	C_6H_5Cl (2–3)	8.05
Cyclohexane (av.: cis, trans)	3.73, 8.07	$C_6H_5OCH_3$ (2–3)	8.30
Cyclohexane (ax–ax)	12.5	$C_6H_5NO_2$ (2–3)	8.36
Cyclohexane (eq–eq and ax–eq)	3.7	$C_6H_5N(CH_3)_2$ (2–3)	8.40
Piperidine (av. $\alpha-\beta$: cis, trans)	3.77, 7.88	Naphthalene (1–2, 2–3)	8.28, 6.85
Oxane (av. $\alpha-\beta$: cis, trans)	3.87, 7.41	Furan (2–3, 3–4)	1.75, 3.3
Cyclohexanone (av. $\alpha-\beta$: cis, trans)	5.01, 8.61	Pyrrole (2–3, 3–4)	2.6, 3.4
Cyclohexene (3–4: cis, trans)	2.95, 8.94	Pyridine (2–3, 3–4)	4.88, 7.67

TABLE 4-4 Carbon Couplings Other Than $^1J(^{13}C{-}^1H)$ (Hz)

CH_3CH_3	−4.8	CH_3CH_3	34.6
CH_3CH_2Cl	2.6	CH_3CH_2OH	37.7
$Cl_2CH{-}CHCl_2$	+1.2	CH_3CHO	39.4
Cyclopropane (^2J)	−2.6	$CH_3C{\equiv}N$	56.5
$(CH_3)_2CHCH_2CH(CH_3)_2$	5.	$CH_3CO_2C_2H_5$	58.8
$(CH_3)_2C{=}O$	5.9	$CH_2{=}CH_2$	67.2
$CH_3(CO)H$	26.7	$CH_2{=}CHCN$	74.1
$CH_3CH{=}C(CH_3)_2$	4.8	C_6H_5CN (ipso)	80.3
$CH_2{=}CH_2$	−2.4	$C_6H_5NO_2$ (1,2)	55.4
$CHCl{=}CHCl$ (cis, trans)	16.0, 0.8	$HC{\equiv}CH$	170.6
$CH_2{=}CHBr$ (cis, trans)	−8.5, +7.5	$(CH_3CH_2)_3P$	+14.1
Benzene [^2J(CH), ^3J(CH)]	+1.0, +7.4	$(CH_3CH_2)_4P^+\ Br^-$	−4.3
$CH_3C{\equiv}CH$ (CH_3, $\equiv CH$)	−10.6, +50.8	$(CH_3O)_3P$	+10.05
CF_3CF_3	46.0	$(CH_3O)_3P{=}O$	−5.8
$CH_3(CO)F$	59.7	$(CH_3)_3P{=}S$	+56.1
$Cl_2C{=}CF_2$	44.2	$CH_3(CH_3O)_2P{=}O$	+142.2

TABLE 4-5 Nitrogen-15 Couplings beyond One Bond (Hz)

CH_3NH_2	−1.0	$CH_3CH_2CH_2NH_2$	1.2
Pyrrole (HNCH)	−4.52	CH_3CONH_2	9.5
Pyridine (NCH)	−10.76	$CH_3C{\equiv}N$	3.0
Pyridinium (HNCH)	−3.01	Pyridine (NCC)	+2.53
$(CH_3)_2NCHO$ (CH_3, CHO)	+1.1, −15.6	Pyridinium (HNCC)	+2.01
$H{-}C{\equiv}N$	8.7	Aniline (NCC)	−2.68
$H_2N(CO)CH_3$	1.3	Pyrrole (HNCC)	−3.92
Pyrrole (HNCCH)	−5.39	$CH_3CH_2CH_2NH_2$	1.4
Pyridine (NCCH)	−1.53	Pyridine (NCCC)	−3.85
Pyridinium (HNCCH)	−3.98	Pyridinium (HNCCC)	−5.30
$CH_3{-}C{\equiv}N$	−1.7	Aniline (NCCC)	−1.29

Problems

4-1 Characterize the indicated protons as (1) homotopic, enantiotopic, or diastereotopic and (2) magnetically equivalent or nonequivalent.

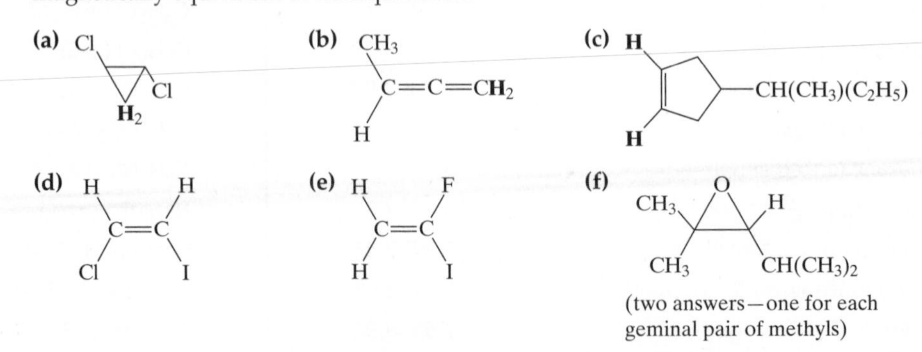

(g) 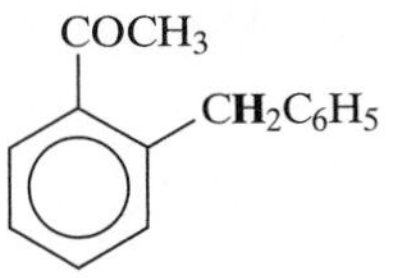 $COCH_3$, $CH_2C_6H_5$

(h) 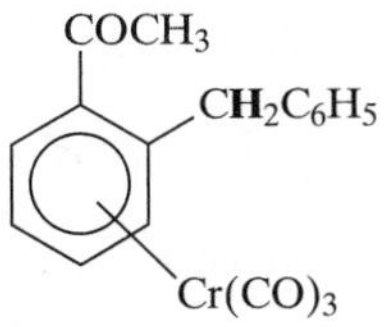 $COCH_3$, $CH_2C_6H_5$, $Cr(CO)_3$

(i) 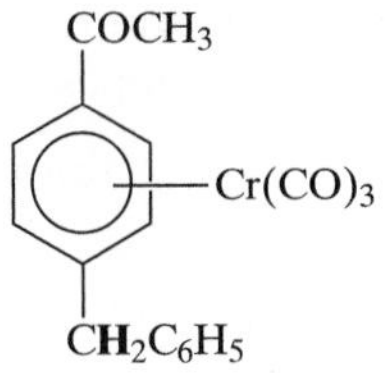 $COCH_3$, $Cr(CO)_3$, $CH_2C_6H_5$

4-2 (a) In the following molecule, are the protons on the double bond homotopic, enantiotopic, or diastereotopic? Explain.

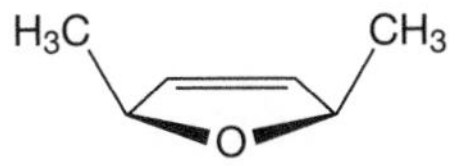

H_3C CH_3 O

(b) Answer as in **(a)** for the following molecule.

H_3C O CH_3

(c) What are the Pople notations for the molecules in **(a)** and **(b)**?

4-3 The 3′ proton of the indene dimer given below exhibits the following 600 MHz spectrum. Construct the tree diagram for this proton. Measure the approximate coupling constants, assign them to proton pairs in the structure, and rationalize their magnitudes in terms of structure and stereochemistry.

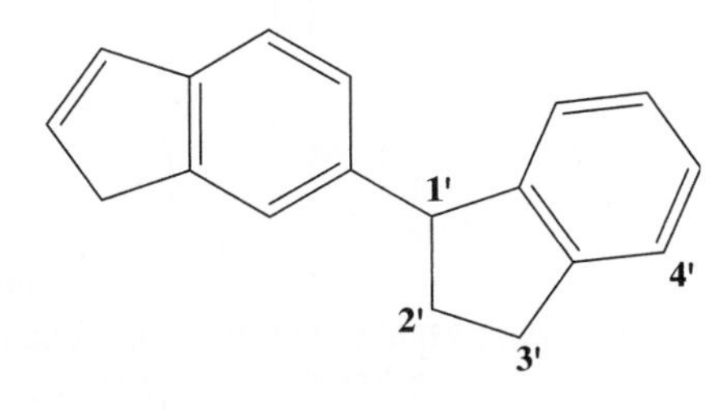

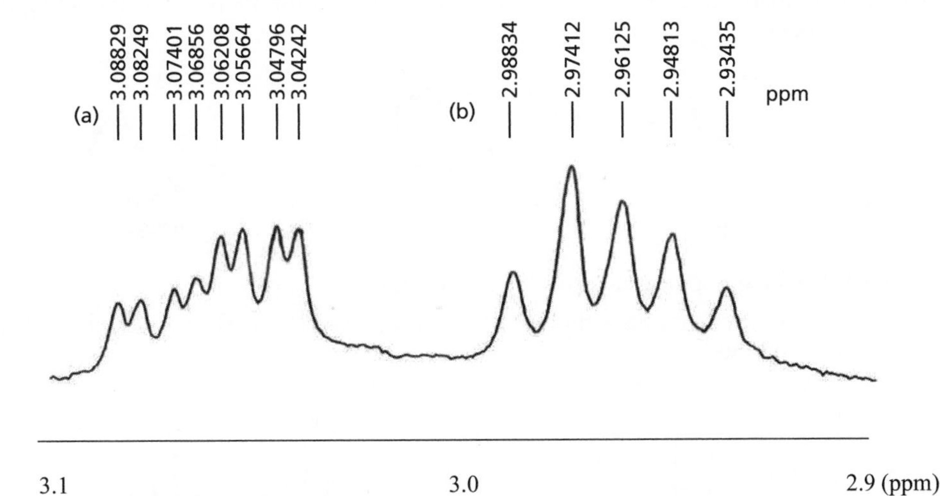

Adapted with permission from P. Spiteller, M. Spiteller, and J. Jovanovich, *Magn. Reson. Chem.*, **40**, 372 (2002).

4-4 What is the spin notation for each of the following molecules (AX, AMX, AA'XX', etc.)? Consider only major isotopes.

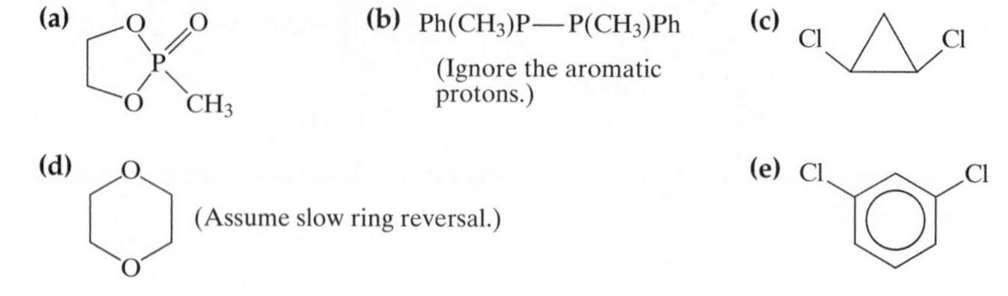

4-5 Write out the rotamers of 2-chloroethanol ($ClCH_2CH_2OH$). What is the spin notation at slow rotation for each rotamer and at fast rotation for the average?

4-6 Consider the following ^1H-decoupled ^{31}P spectrum of the platinum complex with the illustrated structure (the resonances of the anion are omitted). Explain all the peaks and give the spin notation. What should the ^{195}Pt spectrum look like?

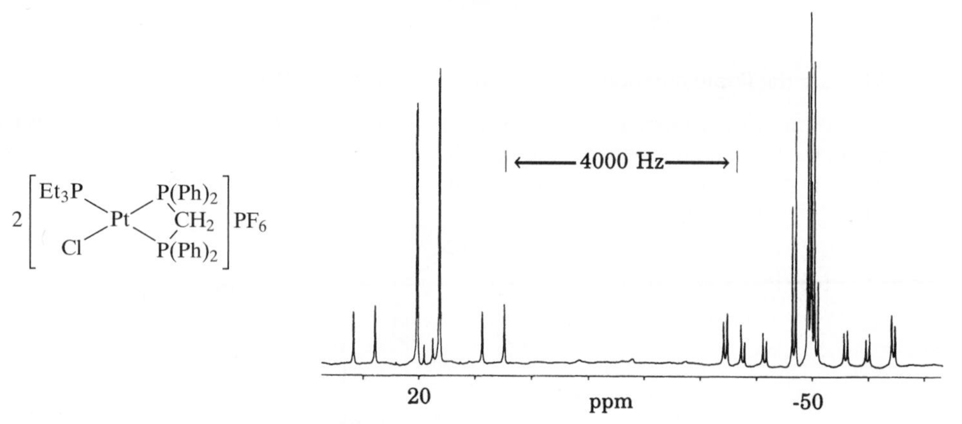

From D. E. Berry, *J. Chem. Educ.*, **71**, 899–902 (1994). Copyright 1994 American Chemical Society. Reprinted by permission of the American Chemical Society.

4-7 Several binary structures between phosphorus and sulfur are possible, including the three shown below.

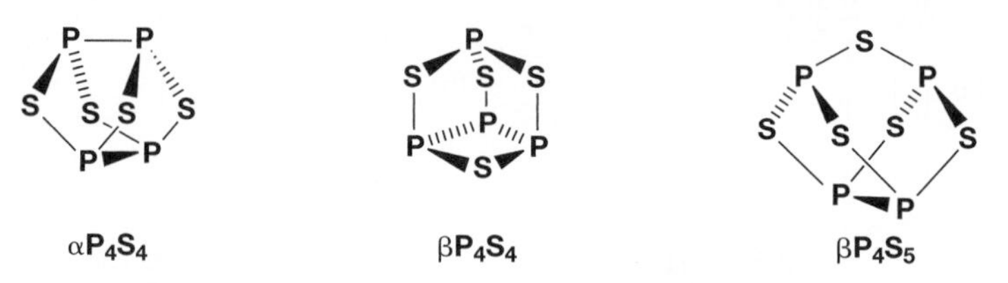

(a) What are the spin systems for these molcules?

(b) The ^{31}P spectra for these three molecules are given below. Which is for which? Explain all the splittings.

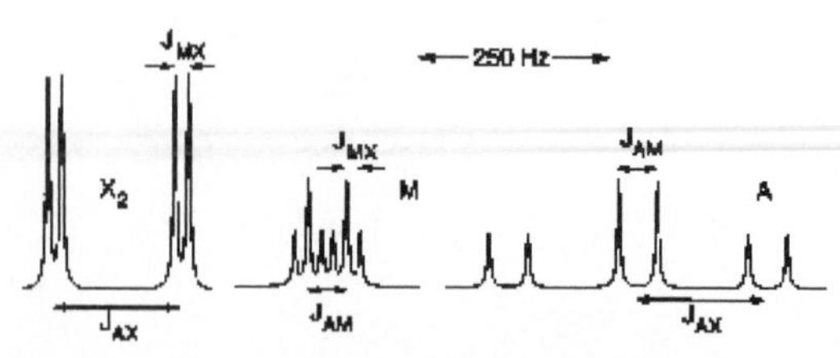

Reproduced by permission of Oxford University Press. From P. J. Hore, *Nuclear Magnetic Resonance*, Oxford, UK: Oxford University Press, 1993, p. 30.

4-8 Construct the stick diagram for the CH resonance of the molecule $BrCH_2CHCH_3CH_2OH$, in which $^3J(CH—CH_3) = 6$ Hz and $^3J(CH—CH_2) = 7$ Hz (for either methylene group).

4-9 The following is the ^1H spectrum of 1,2-dichlorobenzene at **(a)** 90 and **(b)** 750 MHz. Examine each spectrum separately. Is it first or second order? Explain. What effect does the higher field have on the structure of the spectrum?

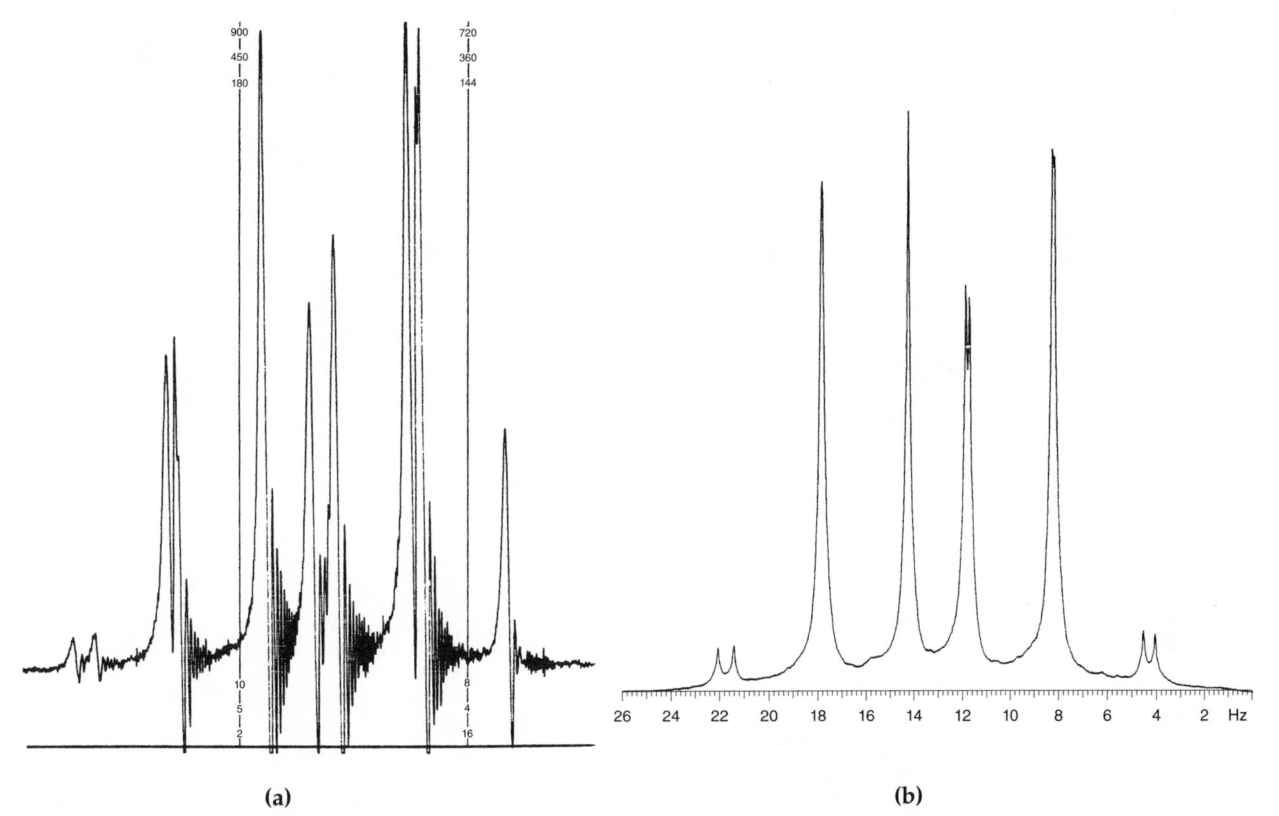

(a) **(b)**

4-10 Eliminating four moles of HBr from the molecule below should give the indicated cyclopropane. The $^1J(^{13}C—^1H)$ for the bridge CH_2 group in the isolated product was measured to be 142 Hz. Explain in terms of product structures.

Br Br Br Br $\xrightarrow[\text{EtOH}]{\text{KOH}}$ H H

4-11 There are two isomers of thiane 1-oxide, **(a)** and **(b)**. The observed geminal coupling constant between the α protons is −13.7 Hz in one isomer and −11.7 Hz in the other. Which coupling belongs to which isomer and why?

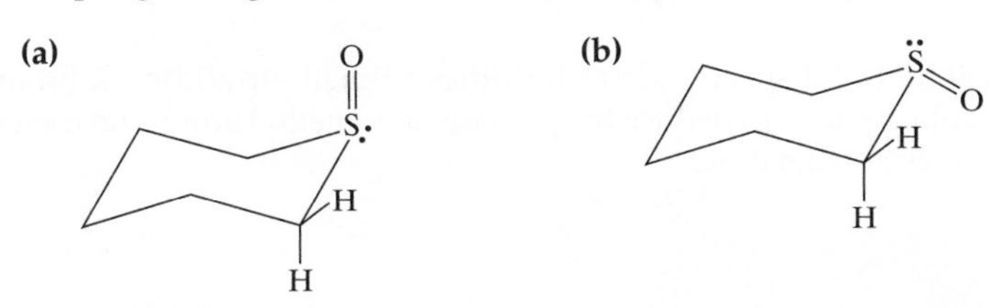

4-12 The ^1H spectrum of 1,3-dioxane (below) at slow ring reversal contains three multiplets with the following geminal couplings: −6.1, −11.2, and −12.9 Hz. Without reference to any chemical shift data, assign the resonances.

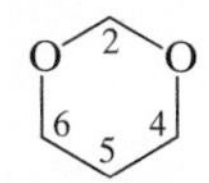

4-13 Does the angular methyl group in *trans*-decalins **(a)** or in *cis*-decalins **(b)** have the larger linewidth? Explain?

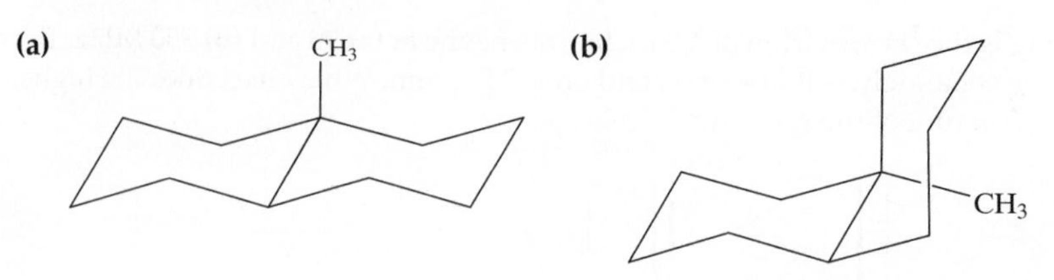

4-14 In cycloheptatriene **(a)**, J_{23} is 5.3 Hz, whereas in its bistrifluoromethyl derivative **(b)**, J_{23} is 6.9 Hz. Explain.

4-15 Explain the following couplings in terms of structure and mechanism:

(a) O H O O H

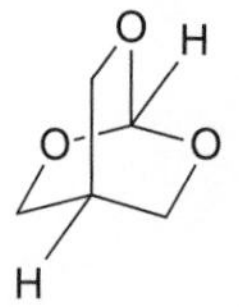

$^5J = 1.7$ Hz

(b) F F H3C

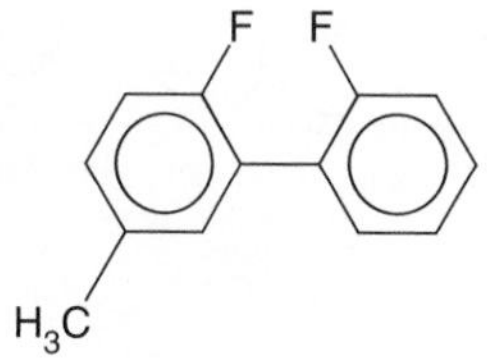

$^5J = 16.5$ Hz

(c) F F H3C

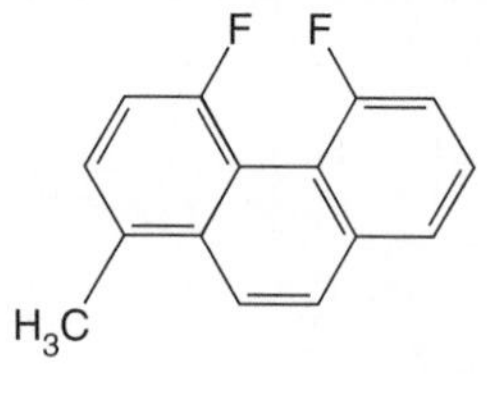

$^5J = 170$ Hz

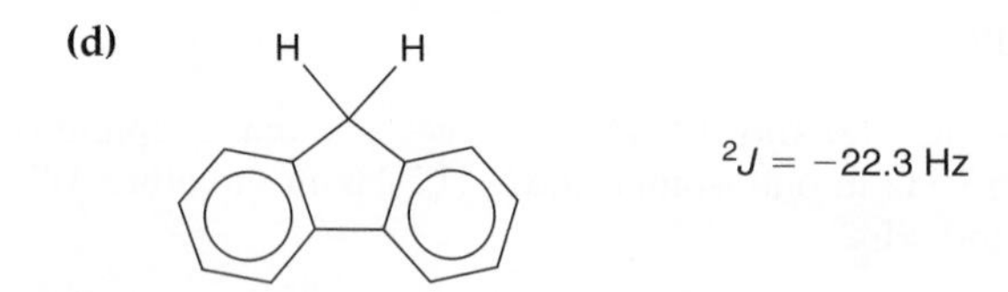

4-16 The following four 300 MHz ^{1}H spectra are of lutidines (dimethylpyridines). From the chemical shifts and coupling patterns, deduce the placement of methyl groups on each molecule. Assume the spacings are first order.

(a)

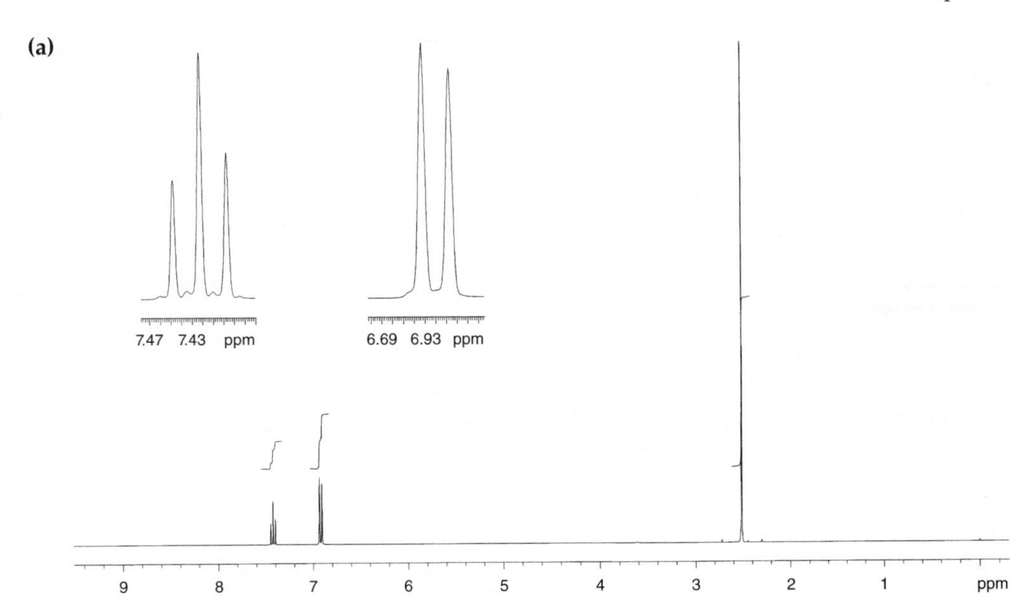

(b)

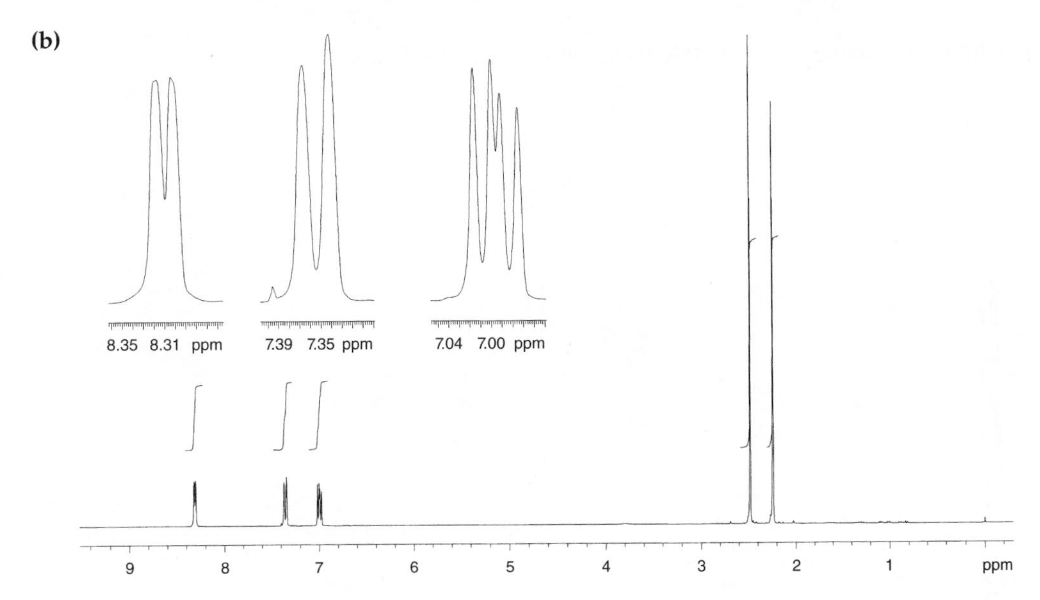

(c)

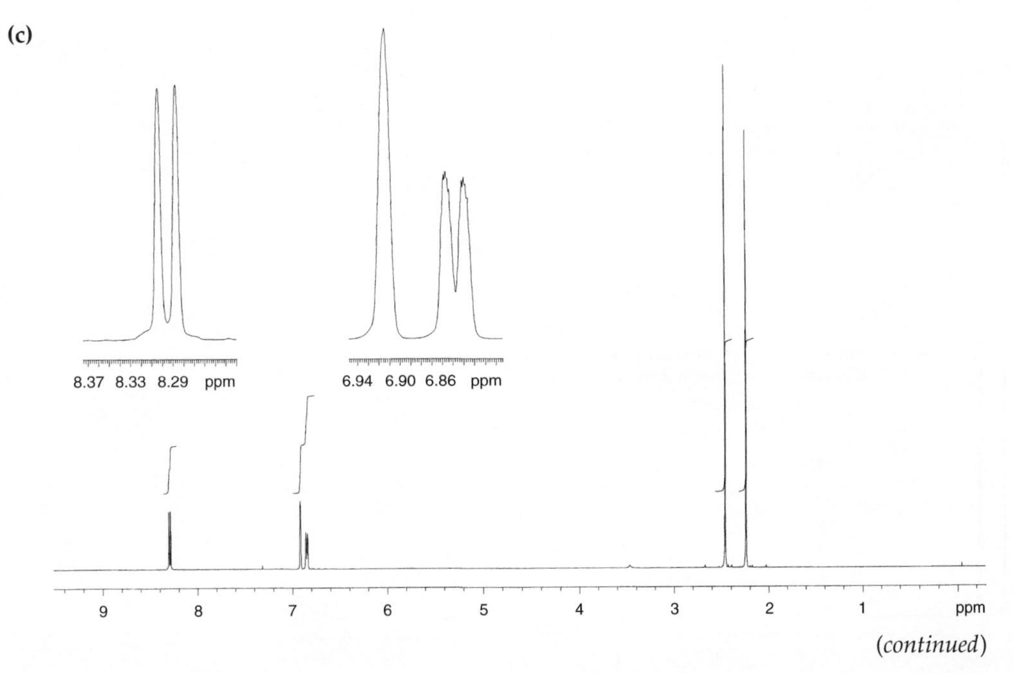

(*continued*)

(d)

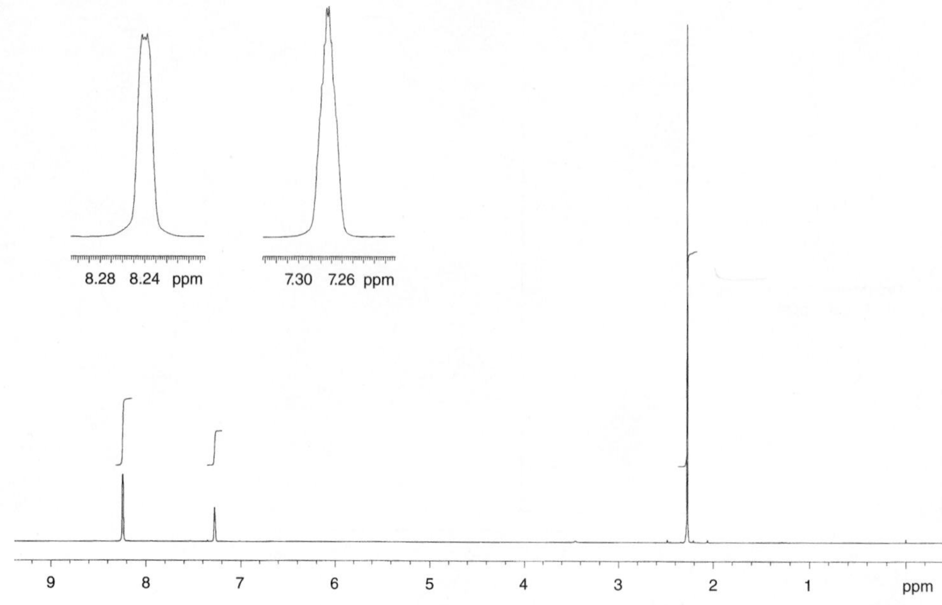

4-17 Proceed as in Problem **4-16** with the following four 300 MHz ^{1}H spectra of dichlorophenols.

(a)

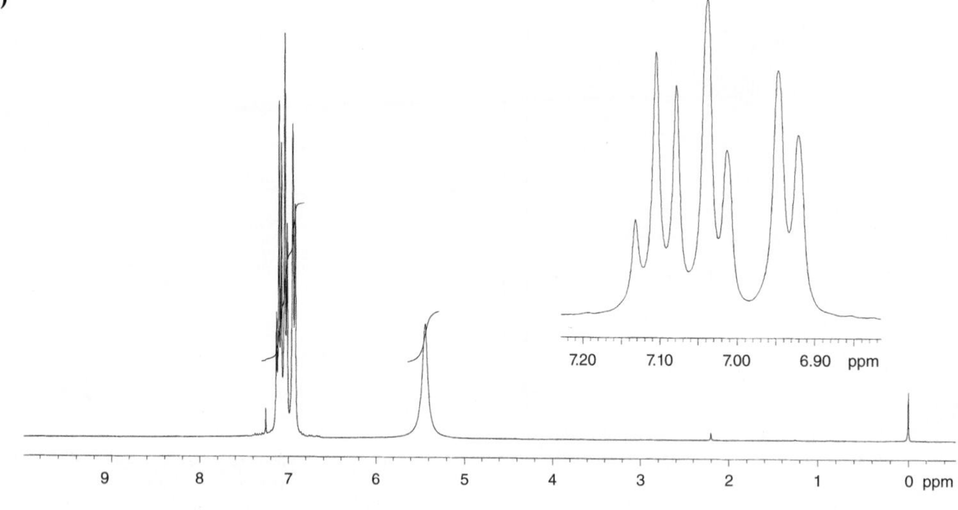

(b)

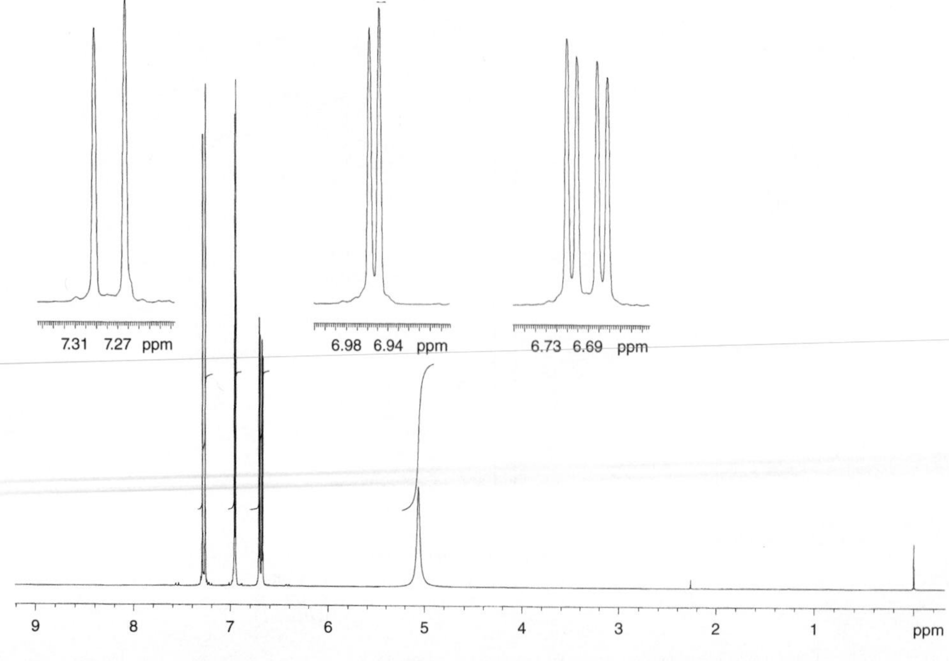

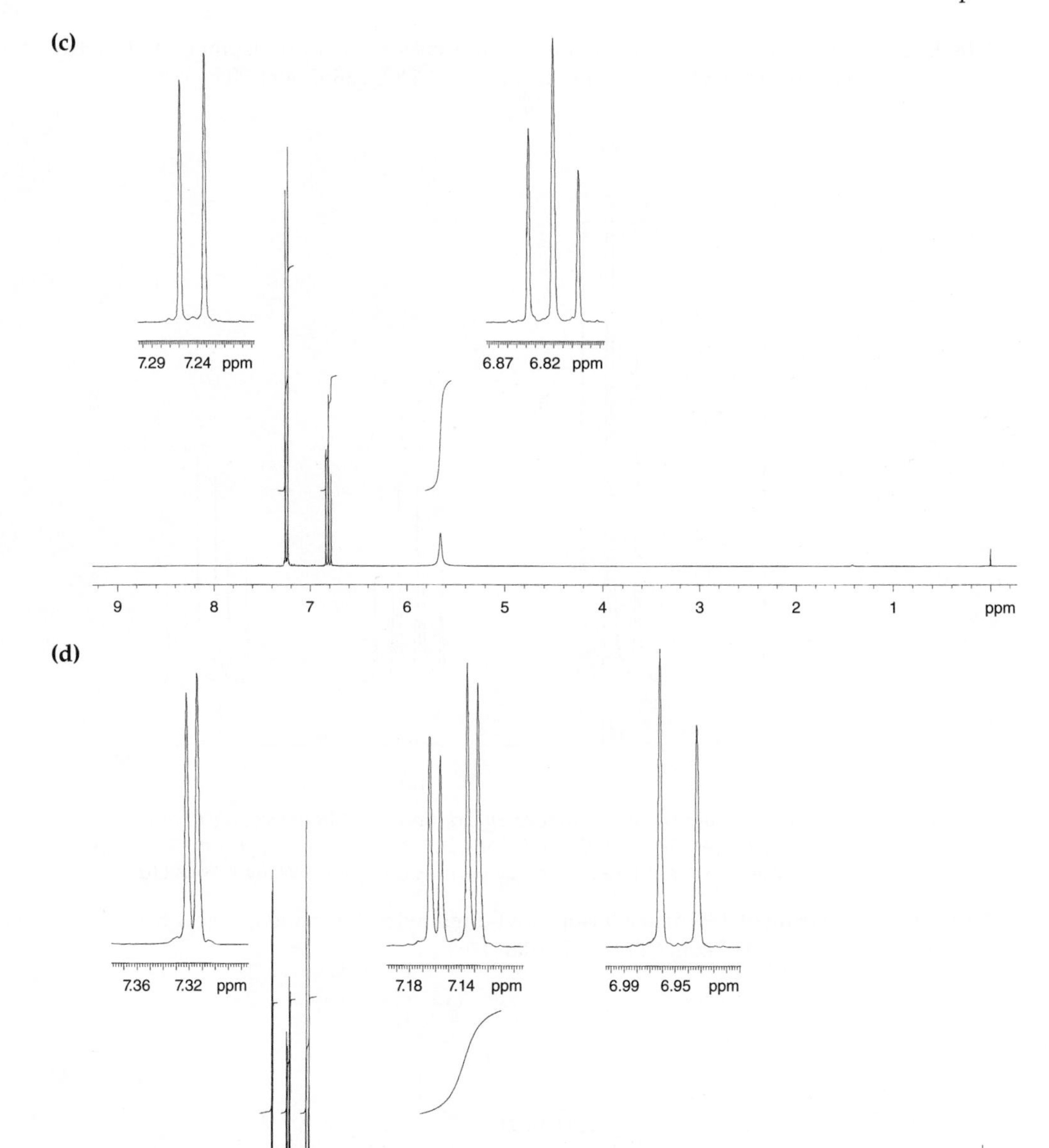

(c)
7.29 7.24 ppm
6.87 6.82 ppm
9
8
7
6
5
4
3
2
1
ppm
(d)
7.36 7.32 ppm
7.18 7.14 ppm
6.99 6.95 ppm
9
8
7
6
5
4
3
2
1
ppm

4-18 Is the 400 MHz ^1H spectrum below of the cis or the trans isomer of dimethyl 1,2-cyclopropanedicarboxylate? Explain. *Conditions*: $CDCl_3$, 25°C, 400 MHz.

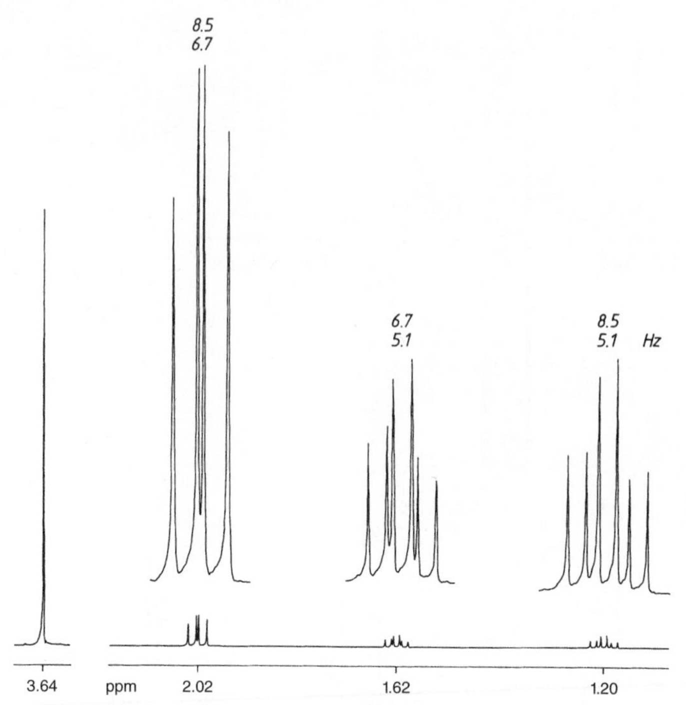

From E. Breitmaier, *Structure Elucidation by NMR in Organic Chemistry*, Chichester, UK: John Wiley & Sons Ltd., 1993, p. 71.

4-19 The ^1H spectrum of 2-hydroxy-5-isopropyl-2-methylcyclohexanone below has a $^3J_{56}$ = 3 Hz in benzene-d_6 but 11 Hz in CD_3OD. Explain.

O, OH, CH_3, $(CH_3)_2CH$

4-20 Analyze the following ^1H spectrum of the illustrated thujic ester. The CH_3 resonances are not shown. Assign the resonances to specific protons and give very approximate coupling constants. Explain your chemical shift assignments.

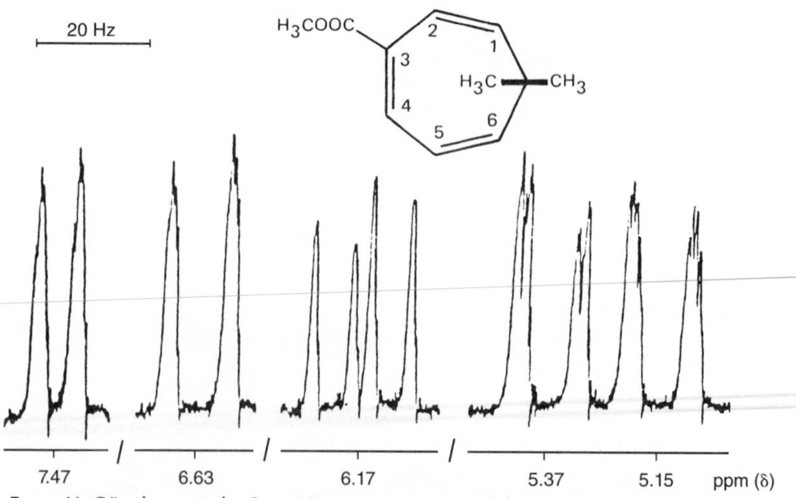

From H. Günther, et al., *Org. Magn. Reson.*, **6**, 388 (1974).

4-21 Deduce the structure (with relative stereochemistry) of the compound $C_6H_{12}O_6$ having the following 300 MHz ^1H NMR spectrum in D_2O [the peaks at δ 2.9 are from the reference, 3-(trimethylsilyl)propionic acid]. Hydroxyl resonances are not shown. The triplet (a) at δ 4.04 has integral 1 and J = 2.8 Hz. The second-order triplet (b) at δ 3.61 has integral 2 and J = 9.6 Hz. The second-order doublet of doublets (c) at δ 3.52 has integral 2 and J = 2.8, 9.6 Hz. The triplet (d) at δ 3.26 has integral 1 and J = 9.6 Hz. The ^{13}C NMR spectrum shows four resonances, all between δ 71 and 75.

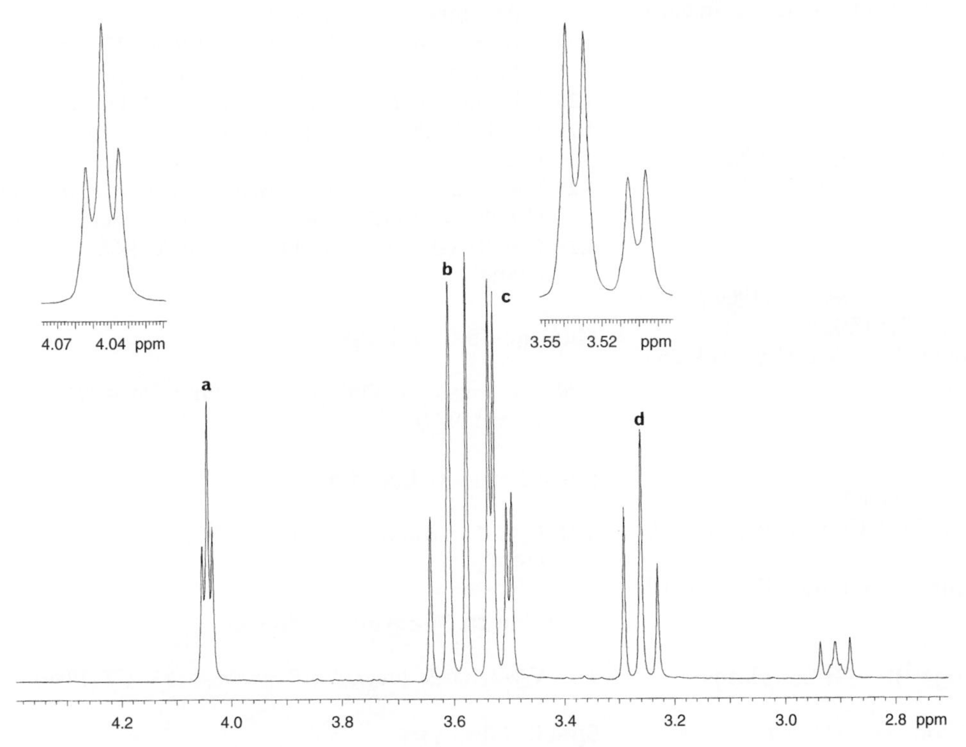

4-22 The covalent and oligomeric nature of organolithium compounds has been demonstrated by examining the spectra of compounds fully labeled with ^{13}C or ^6Li.

(a) The illustrated $^7Li\{^1H\}$ spectrum* of $[Li^{13}CMe_3]_x$ is a 1:3:3:1 quartet with $^1J(^7Li–^{13}C)$ = 14.3 Hz. What can you conclude about the number of nearest neighbor *tert*-butyl groups to lithium in solution? Explain.

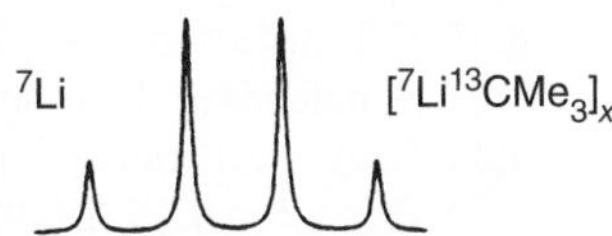

(b) The $^{13}C\{^1H\}$ spectrum of $[^6Li^{13}CMe_3]_x$ at −88°C in cyclopentene is a 1:3:6:7:6:3:1 septet (recall that ^6Li has a spin of 1), with $^1J(^6Li–^{13}C)$ = 5.4 Hz. How many nearest neighbor lithiums are indicated? Explain.

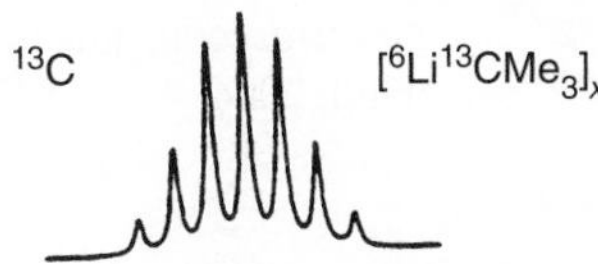

(c) Suggest a structure for *tert*-BuLi under these conditions. Explain.

(d) Above −5°C, the septet is replaced by a nonet (nine lines) with ^1J = 4.1 Hz. Explain in terms of your structure.

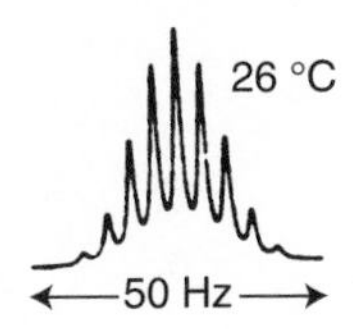

---

*From R. D. Thomas, M. T. Clarke, R. M. Jensen, and T. C. Young, *Organometallics*, **5**, 1851 (1986).

Bibliography

Coupling (general)

4.1 I. Ando and G. A. Webb, *Theory of NMR Parameters,* London: Academic Press, 1983.
4.2 *Nuclear Magnetic Resonance,* A Specialist Periodical Report, London: The Chemical Society, reviewed in each issue.

First-Order Spectral Analysis

4.3 T. R. Hoye, P. R. Hanson, and J. R. Vyvyan, *J. Org. Chem.,* **59,** 4096–4103 (1994).

Magnetic Equivalence

4.4 K. Mislow and M. Raban, *Top. Stereochem.,* **1,** 1 (1966).
4.5 W. B. Jennings, *Chem. Rev.,* **75,** 307 (1975).
4.6 W. H. Pirkle and D. J. Hoover, *Top. Stereochem.,* **13,** 263 (1982).

One-Bond Couplings

4.7 W. McFarlane, *Quart. Rev.,* **23,** 187 (1969).
4.8 C. J. Jameson and H. S. Gutowsky, *J. Chem. Phys.,* **51,** 2790 (1969).
4.9 J. H. Goldstein, V. S. Watts, and L. S. Rattet, *Progr. NMR Spectrosc.,* **8,** 103 (1971).

Geminal, Vicinal, and Long-Range $^1H—^1H$ Couplings

4.10 S. Sternhell, *Rev. Pure Appl. Chem.,* **14,** 15 (1964).
4.11 A. A. Bothner-By, *Advan. Magn. Reson.,* **1,** 195 (1965).
4.12 M. Barfield and B. Charkrabarti, *Chem. Rev.,* **69,** 757 (1969).
4.13 S. Sternhell, *Quart. Rev.,* **23,** 236 (1969).
4.14 V. F. Bystrov, *Russ. Chem. Rev.,* **41,** 281 (1972).
4.15 J. Hilton and L. H. Sutcliffe, *Progr. NMR Spectrosc.,* **10,** 27 (1975).
4.16 M. Barfield, R. J. Spear, and S. Sternhell, *Chem. Rev.,* **76,** 593 (1976).

Carbon-13 Couplings

4.17 J. B. Stothers, *Carbon-13 NMR Spectroscopy,* New York: Academic Press, 1973.
4.18 J. L. Marshall, D. E. Müller, S. A. Conn, R. Seiwell, and A. M. Ihrig, *Acc. Chem. Res.,* **7,** 333 (1974).
4.19 D. F. Ewing, *Ann. Rep. NMR Spectrosc.,* **6A,** 389 (1975).
4.20 R. E. Wasylishen, *Ann. Rep. NMR Spectrosc.,* **7,** 118 (1977).
4.21 P. E. Hansen, *Org. Magn. Reson.,* **11,** 215 (1978).
4.22 V. Wray, *Progr. NMR Spectrosc.,* **13,** 177 (1979).
4.23 G. C. Levy, R. L. Lichter, and G. L. Nelson, *Carbon-13 Nuclear Magnetic Resonance Spectroscopy,* 2nd ed., New York: Wiley–Interscience, 1980.
4.24 P. E. Hansen, *Ann. Rep. NMR Spectrosc.,* **11A,** 65 (1981).
4.25 P. E. Hansen, *Org. Magn. Reson.,* **15,** 102 (1981).
4.26 P. E. Hansen, *Progr. NMR Spectrosc.,* **14,** 175 (1981).
4.27 W. H. Pirkle and D. J. Hoover, *Top. Stereochem.,* **13,** 263 (1982).
4.28 J. L. Marshall, *Carbon–Carbon and Carbon–Proton NMR Couplings,* Deerfield Beach, FL: Verlag Chemie, 1983.
4.29 L. B. Krivdin and E. W. Della, *Progr. NMR Spectrosc.,* **23,** 301 (1991).

Fluorine-19 Couplings

4.30 J. M. Emsley, L. Phillips, and V. Wray, *Progr. NMR Spectrosc.,* **10,** 82 (1977).

Phosphorus-31 Couplings

4.31 E. G. Finer and R. K. Harris, *Progr. NMR Spectrosc.,* **6,** 61 (1970).

Coupling to Quadrupolar Nuclei

4.32 V. Mlynárik, *Progr. NMR Spectrosc.,* **18,** 277 (1986).

Spectral Analysis

4.33 J. D. Roberts, *An Introduction to the Analysis of Spin–Spin Splitting in High-Resolution Nuclear Magnetic Resonance Spectra,* New York: W.A. Benjamin, 1961.
4.34 K. B. Wiberg and B. J. Nist, *The Interpretation of NMR Spectra,* New York: W.A. Benjamin, 1962.
4.35 R. J. Abraham, *The Analysis of High Resolution NMR Spectra,* Amsterdam: Elsevier Science Inc., 1971.
4.36 R. A. Hoffman, S. Forsén, and B. Gestblom, *NMR Basic Princ. Progr.,* **5,** 1 (1971).
4.37 C. W. Haigh, *Ann. Rep. NMR Spectrosc.,* **4,** 311 (1971).
4.38 P. Diehl, H. Kellerhals, and E. Lustig, *NMR Basic Princ. Progr.,* **6,** 1 (1972).
4.39 J. B. Lambert and E. P. Mazzola, *Nuclear Magnetic Resonance Spectroscopy,* Upper Saddle River, NJ: Pearson Prentice Hall, 2004, Appendices 3 and 4.

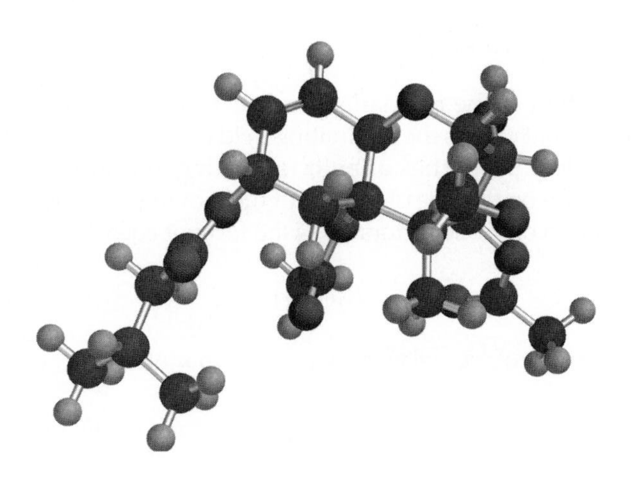

CHAPTER 5

Further Topics in One-Dimensional NMR

Although the chemical shift and the coupling constant are the two fundamental measurable quantities in NMR spectroscopy, several other phenomena may be studied in a single NMR time dimension. In this chapter, we first examine the processes of spin–lattice and spin–spin relaxation, whereby a system moves toward spin equilibrium (Section 2-4). Relaxation times or rates provide another important measurable quantity related to both structural and dynamic factors. Second, we explore in greater detail structural changes that occur on the NMR time scale (Section 2-8). The temporal dependence of chemical shifts and coupling constants influences both line shapes and intensities and can be used to generate rate constants for reactions. Third, we describe the family of experiments that utilize a second irradiation frequency, B_2. Double irradiation can simplify spectra, perturb intensities, and provide information about structure and rate processes. Finally, we expand on the technique of using multiple pulses—often of varied duration rather than only a single 90° pulse, sometimes separated by specific time periods, and even of nonrectilinear shape—to improve sensitivity, simplify spectral patterns, measure relaxation times and coupling constants, draw structural conclusions, and improve the accuracy of pulse timing and definition.

5-1 SPIN–LATTICE AND SPIN–SPIN RELAXATION

Application of the B_1 field at the resonance frequency results in energy absorption and the conversion of some $+\frac{1}{2}$ spins into $-\frac{1}{2}$ spins, so that magnetization in the z direction (M_z) decreases. Spin–lattice, or longitudinal, relaxation returns the system to equilibrium along the z axis, with time constant T_1 and rate constant R_1 ($= 1/T_1$). Such relaxation occurs because of the presence of natural magnetic fields in the sample that fluctuate at the Larmor frequency. Because of the frequency match, excess spin energy can flow into the molecular surroundings, sometimes called the *lattice*, and $-\frac{1}{2}$ spins can return to the $+\frac{1}{2}$ state.

Causes of Relaxation. The major source of these magnetic fields is magnetic nuclei in motion. Like the classic model of a charge moving in a circle, a magnetic dipole in motion creates a magnetic field, whose frequency depends on the rate of motion and on the magnetic moment of the dipole. For appropriate values of these parameters, the resulting magnetic field can fluctuate at the same frequency as the resonance (Larmor) frequency of the nucleus in question, permitting energy to flow from excited spins to the lattice. Such a process is called *dipole–dipole relaxation* [T_1(DD)], because it involves interaction of the resonating nuclear magnetic dipole with the dipole of the nucleus in motion that causes the fluctuating field of the lattice. The resulting relaxation time depends on nuclear properties of both resonating and moving nuclei, on the distance between them, and on the rate of motion of the moving nucleus. Mathematically, the dependence of relaxation on these factors takes the form of eq. 5-1a

$$R_1(\text{DD}) = \frac{1}{T_1(\text{DD})} = n\gamma_{\text{C}}^2\gamma_{\text{H}}^2\hbar^2 r_{\text{CH}}^{-6}\tau_{\text{c}} \qquad \textbf{(5-1a)}$$

for the case of ^{13}C relaxed by protons in motion or the form of eq. 5-1b

$$\frac{1}{T_1(\text{DD})} = \frac{3}{2}n\gamma_{\text{H}}^4\hbar^2 r_{\text{HH}}^{-6}\tau_{\text{c}} \qquad \textbf{(5-1b)}$$

for protons relaxed by protons. As usual, the nuclear properties are represented by the gyromagnetic ratios. The symbol n stands for the number of protons that are nearest neighbors to the resonating nucleus and hence are most effective at relaxing it. The rapid falloff with distance is indicated by the inverse sixth power of the distance r_{CH} (r_{HH}) to the nearest-neighbor hydrogen(s). The motional properties of the hydrogens are described by the effective correlation time τ_c, which is the time required for the molecule to rotate one radian and is typically in the nanosecond-to-picosecond range for organic molecules in solution.

Thus, carbon relaxation is faster (and the relaxation time is shorter) when there are more attached protons, when the internuclear C—H distance is less, and when rotation in solution decreases. A quaternary carbon has a long relaxation time because it lacks an attached proton and because the distance r_{CH} to other protons is large. The ratio of the carbon relaxation time of methinyl to methylene to methyl is 6:3:2 (equivalent to $1:\frac{1}{2}:\frac{1}{3}$), due to differences in the number of attached protons, other things being equal. Because the rate of molecular tumbling in solution slows as molecular size increases, larger molecules relax more rapidly. Thus, cholesteryl chloride relaxes more rapidly than phenanthrene, which relaxes more rapidly than benzene. Eq. 5-1 is an approximation to a more complete equation and represents what is called the *extreme narrowing limit* for smaller molecules. Because the frequency of motion of the moving nuclear magnet must match the resonance frequency of the excited nuclear magnet, dipolar relaxation becomes ineffective for both rapidly moving small molecules and slowly moving large molecules. Many molecules of interest to biochemists fall into the latter category, to which eq. 5-1 does not apply. Rapid internal rotation of methyl groups in small molecules also can reduce the effectiveness of dipole–dipole relaxation. The optimal correlation times (τ_c) for dipolar relaxation lie in the range of about 10^{-7} to 10^{-11} s (the inverse of the resonance frequency). Because the resonance frequency depends on the value of B_0, this range also depends on B_0.

When dipolar relaxation is slow, other mechanisms of relaxation become important. Fluctuating magnetic fields also can arise from (1) interruption of the motion of rapidly tumbling small molecules or rapidly rotating groups within a molecule (spin rotation relaxation), (2) tumbling of molecules with anisotropic chemical shielding at high fields, (3) scalar coupling constants that fluctuate through chemical exchange or through quadrupolar interactions, (4) tumbling of paramagnetic molecules (unpaired electrons have very large magnetic dipoles), and (5) tumbling of quadrupolar nuclei. In the absence of quadrupolar nuclei or paramagnetic species, these alternative mechanisms often are unimportant. A major exception is the relaxation of methyl (CH_3) and trifluoromethyl carbons by spin rotation. Because at higher temperatures relaxation by dipolar interactions becomes less effective but relaxation by spin rotation becomes more effective, these mechanisms may be distinguished by measuring T_1 at multiple temperatures.

Measurement of Relaxation Time. The actual value of T_1 must be known at least approximately in order to decide how long to wait between pulses for the system to return to equilibrium (the delay time). In addition, T_1(DD) offers both structural information, because of its dependence on r_{CH}, and dynamic information, because of its dependence on τ_c. For these reasons, convenient methods have been developed for measuring T_1, the commonest of which is called *inversion recovery*. The strategy is to

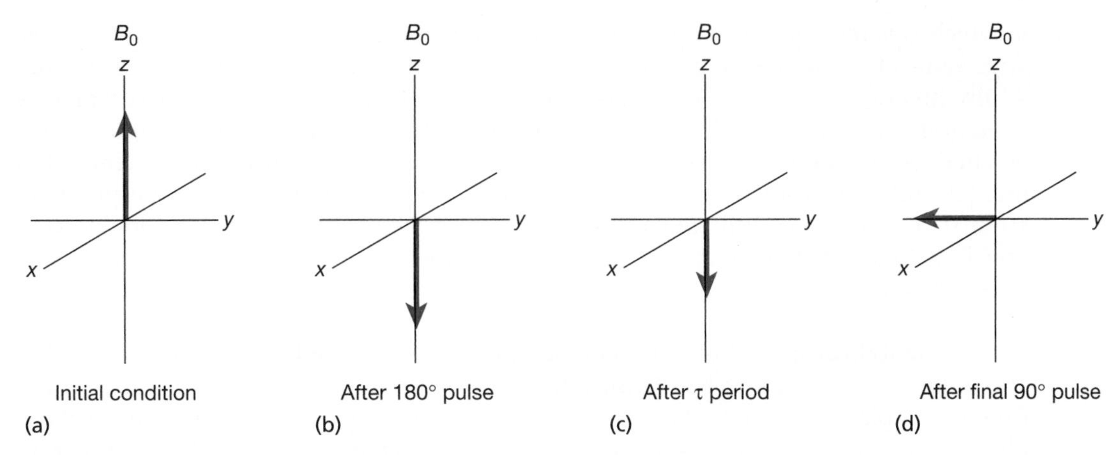

FIGURE 5-1 The inversion recovery experiment.

create a nonequilibrium distribution of spins and then to follow their return to equilibrium as a first-order rate process. Inverting the spins through the application of a 180° pulse creates a maximum deviation from equilibrium (Figure 5-1b). If a very short amount of time τ is allowed to pass (Figure 5-1c) and a 90° pulse is applied to move the spins into the xy plane for observation, the nuclear magnets are aligned along the $-y$ axis (Figure 5-1d) and an inverted peak is obtained. During time τ, some T_1 relaxation occurs (Figure 5-1b and c). The z magnetization at the end of time τ (Figure 5-1c) is smaller than at the beginning (Figure 5-1b). Consequently, the peak produced after the 90° pulse is smaller than if the 180° and 90° pulses had been combined initially as a 270° pulse, that is, if $\tau = 0$. The inversion recovery pulse sequence is summarized as 180° $-$ τ $-$ 90° $-$ Acquire and is an example of a simple multipulse sequence.

Further such experiments with increasingly longer values of τ result in greater relaxation between the 180° and 90° pulses. After the 90° pulse (Figure 5-1d), the resulting peak evolves from negative through zero to positive as τ increases, until complete relaxation occurs when τ is very long. Figure 5-2 shows a stack of such experiments for the carbons of chlorobenzene. Because the carbon ipso to chlorine (C-1) has no directly attached proton, much longer values of τ are needed for the inverted peak to turn over. Relaxation for C-1 is not complete even by τ = 80 s. The intensity I measured at a series of times τ follows an exponential decay according to the first-order kinetics of eq. 5-2,

$$I = I_0(1 - 2e^{-\tau/T_1}) \quad \textbf{(5-2)}$$

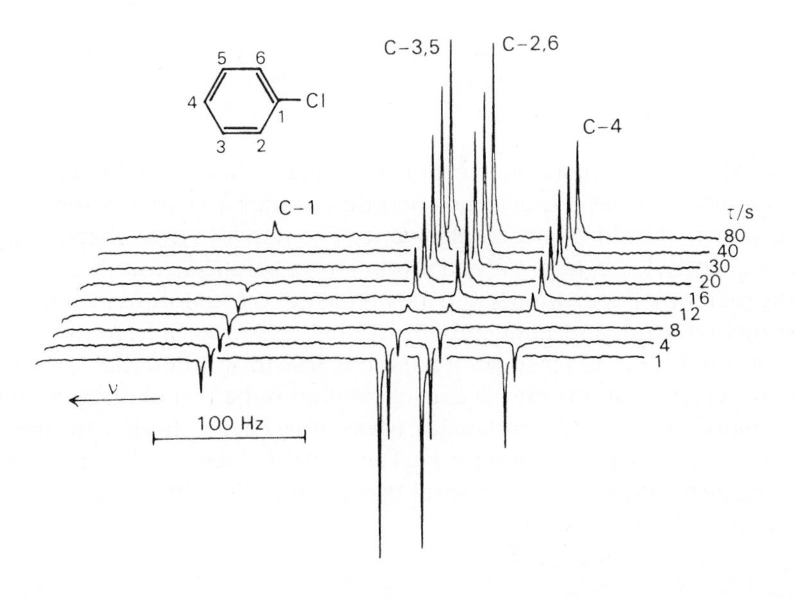

FIGURE 5-2 A stack plot for the inversion recovery experiment of the ^{13}C resonances of chlorobenzene at 25 MHz. The time τ in the pulse sequence 180° $-$ τ $-$ 90° is given in seconds at the right. (From R. K. Harris, *Nuclear Magnetic Resonance Spectroscopy*, London: Pitman Publishing, Ltd., 1983, p. 82. Reproduced with permission of Addison Wesley Longman, Ltd.)

in which I_0 is the equilibrium intensity (measured, for example, initially or after a very large value of τ) and the factor of two arises because magnetization recovery begins from a fully inverted condition. A plot of the natural logarithm of $(I_0 - I)$ versus τ gives a straight line with a slope of $-1/T_1$. An estimate for the spin–lattice relaxation time may be obtained through knowledge of the time τ at which the intensity passes through a null [τ(null) = 8 s for C-4 in Figure 5-2]. At this time, $I = 0$, so that T_1 corresponds to $\tau(\text{null})/\ln 2$, or $1.443\tau(\text{null})$. Such an estimate might be useful, for instance, in deciding how long to wait between repetitive pulses, but it should never be considered a rigorous measurement of T_1.

Transverse Relaxation. Relaxation in the *xy* plane, or spin–spin (transverse) relaxation (T_2), might be expected to be identical to T_1, because movement of the magnetization from the *xy* plane back onto the *z* axis restores *z* magnetization at the same rate as it depletes *xy* magnetization. There are, however, other mechanisms of *xy* relaxation that do not affect *z* magnetization. We already saw in Section 2-4 that inhomogeneity of the B_0 magnetic field randomizes phases in the *xy* plane and hastens *xy* relaxation. As a result, T_2 is expected to be less than or equal to T_1. In addition, *xy* (T_2) relaxation can occur when two nuclei mutually exchange their spins, one going from $+\frac{1}{2}$ to $-\frac{1}{2}$ and the other from $-\frac{1}{2}$ to $+\frac{1}{2}$. This spin–spin, double-flip, or flip-flop mechanism is most significant in large molecules. The process can result in *spin diffusion*. The excitation of a specific proton changes the magnetization of surrounding protons as flip-flop interactions spread through the molecule. The interpretation of the spectra of large molecules such as proteins must take such a process into consideration.

Structural Ramifications. Proton spin–lattice relaxation times depend on the distance between the resonating nucleus and the nearest-neighbor protons. The closer the neighbors are, the faster is the relaxation and the shorter is T_1. The two isomers **5-1a** and **5-1b** [Bz = Ph(C=O)] may be distinguished by their proton relaxation times. In **5-1a**, H_1 is axial and close to the 3 and 5 axial protons, resulting in a T_1 of 2.0 s. In **5-1b**, H_1 is equatorial and has more distant nearest neighbors, resulting in a T_1 of 4.1 s. In this way, the structure of these anomers may be distinguished. The remaining values of T_1 may be interpreted in a similar fashion. For example, H_2 in isomer **5-1a** has only the H_4 axial proton as a nearest neighbor, so its T_1 is a relatively long 3.6 s. In **5-1b**, H_2 has not only the axial H_4, but also the vicinal H_1, as a nearest neighbor, so T_1 is shorter, 2.1 s.

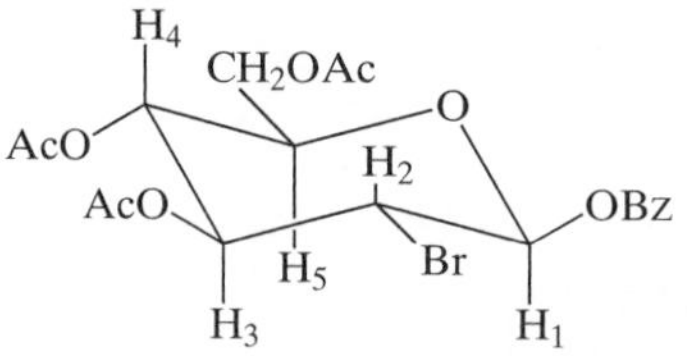

5-1a 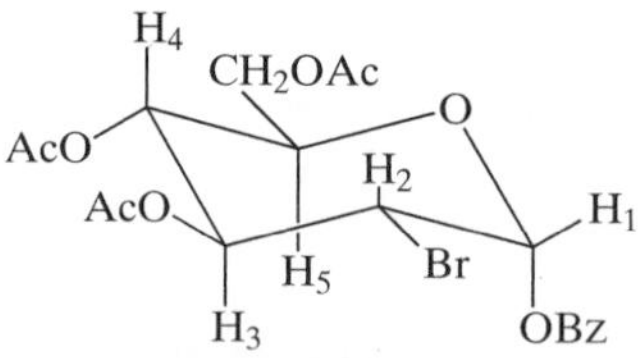 **5-1b**

Anisotropic Motion. When a molecule is rigid and rotates equally well in any direction (isotropically), all the carbon relaxation times (after adjustment for the number of attached protons) should be nearly the same. The nonspherical shape of a molecule, however, frequently leads to preferential rotation in solution around one or more axes (anisotropic rotation). For example, toluene prefers to rotate around the long axis that includes the methyl, ipso, and para carbons, so that less mass is in motion. On average, these carbons (and their attached protons) move less in solution than do the ortho and meta carbons, because atoms on the axis of rotation remain stationary during rotation. The more rapidly moving ortho and meta carbons thus have a shorter effective correlation time τ_c and hence, by eq. 5-1, a longer T_1. The actual values are shown in structure **5-2**. The longer value for the ipso carbon arises because it lacks a directly bonded proton and because r_{CH} in eq. 5-1a is very large.

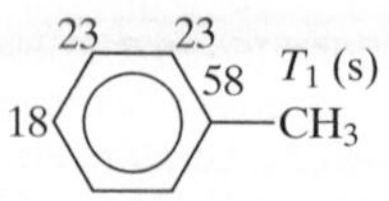

5-2

Segmental Motion. When molecules are not rigid, the more rapidly moving pieces relax more slowly because their τ_c is shorter. Thus, in decane (**5-3**) the methyl carbon

$CH_3CH_2CH_2CH_2CH_2CH_2CH_2CH_2CH_2CH_3$

nT_1 26.1 13.2 11.4 10.0 8.8

5-3

relaxes most slowly, followed by the ethyl carbon, and so on, to the fifth carbon in the middle of the chain. Structure **5-3** gives the values of nT_1 (n is the number of attached protons), so that the figures may be compared for all carbons without considering any substitution patterns. These values reflect the relative rates of motion of each carbon.

Partially Relaxed Spectra. The inversion recovery experiment used to measure T_1 also may be exploited to simplify spectra. In Figure 5-2, the spectrum for $\tau = 40$ s lacks a resonance for the ipso carbon (C-1). Similarly, for a τ of about 10 s, all the other ring carbons are nulled, and only the negative peak for C-1 is obtained. Such *partially relaxed spectra* can be used not only to obtain partial spectra in this fashion but also to eliminate specific peaks. When deuterated water (D_2O) is used as the solvent, the residual HOD peak is undesirable. An inversion recovery experiment can reveal the value of τ for which the water peak is nulled. The rest of the protons will have positive or negative intensities at that τ, depending on whether they relax more rapidly or more slowly than water. The experiment may be refined by applying the 180° pulse selectively only at the resonance position of water. Selection of τ for nulling of this peak then produces a spectrum that lacks the water peak but otherwise is quite normal for the remaining resonances. Such a procedure is an example of *peak suppression* or *solvent suppression*.

Quadrupolar Relaxation. The dominant mode of spin–lattice relaxation for nuclei with spins greater than $\frac{1}{2}$ results from the quadrupolar nature of such nuclei. These nuclei are considered to have an ellipsoidal rather than a spherical shape. When $I = 1$, as for ^{14}N or ^2H, there are three stable orientations in the magnetic field: parallel, orthogonal, and antiparallel, as shown in Figure 5-3. When these ellipsoidal nuclei tumble in solution within an unsymmetrical electron cloud of the molecule, they produce a fluctuating electric field that can bring about relaxation.

The mechanism is different from dipole–dipole relaxation in two ways. First, it does not require a second nucleus in motion; the quadrupolar nucleus creates its own fluctuating field by moving in the unsymmetrical electron cloud. Second, because the mechanism is extremely effective when the quadrupole moment of the nucleus is large, T_1 can become very short (milliseconds or less). In such cases, the uncertainty principle applies, whereby the product of ΔE (the spread of energies of the spin states, as measured by the linewidth $\Delta\nu$) and Δt (the lifetime of the spin state, as measured by the relaxation time) must remain constant ($\Delta E \Delta t \sim$ Planck's constant). Thus, when the relaxation time is very short, the linewidth becomes very large. Nuclei with large quadrupole moments often exhibit very large linewidths—for example, about 20,000 Hz for the ^{35}Cl resonance of CCl_4. The common nuclides ^{17}O and ^{14}N have smaller quadrupolar moments and exhibit sharper resonances, typically tens of hertz. The small quadrupole moment of deuterium results in quite sharp peaks, usually one or a few hertz. The linewidth also depends on the symmetry of the molecule, which controls how unsymmetrical the electron cloud is. Systems with π electrons are more unsymmetrical and

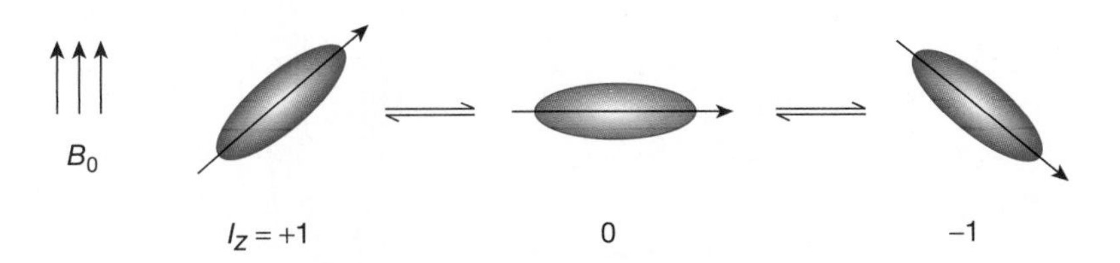

FIGURE 5-3 The spin states for a nucleus with $I = 1$.

give broader lines (as in amides and pyridines for ^{14}N). Spherical or tetrahedral systems have no quadrupolar relaxation, since the electron cloud is symmetrical. Such systems exhibit very sharp linewidths, like those of spin-$\frac{1}{2}$ nuclei (^{14}N in $^{+}NH_4$, ^{6}Li or ^{7}Li in Li^{+}, ^{10}B in $^{-}BH_4$, ^{35}Cl in Cl^{-} or $^{-}ClO_4$, and ^{33}S in SO_4^{2-}).

Very important to the organic chemist is the effect of quadrupolar nuclei on the resonances of nearby protons. When quadrupolar relaxation is extremely rapid, a neighboring nucleus experiences only the average spin environment of the quadrupolar nucleus, so that no spin coupling is observed. Hence, protons in chloromethane produce a sharp singlet, even though ^{35}Cl and ^{37}Cl have spins of $\frac{3}{2}$ and exist in four spin states. Chemists have come to think of the halogens (other than fluorine) as being nonmagnetic, although they appear so only because of their rapid quadrupolar relaxation. At the other extreme, deuterium has a weak quadrupole moment and possesses only s electrons, so that neighboring protons exhibit normal couplings to ^{2}H. Thus, nitromethane with one deuterium (CH_2DNO_2) shows a 1:1:1 triplet, because the protons are influenced by the three spin states (+1, 0, and −1) of deuterium (in analogy to Figure 5-3). Nitromethane with two deuteriums (CHD_2NO_2) shows a 1:2:3:2:1 quintet from coupling to the various combinations of the three spin states (++; +0,0+; +−,00,−+; −0,0−; −−). This quintet is often observed in deuterated solvents such as acetone-d_6, acetonitrile-d_3, or nitromethane-d_3, because incomplete deuteration results in an impurity with a CHD_2 group.

The ^{14}N nucleus falls between these extremes. In highly unsymmetrical cases, such as the interior nitrogen in biuret, $NH_2(CO)$**NH**$(CO)NH_2$, quadrupolar relaxation is rapid enough to produce the average singlet for the attached proton. The protons of the ammonium ion, in contrast, give a sharp 1:1:1 triplet with full coupling between ^{1}H and ^{14}N, since quadrupolar relaxation is absent. When relaxation is at an intermediate rate, it is possible to observe three broadened peaks, one broadened average peak, or broadening to the point of invisibility. Irradiation at the ^{14}N frequency removes the ^{14}N—^{1}H coupling interaction (Section 5-3), so that ^{14}N appears to be nonmagnetic. Figure 5-4 shows the

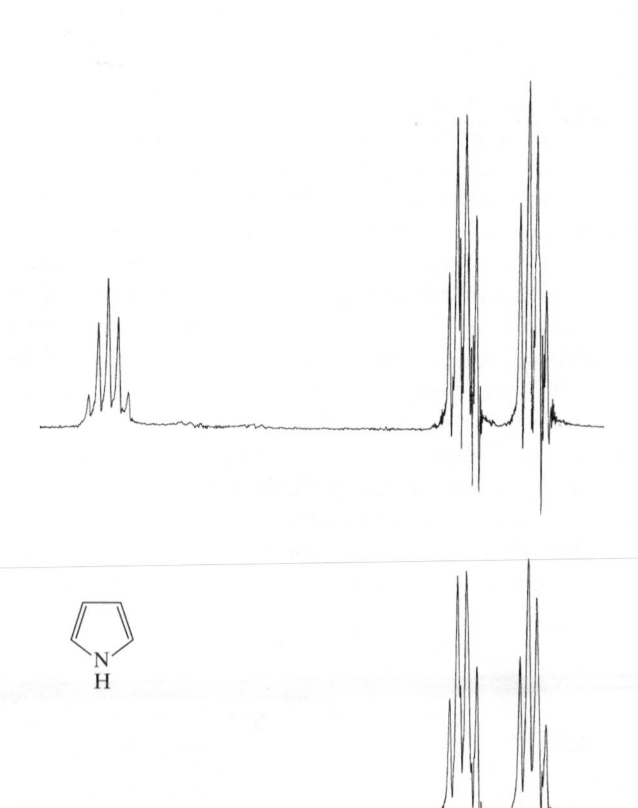

FIGURE 5-4 The 90 MHz proton spectrum of pyrrole with (upper) and without (lower) ^{14}N decoupling.

normal spectrum of pyrrole at the bottom, containing only the AA′BB′ set from the CH protons and no visible NH resonance, because the line is extremely broad. Irradiation at the ^{14}N frequency decouples the NH proton from ^{14}N and results in a deceptively simple quintet NH resonance from coupling to the four CH protons.

5-2 REACTIONS ON THE NMR TIME SCALE

NMR is an excellent tool for following the kinetics of an irreversible reaction traditionally through the disappearance or appearance of peaks over periods of minutes to hours. The spectrum is recorded repeatedly at specific intervals, and rate constants are calculated from changes in peak intensities. Thus, the procedure is a classical kinetic method, performed on the *laboratory time scale*. The molecular changes take place on a time scale much longer than the pulse or acquisition times of the NMR experiment. In addition, and more importantly, NMR has the unique capability for study of the kinetics of reactions that occur at equilibrium and that affect line shapes, usually with activation energies in the range from 4.5 to 25 kcal mol^{-1} (Section 2-8), a range that corresponds to rates in the range from 10^0 to 10^4 s^{-1}. This *NMR time scale* refers to the rough equivalence of the reaction rate in s^{-1} to the frequency spacing in hertz between the exchanging nuclei.

A series of spectra for the interchange of axial and equatorial protons in cyclohexane-d_{11} as a function of temperature is illustrated in Figure 2-31. When the interchange of two such chemical environments occurs much faster than the frequency differences between the two sites, the result is a single peak, reflecting the average environment (*fast exchange*). Keep in mind that these exchanges occur reversibly, and the system remains at equilibrium. When the interchange is slower than the frequency differences, the NMR result is two distinct peaks (*slow exchange*). When the interchange is comparable to the frequency differences, broad peaks typically result. The reaction, then, is said to occur on the NMR time scale. Both fast and slow exchange sometimes may be reached by altering the temperature of the experiment. Intermolecular reactions, such as the acid-catalyzed interchange of protons, also may be studied, as in the case of the hydroxy proton of methanol (Figure 2-30). The following examples are intramolecular.

Hindered Rotation. Normally, rotation around single bonds has a barrier below 5 kcal mol^{-1} and occurs faster than the NMR time scale. Rotation around the double bond of alkenes, on the other hand, has a barrier that is normally above 50 kcal mol^{-1} and is slow on the NMR time scale. There are numerous examples of intermediate bond orders, whose rotation occurs within the NMR time scale. Hindered rotation about the C—N bond in amides such as *N,N*-dimethylformamide (**5-4**) provides a classic example of site exchange. At room temperature, exchange is slow and two methyl resonances are observed, whereas, above 100°C, exchange is fast and a single resonance is observed. The measured barrier is about 22 kcal mol^{-1}.

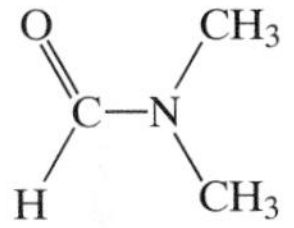

O CH$_3$ C—N H CH$_3$

5-4

Hindered rotation occurs on the NMR time scale for numerous other systems with partial double bonds, including carbamates, thioamides, enamines, nitrosamines, alkyl nitrites, diazoketones, aminoboranes, and aromatic aldehydes. Formal double bonds can exhibit free rotation when alternative resonance structures suggest partial single bonding. The calicene **5-5**, for example, has a barrier to rotation about the central bond of only 20 kcal mol^{-1}.

Steric congestion can raise the barrier about a single bond enough to bring it into the NMR range. Rotation about the single bond in the biphenyl **5-6** is raised to a measurable 13 kcal mol^{-1} by the presence of the ortho substituents, which also provide diastereotopic

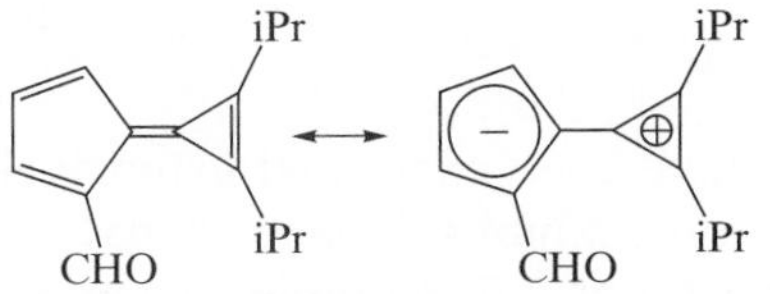

iPr iPr CHO ⟷ iPr iPr CHO

5-5

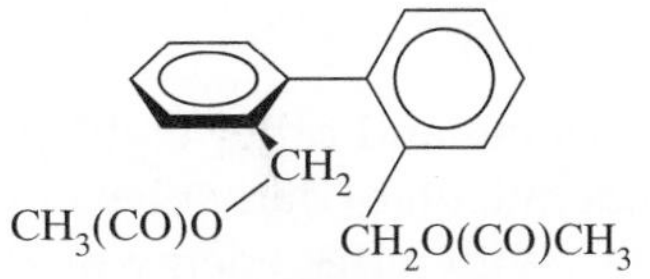

CH$_2$ CH$_3$(CO)O CH$_2$O(CO)CH$_3$

5-6

methylene protons as the dynamic probe. Hindered rotation about an sp^3–sp^3 bond can sometimes be observed when at least one of the carbons is quaternary. Thus, at −150°C, the *tert*-butyl group in *tert*-butylcyclopentane (**5-7**) gives two resonances in the ratio of 2:1, since two of the methyl groups are different from the third (**5-7a**).

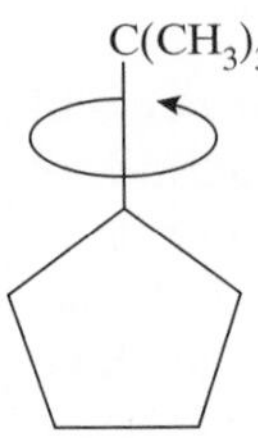

5-7 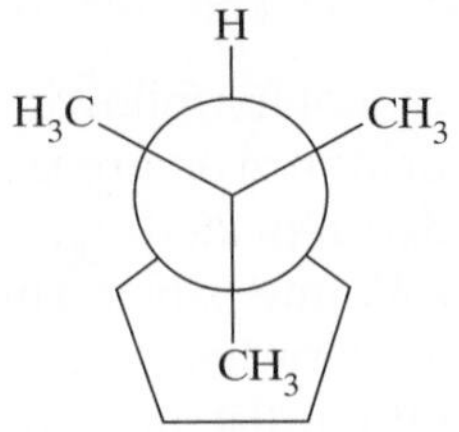 **5-7a**

Hindered rotation has frequently been observed in halogenated alkanes. The increased barrier probably arises from a combination of steric and electrostatic interactions. 2,2,3,3-Tetrachlorobutane (**5-8**) exhibits a 2:1 doublet below −40°C from anti and gauche rotamers that are rotating slowly on the NMR time scale.

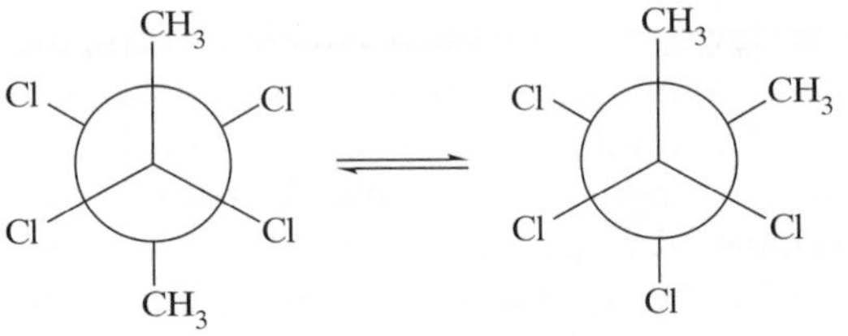

5-8

When both atoms that constitute a single bond possess nonbonding electron pairs, the barrier often is in the observable range. The high barrier may be due to electrostatic interactions or repulsions between lone pairs. For example, the barrier to rotation about the sulfur–sulfur bond in dibenzyl disulfide ($C_6H_5CH_2S{-}SCH_2C_6H_5$) is 7 kcal mol^{-1}. Similar high barriers have been observed in hydrazines (N—N), sulfenamides (S—N), and aminophosphines (N—P).

Ring Reversal. Axial-equatorial interconversion through ring reversal has been studied in a wide variety of systems in addition to cyclohexane, including heterocycles such as piperidine (**5-9**), unsaturated rings such as cyclohexene (**5-10**), fused rings like *cis*-decalin (**5-11**), and rings of other than six members, such as cycloheptatriene (**5-12**).

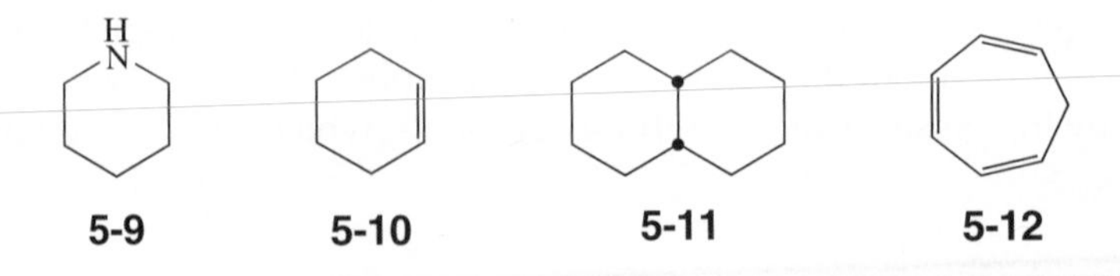

5-9 **5-10** **5-11** **5-12**

Cyclooctane and other eight-membered rings have been examined extensively. The pentadecadeutero derivative of the parent compound exhibits dynamic behavior below −100°C, with a free energy of activation of 7.7 kcal mol^{-1}. The dominate conformation appears to be the boat–chair (**5-13**). Cyclooctatetraene (**5-14**) undergoes a boat–boat ring reversal. The methyl groups on the side chain provide the diastereotopic probe and reveal a barrier of 14.7 kcal mol^{-1}. The favored transition state is a planar form with alternating single and double bonds.

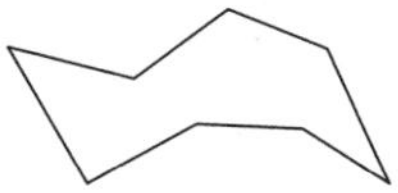

$C(CH_3)_2OH$ $C(CH_3)_2OH$

5-13 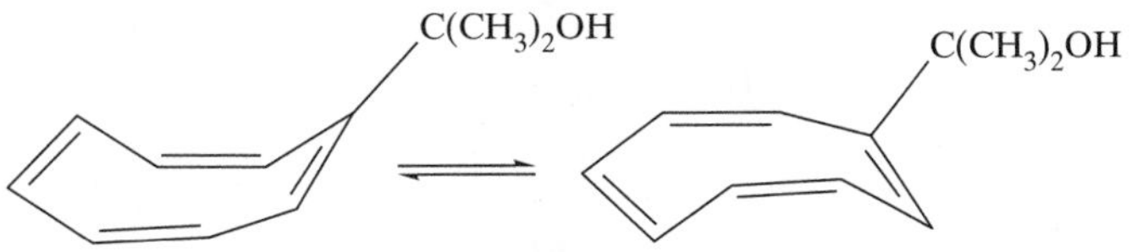 **5-14**

Atomic Inversion. Trisubstituted atoms with a lone pair, such as amines, may undergo the process of pyramidal atomic inversion on the NMR time scale. The resonances of the two methyls in the aziridine **5-15** become equivalent at elevated temperatures through

H_3C CH_3 N H_3C ⇌ H_3C N H_3C CH_3

5-15

rapid nitrogen inversion. This barrier is particularly high (18 kcal mol^{-1}) because of angle strain in the three-membered ring, which is higher in the transition state than in the ground state. The effect is observed to a lesser extent in azetidines (**5-16**, 9 kcal mol^{-1}) and in strained bicyclic systems such as **5-17** (10 kcal mol^{-1}).

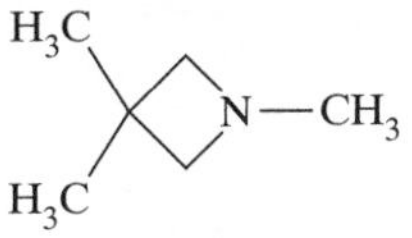

H_3C H_3C N—CH_3 N CH_3

5-16 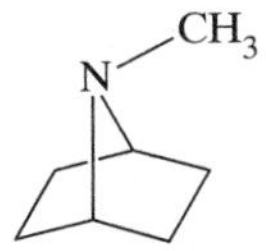 **5-17**

The inversion barrier may be raised when nitrogen is attached to highly electronegative elements. This substitution increases the s character of the ground-state lone pair. Since the transition-state lone pair must remain p-hybridized, the barrier is higher, as in *N*-chloropyrrolidine (**5-18**). When neither ring strain nor electronegative substituents are

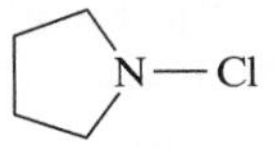

N—Cl N—CH_3

5-18 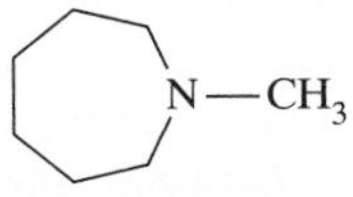 **5-19**

present, barriers are low, but still often measurable, as in *N*-methylazacycloheptane (**5-19**, 7 kcal mol^{-1}) and 2-(diethylamino)propane, $(CH_3CH_2)_2NCH(CH_3)_2$ (6.4 kcal mol^{-1}). In the latter case, the barrier is considered a mix of nitrogen inversion and C—N bond rotation.

Inversion barriers for elements in lower rows of the periodic table generally are above the NMR range. Thus chiral phosphines and sulfoxides are isolable. Barriers must be brought into the observable NMR range by substitution with electropositive elements, as in the diphosphine $CH_3(C_6H_5)P—P(C_6H_5)CH_3$, whose barrier of 26 kcal mol^{-1}

compares with 32 kcal mol^{-1} in $CH_3(C_6H_5)(C_6H_5CH_2)P$. The barrier in phosphole **5-20** is lowered because the transition state is aromatic. Its barrier of 16 kcal mol^{-1} compares with 36 kcal mol^{-1} in a saturated analogue, **5-21**.

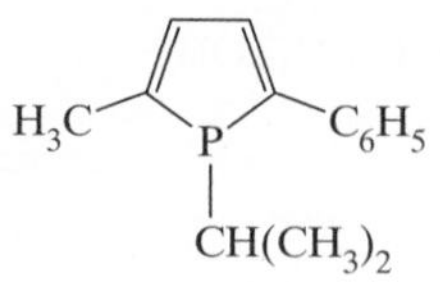

H_3C P C_6H_5 $CH(CH_3)_2$

5-20

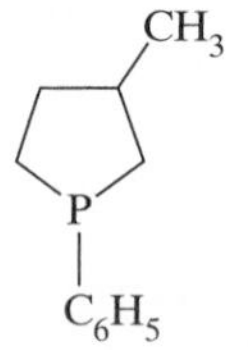

CH_3 P C_6H_5

5-21

Valence Tautomerizations and Bond Shifts. The barriers to many valence tautomerizations fall into the NMR range. A classic example is the Cope rearrangement of 3,4-homotropilidine (**5-22**). At low temperatures, the spectrum has the features expected

5-22

for the five functionally distinct types of protons (disregarding diastereotopic differences). At higher temperatures, the Cope rearrangement becomes fast on the NMR time scale, and only three types of resonances are observed (14 kcal mol^{-1} for the 1,3,5,7-tetramethyl derivative). When a third bridge is added, as in barbaralone (**5-23**), steric

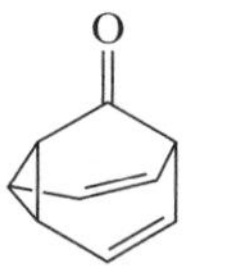

O

5-23

5-24

requirements of the rearrangement are improved, and the barrier is lowered to 9.6 kcal mol^{-1}. When the third bridge is an ethylenic group, the molecule is bullvalene (**5-24**). All three bridges are identical, and a sequence of Cope rearrangements renders all protons (or carbons) equivalent. Indeed, the complex spectrum at room temperature becomes a singlet above 180°C (12.8 kcal mol$^{-1)}$. Molecules that undergo rapid valence tautomerizations often are said to be *fluxional*.

Cyclooctatetraene offers another example of fluxional behavior. In an operation distinct from boat–boat ring reversal depicted in **5-14**, the locations of the single and double bonds are switched via the antiaromatic transition state (**5-25b**). (The transition state to ring reversal in **5-14** has alternating single and double bonds.) The proton adjacent to the substituent is different in the bond-shift isomers **5-25a** and **5-25c**. The barrier to bond switching was determined from the conversion of the proton resonance from two peaks to one (17.1 kcal mol^{-1}). The barrier to bond switching is higher than that to

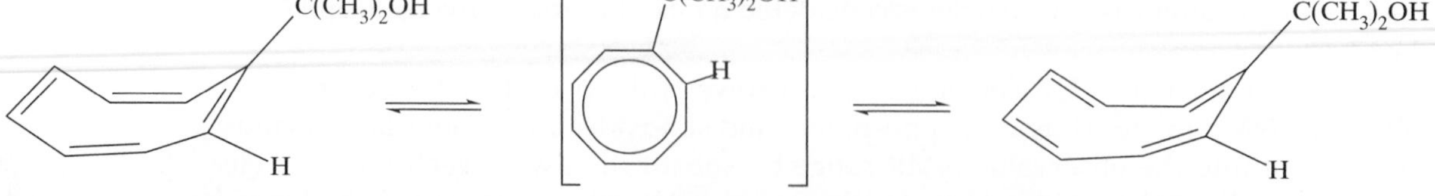

$C(CH_3)_2OH$ H

5-25a

$C(CH_3)_2OH$ H ‡

5-25b

$C(CH_3)_2OH$ H

5-25c

ring reversal because of the antiaromatic destabilization that is present in the equal-bond-length transition state **5-25b**.

Rearrangements of carbocations also may be studied by NMR methods. The norbornyl cation (**5-26**) may undergo 3,2- and 6,2-hydride shifts, as well as Wagner–

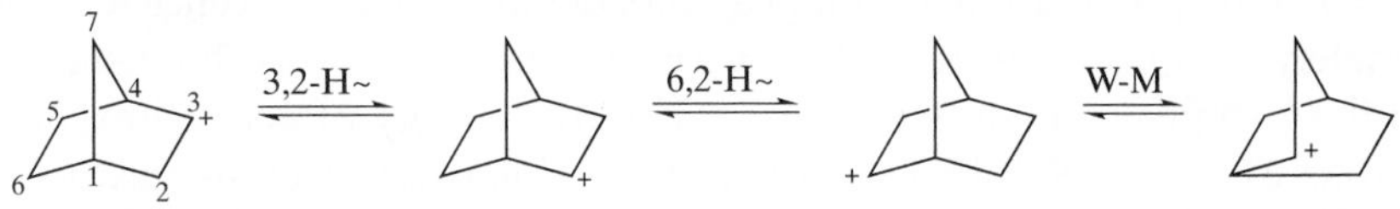

3,2-H~ 6,2-H~ W-M

5-26

Meerwein (W–M) rearrangements. The sum of these processes renders all protons equivalent, so that the complex spectrum below −80°C becomes a singlet at room temperature. The slowed process appears to be the 3,2-hydride shift, whose barrier was measured to be 11 kcal mol^{-1}.

Many examples of fluxional organometallic species have been investigated. Tetra-methylalleneiron tetracarbonyl (**5-27**) exhibits three distinct methyl resonances

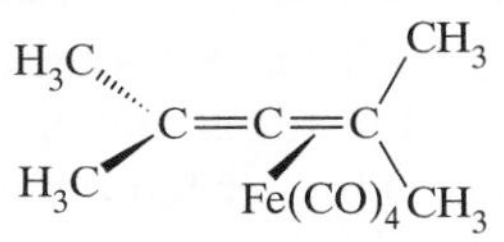

5-27

in the ratio 1:1:2 at −60°C, in agreement with the structure depicted. Above room temperature, however, the spectrum becomes a singlet (9 kcal mol^{-1}) as the $Fe(CO)_4$ unit circulates about the allenic π-electron structure by moving orthogonally from one alkenic unit to the other.

In cyclooctatetreneiron tricarbonyl (**5-28**), the spectrum below −150°C indicates four protons on carbons bound to iron and four on carbons not bound to iron, consistent

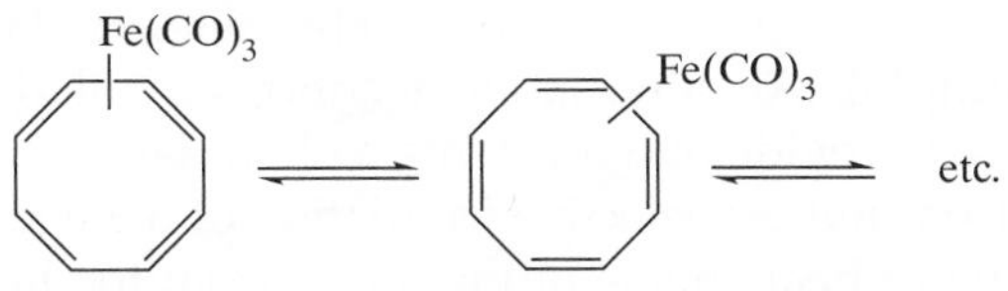

5-28

with the η^4 structure shown. Above −100°C, all the protons converge to a singlet as the iron atom moves around the ring as shown. A bond shift occurs with each 45° movement of the iron atom. Eight such operations result in complete averaging of the ring protons or carbons.

A series of 1,5-sigmatropic shifts occurs in triphenyl-(7-cycloheptatrienyl)tin (**5-29**). At 0°C, the spectrum indicates that bond shifts are slow on the NMR time scale, but, at 100°C, all of the ring protons are equivalent. That the migration is a 1,5 shift to the

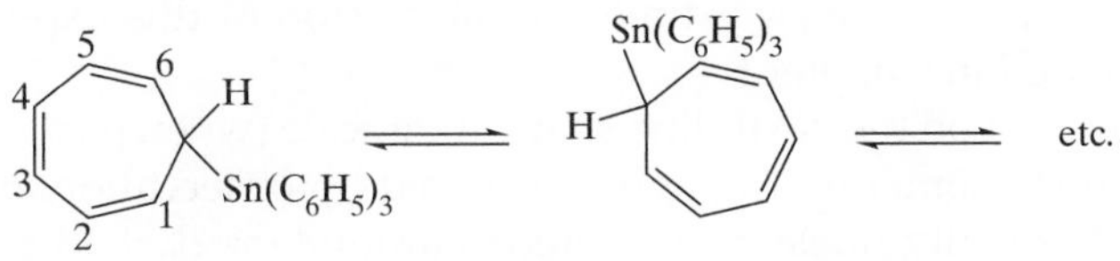

5-29

3 or 4 positions (rather than a 1,2 or 1,3 shift) was demonstrated by double-irradiation experiments (saturation transfer; see shortly).

Quantification. For the simple case of two equally populated sites that do not exhibit coupling (such as cyclohexane-d_{11} in Figure 2-31 or the amide **5-4**), the rate constant (k_c) at the point of maximum peak broadening (the coalescence temperature T_c, approximately −60°C in Figure 2-31) is $\pi\Delta\nu/\sqrt{2}$, in which $\Delta\nu$ is the distance in hertz between the two peaks at slow exchange. The free energy of activation then may be calculated as $\Delta G_c^{\ddagger} = 2.3RT_c[10.32 + \log(T_c/k_c)]$. This result is extremely accurate and certainly easy to obtain, but the equation is limited in its application. For the two-site exchange between coupled nuclei, the rate constant at T_c is $\pi(\Delta\nu^2 + 6J^2)^{1/2}/\sqrt{2}$.

To include unequal populations, more complex coupling patterns, and more than two exchange sites, it is necessary to use computer programs such as DNMR3, which can simulate the entire line shape at several temperatures. Such a procedure generates Arrhenius plots from which enthalpic and entropic activation parameters may be obtained. The procedure is more elegant and more comprehensive, but it is more susceptible to systematic errors involving inherent linewidths and peak spacings than is the coalescence temperature method. Consequently, it is always a good idea to use both line-shape fitting and coalescence temperature methods, when possible, as an internal check.

The proportionality between k_c and $\Delta\nu$ ($k_c = \pi\Delta\nu/\sqrt{2}$) means that the rate constant is dependent on the field strength (B_0). Thus, a change in field from 300 to 600 MHz alters the rate constant at T_c. The practical result is that T_c changes. Since the slow exchange peaks are farther apart at 600 MHz, a higher temperature is required to achieve coalescence than at 300 MHz. At a given field strength, two nuclides such as ^{1}H and ^{13}C have different values of $\Delta\nu$ for analogous functionalities and achieve coalescence at different temperatures. Since $\Delta\nu$ is usually larger for ^{13}C than for ^{1}H, the ^{13}C coalescence temperature often is much higher than the ^{1}H coalescence temperature—for example, for the methyl carbons and hydrogens of *N,N*-dimethylformamide (**5-4**)—even though a single rate process is involved.

Magnetization Transfer and Spin Locking. Alternative procedures not requiring peak coalescence have been developed to expand the kinetic dynamic range of NMR spectroscopy. In many cases, coalescence and fast exchange are never attained. The system may exchange too slowly on the NMR time scale at the highest available temperatures (as determined by the temperature range of the spectrometer, volatility of the solvent, or stability of the sample). An alternative technique, called *saturation transfer* or *magnetization transfer*, can provide rate constants without peak coalescence, that is, at the slow exchange limit. Continuous, selective irradiation of one slow-exchange peak may partially saturate the other peak. Some of the nuclei from the first site turn into nuclei of the second type by the exchange process. The intensity of the second peak then is reduced because the newly transformed nuclei already had been saturated in their previous identity. This reduction in intensity is related to the rate constant of interchange and the relaxation time. Saturation transfer is observed for rates in the range from 10^{-3} to 10^1 s^{-1}, which extends the NMR range based on line-shape coalescence (10^0 to 10^4 s^{-1}) on the slow-exchange end by about three orders of magnitude. In addition to expanding the dynamic range of NMR kinetics, this method permits the easy identification of exchanging partners. For example, in the cycloheptatrienyltin **5-29**, saturation of the 7 proton resonance (geminal to tin) at −10°C (below the coalescence temperature) brings about a decrease in the intensity of the 3,4 proton resonance, indicative of a 1,5 shift. A 1,2 shift would have saturated the 1,6 resonance, and a 1,3 shift would have saturated the 2,5 resonance. The two-dimensional version of this experiment is termed EXSY and is discussed in Chapter 6.

Rates that are fast on the NMR line-shape time scale (when peaks fail to decoalesce at high temperatures) sometimes may be measured by observation at a different resonance frequency. Normally, nuclear spins precess around the B_0 field at their Larmor frequency. Application of the usual 90° pulse in the x direction places the spins in the xy

plane, along the y axis (Figure 2-15a). Continuous B_1 irradiation along the y axis (not a pulse) forces magnetization to precess around that axis (called *spin locking*, as in the cross-polarization experiment of Section 2-9). The spins are said to be locked onto the y axis. Because the spins are precessing at a lower frequency (γB_1, rather than γB_0), they are sensitive to a different range of rate processes, one corresponding to about 10^2 to 10^6 s^{-1}, which extends the NMR range on the fast-exchange end by about two orders of magnitude. Rates are obtained by comparing the relaxation time when the system is spin locked ($T_{1\rho}$) with the usual spin–lattice relaxation time (T_1) and analyzing any differences.

Through line-shape, saturation transfer, and spin-lock methods, the entire range of rates accessible to NMR is about 10^{-3} to 10^6 s^{-1}. Thus, NMR has become an important method for studying the kinetics of reactions at equilibrium over a very large dynamic range.

5-3 MULTIPLE RESONANCE

Special effects may be routinely and elegantly created by using sources of radiofrequency energy in addition to the observation frequency ($\nu_1 = \gamma\!\!\!/ B_1$) ($\gamma\!\!\!/ = \gamma/2\pi$). The technique is called *multiple irradiation* or *multiple resonance* and requires the presence of a second transmitter coil in the sample probe to provide the new irradiating frequency $\nu_2 = \gamma\!\!\!/ B_2$. When the second frequency is applied, the experiment, which is widely available on modern spectrometers, is termed *double resonance* or *double irradiation*. Less often, a third frequency $\nu_3 = \gamma\!\!\!/ B_3$ also is provided, to create a *triple-resonance* experiment. We already have seen several examples of double irradiation experiments, including the removal of proton couplings from ^{13}C (Figure 2-25), the elimination of solvent peaks by peak suppression (Section 5-1), the sharpening of NH resonances by irradiation of ^{14}N (Figure 5-4), and the study of rate processes by saturation transfer (Section 5-2).

Spin Decoupling. One of the oldest and most generally applicable double-resonance experiments is the irradiation of one proton resonance (H_X) and observation of the effects on the AX coupling (J_{AX}) present in another proton resonance (H_A). The traditional and intuitive explanation for the resulting spectral simplification, known as *spin decoupling*, is that the irradiation shuttles the X protons between the $+\frac{1}{2}$ and $-\frac{1}{2}$ spin states so rapidly that the A protons no longer have a distinguishable independent existence. As a result, the A resonance collapses to a singlet. This explanation, however, is inadequate in that it fails to account for phenomena at weak decoupling fields (spin tickling) and even some phenomena at very strong decoupling fields.

The actual experiment involves getting the coupled nuclei to precess about orthogonal axes. The magnitude of the coupling interaction between two spins is expressed by the scalar, or dot, product between their magnetic moments and is proportional to the expression $J\boldsymbol{\mu}_1 \cdot \boldsymbol{\mu}_2 = J\mu_1\mu_2 \cos\phi$. The quantity ϕ is the angle between the vectors (the axes of precession of the nuclei). So long as both sets of nuclei precess around the same (z) axis, ϕ is zero, $\cos 0° = 1$, and full coupling is observed. The geometrical relationship between the spins may be altered by subjecting one of them to a B_2 field. Imagine observing ^{13}C nuclei as they precess around the z axis at the frequency B_1. When the attached protons are subjected to a strong B_2 field along the x axis, they will precess around that axis. The angle ϕ between the ^{13}C and ^{1}H nuclear vectors, then, is 90°, as they respectively precess around the z and x axes. As a result, their spin–spin interaction goes to zero because the dot product is zero ($\cos 90° = 0$). The nuclei are then said to be decoupled.

Spin decoupling has been useful in identifying coupled pairs of nuclei. Figure 5-5 provides such an example for the molecule ethyl *trans*-crotonate (ethyl *trans*-but-2-enoate). The alkenic protons split each other, and both are split by the allylic methyl group to form an ABX_3 spin system. Irradiation at the methyl resonance frequency produces the upper spectrum in the inset for the alkenic protons, which have become a simple AB quartet. A more complex example is illustrated in Figure 5-6. The bicyclic sugar mannosan triacetate, whose structure is given on the left of the figure, has a nearly

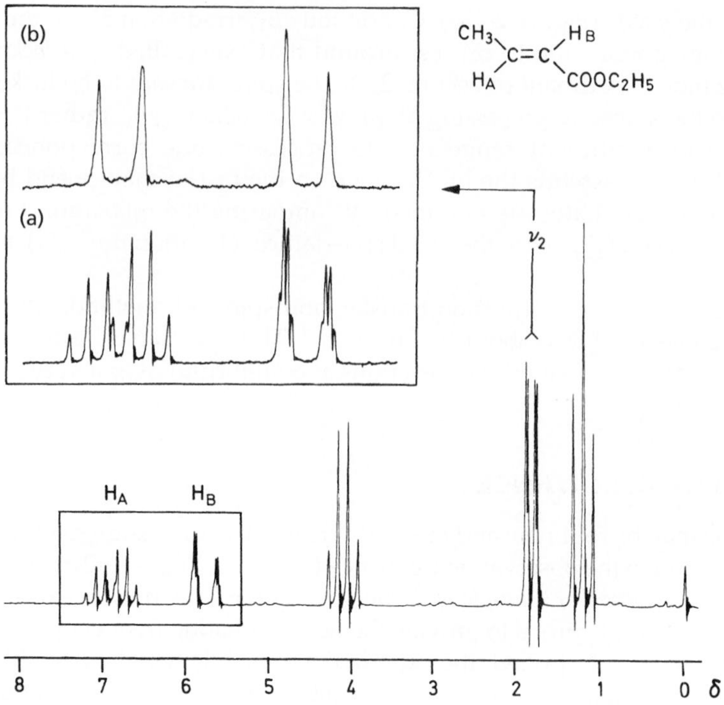

FIGURE 5-5 The ^{1}H spectrum of ethyl *trans*-crotonate. The inset contains an expansion of the alkenic range (a) without and (b) with decoupling of the methyl resonance at δ 1.8. (Reproduced with permission from H. Günther, *NMR Spectroscopy*, 2nd ed., Chichester, UK: John Wiley & Sons, Ltd.,1992, p. 46.)

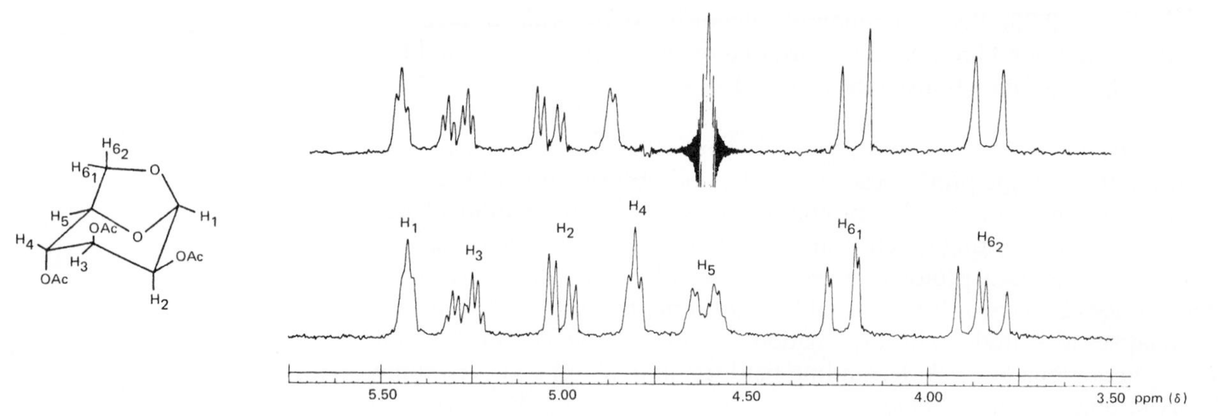

FIGURE 5-6 The 100 MHz ^{1}H spectrum of mannosan triacetate in $CDCl_3$ without decoupling (lower) and with double irradiation at δ 4.62 (upper). (Reproduced with the permission of Varian Associates.)

first-order spectrum with numerous coupling partners. Irradiation of H_5 (δ 4.62) produces simplification of the resonances of its vicinal partners H_4, $H_{6/1}$, and $H_{6/2}$, as well as its long-range zigzag partner H_3.

Difference Decoupling. With complex molecules, it is useful to record the difference between coupled and decoupled spectra. Features that are not affected by decoupling are subtracted out and do not appear. Figure 5-7 shows the ^{1}H spectrum of 1-dehydrotestosterone. The complex region between δ 0.9 and 1.1 contains the resonances of four protons. A comparison of the coupled (Figure 5-7a in the inset) and decoupled (Figure 5-7b) spectra from irradiation of the 6α resonance shows little change as the result of double irradiation. The *difference decoupling spectrum* (Figure 5-7c) is the result of subtracting (a) from (b). The unaffected overlapping peaks are gone. The original resonances of the affected protons are observed as negative peaks with coupling, and the simpler decoupled resonances of the same protons are present as positive peaks. The resonances must be due to the 7α protons. The procedure provides coupling relationships when spectral overlap is a serious problem. This and other simple spin-decoupling experiments have been entirely superseded by two-dimensional experiments (Chapter 6).

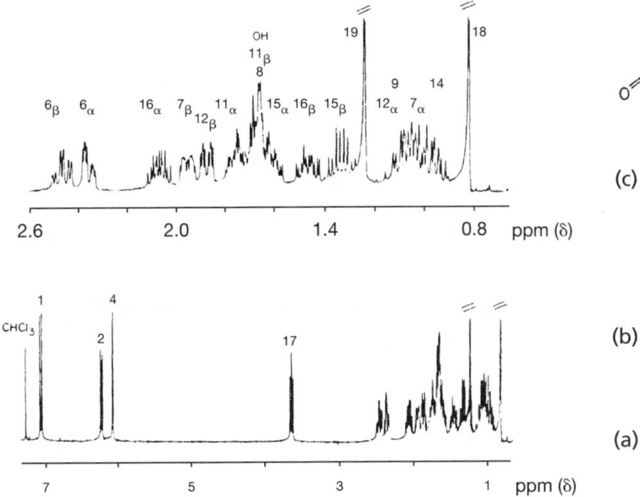

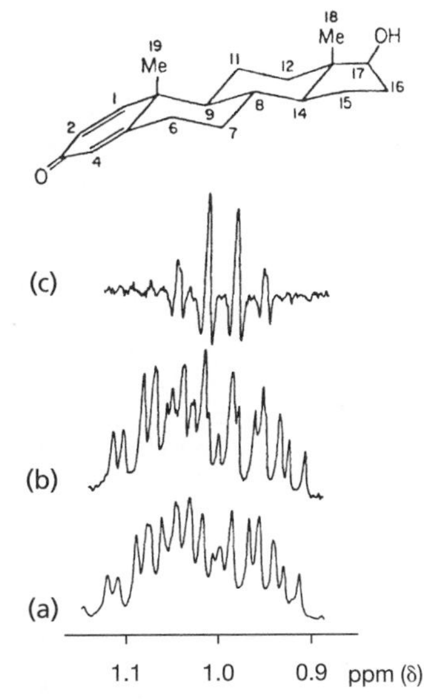

FIGURE 5-7 The 400 MHz ^{1}H spectrum of 1-dehydrotestosterone. The complete spectrum and an expansion of the low-frequency region are given on the left. On the right are shown (a) the coupled spectrum for the δ 0.9–1.1 region; (b) the same region, decoupled from the 6α proton; and (c) the difference spectrum obtained by subtracting (b) from (a). (Reproduced with permission from L. D. Hall and J. K. M. Sanders, *J. Am. Chem. Soc.*, **102**, 5703 [1980]. Copyright 1980 American Chemical Society.)

Classes of Multiple Resonance Experiments. Experiments in which both the irradiated and the observed nuclei are protons are called *homonuclear double-resonance* experiments and are represented by the notation ^{1}H{^{1}H}. The irradiated nucleus is denoted by braces. When the observed and irradiated nuclei are different nuclides, as in proton-decoupled ^{13}C spectra, the experiment is a *heteronuclear double-resonance* experiment and is denoted, for example, ^{13}C{^{1}H}, or ^{13}C{^{1}H}{^{31}P} for a triple-resonance experiment.

Double-resonance experiments also may be classified according to the intensity or bandwidth of the irradiating frequency. If irradiation is intended to cover only a portion of the resonance frequencies, the technique is known as *selective irradiation* or *selective decoupling*. The decoupling shown in Figures 5-5 and 5-6, the peak suppression described in Section 5-1, and the magnetization transfer discussed in Section 5-2 are three examples of selective double irradiation. In the two decoupling experiments, only couplings to the selectively irradiated proton are removed. Nonirradiated resonances can exhibit a small movement in frequency, called the *Bloch–Siegert shift*, which is related to the intensity of the B_2 field and the distance between the observed and irradiated frequencies. An examination of Figure 5-6 reveals several such shifts, found by comparing the relative positions of the resonances in the upper and lower spectra. When all frequencies of a specific nuclide are irradiated, the experiment is termed *nonselective irradiation* or *broadband decoupling*. Figure 2-25 illustrates the ^{13}C spectrum of 3-hydroxybutyric acid both with and without broadband proton double irradiation. The invention of this technique was instrumental in the development of ^{13}C NMR spectroscopy as a routine tool. To cover all the ^{1}H frequencies, B_2 was modulated with white noise, so the technique often was called *noise decoupling*.

Off-Resonance Decoupling. The broadband decoupling experiment removes coupling patterns that could indicate the number of protons attached to a given carbon atom. The *off-resonance decoupling* method was developed to retain this information and still provide some of the advantages of the decoupling experiment. Irradiation above or below the usual 10-ppm range of ^{1}H frequencies leaves residual coupling given by the approximate formula $J_{res} = 2\pi J(\Delta\nu)/\gamma B_2$, in which J is the normal coupling, γ is the gyromagnetic ratio of the irradiated nucleus, and $\Delta\nu$ is the difference between the decoupler frequency and the resonance frequency of a proton coupled to a specific carbon. Because carbon multiplicities remain intact, this technique is useful for determining, with minimal peak overlap, whether carbons are methyl (quartet), methylene (triplet), methine (doublet), or quaternary (singlet). If methylene protons are diastereotopic, methylene carbons can appear as two doublets. The outer peaks of the off-resonance decoupled triplets and quartets usually are weaker than one might expect from the binomial coefficients. As a result, doublets and quartets sometimes are difficult to distinguish. Figure 5-8 shows the spectrum of vinyl acetate with full

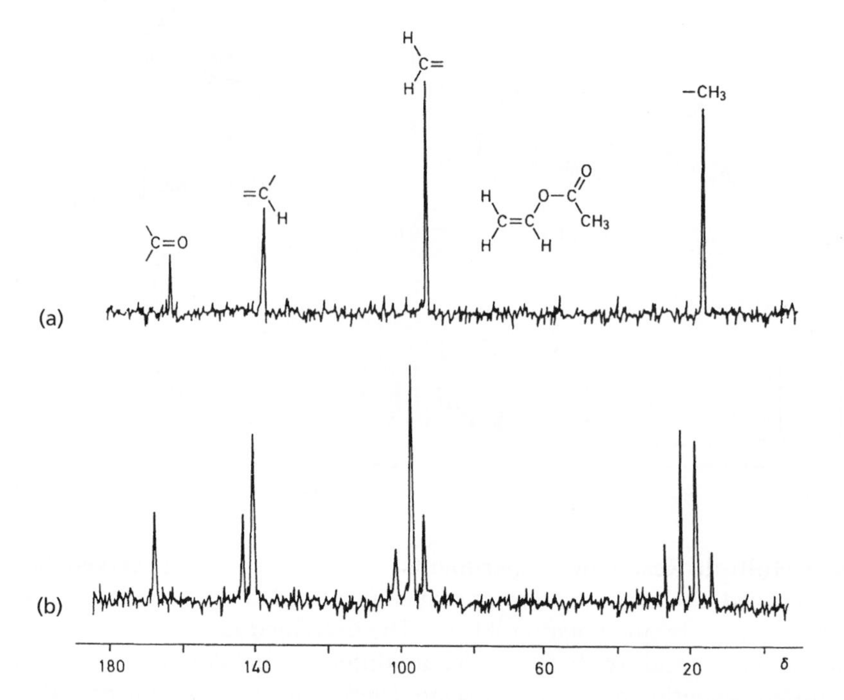

FIGURE 5-8 The ^{13}C spectrum of vinyl acetate (a) with complete decoupling of the protons and (b) with off-resonance decoupling of the protons. (Reproduced with permission from H. Günther, *NMR Spectroscopy*, 2nd ed., Chichester, UK: John Wiley & Sons, Ltd., 1992, p. 270.)

decoupling and with off-resonance decoupling. In complex molecules, peak overlap and ambiguities with regard to quartets often make assignments by this technique difficult. Therefore, it has been superseded by the editing experiments described in Section 5-5.

In early spin-decoupling experiments, the irradiation frequency was left on continuously while the experimenter observed the resonating nuclei. There are two significant problems with this method. First, the application of rf energy at the decoupling frequency generates heat. As B_0 fields increased from 60 to 900 MHz, higher decoupling intensities were required. The resultant heating was unacceptable for biological samples and for many delicate organic or inorganic samples. Second, with higher field strengths, it became increasingly more difficult for B_2 to cover the entire range of ^{1}H frequencies, which had been about 600 Hz at 60 MHz, but became 5000 Hz at 500 MHz.

To overcome these problems of heteronuclear decoupling, modern methods replaced continuous irradiation with a series of pulses that eliminate the effects of coupling. In a ^{13}C{^{1}H} experiment (Figure 5-9 for two spins, ^{13}C—^{1}H), a 90° B_1 pulse applied to the observed ^{13}C nuclei along the x direction moves magnetization from carbon coupled to either spin-up or spin-down protons into the xy plane along the y axis [(Figure 5-9a) → (Figure 5-9b)]. The reference frequency is considered to coincide with the y axis and be midway between the frequencies of the carbons associated with the spin-up (β) and spin-down (α) protons. The two carbon vectors then diverge in the xy plane after the 90° pulse, one becoming faster and the other slower than the carrier frequency (Figure 5-9c). After time τ, a 180° proton pulse (the B_2 of the decoupling experiment) switches the locations of the vectors. The slower-moving vector that was dropping behind the carrier frequency now is replaced by the faster-moving vector (and the faster-moving vector by the slower-moving vector), so that both carbon vectors start to move back toward the y axis (Figure 5-9d). After an equal second period τ, the two vectors coincide on the y axis, only one frequency or peak occurs, and coupling to the protons disappears (Figure 5-9e). The process is repeated during acquisition at a rate (in hertz) that is faster than the coupling constant, so that the effects of coupling are removed. In this way, decoupling can be achieved with short pulses during acquisition rather than with a continuous, high-intensity field during the entire experiment. In practice, the method is limited because the 180° pulse must be very accurate and because the B_2 field is inhomogeneous. Refinements of this experiment have been achieved by replacing the 180° pulse

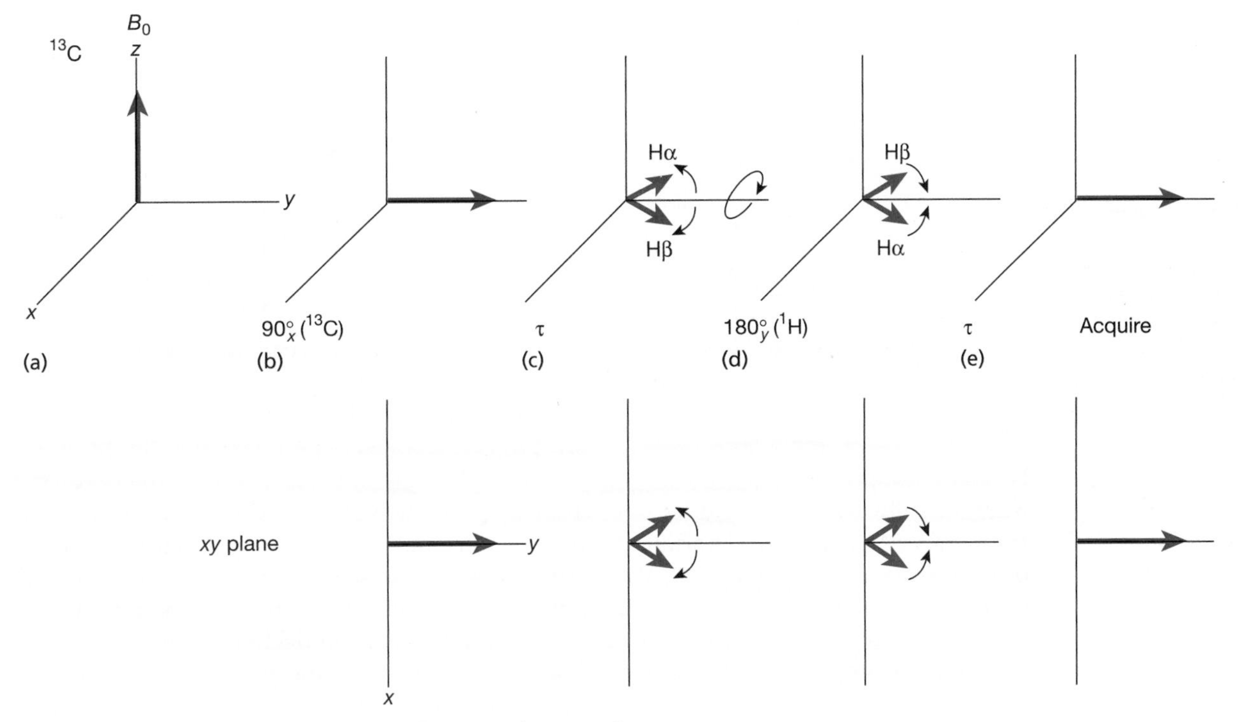

FIGURE 5-9 Pulse sequence to remove heteronuclear coupling.

with several pulses (*composite pulses*) and by cycling their order (*phase cycling*) so as to cancel out the inaccuracies (Section 5-8). Some successful such methods include MLEV-16 (for *Malcolm LEVitt*, the name of one of the developers, with 16 phase cycles) and, in particular, WALTZ-16, which achieves full decoupling across a much wider range than the original continuous method and with a fraction of the power. The source of the name WALTZ is provided in Section 5-8.

5-4 THE NUCLEAR OVERHAUSER EFFECT

Dipole–dipole relaxation occurs when two nuclei are located close together and are moving at an appropriate relative rate (Section 5-1). Irradiation of one of these nuclei with a B_2 field alters the Boltzmann population distribution of the other nucleus and therefore perturbs the intensity of its resonance. No J coupling need be present between the nuclei. The original phenomenon was discovered by Overhauser, but between nuclei and unpaired electrons. The Overhauser effect when both spins are of nuclei was observed first by Anet and Bourne and is of more interest to the chemist. It has great structural utility, because the dipole–dipole mechanism for relaxation depends on the distance between the two spins (eq. 5-1).

Origin. The origin of the *Nuclear Overhauser Effect* (NOE) is illustrated in Figure 5-10. On the left are the states for two spins (A and X) in the absence of double irradiation. Effects of J are irrelevant and are ignored. The diagram represents an expansion of Figure 2-4a for one spin, with β standing for $+\frac{1}{2}$ and α for $-\frac{1}{2}$. There are four spin states: when both spins are β, when the first (A) is β while the second (X) is α, when the first is α while the second is β, and when both are α. There are two A-type transitions (when the A spin flips from β to α)—for example, $\beta\beta$ to $\alpha\beta$—and there are two X-type transitions (when the X spin flips from β to α)—for example, $\alpha\beta$ to $\alpha\alpha$. When $J = 0$, the two A transitions coincide, as do the two X transitions. Because chemical shifts are very small in comparison with the Larmor frequency, the $\alpha\beta$ and $\beta\alpha$ states are almost degenerate. Their difference has been exaggerated to emphasize the different chemical shifts.

αα

A

βα

X

αβ

X

A

ββ

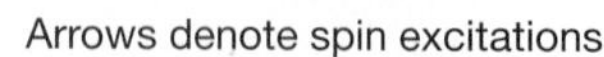

Arrows denote spin excitations

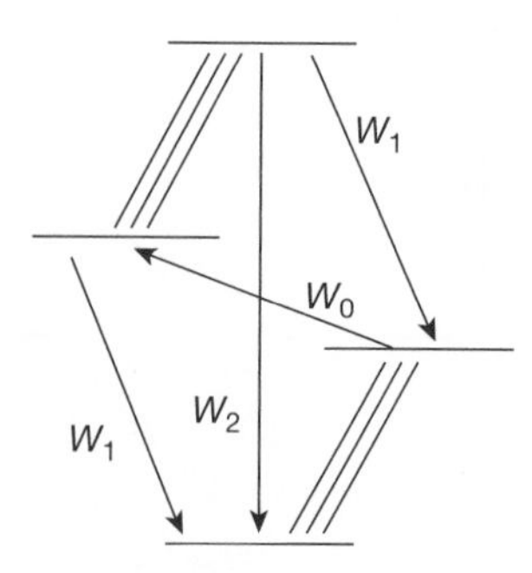

Arrows denote spin relaxation

FIGURE 5-10 Left: The spin states for a normal two-spin (AX) system. Right: The spin states for an AX system when the frequency of A is doubly irradiated.

The normal intensities of the A and X resonances are determined by the difference between the populations of the upper and lower spin states in a spin transition—for instance, between $\alpha\beta$ and $\alpha\alpha$ for one X transition. In the NOE experiment, one resonance frequency (A) is doubly irradiated, and intensity perturbations are monitored at the other resonance frequency (X). When the A resonance is irradiated, as is represented by the multiple parallel lines in the right-hand diagram of Figure 5-10, the population difference between the spin states connecting an A transition decreases through partial saturation. Compared with the normal situation on the left, the populations of $\alpha\alpha$ and $\alpha\beta$ (the upper states) have increased, while those of $\beta\alpha$ and $\beta\beta$ (the lower states) have decreased. Dipolar relaxation from $\alpha\alpha$ to $\beta\beta$, labeled W_2 in the figure, can help restore the system to equilibrium. The new equilibrium present with A irradiation thus can carry spins along the route $\beta\alpha \rightarrow \alpha\alpha \rightarrow \beta\beta \rightarrow \alpha\beta$, depleting $\beta\alpha$ and augmenting $\alpha\beta$. Depleting $\beta\alpha$ enhances the population difference for one X transition ($\beta\beta \rightarrow \beta\alpha$), while augmenting $\alpha\beta$ enhances the population difference for the other X transition ($\alpha\beta \rightarrow \alpha\alpha$). This enhanced polarization of nuclear spin states means that the X intensity is higher in the new equilibrium during the irradiation of A. To a first approximation, the populations of $\alpha\alpha$ and $\beta\beta$ are constant, as they are simultaneously augmented by one transition and depleted by another. Normal relaxation of X nuclei, labeled W_1 in the figure ($\alpha\alpha \rightarrow \alpha\beta$ or $\beta\alpha \rightarrow \beta\beta$), does not alter the X intensity. These processes are unchanged from the diagram on the left.

Relaxation from $\alpha\beta$ to $\beta\alpha$ (called W_0 in Figure 5-10) also can move the irradiated system back toward equilibrium. This relaxation mechanism, however, would result in a *decrease* in intensity of X, since it depletes $\beta\beta$ and augments $\alpha\alpha$ ($\beta\beta \rightarrow \alpha\beta \rightarrow \beta\alpha \rightarrow \alpha\alpha$), with $\alpha\beta$ and $\beta\alpha$ constant. For liquids and relatively small molecules, $W_0 \ll W_2$, so that enhanced intensities are expected. The frequencies of W_2 are in the megahertz range (represented by the large distance between the $\beta\beta$ and $\alpha\alpha$ levels in the figure), whereas those of W_0 are much smaller, in the kilohertz or hertz range (represented by the small, but exaggerated, distance between the $\alpha\beta$ and $\beta\alpha$ levels). Small molecules tumbling in solution produce fields in the megahertz range and hence can provide W_2 relaxation. In contrast, large molecules tumbling in the hertz or kilohertz range can provide W_0 relaxation.

Observation. Double irradiation of A in molecules of molecular weight up to 1000–3000 daltons thus enhances the X intensity, provided that the two nuclei are close enough for W_2 relaxation to dominate (less than about 5 Å). This circumstance corresponds to what we previously called the extreme narrowing limit. For larger molecules—certainly those with molecular weights over 5000—W_0 dominates, and reductions in peak intensity or inverse peaks occur. At some intermediate size (1000–3000), the effect disappears as the crossover between regimes occurs. The change in intensity [denoted by the Greek letter η (eta)] thus depends on the difference between the W_2 and W_0 relaxation rates, in comparison with the total relaxation rates, as given by eq. 5-3.

$$\eta = \frac{\gamma_{\text{irr}}}{\gamma_{\text{obs}}}\left(\frac{W_2 - W_0}{W_0 + 2W_1 + W_2}\right) \tag{5-3}$$

(With two modes, W_1 is doubled.) The effect is observed by comparing intensities I in the presence of double irradiation with those I_0 in its absence via eq. 5-4.

$$\eta = \frac{(I - I_0)}{I_0} \quad \textbf{(5-4)}$$

For small molecules (the extreme narrowing limit), the maximum increment in intensity, η_{max}, is $\gamma_{irr}/2\gamma_{obs}$, so that an initial intensity of unity ($I_0 = 1.0$) increases up to $(1 + \eta_{max})$. [In our example, A was irradiated ("irr") and X observed ("obs").] The maximum enhanced intensity, obtained by rearrangement of eq. 5-4, is given by eq. 5-5.

$$I_{max}(\text{NOE}) = I_0\left(1 + \frac{\gamma_{irr}}{2\gamma_{obs}}\right) \quad \textbf{(5-5)}$$

The increase is almost always less than the maximum, because nondipolar relaxation mechanisms are present and because the observed nucleus is relaxed by nuclei other than the irradiated nucleus.

Whenever the two nuclei are the same nuclide, for example, both protons, the gyromagnetic ratios in eq. 5-5 cancel, η_{max} becomes 0.5, and the maximum intensity enhancement $(1 + \eta_{max})$ is a factor of 1.5, or 50%. For the common case of broadband ^{1}H irradiation with observation of ^{13}C, [^{13}C{^{1}H}], η_{max} is 1.988, so the enhancement is a factor of up to 2.988, or about 200%. Other maximum Overhauser enhancement factors $(1 + \eta_{max})$ include 2.24 for ^{31}P{^{1}H}, 3.33 for ^{195}Pt{^{1}H}, and 3.39 for ^{207}Pb{^{1}H}.

Certain nuclei have negative gyromagnetic ratios, so that, in the extreme narrowing limit, η_{max} becomes negative and a negative peak can result. For irradiation of ^{1}H and observation of ^{15}N [^{15}N{^{1}H}], η_{max} is -4.94. The maximum negative intensity is thus 3.94 times that of the original peak, or an increase of 294% $[(3.94 - 1.00) \times 100]$, but as an inverse peak. If dipolar relaxation is only partial, the ^{15}N{^{1}H} NOE can result in decreased intensity or even a completely nulled resonance. Silicon-29 also has a negative gyromagnetic ratio, so similar complications ensue. For the ^{29}Si{^{1}H} experiment, $\eta_{max} = -2.52$. The maximum enhancement factor $(1 + \eta_{max})$ is then -1.52, which results in an inverted intensity with an increase of 52% over the unirradiated case. For ^{119}Sn{^{1}H} ($\eta_{max} = -1.34$), there is actually a net loss in intensity. The maximum enhancement factor is -0.34, representing a 66% loss in intensity of the negative peak, compared with the peak at the unirradiated position. The NOE is entirely independent of spectral changes that arise from the collapse of spin multiplets through spin decoupling. The NOE does not require that nuclei A and X be spin coupled—only that they be mutually relaxed through a dipolar mechanism.

Large molecules, such as proteins or nucleic acids, with molecular weights higher than about 3000–5000 are dominated by W_0 relaxation. Since the other terms (W_2 and W_1) in eq. 5-3 are small, the value of η_{max} becomes -1 for the homonuclear proton case. Such a situation can result in a loss of signal. Consequently, for large molecules, transient rather than steady-state NOEs often are studied. For example, the buildup (or loss) of signal from the NOE can provide interproton distances. By observing many such relationships, the structures of large biomolecules may be determined quantitatively in a process that rivals X-ray crystallography, but applies to the liquid state (Nobel Prize, 2002).

At the crossover between the extreme narrowing and the large-molecule limits, it is possible that W_2 and W_0 are comparable in magnitude, so that, by eq. 5-3, the NOE goes to zero. The spectroscopist may improve the situation somewhat by changing the solvent or the temperature in order to alter τ_c. Viscosity, in addition to molecular size, can affect the tumbling rates and hence the rate of dipolar relaxation. Thus, viscous media can lower nuclear Overhauser enhancements.

In the traditional NOE experiment, the spectrum is recorded twice, with and without the NOE. Figure 5-11 illustrates the relative timing for the heteronuclear case of a ^{13}C pulse (B_1), a ^{1}H double irradiation field (B_2), and acquisition of the ^{13}C signal (not to scale) in

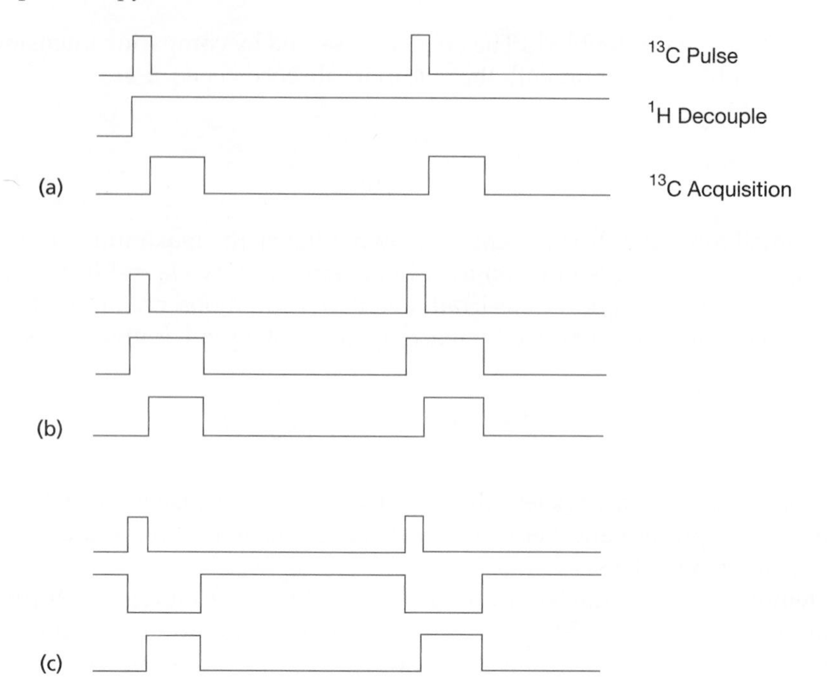

FIGURE 5-11 (a) Observation of ^{13}C with continuous double irradiation of ^{1}H (decoupling and NOE). (b) Double irradiation applied during acquisition, but gated off during the wait period (decoupling, no NOE). (c) Double irradiation applied only during the wait period (NOE, no decoupling). The pulse widths are not to scale. The scheme is shown for two cycles.

order to carry out these two experiments. In the original experiment with continuous broadband decoupling (Figure 5-11a), the B_2 field is turned on and left on. It must be on during acquisition to ensure decoupling, but also during the recovery time, when relaxation occurs and the NOE builds up. The power can be lower during times other than acquisition (*power-gated decoupling*). This experiment results in both decoupling and the Overhauser effect and provides the quantity I in eq. 5-4. By gating the decoupler off during the recovery period, as in Figure 5-11b, but keeping it on during acquisition, the spectroscopist obtains decoupling, but no NOE, providing the quantity I_0 in eq. 5-4. Without irradiation during the recovery period, there is insufficient time for the NOE to build up, and unperturbed intensities are obtained. In practice, the double-resonance frequency is not actually turned off, but is moved far off resonance. A comparison of the intensity I in experiment (a) with the intensity I_0 in experiment (b) provides the NOE via eq. 5-4. Figure 5-11c illustrates an alternative procedure, in which the B_2 field is gated off during acquisition, but is on during the recovery period. Such an experiment provides no decoupling, but generates the NOE, so it is useful for measuring ^{1}H—^{13}C couplings with enhanced intensity.

Difference NOE. For the homonuclear proton NOE experiment (^{1}H{^{1}H}) that parallels experiments (a) and (b) in Figure 5-11, it has traditionally been supposed that the NOE (in percentage, 100η) must exceed about 5% to be accepted as experimentally significant. The *difference NOE experiment*, however, can measure enhancements reliably to below 1%. In this procedure, spectra obtained by the methods analogous to those in Figure 5-11a and b are alternatively recorded and subtracted. Unaffected resonances disappear, and NOEs are signified by residual peaks.

Figure 5-12 illustrates the difference NOE spectrum for a portion of the ^{1}H spectrum of progesterone (**5-30**), in which the resonance of the 19 methyl group has been irradiated (*arrow*). The unirradiated spectrum is given at the bottom, the difference

5-30

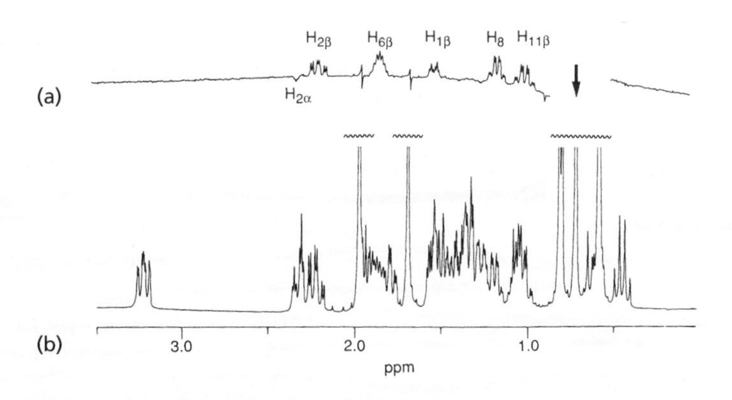

FIGURE 5-12 The 400 MHz ^{1}H spectrum (in part) of progesterone (a) without double irradiation and (b) with irradiation of the CH_3–19 resonance displayed as a difference spectrum. (Reproduced with permission from J. K. M. Sanders and B. K. Hunter, *Modern NMR Spectroscopy*, 2nd ed., Oxford, UK: Oxford University Press, 1993, p. 191.)

spectrum at the top. Enhancements are seen by difference for five nearby protons. In general, for molecules in the extreme narrowing limit, the NOE difference experiment is preferred to the direct experiment. Proton $H_{2\alpha}$ (the equatorial 2 proton) is not close to the 19 methyl group, but its resonances show a small negative NOE. This finding is the result of a three-spin effect. (A is relaxed by B and B by C.) Irradiation at A increases the Boltzmann population for B and enhances the intensity of B. By spin diffusion (Section 5-1), this enhanced intensity of B has the opposite effect on C, decreasing the Boltzmann population and the intensity. As a result, C appears as a negative peak in the difference NOE spectrum. In this example, A is Me-19, B is $H_{2\beta}$, and C is $H_{2\alpha}$. The process occurs most commonly with very large molecules.

Applications. The NOE experiment has three distinct uses. For heteronuclear examples, the foremost use is the increase in sensitivity, which combines with the collapse of multiplets through decoupling to provide the standard ^{13}C spectrum composed of a singlet for each carbon. Because most carbons are relaxed almost entirely by their attached protons, the NOE commonly attains a maximum value of about 200%. Quaternary carbons, with more distant nearest neighbors, do not enjoy this large enhancement.

Second, interpreting ^{13}C spin–lattice relaxation routinely requires a quantitative assessment of the dipolar component, T_1(DD). Because the NOE results from dipolar relaxation, its size is related to the dipolar percentage of overall relaxation. If the maximum, or full, NOE for ^{13}C{^{1}H} of 200% is observed, then $T_1(\text{obs}) = T_1(\text{DD})$. When other relaxation mechanisms contribute to ^{13}C relaxation, the enhancement is less than 200%. The dipolar relaxation for ^{13}C{^{1}H} may be calculated from the expression $T_1(\text{DD}) = \eta\, T_1(\text{obs})/1.988$, in which η is the observed NOE and 1.988 is the maximum NOE (η_{max}). It is possible then to discuss T_1(DD) in terms of structure, according to eq. 5-1.

In the third application, the dependence of the NOE on internuclear distances can be exploited to determine structure, stereochemistry, and conformation. Enhancements are expected when nuclei are close together. The adenosine derivative **5-31** (2′,3′-isopropylidene adenosine) can exist in the conformation shown, with the purine ring lying over the sugar ring (syn), or in an extended form, with the proton on C8 lying over the sugar ring (anti). Saturation of the H1′ resonance brings about a 23% enhancement of the H8 resonance, and saturation of H2′ produces an enhancement of H8 of 5% or less. Thus, H8 must be positioned most closely to H1′, as in the syn form shown. Structural and stereochemical distinctions frequently are made possible by determining the relative orientations of protons. The synthetic penicillin derivative **5-32** could have the spiro sulfur heterocycle

5-31

5-32a

5-32b

oriented either as shown in (a) or with the sulfur atom and $(CH)_{10}$ switched, as in (b). Irradiation of the methyl protons brings about an enhancement of H10 as well as of H3 and clearly demonstrates that the stereochemistry is as shown in **5-32a**.

Limitations. Despite the considerable advantages of the NOE experiment, its limitations must be appreciated. First, three-spin effects, or spin diffusion, may cause misleading intensity perturbations when the third spin is not close to the irradiated nucleus ($H_{2\alpha}$ in Figure 5-12). Second, the size of the molecule can cause NOE effects that are positive, negative, or null. Third, nuclei with negative gyromagnetic ratios can give diminished positive peaks, no peak, or negative peaks with diminished or enhanced intensity. Fourth, chemical exchange can cause an intensity perturbation analogous to the three-spin effect. Irradiation of a nucleus can lead to intensity changes at another nucleus, which can alter its chemical identity through a dynamic exchange such as a bond rotation or valence tautomerization. The NOE can then be observed for the product nucleus, provided chemical exchange is faster than relaxation of the NOE effects. Fifth, unintentional paramagnetic impurities can alter the NOE through intermolecular dipole–dipole relaxation. All these considerations must be taken into account in interpreting NOE experiments. Despite its limitations, the NOE is a very important tool for enhancing intensities and elucidating structures.

5-5 SPECTRAL EDITING

For deducing the structure of organic molecules, one of the most useful pieces of information is a compilation of the substitution pattern of all the carbons—that is, a census of which carbons are methyl, methylene, methine, or quaternary. We have already seen (Figure 5-8) that the off-resonance decoupling procedure provides such information, although with less than ideal results. Through the choice of appropriate pulses and timing, the chemist may accomplish the same task by eliminating some of the resonances from the spectrum or by altering their polarization. Such an experiment is called *spectral editing* and includes solvent suppression, for example.

The Spin–Echo Experiment. Most spectral editing procedures are based on the *spin–echo* experiment devised by Hahn, Carr, Purcell, Meiboom, and Gill in the 1950s, largely to measure spin–spin relaxation times (T_2). An example of this experiment was given in Figure 5-9, in which a 180° pulse brought vectors from spin–spin interactions back together on the y axis as an echo. Such a procedure also refocuses dispersion in the chemical shift caused by magnetic inhomogeneity in the following fashion. As shown in Figure 5-13, in the absence of J, a resonance (Figure 5-13b) fans out over a range of frequencies (Figure 5-13c), because not every nucleus of a given type has exactly the same resonance frequency in an inhomogeneous field. The 180° pulse refocuses all the magnetization back onto the y axis after time 2τ, as in Figure 5-13e. Chemical-shift differences also may be eliminated in this fashion. Repetition of the 180° pulse every 2τ produces a

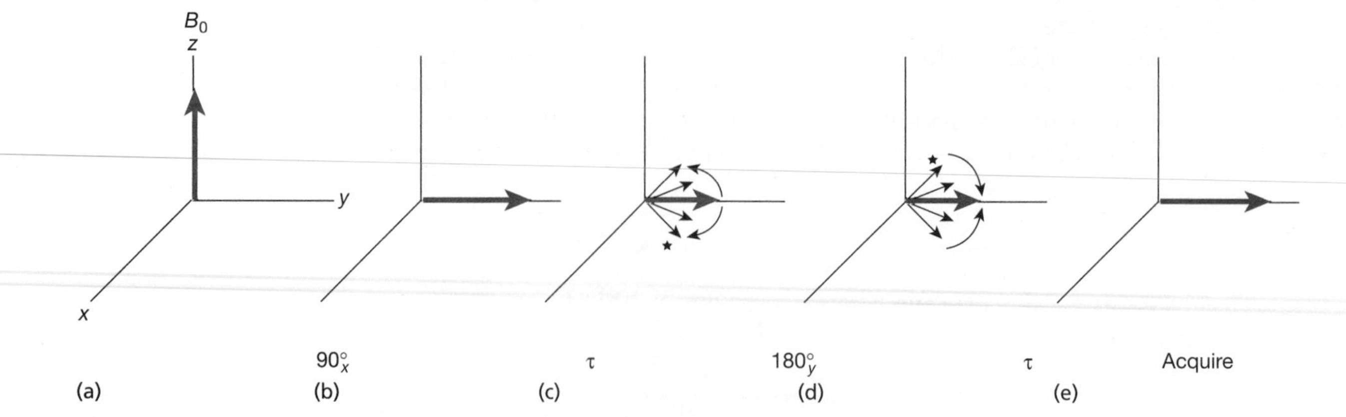

FIGURE 5-13 Spin–echo experiment to eliminate the effects of B_1 inhomogeneity.

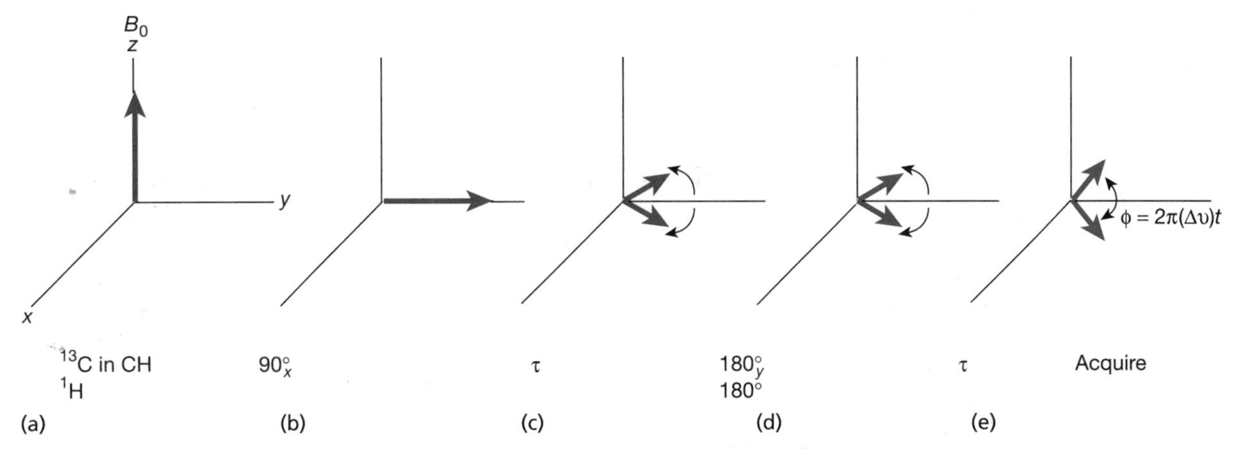

FIGURE 5-14 Pulse sequence that allows spin vectors to evolve to an arbitrary frequency separation ϕ.

train of peaks whose intensities die off with time constant T_2. This relaxation time provides a measure of spin–spin interactions alone, free from the usually dominating effects of field inhomogeneity. The notation T_2^* sometimes is used to denote transverse relaxation that includes the effects of inhomogeneity.

The Attached Proton Test. Although developed to measure T_2, this pulse sequence is able to improve resolution or eliminate coupling constants or chemical shifts after a single cycle. Moreover, it may be modified to achieve other effects. To obtain information about how many protons are attached to a carbon, the coupling information must be manipulated in a fashion different from that used, for example, in Figure 5-9. This is a double-resonance procedure, with pulses applied at both ^{13}C (B_1) and ^{1}H (B_2) frequencies (Figure 5-14 for a methine group, ^{13}C—^{1}H, with the reference frequency set at the ^{13}C resonance). The protons are subjected to a 180° pulse (B_2) at the same time that the carbons are subjected to their 180° pulse (B_1). The two pulses cancel each other, and the vectors from spin–spin coupling continue to diverge, as in Figure 5-14d. The cancelation occurs in the following fashion. Just as the ^{13}C spins are rotated by the 180° ^{13}C pulse between (c) and (d), the signs of the ^{1}H spins are reversed by the 180° ^{1}H pulse. At point (c), the $+\frac{1}{2}$ protons are precessing around the $+z$ axis and the $-\frac{1}{2}$ protons around the $-z$ axis, as in Figure 2-10. The 180° ^{1}H pulse (around either the x or the y axis) switches these identities. The nuclei that were precessing around the $+z$ axis ($+\frac{1}{2}$) are now precessing around the $-z$ axis ($-\frac{1}{2}$) and vice versa. Consequently, the identities of the protons have all been switched. Consider, for example, the faster-moving ^{13}C vector, which may have been associated with the $+\frac{1}{2}$ protons (Hβ). After the 180° ^{13}C rotation, the vector would start catching up to the y axis in the absence of the 180° ^{1}H pulse (as was the case in Figure 5-9d). In the presence of the pulse, however, this vector is now associated with $-\frac{1}{2}$ protons (Hα) and hence is still dropping behind the carrier frequency, as shown in Figure 5-14d. Thus, the effects of the two 180° pulses (^{13}C and ^{1}H) on the vectors derived from coupling cancel out, but those on inhomogeneity do not. The net effect is to achieve an improvement in homogeneity, while at the same time controlling the angle of divergence between the vectors that arise from spin–spin splitting. After the second τ period (total time $t = 2\tau$), these vectors have further diverged to an arbitrary angle ϕ, which is dependent on the difference in their frequencies ($\Delta\nu = J$) and on the total time since the initial 90° pulse, that is, $\phi = (\Delta\omega)t = 2\pi(\Delta\nu)t = 2\pi J(2\tau) = 4\pi J\tau$.

As an aside, in a homonuclear decoupling experiment such as ^{1}H{^{1}H}, a 180° pulse that follows the initial 90° pulse by a time τ has the same effect as the pair of 180° pulses in Figure 5-14. The homonuclear sequence 90°$-\tau-$180°$-\tau-$ Acquire results in refocusing of field inhomogeneities, but continued divergence of the two vectors. The 180° nonselective pulse not only rotates the directions of the vectors for the observed nucleus in the manner

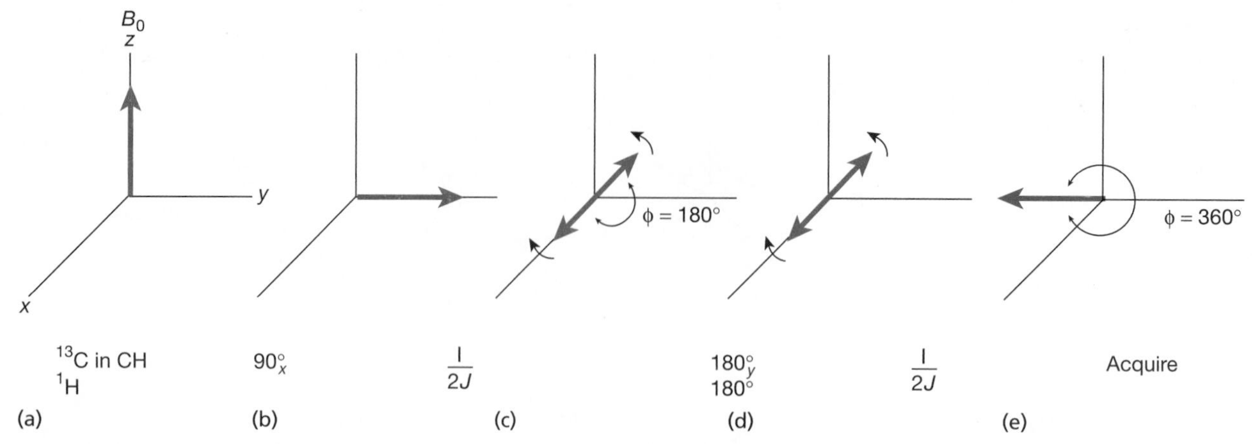

FIGURE 5-15 Pulse sequence for spectral editing of a methine (CH) resonance.

of Figure 5-14c, but also rotates all of the spins of the irradiated nucleus from above the xy plane to below it and vice versa, thus flipping the spins. For example, the pulse rotates the faster-moving vector for the observed nucleus around the y axis. Because of the switch of spins of the irradiated nucleus, it becomes the more slowly moving vector and hence continues to move away from the y axis. After time 2τ, the angle between the vectors is $\phi = 2\pi(\Delta\nu)(2\tau)$.

Returning to the spectral editing experiment begun in Figure 5-14, let us set the time τ to the specific value of $[2J(^{13}\text{C—}^{1}\text{H})]^{-1}$ [J is the coupling between the carbon and hydrogen in the methine group (Figure 5-15); the vectors diverge during one period τ until they are 180° apart, as in Figure 5-15d], since $\phi = 2\pi J(2J)^{-1} = \pi$. After the full pulse sequence ($\tau = 2\pi$), the angle between the vectors is $4\pi J(2J)^{-1}$, or 2π, as in Figure 5-15e. If the spectrum is sampled at this time, the result is a negative singlet, because the spins are all aligned along the negative y direction.

If the same experiment is carried out for a carbon attached to two protons (CH_2, Figure 5-16), the middle peak of the triplet remains on the y axis (coincident with the reference frequency, like the $+y$ vector in Figure 5-16c), and the diverging peaks now differ by $\Delta\nu = 2J$ (the distance between the outer peaks of the triplet). The value of $\phi = 2\pi(\Delta\nu)t$ after τ, then, is $2\pi(2J)(\tau)$, so that, for $\tau = (2J)^{-1}$, the angle is $4\pi J(2J)^{-1}$, or 2π, as for the $-y$ vector in Figure 5-16c and d. After 2τ, $\phi = 4\pi$, so that both vectors are coincident with the positive y axis, as in Figure 5-16e. Consequently, we get a positive peak for methylene carbons and a negative peak for methine protons. Quaternary carbons, of course, always give a positive peak, because, being unsplit, they remain on the positive y axis throughout these pulses. The value of $\Delta\nu$ for the four peaks of a methyl carbon is either J (for the middle two peaks) or $3J$ (for the outer two peaks), which results in refocusing all vectors onto the negative y axis after 2τ and hence produces a negative peak.

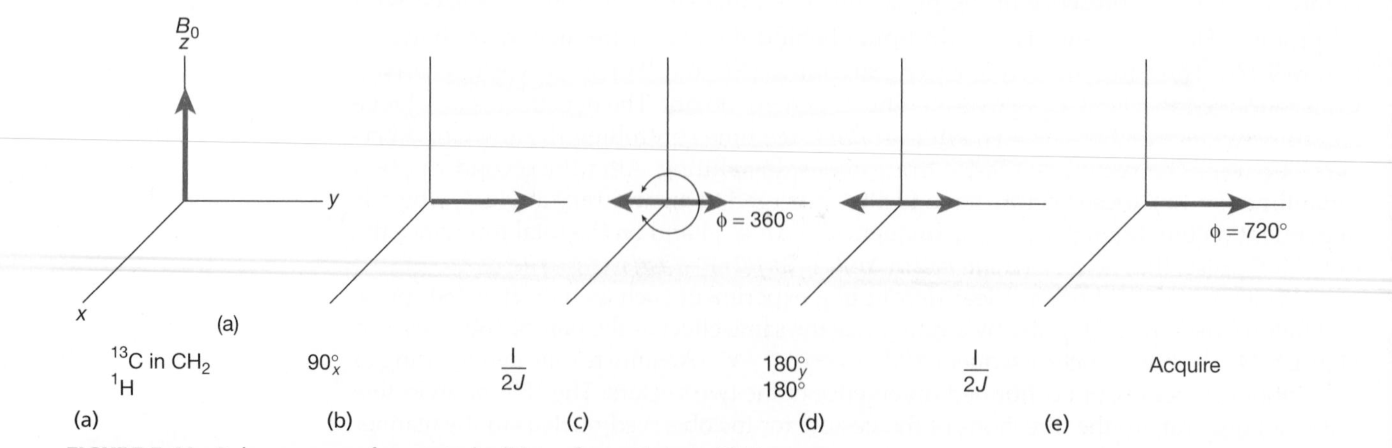

FIGURE 5-16 Pulse sequence for spectral editing of a methylene (CH_2) resonance.

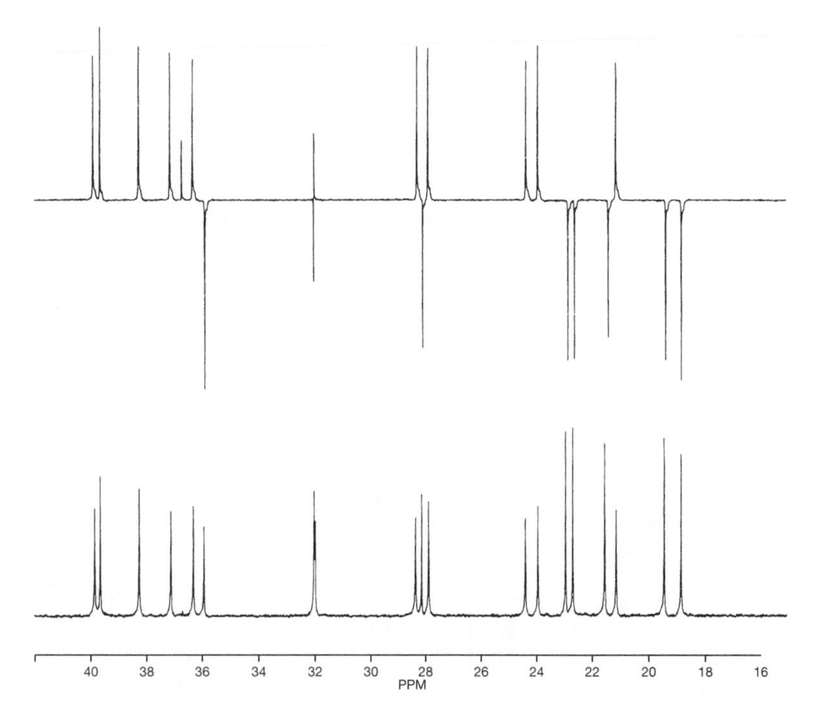

FIGURE 5-17
Lower: The normal proton-decoupled ^{13}C spectrum of cholesteryl acetate. Upper: The attached proton test (APT), phased so that CH_2 and quaternary carbons are positive and CH and CH_3 carbons are negative. (Reproduced with permission from A. E. Derome, *Modern NMR Techniques for Chemical Research*, Oxford, UK: Pergamon Press, 1987, p. 261.)

Figure 5-17 illustrates the result of the complete editing experiment for cholesteryl acetate, which gives negative peaks for CH and CH_3 resonances and positive peaks for C and CH_2. Proton irradiation during acquisition provides decoupling. This experiment affords a visual identification of the substitution pattern of all carbons, and has been called *J modulation* or the ***A****ttached* ***P****roton* ***T****est* (APT). It exists in many variants.

The DEPT Sequence. The procedure illustrated in Figure 5-17 does not distinguish between methine and methyl carbons, so alternative editing procedures have been developed that can provide separate spectra for each substitution pattern. Figure 5-18 illustrates the full set of spectra for the trisaccharide gentamicin, using the DEPT pulse sequence (defined in greater detail in the next section). The DEPT experiment often is presented alternatively as three, rather than four, spectra: the fully decoupled spectrum with all carbons as positive singlets, a spectrum with only CH carbons as positive singlets, and a spectrum with CH_3 and CH carbons positive and CH_2 carbons negative (quaternary carbons then are identified by difference from the complete spectrum). The various DEPT experiments probably are the most commonly used experiments today for ascertaining carbon substitution patterns, because (1) they depend less on the exact value of J than does the aforementioned APT experiment, (2) they provide signal enhancement (Section 5-6), and (3) they easily distinguish CH and CH_3 groups. An edited ^{13}C spectrum is a standard, and sometimes necessary, part of the structural analysis of complex organic molecules.

5-6 SENSITIVITY ENHANCEMENT

Some important nuclei, including ^{13}C and ^{15}N, have low natural abundances and sensitivities. Pulse sequences have been devised to improve the observability of these nuclei when they are coupled to another nucleus of high receptivity, usually a proton. Pulses are applied in such a way that the favorable population of the sensitive nucleus S is transferred to the insensitive nucleus I.

The INEPT Sequence. A common sequence developed by Freeman for this purpose is called INEPT, for ***I****nsensitive* ***N****uclei* ***E****nhanced by* ***P****olarization* ***T****ransfer*, as follows:

$$^{1}H(S) \quad 90^\circ_x - \frac{1}{4J} - 180^\circ_y - \frac{1}{4J} - 90^\circ_y$$

$$^{13}C(I) \quad 180^\circ - \frac{1}{4J} - 90^\circ_x - \text{Acquire}$$

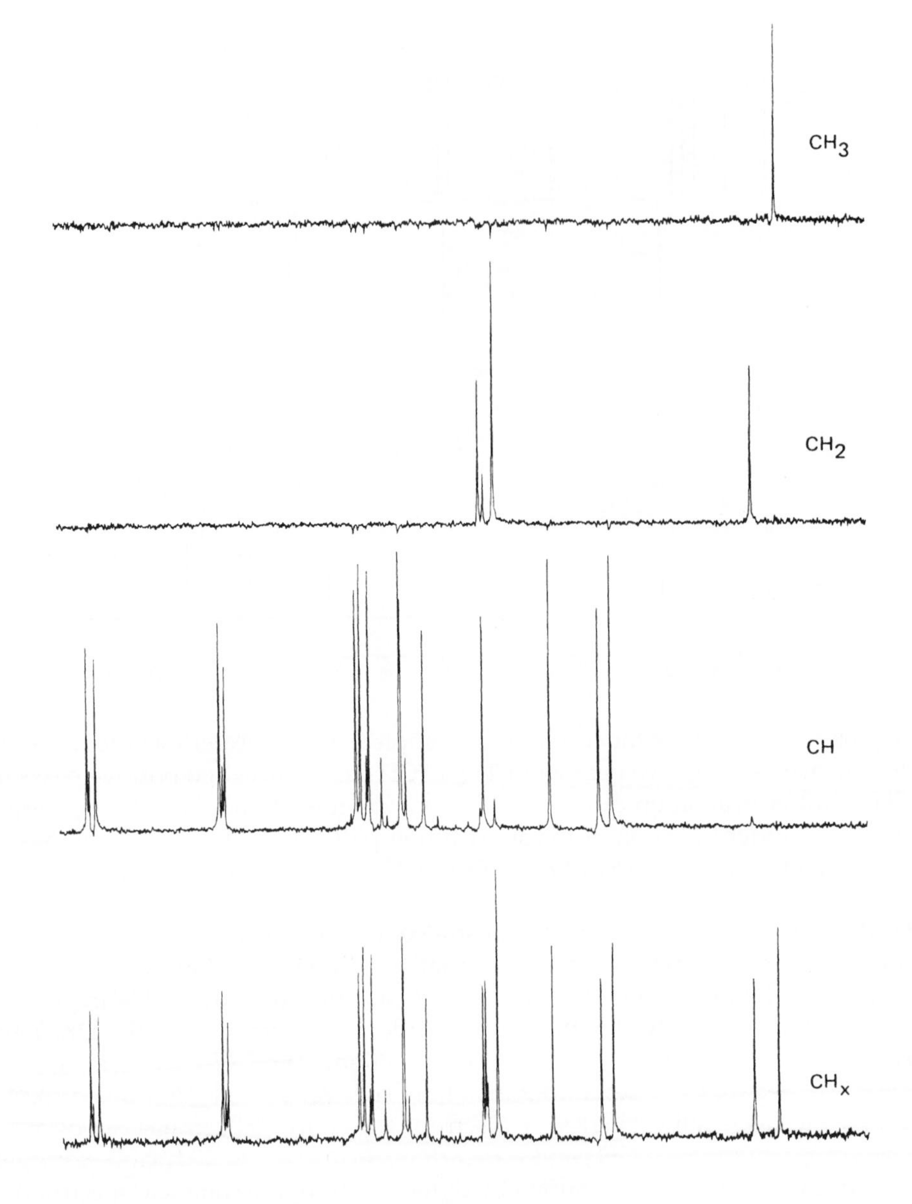

FIGURE 5-18 Spectral editing of the 75.6 MHz ^{13}C spectrum of the trisaccharide gentamicin by the DEPT sequence. The bottom spectrum contains resonances of all carbons with attached protons, and the ascending spectra are respectively of the methine, methylene, and methyl carbons. (Courtesy of Bruker Instruments, Inc.)

The pulses are closely related to the spin–echo experiment in Figure 5-14, with $\tau = (4J)^{-1}$ chosen to leave the ^1H and ^{13}C spin vectors 180° apart, or *antiphase*, after 2τ ($\phi = 2\pi J \cdot 2 \cdot (4J)^{-1} = \pi$). The additional 90° pulses after 2τ are necessary to place the vectors on the appropriate axes.

The results of the pulses are illustrated in Figure 5-19 for the case of two spins, for example, $^{13}C—^1H$. The first set of pulses is applied to the sensitive nucleus (^1H) to prepare it in the antiphase arrangement. The first 90° pulse moves the proton magnetization into the xy plane (Figure 5-19b). The simultaneous 180° pulses on both the proton and the carbon nuclei remove the effects of inhomogeneity, but allow the proton vectors to continue to diverge, as in Figure 5-14. After $\tau = (4J)^{-1}$, the protons are 90° apart (Figure 5-19c), and, after the second $(4J)^{-1}$ period, they are 180° apart (Figure 5-19d). The 90° pulse along the y direction rotates the proton vectors back onto the z axis (Figure 5-19e). Whereas in (a) the protons associated with both carbon spin up and carbon spin down are pointed in the $+z$ direction, in (b) the protons associated with carbon spin $+\frac{1}{2}$, or β, are pointed along the $+z$, but the protons associated with carbon spin $-\frac{1}{2}$, or α, are pointed along the $-z$ direction (or the reverse, depending on the sign of the $^{13}C—^1H$ coupling constant). This situation is termed *antiphase*.

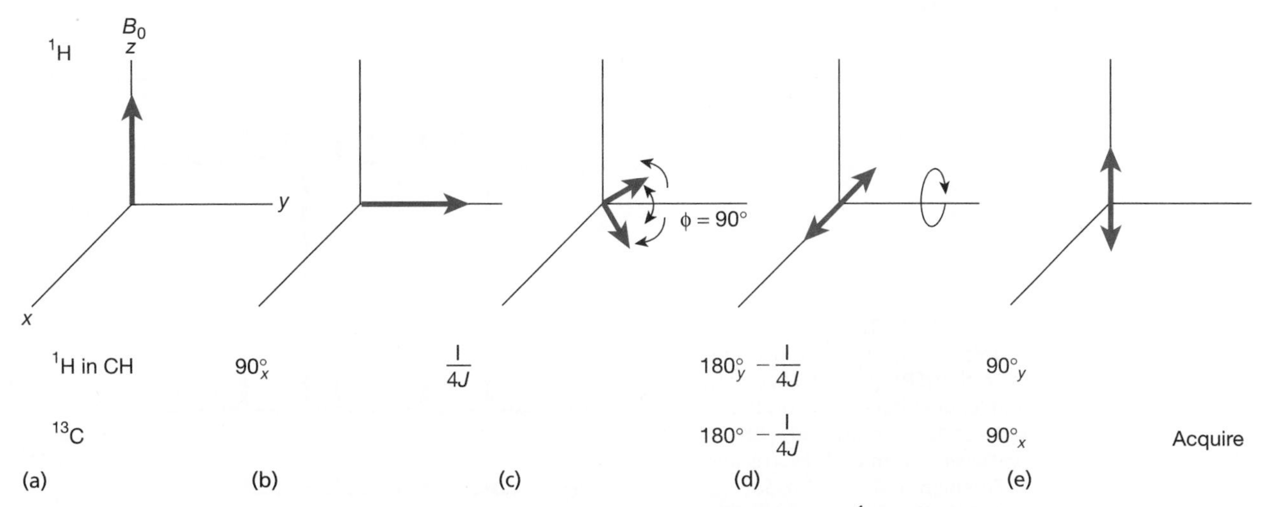

FIGURE 5-19 Pulse sequence for the INEPT experiment, showing the effects on the ^{1}H spin vectors.

The spin energy diagram after these proton pulses is compared with that for the normal two-spin system at the beginning of the sequence in Figure 5-20. The normal diagram on the left shows that the Boltzmann distributions result in more intense ^{1}H resonances ($\beta\alpha \rightarrow \alpha\alpha$ and $\beta\beta \rightarrow \alpha\beta$) than ^{13}C resonances ($\alpha\beta \rightarrow \alpha\alpha$ and $\beta\beta \rightarrow \beta\alpha$), as represented by the greater vertical length of the arrows for the ^{1}H transitions than for the ^{13}C transitions; each arrow goes from a lower to a higher state and hence represents absorption (a positive peak).

The antiphase INEPT arrangement of ^{1}H spin vectors on the right of Figure 5-20 means that two ^{1}H energy levels ($\alpha\alpha$ and $\beta\alpha$) are interchanged, so that the ^{1}H spin flip ($\beta\alpha \rightarrow \alpha\alpha$) gives a negative peak, while the other spin flip ($\beta\beta \rightarrow \alpha\beta$) still gives a positive peak. Thus, one ^{1}H signal is positive and the other is negative (antiphase). An examination of the carbon transitions in the INEPT diagram indicates that their Boltzmann distributions have increased to proton-like proportions (look at the vertical lengths of the arrows, although the representation is not proportional). In this fashion, protons have transferred polarization to carbons. According to the spin energy diagram at the right of the figure, the carbon vectors also are antiphase, since the carbon transition associated with proton spins $+\frac{1}{2}$, or β, is absorptive and must be pointed along the $+z$ direction, whereas the carbon transition associated with proton spins $-\frac{1}{2}$, or α, is emissive and must be pointed along the $-z$ direction. The situation for carbons is identical to that for protons in Figure 5-19e. The final carbon pulse, which is 90° along the x axis, places the antiphase vectors along the y direction for observation. Because of the antiphase relationship, one carbon transition ($\beta\beta$ to $\beta\alpha$) is positive (absorption) and one ($\alpha\beta$ to $\alpha\alpha$) is negative (emission).

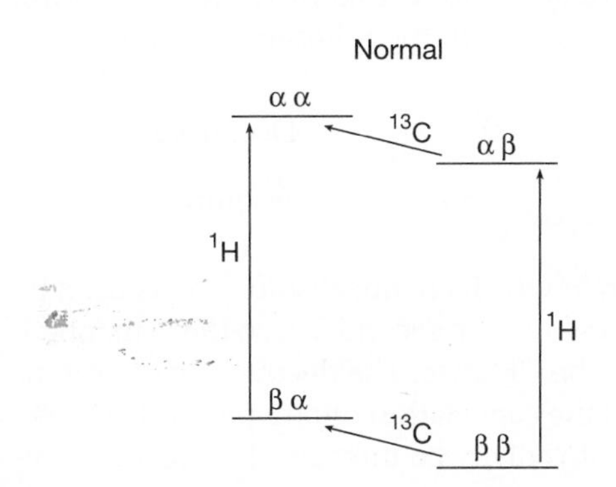

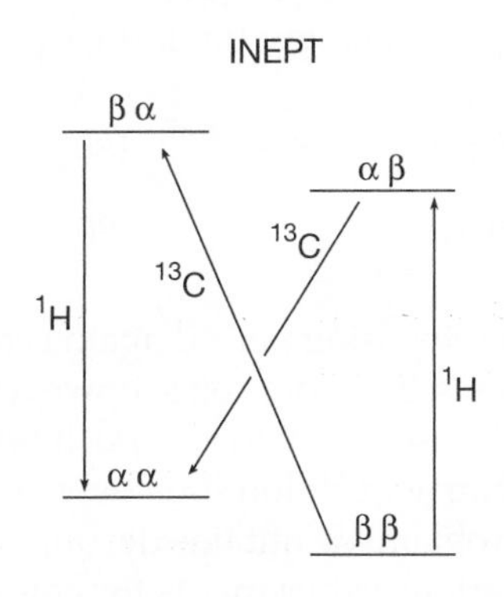

FIGURE 5-20 Spin states for a two-spin (^{13}C—^{1}H) system, normally (left) and after the INEPT pulse sequence (right).

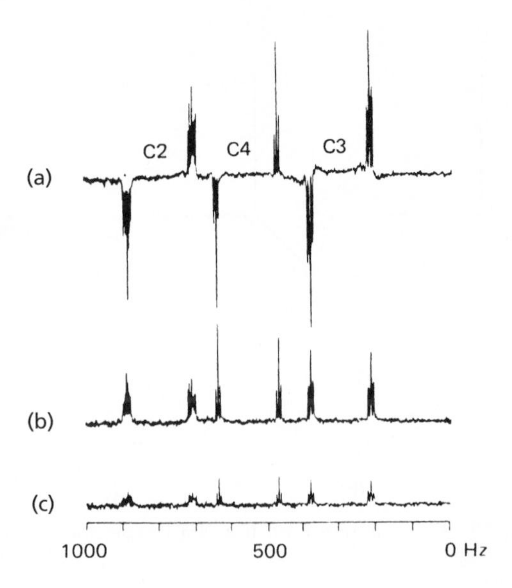

FIGURE 5-21 The proton-coupled ^{13}C spectrum of pyridine, (a) with INEPT, (b) with NOE only, and (c) unenhanced, all on the same scale. (Reproduced with permission from G. A. Morris and R. Freeman, *J. Am. Chem. Soc.*, **101**, 760 (1979). Copyright 1979 American Chemical Society.)

The INEPT sequence results in enhanced signals for the insensitive I nuclei, half of which give negative and half positive peaks for a CH group, such as pyridine at the top of Figure 5-21. For comparison, the figure includes the normal spectrum at the bottom and the spectrum in the middle with gated irradiation in order to obtain the NOE without decoupling. The INEPT spectrum clearly achieves a greater enhancement of sensitivity than does that produced with NOE alone. The maximum increment in intensity is $(1 + |\gamma_S/\gamma_I|)$ (absolute value of γ_S/γ_I) for INEPT, but that for the NOE $(1 + \eta_{max})$ is only $(1 + \gamma_S/2\gamma_I)$ (eq. 5-5) and can be positive or negative. The maximum enhanced intensity available from the INEPT experiment, analogous to that obtained from eq. 5-5 for the NOE experiment, is given by eq. 5-6.

$$I_{max}(\text{INEPT}) = I_0\left|\frac{\gamma_{irr}}{\gamma_{obs}}\right| \tag{5-6}$$

For $^{13}C\{^1H\}$, maximum increased intensities $(I/I_0 = 1 + \eta_{max})$ are 3.98 for INEPT and 2.99 for NOE. When the gyromagnetic ratio of the insensitive nucleus is negative, INEPT has an even greater advantage because of the subtractive factor present in the NOE expression. For $^{15}N\{^1H\}$, the INEPT and NOE factors are 9.87 and −3.94, respectively; for $^{29}Si\{^1H\}$, 5.03 and −1.52; and for $^{119}Sn\{^1H\}$, 2.68 and −0.34. Clearly, INEPT is significantly more effective in each case and is always positive.

Refocused INEPT. There is one apparent drawback to the INEPT experiment. Decoupling of the −1:1 pattern for each CH resonance would lead to precise cancelation and hence a null signal. As methylene triplets give −1:0:1 INEPT intensities and methyl quartets give −1:−1:1:1 intensities, both also would give null signals on decoupling. The *refocused* INEPT pulse sequence was designed to get around this problem and permit decoupling by repeating the INEPT pulses a second time in the following fashion:

$$\begin{array}{lllllllll} ^1\text{H(S)} & 90^\circ_x - \frac{1}{4J} - & 180^\circ_y - \frac{1}{4J} - & 90^\circ_y - \frac{1}{4J} - & 180^\circ - \frac{1}{4J} - \text{Decouple} \\ ^{13}\text{C(I)} & & 180^\circ - \frac{1}{4J} - & 90^\circ_x - \frac{1}{4J} - & 180^\circ_y - \frac{1}{4J} - \text{Acquire} \end{array}$$

The second refocusing period again is a spin echo in which chemical shifts are focused by the 180° pulses. The spin roles, however, are reversed in the second set, so that I magnetization is refocused back to two positive peaks for the CH case. The decoupling of protons during carbon acquisition thus does not result in the cancelation of any peaks. The spectrum that is obtained contains decoupled peaks with enhanced intensity. Figure 5-22 compares the various experiments for chloroform.

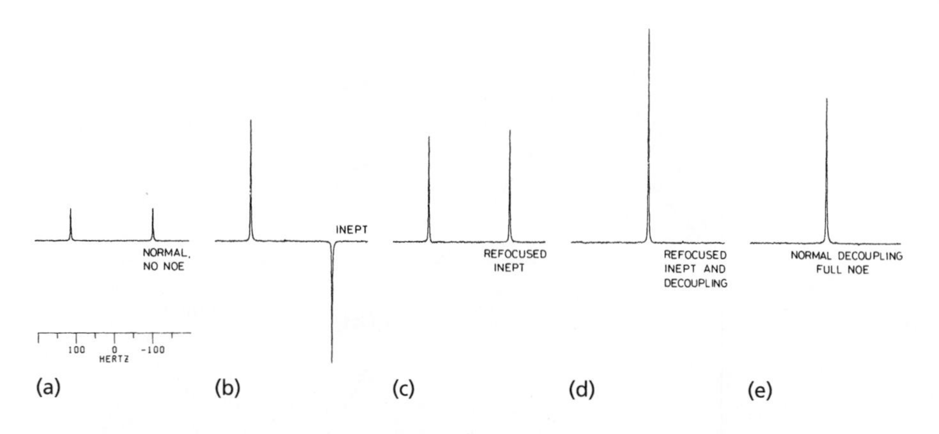

FIGURE 5-22 The ^{13}C spectrum of chloroform (a) without double irradiation, (b) with ^1H irradiation to achieve the INEPT enhancement, (c) with ^1H irradiation to achieve refocused INEPT enhancement, (d) with ^1H irradiation to achieve refocused INEPT enhancement and decoupling, and (e) with normal decoupling to achieve only the NOE. (Reproduced with permission from A. E. Derome, *Modern NMR Techniques for Chemistry Research*, Oxford, UK: Pergamon Press, 1987, p. 137.)

Spectral Editing with Refocused INEPT. The value of $(2J)^{-1}$ for the total period between the last 90° pulse and acquisition (sometimes called Δ_2 to distinguish it from the period Δ_1 or 2τ, the first and last 90° pulses) is appropriate only for the methine fragment CH. For methylene (CH_2) and methyl (CH_3) groups, the vectors do not refocus, so that decoupling would still result in a canceled signal. Alternative values of Δ_2, however, can lead to improved refocusing, with Δ_2 for an arbitrary CH_n fragment given by eq. 5-7.

$$\Delta_2 = (1/\pi J)\sin^{-1}\left(\frac{1}{\sqrt{n}}\right) \tag{5-7}$$

The respective optimum values of Δ_2 for CH, CH_2, and CH_3 are $(2J)^{-1}$, $(4J)^{-1}$, and $\approx (5J)^{-1}$. Thus $\approx (3.3J)^{-1}$ represents a compromise value that yields enhanced, but not optimal, intensities for all substitution patterns under decoupling conditions. In the absence of decoupling, phase differences within a collection of CH, CH_2, and CH_3 resonances would result in peak distortions.

Because the choice of $\Delta_2 = (2J)^{-1}$ in the decoupled, refocused INEPT experiment leads to completely refocused doublets, but antiphase triplets and quartets, this particular experiment with decoupling produces a subspectrum that contains only methinyl resonances. Values of Δ_2 also can be selected to optimize the intensities of methylene and methyl resonances. The idea can be depicted graphically by defining an imaginary angle $\theta = \pi J \Delta_2$. Signal intensities then are found to be proportional to $\sin\theta$ for CH, $\sin(2\theta)$ (or $2\sin\theta\cos\theta$) for CH_2, and $3\sin\theta\cos^2\theta$ for CH_3. Thus, when $\theta = \pi/2$ [and $\Delta_2 = (2J)^{-1}$], the CH signal is optimized and the other signals go to zero. For all other values of θ, the spectrum contains varying proportions of all substitution types. Figure 5-23 illustrates

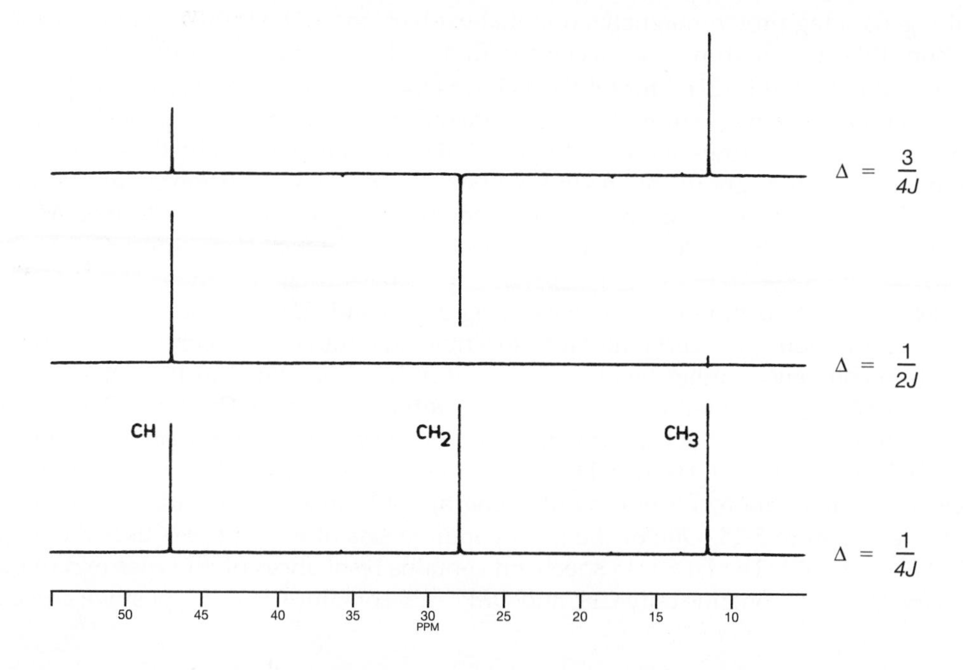

FIGURE 5-23 Intensities of the carbon resonances of an imaginary molecule containing one CH, one CH_2, and one CH_3 under varying values of Δ_2 in the refocused INEPT experiment. (Reproduced with permission from A. E. Derome, *Modern NMR Techniques for Chemical Research*, Oxford, UK: Pergamon Press, 1987, p. 143.)

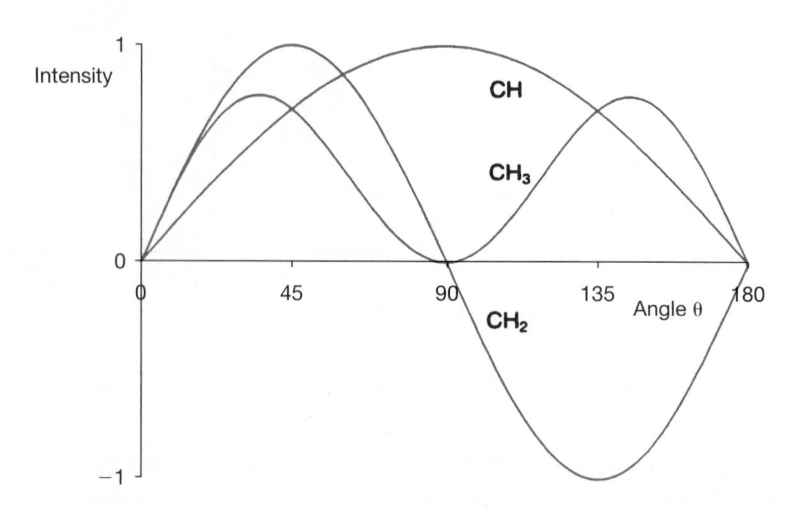

FIGURE 5-24 Variation of signal intensities for CH, CH_2, and CH_3 as a function of $\theta = \pi J \Delta_2$ in the refocused INEPT experiment. (Reproduced from T. D. W. Claridge, *High-Resolution NMR Techniques in Organic Chemistry*, Amsterdam: Pergamon Press, 1999, p. 138.)

the experiment for three values of θ ($\pi/4$, $\pi/2$, and $3\pi/4$) that correspond to $\Delta_2 = 3(4J)^{-1}$, $(2J)^{-1}$, and $(4J)^{-1}$, respectively. Linear combinations of these spectra can lead to edited spectra that contain only methylene or only methyl resonances. Figure 5-24 is a plot of the signal intensities for the three types of carbon as a function of the angle $\theta = \pi J \Delta_2$. The spectra shown were taken as cuts at θ = 135°, 90°, and 45°.

DEPT Revisited. A comparison of the preceding three INEPT spectra allows the multiplicity of all protonated carbon resonances to be determined, albeit with intensities that are not optimized. Section 5-5 describes the APT, which does not distinguish CH from CH_3. The DEPT sequence provides an editing technique that suffers from the drawbacks of neither of these other methods and moreover is less sensitive to experimental imperfections, such as the exact value of J. Already mentioned as the method of choice for spectral editing, DEPT (***D**istortionless **E**nhancement by **P**olarization **T**ransfer*) is similar to refocused INEPT. There are a pair of τ periods $[=(2J)^{-1}]$ followed in the proton channel by a single variable pulse θ, an angle corresponding to that previously defined ($\pi J \Delta_2$, in which Δ_2 corresponds to two τ periods):

$$\text{S} \quad 90^\circ_x - \tfrac{1}{2J} - 180^\circ_y - \tfrac{1}{2J} - \theta_y \quad - \tfrac{1}{2J} - \text{Decouple}$$
$$\text{I} \quad 90^\circ_x - \tfrac{1}{2J} - 180^\circ_x - \tfrac{1}{2J} - \text{Acquire}$$

The DEPT and refocused INEPT sequences begin in a similar fashion, with the 90° ^{1}H (S) pulse generating proton magnetization that evolves under the influence of coupling to carbon. Whereas the time τ between the first and second proton pulses is $(4J)^{-1}$ for refocused INEPT, it is $(2J)^{-1}$ for DEPT, as it was for APT. The second (180°) ^{1}H pulse refocuses proton chemical shifts. The simultaneous initial 90° ^{13}C pulse generates carbon magnetization and brings about a situation that cannot be followed by the vector model we have used throughout this textbook. As both proton and carbon magnetizations, linked by the C–H coupling, are evolving together, the phenomenon is termed ***M**ultiple **Q**uantum **C**oherence* (MQC), or, more specifically, ***H**eteronuclear **M**ultiple **Q**uantum **C**oherence* (HMQC). In essence, the proton and carbon magnetizations have become pooled. The MQC continues to evolve during the second $(2J)^{-1}$ period. The final proton pulse, of duration θ, converts the MQC to single quantum carbon coherence. (Multiple quantum coherence cannot be observed, as it induces no signal in the detection coil, so it must be transformed back into *single quantum coherence*.) The final $(2J)^{-1}$ period allows the development of carbon magnetization, with a dependence on the number of attached protons (CH, CH_2, or CH_3) determined by the value of θ. As with refocused INEPT, the modulation of θ, now a pulse length, results in a series of edited spectra such as those in Figure 5-18. One of the most common sets of experiments uses the angles 45°, 90°, and 135°. The DEPT-45 spectrum contains resonances of all types except quaternary, DEPT-90 contains only CH, and DEPT-135 contains CH/CH_3 positive and CH_2

negative, analogous to the plot in Figure 5-24, readily permitting an assignment of each type of substitution. Spectral subtraction with some loss of signal is required to obtain the fully edited spectra illustrated in Figure 5-18. The term "distortionless" was applied because the initial set of pulses (up to the first 2τ) results not in a combination of positive and negative peaks, but rather in positive 1:1 doublets, 1:2:1 triplets, and 1:3:3:1 quartets in the absence of decoupling.

The INEPT and DEPT sequences assume that coupling between the I and S nuclei is dominant, so that other couplings must be negligibly small. For one-bond $^{13}C—^{1}H$ couplings, this assumption holds, as all $^{1}H—^{1}H$ couplings are much smaller. If polarization is to be transferred from two- or three-bond $^{13}C—^{1}H$ couplings, however, the homonuclear couplings no longer are small in comparison. This situation is more likely to occur when attempts are being made to transfer polarization from protons to silicon, nitrogen, or phosphorus. Because Si—H, N—H, and P—H bonds are relatively uncommon (compared with C—H), recourse must be made to longer-range coupling constants, with attendant difficulties.

5-7 CARBON CONNECTIVITY

The one-bond $^{13}C—^{13}C$ coupling potentially contains a wealth of structural information, as it indicates and characterizes carbon–carbon linkage. Unfortunately, only 1 in about 10,000 pairs of carbon atoms contains two ^{13}C atoms and hence displays a $^{13}C—^{13}C$ coupling in the ^{13}C spectrum. These resonances can be detected as very low intensity satellites on either side of the centerband that is derived from molecules containing only isolated ^{13}C atoms. For bonded pairs of ^{13}C atoms, ^{1}J is about 30–50 Hz, and the satellites are separated from the centerband by half that amount. Coupling also may be present over two or three bonds (^{2}J, ^{3}J) in the range of about 0–15 Hz. Not only are these satellites low in intensity and possibly obscured by the centerband, but, in addition, spinning sidebands, impurities, and other resonances may get in the way.

The pulse sequence INADEQUATE (*Incredible Natural Abundance DoublE QUAntum Transfer Experiment*) was developed by Freeman to suppress the usual (single quantum) resonances and exhibit only the satellite (double quantum) resonances. The pulse sequence is $90^\circ_x - \tau - 180^\circ_y - \tau - 90^\circ_x - \Delta - 90^\circ_\phi$. The homonuclear 180° pulse refocuses field inhomogeneities, but allows the vectors from different $^{13}C—^{13}C$ coupling arrangements to continue to diverge (Section 5-5). If the carrier frequency coincides with the centerband of a carbon resonance, the centerband spins remain on the y axis after the first 90° pulse. The delay time τ is set to $(4J)^{-1}$, so that the vectors for the two satellites from the coupled $^{13}C—^{13}C$ system diverge by 180° after $2\tau\ [=2\pi(\Delta\nu)t = 2\pi J(2/4J) = \pi]$ and lie respectively on the $+x$ and $-x$ axes. The second 90°_x pulse then rotates the centerband spins to the $-z$ axis, but leaves the satellite spins aligned along the x axis. Thus, the centerband signal is not available for detection in the xy plane, but the satellites are.

This pulse sequence is another example of multiple quantum coherence. After the second 90° pulse, the coupled pairs of ^{13}C nuclei evolve together. (Note that each ^{13}C pair is an isolated AX system because of the natural abundance of ^{13}C.) During the period Δ, homonuclear double quantum coherence evolves as the sum of the Larmor frequencies of the two coupled spins. In the two-dimensional variant (Chapter 6), the constant period Δ becomes a variable period. The final 90° pulse reconverts multiple to single quantum coherence for observation. The phase of the final 90° pulse (90°_ϕ) is cycled through a series of directions represented by ϕ ($+x, +y, -x, -y$). The vector diagrams used throughout this book illustrate only the coherence of the spins of a single nucleus. Spins in the xy plane are said to be coherent when they have an ordered relationship between their phases, so that they all precess around one axis and can be depicted by a vector along that direction. Spins rotating with random phase in the xy plane are said to be incoherent. The simultaneous coherence of two spins, as created in the INADEQUATE experiment, is not well depicted by the vector diagrams, so the reason for the final ^{13}C 90° pulse is not well represented.

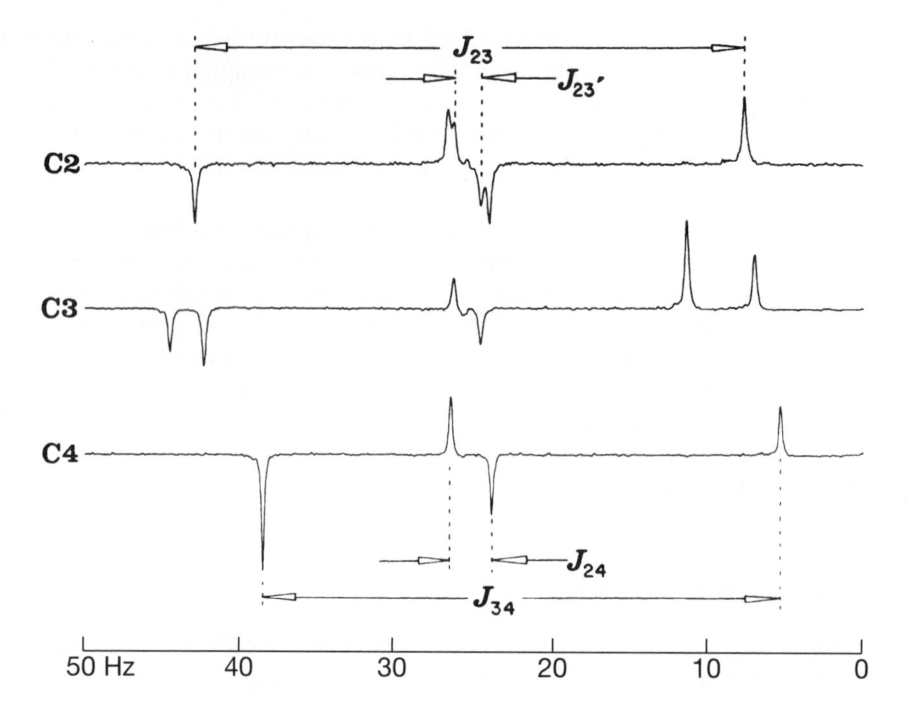

FIGURE 5-25 The one-dimensional INADEQUATE spectrum for the carbons of piperidine. (Reproduced with permission from A. Bax, R. Freeman, and S. P. Kempsell, *J. Am. Chem. Soc.*, **102**, 4849 [1980]. Copyright 1980 American Chemical Society.)

Figure 5-25 contains the INADEQUATE spectrum for piperidine. The double-quantum (satellite) peaks are antiphase, so each $^{13}C—^{13}C$ coupling constant is represented by a pair of peaks—one up, one down (+1:−1). The spectrum for C-4 of piperidine thus contains two such doublets: a large one for $^1J_{34}$ and a small one for $^2J_{24}$. For C-3, there are two large doublets, because the one-bond couplings $^1J_{23}$ and $^1J_{34}$ to the adjacent carbons are slightly different. There also is a small $^3J_{23'}$ between C-2 and the nonadjacent C-3. The spectrum for C-2 shows $^1J_{23}$, $^2J_{24}$, and $^3J_{23'}$.

Although more distant couplings are observable, the most important are the one-bond couplings, which vary slightly for every carbon–carbon bond. Thus, a match of $^1J(^{13}C—^{13}C)$ for any two carbons strongly suggests that they are bonded to each other. Even in complex molecules, there is sufficient variability of couplings that INADEQUATE can be used to map the complete connectivity of the carbon framework, provided that it is not broken by a heteroatom. The major drawback to the INADEQUATE experiment is its extremely low sensitivity, as it uses only 0.01% of the carbons in the molecule. The two-dimensional version is discussed in Section 6-4.

5-8 PHASE CYCLING, COMPOSITE PULSES, AND SHAPED PULSES

We have used 90° and 180° pulses extensively to carry out a variety of experiments. In each case, it is important that the length of the pulse provide the desired angle of rotation accurately. Various artifacts can arise because of imperfections in the pulses. Figure 5-26 illustrates the effect on the inversion recovery experiment ($180^\circ_x - \tau - 90^\circ_x$, Figure 5-1) used to determine T_1, but with the initial inverting pulse not quite 180°. The magnetization after the pulse is slightly off the z axis (Figure 5-26b), so there is a small amount of transverse (xy) magnetization present at the start of the τ period. (Only the y component is shown in Figure 5-26b.) After the period τ, the z magnetization has decreased through T_1 relaxation, and the component of magnetization in the xy plane caused by the pulse imperfection persists (Figure 5-26c). Following the final 90° pulse, the z magnetization is moved into the xy plane for detection (Figure 5-26d). The pulse imperfection in the drawing causes a reduction in intensity, but the spectral phase also can be altered. Almost certainly, there would be errors in the 90° pulse as well, but these are not under consideration here.

Such errors may be eliminated largely by alternating the relative phase of the 180° pulse. The result of an inversion in which the 180° rotation is carried out counterclockwise instead of clockwise about the x axis ($-x$, or −180°) is illustrated in Figure 5-26f.

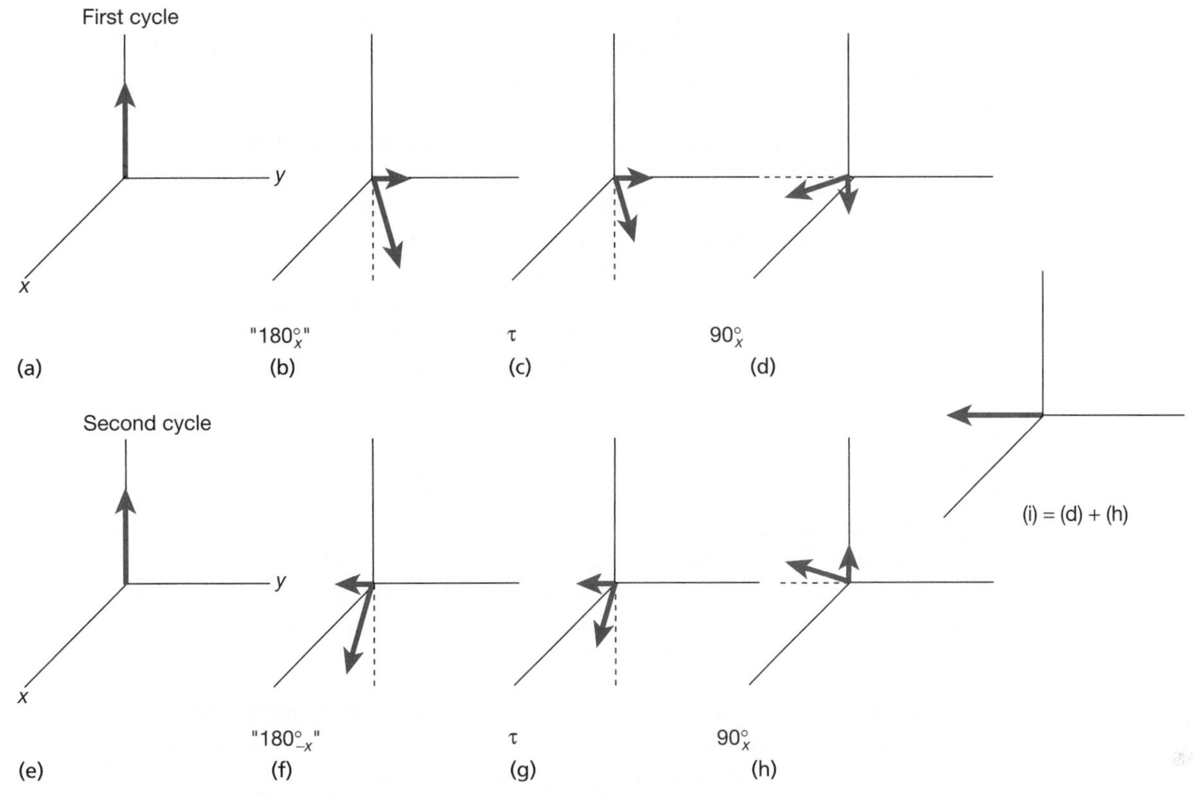

FIGURE 5-26 Phase cycling in the inversion recovery experiment.

The unwanted transverse magnetization now appears along the $-y$ axis. After time τ (Figure 5-26g) and the final 90° pulse (Figure 5-26h), the imperfection is still present, but now has the opposite effect on the z magnetization from that in Figure 5-26d. When the two results are added, as in Figure 5-26i, the effect of the imperfection cancels out. The pulse therefore is alternated between x and $-x$. Such a procedure is called *phase cycling*, a technique that permeates modern NMR spectroscopy.

Phase cycling has improved procedures for broadband heteronuclear decoupling. As described in Section 5-3, modern methods use repeated 180° pulses rather than continuous irradiation. Imperfections in the 180° pulse would accumulate and render the method unworkable. Consequently, phase-cycling procedures have been developed to cancel out the imperfections. The most successful to date is the WALTZ method of Freeman, which uses the sequence 90°_x, 180°_{-x}, 270°_x in place of the 180° pulse $(90 - 180 + 270 = 180)$, with significant cancelation of imperfections. The expanded WALTZ-16 sequence cycles through various orders of the simple pulses and achieves a very effective decoupling result. The sequence 90°, −180°, 270° was given the shorthand notation $1\,\bar{2}\,3$ (1 for 90°, $\bar{2}$ for −180° in which the bar denotes the negative direction, and 3 for 270°). The implied rhythm of the sequence suggested its name.

A third example of phase cycling is used to place the reference frequency in the middle of the spectrum, instead of off to one side. As described heretofore, the NMR experiment is sensitive only to the difference $\Delta\omega$ between a signal and the reference frequency. This situation necessitates placing the reference frequency to one side of all the resonances, so that there is no confusion of two signals that are respectively at a higher and a lower frequency than the reference frequency by exactly the same amount ($+\Delta\omega$ and $-\Delta\omega$). Such sideband detection, however, always contains signals from noise on the signal-free side of the reference. Placement of the reference in the middle of the spectrum avoids this unnecessary noise, but requires a method for distinguishing between signals with $+\Delta\omega$ and those with $-\Delta\omega$. *Quadrature detection* accomplishes this task by splitting the signal in two and detecting it twice, using reference signals with the same frequency, but 90° out of phase. Signals with the same absolute value of $\Delta\omega$, but opposite signs, are distinguished in the experiment (in terms of obtaining θ by knowing both $\sin\theta$ and $\cos\theta$, which are 90° out of phase).

Systematic errors, however, can arise if the two reference frequencies are not exactly 90° out of phase. The resulting signal artifacts, called *quad images*, can appear as low-intensity peaks. The CYCLOPS (***CYCL**ically **O**rdered **P**hase **S**election*) sequence involves four steps that move the 90° pulse and the axis of detection from $+x$ to $+y$ to $-x$ to $-y$ and change the way the two receiver channels are added, with the result that imperfections in the phase difference cancel out.

Phase cycling not only can remove artifacts from pulse or phase imperfections, but also can assist in the selection of coherence pathways. The inversion recovery experiment can be described with a slightly different vocabulary to illustrate this process. When spins are aligned entirely along the z axis, the order of coherence is said to be zero (phases around the xy plane are random). An exact 90° pulse creates maximum single quantum coherence by lining up the spins along, for example, the y direction. Phase cycling in the inversion recovery experiment (Figure 5-26) removes undesired single quantum coherence (transverse or xy magnetization) and leaves coherence of order zero until the end of the τ period, at which time the final 90° pulse creates single quantum coherence. In this way, phase cycling selects the desired degree of coherence. Double quantum coherence, involving the relationship between two spins, is not well illustrated in these vector diagrams. The INADEQUATE experiment involves the selection of double over single quantum coherence (elimination of the centerband and retention of the satellites), in part through phase cycling in the final 90° pulses, whose subscript ϕ refers to a sequence of pulses with different phases.

Composite Pulses. Imperfections in pulses also may be corrected by using *composite pulses* instead of single pulses. The 180° pulse that inverts longitudinal magnetization for the measurement of T_1 or other purposes may be replaced by the series 90°_x, 180°_y, 90°_x, which results in the same net 180° pulse angle, but reduces the error from as much as 20% to as little as 1%. As Figure 5-27 shows, the 180° pulse compensates for whatever imperfection existed in the 90° pulse. Normally, 180° is taken as double the optimized 90° pulse, so errors in one are present in the other. The three components of the WALTZ-16 method (90°_x, 180°_{-x}, 270°_x) also constitute a composite pulse for 180°_x.

Shaped Pulses. For the most part, pulses have been generated by applying rf energy equally over the entire frequency range, with a short duration on the order of microseconds. Such excitations are sometimes referred to as *hard pulses*, in distinction to pulses that require selective excitation, that is, excitation over a restricted frequency range. Selective excitation has been mentioned on several occasions. It is useful, for example, in the saturation transfer experiment (Section 5-2) and in the suppression of specific unwanted peaks (Section 5-1). Frequency selection within two-dimensional spectra (Chapter 6) results in a reduction in dimensionality, so that effects at a single frequency can be examined in detail. (A one-dimensional cut of a two-dimensional spectrum offers the twin advantages of reduced experimental time and decreased storage needs.)

The procedure for producing a selective pulse is to reduce the rf power (B_1) so that the effective frequency range also is reduced. To counter the reduction in power and still achieve the required tip angle, the duration of the pulse is increased, typically into the millisecond range. The simplest such *soft pulses* would have rectangular shapes, that is, from zero intensity instantly up to full intensity for a period of milliseconds and then back to zero intensity, similar to the shapes of the hard pulses in our vector diagrams. Unfortunately, such a pulse shape generates wiggles, or feet, on the signal (Figure 5-28a).

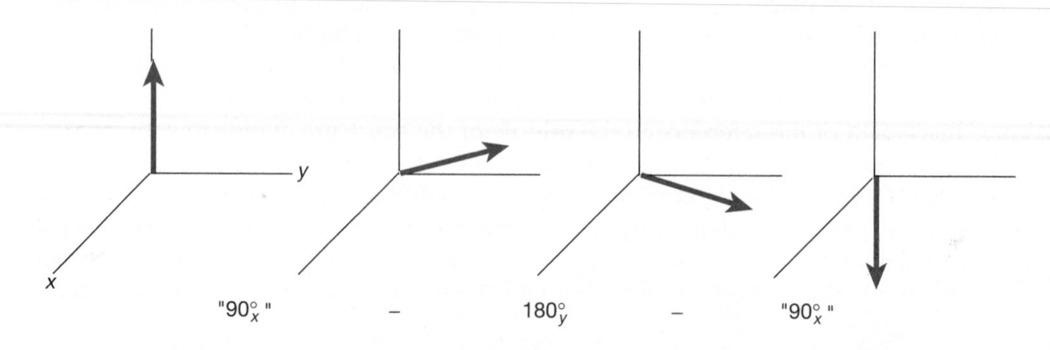

FIGURE 5-27 A composite pulse equivalent to a single 180° pulse.

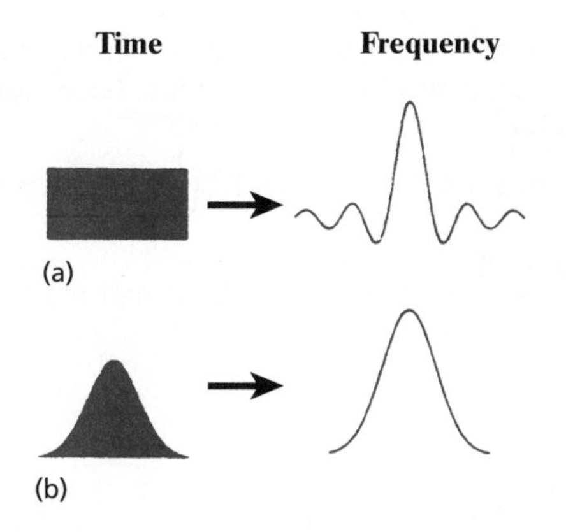

FIGURE 5-28 (a) The result of Fourier transformation of a low-power rectangular pulse. (b) The result for a shaped Gaussian pulse. (Reproduced from T. D. W. Claridge, *High-Resolution NMR Techniques in Organic Chemistry*, Amsterdam: Pergamon Press, 1999, p. 349.)

By a process called *apodization* (meaning "no-feet-ization"), these wiggles may be removed by smoothing off the edges of the peaks (Figure 5-28b).

Such excitations have been called *shaped pulses*, and a considerable effort has been expended to optimize their shapes. A simple Gaussian shape is a considerable improvement over a rectangular pulse, but is not entirely effective in achieving an optimal peak shape. The use of more elaborate mathematical functions improves the shape of the signal, although with increasing loss of intensity. The BURP (***B****and selective*, ***U****niform* ***R****esponse*, ***P****ure phase*) family utilizes an exponentially dependent sinusoidal series of Gaussians with considerable success in a variety of situations (EBURP for 90°, REBURP for 180°).

An early alternative to soft pulses was the DANTE (***D****elays* ***A****lternating with* ***N****utation for* ***T****ailored* ***E****xcitation*) experiment, which used a sequence of short, hard pulses of angle $\alpha = 90°$, followed by a fixed delay τ to achieve selective excitation. [Thus, the pulse sequence is $(\alpha - \tau)_n$.] Nuclei that are on resonance are driven eventually to the y axis and hence are selected, whereas those more removed from the frequency range are not affected. The sequence of hard pulses can achieve a result similar to that of soft pulses and even can be shaped by modulating the duration of the pulse lengths, but DANTE pulses lead to spectral artifacts not created by soft pulses, such as unwanted sidebands.

Problems

5-1 Give the spectral notation (AB, ABX, etc.) for the following substituted ethanes, first at slow C—C rotation, then at fast rotation. Draw all stable conformations. The spectral notation for each frozen form gives the slow-rotation answer. Then imagine free rotation about the C—C bond. The identity of certain protons may average for the fast-rotation answer.

(a) CH_3CCl_3
(b) CH_3CHCl_2
(c) CH_3CH_2Cl
(d) $CHCl_2CH_2Cl$

5-2 Ring reversal in 7-methoxy-7,12-dihydropleiadene (see accompanying figure) can be frozen out at −20°C. Two conformations are observed, in the ratio 2:1. When the high-frequency (low-field) part of the 12-CH_2 AB quartet in the minor isomer is doubly irradiated, the intensity of the 7-methine proton is enhanced by 27%. Double irradiation of the same proton in the major isomer has no effect on the spectrum. What are the two conformational isomers and which is more abundant?

H H
12
7
H OCH_3

5-3 Permethyltitanocene reacts with an excess of nitrogen below −10°C to form a 1:1 complex:

$$[C_5(CH_3)_5]_2Ti + N_2 \rightleftharpoons [C_5(CH_3)_5]_2TiN_2$$

The methyl resonance of the complex is a sharp singlet above −50°C. Below −72°C, the resonance splits reversibly into two peaks of not quite the same intensity. If the nitrogen molecule is doubly labeled with ^{15}N, the ^{1}H-decoupled ^{15}N spectrum contains a singlet and an AX quartet [$J(^{15}N—^{15}N) = 7$ Hz] of not quite the same overall intensity at low temperatures. Explain these observations in terms of structures.

5-4 **(a)** The resonance of the methylene protons of $C_6H_5CH_2SCHClC_6H_5$ in $CDCl_3$ is an AB quartet at room temperature. Why?

(b) The AB spectrum coalesces at high temperatures to an A_2 singlet with a $\Delta G^{\ddagger}$ of 15.5 kcal mol^{-1}. The rate is independent of concentration in the range 0.0190–0.267 M. Explain in terms of a mechanism.

5-5 The ^{1}H spectrum of the following molecule contains resonances from two isomers at room temperature.

H
2′

(a) The spectrum of isomer A contains the resonance of H2′ at δ 6.42, and that of isomer B contains the same resonance in the region δ 7–8. What can you say about the conformations of A and B?

(b) The sample crystallizes only as isomer A. Dissolution of these crystals, however, produces the spectra of both isomers, A and B. What can you say about the barrier for the equilibrium A $\rightleftharpoons$ B? When CH_3 on the double bond is replaced by —CMe_2OH, crystallization still produces only A, but redissolution of the crystals also yields A only and none of B. What can you say about the A $\rightleftharpoons$ B barrier in this compound?

5-6 No coupling is observed between CH_3 and ^{14}N in acetonitrile ($CH_3—C\equiv N$), but there is a coupling in the corresponding isonitrile ($CH_3—N\equiv C$). Explain. This is not a distance effect. The phenomenon is general for nitriles and isonitriles.

5-7 Comment on the following ^{14}N linewidths:

$^{+}NMe_4$	<0.5 Hz		
$MeNO_2$	14	HN (pyrrole ring)	172
Me_3N	77	Aniline ($C_6H_5NH_2$)	1300

5-8 The inversion-recovery ($180° - \tau - 90°$) spectral stack for the aromatic carbons of *m*-xylene is given below. Assign the resonances and explain the order of T_1 (look at the nulls).

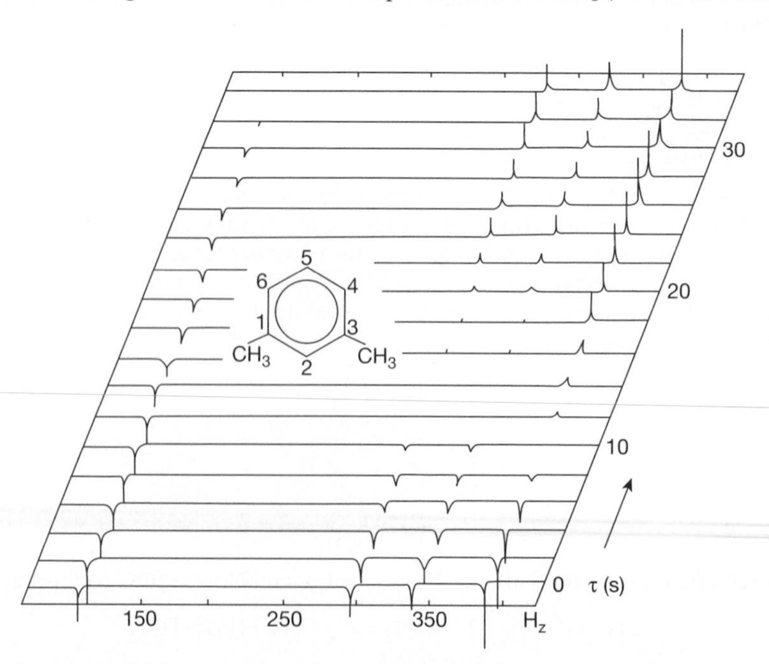

Reprinted with permission of R. Freeman from W. Bremser, H. D. W. Hill, and R. Freeman, *Messtechnik.*, **79**, 14 (1971).

5-9 1-Decanol has the following carbon T_1 values (in s). Explain the order.

CH_3—	CH_2—	CH_2—	CH_2—	$(CH_2)_5$—	CH_2—OH
3.1	2.2	1.6	1.1	0.8-0.83	0.65

5-10 In ribo-C-nucleosides, the base is attached to C1' by carbon. The α and β forms (C1' epimers) may be distinguished by T_1 studies.

(a) Consider the following proton T_1 (s) data.

	H1'	H3'	H5'	H5''
Isomer 1	1.60	1.31	0.45	0.45
Isomer 2	3.33	1.37	0.40	0.40

Which isomer (1 or 2) is α and which is β? Why are the T_1 values for H1' different for isomers 1 and 2, but the values for H3', H5', and H5'' are about the same? Why are the T_1 values for H5' and H5'' smaller than the other values? Use the equation for dipolar relaxation (eq. 5-1) in your reasoning.

(b) Suggest another (not T_1) NMR method for distinguishing these α and β forms.

5-11 **(a)** Consider the following molecule **A**, in which rotation is rapid around all C—C bonds.

What is the spin system when R = R'? Are the protons within a single methylene group homotopic, enantiotopic, diastereotopic, or magnetically nonequivalent? More than one category may apply.

(b) Answer the same question when R ≠ R'.

(c) The R and R' groups were chosen to be a donor (D, 9-anthracyl) or an acceptor (A, 3,5-dinitrophenyl). Three molecules can be constructed, in which both R and R' are D (D—D), both R and R' are A (A–A), and R = D when R' = A (D–A). The ^{1}H spectra of these three molecules, as well as the spectra of the model compounds containing only a single A or D

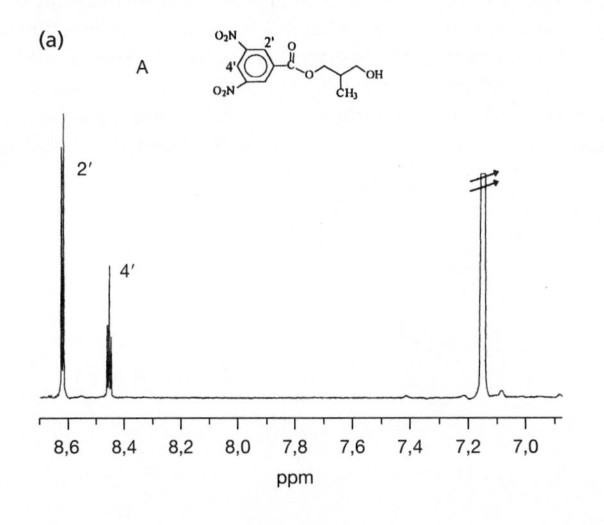

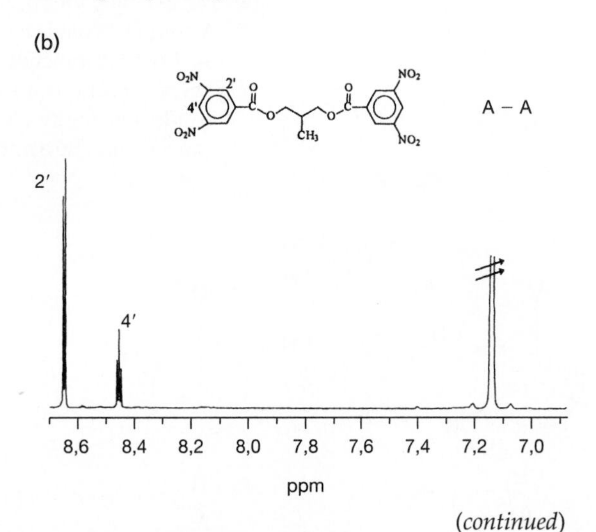

(continued)

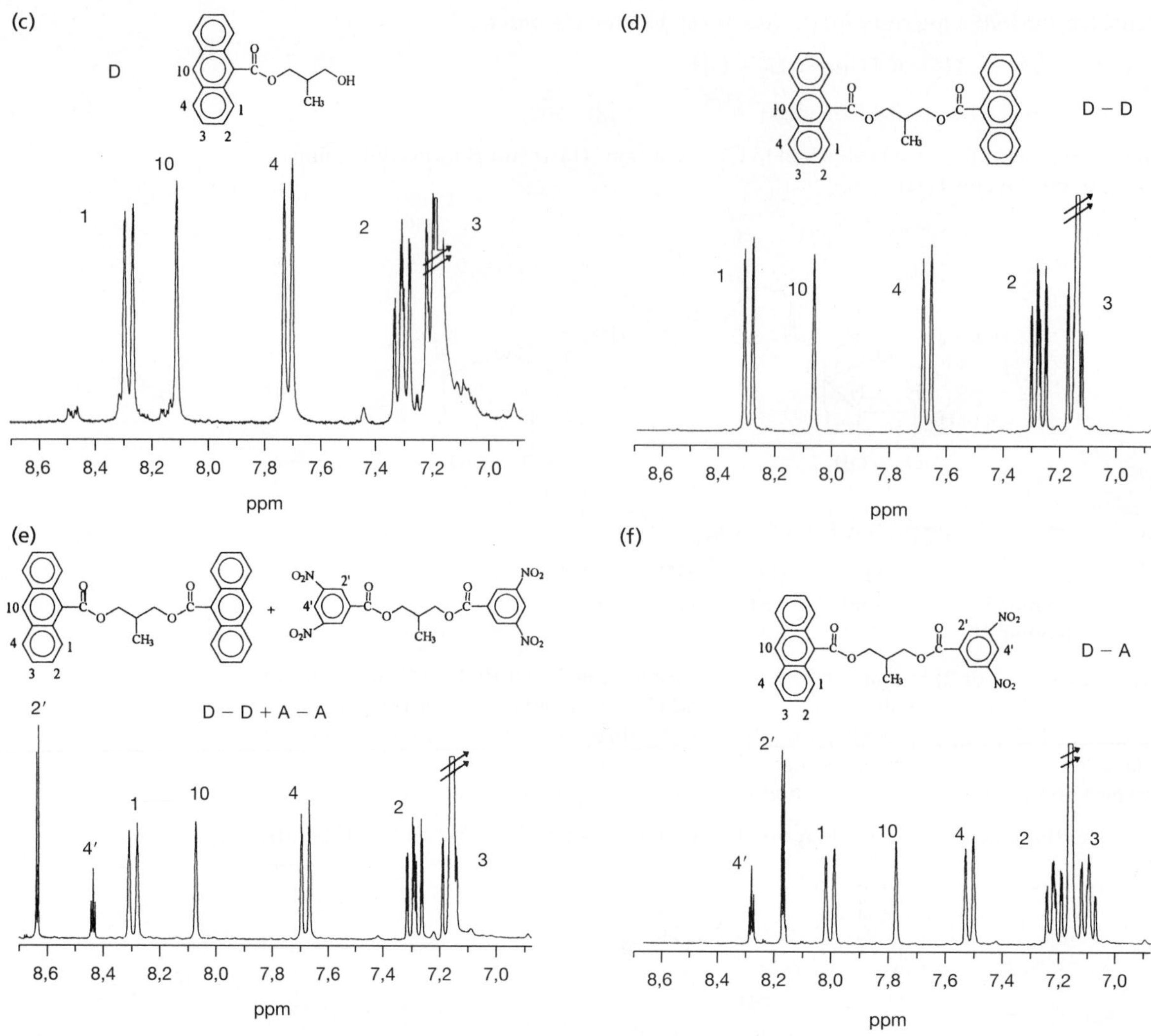

Reproduced with permission from N. J. Heaton, P. Bello, B. Herradón, A. del Campo, and J. Jiménez-Barbero, *J. Am. Chem. Soc.*, **120**, 9636 [1998]. Copyright 1998 American Chemical Society.

and the spectrum of a solution containing equal amounts of D–D and A–A, are given above. Explain the splitting patterns of the A–A molecule in spectrum (b) and of the D–D molecule in spectrum (d). All spectra were measured in C_6D_6.

(d) Explain why there is no difference between the aromatic resonances of A–A and of A in (**a**), nor between D–D and D in (**c**).

(e) Explain why the spectrum of D–A in (f), however, is quite different from those of A–A, D–D, A, and D. What is the purpose of spectrum (e)? What mechanism(s) of interaction between the D and A moieties is (are) eliminated from consideration by these observations?

(f) There are one trans (anti) (**B**) and two gauche (**C**, **D**) conformations around the C–C bonds. For the two C–C bonds, there can be trans–trans and various gauche–trans and gauche–gauche arrangements. The trans–trans conformer, for example, resembles **E**.

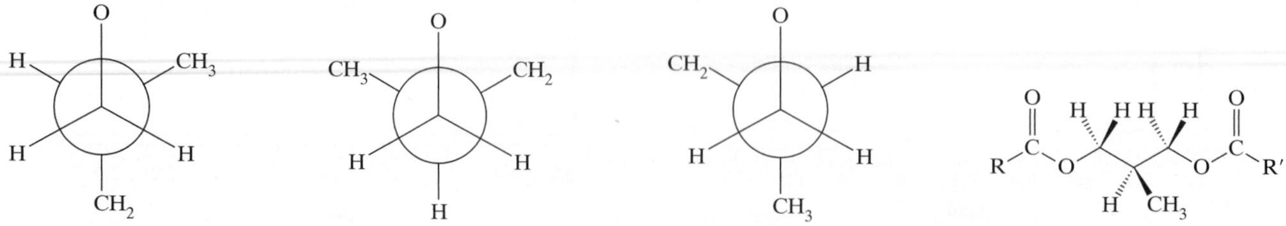

B **C** **D** **E**

The bonds labeled 1 and 2 in the following table are different for D–A, but the same for A–A or D–D.

	J(AX), Hz	*J*(BX), Hz
A–A	6.52	5.62
D–D	6.46	5.55
(A–D)-1	3.60	8.21
(A–D)-2	4.09	8.23

Rotation is fast on the NMR time scale. Couplings were measured at 298 K in C_6D_6 between CHMe (H_X) and CH_2 (H_A and H_B). What conformational conclusion may be drawn from these numbers? Explain.

(g) NOE experiments were carried out on D–A. Irradiation of H10 [see the spectrum in (**c**)] enhanced the resonances of H4, H2′, and H4′. Irradiation of H4′ enhanced the resonances of H1, H4, and H10. Explain.

5-12 **(a)** Trimethylsilylation of *N*-(triisopropylsilyl)indole (see structure) gave a single product in which the ^{1}H spectrum contained four doublets and one doublet of doublets (ignoring long-range couplings). What structures are compatible and incompatible with these observations? Explain.

N

$Si(iPr)_3$

(b) Double irradiation of the trimethylsilyl ^{1}H resonance increased the intensity of two of the doublets. Irradiation of the isopropyl septet increased the intensity of the other two doublets. What is the structure of the product? Explain.

5-13 The 75 MHz ^{13}C spectrum of the drug *N*-propyl-3,4-methylenedioxyamphetamine hydrochloride is given below with WALTZ-16 decoupling (lower) and with the attached proton test (APT, upper). Assign the carbons in the full spectrum, using the edited spectrum and your knowledge of α, β, and γ substituent effects.

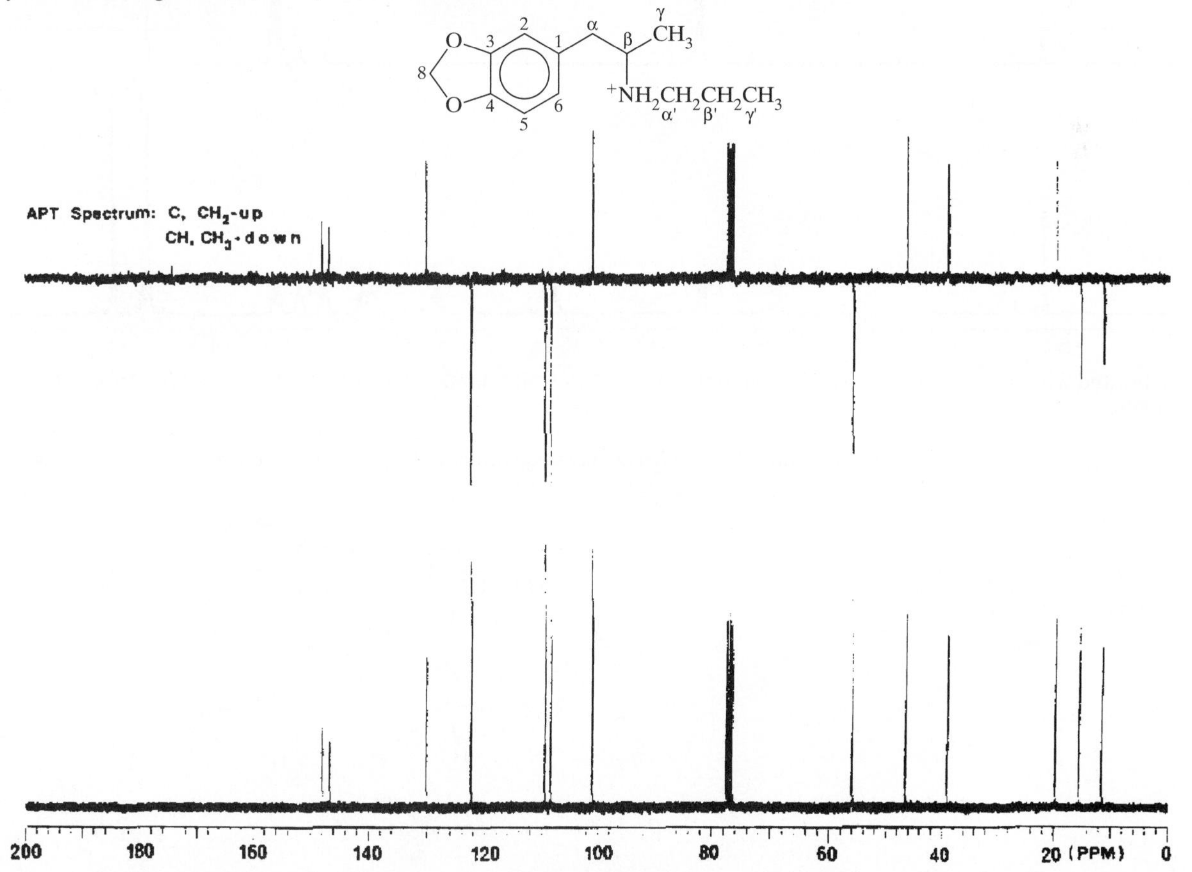

Reprinted with permission of T. A. Dal Cason, J. A. Meyers, and D. C. Lankin, *Forensic Science International*, **86**, 19 (1997).

5-14 Hydroformylation of myrtenol (**A**) was supposed to give the aldehyde **B** but instead produced another product **C** with the bicyclic portion of the molecule entirely intact. The 400 MHz ^{1}H and various difference NOE spectra are provided below, with the resonances from the protons in the bicycle assigned (the multiplet at δ 2.2 comes from H7$_{eq}$ and H4$_a$). The ^{13}C spectrum contained nine peaks in the region δ 20–45, plus peaks at δ 70.06 and 105.31. See structure **D** for the numbering system. The target of NOE irradiation is indicated in each spectrum by a deep valley. Proton 4b also showed a strong NOE with irradiation of the proton at δ 5.4. Use all this information to prove the structure of the product, including stereochemistry.

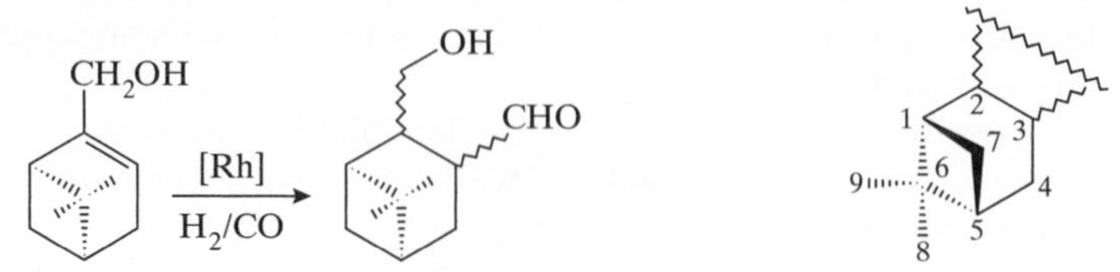

CH$_2$OH [Rh] H$_2$/CO OH CHO

A **B** **D**

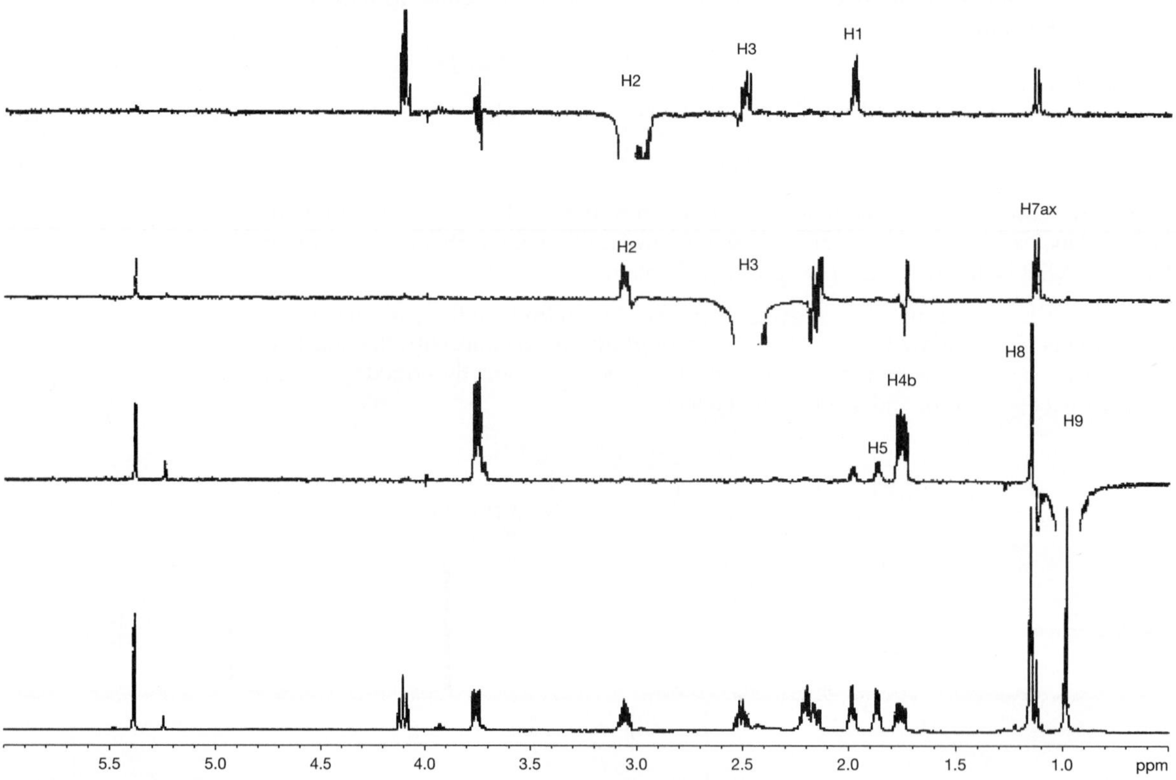

H2 H3 H1

H2 H3 H7ax

H8 H4b H5 H9

5.5 5.0 4.5 4.0 3.5 3.0 2.5 2.0 1.5 1.0 ppm

5-15 The following 500 MHz ^{1}H spectrum is of the taxane structure (**A**) illustrated below.

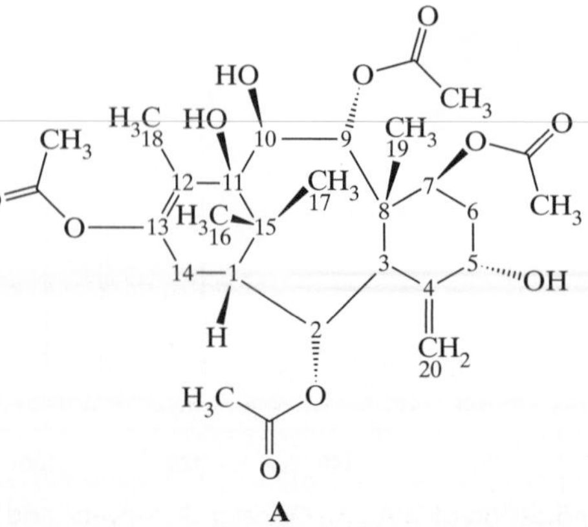

A

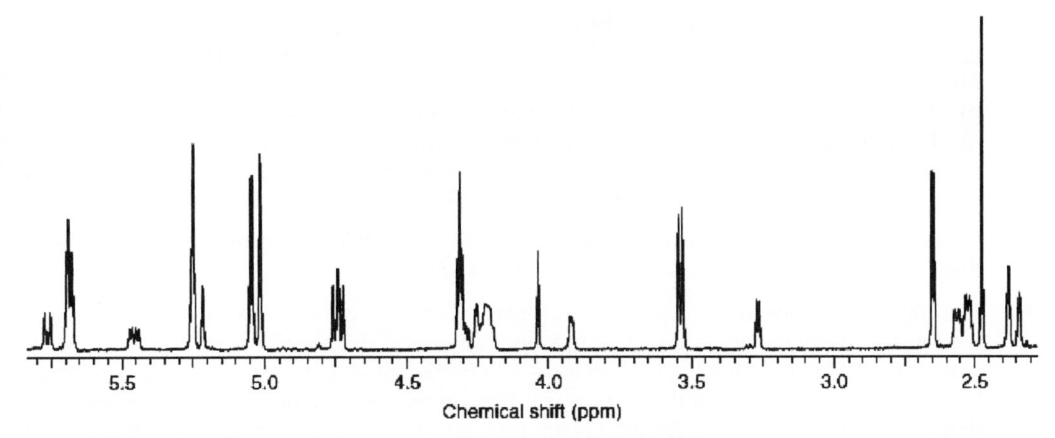

Reprinted with permission of S. Sirol, J.-P. Gorricon, P. Kalck, P. M. Nieto, and G. Commenges, *Magn. Reson. Chem.*, **43**, 799 (2005).

Note the following important spectral characteristics.

1. The methyl resonances and those of H1 and H6 are off scale to low frequency.
2. The middle ring has eight members. Do not expect cyclohexane-like couplings.
3. The OH protons at δ 2.47 and 2.65 are exchanging slowly on the NMR time scale, so that you may expect to see vicinal couplings.
4. H1 resonates at δ 1.89. By spin decoupling, it is coupled to the proton at δ 5.68.
5. The protons at δ 2.36 and 2.55 are coupled to each other ($J = 19$ Hz).
6. The OH proton at position 5 is a doublet off scale to low frequency.

Assign peaks to H2, H3, H5, H7, H9, H10, H14 (two protons), H20 (two protons), OH at position 10, and OH at position 11.

5-16 Glycidol has the following 300 MHz ^{1}H spectrum.

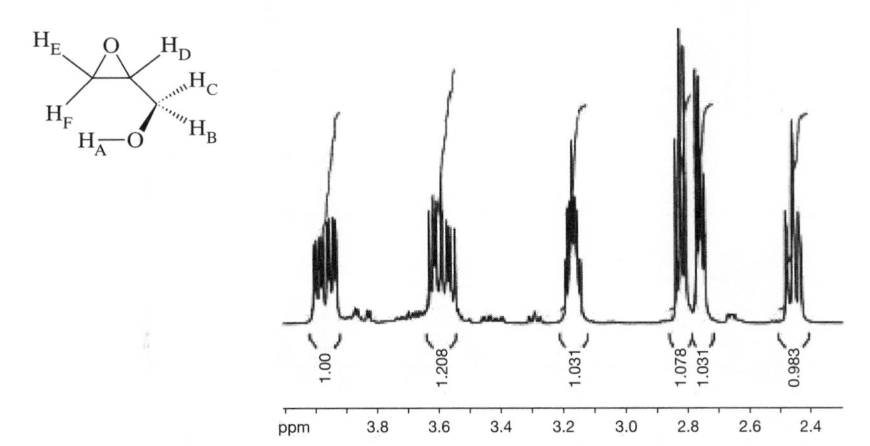

Reproduced with permision from E. Helms, N. Arpaia, and M. Widener, *J. Educ. Chem.*, **84**, 1329 (2007).

The positions of the ^{1}H resonances are as follows: δ 2.53, 2.76, 2.82, 3.16, 3.60, 3.95. The ^{13}C spectrum contains peaks at δ 44.9, 52.8, and 62.5. This experiment was carried out in extremely dry $CDCl_3$, so that coupling of the hydroxy proton H_A appears in the spectrum (no fast exchange or broadening). From the following observations, assign all the ^{1}H and ^{13}C peaks.

(a) The peak at δ 2.53 disappeared on shaking the sample with D_2O, and the peaks at δ 3.60 and 3.95 broadened.

(b) The DEPT experiment showed that the carbon at δ 52.8 is methinyl.

(c) Heteronuclear $^1H\{^{13}C\}$ (actually 2D HETCOR, Chapter 6) double irradiation of the ^{13}C resonance at δ 52.8 decoupled the proton at δ 3.16.

(d) Homonuclear $^1H\{^1H\}$ (actually 2D COSY, Chapter 6) of the ^{1}H resonance at δ 2.53 decoupled the peaks at δ 3.60 and 3.95.

(e) Heteronuclear $^1H\{^{13}C\}$ (actually 2D HETCOR, Chapter 6) double irradiation of the ^{13}C resonance at δ 62.5 decoupled the protons at δ 3.60 and 3.95.

(f) The resonance at δ 2.82 is a triplet with $J = 5.0$ Hz.
(g) The resonance at δ 2.76 is a doublet of doublets with $J = 2.7$ and 5.0 Hz.
(h) The resonances at δ 3.60 and 3.95 share a coupling of 12.7 Hz.
(i) Theoretical calculations indicated that the favored conformation is as shown above.
(j) The resonance at δ 3.60 has a slightly larger coupling with the resonance at δ 2.53 than does the resonance at δ 3.95, as measured from the ddd spacings centered at δ 3.60 and 3.95 (not clear from the broaden triplet at δ 2.53).

In your answer, interpret the couplings of 5.0 and 12.7 Hz. If any protons are enantiotopic or diastereotopic, identify them as such and explain why.

5-17 The 500 MHz ^{1}H spectra and selective nuclear Overhauser experiments given below are of the two γ-butyrolactones **A** and **B**, a functionality found in about 10% of all natural products. The lactone ring may have either a cis- or a trans-fusion to the attached ring. The structures are illustrated without that stereochemistry. In the spectra, the Greek letter α indicates that the proton is down and β that it is up. Each set of spectra contains an entirely assigned ^{1}H spectrum followed by a series of 1D difference NOE experiments. Assign the stereochemistry of each molecule.

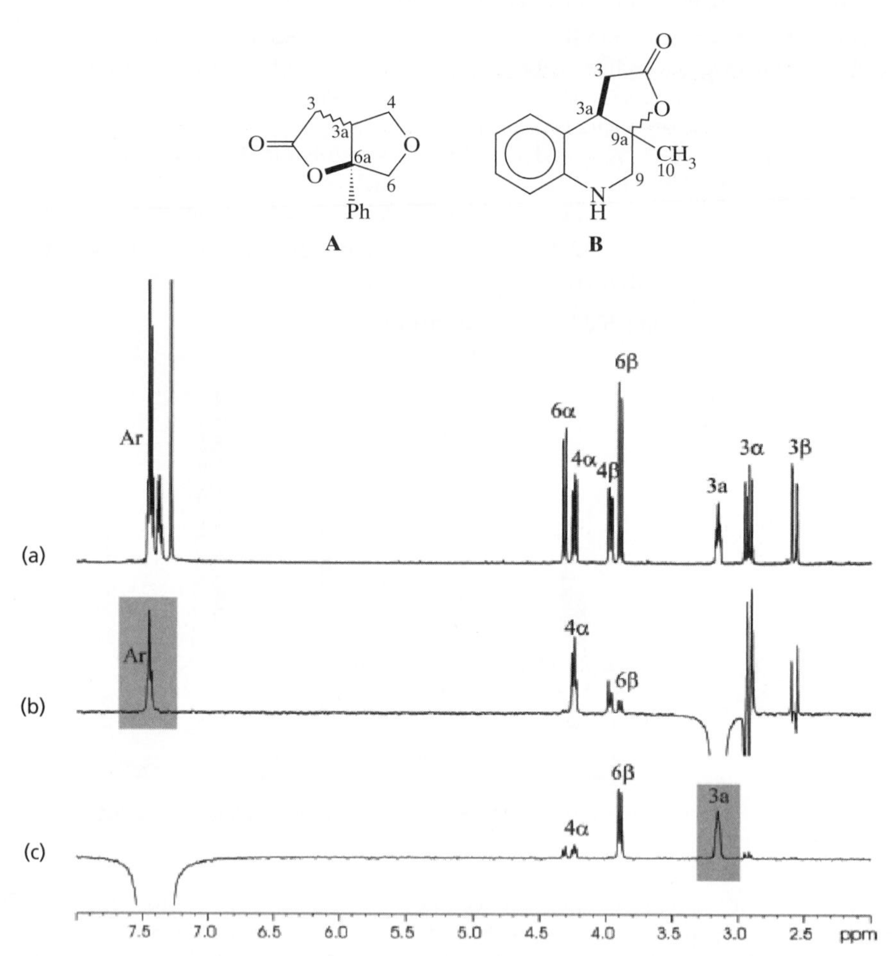

Selective NOEs of **A**: (a) normal ^{1}H spectrum, (b) irradiation of H3a, (c) irradiation of the aromatic protons.

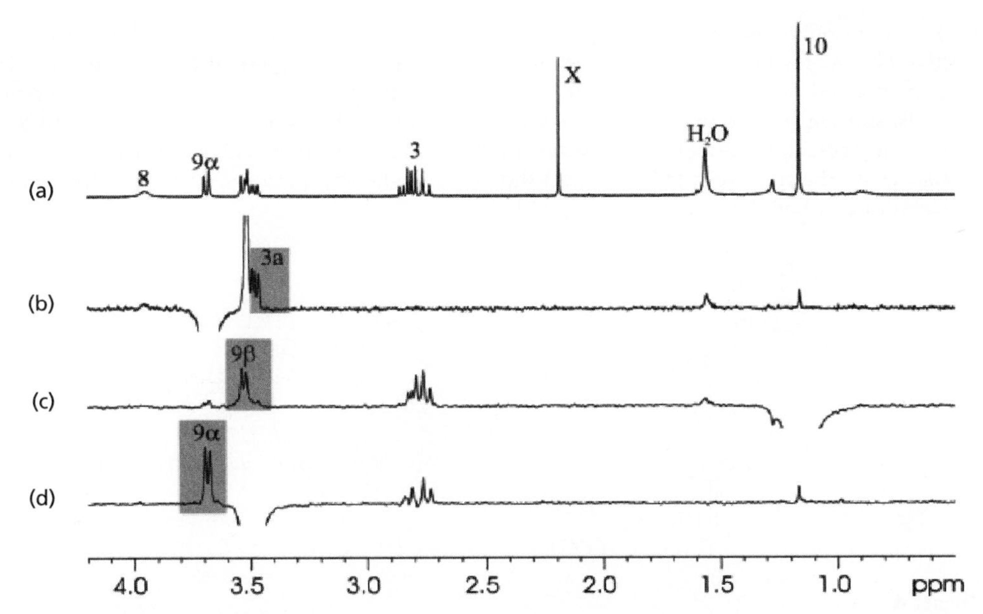

Selective NOEs of **B**: (a) normal ^1H spectrum, (b) irradiation of H9α, (c) irradiation of H10, (d) irradiation of H3a.

The spectra for problem **5-17** are reproduced with permission from X. Xie, S. Tschan, and F. Glorius, *Magn. Reson. Chem.*, **45**, 384–385 (2007).

5-18 The cannabinoid receptor that gives marijuana its psychopharmacological properties also serves as the receptor for the molecule anandamide (from the Sanskrit for *bliss*), claimed to be present in small amounts in chocolate. From the ^{13}C spectrum given below, assign as many peaks as possible. The large peak at δ 40 is from the solvent, DMSO-d_6 , and the peak at δ 25 is the superposition of four peaks. Relaxation times (T_1 in s) were measured for the saturated carbons (except for the four superimposed peaks at δ 25): δ 60.8 (0.74 s), 42.3 (0.65), 35.7 (0.58), 31.8 (1.8), 29.6 (1.5), 27.5 (1.5), 27.1 (0.68), 22.9 (2.6), 14.8 (3.5). These values will be useful in distinguishing C16–C20.

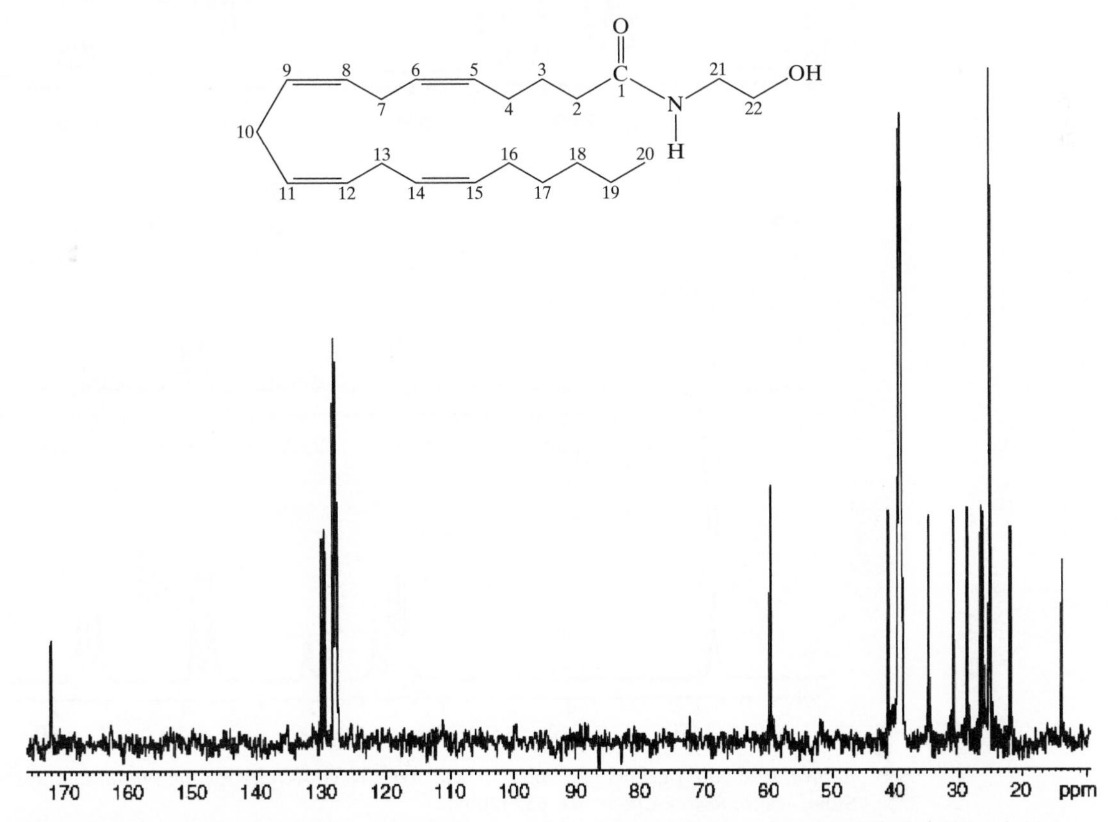

Reproduced with permission from G. Bonechi, A. Brizzi, V. Brizzi, M. Francioli, A. Donati, and C. Rossi, *Magn. Reson. Chem.*, **39**, 433–535 (2001).

5-19 The two isomers below were prepared in a synthetic project. The 300 MHz ^{1}H spectra are given below for the two isomers with the label (a) (aromatic region only). The R groups are aliphatic and off scale, as are the NH resonances. The spectra labeled (b) are the difference NOE spectra for the two isomers. Assign the isomers. Comment on chemical shift and coupling constants as well as the NOEs. In the process, assign the H2, H3, H5, and H6 resonances on the trisubstituted rings and the ortho, meta, and para resonances on the monosubstituted phenyl groups.

NH
4
5
3
6
2
S
1
NHR
CO_2H

CO_2H
NH
4
S
5
3
6
2
1
NHR

Isomer **A**

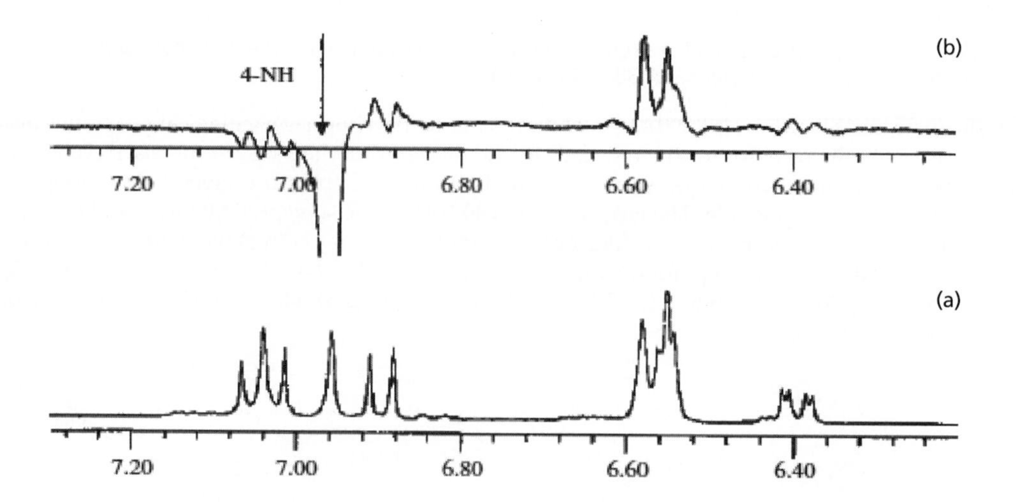

Isomer **B**

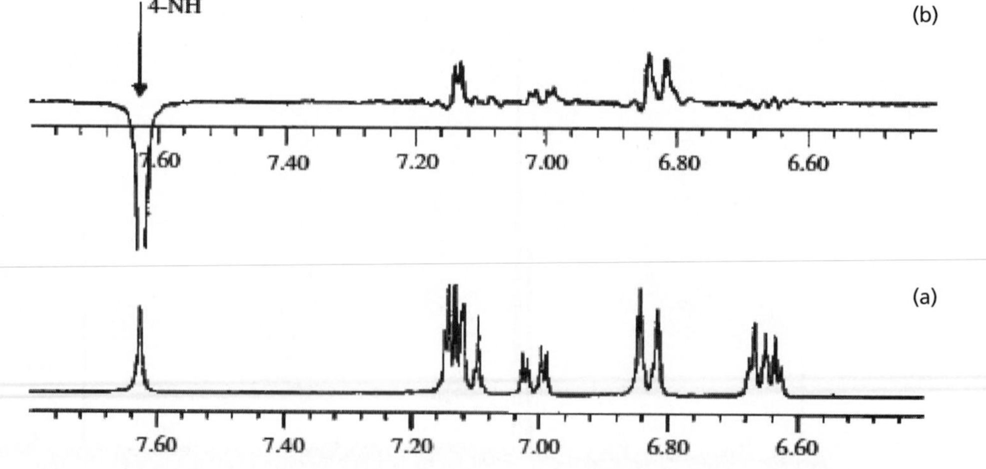

Reproduced with permission from A. R. Katritzky, N. G. Akhmedov, M. Wang, C. J. Rostek, and P. J. Steel, *Magn. Reson. Chem.*, **42**, 652 (2004).

5-20 Draw out the spin vectors in the rotating coordinate system for each step in the following pulse sequence: $90^\circ_x - \left(\frac{1}{4J}\right) - 180^\circ_x - \left(\frac{1}{4J}\right) -$ acquire (all pulses are applied to the observed nucleus). Imagine that your observed nucleus is a ^{13}C atom attached to two protons, with the reference frequency fixed at the Larmor frequency of the carbon. What conclusion can you draw about the appearance of the observed peak at the end of the pulse sequence? Empty coordinate systems are provided for your drawings.

Equilibrium state

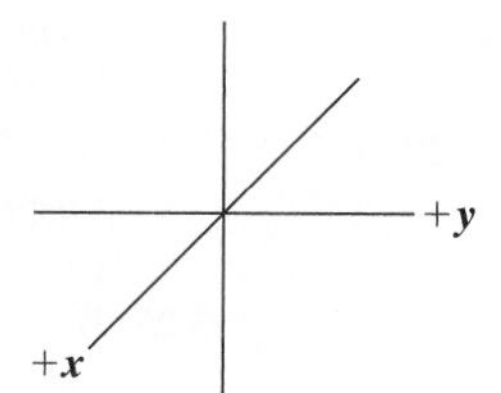

After the 90°_x pulse

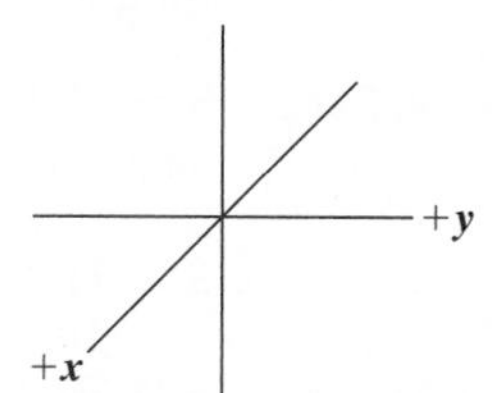

After the $\frac{1}{4J}$ fixed delay

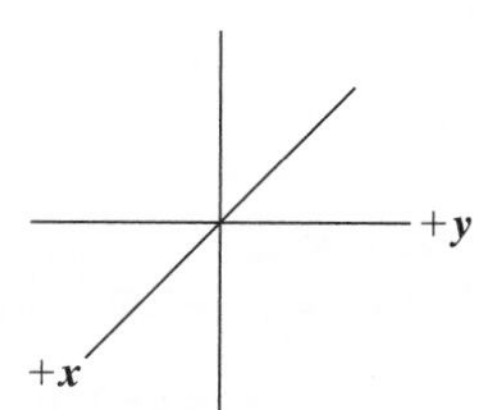

After the 180°_x

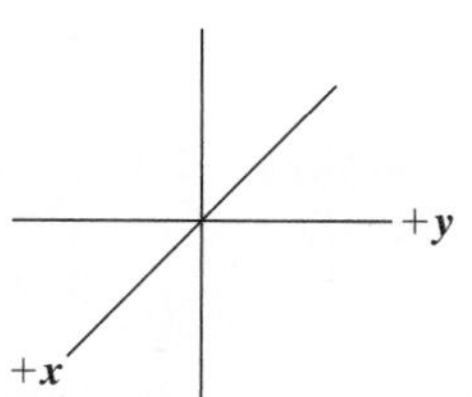

After the second $\frac{1}{4J}$ fixed delay

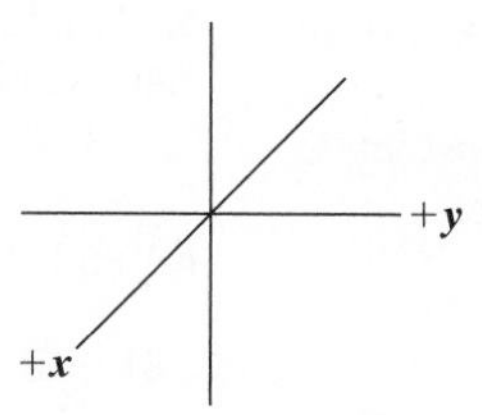

Bibliography

Relaxation Phenomena

General

5.1 D. A. Wright, D. E. Axelson, and G. C. Levy, *Magn. Reson. Rev.*, **3**, 103 (1979).
5.2 G. H. Weiss and J. A. Ferretti, *Progr. NMR Spectrosc.*, **20**, 317 (1988).
5.3 D. J. Wink, *J. Chem. Educ.*, **66**, 810 (1989).

Carbon-13 Relaxation

5.4 J. R. Lyerla, Jr., and G. C. Levy, *Top. Carbon-13 Spectrosc.*, **1**, 79 (1974).
5.5 F. W. Wehrli, *Top. Carbon-13 Spectrosc.*, **2**, 343 (1976).
5.6 D. J. Craik and G. C. Levy, *Top. Carbon-13 Spectrosc.*, **4**, 241 (1983).

Nuclear Overhauser Effect

5.7 J. H. Noggle and R. E. Schirmer, *The Nuclear Overhauser Effect*, New York: Academic Press, 1971.
5.8 R. A. Bell and J. K. Saunders, *Top. Stereochem.*, **7**, 1 (1973).
5.9 J. K. Saunders and J. W. Easton, *Determ. Org. Struct. Phys. Meth.*, **6**, 271 (1976).
5.10 K. E. Kövér and G. Batta, *Progr. NMR Spectrosc.*, **19**, 223 (1987).
5.11 D. Neuhaus and M. P. Williamson, *The Nuclear Overhauser Effect in Structural and Conformational Analysis*, 2nd ed., New York: Wiley-VCH, 2000.

Reactions on the NMR Time Scale

General

5.12 G. Binsch, *Top. Stereochem.*, **3**, 97 (1968).
5.13 L. M. Jackman and F. A. Cotton (eds.), *Dynamic Nuclear Magnetic Resonance Spectroscopy*, New York: Academic Press, 1975.
5.14 A. Steigel, *NMR Basic Princ. Progr.*, **15**, 1 (1978).
5.15 J. I. Kaplan and G. Fraenkel, *NMR of Chemically Exchanging Systems*, New York: Academic Press, 1980.
5.16 G. Binsch and H. Kessler, *Angew. Chem., Int. Ed. Engl.*, **19**, 411 (1980).
5.17 J. Sändstrom, *Dynamic NMR Spectroscopy*. London: Academic Press, 1982.
5.18 M. Ōki (ed.), *Applications of Dynamic NMR Spectroscopy to Organic Chemistry*, Deerfield Beach, FL: VCH, 1985.

Carbon-13 Applications

5.19 *Progr. NMR Spectrosc.*, **11**, 95 (1977).

Hindered Rotation

5.20 H. Kessler, *Angew. Chem., Int. Ed. Engl.*, **9**, 219 (1970).
5.21 W. E. Stewart and T. H. Siddall, *Chem. Rev.*, **70**, 517 (1970).
5.22 M. Ōki, *Top. Stereochem.*, **14**, 1 (1983).
5.23 M. L. Martin, X. Y. Sun, and G. J. Martin, *Ann. Rep. NMR Spectrosc.*, **16**, 187 (1985).
5.24 C. H. Bushweller in J. B. Lambert and Y. Takeuchi (eds.), in *Acyclic Organonitrogen Stereodynamics*, New York: VCH, 1992, pp. 1–55.
5.25 M. Raban and D. Kost, in J. B. Lambert and Y. Takeuchi (eds.), *Acyclic Organonitrogen Stereodynamics*, New York: VCH, 1992, pp. 57–88.
5.26 S. F. Nelsen, in J. B. Lambert and Y. Takeuchi (eds.), *Acyclic Organonitrogen Stereodynamics*, New York: VCH, 1992, pp. 89–121.
5.27 B. M. Pinto, in J. B. Lambert and Y. Takeuchi (eds.), *Acyclic Organonitrogen Stereodynamics*, New York: VCH, 1992, pp. 149–175.

Ring Reversal and Cyclic Systems

5.28 H. Booth, *Progr. NMR Spectrosc.*, **5**, 149 (1969).
5.29 J. B. Lambert and S. I. Featherman, *Chem. Rev.*, **75**, 611 (1975).
5.30 H. Günther and G. Jikeli, *Angew. Chem., Int. Ed. Engl.*, **16**, 599 (1977).
5.31 E. L. Eliel and K. M. Pietrusiewicz, *Top. Carbon-13 Spectrosc.*, **3**, 171 (1979).
5.32 F. G. Riddell, *The Conformational Analysis of Heterocyclic Compounds*, London: Academic Press, 1980.
5.33 A. P. Marchand, *Stereochemical Applications of NMR Studies in Rigid Bicyclic Systems*, Deerfield Beach, FL: VCH, 1982.

Atomic Inversion

5.34 J. B. Lambert, *Top. Stereochem.*, **6**, 19 (1971).
5.35 A. Rauk, L. C. Allen, and K. Mislow, *Angew. Chem., Int. Ed. Engl.*, **9**, 400 (1970).
5.36 W. B. Jennings and D. R. Boyd, in J. B. Lambert and Y. Takeuchi (eds.), *Cyclic Organonitrogen Stereodynamics*, New York: VCH, 1992, pp. 105–158.
5.37 J. J. Delpuech, in J. B. Lambert and Y. Takeuchi (eds.), *Cyclic Organonitrogen Stereodynamics*, New York: VCH, 1992, pp. 169–252.

Organometallics

5.38 K. Vrieze and P. W. N. M. Vanleeuwen, *Progr. Inorg. Chem.*, **14**, 1 (1971).
5.39 B. E. Mann, *Ann. Rep. NMR Spectrosc.*, **12**, 263 (1982).
5.40 K. G. Orrell and V. Šik, *Ann. Rep. NMR Spectrosc.*, **19**, 79 (1987).

Rates from Relaxation Times

5.41 J. B. Lambert, R. J. Nienhuis, and J. W. Keepers, *Angew. Chem., Int. Ed. Engl.*, **20**, 487 (1981).

Multiple Irradiation

General

5.42 R. A. Hoffman and S. Forsén, *Progr. NMR Spectrosc.*, **1**, 15 (1966).
5.43 V. J. Kowalewski, *Progr. NMR Spectrosc.*, **5**, 1 (1969).

5.44 W. McFarlane, *Determ. Org. Struct. Phys. Meth.*, **4**, 150 (1971).
5.45 W. von Philipsborn, *Angew. Chem., Int. Ed. Engl.*, **10**, 472 (1971).
5.46 R. L. Micher, *Magn. Reson. Rev.*, **1**, 225 (1972).
5.47 L. R. Dalton, *Magn. Reson. Rev.*, **1**, 301 (1972).
5.48 W. McFarlane and D. S. Rycroft, *Ann. Rep. NMR Spectrosc.*, **16**, 293 (1985).

Difference Spectroscopy

5.49 J. K. M. Sanders and J. D. Merck, *Progr. NMR Spectrosc.*, **15**, 353 (1982).

Broadband Decoupling

5.50 M. H. Levitt, R. Freeman, and T. Frenkiel, *Advan. Magn. Reson.*, **11**, 47 (1983).
5.51 A. J. Shaka and J. Keeler, *Progr. NMR Spectrosc.*, **19**, 47 (1987).

One-Dimensional Multipulse Methods

General

5.52 R. Benn and H. Günther, *Angew. Chem., Int. Ed. Engl.*, **22**, 350 (1983).
5.53 C. J. Turner, *Progr. NMR Spectrosc.*, **16**, 311 (1984).
5.54 G. A. Morris, *Magn. Reson. Chem.*, **24**, 371 (1986).
5.55 D. L. Turner, *Ann. Rep. NMR Spectrosc.*, **21**, 161 (1989).
5.56 K. Nakanishi, *One-Dimensional and Two-Dimensional NMR Spectra by Modern Pulse Techniques*, Mill Valley, CA: University Science Books, 1990.
5.57 R. R. Ernst and G. Bodenhausen, *Principles of Nuclear Magnetic Resonance in One and Two Dimensions*, Oxford, UK: Oxford University Press, 1990.

Composite Pulses

5.58 M. H. Levitt, *Progr. NMR Spectrosc.*, **18**, 61 (1986).

Multiple Quantum Methods

5.59 G. Bodenhausen, *Progr. NMR Spectrosc.*, **14**, 137 (1986).
5.60 T. J. Norwood, *Progr. NMR Spectrosc.*, **24**, 295 (1992).

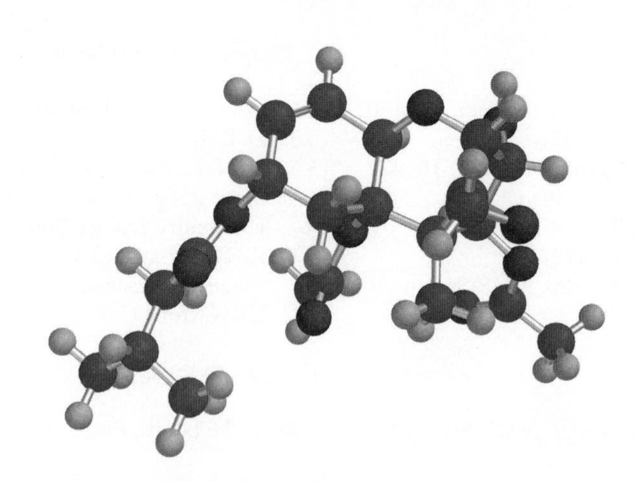

Two-Dimensional NMR Spectroscopy

NMR spectroscopy always has been multidimensional. In addition to frequency and intensity, which serve as the coordinates of the standard one-dimensional (1D) spectrum, reaction rates and relaxation times have provided further dimensions, often presented in the form of stacked plots (Figures 2-31 and 5-2). The second dimension of modern NMR spectroscopy, however, refers to an additional frequency axis. This concept was suggested first in a lecture by Jean Jeener in 1971 and reached wide application in the 1980s, when instrumentation caught up with theory. Usually, we can think of the first frequency dimension as the traditional characterization of nuclei in terms of chemical shifts and couplings. By introducing a second frequency dimension, we examine magnetic interactions between nuclei through structural connectivity, spatial proximity, or kinetic interchange.

6-1 PROTON–PROTON CORRELATION THROUGH *J* COUPLING

In the single-pulse experiments considered up to this point, a 90° pulse is followed by a period during which the free-induction decay is acquired (Figure 6-1a). Fourier transformation of the time-dependent magnetic information into a frequency dimension provides the familiar spectrum of δ values, henceforth called a 1D spectrum.

If the 90° acquisition pulse is preceded by another 90° pulse (Figure 6-1b), useful relationships between spins can evolve. Figure 6-2 illustrates what happens in terms of magnetization vectors. Consider a sample that contains only one type of nucleus without any coupling partners, for example, the ^{1}H spectrum of chloroform or tetramethylsilane. Although the final result of this particular experiment may seem trivial or even pointless at first, it will take on fuller meaning when we introduce relationships with other nuclei. The isolated nucleus of Figure 6-2 begins with the net magnetization **M** aligned along the z axis (Figure 6-2a). The magnetization realigns along the y axis after application of the 90° pulse (Figure 6-2b). If the coordinate system rotates at the reference

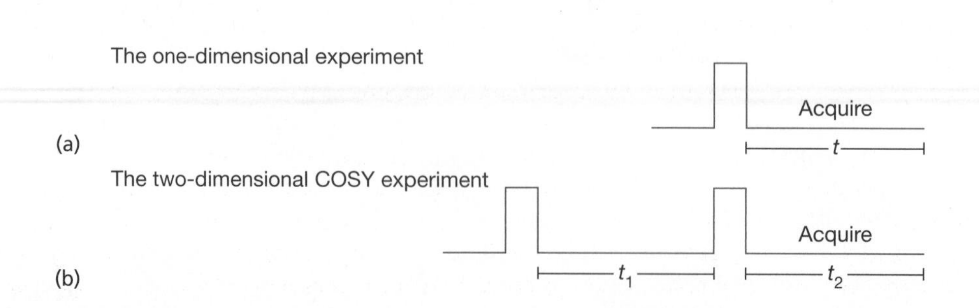

FIGURE 6-1 The pulse arrangements for a single cycle of one-dimensional (1D) NMR spectroscopy (top) and for two-dimensional (2D) NMR correlation spectroscopy (COSY) (bottom). In this diagram, each pulse is 90°. Data are acquired during the time t in the 1D experiment and during t_2 in the 2D experiment.

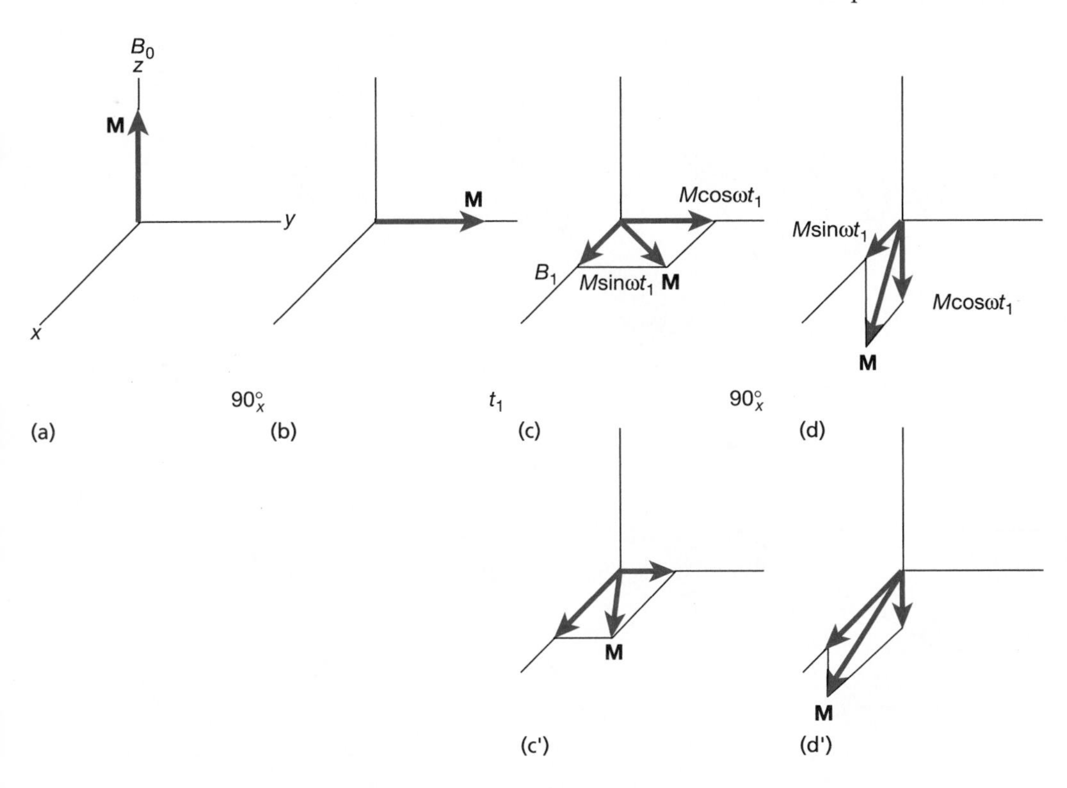

FIGURE 6-2 The pulse sequence for the COSY experiment. For (c'), the magnetization **M** is allowed to evolve a longer time from (b) than in the case of (c) before the final 90°_x pulse is applied to give (d').

frequency and the nucleus resonates at a slightly higher frequency, the spin vector picture begins to evolve. After a short amount of time, the vector **M** moves to a new position in the xy plane, for example, in Figure 6-2c. We ignore longitudinal relaxation (T_1) to simplify the drawings. The evolving magnetization vector may be decomposed geometrically into a y component ($M_y = M \cos \omega t_1$) and an x component ($M_x = M \sin \omega t_1$), in which ω is the difference between the frequency of the reference and that of the resonating nucleus and t_1 is the time elapsed since the 90° pulse.

If, at this point, the second 90° pulse of Figure 6-1b is applied, again along the x axis, the result is different for the two magnetization components illustrated (Figure 6-2d). The x component is unaffected, but the y component is transferred to the negative z axis. If magnetization is detected in the xy plane, only M_x remains. This quantity appears as a free induction decay (FID) during the time t_2 after the second pulse. Fourier transformation of the FID as a function of t_2 yields a signal at the resonance frequency (ν_A). The intensity of this signal is determined by the length of the time period t_1 ($M \sin \omega t_1$). The subscripts are necessary to distinguish the evolution period t_1 from the acquisition period t_2 (Figure 6-1b). If t_1 is relatively short, M_x ($= M \sin \omega t_1$) is small, little x magnetization has developed (Figure 6-2d), and the resulting peak is small. A slightly longer value of t_1 yields a larger x component (Figure 6-2c' and d') as $M \sin \omega t_1$ grows. Note that the spin population (the z, or longitudinal, magnetization) is inverted in Figure 6-2d and d'.

Figure 6-3 shows the result of a whole series of such experiments, with a buildup of M_x ($= M \sin \omega t_1$) as t_1 increases, reaching a maximum when the spin vector **M** is lined up along the x axis. The peak height then decreases as the vector moves to the left of the x axis, reaching zero intensity when it is lined up along the negative y axis. As it passes behind the y axis, the intensity becomes negative, attaining a negative maximum when the vector is aligned along the $-x$ axis. This negative maximum would be slightly smaller than the initial positive maximum, because of T_2 relaxation. It is clear from the figure that this family of experiments generates a sine curve when M_x is plotted as a function of t_1. Only a cycle and a half are illustrated. Frequency (obtained from the Fourier transformation of t_2 to give ν_2) is along the horizontal dimension, and time t_1 is along the vertical dimension. The set of data generated from stepping t_1 in this fashion in fact constitutes an FID that also may be Fourier transformed. Because the frequency ω represented by the sine curve in t_1 is the same as the frequency from the initial Fourier transformation in t_2, the result of the second Fourier transformation is a single peak at

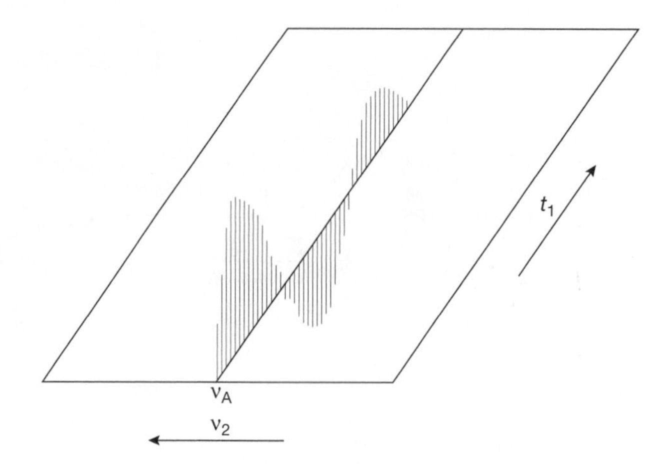

FIGURE 6-3 The solid line slanting upward at frequency ν_A on the horizontal axis serves as the baseline for a series of ^1H spectra of chloroform, according to the COSY pulse sequence for a series of values of t_1. Each peak results from one cycle of 90° – t_1 – 90° followed by Fourier transformation during t_2 (Figure 6-1) to give frequency ν_A on the axis labeled ν_2 (corresponding to the time domain t_2). The period t_1 is ramped up after each cycle. Fourier transformation in the t_1 dimension has not been carried out.

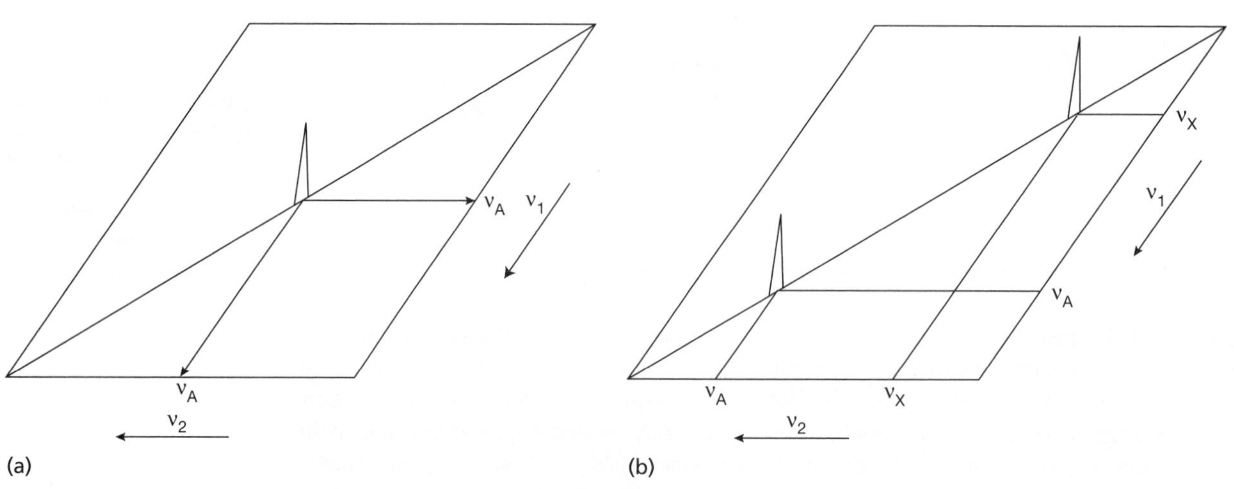

FIGURE 6-4 (a) The result of the COSY experiment after double Fourier transformation for a single isolated nucleus such as that in Figure 6-3. (b) The result of the COSY experiment for two uncoupled nuclei.

the coordinates (ν_A, ν_A) when plotted in two frequency dimensions (Figure 6-4a). This is the trivial result previously alluded to.

The utility of the experiment just discussed becomes evident when two coupled nuclei are treated in the indicated fashion. Two uncoupled nuclei yield the trivial result of two peaks, at (ν_A, ν_A) and (ν_X, ν_X), respectively, as in Figure 6-4b. These peaks are necessarily on the diagonal of the two-dimensional (2D) representation. Profound complications arise when the two nuclei are coupled. Figure 6-5 illustrates the possible spin states for nuclei A and X, as, for example, the alkenic protons in β-chloroacrylic acid, $ClCH{=}CHCO_2H$. The 1D AX spectrum contains four peaks, due to scalar (J) coupling. The diagram is intended to indicate the four different frequencies, from the highest (A1) to the lowest (X2). It is useful first to consider population perturbations during an old-style, 1D selective-decoupling CW experiment. Irradiation, for example, of only transition A1 tends to bring the $\alpha\beta$ and $\beta\beta$ states closer together in population. Consequently, there is a direct effect on the intensities of the connected transitions X1 and X2, which propagates as a secondary effect on the intensity of the A2 transition. With respect to A1, X2 is called a *progressive transition* (a transition that goes on to a higher spin state), X1 a *regressive transition,* and A2 a *parallel transition.*

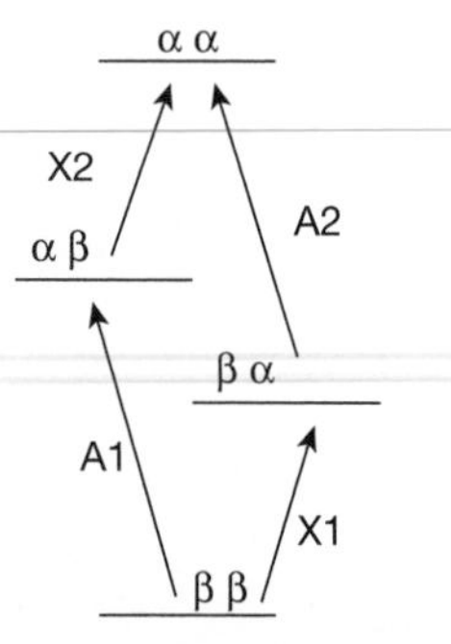

FIGURE 6-5 The energy diagram for an AX spin system.

In the pulse experiment, energy absorption at frequency A1 has similar effects, which bring about magnetization or population transfer. The first 90° pulse serves to label all the magnetization with the 1D frequencies during period t_1: A1, A2, X1, and X2. Figure 6-2d and d' represent a starting point at which the second 90° pulse of Figure 6-1b is applied. Let us set the reference frequency of the rotating frame at frequency A1. The magnetization analogous to that in Figure 6-2 had the frequency A1 during the period t_1. The second 90° pulse transfers magnetization within the xy plane, whereby it can

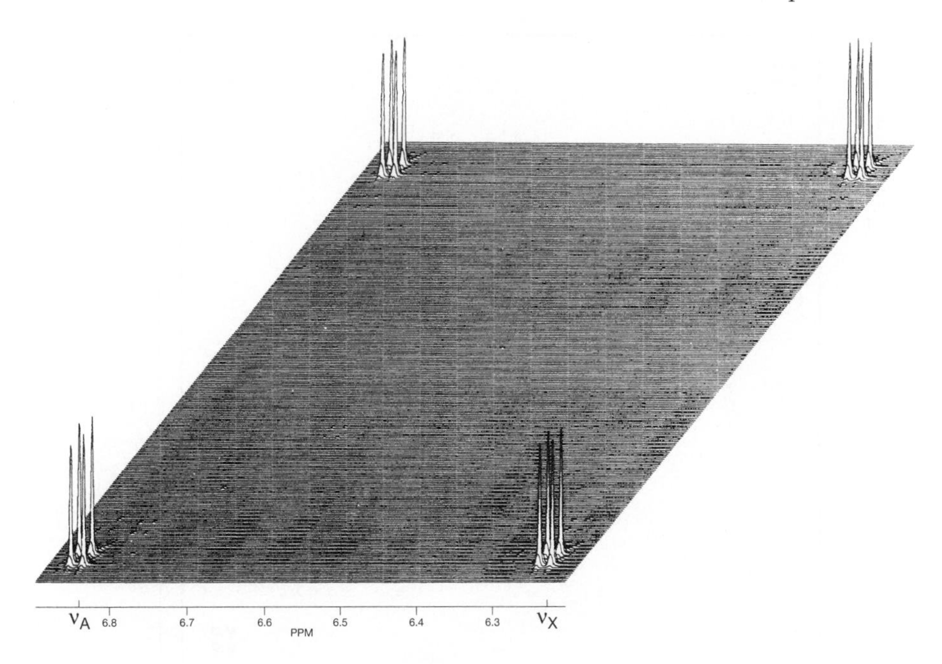

FIGURE 6-6 The stacked representation of the COSY experiment for the two coupled nuclei of β-chloroacrylic acid. (Reproduced from A. E. Derome, *Modern NMR Techniques for Chemistry Research*, Oxford, UK: Pergamon Press, 1987, p. 189.)

precess at any of the allowed frequencies: A1, A2, X1, or X2. Thus the magnetization that had frequency A1 during t_1 can be transferred to A2, X1, or X2 during t_2, and some magnetization remains at A1 during the second time period. Magnetization with frequency A1 during both time periods appears on the diagonal in the 2D representation analogous to Figure 6-4b, at position (A1, A1). The transferred magnetization, however, appears as *cross peaks* off the diagonal, at positions (A1, A2), (A1, X1), and (A1, X2). Each of the four resonances of Figure 6-5 undergoes analogous operations to generate three more diagonal peaks at (A2, A2), (X1, X1), and (X2, X2), plus nine more off-diagonal peaks at, for example, (A2, X1) and (X2, A1).

The 16 peaks (4 on the diagonal and 12 off the diagonal) from the two-dimensional experiment are illustrated in Figure 6-6 for β-chloroacrylic acid (the carboxyl resonance is omitted). The *stacked representation* contains several hundred complete 1D experiments, each representing a different value of t_1. The closely packed horizontal lines are barely distinguishable. On the diagonal from the lower left to the upper right (as drawn into the plots in Figure 6-4) are the four peaks that arise directly from resonance without magnetization transfer, that is, the components of magnetization that possess the same frequency in t_1 and t_2, such as (A1, A1). These four peaks along the diagonal constitute the normal four-peak 1D spectrum. All the peaks off the diagonal represent magnetization transfer by scalar (J) coupling. For example, a transfer between the parallel transitions A1 and A2 is found as symmetrical peaks just above and below the diagonal at the lower left. One peak represents a transfer from A1 to A2, the other from A2 to A1. Because of this reciprocal relationship, all off-diagonal peaks appear in pairs reflected across the diagonal. Normally, the off-diagonal peaks between parallel transitions are more of a nuisance than useful, and they can be reduced or deleted by special techniques. The important information results from magnetization transfer between the A and the X nuclei, whose peaks are the clusters in the upper left and lower right of Figure 6-6, representing all the possible transfers between the A and X transitions: A1 to X1, X2 to A1, and so on—eight in total, including the mirror-image pairs (A to X and X to A) on either side of the diagonal.

Figure 6-7 is the alternative *contour representation* of the same data, in which the distracting baselines are removed and only the peak bases remain, as if the spectator is viewing the spectrum from directly above it. By convention, the original diagonal is from lower left to upper right. The Jeener experiment commonly is given the quasi acronym COSY, for ***CO****rrelation* ***S****pectroscop****Y***. Since most 2D experiments involve spectral correlations, the name is not apt. Alternative terms, such as 90° COSY, COSY90, H,H-COSY, or homonuclear HCOSY, have gone by the wayside through public

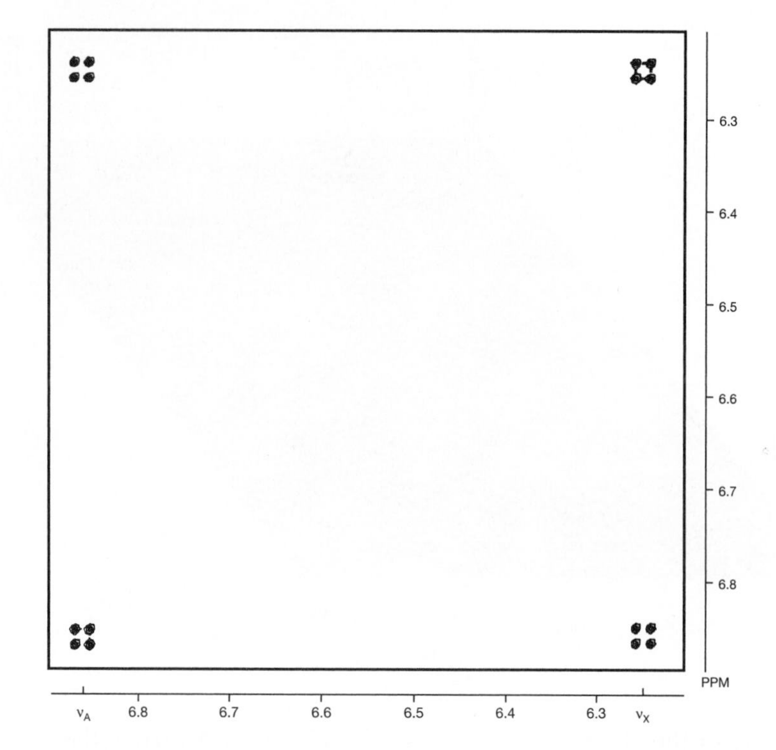

FIGURE 6-7 The contour representation of the COSY experiment for two coupled nuclei of β-chloroacrylic acid. (Reproduced from A. E. Derome, *Modern NMR Techniques for Chemistry Research,* Oxford, UK: Pergamon Press, 1987, p. 191.)

acceptance of the general term COSY. The experiment itself has become an essential part of the analysis of complex proton spectra.

Figure 6-8 is the COSY experiment for the indicated annulene. The 1D spectrum is shown along both the horizontal and the vertical axes, and the resonances are labeled α, β, and A through F. The aromatic protons that are ortho and meta to the ring fusion provide an isolated spin system, and their coupling is represented by the cross peak labeled α,β. The presence of a cross peak normally indicates that the protons giving the connected resonances on the diagonal are geminally (^2J) or vicinally (^3J) coupled. Long-range

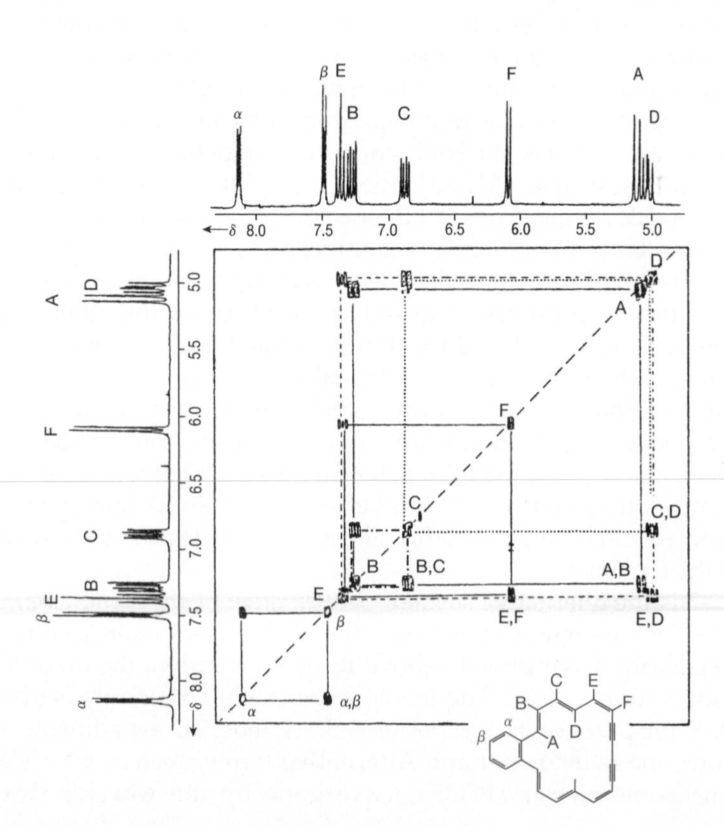

FIGURE 6-8 The COSY experiment for the illustrated annulene. (Reproduced from R. Benn and H. Günther, *Angew. Chem., Int. Ed. Engl.,* **22**, 350 [1983].)

couplings usually do not provide significant cross peaks. Exceptions, however, can be expected, since long-range couplings can be large (Section 4-7).

The COSY analysis of the remainder of the spectrum in Figure 6-8 provides the remaining peak assignments and confirms the structure. Only protons A and F are split by a single neighbor, to give doublets. The trans coupling of A with B should be larger than the cis coupling between E and F. The two doublets then may be assigned as F (with the smaller coupling) at δ 6.1 and A (with the larger coupling) at δ 5.2. It usually is essential in a COSY analysis to be able to make an initial assignment through traditional considerations of chemical shifts and coupling constants (Chapters 3 and 4). The COSY analysis then consists of moving from this known diagonal peak to a cross peak, and then back to the diagonal for the assignment of a new peak. Only A and F have single cross peaks (one coupling partner). All remaining resonances in the large ring have two cross peaks (two coupling partners), which provide the means for assignment. We can start with either A or F. Dropping down from A leads to the cross peak A,B, and horizontal movement to the left leads to a diagonal peak and assignment of proton B. This horizontal path passes through another cross peak, which must be between B and its other coupling partner, C. Moving up from B,C then leads to a diagonal peak and assignment of proton C. Horizontal movement to the right leads to the cross peak C,D and returning upward to the diagonal assigns proton D. Dropping back down from D and passing through C,D leads to the other cross peak from D, labeled E,D. Returning to the diagonal to the left assigns proton E and passes through the other cross peak from E, labeled E,F. Returning upward from E,F to the diagonal completes the assignment with proton F.

A group at IBM has provided a useful example of a more complex COSY analysis with the tripeptide Pro–Leu–Gly (**6-1**). The three carbonyl groups disrupt vicinal connec-

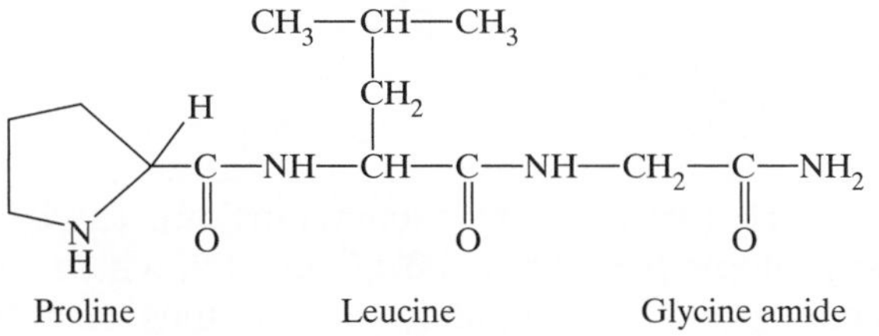

CH_3—CH—CH_3

CH_2

H

C(=O)—NH—CH—C(=O)—NH—CH_2—C(=O)—NH_2

N
H

Proline Leucine Glycine amide

6-1

tivities, so the molecule consists of four independent spin systems: proline, leucine, glycine, and the terminal amide. The 1D ^{1}H spectrum is given at the top of Figure 6-9 without any assignments. The high-frequency (low-field) peaks at δ 7.0–8.3 are from the protons on nitrogen, as there are no aromatic hydrogens, and the broad peak at δ 3.3 is from the solvent HOD. The bottom portion of Figure 6-9 contains the COSY spectrum with connectivities drawn in for the amide resonances. The nonequivalent terminal NH_2 resonances are assigned immediately as δ 7.0 and 7.2 because they have no external connectivities and hence no cross peaks other than between themselves. The Gly NH proton is assigned at δ 8.2 because it is a triplet (next to a CH_2) and has only the single connectivity with the CH_2 group at δ 3.6 (completing the Gly portion of the spectrum). The remaining NH resonance, at δ 8.1, from Leu is a doublet (next to a CH) and has a cross peak with the resonance at δ 4.3, which has other connectivities. There is no third NH resonance, so the Pro NH must be quadrupolar broadened, or is exchanging with HOD.

The top spectrum of Figure 6-10 completes the COSY analysis of the Leu portion and confirms the fact that the NH resonating at δ 8.1 is part of Leu rather than Pro. The expected Leu connectivity is NH $\rightarrow$ CH $\rightarrow$ CH_2 $\rightarrow$ CH $\rightarrow$ CH_3. Cross peaks with the following connectivities (starting with NH) are observed: δ 8.1(d) $\rightarrow$ 4.3(q or dd) $\rightarrow$ 1.5(m) $\rightarrow$ 0.9(dd). Apparently, two of the proton resonances coincide, most likely those from CH_2 and the isopropyl CH. The CH_3 resonance, as expected, is at the lowest frequency and cannot be from any Pro group. Its higher multiplicity (dd, Figure 6-9, top) arises because the two methyl groups are diastereotopic due to the chiral center to which the butyl group is attached.

The lower spectrum of Figure 6-10 shows the Pro connectivity. The highest-frequency resonance (δ 3.7) should be from the CH group adjacent (α) to the carbonyl group. The entire resonance at δ 3.7 is an overlap of this Pro CH (the higher-frequency

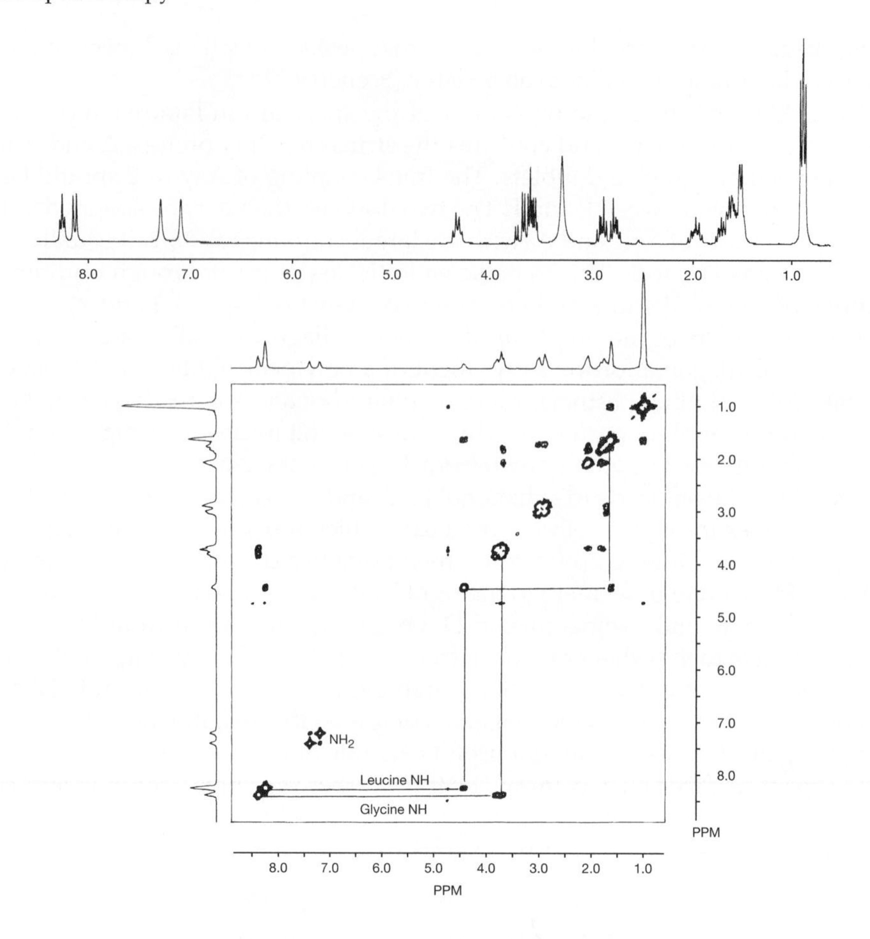

FIGURE 6-9 Top: The 300 MHz ^{1}H spectrum of the tripeptide Pro–Leu–Gly in DMSO. Bottom: COSY spectrum of Pro–Leu–Gly with connectivities of the NH protons. (Courtesy of IBM Instruments, Inc.)

portion) with the Gly CH_2 (the lower-frequency portion). The Pro CH has two cross peaks with the diastereotopic β protons at δ 1.7 and 1.9, which are mutually coupled and have their own cross peak. Unfortunately, the γ protons are nearby (δ 1.6), but their cross peak with the δ protons at δ 2.8 completes the assignment of the spectrum. The fully assigned 1D spectrum and structure are given in Figure 6-11.

False peaks and lack of symmetry around the diagonal can arise in the COSY experiment for several reasons. First, differences in digital resolution in the two periods, t_1 and t_2, may prevent perfect symmetry. Second, incorrect pulse lengths or, third, incomplete transverse relaxation during the delay time can create false cross peaks. Fourth, there may be effects from longitudinal relaxation. Any magnetization in the z direction does not precess during t_1 and therefore is rotated by the second 90° pulse into the position recognized as $\nu_1 = 0$ (the position of the reference frequency). Signals thus occur at $\nu_1 = 0$ and at any value of ν_2 associated with a resonance, resulting in a stream of lines, called *axial peaks* or t_1 *noise* depending on their positions, in the 2D plot. Fifth and finally, folding can occur in two dimensions and can give rise to off-position diagonal peaks and even cross peaks. All these artifacts may be minimized by optimizing pulse lengths, by allowing sufficient time for transverse relaxation, by using phase cycling, or by employing symmetrization. Axial peaks may be suppressed largely by alternating +90° and −90° for the second pulse, thus canceling z magnetization. The more complex CYCLOPS procedure suppresses axial peaks and eliminates other artifacts, such as quad images (Section 5-8). *Symmetrization* is a procedure for imposing bilateral symmetry around the diagonal. Most artifacts are conveniently eliminated by this procedure, but not all. For example, if two resonances have streams of t_1 noise, a point on one stream can occur at the precise mirror position (with respect to the diagonal) of a point on the other stream. Although the streams are largely eliminated, the two peaks at the symmetrical positions are retained and appear as handsome cross peaks. Usually, common sense can reject them.

Many variants of the standard COSY experiment either improve on its basic aims or provide new information. We shall consider several of them without appreciable attention to the details of the pulse sequences.

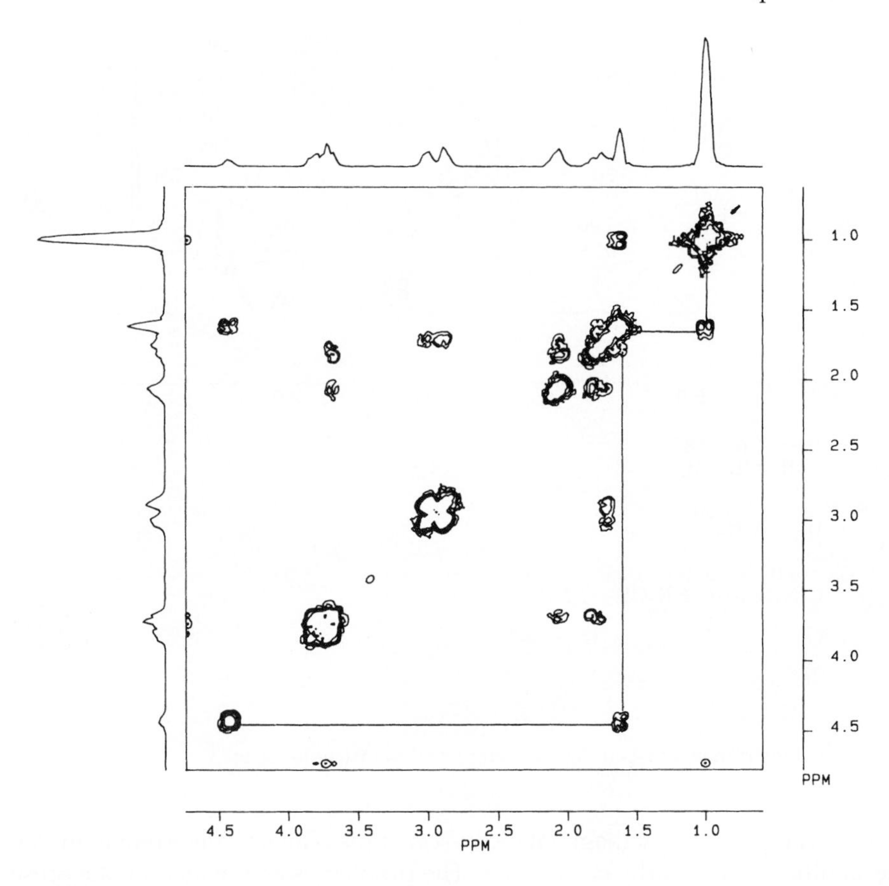

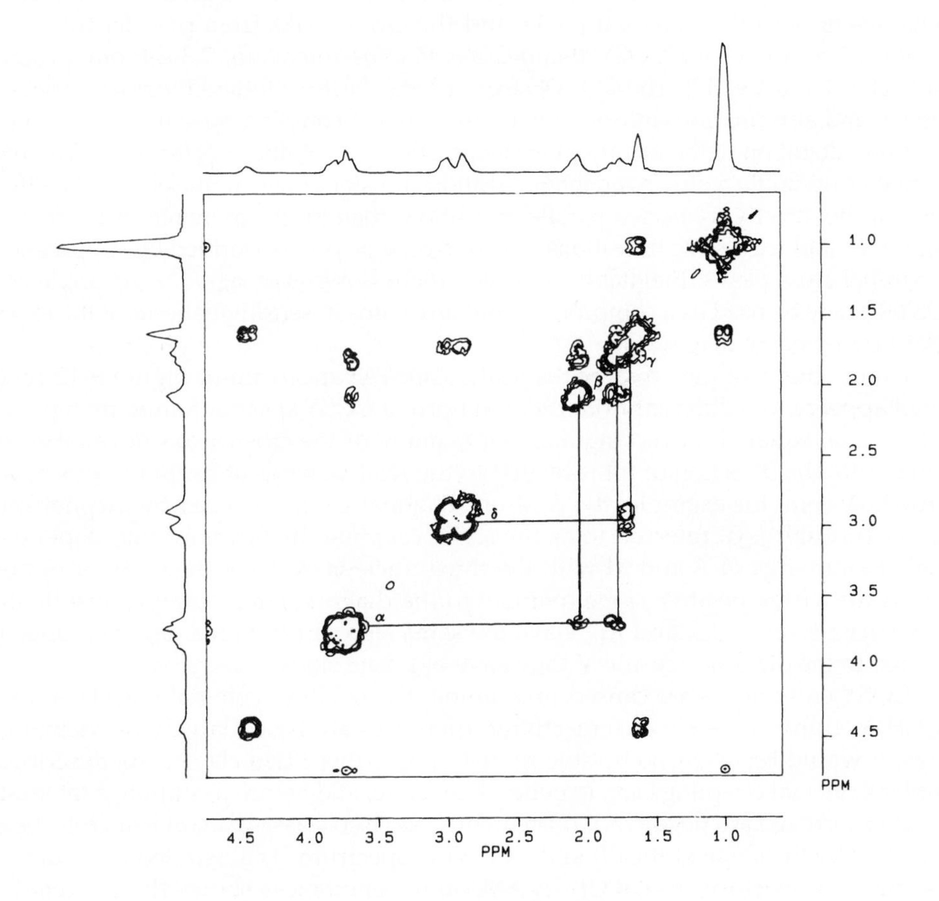

FIGURE 6-10 Top: Connectivity within the leucine portion of Pro–Leu–Gly by COSY. Bottom: Connectivity within the proline portion of Pro–Leu–Gly by COSY. (Courtesy of IBM Instruments, Inc.)

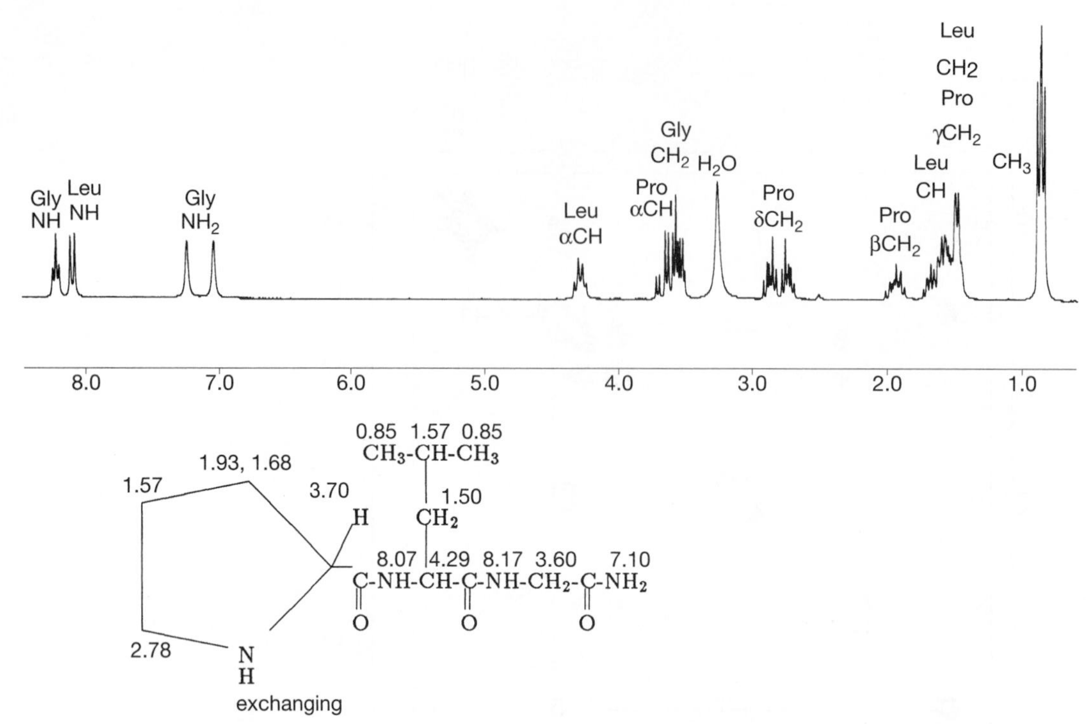

FIGURE 6-11 The fully assigned ^1H spectrum of Pro–Leu–Gly. (Courtesy of IBM Instruments, Inc.)

COSY45. The large size of diagonal peaks sometimes can be a deterrent to understanding the significance of nearby cross peaks. The problem is aggravated by the presence of cross peaks from parallel transitions (Figure 6-5). The COSY45 experiment reduces the intensities of both the diagonal peaks and the cross peaks from parallel transitions. Figure 6-12 compares the COSY90 and COSY45 experiments for 2,3-dibromopropionic acid ($CH_2BrCHBrCO_2H$). The COSY45 experiment clarifies cluttered regions close to the diagonal and also provides information on the signs of coupling constants. The name derives from alteration of the second pulse length: $90° - t_1 - 45° - t_2$ (acquire). The use of the smaller tip angle restricts the magnetization transfer between nuclei, but the effect is larger for the directly connected parallel transitions than for the more remotely connected progressive and regressive transitions. The diagonal peaks are clarified by suppression of the parallel cross peaks. Inevitably, however, there is a loss of signal. A tip angle of 60° (COSY60) may be used as a compromise, but any gains in sensitivity occur at the expense of clarification of the diagonal.

Examination of the cross peaks in the COSY45 spectrum of Figure 6-12 reveals overall appearances different from those in normal COSY spectra. Rather than possessing the usual squarish or rectangular shape, many of the cross peaks have taken on a decided tilt. The direction of tilt is related to the relative signs of coupling constants. In an AMX system, for example, the A,M off-diagonal peak is caused by magnetization transfer through J_{AM}, referred to as the *active coupling*. Its tilt, however, depends on whether couplings of A and M with the third nucleus, X, have the same or opposite signs. A tilt with a positive slope (parallel to the diagonal), for example, results if the two *passive couplings*, J_{AX} and J_{MX}, have the same sign. A tilt with a negative slope (orthogonal to the diagonal) results if they have opposite signs.

COSY cross peaks are caused predominantly by either geminal (HCH) or vicinal (HCCH) couplings. Because connectivity inferences are based largely on vicinal couplings, it would be useful to be able to distinguish these two classes. As described in Chapter 4, vicinal couplings are, in general, positive, and geminal couplings (at least on saturated carbons) are negative. Consequently, the two classes can, in principle, be distinguished by the slope of the tilt in the COSY45 spectrum, as illustrated in Figure 6-12. This spectrum is closer to ABX than AMX but nonetheless shows the expected off-diagonal COSY peaks. The resonances of the diastereotopic CH_2 protons are found at δ 3.67 and 3.89, and the resonance of the methine proton is at δ 4.49, shifted to higher

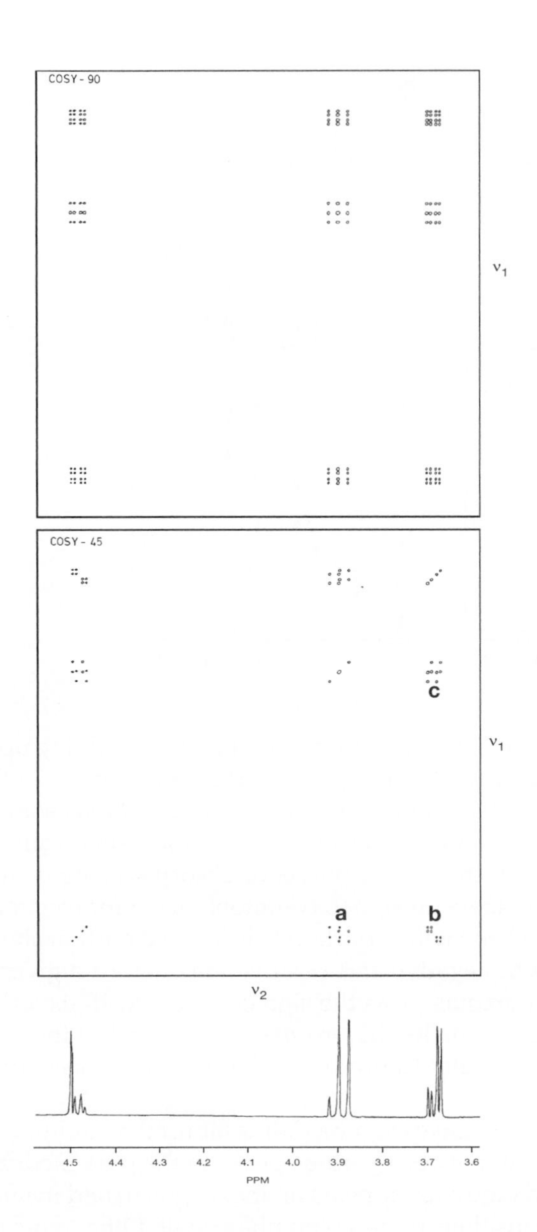

FIGURE 6-12 The COSY90 (top) and COSY45 (bottom) spectra of 2,3-dibromopropionic acid. The 1D spectrum at the bottom is a cross section through the COSY45 spectrum. (Reproduced from A. E. Derome, *Modern NMR Techniques for Chemistry Research*, Oxford, UK: Pergamon Press, 1987, p. 228.)

frequency (lower field) by attachment of the carbon to two electron-withdrawing substituents (Br and CO_2H). The off-diagonal peaks that have been labeled **a** and **b** result from the active coupling of either methylene proton with the methine proton: a positive, vicinal coupling. The passive couplings for these cross peaks are the geminal coupling to the diastereotopic partner and the vicinal coupling to the other methylene proton. As these couplings have opposite sign, the tilt has a negative slope. The off-diagonal peak labeled **c** results from active coupling between the diastereotopic methylene protons. The passive couplings for this cross peak are the two vicinal couplings between the diastereotopic protons and their vicinal neighbor. Because both passive couplings are positive, the tilt has a positive slope. In this fashion, **c** is spotted as a cross peak between geminal protons.

Long-Range COSY (LRCOSY or Delayed COSY). The normal assumption in the COSY experiment is that two- or three-bond (geminal or vicinal) couplings provide the dominant magnetization transfer to create cross peaks. Information from longer-range couplings, however, also can be useful. Introducing fixed delays Δ during the evolution and detection periods [$90° - t_1 - \Delta - 90° - \Delta - t_2$ (acquire)] enhances magnetization transfer from small couplings at the expense of large couplings. Figure 6-13 compares the COSY and LRCOSY experiments for a polynuclear aromatic compound. In the COSY spectrum, cross peaks occur only between ortho neighbors (1,2 and 3,4). In the LRCOSY spectrum, additional cross peaks arise between peri neighbors (5,6 and 4,6). Information on the connectivity between fused aromatic rings thus becomes available in the LRCOSY case.

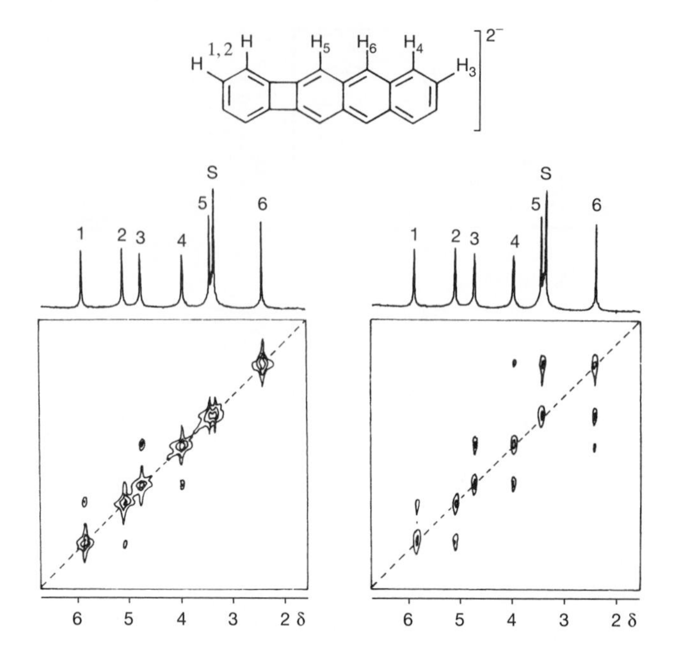

FIGURE 6-13 The 400 MHz COSY (left) and LRCOSY (right) spectra of naphthobiphenylene dianion. (The signal S is from solvent.) (Reproduced from H. Günther, *NMR Spectroscopy,* 2nd ed., Chichester, UK: John Wiley & Sons, Ltd., 1995, p. 300.)

Phase-Sensitive COSY (ϕ-COSY). Fourier transformation involves building up a signal from the sum of sine and cosine curves. Every point in the spectrum has both sine and cosine contributions, which are 90° out of phase. These contributions sometimes are called, respectively, the imaginary and real terms and lead mathematically to the *dispersion-mode* and *absorption-mode* spectra. An in-phase, or absorption, signal has the familiar form of a positive peak. A dispersion signal, commonly used for electron spin resonance spectra, has a sideways S shape with a portion below and a portion above the baseline. Such a signal produces both negative and positive maxima for a given peak and a value of zero at the resonance frequency as the sign changes. NMR experiments normally are tuned to the absorption mode by the process of phasing, but the two signals also may be combined mathematically to produce what is called a *magnitude,* or *absolute-value, spectrum.*

Many of the COSY experiments we have examined thus far used magnitude representations because of phase differences between various peaks in the pure modes. Both magnetization that is not transferred (and thus appears on the diagonal) and magnetization that is transferred to parallel transitions undergo no phase shift. Other cross peaks, however, exhibit phase shifts. A transfer between progressive transitions (A1 to X2 in Figure 6-5) shifts the phase −90°, and a transfer between regressive transitions (A1 to X1) shifts it +90°. Because absorption and dispersion modes differ by 90°, phasing the diagonal peak to absorption results in dispersive cross peaks, or vice versa. Moreover, cross peaks from progressive and regressive transitions are always out of phase by 180°. (If one cross peak represents positive absorption, then the other represents negative absorption; or if one cross peak begins a dispersive signal negatively, the other begins positively.) The magnitude, or absolute-value, mode is used to eliminate all phase differences and produce absorption-like peaks. The resulting peaks tend to be broad and often are distorted. In small molecules with little peak overlap, there may be no problem with the use of magnitude spectra, but larger molecules such as proteins, polysaccharides, or polynucleotides may produce unacceptable overlap. The phase-sensitive COSY experiment then can tune the cross peaks to a pure absorption (real) mode. This experiment not only provides enhanced resolution, but also enables coupling constants to be read more easily from the cross peaks when the data are highly digitized.

Because the 2D method involves two time domains, the transformations in both t_1 and t_2 generate real and imaginary components. As a result, the phase-sensitive 2D signal has four modes rather than two. These phase modes, or quadrants, correspond to both frequency signals being real, both being imaginary, or one being real while the other is imaginary. Figure 6-14 illustrates the four modes. The real–real (RR) mode produces the familiar peak with a contour shaped like a four-pointed star at the base.

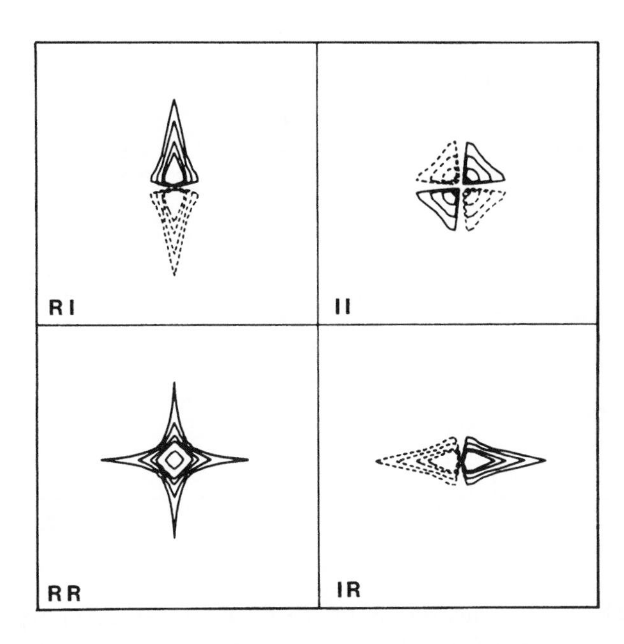

FIGURE 6-14 The four types of 2D phase quadrants, corresponding to frequency modes that are real–real (RR), imaginary–real (IR), real–imaginary (RI), and imaginary–imaginary (II). (Reproduced from A. E. Derome, *Modern NMR Techniques for Chemistry Research,* Oxford, UK: Pergamon Press, 1987, p. 207.)

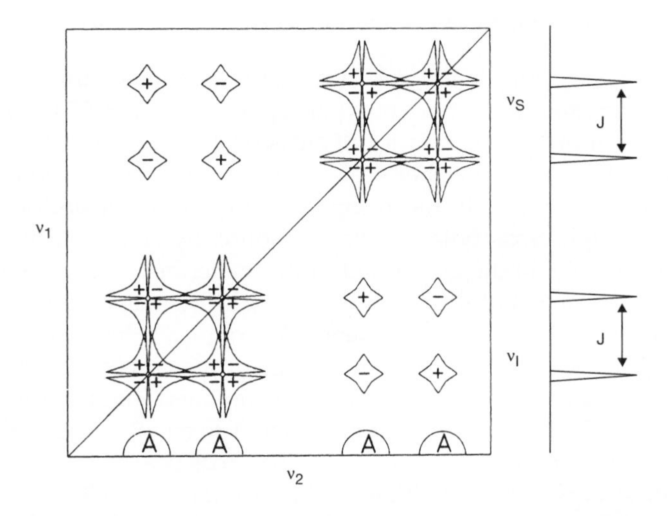

FIGURE 6-15 Phase-sensitive COSY diagram for two spins, with the diagonal peaks in dispersion mode and the cross peaks in antiphase absorption mode. The 1D spectrum is on the right. (Reproduced from F. J. M. van de Ven, *Multidimensional NMR in Liquids,* New York: VCH, 1995, p. 171.)

Figure 6-15 illustrates what the COSY spectrum for two spins looks like when the diagonal and parallel components are tuned dispersively (both imaginary) and the progressive and regressive cross peaks are tuned absorptively (both RR, but 180° out of phase). This common phase-sensitive representation provides straightforward identification of the cross peaks derived from coupling and hence determines *J*.

Multiple Quantum Filtration. The 1D INADEQUATE pulse sequence suppresses the centerband singlet in order to measure $^{13}C-^{13}C$ couplings from the satellites (Section 5-7). The procedure involves creating double quantum coherences (Section 5-8). A similar procedure may be used in 2D ^1H spectra to suppress singlets, which are single quantum coherences. Such singlets may arise from solvent or from uncoupled methyl resonances, both of which can constitute major impediments in locating highly split resonances in a complex spectrum. In a *Double Quantum Filtered* ***COSY*** (DQF–COSY) experiment, an extra 90° pulse is added after the second 90° COSY pulse, and phase cycling converts multiple quantum coherences into observable magnetizations. The resulting 2D spectrum lacks all singlets along the diagonal. For example, the spectrum of lysine, $^+NH_3CH(CH_2CH_2CH_2CH_2NH_2)CO_2^-$, on the left of Figure 6-16 has no solvent (HOD) peak, which was suppressed as a single quantum coherence. An important feature of the phase-sensitive DQF–COSY experiment is that double quantum filtration allows both diagonal and cross peaks to be tuned into pure absorption at the same time. This feature reduces the size of all the diagonal signals and permits cross peaks close to the diagonal to be observed. The only disadvantage of DQF–COSY is a reduction in

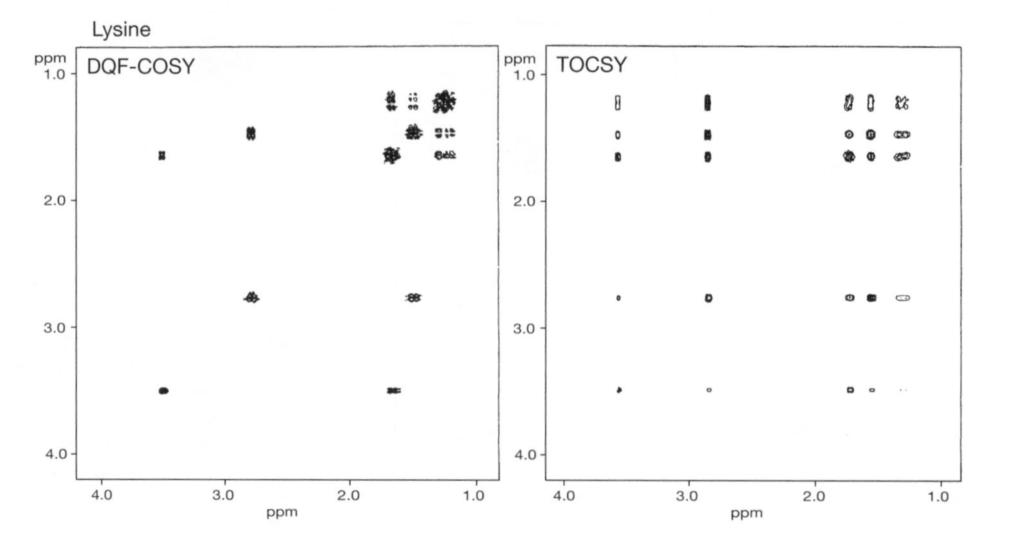

FIGURE 6-16 The DQF–COSY and TOCSY spectra of lysine. (Reproduced from J. N. S. Evans, *Biomolecular NMR Spectroscopy,* Oxford, UK: Oxford University Press, 1995, p. 428.)

sensitivity by a factor of two. The *Triple Quantum Filtered* ***COSY*** (TQF–COSY) experiment removes both singlets and AB or AX quartets, providing even greater spectral simplification. It is rarely used because of a concomitant increased loss of sensitivity.

TOtal Correlation Spectroscop Y (TOCSY). In the standard COSY experiment, the connectivity within an entire spin system, such as that in a butyl group ($CH_3CH_2CH_2CH_2$—), must be mapped out from proton to proton via a series of cross peaks. By spin locking the protons during the second COSY pulse, the chemical shifts of all the protons may be brought essentially into equivalence. Recall that resonance frequencies of protons and carbons are made equal through cross polarization for solids by achieving the Hartmann–Hahn condition (Section 2-9). In the 2D variant of this experiment, the initial 90° pulse and the period t_1 occur as usual, but the second pulse locks the magnetization along the y axis so that all protons have the spin lock frequency. All coupled spins within a spin system then become closely coupled to each other, and magnetization is transferred from one spin to all the other members, even in the absence of J couplings. The spectrum on the right of Figure 6-16 shows the TOCSY experiment for lysine. The methylene group at the lowest frequency (upper right corner) exhibits four TOCSY cross peaks, one with each of the other three methylene groups and one with the methine proton. The TOCSY experiment, a variation of which is called the ***HO****monuclear* ***HA****rtmann-****HA****hn* or HOHAHA, experiment, has particular advantages for large molecules, including enhanced sensitivity and, if desired, the phasing of both diagonal and cross peaks to the absorption mode. The process of identifying resonances within specific amino acid or nucleotide residues is considerably simplified by this procedure. Each residue can be expected to exhibit cross peaks among all its protons and none with protons of other residues.

Relayed COSY. An alternative, but less general, method for displaying extended levels of connectivity is provided by ***R****elayed* ***C****oherence* ***T****ransfer* (RCT). The typical COSY experiment for an AMX system with $J_{AX} = 0$ produces cross peaks between A and M and between M and X. It is not unusual for the key diagonal peak for M to be coincident with a resonance from another spin system, making it difficult to follow the connectivity path. (Recall, for example, the Leu portion of the COSY spectrum of Pro–Leu–Gly in Figure 6-10.) The RCT experiment generates a cross peak between the A,M and M,X cross peaks, eliminating the ambiguity. The result is shown diagrammatically in Figure 6-17 for AMX and A'M'X' systems whose M and M′ resonances coincide. The COSY experiment contains the expected four cross peaks. The RCT experiment contains two additional cross peaks, connecting the cross peaks of the individual spin systems. The two new cross peaks are labeled (A,M,X) and (A',M',X'). The connectivity of AMX and of A'M'X' is then rendered unambiguous.

***J*-Resolved Spectroscopy.** In Chapter 5, we saw how the spin–echo experiment can isolate or remove characteristics of chemical shifts or coupling constants. Spin echoes may be used in two dimensions to generate one frequency dimension representing chemical shifts and

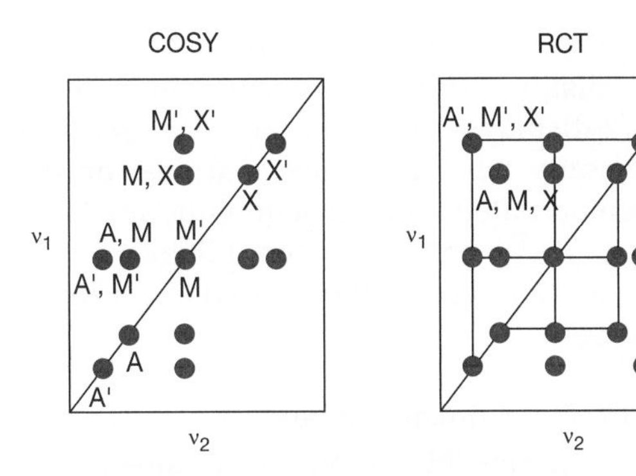

FIGURE 6-17 Diagram of COSY and relayed coherence transfer (RCT) experiments for two three-spin systems (AMX and A'M'X') whose M and M' portions overlap. (Adapted from F. J. M. van de Ven, *Multidimensional NMR in Liquids*, New York: VCH, 1995, p. 233.)

another representing coupling constants. The sequence $90° - \frac{1}{2}t_1 - 180° - \frac{1}{2}t_1 - t_2$, for example, uses the 180° pulse to refocus chemical shifts during t_1. The result for a glucose derivative is shown in Figure 6-18. The ^{1}H frequencies are found on the horizontal axis (ν_2 in δ), with the normal 1D spectrum displayed at the top (Figure 6-18a). The vertical axis ("f_1" = ν_1) contains only proton–proton coupled multiplets, each centered about a zero frequency point; that is, all multiplets occur at the same chemical shift in $\nu_1 = 0$. Thus, the multiplet at the highest frequency (lowest field) from H-3 is a quartet, seen with further splitting when viewed from the vertical axis. By taking a projection at an angle (45°) that causes each of the members of the individual multiplets to overlap when viewed from the horizontal axis, as at the bottom (b), a display is obtained that, in essence, is a proton–proton decoupled proton spectrum. Resonances devoid of any couplings are present at each frequency. This projection is a novel way to examine ^{1}H spectra, although it has not seen widespread use because it reveals no connectivities.

The pulse sequence, as a variant of the spin–echo experiment, also refocuses the spread of frequencies caused by field inhomogeneity, so that some improvement in resolution is obtained. The inset at the lower right of Figure 6-18 shows the normal 1D spectra of H-4 and H-5 at the top (Figure 6-18c and e) and the unrotated projection of the 2D *J*-resolved spectra at the bottom [Figure 6-18d and f, extracted from the projected spectrum (Figure 6-18a) at the top of the 2D display]. The much higher resolution of the 2D resonances is clearly evident. Thus, the procedure is an effective way to measure *J* accurately, particularly when *J* is poorly resolved in the 1D spectrum. The experiment fails for closely coupled nuclei (second-order spectra).

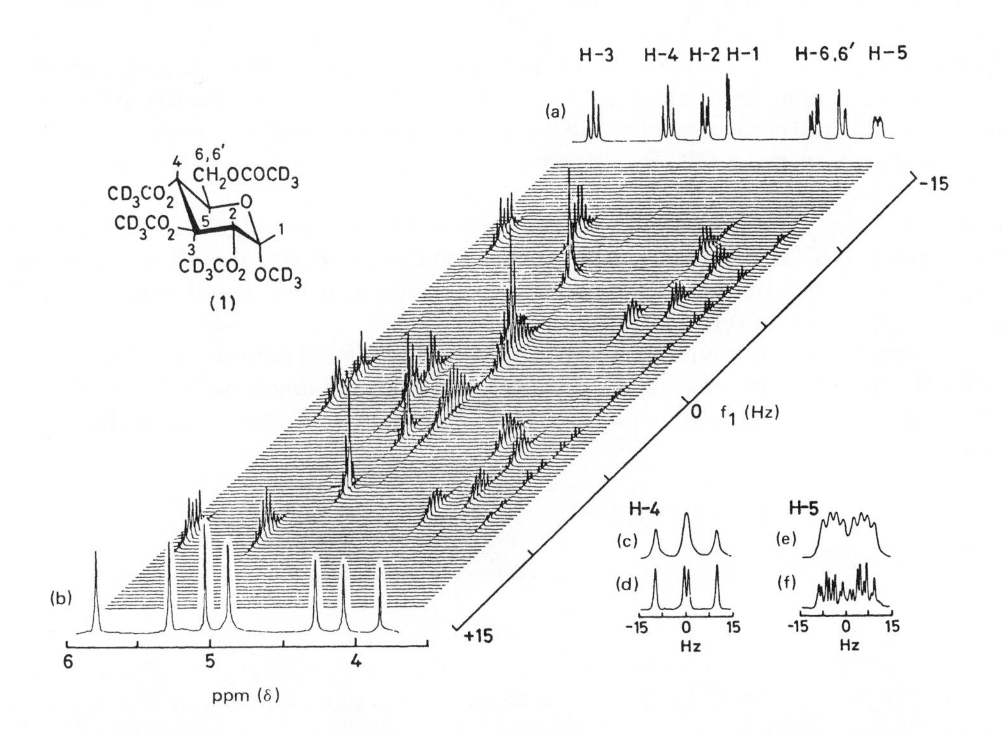

FIGURE 6-18 The 270 MHz 2D *J*-resolved ^{1}H spectrum of 2,3,4,6-tetrakis-*O*-trideuteroacetyl-α-D-glucopyranoside. (Reproduced from L. D. Hall, S. Sukumar, and G. R. Sullivan, *J. Chem. Soc., Chem. Commun.*, 292 [1979]).

In addition to resolving small couplings that may be absent in the 1D spectrum, the J-resolved procedure can be used to distinguish homonuclear from heteronuclear couplings. The vertical axis in the figure displays couplings only between spins that were affected by the 180° pulse. Hence, only $^1H-^1H$ couplings appear on that axis. Couplings to heteronuclei are not phase modulated and, consequently, appear as spacings along the horizontal axis. In this way, $^1H-^{19}F$ and $^1H-^{31}P$ couplings may be distinguished from $^1H-^1H$ couplings, which are removed in the rotated spectrum such as that shown in (b) at the bottom of the figure.

COSY for Other Nuclides. The basic COSY experiment can be carried out for any spin-$\frac{1}{2}$ nucleus that is 100% abundant. In addition to $^1H-^1H$, the procedure thus is applicable to $^{19}F-^{19}F$ (F,F-COSY) and to $^{31}P-^{31}P$ (P,P-COSY), respectively, in organofluorine and organophosphorus compounds. When the nuclide is less than 100% abundant, as for Li,Li-COSY, the uncoupled centerband must be separated from the coupled satellites, usually by the 2D INADEQUATE procedure (Section 6-4).

6-2 PROTON–HETERONUCLEUS CORRELATION

Cross peaks in the standard COSY experiment are generated through magnetization transfer that arises from scalar (J) coupling between protons. Coupling from a proton to a different nuclide, such as carbon-13, should be able to generate a similar response. Analogous cross peaks, then, would provide very useful information about which carbons are bonded to which protons. Thus, the assignment of a proton resonance would automatically lead to the assignment of the resonance of the carbon to which it is bonded, and vice versa. This field has seen considerable development recently, and there now are several pulse sequences commonly used to explore connectivity between protons and carbon or other heteronuclides.

HETCOR. The simplest 2D sequence that includes magnetization transfer proton to carbon takes the form given in Figure 6-19. The pulse sequence is reminiscent of the 1D INEPT sequence, and manipulation of magnetization is much the same as in Figure 5-19, but without the 180° pulse. The initial 90° ^1H pulse generates y magnetization. For the simplest case, in which one carbon is bonded to one proton (as in $CHCl_3$), ^1H magnetization evolves during the period t_1 according to its Larmor frequency. Two ^1H vectors diverge due to coupling with ^{13}C. The second 90° ^1H pulse generates nonequilibrium z magnetization that is transferred to ^{13}C in the manner of the INEPT experiment of Figure 5-19. The single 90° ^{13}C pulse then provides the ^{13}C free-induction decay that is acquired during t_2. The 2D spectrum then has one axis in ^1H frequencies (ν_1) and one in ^{13}C frequencies (ν_2).

For the simple heteronuclear AX case, the 2D spectrum contains two peaks when it is projected onto either the ^1H or the ^{13}C axis (the A and X portions, respectively, of the AX spectrum), as in the INEPT experiment. Thus, the ^1H resonances that are detected correspond to the ^{13}C satellites of the usual ^1H spectrum. The 2D display contains four peaks: two along a diagonal and two symmetrically off the diagonal. Moreover, the peaks are in antiphase for each nuclide, since the INEPT spectrum without decoupling generates one peak up and one peak down. Decoupling would result in summing the peaks algebraically to zero.

To bring about decoupling, another pulse and two fixed periods are added (Figure 6-20). The first (90°_x) ^1H pulse allows chemical shifts and coupling constants to evolve during t_1. The 180° ^{13}C pulse refocuses the H–C coupling constants in the ^1H dimension,

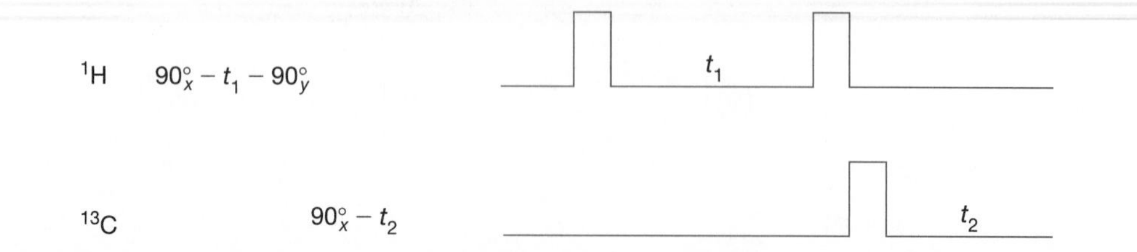

FIGURE 6-19 The pulse sequence for the HETCOR experiment.

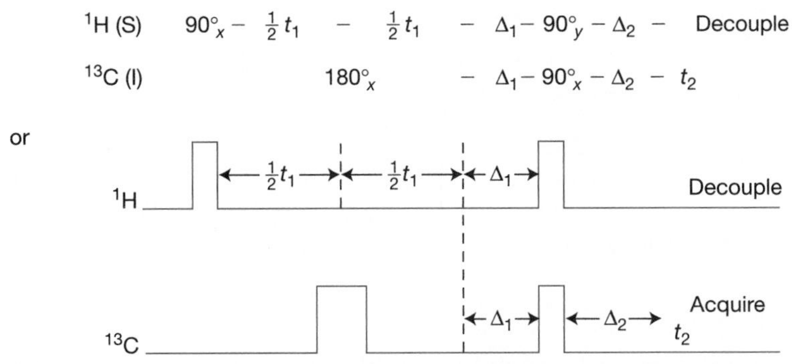

^{1}H (S) $\quad 90°_x - \frac{1}{2}t_1 \quad - \quad \frac{1}{2}t_1 \quad - \Delta_1 - 90°_y - \Delta_2 -$ Decouple

^{13}C (I) $\quad 180°_x \quad - \Delta_1 - 90°_x - \Delta_2 - t_2$

or

^{1}H $\quad \leftarrow \frac{1}{2}t_1 \rightarrow \leftarrow \frac{1}{2}t_1 \rightarrow \leftarrow \Delta_1 \rightarrow$ Decouple

^{13}C $\quad \leftarrow \Delta_1 \rightarrow \quad \leftarrow \Delta_2 \rightarrow \; t_2$ Acquire

FIGURE 6-20 The pulse sequence for the HETCOR experiment with decoupling.

thereby decoupling ^{1}H from ^{13}C. The fixed time Δ_1 allows the ^{1}H vectors to obtain the antiphase (180° out of phase) relationship illustrated in Figure 5-19d. The second ($90°_y$) ^{1}H pulse moves the vectors in antiphase relationship onto the z axis and enables polarization to be transferred to ^{13}C, also in the antiphase relationship. The 90° ^{13}C pulse is for observation. The second fixed time, Δ_2, restores phase alignment and permits ^{13}C to be decoupled from ^{1}H during ^{13}C acquisition.

Figure 6-21 illustrates this experiment for an adamantane derivative. The ^{1}H frequencies are on the vertical axis, and the ^{13}C frequencies are on the horizontal axis. The respective 1D spectra are illustrated on the left and at the top. The 2D spectrum is composed only of cross peaks, each one relating a carbon to its directly bonded proton(s). There are no diagonal peaks (and no mirror symmetry associated with a diagonal), because two different nuclides are represented on the frequency dimensions. Quaternary carbons are invisible to the technique, as the fixed times Δ_1 and Δ_2 normally are set to values for one-bond couplings. This experiment often is a necessary component in the complete assignment of ^{1}H and ^{13}C resonances. Its name, ***HET**eronuclear chemical-shift* ***COR**relation*, is abbreviated as HETCOR, but other acronyms—for example HSC, for ***H**eteronuclear **S**hift **C**orrelation*, and H,C-COSY—have been used. The method may be applied to protons coupled to many other nuclei, such as ^{15}N, ^{29}Si, and ^{31}P, as well as ^{13}C.

Figure 6-21 illustrates two advantages of the HETCOR experiment: (1) The correlation between protons and carbons means that spectral assignments for one nuclide

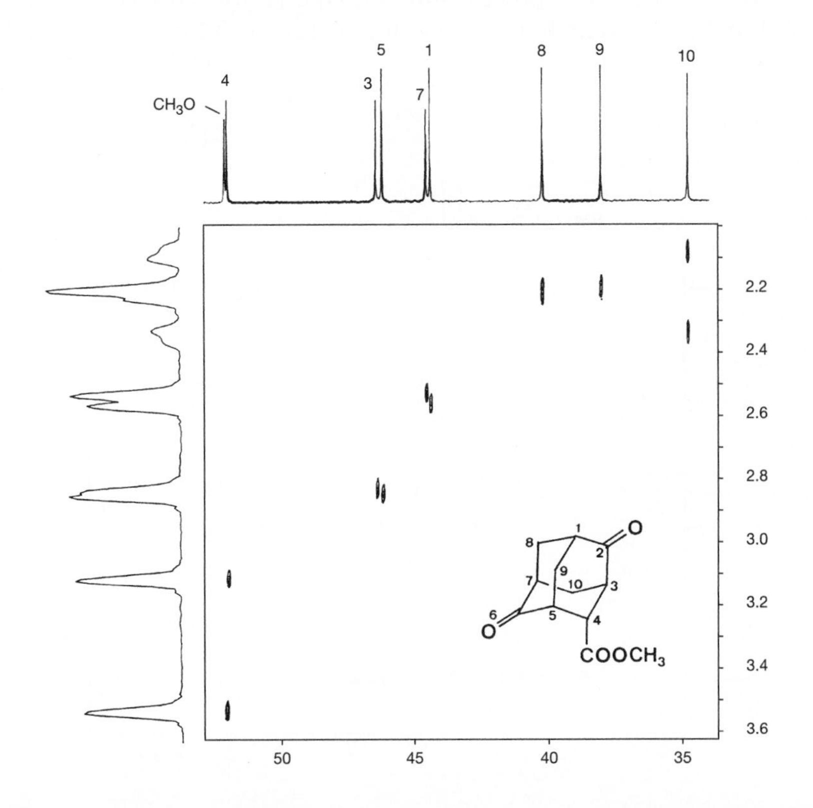

FIGURE 6-21 The HETCOR spectrum of 4-(methoxycarbonyl)adamantane-2,6-dione. (Reproduced from H. Duddeck and W. Dietrich, *Structure Elucidation by Modern NMR*, Darmstadt, Germany: Steinkopff Verlag, 1989, p. 22.)

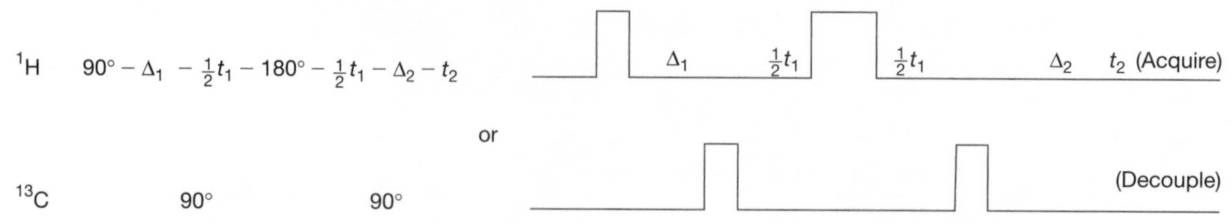

FIGURE 6-22 The pulse sequence for the HMQC experiment.

automatically lead to spectral assignments for the other. Thus, ^{13}C assignments can assist in ^{1}H assignments, and vice versa. (2) Overlapping proton resonances often can be dispersed in the carbon dimension. Even at very high fields, proton resonances can overlap; consider, for example, those of H8 and H9, which coincide at δ 2.2 in the figure. The presence of the two cross peaks from C8 (δ 40) and C9 (δ 38) reveals the spectral overlap in the proton dimension. Similar considerations apply to H3 and H5.

COSY cross peaks can arise from either geminal (HCH) or vicinal (HCCH) connectivities, so that ambiguities can be present. Geminally related protons that are diastereotopic are attached to a common carbon and hence have HETCOR connectivities to a single ^{13}C frequency, whereas vicinally related protons are attached to different carbons and thus have HETCOR connectivities to two different ^{13}C frequencies. This advantage is not illustrated in the figure, but is useful for distinguishing geminal from vicinal relationships in COSY spectra.

HMQC. A major drawback to the HETCOR experiment is the low sensitivity that results from detection of the X nucleus (usually ^{13}C). The HMQC (***H****eteronuclear correlation through* ***M****ultiple* ***Q****uantum* ***C****oherence*) experiment uses *inverse detection*, whereby ^{13}C responses are observed in the ^{1}H spectrum. The pulse sequence is given in Figure 6-22 and represents a transfer of coherence rather than of polarization. The initial ^{1}H magnetization from the 90° pulse becomes antiphase during the fixed period Δ_1 through the ^{1}H–^{13}C coupling constant. Multiple quantum coherence then is created by the first ^{13}C pulse. The remainder of the sequence is designed to select double or higher quantum coherence (from the ^{13}C satellites in the ^{1}H spectrum) over single quantum coherence (from the ^{1}H centerbands), in a process similar to the 1D INADEQUATE experiment in Section 5-7. The 2D representation, as shown in Figure 6-23 for camphor, still includes

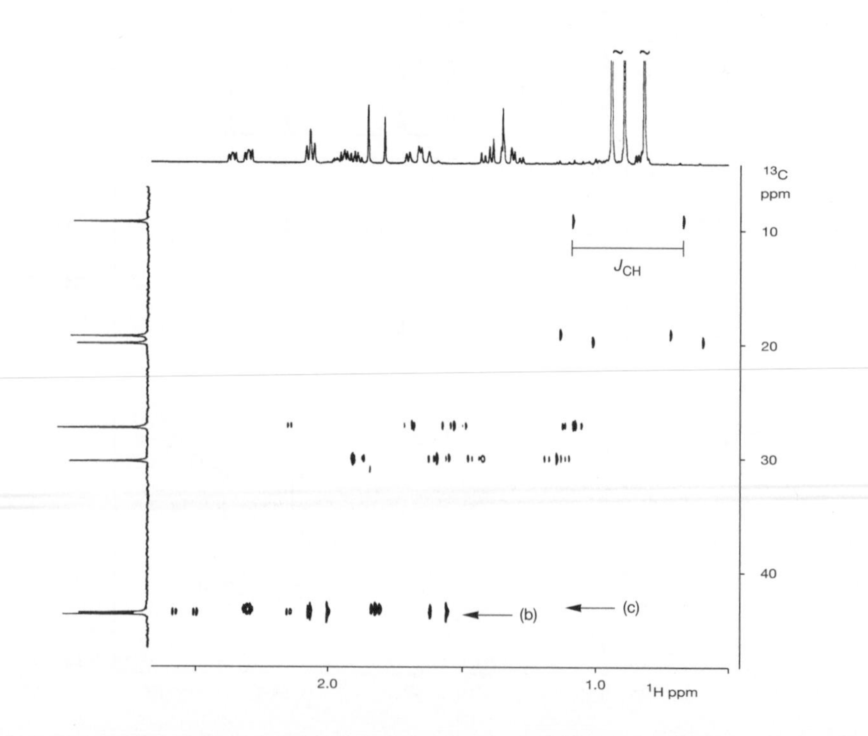

FIGURE 6-23 The HMQC spectrum of camphor, with ^{1}H–^{13}C couplings retained in the ^{1}H dimension. (Reproduced from J. K. M. Sanders and B. K. Hunter, *Modern NMR Spectroscopy,* 2nd ed., Oxford, UK: Oxford University Press, 1993, p. 111.)

the ^{1}H–^{13}C coupling information in the ^{1}H dimension, although ^{13}C irradiation can be applied during the ^{1}H t_2 acquisition period to provide decoupling. The major difference between HETCOR and HMQC is that the acquisition period t_2 is at ^{13}C frequencies in the former experiment (Figure 6-20), but at ^{1}H frequencies in the latter (Figure 6-22). Consequently, the HMQC experiment is much more sensitive. Like HETCOR, HMQC can be used with heteronuclei other than ^{13}C, of which the experiment involving ^{15}N is the most common.

BIRD–HMQC. The most difficult aspect of implementing the HMQC experiment is the suppression of signals from protons attached to ^{12}C (the centerband or single quantum coherences) in favor of the protons attached to ^{13}C (the satellites or double quantum coherences). The use of pulsed field gradients (Section 6-6) is the most effective technique, but many spectrometers still lack the hardware required for their generation. Fortunately, there is an effective alternative for the suppression of centerbands by means of the BIRD (***BI**linear **R**otation **D**ecoupling*) sequence, which is outlined by the vector notation in Figure 6-24. Two sets of vectors are followed, one for the protons attached to ^{12}C and one for the protons attached to ^{13}C. The initial 90° proton pulse (Figure 6-24a and a') along the x direction moves all magnetization onto the y axis (Figure 6-24b and b'). (Keep in mind that eventually the inverse detection HMQC pulses will be applied, with ultimate detection in the proton channel.) Protons on ^{12}C are unsplit, so their magnetization evolves as depicted in the upper set of vector diagrams. After the delay period $(2J)^{-1}$, these vectors have reached some arbitrary angle with respect to the y axis (Figure 6-24c), according to their individual Larmor frequencies. (Only one such frequency is illustrated.) By contrast, protons on a ^{13}C evolve as two vectors (Figure 6-24c'), separated by the frequency of the one-bond coupling constant ($\Delta\omega = 2\pi\Delta\nu = 2\pi J$), as illustrated in the lower set of vector diagrams. (The centers of the diagrams are maintained on the y axis for viewing simplicity, but the chemical shifts of the vectors evolve according to their Larmor frequencies.) After the delay period $(2J)^{-1}$, these vectors are separated by a 180° angle [Figure 6-24c', Section 5-5, $\phi = 2\pi(\Delta\nu)t = 2\pi J(2J)^{-1} = \pi$]. At this time, a 180° pulse is applied to the ^{1}H channel along the y axis, and a 180° pulse is applied to the ^{13}C channel. The ^{1}H pulse rotates the vector for the protons attached to ^{12}C about the y axis (Figure 6-24d), so that, after another period $(2J)^{-1}$, that vector converges onto the y axis to create a spin echo (Figure 6-24e). The simultaneous ^{13}C pulse, however, switches the spin identities of the coupled partners for the protons attached to ^{13}C, so that the vectors continue to diverge, as was discussed with regard to Figure 5-14. After the second period $(2J)^{-1}$, these vectors converge onto the $-y$ axis (Figure 6-24e', $\phi = 360°$). Application of the second 90° pulse along the x axis moves all the proton magnetization back to the z axis. Because they had opposite phases in Figure 6-24e and e', the protons attached to ^{12}C now point along the $-z$ axis (Figure 6-24f), whereas the protons attached to ^{13}C point along the $+z$ axis (Figure 6-24f'). The experiment then requires a delay time τ for relaxation. The protons attached to ^{13}C are already essentially at equilibrium and remain unaffected (Figure 6-24g'), but those attached to ^{12}C are upside down and hence begin to relax back to equilibrium. The HMQC pulses are applied after a time τ that brings the magnetization for the protons attached to ^{12}C exactly to zero (Figure 6-24g). At this point, BIRD has suppressed the single quantum coherences (^{12}C–^{1}H, Figure 6-24g) and selected the multiple quantum coherences (^{13}C–^{1}H, Figure 6-24g').

The overall experiment thus becomes BIRD–τ–HMQC–DT, in which τ is chosen to null the single quantum coherences and DT is the normal delay time between pulse repetitions. DT includes the time taken to acquire the signal, as well as a recycle time during which the signal is regenerated through relaxation. The details of the experiment require some knowledge of the relaxation times T_1 and an appropriate choice of the intervals τ and DT. The optimized result can provide effective suppression of the unwanted signals for small molecules. Larger molecules, however, can undergo a negative nuclear Overhauser effect (Section 6-3) from the inverted signals, with loss of sensitivity.

HSQC. The *Heteronuclear Single Quantum Correlation* (HSQC) experiment is an alternative to HMQC that accomplishes a similar objective. The experiment generates, via an

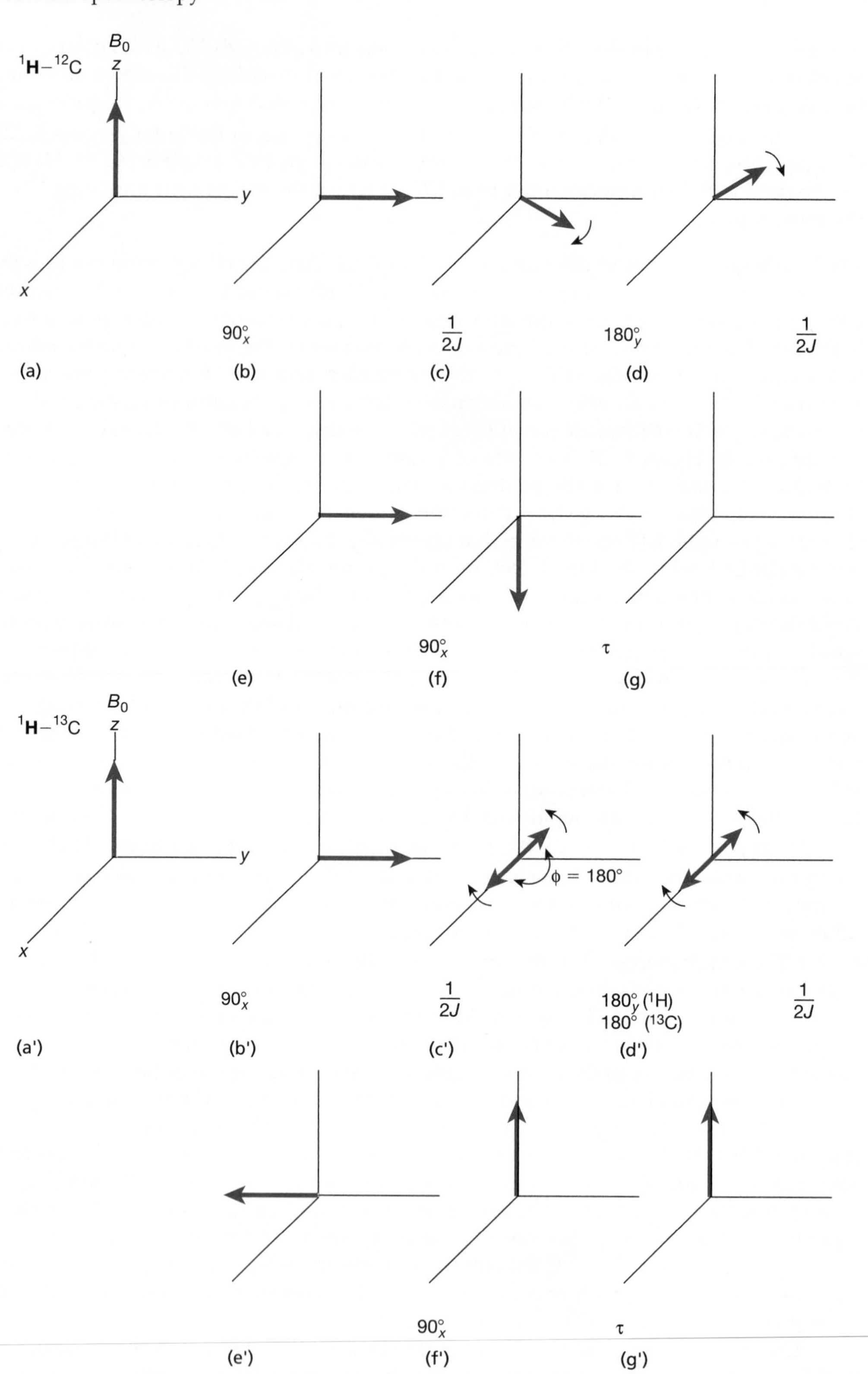

FIGURE 6-24 The BIRD sequence for the selection of signals from protons attached to ^{13}C over those from protons attached to ^{12}C.

INEPT sequence, single quantum ^{13}C (or ^{15}N) coherence, which evolves and then is transferred back to the proton frequency by a second INEPT sequence, this time in reverse. The main difference from the HMQC result is that HSQC spectra do not contain ^{1}H–^{1}H couplings in the ^{13}C (ν_1) dimension. As a result, HSQC cross peaks tend to have improved resolution over analogous HMQC cross peaks. HSQC is preferred when there is considerable spectral overlap.

COLOC. To focus on longer-range H–C couplings, the fixed times Δ_1 and Δ_2 of HETCOR can be lengthened accordingly. Loss of magnetization due to transverse relaxation then reduces sensitivity significantly. The COLOC pulse sequence (***CO****rrelation spectroscopy via* ***LO****ng-range* ***C****oupling*) avoids this problem by incorporating the ^{1}H evolution period t_1 inside the Δ_1 delay period. Figure 6-25 shows the COLOC spectrum for vanillin. The circled cross peaks are residues from one-bond couplings. The only long-range coupling of the methoxy group is with C-3, which indicates that methoxy is connected at that point. Other long-range couplings, however, also are seen, for example, between C-1 and H-5, C-3 and H-5, and C-2 and H-7.

The principal disadvantage of the COLOC sequence lies in the fixed nature of the evolution period. In such a pulse sequence, C–H correlations are diminished or even absent when two- and three-bond ^{1}H–^{13}C couplings are of a magnitude similar to that of ^{1}H–^{1}H couplings within a molecular fragment. This situation occurs quite commonly (Chapter 4). The FLOCK sequence (so named because it contains three BIRD sequences, Figure 6-24) contains a variable evolution time, in which t_1 becomes progressively larger, and avoids the potential loss of C–H correlations. Although the experiment detects carbon, it is quite useful when there is signal overlap in the ^{1}H spectrum.

HMBC. Correlations through longer-range H–C couplings offer at least two potential advantages. (1) Heteronuclei that lack an attached proton are invisible to the HETCOR/HMQC/HSQC family of experiments. Thus, carbonyl groups and tetrasubstituted carbons cannot be studied. Such carbons, however, are likely to show coupling to protons two or more bonds away. (2) A single carbon can be correlated with several neighboring protons. Hence, connectivities over heteroatoms or carbonyl groups can help define larger groupings of atoms, thereby complementing information from COSY. The COLOC and FLOCK sequences, however, suffer from the same drawback as HETCOR, namely, low sensitivity arising from direct observation of a nuclide (^{13}C, ^{15}N, and so on) with a low gyromagnetic ratio.

Proton-detected *Heteronuclear Multiple Bond Correlation* (HMBC) is designed to provide correlations between protons and heteronuclei such as carbon or nitrogen by a pulse

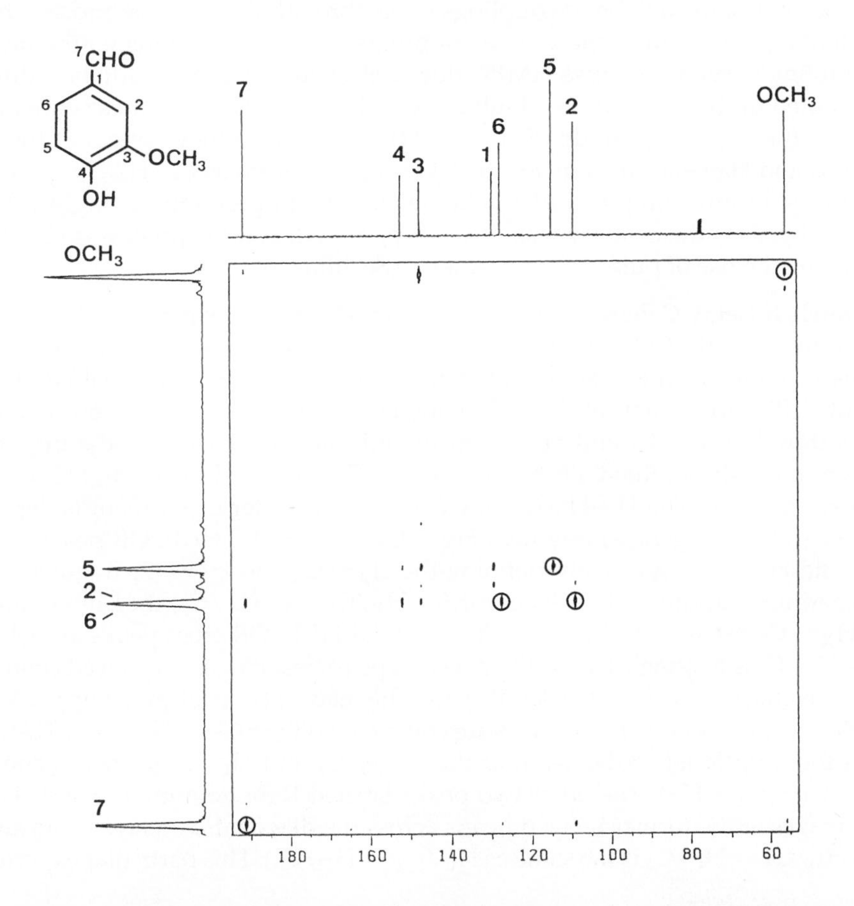

FIGURE 6-25 The COLOC spectrum of vanillin. (Reproduced from H. Duddeck and W. Dietrich, *Structure Elucidation by Modern NMR,* Darmstadt, Germany: Steinkopff Verlag, 1989, p. 24.)

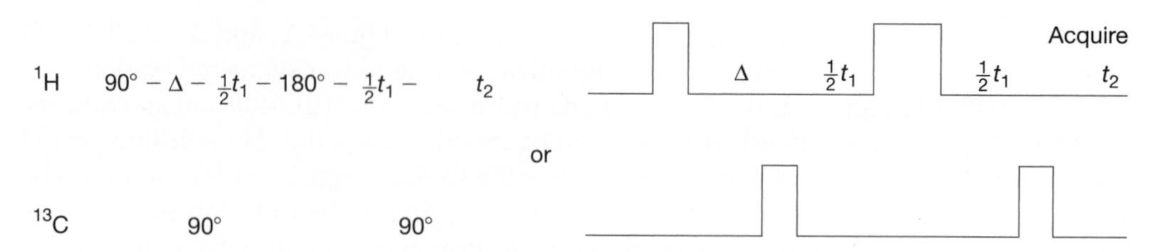

FIGURE 6-26 The pulse sequence for the HMBC experiment.

sequence similar to that used in HMQC (Figure 6-26). A second delay time (Δ_2) is no longer necessary, because H–C couplings are not intended to be removed. The major difference from HMQC is the duration of the initial delay time Δ, which is lengthened in order to select for the smaller couplings over two or more bonds. These couplings are usually in the range 0–15 Hz [Section 4-5 for ^2J(HCC) and Section 4-6 for ^3J(HCCC)]. A typical delay Δ of $(^2J)^{-1}$ then corresponds to the range 60–200 ms ($J = 2.5$–8.3 Hz), in comparison with 24 ms for HMQC. Shorter delay times tend to improve sensitivity (there is less loss of signal through relaxation), but longer times may be entirely acceptable for small molecules with relatively long relaxation times. Longer delay times occasionally permit observation of connectivities through four-bond couplings, such as ^4J(HCCCC), or even five (Section 4-7). Interpreting the magnitudes of H–C couplings over two to five bonds entails all the subtleties of analogously interpreting H–H couplings and includes consideration of inductive effects, zigzag pathways, π bonding, Karplus stereochemical considerations, and so on (Sections 4-6 and 4-7). Methyl groups often exhibit the most intense HMBC correlations, because of the multiplicative effect of simultaneous detection by three protons and because free rotation provides almost no stereochemical restrictions that can reduce couplings.

Figure 6-27 provides the HMBC spectrum of the illustrated heterocycle. As with HETCOR, the spectrum contains only cross peaks. The carbonyl carbon C8 apparently couples with H2, H5, H6, H6', H7, and H7', as it exhibits cross peaks with all these protons, representing all possible two- and three-bond couplings. Three further points are worth noting. First, the proton pairs on C4, C6, and C7 are diastereotopic. Separate peaks are observed for each such connectivity, for example, that between C8-H7 and C8-H7'. Second, some one-bond couplings break through the selection process, because of accidental phase coincidence. Such couplings show up as doublets (the one-bond H–C coupling survives because HMBC does not include carbon irradiation during t_2) and are noted in the spectrum by double-headed arrows. The cross peaks are obvious, because of the large magnitude of $^1J(^{13}C-^1H)$ and because they appear at the coincidence of C and H chemical shifts for the CH fragment, such as C2/H2. They can be filtered out by a more complex pulse scheme. Third, suppression of single coherence ($^1H-^{12}C$) signals is the primary task of the experiment, and the process is significantly enhanced by the use of pulsed field gradients (Section 6-6).

Heteronuclear Relay Coherence Transfer. The relay COSY experiment (RCT) may be adapted to the HETCOR context. The result of such an experiment is shown in Figure 6-28 for the illustrated acetal of acrolein, along with the normal HETCOR experiment ("HSC" in the figure). The following HETCOR cross peaks are present: H_C/C_2 at the bottom left, H_B/C_1 and H_A/C_1 in the middle, and H_D/C_3 at the upper right. This experiment shows that C_3 is bonded to H_D, C_2 is bonded to H_C, and C_1 is bonded to both H_A and H_B. The H—H—C relay coherence transfer experiment is depicted at the upper right. In a general fragment $H_X—C_X—C_Y—H_Y$, HETCOR peaks occur for H_X/C_X and H_Y/C_Y, respectively, defining the $H_X—C_X$ and $H_Y—C_Y$ fragments. In the RCT experiment, additional peaks occur for H_X/C_Y and H_Y/C_X, thereby defining the larger $H_X—C_X—C_Y—H_Y$ fragment. The normal HETCOR cross peaks are labeled N (that for H_C/C_2 is missing). The additional cross peaks resulting from relayed connectivity between given CH pieces are labeled R. Thus, the relayed (R) peak at the upper left indicates that the H_D/C_3 and H_C/C_2 pairs are connected ($H_D—C_3—C_2—H_C$). The relayed peak in the middle left indicates that the $H_{A/B}/C_1$ and H_C/C_2 pairs are connected ($H_{A/B}—C_1—C_2—H_C$). The other two peaks labeled R (bottom middle and right) are mirror images of the former relayed peaks across the diagonal, as relay occurs symmetrically ($H_D/C_3 \rightarrow H_C/C_2$ is the same as $H_C/C_2 \rightarrow H_D/C_3$). This particular experiment is

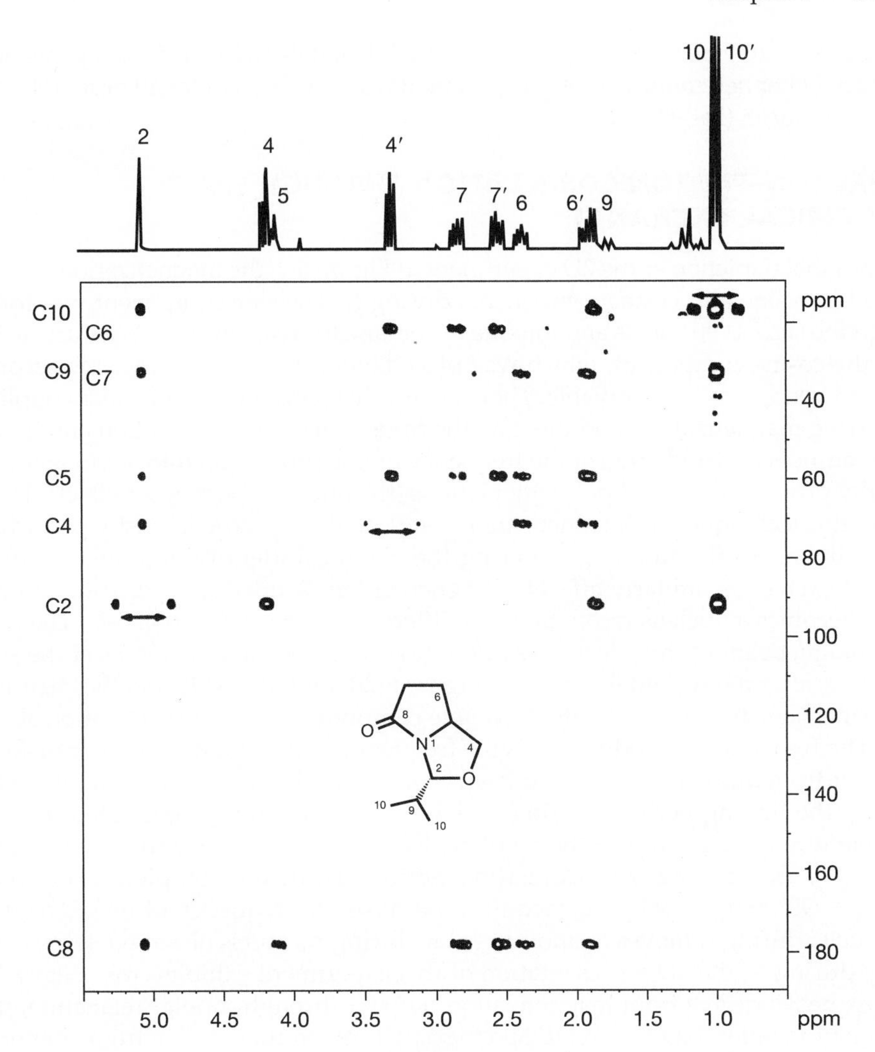

FIGURE 6-27 The HMBC spectrum of the illustrated heterocycle. The 1D ^{1}H spectrum is given at the top. The vertical axis corresponds to ^{13}C frequencies, the horizontal axis to ^{1}H frequencies. One-bond correlations are indicated by peaks connected by double-headed arrows. (Reproduced from T. D. W. Claridge, *High-Resolution NMR Techniques in Organic Chemistry*, Oxford, UK: Pergamon Press, 1999, p. 245.)

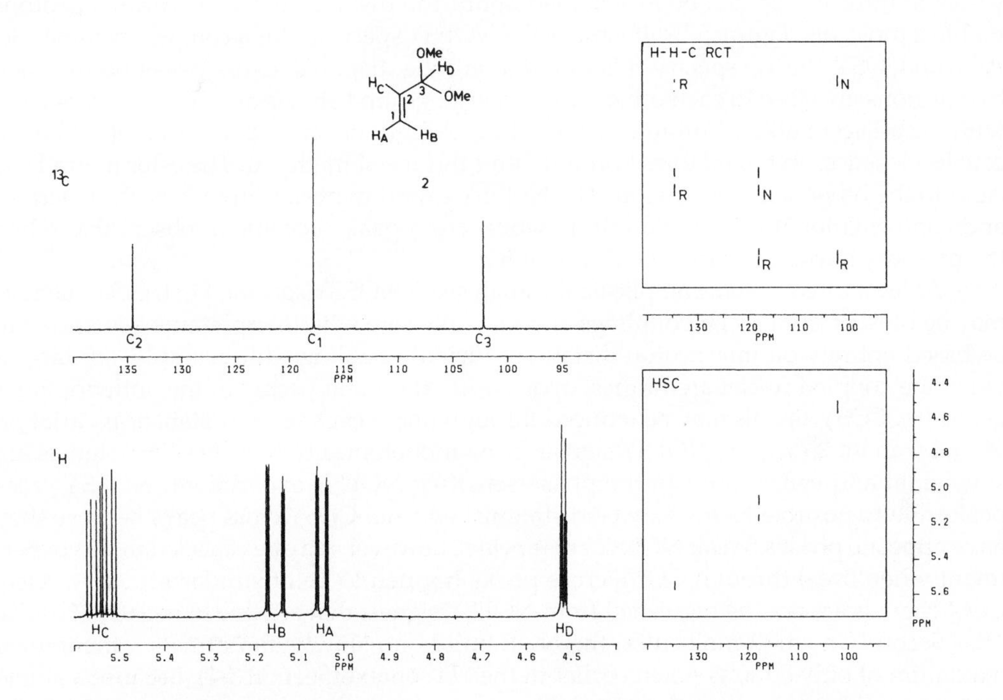

FIGURE 6-28 Left: The ^{13}C and ^{1}H spectra of the dimethyl acetal of acrolein. Right: The normal HETCOR (labeled HSC, bottom) and the H–H–C relay coherence transfer (top) experiments for the same molecule. (Reproduced from A. E. Derome, *Modern NMR Techniques for Chemistry Research*, Oxford, UK: Pergamon Press, 1987, p. 257.)

called H–H–C RCT because the pulses involve ^{1}H signals twice and ^{13}C signals once, in that order. Other heteronuclear relay experiments can involve a different order (H—C—H) or different nuclei (H—H—N).

6-3 PROTON–PROTON CORRELATION THROUGH SPACE OR CHEMICAL EXCHANGE

In the original depiction of the 2D experiment in Figure 6-2, the magnetization vector was resolved into sine and cosine components during t_1. The sine component was followed through the second 90° pulse and into the t_2 domain, to create the COSY sequence. We ignored the cosine component, which was placed along the $-z$ axis after the second 90° pulse and hence was unobservable. There are mechanisms other than scalar coupling for transferring magnetization, and these methods can affect z magnetization and hence the cosine component. Irradiation at the frequency of one proton can transfer magnetization to nearby protons through dipolar interactions (the nuclear Overhauser effect). The clear effect of this technique on z magnetization is reflected in spectral intensity perturbations in one dimension (Section 5-4). Altering the chemical identity of a nucleus through chemical exchange similarly affects z magnetization. A nucleus resonating at one frequency becomes a nucleus resonating at a different frequency (Section 5-2). The population (z magnetization) thus decreases at the first frequency and increases at the second.

Thus, after the second 90° pulse in Figure 6-2d, both the NOE and the chemical exchange mechanisms can modulate the cosine component of the magnetization along the z axis. The frequency of modulation is the frequency of the magnetization transfer partner, either from dipolar relaxation or from chemical exchange. After a suitable fixed period (τ_m, the mixing period), during which this modulation is optimized, the cosine component may be moved to the xy plane by a third 90° pulse and may be detected along the y axis during a t_2 acquisition period. Thus, the complete experiment is $90° - t_1 - 90° - \tau_m - 90° - t_2$ (acquire). Because the frequency of magnetization of some nuclei during t_1 moves to another value during τ_m and is observed at the new frequency during t_2, the 2D representation of this experiment exhibits cross peaks. When the cross peaks derive from magnetization transfer through dipolar relaxation, the 2D experiment is called NOESY (***NOE*** *Spectroscop****Y***). When they derive from chemical exchange, the experiment is called EXSY (***EX****change* *Spectroscop****Y***).

The duration of the fixed time τ_m depends on the relaxation time T_1, the rate of chemical exchange, and the rate of NOE buildup. In the case of the NOESY experiment, valuable information can be ascertained about the distance between various protons within a molecule. Figure 6-29 illustrates the NOESY spectrum for a complex heterocycle. As with COSY, the 1D spectrum is found along the diagonal. Cross peaks occur when two protons are close to each other. Thus, methyl group **l** shows an expected cross peak with the adjacent alkenic proton **a** (upper left). Additional cross peaks of methyl **l** indicate its closeness to the methinyl proton **f** and the acetal methyl **n**. The ester methyl **e** is close to the other acetal methyl **m**. The NOESY experiment can provide both structural and conformational information. In practice, cross peaks become unobservable when the proton–proton distance exceeds about 5 Å.

At least three factors complicate the analysis of NOESY spectra. First, COSY signals may be present from scalar couplings and may interfere with interpretations intended to be based entirely on interproton distances. Vicinal couplings, for example, are largest when the coupled nuclei are farthest apart, a situation that occurs in the antiperiplanar geometry. COSY signals may be reduced through phase cycling or by statistical variation of τ_m by about 20%. (The NOESY signals grow monotonically, but the COSY signals are sinusoidal and cancel out.) In the phase-sensitive NOESY experiment, NOESY cross peaks due to positive NOEs may be distinguished from COSY cross peaks because they have opposite phases. Weak NOESY cross peaks, however, may be canceled in this experiment when breakthrough COSY cross peaks happen to be of similar intensity. Also, COSY signals are not distinguished from NOESY signals that are due to negative NOE's.

Second, in small molecules, the NOE builds up slowly and attains a theoretical maximum of only 50%, as noted earlier in the 1D context (Section 5-4). Because a single proton may be relaxed by several neighboring protons, the actual maximum normally

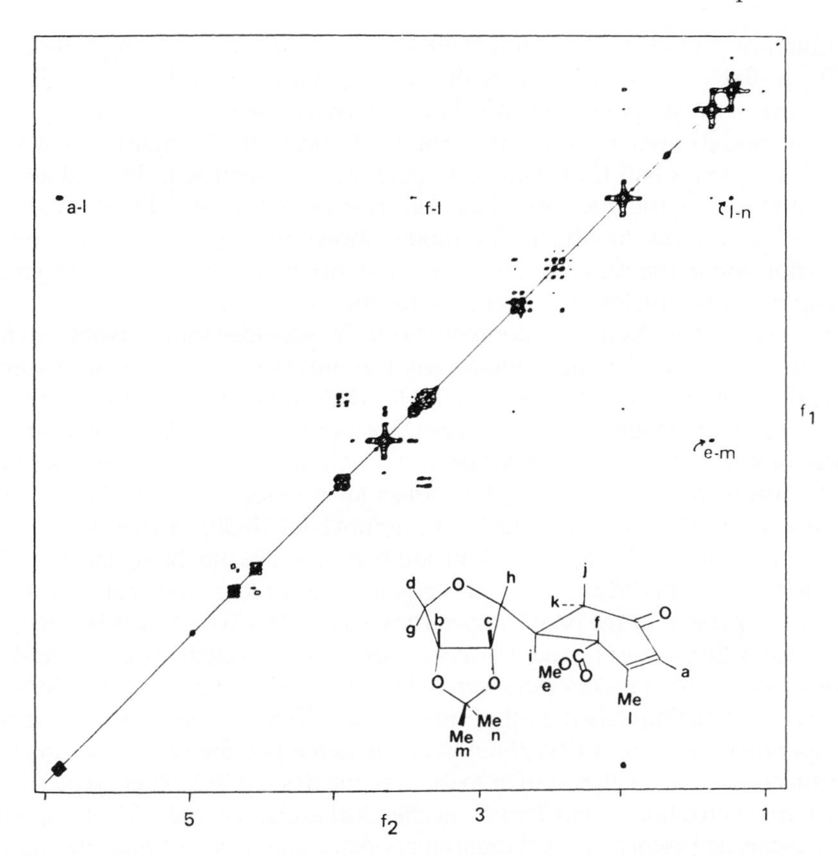

FIGURE 6-29 The ^{1}H NOESY spectrum for the indicated compound. (Courtesy of Bruker Instruments, Inc.)

is much less than 50%. (Of course, the same problem exists in the 1D NOE experiment.) Moreover, as the molecular size increases and behavior departs from the extreme narrowing limit, the maximum NOE decreases to zero and becomes negative. Thus, particularly for medium-sized molecules, the NOESY experiment may fail. For larger molecules, whose relaxation is dominated by the W_0 term, not only is the maximum NOE −100% rather than +50%, but also the NOE buildup occurs more rapidly. The NOESY experiment thus has been of particular utility in the analysis of the structure and conformation of large molecules such as proteins and polynucleotides.

Third, in addition to its transfer directly from one proton to an adjacent proton, magnetization may be transferred by spin diffusion. In this mechanism, already described for the 1D experiment (Section 5-4), magnetization is transferred through the NOE from one spin to a nearby second spin and then from the second to a third spin that is close to the second spin, but not necessarily to the first one. These multistep transfers can produce NOESY cross peaks between protons that are not close together. Spin diffusion can occur even through two or more intermediate spins, but the process becomes increasingly less efficient. The direct transfer of magnetization and its transfer by spin diffusion sometimes may be distinguished by examining the NOE buildup rate, as illustrated in Figure 6-30. In

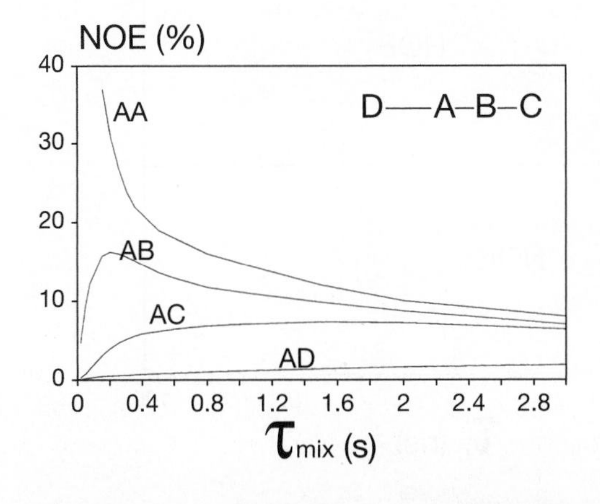

FIGURE 6-30 Peak intensities calculated for a hypothetical NOESY experiment involving four nuclei, D—A–B–C, with D 4 Å from A and B 2 Å from A and C. The curve labeled AA is for the diagonal peak, and the remaining curves are for the various cross peaks. (Reproduced from F. J. M. van de Ven, *Multidimensional NMR in Liquids,* New York: VCH, 1995, p. 188.)

this hypothetical plot of the NOE intensity as a function of the mixing time τ_m for the system D—A–B–C, AA is the intensity of the diagonal peak, and the other lines represent intensities of cross peaks. The NOE between two close protons (A and B, separated by 2 Å in the model) rises most rapidly. Protons A and D are 4 Å apart and show a very small NOE. Protons A and C also are 4 Å apart (2 Å between A and B and another 2 Å between B and C), but the intensity of the AC cross peak rises steadily through spin diffusion with B as the intermediary. The model shows that spin diffusion provides the major contribution to the AC cross peak for a distance of 4 Å, but, for the AB cross peak, the buildup of spin diffusion is slower than for the direct transfer.

The rotating-frame NOESY experiment (ROESY) provides some advantages for small and medium-sized, as well as large, molecules. The pulse sequence for ROESY (previously called CAMELSPIN) is similar to TOCSY or HOHAHA, although the period of spin locking is chosen to optimize magnetization transfer through the NOE (via dipolar interactions) rather than through scalar couplings. Whereas the NOE decreases to zero and becomes negative as the mean correlation time τ_c for molecular rotation increases (larger molecules move more slowly), in the rotating frame the maximum NOE (ROE) remains positive and even increases from 50% to 67.5% (Figure 6-31). In addition to enhancing the signal, the ROESY experiment decreases spin diffusion, offering advantages for large molecules. Just as COSY artifacts may be present in the NOESY spectrum, so can TOCSY artifacts be present in the ROESY spectrum. Steps must be taken to remove them, as provided by the T-ROESY variant. The use of a weak static spin-lock pulse can reduce the TOCSY peaks. Positive ROESY cross peaks also can be distinguished easily from negative TOCSY peaks in a phase-sensitive ROESY experiment. As with COSY/NOESY, there is the possibility of canceling signals if TOCSY artifacts have intensities similar to those of the desired ROESY cross peaks.

When magnetization is transferred via chemical exchange in the EXSY experiment, it may be necessary to perform several preliminary experiments to optimize the value of τ_m, which should be approximately $1/k$. Figure 6-32 illustrates the EXSY experiment from an early example by Ernst, in which the diagonal peaks run nontraditionally from upper left to lower right. At fast exchange, the 1D ^{1}H spectrum of the heptamethylbenzenium ion contains only one methyl resonance, as the methyl group moves around the ring. At slow exchange, there are distinct resonances for the four types of methyls labeled on the left in the figure. The EXSY experiment shows which methyls interchange with which. One can imagine 1,2, 1,3, or 1,4 shifts, but the EXSY experiment agrees only with the 1,2 mechanism. Each off-diagonal peak indicates magnetization transfer between two diagonal peaks. Thus, the 1 methyls have a cross peak only with (and hence exchange only with) the 2 methyls, the 2 methyls exchange with the 1 and 3 methyls, the 3 methyls exchange with the 2 and 4 methyls, and the 4 methyls exchange only with the 3 methyls. This is the pattern expected for 1,2 shifts.

The intensities of the cross peaks depend on the rate constant for exchange. For the case of exchange between equally populated sites lacking spin–spin coupling, such as the two methyls of *N,N*-dimethylformamide [H(CO)N(CH_3)$_2$], the rate constant k is

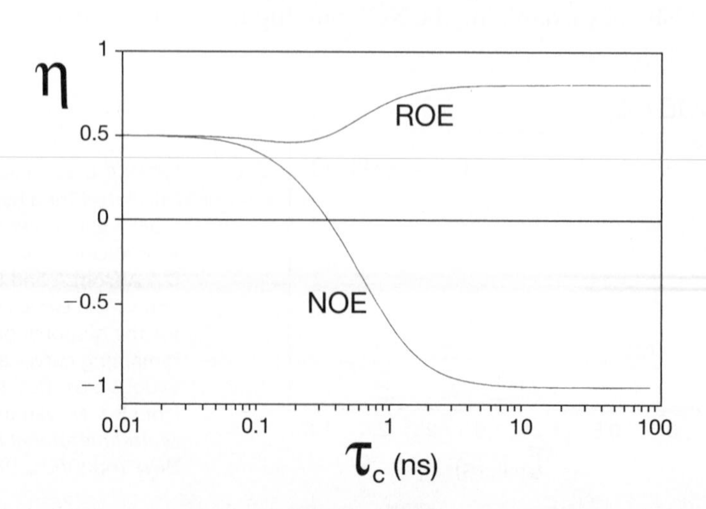

FIGURE 6-31 The enhancement factor η as a function of the effective correlation time τ_c for the standard nuclear Overhauser experiment (NOE) and for the spin-lock, or rotating-coordinate, variant (ROE). The curves were calculated for an interproton distance of 2.0 Å and a spectrometer frequency of 500 MHz. (Reproduced from F. J. M. van de Ven, *Multidimensional NMR in Liquids,* New York: VCH, 1995, p. 251.)

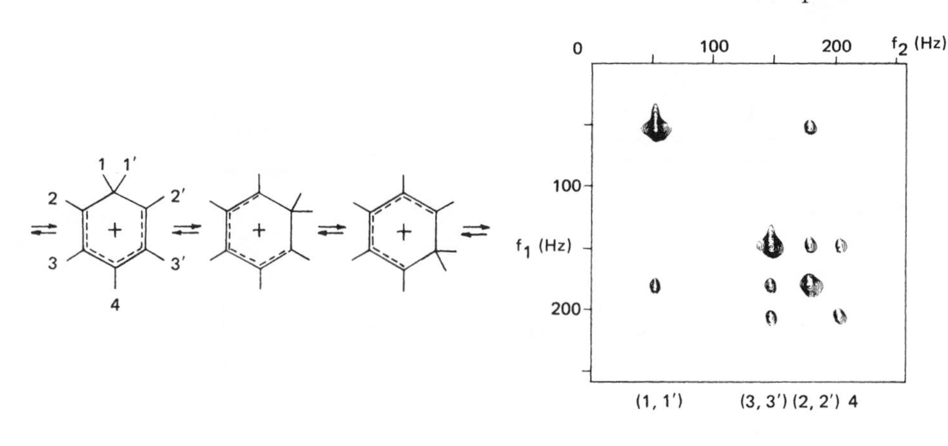

FIGURE 6-32 The ^{1}H EXSY spectrum for the heptamethylbenzenium ion. (Reproduced with permission from R. H. Meier and R. R. Ernst, *J. Am. Chem. Soc.*, **101,** 6441 [1979]. Copyright 1979 American Chemical Society.)

related to the mixing time τ_m, the intensity of the cross peak I_c, and the intensity of the diagonal peaks I_d by the formula of eq. 6-1.

$$\frac{I_d}{I_c} \sim \frac{(1 - k\tau_m)}{k\tau_m} \tag{6-1}$$

Rearranging this expression algebraically gives the rate constant k from eq. 6-2.

$$k \sim \frac{1}{[\tau_m(I_d/I_c + 1)]} \tag{6-2}$$

Since the pulse sequence is the same for EXSY and NOESY, cross peaks in the NOESY (or ROESY) experiment might be mistaken for EXSY cross peaks, or vice versa. They can be distinguished in the phase-sensitive experiment, since EXSY and NOESY/ROESY peaks have opposite phases.

6-4 CARBON–CARBON CORRELATION

The 1D INADEQUATE experiment provides a method for measuring ^{13}C–^{13}C coupling constants and for determining carbon–carbon connectivity by establishing coupling magnitudes that are common to two carbon atoms (Section 5-7). In practice, application of the method to solving connectivity problems is complicated not only by the inherently low sensitivity of detecting two dilute nuclei but also by the similarity of many ^{13}C–^{13}C couplings. Duddeck and Dietrich have pointed out that all the one-bond carbon–carbon couplings in cyclooctanol fall into the narrow region from 34.2 to 34.5 Hz, except for C_1—C_2, which is 37.5 Hz. The second problem may be largely alleviated by translating the experiment into two dimensions. The original INADEQUATE experiment (Section 5-7) can be adapted directly to two dimensions by incrementing the fixed time Δ as the t_1 domain: $90^\circ_x - 1/4J_{CC} - 180^\circ_y - 1/4J_{CC} - 90^\circ_x - t_1 - 90^\circ_\phi - t_2$ (acquire).

The period t_1 is used to encode the double quantum frequency domain. The resulting 2D display contains a horizontal axis in ν_2 (the normal ^{13}C frequencies) and a vertical axis that is a double quantum domain represented by the sum of the frequencies of coupled ^{13}C nuclei ($\nu_1 = \nu_A + \nu_X$). The latter frequencies are referenced to a transmitter frequency at zero.

Figure 6-33 illustrates the 2D INADEQUATE spectrum of menthol (**6-2**). The experiment also has been called C–C–COSY, as the cross peaks represent connectivity between

7 CH_3, 1, 2, 3, 4, 5, 6, OH, 8, 9 H_3C, 10 CH_3

6-2

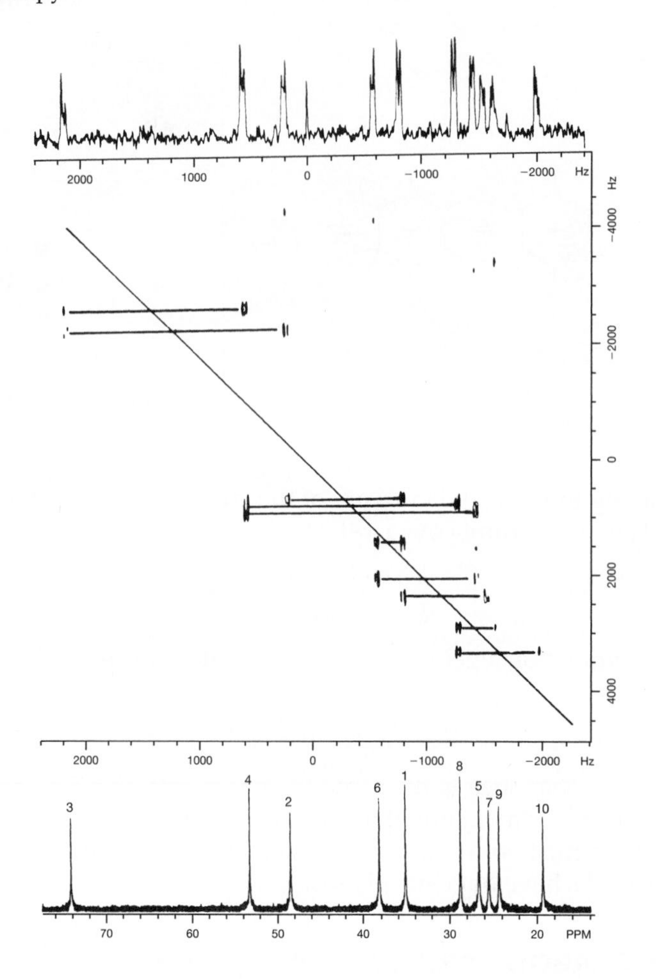

FIGURE 6-33 The 2D INADEQUATE spectrum of menthol, with the ^{1}H-decoupled ^{13}C spectrum. (Reproduced from G. E. Martin and A. S. Zektzer, *Two-Dimensional NMR Methods for Establishing Molecular Connectivities*, New York: VCH, 1988, p. 362.)

two carbons. There are no diagonal peaks (which would arise from ^{13}C nuclei with ^{12}C neighbors), because the experiment removes single quantum signals. The diagonal usually is drawn in, as in the figure. The normal proton-decoupled ^{13}C spectrum is shown at the bottom. At the top, the 2D procedure permits recovery of the carbon-coupled ^{13}C spectrum through a projection of the ν_2 dimension.

To obtain connectivity from a 2D INADEQUATE experiment, a single assignment is made and the remainder of the structure is mapped, much as with COSY. Only a gap caused by the presence of a heteroatom, C—X—C, prevents mapping the entire skeleton. For menthol, the oxygen-substituted C-3 resonates at the highest frequency (far left) and serves as the starting point. Horizontal lines are drawn between coupled carbons in the 2D spectrum, passing through the diagonal at their midpoints. There are two cross peaks at the C-3 frequency, corresponding to connectivities to C-2 and to C-4. Of these, the secondary C-2 should be at lower frequency (higher field). The connectivity then may be mapped in the following fashion: C-2 → C-1 → C-6 (and from C-1 to the C-7 methyl) → C-5 → C-4 → C-8 (and from C-4 to the original C-3) → C-9 and C-10. The 2D INADEQUATE procedure also is applicable to concatenations of other coupled, dilute nuclides, such as ^{29}Si/^{29}Si, ^{11}B/^{11}B, and ^{6}Li/^{6}Li in organosilicon, organoboron, and organolithium systems, respectively.

The major disadvantage to this experiment is its extremely low sensitivity. A variation is INEPT-INADEQUATE (sometimes called ADEQUATE), which uses proton observation and pulsed field gradients over pathways such as H—C—C. The resulting spectrum resembles that of 2D INADEQUATE, but can contain peaks only when at least one of the paired carbons has an attached proton. Although such technical refinements ameliorate the problem of sensitivity, this family of experiments has not been widely used.

6-5 HIGHER DIMENSIONS

The enormous complexity of spectra of large biomolecules such as proteins, polynucleotides, and polysaccharides has led to the development of three-dimensional (3D) and four-dimensional (4D) experiments. Two independently incremented evolution periods (t_1 and t_2), in conjunction with three separate Fourier transformations and of an acquisition period t_3, result in a cube of data with three frequency coordinates.

Figure 6-34, from a study by van de Ven, illustrates the complexity of the 2D NOESY spectrum of a DNA-binding protein of phage Pf3, consisting of 78 amino acids. The vertical line at δ 9.35 highlights the problems at a single resonance position in the NH region. The NH proton in a given peptide unit —CHR'—CO—NH—CHR—CO could have one cross peak with its own CHR proton and another with the neighboring CHR′ protein, but the NOESY spectrum contains more than a dozen cross peaks at the one frequency of δ 9.35. Thus, more than one NH must be generating cross peaks at that frequency.

The nitrogen HMQC experiment provides information about the connectivity of nitrogens and their attached protons. For proteins, the use of HMQC normally requires isotopic enrichment of ^{15}N, which is obtained by growing an organism in a medium containing a single nitrogen source, such as $^{15}NH_4Cl$. (Similarly, ^{13}C enrichment may be obtained from a medium containing ^{13}C-labeled glucose.) The normal 2D HMQC spectrum (^{15}N vs. ^{1}H) for this same protein is given in Figure 6-35, and two connectivities are seen at the ^{1}H frequency of δ 9.35 (vertical line). Accordingly, there are two NH resonances (or more if there are coincidences) at δ 9.35.

The 3D experiment takes the 2D experiments in Figures 6-34 and 6-35 into an additional dimension. The 3D procedure illustrated in Figure 6-36 labels each NOESY peak with the ^{15}N frequency through the HMQC method, thus combining NOESY and HMQC data. The pulses and time delays constitute the standard ^{1}H NOESY sequence

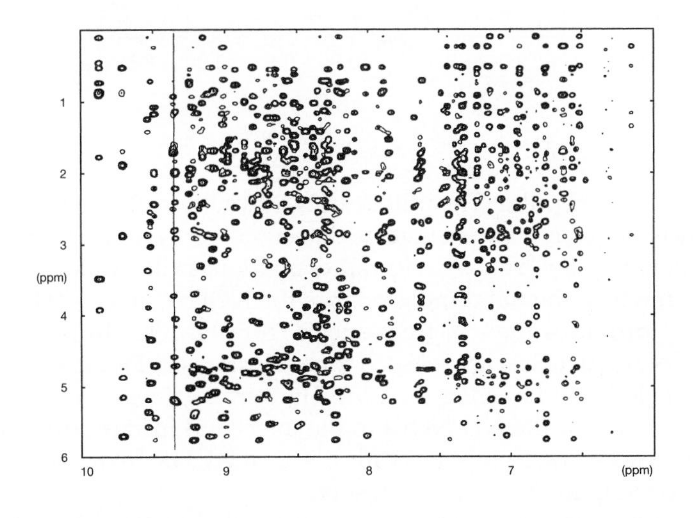

FIGURE 6-34 A portion of the NOESY spectrum of a DNA-binding protein of phage Pf3 containing cross peaks between NH and aliphatic protons. (Reproduced from F. J. M. van de Ven, *Multidimensional NMR in Liquids*, New York: VCH, 1995, p. 296.)

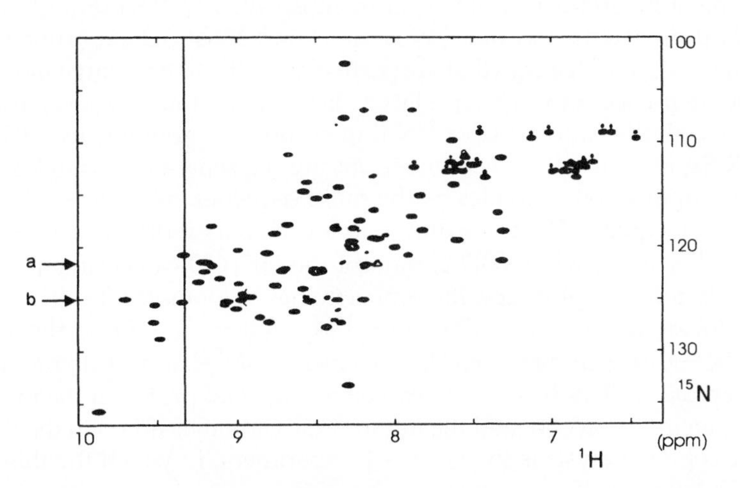

FIGURE 6-35 The ^{1}H/^{15}N HMQC spectrum of ^{15}N-labeled Pf3. The arrows are explained in Figure 6-38. (Reproduced from F. J. M. van de Ven, *Multidimensional NMR in Liquids*, New York: VCH, 1995, p. 297.)

FIGURE 6-36 The pulse sequence for the 3D NOESY/HMQC experiment.

^{1}H	90° − t_1 − 90° − τ_m − 90° − 180° − t_3 (Acquire)
^{15}N	180° − Δ − 90° − t_2 − 90° − Δ − Decouple

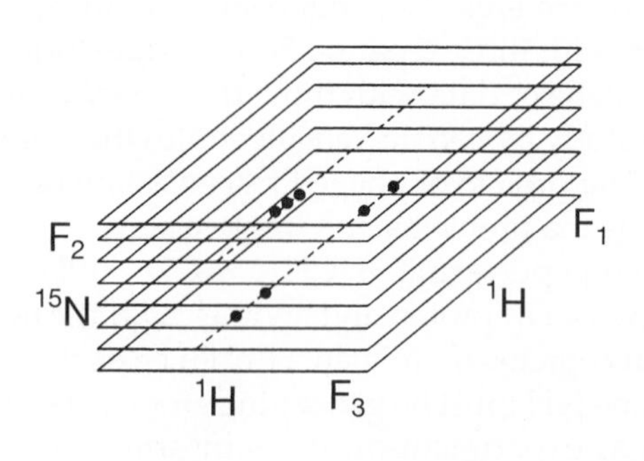

FIGURE 6-37 Diagram of the 3D NOESY–(^{1}H/^{15}N)HMQC spectrum of Pf3 in three frequency dimensions: F_1, F_2, and F_3 (ν_1, ν_2, and ν_3). (Reproduced from F. J. M. van de Ven, *Multidimensional NMR in Liquids,* New York: VCH, 1995, p. 299.)

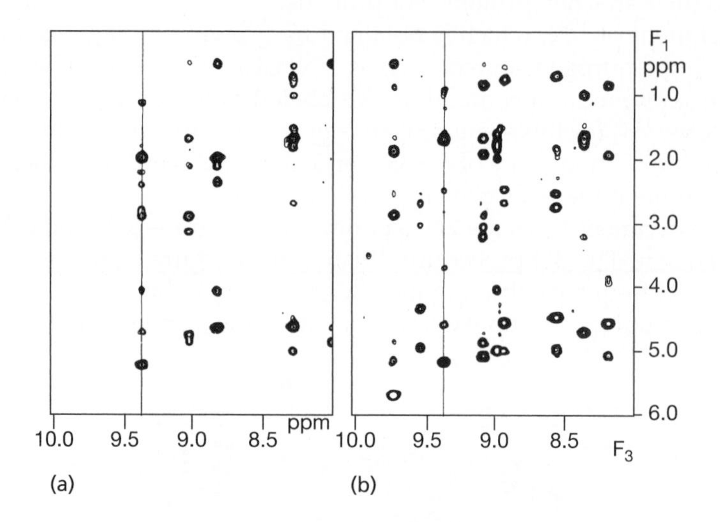

FIGURE 6-38 Two $\nu_1 - \nu_3$ ($F_1 - F_3$) planes taken from the 3D NOESY–(^{1}H/^{15}N)HMQC spectrum of Pf3, corresponding to the frequencies indicated by arrows in Figure 6-35. (Reproduced from F. J. M. van de Ven, *Multidimensional NMR in Liquids,* New York: VCH, 1995, p. 300.)

through the third 90° ^{1}H pulse. The pulses and delays thereafter make up the standard HMQC sequence, which ends with inverse detection of ^{15}N at the ^{1}H frequencies in t_3. The totality of data requires a cube for its representation, as illustrated diagrammatically in Figure 6-37, in which the flat dimensions are the NOESY data in ^{1}H frequencies and the vertical axis provides the ^{15}N frequencies from HMQC. In practice, horizontal planes (single ^{15}N frequencies) are selected for analysis, as in Figure 6-38 for δ 120.7 and 124.9 (arrows labeled a and b in Figure 6-35). The vertical lines at δ 9.35 each show two dominant cross peaks for the NH NOEs to the inter- and intraresidue CHR. Note that both ^{15}N frequencies show a cross peak for δ 9.35 at a CHR frequency of δ 5.2, so that the question of overlap in Figure 6-35 is resolved.

This type of heteronuclear 3D experiment is called NOESY–HMQC (Figure 6-37, with two ^{1}H dimensions and one ^{15}N dimension.) Most 3D experiments use high-sensitivity methods and displays that are particularly effective for large molecules. Thus, COSY is not often used, but TOCSY–HMQC is a useful method for separating ^{1}H–^{1}H coupling connectivities into a ^{13}C or ^{15}N dimension. The homonuclear 3D experiment NOESY–TOCSY, in which all three dimensions are ^{1}H, separates through-space connectivities from coupling connectivities by the pulse sequence 90° − t_1 − 90° − τ_m − t_2 − (spin lock) − t_3 (acquire). The three dimensions each may represent a different nuclide, as in ^{1}H/^{13}C/^{15}N, to provide a 3D variant of the HETCOR experiment. The nuclides usually are selected to explore specific connectivities in biomolecules. The H—N—CO experiment looks at the connection ^{1}H—^{15}N—(C)—^{13}C=O in the peptide unit —NH—CHR—CO— and requires double labeling of ^{15}N and ^{13}C to provide sufficient sensitivity in proteins. The 3D cross peaks connect the HN proton in the first dimension, the HN nitrogen in the second, and the intraresidue carbonyl carbon in the third. An analogue in nucleotide analysis is the H—C—P experiment, in which the third dimension is ^{31}P. Numerous variations of these triple-resonance experiments exist. In particularly complex cases, a fourth time domain t_4 may be introduced to produce 4D experiments.

When through-bond connectivity experiments are combined with the spatial information from buildup rates of NOESY cross peaks, proton–proton distances can be obtained by comparison with known bond lengths. The result can be a complete 3D structure of large molecules. Such solution-phase structures complement solid-phase information from X-ray crystallography. In this way, NMR spectroscopy has become a structural tool for obtaining detailed molecular geometries of complex molecules in solution.

6-6 PULSED FIELD GRADIENTS

Field inhomogeneity has been mentioned as the primary contributor to transverse (xy) relaxation (T_2) (Sections 2-4 and 5-1). Transverse magnetization arises because the phases of individual magnetic vectors become coherent (Figure 2-11) rather than random (Figure 2-10). In a perfectly homogeneous field, this coherence would relax only through spin–spin interactions. In an inhomogeneous field, however, the existence of slightly unequal Larmor frequencies permits vectors to move faster or more slowly than the average, thereby randomizing their phases and destroying transverse magnetization, as described in Section 2-4.

There are several situations in which transverse magnetization is unwanted and may be eliminated by the application of a ***Pulsed Field Gradient*** (PFG), also called a *gradient pulse*. An example of a PFG is presented in Figure 6-39. The gradient is along the direction (z) of the B_0 field. When a PFG is applied, nuclei with different positions in the sample (different z coordinates) resonate at different frequencies. Such spatial encoding of frequency information is the fundamental principle of ***Magnetic Resonance Imaging*** (MRI). In the present context, PFGs may be viewed as a method for inducing transverse relaxation very quickly by a rapid dephasing of the spins. In a typical 2D pulse sequence, a delay time is necessary between repetitions of the pulse sequence in order for relaxation to occur. If repetition occurs before transverse magnetization has relaxed to zero, sensitivity is reduced and artifacts may occur in the 2D spectrum. Consequently, the application of a PFG at the beginning of the sequence reduces or avoids these problems.

For the NOESY experiment, the following pulse sequence may be used: $G1 - 90° - t_1 - 90° - \tau_m, G2 - 90° - t_2$ (acquire). The first PFG (G1) destroys residual transverse magnetization from previous pulses by dephasing the magnetic vectors. The second PFG (G2) is applied during the mixing period τ_m. Only the effects on longitudinal (z) magnetization are of interest during this period. The second PFG helps to eliminate false cross peaks that can arise from pre-existing transverse magnetization.

PFGs also may be used to remove unwanted resonances. One of the most successful methods for solvent suppression destroys solvent magnetization by a PFG, but retains all other resonances through the sequence G–S–G–t(acquire), which is reminiscent of solvent suppression through partial relaxation (Section 5-1). In this sequence (called WATERGATE for ***WATER*** *suppression by* ***G****r****A****dient-****T****ailored* ***E****xcitation*), the S pulse is chosen to invert all resonances except the solvent peak, which is left at zero magnetization, as in partially relaxed spectra (Section 5-1). The two identical gradients mimic a spin–echo process, whereby the dephasing of the first PFG is undone by the second PFG through the process of *rephasing*. Normally, the result is a *gradient echo*. Such an

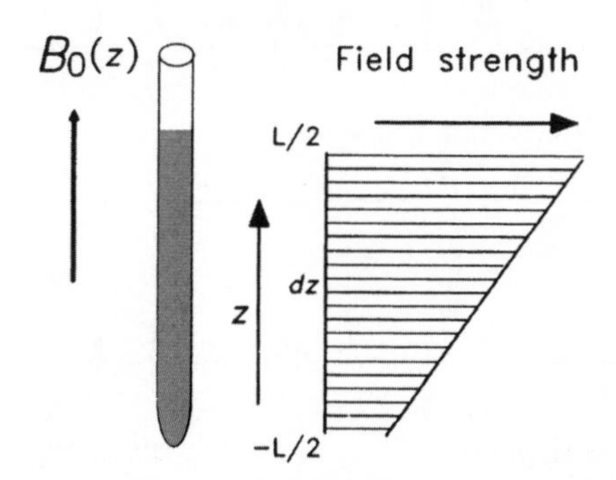

FIGURE 6-39 Diagram of a B_0 field gradient along the z direction. (Reproduced from F. J. M. van de Ven, *Multidimensional NMR in Liquids*, New York: VCH, 1995, p. 212.)

echo occurs for all resonances except solvent. Because the middle pulse eliminates solvent magnetization, the final PFG cannot rephase solvent resonance. All other resonances, however, are rephased by the second gradient pulse.

In addition to dephasing transverse magnetization, PFGs are used to select a coherence order. The use of phase cycling to select a coherence order inevitably involves multiple scans, by which pulse sequences move through 4, 16, or 64 variations with switching, for example, of x, $-x$, y, and $-y$. A full exploitation of phase cycling thus is time consuming. The development of zero, single, or double quantum coherence depends on the rate of various dephasing processes. Proper use of PFGs permits the selection of a coherence order without the repetitive scans of phase cycling. For example, in the inverse-detection HMQC experiment, the single quantum coherence signal for protons attached to ^{12}C (or ^{14}N) must be suppressed while selecting the multiple quantum coherence signal for protons attached to ^{13}C (or ^{15}N). Through phase cycling, this selection is achieved by measuring the difference between two strong signals. The PFG method selects and measures the small difference signal directly in a single scan.

PFG procedures have been developed to implement most of the 1D and 2D experiments discussed in Chapters 5 and 6. In particular, the procedures may be used for INADEQUATE, all common 2D experiments, and spectral editing. A PFG combination of DEPT and HMQC results in editing of proton spectra according to carbon substitution patterns. A PFG-based multiple quantum filtration leads to evolution of double, triple, or quadruple quantum coherence providing proton spectra containing only CH, CH_2, CH_3 resonances, respectively. Broadband decoupling removes coupling to ^{13}C, while proton–proton couplings remain. Figure 6-40 illustrates the result for brucine (**6-3**).

PFGs may be used to optimize the NOE experiment. Although the 2D NOESY experiment is useful in analyzing spatial relationships in large molecules such as proteins, the

6-3

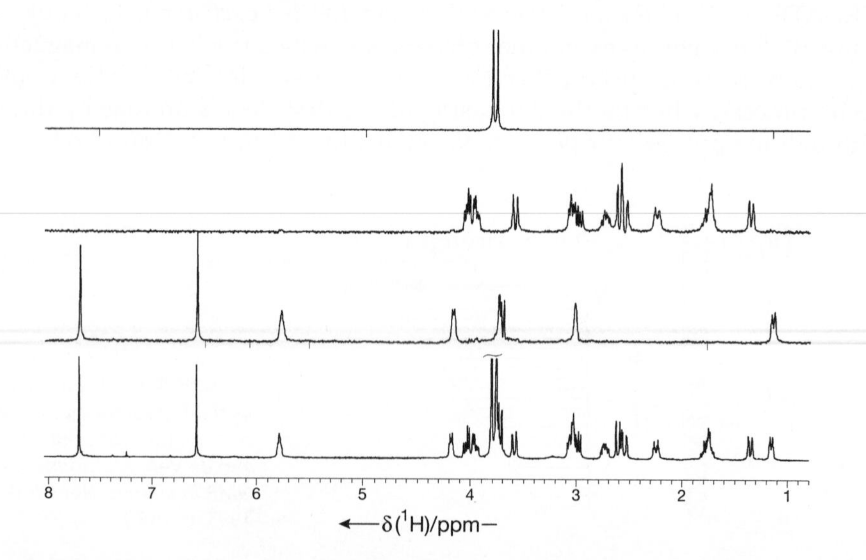

FIGURE 6-40 Editing of the ^1H spectrum of brucine (**6-3**) into subspectra for CH_3, CH_2, and CH (top to bottom). The complete ^1H spectrum is given at the bottom. (Reproduced with permission from T. Parella, F. Sánchez-Ferrando, and A. Virgili, *J. Magn. Reson. A*, **117**, 80 [1995].)

enhancements are weaker for smaller molecules. The 1D NOE experiment thus may be more appropriate for small molecules. The difference experiment described in Section 5-4 was instrumental in lowering the limit for ^{1}H NOEs from about 5% to about 1%. The experiment, however, suffers from problems arising from incomplete subtraction between the irradiated and unirradiated spectra. Such *subtraction artifacts* limit the difference NOE method. The use of pairs of PFGs can yield Overhauser enhancements without difference methods. This procedure has been called *excitation sculpting* and involves a pulse sequence of the general type G1–S–G1–G2–S–G2, in which S represents a selective inversion pulse or sequence of pulses (to produce, for example, the NOE). For the selected spins, the identical G1 pulses act in opposition and hence refocus the magnetization to produce a gradient echo. Spins outside the selected frequency range absorb the cumulative effect of G1 + G1 and are fully dephased. Hence, after a single gradient echo, all resonances have been eliminated, except those in the selected range of the pulse S. Because S may not have ideal phase properties, the gradient echo is repeated a second time (G2/G2) with a gradient of different magnitude, to avoid accidental refocusing of unwanted dephased magnetization. The resulting sequence has been called the *Double **PFG** Spin Echo* (DPFGSE) experiment. In practice, the sequence is preceded by a nonselective 90° pulse. All nuclei move into the *xy* plane, and the gradient pulses then dephase all resonances except those selected by the S pulse.

If the double PFGs are followed by a mixing time τ_m that permits the development of the NOEs that arise due to irradiation by the pulse S, the only resonances that develop are those coming from the NOE. All others have been dephased. The sequence then continues with another PFG to eliminate transverse magnetization and finishes with a 90° pulse for acquisition. The complete pulse sequence is thus 90°–G1–S–G1–G2–S–G2–90°–τ_m–G3–90°–*t*(acquire), in which all 90° pulses are along the *x* axis, S is a selective pulse for the target (irradiated) nucleus, τ_m is set to optimize the NOEs, and G3 eliminates transverse magnetization. Figure 6-41 shows the results of this experiment with 11β-hydroxyprogesterone (compare Figure 5-12). The bottom spectrum (a) contains the unirradiated spectrum, and the ascending spectra contain the DPFGSE NOE experiments on a series of target nuclei. These spectra are taken directly, not by a difference technique. Only Overhauser-enhanced resonances result. The technique easily extends the limit for ^{1}H NOEs to 0.1% and has been used to observe enhancements as small as 0.02%.

Excitation sculpting also can be used for solvent suppression in the DPFGSE version of WATERGATE [G1–S–G1–G2–S–G2–*t*(acquire)], in which the solvent peak is selected for dephasing during S and all other resonances are refocused. Figure 6-42 illustrates the removal of the solvent resonance for 2 mM sucrose in 9:1 H_2O/D_2O.

6-7 DIFFUSION ORDERED SPECTROSCOPY

Entirely different from all previously described experiments is a family based on diffusion properties. Magnetic properties of molecules have a fundamental connection with their motional properties. For example, most relaxation mechanisms arise from rotational motion that generates fluctuating magnetic fields (Section 5-1). In an inhomogeneous field such as provided by gradients (Figure 6-39), translational motion can move a molecule from one value of the magnetic field to another one. Consequently, the diffusional coefficient of a molecule determines in part how a molecule responds to magnetic field gradients. Molecular diffusion depends on many properties of the environment, such as temperature and viscosity, which normally are constant within a liquid. Thus, differences in diffusional properties between molecules are primarily a function of molecular size and shape. Multiple components in solution would diffuse at different rates and respond to the gradient in different fashions. Methods have been developed to exploit these differences in order to use NMR to separate components according to their diffusion properties. Such experiments as a group have been termed *Diffusion **O**rdered Spectroscop**Y*** or DOSY.

The intensity I_A of the signal for component A in such a mixture can be expressed as eq. 6-3,

$$I_A = I_A(0)\exp(-D_A Z) \qquad \textbf{(6-3)}$$

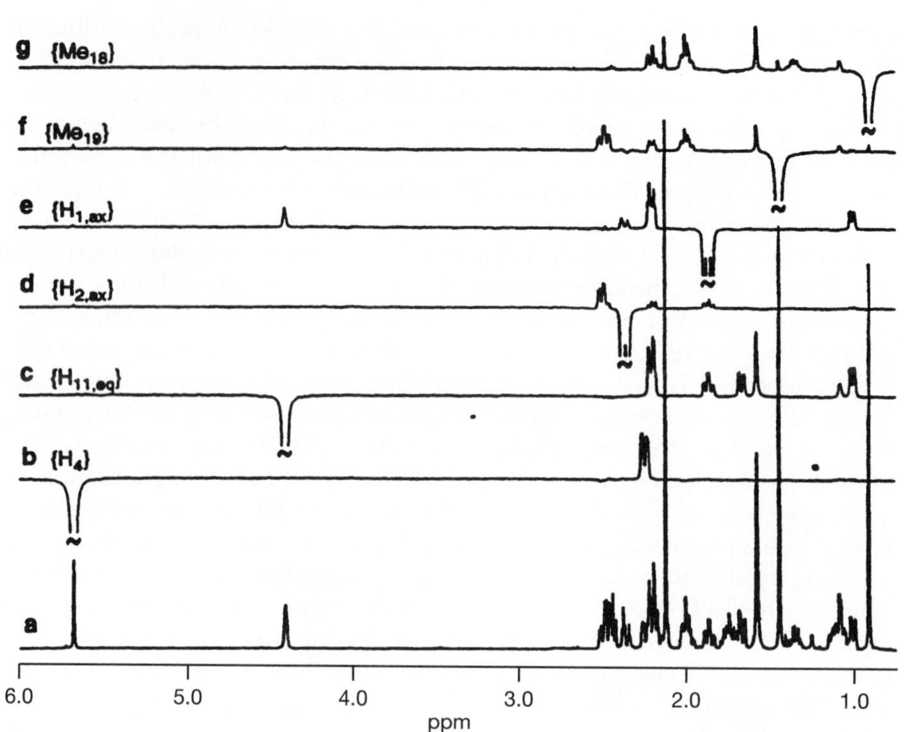

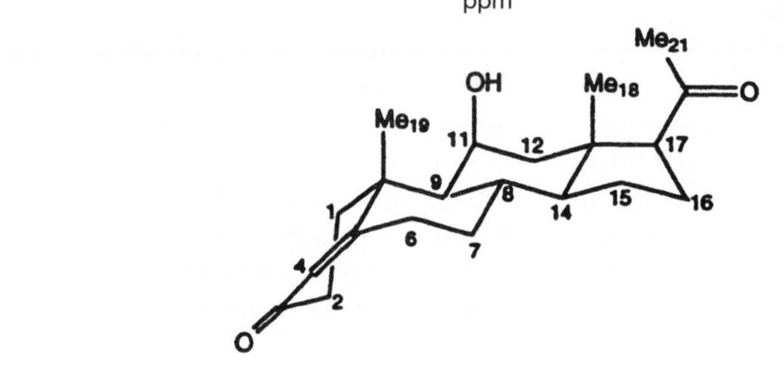

FIGURE 6-41 The double pulse field gradient spin echo (DPFGSE) NOE experiment for 11β-hydroxyprogesterone: (a) the unirradiated spectrum; (b)–(g) spectra with irradiation at selected frequencies. [Reproduced with permission from K. Stott, J. Keeler, Q. N. Van, and A. J. Shaka, *J. Magn. Reson. A*, **125**, 322 (1997).]

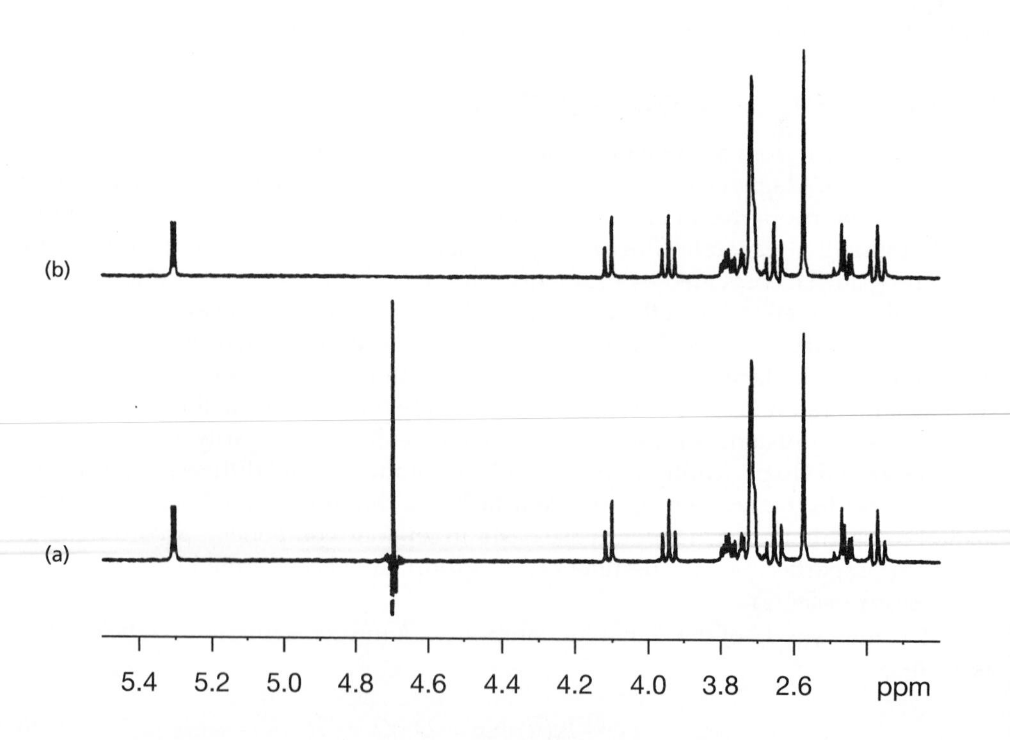

FIGURE 6-42 (a) Water suppression with excitation sculpting on 2 mM sucrose in 9:1 H_2O/D_2O. (b) The residual solvent peak has been eliminated by further processing. (Reproduced from T. D. W. Claridge, *High-Resolution NMR Techniques in Organic Chemistry*, Amsterdam: Pergamon Press, 1999, p. 365.)

in which $I_A(0)$ is the intensity of the signal at zero gradient (the normal 1D signal), I_A is the intensity of the signal in the presence of a gradient of formula Z, and D_A is the diffusion coefficient of component A. The intensities of other components of the mixture differ only as the result of their diffusion coefficients D_i, as all components experience the same gradient Z. There are many different DOSY experiments, and the art of the experiment is the nature of the function Z. A series of experiments is recorded with variation of the amplitude of the gradient: the stronger the gradient, the smaller the signal intensity. After Fourier transformation of the FID, the experiment generates a series of peak intensities as a function of the gradient Z. A 2D array is created similar to that in Figure 6-3, with two important differences. First, although the horizontal axis still is the normal 1D frequency dimension, the vertical axis is the gradient amplitude. Second, instead of generating a sinusoidal array of intensities along the y axis, the intensities decrease exponentially. This unrecorded intermediate stage analogous to Figure 6-3 would have a series of peaks, one series for each component along the x axis, whose intensities decay from the bottom of the spectrum (zero gradient) to a diminishingly small value at the top (large gradient).

Because these simple exponential decays are not sinusoidal, the final step is not a second Fourier transformation but rather a process that has been termed a *DOSY transformation*, whereby the data are transformed into a diffusion domain instead of a frequency domain. There are numerous mathematical procedures for carrying out the DOSY transformation, including exponential fits, maximum entropy, and multivariate analysis. The result resembles a traditional 2D display, in which the horizontal axis is the normal ^{1}H frequency display and the vertical axis is the diffusion constant. Figure 6-43 illustrates the result for the DOSY experiment of a three-component system that contains the solvent tetrahydrofuran (THF), dioctyl phthalate (DOP), and the polymer polyvinyl chloride (PVC). The normal 1D spectrum is recorded at the top, and the 2D spectrum contains cross peaks located according to frequency along the horizontal axis and diffusion along the vertical axis. Each component diffuses at a different rate and consequently provides a series of cross peaks at different levels on the diffusion axis. More rapidly diffusing components are found lower on the plot. It is clear from the vertical positions that THF diffuses most rapidly and PVC most slowly, as expected for their molecular sizes. The cross peaks in the middle are identified as those from DOP by the presence of aromatic resonances. Each horizontal cut may be projected onto the vertical axis to create the equivalent of a chromatogram, in which each component is represented by a single peak on the gradient axis according to its diffusion coefficient, with an intensity appropriate to its molar representation in the mixture. Ideally, each distinct component of a mixture would generate a separate spectrum on the vertical axis. In essence, DOSY results in a chromatographic-like separation, with the important differences that there is no physical separation of components and no sample preparation beyond normal NMR sample preparation. The ability for NMR to separate components according to molecular size has led to a host of new applications.

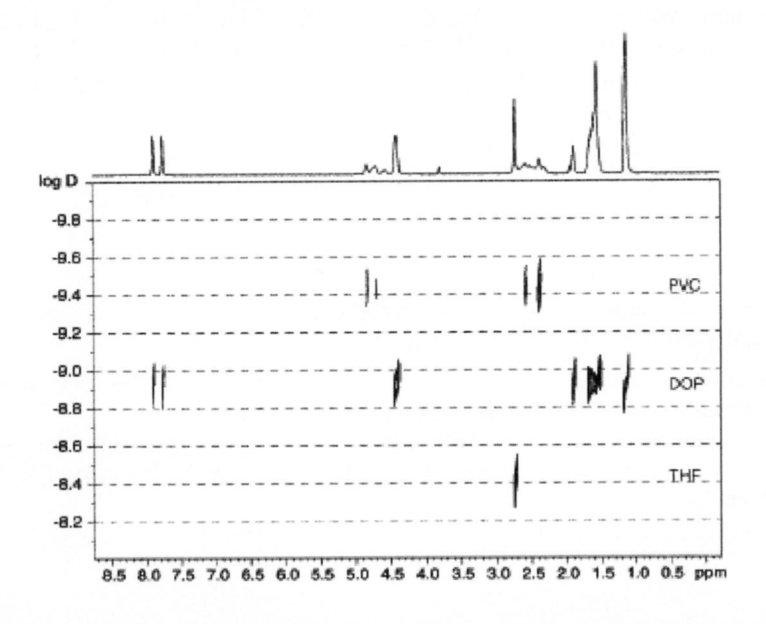

FIGURE 6-43 The DOSY spectrum of a mixture of polyvinyl chloride (PVC), dioctyl phthalate (DOP), and tetrahydrofuran (THF). (Reproduced with permission from S. Ahn, E.-H. Kim, and C. Lee, *Bull. Korean Chem. Soc.*, **26**, 332 [2005].)

6-8 SUMMARY OF TWO-DIMENSIONAL METHODS

A bewildering array of 2D methods is available to the NMR spectroscopist today. This chapter has described a number of the most widely used such experiments. A routine structural assignment begins with recording the 1D ^{1}H and ^{13}C spectra. Many resonances may be assigned according to the principles outlined in Chapters 3 and 4 on chemical shifts and coupling constants. Normally, recourse is made to 2D methods only if this traditional approach is insufficient. Some type of spectral editing for determining the number of protons attached to each carbon, such as DEPT, is helpful in completing the ^{13}C assignments (Section 5-5). The HETCOR or HMQC experiment then provides correlations between the ^{13}C and ^{1}H resonances, possibly with completion of the ^{1}H assignments.

Further 2D methods are necessary if the structure is not deduced in the process of making spectral assignments for hypothetical or expected structures. The COSY experiment lays the groundwork for building structures through ^{1}H–^{1}H connectivities based on J couplings. For small molecules, there may not be enough vicinal or geminal couplings for the method to be useful. For molecules of medium complexity, COSY may be sufficient to provide the entire structure by confirming expected ^{1}H–^{1}H connectivities based on vicinal and geminal couplings. The analogous experiment based on long-range couplings (LRCOSY) may be necessary to assign connections between molecular pieces that do not involve vicinal protons, as, for example, between two rings, for substituents on a ring, or over a heteroatom or carbonyl group.

Additional 2D experiments may be necessary for larger molecules. As peaks accumulate along the diagonal, the COSY45 or DQF-COSY experiment may be used to simplify that region and uncover cross peaks that are close to the diagonal. For even larger molecules, the TOCSY or relayed COSY experiments may be necessary. Further connectivities between protons and carbons may be explored through longer-range couplings (HMBC) in order to define structural regions around quaternary carbons. If ^{1}H–^{1}H couplings need to be measured, either the J-resolved method or DQF-COSY may be carried out.

The NOESY experiment provides information about the proximity of protons and hence is used primarily for distinguishing structures that have clear stereochemical differences. For larger molecules, the ROESY experiment may offer some advantages because of its lower tendency to exhibit spin diffusion. The related EXSY experiment is used only when chemical exchange is being investigated, to supplement information from one dimension.

The 2D INADEQUATE or INEPT–INADEQUATE experiment requires additional spectrometer time. It is usually an experiment of last resort, although specific structures may be particularly amenable to this technique, as, for example, when there are several quaternary carbons that prevent COSY analysis.

Kupĉe and Freeman have developed an experiment to implement many of these methods with one super sequence, which they call PANACEA, for *Parallel Acquisition Nuclear magnetic resonance All-in-one Combination of Experimental Application*. This single experiment generates the 1D ^{13}C, 2D INADEQUATE, HSQC, and HMBC spectra, in a run of about nine hours on 10 mg of material of several hundred dalton molecular weight.

Problems

For a large selection of relatively straightforward 2D spectra, see Problem **6-1**. *The remaining problems involve molecules of medium to high complexity, although none is so complex as to require 3D methods.*

6-1 The spectra in this problem provide relatively routine practice. Assign the structure for each unknown from the 300 MHz ^{1}H, 75 MHz ^{13}C, COSY, HETCOR, and, in some cases, DEPT spectra. In the DEPT spectra, methine and methyl carbons give positive peaks and methylene carbons negative peaks at the top. Only methine carbons are in the middle, and a full spectrum of all carbons is at the bottom.

(a) $C_{11}H_{16}$

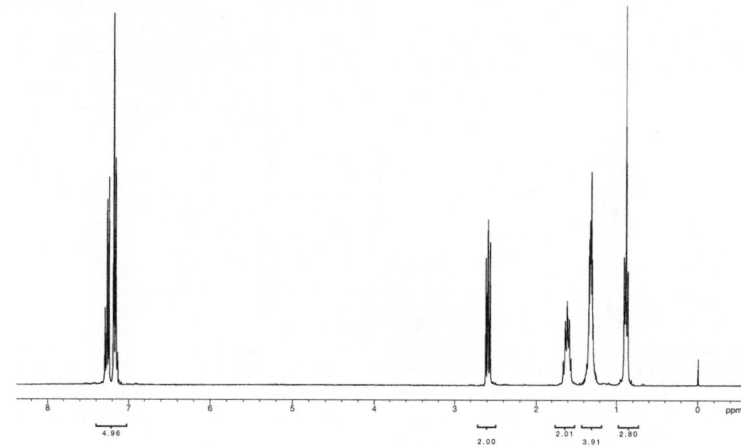

DEPT spectra

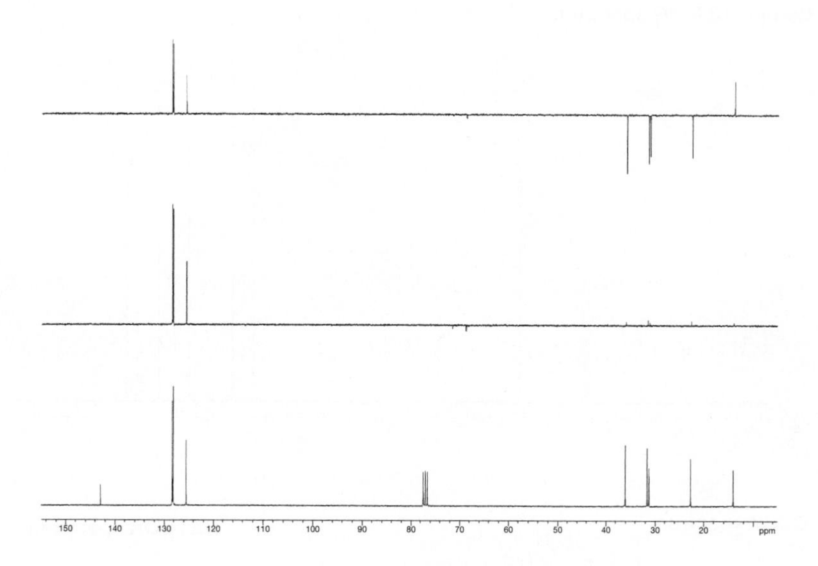

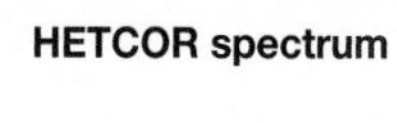

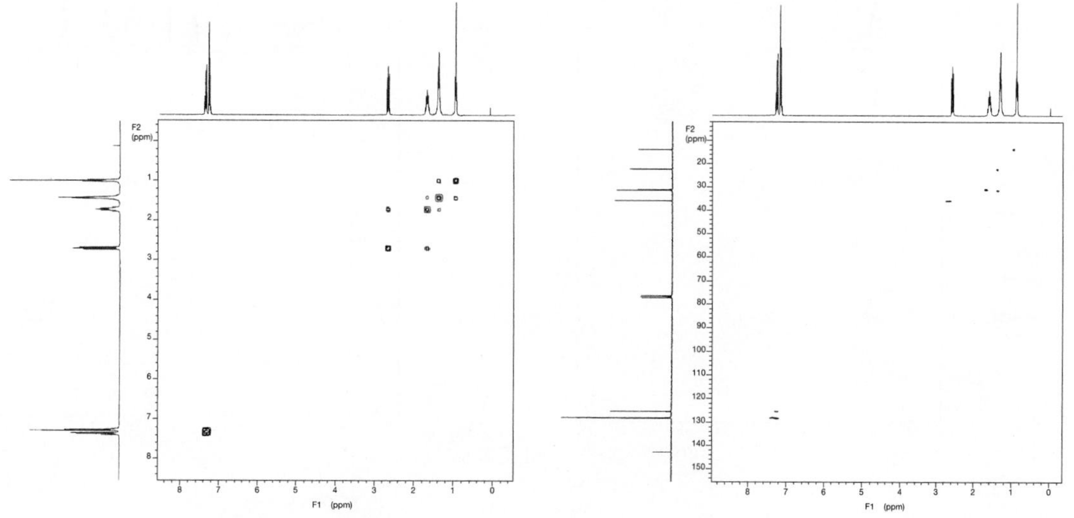

(b) C_8H_{14}

Proton NMR spectrum (CDCl$_3$)

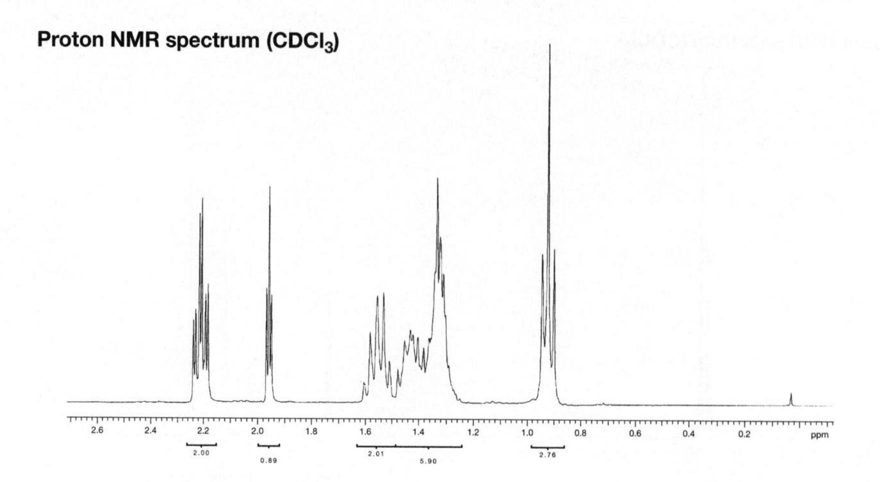

Carbon-13 NMR spectrum

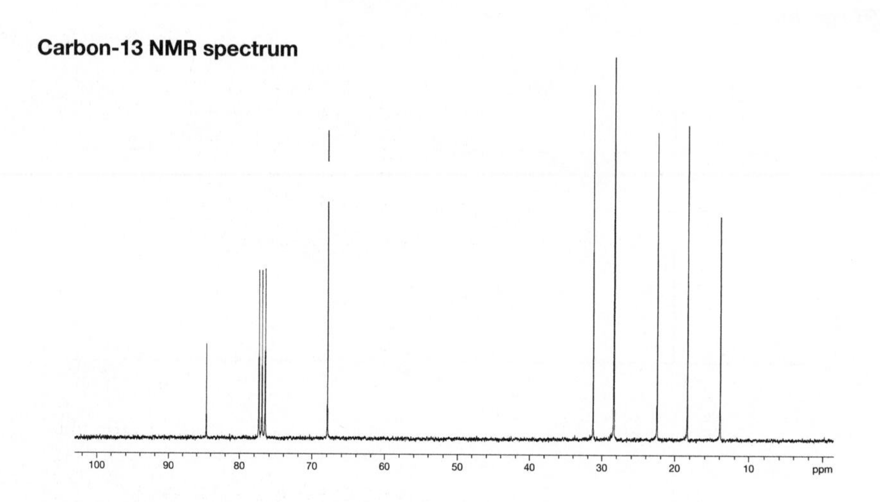

COSY spectrum **HETCOR spectrum**

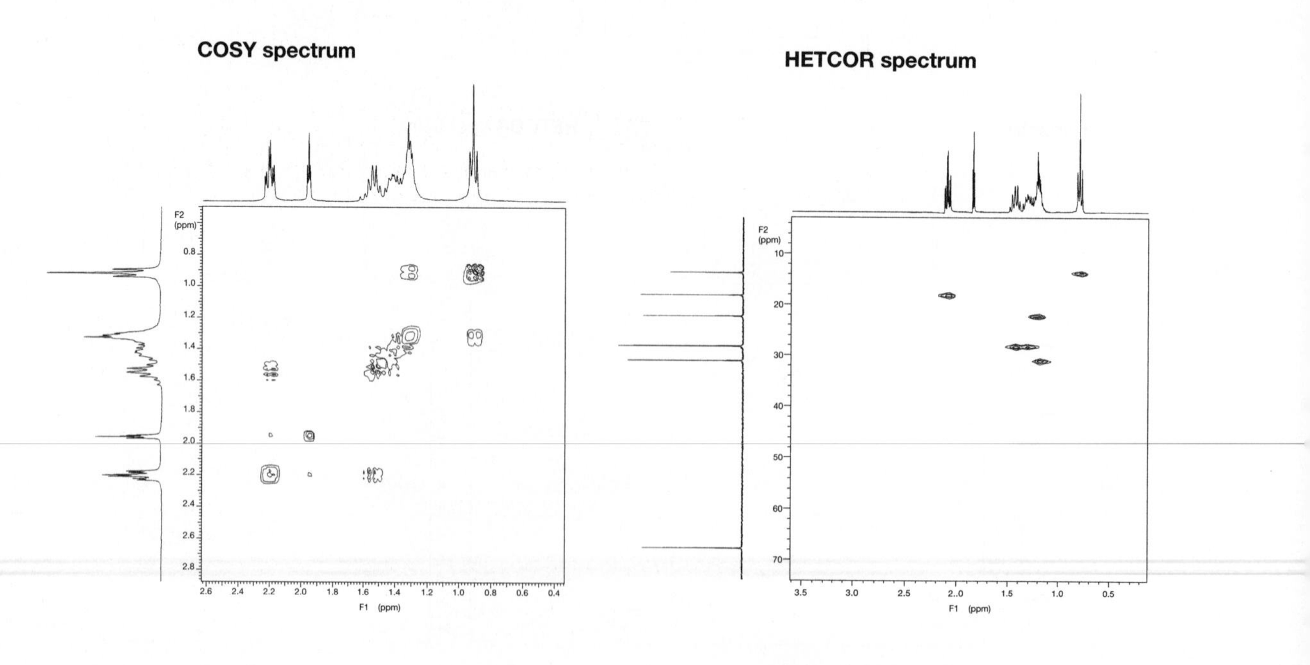

(c) C_5H_9Cl

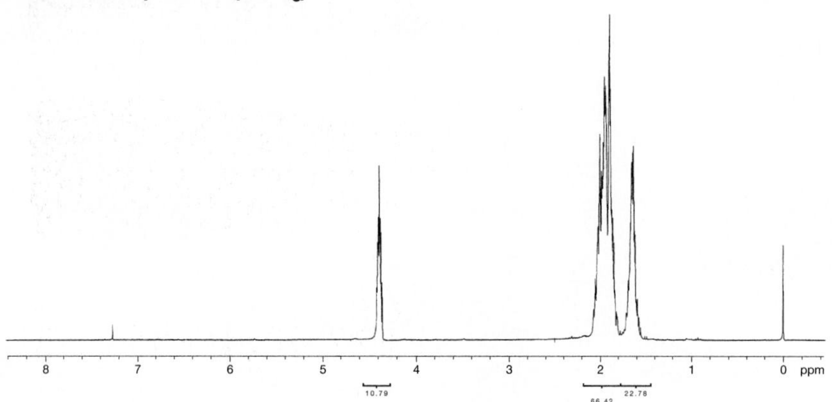

Carbon-13 spectrum

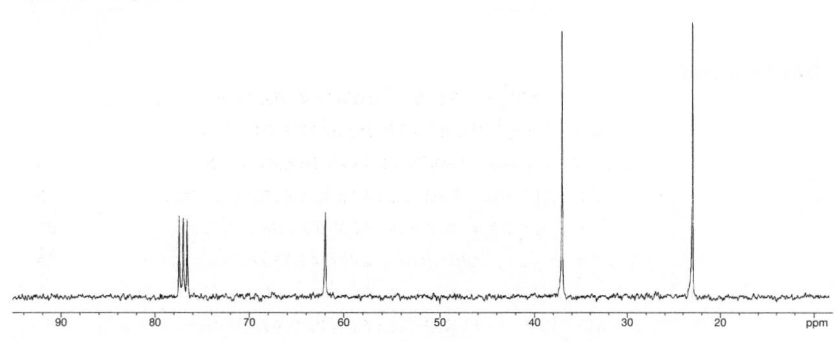

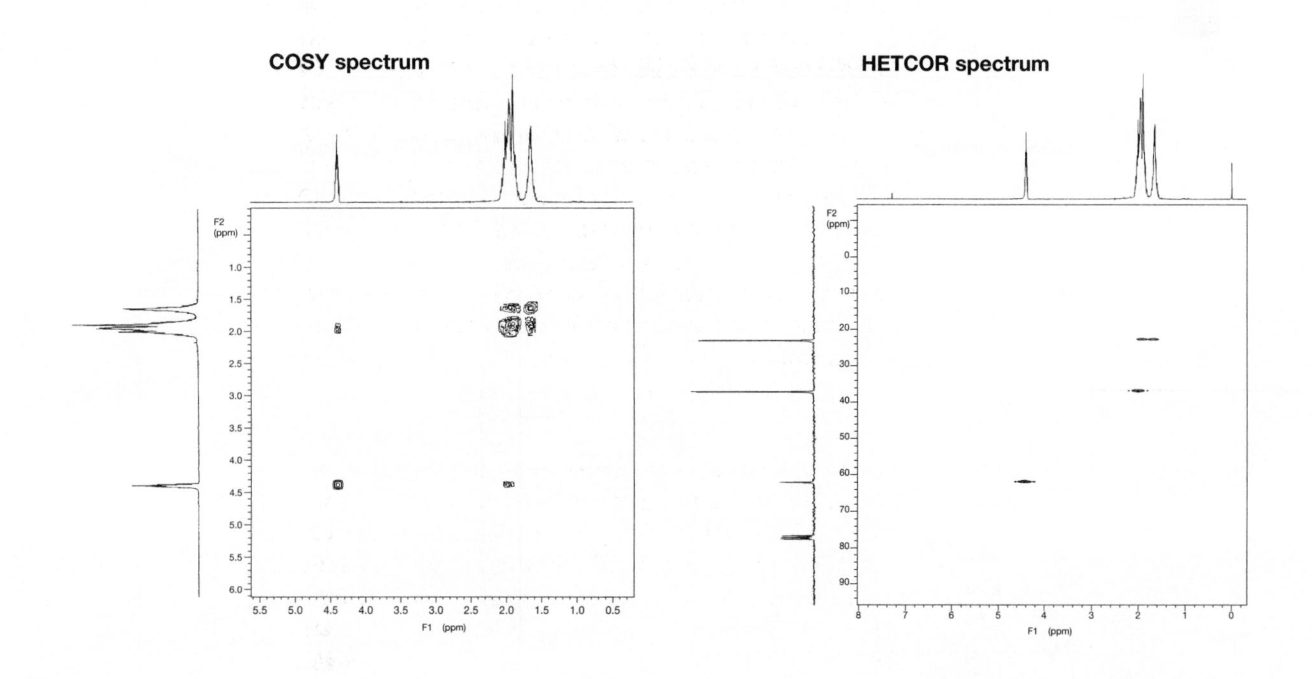

(d) $C_7H_8O_2$

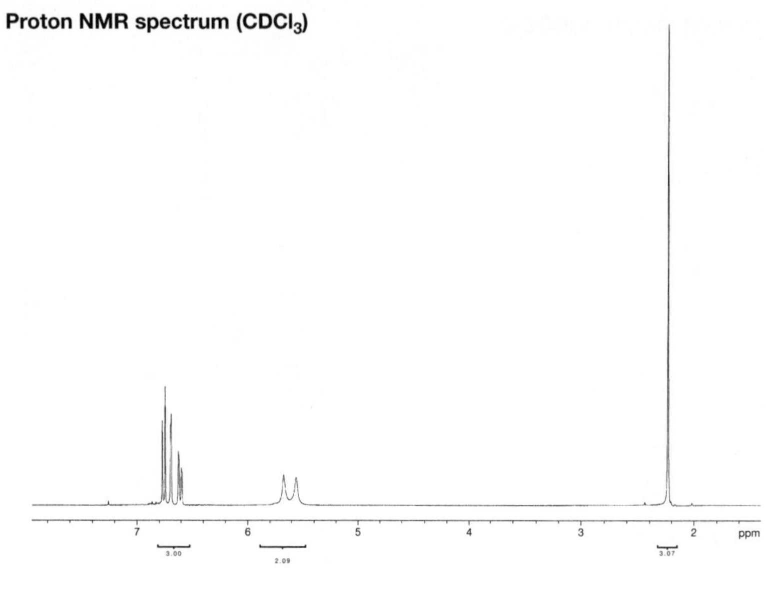
Proton NMR spectrum (CDCl3)
7
6
5
4
3
2
ppm
3.00
2.09
3.07

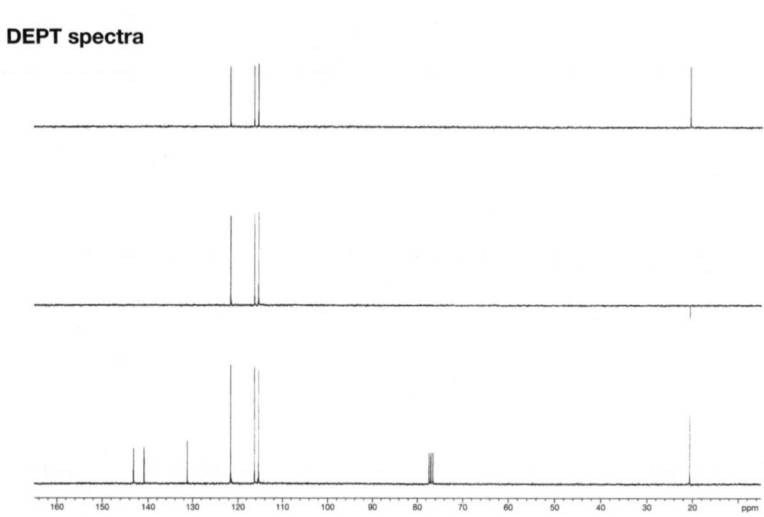
DEPT spectra
160
150
140
130
120
110
100
90
80
70
60
50
40
30
20
ppm

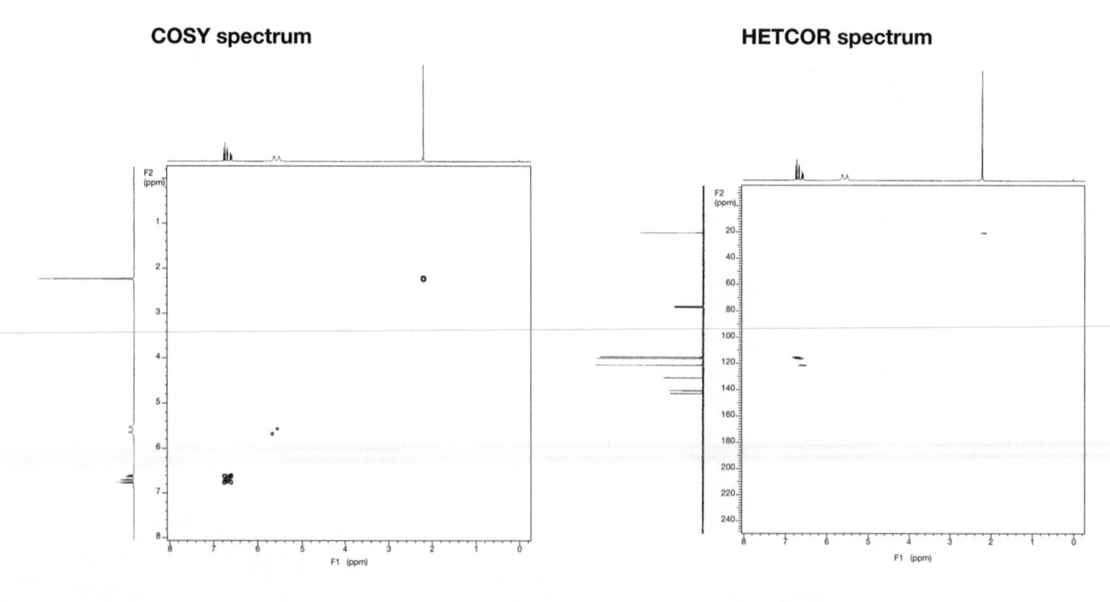
COSY spectrum
HETCOR spectrum
F2 (ppm)
F1 (ppm)

(e) $C_6H_{10}O$

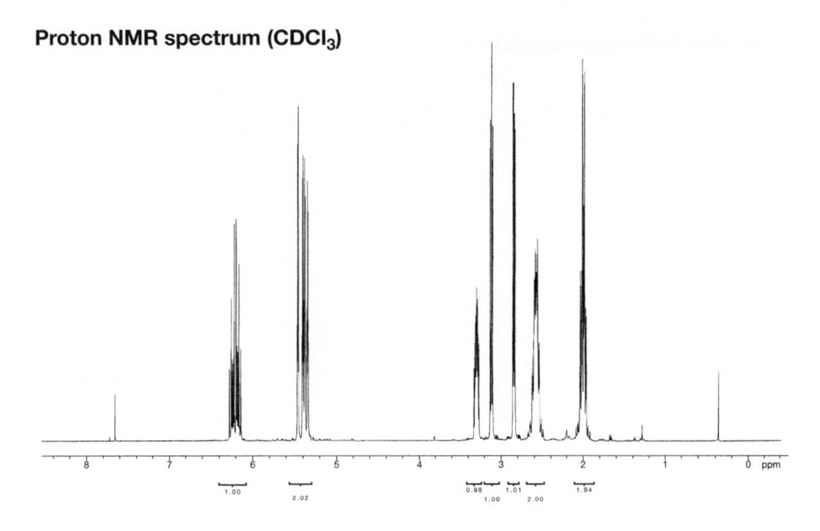
Proton NMR spectrum (CDCl3)
8
7
6
5
4
3
2
1
0 ppm
1.00
2.02
0.98
1.00
1.01
2.00
1.94

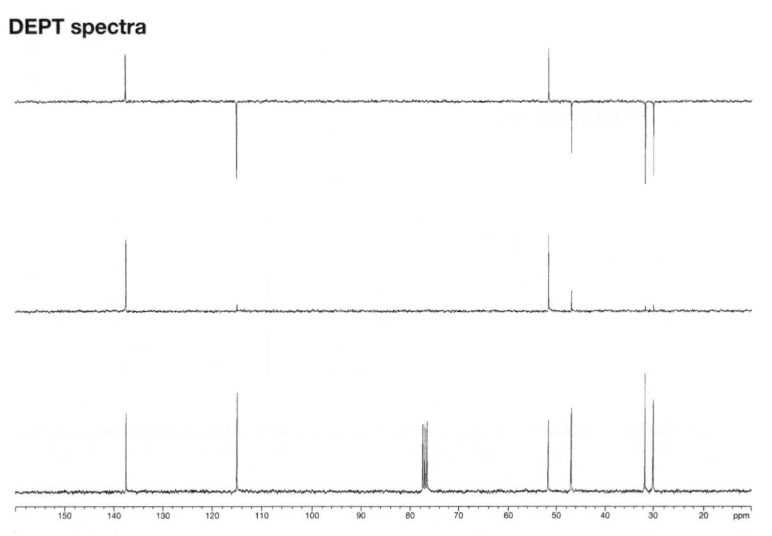
DEPT spectra
150
140
130
120
110
100
90
80
70
60
50
40
30
20
ppm

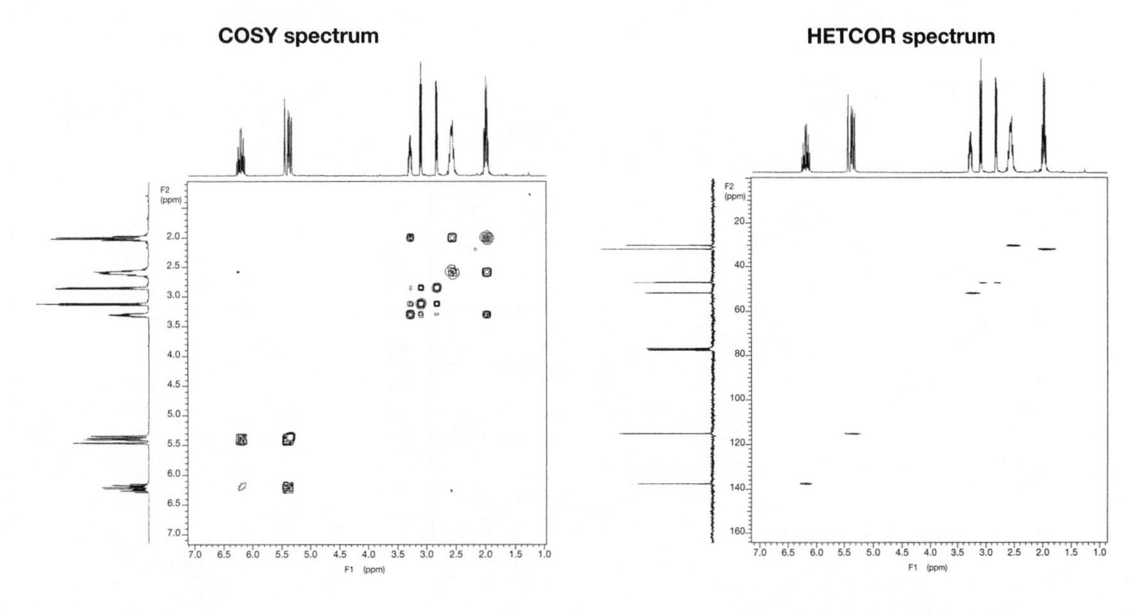
COSY spectrum
HETCOR spectrum
F2 (ppm)
F1 (ppm)

(f) $C_7H_8O_2$

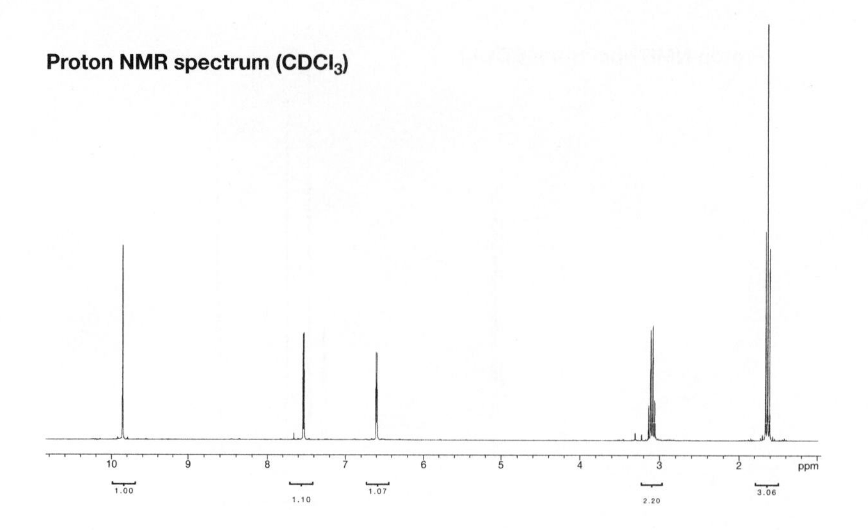
Proton NMR spectrum (CDCl3)
ppm
1.00
1.10
1.07
2.20
3.06

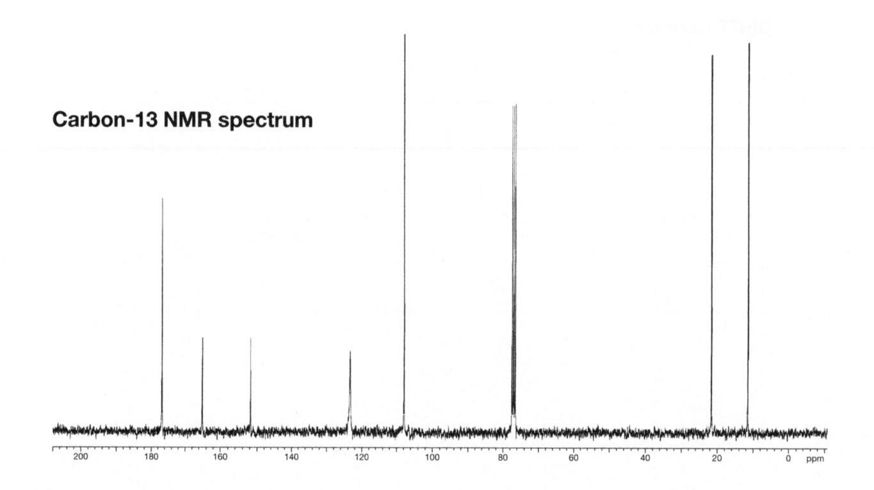
Carbon-13 NMR spectrum
ppm

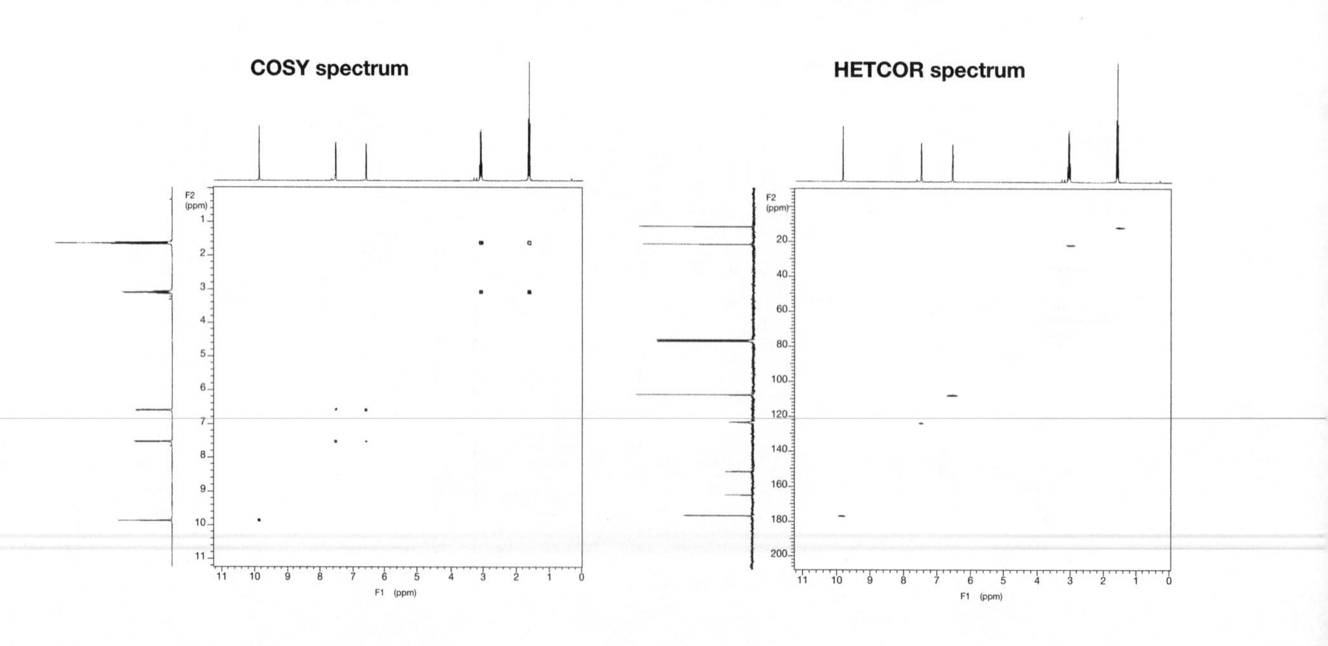
COSY spectrum
HETCOR spectrum
F2 (ppm)
F1 (ppm)

(g) C_6H_7NO

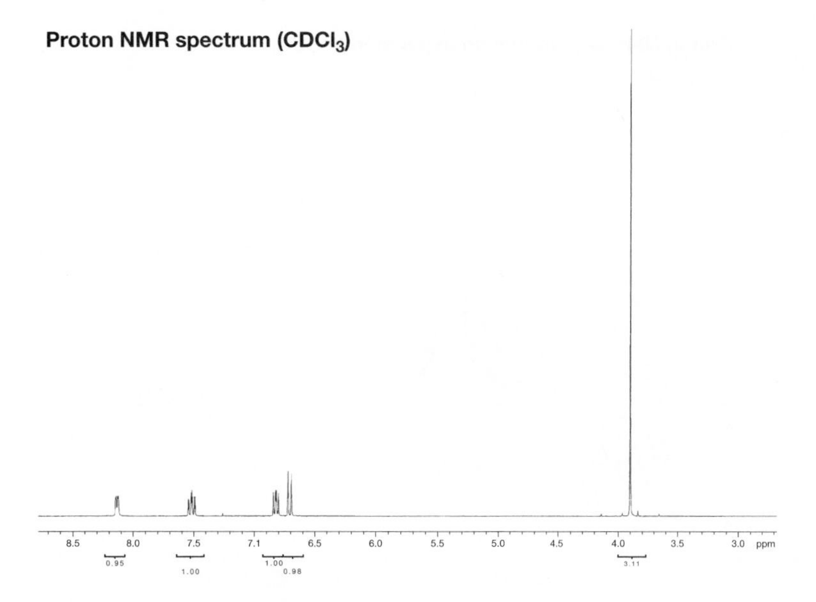
Proton NMR spectrum ($CDCl_3$)
8.5 8.0 7.5 7.1 6.5 6.0 5.5 5.0 4.5 4.0 3.5 3.0 ppm
0.95
1.00
1.00
0.98
3.11

Carbon-13 NMR spectrum

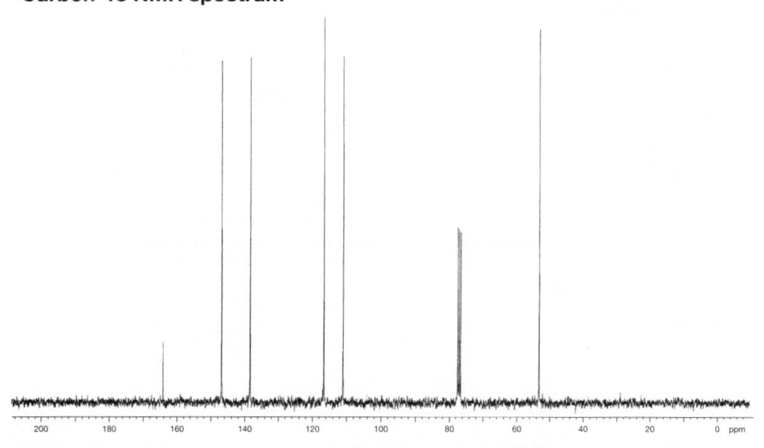
200 180 160 140 120 100 80 60 40 20 0 ppm

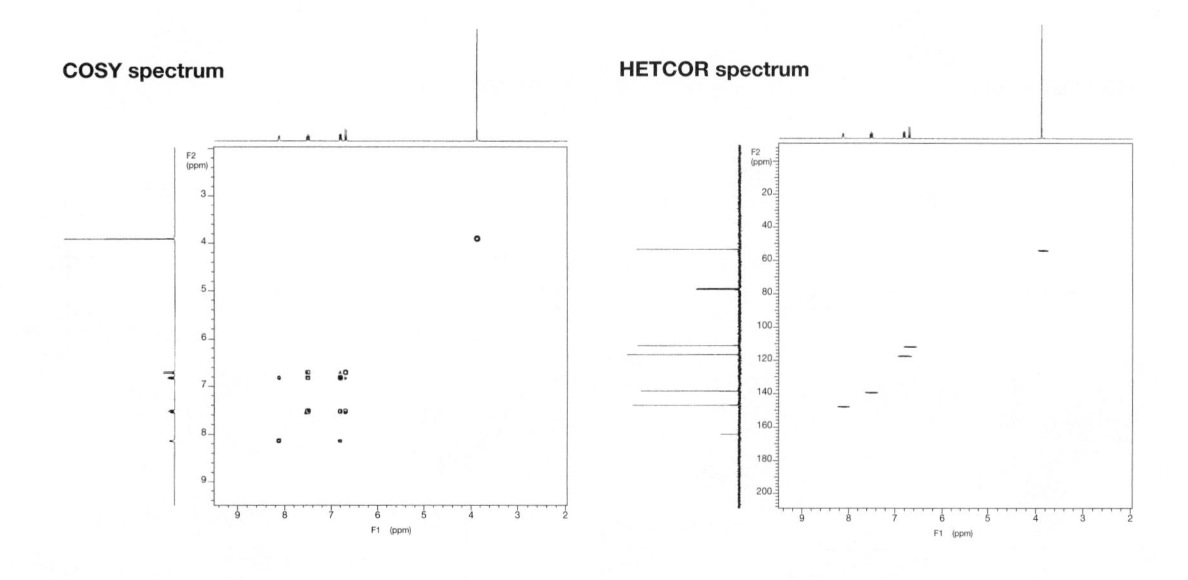
COSY spectrum
F2 (ppm)
3 4 5 6 7 8 9
9 8 7 6 5 4 3 2
F1 (ppm)
HETCOR spectrum
F2 (ppm)
20 40 60 80 100 120 140 160 180 200
9 8 7 6 5 4 3 2
F1 (ppm)

(h) $C_{10}H_{16}O$

Proton NMR spectrum with expansion ($CDCl_3$)

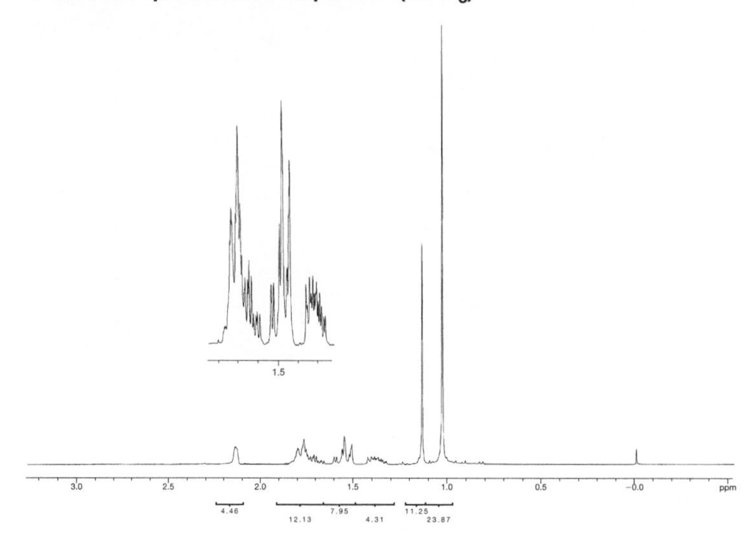

DEPT spectra

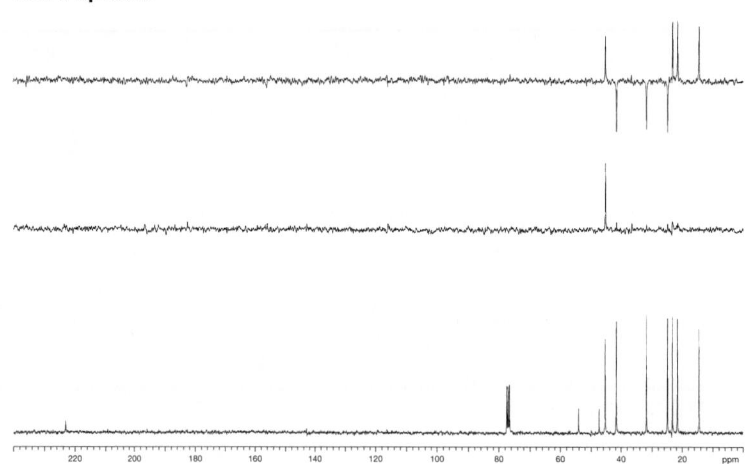

COSY spectrum **HETCOR spectrum**

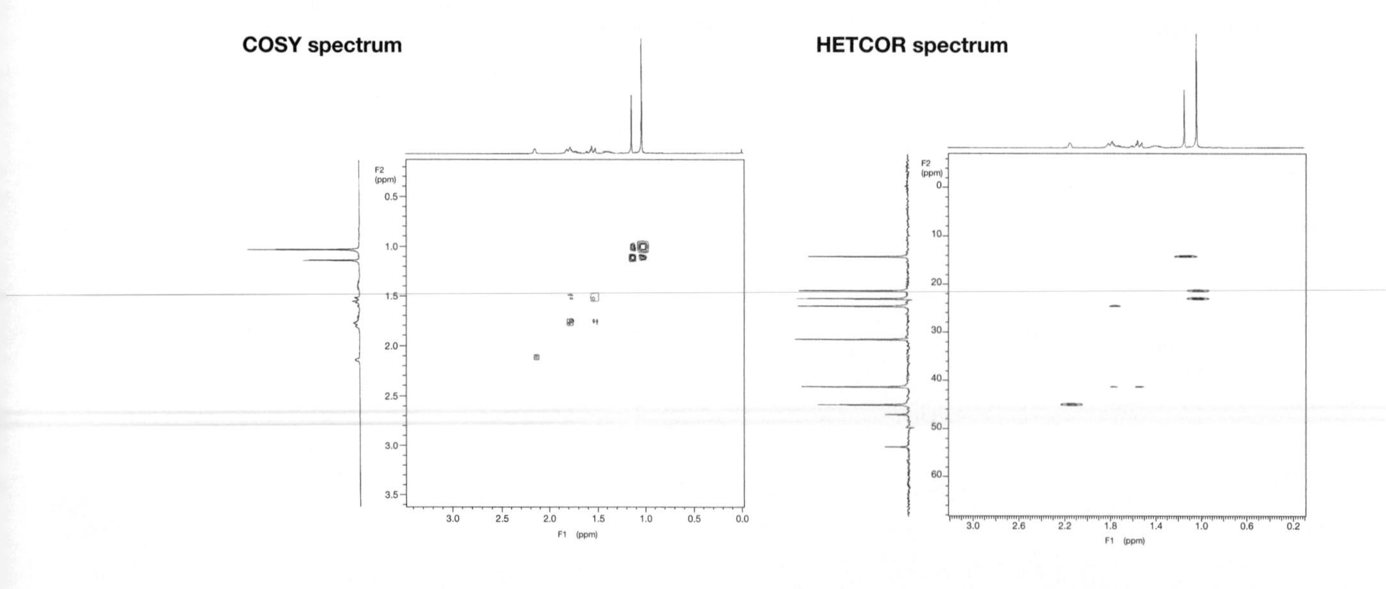

(i) $C_{10}H_{16}O$

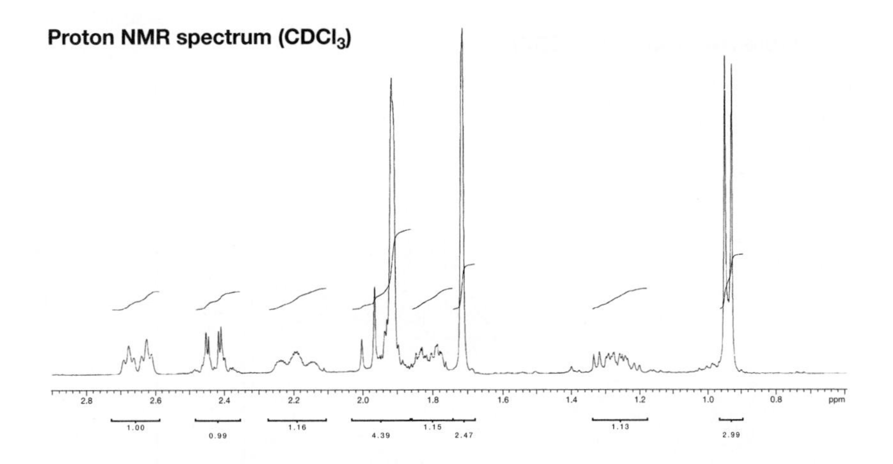
Proton NMR spectrum (CDCl3)
ppm
1.00
0.99
1.16
4.39
1.15
2.47
1.13
2.99

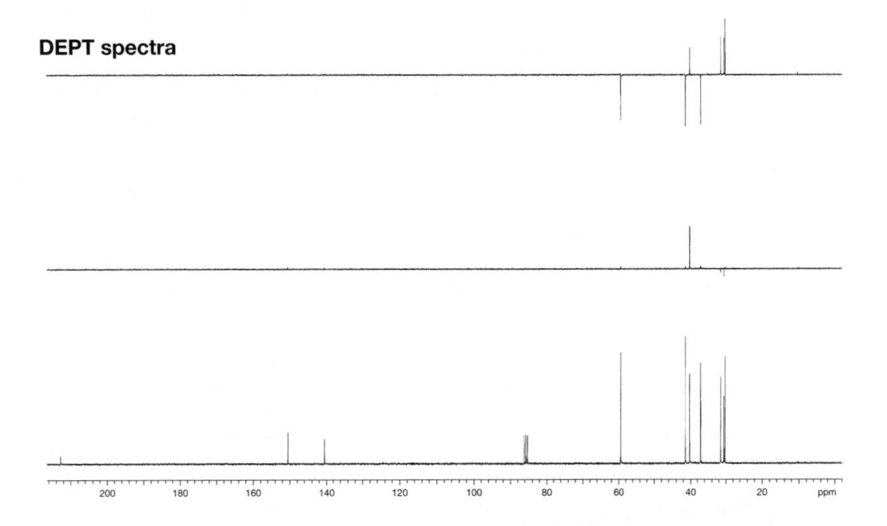
DEPT spectra
ppm

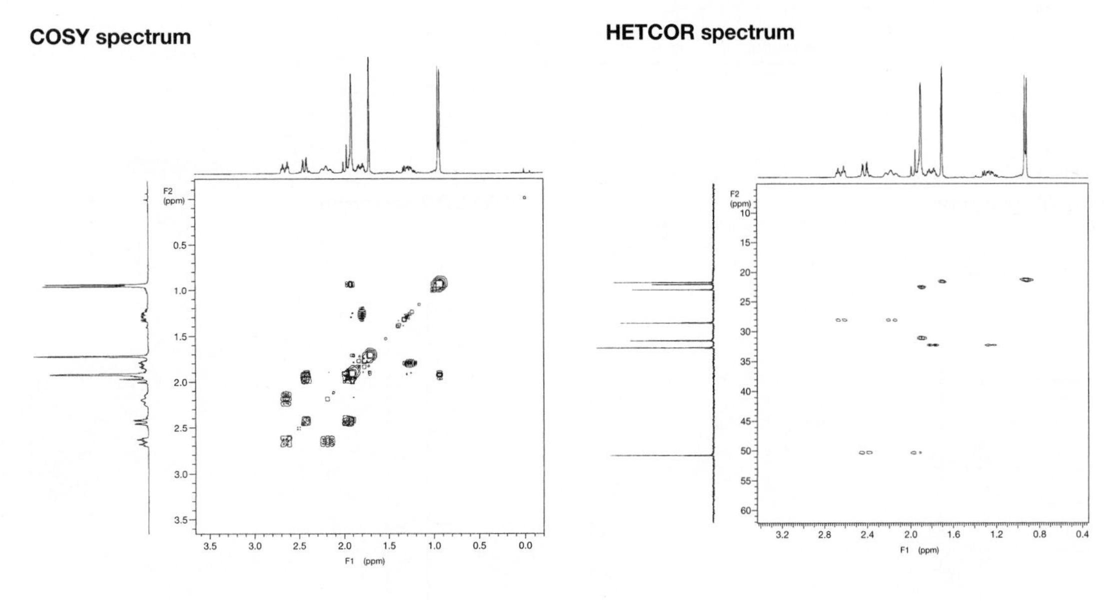
COSY spectrum
HETCOR spectrum
F2 (ppm)
F1 (ppm)

(j) $C_{15}H_{18}$

Proton NMR spectrum (CDCl$_3$)

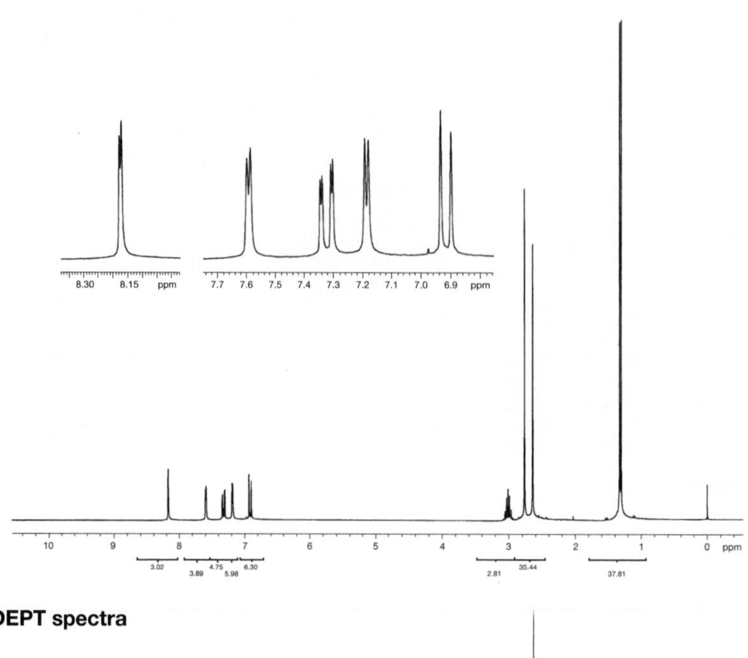

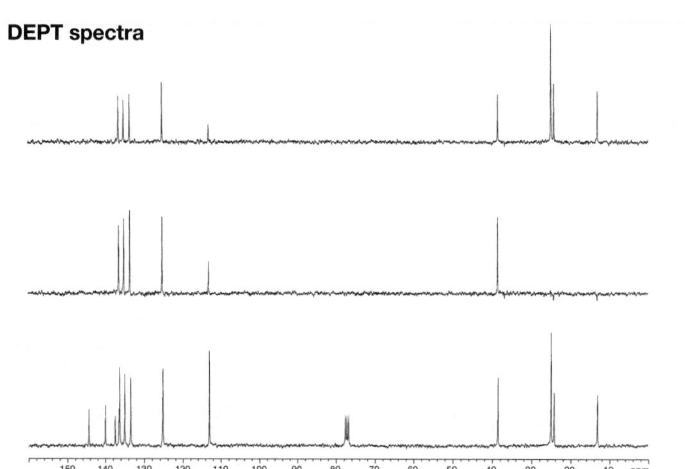

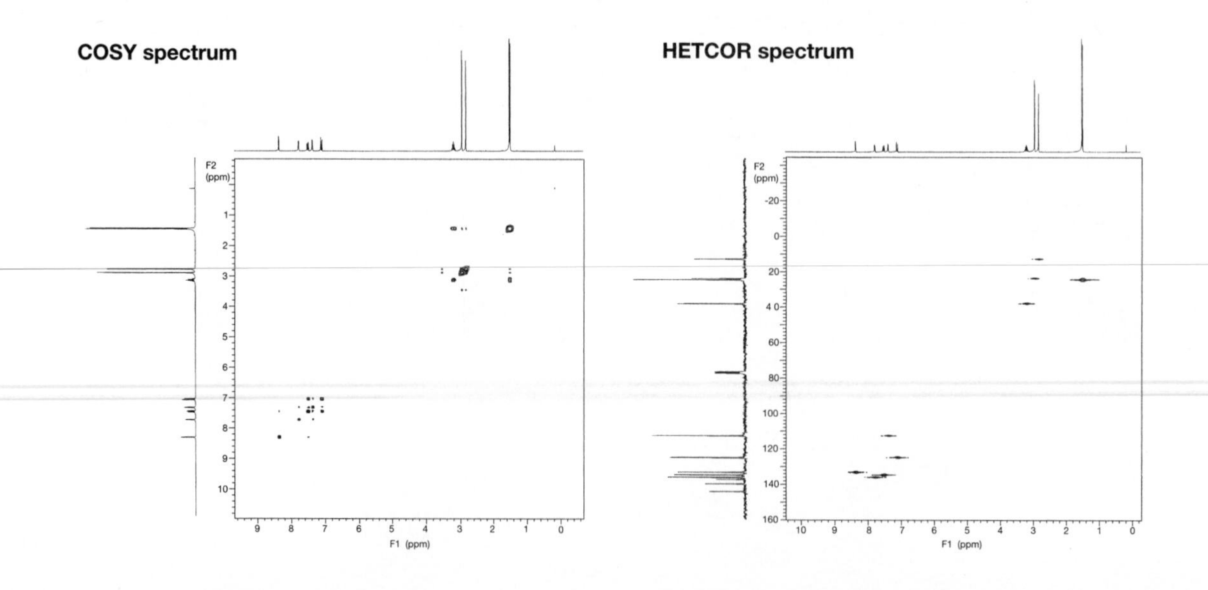

6-2 The 300 MHz COSY spectrum below is of a molecule with the formula $C_{14}H_{20}O_2$. The 1D spectrum is given on either edge. In addition to the illustrated resonances, the ^1H spectrum contains a broad singlet at δ 7.3 with integral 5. What is the structure? Show your reasoning.

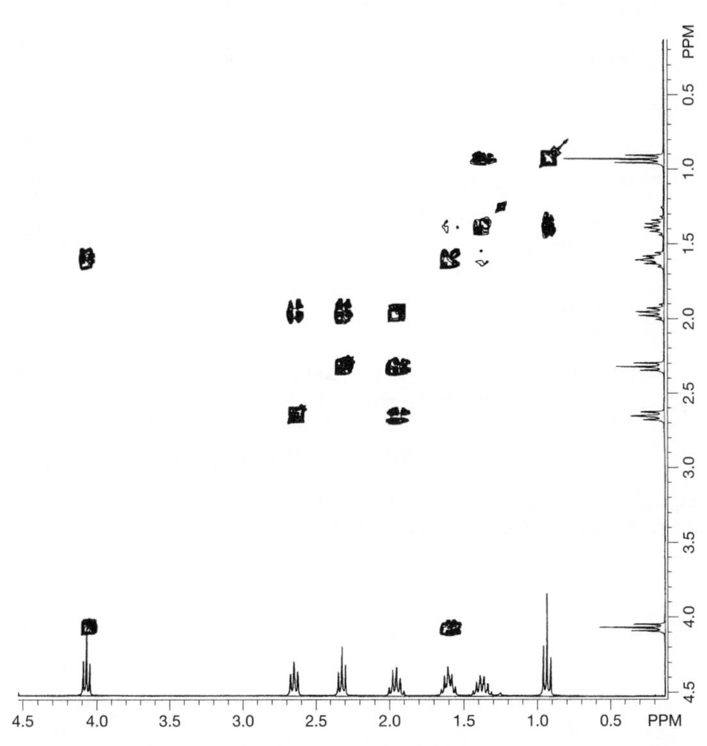

Reproduced with permission from S. E. Branz, R. G. Miele, R. K. Okuda, and D. A. Straus, *J. Chem. Educ.*, **72**, 659–661 (1995). Copyright 1995 American Chemical Society.

6-3 Below are the 1D ^1H and 2D COSY spectra of quinoline, whose structure is shown.

(a) Assign what resonances you can from the 1D spectrum. Explain each assignment in terms of chemical shifts and coupling constants.

(b) Assign all remaining protons from the COSY spectrum, and explain the 1D splitting patterns of these protons.

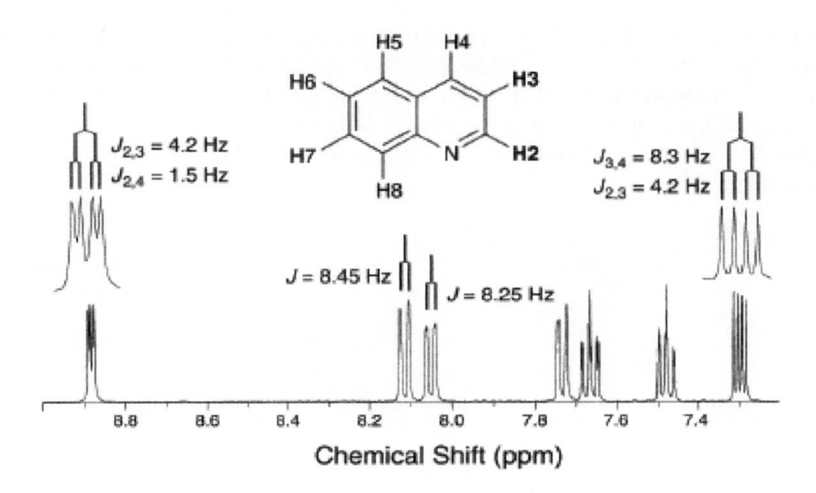

(continued)

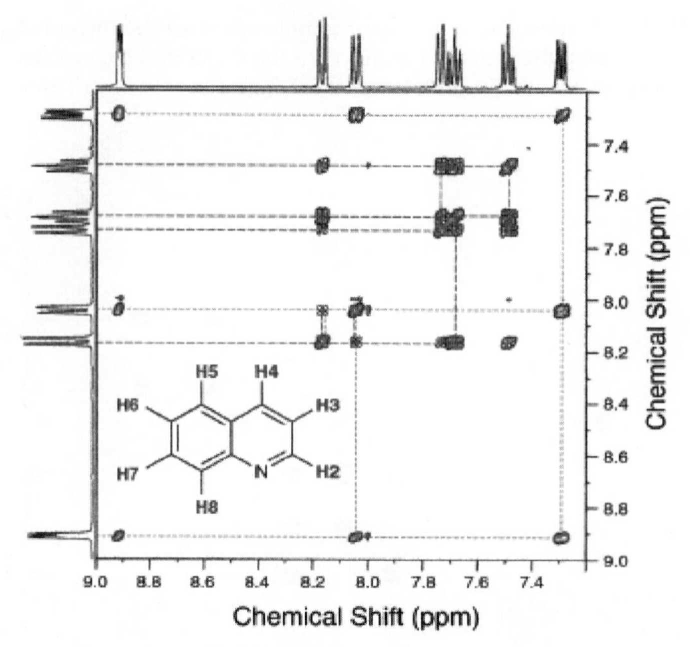

Reproduced with permission from P. J. Seaton, R. T. Williamson, A. Mitra, and A. Assarpour, *J. Chem. Educ.*, **79**, 107 (2002).

6-4 Trimerization of indole-5-carboxylic acid gives one of the following two isomers:

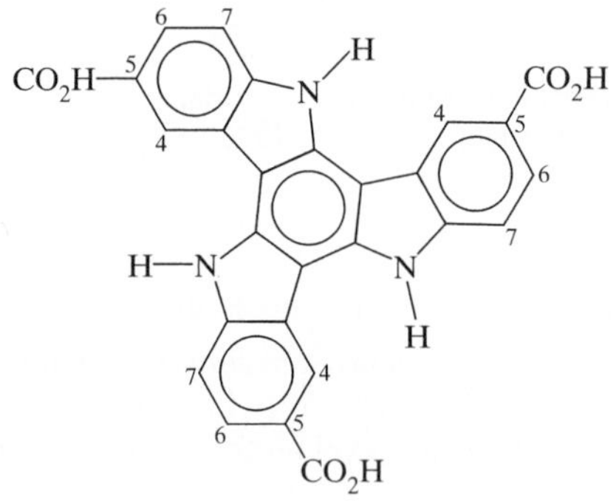

I

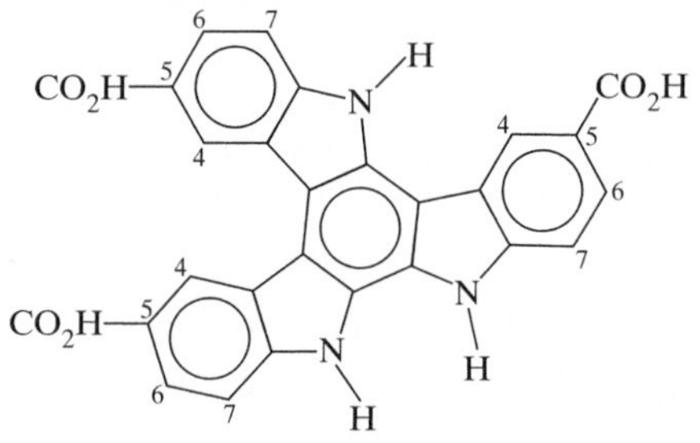

II

Shown below are the 360.1 MHz ^1H spectrum of the product and the COSY spectrum (with a blowup of the δ 7.8–8.2 region) in DMSO-d_6. The signal marked with an asterisk is an impurity. The ^1H signals are labeled A through N. Signals A to D were removed with the addition of D_2O. Signal A is a broad peak at the base of B. In the 1D NOE experiment, irradiation of B affected F/G and N, irradiation of C and D affected M and L, respectively, and irradiation of F/G affected B.

(a) From the overall appearance of the spectrum, is the trimer I or II? Explain.

(b) Using peak multiplicities, the NOE experiments, and the COSY spectrum, assign all the resonances. Discuss your reasoning in a step-by-step fashion. You should end up with an assignment of peaks A through N to specific protons.

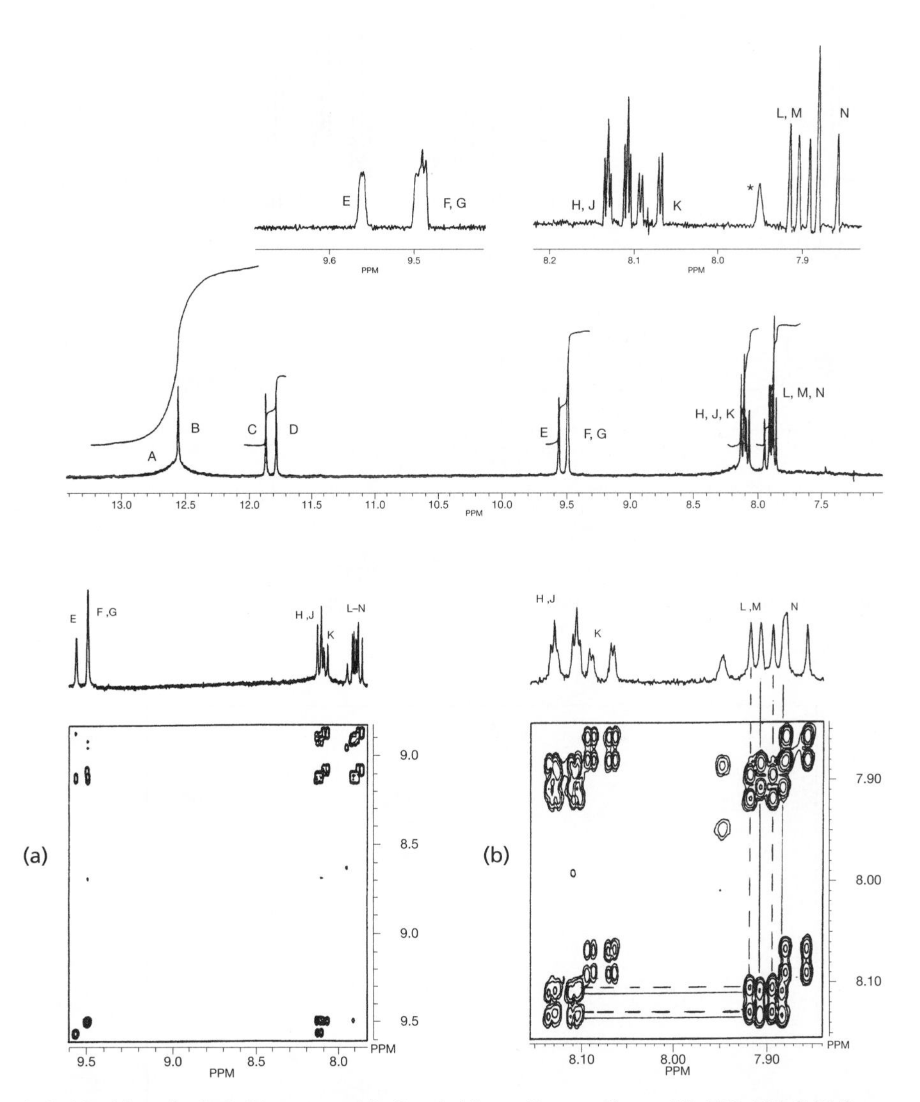

J. G. Mackintosh, A. R. Mount, and D. Reed, *Magn. Reson. Chem.*, **32**, 559–560 (1994). Reproduced with permission from John Wiley & Sons, Ltd.

6-5 The following 500 MHz ^{1}H spectrum is of the illustrated sugar derivative (extracted from a bean). Hydroxy resonances are omitted.

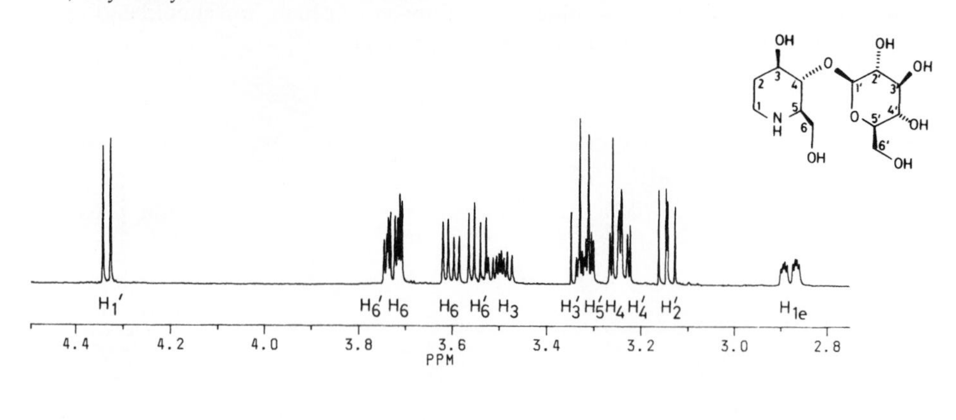

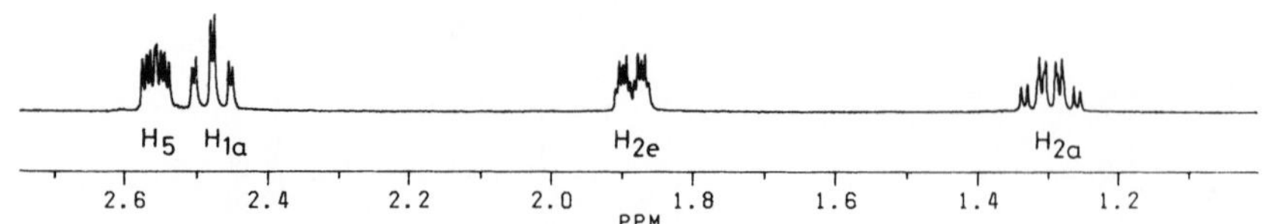

(a) The complete COSY spectrum, including the assignment of one resonance, is as follows. Complete the assignment for protons in the sugar ring.

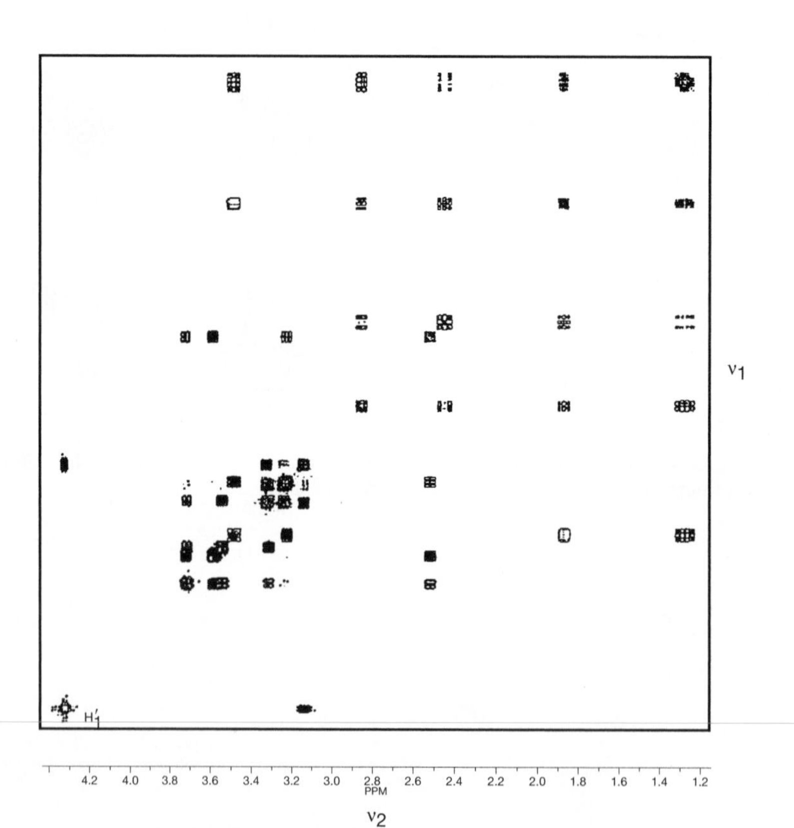

A. E. Derome, *Modern NMR Techniques for Chemistry Research,* Oxford, UK: Pergamon Press, 1987, p. 257.

(b) The following is a slightly expanded version of the low-frequency portion of the COSY spectrum, again with the assignment of one resonance. Complete the assignment for the protons of the piperidine ring. First, it is advisable to draw out the chair conformation.

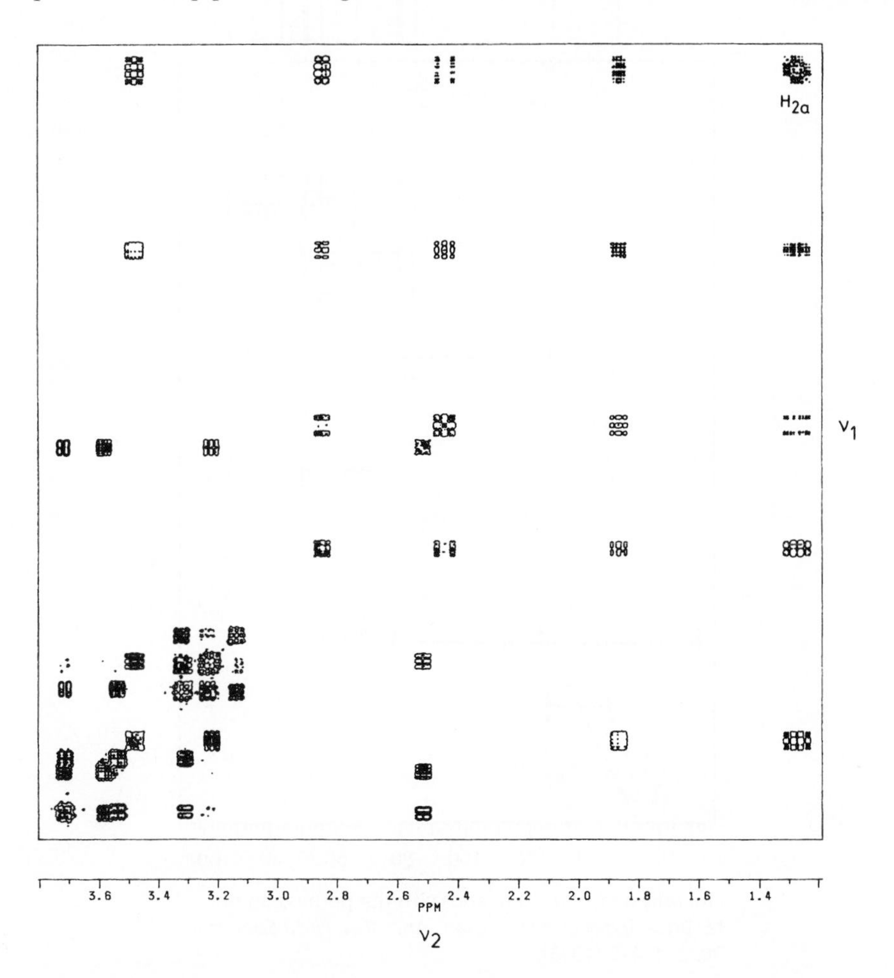

6-6 With only the following 2D INADEQUATE spectrum, derive the structure for the skeleton of isomontanolide. There are two overlapping resonances at δ 78, so you must work around ambiguities at that chemical shift. Also, there are substituents on the carbons in the range δ 60–80 that are not defined by the experiment.

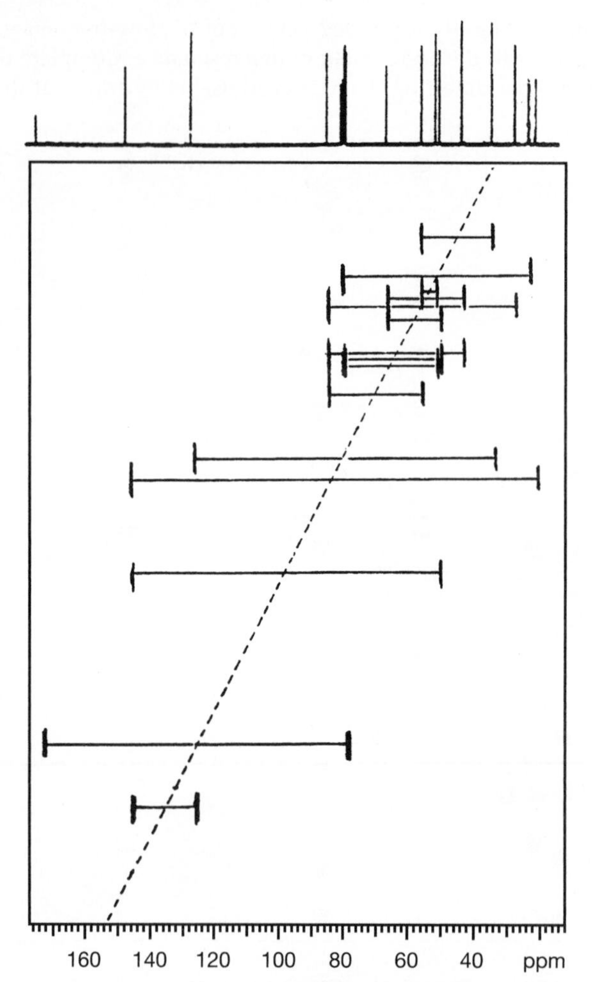

Liberally adapted from, and with the permission of, M. Budesinsky and D. Saman, *Ann. Rev. NMR Spectrosc.*, **30**, 231–475 (1995).

6-7 The molecule with structure **A** gave the following DQF–COSY (sugar portion only) in DMSO-d_6.

(a) Assign the sugar protons as fully as possible, and explain their positions.

(b) Assign the ^{1}H resonance at δ 12.56, and explain its high-frequency position.

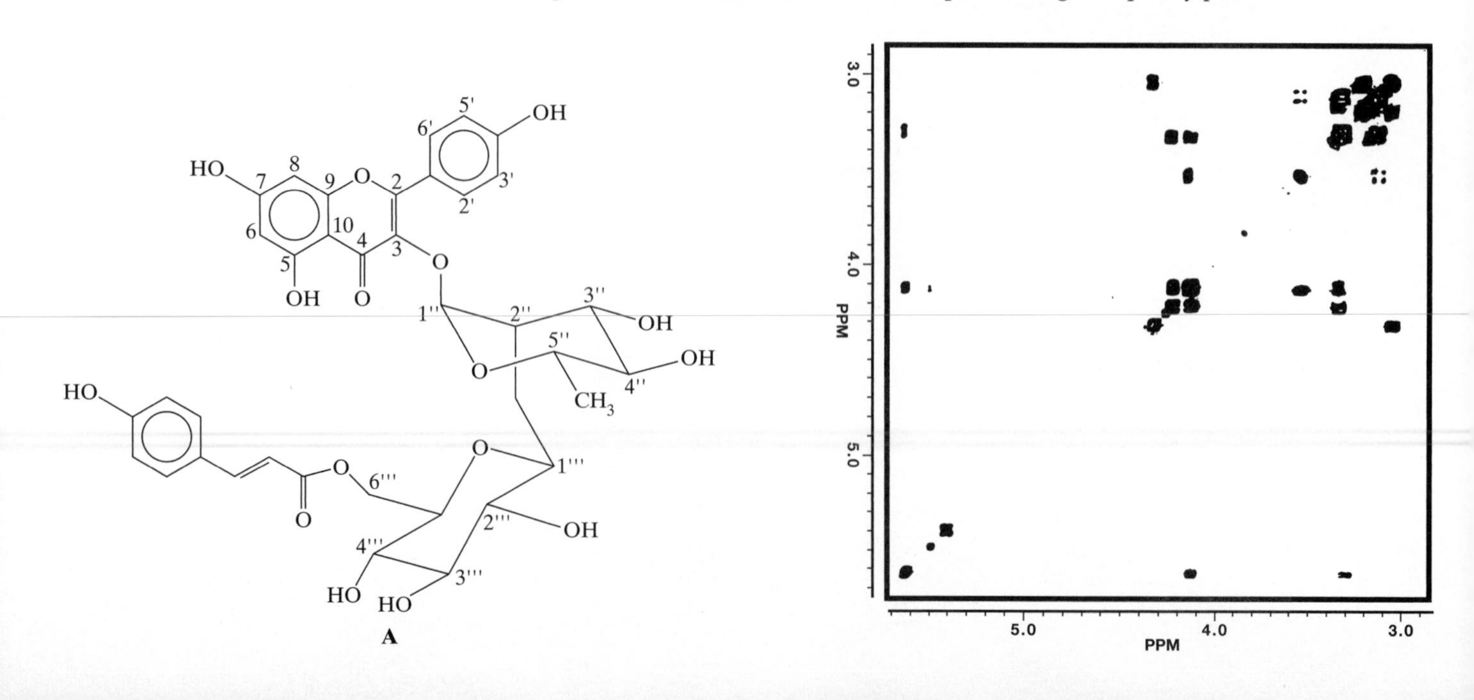

(c) The ^{1}H spectrum exhibited vicinal couplings of ^3J = 9.4 Hz for H—O—C(3'')—H and of ^3J = 3.1 Hz for H—O—C(2''')—H. Explain in terms of medium effects and conformations.

6-8 The isomeric 4,5-cyclopropanocholestan-3-ols **A** and **B** are expected to be inhibitors of cholesterol oxidase. (Protons pointing up are labeled β and protons pointing down are α.) The undepicted remainder of the structure is the cholestane skeleton and is not relevant to this problem. The DFQ–COSY and NOESY spectra for both isomers are given below.

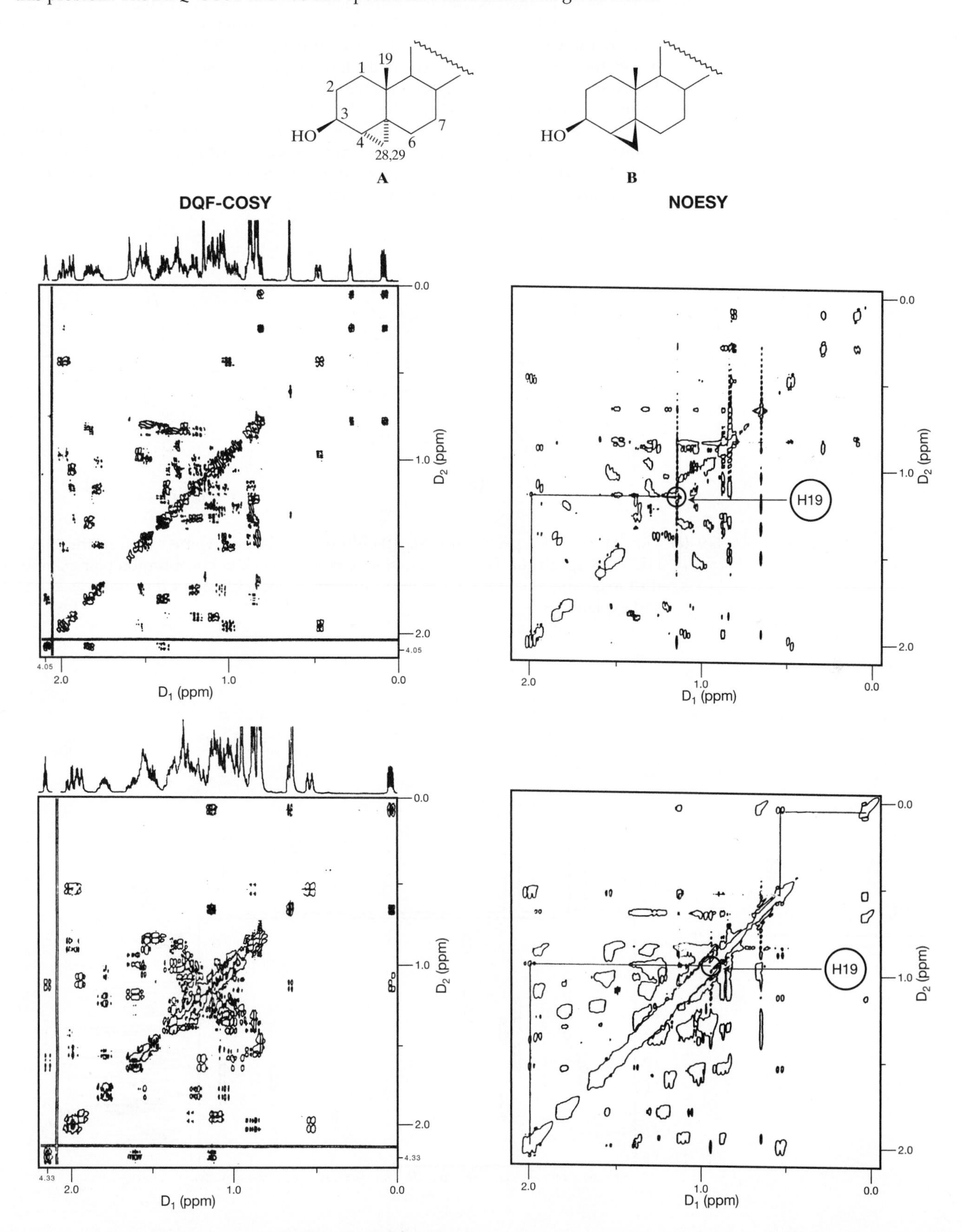

(a) Starting with H(3α) in the DQF–COSY spectrum of one isomer, assign the cyclopropane protons (H4, H28, H29). [This approach works only for one isomer; see part **(b)**.] Note the discontinuity in the DQF–COSY axes between δ 2 and 4. Assign cross peaks for H4/H28, H4/H29, and H28/H29.

(b) The approach fails for the other isomer. Why? By analogy with your assignments in **(a)**, assign the cyclopropane resonances and cross peaks anyway.

(c) In both isomers, there is a third low-frequency peak in addition to the cyclopropane resonances (δ 0.5). It has a large J = 13.5 Hz, with a partner at about δ 2 in both isomers. Identify the DFQ–COSY cross peak between the J = 13.5 Hz partners in both isomers. Assign these resonances and explain the low-frequency position for the one partner.

(d) The NOESY spectra for both isomers reveal that H19 (the methyl group) is close to the δ 2 proton, from **(c)**. The proton to which the δ 2 proton is coupled (J = 13.5 Hz) shows a NOESY cross peak with one of the other low-frequency cyclopropane protons in only one isomer. These cross peaks are indicated on the spectra. With your peak assignments and these NOESY data, assign the isomers to the spectra. The expected conformations are given below. Show the NOESY relationships.

1a **1b**

6-9 **(a)** The ^{1}H and ^{13}C spectra of *trans*-10-chlorodecal-2-one (**A**) are given in the following HETCOR spectrum. The conformation is rigid, so each geminal proton pair exhibits distinct axial and equatorial resonances. Identify the C6, C7, and C8 resonances *as a group*, and explain.

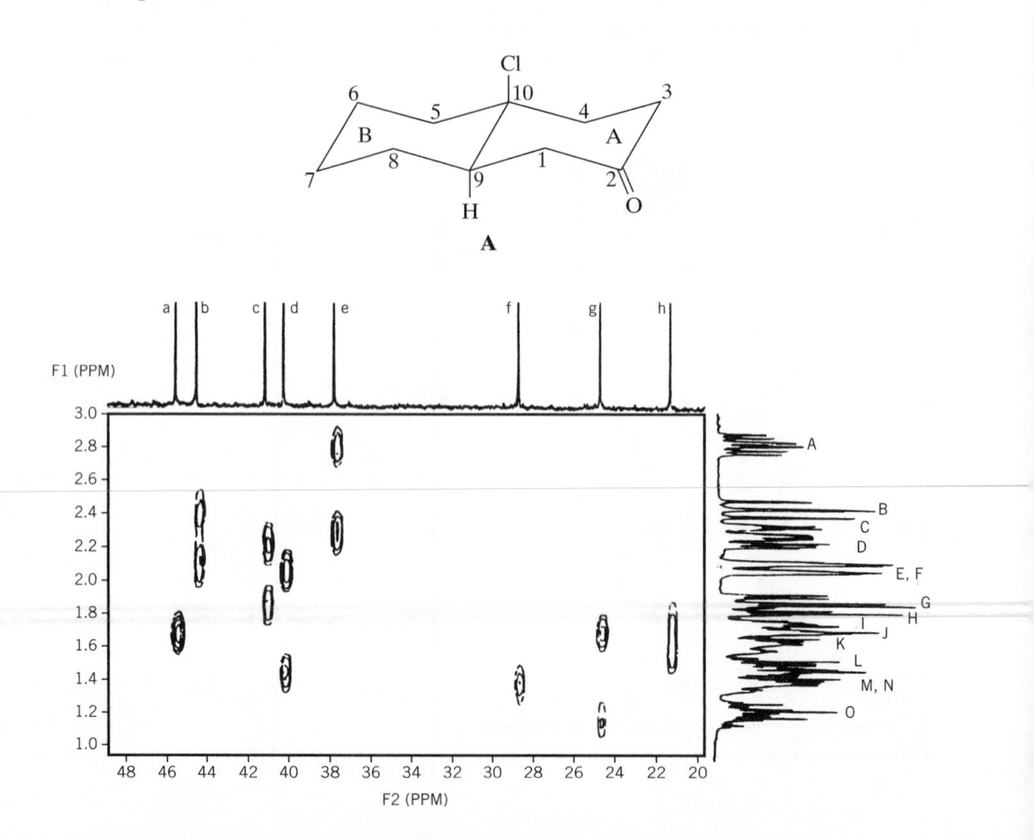

(b) Now assign H9 and C9. Explain your logic.

(c) What ^{1}H signals are expected to occur at the highest frequency (lowest field)? Explain.

(d) From the following COSY spectrum and the previous HETCOR spectrum, assign all the protons and carbons in the A ring. *Caution*! In this spectrum, the normal axial–equatorial relationship is reversed, as H1a and H3a respectively occur at a *higher* frequency than H1e and H3e.

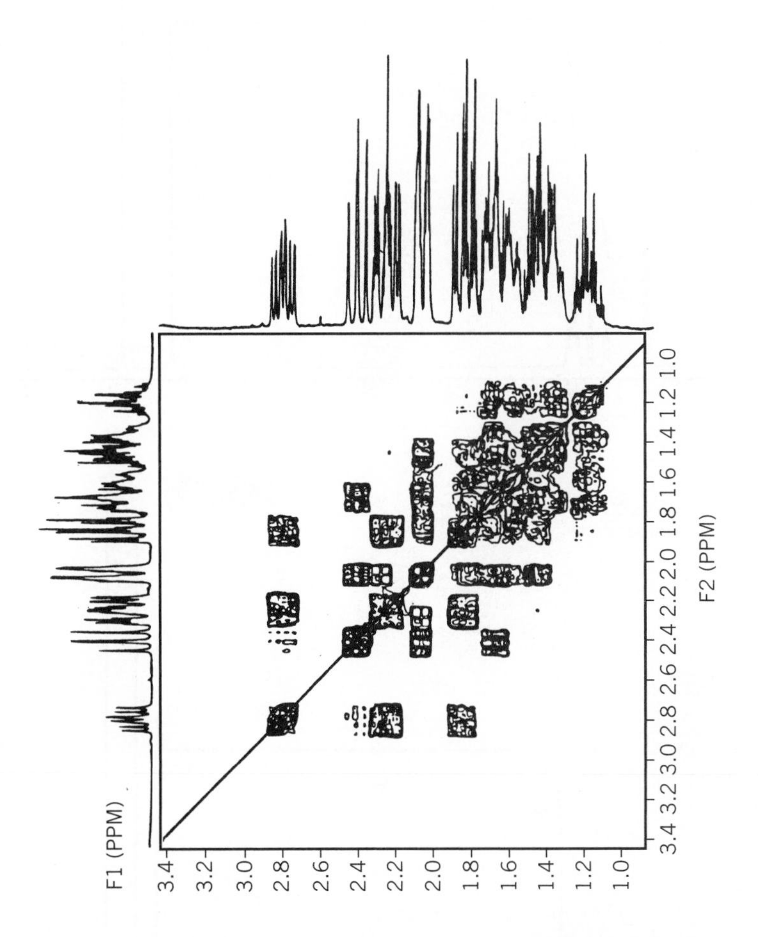

6-10 **(a)** The controlled substance methaqualone has the formula $C_{16}H_{14}N_2O$ (MW 250). A closely related analogue with MW 264 started to appear as a replacement in the illegal drug market. From their infrared spectra, both molecules contain a carbonyl group (C=O, at 1705 cm^{-1}). What structural fragments can you deduce from the 300 MHz ^{1}H spectra of methaqualone and its analogue given on the next page.

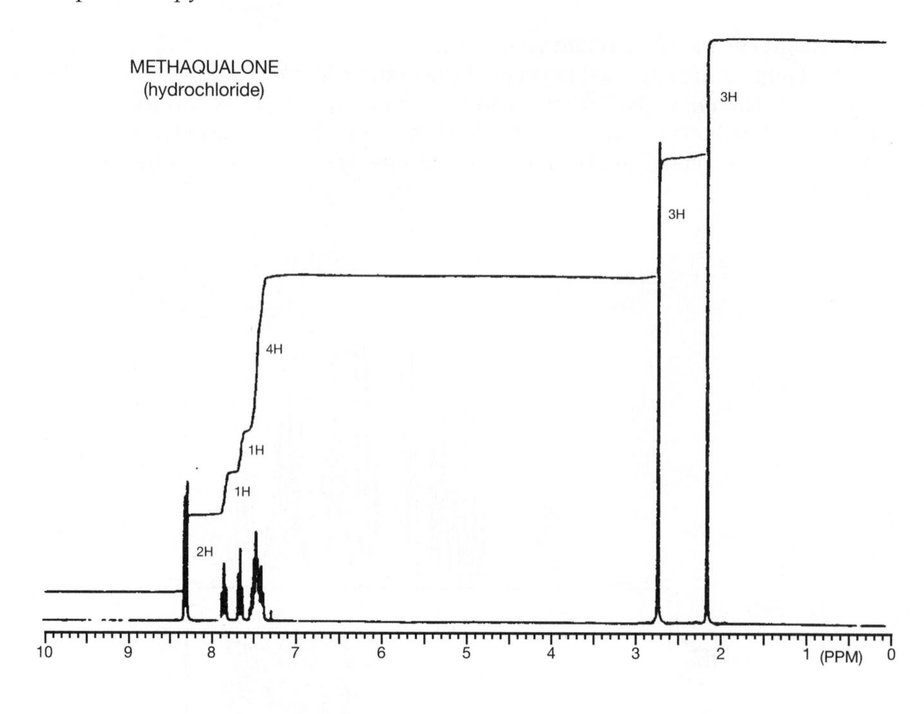

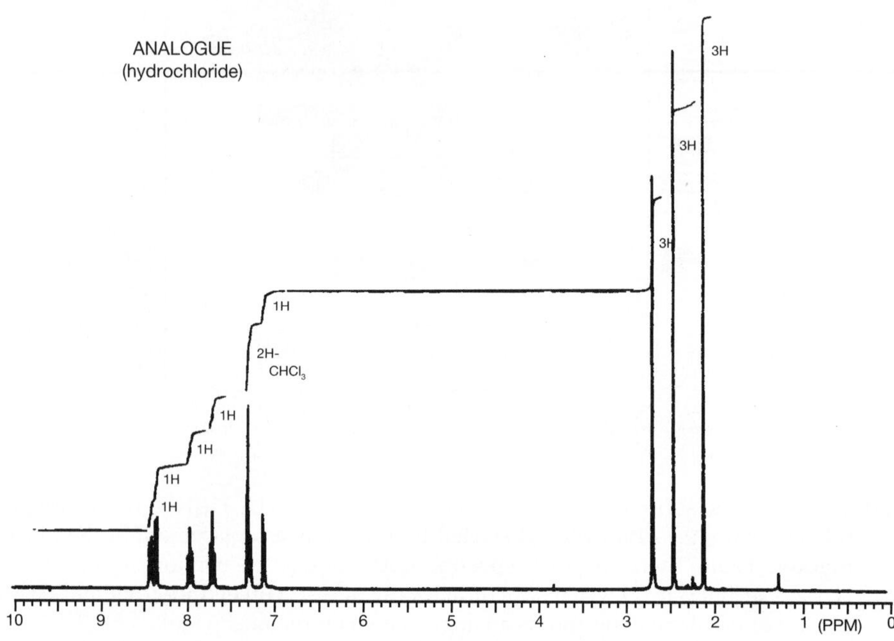

(b) From the following expansion of the high-frequency portion of the spectrum of the analogue, deduce the substructures responsible for these resonances.

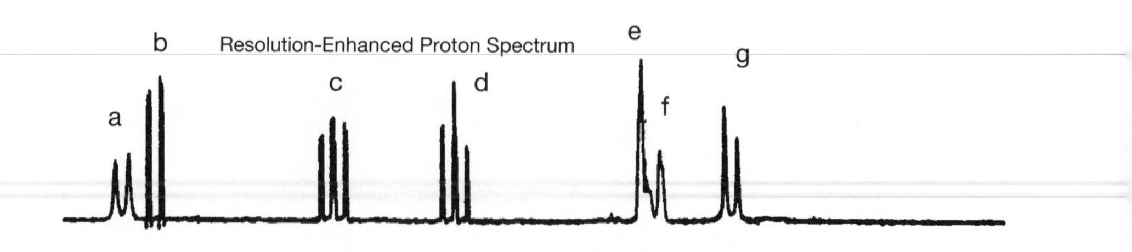

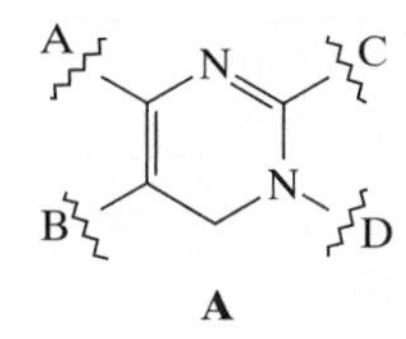

(c) Both the parent and the analogue contain a pyrimidine ring (**A**) that is unsaturated and substituted at all positions. Now assemble the entire analogue molecule.

(d) Assign all the cross peaks in the following COSY spectrum for the analogue molecule.

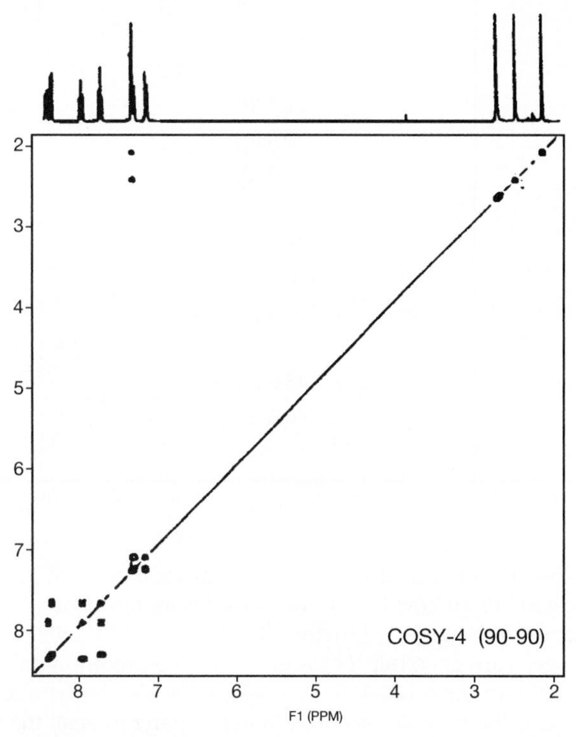

(e) From the following NOESY spectrum, complete any unresolved aspects of the structure of the analogue molecule.

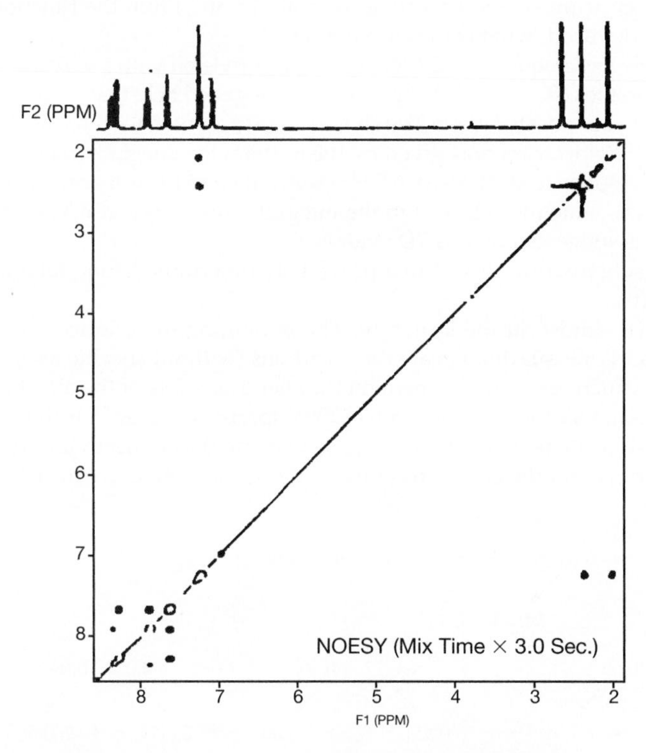

(f) From the structure of the analogue molecule and earlier spectral data, deduce the structure of methaqualone.

6-11 Upjohn scientists isolated a potent inhibitor of the cholesteryl ester transfer protein U-106305. The high resolution mass spectrum indicated that the formula was $C_{28}H_{41}NO$, and the ^{13}C spectrum had 27 distinct resonances (including one pair of equivalent carbons at δ 20.02). DEPT spectra indicated the multiplicities given in the following table.

^{13}C Chemical Shift and Multiplicities	^{1}H Chemical Shift and Coupling Constants	^{13}C Chemical Shift and Multiplicities	^{1}H Chemical Shift and Coupling Constants
7.6 (T)	0.07, 0.09 (dt: 8.43, 4.85)	20.0 (D)	1.00 (m)
7.6 (T)	0.12, 0.16 (dt: 8.39, 4.90)	20.02 (Q)	0.90 (d: 6.8)
8.0 (T)	0.08 (not first order)	20.7 (D)	1.29 (m)
11.4 (T)	0.32, 0.34 (dt: 8.20, 4.77)	21.8 (D)	0.68 (m)
13.4 (T)	0.65 (dt: 8.59, 4.87)	22.4 (D)	0.94 (m)
14.8 (T)	0.34, 0.43 (dt: 8.33, 4.60)	24.0 (D)	1.01 (m)
14.8 (D)	0.63 (m)	28.5 (D)	1.77 (h: 6.8)
17.9 (D)	0.57 (dq: 13.27, 4.93)	46.7 (T)	3.20 (d: 6.8)
18.0 (D)	0.58 (m)	120.0 (D)	5.91 (d: 15.2)
18.2 (D)	0.49 (m)	130.4 (D)	4.98 (dd: 15.5, 7.2)
18.2 (D)	0.51 (m)	131.0 (D)	4.98 (dd: 15.5, 7.7)
18.4 (D)	0.53 (m)	148.8 (D)	6.24 (dd: 15.2, 9.8)
18.41 (Q)	1.02 (d: 6.0)	166.0 (S)	
18.8 (D)	0.60 (m)		

(a) The infrared spectrum showed intense bands at 1630 and 1558 cm^{-1}. What functional group is suggested? What NMR peak confirms this assignment?

(b) What substructures are suggested by the ^{13}C peaks at δ 120–150?

(c) The HETCOR spectrum gave full ^{1}H assignments (see preceding table). From the ^{1}H resonances correlated with the δ 120–150 ^{13}C peaks, what else can you tell about the substructures from **(b)**? Use the magnitudes of J (values in parentheses), the values of the four ^{1}H and ^{13}C chemical shifts, and the proton multiplicities. In particular, note that δ 120 is d rather than dd.

(d) The UV–vis spectrum showed a strong band at 215 nm. From the functional groups you have already deduced, what is the chromophore?

(e) Examine the six low-frequency ^{13}C triplets. Each is correlated with a very low frequency (high field) pair of protons ($\delta < 0.7$). What grouping is suggested here that is present six times?

(f) Now count up your unsaturations. You should have accounted for them all. Enumerate them.

(g) The DQF–COSY spectrum was given by the authors for one substructure as follows: the integral-6 ^{1}H resonance at δ 0.90 (d: 6.8 Hz) was linked to the integral-1 resonance at δ 1.77 (heptet: 6.8 Hz), which was linked to the integral-2 resonance at δ 3.20 (d: 6.8 Hz). What substructure is suggested by this 2D evidence?

(h) How is this substructure linked to a previously determined functionality? Look at the chemical shifts.

(i) You now have almost all the structure. The remaining unassigned ^{13}C resonances are 12 doublets and one quartet. Locate these carbons (without specific assignment) on your previous substructures, and comment on the chemical shifts of the attached protons.

(j) These protons are all found in the DQF–COSY spectrum, except for the δ 1.29 resonance. Even at 600 MHz, there is severe overlap, so here are the connectivities derived from this experiment. Write out the entire structure, without some stereochemistries.

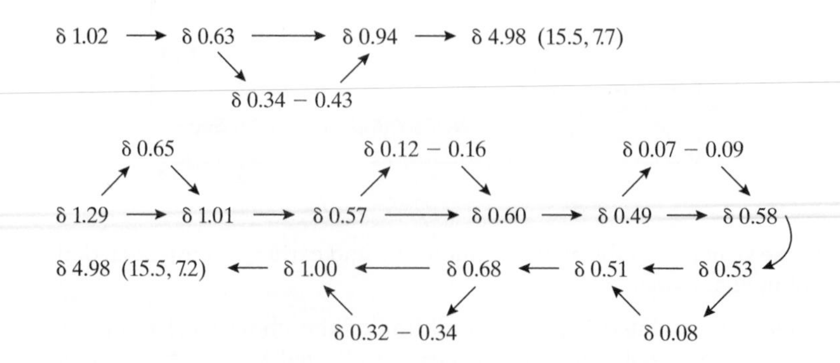

(k) Look at the available multiplicities and *J* values for the protons associated with the six high-frequency triplets. All are dt with *J* either ~8.4 or ~4.8 Hz. Give the full structure with all stereochemistries.

Bibliography

See also general texts referenced in previous chapters.

General

6.1 A. Bax, *Two-Dimensional Nuclear Magnetic Resonance in Liquids*, Boston: D. Reidel Publishing, 1982.
6.2 R. R. Ernst, G. Bodenhausen, and A. Wokaun, *Principles of Nuclear Magnetic Resonance in One and Two Dimensions*, Oxford, UK: Oxford University Press, 1987.
6.3 G. E. Martin and A. S. Zektzer, *Two-Dimensional NMR Methods for Establishing Molecular Connectivity*, New York: VCH, 1988.
6.4 J. Schraml and J. M. Bellama, *Two-Dimensional NMR Spectroscopy*, New York: Wiley-Interscience, 1988.
6.5 K. Nakanishi, *One-Dimensional and Two-Dimensional NMR Spectra by Modern Pulse Techniques*, Mill Valley, CA: University Science Books, 1990.
6.6 R. R. Croasmun and R. M. K. Carlson (eds.), *Two-Dimensional NMR Spectroscopy*, 2nd ed., New York: VCH, 1994.
6.7 F. J. M. van de Ven, *Multidimensional NMR in Liquids*, New York: VCH, 1995.
6.8 J. N. S. Evans, *Biomolecular NMR Spectroscopy*, Oxford, UK: Oxford University Press, 1995.
6.9 Atta-ur-Rahman and M. I. Choudhary, *Solving Problems with NMR Spectroscopy*, San Diego: Academic Press, 1996.
6.10 H. Friebolin, *Basic One- and Two-Dimensional NMR Spectroscopy*, 4th ed., Weinheim, Germany: Wiley-VCH, 2005.
6.11 J. H. Simpson, *Organic Structure Determination Using 2-D NMR Spectroscopy*, San Diego: Academic Press, 2008.

COSY

6.12 A. Kumar, *Bull. Magn. Reson.*, **10**, 96–118 (1988).

Long-Range HETCOR

6.13 G. E. Martin and A. S. Zektzer, *Magn. Reson. Chem.*, **26**, 631–652 (1990).

HMBC

6.14 G. E. Martin, *Ann. Rev. NMR Spectrosc.*, **46**, 36–100 (2002).

HOESY

6.15 M. Yemloul, S. Bouguet-Bonnet, L. Aï cha Ba, G. Kirsch, and D. Canet, *Magn. Reson. Chem.*, **46**, 939–942 (2008).

EXSY

6.16 K. G. Orrell, V. Ŝik, and D. Stephenson, *Progr. NMR Spectrosc.*, **22**, 141 (1990).
6.17 C. L. Perrin and T. J. Dwyer, *Chem. Rev.*, **90**, 935–967 (1990).

2D INADEQUATE

6.18 J. Buddrus and J. Lambert, *Magn. Reson. Chem.*, **40**, 3–23 (2002).

Multiple Quantum Methods

6.19 T. J. Norwood, *Progr. NMR Spectrosc.*, **24**, 295–375 (1992).

Pulsed Field Gradients

6.20 W. S. Price, *Ann. Rev. NMR Spectrosc.*, **32**, 51–142 (1996).
6.21 T. Parella, *Magn. Reson. Chem.*, **34**, 329–347 (1996).
6.22 K. Stott, J. Keeler, Q.-N. Yan, and A. J. Shaka, *J. Magn. Reson. A*, **125**, 302–324 (1997).

DOSY

6.23 C. S. Johnson, *Progr. NMR Spectrosc.*, **34**, 203–256 (1999).
6.24 Y. Cohen, L. Avram, and L. Frish, *Angew. Chem., Int. Ed. Engl.*, **44**, 520–554 (2005).

PANACEA

6.25 E. Kupĉe and R. Freeman, *J. Am. Chem. Soc.*, **130**, 10788–10792 (2008).

PART II

Mass Spectrometry

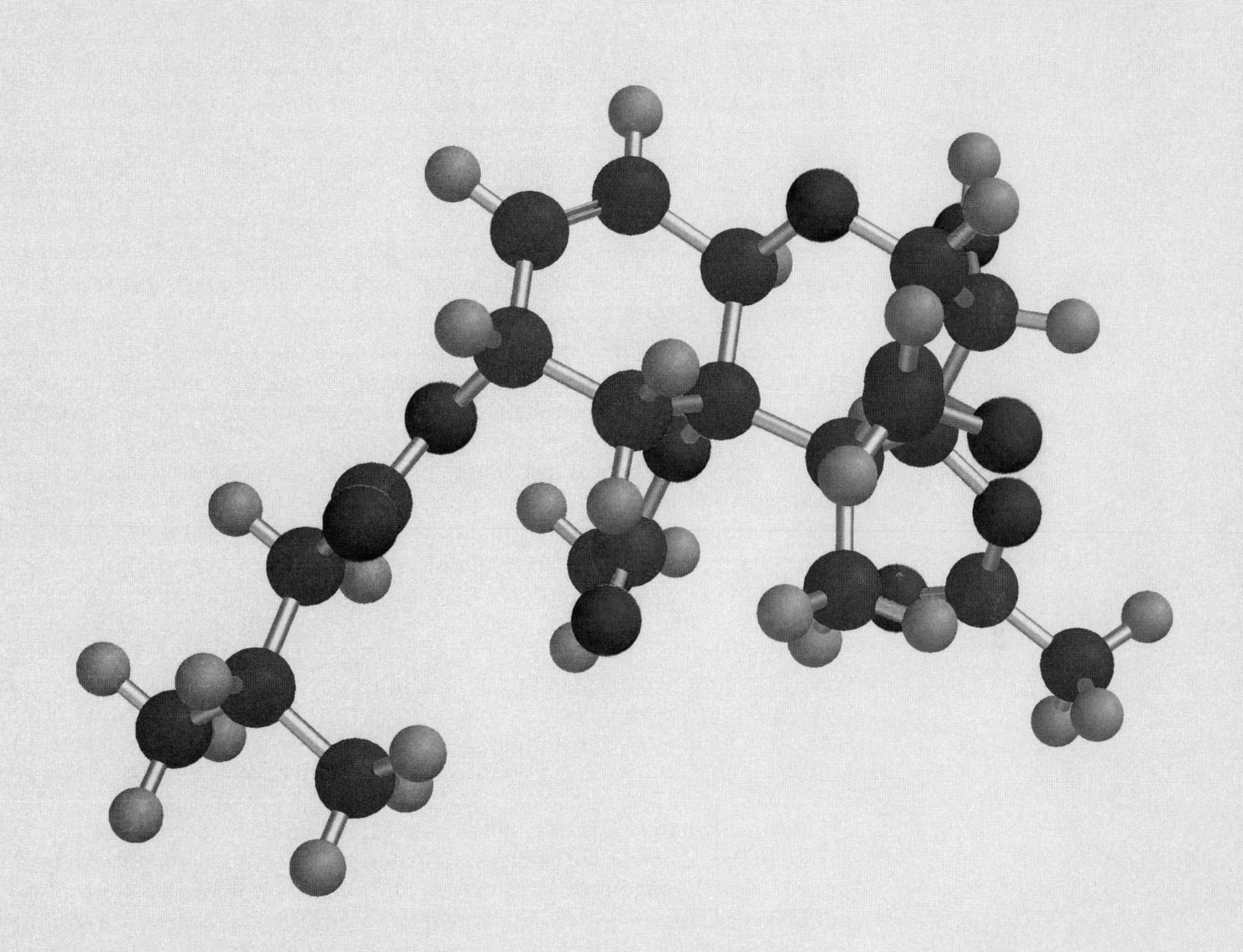

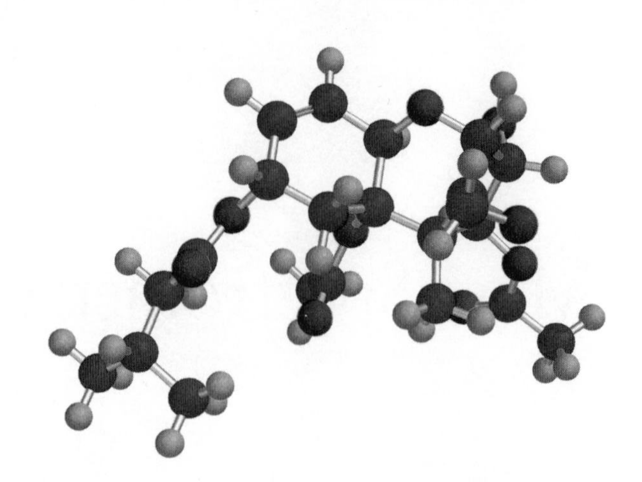

Mass Spectrometry: Instrumentation

7-1 INTRODUCTION

Mass spectrometry is probably today's most rapidly developing methodology in the collection of organic spectroscopic methods, and it is difficult to keep pace with the myriad technical innovations that occur each year. A 10-year-old mass spectrometer can easily be viewed as obsolete or at least greatly lacking the full capabilities of a new machine. Advances have been made in how ions are formed, how their masses are determined, and how they are detected. Wholly new types of instruments routinely come to market, and many hybrid instruments have been developed by linking together existing technologies. Despite all these changes and technological development, the basic data generated by mass spectrometry remains remarkably simple, given the power of the technique. All one obtains is an indicator of the molecular weight of a species of interest. How is it possible to tackle such a wide variety of analytical problems with such seemingly simple data? The answer lies in the fact that mass spectrometry is one of the few general techniques that permits manipulation of individual molecules and, as a result, links molecular weight information from one molecule to the weights of its reaction/fragmentation products. Still, this apparently limited amount of information may seem inadequate for solving complex problems, but two other factors play important roles. First and most importantly, there is great regularity in the reaction/fragmentation behavior of organic ions, and, consequently, very reliable structural predictions can be made on the basis of patterns in ion fragmentation. Second, molecular structure can be probed in a variety of different experiments and assignments based on the consensus of multiple spectra. In the case of a protein, potentially thousands of spectra might be used in conjunction with powerful bioinformatics software to confirm its identity. In the following chapters, we will explore how mass spectrometry can be used to solve a variety of structural problems in modern organic and bioorganic chemistry.

How Does It Work? Mass spectrometry involves three basic steps: (1) formation of ions from an analyte (ionization); (2) separation of the ions, in space or time, based on their mass-to-charge ratio (mass analysis); and (3) detection of mass-sorted ions. The first step is necessary because only ions can be manipulated by electric and magnetic fields, which are used in all mass analyzers to sort ions by their mass-to-charge ratio. In the second step, the application of a variety of ion physics can be used to separate a collection of ions of nearly unlimited complexity into its individual mass-to-charge components. Ions can be raced down a vacuum tube and sorted from fastest to slowest, or they can be trapped in stable orbits and separated by differences in their orbital periods (frequency). Third and finally, the ions need to be converted to an electric current or potential for detection. This process allows for massive amplification and is at the heart of the sensitivity of the technique.

What Kind of Problems Can It Solve? The most obvious problem that can be addressed by mass spectrometry is the verification of a chemical structure. Obtaining the molecular weight of an analyte can confirm that it has the proper molecular formula for a proposed structure. Such data are routinely part of the evidence used to verify the structures of new synthetic species. Mass spectrometry, however, can provide much more structural information than just a molecular weight. Through a variety of means, it is possible to fragment analyte ions in mass spectrometers. On the basis of past experience or comparison to authentic samples, it is possible to predict how structures fragment and what types of fragment ions are produced. The combination of an accurate match on the molecular weight of a species as well as a plausible fragmentation pattern is excellent evidence for a proposed structure. This approach can be multiplexed and used to investigate structures of very complex species, such as proteins. Along with determining structures, mass spectrometry can provide information about the physical properties of analytes. For example, data from mass spectrometry have been widely used to estimate the heats of formation of a wide range of reactive intermediates and provide key information for the accurate determination of bond dissociation energies. Finally, mass spectrometry can be used as a fine analytical tool for quantifying the concentrations of analytes in potentially complex mixtures. We explore all of these applications in the next few chapters.

7-2 IONIZATION

Because mass spectrometry can be applied only to charged species, all mass spectrometry experiments must begin with an ionization process. In early work, the ability to ionize an analyte was potentially a major problem and greatly limited the range of species that could be analyzed by mass spectrometry. Over the past few decades, the development of a wide variety of novel ionization methods has greatly enhanced the generality of mass spectrometry and has allowed for the analysis of virtually any molecular species by mass spectrometry.

The choice of an ionization method depends greatly on the nature of the analyte as well as the type of data one wishes to obtain. Part of this decision involves the properties of the ionic species to be used in the experiment. Techniques are available to produce cations, radical cations, anions, and radical anions from neutral substrates. The charge state (cation or anion) and electronic state (radical or closed-shell) play critical roles in the way ions fragment in a mass spectrometry experiment and therefore have an important impact on the information content of the experiment. Another component of the decision-making process is the physical state of the analyte (gas, liquid, or solid). Specific techniques have been developed to handle analytes in each of these states.

There are four major approaches to the ionization of an analyte. The earliest, and still the most common for small organic compounds, is electron ionization (EI), referred to in the past as electron impact. The major ionization path is the formation of the radical cation of the analyte. Chemical ionization (CI) was an outgrowth of EI and generally involves the gas phase reaction of the analyte with an ion produced either directly or indirectly by EI. The reactant ion can be a cation or an anion, and therefore CI can produce either positive or negative ions. Both of these techniques are gas-phase processes and therefore are limited to analytes with a reasonable vapor pressure. Desorption methods were developed to permit the ionization of solids. They involve a variety of mechanisms for producing charged species. Depending on the conditions, the analyte can be observed as a cation, as an anion, or as a complex with a permanent ion such as Na^+. Finally, spray methods have been developed to handle analytes in liquid solutions. In these methods, the analyte is typically already ionized in solution or forms a complex with an ionized species in the solution. A unique feature of the spray methods is their ability to routinely produce species with multiple charges. These ionization methods are summarized in Table 7-1.

TABLE 7-1 Ionization Methods

Type	Examples	Analyte State	Typical Ion Types	Fragmentation
Electron ionization	EI	Gas	$M^{+\cdot}$	Extensive
Chemical ionization	CI	Gas	$(M+H)^+$, $(M-H)^-$	Limited
Desorption	MALDI, FAB, LSIMS	Solid	$(M+H)^+$, $(M-H)^-$, $(M+Cat)^+$	Limited
Spray	ESI, APCI	Solution	$(M+nH)^{n+}$, $(M-nH)^{n-}$, $(M+nCat)^{n+}$	Little

Note: Cat^+ = permanently charged cation such Na^+ or R_4N^+, MALDI = matrix-assisted laser desorption ionization, FAB = fast-atom bombardment, LSIMS = liquid secondary ion mass spectrometry, ESI = electrospray ionization, APCI = atmospheric pressure chemical ionization.

Another aspect of the ionization process is the internal energy of the resulting ion. In electron impact, the ionizing electrons have sufficient energy to ionize the analyte and leave it with large amounts of internal energy (electronic or vibrational), which can lead to extensive fragmentation. The energetics of chemical ionization depend on the nature of the chemical reaction that is employed and can produce ions with significant or little internal energy, which leads to various degrees of fragmentation during the ionization process. Desorption and spray processes are often referred to as "soft" ionization methods, because the ions can be formed with little internal energy and analytes undergo limited or no fragmentation during ionization. In desorption methods, a large amount of energy is deposited into the supporting matrix (the analyte is mixed in a chemical matrix in these methods), but generally little energy is transferred to the analyte itself, so limited fragmentation is observed. In spray methods, ions are formed with almost no excess energy.

Each of the ionization methods plays an important role in the mass spectrometry of various classes of analytes from small hydrocarbons to large biopolymers. In the next sections, each of the ionization methods is described in detail.

7-2a Electron Ionization

Electron ionization, which is sometimes referred to as electron impact ionization, is one of the simplest ionization methods and has been widely used in the analysis of small organic species. Its strengths are its ease of implementation, high degree of reproducibility, and ability to produce both parent and fragment ions. Its major weaknesses are its requirement that the analyte be volatile and the inability to control the extent of fragmentation carefully during the ionization process. The first issue limits it to relatively small species, and the second issue can lead to situations in which none of the analyte survives the ionization process and only fragment ions are observed. Within these constraints, however, EI is a robust, general method of analysis.

Ionization Process. The fact that electron ionization generally leads to the formation of cationic species might seem counterintuitive. Why would the addition of an electron to an analyte produce a cation? The answer is found in the electron affinities of typical organic analytes and the energy of the electrons used in the ionization process. We begin by considering a set of possible outcomes when an electron interacts with an analyte (Figure 7-1). Path (a) might seem the most logical. The ionizing electron is trapped by the analyte to produce a radical anion. One problem with this path is obvious in the data in Table 7-2. Most simple organics do not have significant electron affinities and

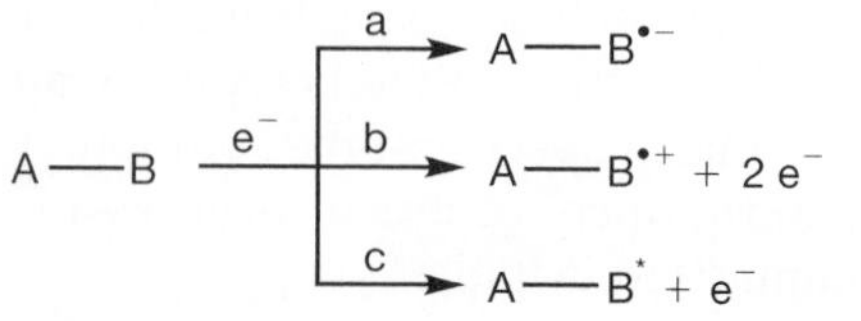

FIGURE 7-1 Potential pathways for the interaction of a fast electron with an organic substrate, A—B. The symbol * indicates that the neutral substrate is electronically excited in the process. Although not indicated explicitly, the radical ions are generally formed in vibrationally excited states.

$$A{-}B \xrightarrow{e^-} \begin{cases} \xrightarrow{a} A{-}B^{\bullet -} \\ \xrightarrow{b} A{-}B^{\bullet +} + 2\,e^- \\ \xrightarrow{c} A{-}B^{*} + e^- \end{cases}$$

TABLE 7-2 Electron Affinities of Typical Organic Species

Functional Group	Example of Analyte	Electron Affinity (eV)
Ketone	Acetone	0.0015
Nitrile	Acetonitrile	0.01
Aldehyde	Benzaldehyde	0.39
Nitro	Nitromethane	0.50
Nitro	Nitrobenzene	1.00
Nitro	Trinitrobenzene	2.63

Data from webbok.nist.gov.

many even have negative electron affinities (if formed, their radical anions are thermodynamically unstable toward electron detachment). Only species with powerful electron-withdrawing groups, such as nitro groups, have significant electron affinities. Another factor is that the ionizing electrons typically have kinetic energies on the order of tens of electron volts (70 eV is standard), and there is little chance for analytes to capture electrons with such excess energy and dissipate it. As a result, path (a) is unlikely, and instead the system follows paths (b) or (c).

In path (b), the fast-moving electron interacts with the electron cloud of the analyte and excites a bound electron to a dissociative state. In simple terms, some of the ionizing electron's kinetic energy is transferred to the bound electron, providing it with enough energy to escape. This is the key pathway in EI. In this process, the resulting radical cation is generally formed in a vibrationally excited state, which can lead to bond cleavage and fragmentation products (see the next section). In path (c), the electron interacts with the electron cloud of the analyte but is not trapped and does not provide sufficient electronic excitation to dissociate an electron and form an ion. This is a nonproductive path. Is it possible for EI to produce significant amounts of negative ions? Although direct capture of a 70 eV electron by an organic substrate is unlikely, note that path (b) produces a second electron, which typically has relatively low kinetic energy. These electrons can be captured by organics with significant electron affinities and produce radical anions. This process is relevant to negative ion chemical ionization (see Section 7-2b).

Internal Energy Deposition and Fragmentation. A hallmark of EI is the production of fragment ions resulting from bond cleavage during the ionization process. Fragmentation provides valuable data that can be used to identify analytes (fingerprinting) or to provide insight into the structural features of an analyte. Given their kinetic energy, it is not surprising that ionizing electrons can induce fragmentation. At typical EI energies (70 eV), an ionizing electron could potentially deliver more than 1500 kcal mol^{-1} of energy to an analyte. Since less than 250 kcal mol^{-1} generally is needed to ionize organic analytes, a large amount of energy is available for internal excitation of the resulting radical cation. Most of this energy is retained in the ionizing electron as kinetic energy, but there are two mechanisms for transferring it to the radical cation. First, the radical cation can be formed directly in an electronically excited state, which relaxes via internal energy conversion to give a vibrationally excited state. Second, because the ionization process occurs on a very short time scale, the radical cation is formed in the geometry of the initial, neutral analyte, which often correlates with a high vibrational state for the resulting radical cation. Either mechanism is capable of producing a radical cation with sufficient internal energy to undergo a bond cleavage process. Fragmentation also is favorable because radical cations typically have weaker bonds than do typical closed-shell organic species. Two examples are given in Figure 7-2. In ethanol, the *molecular ion peak*, which corresponds to the radical cation of ethanol, appears at m/z 46. It is a significant peak, but a fragment at m/z 31 is the largest peak, referred to as the *base peak* of the spectrum (the one used in the normalization to 100% relative intensity). This peak represents a loss of 15 mass units, likely a CH_3 group. There is also a strong signal for m/z 45, which is formally from loss of H^-, but is more likely from loss of $H^\bullet$ from m/z 46, the parent radical cation. The spectrum for 1-decanol has

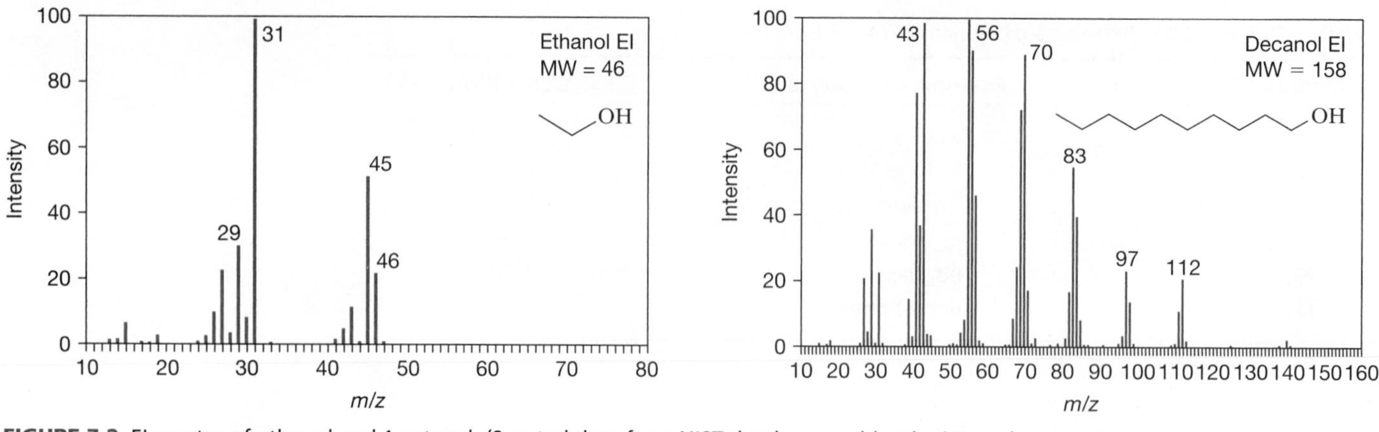

FIGURE 7-2 EI spectra of ethanol and 1-octanol. (Spectral data from NIST database, webbook.nist.gov.)

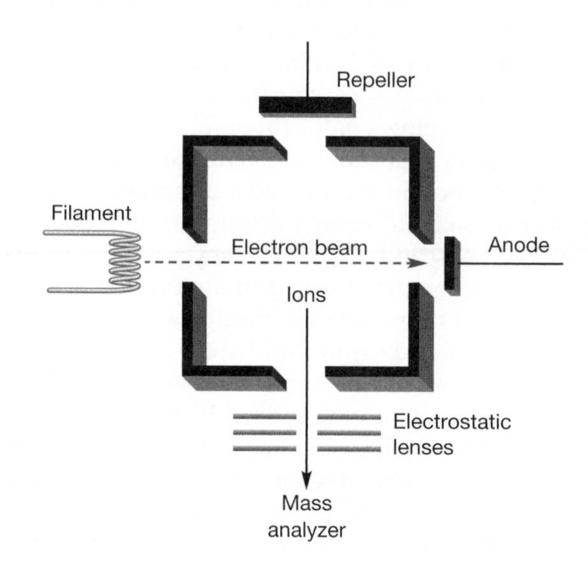

FIGURE 7-3 Schematic of an EI source.

many more fragments. One can see groups of fragments with the same number of carbons, but varying numbers of hydrogens (two to eight carbons). Conspicuously absent is the molecular ion peak expected at m/z 158. Fragmentation processes are discussed in detail in Chapter 8.

Typical EI Sources. One of the attractive features of EI is the simplicity of the ionization source, which generally allows for highly reproducible, trouble-free operation. The heart of an EI source is a heated filament and an ion volume chamber, separated by a fixed potential (Figure 7-3). The analyte is introduced into the source as a vapor at low pressure ($\sim 10^{-5}$ torr) and interacts with the electron beam. Although a range of voltages could be used for accelerating the electrons, 70 eV has become a standard, in part because it reliably produces radical cations that fragment in a reproducible way. Over the years, a large database of 70 eV EI spectra has been developed and extensive libraries of 70 eV EI reference spectra are available (the National Institute of Standards and Technology [NIST] provides access to a free library at webbook.nist.gov). Once formed, the ions are extracted from the source by an electric potential and passed on to the mass analyzer of the system.

7-2b Chemical Ionization

Chemical ionization relies on gas-phase reactions to produce the desired ionization product of the analyte. In early work, it was recognized that gas-phase reactions were occurring in EI sources, and CI evolved as way of turning these processes into a useful, general ionization method. In CI, EI is generally used to produce a reagent ion, which then reacts with the analyte to give an ion suitable for mass analysis. An advantage of CI is that a variety of reagent ions can be formed in most instruments, and therefore the ionization process can be tailored to the analyte. Typically, CI relies on proton transfer or electron transfer reactions to produce the analyte ions. More exotic processes allow analytes to be formed as either closed-shell or radical cations and anions. A second advantage of CI is that the energetics of the ionization process can be controlled by the choice of the reagent ion, and, as a result, there is more control over the level of fragmentation in the ionization process.

Ionization Process. The first step is typically EI of a reagent gas to give a reagent ion or an ion that will be transformed into the reagent ion in a series of gas-phase reactions. During conventional EI, gas-phase reactions are suppressed by employing low pressures (~10^{-5} torr) in the source. In contrast, CI sources operate with much higher reagent gas pressures (~0.5 torr), which foster reactions between the initial radical cations and neutral reagent gas molecules. The process is best described with an illustrative example.

Methane is a very common reagent gas that leads to the formation of CH_5^+ as the reagent ion. Under EI conditions, methane is expected initially to form its radical cation, $CH_4^{+\bullet}$ (eq. 7-1). This radical cation is a strong acid that is capable of protonating methane to give CH_5^+, the ultimate reagent ion (eq. 7-2), which is capable of protonating the analyte (eq. 7-3).

$$CH_4 + e^- \longrightarrow CH_4^{\bullet +} + 2\,e^- \qquad \textbf{(7-1)}$$

$$CH_4^{\bullet +} + CH_4 \longrightarrow CH_5^+ + CH_3^\bullet \qquad \textbf{(7-2)}$$

$$CH_5^+ + M \longrightarrow (M + H)^+ + CH_4 \qquad \textbf{(7-3)}$$

Addition of H^+ to the analyte with the loss of H_2 also is a possibility and leads to $(M - H)^+$ ions. These are referred to as hydride loss ions. Unlike in EI, the analyte is formed as a closed-shell, protonated cation, $(M + H)^+$, rather than a radical cation. This difference can greatly alter the fragmentation behavior of the ionized analyte and, of course, leads to the production of a parent ion whose mass is one unit greater than the molecular weight of the analyte. Additional reactions of the reagent ion with the reagent gas are possible. In the case of methane, ions such as $C_2H_5^+$ and $C_3H_5^+$ can be formed (the former can result from the reaction $CH_3^+ + CH_4 \rightarrow C_2H_5^+ + H_2$). Along with ionization by proton transfer, it is possible for the reagent ion to form an adduct with the analyte. For example, CI with NH_4^+ as the reagent ion can often lead to ammonium complexes of the general structure $(M + NH_4)^+$. Methane is just one of many gases that could be used to produce reagent ions capable of protonating analytes (Table 7-3). The key difference among them is the acidity of the resulting reagent ion. This topic is discussed in detail in the next section.

The production of negative reagent ions capable of deprotonating analytes follows a similar pathway. EI in the presence of buffer gas (reagent gas) leads to the production of slow electrons that can be captured by the reagent gas to give radical anions or to give closed-shell anions via dissociative attachment. As noted above (see Table 7-2), most organics have low electron affinities, so the process is not particularly general and is

TABLE 7-3 Typical Positive Ion Reagent Gases and the Resulting Reagent Ions

Reagent Gas	Reagent Ion(s), RH^+	Proton Affinity (R)[a]	Analyte Ion(s)
H_2	H_3^+	100.9	$(M+H)^+$, $(M-H)^+$
CH_4	CH_5^+, $C_2H_5^+$	129.9, 162.6	$(M+H)^+$, $(M-H)^+$, $(M+C_2H_5)^+$
$(CH_3)_3CH$	$(CH_3)_3C^+$	191.7	$(M+H)^+$, $(M-H)^+$, $(M+C_4H_9)^+$
NH_3	NH_4^+	204.0	$(M+H)^+$, $(M-H)^+$, $(M+NH_4)^+$

[a]Proton affinity is the $-\Delta H_{rxn}$ of the following process: $R + H^+ \Rightarrow RH^+$. Values are in kcal mol^{-1} from webbook.nist.gov.

TABLE 7-4 Typical Negative Ion Reagent Gases and the Resulting Reagent Ions

Reagent Gas	Reagent Ion, R^-	Proton Affinity (R^-)[a]	Analyte Ion
CH_4/N_2O	HO^-	390.3	$(M-H)^-$
CH_4/CH_3ONO	CH_3O^-	382.0	$(M-H)^-$
NF_3	F^-	371.3	$(M-H)^-$

[a]Proton affinity is the $-\Delta H_{rxn}$ of the following process: $R^- + H^+ \Rightarrow RH$. Values are in kcal mol^{-1} from webbook.nist.gov.

limited to reagent gas species that have significant electron affinities or can dissociate to give energetically favorable closed-shell anions. The process with NF_3 is shown in eqs. 7-4 and 7-5.

$$NF_3 + e^- \longrightarrow NF_2^\bullet + F^- \quad \textbf{(7-4)}$$

$$F^- + M \longrightarrow (M-H)^- + HF \quad \textbf{(7-5)}$$

Dissociative electron capture leads to the production of fluoride ions that can deprotonate an analyte to give a closed-shell anion. Anions tend to fragment in very different ways from cations (often simply undergoing electron detachment) and produce parent ions whose mass is one unit less than the molecular weight of the analyte. A variety of reagent ions are used in negative ion CI with a range of gas-phase basicities (Table 7-4).

Finally, CI can be used as an alternative method of forming the radical cations typically generated in EI. Here, the reagent ion acts as a one-electron oxidant and forms the analyte ion by an electron transfer process. As in other CI methods, the reagent ion is generally formed by EI. The advantage of CI in this respect is that it allows one to tune the energetics of the process by varying the ionizing power of the reagent ion (generally measured as the ionization potential of the reagent gas). An example is presented in eqs. 7-6 and 7-7. As with EI, fragmentation can be significant, especially if the electron transfer process is highly exothermic. Typical systems are listed in Table 7-5.

$$Ar + e^- \longrightarrow Ar^{\bullet +} + 2\,e^- \quad \textbf{(7-6)}$$

$$Ar^{\bullet +} + M \longrightarrow M^{\bullet +} + Ar \quad \textbf{(7-7)}$$

Internal Energy Deposition and Fragmentation. Unlike with EI, one has significant control over the energetics of CI processes, so that fragmentation can be limited. In Figure 7-4, an example is given for the ionization of tetrahydrofuran (THF). In the first panel, 70 eV EI is used, whereas in the second panel, CI is used with THF itself as the

TABLE 7-5 Typical Reagent Gases Used for Charge Transfer Ionization

Reagent Gas	Reagent Ion, $R^{+\bullet}$	Ionization Potential (R)[a]	Analyte Ion
He	$He^{+\bullet}$	24.6	$M^{+\bullet}$
Ne	$Ne^{+\bullet}$	21.6	$M^{+\bullet}$
Ar	$Ar^{+\bullet}$	15.8	$M^{+\bullet}$
N_2	$N_2^{+\bullet}$	15.6	$M^{+\bullet}$
Xe	$Xe^{+\bullet}$	12.1	$M^{+\bullet}$
C_6H_6	$C_6H_6^{+\bullet}$	9.2	$M^{+\bullet}$
C_6H_5Cl	$C_6H_5Cl^{+\bullet}$	9.1	$M^{+\bullet}$

[a]Ionization potential in electron volts. Values are from webbook.nist.gov.

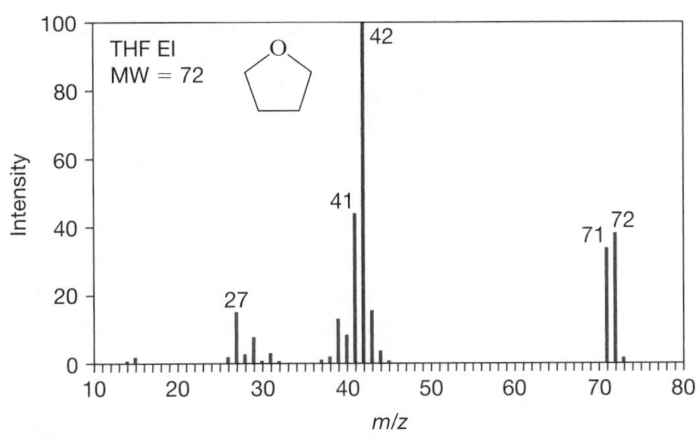

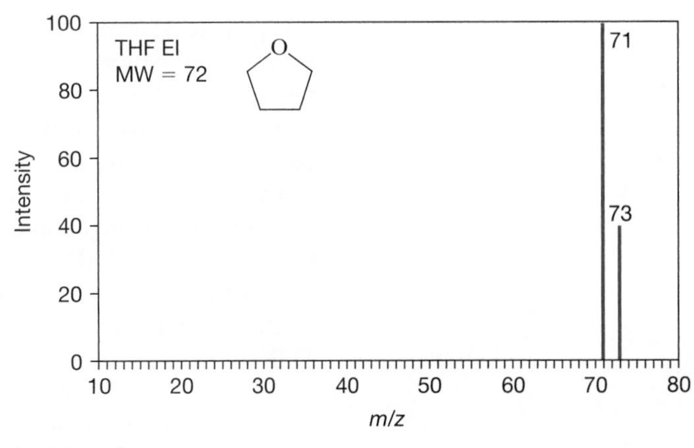

FIGURE 7-4 EI and CI spectra of tetrahydrofuran. (EI data from webbook.nist.gov.)

reagent gas (THF produces primary EI ions that react with more THF). The EI spectrum is dominated by fragment ions, with the base peak being *m/z* 42 (presumably formed by loss of CH_2O from THF). This is a radical cation with the formula $C_3H_6^{+\cdot}$ (potentially ionized propene). One also sees a closed-shell ion at *m/z* 41 that could be the allyl cation ($C_3H_5^+$). Finally, peaks are present at *m/z* 71 and 72. The latter is the molecular ion. The ion at *m/z* 71 formally is the loss of H^- (hydride). As one can see, only a small portion of the ion intensity originates from the molecular ion. The CI spectrum is much simpler and shows only two peaks that are related directly to THF, a protonated ion at *m/z* 73 and the hydride loss ion at *m/z* 71. Both of these represent *quasi-molecular ions*, that is, CI ions indicating the molecular weight of an analyte. Here, the ion at *m/z* 71 is probably the result of protonation, to give *m/z* 73 followed by the loss of H_2.

The key factors to controlling CI are the ionizing properties of the reagent ions, such as their acidity, basicity, or electron affinity. These properties can be illustrated with the CI of THF to give protonated species. In eqs. 7-8 through 7-10,

$$\begin{array}{ccc} & \Delta H = -66\ \text{kcal mol}^{-1} & \\ CH_5^+ + C_4H_8O & \longrightarrow & CH_4 + C_4H_9O^+ \\ \text{PA} = & & \text{PA} = \\ 196\ \text{kcal mol}^{-1} & & 130\ \text{kcal mol}^{-1} \end{array} \tag{7-8}$$

$$\begin{array}{ccc} & \Delta H = -4\ \text{kcal mol}^{-1} & \\ C_4H_9^+ + C_4H_8O & \longrightarrow & C_4H_8 + C_4H_9O^+ \\ \text{PA} = & & \text{PA} = \\ 196\ \text{kcal mol}^{-1} & & 192\ \text{kcal mol}^{-1} \end{array} \tag{7-9}$$

$$\begin{array}{ccc} & \Delta H = +8\ \text{kcal mol}^{-1} & \\ NH_4^+ + C_4H_8O & \longrightarrow & NH_3 + C_4H_9O^+ \\ \text{PA} = & & \text{PA} = \\ 196\ \text{kcal mol}^{-1} & & 204\ \text{kcal mol}^{-1} \end{array} \tag{7-10}$$

the ionization process varies from highly exothermic to endothermic, depending on the nature of the reagent ion. The exothermicity of the reaction can be calculated easily by comparing the proton affinities (PAs) of the analyte and the product neutral from that of the reactant ion (when the analyte has the higher proton affinity, the CI process is exothermic). With CH_5^+, the proton transfer is exothermic by more than 65 kcal mol^{-1}, and the resulting ion, $C_4H_9O^+$, can contain a large amount of vibrational excitation. Of course, not all of this energy is deposited in the product, and some is in the kinetic and vibrational energy of the departing methane. With the *tert*-butyl cation ($C_4H_9^+$), the reaction is less exothermic because the reactant ion is less acidic; that is, C_4H_8 has a higher proton affinity than does CH_4. In this case, the reaction is barely exothermic, and the protonated analyte ion is formed with very little vibrational excitation. Finally, when the ammonium ion,

NH_4^+, is used as the reactant ion, the CI process is endothermic and THF is not ionized to its $(M + 1)^+$ cation under these circumstances. As this simple example illustrates, one potentially can tune the CI process to meet the needs of the analysis. Analyte ions can be formed in highly exothermic processes, favoring fragmentation, or in mildly exothermic processes, favoring the formation of intact parent ions, $(M + 1)^+$. The possibility of selectivity also exists in the CI process. For example, $C_4H_9^+$ is capable of protonating THF but does not protonate alkanes or many simple hydrocarbons. As a result, THF can be selectively ionized by $C_4H_9^+$ out of a background of alkanes or other hydrocarbons. The selectivity of the CI process requires a careful choice of reactant ions. If we were to use NH_4^+ as the reactant ion, a mixture containing THF would give no protonated ions for the THF—only species in the mixture with proton affinities greater than that of NH_3 would give M + 1 ions. In contrast, adducts of the form $[C_4H_8O \cdot NH_4^+]$ might be formed.

Selectivity becomes more important in negative ion CI because only a limited number of functional groups lead to reasonably stable negative ions. This is particularly true for charge transfer because only a few functional groups give organic species with high electron affinities. Such selectivity is the basis of negative ion CI of nitro-based explosive materials.

Typical CI Sources. CI sources are adapted from EI sources but operate at higher pressures to allow for ion/molecule reactions to occur. Generally, a small chamber referred to as an ion volume is incorporated into the EI source vacuum manifold to contain the reagent gas at a modest pressure. The electron beam passes through the ion volume and produces the reagent ions. CI sources generally use higher-energy electrons, about 200 eV rather than 70 eV, in order to obtain the optimum ionization level in the source.

7-2c Desorption Methods: FAB and MALDI

Electron ionization and chemical ionization are widely used by organic chemists for the analysis of small, relatively volatile molecules. These gas-phase ionization techniques, however, are not viable with species lacking a significant vapor pressure. This limitation prohibits large molecules, salts, and highly polar species. Desorption techniques operate on analytes in the condensed phase and therefore do not suffer from limitations related to vapor pressure. Over the years, a number of methods have been developed and then gone out of fashion as more versatile methods replaced them. Rather than document all of them, we present only the two most currently common approaches for organic systems: *fast-atom bombardment* (FAB) and *matrix-assisted laser desorption ionization* (MALDI). The former found wide application in the recent past, and references to studies involving FAB spectra are still common in the current literature, whereas the latter is the most widely used desorption technique today. Both are based on the principle of transferring a large amount of localized energy to a matrix containing the analyte. They are capable of producing ions from large, nonvolatile analytes and tend to yield singly charged, protonated species, that is, $(M + H)^+$ ions, with limited fragmentation.

Ionization Process. In FAB, the analyte is generally mixed with or suspended in a nonvolatile liquid referred to as the matrix. The matrix not only supports the analyte, but also acts as the ionizing agent (see below). Glycerol is a typical matrix material, but others are commonly used, depending on the properties of the analyte. As the name suggests, ionization is caused by a beam of high-energy neutral atoms striking the matrix. This beam is initiated by the ionization and acceleration (5–8 kV) of a noble gas, typically xenon. The preference for xenon is related to its high atomic weight, which leads to greater momentum transfer when striking the FAB target. The beam of Xe cations is directed into a chamber containing a low pressure of the noble gas. As the beam passes through the chamber, there are multiple interactions with the gas, with a high probability that the charge on the high-velocity heavy atoms will be stripped off via charge transfer processes. The interactions involved with the charge-stripping process do not greatly alter the momentum of the atoms in the beam, and, as a result, the ion beam is efficiently converted to a high-velocity neutral xenon beam. The original beam is generally aligned to strike the surface at an angle to deliver a large amount of

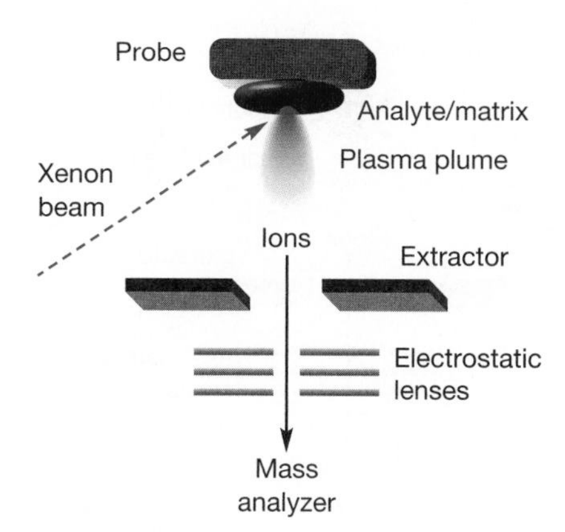

FIGURE 7-5 Schematic of generic FAB source and ionization process.

highly localized energy to the matrix. The energy of the collision heats the surface and causes a plume of material (matrix and analyte) to be sputtered off the surface. Each collision is estimated to drive about 1000 molecules off the surface. A set of electrostatic lenses is used to direct the resulting ions to the mass analyzer of the instrument (Figure 7-5).

There are multiple pathways to ionization. If the analyte is charged or can easily be charged by reaction with the matrix, the possibility exists of preformed ions being sputtered off the surface. In the case of analytes that are not easily charged, a more complicated path is likely. The intense heating from the collision can generate a plasma in the plume, which contains matrix fragments (charged and neutral) as well as slow electrons. The situation is not so different from the plasma found in a CI source. Gas-phase reactions of the analyte with charged species in the plume lead to analyte ions. The process can produce molecular ions ($M^{+\cdot}$), protonated ions $(M + H)^+$, and cluster ions with metal cations $(M + Metal)^+$ if they are present in the matrix. Like CI, FAB is a relatively gentle ionization technique (most of the energy is transferred to the matrix molecules) and, consequently, fragmentation is fairly limited. A key complication of FAB sources is that one observes ionization products from both analyte and matrix. The matrix tends to lead to a large number of signals in the low mass range from ionized matrix molecules and clusters containing them. As a result, this part of the spectrum is of limited value, and FAB is best suited for analytes whose mass is significantly larger than that of the matrix. Overall, FAB spectra tend to be relatively complex, with a variety of background ions due to the matrix.

A similar approach replaces the fast xenon atoms with cesium cations. Here, the technique is referred to as liquid secondary ion mass spectrometry (LSIMS). It provides similar data and is often used today in the place of FAB.

Ionization Process: MALDI. As in FAB, MALDI is based on the principle of depositing energy not directly into the analyte, but instead into a matrix, which serves to ionize the analyte. In MALDI, the energy is provided by excitation of the matrix molecules via a laser pulse. Nitrogen lasers (337 nm) are typically used and matrix species chosen so that they have a reasonable absorbance at this wavelength. Commonly used examples are 2,5-dihydroxybenzoic acid, 3-hydroxypicolinic acid, and α-cyano-4-hydroxycinnamic acid. The mixing ratio of matrix to analyte is generally on the order of 1000:1 or greater. Short, powerful, highly focused (~0.1 mm) laser pulses are used to heat/excite a spot on the analyte/matrix mixture rapidly. The energy leads to evaporation/ablation of the matrix along with the analyte (Figure 7-6). The ionization process is complex and most likely involves a variety of pathways. The matrix material is generally acidic, and protonation of the analyte before evaporation is one pathway to analyte ions. Even matrix materials that are not acidic in their ground state can act as strong acids when electronically excited by the laser pulse. Alkali metal cations can be introduced into the matrix, and cation attachment also can serve as an ionization mechanism. In the plume caused by the laser pulse, a plasma is formed that can allow for gas-phase ionization processes such as proton transfer, charge transfer, and electron capture. Finally, it is possible for the analyte to be directly photoionized by the laser pulse. Depending on the analyte and matrix, MALDI produces protonated ions $(M + H)^+$, molecular ions ($M^{+\cdot}$),

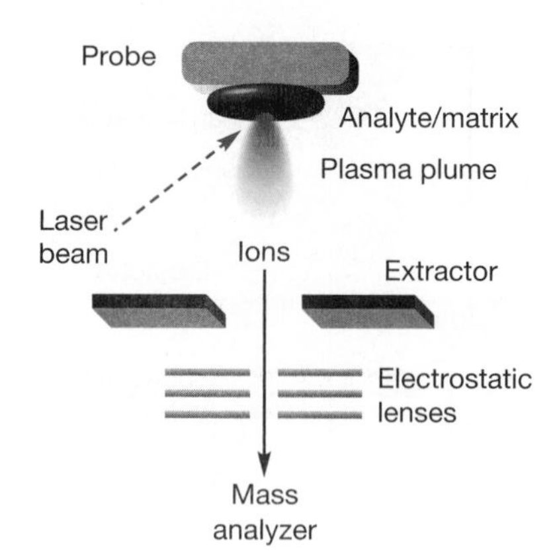

FIGURE 7-6 MALDI ionization process.

or cationized ions $(M + Cat)^+$. Analytes prone to forming negative ions produce deprotonated ions $(M - H)^-$ and molecular anions $(M^{-\cdot})$. As with FAB, interference by ions derived from the matrix is a problem, so the technique is at its best when applied to high molecular weight analytes that will not overlap with the matrix ions. A sample MALDI spectrum is provided in Figure 7-7. This example is a mixture from the radical polymerization of styrene. A silver salt was used in the matrix to provide the charge carrier. Each species is represented by a collection of peaks because silver has multiple isotopes and the compounds are large enough that some molecules are likely to contain one or more carbon-13 isotopes (the impact of isotopes on spectra is discussed in detail in Chapter 9). Note also in the spectrum that at low *m/z*, a significant background signal that can be attributed to matrix ion is present.

Energy Deposition and Fragmentation. A key characteristic of matrix/desorption methods is that the energy used to fuel the ionization process is initially delivered to the matrix molecules. As a result, these methods are referred to as soft ionization methods and have a tendency to produce intact molecular or quasi-molecular ions. The energy imparted to the matrix, however, can be transferred to the analyte, and significant heating can occur in the plume generated by the ionization event. Consequently, some degree of fragmentation is observed, especially with analyte ions that are particularly fragile. The level of fragmentation can be controlled to some extent by the rate at which the energy is delivered to the sample, for example, laser intensity. Because MALDI often is applied to large molecules with many degrees of freedom and is linked to mass analyzers with short ion-residence times (such as time-of-flight), metastable ions can play an important role in fragmentation processes. Metastable ions contain enough energy to dissociate but do so on a time scale that leads to fragmentation after the ion leaves the source. In the context of MALDI, this phenomenon is referred to as post-source decay (PSD) and is discussed in a subsequent section.

Desorption Ionization Sources. These sources are relatively simple and involve a sample plate for the introduction of the analyte/matrix mixture. The neutral beam (FAB), ion beam (LSIMS), or laser (MALDI) is directed at the sample plate, and a series of electrostatic lenses are used to extract the ions and direct them to the mass analyzer. Because the sample plate is enclosed in the vacuum system of the mass spectrometer, a number of novel approaches have been developed to increase the efficiency of sample introduction, including the use of trays with large numbers of sample spots (96 or more) for the rapid analysis of multiple samples off a single plate.

7-2d Electrospray Ionization (ESI)

Like desorption methods, *electrospray ionization* has been widely used to study large species, but it is also suitable for the ionization of low molecular weight compounds. Its ionization process is quite different from the methods previously discussed. As a result, it offers access to other types of species, particularly multiply charged ions. In this method, little energy is transferred to the analyte, so it is truly a very soft ionization

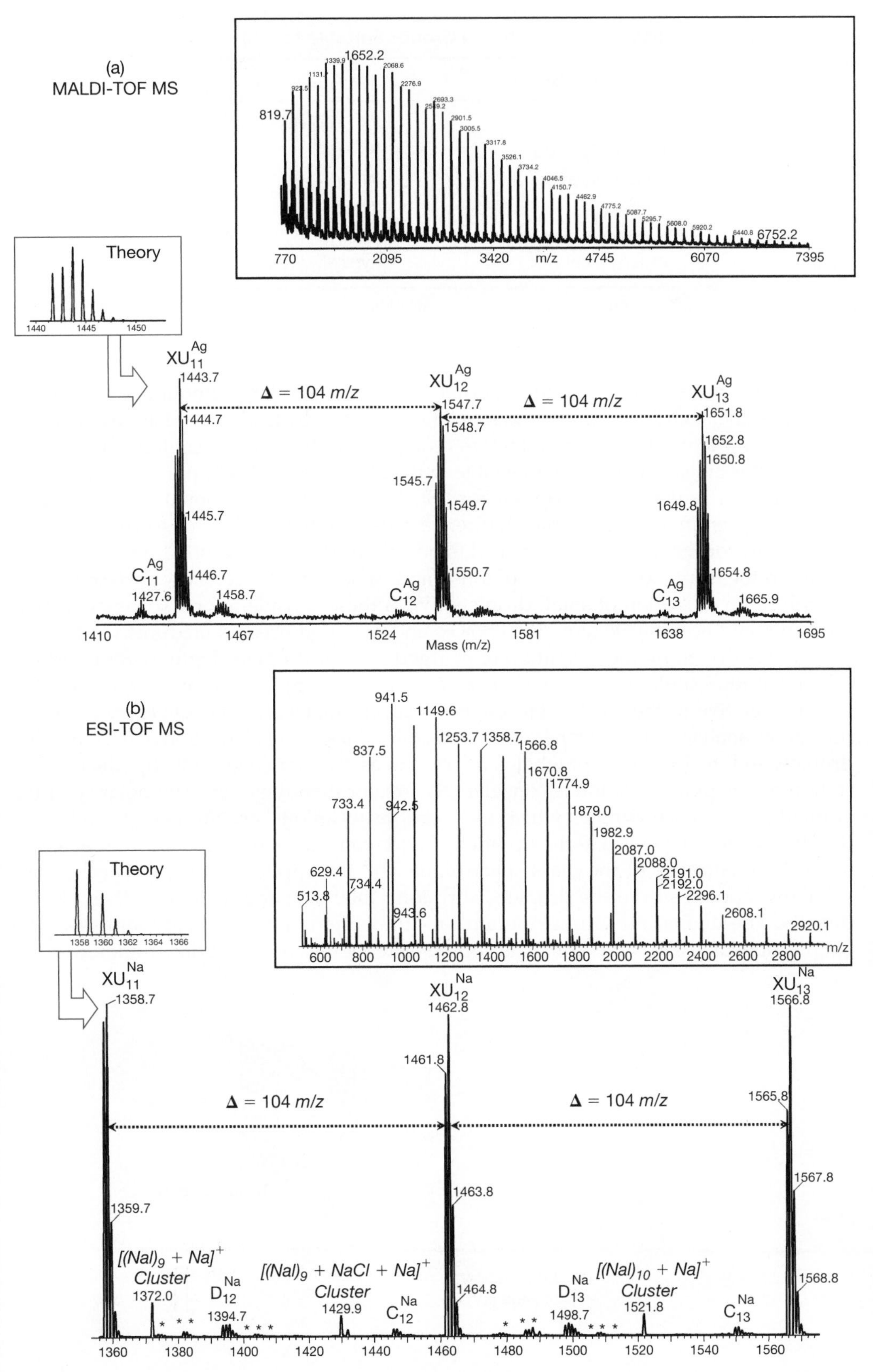

FIGURE 7-7 Spectra of a styrene polymerization mixture. (a) MALDI ionization is employed using a matrix that included silver trifluoroacetate as an additive. The series of XU_n ions consists of the polymers of various sizes (n), with Ag^+ as the charge carrier. The C and D series of ions contain components of the radical initiator. (b) The same sample is ionized using electrospray ionization, with sodium iodide added to the solution to assist in the formation of polymer/Na^+ complexes, that is, with sodium as the charge carrier. Sodium/halide clusters are apparent in the electrospray ionization spectrum. (Reprinted with permission from C. Ladavière, P. Lacroix-Desmazes, F. Delolme, *Macromolecules*, **42**, 70–84 [2009]. Copyright 2009. American Chemical Society.)

method. However, the low energy associated with the ionization process often limits the approach to ions that are preformed in solution (or during the evaporation process associated with ESI). This limitation is not so constrictive as it may seem, in that ESI can generate ions efficiently from ionic species that might be present in very low concentrations in solution.

TABLE 7-6 Functional Groups Suitable for ESI

Functional Group	Analyte Ion[a]
Amines	$(M+H)^+$, $(M+Cat)^+$
Carboxylic acids	$(M-H)^-$
Sulfonic acids	$(M-H)^-$
Sugars	$(M+Cat)^+$
Ketones, aldehydes, ethers, alcohols	$(M+H)^+$, $(M+Cat)^+$ (typically weak)

[a]Cat^+ is usually an alkali cation such as Na^+.

Ionization Process. ESI is unique in that at no point is a large amount of energy deposited in the sample to initiate the ionization process. Instead, ESI generally relies on auto-ionization in solution and therefore requires that the analytes contain a functional group that is at least marginally ionizable in the ESI solvent. This requirement limits the range of analytes that are appropriate for ESI. A partial list of functional groups suitable for ESI is given in Table 7-6. This requirement also limits the range of solvents that are appropriate for ESI. Good solvents need to be volatile (to aid in evaporation), to be polar (to aid in the formation/solubility of ions), and to have a relatively low surface tension (to aid in the formation of small droplets). With these considerations, solvents such as methanol and acetonitrile are good choices and are often used as mixtures with water.

ESI is driven by the formation of charged solvent droplets. Figure 7-8 provides a schematic representation of the electrospray process. Application of a large potential (1–5 kV relative to the entrance to the mass spectrometer) to a needle carrying a low flow of the analyte solution (generally 1–5 $\mu L\ min^{-1}$) leads to polarization of the emerging droplets and the formation of what is referred to as a Taylor cone. At its tip, the concentration of one polarity of ion is considerably greater than the other. The polarity of the applied potential determines which ions, cations or anions, accumulate at the tip. As the charged droplets form at the tip of the Taylor cone, they are driven from the needle toward the entrance of the mass spectrometer by the applied potential. On the way, solvent evaporates from the droplets, shrinking their size and increasing their charge density. The ionization source operates at atmospheric pressure. Ambient gas, or a counterflow of gas in some cases, can compensate for the cooling of the droplet from the

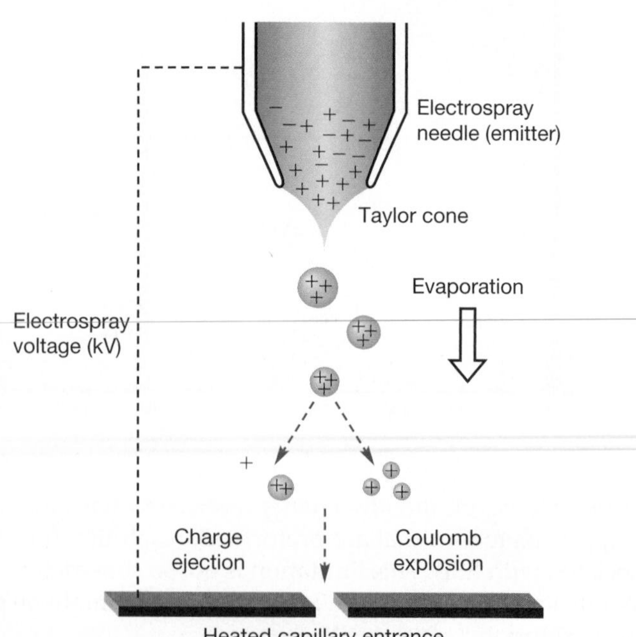

FIGURE 7-8 ESI process.

evaporation process. The droplets generally enter the mass spectrometer via a heated capillary interface, which aids in the evaporation of the charged droplets. At a certain point in the evaporation process, the internal electrostatic repulsion in the droplet overcomes the surface tension and the droplet fragments. The actual mechanism for the formation of gas-phase ions in the ESI process remains somewhat controversial. One explanation is that when the droplet reaches a critical size, internal electrostatic repulsion is reduced by directly ejecting a bare ion from the droplet. Another possibility is that the droplets shrink via evaporation/fragmentation cycles, eventually producing droplets with a single charged species, which desolvates to give a bare ion. This has been called the charged residue model. Some data indicate that both mechanisms can be operative, depending on the nature of the analyte and the solvent. In any case, the electrospray process eventually generates bare ions, often with multiple charges if they contain more than one ionizable site. A sample spectrum is shown in Figure 7-7. This is the same sample that was used as the MALDI example. Here, sodium iodide was added to the solution to form Na^+ adducts of the polymers. Unlike the MALDI spectra, there is little background in the low *m/z* range because no matrix is used in ESI. The spectrum, however, does contain signals for clusters of the NaI additive that can complicate the analysis.

Energy Deposition and Fragmentation. As the above description indicates, the ions and their precursors never obtain high internal energy in the ESI process, so little or no fragmentation is anticipated. In fact, the opposite effect is often seen—ESI can lead to the formation of noncovalent complexes between ions and neutral species or ions of opposite polarity (the complex still must retain a net charge to be observed by mass spectrometry). Such processes can lead to rich spectra from reasonably simple species. For example, consider the ESI of benzoic acid from a methanol solution. Ionization of the carboxylic acid leads to the expected benzoate ion in the negative ion mode $(M-H)^-$. Benzoate, however, makes a fairly strong gas-phase complex with benzoic acid. If the benzoic acid concentration is sufficiently high, complexes of benzoate and benzoic acid are expected $(M_2-H)^-$. If sodium cations also are present in the electrospray solution, a sodium complex of two benzoates, $(M_2 - 2H + Na)^-$, might also be observed. In solutions with high salt contents, large complexes containing several cations and anions can dominate the ESI spectrum and produce very complicated patterns. As a result, ESI is at its best in solutions with relatively low salt concentrations ($<10^{-4}$ M) and offers special challenges in the presence of buffer. With ESI, fragmentation generally requires the application of accelerating potentials after ion formation to cause collisional excitation, as discussed in Chapter 8.

ESI Sources. ESI sources have simple designs, although there is a good deal of variety in the specific components used in the design. The two major components are an electrospray needle (often called an emitter) and a heated capillary that serves as an interface with the mass analyzer. In a typical experiment, the emitter is raised to a relatively high potential (1–5 kV) with respect to the inlet to the heated capillary. The emitter may be a stainless steel needle or, in the case of nanospray systems, a drawn-out silica capillary that is charged either by a liquid contact or by a metal coating. The needle can be aligned in an in-line or an orthogonal arrangement with respect to the heated capillary. This portion of the source is at atmospheric pressure, but, in some cases, nitrogen gas rather than air is used as the atmosphere. The charged droplets, with or without the assistance of a nebulizing gas flow, are drawn to the entrance of the heat capillary by the applied potential as well as by the flow of gas entering the capillary. The heated capillary aids in the evaporation of the droplets and leads to a region at roughly 1 torr directly behind the capillary. From here, ions are directed toward the mass analyzer by electrostatic lenses and generally pass through differential pumping regions with sequentially lower pressures.

7-2e Atmospheric Pressure Chemical Ionization (APCI)

A variation of ESI is referred to as *atmospheric pressure chemical ionization*. The two methods are linked because they share very similar source designs. The key difference is the nature of the analyte. In ESI, the analyte must be able to form a reasonably stable ion in solution. This is a possibility for many species, including most biologically relevant compounds.

A number of important, polar nonvolatile species, however, are incapable of auto-ionization in an ESI source. For these systems, an alternative would be to induce ionization via gas-phase reactions with reagent ions, that is, CI. In practice, the analyte is introduced as a solution into the source chamber via a heated nebulizer. A corona discharge is generated between a needle at a high potential and the source chamber walls. This discharge can ionize either the solvent or the nebulizing gas, and subsequent gas-phase reactions lead to analyte ions. The nature of the ionization process (formation of protonated species or radical ions) depends on the details of the gases and solvents employed. The ions are introduced into the mass spectrometer via the same type of interface that is used with an ESI source. As with other CI methods, there is a tendency to form singly charged ions. Although not used so widely as ESI, APCI fills an important niche in the analysis of mid-sized, nonvolatile species (~500–1500 daltons) that do not contain ionizable groups. It has particularly useful applications in drug discovery and metabolite analysis.

7-2f Hybrid Methods

Recently, success has been achieved in using ESI in desorption ionization schemes. The strategy is to use the ESI source to produce charged droplets that can incorporate analytes that are on surfaces or in the atmosphere. A useful example is desorption electrospray ionization (DESI) from the Graham Cooks lab. Here, an electrospray source with a high nebulizing gas flow is directed at an insulator surface that contains the analyte. The interaction of the electrospray stream with the surface leads to the incorporation of analyte molecules in the charged droplets, which are then captured and directed to the heated capillary of a conventional ESI source. The advantage of the approach is that there are relatively few limits on the nature of the insulator surface, and therefore trace species on objects, such as an envelope, can be analyzed directly without the need to extract them from the object. This is a rapidly growing area, and new implementations are continually emerging.

7-3 MASS ANALYSIS

The heart of any mass spectrometer is the mass analyzer. Over the years, a remarkable number of mass analyzers have been designed on the basis of a variety of physical principles. Each takes advantage of the unique feature of ions, the fact that their motion can be manipulated by the application of either electric or magnetic fields. The ability to control the course of ions allows for their separation in space, time, or frequency of motion. The characteristics of mass analyzers vary greatly, particularly in terms of mass range, resolution, and operating pressures. The choice of the most appropriate mass analyzer for a problem depends on the properties of the analyte, its physical form (i.e., pure or part of a complex mixture), and the desired information. In many cases, more than one analyzer has the capabilities to solve a problem, so that availability can become the ultimate deciding factor. In this section, six major types of analyzers are presented, along with a description of some hybrid instruments.

7-3a Time-of-Flight Mass Analyzers

Principle of Operation. The relatively simple physics of *time-of-flight* (TOF) analyzers is based on the fact that ions in an accelerating potential reach the same kinetic energies, but have different velocities, depending on their masses (eq. 7-11). In such a situation, mass analysis comes down to determining the velocities of ions moving in a field-free region after the initial acceleration, which can be accomplished by measuring how long the ions need to traverse a fixed distance, that is, their time of flight (eqs. 7-11, 7-12, and 7-13, where V is the applied voltage, d is the flight distance, and t is the flight time).

$$E = V \cdot z = 1/2\, mv^2 \tag{7-11}$$

$$v = \left(\frac{2Vz}{m}\right)^{1/2} \tag{7-12}$$

$$m/z = 2Vt^2/d^2 \tag{7-13}$$

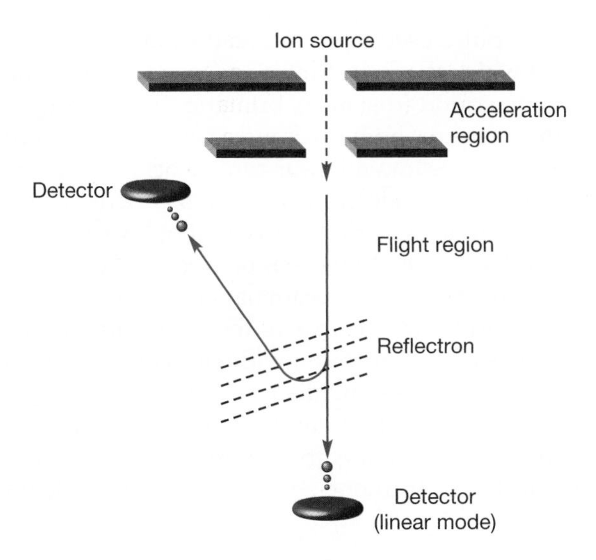

FIGURE 7-9 Schematic of TOF mass analyzer. Ions with lower *m/z* travel faster and arrive first at the detector. Ions can be detected directly (bottom detector) or have their motion reversed with a reflectron and be detected on the upper-left detector.

A direct analogy can be made to the way in which police aircraft track the speed of cars by determining the time it takes them to traverse the distance between two lines painted on the pavement.

A general schematic of a TOF analyzer is given in Figure 7-9. Typically, high accelerating voltages are used (10–30 kV), and ions can reach velocities on the order of 100,000 m s^{-1} in the instrument. Under these conditions, flight times are exceptionally short (tens of microseconds), and therefore spectra can potentially be collected at a very high rate. The high velocities place heavy demands on the electronics of the ion detection system because there is little separation in the arrival times of ions with similar masses. For example, consider ions at *m/z* values of 1000 and 1001 with a nominal flight time of 100 μs. The difference in arrival times would be about 0.05%, or 50 ns. Therefore, high resolution requires extremely fast detection schemes.

Another factor with TOF instrumentation is the initial packet of ions. The experiment as described requires all the ions to be present at the same place, at the same time, and then to be suddenly accelerated to their flight velocity. This requirement immediately suggests an ion source that produces a high-density pulse of ions. As you will recall from the previous section, the one source that would fit this description is that used in the MALDI experiment. Lasers can be designed to provide short, intense pulses that provide the starting packet of ions for the TOF acceleration. It is not a coincidence that the recent explosion in TOF technology followed the development of MALDI sources. Although there are some complications in adapting continuous ion sources like ESI to TOF mass analyzers, hybrid instruments have been developed that allow TOF to be applied to a variety of ion sources (see below).

Analyzer Design. The general design principles are rather simple and amount to an ion source, flight tube, and detector system, but to achieve high resolution, a number of elaborations on this simple design are needed. A key problem is that sources produce ions with a variety of initial kinetic energies, potentially varying on the order of 10 eV. This value is about 0.1% of the acceleration potential and therefore could lead to peak broadening on the order of ±0.1%, an unacceptable reduction in resolution for many applications. To address this issue, several modifications to the basic design have been developed. One common approach is the *reflectron*. In this version of TOF, the flight tube ends with an electrostatic mirror that nearly reverses the path of the ions on their way to the detector (see Figure 7-9). The advantage of the reflectron is that it acts as a focusing device for the ions and reduces some of the peak broadening associated with variations in starting kinetic energies. The mechanism of the correction is rather straightforward. Ions with higher than average kinetic energies penetrate deeper into the reflectron before being reversed and therefore experience a greater path length to the detector. Conversely, the low-energy ions (slower) do not penetrate so deeply into the reflectron and experience a shorter path length. The net effect is to focus the ions of a particular *m/z* ratio and produce narrower ion packets and peak widths. A second approach involves manipulating the voltages and timing of the extraction pulse used to accelerate the ions (*delayed extraction*). The goal is to create a situation in which the fastest-moving ions of a particular m/z experience a slightly shorter acceleration period (distance), bringing their velocity close to the average for the ion packet. A third approach involves *orthogonal acceleration* of the ions. Here, the ions are focused into a narrow beam with near zero velocities on the x and y axes of motion and a moderate velocity (tens of electron volts) on the z axis. The beam also may be trapped in an ion accumulation device prior to acceleration. An acceleration pulse of 10 to 30 kV is rapidly applied orthogonally to the axis of motion of the ions, and a packet of ions is punched out of the beam. Because the beam has almost zero velocity in this direction, the packet remains very narrow on the axis of the acceleration. There is continual broadening along the initial axis of motion, but a wide detector design allows for the simultaneous detection of the ions spreading out along this axis.

As the previous paragraph indicates, there are a variety of modern implementations of TOF analyzers. Efforts to generate analyzers with high resolution and efficient duty cycles have led to highly sophisticated instruments, but all of them operate on the same basic principles.

Capabilities and Limitations. One of the strong suits of TOF is a nearly unlimited mass range, and instruments can routinely operate at m/z of more than 100,000. This attribute is particularly attractive for MALDI because low charge states lead to high m/z ratios. TOF also provides excellent *mass-resolving power* on the order of 1000 to 10,000 or higher. Mass-resolving power refers to the ability of an analyzer to separate the signals of two masses with only marginal peak overlap (~10%). It can be defined as $m_1/(m_1 - m_2)$, in which m_1 and m_2 are the peaks that are being resolved. For example, at a mass-resolving power of 1000, a peak at m/z 1000 has marginal overlap with one at m/z 1001. In addition, the technique offers good mass accuracy (5–50 ppm). Efficiency in the context of mass analyzers is a measure of how well the analyzer's duty cycle matches the ion source capabilities. In the optimum situation, the mass analyzer can process ions as fast as the source produces them, leading to 100% efficiency. This ideal is not always achievable. For example, in the simplest implementation of MALDI-TOF, the source would be idle while the ions are in the flight tube on the way to the detector. As a result, the mass analyzer becomes the bottleneck in the process, and the experiment becomes limited by the duty cycle of the mass analyzer rather than by the ion source. For TOF, the efficiency depends greatly on the instrument design. TOF systems that incorporate an ion accumulation device can have efficiencies near 100% because, while one packet of ions is in the flight tube, another is being collected in the accumulation device for the next mass analysis. Finally, there is no practical way to activate the ions in the mass analyzer, so fragmentation must result from activation before entering the mass analyzer. The implications of this limitation become clearer in Chapter 8.

7-3b Ion Cyclotron Resonance (ICR) Traps: Fourier Transform Mass Spectrometry

Principle of Operation. *Ion cyclotron resonance* instruments also operate on the basis of rather straightforward principles of physics. When ions pass through a magnetic field, their paths are deflected. This deflection can be used to force ions into a stable, cyclic trajectory within the magnetic field, thereby trapping them indefinitely. The radius r of the pathway is correlated with the velocity v of the ion via eq. 7-14,

$$\mathrm{r} = \frac{mv}{z\mathrm{B}} \quad \textbf{(7-14)}$$

$$\omega = \frac{z\mathrm{B}}{m} \quad \textbf{(7-15)}$$

$$\frac{\Delta f}{\Delta m/z} = \mathrm{B} \quad \textbf{(7-16)}$$

in which B is the magnitude of the magnetic field. As one can see, an increase in velocity leads directly to an increase in the radius of the trajectory. Eq. 7-14 can be rearranged by

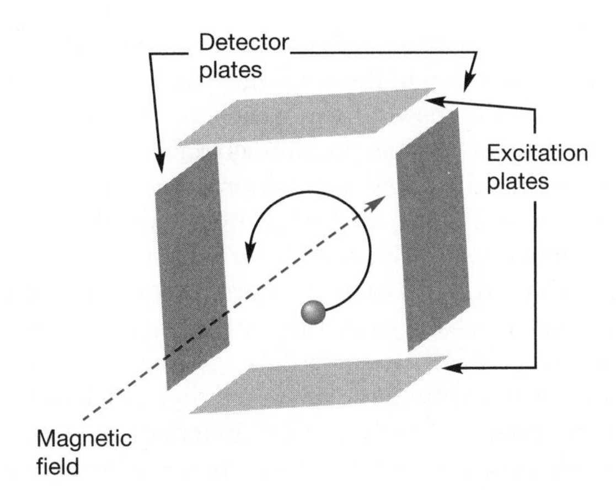

FIGURE 7-10 Schematic of ICR and FT detection. Trapping plates are not shown.

substituting the angular frequency, ω, for the velocity and radius terms, using the simple relationships $\omega = 2\pi f$ and $f = v/2\pi r$, where f is the frequency of rotation in the magnetic field. This manipulation leads to the fundamental relationship of ICR mass spectrometry, eq. 7-15, in which the frequency of motion is related to the m/z ratio. Because there are ways of measuring frequencies with exquisite accuracy, ICR can offer very accurate mass measurements. One should also note from eq. 7-15 that frequencies scale with the magnitude of the magnetic field. This relationship indicates that the difference in frequency for a pair of ions with different m/z ratios is directly proportional to the magnitude of the magnetic field (eq. 7-16). As a result, increasing the magnetic field leads to greater resolving power, that is, a greater difference in the measured quantity, frequency. This same situation occurs in NMR, and ICR instruments typically use the same superconducting magnets that are employed in high-field NMR spectrometers (5–12 T). This is one of several similarities between ICR mass spectrometry and NMR spectroscopy.

A unique aspect of ICR spectrometry is that ion detection is nondestructive. For nearly all of the mass analyzers that are described herein (orbitrap is the only other exception), ion detection is based on discharging the ion on a surface and amplifying the signal via a variety of means. In an ICR experiment, it is possible to measure an image current that results from the ions moving toward and away from a pair of plates in the trapping magnetic field as the ions trace out their cyclic paths (Figure 7-10). As noted above, the angular frequency of the ion is related to its m/z ratio, and the oscillation of the resulting image current has the same frequency, making it a useful measure of the ion's m/z. The image current can be measured only if the motion of the trapped ions is coherent. This characteristic can be accomplished by applying a broadband radiofrequency (rf) pulse that accelerates all the ions into coherent motion and thereby maximizes the resulting image current. In this context, coherent ion motion refers to the fact that ions of a particular m/z move as packets and approach each of the image current plates as a group (i.e., ions of specific m/z have the same frequency and phase in their circular paths). All modern ICR instruments use a Fourier transform (FT) approach to analyze the m/z of the ions trapped in the ICR. This is the second key similarity with NMR and also the reason the method has been dubbed FT-MS. If all the ions are simultaneously excited into coherent motion, the resulting image current is a superposition of all the cyclotron frequencies of the ions in the trap. As a result, an FT of the time-domain oscillations in the image current provides the frequencies and therefore the m/z values of all the ions in the trap. From a single excite/detect/FT cycle, an entire mass spectrum can be obtained. Because the ions are not destroyed in the process, the cycle can be repeated. This repetition is very useful if one is tracking the progress of reactions occurring in the ICR trap.

Analyzer Design. The actual ICR cell is rather simple, with a pair of plates along one axis for exciting the ions into coherent motion and another pair for detecting the image current. Finally, plates on the third axis are used to provide a trapping potential; that is, they keep the ions from drifting along the axis orthogonal to the plane of the cyclotron motion. A key aspect of this type of analyzer is that any deviations in the path or in velocities of the ions lead directly to a reduction in the resolution of the instrument. As a result, collisions must be avoided during the mass analysis cycle of an ICR, and background gas pressures greatly reduce performance. This problem is exacerbated by the fact that the resolution of FT analyses is maximized by long acquisition times. Therefore, high vacuum is a necessity for high resolution in FT-MS, and pressures under 10^{-10} torr are desirable in the analyzer. This requirement creates the need for a robust vacuum system and an efficient ion guide system that allows for ions to be produced at the relatively high pressures used in typical ionization sources—for example, atmospheric pressure for ESI—and then transferred to the analyzer across an enormous pressure gradient. This arrangement is generally accomplished with a differential pumping system, in which the ions pass through regions of decreasing pressure, each with its own dedicated pumping system. These additions greatly increase the cost of FT-MS instruments. When combined with the cost of a superconducting magnet, FT-MS becomes a very expensive methodology with significant maintenance demands. Of course, these drawbacks are balanced by the exceptional high-resolution capabilities that these instruments offer.

Capabilities and Limitations. High resolution and mass accuracy are the key attributes of the ICR method. Mass-resolving powers can reach 10^6 in these instruments, with mass accuracy on the order of a few parts per million. As a result, ICR instrumentation can probe low as well as high *m/z* ions in exquisite detail. As with TOF analyzers, ions cannot be produced and analyzed simultaneously in an ICR. In the simplest implementation, the ion source is not productive during mass analysis in an ICR, but hybrid instruments with ion accumulation devices have been developed to allow ions to be produced continuously and then injected as packets into the ICR. As noted above, the key limitations of the ICR experiment are the need for high vacuum and high magnet field strengths. The former leads to elaborate differential pumping between the ICR and atmospheric pressure ion sources. In addition, it complicates efforts to excite ions collisionally in order to produce fragmentation spectra. The need for a high magnetic field necessitates expensive superconducting magnets with moderately high maintenance costs.

7-3c Quadrupole Ion Trap Mass Analyzers

Principle of Operation. Although based on fundamentally different ion physics, *quadrupole ion traps* (QIT) and ICR traps share many things in common. Both are trapping analyzers with long ion residence times, both exclude unwanted ions by the application of rf pulses, and both use rf pulses to induce fragmentation of trapped ions. QITs, however, use an alternating electric field rather than a magnetic field. Unlike ICR, for which the physics of ion trapping is fairly straightforward, the interaction of the ions with the quadrupole field in an ion trap leads to difficult mathematics and ion pathways that cannot be explicitly defined under the conditions used in practical applications. Under the proper conditions, ions travel in stable trajectories within the ion trap and can be held for extended periods. The Mathieu equation describes conditions that result in stable trajectories. The trapping potential is often likened to a ball resting in the center of a saddle potential (Figure 7-11). If the saddle is static, then the ball immediately begins to roll down one side and be lost. If the saddle is rotated by 90°, however, the downward-sloped potential is replaced by an upward-sloped potential, and the ball is driven back toward the center. Rotation of the saddle is equivalent to alternation of the electric field in the ion trap. If the saddle rotates at the appropriate rate, the ball is locked into a stable motion moving slightly up and down the faces of the saddle and never slides off it. In the same way, ions become trapped in stable trajectories in the quadrupole field created by the application of the alternating electric field. Ions are drawn toward a wall with an attractive potential, but the reversal of the field changes the potential to repulsive and the ions' path is reversed. An interesting aspect of quadrupole ion traps is that the same field can trap positive and negative ions. This has led to some novel applications of the instruments (see Chapter 8). Returning to the saddle model, one would imagine that for a ball of a given mass, the ability to trap it would depend on the shape of the saddle and the frequency of the rotation. Another way of

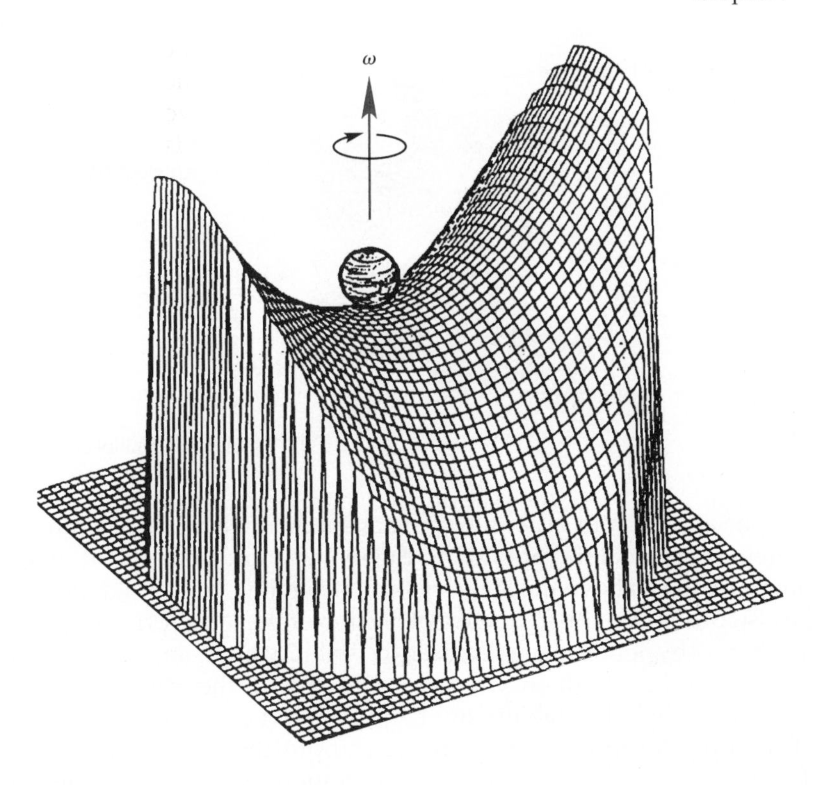

FIGURE 7-11 Saddle-potential analogy of the QIT. Rotating potential provides a counterbalance to any motion away from the center. (Modeled after a figure in Wolfgang Paul's Nobel lecture in 1989.)

looking at it would be that for a given saddle shape and speed of rotation, only some balls could be trapped—some are too heavy (eventually sliding down the saddle) and some too light (thrown off the top of the saddle by gaining too much velocity). The same is true in the ion trap—only a relatively narrow range of *m/z* values can be trapped under a given set of conditions, as defined by the Mathieu equation.

$$q_z = -\frac{8zV}{mr_o^2\omega^2} \qquad \textbf{(7-17)}$$

In commercial instruments, the operative experimental parameter from the Mathieu equation is q_z, which is a function of the dimensions of the ion trap and the magnitude as well as the frequency of the alternating electric field (eq. 7-17, where V is the rf field strength, r_0 is the radius of the trap, and ω is the nominal rf frequency of the trap). The q_z term is inversely proportional to the *m/z* of the ion. Stable trajectories are possible when q_z is between ~0.15 and 0.90. At low values, ions simply drift out of the field (equivalent to sliding off the saddle) and, at high values, they are ejected from the field (equivalent to flying off the top of the saddle). The upper value is an absolute limit, whereas the lower value refers to the point at which ion trapping becomes less efficient. These limits explain the narrow trapping range of QIT instruments. Because q_z is inversely proportional to *m/z*, the low limit of q_z defines the highest mass that can be trapped, and the high limit of q_z defines the lowest mass that can be trapped. If we set the lowest mass at *m/z* 100 ($q_z = 0.90$), the high mass limit becomes *m/z* 600 [the point at which q_z reaches 0.15 ($100 \times 0.90/0.15$)]. This result does not imply that the instrument is limited to *m/z* values up to 600; it only means that if one wants to trap ions with a higher *m/z*, then the low mass cutoff has to be greater. Thus, to get to *m/z* 1200, a low mass cutoff of *m/z* 200 is needed.

Another key difference between QITs and virtually all other common mass analyzers is that instead of striving for high vacuum, one intentionally introduces a small background pressure of an inert gas, generally helium (~10^{-3} torr). In contrast to other instruments, the presence of a background gas pressure actually enhances the resolution of a QIT. The gas dampens the motion of the ions in response to the quadrupole field and tends to keep the ion cloud toward the center of the device. A small ion cloud leads to a narrow peak width because all the ions are approximately the same distance from the detector when they are ejected from the analyzer and detected. Although there

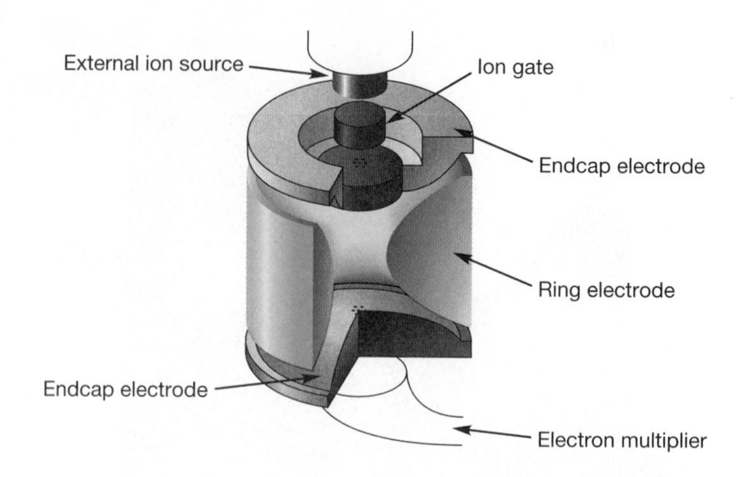

FIGURE 7-12 Schematic of three-dimensional QIT.

is more than one approach to generating a mass spectrum in a QIT, a common way is to ramp the alternating potential applied to the ion trap and, as a result, mass by mass push the ions' q-values above the stability limit (from low *m/z* to high *m/z*). The ramping voltage also pushes up the frequency of motion of the ions. If rf energy is applied at a frequency that corresponds to a point near the stability limit, ions are promptly ejected and detected when their frequency is in resonance with the excitation rf energy, just before they would reach the stability limit and drift out of the ion trap. In QITs, *m/z* is correlated with the trapping voltage needed to put the ion in resonance with this applied rf and is determined on the basis of calibrations with ions of known *m/z*.

Analyzer Design. Traditionally, QITs have adopted a three-dimensional design built around a ring electrode and a pair of end-cap electrodes (Figure 7-12). The end-cap electrodes have a hyperbolic shape, and the alternating electric field is applied to the ring and end-cap electrodes. The device is small, and the radius of the ring is on the order of 0.7 cm. Alternatively, the quadrupole field can be applied in two dimensions, with the ions being trapped between a set of four rods with alternating potentials on adjacent rods. This setup is referred to as a linear ion trap and is akin to quadrupole ion filters (Section 7-3d), except that the ions are trapped along the axis of the rods by a set of offset potentials. A key difference in the designs is that ions are spread along the axis of the linear ion trap rather than concentrated at the center of a three-dimensional trap. The advantage is that more ions can be trapped without incurring significant degradation of the quadrupole field due to ion/ion interactions (referred to as space/charge effects). Because ion/ion interactions occur over relatively long ranges in a vacuum, ion densities must be limited to minimize the contribution of the ion/ion interactions to the net field experienced by a given ion. If not, the quadrupole field is altered by the space/charge effects, and the net result is peak broadening and apparent mass shifts. The ability to trap more ions gives linear ion traps a significant advantage in sensitivity over traditional three-dimensional ion traps. In both types of instruments, it is necessary to control the number of ions in the trap carefully. Automatic gain control systems have been developed to generate optimum ion densities in ion traps. Once trapped in these devices, ions can be detected by ejecting them along one of the field axes through a hole in the end-cap electrode (or slots in a linear quadrupole rod). The ejected ions are typically detected with an electron multiplier system. Because ion traps operate at fairly high pressures (for a mass spectrometer), they are easily interfaced with a variety of ion sources and do not need extensive pumping capabilities, even with atmospheric pressure ion sources.

Capabilities and Limitations. QITs cannot match TOF or ICR instruments in terms of mass-resolving power and mass accuracy. Typical QIT instruments give mass-resolving powers of $\sim 10^4$, with an accuracy of about 100 ppm. These are useful ranges for most applications but preclude high resolution spectra. Along with problems in simultaneously trapping low and high *m/z* ions, there typically is an upper limit on their mass range of about 2000 to 4000, although special modes of operation can significantly extend this limit. The strengths of QITs are their versatility, low cost, and robust nature.

In addition, the relatively high pressure used in QITs makes them easy to link to a variety of ion sources.

7-3d Quadrupole Mass Filters

Principle of Operation. These instruments employ the same general type of electric field as linear QITs, but operate in a beam mode whereby ions of a particular *m/z* range pass through the device with stable trajectories, while other ions strike the rods. In a sense, the ions are trapped in two dimensions but have a significant velocity in the third dimension. These devices can be used simply as ion guides to transfer all ions from one region to another of an instrument or in a mass-selective mode in which ions of only a narrow *m/z* window can pass through. The latter mode of operation turns the device into a mass analyzer because one can ramp the voltage of the quadrupole field and generate a mass spectrum by sequentially allowing ions of increasing (or decreasing) *m/z* to pass through the filter and strike the detector. These devices are best suited for continuous ion sources because a new packet of ions must be used for the analysis of each mass—all the ions except the *m/z* of interest are sacrificed during the detection of ions at a particular *m/z* value.

Analyzer Design. As noted above, a quadrupole mass filter is much the same as a linear QIT, except for the fact that it is a beam device. It is constructed of four rods with alternating potentials on adjacent rods. Ions from the source enter the device along the axis of the rods and under the proper conditions pass through with a stable trajectory. Ions with *m/z* values outside the stability window strike one of the rods and are lost. A schematic is shown in Figure 7-13. The length, diameter, and spacing of the rods depend on the nature of the mass spectrometer and the desired ion throughput. Like QITs, quadrupole mass filters can tolerate relatively high pressures ($\sim 10^{-5}$ torr), but there is no advantage in resolution or performance in increasing the pressure further with a buffer gas. It is possible to link several quadrupole mass filters together to design more powerful mass spectrometers. A very common instrument is the triple quadrupole. Here, the first quadrupole is used as a mass selection device that allows only ions of a particular *m/z* to pass through to the second quadrupole. In most applications, the second quadrupole is used as a collision cell, and ions are accelerated into this region. It is set to act as an ion guide to allow all ions (parent and fragments) to pass through to the third quadrupole. The final quadrupole is used as a mass analyzer and is scanned to generate a mass spectrum of the mixture created in the second quadrupole. Triple quadrupoles are useful instruments because they allow for mass selective fragmentation with a very wide range of collision energies. Moreover, they allow for novel experiments in which the first and third quadrupoles are scanned in unison to track ions with a particular neutral loss during fragmentation (see Chapter 8).

Capabilities and Limitations. In performance, quadrupole mass filters and QITs have much in common. They offer modest mass-resolving powers ($\sim 10^4$) and mass accuracies (100 ppm). Unlike ion traps, they do not have constraints associated with trapping a wide range of *m/z* and can scan directly across their entire mass range. This range, however, is generally constrained to an upper limit of about 2000. One strength

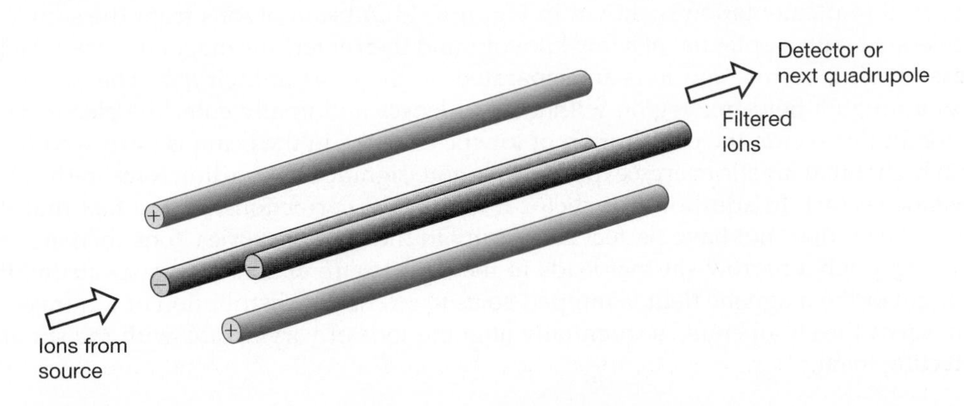

FIGURE 7-13 Schematic of quadrupole mass filter.

of these devices is their very wide *dynamic response* range, which represents the range of ion intensities that can be accurately recorded. With a dynamic range of $\sim 10^6$, quadrupole ion filters are superior instruments for quantitation, being able to quantify ions simultaneously with very high and low intensities. This range is several orders of magnitude greater than that of trapping devices like QITs and ICRs. Their efficiency is limited because only ions of one m/z from the source are analyzed at a time while the rest of the ions from the source are discarded. Finally, quadrupole mass filters tend to be rugged and relatively inexpensive because they rely only on electric fields and do not require high vacuum systems. They are easily combined with a variety of continuous ion sources.

7-3e Sector Mass Analyzers

Principle of Operation. For many years, *sector mass spectrometers* were the standard for high-performance mass spectrometry. Although these instruments are still widely used, they are being replaced by TOF, FT-ICR, and orbitrap (see Section 7-3f) instruments, which have some of the same performance characteristics but are more versatile and less cumbersome. Typically, the instruments consist of two sectors: a magnetic sector that separates ions on the basis of momentum and an electrostatic sector that separates ions on the basis of their kinetic energy. Only the magnetic sector is needed to have a mass analyzer, but greater resolution is attained by including the electrostatic sector. The basic physics of the magnetic sector is similar to that of ICR. The action of the magnetic field perpendicular to the trajectory of the ions bends their pathway. In an ICR, this effect is used to create a stable, circular pathway. In a magnetic sector, the ion pathways are simply altered by the magnetic field—ions have varying arcs, dependent on their m/z, as they pass through the magnetic sector. The net result is a spatial separation of the ions on the basis of their m/z. This process is akin to light passing through a prism leading to spatial separation of the wavelengths of light. If this explanation sounds familiar, it is because introductory organic texts usually use sector instruments and this analogy to explain mass spectrometry. So what role is left for the electrostatic sector? Here, the ions pass through an arc in which the outer wall has a repulsive potential and the inner wall an attractive potential. This setup also causes a bend in the trajectories of the ions. The electrostatic sector is not directly used to separate the ions by m/z, but instead is employed to even out the distribution of ion kinetic energies. Recall that the magnetic sector separates ions on the basis of momentum and this condition can be correlated with m/z only if the ions have the same kinetic energy. When combined, the two sectors provide a mass spectrometer with exceptional mass-resolving power, in part because the sectors work together to provide a high degree of ion beam focusing.

Analyzer Design. There have been a wide variety of implementations of sector instrument designs, in terms of the geometries of the sectors and the order of the magnetic and electrostatic sectors, that is, which sector is encountered first by the beam. One common feature of all sector instruments is the need for high vacuum, because the mass analysis is dependent on ions that pass through the instrument on trajectories unaffected by collisions with background gas. For the present purposes, there is no need to provide an exhaustive discussion of the various ways that sector instruments can be operated. A typical implementation is shown in Figure 7-14. A beam of ions from the source is accelerated with a potential of a few kilovolts and then enters the magnetic sector. Here, mass filtering occurs, and ions are separated on the basis of their m/z. The ions then pass through a field-free region with a set of lenses and finally enter the electrostatic sector. In this sector, the distribution of kinetic energies in the beam is narrowed (ions with high initial kinetic energy experience a more significant retarding force in the electrostatic sector). In addition, the sector makes some corrections for the fact that the source beam does not have perfect colinearity in the ion trajectories. Ions then are collected through a narrow slit that leads to the detector. To obtain a mass spectrum, the strength of the magnetic field is ramped so as to sweep the distribution of ion trajectories across the slit opening, sequentially aligning ions of varying m/z with the slit and detecting them.

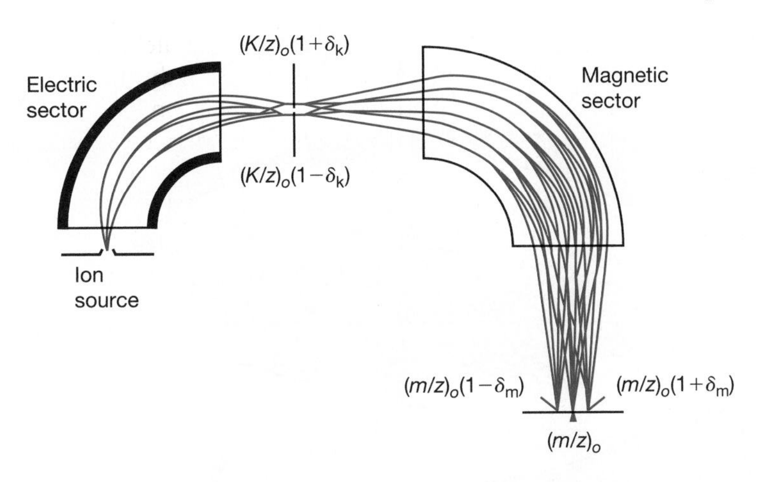

FIGURE 7-14 Double-focusing mass spectrometer of Nier–Johnson geometry. The magnetic field is normal to the plane, and the electric field lies in the plane. Ions of three different mass-to-charge ratios (δ_m) and a range of angles emerge from an ion source with a spread in kinetic energies and are focused for angle and kinetic energy but dispersed for *m/z* ratio, at the point detector. The electric and magnetic sectors both disperse ions according to their momentum-to-charge (*K/z*) ratios. (Based on R. G. Cooks et al., *Encyclopedia for Applied Physics*, vol. 19, New York: VCH Publishers, 1997, p. 289.)

Capabilities and Limitations. Sector instruments offer high performance in terms of mass accuracy (1–5 ppm) and are most often used in analytical labs today for that purpose. They also have good mass-resolving power (up to 10^5) and an exceptional dynamic range (up to 10^8 variation in ion abundances). Their mass ranges are not so great as those of TOF and ICR, and are generally limited to *m/z* of about 10,000. A major limitation is their efficiency because, during a scan, only ions passing through the slit are detected, while ions of all other masses are discarded. This issue is particularly important in restricting the applicability of sector instruments for liquid chromatography/mass spectrometry (LC/MS) or gas chromatography/mass spectrometry (GC/MS) analysis, for which efficient and rapid spectral analysis is needed. The need for high vacuum, the high initial cost, and the sheer size of sector instruments also have made them less attractive in comparison with more modern mass analyzers.

7-3f Orbitrap Analyzers

Principle of Operation. The orbitrap is a relatively new type of mass analyzer. Like QITs, it depends on an electric field to trap ions rather than the magnetic fields used in ICRs. However, like ICRs, it uses an image current for detection of ion oscillation frequencies in the device. This combination is very attractive in that it eliminates the complications of maintaining high magnetic fields, but offers the sensitivity and accuracy of FT analysis of the ion frequencies. The principle of operation is rather simple, but its practical implementation is difficult. The device consists of a spindle electrode surrounded by a roughly cylindrical electrode (Figure 7-15). The potentials are set such that the ions are attracted to the spindle electrode (unlike with QITs, a fixed rf potential is used). For a stable trajectory, the attraction to the spindle electrode must be balanced by the centrifugal forces of the ions orbiting around the spindle. This arrangement is the origin of the name of the device. One can make analogies to planetary motion, although the nature of the field is significantly different from that of a gravitational field. The attraction to the spindle electrode traps the ions in two dimensions, and repulsive potentials at end-cap electrodes trap the ions along the axis of the spindle (*z* axis). In this arrangement, the ions undergo three types of motion: (1) cycling around the spindle electrode, (2) radial oscillations between smaller and larger orbits, and (3) oscillations along the axis of the spindle. Only the last one is independent of the initial conditions of the ion trapping and therefore is the preferred motion for ion detection. The frequency of the oscillation along the axis of the spindle provides the measure of the *m/z* of the ions, and FT methods can be used for a collection of ions.

Analyzer Design. More so than with most other analyzers, the trapping capabilities of orbitrap analyzers are more closely linked to the physical attributes of the device rather than to applied magnetic or rf fields. As a result, the machining of the device is critical to its performance. In addition, the introduction of ions to the orbitrap is challenging because stable trajectories in the trap can be achieved only with a fairly narrow set of starting ion trajectories. In the commercial example of an orbitrap, the device is

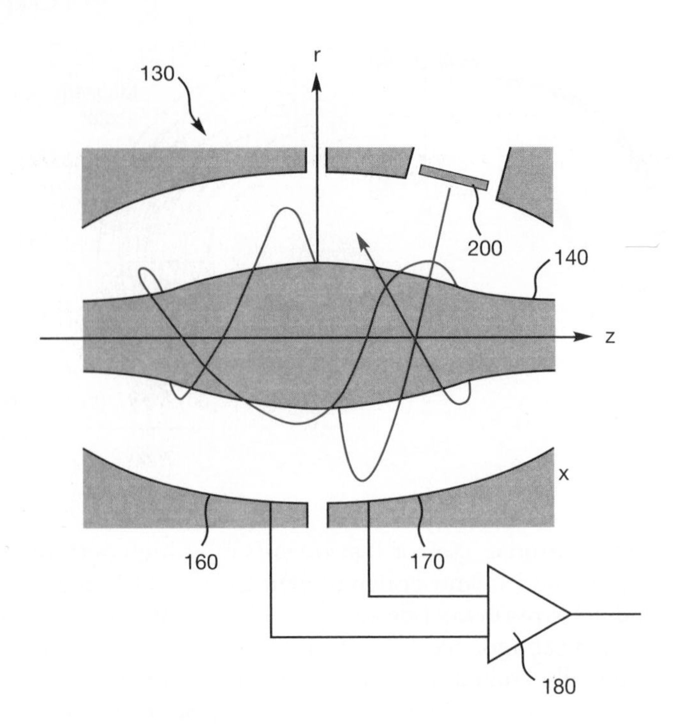

FIGURE 7-15 Original schematic of orbitrap analyzer from U.S. Patent 6995364. Spindle is labeled 140. Split outer electrodes are labeled 160 and 170.

linked to two linear ion traps. The first is used for ion accumulation and fragmentation (see Section 7-3g, on hybrid instruments). The second has a novel semicircular shape and is used to concentrate the ion packets for injection through a narrow port into the orbitrap. As ions are introduced into the orbitrap, the voltage on the spindle is ramped to trap the ions in the field and is then relaxed to a static potential once the ions are trapped. The outer electrode is split, and oscillations along the axis of the spindle lead to an image current across the two sections of the outer electrode. This image current is detected and converted from the time domain to the frequency domain by an FT. The frequencies provide the measure of the *m/z* values of the ions in the orbitrap.

Capabilities and Limitations. The orbitrap was developed to compete with FT-ICR instruments in terms of capabilities. Orbitraps have high mass accuracy in the range of 1 to 2 ppm and mass-resolving powers in the 10^5 range. They also have high mass capabilities, which reach up to about *m/z* 50,000. In each of these aspects, orbitraps are competitive with FT-ICRs. The orbitrap, however, is less versatile in terms of manipulating ions and is not useful for carrying out reactions or inducing fragmentation. As noted above, in commercial systems, this drawback is solved by linking the orbitrap to other mass analyzers that are capable of handling these tasks. As a result, orbitraps are inherently hybrid instruments (see below) and rely on a combination of technologies to produce a versatile mass spectrometer. Because they do not use high magnetic fields, orbitraps do not have high maintenance costs, but the need to link them to other mass analyzers as well as the precision machining makes them relatively expensive commercial instruments.

7-3g Hybrid Instruments

In recent years, there has been a growing tendency to expand the capabilities of mass spectrometers by incorporating two types of mass analysis devices. The goal is to combine the qualities of the two devices and create an instrument with exceptional capabilities. One common approach is to link a QIT with another mass analyzer, particularly one with high mass-resolving power. The ion trap is used to accumulate ions and potentially to activate mass-selected ions collisionally to give fragment ions (MS/MS scans, Chapter 8). The ion trap is very well suited for both of these tasks, but does not offer high resolution or mass accuracy. This deficiency can be rectified by transferring the ions from the ion trap to an FT-ICR, an orbitrap, or another high resolution mass analyzer, and with that device, generate the ultimate mass spectrum. This combination

TABLE 7-7 Capabilities of Typical Mass Analyzers

Analyzer	*m/z* range	Resolving Power	Mass Accuracy	Dynamic Range
TOF	10^6	10^4	5–50 ppm	10^4
ICR	10^5	10^6	$<$5 ppm	10^4
QIT	10^4	10^4	100 ppm	10^4
Quadupole mass filter	10^4	10^4	100 ppm	10^6
Magnetic/electrostatic sector	10^4	10^5	1–5 ppm	10^8
Orbitrap	10^4	10^5	1–2 ppm	10^4

provides an instrument with fast, versatile ion accumulation and fragmentation characteristics (from the ion trap), as well as high mass-resolving power. Moreover, the combination can offer superior efficiency because each mass analyzer can act independently. For example, the ion trap can be accumulating and fragmenting ions while the second mass analyzer is generating the mass spectrum from the previous packet of ions. The addition of the second analyzer does not slow the work flow, and it actually leads to some work acceleration because it allows for tasks to be shared by the two devices. Another very popular combination brings together quadrupole ion filters and a TOF analyzer. Here, the quadrupole ion filters are used to mass-select ions and fragment them in a collision cell (much like in a triple quadrupole). The beam is then directed to a TOF analyzer and orthogonally accelerated down the flight tube. As with most hybrid instruments, special ion optics are used to manipulate the ions in the transition from quadrupole to the TOF analyzer and to correct for issues related to the initial velocity of the ion beam. Development of hybrids is a common strategy in modern instrumentation, and one expects to see more of them in the future.

7-3h Summary of Mass Analyzers

In the previous sections, a number of mass analyzers have been described. Each has a set of strengths, and several are generally suitable for addressing a particular research problem. In Table 7-7, a brief summary of the capabilities of various mass analyzers is presented. The values in the table are generic, and often there are specialized implementations of the mass analyzer that might significantly exceed those listed in the table.

7-4 SAMPLE PREPARATION

A key determining factor in the choice of an ionization method is the properties of the sample, whether it be a liquid or solid, volatile or nonvolatile, pure or a mixture. The properties also play an important role in the preparation of the sample for analysis. Pure samples can be introduced in a variety of straightforward ways, depending on the type of ionization source, as discussed in the previous sections on ionization sources. In contrast, complex mixtures generally require some type of separation step as a part of the sample analysis, and chromatography is commonly linked to mass spectrometry. These approaches lead to the well-known tandem techniques of GC/MS and LC/MS.

7-4a GC/MS

For volatile samples, GC is the most effective separation method prior to MS. The GC is easily interfaced with EI or CI sources because the gas stream coming off a capillary column can be directed into the inlet of the EI source. In this way, compounds are sequentially ionized as they are chromatographically separated. Relatively high substrate pressures occur as they elute, and the sensitivity of GC/MS analyses is quite high. Because EI is a general ionization method, substrates generally do not need to be modified before analysis if they are sufficiently volatile to pass through the GC. For samples of relatively low volatility, derivatization is a common strategy. Derivatization schemes involve globally converting a functional group to one that offers greater volatility. A common approach is the conversion of an alcohol to a silyl ether. The reaction of ROH with reagents such as $(CH_3)_3SiCl$ is fast and quantitative, leading to $ROSi(CH_3)_3$ products, which are less polar, lack hydrogen bond donors, and consequently are much more volatile. For example, trimethylsilylation of sugars like glucose is predicted to increase their vapor pressure by orders of magnitude. This approach can greatly expand the range of compounds that can be analyzed by GC/MS. For the analysis of small, only moderately complex organic compounds, GC/MS on an instrument with an EI or CI source is the method of choice.

7-4b LC/MS

For samples with low volatility, LC is a good choice for the separation phase of the experiment. In this case, the analyte is part of the liquid eluent from the LC, and atmospheric pressure ionization sources are generally used. If the analyte is naturally charged, ESI is used and, if not, APCI can be used. These sources require very low flows of eluent (μl min^{-1} to nL min^{-1}), and interfaces often incorporate a split-flow system to reduce the quantity of eluent entering the source.

7-4c Data from GC/MS and LC/MS Systems

One is accustomed to mass spectrometry experiments that produce a single spectrum for the analyte. When the spectrometer is linked to chromatography, however, 100 s to 1000 s of spectra are recorded during the chromatographic run. In analyzing the data, it is best to think of the mass spectrometer as a detector on the GC or LC system, and the total ion count (TIC) is often employed as the y axis in chromatograms produced by GC/MS and LC/MS instruments. What differentiates the data from those obtained on a typical GC is that they are multidimensional in nature. At every time point, one is not limited to a single intensity, but, instead, signal intensities are available for every peak in an entire mass spectrum. As a result, chromatograms can be created using the intensity of any peak in the collection of spectra. In many applications, this approach is used to identify species in a sample by the combination of the retention time (RT) and parent mass. A more powerful approach is to add yet another dimension to the analysis, namely, fragmentation spectra. As spectra are collected during a chromatographic run, the mass spectrometer can be programmed to record not only full spectra but also fragmentation spectra for parent ions identified in a full mass spectrum (fragmentation spectra are discussed in detail in the following chapter). In this way, three levels of data are obtained on each unknown analyte in a sample: the RT, its parent mass, and the fragments that it forms. This level of characterization provides the gold standard of analyte identification.

7-4d Gel Separations

In the field of proteomics, it has become common practice to incorporate gel separations of proteins into the work flow. Typical SDS-PAGE (sodium dodecyl sulfate polyacrylamide gel electrophoresis) gels can be used to separate and greatly reduce the complexity of protein mixtures. Spots from the one-dimensional or two-dimensional gels can be excised and then further processed, which generally involves treatment with trypsin to cleave the protein into peptides. The mixture of peptides from each spot can be analyzed by mass spectrometers with either a MALDI or an ESI source. In the latter case, an LC is generally the interface, and, as a result, there are multiple levels of separation before the mass spectrometric analysis. This approach is exceptionally valuable in the analysis of samples that could potentially contain thousands of analytes, such as proteins in a cell lysate.

Worked Problems

The following four problems serve as worked illustrations. Additional problems in the following section are left for the reader to work.

7-1 Benzyl methyl ether has a proton affinity of 195.2 kcal mol^{-1}. Suggest appropriate positive ion CI gases that might give (i) limited fragmentation and (ii) extensive fragmentation.

Answer. If one wants to limit fragmentation, then the conjugate base of the reagent ion should have a proton affinity (PA) slightly smaller than the analyte. In this case, isobutane is a good choice. It produces the *t*-butyl cation, whose conjugate base, isobutene, has a PA of 191.7 kcal mol^{-1} (Table 7-3). This leads to the following proton transfer reaction:

$$(CH_3)_3C^+ + C_6H_5CH_2OCH_3 \rightarrow (CH_3)_2C{=}CH_2 + C_6H_5CH_2O(H)CH_3{}^+$$

$$\Delta H_{\text{reaction}} = \Delta H_f((CH_3)_2C{=}CH_2) - \Delta H_f((CH_3)_3C^+) + \Delta H_f(C_6H_5CH_2O(H)CH_3{}^+) - \Delta H_f(C_6H_5CH_2OCH_3)$$

but by definition $PA((M) = \Delta H_f(MH^+) - \Delta H_f(M)$

$$\therefore \Delta H_{\text{reaction}} = PA((CH_3)_2C{=}CH_2) - PA(C_6H_5CH_2OCH_3) = 191.7 - 195.2 \text{ kcal mol}^{-1} = -3.5 \text{ kcal mol}^{-1}$$

The reaction is exothermic by only 3.5 kcal mol^{-1} (195.2–191.7). Under these circumstances, little fragmentation is expected. To maximize fragmentation, the proton transfer reaction needs to be much more exothermic, which requires a reagent ion with a less basic conjugate base. A natural choice is methane. Here the proton transfer reaction

$$CH_5{}^+ + C_6H_5CH_2OCH_3 \rightarrow CH_4 + C_6H_5CH_2O(H)CH_3{}^+$$

is exothermic by 65.3 kcal mol^{-1} (129.9–195.2).

7-2 1,2,3,4-Tetrafluorobenzene is ionized by charge exchange with argon. If the ionization energy of the organic molecule is 9.6 eV and that of argon is 15.8 eV, what is the internal energy of the molecular ion?

Answer. Charge exchange proceeds as follows:

$$Ar^{+\cdot} + C_6F_4H_2 \rightarrow Ar + C_6F_4H_2{}^{+\cdot}$$

$$\Delta H_{\text{reaction}} = \Delta H_f(Ar) - \Delta H_f(Ar^{+\cdot}) + \Delta H_f(C_6F_4H_2{}^{+\cdot}) - \Delta H_f(C_6F_4H_2)$$

but by definition $IE(M) = \Delta H_f(M^{+\cdot}) - \Delta H_f(M)$

$$\therefore \Delta H_{\text{reaction}} = IE(C_6F_4H_2) - IE(Ar) = 9.6 - 15.8 \text{ eV} = -6.2 \text{ eV}$$

This 6.2 eV appears as internal energy of $C_6F_4H_2{}^{+\cdot}$ and leads to a spectrum that has considerable fragmentation (not shown). The base peak in this case is due to CF_2 loss, common in fluorinated compounds, whereas losses of CHF and CHF_2 also give intense peaks. The molecular ion abundance is only 22% of the base peak.

7-3 Predict the results of attempted negative chemical ionization of nitrobenzene and biacetyl (electron affinities of 0.4 and 1.1 eV) when the experiment is done under conditions that promote electron attachment (e^- reagent).

Answer. Electron attachment is often successful when the compound has a positive electron affinity. Both compounds should yield molecular anions, M^-.

7-4 The CI (methane) mass spectra of alkanes show prominent $(M - H)^+$ ions in the molecular ion region. Explain this observation.

Answer. The major methane reagent ion is $CH_5{}^+$ (Table 7-3). The reaction $RH + CH_5{}^+ \Rightarrow RH_2{}^+ + CH_4$ is only slightly exothermic for alkanes (it is thermoneutral for $R = CH_3$). Hence $(R - H)^+$ does *not* arise by fragmentation of the usual molecular ion, $(M + H)^+$. Rather, other ion-molecular reactions can compete with protonation. In particular, hydride abstraction occurs readily for alkanes.

$$C_2H_5{}^+ + RH \rightarrow C_2H_6 + R^+$$

Problems

7-1 For each of the following compounds, calculate the reaction enthalpy for a proton transfer from $(CH_3)_3C^+$, the reagent ion from isobutane. You will need to consult webbook.nist.gov for proton affinities.

(a) *t*-butyl methyl ether
(b) isobutylamine
(c) naphthalene
(d) methyl benzoate
(e) bromobenzene

7-2 For each of the following compounds, calculate the reaction enthalpy for an electron transfer to Ar^+, the reagent ion from argon. You will need to consult webbook.nist.gov for ionization potentials.

(a) *t*-butyl methyl ether
(b) isobutylamine
(c) naphthalene
(d) methyl benzoate
(e) bromobenzene

7-3 For each of the following compounds, calculate the reaction enthalpy for a proton transfer to F^-, the reagent ion from NF_3. You will need to consult webbook.nist.gov for protonation enthalpies of the conjugate bases of the analytes.

(a) acetone
(b) phenylacetylene
(c) chloroform
(d) 4-nitrotoluene
(e) benzyl alcohol

7-4 For each of the following situations, suggest instruments (ionization source and analyzer) that would be able to solve the problem. In most cases, more than one answer is correct. Briefly justify your answers.

(a) You need to analyze a complex mixture of hundreds of peptides with masses in the range of 500–2000 daltons. You need a mass accuracy of ± 1 dalton for your application.
(b) You need to analyze the head-space vapor over an aqueous solution. You are looking for a specific set of small organic compounds (<300 daltons) and anticipate very large variations in concentration ($>10^4$).
(c) You need to analyze a mixture of polymers with high molecular weights (1000–50,000 daltons). They are all from the same preparation and your goal is to determine the molecular weight distribution. The material has limited polarity and no functional groups that ionize readily in solution.

Bibliography

7.1 Chemical ionization: A. G. Harrison, *Chemical Ionization Mass Spectrometry*, 2nd ed., Boca Raton, FL: CRC Press, 1992; M. L. Vestal, *Chem. Rev.*, **101**, 361 (2001).

7.2 Fast-atom bombardment: P. A. Demirev, *Mass Spectrom. Rev.*, **14**, 279 (1995).

7.3 Matrix-assisted laser desorption: F. Hillenkamp and P. Peter-Katalinic, *MALDI MS: A Practical Guide to Instrumentation, Methods and Applications*, Weinheim, Germany: Wiley-VCH, 2007; M. Karas and R. Kruger, *Chem. Rev.*, **103**, 427 (2003); K. Dreisewerd, *Chem. Rev.*, **103**, 395 (2003); R. Knochenmuss and R. Zenobi, *Chem. Rev.*, **103**, 441 (2003); M. Karas, U. Bahr, and U. Giessmann, *Mass Spectrom. Rev.*, **10**, 335 (1991).

7.4 Electrospray ionization: R. B. Cole, *Electrospray Ionization Mass Spectrometry*, New York: Wiley, 1997; J. B. Fenn, M. Mann, C. K. Meng, S. F. Wong, and C. M. Whitehouse, *Mass Spectrom. Rev.*, **9**, 37 (1990); N. B. Cech and C. G. Enke, *Mass Spectrom. Rev.*, **20**, 362 (2001); J. A. Loo, *Mass Spectrom. Rev.*, **16**, 1 (1997).

7.5 Atmospheric pressure chemical ionization: M. L. Vestal, *Chem. Rev.*, **101**, 361 (2001).

7.6 Desorption electrospray ionization: R. G. Cooks, Z. Ouyang, Z. Takats, and J. M. Wiseman, *Science*, **311**, 1566 (2006).

7.7 Time-of-flight: M. Guilhaus, D. Selby, and V. Mlynski, *Mass Spectrom. Rev.*, **19**, 65 (2000); H. Wollnik, *Mass*

Spectrom. Rev., **12**, 89 (1993); N. Mirsaleh-Kohan, W. D. Robertson, and R. N. Compton, *Mass Spectrom. Rev.*, **27**, 237 (2008).

7.8 Ion cyclotron resonance: A. G. Marshall, C. L. Hendrickson, and G. S. Jackson, *Mass Spectrom. Rev.*, **17**, 1 (1998).

7.9 Quadrupole ion traps: R. E. March, *Quadrupole Ion Mass Spectrometry*, Hoboken, NJ: John Wiley, 2005; D. J. Douglas, A. J. Frank, and D. M. Mao, *Mass Spectrom. Rev.*, **24**, 1 (2005); S. A. McLuckey and J. M. Wells, *Chem. Rev.*, **101**, 571 (2001).

7.10 GC/MS: M. McMaster, *GC/MS: A Practical User's Guide*, Hoboken, NJ: John Wiley, 2008; H.-J. Hübschmann, *Handbook of GC/MS: Fundamentals and Applications*, Weinheim, Germany: Wiley-VCH, 2007.

7.11 LC/MS: M. McMaster, *LC/MS: A Practical Users Guide*, Hoboken, NJ: John Wiley, 2005; R. E. Ardrey, *Liquid Chromatography Mass Spectrometry: An Introduction*, Chichester, England: John Wiley, 2003.

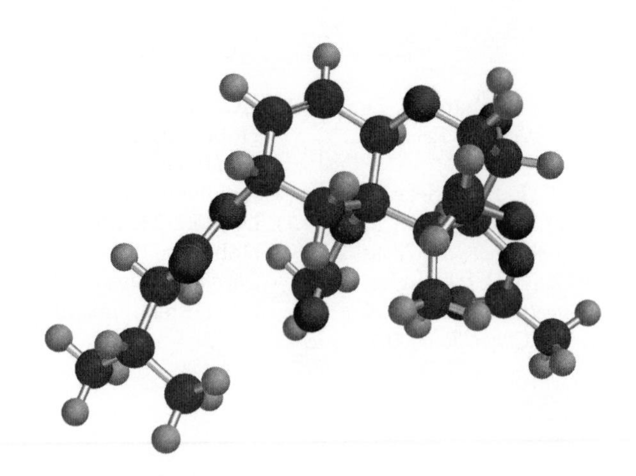

Fragmentation and Ion Activation Methods

8-1 INTRODUCTION

In the previous chapter, the instrumentation of modern mass spectrometry was described, with a focus on the tools needed to obtain the mass of a species derived from a particular analyte. If mass spectrometry were a tool whose sole purpose was the determination of the molecular weights of analytes, its scope would be quite limited and the methodology would be relegated to providing just a small component of the evidence needed for the verification of an analyte structure. Fortunately, mass spectrometry is capable of providing significantly more information because it also can probe the fragmentation patterns of the ions derived from analytes. Fragmentation is a natural outcome of electron ionization (EI), but also occurs with other ionization methods. If an ion is first selected by its *m/z* and then forced to fragment, the technique is referred to as *tandem mass spectrometry* or *MS/MS*, indicating that one mass analysis cycle was used to select the ion of interest and a second to identify the fragmentation products. This approach is generally adopted when a soft ionization technique like electrospray ionization (ESI) or matrix-assisted laser desorption (MALDI) is used. In any implementation, fragmentation greatly enhances the information content derived from mass spectrometry because a single ion can lead to a dozen or more fragments, each potentially containing useful structural data about the analyte. The value of these data is derived from the fact that organic ions tend to fragment in consistent ways, so it is possible to make reliable structural predictions on the basis of fragmentation patterns. Moreover, fragmentation patterns are generally unique (though not always) so that even isomeric species can be distinguished despite having identical molecular weights. Overall, fragmentation data can provide the pieces of the puzzle needed for identifying the structure of an analyte or, in other applications, simply a fingerprint for matching an analyte with a known species. In either case, it is the typical richness of fragmentation spectra that makes the identification possible.

In this chapter, we examine general patterns in the fragmentation of organic ions. In many cases, the mechanisms of these fragmentations should be familiar to the reader, from previous experience in organic chemistry. Like most processes, they usually are driven by energetic considerations, often involving known thermochemistry. We begin with some general principles of ion excitation, consider the fragmentation of common functional groups as both molecular cations (i.e., radical cations) and protonated species, and finally survey the ways that fragmentation methods are incorporated into modern mass spectrometry.

8-2 ION ENERGETICS

One might ask why fragmentation is such an integral part of mass spectrometry. The answer comes from the energetics of ions in a low-pressure environment. In the absence of cooling collisions with a background gas, excited ions retain the energy from activation processes for relatively long periods. Although emission of a photon or collision with a trace gas is possible, bond cleavage leading to fragmentation can be the dominant pathway for releasing the excess internal energy. This observation is true, in part, because ions, particularly radical ions, often have weak bonds and low barriers to fragmentation. Fragmentation can occur promptly after activation or be delayed and occur later in the mass spectrometry experiment if the process is slow. This gas-phase situation is completely different from the condensed phase, for which interactions with the solvent or matrix rapidly allow for the dispersal of excess energy and thermalization of a species to the temperature of the environment. The key issue here is that mass spectrometry easily allows large quantities of energy to be transferred to and retained in individual molecules.

In virtually all fragmentation events, the analyte ion must be vibrationally or electronically excited to initiate the process. In the case of EI, the excitation is a result of dislodging the electron. Given the high energy of a 70 eV electron ($>$1600 kcal mol^{-1}), significant electronic and vibrational excitations are expected, leading to ions with extensive internal energy. This energy fuels the bond cleavage processes. The origins of the initial vibrational excitations can, in part, be linked to the time scale of the EI process. The electron interaction with the analyte is in the femtosecond range, whereas vibrations are on the picosecond time frame. As a result, the initial product of the electron ionization (typically a radical cation) is formed with the structure (geometry) of the starting analyte. In other words, the *Franck-Condon principle* applies to the ionization. Because radical cations often prefer significantly different geometries than do neutral species, the net effect is the production of the radical cation in a vibrationally excited state (geometric distortions are equivalent to vibrational excitation).

Alternatively, the energy can be delivered to the ion after its formation. The most common approach in this case is to use electric fields to increase the kinetic energy of the analyte ion and convert that energy to vibrational (or possibly electronic) excitations via collisions with inert target atoms or molecules such as helium, argon, or nitrogen. This process is referred to as *collision-induced dissociation* (CID) or *collision-activated dissociation* (CAD). The activation can be provided by a single, high-energy collision or by multiple, low-energy collisions. Ions also can be activated by photons, and methods have been developed that use infrared (IR), visible, or ultraviolet (UV) radiation to force them into excited states that are capable of undergoing fragmentation processes. One of the more common implementations is *infrared multiphoton dissociation* (IRMPD).

8-2a Competing Pathways: Ion Stability

Although fragmentation might be viewed as an idiosyncratic aspect of mass spectrometry, it is guided by the same principles as any organic reaction, and the regular patterns seen in fragmentation pathways are driven by fundamental reaction mechanisms. As in condensed-phase chemistry, thermodynamics plays a large role and dominant pathways often are those that are favorable energetically, that is, least endothermic. The major differences are that the high energies deposited in the ions allow for highly endothermic reactions and many competing pathways (little selectivity).

One could calculate reaction enthalpies for all the competing processes, either on the basis of literature values or computational modeling, and use these data to predict the dominant fragmentation pathways. This approach can be tedious and, for complex species, require making a number of assumptions based on analogies to simpler species with known thermochemistry. Fortunately, the situation is somewhat simplified for positive ions because the stabilities of cations, particularly carbocations, generally are much more sensitive to structural effects and substituent effects than are neutral species, including radicals. As a result, identifying the pathway that produces the most

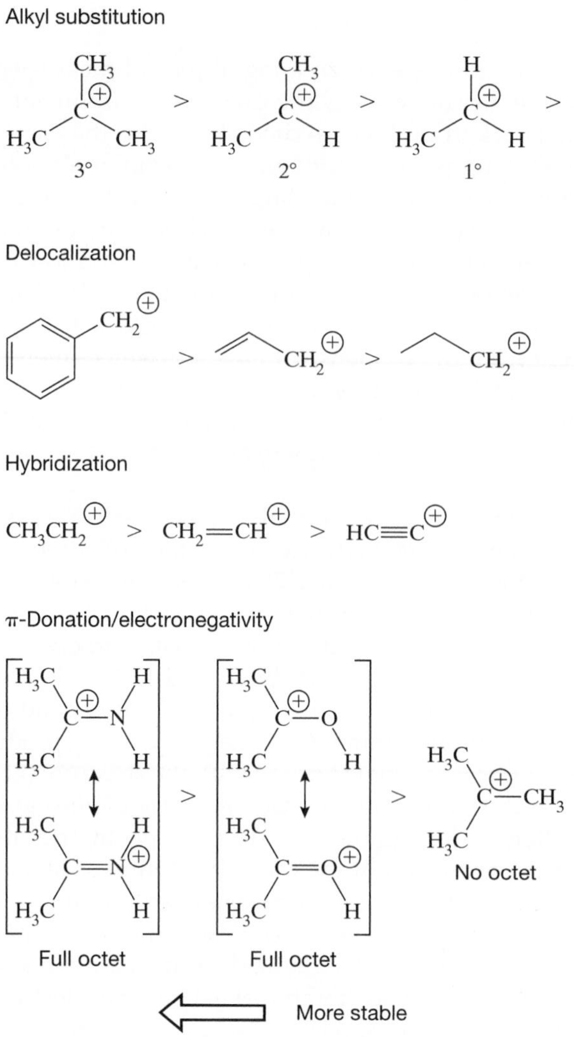

Alkyl substitution

3° > 2° > 1° >

Delocalization

Hybridization

$CH_3CH_2^{\oplus} > CH_2{=}CH^{\oplus} > HC{\equiv}C^{\oplus}$

π-Donation/electronegativity

Full octet > Full octet > No octet

More stable

FIGURE 8-1 General stability patterns for carbocations.

stable cation in a fragmentation process is often sufficient to predict the preferred products. In this way, one can generally use chemical intuition to predict or rationalize logical fragmentation pathways. Here, it is worthwhile to review a few trends in organic cation stability (Figure 8-1). Quantitative data are presented in Table 8-1 in terms of hydride ion affinities (HIAs; lower values indicate more stable cations). With this information it is easy to make some simple predictions about ion fragmentation patterns.

Consider the ionization of propene by EI. Three potential fragmentation pathways are given in eqs. 8-1 to 8-3. From Table 8-1, it is clear that the most stable cation from the three pathways is the allyl cation (eq. 8-3). Not surprisingly, the dominant pathway for

$$CH_2{=}CHCH_3^{\bullet +} \rightarrow CH_2{=}CH^+ + CH_3\bullet \quad \textbf{(8-1)}$$

$$CH_2{=}CHCH_3^{\bullet +} \rightarrow CH_2{=}CH\bullet + CH_3^+ \quad \textbf{(8-2)}$$

$$CH_2{=}CHCH_3^{\bullet +} \rightarrow CH_2{=}CHCH_2^+ + H\bullet \quad \textbf{(8-3)}$$

fragmentation is loss of a hydrogen atom to give the allyl cation (*m/z* 41). A small signal is seen for the vinyl cation ($CH_2{=}CH^+$, *m/z* 27), and there is little intensity from the methyl cation (Figure 8-2). Secondary fragmentations also are evident in the spectrum. Although this approach is only approximate (we did not consider the stability of the neutral products) and does not take into account the kinetics of the fragmentations (some pathways might have large barriers), it offers a very rapid way of evaluating potential pathways. In the next sections, we look at some well-established guiding principles for predicting likely fragmentation pathways.

TABLE 8-1 Hydride Ion Affinities of Carbocations

Cation	HIA (kcal mol^{-1})	Cation	HIA (kcal mol^{-1})
CH_3^+	314	$CH_3OCH_2^+$	234
$CH_3CH_2^+$	271	$(CH_3)_2NCH_2^+$	197
$(CH_3)_2CH^+$	250	$CH_3SCH_2^+$	236
$(CH_3)_3C^+$	237	FCH_2^+	291
$CH_2{=}CH{-}CH_2^+$	256	$ClCH_2^+$	282
$Ph{-}CH_2^+$	239	$BrCH_2^+$	266
$CH_2{=}CH^+$	288	HCC^+	381

Values from webbook.nist.gov or D. A. Shea and R. J. J. M. Steenvoorden, *J. Chem. Phys A*, **101**, 9728–9731 (1997).

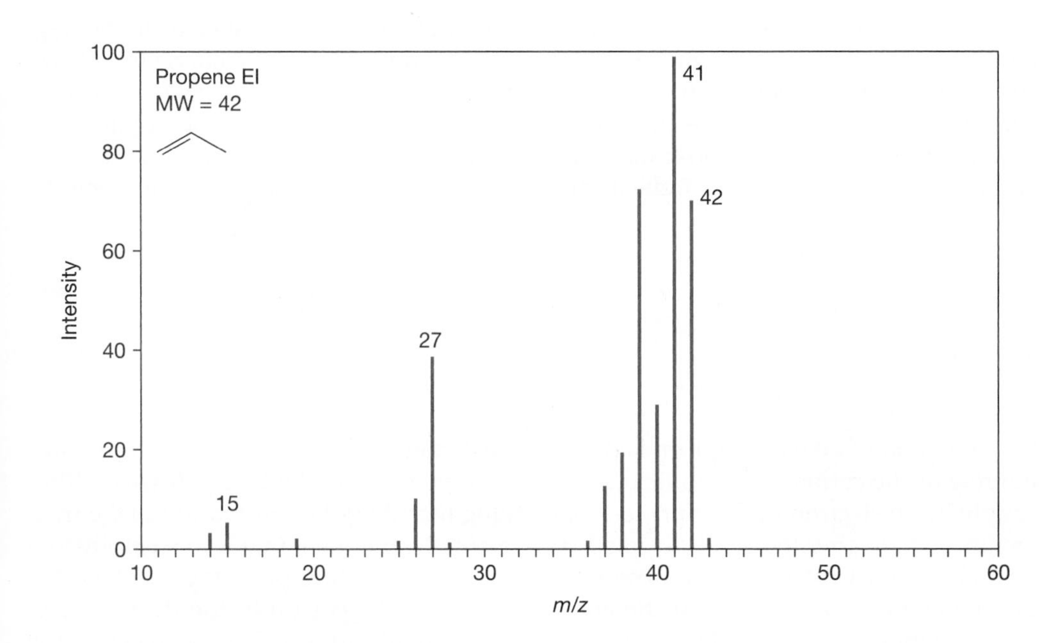

FIGURE 8-2 EI spectrum of propene. (EI data from webbook.nist.gov.)

8-2b Competing Pathways: General Principles

Preferences based on the enthalpy of the fragmentation can be couched in a number of different terms, each of which provides a guide to predicting likely fragmentations. One is the *odd/even-electron* rule. Here, an even-electron ion refers to a closed-shell species, that is, one with no unpaired electrons, and an odd-electron ion refers to a free radical with an unpaired electron. Recall that EI often initially produces radical cations, which, of course, are odd-electron species. When an odd-electron species fragments, there are two common outcomes: (1) formation of a new radical cation with loss of a closed-shell neutral species or (2) formation of a closed-shell, even-electron cation with loss of a radical species. The latter result tends to be more common because it generally does not require a rearrangement and therefore has favorable kinetics. In each case, the mechanisms are radical in character, and the total number of unpaired electrons does not change. During CI, ESI, or MALDI, closed-shell, even-electron species often are the initial ionization product (usually protonated analytes). In this case, the fragmentation mechanisms are generally closed-shell processes, often related to acid-catalyzed mechanisms for cations, and, as a result, the products are almost always closed-shell, even-electron ions and neutrals. The general rule is *odd-electron species are formed only from the fragmentation of an odd-electron ion.* The principle behind the rule is clearly thermodynamic. The fragmentation of an even-electron species to give two odd-electron species, both with an atom bearing an unpaired electron, is energetically unfavorable. Consider the fragmentation of $C_2H_5^+$, an even-electron species that can be viewed as either the ethyl cation or protonated ethene. A logical neutral loss to give an even-electron species would be expulsion of H_2 to give $C_2H_3^+$ (eq. 8-4). Alternatively, it could fragment to give the radical cation of ethene, $C_2H_4^+$ and a hydrogen atom (eq. 8-5). The former process is significantly less endothermic. As with most rules in mass spectrometry, this one is only a guideline and exceptions are numerous. Because the origin of the rule is thermodynamic, there can be specific cases for which formation of odd-electron species from an even-electron ion might be a favorable path because it leads to some special stabilization in one of the products, such as the formation of an aromatic ring.

$$C_2H_5^+ \xrightarrow{\Delta H = +51 \text{ kcal mol}^{-1}} C_2H_3^+ + H_2 \quad (8\text{-}4)$$

$$C_2H_5^+ \xrightarrow[\Delta H = +92 \text{ kcal mol}^{-1}]{} C_2H_4^{\bullet+} + H\bullet \quad (8\text{-}5)$$

The second general principle is *Stevenson's rule*, which is rather specific to the fragmentation of radical cations and provides a direct comparison of the energetics of two competing fragmentation pathways. It is, in fact, a variation on the ideas presented in Section 8-2a. In radical cations, fragmentations often lead to two competing pathways that differ only by an electron transfer process. This rule is best illustrated with an example (eqs. 8-6 and 8-7). The radical cation of 2-methylhexane can fragment along the

IP = 8.02 eV (8-6)

IP = 7.37 eV (8-7)

C_2—C_3 bond to give an isopropyl and an *n*-butyl fragment, either of which could take the role of the cation or the radical species. As noted above, the two pathways differ simply by an electron transfer process, something that should occur readily as the fragments separate after the bond cleavage. As a result, the fragment that is most easily ionized is the most likely one to form a cation in the competing pathways. In fact, the difference in the enthalpies of the competing pathways is simply the difference in the ionization potentials (IPs) of the two possible neutral products. In this case, the isopropyl group has the lower IP, that is, most easily ionized, and therefore we predict that the pathway in eq. 8-6 dominates because it is less endothermic by 0.65 eV (15 kcal/mol), the difference in the IPs of the isopropyl and *n*-butyl groups. A spectrum is shown in Figure 8-3. The signal for the isopropyl cation, *m/z* 43, is more intense than

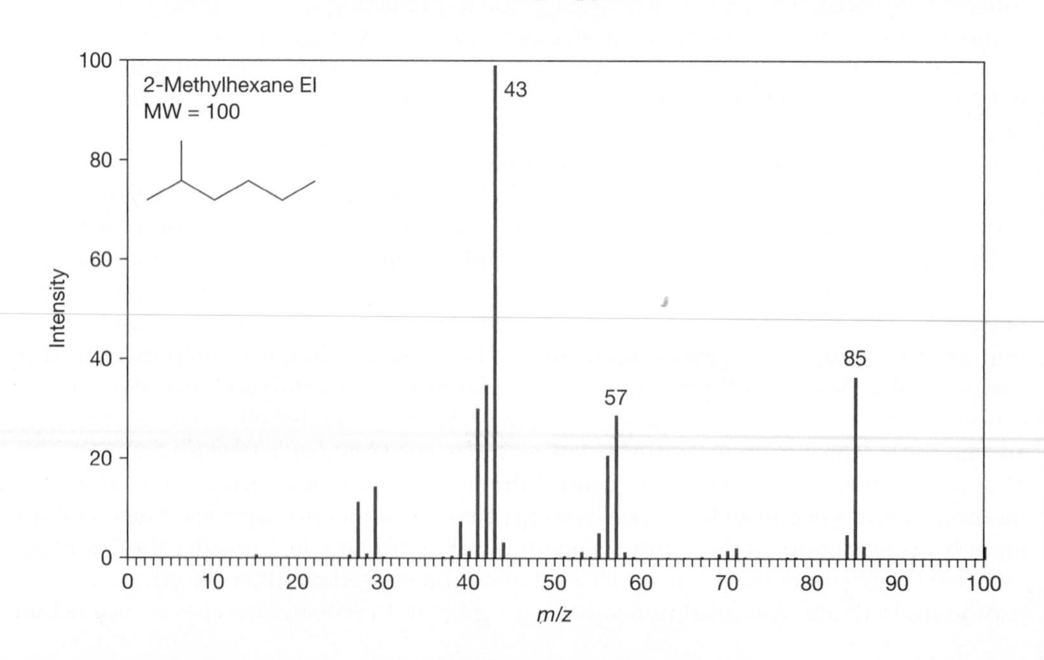

FIGURE 8-3 EI spectrum of 2-methylhexane. (EI data from webbook.nist.gov.)

that for the butyl cation, *m/z* 57 (because rearrangements occur in carbocations, it is possible that this ion eventually becomes the *tert*-buyl cation). In addition, a strong signal is seen for methyl loss (*m/z* 85), presumably to give the 2-hexyl cation. You will note that the same prediction would have been made simply on the basis of the estimated stabilities of the carbocations, 2° cations being more stable than 1° cations, using the logic from Section 8-2a. Fragmentations of radical cations to give another radical cation and a closed-shell neutral species also obey this rule. In short, Stevenson's rule can be stated as follows: *in fragmentations of a radical cation, there is a preference for producing the radical with the higher IP and therefore the cation which is associated with the radical with the lower IP.*

8-2c Competing Pathways: Charge Localization

It is also worth considering the impacts of charge localization on the sites of fragmentation in a radical cation. In the ionization process, electrons are likely lost from the highest-energy orbitals, which often are localized to some extent at functional groups with π bonds or lone pairs. In early organic mass spectrometry, Djerassi suggested that localization of the charge and radical center could act as a driving force in directing the location of bond cleavages during fragmentation. In this view, the critical factor in deciding which bond will cleave is related to a property of the radical cation (site of charge/radical localization) rather than a property of the fragmentation products (cation/radical stability). In most cases, both points of view suggest the same cleavage site because similar factors are at play in stabilizing the charge site in the initial radical cation and the cationic products of fragmentation. Nonetheless, there are certain situations in which the charge localization model seems to provide more insight into the preferred fragmentation pathways. This is particularly true in carbonyl-containing species in which fragmentations are often directed by the ionized carbonyl group.

8-2d Competing Pathways: Internal Energies

In condensed-phase reactions, temperature plays an important role in balancing the competition between pathways. Higher-energy reaction pathways become more competitive at higher temperatures because the impact of differences in activation energies is scaled by $1/T$ in the Arrhenius equation or transition state theory. By the same token, reactions tend to be more selective at lower temperatures. In the low-pressure environments of mass spectrometers it is difficult to assign temperatures (there are too few collisions to equilibrate translational, vibrational, and rotational energies), so gas-phase ion populations are generally characterized in terms of *internal energy distributions*. Nonetheless, the same principles hold. When an ion population has high internal energies, many pathways become viable and competitive, despite having significantly different activation barriers. For this reason, activation processes that lead to high internal energies, such as EI, often give dozens of significant fragments. In contrast, activation processes that lead to relatively low internal energies, such as multicollisional activation in an ion trap, tend to give fewer fragments because only enough energy is available in these ions to overcome fragmentation pathways with the lowest barriers. As a result, the complexity of fragmentation spectra is intimately linked to the magnitude of the energy deposited in the activation process. This fact must be considered when comparing spectra from instruments employing different activation mechanisms.

8-2e Competing Pathways: Typical Fragments

In the analysis of mass spectra, it is very helpful to know the masses of typical fragments, neutral and charged, that might be generated. In this way, it is possible to make predictions quickly about pathways and develop suggestions of likely structural elements in an analyte. For example, whenever one sees a loss of 15 mass units from an ion, the immediate assumption is that a CH_3 group has been lost. This observation suggests that the analyte contains a methyl group and that the methyl group is in a location such that its loss leads to a reasonably stable ion. Although it is not necessary to memorize all of the masses in Tables 8-2 and 8-3, knowledge of at least the most common ones makes the analysis of fragmentation spectra much more efficient. One helpful coincidence is that

TABLE 8-2 Possible Formulas of Common Closed-Shell, Cationic Fragment Ions

Mass	Ion	Mass	Ion
29	$C_2H_5^+$, HCO^+	55	$C_4H_7^+$
30	$CH_2NH_2^+$	57	$C_4H_9^+$, $C_2H_5CO^+$
31	CH_2OH^+	59	CH_3OCO^+, $C_2H_5OCH_2^+$
39	$C_3H_3^+$	61	$CH_3SCH_2^+$
41	$C_3H_5^+$	69	$C_5H_9^+$, CF_3^+
43	$C_3H_7^+$, CH_3CO^+	71	$C_5H_{11}^+$, $C_3H_7CO^+$
44	$CH_3CHNH_2^+$	77	$C_6H_5^+$ (phenyl), $C_2H_5OCO^+$
45	HCO_2^+, $CH_3OCH_2^+$	79	$C_6H_7^+$
46	NO_2^+	85	$C_6H_{13}^+$, $C_4H_9CO^+$
47	CH_2SH^+	91	$C_7H_7^+$ (benzyl, tropyl)
51	$C_4H_3^+$	99	$C_7H_{15}^+$, $C_5H_{11}CO^+$
53	$C_4H_5^+$	105	$C_8H_9^+$, $C_6H_5CO^+$

TABLE 8-3 Possible Formulas of Common Neutral Fragments

Mass	Fragment[a]	Mass	Fragment[a]
15	*CH_3*	31	*CH_3O*
16	CH_4, *NH_2*	32	CH_3OH, N_2H_4
17	NH_3, *OH*	33	*HS*
18	H_2O	35, 37	*Cl*
26	C_2H_2	36, 38	HCl
27	HCN	42	C_3H_6, CH_2CO
28	C_2H_4, CO, N_2	43	*C_3H_7*
29	*C_2H_5*	44	C_3H_8, CO_2, CH_3CHO, N_2O
30	C_2H_6, CH_2O, *NO*	46	*NO_2*, C_2H_6O

[a]Odd-electron species are in italics.

for analytes containing common elements other than nitrogen—that is C, H, O, F, Si, P, S, Cl, and Br—the molecular weight is an even number. As a result, radical cations of these analytes have even *m/z* values (at least for singly charged species), and closed-shell fragments (even-electron) have odd *m/z* values. Consequently, one can easily distinguish between the two possibilities in a spectrum. When an analyte contains a single nitrogen (or an odd number of them), however, the pattern is reversed, so that radical cations give odd *m/z* values and closed-shell cations give even *m/z* values. On the other hand, when the analyte contains an even number of nitrogens (0, 2, 4, 6 nitrogens), the normal pattern is observed. This has been called the *nitrogen rule*.

8-3 FRAGMENTATION PATTERNS OF FUNCTIONAL GROUPS

Just as organic chemistry texts can be conveniently organized along a functional group approach, the same can be done with fragmentation patterns. The features of functional groups allow specific mechanisms to operate, and these mechanisms can be generalized with reasonable success to all compounds bearing that functional group. This regularity of behavior is one of the key factors that allow one to link chemical structures with fragmentation patterns. In the following sections, typical fragmentation patterns of several common functional groups are presented. The emphasis is on radical cations derived from EI, but with appropriate functional groups, the fragmentation behavior of protonated species also is considered.

8-4 HYDROCARBONS

8-4a Alkanes and Alkenes

The fragmentation patterns of radical cations derived from alkanes generally involve sequential C—C bond cleavages along with hydrogen atom or molecule expulsions and often are rather complex. As discussed above, there is a tendency to produce the most stable cations, but in the absence of exceptionally stable fragmentation products, many product ions are usually formed. A comparison of the EI spectra of octane and 2,2-dimethylhexane is instructive (Figure 8-4). The former shows a small molecular ion (*m/z* 114) followed by a series of alkyl cations with six (*m/z* 85), five (*m/z* 71), four (*m/z* 57), three (*m/z* 43), and two carbons (*m/z* 29). Another way of interpreting these signals is by their neutral losses (see Table 8-3). They correspond to loss of 29 (ethyl), 43 (propyl), 57 (butyl), 71 (pentyl), and 85 (hexyl), respectively. Except for *m/z* 43, all of these signals have roughly equal intensities. In addition, peaks are seen for those ions missing a hydrogen atom or molecule, such as *m/z* 42 and 41, so that the spectrum contains a small cluster of ions for each carbon number. For 2,2-dimethylhexane, the spectrum is dominated by *m/z* 57, presumably the very stable *tert*-butyl cation. The molecular ion is missing because very facile fragmentation pathways are available. A peak at *m/z* 99 corresponds to loss of CH_3, to give another 3° cation, the 2-methyl-2-hexyl cation. With respect to three-carbon species, 2,2-dimethylhexane gives the largest signal for *m/z* 41, the allyl cation. This species is most likely the result of a secondary fragmentation of the *tert*-butyl cation. Another diagnostic aspect of this spectrum is the lack of signals at *m/z* 71 and 85, products that would require bond cleavage to give 1° cations initially. As you can see, the fragmentations are driven mainly by thermodynamic considerations, but the spectra are complicated by a large number of pathways and potential secondary fragmentations such as the loss of hydrogen atoms and molecules.

Although it is easy to distinguish between the spectra for the isomeric molecules octane and 2,2-dimethylhexane, would it be possible to identify these compounds solely based on their EI mass spectra? In Figure 8-5, spectra are given for two additional isomers of C_8H_{18}: 3,3-dimethylhexane and 2,2,3-trimethylpentane. They differ to some extent from those shown in Figure 8-4, but many similarities can be noted, particularly when comparing octane with 3,3-dimethylhexane or 2,2-dimethylhexane with 2,2,3-trimethylpentane. Could we look at any one of these spectra alone and confidently assign a structure? The answer to that question is probably no. The only obvious differences between the spectra in the left panels of Figures 8-4 and 8-5 are the peaks at *m/z* 99 and 114. One could easily look at the spectrum for 3,3-dimethylhexane and accidentally assign it to octane because they share many of the same peaks. As Figures 8-4 and 8-5 illustrate, there are many, sometimes subtle differences in the spectra of isomers. Consequently, mass spectra are often used to assign spectra on the basis of comparisons, by matching unknown spectra with those in a database of known compounds. These differences, though often subtle, generally are logical. For example, the

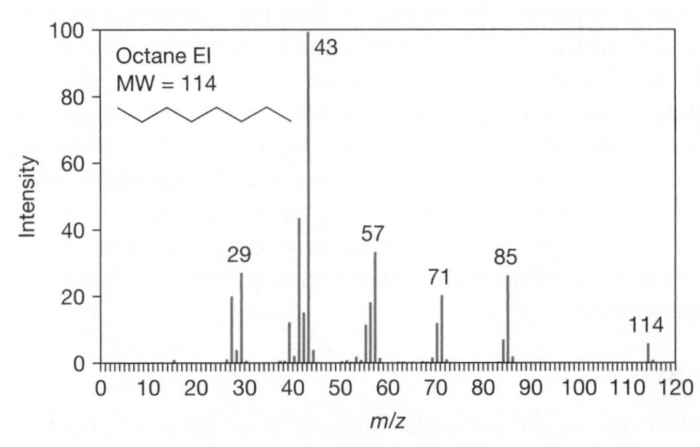

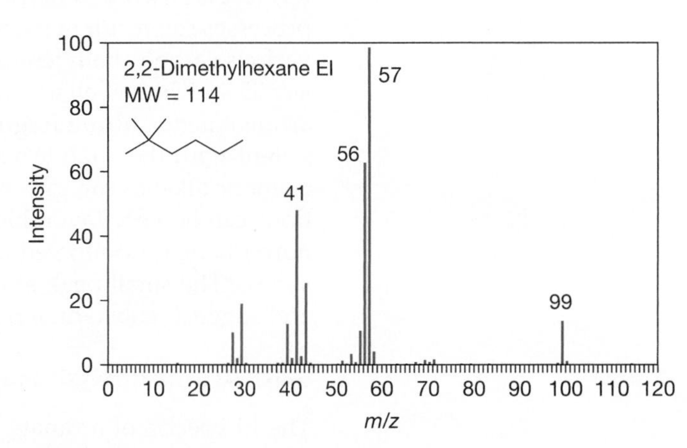

FIGURE 8-4 EI spectra of octane and 2,2-dimethylhexane. (Data from webbook.nist.gov.)

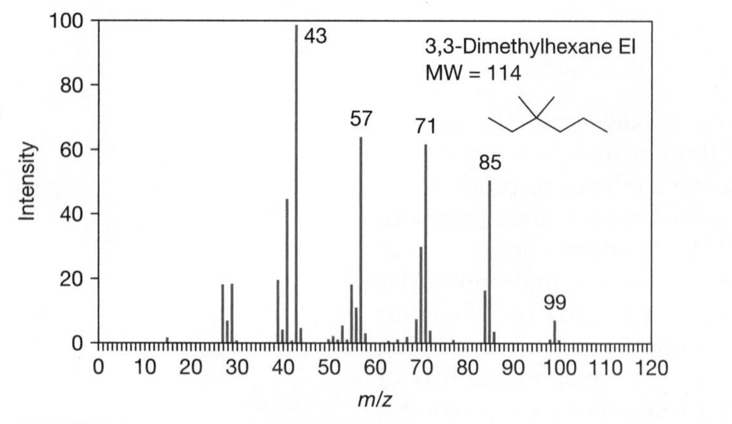

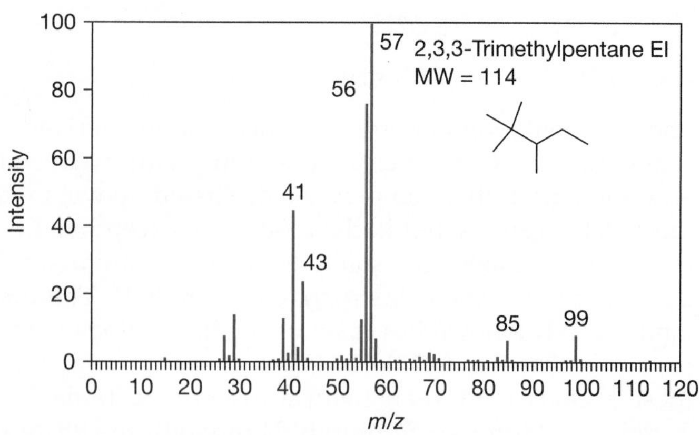

FIGURE 8-5 EI spectra of 3,3-dimethylhexane and 2,2,3-trimethylpentane. (Data from webbook.nist.gov.)

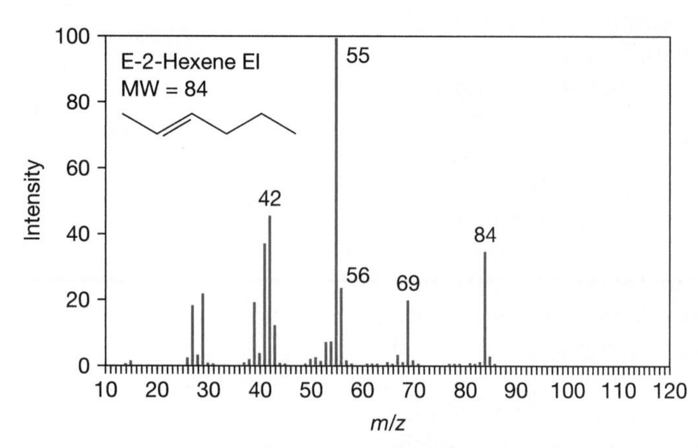

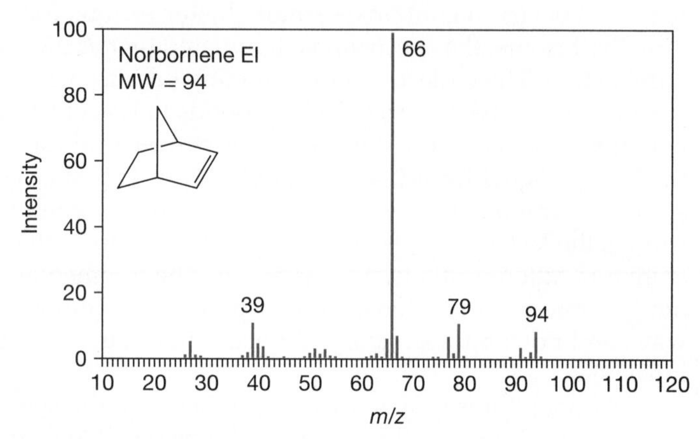

FIGURE 8-6 EI spectra of E-2-hexene and norbornene. (Data from webbook.nist.gov.)

absence of the molecular ion peak for 3,3-dimethylhexane (*m/z* 114) is related to the fact that the molecule is able to fragment easily to 3° cations. The presence of *m/z* 99 in this spectrum makes sense because methyl loss can give a 3° cation. The spectrum of 2,2,3-trimethylpentane differs from that of 2,2-dimethylhexane by the presence of a peak at *m/z* 85 (loss of ethyl). Unlike the case with 2,2-dimethylhexane, loss of ethyl from 2,2,3-trimethylpentane gives a stabilized cation (2°).

Alkenes tend to lose alkyl groups and often produce allylic cations. This can be seen in the spectrum for E-2-hexene in Figure 8-6, in which loss of ethyl (29 mass units) gives the base peak at *m/z* 55, likely the 2-buten-1-yl allylic cation. In Figure 8-2, propene is seen to undergo a related process, hydrogen atom loss, to give the allyl cation. Competing with this process is C—C cleavage accompanied by a hydrogen transfer, to give two alkene products, one of which is an ionized, radical cation. These processes can result in pairs of peaks, separated by one mass unit, and corresponding to various carbon chain lengths. For example, in E-2-hexene, significant peaks are seen at *m/z* 55 and 56 as well as *m/z* 41 and 42. However, alkenes tend to undergo extensive rearrangements during fragmentation and often give product ions that do not seem consistent with the analyte's structure. As one might expect under these circumstances, isomeric alkenes can give very similar spectra. In cyclic alkenes, retro Diels–Alder reactions can be very favorable with appropriate analytes. In Figure 8-6, the spectrum for norbornene is dominated by the loss of ethene, resulting in ionized cyclopentadiene at *m/z* 66. The small peak at *m/z* 79 is for methyl loss, to likely give protonated benzene, and suggests substantial rearrangement.

8-4b Aromatic Hydrocarbons

The EI spectra of aromatic hydrocarbons are generally simpler than those of alkanes. Bond strengths typically are greater, which disfavors bond cleavages, and there can be

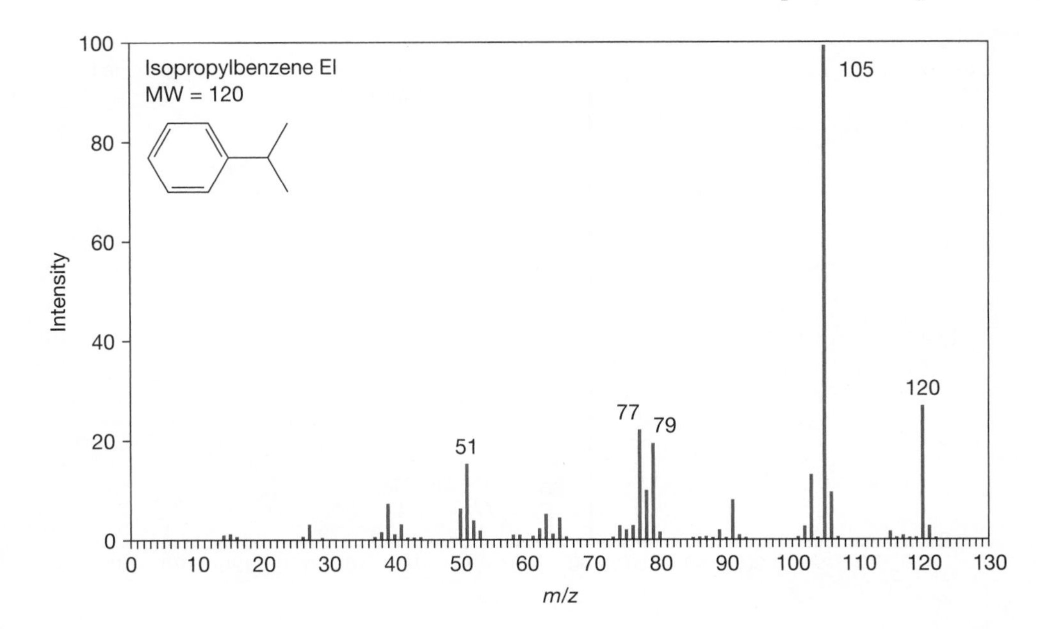

FIGURE 8-7 EI spectrum of isopropylbenzene. (EI data from webbook.nist.gov.)

large disparities in the stabilities of fragment ions, so fewer pathways are available to the analyte ions. The EI spectrum of isopropyl benzene is a good example (Figure 8-7). The base peak is *m/z* 105, which represents loss of methyl, to initially give a stable benzylic cation that eventually rearranges to an even more stable tropylium (cycloheptatrienylium) ion. This is a common rearrangement and, in the text, cyclic $C_7H_7^+$ species are referred to as benzyl/tropyl cations. Aside from this ion, there is a significant molecular ion, *m/z* 120, and collections of ions around *m/z* 77–79 and *m/z* 50–53. Each of these is common in benzene derivatives. First, the relatively high stability of aromatic analytes allows molecular ions to survive. Second, the benzene core tends to remain intact despite extensive fragmentation of substituents, so the phenyl cation (*m/z* 77) and protonated benzene (*m/z* 79) are common product ions. Finally, if the benzene ring breaks down, four-carbon ring-opening products with low hydrogen counts (mostly $C_4H_3^+$) are characteristic, in part because not many ions are typically found in this part of the *m/z* scale. Analogous behavior is seen in polycyclic aromatic hydrocarbons.

Summary for Hydrocarbons

1. Alkanes undergo C—C cleavage generally to produce the most stable carbocation. Molecular ion peaks are sometimes very weak.
2. Aromatic hydrocarbons also undergo C—C cleavage to give the most stable carbocation. Molecular ion signals are sometimes very strong. Benzyl/tropyl and phenyl cations are common.

8-5 AMINES

As noted earlier, nitrogen-containing species follow a different pattern in terms of odd- and even-numbered ion masses. The radical cations give odd-numbered *m/z* values, and closed-shell species give even-numbered *m/z* values. Realizing this difference is important in the analysis of amine mass spectra.

8-5a Radical Cations of Amines

As noted earlier, atoms with lone pairs can act as powerful stabilizing groups for adjacent carbocations. As a result, the radical cations from amines are prone to α-cleavage processes, by which a C—C bond that is one removed from the nitrogen center cleaves to

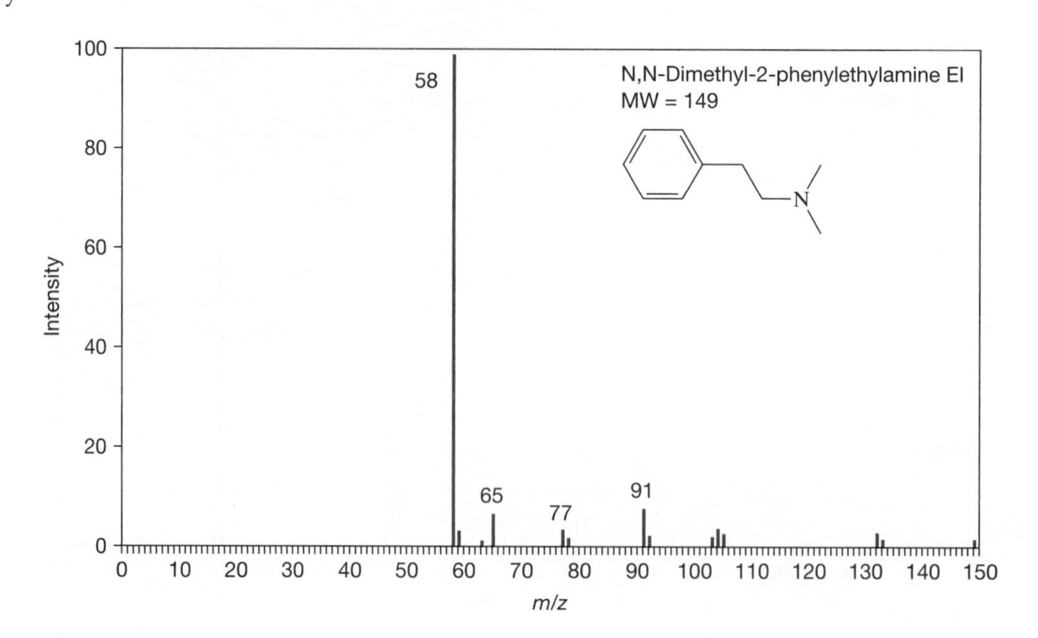

FIGURE 8-8 EI spectrum of *N,N*-dimethyl-2-phenylethylamine. (EI data from webbook.nist.gov.)

give an immonium cation (eq. 8-8). This process is governed by Stevenson's rule, but the immonium cation inevitably is favored over an alkyl cation. As an example, the EI

$$R_2\dot{N}^{+}\text{—C—C—} \longrightarrow \left[R_2\ddot{N}\text{—}\overset{+}{C} \longleftrightarrow R_2\overset{+}{N}\text{=C}\right] + \cdot\text{C} \tag{8-8}$$

spectrum for *N,N*-dimethyl-2-phenylethylamine is very simple and is dominated by a peak at *m/z* 58, the α-cleavage product (Figure 8-8). This product is closed shell with an even number of electrons and moreover has a full octet of electrons. In contrast, the even-electron products of alkane fragmentation have only a sextet. Note that the alternative product of this cleavage, the benzyl/tropyl cation at *m/z* 91, is quite small, despite the high stability of this hydrocarbon cation—the advantage of forming an octet-satisfied immonium cation is large. In addition, the common phenyl fragmentation products at *m/z* 77–79 also lack octets and are less competitive than the α-cleavage process. Once formed, the α-cleavage product is a closed-shell, even-electron species and undergoes secondary fragmentations similar to those of protonated amine species, such as loss of an alkene (eq. 8-9). This is an example of an ene reaction (although a stepwise

(8-9)

process involving C—N cleavage followed by proton transfer is also possible). As a result, α-cleavage can give a series of ions that differ in mass by that of an alkene, for example, 28, 42, and 56.

The situation with aromatic amines such as anilines is fairly similar. For example, *N,N*-diethylaniline gives an α-cleavage product ion at *m/z* 134 (loss of CH_3) as well as a secondary fragmentation ion at *m/z* 106, resulting from loss of ethene from the base peak ion (Figure 8-9). As expected for a benzene derivative, there is a peak at *m/z* 77 for the phenyl cation. Pyridine derivatives act much like aromatic hydrocarbons, and their spectra are dominated by cleavages to give stabilized cations. Figure 8-10 shows the spectrum for 2-isopropylpyridine, with the molecular ion at $m/z = 121$. As expected, hydrogen atom loss ($m/z = 120$) and methyl loss (*m/z* 106) dominate and are the result of initially forming a benzylic-type cation. There also is a peak at *m/z* 78 corresponding to a phenyl-like cation (isopropyl loss). The peak at *m/z* 93 involves the net loss of C_2H_4 and is most likely the result of a fairly complex rearrangement process.

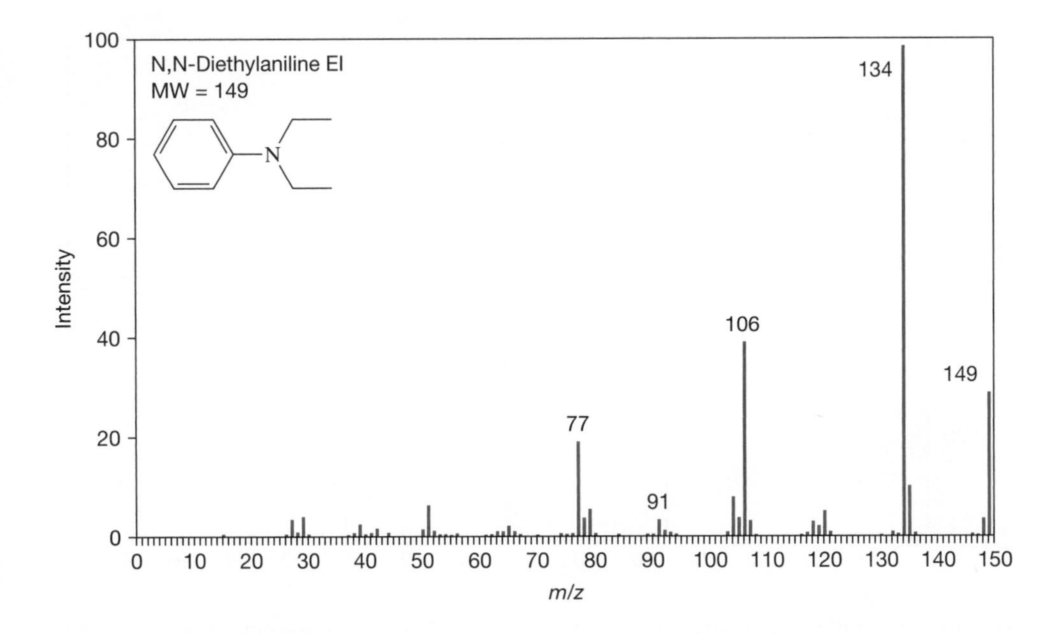

FIGURE 8-9 EI spectrum of *N,N*-diethylaniline. (Data from webbook.nist.gov.)

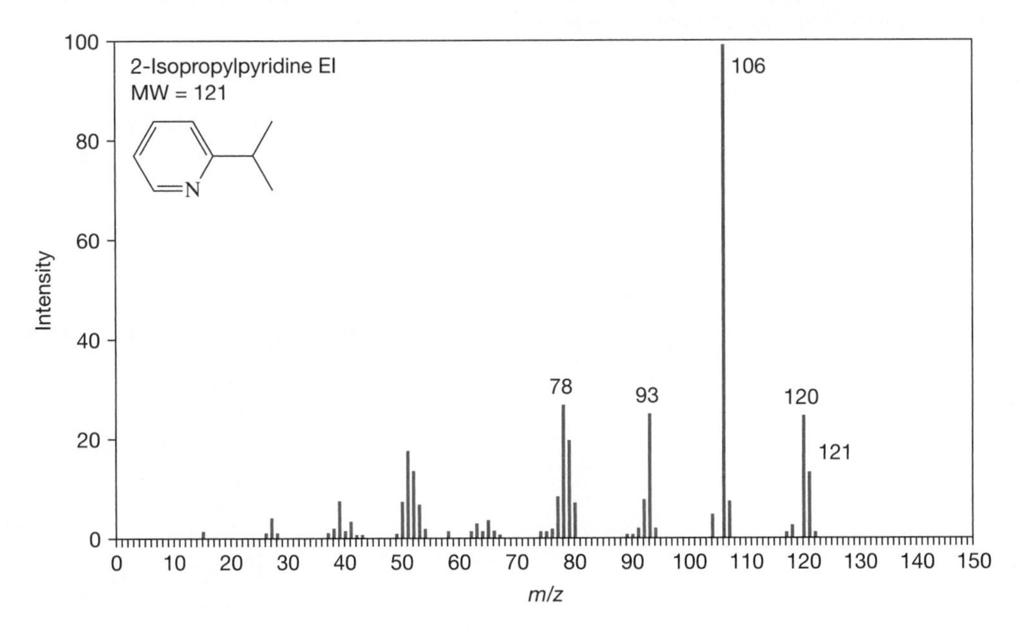

FIGURE 8-10 EI spectrum of 2-isopropylpyridine. (Data from webbook.nist.gov.)

8-5b Protonated Amines

Given the rather high basicity of amines, it is no surprise that they readily form protonated species during processes like CI, ESI, or MALDI. As noted in the previous chapter, ESI and MALDI often produce analyte ions with little or no fragmentation because little energy is deposited during the ionization process. To gain structural information, ions from ESI and MALDI are forced to undergo fragmentation by the subsequent addition of energy by either collisions or photolysis. Methods for postionization activation are discussed at the end of this chapter. Fragmentation of protonated amines is fairly straightforward and is dominated by S_N1-type processes, whereby a C—N bond cleaves to give a carbocation and a neutral amine as the leaving group. As these products are formed, there is the possibility of further reactions within the initial product complex, in particular, proton transfer from the carbocation to the amine, to give a new ammonium ion and an alkene. Because amines generally have higher proton affinities than do alkenes, this process tends to be exothermic; as a result, alkene loss is one of the most common pathways in the fragmentation of protonated amines. As an example, consider the fragmentation of protonated triethylamine (Figure 8-11). The spectrum exhibits the parent ion at *m/z* 102 (one unit greater than the molecular weight) and

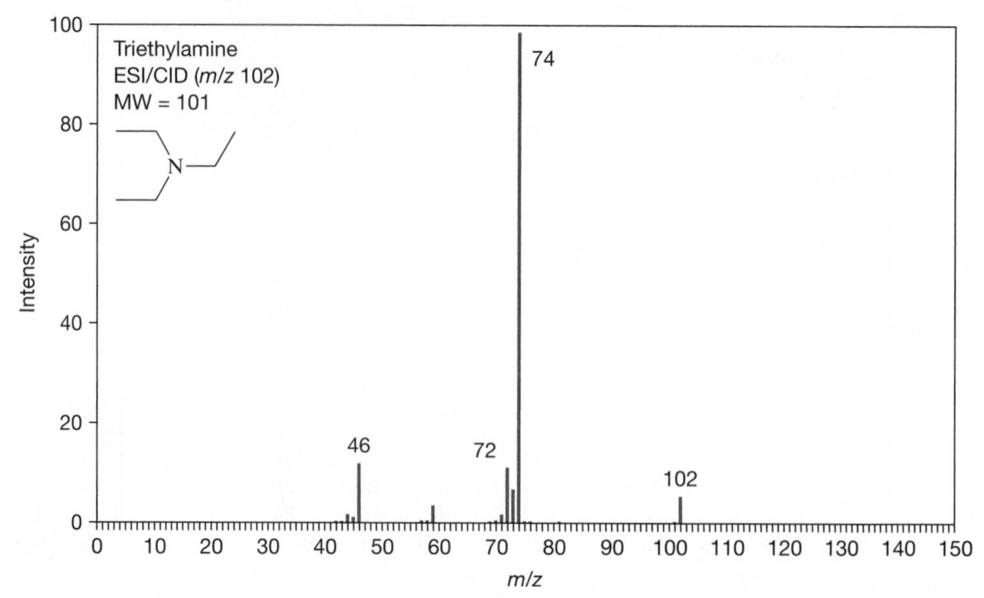

FIGURE 8-11 CID spectrum of protonated triethylamine. The ion was generated by ESI, and *m/z* 102 was selected for CID in a quadrupole ion trap.

fragments at *m/z* 74 (loss of C_2H_4) and *m/z* 46 (loss of two C_2H_4). Mechanisms are shown in eqs. 8-10 and 8-11. This sequence of fragmentations is very common in amines.

$$\underset{m/z\ 102}{(CH_3CH_2)_3\overset{+}{N}{-}H} \longrightarrow CH_2CH_2^{+}\ \cdots\ H{-}\ddot{N}(CH_2CH_3)_2{-}H \longrightarrow CH_2{=}CH_2 + \underset{m/z\ 74}{(CH_3CH_2)_2\overset{+}{N}H{-}H} \tag{8-10}$$

$$\underset{m/z\ 74}{(CH_3CH_2)_2\overset{+}{N}H{-}H} \longrightarrow \underset{m/z\ 46}{CH_3CH_2\overset{+}{N}H(H){-}H} + CH_2{=}CH_2 \tag{8-11}$$

In addition, loss of alkanes is sometimes seen. One reasonable mechanism is hydride rather than proton transfer within the initial carbocation/amine fragmentation complex (eq. 8-12). This process allows for the formation of a nitrogen-stabilized cation (immonium) ion of the same type seen in the fragmentation of the radical cations of amines.

$$CH_3CH_2^{+}\ \ CH_3CH(H)(CH_3CH_2)\ddot{N}{-}H \longrightarrow CH_3CH_3 + \underset{m/z\ 72}{CH_3CH{=}\overset{+}{N}(CH_2CH_3){-}H} \tag{8-12}$$

Finally, if the initial carbocation formed by C—N cleavage is particularly stable, it can escape the complex and lead to the formation of an alkyl cation, such as the ethyl cation from protonated triethylamine, although it is not competitive in this case because of its low stability.

Summary for Amines

Radical Cations

1. Weak molecular ions unless it is an aromatic amine.
2. α-Cleavage to produce immonium ions. Subsequent cleavages generally involve loss of an alkene, via C—C cleavage of another α bond (eq. 8-9).

Protonated Amines

1. Alkene loss via C—N cleavage followed by proton transfer (eq. 8-10). Process potentially can be repeated on each alkyl chain (eq. 8-11).
2. Alkane loss via C—N cleavage followed by hydride transfer to produce immonium ion (eq. 8-12). Subsequent losses are those of an immonium ion, that is, alkene loss by C—C cleavage of the α bond.

8-6 ALCOHOLS, ETHERS, AND PHENOLS

8-6a Radical Cations of Alcohols, Ethers, and Phenols

The fragmentation patterns of alcohols and ethers share many of the same pathways as amines because they offer the possibility of forming heteroatom-stabilized carbocations. As a result, an α-cleavage process analogous to that in eq. 8-8 also is common for these species. The greater electronegativity of oxygen, however, makes the hetero-substituted carbocations somewhat less favorable, and, consequently, formation of unstabilized carbocations becomes more competitive. As an example, compare the EI spectrum of methyl 2-phenylethyl ether (Figure 8-12) with that of *N,N*-dimethyl-2-phenylethylamine (see Figure 8-8). In the ether, there are many significant fragment ions, spanning species with four to eight carbons. The richness of the spectrum is driven by the fact that none of the ions is exceptionally stable. This situation is also indicated by the relatively large peak for the molecular ion at *m/z* 136. The base peak is a benzyl/tropyl ion at *m/z* 91, which is the result of an α-cleavage process in which the charge is retained on the hydrocarbon and the hetero-substituted fragment becomes the radical, $^{\bullet}CH_2OCH_3$. There also is a significant peak at *m/z* 45 for α-cleavage to give an oxygen-stabilized ion, $^{+}CH_2OCH_3$. Typical benzene derivative peaks such as at *m/z* 51, 65, and 77 also are seen in the spectrum. The peak at *m/z* 104 is interesting in that it represents loss of a closed-shell species, CH_3OH, to give the radical cation of styrene as a fragment ion. Ethers are prone to undergo C—O cleavage to produce a carbocation and an alkoxy radical. In this

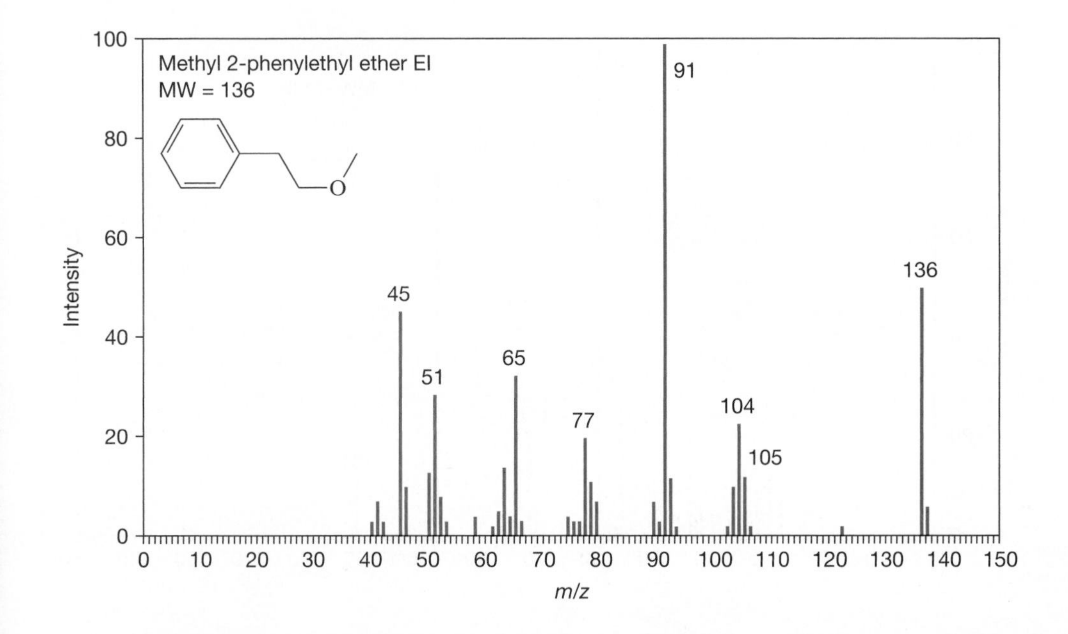

FIGURE 8-12 EI spectrum of methyl 2-phenylethyl ether. (Data from webbook.nist.gov.)

case, hydrogen atom transfer follows to produce methanol (eq. 8-13), presumably because styrene gives a relatively stable radical cation (there is also a signal at m/z 105 for

m/z 136 $\longrightarrow$ $\longrightarrow$ m/z 104 $+$ HO: **(8-13)**

simple C—O cleavage to give the 2-phenylethyl cation initially). The EI spectrum of ethyl *tert*-pentyl ether also illustrates how rearrangements and secondary fragmentations can play a large role with the radical cations of ethers (Figure 8-13). In this case, there is no peak for the molecular ion at m/z 116. Significant signals are seen at m/z 101 and 87 for loss of CH_3 and CH_3CH_2, respectively. Both of these ions are α-cleavage products, most likely on the *tert*-pentyl fragment and result in oxygen-stabilized, 3° carbocations (eq. 8-14 and 8-15). One might ascribe loss of CH_3CH_2 to a C—O cleavage of the ether bond,

m/z 116 $\longrightarrow$ m/z 101 $+ CH_3\cdot$ **(8-14)**

m/z 116 $\longrightarrow$ m/z 87 $+ CH_3CH_2\cdot$ **(8-15)**

but the resulting cation is less stable. There is a peak at m/z 71, however, that corresponds to C—O cleavage with loss of $CH_3CH_2O\cdot$ (45 mass units), to give the very stable *tert*-pentyl cation. The ions at m/z 73 and 59 are probably the result of secondary fragmentations of the α-cleavage products, and both involve loss of $CH_2{=}CH_2$ via a process similar to that seen in amines (eq. 8-16).

m/z 87 $\longrightarrow$ $\longrightarrow$ m/z 59 $+ CH_2{=}CH_2$ **(8-16)**

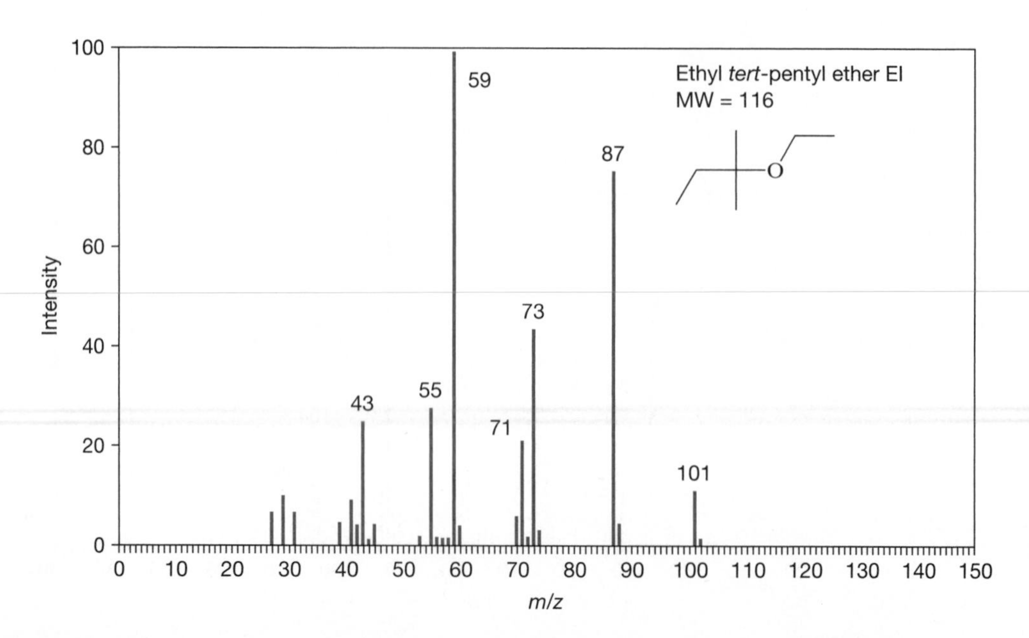

FIGURE 8-13 EI spectrum of ethyl *tert*-pentyl ether. (Data from webbook.nist.gov.)

As one might expect, the fragmentation pathways of alcohols show many of the same features as those of ethers. As an example, the spectrum of 3,3-dimethyl-2-butanol is shown in Figure 8-14. The molecular ion is at *m/z* 102 and gives α-cleavages to produce ions at *m/z* 87 (methyl loss) and *m/z* 45 (*tert*-butyl loss). The *tert*-butyl loss pathway is also responsible for the ions at *m/z* 56 and 57 (eq. 8-17). The latter is simply the

:OH (·+) *m/z* 102 → *m/z* 45 (:OH, + ·) → *m/z* 56 (·+) + :OH ; :OH (·+) *m/z* 102 → *m/z* 57 (+ + ·:OH) **(8-17)**

tert-butyl cation and can be explained by Stevenson's rule because the 1-hydroxyethyl radical has a higher IP than does the *tert*-butyl radical. As a result, the alkyl cation is favored over the oxygen-stabilized cation in this α-cleavage. The ion at *m/z* 56, the radical cation of isobutene, can be rationalized by a hydrogen atom transfer within the initial complex of the α-cleavage products. Finally, the ion at *m/z* 69 is likely the result of dehydration of *m/z* 87 and involves a rearrangement. Loss of H_2O from alcohols is common and often part of a potentially complex rearrangement process. Figure 8-15 shows the spectrum of cyclohexanol. Species in which the functional group is part of a ring or directly attached to a ring present special problems in the interpretation of fragmentation spectra because bond cleavages often result in ring opening rather fragmentation. As a result, observed fragment ions involve at least two cleavage events and sometimes offer little insight into the structure of the analyte. For cyclohexanol, there are three prominent peaks in the EI spectrum. The first, at *m/z* 82, is from loss of H_2O, to give the radical cation of the dehydration product, cyclohexene. This process is common, particularly in cyclic alcohols or acyclic alcohols that contain potential radical-stabilizing groups. Another prominent peak is *m/z* 57, which corresponds to loss of 43 mass units. This mass could be indicative of the loss of C_3H_7 or C_2H_3O. At first glance, neither of these seems to be a likely neutral fragment from cyclohexanol. As is often the case, the best way to approach an unexpected reaction product is to begin by considering likely reaction pathways. For the radical cation of an alcohol, we know that α-cleavage is a very common pathway. In cyclohexanol this process leads to ring opening, to give a radical cation in which the radical and the charge are localized on different atoms

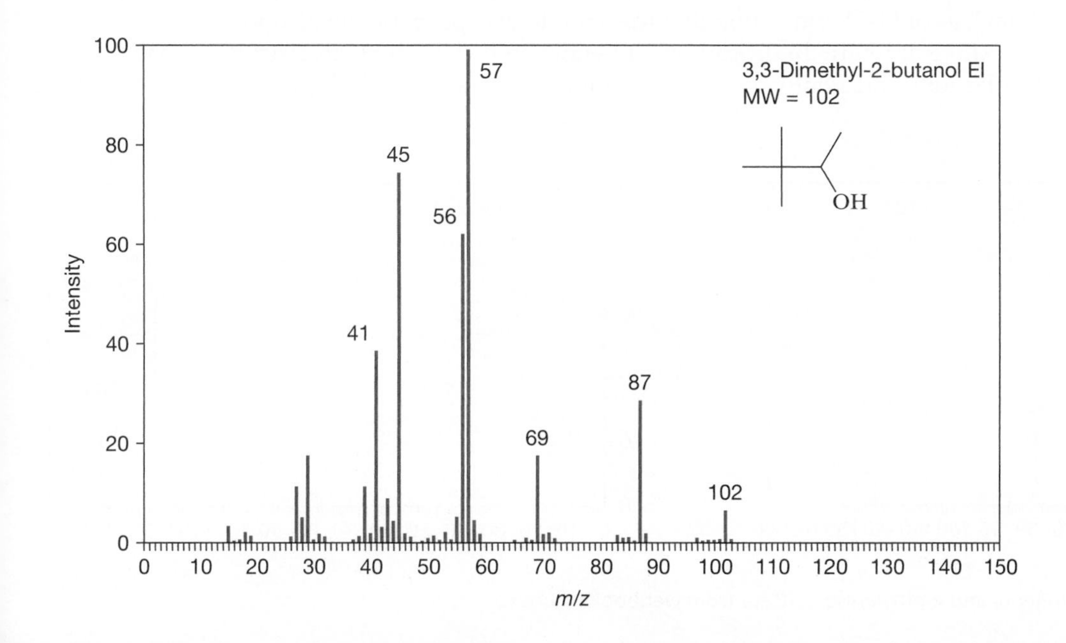

FIGURE 8-14 EI spectrum of 3,3-dimethyl-2-butanol. (Data from webbook.nist.gov.)

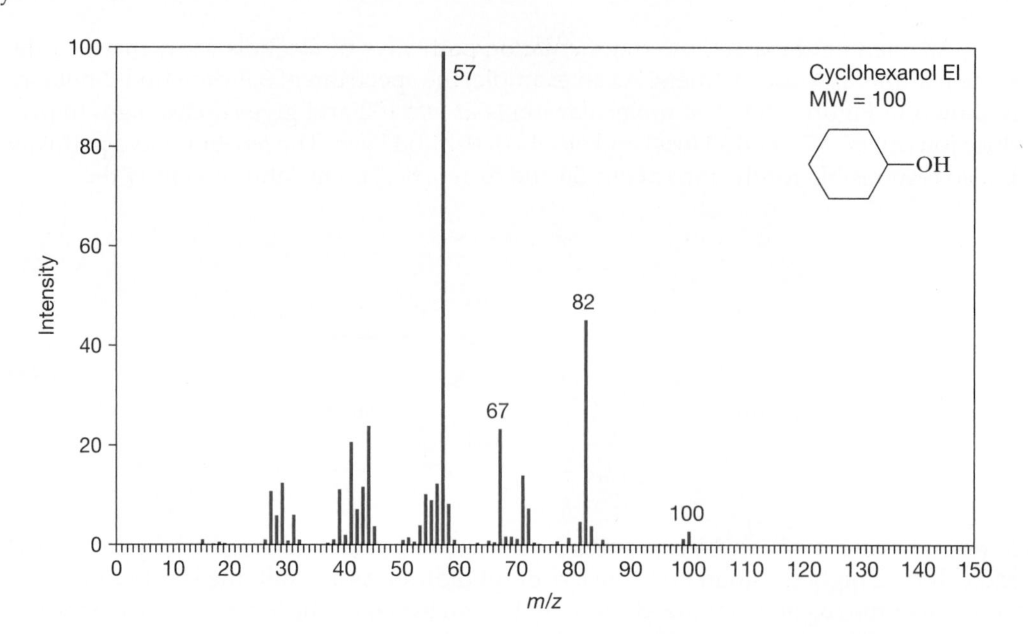

FIGURE 8-15 EI spectrum of cyclohexanol. (Data from webbook.nist.gov.)

(eq. 8-18). Such odd-electron ions are common in mass spectrometry and are referred to as *distonic ions*. Alkyl radicals, in the gas phase or solution, are known to undergo

m/z 100 → ... → ... → $H_2\dot{C}$ + m/z 57 **(8-18)**

intramolecular hydrogen atom transfers when the process leads to a more stable radical. In this case, a delocalized radical can be produced by the hydrogen atom transfer, and the process is favorable. Finally, alkyl radicals are also prone to the formation of an alkene by the expulsion of alkyl radical. The net result is production of m/z 57, protonated acrolein. The third significant peak in the spectrum is at m/z 67, which must result from loss of CH_5O in some form from the parent ion. The most logical assumption is that the fragmentation involves loss of CH_3 and H_2O, but CH_4 and HO are another possibility. In either case, it is difficult to see how these products could be formed without a good deal of rearrangement. Overall, this spectrum illustrates how fragmentation spectra can sometimes offer little structural information. Here, the most useful clue came from loss of H_2O, indicating that the analyte was probably an alcohol.

Like aromatic hydrocarbons, phenols and aryl ethers give strong signals for molecular ions and relatively simple fragmentation spectra. In Figure 8-16, EI spectra are

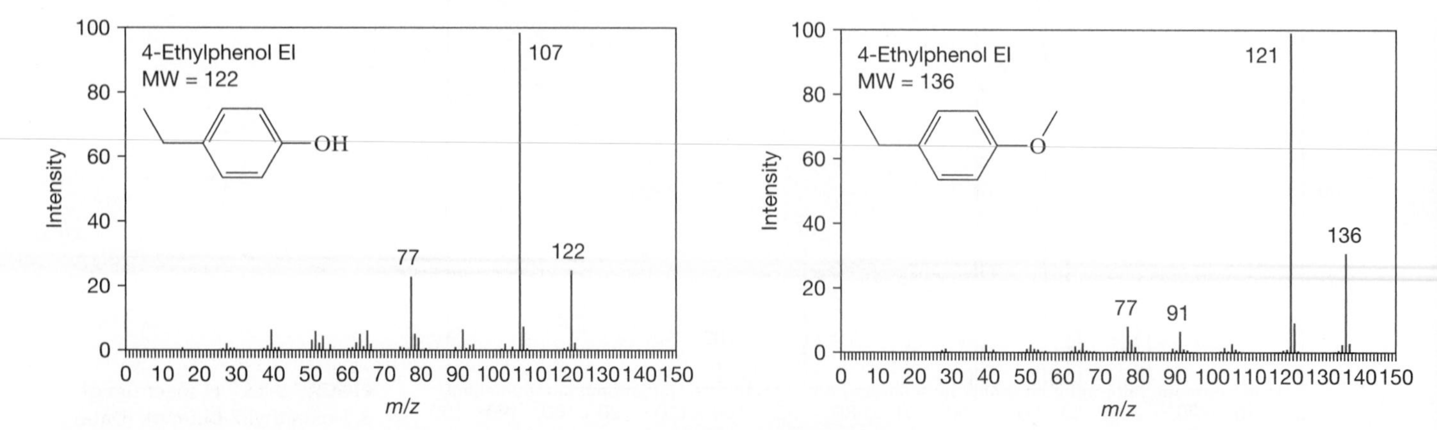

FIGURE 8-16 EI spectra of 4-ethylphenol and 4-ethylanisole. (Data from webbook.nist.gov.)

given for 4-ethylphenol and 4-ethylanisole. In both cases, large molecular ion signals are observed, and loss of methyl is the most important fragmentation pathway. This process initially produces a benzyl cation, which derives significant stabilization from the oxygen group at the para position of the ring. In addition, typical benzene-type fragments are seen in the spectra. When the aryl ether has a more complicated group than methyl, fragmentations of the alkyl group are common. For example, ethyl phenyl ethers give large signals for loss of ethene (eq. 8-19). Similar behavior is seen in higher homologues.

$$C_6H_5\text{–}\dot{O}^{+}\text{–}CH_2CH_3 \longrightarrow C_6H_5\text{–}\dot{O}H^{+} + CH_2{=}CH_2 \qquad \textbf{(8-19)}$$

8-6b Protonated Alcohols and Ethers

Because of their lower basicity, it is more difficult to form protonated alcohols and ethers by ESI, but they are readily formed in MALDI and CI processes. As one might expect from analogies to condensed-phase chemistry, loss of H_2O or ROH is a common fragmentation pathway when it leads to a reasonably stable carbocation. As with amines, proton transfer can occur after C—O cleavage of a protonated alcohol or ether. This process leads to loss of a neutral alkene and the formation of H_3O^+ or ROH_2^+ rather than a carbocation (eq. 8-20). Finally, ethers also are capable of undergoing a hydride transfer in the alcohol/alkyl cation complex, resulting in a protonated aldehyde or ketone and an alkane (eq. 8-21). As an example, consider the chemical ionization spectrum

$$CH_3CH_2CH_2\text{–}O^{+}(H)\text{–}R \nearrow CH_3CH_2\overset{+}{C}H_2 + ROH \longrightarrow CH_3CH{=}CH_2 + RO\overset{+}{H}_2 \qquad \textbf{(8-20)}$$

$$\searrow CH_3CH_2CH_2\text{–}\ddot{O}\text{–}H + R^{+} \longrightarrow CH_3CH_2CH{=}\overset{+}{O}\text{–}H + RH \qquad \textbf{(8-21)}$$

of 1-decanol (Figure 8-17). Because it was generated by chemical ionization, it actually contains the fragment ions of two species. The protonated species would appear at *m/z* 159, but it readily fragments by loss of H_2O to give *m/z* 141 as the base peak. This alkyl cation, 1-decyl, undergoes typical fragmentations to give a series of smaller alkyl cations at *m/z* 99, 85, 71, and 57. The other species initially formed by CI is a hydride abstraction product at *m/z* 157, which can undergo further fragmentation with loss of H_2O,

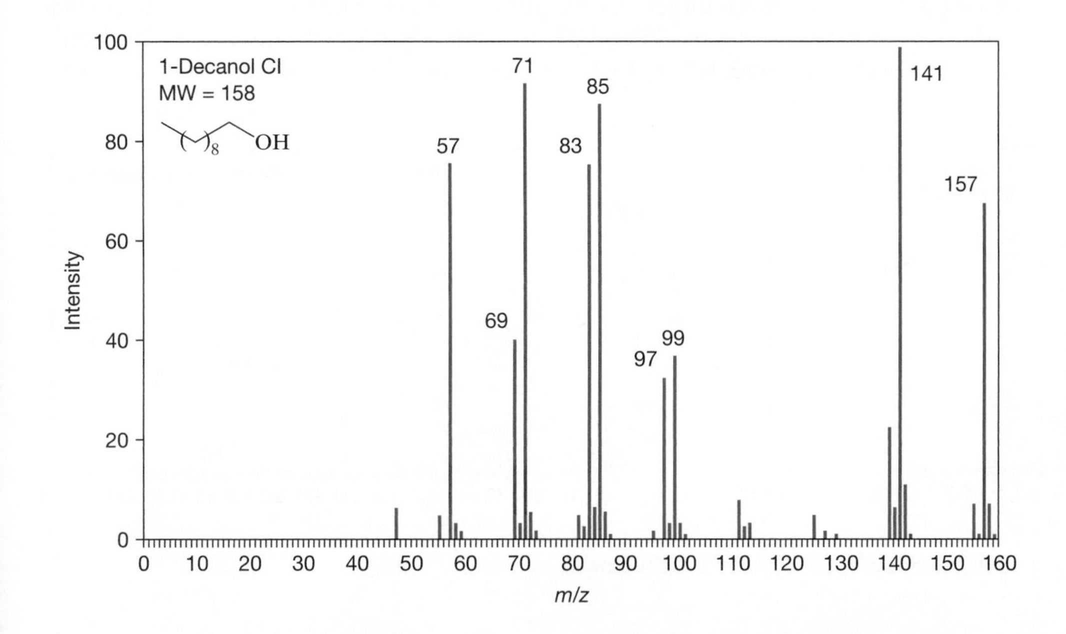

FIGURE 8-17 Chemical ionization mass spectrum of 1-decanol. (From M. S. B. Munson and F. H. Field, *J. Am. Chem. Soc.*, **88**, 2621–2630 [1966].)

followed by additional fragmentation to produce a series of unsaturated cations at *m/z* 97, 83, and 69. The presence of multiple parent ions in fragmentation spectra is a common complication of CI, but can be dealt with in instruments capable of isolating sets of ions before fragmentation (see Section 8-11). Comparison of this spectrum with the EI spectrum of 1-decanol (Figure 7-2 in Chapter 7) provides a dramatic example of how greatly EI and CI spectra can differ for a given compound.

Summary for Alcohols and Ethers

Radical Cations

1. Weak molecular ions unless it is an aromatic ether or phenol.
2. α-Cleavage to give oxygen-stabilized carbocations (oxonium ions). In ethers, subsequent C—O cleavage to give a carbonyl/alkyl cation pair leads to alkene loss via a proton transfer (eq. 8-16).
3. Because the oxonium ions are not exceptionally stable, other processes can occur as part of α-cleavage. If the other component in the α-cleavage gives an especially stable carbocation, for example, benzyl/tropyl, the charge can be retained on this group with loss of a neutral α-alkoxy radical. Hydrogen atom transfer can occur after α-cleavage to give the radical cation of an alkene and a neutral alcohol or ether (eq. 8-17).
4. C—O cleavage with loss of an alkoxy radical to give an odd-numbered cation or loss of ROH (or H_2O) to give an even-numbered, radical cation (eq. 8-13).

Protonated Alcohols and Ethers

1. Cleavage of the C—O bond to give an alkyl cation and an alcohol or H_2O.
2. After C—O cleavage, proton transfer to give a protonated alcohol with loss of a neutral alkene (eq. 8-20).
4. After C—O cleavage, hydride transfer to give a protonated aldehyde or ketone with loss of a neutral alkane (eq. 8-21).

8-7 ALKYL AND ARYL HALIDES

Halogens are not nearly as effective as oxygen or nitrogen in stabilizing an adjacent carbocation (see Table 8-1), so we anticipate that α-cleavage to give a halo-substituted ion plays a less significant role in the radical cations of alkyl and aryl halides. Instead, typical losses are the halogen atom to give a carbocation, hydrohalogen loss to give the radical cation of an alkene, and α-cleavage to give a carbocation (if it is relatively stable) along with a halo-stabilized radical. The spectrum for 1-chloro-3-methylbutane offers a good example (Figure 8-18). There is no molecular ion, and chlorine is absent in all significant peaks. The nominal molecular weight is listed as 106/108 in the figure because chlorine presents two isotopes, 35 and 37, in relatively high abundance, so molecular ions are expected at two *m/z* values. The impact of isotopes on mass spectra is

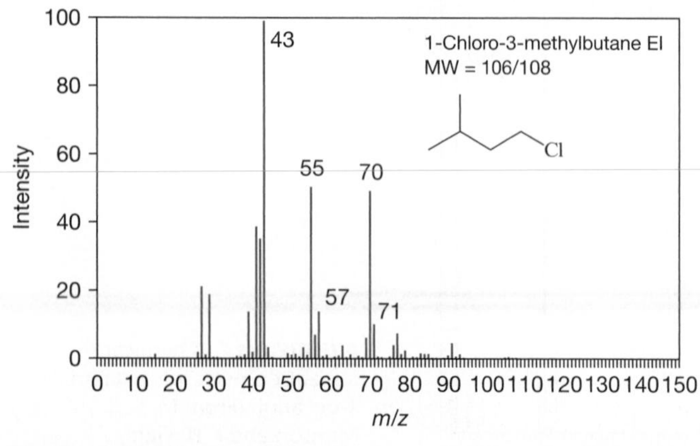

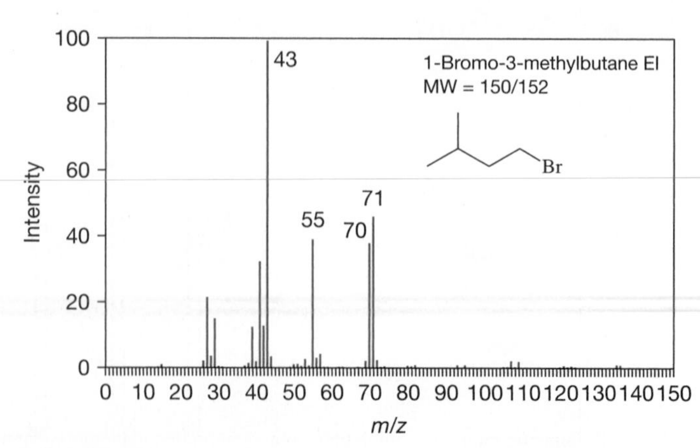

FIGURE 8-18 EI spectra of 1-chloro-3-methylbutane and 1-bromo-3-methylbutane. (Data from webbook.nist.gov.)

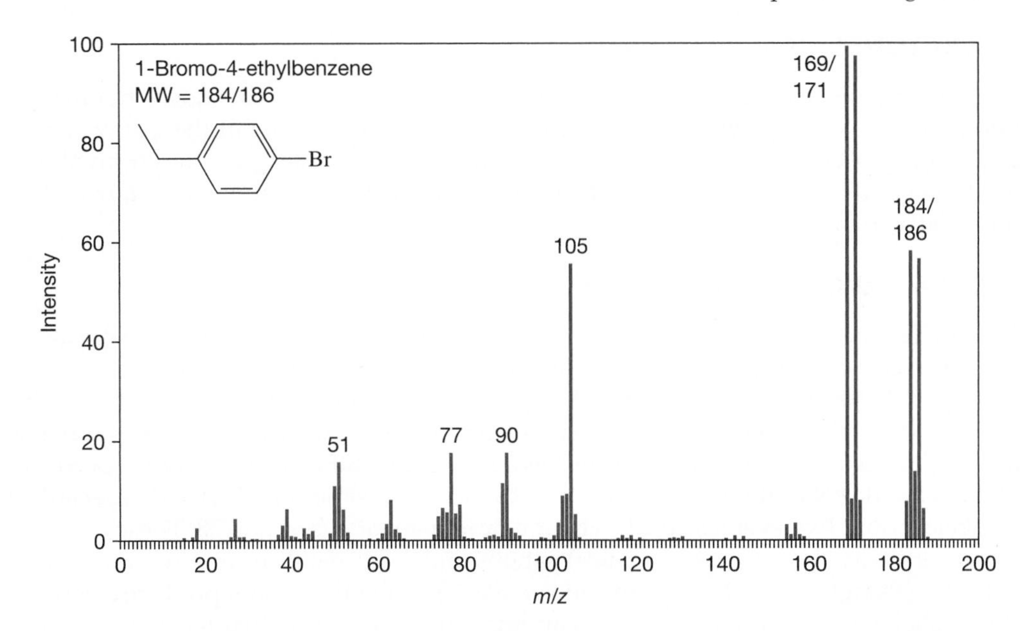

FIGURE 8-19 EI spectrum of 1-bromo-4-ethylbenzene. (Data from webbook.nist.gov.)

discussed in detail in the next chapter. Direct loss of Cl is seen at *m/z* 71, but a more intense peak is seen for loss of HCl to give a radical cation of C_5H_{10}, *m/z* 70. The peak at *m/z* 55 is likely a secondary fragmentation of the $C_5H_{10}^{+}$ by loss of methyl to give an allylic cation. The base peak is the propyl cation, which could result from loss of Cl followed by loss of $CH_2{=}CH_2$ from the molecular ion. These processes are outlined in eqs. 8-22 and 8-23.

m/z 106/108 → CH_2^+ (*m/z* 71) + :Cl· → $H_3C{-}CH^+{-}CH_3$ (*m/z* 43) + $CH_2{=}CH_2$ **(8-22)**

m/z 106/108 → alkene radical cation (*m/z* 70) + HCl → allylic cation (*m/z* 55) + $CH_3\cdot$ **(8-23)**

Figure 8-18 also shows the spectrum for the analogous bromide. Bromine, like chlorine, has two common isotopes, so the molecular ion would appear at both *m/z* 150 and 152. The spectra are quite similar, but one can see that compared with the chlorine system, bromides are more likely to lose X radicals (to give *m/z* 71) rather than HX (to give *m/z* 70). The spectrum for 1-bromo-4-ethylbenzene provides an example of an aryl halide (Figure 8-19). As is typical of aromatic species, the molecular ion peaks are strong. The base peak comes from a fragmentation unrelated to the bromine atom and corresponds to methyl loss, to give a benzyl/tropyl cation at *m/z* 169/171. A key and expected difference between the spectra in Figures 8-17 and 8-18 is that the aryl halide is much less likely to undergo C—X cleavage and X is retained in a major fragment ion. Nonetheless, C—Br cleavage is observed and leads to *m/z* 105.

Summary for Alkyl and Aryl Halides

Radical Cations

1. Weak molecular ions unless it is an aromatic species.
2. Cleavage of the C—X bond to give an alkyl cation with loss of a halogen atom (eq. 8-22).
3. Loss of H—X to give the radical cation of an alkene (eq. 8-23).

8-8 KETONES AND ALDEHYDES

The presence of a carbonyl group in an analyte offers some very diagnostic fragmentation pathways. First, acyl cations [$R(CO)^+$] are reasonably stable in the gas phase, so direct fragmentation at the carbonyl is common. Second, a fragmentation involving multiple bonding changes, the *McLafferty rearrangement*, is a highly characteristic feature of all carbonyl species.

8-8a Radical Cations of Ketones

The key features of the EI spectra of ketones are (1) α $C—C$ cleavages at the carbonyl, to give an acyl cation and an alkyl radical, or (2) distonic radical cations via a single or a double McLafferty rearrangement. The spectrum for 4-octanone provides a very instructive example of the fragmentation pathways of ketone radical cations. As is often is the case for ketones, the molecular ion (*m/z* 128) is a significant peak in the spectrum. There is a series of odd *m/z* peaks that correspond to the expected alkyl and acyl cations (see below), but two even *m/z* peaks also are prominent at *m/z* 86 and 58. These are noteworthy because they represent radical cations and must result from a rearrangement pathway (a single bond cleavage in a molecular ion generally cannot produce a neutral, closed-shell species). Their presence is an immediate indication that the analyte might be a carbonyl species. We first consider the expected peaks from simple cleavage at the carbonyl carbon (eqs. 8-24 and 8-25). Because acyl radicals almost always have

m/z 128 — α-cleavage → $\dot{C}H_2$ (propyl radical) + acyl cation (*m/z* 71) → CH_2^+ (*m/z* 43) **(8-24)**

m/z 128 — α-cleavage → $\dot{C}H_2$ (ethyl... radical) + acyl cation (*m/z* 85) → $^+CH_2$ (*m/z* 57) **(8-25)**

lower ionization energies than alkyl radicals, the charge is generally retained on the acyl group by Stevenson's rule. Cleavage along the $C_3—C_4$ bond leads to an acyl cation at *m/z* 85, which can further fragment by loss of CO to give the butyl cation at *m/z* 57. Alternatively, cleavage along the $C_4—C_5$ bond leads to an acyl cation at *m/z* 71, which can further fragment by loss of CO to give the propyl cation at *m/z* 43. Together, these pathways account for the major odd *m/z* species in the spectrum. To account for *m/z* 86 and 58, McLafferty rearrangements are needed. The ion at *m/z* 86 results from loss of 42 mass units, which could be interpreted as loss of either C_3H_6 or $CH_2C{=}O$. The former seems more likely because 3-carbon chains are present in the molecule, whereas the ketene unit would have to be excised from the center of octanone. Although it might be tempting to relate this product to the pathway in eq. 8-25 (possibly by a hydrogen atom transfer from the propyl radical), it actually results from a fragmentation involving the butyl substituent on the ketone. The ionized ketone carbonyl is a good hydrogen atom abstractor and can very readily remove a hydrogen from the γ-carbon of the alkyl chain via a six-membered ring transition state. This process results in a distonic radical cation, in which the charge is localized on the carbonyl group and the unpaired electron on the γ-carbon. The alkyl radical loses an alkene (propene in this case) to give the delocalized radical cation of an enol (eq 8-26). In early work, it was assumed that the McLafferty

m/z 128 → *m/z* 128 → [enol radical cation ↔ resonance form] (*m/z* 86) + $H_2C{=}CH—CH_3$ **(8-26)**

rearrangement was a concerted process and that the first two steps occurred simultaneously to give the ionized enol directly, without the intermediate distonic ion. More recently, evidence seems to suggest that it is stepwise. Timing of the events is not a key issue in terms of spectral interpretation, however, because either mechanism, stepwise or concerted, gives the same product. The gap between the ions at *m/z* 86 and 58 is 28 mass units, which could correspond to either $CH_2{=}CH_2$ or CO. Again, the structure of the *m/z* 86 ion in eq. 8-26 suggests that alkene loss is more likely. In fact, the mechanism is identical to that for the first McLafferty rearrangement. It is most easily presented starting with the enol resonance form of the radical cation, which can undergo a hydrogen atom transfer from the γ-carbon to the enol oxygen, resulting in a distonic radical cation, which subsequently expels $CH_2{=}CH_2$ to form yet another distonic ion (eq. 8-27).

m/z 86 → *m/z* 86 → $H_2C{=}CH_2$ + *m/z* 58 **(8-27)**

The product ion is stabilized by delocalization in the allyl radical component of the ion. In 4-octanone (Figure 8-20), both alkyl chains contain a γ-hydrogen, but there is almost no evidence of an initial McLafferty rearrangement on the propyl chain, which would lead to the loss of $CH_2{=}CH_2$, to give an ion at *m/z* 100 (eq. 8-29). The key difference can be seen in a comparison of eqs. 8-28 and 8-29. In the former, a 2° radical is formed in

m/z 128 —Fast→ 2° radical → *m/z* 86 + $H_2C{=}C(H)CH_3$ **(8-28)**

m/z 128 —Slow→ 1° radical → $H_2C{=}CH_2$ + *m/z* 100 **(8-29)**

the first step, whereas production of the ion at *m/z* 100 requires the initial formation of a less stable 1° radical. The difference in rates is great enough that only the pathway via the 2° radical is observed to a significant extent. The McLafferty rearrangement can pass through a 1° radical (eq. 8-27), but it is less favorable and rarely seen as the *initial* fragmentation step of a ketone (eq. 8-27 is a secondary fragmentation).

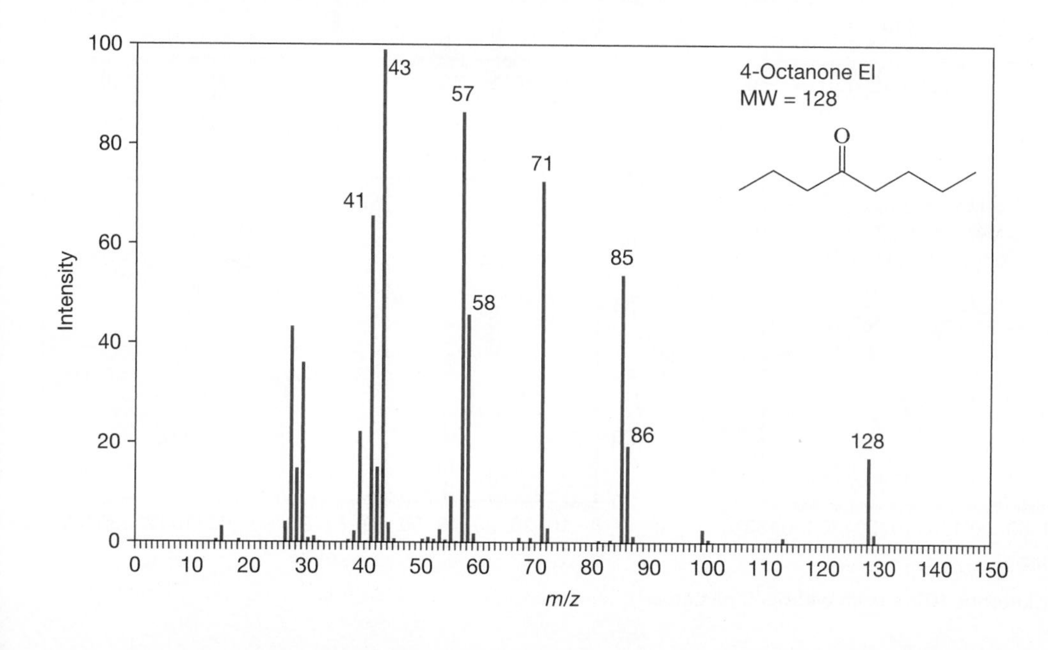

FIGURE 8-20 EI spectrum of 4-octanone. (Data from webbook.nist.gov.)

In Figure 8-21, spectra are presented for a series of ketones with the formula $C_7H_{14}O$, illustrating how the combination of α-cleavages and McLafferty rearrangements can provide very useful clues in determining chemical structures. With 2-heptanone, a characteristic McLafferty ion at *m/z* 58 indicates that it is a methyl ketone with no substitution at the α-carbon of the longer alkyl chain. Otherwise, a higher homologue of the McLafferty ion would be expected. The complementary α-cleavage ions at *m/z* 43 (acetyl cation) and 71 (pentyl cation) confirm this conclusion. These observations, however, are not enough to assign a structure definitively. For example, the same pattern is expected for 5-methyl-2-hexanone, because it is also a methyl ketone without an α substituent and would give ions of the same *m/z* as for α-cleavages (the difference being that an isomeric pentyl cation is produced). As expected, the spectra for 2-heptanone and 5-methyl-2-hexanone are very similar and differ in subtle ways that could not be used to easily distinguish them. In contrast, 3-methyl-2-hexanone is quite different, because it bears a substituent at the α-carbon of the ketone, and, as a result, the McLafferty ion is shifted up in mass to *m/z* 72. One still sees the expected complementary α-cleavage ions at *m/z* 43 and 71, which indicate a methyl ketone, that is, formation of the acetyl cation. For comparison, the spectrum of 3-heptanone also is shown in Figure 8-21. The McLafferty ion also occurs at *m/z* 72, which is consistent with an ethyl ketone or a ketone with an α-methyl group. The key differences are observed in the α-cleavages. In this case, three strong signals are seen for them. Cleavage between C_2 and C_3 yields *m/z* 85 ($C_4H_9CO^+$), which can lose CO to give *m/z* 57. The other α-cleavage, between C_3 and C_4, yields *m/z* 57 ($C_2H_5CO^+$), which can lose CO to give *m/z* 29. In this case, *m/z* 57 can be formed in two ways by α-cleavages, but the ions differ in formula and could be distinguished by high resolution spectra, as discussed in Chapter 9. In any case, α-cleavages provide a clear means of distinguishing between 3-methyl-2-hexanone and 3-heptanone, despite the similarity in the McLafferty ions. It should be noted that the methyl ketones gave significant signals only for α-cleavage along the $C_2—C_3$ bond.

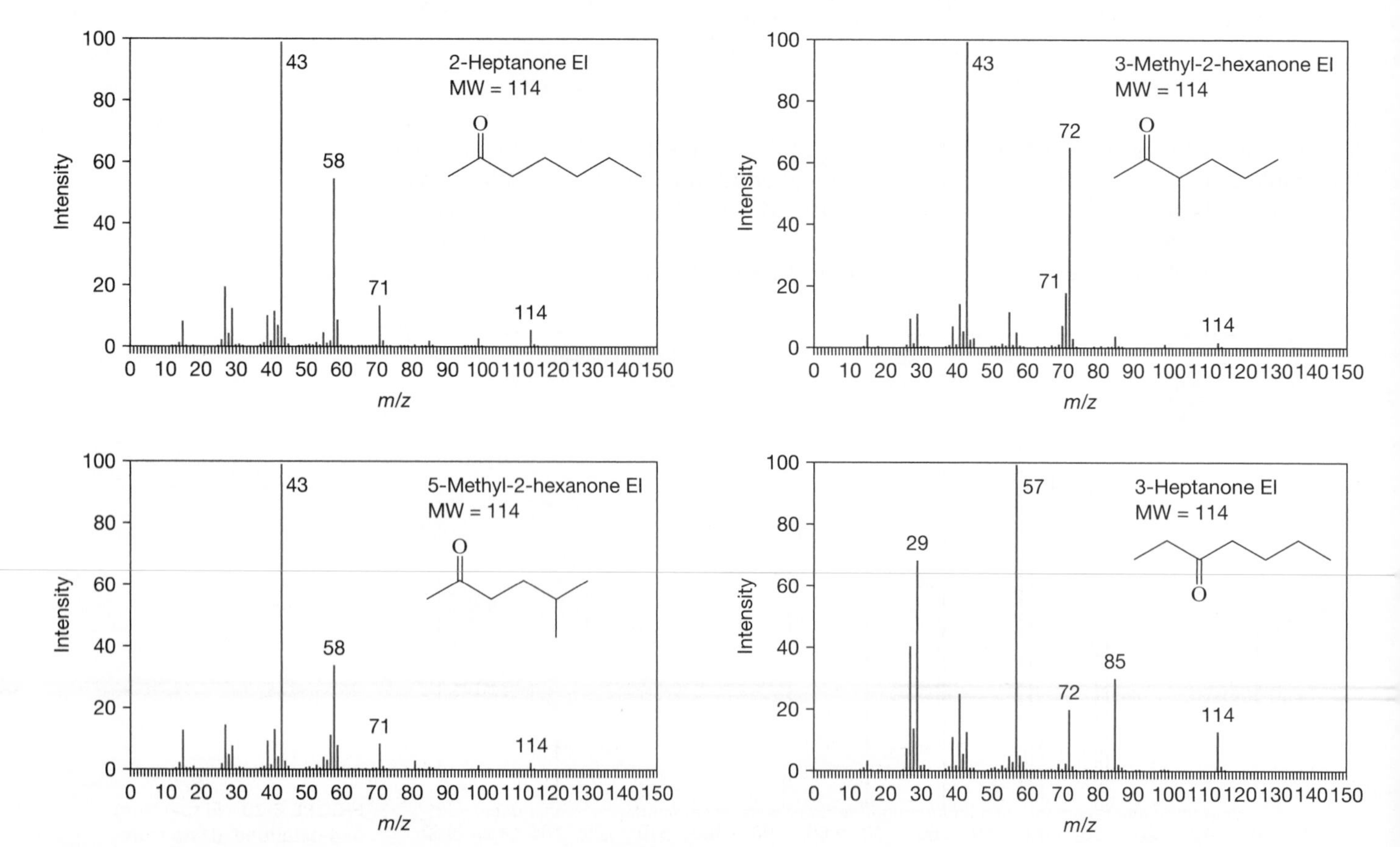

FIGURE 8-21 EI spectra of isomeric ketones. (Data from webbook.nist.gov.)

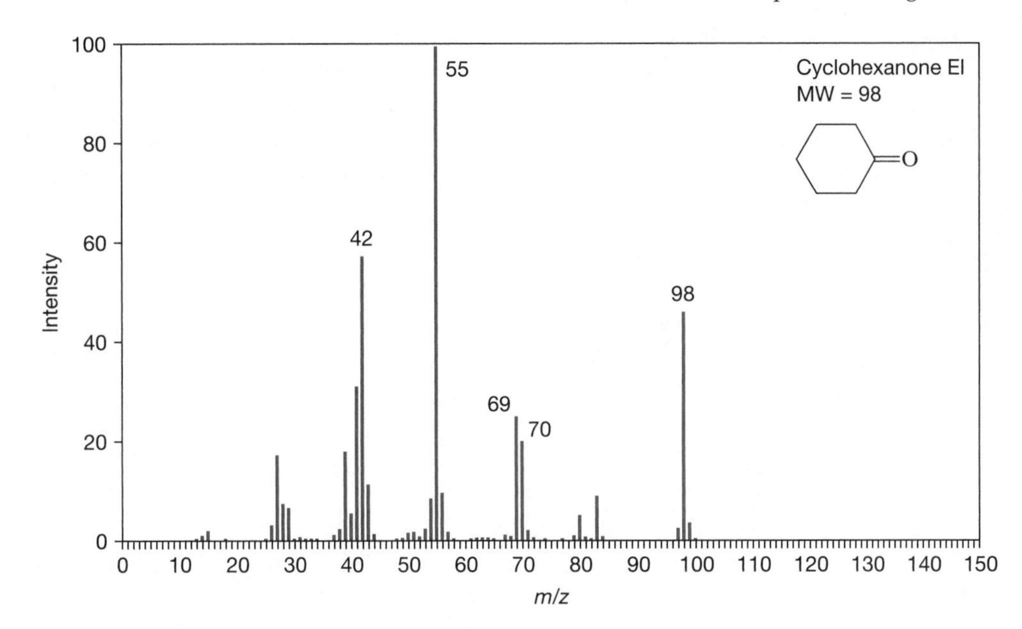

FIGURE 8-22 EI spectrum of cyclohexanone. (Data from webbook.nist.gov.)

This pathway is driven by the fact that cleavage along the $C_1—C_2$ bond would be energetically less favorable because it requires the formation of a methyl radical rather than a 1° radical as a fragment.

In cyclic ketones, such as cyclohexanone, geometric constraints prevent McLafferty rearrangements, and α-cleavages lead to ring opening, and hence not to the formation of fragments. As a result, the EI spectra are based on secondary fragmentations of the parent ion. In Figure 8-22, there is a strong molecular ion peak, which is common for cyclic ketones, and the base peak is *m/z* 55. The most likely path to this ion is α-cleavage, followed by an intramolecular hydrogen atom transfer, to give the radical cation of a ketene derivative, which undergoes a final C–C cleavage to produce an unsaturated acyl cation (eq. 30). This pathway is general, and the *m/z* 55 ion also is seen in the spectra of cyclopentanone and cycloheptanone. Higher homologues of the ion are seen in the spectra of cyclic ketones with alkyl substituents at positions 2 or 3 on the ring. In addition, the spectrum shows strong signals for a pair of radical cation species at *m/z* 70 and 42. They are also easily explained by the initial α-cleavage to give the distonic ion, which alternatively can shed sequential $CH_2{=}CH_2$ units from the alkyl radical chain (eq. 8-31). This process is common for distonic radicals of this general type.

$$\text{cyclohexanone}^{+\cdot}\ (m/z\ 98) \rightarrow \dot{C}H_2\cdots H\overset{+}{C}{=}\ddot{O} \rightarrow \dot{C}H_3\ \overset{+\cdot}{O}\!=\!C\cdots H \rightarrow H_2\dot{C}\!-\!CH_3 + H_2C{=}CH\!-\!\overset{+}{C}{=}\ddot{O}\ (m/z\ 55) \qquad \textbf{(8-30)}$$

$$-\ CH_2{=}CH_2:\quad H_2\dot{C}\!-\!CH_2CH_2\!-\!\overset{+}{C}{=}\ddot{O}\ (m/z\ 70) \xrightarrow{-\ CH_2=CH_2} H_2C{=}C{=}\overset{+\cdot}{\ddot{O}}\ (m/z\ 42) \qquad \textbf{(8-31)}$$

8-8b Radical Cations of Aldehydes

Aldehydes undergo many of the same fragmentations as ketones, although they are limited to a single McLafferty rearrangement. As an example, consider the EI spectrum of octanal, which is shown in Figure 8-23 (for its ketone isomer 4-octanone; see Figure 8-20). At first glance, it is clear that octanal gives a much weaker molecular ion peak and a much more diverse set of fragments. The most noteworthy aspect is the large number of

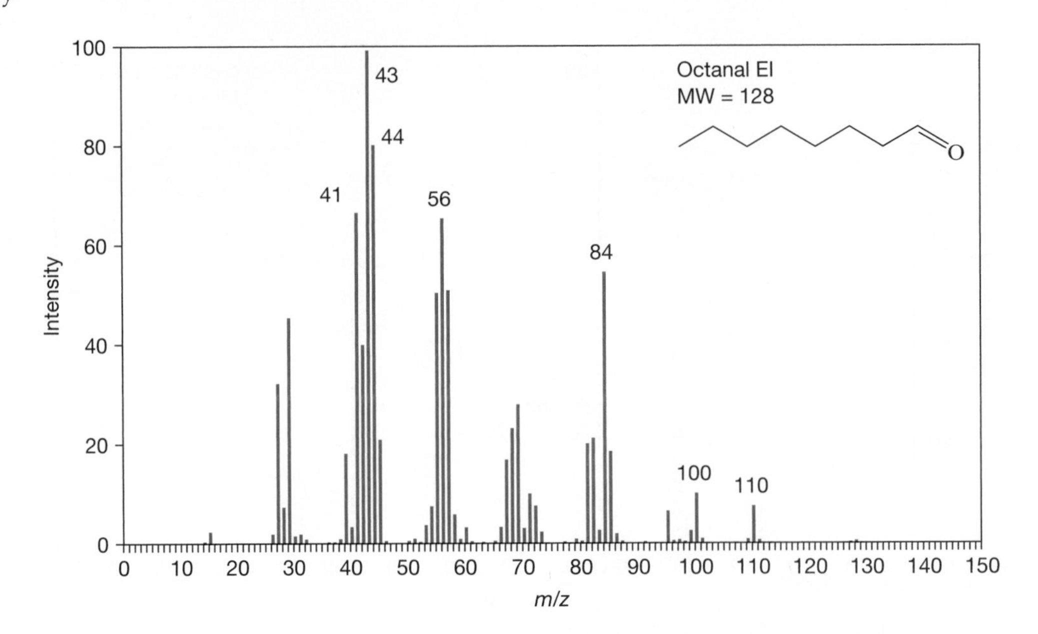

FIGURE 8-23 EI spectrum of octanal. (Data from webbook.nist.gov.)

radical cations (even-numbered *m/z*) in the spectrum, indicating that rearrangements are a common pathway. Loss of H_2O or CO is typical of aldehydes and gives rise to peaks at *m/z* 110 and 100, respectively. There is also a prominent peak at *m/z* 44, which is the result of a McLafferty rearrangement. This ion is particularly characteristic and is seen for all aldehydes that lack substituents at the α-carbon. For substituted aldehydes, the mass of the McLafferty peak is increased by the mass increment of the α-substituent. The ions at *m/z* 84 and 56 are also related to the McLafferty rearrangement. Unlike the enols produced from the McLafferty rearrangement of ketones, those from aldehydes have relatively high ionization energies, which can be greater than those of the complementary alkene fragments. By Stevenson's rule, one then would expect to see the radical cations of the alkene fragments in the spectrum (eq. 8-32). In this case, the ionization

$$m/z\ 128 \longrightarrow \left[\text{CH}_2 + \text{Ö}^{\cdot+}\text{—H}\ (m/z\ 44)\right] \text{ or } \left[\text{CH}_2^{\cdot+}\ (m/z\ 84) + \text{Ö—H}\right] \tag{8-32}$$

energies of 1-hexene and ethenol are similar, and consequently both radical cations are seen (*m/z* 84 and 44). Like other ionized alkenes, 1-hexene can lose $CH_2{=}CH_2$, to give *m/z* 56.

Summary for Aldehydes and Ketones

1. McLafferty rearrangements, single or double, to give distonic ions with characteristic masses of 44, 58, 72, and 86 for alkyl systems (eq. 8-26, 8-27).
2. α-Cleavage to give acyl cations with characteristic masses of 43, 57, 71, and 85 for alkyl systems. These ions often undergo secondary fragmentations, with loss of CO to give the corresponding alkyl cations (eq. 8-24).
3. Aldehydes give McLafferty ions at *m/z* 44 unless they have a substituent at the α-carbon (eq. 8-32)
4. In aldehydes, α-cleavage can also produce alkyl cations directly with the loss of an acyl radical.

8-9 CARBOXYLIC ACIDS AND DERIVATIVES

8-9a Carboxylic Acids

The most characteristic fragmentation pathway of alkyl carboxylic acids is a McLafferty rearrangement, which produces an ion at *m/z* 60 if the carboxylic acid lacks substituents at the α-carbon (eq. 8-33).

$$R\text{–}CH_2\text{–}C(=\overset{+\cdot}{O}:)\text{–}OH \longrightarrow CH_2=C(\text{–}\overset{+\cdot}{O}:\text{–}H)\text{–}OH \quad (m/z\ 60) \qquad \textbf{(8-33)}$$

The other major pathways are loss of alkyl radicals, the hydroxyl radical, or the elements of the carboxyl group (HCO_2). The last two are most common in aryl carboxylic acids. The spectra for two examples are shown in Figure 8-24. In octanoic acid, the base peak is *m/z* 60, the characteristic McLafferty rearrangement product of carboxylic acids. The complementary alkene radical cation (*m/z* 84) also is observed, an indication that the IP of 1-hexene is similar to that of the enol of acetic acid (the peak at *m/z* 85 is probably related by hydrogen atom transfer). In addition, a small molecular ion (*m/z* 144) is observed, as well as a series of peaks for the loss of alkyl groups: ethyl (*m/z* 115), propyl (*m/z* 101), butyl (*m/z* 87), and pentyl (*m/z* 73). Each retains the carboxylic acid group. In the aryl carboxylic acid, there is a more significant molecular ion at *m/z* 150, and the base peak is the loss of the components of CO_2H, to give *m/z* 105. The other losses in the spectrum are typical of aryl species, such as methyl loss to give a benzyl/tropyl cation (*m/z* 135).

In negative ion mode, carboxylates typically fragment by decarboxylations to give carbanions, which can undergo subsequent fragmentations. Simple carbanions, however, have very low (sometimes negative) electron binding energies. If so, an electron can be lost during the fragmentation process, yielding only neutral fragmentation products. In this case, processes like CID diminish the parent ion intensity, but no fragments are observed. This neutralization behavior is unique to negative ions and is relatively common because many negatively charged functional groups have low electron binding energies. This behavior is one of the reasons that negative ions are less often used in analytical mass spectrometry.

8-9b Esters

Esters can give rich spectra with a good deal of information about the carboxylic acid and alkoxy components of the ester. Like other carbonyl species, the McLafferty rearrangement provides characteristic peaks, either from the alkyl chain of the carboxylic acid or from the alkoxy component. Double McLaffertys are also a possibility. Like carboxylic acids, the typical McLafferty ion masses are *m/z* 60, 74, 88, 102, and 116. One

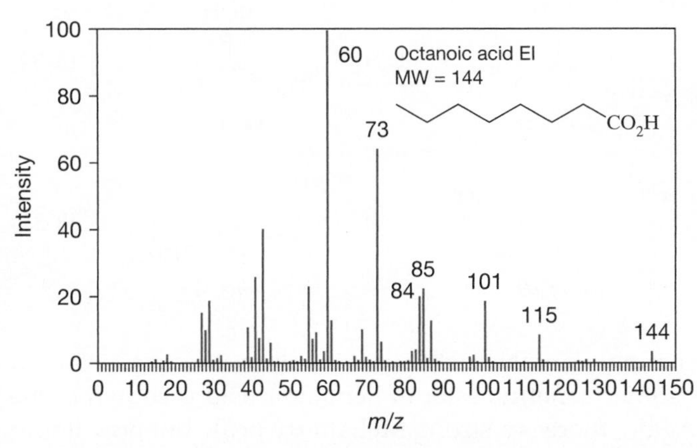

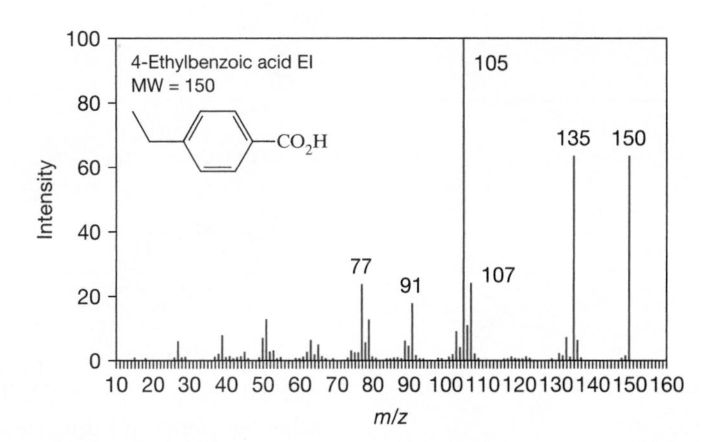

FIGURE 8-24 EI spectra of octanoic and 4-ethylbenzoic acid. (Data from webbook.nist.gov.)

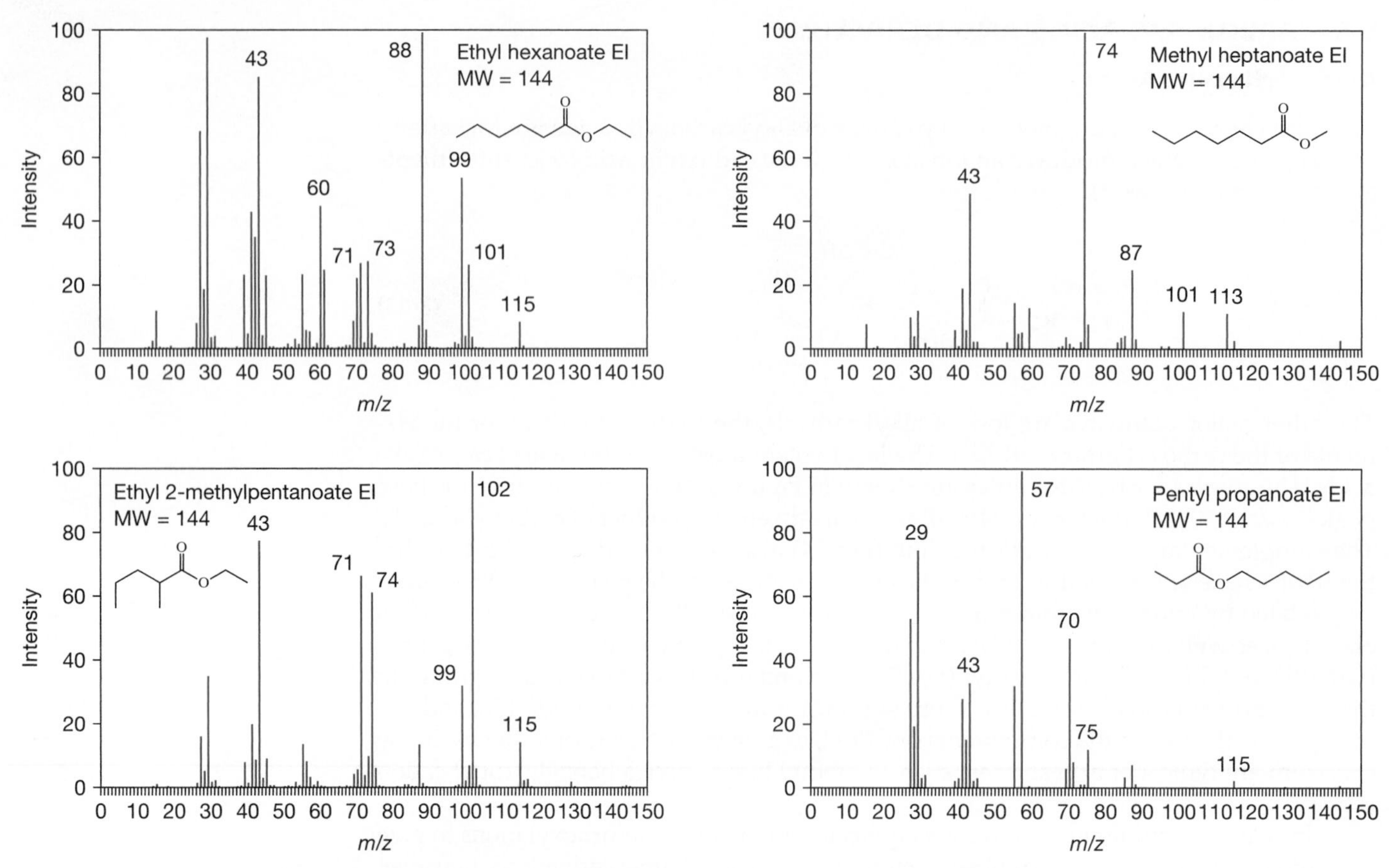

FIGURE 8-25 EI spectra of isomeric esters. (Data from webbook.nist.gov.)

also sees peaks that appear to be α-cleavages on the carboxylic acid as well as loss of the alkoxy group to give an acyl cation. Finally, γ-cleavages are seen on the carboxylic acid component. In Figure 8-25, spectra for a series of isomeric esters illustrate many of the fundamental fragmentations of esters. In the top left panel, ethyl hexanoate gives a complex spectrum with peaks clustered for species with various numbers of carbons. There are strong peaks for a pair of even-numbered ions at *m/z* 88 and *m/z* 60, with the former being the base peak. These peaks represent sequential McLafferty rearrangements. The first is loss of butene from the carboxylic acid component, followed by ethene loss from the alkoxy component, to give the familiar radical cation $H_2CC(OH)_2^{+\bullet}$, which is also often seen in the McLafferty rearrangements of carboxylic acids (eq. 8-34). The other characteristic peak is loss of ethoxyl, to give *m/z* 99. This process generates an acyl cation, which could lose CO to give an alkyl cation at *m/z* 71 (eq. 8-35). In this case,

$$\text{ethyl hexanoate}^{+\bullet}\ (m/z\ 144;\ \gamma,\ \beta,\ \alpha\ \text{marked}) \xrightarrow{-C_4H_8} H_2C{=}C(\ddot{O}H^{+\bullet})OCH_2CH_3\ (m/z\ 88) \xrightarrow{-C_2H_4} H_2C{=}C(\ddot{O}H^{+\bullet})OH\ (m/z\ 60) \qquad \textbf{(8-34)}$$

$$\text{ethyl hexanoate}^{+\bullet}\ (m/z\ 144) \xrightarrow{-C_2H_5O} CH_3(CH_2)_4C{\equiv}\ddot{O}^{+}\ (m/z\ 99) \xrightarrow{-CO} CH_3(CH_2)_3\overset{+}{C}H_2\ (m/z\ 71) \qquad \textbf{(8-35)}$$

however, CO loss is a minor channel. Finally, γ-cleavage involves loss of the propyl radical and gives *m/z* 101. The isomeric methyl ester of heptanoic acid is shown in the adjacent panel of Figure 8-25. Again, there is a strong McLafferty peak, but now it is at *m/z* 74 and represents loss of pentene. This observation immediately suggests that the carboxylic acid component contains at least seven carbons, because in a McLafferty rearrangement at least two carbons (α and carbonyl) are retained when the alkene is lost. The second McLafferty rearrangement is absent because an ethyl or higher ester is required for it. Loss of a methoxy group to give *m/z* 113 suggests a methyl ester. Finally, γ-cleavage (loss of butyl) gives *m/z* 87. Each of these features allows one to distinguish easily between ethyl hexanoate and methyl heptanoate. The third isomer in Figure 8-25 is ethyl 2-methylpentanoate. Again, the McLafferty rearrangements provide the most structural insight. The first rearrangement ion appears at *m/z* 102 and involves loss of propene from the carboxylic acid component. This step is followed by loss of ethene from the alkoxy component, to give *m/z* 74. The fact that the double McLafferty peak is 14 units higher than for unbranched species, that is, *m/z* 60, immediately suggests methylation at the α-carbon. Loss of ethoxyl to give *m/z* 99 is prominent in the spectrum, as well as subsequent CO loss to give *m/z* 71. It is not surprising that this pathway is more prevalent than in ethyl hexanoate because it leads to a more stable, 2° cation. Note that this ion could also be viewed as an α-cleavage product from the carboxylic acid component of the ester. The γ-cleavage gives *m/z* 115 in this case (loss of ethyl). Finally, consider the EI spectrum of pentyl propanoate. The most prominent even-numbered ion is at *m/z* 70. This is not a typical McLafferty ion for an ester or a carboxylic acid. The neutral loss in this case is a McLafferty mass, 74. Here, a McLafferty rearrangement on the alkoxy component of the ester would lead to loss of pentene and formation of the radical cation of propanoic acid. The ionization energy of propanoic acid (10.4 eV) is greater than that of 1-pentene (9.5 eV) and, by Stevenson's rule, the radical cation of pentene at *m/z* 70 is the dominant pathway. Carboxylic acids often have higher IPs than do alkenes, so this outcome is common when the McLafferty rearrangement initially occurs on the alkoxy component of an ester. A McLafferty rearrangement is not possible on the carboxylic acid component in this case (at least a four-carbon chain is needed). The base peak is *m/z* 57, which is an acyl cation resulting from the characteristic alkoxy loss pathway (pentoxy in this case). The ethyl cation at *m/z* 29 is also prominent and can be formed by loss of CO from this acyl cation. As this exercise indicates, the EI spectra of esters produce a useful variety of fragments, and there is often enough information to distinguish between related closely species.

Aryl esters tend to provide larger signals for molecular ions, and the McLafferty rearrangement is possible only on the alkoxy component of the ester. In addition, alkoxy loss followed by CO loss is also common. Ethyl benzoate (Figure 8-26) provides a simple example. The base peak is for the loss of ethoxy to give an acyl cation (*m/z* 105) that can lose CO to give the phenyl cation (*m/z* 77). There is a strong McLafferty for loss of ethene to produce the radical cation of benzoic acid at *m/z* 122.

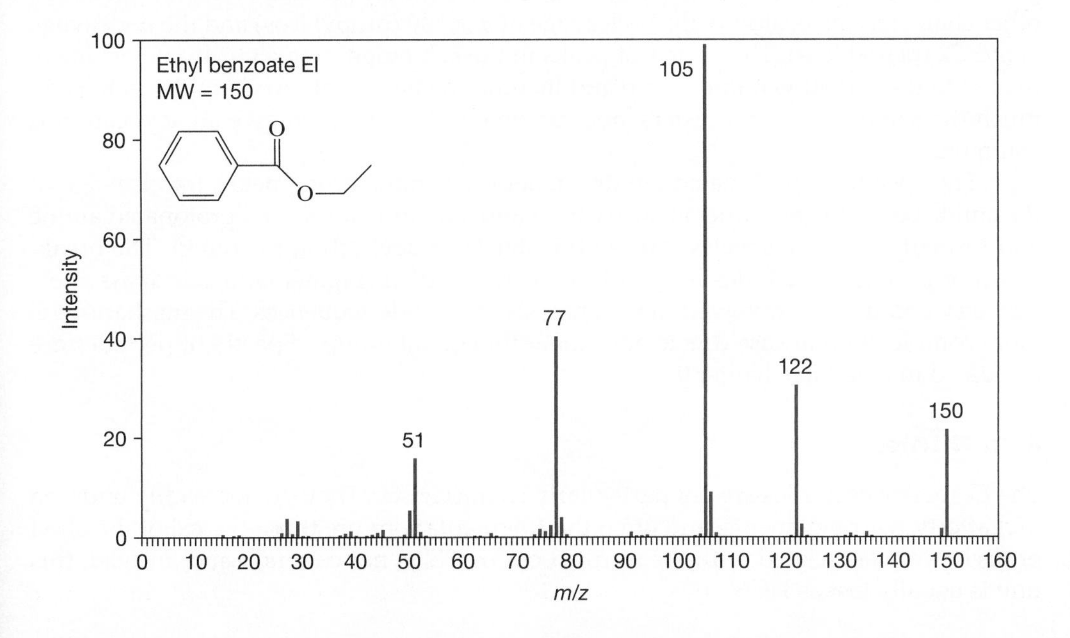

FIGURE 8-26 EI spectrum of ethyl benzoate. (Data from webbook.nist.gov.)

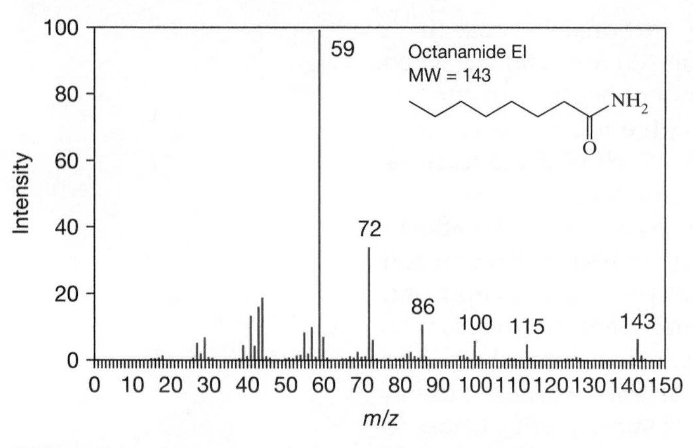

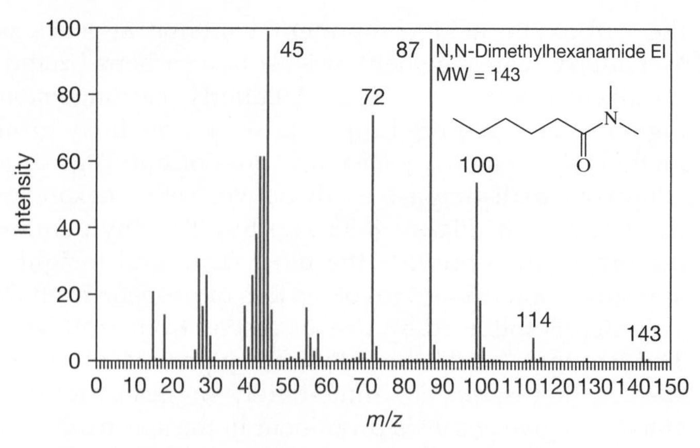

FIGURE 8-27 EI spectra of octanamide and *N,N*-dimethylhexanamide. (Data from webbook.nist.gov.)

Protonated esters are readily formed during CI and undergo characteristic fragmentations. Loss of the alkoxy group as an alcohol is common and leads to an acyl cation. This pathway can also lead to subsequent loss of CO, to give an alkyl cation. Protonated esters also can lose an alkene from the alkoxy component, to give a protonated carboxylic acid, although this process is not formally a McLafferty rearrangement, the neutral mass loss is the same. This type of loss, β-elimination, is a well-known thermal process of esters, and a similar mechanism is likely in the gas phase.

8-9c Amides

Not surprisingly, the fragmentation processes of amides parallel those of esters. A key difference is the presence of a nitrogen atom, which reverses the even/odd-numbered mass rule (odd masses correspond to radical cations when a nitrogen is present). Aside from that difference, the normal mix of McLafferty rearrangements, α-cleavages, and γ-cleavages is expected. The EI spectra for a pair of isomeric amides are shown in Figure 8-27. In octanamide, the molecular ion is an odd-numbered mass (*m/z* 143), and the base peak also is an odd-numbered mass (*m/z* 59), which implies that it is a radical cation if it contains nitrogen. In this case, it does and is the result of a McLafferty rearrangement with loss of hexene to give the radical cation of $CH_2{=}C(OH)NH_2$. This is the typical McLafferty ion for a primary amide. The other prominent peak in the spectrum is *m/z* 72, which is the γ-cleavage product. Because of the nitrogen, this closed-shell species has an even-numbered mass. In *N,N*-dimethylhexanamide, the McLafferty peak is shifted to *m/z* 87 by the addition of the two methyl groups to the nitrogen. The other characteristic peaks are the γ-cleavage at *m/z* 100 (propyl loss) and the α-cleavage at *m/z* 72 (pentyl loss). The cluster of peaks in the *m/z* range from 43 to 45 is most likely related to the dimethylamino group and include the base peak. Aryl amides behave in much the same way as aryl esters, and amide bond cleavage to give an acyl cation is common.

The spectra of protonated amides generally exhibit strong peaks for cleavage of the amide bond, to give either an acyl cation and a neutral amine or a protonated amine and formally a ketene species (proton transfer from acyl cation to amine). The breakdown of protonated amides is one of the better-studied fragmentations in mass spectrometry because it is involved in the analysis of peptide sequences. The mechanism is more complex in that case due to additional functional groups. Spectra of peptides are discussed in detail in Chapter 9.

8-9d Nitriles

The EI spectra of nitriles are not particularly characteristic. They do not readily undergo McLafferty rearrangements, and often the fragmentations are those typical of the alkyl or aryl group associated with the nitrile. Loss of CN is not common and, instead, this unit is usually lost as HCN.

> **Summary for Carboxylic Acids and Derivatives**
>
> 1. McLafferty rearrangements to give distonic ions with characteristic masses are common. Carboxylic acids give *m/z* 60 unless there is a substituent at the α-carbon. Esters are prone to double McLafferty rearrangements, to give ions of *m/z* 60, 74, 88, 102, and so forth. Single McLafferty rearrangements generally occur most readily on the carboxylic acid component of the ester. Amides produce odd-numbered McLafferty ions in the *m/z* series 59, 73, 87, and so on.
> 2. All derivatives are prone to α-cleavages at either side of the carbonyl. In esters this process can lead to alkoxy loss to give an acyl cation, potentially followed by CO loss formally to give the product of α-cleavage on the carboxylic acid component of the ester.
> 3. In addition, γ-cleavage is common on the carboxylic acid components of these species.

8-10 OTHER FUNCTIONAL GROUPS

8-10a Nitro Compounds

The EI spectra of nitroalkanes are usually dominated by the loss of NO_2, a moderately stable radical. The resulting alkyl cations undergo typical secondary fragmentations. Nitroarenes have interesting fragmentation patterns and offer a good illustration of alternative pathways in arenes. In Figure 8-28, the EI spectra of 2- and 4-nitrotoluene provide examples of typical nitroarene fragmentations. In the para isomer, NO_2 loss is common and gives a phenyl cation at *m/z* 91. In addition, there is a very interesting peak at *m/z* 107, which must correspond to loss of NO. This process is an example of a Smiles-type rearrangement and involves a spiro intermediate to give a quinone-like product (eq. 8-36). The common arene fragment, $C_5H_5^+$ (*m/z* 65), also is seen. A very different spectrum is observed for the ortho isomer. The key difference is that the nitro group can interact with the ortho methyl group. The surprising loss of 17 (hydroxyl) is a result of this interaction. Intramolecular hydrogen atom transfer followed by N—OH cleavage yields a nitroso product (eq. 8-37). The NO_2 loss peak (*m/z* 91) also is seen in this

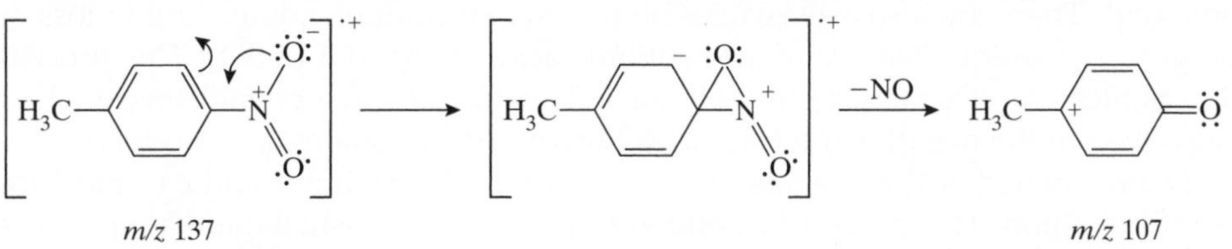

(8-36)

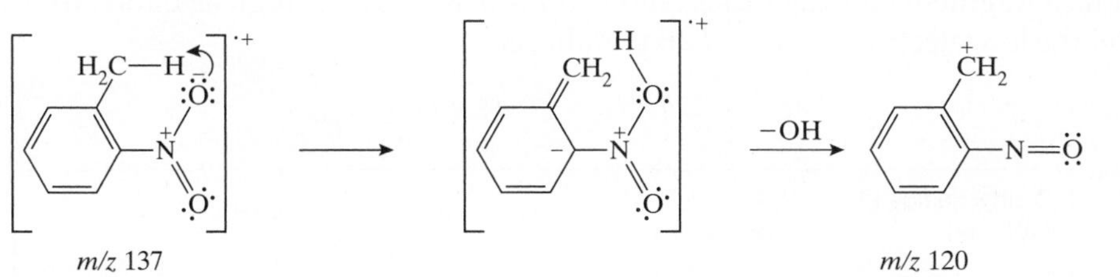

(8-37)

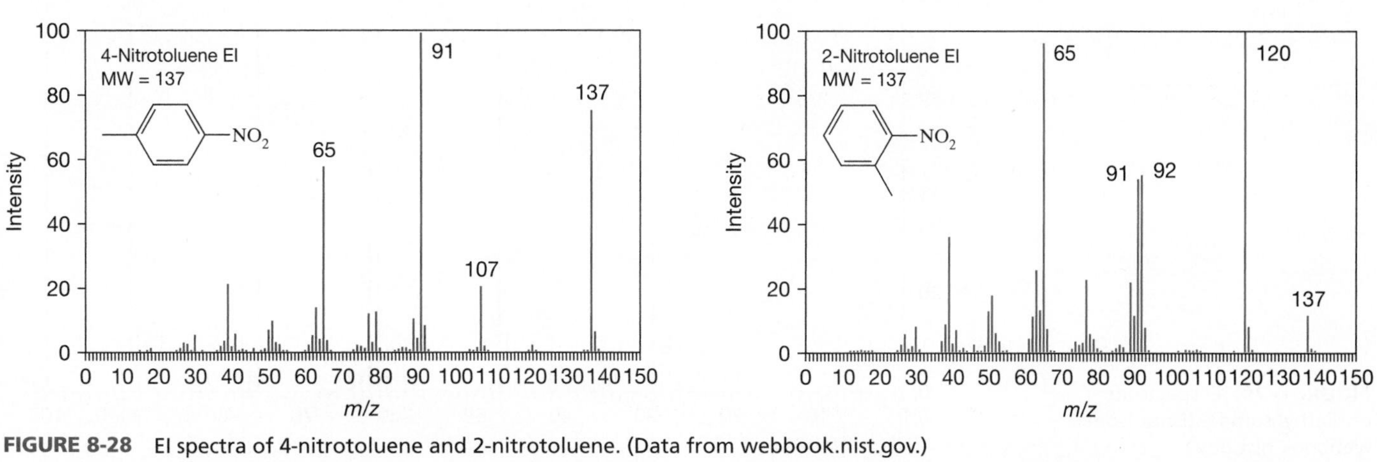

FIGURE 8-28 EI spectra of 4-nitrotoluene and 2-nitrotoluene. (Data from webbook.nist.gov.)

spectrum, along with $C_5H_5^+$ (*m/z* 65). A strong peak is also present at *m/z* 92. It most likely is derived from *m/z* 120 through a complex ring opening/rearrangement with loss of C_2H_4. The spectra are easily distinguished, but this is only because of the *ortho effect*. For example, the meta isomer gives a spectrum that is very similar to that of the para isomer. The ortho effect is also seen with other functional groups, and ortho-substituted alkyl esters and carboxylic acids undergo neutral losses of ROH and H_2O, respectively, by a similar process.

8-10b Sulfur-Containing Functional Groups

Thiols and sulfides tend to behave like alcohols and ethers, but sulfur is better able to stabilize adjacent carbocations than is oxygen, so α-cleavage is a common process. In addition, thiols have a strong tendency to lose H_2S, to produce an even-numbered ion corresponding to the radical cation of an alkene. This ion can dominate the spectrum along with its subsequent fragmentation ions. For example, butanethiol gives a strong peak for loss of H_2S, to give the radical cation of butene, which subsequently fragments to give the allyl cation by loss of methyl. The spectrum for diethylsulfide is given in Figure 8-29 and illustrates common features of sulfides. Methyl loss represents α-cleavage and gives a sulfur-stabilized cation at *m/z* 75. Cleavage at the C—S bond leads to the peaks at *m/z* 61 and 62. The former corresponds to loss of ethyl and the latter to loss of ethene, to give the radical cation of ethanethiol, which can undergo an α-cleavage (methyl loss) to give $^+CH_2SH$ at *m/z* 47 (C—S cleavage followed by hydrogen atom transfer).

Disulfides and sulfoxides often undergo C—S cleavage along with proton transfer (1,2-elimination on the alkyl group) to give even-numbered ions by loss of an alkene, that is, radical cations of RSSH and RSOH. This pathway is not common in sulfones, which generally undergo C—S cleavage to give alkyl cations. Sulfonate esters often cleave at the S—O bond with loss of an alkoxy group. They also can undergo 1,2-elimination on the alkoxy component to produce the radical cation of an alkene with loss of a sulfonic acid. There are also α-cleavages on the alkoxy component, leading to loss of an alkyl group. A unique feature of aryl sulfonic acids is the loss of SO_2. This process again is a Smiles-like rearrangement, as in eq. 8-36, and leads to a phenol species. This loss is also seen in the negative ion fragmentation of aryl sulfonates and produces phenolates. Deprotonated sulfonic acids also cleave at the C—S bond and can produce SO_3^{-}, a radical anion. This process is a rare example of a closed-shell ion that produces two radical fragments during dissociation and is driven by the high electron affinity of SO_3 and the low electron affinity of alkyl radicals.

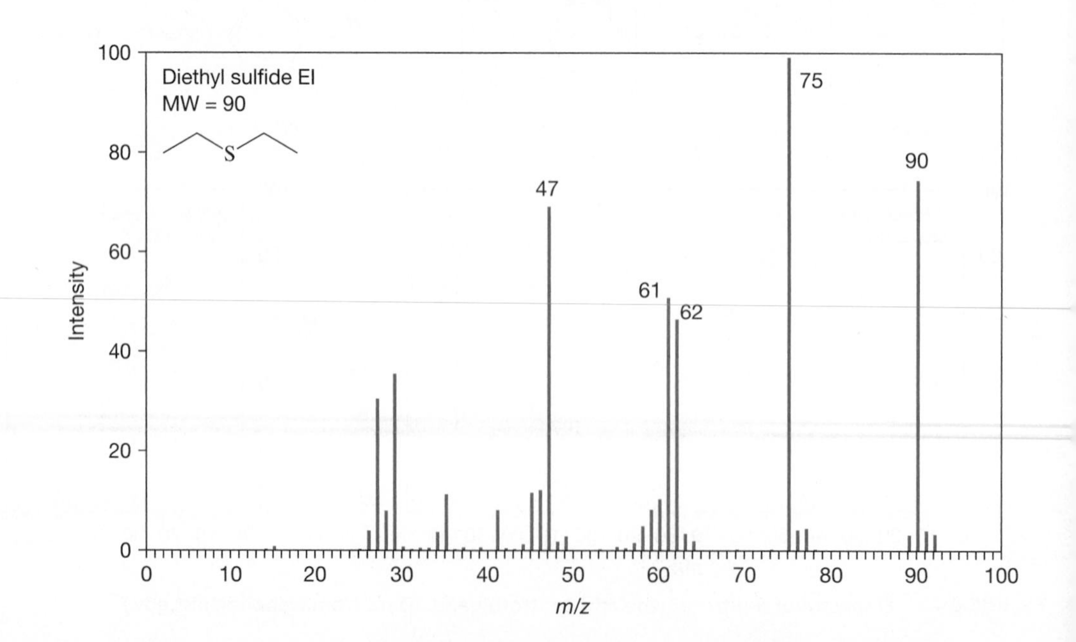

FIGURE 8-29 EI spectrum of diethyl sulfide. (Data from webbook.nist.gov.)

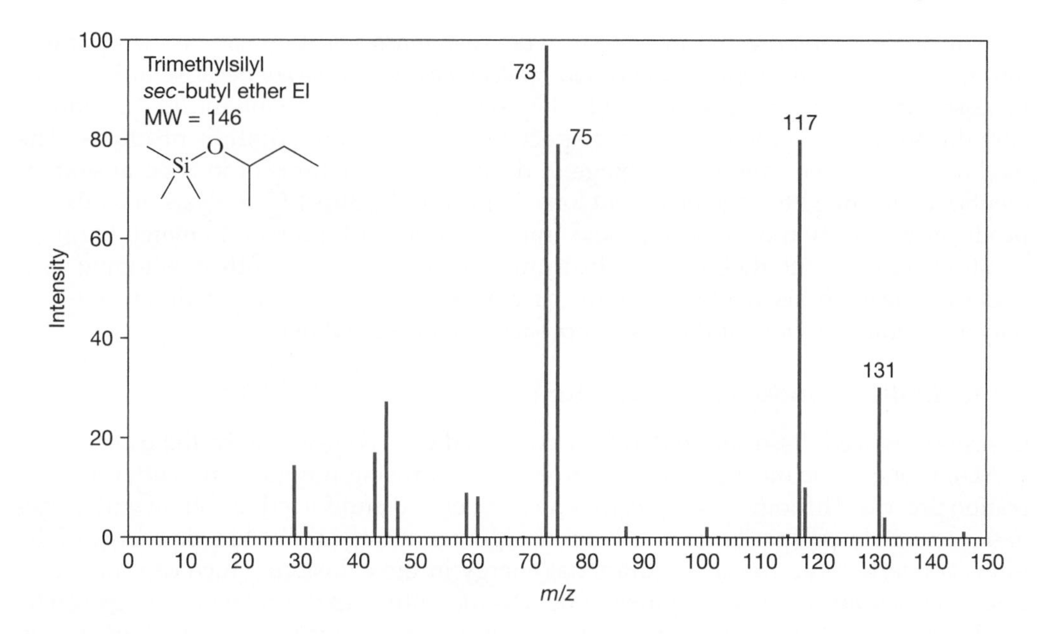

FIGURE 8-30 EI spectrum of trimethylsilyl *sec*-butyl ether. (Data from webbook.nist.gov.)

8-10c Silanes

Silicon supports positive charges fairly well, and the radical cations of alkyl silanes generally fragment by C—Si cleavage to give a R_3Si^+ species. For trimethylsilyl derivatives, this reaction often leads to the production of *m/z* 73, $(CH_3)_3Si^+$. Subsequent fragmentations of silyl cations generally involve the loss of an alkyl chain from silicon as an alkene with the formation of a new Si—H bond (1,2-elimination). The situation for trimethylsilyl ethers is somewhat more complicated, and several pathways are available. The trimethylsilyl cation can be formed by Si—O cleavage. There are two types of α-cleavages. C—C cleavage on the alkyl chain of the ether leads to the usual oxygen-stabilized carbocation. Alternatively, C—Si cleavage leads to an oxygen-stabilized silyl cation. This stabilized silyl cation can undergo a C—O cleavage/proton transfer process (1,2-elimination) to lose an alkene and produce another oxygen-stabilized silicon cation, $(CH_3)_2SiOH^+$, at *m/z* 75. The spectrum of trimethylsilyl *sec*-butyl ether in Figure 8-30 provides a good example (eq. 8-38). Strong peaks are seen for *m/z* 73 and 75, the characteristic

$$(H_3C)_3Si^{+\cdot}\text{—O—}sec\text{-}C_4H_9 \xrightarrow{-CH_3} (H_3C)_2Si^{+}\text{—O—}sec\text{-}C_4H_9 \;(m/z\ 131) \longrightarrow (H_3C)_2Si{=}O^{+}\ \cdots \longrightarrow (H_3C)_2Si^{+}\text{—OH}\ (m/z\ 75) + C_4H_8 \quad (8\text{-}38)$$

trimethylsilyl ether fragments [$(CH_3)_3Si^+$ and $(CH_3)_2SiOH^+$, respectively]. In addition, signals are seen for loss of methyl and ethyl (α-cleavages). The former is probably a mix of C—C and C—Si cleavage products (C—Si cleavage is required to produce *m/z* 75).

8-11 DISSOCIATION METHODS AND TANDEM MASS SPECTROMETRY

Throughout the previous section, much of the emphasis was on spectra obtained by EI, a method that produces ions with excess internal energy and causes extensive fragmentation. Such is not the case with other ionization processes. Although CI can also lead to significant fragmentation, the modern soft ionization techniques described in Chapter 7, ESI and MALDI, do not generally cause extensive fragmentation of the analyte. When we use these methods, do we have to give up the rich structural information found in fragmentation spectra? Thankfully, the answer is no, as there are a variety of techniques available for causing fragmentation after the ionization process. In fact, these methods offer greater control of the process because the amount of energy used to induce the fragmentation can be controlled. It is possible to probe the fragmentation processes of specific analyte ions present in a mixture because mass selection can be incorporated into the process. This approach is called *tandem mass spectrometry* because at least two mass spectrometry scans are involved. The first scan isolates the ion for fragmentation, and the second produces the mass spectrum of the fragmentation products. The method offers an enormous advantage in dealing with mixtures of ions because fragments can be linked to specific parent ions. For example, direct EI analysis of a mixture produces a collection of molecular and fragment ions with no way to match the fragments effectively with their parents. Instrument makers are constantly developing variations on the methods available to induce fragmentation after the ionization process, so only a general overview of the basic approaches is presented here.

8-11a Collision-Induced Dissociation

Collision-induced dissociation (CID), also referred to as collision-activated dissociation (CAD), is one of the most common approaches for causing fragmentation after the ionization process. The logic is simple. Ions are accelerated and forced to collide with target gases, usually inert species such as helium, argon, or nitrogen. The kinetic energy of the ion is converted into internal vibrational energy in the collision, which can fuel bond cleavages and formation of fragment ions. The magnitude of the collision energy can be used to manipulate the extent of fragmentation. The process is illustrated schematically in Figure 8-31.

The way in which CID is implemented has a major effect on the nature of the resulting fragmentation process. Implementation is controlled to some extent by the type of mass analyzer. A comparison of CID in a triple quadrupole and a quadrupole ion trap mass spectrometer gives a good contrast of two extreme cases in the implementation of CID. In a triple quadrupole, the first quadrupole can be used as a mass filter to select the analyte ion of interest for fragmentation. Before the selected analyte ion enters the second quadrupole, an electric field is applied that accelerates the ion in order to induce the fragmentation process. Typically, this process might involve energies from a few volts to 10s of volts (an electron volt is 23.1 kcal mol^{-1} of kinetic energy, but not all of this energy is available for fragmentation; see below). The second quadrupole is set to act as an ion guide—that is, it transmits all ions—and it is used as a collision cell in the process. An inert gas such as argon is present, so that collisions occur as the ions are passing through the second quadrupole. The third quadrupole is used as a mass analyzer and generates the mass spectrum of the fragment species. The key characteristic of the CID process is that it is driven by a few, highly energetic collisions in the second quadrupole. In the limiting case, fragmentation would be caused by a single collision, although this mode of operation is not typical. If the ion has sufficient initial kinetic energy, the collision can produce enough internal energy to cause a variety of fragmentation processes and produce a rich spectrum. This case is akin to what happens during EI. With high internal energy after the collision, processes with either high or low barriers can compete because the ions have more than enough energy to cross any of the barriers. As a result, many fragments can be formed.

In a quadrupole ion trap, a much different series of events leads to CID. After the analyte ion is selected in the ion trap, it is activated by the application of rf energy at the ion's resonance frequency. The target in an ion trap is the helium buffer gas used to enhance the mass analyzer's resolution (~10^{-3} torr). The amount of energy that can be

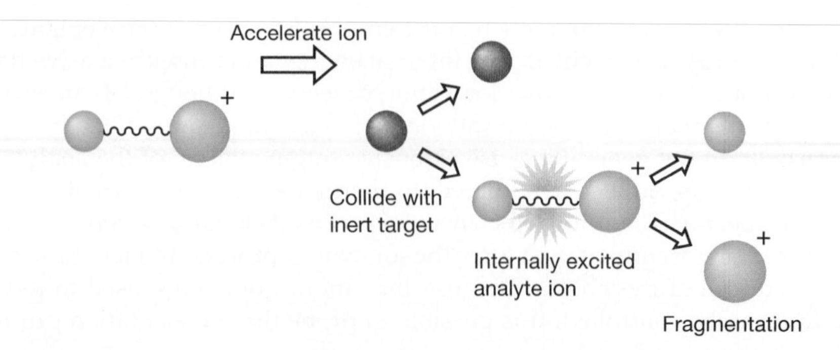

FIGURE 8-31 Schematic of the CID process. An ion is accelerated and allowed to collide with a target atom. The resulting collision internally excites the ion, and bond cleavage leads to fragmentation.

applied in a single collision is limited because the ion must remain trapped and a high-energy rf pulse would lead to its ejection from the ion trap. Under these circumstances, the only way to provide enough internal energy for fragmentation is to apply the rf energy for an extended period (10s of ms) and force the ion to undergo low-energy (0.5–5 V) collisions repeatedly with the helium. In this way, the internal energy of the ion is slowly ramped up to the dissociative limit. What makes this process so different is that as the ions gain sufficient internal energy to cross the lowest dissociative barrier, this fragmentation pathway becomes active and competes with ion activation via collisions. The net result is that ions often fragment only by the lowest energy processes because they dissociate before enough internal energy is accumulated to access the higher energy fragmentation processes. For this reason, multi-collision CID processes tend to produce simpler fragmentation spectra dominated by low-energy dissociation pathways. This characteristic can limit the information content to some extent.

Although the major factor in determining the energy transfer in CID is the acceleration voltage of the parent ion, the nature of the target atom/molecule is also important. Because of conservation of momentum, all of the kinetic energy involved in a collision cannot be converted into internal energy, and a portion of the collision energy is deposited in the translational energies of the products. The center-of-mass collision energy, E_{CM} (the energy calculated in a moving frame of constant momentum) provides a good upper limit for the energy available for internal excitation and fragmentation (eq. 8-39, in

$$E_{CM} = E_{Lab} \frac{m_{target}}{m_{target} + m_{ion}} \tag{8-39}$$

which m_{ion} is the mass of the ion, m_{target} is the mass of the target, and E_{Lab} is the kinetic energy due to the acceleration voltage). From this equation it is clear that E_{CM} is greatest when the target mass is large (as m_{target} goes to infinity, E_{CM} approaches E_{Lab}). For this reason, heavy targets such as argon are popular for CID. Ion traps, however, generally use helium because it is a much more satisfactory gas for enhancing the resolution of an ion trap mass analyzer, a critical consideration. Consequently, CID in an ion trap requires many collisions because E_{CM} is forced to be small due to the requirements of a small acceleration rf (low E_{Lab}) and a light target.

A sample CID sequence is given in Figure 8-32 for the dipeptide Gly-Arg. In this case the protonated ion is formed by electrospray and isolated in an ion trap mass spectrometer. CID then leads to the fragmentations in the second panel. This process is referred to as MS/MS (or MS^2) because the parent ion was isolated in one mass spectrometry scan and fragmented, and then a spectrum was generated in a second mass spectrometry scan. In the third panel, the major fragment ion has been isolated, and, in the fourth panel, it is forced to undergo a second CID process. This experiment is an example of an MS/MS/MS or MS^3 scan. The fragmentations of peptides are discussed in detail in Chapter 9. The sequence in Figure 8-32 illustrates one of the strengths of CID approaches. In instruments capable of MS^n experiments, specific fragmentation pathways can be mapped by isolating the primary fragments and then identifying their secondary products and possibly tertiary products. This information can be invaluable in elucidating

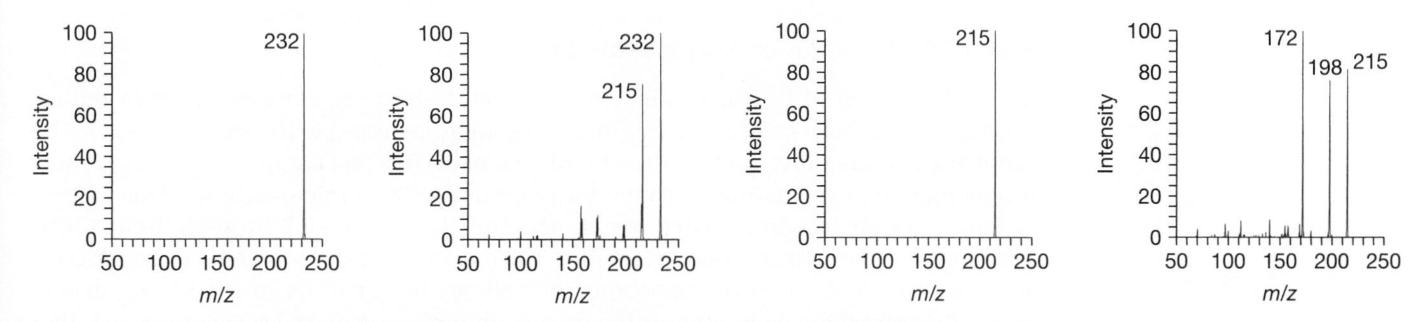

FIGURE 8-32 Fragmentation of Gly-Arg by CID. From left to right, *m/z* 232 (protonated parent) is isolated, *m/z* 232 is fragmented to give mainly *m/z* 215 by loss of NH_3, *m/z* 215 is isolated, and *m/z* 215 is fragmented.

the structures of large compounds and also provides very tight constraints when comparing unknowns with authentic samples. For example, if the chromatographic retention time, parent mass, primary fragment masses, and secondary fragment masses match an authentic sample, an identification can be made with very high certainty.

8-11b Infrared Multi-photon Dissociation

Photoactivation of ions also is possible. Again, the goal is to introduce enough internal energy into the ion to cause fragmentation. The most common implementation involves IR activation of ions with a CO_2 laser and is referred to as infrared multiphoton dissociation (IRMPD). Trapping devices such as ion traps or ion cyclotron mass spectrometers are generally needed because the ions must be held in the photon beam for an extended period. IRMPD is much like low-energy CID in that a single IR photon is insufficient to induce dissociation, so the fragmentation process involves the absorption of many photons until the ion has sufficient energy for bond cleavage. As in low-energy CID, processes with low barriers dominate because fragmentation can occur before the ion absorbs enough photons to access higher-energy pathways.

8-11c Electron Capture/Transfer Dissociation

These two relatively new methods are specifically aimed at multiply charged positive ions from ESI. In *electron capture dissociation* (ECD), a beam of slow-moving electrons is allowed to interact with trapped, multiply charged cations. Electron capture leads to a reduction in the charge, conversion to a radical cation species (usually the analyte ions are closed-shell ions), and release of potentially 100s of kcal mol^{-1} of electron/charge recombination energy. The combination of the bond weakening associated with radical cation formation and the recombination energy is generally sufficient to induce fragmentation. *Electron transfer dissociation* (ETD) is closely related and involves the reaction of the multiply charged cation with a negative ion that has a low electron binding energy (the radical anion of fluoranthene is commonly used). Electron transfer from the anion reduces the multiply charged cation's charge and initiates the same cascade of events as ECD. Both processes require trapping instruments to hold the ions in place while the reduction reactions occur.

The need for these methods is linked to the analysis of large species such as proteins and peptides. When CID is applied to a very large ion, the energy can be dispersed into potentially thousands of vibrational degrees of freedom. In such a case, the probability that a large amount of energy is localized in a particular bond is small, so dissociation can be very inefficient. To offset the dispersal of internal energy, very large collision energies can be necessary, sometimes higher than are practical for the instrumentation. In ECD and ETD, the energy is localized initially at the site of the electron/charge recombination, which also is the site of the bond weakened by the presence of the unpaired electron. As a result, cleavage at that site can be very efficient if it occurs before the energy is dispersed into all the vibrational modes. This is often the case. These methods are providing a powerful complement to CID methods in the analysis of peptides and proteins.

8-11d Dissociation of Metastable Ions

In the discussion of EI spectra, it was assumed that fragmentations occur in the ion source and that both parent and fragment ions are transferred to the mass analyzer. This is not always true. Ions can be formed with extensive internal energy, sufficient to cause fragmentation, but they may survive for potentially 10s of microseconds. This lifetime is long enough in some instruments, specifically sector and time-of-flight (TOF) machines, for the ions to dissociate within the mass analyzer. These ions are referred to as *metastable*. Production of metastable ions allows for a variety of interesting experiments that probe the energetics of the dissociation process. In the present context, they can be used to generate MS/MS-like spectra. If one can mass-select metastable precursor ions, the decay of the metastable ions provides a fragmentation spectrum. Today, the most widely employed application of metastable ions is in MALDI-TOF instruments, and the overall process is called *post-source decay* (PSD). Although instrumentally somewhat complicated, PSD provides valuable structural information that can be interpreted like CID spectra.

Worked Problems

These three problems are worked. Additional problems in the subsequent section are left for the reader to work.

8-1 The EI spectrum of 3-hexanol is given below.

(a) Rationalize the small size of the molecular ion peak at *m/z* 102.
(b) Assign structures to all the labeled species and rationalize their formation.

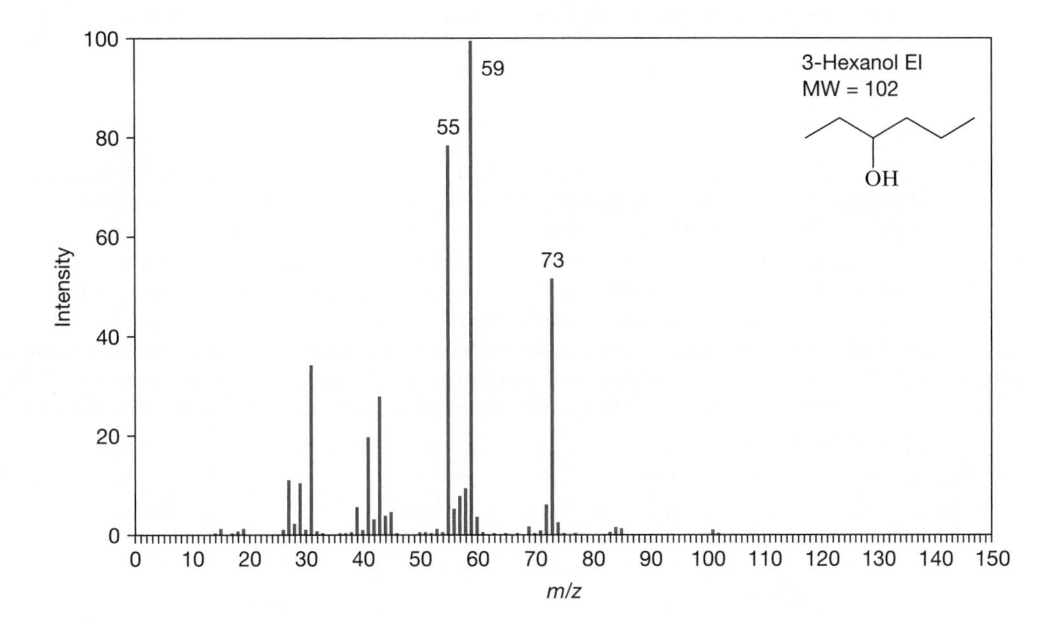

Answer

(a) The molecular ion gives a small peak because the alcohol offers favorable fragmentation pathways. In particular, two α-cleavage paths are available, each giving oxygen-stabilized cations.

(b) As noted, two α-cleavages are possible, leading to the loss of C_2H_5 (29 mass units) or C_3H_7 (43 mass units). They account for *m/z* 73 and 59, respectively. The ion at *m/z* 55 is most likely the dehydration product of *m/z* 73. The fact that there is a peak at *m/z* 41 for the analogous dehydration product of *m/z* 59 gives more confidence to this assignment. The most direct dehydration product would be protonated 1-butyne (a vinylic cation), but complex rearrangements are common in H_2O loss, so formation of a more stable, allylic cation is likely.

H OH *m/z* 73 | H OH *m/z* 59 | H H *m/z* 55

8-2 The EI spectrum of 5-phenyl-2-pentanone is given on the next page.

(a) Assign structures to all the labeled species and rationalize their formation.
(b) Explain why there is no McLafferty ion in the spectrum.

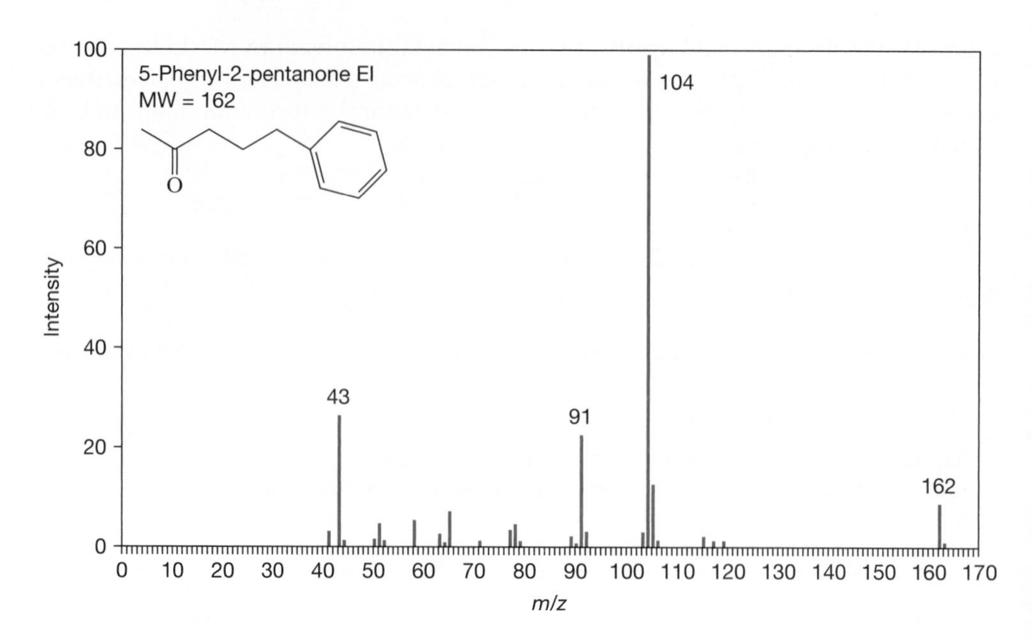

Answer

(a) The peak at m/z 43 represents an α-cleavage and is the acetyl cation. The complementary alkyl cation at m/z 119 is weak, but it could readily lose $CH_2{=}CH_2$, to produce the benzyl/tropyl cation at m/z 91. The most interesting ion is m/z 104. It is an even-numbered ion, but it is not a McLafferty ion. It results from the loss of 58, which, given the structure, must be C_3H_6O. The ion must be the radical cation of styrene. It is formed in a McLafferty process with the loss of 58, a McLafferty mass.

(b) The McLafferty rearrangement produces the enol of acetone and styrene. In a typical McLafferty process, the acetone enol would carry the charge and give an ion at m/z 58. Here, styrene has a lower IP than does the enol of acetone, so it carries the charge by Stevenson's rule.

m/z 43 $\qquad$ m/z 91 $\qquad$ m/z 104

8-3 The EI spectrum of ethyl 2-methylbenzoate is given below.

(a) Assign structures to all the labeled species and rationalize their formation.

(b) Which peaks would you expect to be absent in the EI spectrum of ethyl 3-methylbenzoate?

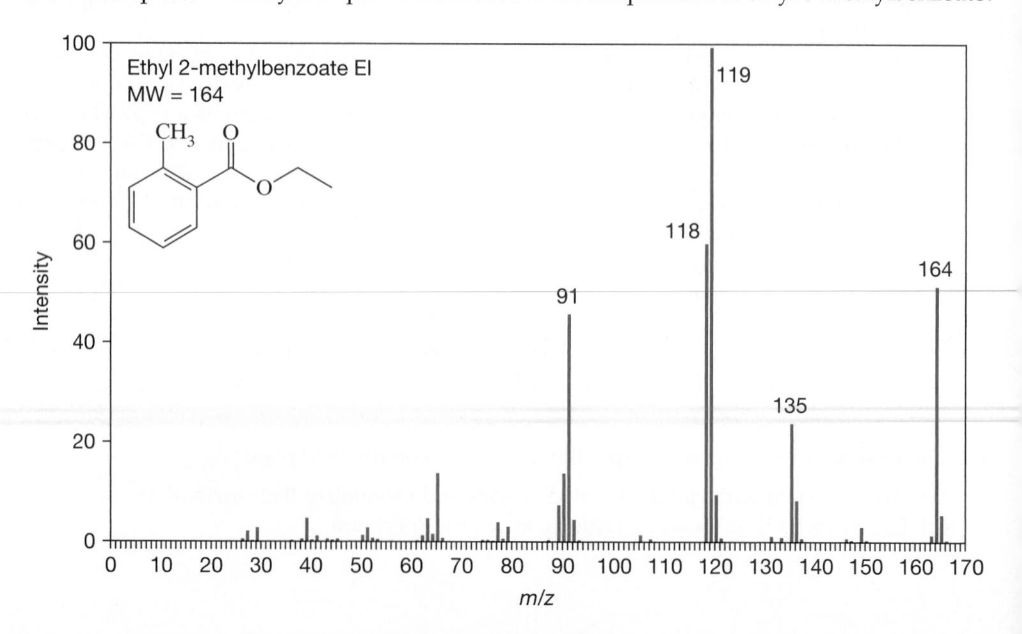

Answer

(a) The easiest peaks to explain are at m/z 119 and m/z 91. The former is loss of ethoxy to give an acyl cation, and the latter is effectively the result of loss of CO from this ion, to give a phenyl cation. The peak at m/z 135 is the result of ethyl loss (29 mass units). It might be tempting to ascribe this ion to a carboxyl cation, i.e., charge on the CO_2 unit. This structure is unlikely because the resulting ion is not stable. The process leading to the ion at m/z 135 is likely from the ortho effect. In the molecular ion, hydrogen atom transfer from the ortho methyl group to the carbonyl oxygen sets the system up for ethyl loss, initially producing a benzylic cation that could collapse to a protonated lactone. Alternatively, hydrogen atom transfer to the ethoxy oxygen gives a distonic ion and allows for loss of ethanol (46 mass units) with formation of m/z 118, a radical cation.

(b) The ions at m/z 135 and 118 require the ortho effect and are not expected in the meta isomer. In that isomer, the ion at m/z 135 is replaced by one at m/z 136. This ion corresponds to loss of $CH_2{=}CH_2$ from the ethyl group. This fragmentation is typical for an ester and involves a hydrogen atom transfer from the β-carbon of the ethyl group to the carbonyl. Its suppresion in the spectrum of the ortho isomer suggests that hydrogen atom transfer from the ortho methyl group to the carbonyl is more facile.

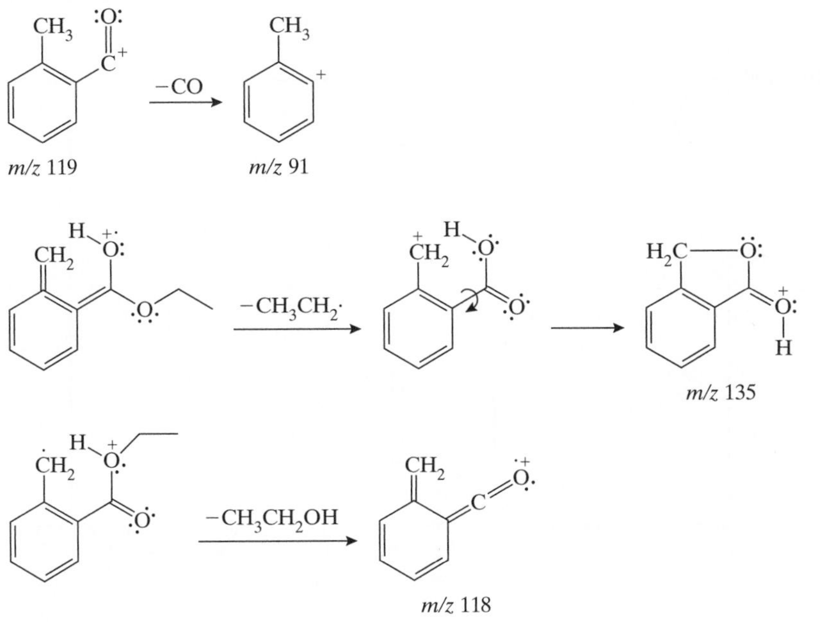

$-CO$

m/z 119 $\quad$ m/z 91

$-CH_3CH_2\cdot$

m/z 135

$-CH_3CH_2OH$

m/z 118

Problems

8-1 For each of the following compounds, predict the product masses of potential single and double (if possible) McLafferty rearrangements.

(a) 1-phenyl-3-heptanone
(b) ethyl 2-methylpentanoic acid
(c) 2,2-dimethyl-6-phenylhexanoic acid
(d) 4,6-dimethyl-5-decanone

8-2 In the EI spectrum of hexanal, strong peaks of nearly equal intensity are seen for m/z 44 and 56.

(a) Give the structures of these ions.
(b) What does the similarity in peak intensities tell you about the electron affinities of the cations responsible for these peaks?

8-3 For each of the following EI spectra, assign structures to all the labeled species and rationalize their formation.

(a)

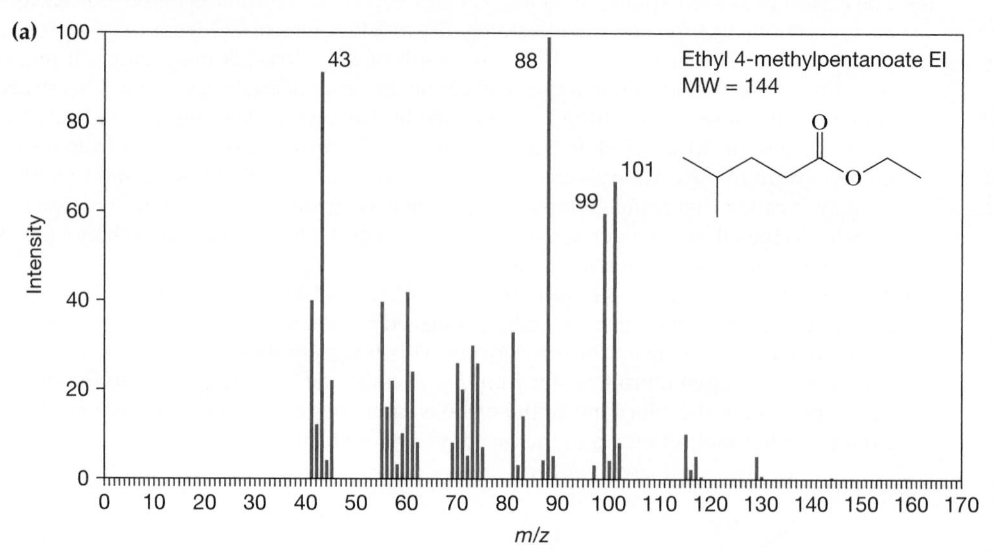

(b)

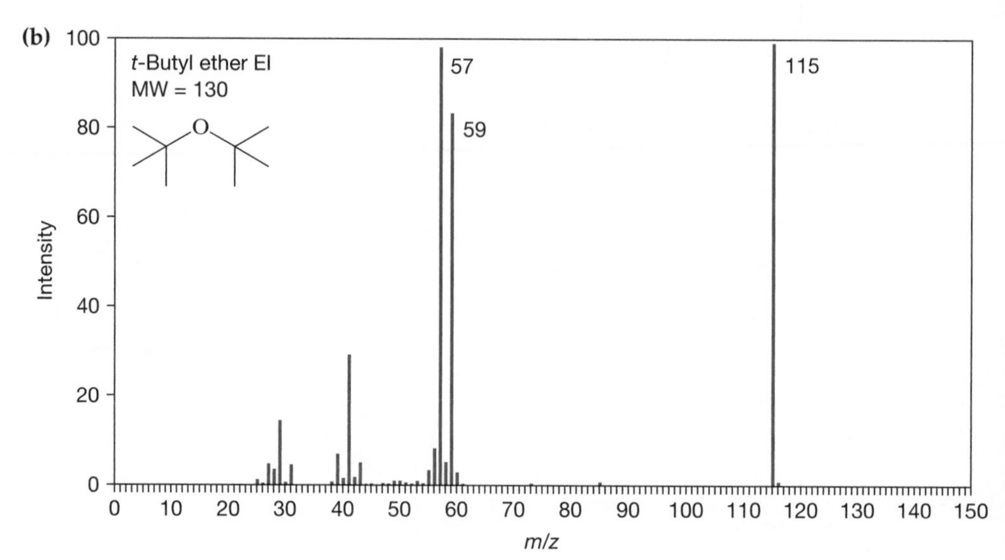

(c)

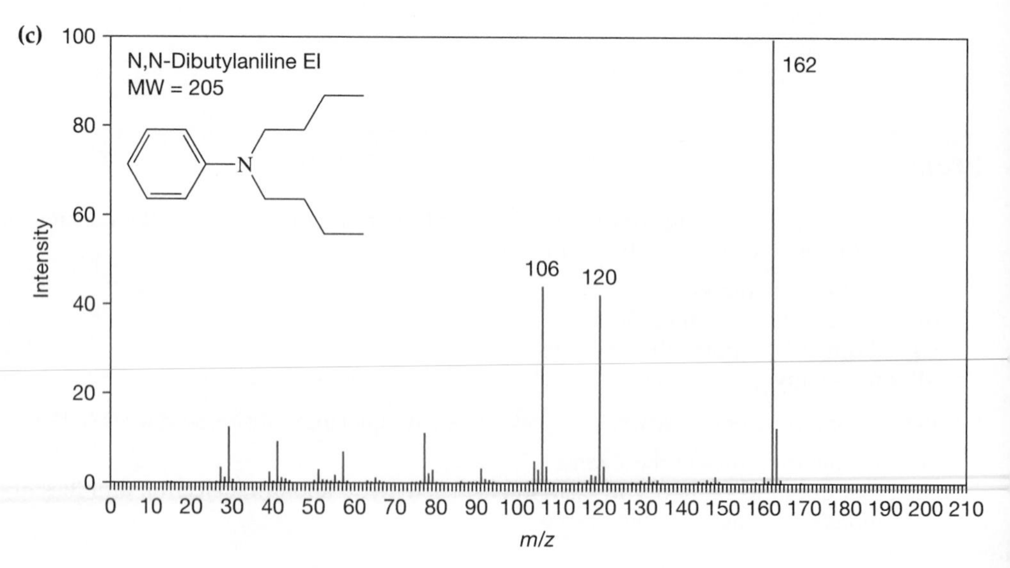

(d)
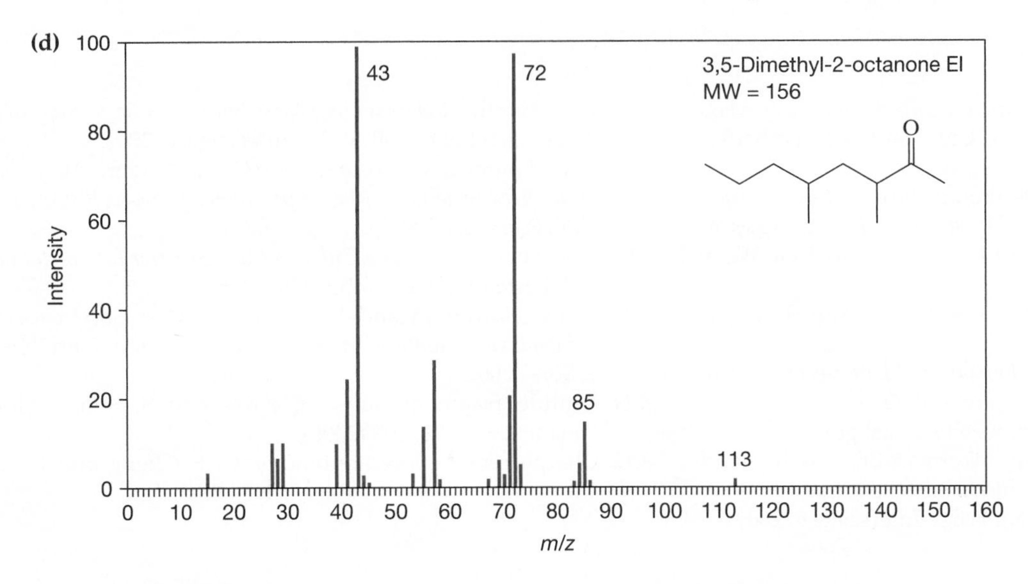

(e)
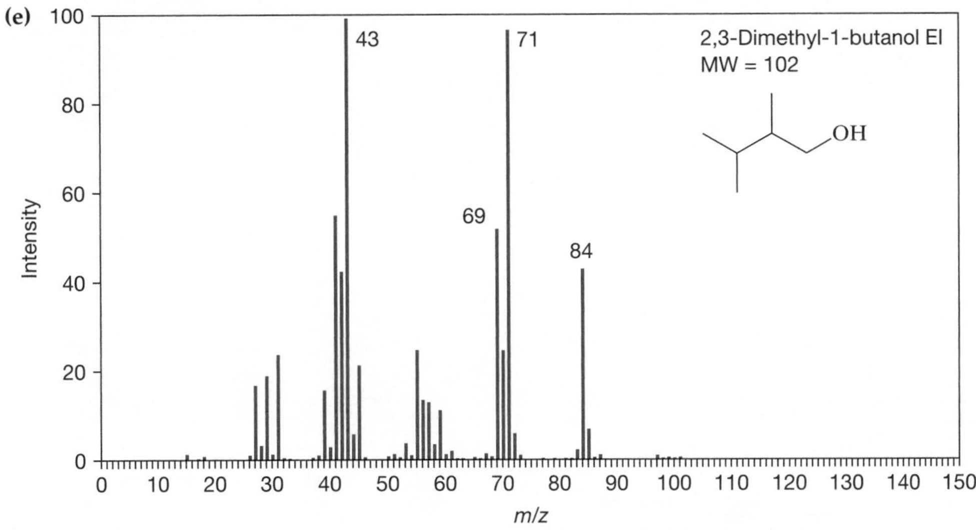

8-4 For each of the following pairs of isomers, suggest at least two significant differences you would expect in their EI spectra.

(a) 2-hexanone and 3-methyl-2-pentanone
(b) ethyl heptanoate and propyl hexanoate
(c) ethyl benzoate and methyl 4-methylbenzoate
(d) *N*-propylbutanamine and *N,N*-diethylpropanamine

8-5 For each of the following pairs, list a difference expected in the fragmentation of the radical cation of the analyte compared with the protonated analyte.

(a) tributylamine
(b) dibutyl ether
(c) butyl benzoate

Bibliography

8.1 F. W. McLafferty and F. Turecek, *Interpretation of Mass Spectra*, 4th ed., Sausalito, CA: University Science Books, 1993.

8.2 J. T. Watson and O. D. Sparkman, *Introduction to Mass Spectrometry: Instrumentation, Applications, and Strategies for Data Interpretation*, 4th ed., Chichester, England: John Wiley & Sons, 2007.

8.3 J. H. Gross, *Mass Spectrometry: A Textbook*, Berlin: Springer-Verlag, 2004.

8.4 A. G. Harrison, *Chemical Ionization Mass Spectrometry*, 2nd ed., Boca Raton, FL: CRC Press, 1992.

8.5 NIST Database of EI Spectra: webbook.nist.gov.

8.6 R. Ekman, J. A. Silberring, Westman-Brinkmalm, and A. Kraj, *Mass Spectrometry: Instrumentation, Interpretation, and Applications,* Hoboken, NJ: Wiley-Interscience, 2009.

8.7 R. M. Smith, *Understanding Mass Spectra. A Basic Approach*, 2nd ed., Hoboken, NJ: Wiley-Interscience, 2004.

8.8 J. L. Holmes, C. Aubry, and P. M. Mayer, *Assigning Structures to Ions in Mass Spectrometry*, Boca Ratan, FL: CRC Press, 2007.

8.9 T. A. Lee, *A Beginner's Guide to Mass Spectral Interpretation*, Chichester, England: John Wiley & Sons, 1998.

8.10 E. De Hoffmann and V. Stroobant, *Mass Spectrometry: Principles and Applications*, Chichester, England: John Wiley & Sons, 2002.

8.11 Peptide fragmentations: B. Paizs and S. Suhai, *Mass Spectrom. Rev.*, **24**, 508 (2005).

8.12 Charge remote fragmentations: C. F. Cheng and M. L. Gross, *Mass Spectrom. Rev.*, **19**, 398 (2000).

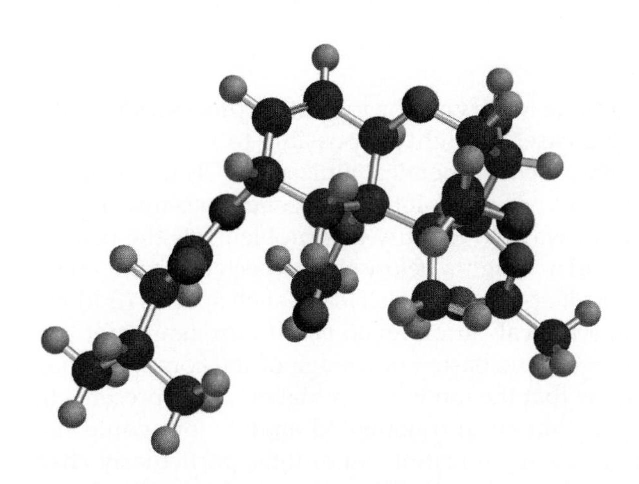

CHAPTER 9

Interpretation of Mass Spectra

9-1 INTRODUCTION

The two previous chapters described instrumentation for mass spectrometry and typical fragmentation patterns of organic functional groups. In this chapter, the focus is on the general factors that go into analyzing mass spectra and extracting critical information. Of course, mass spectrometry is used for a variety of purposes, and the way the data are analyzed is driven by the goals of the particular study. Here, the focus is on the applications of mass spectrometry that are most common in organic chemistry: (1) exact mass measurements to provide evidence for synthetic and natural products, (2) fragmentation analysis to support proposed chemical structures, and (3) fragmentation analysis as a guide to predict an analyte's structure. Each of these topics was touched upon in the previous two chapters, but the specific aim here is to provide practical tools for gaining the needed information from a given mass spectrum.

9-2 ANALYTE MASS ANALYSIS

The most direct type of information from a mass spectrum, it would seem, should be the molecular weight of the analyte. This type of data, however, can be difficult to obtain in some cases. Moreover, the way it is interpreted is closely related to the ionization source and, to a lesser extent, to the type of mass analyzer. The first step is to identify the analyte's peak in the mass spectrum.

9-2a Molecular Ions

The best practical definition of a *molecular ion* is a species that contains an intact analyte molecule and is charged in a way that is characteristic of the ionization source. Typically, the term has been associated with the radical cation of the analyte because this type of species is commonly seen in electron ionization (EI) spectra. However, other ionization sources, particularly soft ionization sources like electrospray ionization (ESI) or matrix-assisted laser desorption ionization (MALDI), rarely give radical cations. Here, the analyte generally appears as a closed-shell species, either protonated to give a cation or deprotonated to give an anion. In some cases, these species have been referred to as *quasi molecular ions* because they provide the necessary information about the molecular weight of the analyte, although the observed *m/z* does not directly match the molecular weight. How does one identify the molecular ion in a mass spectrum? The answer to this question is not altogether satisfying, particularly in samples that possibly contain impurities. Simply put, one generally must have some reasonable prediction of the molecular weight of the analyte to identify its molecular ion in a spectrum. Here are the issues. First, the molecular ion may not appear or may be very weak in spectra from an electron ionization source. Chapter 8 presented many examples of analytes that were too fragile to survive the ionization process, so all of the analyte molecules fragmented. In this case, it might be possible to reduce the ionization energy or apply another method such as chemical ionization (CI) to obtain an intact analyte molecular ion, but how does one know if the molecular ion is missing in an EI spectrum? There are two important ways to identify this problem: (1) the peaks in the spectrum all correspond to molecular weights below that expected for the analyte or (2) the candidate molecular ion peak appears to be a closed-shell species (odd *m/z* if nitrogen is not present) rather than a radical cation (even *m/z* if nitrogen is not present). Drawing either conclusion requires some basic knowledge of the composition of the analyte. Second, one cannot assume that the ion with the highest *m/z* represents the molecular ion—the assumption here is that an unfragmented analyte ion would have the highest molecular weight. Under some ionization conditions, particularly chemical ionization and electrospray ionization, adduct ions can be formed between the analyte and a charge carrier such as an ammonium ion or a sodium cation. If one does not take this possibility into account, that is, if one assumes the charge carrier is a proton, then the molecular weight assignment will involve a significant error. Third, low-concentration impurities in the sample can produce ions that might be mistaken for the molecular ion of the analyte. This possibility is particularly likely in EI because weak molecular ion signals are common and difficult to distinguish from low concentration impurities.

These issues with molecular ions highlight the fact that mass spectrometers are not machines that directly convert analyte samples into molecular weight information. Some level of analysis is required along with background information about the analyte. For this reason, mass spectrometry facilities generally request a proposed structure, a predicted molecular weight, and possible impurities when samples are submitted for analysis.

9-2b Exact Mass and Mass Defects

In general, chemists speak in terms of nominal masses and ignore the fact that, aside from ^{12}C, isotopic weights differ slightly from integer values. The differences are referred to as *mass defects*. These mass defects are further hidden by the fact that chemists generally think in terms of average atomic weights rather the weights of specific isotopes. For example, although ^{12}C is defined as having an atomic weight of exactly 12, carbon has an average atomic weight of 12.0107 because one must take into account the other naturally occurring isotopes, namely ^{13}C, in the atomic weight calculation. Mass defects, however, are a very useful tool in the analysis of organic compounds and are the key reason why high resolution mass spectra are often required by journals as part of the evidence for new synthetic products. It is the measurement of exact masses that allows for the calculation of mass defects and the assignment of elemental compositions from molecular weight information.

One of the relatively unique features of mass spectrometry is that it is truly a molecular technique—the information obtained in the experiment is the summation of data points from individual molecules rather than the average of data points from a collection of molecules. The subtle difference between these two situations is best described with an example. If one were to generate carbon ions (C^+), possibly by EI on a substrate like ethyne, the resulting spectrum would exhibit two significant peaks. One would be at an *m/z* of exactly 12 and the other at an *m/z* of 13.00336. The ^{12}C and ^{13}C ions would produce different, distinct peaks in a ratio of approximately 100:1. In contrast, if one were to weigh exactly one mole of carbon, one would obtain a single atomic mass for the sample of 12.0107, the weighted average of the masses of the ^{12}C and ^{13}C isotopes.

The value of exact mass measurements, however, becomes apparent in the study of molecules because mass defects are not scalar functions of atomic weight and, consequently, are unique to each atom. This fact provides the handle for distinguishing between species with identical nominal masses, but different exact masses. 2-Naphthylethanol ($C_{12}H_{12}O$) and its potential decomposition product from harsh oxidation conditions, naphthoic acid ($C_{11}H_8O_2$), provide a good example (Scheme 9-1). Each has the same nominal mass and would be expected to give a peak at *m/z* 172 in a low resolution EI spectrum, but

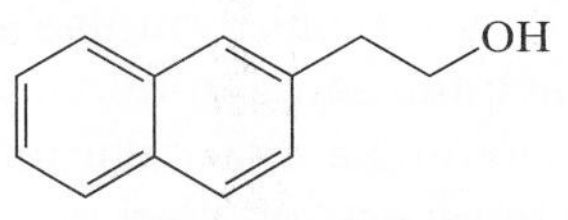

2-(β-naphthyl)ethanol
$C_{12}H_{12}O$
nominal mass = 172
exact mass = 172.08882

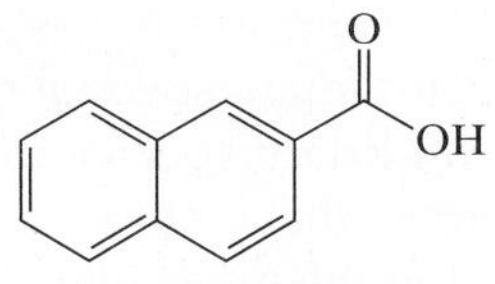

β-naphthoic acid
$C_{11}H_8O_2$
nominal mass = 172
exact mass = 172.05243

Scheme 9-1

they differ in composition. Using the values in Table 9-1, we calculate the mass for the dominant isotopic composition of 2-naphthylethanol as 172.08882 and for naphthoic acid, 172.05243 (these are referred to as monoisotopic masses). These figures represent a difference of about 0.02% (200 ppm) and could easily be distinguished by a mass spectrometer with high mass-resolving power and accuracy, such as a Fourier transform mass spectrometer (FT-MS), sector instrument, or time-of-flight mass spectrometer. In this way, it would be simple to detect if either or both materials were present in a sample. The key difference in the formula is the exchange of a CH_4 unit for an O in going from the alcohol to the carboxylic acid, which amounts to the exchange of 10 protons and 6 neutrons for 8 protons and 8 neutrons, and leads to the observed mass defect (this analysis of the defect is illustrative, but simplistic).

TABLE 9-1 Exact Masses and Abundances for Isotopes of Common Elements

Element	Isotope	Mass	% Abundance
Hydrogen	^{1}H	1.007 825	99.9885
	^{2}H	2.014 101	0.0115
Lithium	^{6}Li	6.015 122	7.59
	^{7}Li	7.016 004	92.41
Boron	^{10}B	10.012 937	19.9
	^{11}B	11.009 305	80.1
Carbon	^{12}C	12.	98.93
	^{13}C	13.003 354	1.07
Nitrogen	^{14}N	14.003 074	99.632
	^{15}N	15.000 108	0.368
Oxygen	^{16}O	15.994 914	99.757
	^{17}O	16.999 131	0.038
	^{18}O	17.999 160	0.205
Fluorine	^{19}F	18.998 403	100
Sodium	^{23}Na	22.989 769	100
Aluminum	^{27}Al	26.981 538	100
Silicon	^{28}Si	27.976 926	92.2297
	^{29}Si	28.976 494	4.6832
	^{30}Si	29.973 770	3.0872
Phosphorus	^{31}P	30.973 761	100
Sulfur	^{32}S	31.972 070	94.93
	^{33}S	32.971 458	0.76
	^{34}S	33.967 866	4.29
Chlorine	^{35}Cl	34.968 852	75.78
	^{37}Cl	36.965 902	24.22
Bromine	^{79}Br	78.918 338	50.69
	^{81}Br	80.916 291	49.31
Iodine	^{127}I	126.904 468	100

Data from physics.nist.gov.

As the previous example suggests, exact mass measurements provide a general approach for converting mass spectral data to molecular formulae and can distinguish between multiple candidate species that have the same nominal mass, but different exact masses. Of course, this method of determining a molecular formula is contingent on correctly identifying the molecular ion in the spectrum (Section 9-2a).

9-2c Isotope Patterns

As noted above, mass spectrometry probes individual molecules, so a signal is seen for each isotopic composition of an analyte. Because nearly all of the elements have more than one naturally occurring isotope, one must anticipate that analyte peaks are split into a set of signals, each representing a different combination of isotopes in the analyte. Carbon, hydrogen, nitrogen, and oxygen, however, have single, dominant isotopes, so small organic species generally give simple isotope patterns. The biggest impact usually comes from ^{13}C, which has a relative abundance of approximately 1% relative to ^{12}C. This percentage is not very significant in a species with a single carbon, but with a 10-carbon species, there is roughly a 10% chance that an analyte molecule contains one ^{13}C (and nine ^{12}C atoms). As a result, the analyte presents two peaks for the molecular ion, separated by one unit and in a ratio of 10:1. From this analysis, it is not surprising that most organic analytes display a signal at M + 1 for the species with one ^{13}C (this result is evident in many of the spectra shown in Chapter 8 and later in this chapter). As the number of carbons rises, the effect of ^{13}C on the spectrum becomes more significant and peaks for species with two and three ^{13}C atoms appear in the spectrum. The EI spectrum for buckminsterfullerene, C_{60}, provides a nice example of the influence of ^{13}C on isotope patterns (Figure 9-1). In this case, the signal for the species with one ^{13}C is nearly as tall as the one for the all ^{12}C species. It is easy to imagine how complicated the isotope pattern would be for a large polymer, such as a protein. With potentially thousands of carbons, these species give rich isotopic envelopes instead of single peaks, and the dominant species have many ^{13}C atoms. In fact, the signal for the species with all ^{12}C becomes vanishingly small in such a large molecule because so very many combinations are possible for distributing the ^{13}C atoms in the molecule (the probability of an all ^{12}C species in a molecule with 1000 carbons is roughly 0.99^{1000}, which equals only 4.3×10^{-5}!).

There are also many common elements that have significant contributions from several isotopes (Table 9-1). The most relevant for organic chemistry are chlorine, bromine, silicon, and sulfur. As illustrated in Chapter 8, the impact of multiple isotopes

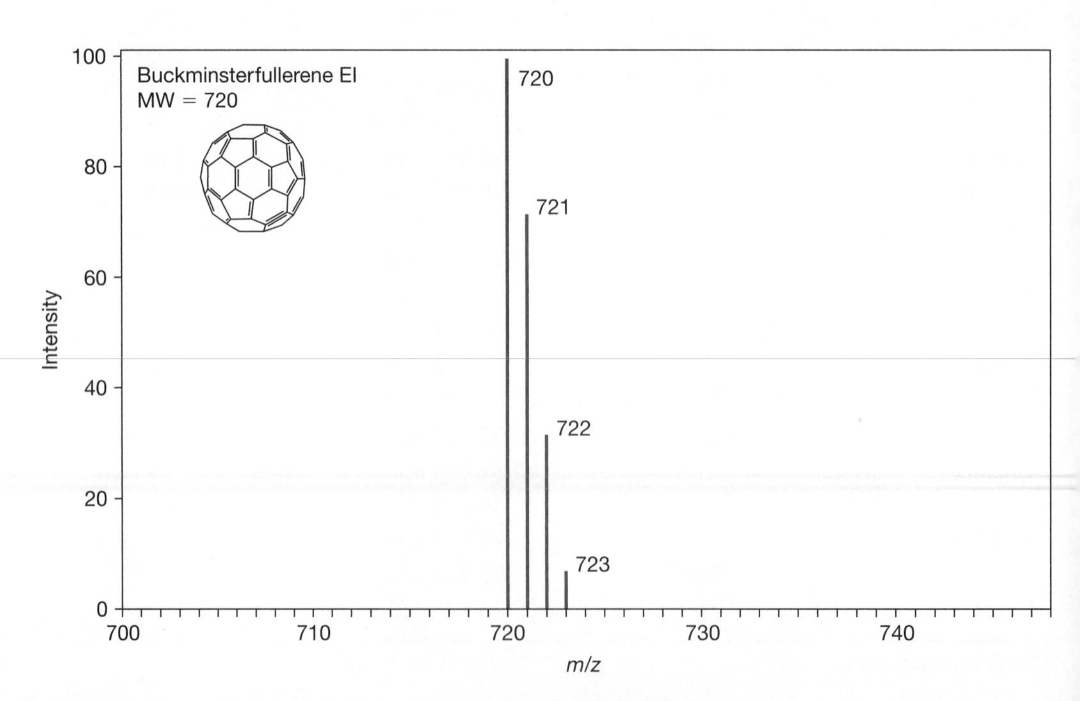

FIGURE 9-1 The EI spectrum of buckminsterfullerene (molecular ion region). (Data from webbook.nist.gov.)

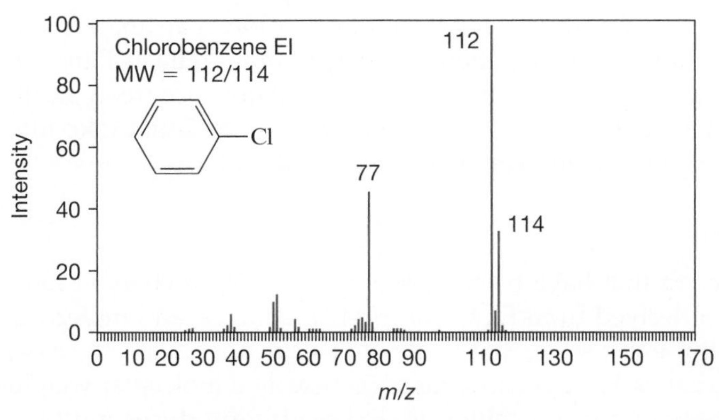

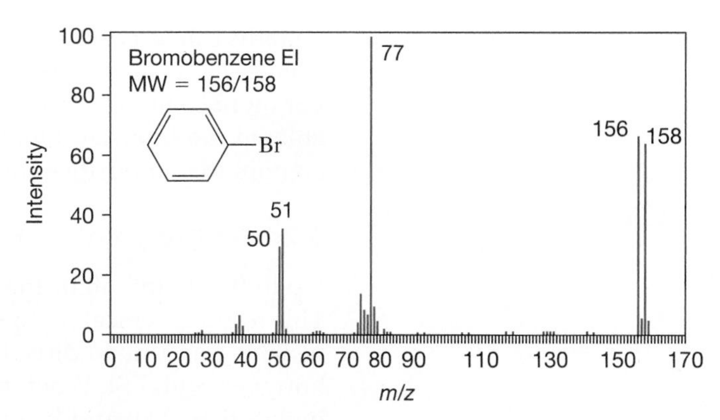

FIGURE 9-2 EI spectra of chlorobenzene and bromobenzene. (Data from webbook.nist.gov.)

is dramatic for species that contain chlorine or bromine. In fact, the isotopes of these atoms provide diagnostic patterns that can be used as a tool for identifying species that contain them. Whenever one sees a pair of peaks separated by two mass units in a ratio of 3:1 or 1:1, the immediate suspicion is that chlorine or bromine is present. In Figure 9-2, spectra are given for chlorobenzene and bromobenzene. The isotope patterns in these spectra clearly suggest the presence of chlorine and bromine, respectively. Notice also the small contributions from ^{13}C at one unit above the dominant molecular ion peaks. The impacts with silicon and sulfur are subtler. In each case, they have small but significant contributions (3%–4%) from an isotope two mass units higher than the dominant isotope. For small organic species, the contribution from ^{13}C is not large enough to have a significant impact at M + 2, so the appearance of a small peak two mass units higher than the dominant species is evidence of silicon or sulfur (silicon-containing species generally also exhibit an unusually large peak at M + 1). Finally, transition metals often give very rich isotope patterns that can sometimes contain contributions from four or more isotopes. The EI spectrum of nickelocene, $Ni(C_5H_5)_2$, provides a good example (Figure 9-3). Here, the dominant molecular ion peak is at *m/z* 188 (from ^{58}Ni), but the isotope pattern spans the region from *m/z* 188 to 194 and encompasses five isotopes of nickel (58, 60, 61, 62, and 64), along with contributions from ^{13}C. The next most significant peak is at *m/z* 190 (from ^{60}Ni). The peak at *m/z* 186 is a fragment peak (from loss of H_2). A similar pattern is seen for the signal associated with the loss of C_5H_5, which is centered at *m/z* 123.

As these spectra demonstrate, isotope patterns can be valuable tools for quickly identifying the presence or absence of elements. For example, if a product is thought to contain bromine, it must have the element's characteristic pair of equally strong isotope peaks separated by two units in the molecular ion. Obviously, knowledge of isotope

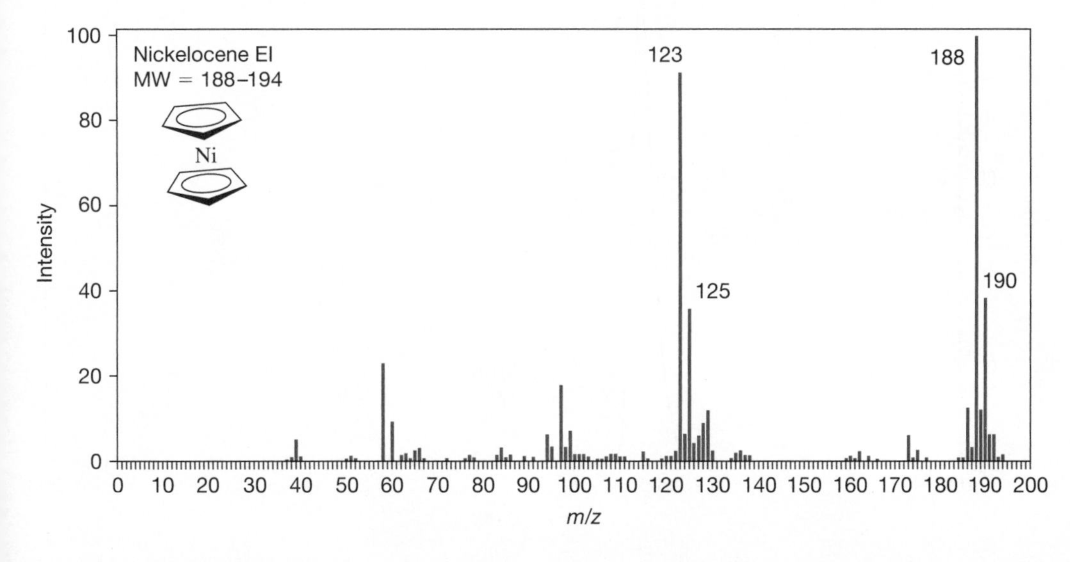

FIGURE 9-3 EI spectrum of nickelocene. (Data from webbook.nist.gov.)

patterns can be critical in some cases for identifying the molecular ion of an analyte. For simple species, it is possible to estimate isotope patterns quickly with the data in Table 9-1. For more complicated species, a number of isotope pattern calculators are freely available on the Internet. Finally, when considering isotope patterns, one must take into account that impurities could be present and can distort expected patterns.

9-2d Multiply Charged Ions

Up to this point, all of the spectra that have been shown involve singly charged ions. This result is typical for spectra derived from EI, CI, or MALDI sources, and molecular weights can be read directly off the m/z scale, that is, $z = 1$. This is not always the case, however, with ESI. When the analyte is sufficiently large (generally a molecular weight higher than 200) and has more than one ionizable site, ESI readily produces multiply charged ions, sometimes appearing simultaneously with different charge states, e.g., +1, +2, +3. Of course, such a result presents some potential problems in assigning molecular weights because the spectrum provides m/z values, not mass, so m can be determined only if z is known. Fortunately, it is generally a straightforward task to identify the charge state of a given ion. The trick is to take advantage of the known spacing between the peaks for a signal's isotope distribution. Because the nominal masses of all isotopes are integers, their signals should be separated by integer mass units. For organics, the presence of ^{13}C ensures that there are species represented in the spectrum that should be separated by one mass unit. Therefore, the mass gaps in the isotopic envelopes of organics are always 1, that is, $\Delta m = 1$ and, as a result, provide a measure of z through eq. 9-1,

$$z = 1/\Delta(m/z) \tag{9-1}$$

in which $\Delta(m/z)$ is the gap between neighboring signals in the isotope pattern. This relationship provides a very simple way to obtain z. If the gap between neighboring signals in the isotope pattern is 0.5, then $z = 2$. If it is 0.33, then $z = 3$. Once z is known, it is easy to convert m/z data into molecular weights. It is important, however, to keep track of the charge carriers. For example, a doubly protonated analyte gives an apparent mass that is two units higher than its molecular weight (the mass of the two added protons). An example is given in Figure 9-4 for a dianion. The molecular weight of the ion is 402, but the signal occurs at m/z 201. The isotope pattern has peaks separated by 0.5 unit, indicating a doubly charged ion. The molecular weight of the ion is 402, but after accounting for the loss of two protons in forming the dianion, a molecular weight of 404 is predicted for the analyte. Of course, the mass analyzer must be capable of resolving the isotopic signals. For small species, this is not a major issue, but for large ones with many charges, it can be a problem. For example, a peptide with a mass of 5000 and five positive charges would require

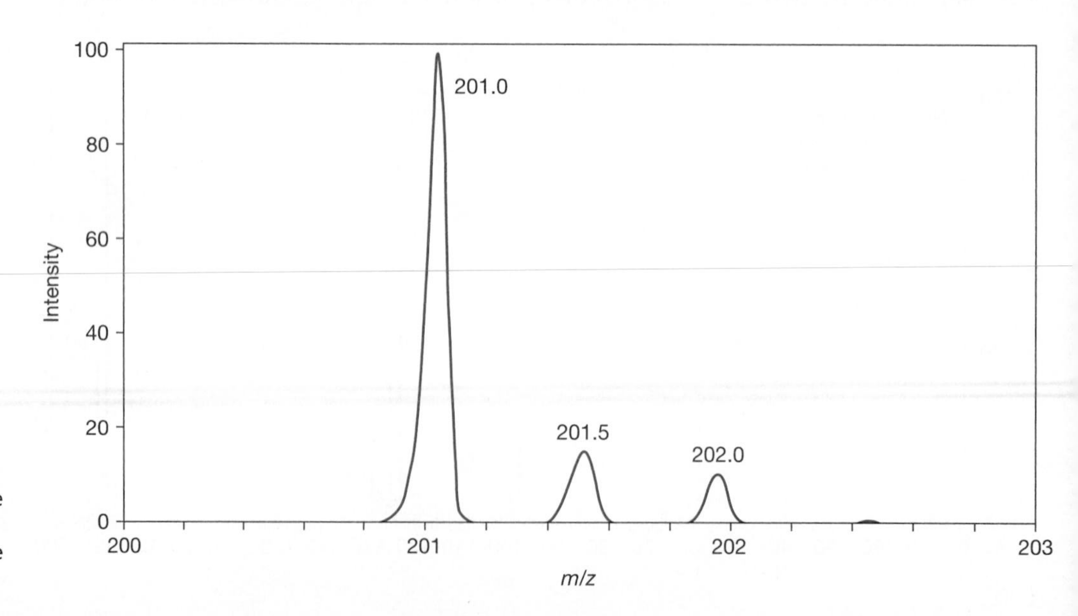

FIGURE 9-4 The ESI spectrum of 2,6-bis(3-sulfatopropoxy)naphthalene in negative ion mode. The presence of two sulfur atoms accounts for the unusually large signal for the second isotope peak.

sufficient resolution to identify signals at m/z 1000 separated by 0.2 unit. This pushes the limits of simple analyzers such as quadrupoles. The situation becomes more difficult with very large analytes like proteins where the gaps can be as little as 0.02 unit between isotope signals (i.e., charge = 50). These problems can be addressed by shifting to a mass analyzer with higher resolution or by taking advantage of additional information in the spectra, such as the gaps between peaks for the analyte in different charge states (remember that ESI sources usually produce multiply charged ions in several charge states).

The number of charges that can be carried by an analyte is a function of the size of the molecule as well as the number of ionizable sites, and can become quite large. For example, proteins often have as many as 40 charges when formed by ESI. The limiting factor is that as each charge is added, the molecule gains additional internal electrostatic repulsion. In a vacuum, this is a critical issue because with a dielectric constant of 1, short-range electrostatic repulsions can lead to enormous energy penalties. As a rough guide, one can estimate the impact using the following relationship between the repulsion and distance: 330 kcal/mol/Å. The limiting factor in forming a multiply charged ion by ESI is that there is always the possibility of proton transfer to or from the ESI solvent. As a result, when the internal electrostatic repulsion becomes too large, the multiply charged ion undergoes a gas-phase proton transfer reaction with the ESI solvent and reduces its overall charge. In practice, large molecules capable of forming multiply charged ions tend to have m/z values of about 500 to 1500, independent of the total size of the molecule. This range apparently provides a reasonable compromise between the inherent basicity advantage of most analytes (relative to the solvent) and the internal electrostatic repulsion from the multiple charges. Of course, typical organic analytes generally are relatively small and are usually limited to forming doubly charged ions. In any case, one needs to take into consideration the possibility of multiply charged ions in an ESI spectrum of any species with an expected molecular weight higher than 200 that has two or more easily ionized sites.

9-3 STRUCTURES FROM SPECTRA

The information from Chapter 8 and Section 9-2 provides a foundation for determining structures from spectra. As noted earlier, typically one needs more data than simply a mass spectrum to determine a structure. Often, one has data about how the analyte was prepared or the results of chemical tests that suggest a particular functional group. The following examples are relatively straightforward, but they offer good illustrations of how mass spectrometric data can be used to unravel structural features in an analyte. More examples are provided in the problems at the end of the chapter.

In the analysis of the spectra of unknown species, the following logical progression is useful. These steps can also be applied when confirming a proposed structure for an analyte.

1. Determine the molecular weight of the analyte.
2. Determine the molecular formula of the analyte, potentially using exact mass data.
3. Identify possible structural features based on neutral fragment losses.
4. Use any available information about the analyte's functional group to search for characteristic fragmentations
5. Attempt to rationalize neutral fragment losses for a proposed structure.

9-3a Structural Analysis 1

In the following example, the sample is from a solvolysis reaction. The expected product has been identified, but an additional spot was identified by thin-layer chromatography and later isolated. The chemical reaction and expected product are shown in eq. 9-2.

Br

H_2O/Acetone

OH

(9-2)

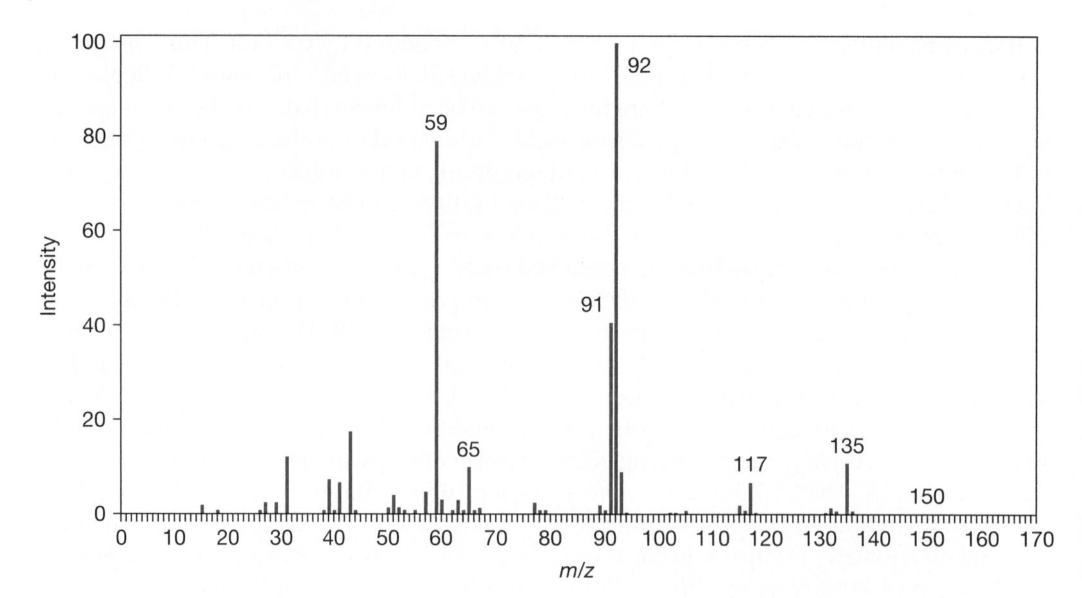

FIGURE 9-5 EI spectrum of analyte in structural analysis 1.

An EI spectrum of the by-product of the reaction is shown in Figure 9-5. The spectrum is from a low resolution mass spectrometer, so it is possible to determine only the nominal mass. In this case, that is sufficient because the reaction conditions in eq. 9-2 provide enough information to elucidate the molecular formula from the nominal mass of the analyte. No nitrogen is in the system, so an even-numbered molecular ion is expected. In addition, there is no evidence of the characteristic bromine isotope pattern. This observation leads to the conclusion that the product is limited to carbon, hydrogen, and oxygen. A small, even-numbered peak is observed at *m/z* 150 and is the molecular ion. The most likely compositions with this mass containing only carbon, hydrogen, and oxygen are $C_{10}H_{14}O$, $C_9H_{10}O_2$, or $C_8H_6O_3$. With a starting composition of $C_{10}H_{13}Br$, the first is very reasonable (swap Br for OH), but the others require considerable rearrangement to give reasonable structures. In fact, the expected product has this formula and therefore the unknown analyte could simply be an isomer of the expected product.

The next step is to consider the fragmentation pattern. The peak at *m/z* 135 indicates loss of methyl (15 mass units), which suggests that the rearranged product contains a methyl group. Aside from this peak, the key peaks in the spectrum are *m/z* 59, 91, and 92. The last two point to a benzyl component in the molecule, particularly *m/z* 92, which is formed by hydrogen atom transfer to a benzyl group. The ion at *m/z* 59 is likely $C_3H_7O^+$ (see Table 8-2). Interestingly, *m/z* 59 and 91 are *complementary ions* (their sum is the total molecular weight). This result suggests that the benzyl group is linked to the C_3H_7O group by a fairly weak bond. The formula indicates a phenyl-substituted alcohol or ether, and therefore α-cleavage is a likely pathway. Scheme 9-2 outlines the

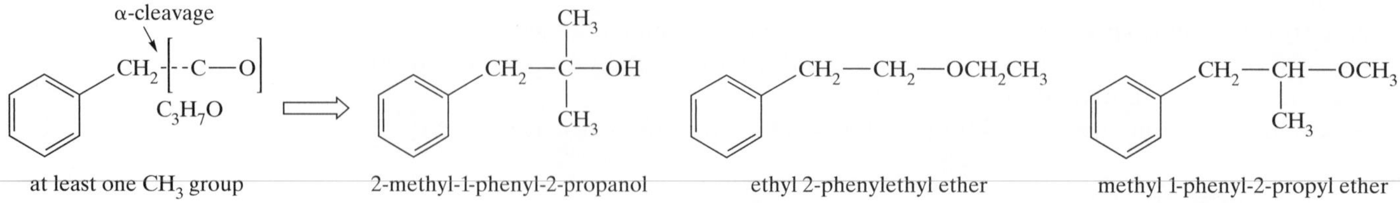

α-cleavage

CH_2–[C–O] C_3H_7O ⟹

CH_2–C(CH_3)$_2$–OH CH_2–CH_2–OCH_2CH_3 CH_2–CH(CH_3)–OCH_3

at least one CH_3 group — 2-methyl-1-phenyl-2-propanol — ethyl 2-phenylethyl ether — methyl 1-phenyl-2-propyl ether

Scheme 9-2

structural possibilities. Given that the unassigned C_3H_7O component must contain C–O and a CH_3 group, one is left with three options. If it is an alcohol, it must be 2-methyl-1-phenyl-2-propanol. If it is an ether, it could be either the ethyl ether of 2-phenylethanol or the methyl ether of 1-phenyl-2-propanol. Other possible fragmentation pathways for each of these options include C–O cleavage, possibly accompanied by hydrogen transfer, to give the losses of HO (or H_2O), CH_3CH_2O (or CH_3CH_2OH), or CH_3O (or CH_3OH), respectively. The spectrum in Figure 9-5 shows only weak peaks for these losses, with the strongest being *m/z* 132 for loss of H_2O. The peak at *m/z* 117, however, is useful in distinguishing between the possibilities. It involves loss of 33, which is likely the combined loss of CH_3 and H_2O. This observation (i.e., loss of H_2O) immediately suggests the alcohol, 2-methyl-1-phenyl-2-propanol, which is the correct assignment for this spectrum.

9-3b Structural Analysis 2

Ozonolysis/reduction of an unknown unsaturated hydrocarbon produces a compound that gives the EI spectrum in Figure 9-6. From this information we can conclude that it is an aldehyde or a ketone. The molecular weight appears to be 162, an even-mass ion, suggesting the presence of zero or two nitrogens. With no nitrogens, the formula would appear to be $C_{11}H_{14}O$. Analysis on an instrument with high mass accuracy gives a molecular weight of 162.10448. This value is consistent with the proposed formula. For aldehydes and ketones, a good first step is to look for McLafferty ions. In this case, there are no significant fragment peaks at even masses, so a McLafferty rearrangement is not a dominant pathway. This result suggests that the chain lengths from the carbonyl group are not adequate to allow for a McLafferty rearrangement. More specifically, the chains are no longer than propyl (or do not contain an appropriate hydrogen). The next step is to look for products from cleavage at the carbonyl. A key to identifying them is that they are often accompanied by a peak 28 mass units lower for the subsequent loss of CO. In this spectrum, *m/z* 71 is an obvious candidate because it is a typical mass for alkyl ketones, and a strong signal is also seen for *m/z* 43, its CO loss product. The peak at *m/z* 91 is the complement of *m/z* 71, and its mass suggests a benzyl group. Recall that, in cleavages at a carbonyl, typically acyl cations are formed, but, if an especially stable cation is possible, the alkyl portion can carry the charge. Scheme 9-3 outlines the

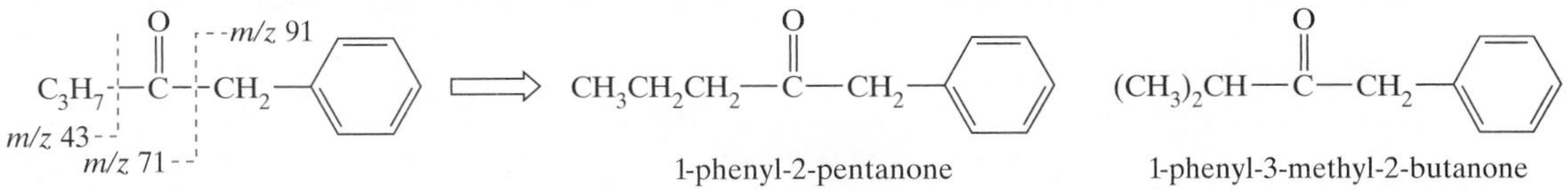

C_3H_7–C(=O)–CH_2–C_6H_5 (cleavage labels: *m/z* 43, *m/z* 71, *m/z* 91) ⟹ $CH_3CH_2CH_2$–C(=O)–CH_2–C_6H_5 and $(CH_3)_2CH$–C(=O)–CH_2–C_6H_5

1-phenyl-2-pentanone   1-phenyl-3-methyl-2-butanone

Scheme 9-3

structural possibilities. The only question remaining is the structure of the C_3H_7 component, *n*-propyl or isopropyl. Both are consistent with a weak or nonexistent McLafferty rearrangement. It is tempting to conclude that it must be isopropyl because of the high intensity of *m/z* 43 (possibly the isopropyl cation), but, as a secondary fragment, this is not a safe inference. The absence of a peak for the loss of CH_3 (*m/z* 147) might be taken as evidence for *n*-propyl, but cleavage of groups from the α-carbon of a carbonyl group is not

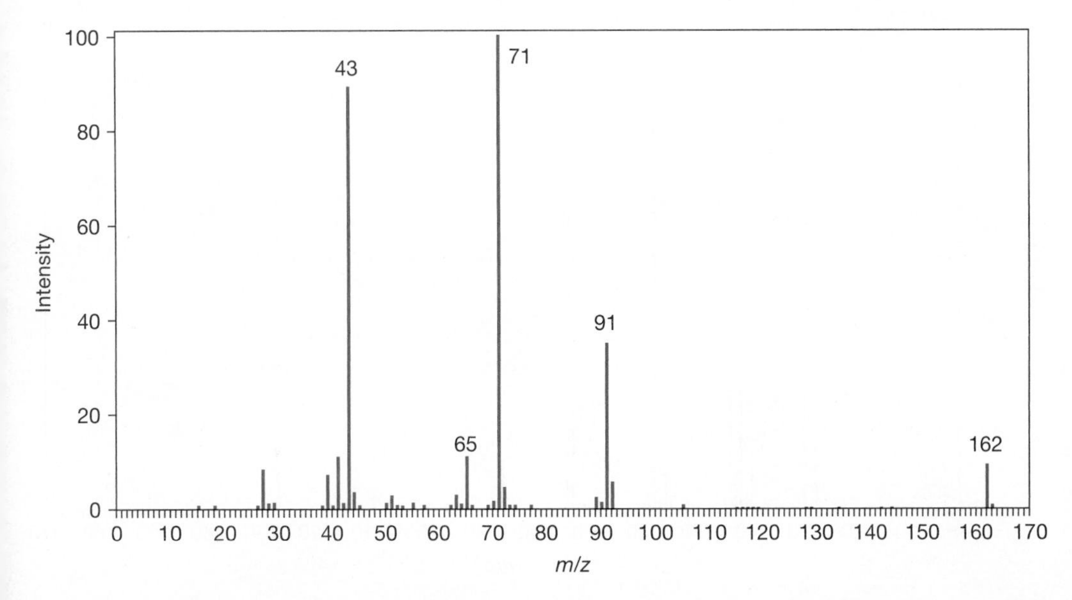

FIGURE 9-6 EI spectrum of analyte in structural analysis 2.

common because the carbonyl destabilizes an adjacent carbocation. In short, there is not enough information to make a decision, and another approach would be needed to distinguish between the options. The spectrum is actually 1-phenyl-2-pentanone.

9-3c Structural Analysis 3

In this case, no information is available about the origin of the analyte in Figure 9-7. It is rarely possible in this situation to define a structure definitively with only an EI spectrum. Nonetheless, it is a worthwhile exercise to deduce as much structural information as possible from the spectrum. The molecular ion is at *m/z* 156 (in this instance, one has to assume that the analyte is producing a significant molecular ion and has not undergone 100% fragmentation). The even mass suggests no nitrogen, and the absence of large isotope peaks rules out chlorine or bromine. Several molecular formulae for this nominal mass are possible, including $C_{12}H_{12}$, $C_{11}H_{24}$, $C_{10}H_{20}O$, $C_{10}H_{17}F$, or $C_9H_{16}O_2$. These formulae lead to a nearly intractable number of structural possibilities, so an accurate mass measurement would be exceptionally useful. First consider the predicted molecular weights for each formula: $C_{12}H_{12}$, 156.093900; $C_{11}H_{24}$, 156.187800; $C_{10}H_{20}O$, 156.151414; $C_{10}H_{17}F$, 156.131428; and $C_9H_{16}O_2$, 156.115028. The differences are significant and generally in the range of about 100 ppm. In this case, the experimental mass of 156.15293 is consistent with $C_{10}H_{20}O$, for which the unsaturation number U (eq. 1-1) is 1. Next consider the fragment losses in Figure 9-7. The peak at *m/z* 127 (loss of 29) points to an ethyl group in the structure. The ion at *m/z* 72 matches the mass of a typical McLafferty ion for a ketone. The presence of oxygen in the formula from the high resolution mass measurement is consistent with a ketone, which then accounts for the single unsaturation. If so, *m/z* 72 indicates an ethyl ketone or a methyl ketone with an additional methyl group at carbon 3 (i.e., the McLafferty ion is a four-carbon species). The

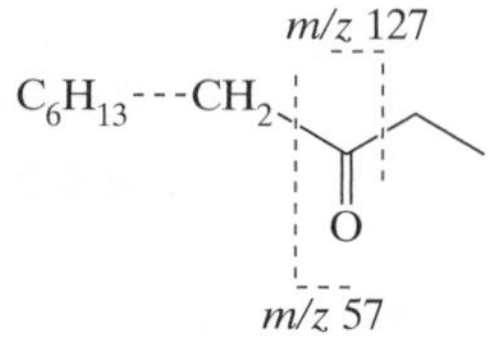

m/z 127

C_6H_{13}---CH_2

O

m/z 57

Scheme 9-4

peak for ethyl loss suggests that the former is more likely (Scheme 9-4). Finally, one would expect cleavage on the opposite side of the carbonyl (loss of C_6H_{13}) to give the

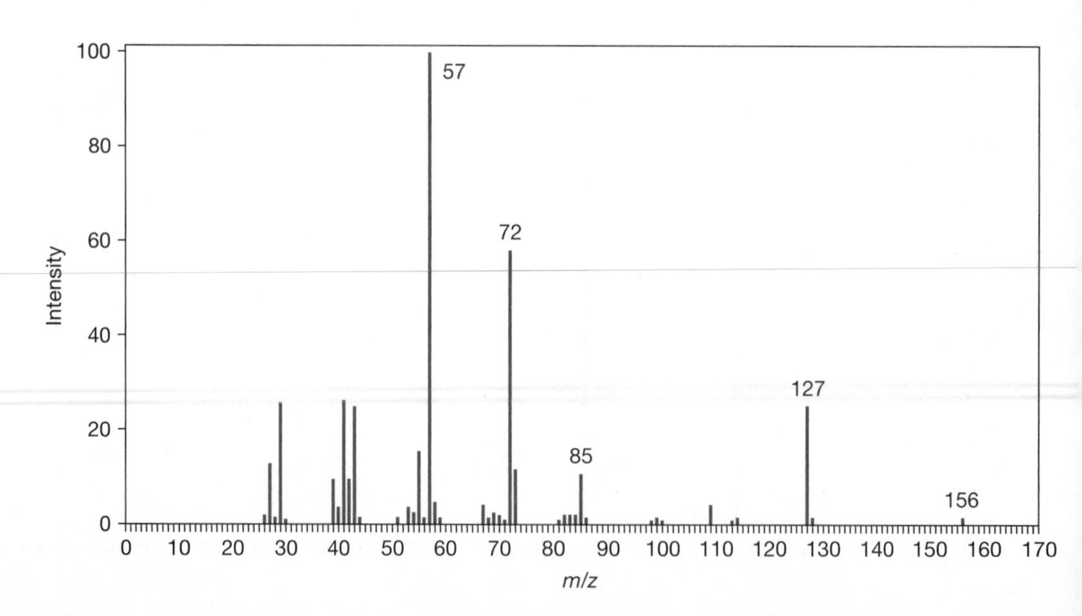

FIGURE 9-7 EI spectrum of the analyte in structural analysis 3.

propionyl cation at *m/z* 57. This is indeed a strong peak in the spectrum and confirms the conclusion that the analyte is an ethyl ketone. The peak at *m/z* 85 is intriguing but actually offers little structural insight. If it is derived from the right side of the structural outline in Scheme 9-4, it must be $(CH_2CH_2C(O)CH_2CH_3)^+$, a somewhat uncommon γ-cleavage product. If it is from the left side, it would be $C_6H_{13}^+$, possibly a stabilized carbocation due to branching. There is little in the spectrum to help differentiate between these possibilities, so the peak at *m/z* 85 does not aid in the analysis. One is left with the information in Scheme 9-4 as the best deduction of the structure. The analyte in this case is relatively simple, 3-decanone, and the C_6H_{13} component in Scheme 9-4 is *n*-hexyl. Despite the temptation to use the peak at *m/z* 85 to argue for branching in the C_6H_{13} unit, it is in fact a linear chain.

9-3d Structural Analysis 4

The spectrum in Figure 9-8 was generated in an ESI source and involves collision-induced dissociation (CID) of *m/z* 144, the base peak in the full spectrum. Because the experiment was done in positive ion mode, it is reasonable to assume that it is a protonated ion. Other cationated species, such as sodiated, are possible, but without added information, protonation would be the first guess. With ESI, there is the possibility of multiply charged ions, but that can be ruled out by examining the isotope pattern in the initial mass spectrum (not shown), which has one-unit spacings between the isotope peaks. In a closed-shell, protonated species, odd-numbered masses are anticipated unless the analyte contains a nitrogen. With an initial mass of 144, it is likely that the analyte contains a nitrogen. The first set of losses is 42 (*m/z* 102) and 44 (*m/z* 100). Because the initial ion is closed shell, we expect major losses to be closed shell as well, rather than radical species. The fact that the product ions are still even numbered suggests that they retain the nitrogen atom. A loss of 42 is likely to be C_3H_6 (propene) or C_2H_2O (ketene; see Table 8-3). Although ketene loss is an option (suggesting a possible acetamide derivative), recall that alkene loss is a common pathway for protonated amines, so propene loss seems more likely. This deduction is reinforced by the fact that alkane loss is also common with protonated amines and the peak at *m/z* 100 could correspond to propane loss. The peak at *m/z* 60 could be from loss of 84, or it could be a secondary fragment. This can easily be determined by an MS^3 experiment. In such an experiment, *m/z* 144 would be fragmented and the product at *m/z* 102 isolated. This ion would then be subjected to CID to identify its fragmentation products. This type of experiment is possible in trapping instruments or a triple-quadrupole. In this case, *m/z* 60 is a secondary fragment derived from *m/z* 102 (spectrum not shown). Therefore,

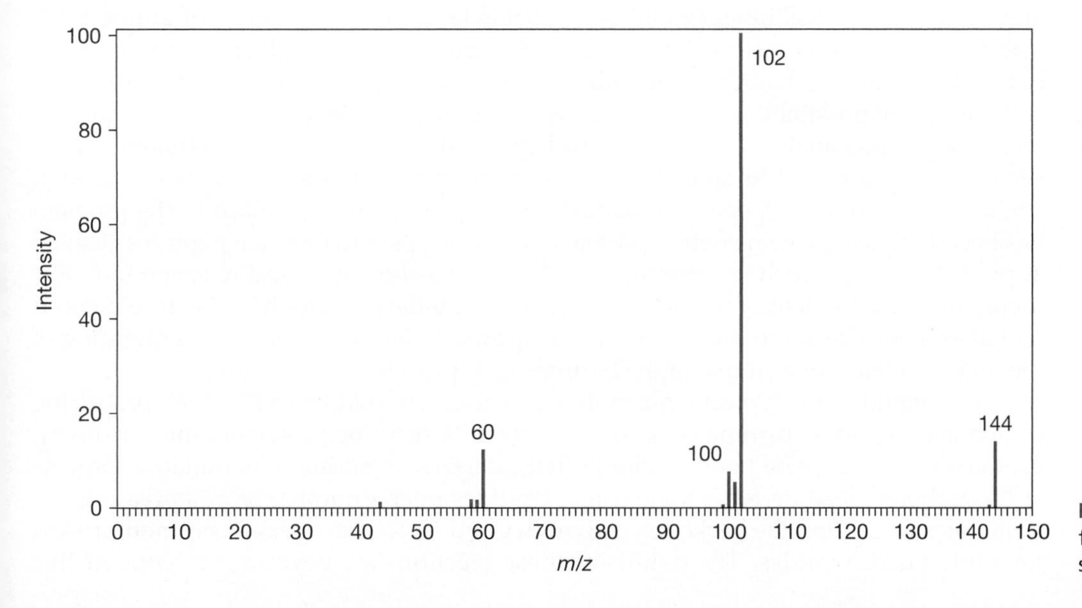

FIGURE 9-8 CID of *m/z* 144 formed by ESI of the analyte in structural analysis 4.

m/z 60 must represent a second loss of 42. Using the same logic, one would conclude it was propene loss. This line of reasoning leads to the partial structure in Scheme 9-5.

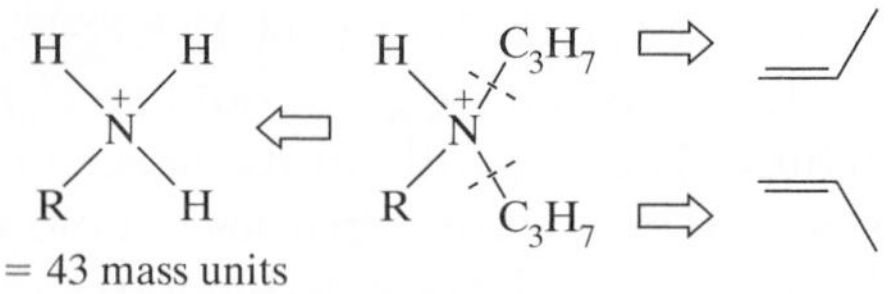

R = 43 mass units

Scheme 9-5

This structure is consistent with a tripropylamine, but there is no information about whether the propyl groups are *n*-propyl or isopropyl. The spectrum in Figure 9-8 is, in fact, derived from tripropylamine, that is, *n*-propyl.

9-4 BIOPOLYMERS

The analysis of spectra from polymeric materials is one of the great success stories in modern mass spectrometry, and makes it an indispensable tool in the analysis of large, complex materials with varying repeating units, particularly biopolymers such as proteins and polynucleotides. As in the analysis of relatively small organic compounds, it is the fragmentation patterns of polymers that reveal their structures, and the pattern is dependent on the polymer breaking down in predictable ways when it is activated. This field of study is rich, and methods for the analysis of many synthetic and biopolymers have been developed. This section, however, focuses only on the analysis of peptides and proteins by mass spectrometry. This type of work is at the foundation of many proteomics studies and provides a good example of how mass spectrometric data, in conjunction with database information, can be used to elucidate the structures of exceptionally large, complex molecules. The same general sequencing strategies are used with other biopolymers such as polynucleotides and polysaccharides. In synthetic polymers, sequencing is generally not the primary goal, and mass spectrometry is a tool for analyzing the size and dispersion of the mixture of polymeric products.

9-4a Peptides and Proteins

In the analysis of peptides, it takes only a few amino acid residues to create an analytical problem with a seemingly intractable number of structural possibilities. For example, a tetrapeptide composed of the common amino acids offers 160,000 (20^4) possible sequences. Because all of the common amino acids have different molecular weights (with the exception of leucine and isoleucine), the number of possibilities is greatly reduced if you can determine the molecular weight of the peptide by mass spectrometry. Nonetheless, at least 24 possibilities still are left (4!, based on sequence of amino acids), and many more possibilities would be available because combinations of amino acids lead to the same combined weight. In short, one must contend with an enormous number of structural possibilities for peptides of a modest size, and proteins offer astronomical numbers of possibilities. How can mass spectrometric data provide insight into such an overwhelming analytical problem? The key is in the regular patterns of fragmentation seen in peptides as well as some very powerful bioinformatics tools that can be used in conjunction with mass spectrometric data. The most common approach to the problem has been to digest proteins with a protease such as trypsin to produce peptides cleaved at predictable sites. Such an approach is referred to as *bottom-up* and relies on bioinformatics methods to identify the protein sequence once the peptides have been sequenced and identified. The alternative is a *top-down* approach that involves direct sequencing of the intact protein. An example of the bottom-up approach is given below.

For peptides, the typical ionization source is either MALDI or ESI, each producing protonated peptides. Turning back to some of the basic principles of organic chemistry, protonation of a peptide bond (amide) often catalyzes an addition/elimination process at the carbonyl, leading to C–N cleavage. Peptides offer a number of potential nucleophilic sites (including the backbone carbonyls), and C–N cleavage is a common process for protonated peptides. The details of these reactions are beyond the scope of this discussion, but the key issue is that this process provides a regular pattern in the fragmentation pathways of peptides. Because peptides can fragment in different ways under different activation conditions, it has been useful to adopt a standard, general nomenclature for peptide fragmentations that was developed by Klaus Biemann (Scheme 9-6).

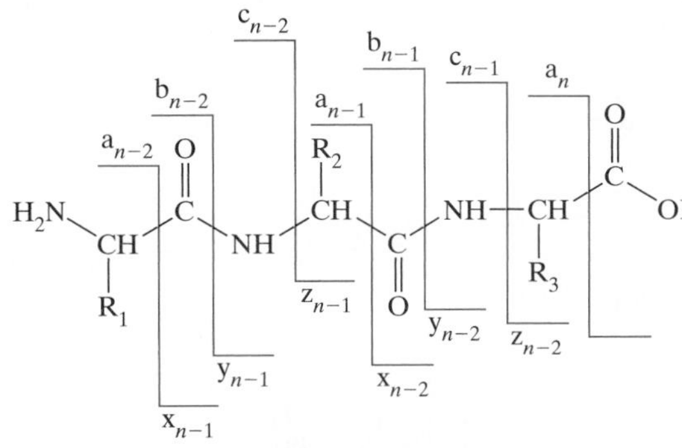

Scheme 9-6

Here, lower-case letters are used to define the site of the cleavage. In the case of C–N cleavage at the carbonyl, b and y ions are formed. The b ions correspond to the fragmentation product from the *N*-terminal side (left side as shown), and the y ions correspond to the fragmentation product from the *C*-terminal side (right as shown). In a singly charged ion, only one of these fragments would carry a charge, but in multiply charged ions, both could be charged and detected. The lower-case numbers indicate the number of amino acid residues in the fragment. For example, a b_5 ion represents the first five amino acid residues in a peptide, whereas a y_5 represents the last five. The y ions are simply truncated peptides, but the b ions are rearranged as a part of the addition/elimination process and often are cyclic species. Examples of structures are given in Scheme 9-7. In the y_3 ion,

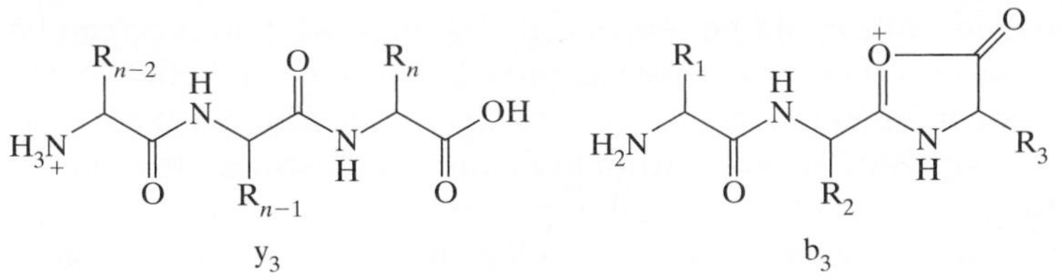

Scheme 9-7

n refers to the total number of residues in the peptide (only the last three are in y_3). In the b_3 ion, the carbonyl of one amide was the nucleophile that attacked the adjacent amide carbonyl. Such an oxazolidinone is believed to be a common structure for b ions. In this scheme, the first three amino acids of the protein are represented. Other fragmentation processes lead to different cleavage points. For example, electron capture dissociation leads to mainly c and z ions. It is not important which type of bond in the peptide backbone cleaves, just that it cleaves reliably under the fragmentation conditions.

The next question is how this fragmentation data can lead to a peptide sequence. In the ideal situation, fragmentation from collision-induced dissociation would lead to a continuous series of b or y ions. The difference in mass between consecutive members of the series would be the mass of one of the amino acids. With the whole series, one could sequentially determine the mass of each of the amino acids and obtain the sequence of the peptide. This process is outlined in Scheme 9-8. For example, the difference in mass

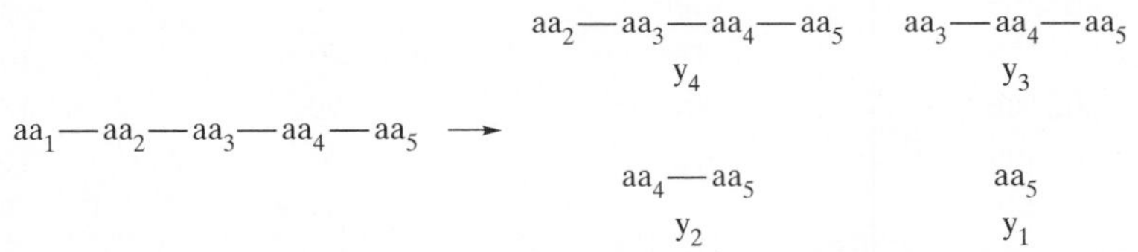

$aa_1—aa_2—aa_3—aa_4—aa_5 \rightarrow$ $aa_2—aa_3—aa_4—aa_5$ (y_4); $aa_3—aa_4—aa_5$ (y_3); $aa_4—aa_5$ (y_2); aa_5 (y_1)

Scheme 9-8

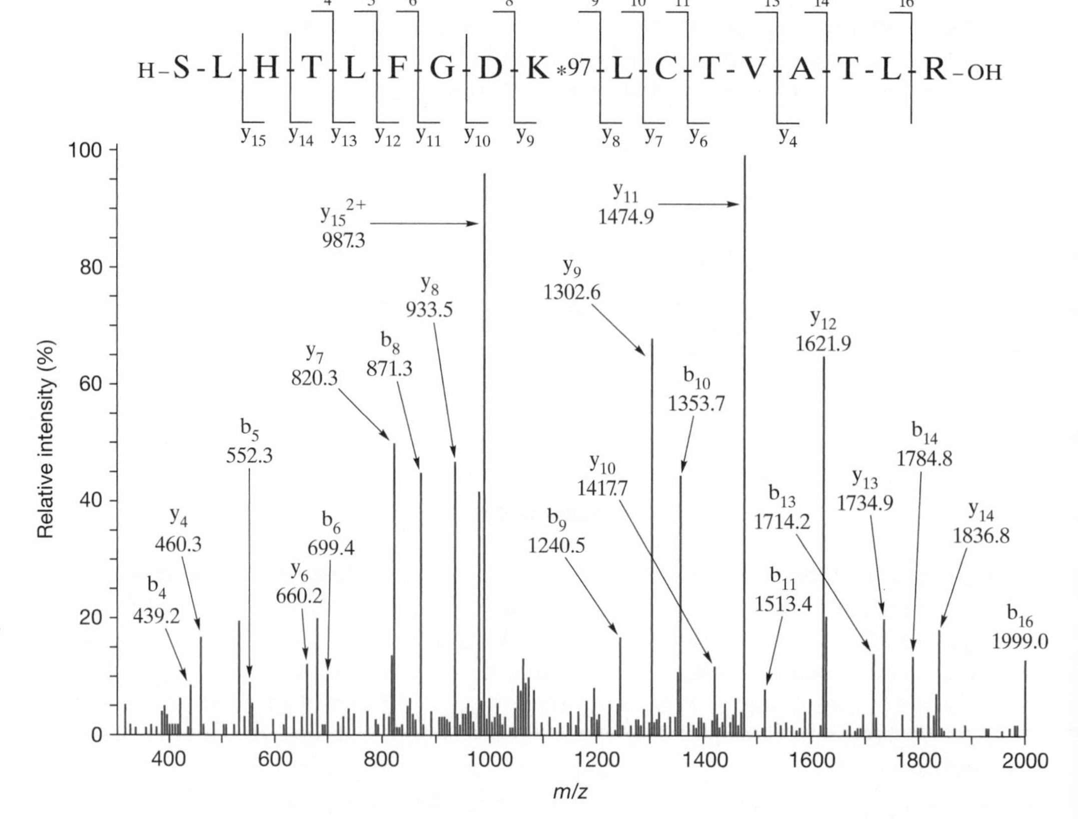

FIGURE 9-9 A peptide derived from the digestion of human serum albumin with trypsin. One-letter amino acid abbreviations are used. The lysine residue (residue 97 in the protein sequence) bears an oxidative modification and has an unconventional weight.

between the full peptide and y_4 is the weight of aa_1, and the difference in mass between y_4 and y_3 is the weight of aa_2. This approach is general and is a way that any polymer with varying repeating units could be sequenced. The process, however, can be much more complicated than this simple example suggests. In particular, if both b and y ions are produced during the fragmentation, how can one determine from the mass whether an ion is from the b or y sequence? The spectrum in Figure 9-9 illustrates this complication. There are a good number of b and y ions in the spectrum, but they are mixed together with no obvious way to distinguish between them. Of course, with the peaks labeled, it is easy to see, for example, that the gap between y_{10} (m/z 1417.7) and y_{11} (m/z 1474.9) corresponds to glycine, which has a residue mass of 57 units (the molecular weight of glycine is 75, but the 18 units of an H_2O component are lost when it is incorporated into a peptide and an amide bond is formed). A full list of amino acid residue masses is given in Table 9-2.

TABLE 9-2 Masses of Amino Acid Residues[a]

Amino Acid	One-Letter Abbreviation	Monoisotopic Residue Mass	Amino Acid	One-Letter Abbreviation	Monoisotopic Residue Mass
Ala	A	71.03711	Leu	L	113.08406
Arg	R	156.10111	Lys	K	128.09496
Asn	N	114.04293	Met	M	131.04049
Asp	D	115.02694	Phe	F	147.06841
Cys	C	103.00919	Pro	P	97.05276
Glu	E	129.04259	Ser	S	87.03203
Gln	Q	128.05858	Thr	T	101.04768
Gly	G	57.02146	Trp	W	186.07931
His	H	137.05891	Tyr	Y	163.06333
Ile	I	113.08406	Val	V	99.06841

[a]The actual amino acid mass is larger by 18.01056, the mass of H_2O lost in forming the peptide bond.

In general, peak labeling depends on the application of sophisticated bioinformatics methods or software designed to identify overlapping sequences in mass lists. In the first approach, fragmentation spectra are predicted for all of the peptides that might potentially be in a sample, and the one that provides the best match to the experimental spectrum is used in the labeling process. This method requires knowledge of the sequences of all the proteins that could be in the digest that produced the peptide of interest. Although this might seem like an extreme requirement, the various genome projects have made this type of data routinely available for a variety of species. It still represents a significant computational challenge, but powerful software such as SEQUEST, MASCOT, and Protein Prospector have been developed and have made this task routine. The spectrum in Figure 9-9 was assigned in this way. The second approach is referred to as *de novo* sequencing and starts with no assumptions other than that the peptide is composed of the common amino acids (some implementations allow for the presence of unknown, atypical amino acids). Here the software seeks natural patterns in the mass data and from this information constructs a sequence. This method is much more dependent on having long stretches of sequential fragments, such as b and y ions, and is more likely to fail than database methods.

9-4b Other Biopolymers

Similar approaches have been developed for the analysis of oligonucleotides and oligosaccharides. The former are relatively straightforward because the repeating units are simple and fragmentation patterns are well established. The situation with oligosaccharides is much more complicated. Here, the varying repeating units often have the same mass (stereoisomers), so they cannot be easily distinguished during fragmentation. In addition, sugars can display a variety of linkage patterns and involve branch points. All of these factors have made the sequence analysis of oligosaccharides a challenge for mass spectrometry, but good headway has been made in establishing expected fragmentation patterns for common structural motifs.

Problems

*The answer is provided for part **(a)** of Problem **9-1** as a guide for subsequent parts.*

9-1 Even species with a relatively simple molecular formula can produce rich isotopic patterns. For each of the following molecular formulae, (i) provide a list of the expected nominal masses for the peaks in the isotopic envelope of the molecular ion, (ii) give the exact masses for all the contributors to the isotopic envelope of the molecular ion, and (iii) predict the dominant *m/z* in the isotopic envelope of the molecular ion. In answering parts (i) and (ii), consider only combinations that would contribute more than 1% to the intensity of the molecular ion.

(a) C_3H_7ClS

Answer First consider each of the elements. With carbon, ^{13}C is important, but at 1% natural abundance and three carbons, only molecules with zero or one ^{13}C are expected to contribute at a level greater than 1%. With seven hydrogens, the deuterium content should be negligible. Chlorine provides significant signals for ^{35}Cl and ^{37}Cl. Finally, sulfur has greater than 1% contributions from ^{32}S and ^{34}S, but not ^{33}S.

(i) Given these isotopes, the lowest mass in the isotopic envelope corresponds to $^{12}C_3{}^1H_7{}^{35}Cl^{32}S$ with a nominal *m/z* of 110. The highest mass in the envelope corresponds to $^{12}C_2{}^{13}C^1H_7{}^{37}Cl^{34}S$ with a nominal *m/z* of 115, but the presence of two low-abundance isotopes (^{13}C and ^{34}S) pushes it below the 1% threshold defined in the problem. Therefore, $^{12}C_3{}^1H_7{}^{37}Cl^{34}S$ is the highest representative with an *m/z* of 114. The presence of a ^{13}C, a shift of one mass unit from ^{12}C, ensures that there are peaks at each intermediate *m/z*, so the envelope includes 110, 111, 112, 113, and 114.

(ii) $^{12}C_3{}^1H_7{}^{35}Cl^{32}S$: 109.99570; $^{12}C_2{}^{13}C^1H_7{}^{35}Cl^{32}S$: 110.99905; $^{12}C_3{}^1H_7{}^{37}Cl^{32}S$: 111.99275; $^{12}C_3{}^1H_7{}^{35}Cl^{34}S$: 111.99149; $^{12}C_2{}^{13}C^1H_7{}^{37}Cl^{32}S$: 112.99610; $^{12}C_2{}^{13}C^1H_7{}^{37}Cl^{34}S$: 112.99485; and $^{12}C_3{}^1H_7{}^{37}Cl^{34}S$: 113.98854. Notice that the nominal peaks at *m/z* 112 and 113 have two components with different mass defects.

(iii) In this case it is straightforward to determine that $^{12}C_3{}^{1}H_7{}^{35}Cl^{32}S$ is the dominant species because it contains each of the most abundant isotopes and there are not enough of any of the atoms for statistical distributions to have a significant effect.

(b) CH_2BrCl

(c) $C_{10}H_{10}Fe$

(d) $C_3H_9O_3$

9-2 The spectrum below was derived at relatively low resolution from ESI of a basic analyte under positive ion conditions. The isotope envelope is shown for a major peak in the electrospray spectrum (no fragmentation has occurred). Assume that the species contains only C, H, and N and assign the molecular formula.

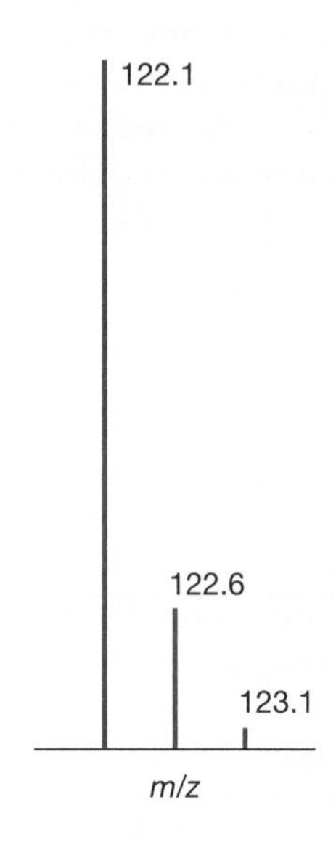

9-3 Derive as much structural information as possible from each spectrum. Assume that it is a 70 eV EI spectrum.

(a) In the epoxidation of 1-pentene, the product mixture contained a small amount of the compound responsible for the following spectrum, in which exact mass data are given for the molecular ion.

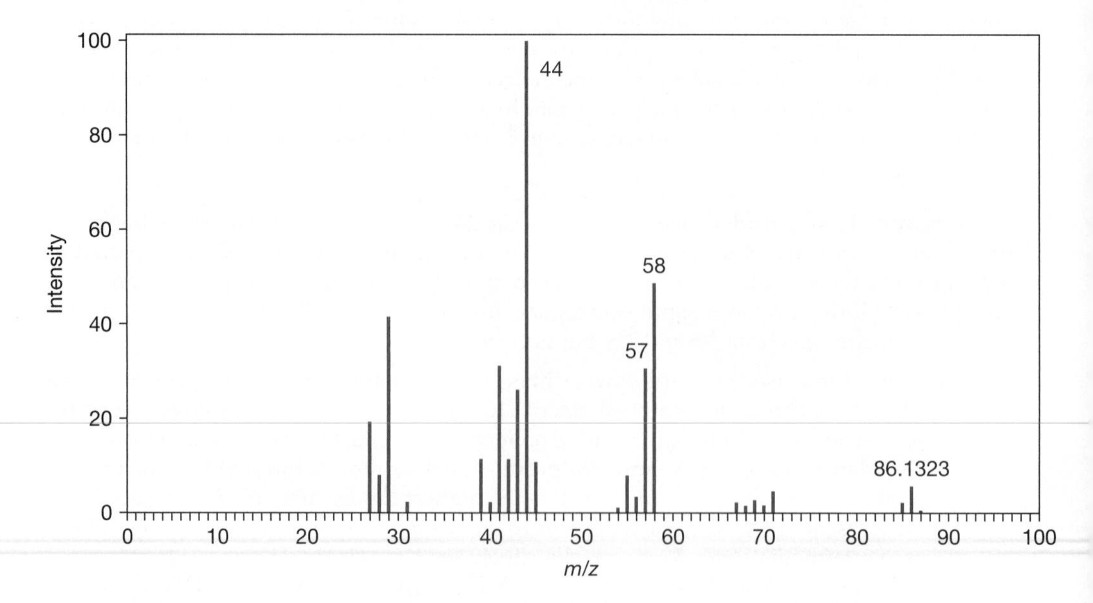

(b) In the hydrolysis of pentanenitrile, the compound responsible for the following spectrum (with exact mass data for the molecular ion) was a significant contributor to the product mixture.

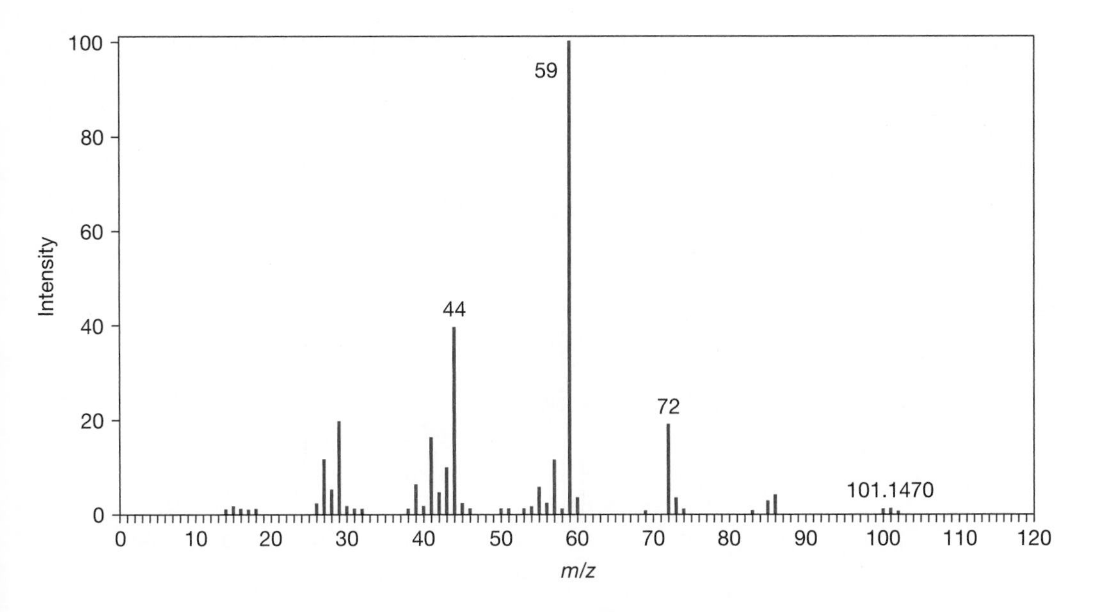

(c) In the oxidation of 2-hexanol, unusually harsh conditions were used. The compound responsible for the following spectrum (with exact mass data for the molecular ion) was a significant contributor to the product mixture.

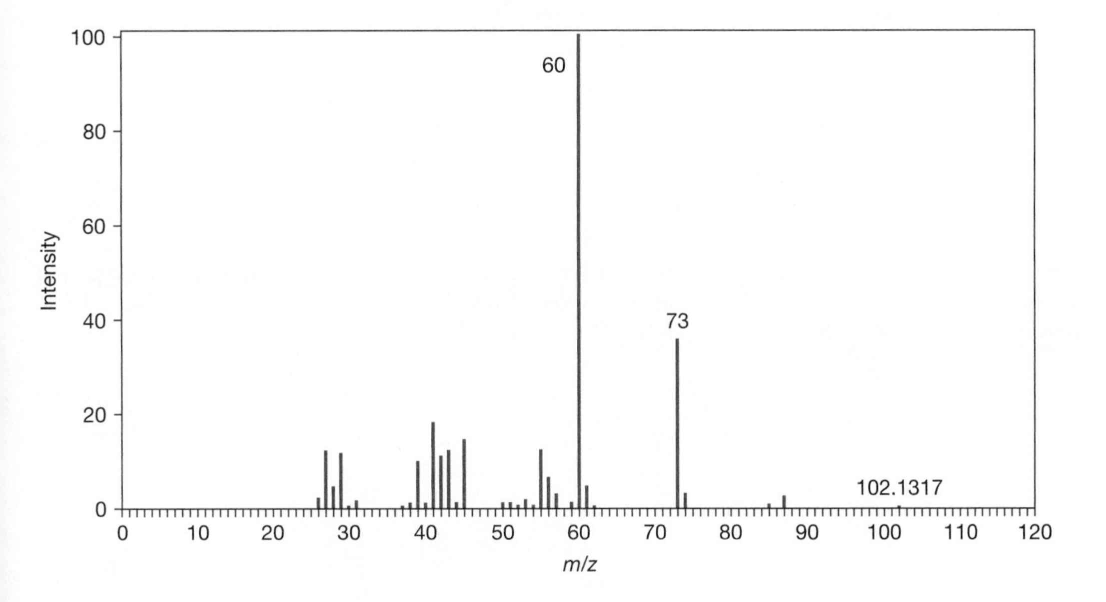

(d) The following EI spectrum (with exact mass data for the molecular ion) is from a headspace analysis of an aqueous solution. It is expected that the solution contains one volatile species from the decomposition of a phase-transfer catalyst.

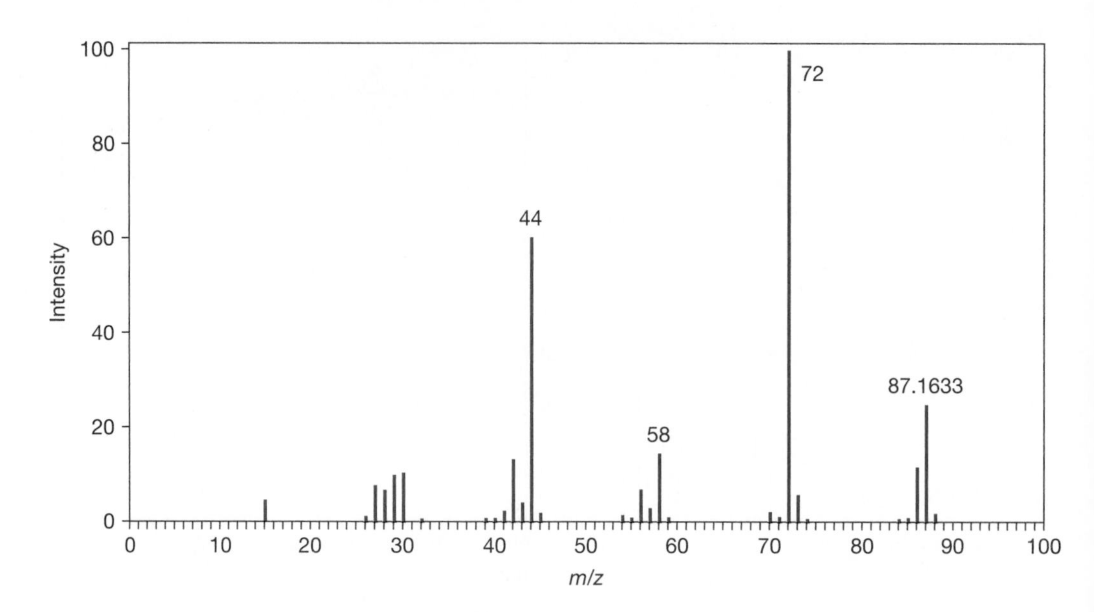

9-4 A peptide ion is formed by ESI under positive ion conditions. The major ion in the spectrum has a mass of 763.3403. It produces the following series of y ions. Determine the sequence.

y_1	90.055	y_4	450.2129
y_2	246.1561	y_5	579.2555
y_3	347.2037	y_6	676.3083

Bibliography

Also see Chapter 8 bibliography.

9.1 C. Dass, *Fundamentals of Contemporary Mass Spectrometry*, Hoboken, NJ: John Wiley & Sons, 2007.

9.2 M. S. Lipton and L. Paŷa-Tolic (eds.), *Methods in Molecular Biology*, **492**, 1 (2008).

9.3 M. Kinter and N. E. Sherman, *Protein Sequencing and Identification Using Tandem Mass Spectrometry*, New York: John Wiley & Sons, 2000.

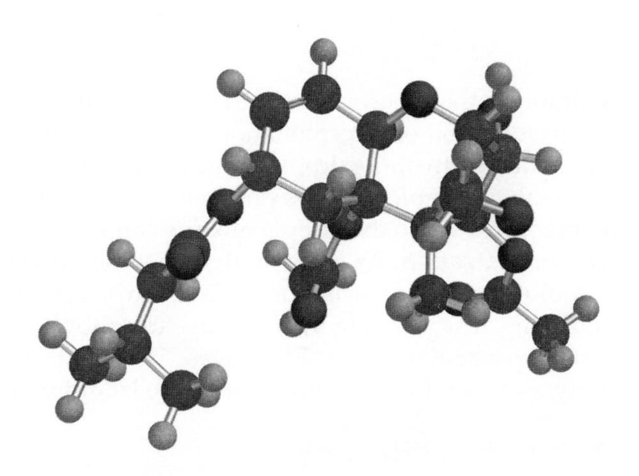

Quantitative Applications

10-1 INTRODUCTION

Along with identification of analytes, mass spectrometry can be used for a variety of quantitative applications. These range form the direct determination of analyte concentrations to the measurement of thermodynamic quantities such as acidities and basicities. A key aspect of any quantitative measurement by mass spectrometry is accounting for effects related to the variable efficiency of ionization and analyte detection. Response effects in mass spectrometry are more complicated than those found in typical optical spectroscopy (or proton NMR), and dependent on many factors, including the mass, basicity/acidity, hydrophobicity, and structural stability of the analyte. In this chapter, the general features of quantitative mass spectrometry are described briefly, and some specialized applications focused on determining the physical properties of gaseous ions and neutrals are introduced.

10-2 FACTORS CONTROLLING MASS SPECTROMETER SIGNAL INTENSITIES

10-2a Ionization Efficiency

In the majority of spectroscopic methods described in this book, the measurements rely on well-defined processes such as the absorption or emission of a photon. These events have a solid physical foundation, can be modeled theoretically, and generally can be initiated in a very reproducible manner. In contrast, ionization methods tend to be based on an aggregate of poorly defined events, and efficiencies depend on a whole array of factors, many idiosyncratic to the design properties of the ion source. The most reproducible of the ionization sources tends to be electron ionization. If the accelerating voltage, flux, and analyte pressure are carefully controlled, there is no reason why the source would not consistently produce ions at a constant rate. Even if these variables are managed (a process that is not trivial), the ions from the source need to be transferred to the mass analyzer, potentially isolated, and then detected in a highly predictable and reproducible manner. These requirements are difficult to meet, particularly in comparisons across different experimental platforms. Other sources such as MALDI and ESI rely on a collection of reactions in the condensed phase, plasma, or gas phase before the ultimate analyte ion is formed. Again, careful control of conditions can lead to a good deal of reproducibility in these ion sources, but it would be difficult to obtain truly quantitative measures of concentration from their absolute ion intensities.

Another issue is ion suppression. In all ionization sources, a certain number of charges are created, and the molecules within the source compete for the charge carriers, for example, electrons, protons, or basic anions. Consequently, the intensity of a particular analyte can be dependent on the other species in the sample or the elution mixture. This problem is very common in ESI. If samples have significant inorganic salt concentrations, these permanently charged species, rather than the analyte (which may require protonation or deprotonation to be charged), can act as the dominant charge carriers in the electrospray droplets and potentially suppress the signal of the less easily ionized analyte completely. This possibility is not limited to ESI. Matrix effects are seen in all ion sources when the sample mixture contains a species that can compete with the analyte for ionization. As a result, it is generally difficult to correlate concentrations with absolute source intensities in mass spectrometry. Nonetheless, with careful control of conditions, it is possible to create useful calibration curves based on absolute ion intensities, particularly with EI sources. These curves, however, need to be updated regularly, and therefore the approach can be labor intensive.

10-2b Detection Efficiency

If an ionization source had perfect reproducibility and gave intensities proportional to analyte concentrations, the next challenge would be to transfer the ions to the mass analyzer and detect them without any bias. Detection efficiency, defined in this way, is based to a great extent on the properties of the mass analyzer, as discussed in detail in Chapter 7. A key parameter in quantitation is dynamic range. This quantity is a measure of the variation in signal intensities that can be determined concurrently (or nearly so) in a mass analyzer (see Table 7-7). What might limit the dynamic range of a mass analyzer? Consider a quadrupole ion trap. These instruments generally trap 100,000 or fewer ions during a scan. Under these circumstances, it is difficult to measure a ratio of two species larger than 1000:1 because at greater ratios there would be too few ions of the low concentration species for accurate counting statistics. As a result, quadrupole ion traps tend to have small dynamic ranges and are limited to making comparisons between species with relatively similar concentrations. At another extreme are quadrupole mass filters. These instruments, as well as triple quadrupoles (their tandem mass spectrometer equivalent), have large dynamic ranges because they are capable of transmitting a very large number of ions and therefore can obtain good statistics for species with both high and low relative concentrations. This property is one of the reasons that triple quadrupoles are popular instruments for quantitation. Good dynamic ranges also are seen in sector instruments and, to a lesser extent, in TOF instruments. Because FT-ICR is a trapping technique with limited ion storage capabilities, it suffers from the same issues in dynamic range as quadrupole ion traps. Thus quantitation in mass spectrometry is heavily dependent on the nature of the mass analyzer.

Another issue is mass discrimination. Although mass analyzer/detection systems are designed with the goal of offering equal response across a wide m/z range, this equality generally is not realized in practical applications. As a result, an equal number of ions in the mass analyzer at two different m/z values do not typically give equal peak heights. Often, if the m/z values are close, the effect is small, but it can be quite significant for widely separated peaks. Again, this problem can be addressed with calibration curves and internal standards.

10-3 INTERNAL STANDARDS AND RELATIVE INTENSITIES

Despite the limitations stated above, mass spectrometry can be used effectively in many situations in which quantitative data are desired. The key is to account for the variability in source conditions and choose conditions that suit the dynamic range of the mass analyzer. The first of these issues is best handled by the use of internal standards and the measurement of relative rather than absolute signal intensities. Because the analyte and internal standard are simultaneously ionized, variations in ion production from the source should not affect the ratio of analyte ions to those derived from the internal standard. In addition, the concentration of the internal standard can be adjusted to limit the dynamic range needed for the analysis. Matrix effects are still important because the analyte and internal standard could experience different levels of ion suppression from the matrix if they have significantly different physical properties. For this reason, isotopically labeled analytes often are used as internal standards in quantitative analyses. They match the properties of the analyte almost perfectly, yet can be distinguished easily

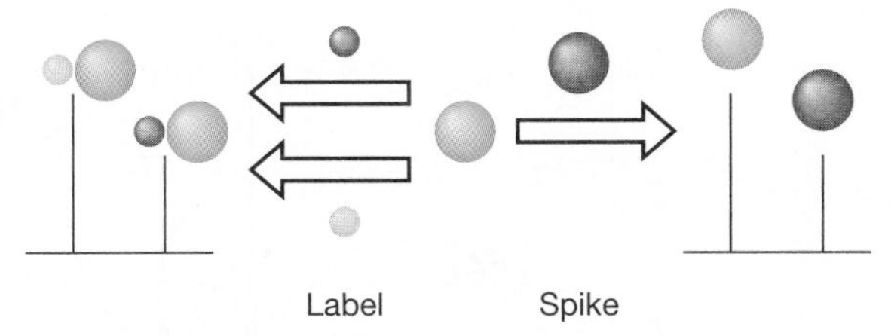

FIGURE 10-1 Strategies for using isotopic internal standards. The sample can be spiked with a known amount of an isotopically labeled analog. Alternatively, samples from two different reaction conditions (control and modified condition) can be labeled with isotopic tag species and combined—the difference in intensities quantifies the effect of the modified condition on the analyte concentration. This label approach is often used to compare samples from two different biological conditions.

by their mass differences. In any case, it is generally best to establish calibration curves with matrix conditions similar to those expected in the actual analysis. With a calibration curve, it is not necessary for the detection efficiency of the analyte ion and the internal standard to be identical, but any differences in detection efficiency must be reproducible.

Another issue with internal standards arises when chromatography is part of the mass spectrometry experiment. During chromatography, the internal standard is separated from the analyte; consequently, they are ionized and detected at different time points, and therefore the analysis becomes dependent on a constant ionization rate from the source. Matrix effects also become an issue because the analyte and internal standard are ionized out of different mixtures. Using isotopically labeled analytes as the internal standard can lead to co-elution and eliminate these problems; however, co-elution is not necessarily the case with deuterium labels because deuteration slightly affects retention times, but ^{13}C and ^{15}N labels generally have little effect on the retention time and truly co-elute with the analyte. In any case the internal standard can be introduced into the experiment in a great number of ways, such as spiking the solution after chromatographic separation (Figure 10-1). The breadth of available quantitation schemes goes well beyond the scope of this text and, generally, an accurate mass spectrometric quantitation methodology can be developed for nearly any analyte.

The analysis of 5-nitroimidazole–based antibiotics in food products provides a simple, illustrative example of a typical approach. The target compounds are the drugs as well as their metabolites, which are outlined in Scheme 10-1. The last three species listed in

DMZ: R_1 = Me, R_2 = Me
RNZ: R_1 = Me, R_2 = $CH_2OC(O)NH_2$
MNZ: R_1 = CH_2CH_2OH, R_2 = Me
IPZ: R_1 = Me, R_2 = iPr
DMZOH: R_1 = Me, R_2 = CH_2OH
MNZOH: R_1 = CH_2OH, R_2 = CH_2OH
IPZOH: R_1 = Me, R_2 = $COH(CH_3)_2$

Scheme 10-1

Scheme 10-1 are metabolites. Each of these compounds is amenable to LC/MS analysis, and the study was completed using a triple quadrupole instrument. For the species with a CH_3 group at R_1, the corresponding CD_3 species were prepared and used as internal standards. For the compounds without a CH_3 group at R_1, deuterated DMZOH was used as the standard. The first step in the quantitation scheme is identifying the masses to be used for the analysis. For confident identification, fragment ions from MS/MS spectra are used, rather than the protonated analytes themselves. This experiment requires matching both the analyte and the fragment masses simultaneously, so coincident mass conflicts with other species in the sample are much less likely. The collision-induced dissociation (CID) spectra of two of the protonated analytes are shown in Figure 10-2. In each case, two product ions were chosen and their yields used for quantitation. This approach is referred to as *multiple reaction monitoring* (MRM). By using two product ions, a more reliable and robust quantification scheme is obtained (Figure 10-3). Once the quantitation masses are chosen, calibration curves can be generated by spiking the sample matrix (in this case a food product) with the analytes and standards, and then comparing the response of the analytes

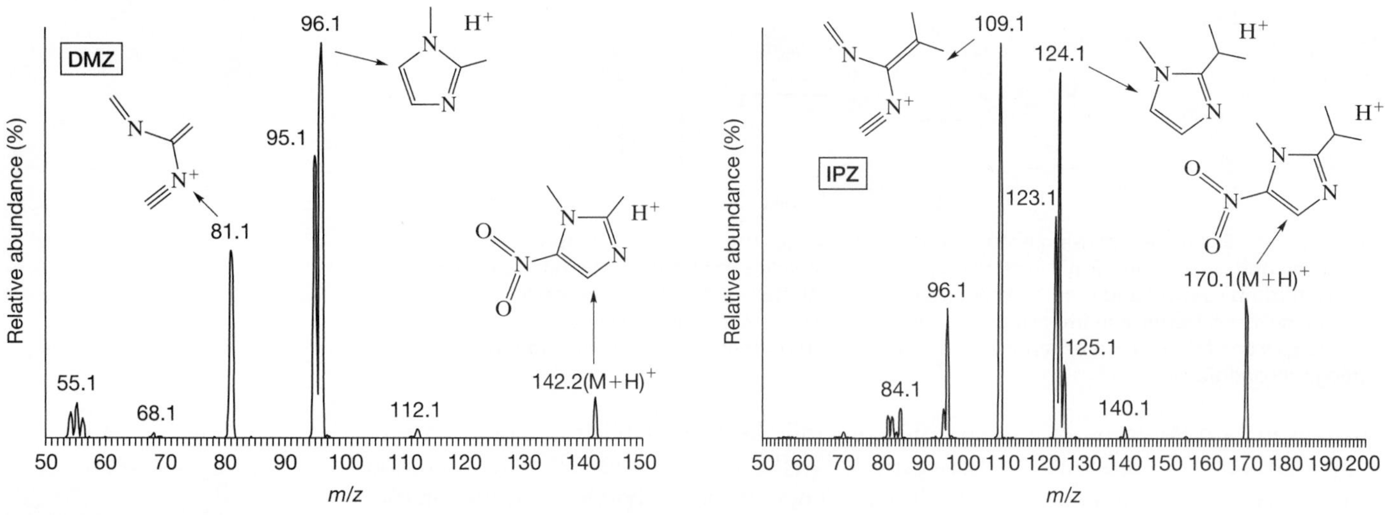

FIGURE 10-2 CID spectra of protonated nitroimidazole analytes. (Reprinted with permission from P. Mottier, I. Hure, E. Gremand, and P. A. Guy, *J. Agric. Food Chem.*, **54**, 2018 [2006]. Copyright 2006 American Chemical Society.)

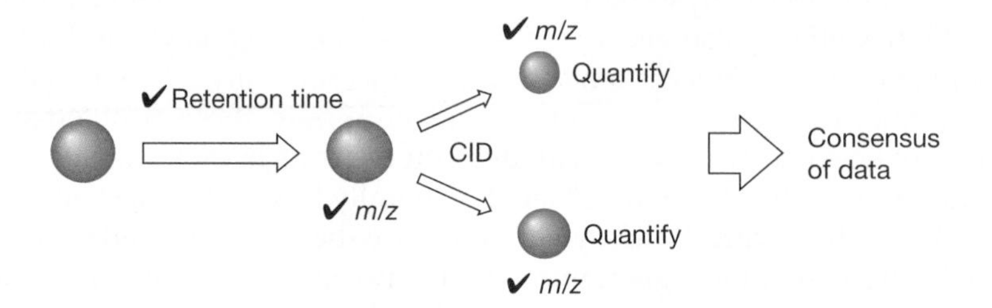

FIGURE 10-3 Multiple reaction monitoring scheme. Matching the retention time, parent's *m/z*, and CID products' *m/z* gives high confidence in the ions designated for quantification. The use of multiple ions for quantification allows for an internal check of consistency.

versus the internal standard. The MRM chromatogram can be recorded and the signal intensities integrated for the CID product ions from the various analytes in the study. In such a run, full mass spectra are recorded. When a signal for a protonated analyte is detected, CID spectra are recorded for it. The sequence cycles back to obtain more full spectra and again completes a CID scan whenever a signal is detected at the *m/z* of an analyte ion. This process is referred to as a *data-dependent scan sequence* (Figure 10-4). A set of MRM chromatograms for these analytes is given in Figure 10-5 (target masses used for MS/MS quantification are listed above spectra). In this system, the precision of the measurement at the parts per million level was better than 10% when an isotopically labeled standard was used, but rose to over 20% in cases in which an analogue was used—that is, the cases when R_1 was not CH_3—so that DMZOH was used as a surrogate label. Although this level of accuracy at such low concentrations is not generally required in the analysis of mixtures from organic syntheses, it is important in drug studies.

The popularity of using isotopically labeled internal standards has led to the commercial production of a wide range of common species with ^{2}H, ^{13}C, or ^{15}N labels, including many for common drugs, their metabolites, biochemically relevant species, environmental toxins, and simple organic materials. With more complex species such as peptides, alternative approaches have been designed because it is not practical to synthesize all of the potential peptides that might be useful in the analysis of a protein sample. In these cases, methods have been developed to derivatize the analyte with an isotopically labeled group (Figure 10-1). Here, a control sample and the analyte sample can be derivatized with reagents bearing different isotopic labels, and the observed ratio of peak heights can be used for quantification relative to the control sample. In the most sophisticated applications, different isobaric isotopomers (species with the same mix of isotopes, but in different structural locations) of the derivatization reagent can be designed such that they fragment under CID conditions to give products with different masses. In this way, the derivatized mixture (sample and control) give species that co-elute and give parent ions with identical masses. When isolated and fragmented, however, they give product ions with different

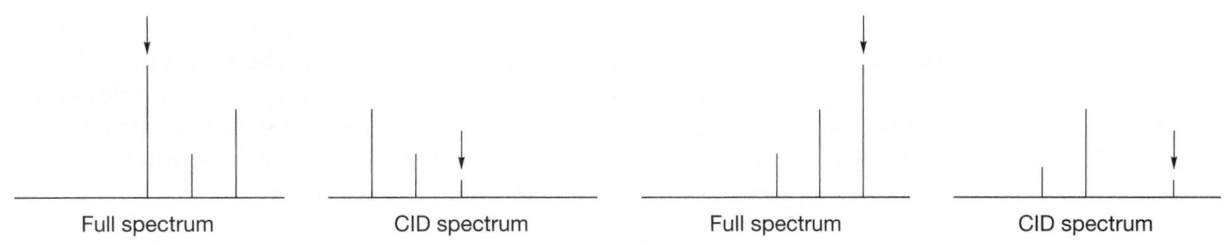

FIGURE 10-4 Data-dependent scan sequence. In the first full mass spectrum, the most intense peak is chosen (indicated by arrow). It is subjected to CID to produce the subsequent spectrum. Another full mass spectrum is taken at a slightly later retention time, and again the most intense peak is chosen for CID and produces the subsequent spectrum. Many variations are possible. Peaks chosen for CID can be restricted to those on a preset mass list, or peaks already subjected to CID can be excluded to avoid redundant data (the next most intense peak would be chosen instead).

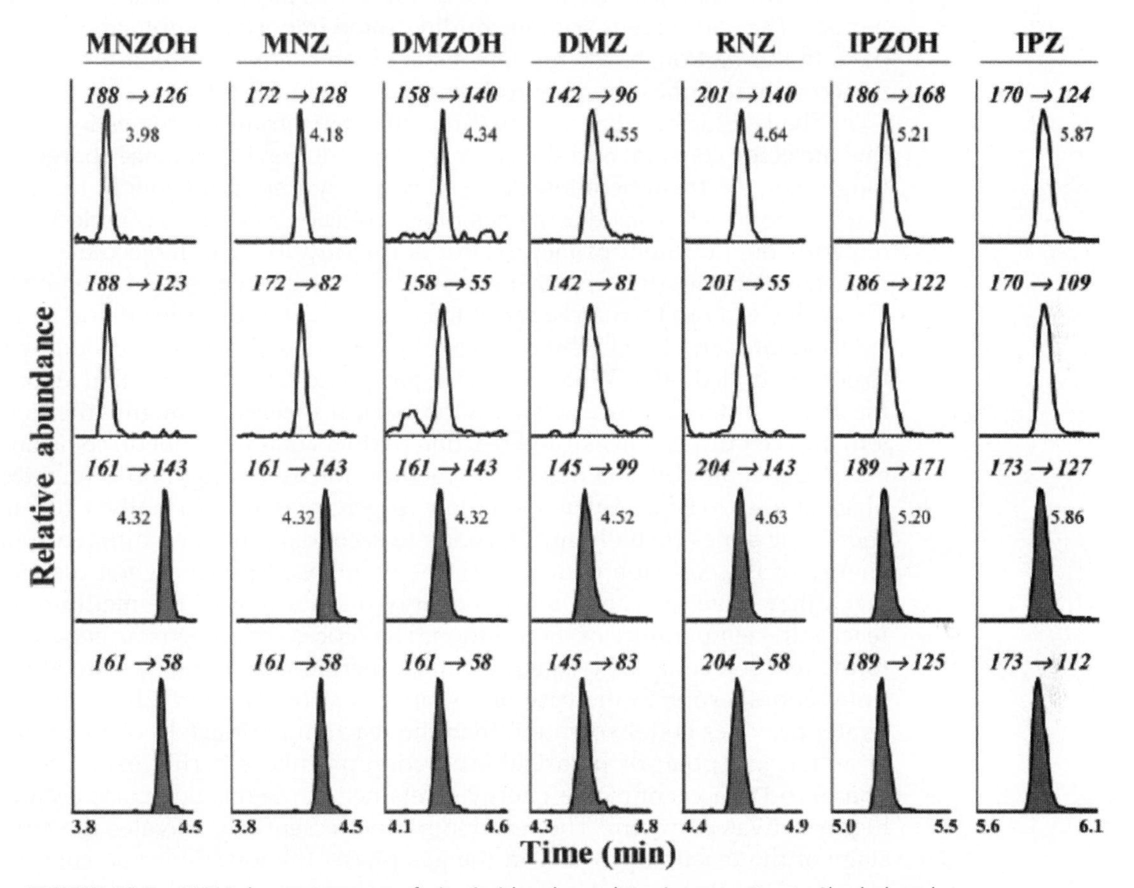

FIGURE 10-5 MRM chromatograms of nitroimidazole analytes in egg extracts. Shaded peaks represent the internal standards. The parent and product ion masses are indicated above each peak. (Reprinted with permission from P. Mottier, I. Hure, E. Gremand, and P. A. Guy, *J. Agric. Food Chem.*, **54**, 2018 [2006]. Copyright 2006 American Chemical Society.)

masses, and the peak ratio provides a measure of the relative concentrations. This approach can be very accurate because many of the uncertainties in ionization and detection are eliminated, given that the analyte and control have the same chromatographic, ionization, and fragmentation properties as well as the same mass—the isotopic difference is apparent only in the CID process.

10-4 MEASURING THERMODYNAMIC PROPERTIES WITH MASS SPECTROMETRY

Although mass spectrometry generally is viewed as a tool to obtain structural information, it also has been used extensively to determine the thermodynamic properties of ions and neutral species. These properties range from the obvious ones, such as gas-phase acidities, basicities, and ionization potentials, to some surprising ones, like the bond dissociation energies of neutral compounds. The following subsections illustrate several common applications of mass spectrometry in thermodynamic measurements. Many require specialized instrumentation, but some can be completed using commercial instruments with no modifications. The examples provide an overview of the broad capabilities of mass spectrometry in organic chemistry.

10-4a Gas-Phase Ion Chemistry

In most applications of mass spectrometry, the goal is to restrict ion chemistry to the source region and avoid gas-phase reactions in the mass analyzer aside from CID processes. This is one of the reasons that moderate to high vacuum is needed for the manifold housing the mass analyzer. It is possible, however, to employ the mass analyzer as a versatile chemical reactor and use it to probe gas-phase reactivity and thermodynamics. The subsequent sections outline some important applications of this type of data. In this section, a few fundamentals of ion chemistry are outlined, and contrasts with condensed-phase and neutral chemistry are highlighted.

The key factors that set apart ion chemistry from processes in solution are the low dielectric constant and the absence of counterions to balance charges. As a result, long-range electrostatic interactions play a very important role in reactivity. A good starting point is to consider the gas-phase collision rate for ion/molecule reactions. It depends on a number of factors, but is roughly 10^{-9} cm^3 molecule^{-1} sec^{-1}. Put into more familiar units, it is approximately 10^{12} M^{-1} sec^{-1}, a remarkably high rate constant. This value is more than an order of magnitude higher than gas-phase neutral/neutral collision rates and two to three orders of magnitude higher than condensed-phase diffusion controlled rates. What makes the gas-phase rate so high is that at long distances, the charge of the ion interacts with the neutral species (through either ion/dipole or ion/induced-dipole forces), and reaction partners that were not on collision trajectories are brought onto collision trajectories by these forces. The high collision rates allow gas-phase studies to be completed with low reagent pressures (usually $<10^{-5}$ torr) and on short time scales (usually milliseconds to seconds). The next difference involves the energy of the collision complex. In the condensed phase, when reaction partners meet, they have the same internal energy distribution as the medium; that is, they react at the temperature of the solution. The process of going from separated reactants to the reaction complex is usually almost thermoneutral in solution, although sometimes endothermic in the case of an ion in a strongly stabilizing solvent like water (water provides better solvation than the reaction partner). In contrast, the approach of an ion to a polar or polarizable reaction partner is highly exothermic in the gas phase, and in a vacuum this energy is retained in the reaction complex and leads to a highly activated system. The reaction partner essentially solvates the ion in this first stage of the reaction process. In the gas phase, the ion/molecule complex typically contains 10 to 25 kcal mol^{-1} of excess internal energy, which can be used to fuel subsequent reactions. This situation is outlined in Figure 10-6 and has been referred to as a *Brauman double-well potential*. In the particular case of Figure 10-6, the transition state is lower in energy than the separated reactants, so every collision leads to a complex

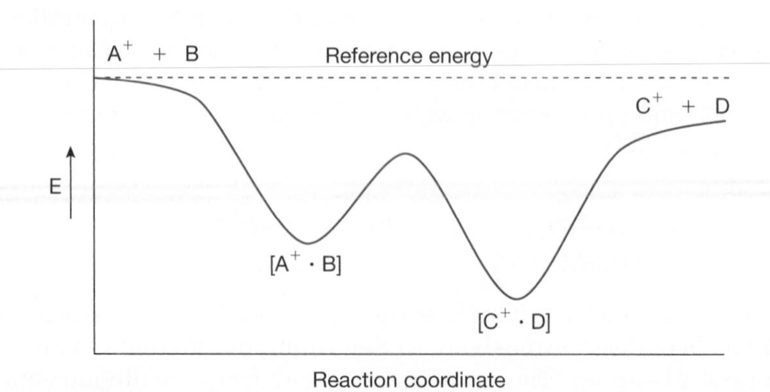

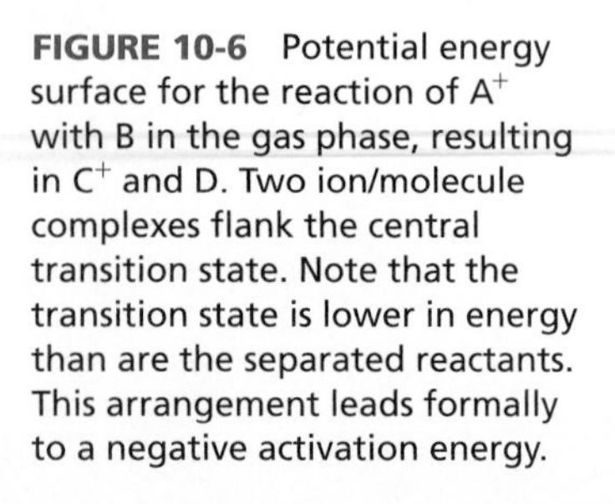

FIGURE 10-6 Potential energy surface for the reaction of A^+ with B in the gas phase, resulting in C^+ and D. Two ion/molecule complexes flank the central transition state. Note that the transition state is lower in energy than are the separated reactants. This arrangement leads formally to a negative activation energy.

with sufficient energy to cross the barrier, and a very efficient reaction is expected, although reaction dynamics effects can slow the reaction below the collision-controlled limit. In the analogous condensed-phase reaction, the reactants and reaction complex would be at roughly the same energy (the energy of the reaction complex in Figure 10-6) and face a significant barrier to crossing the transition state. In gas-phase reactions without a barrier, for example, proton transfers between heteroatoms such as N and O, the surface collapses to a single well for the hydrogen-bonded complex. A final difference is that throughout its lifetime, the gas-phase reaction complex may not and generally does not experience any collisions. This situation is a result of the low pressures typically employed in mass spectrometers and the short lifetimes of the reaction complexes. Consequently, the complexes generally retain the reaction energy and must dissociate to release their excess energy (emission of a photon is another possibility, but it is not particularly common in small systems). Therefore, addition reactions are unusual in the gas phase (with the exception of very large molecules), and addition products either fragment back to reactants or proceed to a dissociation product. For example, the ion/molecule complexes in Figure 10-6 are the most stable species on the potential energy surface, but usually are not observed because they have the internal energy of the reactants, that is, the reference energy, and dissociation to reactants or products is energetically viable and strongly favored by entropy.

Although this discussion has focused on the differences between gas-phase ion chemistry and condensed-phase chemistry, there are a great number of similarities. In general, organic chemistry in the gas phase has many parallels with what is seen in solution. The following sections touch mainly on thermochemical measurements, but, as this discussion suggests, organic reaction chemistry also can be probed by mass spectrometry.

10-4b Gas-Phase Acidities and Basicities

In the condensed phase, one of the more important characteristics of organic compounds is their acid/base properties. As one might expect, these properties are highly dependent on solvation, and therefore gas-phase studies provide a picture of the analyte's intrinsic acidity or basicity. How, then, can a mass spectrometer be used to determine the acid/base properties of an analyte? Although self-ionization, for example, $AH \rightarrow A^- + H^+$, is not possible in a low-dielectric medium such as a vacuum, relative acidities and basicities can be determined by measuring equilibrium constants. This approach is the same as that used in solution with species that are too weakly acidic or basic to undergo self-ionization. Establishing an equilibrium requires time, so these applications require instruments that can trap ions for an extended period, such as FT-ICRs and quadrupole ion traps. The experiment is rather straightforward. A low background pressure of the compounds to be compared is introduced into the mass analyzer's vacuum manifold. This capability generally is not a design feature of commercial mass spectrometers and requires some modification to the system. One of the ions of interest is generated by the ionization source and is isolated in the mass analyzer. At this point, gas-phase reactions can occur, and an equilibrium mixture of ions is produced. A measurement of relative gas-phase acidity is exemplified in eq. 10-1. The equilibrium constant

$$AH + B^- \rightleftharpoons A^- + BH \qquad \textbf{(10-1)}$$

can be calculated from the intensities of the ions, $I(A^-)$ and $I(B^-)$, and the pressures of the two analytes, $p(AH)$ and $p(BH)$, according to eq. 10-2. Because the ion concentrations in

$$K = \frac{I(A^-)\,p(BH)}{I(B^-)\,p(AH)} \qquad \textbf{(10-2)}$$

mass spectrometers are always very small compared with the pressures of neutral reagents (usually by a factor of 10^5 or more), the pressures of the neutral reagents can be treated as a constant, that is, the pressure introduced into the instrument.

In the gas phase, acidities and basicities generally are defined in terms of enthalpy changes rather than equilibrium constants, that is, pK_a, because self-ionization constants in the gas phase have very low values, which are awkward to present in terms of pK_a. The most common quantities used in the gas phase are the enthalpy required for self-ionization (ΔH_{acid}, eq 10-3) and the proton affinity (PA, eq. 10-4), the enthalpy released

$$AH \xrightarrow{\Delta H_{acid}} A^- + H^+ \quad (10\text{-}3)$$

$$A + H^+ \xrightarrow{PA = -\Delta H} AH^+ \quad (10\text{-}4)$$

in protonating a species. These equations describe the same process—deprotonation/protonation—and are numerically equivalent, although ΔH_{acid} is usually used to characterize neutral species that produce anions by dissociation and PA is used to characterize the protonation of a neutral species.

In Table 10-1, some typical values of ΔH_{acid} and PA are given for common compounds and functional groups. You will notice that the gas-phase trends in acidity/basicity parallel those seen in solution, for the most part, but the absolute values are very different. For example, H_2O has a ΔH_{acid} value of 390 kcal mol^{-1}, which corresponds to a pK_a of more than 280! The other difference is that delocalization and polarizability play a much greater role in the gas phase. For example, phenol is nearly as acidic as acetic acid and, surprisingly, toluene is more acidic than water. This difference is driven by the low dielectric of a vacuum and, in the case of H_2O, the absence of hydrogen bonding to stabilize the anion. Similar, but less dramatic, effects are seen when acidities are measured in aprotic solvents like DMSO. The effect on the bases is the same. Larger, more polarizable species such as pyridine exhibit enhanced basicity compared with the condensed phase.

In the experiment described above, a gas-phase equilibrium is established, which requires that each of the reagents be volatile enough to provide a measurable and sufficient pressure such that equilibrium can be reached within the trapping time of the instrument (at least 10^{-8}–10^{-9} torr). This condition is possible for many simple species, but not for large, highly polar species, such as peptides or organometallic species. To study the properties of nonvolatile analytes, Graham Cooks and co-workers developed a method centered on the dissociation of gas-phase complexes. These can be produced by ESI and therefore have no constraints related to the volatility of the substrate. It has been called the *Cooks Kinetic Method* because rather than establishing an equilibrium, the experiment characterizes the relative kinetics of cluster dissociation. The method is outlined in Scheme 10-2 for a proton affinity measurement. By ESI, a proton-bound

$$[B_1\text{----}H\text{----}B_2]^+ \xrightarrow{CID} \begin{cases} \xrightarrow{k_1} B_1H^+ + B_2 \\ \quad\quad \updownarrow K \\ \xrightarrow{k_2} B_1 + B_2H^+ \end{cases}$$

Scheme 10-2

TABLE 10-1 ΔH_{acid} and Proton Affinities of Various Organic Compounds

Compound	ΔH_{acid} (kcal mol^{-1})	Compound	PA (kcal mol^{-1})
HCl	333	H_2O	165
CH_3CO_2H	349	$CH_3C(O)CH_3$	194
Ph–OH	350	NH_3	204
$CH_3C(O)CH_3$	369	$Ph–NH_2$	211
$Ph–CH_3$	381	CH_3NH_2	215
CH_3OH	382	Pyridine	222
H_2O	390	$(CH_3CH_2)_3N$	235
CH_4	417	$(CH_3)_2NCH_2CH_2N(CH_3)_2$ (TMEDA)	242

Values from webbook.nist.gov.

complex of two bases is formed with a net positive charge. Under CID conditions, a hydrogen bond is cleaved to produce a protonated base as well as a free base. The key assumption is that the branching ratio of the two pathways is a measure of the equilibrium constant, K, between the products (eq. 10-5). It is a very sensible assumption, given

$$\frac{k_2}{k_1} = K \tag{10-5}$$

that the preference for forming the CID products should be heavily dependent on their stability, and, in fact, the complex is the key intermediate in the proton transfer reaction involved in the equilibrium. However, the relationship in eq. 10-5 is only an approximation because it requires that the processes associated with k_1 and k_2 have no barriers aside from their inherent endothermicity. This approximation is good for small molecules, but is not so safe for large species that may undergo extensive conformational changes in the dissociation process. Another issue is that the dissociation occurs in an activated complex, so the temperature associated with eq. 10-5 is not rigorously defined (of course, the temperature is needed to convert K to an energetic term); however, protocols have been developed for estimating effective temperatures in the dissociation processes.

Overall, the Cooks Kinetic Method has proved to be a robust approach for determining gas-phase equilibrium constants. Another advantage of the approach is that it can be implemented on any ESI instrument that is capable of CID because it does not require the introduction of reagent gases. As a result, these types of experiments can be carried out in most general mass spectrometry facilities. The approach can be applied to other equilibrium processes, such as electron transfer reactions, provided the following requirements are met: (1) an appropriate complex can be formed between the species of interest; (2) the complex fragments mainly into the desired reaction partners; and (3) there is no barrier, aside from endothermicity, to the dissociation process. Figure 10-7 shows a spectrum for CID of a proton-bound complex of two pyridine-derived bases. The greater intensity for protonated 3-chloropyridine indicates that it is slightly more basic than 3-fluoropyridine.

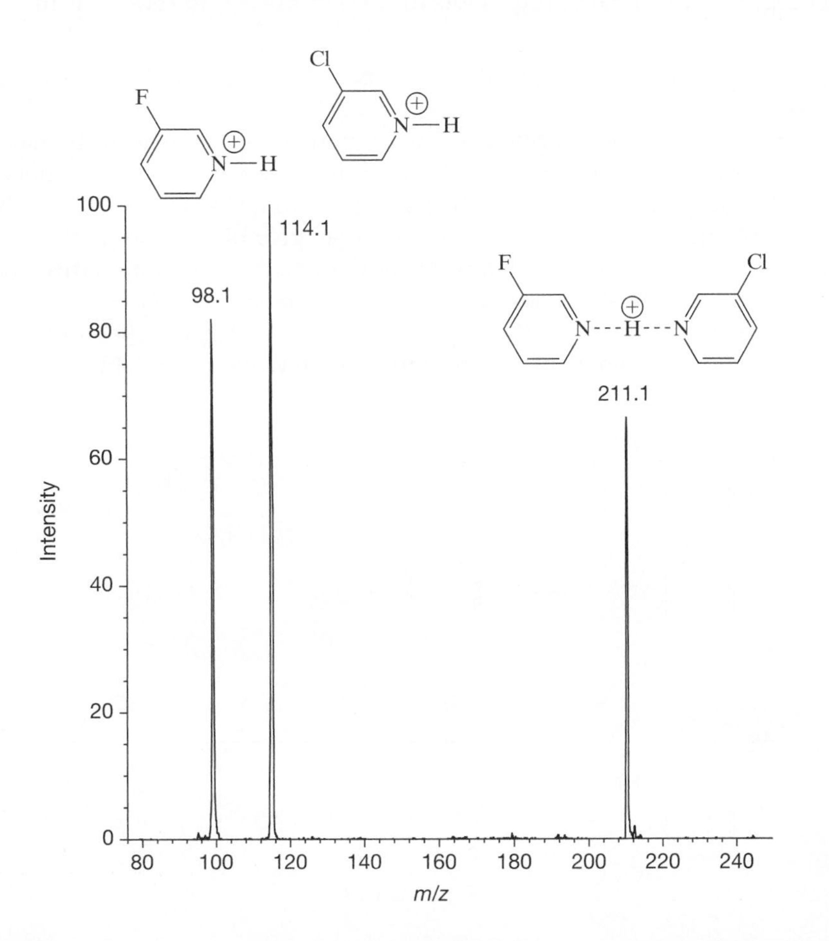

FIGURE 10-7 CID of the proton-bound complexes of 3-fluoropyridine and 3-chloropyridine. The peaks at *m/z* 98.1 and 114.1 correspond to protonated 3-fluoropyridine and 3-chloropyridine, respectively. The peak at *m/z* 211.1 is the complex. Only the ^{35}Cl-containing species were isolated and fragmented.

10-4c Bond Dissociation Energies

It might be surprising that the most accurate measurements of bond dissociation energies (BDEs), a process involving neutral species, often come from experiments with ions. This application highlights the versatility of mass spectrometry as well as the intrinsic benefits of working with species that can be manipulated by magnetic and electric fields. Here, the BDE is recast into a series of steps, heterolytic dissociation followed by processes that lead to radicals. The key relationship is shown in Scheme 10-3, in

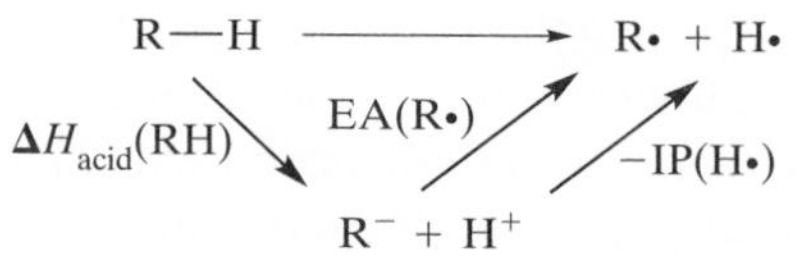

R—H ⟶ R• + H•

ΔH_{acid}(RH) ↘ EA(R•) ↗ −IP(H•) ↗

$R^- + H^+$

Scheme 10-3

which EA is the electron affinity of the radical species and IP is the ionization potential of the hydrogen atom. The relationship can be expressed conveniently in the form of eq. 10-6.

$$\mathrm{BDE(R{-}H)} = \Delta H_{acid}\mathrm{(RH)} + \mathrm{EA(R\bullet)} - \mathrm{IP(H\bullet)} \tag{10-6}$$

As outlined above, there are good ways to determine ΔH_{acid}, and under favorable conditions it can be obtained with uncertainties at the ±1 kcal mol^{-1} level. The electron affinity of a radical can be determined by anionic photoelectron spectroscopy. This type of mass spectrometry experiment uses a typical ion source to generate anions. The desired *m/z* species is isolated and photodetached with a laser pulse, to give the radical species and an electron. If the kinetic energy distribution of the detached electrons is determined, the electron binding energy of the anion (equivalent to the EA of the corresponding radical) can be calculated from the energy of the laser photon. This result is expressed in eq. 10-7, in which $h\nu$ represents the photon energy and KE_{max} is the maximum

$$\mathrm{EA(R\bullet)} = h\nu - KE_{max} \tag{10-7}$$

kinetic energy of the detached photon. The photon energy and electron energy can be measured with very high accuracy, so the resulting EA values have low uncertainties. Of course, the ionization potential of the hydrogen atom is known with exquisite accuracy. In general, the cycle shown in Scheme 10-3 can give BDE values with uncertainties of 1 kcal mol^{-1} or less, which is usually better than methods based on directly probing the reactivity of the radical species, that is, radical kinetics.

A fundamental example of the approach is given in Scheme 10-4 for benzene. The ΔH_{acid} of benzene has been measured by a number of workers, and the consensus value

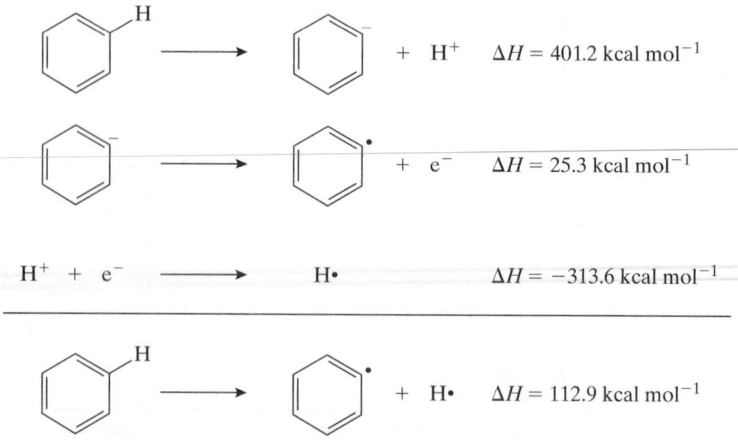

C_6H_5–H ⟶ $C_6H_5^-$ + H^+ $\Delta H = 401.2$ kcal mol^{-1}

$C_6H_5^-$ ⟶ $C_6H_5\bullet$ + e^- $\Delta H = 25.3$ kcal mol^{-1}

$H^+ + e^-$ ⟶ H• $\Delta H = -313.6$ kcal mol^{-1}

C_6H_5–H ⟶ $C_6H_5\bullet$ + H• $\Delta H = 112.9$ kcal mol^{-1}

Scheme 10-4

is 401.2 ± 0.4 kcal mol^{-1}. The photoelectron spectrum of the phenyl anion suggests an electron binding energy of 25.3 ± 0.1 kcal mol^{-1}. The ionization potential of the hydrogen atom is 313.6 kcal mol^{-1}. Taken together (see Scheme 10-3), these data indicate a C–H bond dissociation energy of 112.9 kcal mol^{-1} for benzene, with an uncertainty of only 0.5 kcal mol^{-1}. This value is much more accurate than that obtained with approaches based on neutral chemistry. Determining BDEs via the cycle in Scheme 10-3 provides a nice illustration of the extreme generality of mass spectrometric approaches.

10-4d Relative Condensed-Phase Association Constants

Electrospray ionization allows ions, in some sense, to be extracted directly from solution and introduced into a mass spectrometer. As discussed in Chapter 7, the process is complicated, and there is not complete consensus on the actual physical process that leads to the bare ions. Nonetheless, considerable interest has been shown in the relationship between the distribution of condensed-phase ionic species and those observed in ESI spectra. If ESI could provide a snapshot of ionic species in solution, it would be a powerful technique for quantitatively characterizing association reactions in solution and consequently probing many important binding processes in chemistry and biochemistry. For example, it could be used to measure the relative binding constants of a set of potential drugs with receptor sites in enzymes. In this way, mass spectrometry could replace many cumbersome assay techniques and give rapid measures of binding constants in solution. The required relationship is given in eq. 10-8, in which R(S) represents the concentration

$$\frac{R(S_1)_{solution}}{R(S_2)_{solution}} = \frac{I[R(S_1)]_{ESI}}{I[R(S_2)]_{ESI}} \qquad \textbf{(10-8)}$$

of a receptor/substrate complex in solution and $I[E(S)]_{ESI}$ is the intensity of the corresponding complex in the ESI spectrum.

The key question is whether the relationship expressed in eq. 10-8 is quantitatively reliable in general. Over the past decade, this issue has been contentious and controversial. As is often the case, the best way to probe the reliability of a proposed relationship is to consider the inherent assumptions in the relationship. In this case, the required assumptions are the following: (1) the relative concentrations of the competing receptor/substrate complexes are unaffected by the rapid solvent evaporation process, and (2) when the droplets reach the stage at which ion formation occurs, both of the competing complexes have equal ionization efficiencies. Both of these assumptions are difficult. During the ESI evaporation process, one expects a rapid change in the concentrations of the receptor and substrate in the droplet. This event will be controlled partially by the relative volatilities and solubilities of the substrates. For eq. 10-8 to be valid, however, the rapid evaporation and concentration changes during ESI must not perturb the receptor/substrate equilibria. This scenario is most likely if the reactions involved in the equilibrium are slow relative to the ESI process or if the substrates have very similar physical properties. The second assumption is reasonable if the substrates are relatively similar and if the receptor species is relatively large such that the receptor properties dominate the complex. A problem still exists, however, if there is differential partitioning of the species with respect to the bulk and surface of the electrospray droplet, because ionization occurs at the droplet surface. This effect can be important and, in some cases, has been shown to give results for which binding preferences are reversed between gas-phase and condensed-phase measurements. Nonetheless, it appears that the assumptions hold in a variety of cases, and there has been success in using ESI to monitor condensed-phase equilibria. Given the high speed of mass spectrometric analyses, this approach offers great potential in high-throughput screening of drug candidates and potential enzyme inhibitors. Careful validation experiments, however, are required because there are many counterexamples in which the key assumptions outlined above fail.

Problems

10-1 Your goal is to develop a quantitative assay for *N*-methylephedrine, using LC/MS. *N*-methylephedrine is part of a complex biological mixture. Suggest a practical internal standard methodology for accurate quantification.

10-2 Describe a practical strategy for determining the gas-phase proton affinity of acridine.

10-3 The gas-phase ΔH_{acid} of ethyne is 378.4 kcal mol^{-1} and the EA of ethynyl radical is 2.969 eV. Using these data, calculate the BDE of the C–H bond in ethyne.

10-4 Using the data in Figure 10-7, estimate the difference in proton affinity between 3-fluoropyridine and 3-chloropyridine. In the calculation, assume that the effective temperature of the experiment is 350 K and that, for the equilibrium process outlined in eq. 10-5, $\Delta S = 0$.

10-5 What are the potential difficulties in the following experiment? Equal molar quantities of 4-methoxyaniline and ethylamine are dissolved in a methanol/water mixture, along with a macrocyclic receptor that targets ammonium ions. The solution is prepared with appropriate concentrations for ESI mass spectrometry, and a spectrum is obtained that has signals for the complexes of protonated 4-methoxyaniline and ethylamine with the macrocyclic receptor. The relative binding constants of the receptor for protonated 4-methoxyaniline and ethylamine are calculated from the signal intensities of their complexes.

Bibliography

10.1 P. Traldi, F. Magno, I. Lavagnini, and R. Seraglia, *Quantitative Applications of Mass Spectrometry*, Chichester, England: John Wiley & Sons, 2006.

10.2 W. L. Buddle, *Analytical Mass Spectrometry: Strategies for Environmental and Related Applications*, New York: Oxford University Press, 2001.

10.3 R. A. Boyd, C. Basic, and R. A. Bethem, *Trace Quantitative Analysis by Mass Spectrometry*, Chichester, England: John Wiley & Sons, 2008.

10.4 Protein and peptide quantitation: J. Lill, *Mass Spectrom. Rev.* **22**, 182 (2003).

10.5 Drug quantitation: G. Hopfgartner, and E. Bourgogne, *Mass Spectrom. Rev.* **22**, 195 (2003).

10.6 Gas-phase acid/base measurements: G. Bouchoux, *Mass Spectrom. Rev.* **26**, 775 (2007).

10.7 Gas-phase ion chemistry: S. Gronert, *Chem. Rev.* **101**, 329 (2001).

10.8 Cooks Kinetic Method: R. G. Cooks, J. S. Patrick, T. Kotiaho, and S. A. McLuckey, *Mass Spectrom. Rev.* **13**, 287 (1994); P. B. Armentrout, *J. Amer. Soc. Mass Spectrom.* **11**, 371 (2000).

10.9 Bond dissociation energies: S. J. Blanksby and G. B. Ellison, *Acct. Chem. Res.* **36**, 255 (2003); J. Berkowitz, G. B. Ellison, and D. Gutman, *J. Phys. Chem.* **98**, 2744 (1994).

PART III

Vibrational Spectroscopy

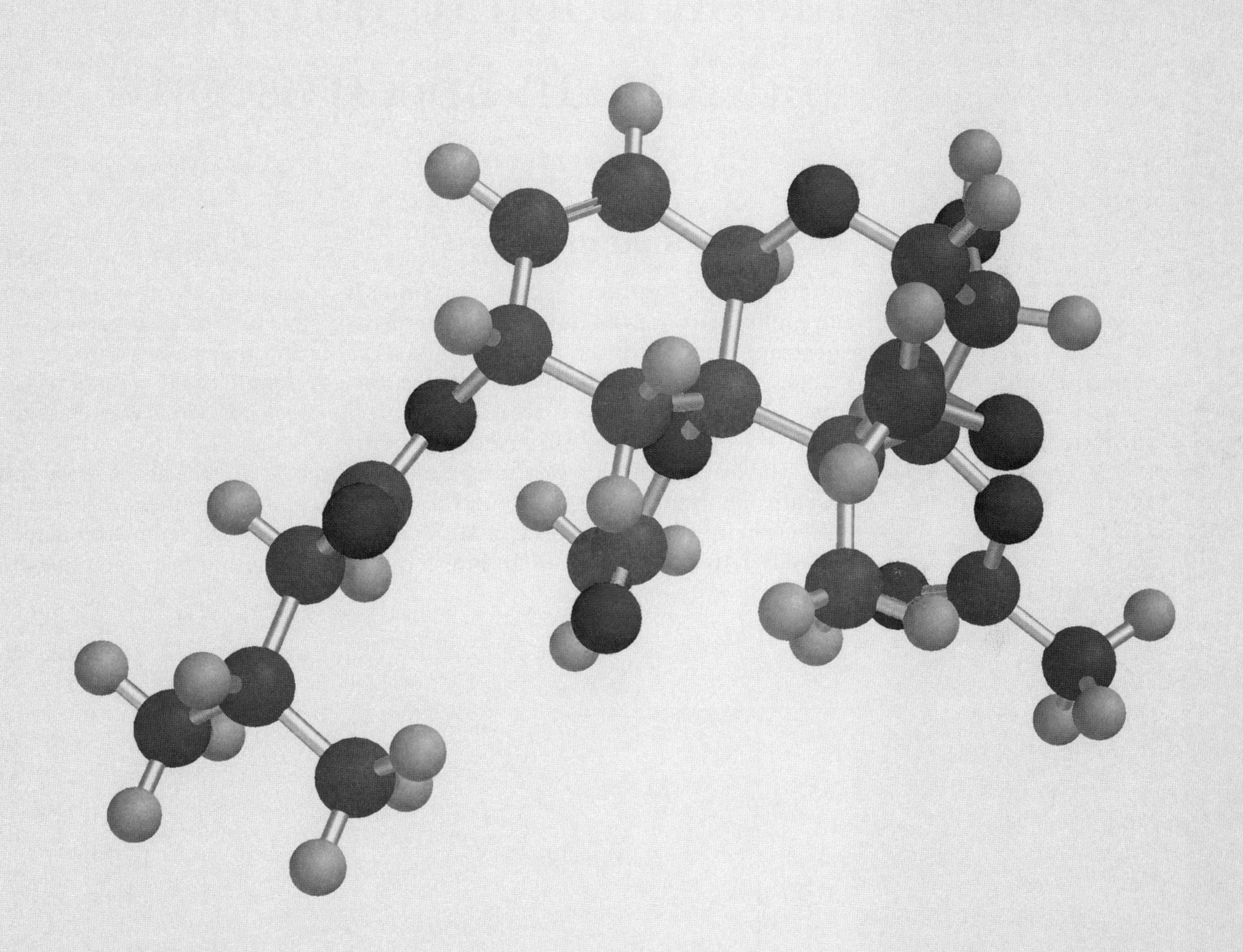

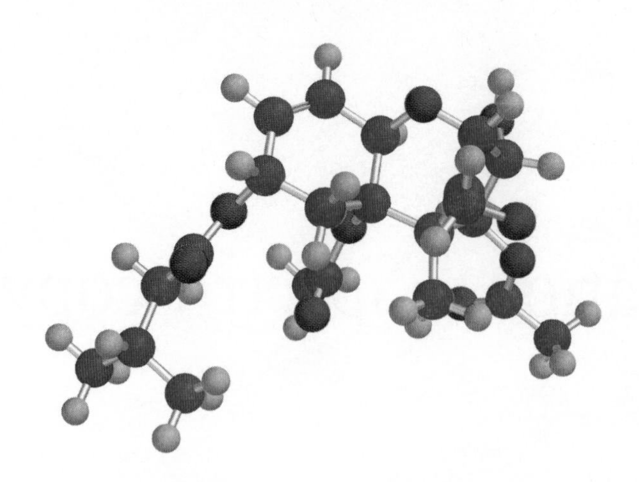

Introduction to Infrared and Raman Spectroscopy

11-1 INTRODUCTION

Infrared (IR) absorption spectroscopy provides a rapid and simple method for obtaining preliminary information on the identity or structure of an organic molecule. The spectrum is a plot of the percentage of IR radiation that passes through the sample (% transmission) versus some function of the wavelength of the radiation (Figure 11-1). The positions and relative sizes of the absorption peaks or bands give clues to the structure of the molecule, as indicated on the figure.

The instrument that produces the spectrum is known as an IR spectrophotometer (usually shortened to spectrometer). Modern IR spectrometers are based on the Michelson interferometer (Section 11-5b). The absorption spectrum is obtained by means of Fourier transformation of an interferogram. Hence the instruments are known as

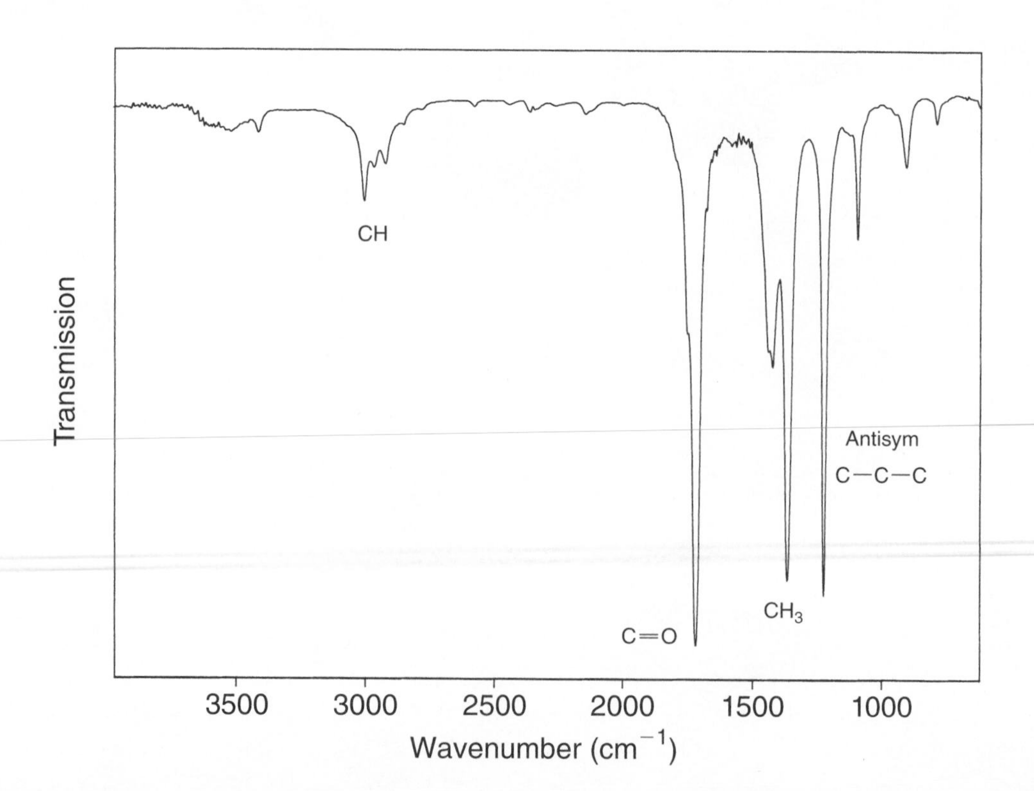

FIGURE 11-1 The IR spectrum of a thin film of acetone $(CH_3)_2CO$ between two KBr plates.

Fourier transform infrared (FT–IR) spectrometers. Earlier instruments known as dispersive IR spectrometers are based on monochromators that disperse the radiation from an IR source into its component wavelengths (Section 11-5a). The spectrum is obtained by measuring the amount of radiation absorbed by a sample as the wavelength is varied. Although few dispersive IR spectrometers are still in use today, the early literature and some spectral libraries are based on spectra recorded on these instruments.

Raman spectroscopy provides information complementary to that obtained from IR spectroscopy. This technique is discussed in Sections 11-3, 11-7, and 11-8. Again, there are two types of instruments: FT and dispersive.

Information on the structure of a molecule is obtained from a detailed study of those bands in the spectrum that are characteristic of certain functional groups. Some such bands are indicated on Figure 11-1 for the molecule acetone. Characteristic group frequencies are discussed in detail in Chapter 12.

11-2 VIBRATIONS OF MOLECULES

Infrared and Raman spectroscopies give information on molecular structure through the frequencies of the normal modes of vibration of the molecule. A *normal mode of vibration* is one in which each atom executes a simple harmonic oscillation about its equilibrium position. All atoms move in phase with the same frequency while the center of gravity of the molecule is stationary. A model of the molecule can be imagined, with balls representing atoms and springs representing bonds. Vibrations of the model involve stretching and bending the springs, together with motions of the balls.

According to classical mechanics, the frequency of vibration ν (s^{-1}) of two balls of mass m (kg) connected by a spring with force constant k ($N\,m^{-1}$) can be calculated from eq. 11-1.

$$\nu = \frac{1}{\pi}\sqrt{\frac{k}{2m}} \qquad \textbf{(11-1)}$$

The force constant k is a measure of the resistance to stretching of the spring. The force needed to displace a mass m by a distance x is $F = kx$.

There are $3N - 6$ normal modes of vibration of a molecule (N is the number of atoms). Each atom has three degrees of motional freedom, which can be thought of as motions in the x, y, and z directions. Thus N atoms have $3N$ independent motions. When the atoms are connected together in a molecule, however, the motions are no longer independent. Three motions become translations of the molecule, whereby all atoms move simultaneously in the x, y, or z directions. Another three are rotations, whereby all atoms rotate in phase about the x, y, or z axes. There remain $3N - 6$ motions, in which internuclear distances and bond angles change, but the center of gravity of the molecule does not move. This number is increased by one (to $3N - 5$) when the molecule is linear.

The methods of classical mechanics used to calculate the frequencies and forms of the normal modes of vibration of a ball-and-spring model also apply to molecules. Appropriate atomic masses, molecular dimensions, and force constants, however, must be used, and certain rules of quantum mechanics must be applied. This subject is discussed further in Section 12-1.

11-3 INFRARED AND RAMAN SPECTRA

An IR spectrum is obtained when a sample absorbs radiation in the region of the electromagnetic spectrum known as the infrared (see Figure 1-2 in Section 1-2). The expression *absorption band* is used to denote a feature observed in the spectrum. If the absorption band is relatively narrow and sharp, the word *peak* is used.

In IR absorption, energy is transferred from the incident radiation to the molecule, and a quantum mechanical transition occurs between two vibrational energy levels, from E_1 to E_2. The difference in energy (joules) between the two vibrational energy levels is related directly to the frequency ν (s^{-1}) of the electromagnetic radiation, as shown in eq. 11-2,

$$\Delta E = E_2 - E_1 = h\nu \tag{11-2}$$

in which h is Planck's constant (6.624×10^{-34} J·s). The frequency of vibration of the molecule corresponds directly to the frequency of IR radiation absorbed.

The most important transitions are from the ground state (all vibrational quantum numbers $\nu_i = 0$) to the first excited levels ($\nu_i = 1$). These allowed transitions are known as *fundamentals*. They usually give rise to strong absorption bands in the infrared. A transition from the ground state to a level with one $\nu_i = 2$ is known as an *overtone*. A transition to a level for which $\nu_i = 1$ and $\nu_j = 1$ (i and j are two different vibrations) is known as a *combination*. Overtones and combinations are forbidden by the simple harmonic oscillator theory of molecular vibrations, but they become weakly allowed when anharmonicity is taken into account. Many weak absorptions in an IR spectrum can be attributed to overtones or combinations. The substance for study is usually in the form of a solid, liquid, or solution, but gas-phase spectra also can be obtained.

A Raman spectrum is produced by a *scattering* process. Monochromatic incident radiation from a laser is scattered by the sample. In earlier dispersive instruments, the scattered (visible) light is usually observed instrumentally in a direction at 90° to the incident radiation. In more common FT Raman instruments, the scattered light is usually collected at 180° to the incident laser beam.

Raman spectra result from inelastic collisions of photons with molecules. In an inelastic collision, some energy is transferred either from the photon to the molecule or from the molecule to the photon, as illustrated in Figure 11-2. In the former case, the molecule is left in a higher vibrational energy level (giving rise to the so-called *Stokes lines*). In the latter case, the molecule must already be in an excited state, so that it can return to a lower state after giving up energy to the photon (giving rise to the so-called *anti-Stokes lines*). Since most molecules are in their ground vibrational state at normal temperatures, only the Stokes lines are important, and it is these that constitute the Raman spectrum of interest. A typical Raman spectrum is shown in Figure 11-3 for acetone. The zero on the abscissa corresponds to the frequency of the laser line (ν_L) used to excite the spectrum. The positions of the peaks correspond to differences between ν_L and the observed scattered frequencies (ν_{obs}).

Vibrations of certain functional groups, such as OH, NH, CH_3, C=O, and C_6H_5, always give rise to bands in the IR and Raman spectra within well-defined frequency ranges regardless of the molecule containing the functional group. The exact position of the *group frequency* within the range gives further information on the environment of the functional group. For example, we might take the carbonyl stretching band ($\nu_{C=O}$) of simple aliphatic aldehydes or ketones as the standard (ca. 1730 cm^{-1}). Carboxylic acid (monomers), acid chlorides, and esters have their $\nu_{C=O}$ bands at higher frequencies,

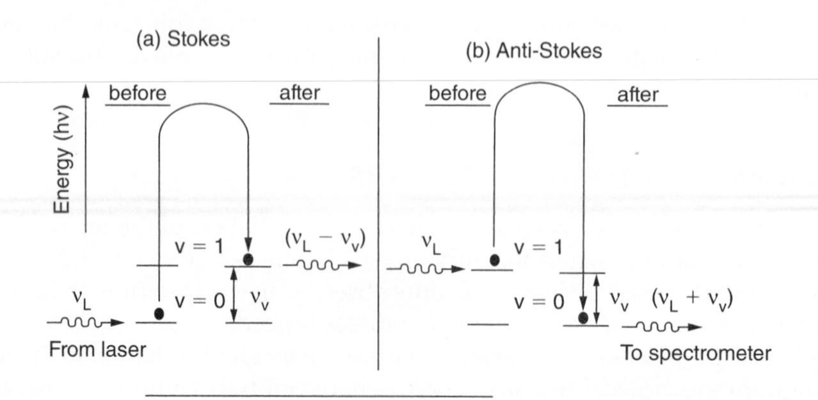

FIGURE 11-2 The mechanism of Raman scattering.

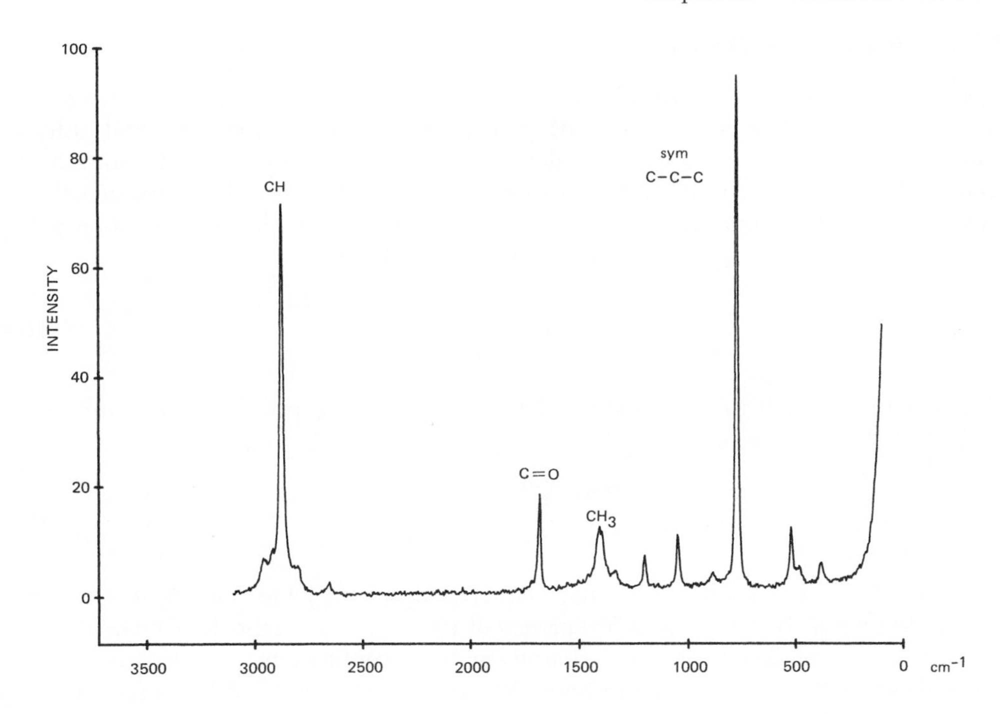

FIGURE 11-3 The Raman spectrum of acetone.

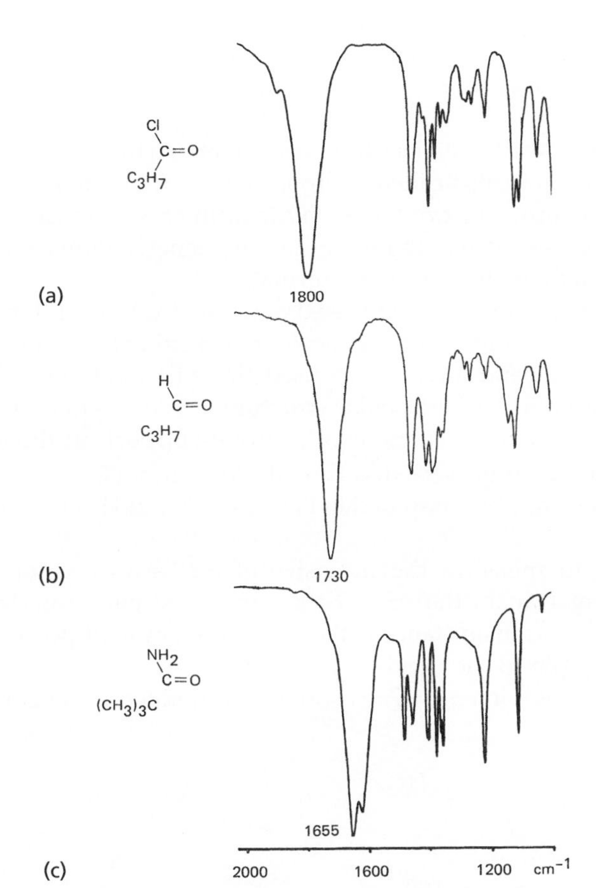

FIGURE 11-4 The carbonyl stretching bands of (a) butyryl chloride, (b) butyraldehyde, and (c) trimethylacetamide. (Reproduced with permission of Aldrich Chemical Co., Inc., from *The Aldrich Library of FT–IR Spectra.*)

whereas amides and aromatic ketones have lower carbonyl stretching frequencies (Figure 11-4).

The observation of a band in the spectrum within an appropriate group frequency range can indicate the presence of one *or more* different functional groups in the molecule, because there is considerable overlap of the ranges for many functional groups. It is therefore necessary to examine other regions of the spectrum for confirmation of a particular group. Examples of this procedure are given in Chapter 12.

11-4 UNITS AND NOTATION

As Figure 11-1 shows, a spectrum is recorded graphically with some measure of the frequency as the abscissa and the amount of absorption as the ordinate. Several units are used for both ordinate and abscissa scales. The unit of frequency ν is s^{-1} (vibrations per second). For molecular vibrations this number is very large (on the order of $10^{13}\,s^{-1}$) and inconvenient. A more convenient unit, $\overline{\nu}$ (the wavenumber in cm^{-1}), is obtained by dividing the frequency by c, the velocity of light (eq. 11-3).

$$\overline{\nu} = \frac{\nu}{c} \tag{11-3}$$

Thus a vibration with frequency $3 \times 10^{13}\,s^{-1}$ has a corresponding wavenumber of $1000\,cm^{-1}$ as shown in eq. 11-4.

$$\frac{3 \times 10^{13}\,s^{-1}}{3 \times 10^{10}\,cm\,s^{-1}} = 10^{3}\,cm^{-1} \tag{11-4}$$

Although the wavenumber is the frequency divided by the velocity, it is common practice to refer to $1000\,cm^{-1}$ as a "frequency of $1000\,cm^{-1}$" with the division by c understood. The three accepted ways of denoting cm^{-1} are "centimeters to the minus one," "reciprocal centimeters," or "wavenumbers." Wavelength (λ) is related to frequency (ν) or wavenumber ($\overline{\nu}$) as given in eq. 11-5.

$$\frac{1}{\lambda} = \frac{\nu}{c} = \overline{\nu} \tag{11-5}$$

IR spectroscopists formerly expressed wavelength in micrometers (10^{-6} m) for the abscissa scale in their spectra. This unit, also called microns, may be found in many older texts and papers on IR spectroscopy. Most of the earlier instruments produced a spectrum whose abscissa was linear in wavelength rather than frequency. All modern IR instruments record spectra with a linear wavenumber format.

The positions of Raman lines cannot be expressed in units of wavelength because the lines are measured as frequency shifts from the incident or exciting laser line. Hence the wavelength of a Raman line depends on the laser used. Since IR and Raman spectra are used together to give information on molecular structure, it is convenient to use a common unit, the wavenumber (cm^{-1}). There also is strong support on theoretical grounds for using cm^{-1}, since this unit is related directly to energy ($E = h\nu = hc\,\overline{\nu}$). The IR and Raman presentations may be compared in Figures 11-1 and 11-3 for the spectra of acetone.

Several units are used to measure the intensity of an IR absorption peak. Transmittance (T) and percent transmittance ($\%T$) are the most common, but absorbance (A) also is encountered. Transmittance is the ratio of the radiant power or intensity (I) transmitted by a sample to the incident intensity (I_0), as shown in eq. 11-6, and percent transmittance, as shown in eq. 11-7. Absorbance is defined in several ways in eq. 11-8.

$$T = \frac{I}{I_0} \tag{11-6}$$

$$\%T = 100\frac{I}{I_0} \tag{11-7}$$

$$A = \log_{10}\frac{I_0}{I} = \log_{10}\frac{1}{T} = \log_{10}\frac{100}{\%T} \tag{11-8}$$

In solution spectra, the intensity of absorption can be related to the concentration and the pathlength by the Beer–Lambert–Bouguer law, eq. 11-9,

$$A = \epsilon cl \tag{11-9}$$

in which c is the concentration in mol L^{-1} and l is the pathlength in centimeters. The constant ϵ is the molar absorption coefficient, with units L mol^{-1} cm^{-1}. Thus ϵ is the absorbance produced by a solution of concentration 1.0 M in a cell with a pathlength of 1.0 cm. Other names for ϵ include extinction coefficient and molar absorptivity.

Intensities in Raman spectra are much less quantitative than in IR spectra because the height of a peak depends on factors such as the laser power, the wavelength of the exciting radiation, the detector, and the amplification system used. Thus quantitative results can be obtained only if an internal standard is used to determine the amount of sample actually in the laser beam and giving rise to scattering.

The intensity of a Raman line is a *linear* function of concentration, whereas the intensity of an IR absorption band is a logarithmic function of concentration (eq. 11-8). Thus, doubling the concentration of a solution should double the intensities of all Raman lines for identical instrumental settings, whereas the apparent effect on the IR peak heights depends on the peak. For example, a weak IR band appears to be affected much more than a strong band, since doubling the concentration almost doubles the intensity of a weak band but might change that of a strong band only about 10%. Caution must be exercised in discussing both IR and Raman band intensities. For present purposes, the normal IR range is taken to be 4000 to 400 cm^{-1}. Some IR spectrometers, however, cover a somewhat wider range, overlapping the far IR region to 200 cm^{-1}. The region below 200 cm^{-1} is not readily accessible by IR spectroscopy, but vibrational spectra can be obtained in this region by Raman spectroscopy (Section 11-7).

11-5 INFRARED SPECTROSCOPY: DISPERSIVE AND FOURIER TRANSFORM SPECTRA

Infrared spectra can be obtained by either dispersive or interferometric methods. Here an analogy may be drawn to NMR spectroscopy. In Part I it was seen that an NMR spectrum can be obtained by either continuous wave or FT methods. Analogously for IR spectra, dispersive instruments record the spectrum in the frequency domain, whereas interferometers record the spectrum in the time domain. The latter result is an interferogram, which must be converted to the frequency domain by means of a Fourier transformation to obtain the infrared spectrum.

11-5a Dispersive Infrared Spectrometers

Although instrument manufacturers no longer manufacture dispersive IR spectrometers, these instruments are still in use and are important in the historical development of IR spectroscopy. A dispersive IR spectrometer consists of three basic parts: (1) a source of continuous IR radiation, such as a coil of wire heated to high temperatures, (2) a monochromator to disperse the radiation into its spectrum, and (3) a sensitive thermocouple to detect IR radiation. The sample is placed usually between the source and the monochromator. The monochromator contains a grating that disperses the continuous radiation into its spectrum of monochromatic components. A mechanical scanning device passes the component frequencies sequentially and continuously to the detector. In this way, the detector senses which frequencies have been absorbed or partially absorbed by the sample and which frequencies have been unaffected.

The ability of the instrument to distinguish between absorptions at closely similar frequencies is known as the *resolution*. For most applications, a resolution of 4 cm^{-1} is adequate.

11-5b Fourier Transform Infrared Spectrometers

FT–IR spectrometers are based on the Michelson interferometer. This instrument (Figure 11-5) consists of two plane mirrors, M1 and M2, mounted at 90° to each other and a semireflecting beam splitter (BS). One of the mirrors (M1) is fixed; the other (M2) can be moved very precisely and reproducibly through a distance (δ) of a few millimeters. The beam splitter transmits 50% of the incident radiation to one mirror and reflects 50% to the other. After reflection at M1, 50% of the radiation travels back through the BS and

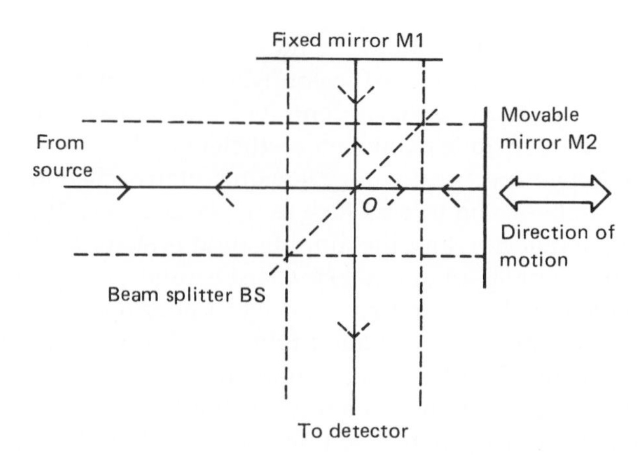

FIGURE 11-5 Schematic diagram of a Michelson interferometer.

recombines with 50% of the radiation returned from mirror M2 and reflected by the BS. The optical path difference between the beams is known as the retardation x ($x = 2\delta$); unless $x = 0$, the recombined beams interfere. With this arrangement, 50% of the radiation returns to the source and the other 50% passes through the sample to a detector. Interferometers cannot use thermocouples as sources of IR irradiation, as the scan time is very short and the thermocouple response times are very long. One type of IR detector with a response time fast enough for an interferometer is the pyroelectric bolometer. The beam splitter is usually a very thin layer (typically 0.4 μm) of germanium deposited on an optically flat KBr plate.

To understand how an interferometer can be used to measure an IR spectrum, consider first a monochromatic beam of radiation of wavelength λ passing through the instrument. When $x = 0$ or $n\lambda$ (n is an integer), the recombined beams are exactly in phase, so the signal at the detector is a maximum. As mirror M2 is moved, the beams interfere, and the signal falls to zero when $x = \lambda/2$. As mirror M2 continues to move at a constant velocity, the signal intensity $I(x)$ varies according to a cosine function (eq. 11-10),

$$I(x) = 0.5\, I(\nu) \cos 2\pi\nu x \qquad \textbf{(11-10)}$$

in which $I(\nu)$ is the intensity of the source at frequency ν. The factor of 0.5 in eq. 11-10 occurs because only one half of the incident radiation reaches the detector. The other half is reflected back toward the source. A plot of $I(x)$ vs. x is known as an interferogram.

When a continuous source of IR radiation is used, an infinite number of wavelengths pass simultaneously through the interferometer, and only when $x = 0$ are all wavelengths in phase. At any other position of mirror M2, a very complex interference pattern results, giving rise to an interferogram like that in Figure 11-6a. To obtain this interferogram, the mirror M2 is moved from a negative x through the position of zero path difference to positive x. At $x = 0$, all wavelengths interfere constructively and produce the very strong central signal. The intensity $I(x)$ of the radiation reaching the detector at any other retardation is the sum of the intensities of all the interfering wavelengths at this mirror position.

A sample placed between the interferometer and the detector reduces the intensity of radiation at any frequency at which the sample absorbs. Thus the IR absorption spectrum is contained in the resulting interferogram. Figure 11-6b shows the interferogram of Figure 11-6a modified by the absorption of a 0.05-mm-thick polystyrene film placed between the interferometer and the detector. There is clearly a difference in the interferograms of Figures 11-6a and 11-6b, but to obtain quantitative information about the absorption from the interferograms, we need the data in the frequency domain form $I(\nu)$ versus ν. To obtain this information, a Fourier transformation is performed on both interferograms, as with FT–NMR. A FT–IR spectrum of the polystyrene sample used to obtain the interferogram of Figure 11-6b is shown in Figure 11-7. The two main advantages of interferometers over dispersive spectrometers for IR spectroscopy are speed and sensitivity.

The resolution of a FT–IR spectrometer depends on the reciprocal of the retardation x. Hence, for a resolution of 4 cm^{-1}, the moving mirror must move 0.25 cm.

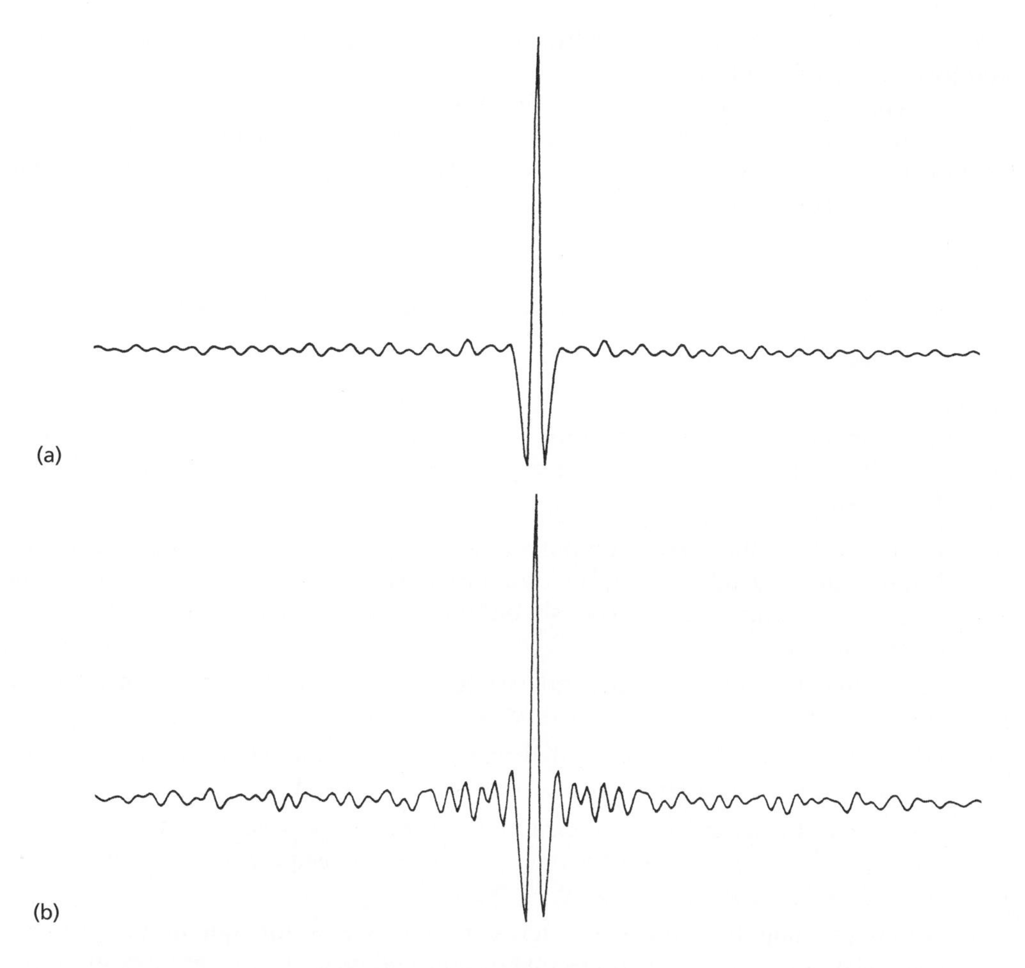

FIGURE 11-6 Interferograms from a continuous source: (a) with no sample; (b) with a 0.05 mm film of polystyrene placed between the interferometer and the detector.

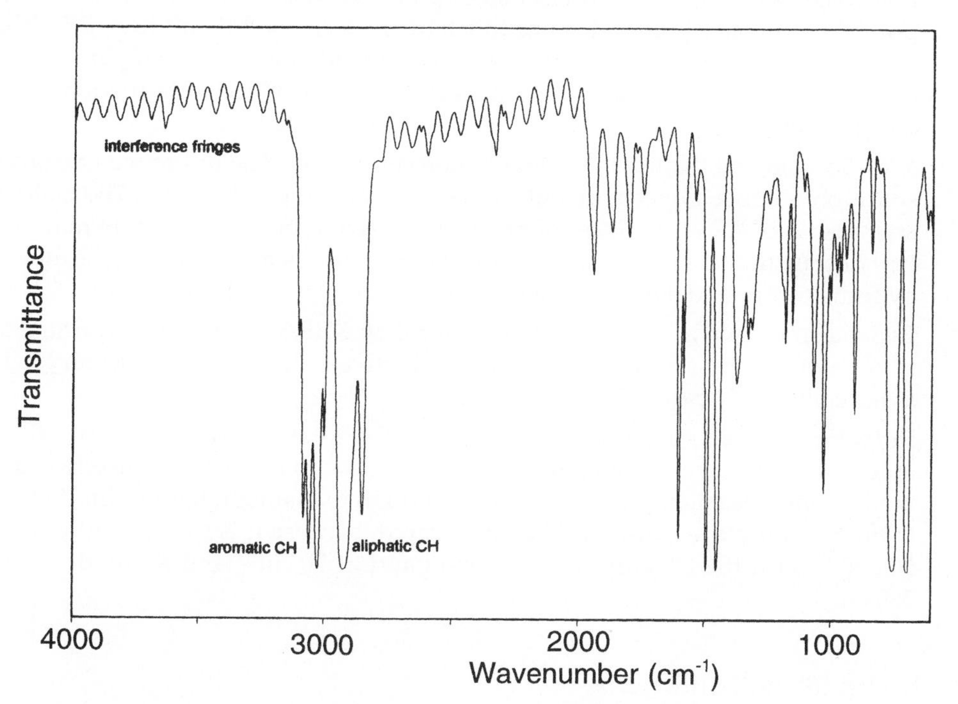

FIGURE 11-7 The FT–IR spectrum of a 0.05 mm film of polystyrene obtained from the interferogram shown in Fig. 11-6b.

The FT is computed from the digitized values of the interferogram sampled at equal intervals (Δx) of retardation. The spectral range covered is inversely proportional to the sampling interval. For a spectral range of 4000 to 400 cm^{-1}, x is on the order of 10^{-4} cm. These sampling points must be separated by precisely equal intervals; otherwise, noise is introduced into the computed spectrum. This precision is possible when sampling points are triggered by the zero crossings of an interferogram generated by a helium-neon laser.

As with NMR spectroscopy, co-addition of several scans results in an improved signal-to-noise ratio (S/N). It should be noted that the spectrum cannot be plotted until all of the interferograms have been recorded and averaged and the Fourier transformation has been carried out.

11-6 SAMPLING METHODS FOR INFRARED TRANSMISSION SPECTRA

11-6a Support Materials

In all transmission studies except the pressed-pellet method, discs or plates transparent to IR radiation are needed to support the sample. Several IR transmitting materials are in common use. Most can be purchased as polished windows or as sawn crystal blanks ready for polishing.

NaCl is perhaps the most commonly used material. The rock salt region is the traditional range (4000–650 cm^{-1}) studied by earlier workers. This material is the least expensive and is easy to polish, but it fractures easily and is transparent only as far as 650 cm^{-1}.

KCl is almost as cheap as NaCl and easy to polish, but it fractures when subjected to stress. It is transparent to 500 cm^{-1}.

KBr is another traditional material. It costs about 50% more than NaCl or KCl. It is easy to polish but is fragile. KBr is transparent to 400 cm^{-1}.

CsI does not fracture under mechanical or thermal stress. It is, however, very water soluble, very soft, and tricky to polish. Its cost is quite high (about 10 times that of NaCl), but it is transparent to 150 cm^{-1}.

CaF_2 is very slightly soluble in water, so it may be used for aqueous or D_2O solutions. The cost is about five times that of polished NaCl. CaF_2 windows should be purchased already polished, since this hard material is difficult to polish. These windows normally do not need repolishing. Disadvantages of CaF_2 are that it is transparent only to 1100 cm^{-1} and it fractures when subjected to thermal or mechanical stress.

KRS-5 is a mixed thallium bromide iodide compound. The bright red crystals are sparingly soluble in water, do not cleave, and transmit to 200 cm^{-1}. The main disadvantages of KRS-5 are its high price and a high refractive index that may cause loss of transmitted energy by scattering. KRS-5 also is toxic and may be attacked by some compounds in an alkaline solution.

ZnSe (also known as IRTRAN 4) is harder than KRS-5, is insoluble in water, and is not attacked by alkalis. It has a high refractive index, is very toxic, and, when attacked by acids, produces the unpleasant gas H_2Se.

Polyethylene and polytetrafluoroethylene (PTFE) are used in low-cost, disposable cells. They are not usually used as window materials, because two disks made from these materials are not rigid enough to maintain liquid films between them. Polyethylene is used in the far infrared, because it is transparent to below 50 cm^{-1}, with the exception of a sharp band at 70 cm^{-1} and a broader one at 380 cm^{-1}.

11-6b Liquids and Solutions

Probably the easiest method to obtain a qualitative IR spectrum of a liquid is to place one drop of the liquid onto a disc of NaCl, KBr, and so forth, cover the drop with a second disc, and mount the pair in a holder. Teflon spacers may be used to give various pathlengths. Fixed-pathlength, sealed cells also are available. These usually have amalgamated silver or lead spacers. The cells are filled, emptied, or flushed by means of a syringe through conventional Luer (twist and lock) ports. Teflon stoppers are used to

TABLE 11-1 Useful Solvents for Infrared Solution Spectra

Solvent	Useful Regions (cm^{-1})	Typical Pathlength (mm)
CCl_4	All except 850–700	0.5
$CHCl_3$	All except 1250–1175 and below 820	0.25
Benzene	All above 750 except 3100–3000	0.1
CH_2Cl_2	All above 820 except 1300–1200	0.2
Acetone	2800–1850 and below 1100	0.1
Acetonitrile	All except 2300–2200 and 1600–1300	0.1
Cyclohexane	Below 2600	0.1
N,*N*-Dimethylformamide	2750–1750 and below 1050	0.05
Diethyl ether	All except 3000–2700 and 1200–1050	0.05
Heptane and hexane	All except 3000–2800 and 1500–1400	0.2
Dimethyl sulfoxide	All except 1100–900	0.05

close the ports. For the far-IR range, polyethylene cells are available with various pathlengths but are easily contaminated.

Commercially available, disposable IR Cards consist of microporous polyethylene or PTFE films mounted in cards. They may be used for analysis of liquids or solids soluble in organic solvents. The cards have a circular aperture containing a thin film of the microporous material. A blank card is used to obtain a background spectrum. A small quantity (~50 μl) of a liquid is applied directly to the IR Card film. Solids soluble in organic solvents are applied as solutions, and the solvent is allowed to evaporate. The card then is placed in the sample compartment of an FT–IR spectrometer, and the spectrum is recorded. The polyethylene card may be used from 4000 to 400 cm^{-1}, except for the region between 3000 and 2800 cm^{-1}. The PTFE card has a more limited range from 4000 to 1300 cm^{-1}, but it can withstand temperatures up to 200°C.

Although it is often most convenient to record the IR spectrum of a compound in solution, some of the best solvents have very strong IR absorption bands that obscure parts of the spectrum of the compound. Water, for example, absorbs strongly throughout the spectrum and is rarely used in routine IR work. Water, however, can be useful for examination of compounds such as sugars, amino acids, and compounds of biochemical interest, although special window materials such as CaF_2 or KRS-5 must be used. FT–IR instruments have the capability to signal-average over a large number of scans, so that a spectrum can be obtained even when a sample absorbs very strongly and very little radiation passes through it. FT–IR spectrometers can produce a spectrum from surprisingly little transmission (as low as 0.1% T).

In dispersive instruments, weak to medium solvent absorption can be removed from the spectrum by using a pair of matched cells with the solvent in the reference beam and the solution in the sample beam. With FT–IR instruments, a spectrum of the solvent can be subtracted from the spectrum of the solution. It should be noted, however, that, while recording a spectrum in a region in which the solvent absorbs strongly, essentially no energy passes to the detector, and the S/N ratio is very poor. Only those regions of the spectrum in which the solvent does not absorb strongly can be used, so that several solvents may be needed to obtain a complete spectrum. A short list of solvents and their useful regions is given in Table 11-1.

11-6c Solids

Solid samples are handled in the form of either mulls or pressed discs. To make a mull, a small amount of the sample is ground in an agate or mullite mortar. Then a drop of mulling material, usually a pure, colorless paraffin oil, is added and the grinding continued. The mixture should have the consistency of a thin paste. It is transferred to a window of NaCl, KBr, and so on, and covered with a second window. A thin film is produced by gentle pressure with a slight rotating movement. The two plates with the mull between are placed in a cell holder, and the spectrum is recorded. There are strong bands at 2900, 1470, and 1370 cm^{-1} and a weak band at 720 cm^{-1} from the paraffin oil. If these bands are stronger than the peaks from the sample, then more sample and less oil are required. If the sample peaks are too strong, the two windows can be squeezed more closely together or a small drop of oil can be added. The user must experiment with the mull to obtain the best results.

When the region near 2900 cm^{-1} is important, another mulling material, usually a chlorofluorocarbon oil, must be used. This material is completely opaque below 1400 cm^{-1}, and it also has a band at 1650 cm^{-1}. Another compound useful for mulls, when the 2900 cm^{-1} region is to be studied, is hexachlorobutadiene. This compound has no absorptions above 1650 cm^{-1} and has a useful window between 1500 and 1250 cm^{-1}.

Another sampling method for solid compounds is the pressed pellet technique. In an agate or mullite mortar, a few milligrams of the sample are ground together with about 100 times the quantity of a matrix material that is transparent in the infrared. The usual material is KBr, although other compounds such as CsI, TlBr, and polyethylene are used in special circumstances (see below). The finely ground powder is introduced into a stainless steel die, usually 13 mm in diameter, which is evacuated for a few minutes with a vacuum pump to remove air from between the particles. The powder then is pressed into a disc between polished stainless steel anvils at a pressure of about 30 tons in.$^{-2}$. Other devices, such as a hand-held press, are available for making smaller (7, 3, and even 1 mm) KBr pellets.

A well-made KBr pellet has 80% to 90% transmittance in regions below 3000 cm^{-1} in which the sample itself does not absorb. Between 4000 and 3000 cm^{-1}, the transmission sometimes is low due to scattering effects. The amount of scattering by mulls and pellets depends on the relative refractive indices of the sample and the matrix or mulling material. When the refractive indices are similar, large particles cause serious scattering at high wavenumbers. At lower wavenumbers, the particle size becomes less important. The particle size should be less than about 20 μm for good pellets. This size usually can be achieved by hand grinding in a hard mortar, as discussed above.

Matrices other than KBr may be used. CsI is useful when spectra down to 200 cm^{-1} are required. One disadvantage of CsI is that it is a very hygroscopic material. Lower pressures are needed for CsI pellet formation than for KBr. TlBr is used when materials of high refractive index are studied. It has a refractive index of 2.3 in the IR region and is transparent to 230 cm^{-1}. It is not hygroscopic and can be used in conditions of high humidity, but is toxic. Powdered polyethylene has been used for making pellets for far-IR spectra because, apart from a band at 80 cm^{-1}, its spectrum between 400 and 10 cm^{-1} is free of absorption.

In addition to the scattering problems discussed above, the pressed-disc method has other disadvantages. Changes may occur in the sample during the grinding and pressing processes, and the sample may react with absorbed water or even with the matrix material.

11-7 RAMAN SPECTROSCOPY

Raman spectrometers give information on vibrational frequencies in the range covered by both mid- and far-IR spectrometers (4000–10 cm^{-1}). Samples in the form of liquids, solutions, powders, and single crystals can be handled by standard sampling techniques. Part of the low-frequency end of the spectrum is always obscured by Rayleigh scattering of the exciting radiation. Rayleigh scattering is due to elastic scattering of the laser photons by the molecules of the sample. No energy is exchanged, and the scattered photons emerge from the sample with the same frequency as that of the laser line. This process is several orders of magnitude more probable than the inelastic (Raman) scattering and gives rise to an extremely strong band centered at 0 cm^{-1} in the Raman spectrum. The width of the Rayleigh line, as it is called, increases from solids to pure liquids to solutions. The edge of the Rayleigh line is clearly evident in the Raman spectrum of liquid acetone in Figure 11-3.

Experienced users of Raman spectroscopy can obtain the same kind of structural information as the IR spectroscopist. They use a somewhat different set of group frequencies, however, because vibrations that give rise to strong, characteristic IR absorption are often weak in the Raman spectrum. The converse is also true, as can be seen by comparing Figures 11-3 and 11-1. Raman spectra complement IR spectra, and the two techniques used together provide important and often unique information for organic structure determination.

Raman spectroscopy has certain advantages over IR. One is that simpler spectra are more easily observed in the Raman because of the absence of overtone or combination bands, which are an order of magnitude weaker in the Raman than in the infrared. A second advantage is the wider choice of solvents for solution spectra. In particular, water may be used routinely, and other solvents have more clear regions in the Raman than in the infrared. Information on the lower-frequency region 200–50 cm^{-1}, corresponding to the far infrared, is obtained easily.

There are, however, certain disadvantages of Raman spectroscopy. One is the inherent weakness of Raman spectra, which often can be masked by background scattering from small suspended particles in liquid or solution samples. Absorption of the laser radiation can cause heating of liquid samples and charring of solid samples. If the compound to be studied fluoresces under visible light, the Raman spectrum is totally obscured. Fluorescence is often a serious problem even when the sample is not fluorescent itself, because a trace of fluorescent impurity can obscure the Raman spectrum. The problem of fluorescence is reduced when the laser excitation is in the red or the near infrared. It is sometimes difficult to get good spectra from solids unless they are crystalline. Thus in practice one often cannot obtain a Raman spectrum of a sample, whereas an IR spectrum always can be recorded. Another disadvantage of Raman spectroscopy is the high cost of the instruments.

Resonance Raman spectroscopy can provide enhanced sensitivity when the incident laser energy is tuned to coincide with that of an electronic transition of the sample (Part IV). Vibrational modes from that transition exhibit considerably enhanced Raman intensity, up to a factor of 10^6, much larger than those from other vibrations. *Surface-enhanced Raman spectroscopy* (SERS) can yield enhancement factors up to 10^{10} and is sufficiently sensitive to observe spectra of single molecules. In conjunction with resonance effects, *surface-enhanced resonance Raman spectroscopy* (SERRS) can achieve enhancement factors of 10^{14}. The SERS effect derives from the ability of some surfaces to create an enhanced electric field.

11-8 RAMAN SAMPLING METHODS

Sampling methods for Raman spectroscopy are usually simpler than those for IR. In principle, the sample simply is placed in the laser beam in front of the entrance slit of the monochromator of a dispersive Raman instrument, or at the sample position of an FT–Raman spectrometer, and a laser beam is directed to the sample. In either case, the sample itself then becomes the source of radiation passing into the spectrometer. Focusing the laser beam increases the Raman intensity but also may damage the sample. Sample damage can be prevented by reducing the laser power. In the case of liquids, solutions, and gases, Raman scattering may be increased by passing the laser beam through the sample several times by means of mirrors. Because of the simplicity and efficiency of sampling, Raman spectroscopy has become an analytical method of choice for the examination of materials such as art objects that are too valuable for removal of a sample.

11-8a Liquids and Solutions

A small-volume quartz cell (5 mm × 5 mm × 5 cm), such as that used for UV absorption work (Part IV), can be used as a Raman liquid sample container. Although quartz is preferable because it has low fluorescence, cells made from ordinary Pyrex glass are usually satisfactory.

Raman spectroscopy is ideally suited for microsampling. The laser beam can be focused to a very small area, and samples contained in melting-point capillary tubes yield virtually the same spectra as much larger samples. Most instrument manufacturers now offer microscope attachments for their FT–Raman spectrometers.

Techniques for handling solutions are essentially the same as those for pure liquids. Solvents should be clean, pure, redistilled, and filtered to avoid problems from suspended particles or fluorescence. As we saw in Section 11-6a, most of the common solvents have extensive regions of absorption in the infrared. Although this is not the case for the Raman, there is still the drawback that the intensities obtained from dilute solutions cannot be increased by having a longer pathlength. There is a lower limit to concentrations that can be used, normally 1% to 5% by weight (about 0.1 M).

Of course, every solvent has a Raman spectrum, but Raman lines are usually much narrower than IR absorption bands. In addition, overtones and combinations are very much weaker in the Raman. Hence, there are many completely clear regions in the Raman spectra of all the usual solvents. Water is an excellent solvent for Raman solution studies. This property is extremely important when examining compounds of biological interest, since their natural environment is usually aqueous solution.

11-8b Solids

Polycrystalline powders are best handled in capillary tubes or tamped into a hole in the end of a metal rod. Irregular pieces of solid materials may be glued to a support rod and held in the laser beam. Powders may be pressed into pellets and examined in the same way. High background scatter often is observed, especially from amorphous materials.

A crystal 1 or 2 mm long can be mounted on a goniometer head in the same way that crystals are mounted for X-ray examination. Single crystals can be carefully positioned so that the laser beam passes along one of the axes of the crystal. In this case, when a polarization analyzer is used (Section 11-9), important information concerning the symmetry of the normal vibrations of the molecule can be obtained.

11-9 DEPOLARIZATION MEASUREMENTS

Raman spectra can be used to classify molecular vibrations as either totally symmetric or non-totally symmetric, with respect to the elements of symmetry that the molecule possesses. This distinction is accomplished by means of depolarization measurements, which can be made using a polarization analyzer. Raman scattering is produced by the interaction between the radiation from a plane-polarized laser beam and the change in the polarizability of the molecule associated with the molecular vibrations. As a result of this interaction, some of the scattered light is no longer plane polarized. Hence there is a difference in intensity between the light transmitted through the analyzer when it is oriented parallel to the plane of polarization of the laser beam ($I_{\parallel}$) and when it is turned through 90° into the perpendicular orientation ($I_{\perp}$). The ratio $I_{\perp}/I_{\parallel}$ is known as the *depolarization ratio* ρ, with a theoretical maximum value of 0.75. A band with ρ less than 0.75 is said to be polarized, and a band with ρ exactly equal to 0.75 is said to be depolarized.

Some molecular vibrations are symmetric with respect to all elements of symmetry that the molecule possesses. These totally symmetric modes of vibration give rise to polarized Raman bands (ρ between 0 and 0.75). Other vibrations are antisymmetric with respect to some of the symmetry elements. These non-totally symmetric vibrations give rise to depolarized Raman bands ($\rho = 0.75$). Hence depolarization ratios can be used to identify the totally symmetric vibrations of a molecule.

To make depolarization measurements, the spectrum is scanned twice with identical instrument settings. The only difference is the orientation of the analyzer (parallel or perpendicular). Peak heights (I) and baselines (I_0) are noted for each orientation, and depolarization ratios are calculated from eq. 11-11.

$$\rho = \frac{(I - I_0)_{\perp}}{(I - I_0)_{\parallel}} \tag{11-11}$$

It should be noted that depolarization ratios cannot be measured for powdered solid samples.

Depolarization measurements may be used to separate two overlapping bands, when one is due to a totally symmetric vibration. The intensity of a band from a totally symmetric vibration is usually reduced drastically on rotating the analyzer from parallel to perpendicular orientation, whereas the intensity of a band from a non-totally symmetric vibration is reduced by only 25%. Hence the band from the non-totally symmetric mode predominates in the perpendicular spectrum.

11-10 INFRARED REFLECTION SPECTROSCOPY

When a ray of light strikes an interface between two nonabsorbing materials of different refractive index (n_1 and n_2), the light is partially transmitted and partially reflected. This property is used in the manufacture of beam splitters for FT–IR spectrometers. When light enters material 2 from material 1 with n_1 less than n_2 (for example, air to glass), the result is *external reflection*. For external reflection, the reflectivity can never be 100% and is usually much less. In the case when n_1 is greater than n_2, we have *internal reflection*, and in this case the reflection is total when the angle of incidence is between the critical angle and 90° (grazing incidence). Both internal and external reflection can be used to obtain spectra.

Total internal reflection can be observed in a glass of water. When the inside of the glass is viewed through the water surface, it appears to be completely silvered and opaque. When the outside of the glass is touched with a finger, however, details of the ridges and whorls on the skin are clearly seen, but the silvered effect remains between these features. Total reflection is destroyed where the skin actually makes contact with the glass. This result can be explained by penetration of the electromagnetic field of the light into the rarer medium (n_2 smaller refractive index) by a fraction of a wavelength, as illustrated in Figure 11-8.

When the light is in the IR region and the rarer medium is a compound that absorbs IR radiation, the penetrating radiation field can interact by means of an absorbing mechanism to produce an attenuation of the total (internal) reflection. This interaction can be described in terms of an effective thickness, which corresponds to a sample thickness in a normal absorption process. This phenomenon is most commonly termed *attenuated total reflection* (ATR).

The technique of internal reflection spectroscopy consists of recording the wavelength dependence of the reflectivity, through the usual IR region, at the interface between a sample compound and a material that has a higher refractive index but is transparent to the infrared radiation. The radiation approaches and leaves the interface through the denser medium. In the simplest experimental arrangement, the material of high refractive index is in the form of a prism.

Multiple internal reflection (MIR) can be achieved using a crystal of the form shown in Figure 11-9. Multiple reflections can enhance the spectra of weakly absorbing samples. The angle of incidence is fixed in this method, and more than one plate may be needed to study a variety of compounds. The thinner the plate, the more internal reflections there are. A typical plate is 2 mm thick, 5 cm long, and 2 cm high, with the ends cut at 45°.

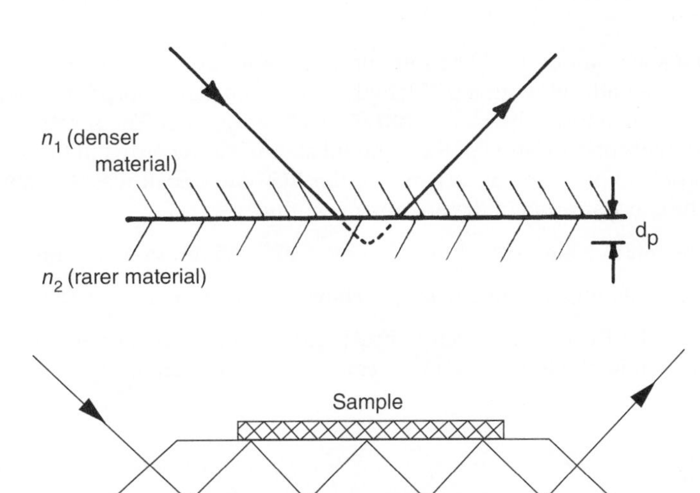

FIGURE 11-8 The path of a ray of light in total internal reflection. The ray penetrates a fraction of a wavelength (d_p) beyond the reflecting surface into the medium of the rarer refractive index, n_2.

FIGURE 11-9 Multiple internal reflection (MIR).

This plate gives about 25 reflections. The ends of the plates may be cut at various angles, with 30°, 45°, or 60° being most commonly used. A large energy loss of 50% or more occurs in the sample beam.

The depth of penetration of the infrared radiation into the sample depends on the wavelength, the refractive indices of sample and crystal, and the angle of incidence of the reflected ray at the interface. The most commonly used crystal material is KRS-5 (Section 11-6). Other materials such as IRTRAN-2 or germanium also are used.

Good spectra comparable to IR transmission spectra often can be obtained using ATR methods, if proper choice is made of crystal material and angle of incidence. Differences may be observed between IR transmission and ATR spectra. In general, bands are shifted slightly to lower frequencies in ATR, and the high-frequency bands are weaker in ATR than in transmission IR experiments. Although the refractive index of the internal reflection element must be greater than that of the sample, the best spectra are obtained when the difference in refractive indices is small.

The single-reflection diamond ATR accessory is useful for analysis of hard solid materials, powders, and single fibers. Samples are crushed between the diamond ATR element and a sapphire anvil. For 45° incidence at 1000 cm^{-1}, with a sample of refractive index of 1.5, a depth of penetration of 2.0 μm and an effective pathlength of 4.4 μm are obtained. These characteristics are sufficient to obtain good FT–IR spectra from most materials.

The ATR method is extremely useful for studying films, coatings, and surfaces in general and for samples that are too thick or too strongly absorbing for transmission methods. In many cases, no sample preparation is necessary, as the sample is just laid onto the surface of the crystal.

External reflection can be subdivided into two categories. Some of the reflected radiation from a sample travels along the theoretical path that would be followed if the surface were perfectly flat. This portion is called *specular reflection*. The rest of the reflected radiation is scattered in all directions. This portion is called *diffuse reflection*. IR spectra can be obtained from both kinds of reflection. For specular reflection spectra, simple reflectance attachments are available from the manufacturers of spectrometer accessories.

Powders and solid samples with rough surfaces give mainly diffuse reflection, so the reflected radiation must be collected over a wide solid angle. Accessories that enable *diffuse reflectance FT–IR spectra* (DRIFTS) to be recorded are available from several manufacturers.

Problems

11-1 Calculate the vibrational frequency (s^{-1}) of a system consisting of two balls of mass 1.66×10^{-27} kg connected by a spring with force constant 510 N m^{-1}.

11-2 A strong IR absorption band has a peak maximum at 7% *T*. What is the absorbance of the peak?

11-3 The infrared spectrum of a 0.20 M solution of a compound in CCl_4 solution was recorded using a variable path cell with a pathlength of 0.50 mm. An absorption band was observed at 1730 cm^{-1}, with a peak absorbance of 1.05 absorbance units. The spectrum of a CCl_4 solution of unknown concentration of the same substance was recorded in the same cell, but at a pathlength of 0.15 mm. The absorbance of the 1730 cm^{-1} peak was 0.65 absorbance units. Calculate the concentration of the unknown solution.

11-4 Calculate the relative intensity of light scattered at 514.5 nm to that scattered at 1064.8 nm.

11-5 Enumerate the advantages of FT over dispersive IR spectroscopy.

11-6 Give reasons why FT–Raman spectroscopy is carried out in the near-IR (10,000–6000 cm^{-1}) rather than the mid-IR (4000–400 cm^{-1}) region of the spectrum.

Bibliography

General Infrared Spectroscopy

11.1 J. R. Durig (ed.), *Analytical Applications of FT–IR to Molecular and Biological Systems,* Norwell, MA: Reidel Publishing Co., 1980.

11.2 A. E. Martin, *Infrared Interferometric Spectrometers,* vol. 8 of J. R. Durig (ed.), *Vibrational Spectra and Structure,* Amsterdam: Elsevier Science, 1980.

11.3 P. R. Griffiths and J. A. de Haseth, *Fourier Transform Infrared Spectrometry,* New York: John Wiley & Sons, 1986.

11.4 H. A. Willis, J. H. Van Der Maas, and R. G. J. Miller (eds.), *Laboratory Methods in Vibrational Spectroscopy,* 3rd ed., Chichester, UK: John Wiley & Sons Ltd., 1987.

11.5 J. Workman and C. Weyer, *Practical Guide to Interpretive Near-Infrared Spectroscopy,* Boca Raton, FL: CRC Press, Taylor and Frances Group, 2008.

General Raman Spectroscopy

11.6 D. A. Long, *Raman Spectroscopy,* London: McGraw-Hill Book Co., 1977.

11.7 P. Hendra, C. Jones, and G. Warnes, *Fourier Transform Raman Spectroscopy,* Chichester, UK: Ellis Horwood, 1991.

11.8 D. B. Chase and J. F. Rabolt (eds.), *Fourier Transform Raman Spectroscopy,* New York: Academic Press, 1994.

11.9 J. R. Ferraro and K. Nakamoto, *Introductory Raman Spectroscopy,* New York: Academic Press, 1994.

11.10 E. Smith and G. Dent, *Modern Raman Spectroscopy: A Practical Approach,* Hoboken, NJ: John Wiley & Sons, 2005.

11.11 SERS: C.L. Haynes, C.R. Yonzon, X. Zhang, and R. P. Van Duyne, *J. Raman Spectrosc.,* **36,** 471 (2005).

Reflection Spectroscopy

11.12 N. J. Harrick, *Internal Reflection Spectroscopy,* New York: Interscience, 1967; *Review and Supplement,* Ossining, NY: Harrick Scientific Corp., 1985.

11.13 DRIFTS: *Optical Spectroscopy: Sampling Techniques Manual,* Ossining, NY: Harrick Scientific Corp., 1987.

11.14 M. W. Urban, *Attenuated Total Reflectance Spectroscopy of Polymers,* Washington, DC: American Chemical Society, 1996.

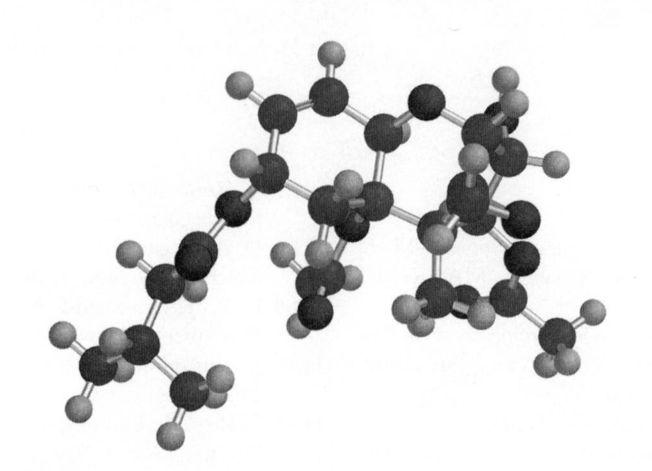

Group Frequencies

12-1 INTRODUCTION

Although the subject of group frequencies is essentially empirical in nature, it has a sound theoretical basis. Infrared (IR) and Raman spectra of a large number of compounds containing a particular functional group, such as carbonyl, amino, phenyl, nitro, and so forth, have certain features that appear at generally the same frequency for every compound containing the group. The similarity of frequency results from the approximately constant values of the stretching force constant in different molecules (eq. 11-1). It is reasonable, then, to associate these spectral features with the functional group, provided that a sufficiently large number of different compounds containing the group have been studied. For example, the IR spectrum of any compound that contains a C=O group has a strong band between 1800 and 1650 cm^{-1}. Compounds containing $—NH_2$ groups have two IR bands between 3400 and 3300 cm^{-1}. Nitro groups are characterized by IR and Raman bands near 1550 and 1350 cm^{-1}. These are just three examples of the many characteristic frequencies of chemical groups observed in IR and Raman spectra.

Pairs of atoms joined by bonds in a molecule can be treated theoretically as diatomic molecules. This simple approach gives surprisingly good results when one of the atoms of the pair is a light atom that is not bonded to any other atom, for example, C—H and N—H in CH_3NH_2 or C—H and C=O in $(CH_3)_2CO$. The stretching frequencies in cm^{-1} of these diatomic groups may be calculated from eq. 12-1,

$$\nu(\text{cm}^{-1}) = 130.3\sqrt{\frac{k}{\mu}} \qquad \textbf{(12-1)}$$

in which k is the force constant (N m^{-1}) and μ is the reduced mass, $m_1m_2/(m_1 + m_2)$ in atomic mass units. The numerical constant $130.3 = 1/2\pi c\sqrt{N \times 10^{-1}}$ (N is Avogadro's number, 6.022×10^{23}, and c is the velocity of light, 2.998×10^8 m s^{-1}). In older texts and papers, force constants have units of mdyn A^{-1} (10^5 dyn cm^{-1}) and c has units of cm s^{-1}. When these units are used, the numerical constant of eq. 12-1 becomes 1303.

When two or more identical functional groups are present in a molecule, one might expect to observe a band for each group at a similar frequency in the IR spectrum. These bands, however, often are not resolved. When the groups are attached to the same carbon atom or to two adjacent atoms, the frequencies may be spread over a few hundred wavenumbers by strong interactions. For example, in the ethylene molecule (C_2H_4), there are four CH groups, and one might expect to find four frequencies near 3000 cm^{-1} due to C—H stretching vibrations. The CH groups are attached to the same or adjacent carbon atoms, and the four observed CH stretching frequencies are 3270, 3105, 3020, and 2990 cm^{-1}.

12-2 FACTORS AFFECTING GROUP FREQUENCIES

12-2a Symmetry

The vast majority of organic molecules have little or no symmetry. Nevertheless, some knowledge of symmetry can be of considerable help in understanding the factors that affect intensities of group frequencies.

For a vibration to give rise to absorption of IR radiation (to be active), it must be associated with an oscillating electric dipole. For a vibration to be Raman active, it must give rise to a change in the polarizability of the molecule. This change in turn gives rise to an induced dipole through interaction with the electric field of the incident laser radiation. Some vibrations are inactive in the IR or Raman, usually as a consequence of symmetry.

A molecule with a center of symmetry has no permanent dipole moment, and a vibration that is symmetric with respect to the center of symmetry (a symmetric mode) does not generate an oscillating dipole. This vibration is inactive (does not absorb) in the infrared. A vibration that is antisymmetric with respect to the center of symmetry (an antisymmetric mode), however, produces a transient oscillating dipole moment that interacts with the electric field of the radiation. This vibration is active (absorbs) in the infrared. In contrast, symmetric modes are always active (give rise to scattering) in the Raman spectrum, because they produce a change in the polarizability of the molecule, but antisymmetric modes are not active in the Raman. This *mutual exclusion rule* between IR and Raman activity is illustrated by the spectra of *trans*-1,2-dichloroethene (Figure 12-1). The C=C stretching vibration of *trans*-1,2-dichloroethene is not observed in the infrared, but is seen at 1580 cm^{-1} in the Raman spectrum. The C—Cl stretches give rise to two vibrational modes, a symmetric mode observed in the Raman spectrum at 840 cm^{-1} and an antisymmetric mode in the infrared at 895 cm^{-1}. Similarly, the two CH bending modes are observed at 1200 cm^{-1} (IR) and 1270 cm^{-1} (Raman).

The C≡C stretching mode provides another example of the effect of a center of symmetry. For methylacetylene ($CH_3C{\equiv}CH$) the vibration is both IR and Raman active, with bands at 2150 cm^{-1}. On the other hand, in dimethylacetylene ($CH_3C{\equiv}CCH_3$), which has a center of symmetry, the exclusion rule applies. A strong C≡C stretching band is found in the Raman spectrum near 2150 cm^{-1}, but no band is observed in the IR spectrum at this frequency.

Some molecules have threefold or higher axes of symmetry, as well as mirror planes or other symmetry elements. Such molecules are said to have high symmetry

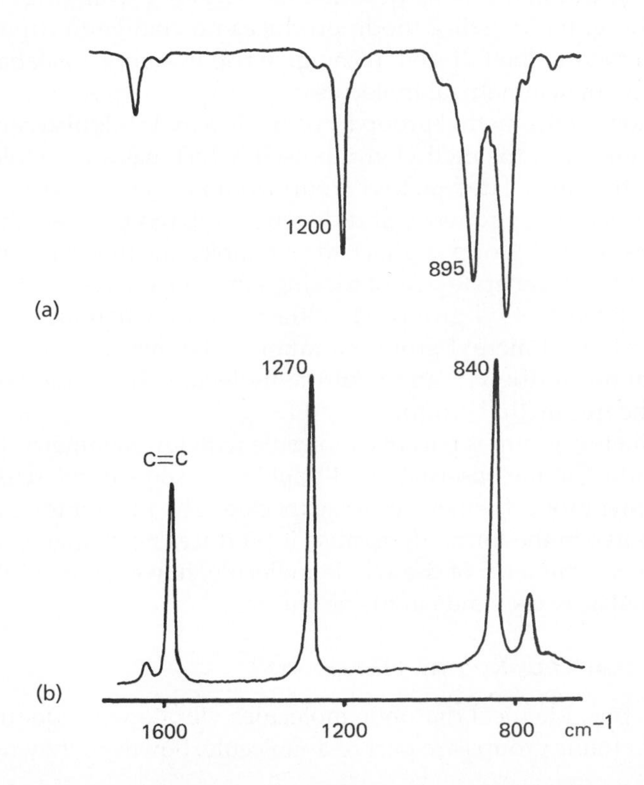

FIGURE 12-1 Portions of the (a) infrared and (b) Raman spectra of *trans*-1,2-dichloroethene.

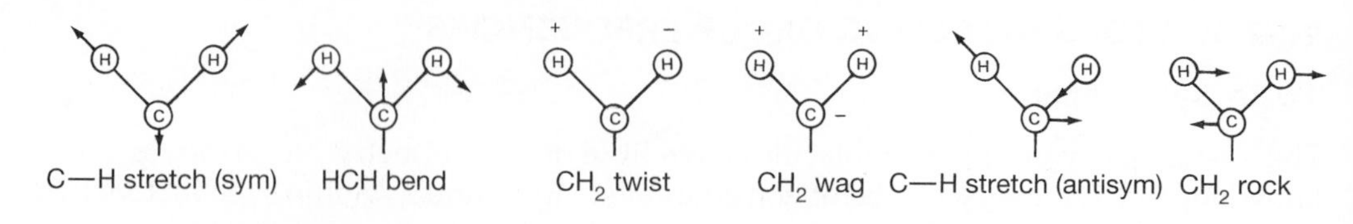

FIGURE 12-2 The vibrations of a CH_2 group. The arrows show the direction of motion of atoms in the plane of the CH_2 group, while the + and − signs denote motion above and below the plane, respectively.

and to have simple IR and Raman spectra. For example, the benzene molecule has a sixfold axis of symmetry perpendicular to the plane of the ring and many other symmetry elements. It has 12 atoms and therefore $(3N - 6) = 30$ normal modes of vibration. The first effect of the high symmetry is that 10 pairs of these vibrations have identical frequencies (degenerate modes), leaving 20 different normal frequencies. The second effect of the high symmetry is to reduce the number of modes for which there is a change in dipole moment (IR active) or a change in polarizability (Raman active). In fact, the IR spectrum of benzene contains only four bands from fundamentals (defined in Section 11-3), and the Raman spectrum contains only six.

The symmetry of the benzene molecule is reduced by substitution, as in 1,3,5-trichlorobenzene, in which the sixfold axis is replaced by a threefold axis. For this molecule, the number of IR and Raman active modes is greater than for benzene, but there still are some degenerate modes and the spectra are relatively simple. When the symmetry is further lowered, as in 1-chloro-2-bromobenzene, which has only the plane of the benzene ring as a symmetry element, all 30 normal modes are active in both IR and Raman. Because of the symmetry of the benzene ring itself, however, some of these vibrations appear only very weakly and are hard to distinguish from the weak bands from overtones and combinations.

In larger, more complicated molecules, local symmetry may exist for a homonuclear diatomic group such as C=C or S—S, so that the IR absorption from the group vibration may be weak or absent. In such cases, the Raman spectrum can confirm the presence (or absence) of the functional group.

The vibrations of a methylene group (CH_2) can be described in terms of the local symmetry of the group, which has a twofold axis and two planes of symmetry. Figure 12-2 shows the vibrations associated with a CH_2 group within a molecule. An isolated CH_2 group has three modes: the symmetric and antisymmetric (with respect to the twofold axis) stretching, and the bending (scissors) vibrations. When the group is part of a larger molecule, three additional modes described as twisting, wagging, and rocking are produced. Of these, the twisting mode produces no change in dipole moment and hence is not observed in the infrared. It can give rise to a very weak band, however, in the spectrum of an unsymmetrical molecule.

The vibrations of the methyl group ($—CH_3$) in a molecule also can be described in terms of the symmetry of the methyl group itself, which has a threefold axis and three planes of symmetry. An isolated methyl group would have $3N - 6 = 6$ normal modes of vibration comprising symmetric and degenerate pairs of stretching and bending modes. When the methyl group is attached to a molecule, three new modes appear: a torsional mode and a degenerate pair of rocking vibrations. These motions would be rotations in the isolated methyl group. Thus there are four regions of the spectrum in which we expect to find methyl group vibrations. The methyl group also contributes three skeletal modes to the vibrations of the molecule. These modes correspond to translations of the free methyl group.

When the methyl group is part of a molecule with low symmetry, the degeneracies are removed, leading to the observation of doublets in some of the regions of the spectrum where methyl group frequencies are expected. The methyl torsional mode is expected to be inactive in the infrared, because it produces no change in dipole moment. Because of the low symmetry of the whole molecule, however, methyl torsions sometimes are observed as weak bands in the far infrared.

12-2b Mechanical Coupling of Vibrations

Two completely free, identical diatomic molecules vibrate with identical frequencies. When the two diatomic groups are part of a molecule, however, they no longer vibrate independently of each other because the vibration of one group causes displacements of the other atoms in the molecule. These displacements are transmitted through the molecule and interact with the vibration of the second group. The resulting vibrations appear as in-phase and out-of-phase combinations of the two diatomic vibrations. When the groups are widely separated in the molecule, the coupling is very small and the two frequencies may not be resolved. The two C—H stretching modes in acetylene (H—C≡C—H) are observed at 3375 cm^{-1} in the Raman spectrum (in phase) and 3280 cm^{-1} in the infrared (out of phase). In diacetylene (H—C≡C—C≡C—H) the two C—H stretching vibrations have closer frequencies, 3330 and 3295 cm^{-1}.

Mechanical coupling occurs in two C=C groups coupled through a common atom, as in the allene molecule, CH_2=C=CH_2. In the absence of strong coupling, one might expect to observe a band in the IR spectrum near 1600 cm^{-1} from the out-of-phase (antisymmetric) vibrations of the C=C groups and a band in the Raman spectrum from the in-phase (symmetric) modes at a similar frequency. For the 1,3-butadiene molecule (CH_2=CH—CH=CH_2) these bands are, in fact, observed at 1640 cm^{-1} in the infrared and 1600 cm^{-1} in the Raman. For allene, however, the observed frequencies are near 1960 and 1070 cm^{-1}. This result can be understood in terms of mechanical coupling of the two C=C group vibrations. When such coupling occurs, it is usually found that the higher-frequency mode is the antisymmetric vibration and the lower frequency mode is the symmetric vibration.

It is possible for coupling to occur between dissimilar modes such as stretching and bending vibrations when the frequencies of the vibrations are similar and the two groups involved are adjacent in the molecule. An important example is found in secondary amides, in which the C—N stretching vibration is of a similar frequency to that of the NH bending mode. Interaction of these two vibrations gives rise to two bands in the spectrum, one at a higher and one at a lower frequency than the uncoupled frequencies. These bands are known as Amide II and Amide III bands. (The *Amide I band* is the C=O stretching mode.)

Chains of singly bonded (saturated) carbon atoms, of course, are not linear, so the simple model used for the allene molecule must be modified. In addition, we ignored the bending of the C=C=C group in allene, which cannot couple with the stretching modes, because it takes place at right angles to the stretching vibrations. Mechanical coupling always occurs between C—C single bonds in an organic molecule, and so there is no simple C—C group stretching frequency. One can expect that there always will be several bands in the IR and Raman spectra in the 1200–800 cm^{-1} range in compounds containing saturated carbon chains. Certain branched chain structures, such as the *tert*-butyl group, $(CH_3)_3C$—, and the isopropyl group, $(CH_3)_2CH$—, do have characteristic group frequencies involving the coupled C—C stretching vibrations. These systems are discussed further in later sections.

In many molecules, mechanical coupling of the group vibrations is so widespread that few, if any, frequencies can be assigned solely to functional groups. Many such examples are found in aliphatic fluorine compounds, in which the CF and CC stretching modes are coupled with each other and with FCF and CCF bending vibrations. The presence of fluorine can be deduced from several very strong IR bands in the region between 1400 and 900 cm^{-1}.

12-2c Fermi Resonance

A special case of mechanical coupling, known as *Fermi resonance*, results from coupling of a fundamental vibration with an overtone or combination. Such interactions can shift group frequencies and introduce extra bands. For a polyatomic molecule, there are $3N - 6$ energy levels for which only one vibrational quantum number (ν_i) is 1 and all the rest are zero. In addition to these *fundamentals*, there are the levels for which one ν_i is 2, 3, and so on (*overtones*), or for which more than one ν_i is nonzero (*combinations*). Therefore, a very large number of vibrational energy levels exist, and it quite often happens that the energy of an overtone or combination level is very close to that of a fundamental. This situation is termed *accidental degeneracy*, and an interaction known as Fermi resonance can occur between these levels, provided that the symmetries of the levels are the same. Since most organic molecules have no symmetry, all levels have the same symmetry and Fermi resonance effects occur frequently in vibrational spectra.

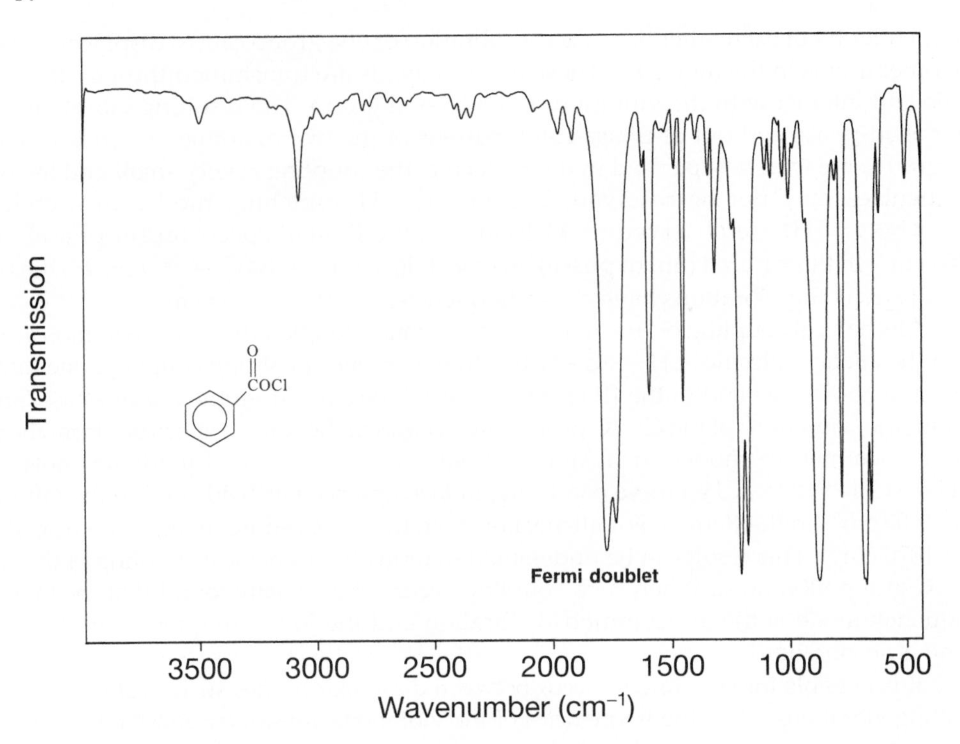

FIGURE 12-3 The IR spectrum of benzoyl chloride showing the Fermi doublet at 1760–1720 cm^{-1}.

Normally, an overtone or combination band is very weak in comparison with a fundamental, because these transitions are not allowed. When Fermi resonance occurs, however, there is a sharing of intensity, and the overtone can be quite strong. The result is the same as that produced by two identical groups in the molecule. As an example, two peaks are observed in the carbonyl stretching band of benzoyl chloride, near 1760 and 1720 cm^{-1} (Figure 12-3). If this were an unknown compound, one might be tempted to suggest that there were two nonadjacent carbonyl groups in the molecule. The lower-frequency band, however, is due to the first overtone of the CH out-of-plane bending mode at 865 cm^{-1} in Fermi resonance with the C=O stretching fundamental.

Numerous other well-characterized examples of Fermi resonance are known. The N—H stretching mode of the —CO—NH— group in polyamides, peptides, proteins, and so on, appears as two bands near 3300 and 3205 cm^{-1}. The N—H stretching fundamental and the overtone of the N—H deformation mode near 1550 cm^{-1} combine through Fermi resonance to produce the two observed bands. The CH stretching region of the —CHO group in aldehydes provides another example of Fermi resonance. Two bands often are observed near 2900 and 2700 cm^{-1} in the IR spectra of aldehydes (see Figure 12-12 in Section 12-6f). This doubling is attributed to Fermi resonance between the overtone of the C—H deformation mode, which would have a frequency near 2 × 1400 cm^{-1}, and the C—H stretching mode, which also would occur near 2800 cm^{-1} in the absence of Fermi resonance.

12-2d Hydrogen Bonding

Hydrogen bonding (written X—H···Y) occurs between a hydrogen atom bonded to an electronegative element X, as in OH or NH, and an atom Y possessing one or more nonbonding electron pairs, usually O or N. The main effects on the IR and Raman spectra are broadening of bands in the spectra and shifts of group frequencies. X—H stretching frequencies are lowered by hydrogen bonding, and X—H bending frequencies are raised. Hydrogen bonding also affects the frequencies of the acceptor group, but the frequency shifts are less than those of the X—H group. Solvents such as CCl_4, which do not interact with the solute, can reduce the extent of hydrogen bonding and even eliminate the effect in very dilute solutions.

Hydrogen bonding manifests itself in very broad OH and NH stretching bands at frequencies considerably lower than those of the unbonded groups. Changes in the intensity of these bands can be brought about by changes in temperature and concentration, both of which affect the degree of hydrogen bonding. A very broad band centered near 3100 cm^{-1} in the spectrum of a pure carboxylic acid is due mainly to OH stretching of hydrogen-bonded carboxylic acid oligomers. In solutions of carboxylic acids in non–hydrogen bonding solvents such as CCl_4, the presence of monomer, dimer, and oligomeric species can be identified by bands in the carbonyl stretching region of the IR spectrum. Monomer–dimer equilibria have been studied for a variety of molecules such as phenols, alcohols, and compounds that self-associate to form hydrogen-bonded species.

12-2e Ring Strain

The effect of ring strain on group frequencies is quite interesting and useful for diagnostic purposes. As an example, consider the series of alicylic ketones: cyclohexanone, cyclopentanone, and cyclobutanone. The IR spectra are given in Figure 12-4. The observed carbonyl stretching frequencies are 1714, 1746, and 1783 cm^{-1}. The increase in $\nu_{C=O}$ is attributed to mechanical interaction with the adjacent coplanar C—C single bonds, which changes as the double bond–single bond angle changes.

The increase in frequency with increasing angle strain is generally observed for double bonds directly attached (exocyclic) to rings. Frequency changes similar to those observed for cyclic ketones are found in the series of compounds methylenecyclohexane, methylenecyclopentane, and methylenecyclobutane, in which a $C{=}CH_2$ group replaces the C=O group. The observed C=C stretching frequencies are 1649, 1656, and 1677 cm^{-1}, respectively. In contrast, when the double bond is endocyclic, a decrease in the ring angle causes a *lowering* of the C=C stretching frequency. The observed frequencies for cyclohexene, cyclopentene, and cyclobutene are 1650, 1615, and 1565 cm^{-1}, respectively.

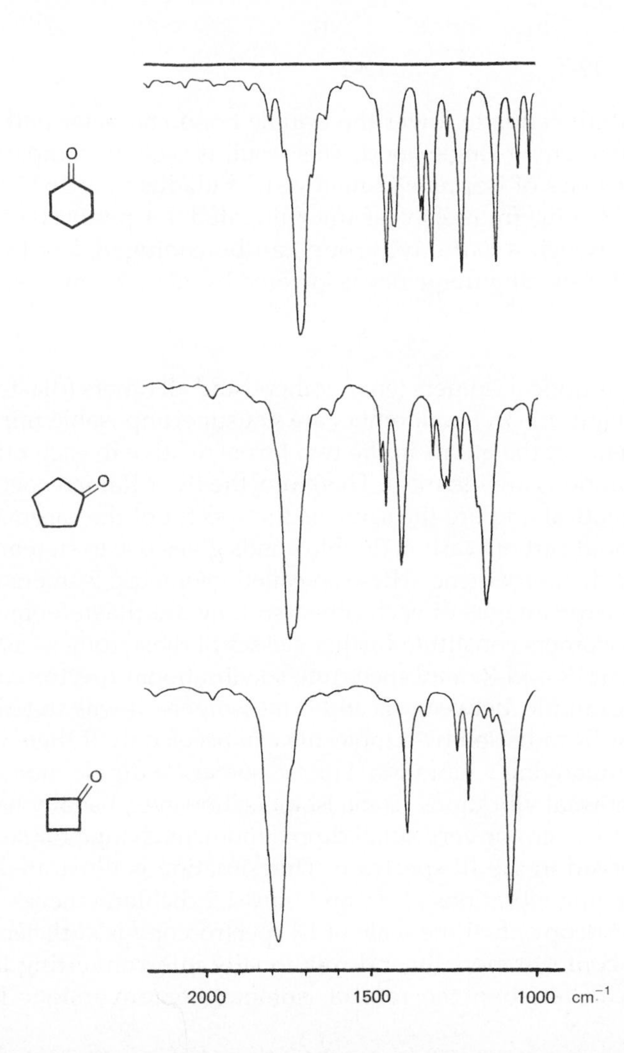

FIGURE 12-4 Portions of the IR spectra of three cyclic ketones showing the influence of increasing ring strain on the C=O stretching frequency. (Reproduced with permission of Aldrich Chemical Co., Inc., from *The Aldrich Library of FT–IR Spectra*.)

Unlike C=O and C=C groups, the P=O, S=O, and SO_2 groups have stretching frequencies that are little affected by being part of a strained ring, because the double bond in these groups is not coplanar with the two attached single bonds.

12-2f Electronic Effects

Effects arising from the change in the distribution of electrons in a molecule produced by a substituent atom or group often can be detected in the vibrational spectrum. There are several mechanisms, such as polar (also called "inductive") and resonance effects, that can be used to explain observed shifts and intensity changes in a qualitative way. These effects involve changes in electron distribution in a molecule and cause changes in the force constants that are, in turn, responsible for changes in group frequencies. Polar and resonance effects have been used successfully to explain the shifts observed in C=O stretching frequencies produced by various substituent groups in compounds such as acid chlorides and amides. High C=O stretching frequencies usually are attributed to polar effects, and low frequencies arise when delocalized structures are possible. For example, in acid chlorides the C=O frequency is near 1800 cm^{-1}, which is high compared with that observed for aldehydes or ketones (1730 cm^{-1}). On the other hand, in amides the carbonyl frequency is lower (near 1650 cm^{-1}). An example of this behavior was shown earlier in Figure 11-4. In acid chlorides, the electronegative chlorine atom adjacent to the carbonyl group increases electron density in the double bond, raises the C=O stretching force constant, and causes an increase in frequency, whereas in amides the delocalized electronic structure (**12-1** ↔ **12-2**) lowers the force constant and leads to a decrease of the C=O stretching frequency.

O R C NH_2 ↔ O^- R C NH_2^+

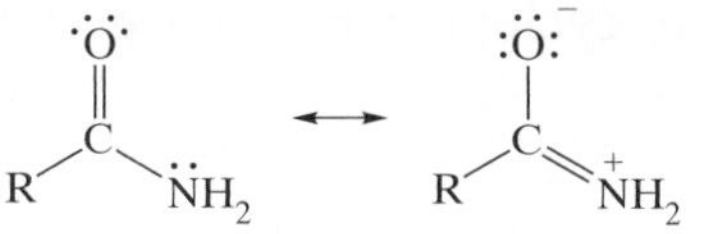

12-1 **12-2**

Conjugation of double bonds tends to lower the double bond character and increase the bond order of the intervening single bond. This result is seen by comparing the two C=C stretching frequencies of isoprene (2-methyl-1,3-butadiene) at 1637 and 1604 cm^{-1} with the C=C stretching frequency of unconjugated 1,4-pentadiene at 1644 cm^{-1}. For compounds in which a carbonyl group can be conjugated with an alkenic double bond, the C=O stretching frequency is lowered by 20 to 30 cm^{-1}.

12-2g Stereoisomerism

Stereoisomers can be classified as optical isomers (enantiomers) and all others (diastereomers) (see the isomer tree in Figure 1-3). Enantiomers are nonsuperimposable mirror images. Because the arrangements of the atoms in the two forms relative to each other are the same, the molecular vibrations are identical. Therefore, the IR or Raman spectra of a pair of enantiomers are identical and are the same as the spectra of the racemate.

The absence of rotation about carbon–carbon double bonds gives rise to stereoisomers, such as *cis*- and *trans*-1,2-dichloroethene. These so-called geometric isomers are neither superimposable nor mirror images of each other, so they are diastereomers. Rotational and conformational isomers constitute further classes of diasteromers. Each isomer has a completely different IR and Raman spectrum, so vibrational spectroscopy is useful in distinguishing, for example, between cis and trans isomers. It was stated in Section 12-2a that absorption of IR radiation by a molecule can occur only if there is a change of dipole moment accompanying a vibration. For cis isomers, a dipole moment change occurs for most of the normal vibrations. Trans isomers, however, usually have higher symmetry, which leads to a zero or very small dipole moment change for some vibrations, which are not observed in the IR spectrum. This situation is illustrated in Figure 12-5 for the C—Cl stretching vibrations of *cis*- and *trans*-1,2-dichloroethene.

In contrast to NMR spectroscopy, the time scale of IR spectroscopy is sufficiently fast to give distinct bands for conformationally and rotationally interconverting isomers. Vibrations occur much faster than the rate of isomer interconversion. The

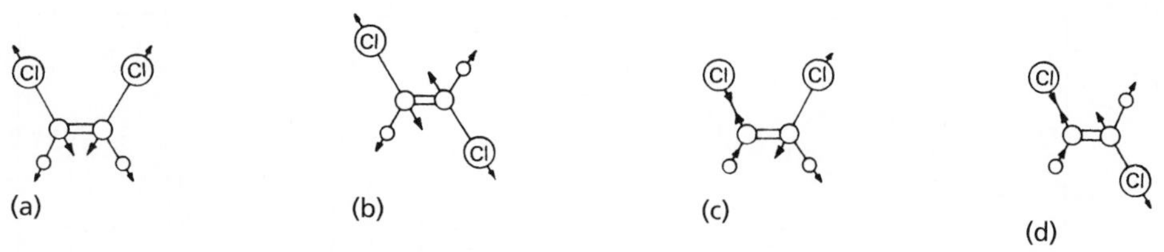

(a) (b) (c) (d)

FIGURE 12-5 The C—Cl stretching modes of *cis*- and *trans*-1,2-dichloroethene: (a) cis (symmetric), the dipole moment changes (observed in both IR and Raman at 710 cm^{-1}); (b) trans (symmetric), no change in dipole moment (not observed in IR but observed in Raman at 840 cm^{-1}; see Figure 12-1); (c) cis (antisymmetric), the dipole moment changes (observed in both IR and Raman at 840 cm^{-1}); (d) trans (antisymmetric), the dipole moment changes, but there is no change in polarizability (observed only in IR at 895 cm^{-1}; see Figure 12-1).

axial–equatorial conformations of cyclohexane derivatives, for example, interconvert rapidly on the NMR time scale at room temperature. In contrast, for α-chloro-substituted cyclohexanones, two distinct carbonyl stretching frequencies can be observed, as ring reversal is slower than the rate of vibration. One band is found near 1745 cm^{-1} from the equatorial conformation, **12-3**, and a second band near 1725 cm^{-1} from the axial isomer, **12-4**. The relative proportions of axial and equatorial forms change with phase, temperature, and solvent. Such changes can be followed readily in the vibrational spectra. In cyclohexanols, the equatorial C—OH stretching frequency is 1050 to 1030 cm^{-1}, while in the axial conformation the frequency is 10 to 30 cm^{-1} lower.

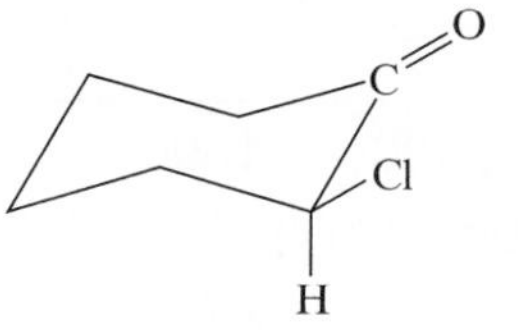

equatorial
$\nu_{CO} = 1745\ cm^{-1}$
12-3

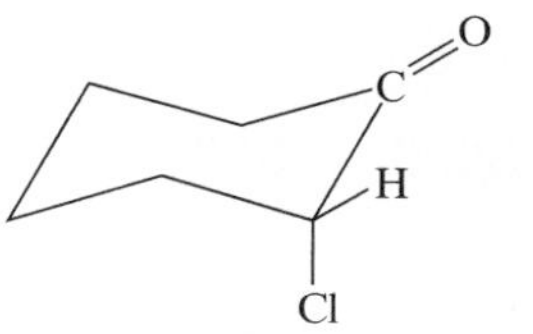

axial
$\nu_{CO} = 1725\ cm^{-1}$
12-4

Ortho-halogenated benzoic acids also show two carbonyl stretching frequencies, from the two rotational isomers such as **12-5** and **12-6**, which could be described as *cis*

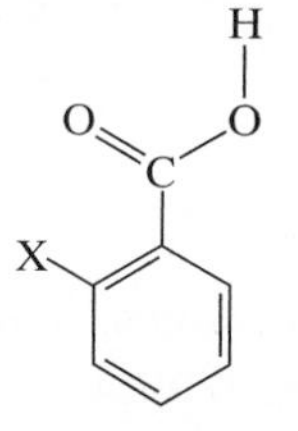

cis
12-5

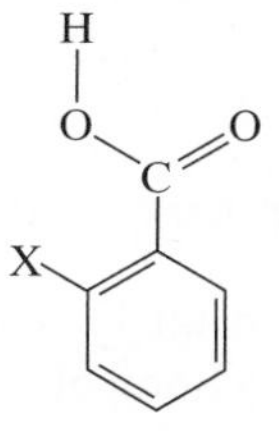

trans
12-6

and *trans* with respect to the halogen and C=O groups. Vinyl ethers show a doublet for the C=C stretching mode at 1640 to 1620 and 1620 to 1610 cm^{-1}. These bands correspond to rotational isomers about the C—O bond.

12-2h Tautomerism

Tautomerism is a special case of constitutional isomerism in which the isomers are readily interconvertible under ambient conditions, usually by a hydrogen shift. IR spectroscopy offers a useful means of distinguishing between possible tautomeric structures. A simple example is found in β-diketones. The keto form, **12-7**, has two C=O groups, which have separate stretching frequencies. A doublet is often observed in the usual ketone carbonyl stretching region, near 1730 cm^{-1}. The enol form, **12-8**, on the

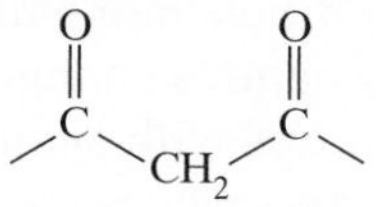

keto
12-7

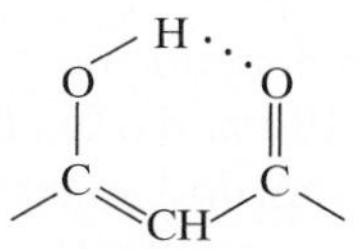

enol
12-8

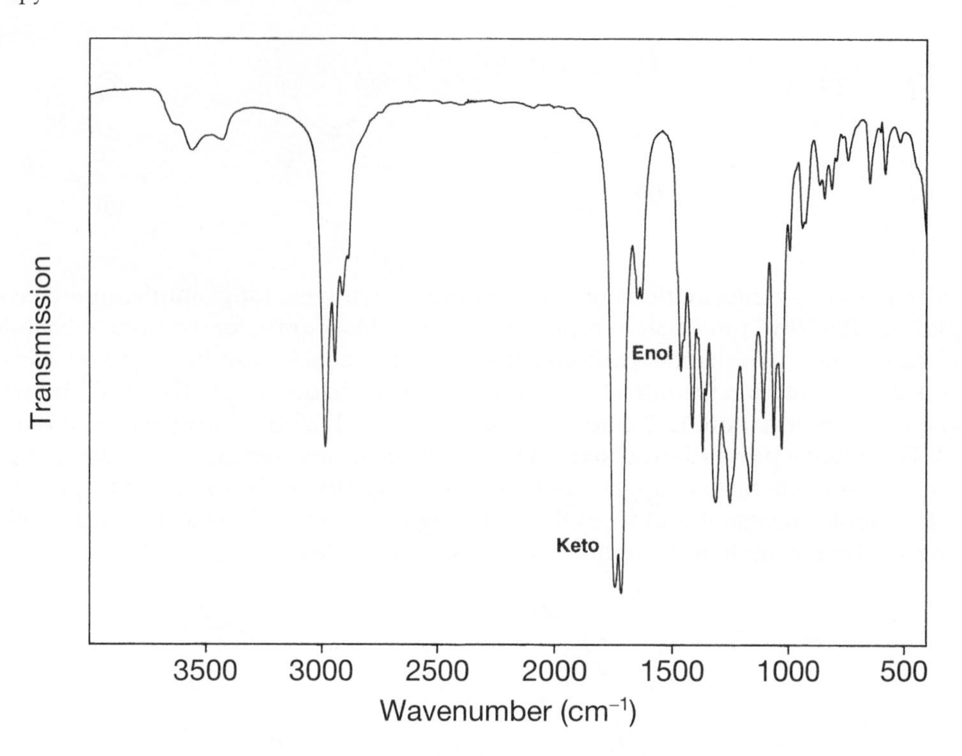

FIGURE 12-6 The IR spectrum of ethyl propionylacetate.

other hand, has only one carbonyl group. The frequency of the carbonyl group is lowered by 80 to 100 cm^{-1} as the result of hydrogen bonding and conjugation. This structure also has an alkenic double bond that should give a band between 1650 and 1600 cm^{-1}. The C=O and C=C vibrations then may appear as two overlapping bands with closely spaced peaks. An example of such tautomerism is shown in Figure 12-6. This particular compound, ethyl propionylacetate, clearly shows both keto and enol forms.

12-3 INFRARED GROUP FREQUENCIES

Group frequencies fall within fairly restricted ranges, regardless of the compound in which the group is found. Mechanical coupling, symmetry, or other effects discussed in the previous sections may occasionally cause even a good group frequency to misbehave. An alphabetical list of functional groups with frequency ranges and intensities is given in Table 12-1. This table is by no means comprehensive. To make full use of the group frequency method for structure determination, the references cited at the end of this chapter should be consulted.

12-4 RAMAN GROUP FREQUENCIES

Depending on the symmetry of a molecule, its vibrational frequencies may give rise to IR absorption, Raman scattering, or both. In the last case, observed wavenumbers are numerically the same in Raman and IR, but the intensities often are quite different. In most cases, information obtained from the Raman spectrum duplicates that obtained from the infrared. In some cases, however, the Raman spectrum provides additional information, especially in the low-frequency region in which far-IR spectra may not be available.

Since IR absorption depends on change of dipole moment, we expect polar bonds or groups to give strong IR bands. On the other hand, a change in polarizability is necessary for Raman scattering, and so bonds or groups with symmetrical charge distributions are expected to give rise to strong Raman lines. Some of the most important Raman group frequencies are found for the C=C, N=N, C≡C, and S—S stretching modes. Some Raman group frequencies are given in Table 12-2.

TABLE 12-1 Characteristic IR Frequencies of Some Functional Groups and Classes of Compounds

Group or Class	Frequency Ranges (cm^{-1}) and Intensities[a]	Assignment and Remarks
Acid halides R—C(=O)—X		
Aliphatic	1810–1790 (s)	C=O stretch; fluorides 50 cm^{-1} higher
	965–920 (m)	C—C stretch
	440–420 (s)	Cl—C=O in-plane deformation
Aromatic	1785–1765 (s)	C=O stretch; also a weaker band (1750–1735 cm^{-1}) due to Fermi resonance
	890–850 (s)	C—C stretch (Ar—C) or C—Cl stretch
Alcohols		
Primary —CH_2OH	3640–3630 (s)	OH stretch, dil CCl_4 soln
	1060–1030 (s)	C—OH stretch; lowered by unsaturation
Secondary —CHROH	3630–3620 (s)	OH stretch, dil CCl_4 soln
	1120–1080 (s)	C—OH stretch; lower when R is a branched chain or cyclic
Tertiary —CR_2OH	3620–3610 (s)	OH stretch, dil CCl_4 soln
	1160–1120 (s)	C—OH stretch; lower when R is branched
General —OH	3350–3250 (s)	OH stretch; broad band in pure solids or liquids
	1440–1260 (m–s, br)	C—OH in-plane bend
	700–600 (m–s, br)	C—OH out-of-plane deformation
Aldehydes R—C(=O)—H	2830–2810 (m) } 2740–2720 (m) }	Fermi doublet; CH stretch with overtone of CH bend
	1725–1695 (vs)	C=O stretch; slightly higher in CCl_4 soln
	1440–1320 (s)	H—C=O bend in aliphatic aldehydes
	695–635 (s)	C—C—CHO bend
	565–520 (s)	C—C=O bend
Alkenes		
Monosubst —CH=CH_2	—	*See* Vinyl
Disubst —CH=CH— , >C=CH_2	—	*See* Vinylene and vinylidene
Trisubst >C=CH—	3050–3000 (w)	CH stretch
	1690–1655 (w-m)	C=C stretch
	850–790 (m)	CH out-of-plane bending
Tetrasubst >C=C<	1690–1670 (w)	C=C stretch, may be absent for symmetrical compounds
Alkyl	2980–2850 (m)	CH stretch, several bands
	1470–1450 (m)	CH_2 deformation
	1400–1360 (m)	CH_3 deformation
	740–720 (w)	CH_2 rocking
Alkynes RC≡C—H	3300–3250 (m–s)	Terminal ≡C—H stretch
	2250–2100 (w–m)	C≡C, frequency raised by conjugation
	680–580 (s)	C≡CH bend
Amides		
Primary —$CONH_2$	3540–3520 (m)	NH_2 stretch (dil solns); bands shift to
	3400–3380 (m) }	3360–3340 and 3200–3180 in solid
	1680–1660 (vs) }	C=O stretch (Amide I band)
	1650–1610 (m)	NH_2 deformation; sometimes appears as a shoulder (Amide II band)

[a]*Note*: s = strong; m = medium; w = weak; v = very; br = broad.

(*continued*)

TABLE 12-1 Characteristic IR Frequencies of Some Functional Groups and Classes of Compounds (*continued*)

Group or Class	Frequency Ranges (cm^{-1}) and Intensities[a]	Assignment and Remarks
	1420–1400 (m–s)	C—N stretch (Amide III band)
Secondary —CONHR	3440–3420 (m)	NH stretch (dil soln); shifts to 3300–3280 in pure liquid or solid
	1680–1640 (vs)	C=O stretch (Amide I band)
	1560–1530 (vs)	NH bend (Amide II band)
	1310–1290 (m)	C—N stretch
Tertiary —$CONR_2$	1670–1640 (vs)	C=O stretch
Genera l—$CONR_2$	630–570 (s)	N—C=O bend
	615–535 (s)	C=O out-of-plane bend
	520–430 (m–s)	C—C=O bend
Amines		
Primary —NH_2	3460–3280 (m)	NH stretch; broad band, may have some structure
	2830–2810 (m)	CH stretch
	1650–1590 (s)	NH_2 deformation
Secondary —NHR	3350–3300 (vw)	NH stretch
	1190–1130 (m)	C—N stretch
	740–700 (m)	NH deformation
	450–400 (w, br)	C—N—C bend
Tertiary —NR_2	510–480 (s)	C—N—C bend
Amine hydrohalides $RNH_3^+X^-$	2800–2300 (m–s)	—NH_3^+ stretch, several peaks
$R'\overset{+}{N}H_2R\ X^-$	1600–1500 (m)	NH deformation (one or two bands)
Amino acids —C(NH_2)—COOH	3200–3000 (s)	H-bonded NH_2 and OH stretch; ν broad band in solid state
(or —$C\overset{+}{N}H_3COO^-$)	1600–1590 (s)	COO^- antisym stretch
	1550–1480 (m–s)	—NH_3^+ deformation
	1425–1390 (w–m)	COO^- sym stretch
	560–500 (s)	COO^- rocking
Ammonium NH_4^+	3350–3050 (vs)	NH stretch; broad band
	1430–1390 (s)	NH_2 deformation; sharp peak
Anhydrides —CO—O—CO—	1850–1780 (variable)	Antisym C=O stretch
	1770–1710 (m–s)	Sym C=O stretch
	1220–1180 (vs)	C—O—C stretch (higher in cyclic anhydrides)
Aromatic compounds	3100–3000 (m)	CH stretch, several peaks
	2000–1660 (w)	Overtone and combination bands
	1630–1430 (variable)	Aromatic ring stretching (four bands)
	900–650 (s)	Out-of-plane CH deformations (one or two bands depending on substitution)
	580–420 (m–s)	Ring deformations (two bands)
Azides —$N=\overset{+}{N}=N^-$	2160–2080 (s)	N=N=N stretch
Bromo —C—Br	650–500 (m)	C—Br stretch
***tert*-Butyl** $(CH_3)_3C$—	2980–2850 (m)	CH stretch; several bands
	1400–1370 (m) 1380–1360 (s)	CH_3 deformations
Carbodiimides —N=C=N—	2150–2100 (vs)	N=C=N antisym stretch

[a]*Note*: s = strong; m = medium; w = weak; v = very; br = broad.

Group or Class	Frequency Ranges (cm^{-1}) and Intensities[a]	Assignment and Remarks
Carbonyl $\backslash C{=}O /$	1870–1650 (vs, br)	C=O stretch
Carboxylic acids $R{-}C({=}O){-}OH$	3550–3500 (s)	OH stretch (monomer, dil soln)
	3300–2400 (s, v br)	H-bonded OH stretch (solid and liquid states)
	1800–1740 (s)	C=O stretch of monomer (dil soln)
	1710–1680 (vs)	C=O stretch of dimer (solid and liquid states)
	960–910 (s)	C—OH deformation
	700–590 (s)	O—C=O bend
	550–465 (s)	C—C=O bend
Chloro —C—Cl	850–550 (m)	C—Cl stretch
Cycloalkanes	580–430 (s)	Ring deformation
Diazonium salts $-\overset{+}{N}{\equiv}N$	2300–2240 (s)	N≡N stretch
Esters $R{-}C({=}O){-}OR'$	1765–1720 (vs)	C=O stretch
	1290–1180 (vs)	C—O—C antisym stretch
	645–575 (s)	O—C—O bend
Ethers —C—O—C—	1280–1220 (s)	C—O—C stretch in alkyl aryl ethers
	1140–1110 (vs)	C—O—C stretch in dialkyl ethers
	1275–1200 (vs)	C—O—C stretch in vinyl ethers
	1250–1170 (s)	C—O—C stretch in cyclic ethers
	1050–1000 (s)	R(alkyl)—C—O stretch in alkyl aryl ethers
Fluoroalkyl $-CF_3$, $-CF_2-$, etc.	1400–1000 (vs)	C—F stretch
Isocyanates —N=C=O	2280–2260 (vs)	N=C=O stretch
Isothiocyanates —N=C=S	2140–2040 (vs, br)	C=N=S antisym stretch
Ketones $R(R')C{=}O$	1725–1705 (vs)	C=O stretch in saturated aliphatic ketones
	1700–1650 (vs)	C=O stretch in aromatic ketones
	1705–1665 (s) } 1650–1580 (m) }	C=O and C=C stretching in α, β-unsaturated ketones
Lactones (R ring with CH_2—C(=O)—O—CH_2)	1850–1830 (s)	C=O stretch in β-lactones
	1780–1770 (s)	C=O stretch in γ-lactones
	1750–1730 (s)	C=O stretch in δ-lactones
Methyl $-CH_3$	2970–2850 (s)	CH stretch in $C{-}CH_3$ compounds
	2835–2815 (s)	CH stretch in methyl ethers ($O{-}CH_3$)
	2820–2780 (s)	CH stretch in $N{-}CH_3$ compounds
	1470–1440 (m)	CH_3 antisym deformation
	1390–1370 (m–s)	CH_3 sym deformation
Methylene $-CH_2-$	2940–2920 (m) } 2860–2850 (m) }	CH stretches in alkanes
	3090–3070 (m) } 3020–2980 (m) }	CH stretches in alkenes
	1470–1450 (m)	CH_2 deformation
Naphthalenes	645–615 (m–s) } 545–520 (s) }	In-plane ring bending
	490–465 (variable)	Out-of-plane ring bending

[a]*Note*: s = strong; m = medium; w = weak; v = very; br = broad.

(continued)

TABLE 12-1 Characteristic IR Frequencies of Some Functional Groups and Classes of Compounds (*continued*)

Group or Class	Frequency Ranges (cm^{-1}) and Intensities[a]	Assignment and Remarks
Nitriles —C≡N	2260–2240 (w)	C≡N stretch in aliphatic nitriles
	2240–2220 (m)	C≡N stretch in aromatic nitriles
	580–530 (m–s)	C—C—CN bend
Nitro —NO_2	1570–1550 (vs) } 1380–1360 (vs) }	NO_2 stretches in aliphatic nitro compounds
	1480–1460 (vs) } 1360–1320 (vs) }	NO_2 stretches in aromatic nitro compounds
	920–830 (m)	C—N stretch
	650–600 (s)	NO_2 bend in aliphatic compounds
	580–520 (m)	NO_2 bend in aromatic compounds
	530–470 (m–s)	NO_2 rocking
Oximes =NOH	3600–3590 (vs)	OH stretch (dil soln)
	3260–3240 (vs)	OH stretch (solids)
	1680–1620 (w)	C=N stretch; strong in Raman
Phenols Ar—OH	720–600 (s, br)	O—H out-of-plane deformation
	450–375 (w)	C—OH deformation
Phenyl C_6H_5—	3100–3000 (w–m)	CH stretch
	2000–1700 (w)	Four weak bands; overtones and combinations
	1625–1430 (m–s)	Aromatic C=C stretches (four bands)
	1250–1025 (m–s)	CH in-plane bending (five bands)
	770–730 (vs)	CH out-of-plane bending
	710–690 (vs)	Ring deformation
	560–420 (m–s)	Ring deformation
Phosphates $(RO)_3P{=}O$		
R = alkyl	1285–1255 (vs)	P=O stretch
	1050–990 (vs)	P—O—C stretch
R = aryl	1315–1290 (vs)	P=O stretch
	1240–1190 (vs)	P—O—C stretch
Phosphines —PH_2, —PH	2410–2280 (m)	P—H stretch
	1100–1040 (w–m)	P—H deformation
	700–650 (m–s)	P—C stretch
Pyridyl —C_5H_4N	3080–3020 (m)	CH stretch
	1620–1580 (vs) } 1590–1560 (vs) }	C—C and C—N stretches
	840–720 (s)	CH out-of-plane deformation (one or two bands, depending on substitution)
	635–605 (m–s)	In-plane ring bending
Silanes —SiH_3	2160–2110 (m)	Si—H stretch
—SiH_2—	950–800 (s)	Si—H deformation
Silanes (fully substituted)	1280–1250 (m–s)	Si—C stretch
	1110–1050 (vs)	Si—O—C stretch (aliphatic)
	840–800 (m)	Si—O—C deformation
Sulfates R—O—SO_2—O—R	1140–1350 (s) } 1230–1150 (s) }	S—O stretches in covalent sulfates
Sulfate salts	1260–1210 (vs)	S=O stretches in alkyl sulfate salts
R—O—SO_3^- M^+		
(M = Na^+, K^+, etc.)	810–770 (s)	C—O—S stretch
Sulfides C—S—	710–570 (m)	C—S stretch

[a]*Note*: s = strong; m = medium; w = weak; v = very; br = broad.

Group or Class	Frequency Ranges (cm^{-1}) and Intensities[a]	Assignment and Remarks
Sulfones —SO_2—	1360–1290 (vs)	SO_2 antisym stretch
	1170–1120 (vs)	SO_2 sym stretch
	610–545 (m–s)	SO_2 scissor mode
Sulfonic acids —SO_2OH	1250–1150 (vs, br)	S=O stretch
Sulfoxides >S=O	1060–1030 (s, br)	S=O stretch
Thiocyanates —S—C≡N	2175–2160 (m)	C≡N stretch
	650–600 (w)	S—CN stretch
	405–400 (s)	S—C≡N bend
Thiols —S—H	2590–2560 (w)	S—H stretch
	700–550 (w)	C—S stretch
Triazines $C_3N_2Y_3$	1600–1500 (vs)	Ring stretching
1,3,4,5-trisubst	1380–1350 (vs)	Ring stretching
	820–800 (s)	CH out-of-plane deformation
Vinyl —CH=CH_2	3095–3080 (m)	=CH_2 stretching
	3030–2980 (w–m)	=CH stretching
	1850–1800 (w–m)	Overtone of CH_2 out-of-plane wagging
	1645–1615 (m–s)	C=C stretch
	1000–950 (s)	CH out-of-plane deformation
	950–900 (vs)	CH_2 out-of-plane wagging
Vinylene —CH=CH—	3040–3010 (m)	=CH_2 stretching
	1665–1635 (w–m)	C=C stretch (cis isomer)
	1675–1665 (w–m)	C=C stretch (trans isomer)
	980–955 (s)	CH out-of-plane deformation (cis isomer)
	730–665 (s)	CH out-of-plane deformation (trans isomer)
Vinylidene >C=CH_2	3095–3075 (m)	=CH_2 stretching
	1665–1620 (w–m)	C=C stretch
	895–885 (s)	CH_2 out-of-plane wagging

[a]*Note*: s = strong; m = medium; w = weak; v = very; br = broad.

TABLE 12-2 Characteristic Raman Frequencies of Some Functional Groups

Group or Class	Frequency Ranges (cm^{-1}) and Intensities[a]	Assignment and Remarks
Acetylenes ≡CH	3340–3270 (s)	CH stretch
(alkynes) R—C≡C—R	2300–2190 (s)	C≡C stretch in disubstituted acetylenes, sometimes two bands (Fermi doublet)
R—C≡CH	2140–2100 (s)	C≡C stretch in monoalkyl acetylenes
	650–600 (m)	C—C≡CH deformation
Acid chlorides R—C(=O)Cl	1800–1790 (s)	C=O stretch
Alcohols R—OH	3400–3300 (vw)	OH stretch; broad band
	1450–1350 (m)	OH in-plane bend
	1150–1050 (m–s)	C—O antisym stretch
	970–800 (s)	C—C—O sym stretch

[a]*Note*: s = strong; m = medium; w = weak; v = very; br = broad.

(continued)

TABLE 12-2 Characteristic Raman Frequencies of Some Functional Groups (*continued*)

Group or Class	Frequency Ranges (cm^{-1}) and Intensities[a]	Assignment and Remarks
Aldehydes R—C(=O)—H	1730–1700 (m)	C=O stretch
***n*-Alkanes**	2980–2800 (vs)	CH stretch
	1475–1450 (s)	CH_3 antisym deformation
	1350–1300 (m–s)	CH_2 bend
	340–230 (s)	—C—C—C— bend
Alkenes	3090–3010 (m)	CH stretch
	1675–1600 (m–s)	C=C stretch, stronger than IR
	1450–1200 (vs)	CH in-plane deformation
cis alkenes R'CH=CHR	590–570 (m)	
	420–400 (m)	Skeletal deformations
	310–290 (m)	
trans alkenes R'CH=CHR	500–480 (m)	Skeletal deformations
	220–200 (m)	
Terminal alkenes RCH=CH_2	500–480 (m)	
RR'C=CH_2	440–390 (m)	Skeletal deformations
	270–250 (m)	
Allenes C=C=C	2000–1960 (s)	—C=C=C— antisym stretch
	1080–1060 (vs)	—C=C=C— sym stretch
Amides		
Primary —$CONH_2$	3540–3520 (w)	NH_2 antisym stretch (dil soln)
	3400–3380 (w)	NH_2 sym stretch (dil soln)
	1680–1660 (m)	C=O stretch (Amide I band)
	1420–1400 (s)	C—N stretch (Amide III band)
Secondary —CONHR	3440–3420 (s)	NH stretch (dil soln)
	1680–1640 (w)	Amide I band
	1310–1280 (s)	Amide III band
Tertiary —$CONR_2$	1670–1640 (m)	Amide I band
Amines, aliphatic		
Primary —RNH_2	3550–3330 (m)	NH_2 antisym stretch
	3450–3240 (m)	NH_2 sym stretch
	1090–1070 (m)	C—N stretch
Secondary R'NHR	3350–3300 (w)	NH stretch
	1190–1130 (m)	C—N stretch
Amines, aromatic	1380–1250 (s)	C—N stretch
Amino acids —CNH_2COOH	1600–1590 (w)	OCO antisym stretch
or (—$C\overset{+}{N}H_3COO^-$)	1400–1350 (vs)	OCO sym stretch
	900–850 (vs)	C—C—N sym stretch
Anhydrides —CO—O—CO—	1850–1780 (w–m)	C=O antisym stretch
	1770–1710 (m)	C=O sym stretch
Aromatic compounds	3070–3020 (s)	CH stretch
	1620–1580 (m–s)	C=C stretch; may be weak in IR
	1045–1015 (m)	CH in-plane bend
	1010–990 (vs)	Ring breathing (absent in 1,2- and 1,4-disubstituted compounds)
	900–650 (m)	CH out-of-plane deformation (one or two bands)

[a]*Note*: s = strong; m = medium; w = weak; v = very; br = broad.

Group or Class	Frequency Ranges (cm^{-1}) and Intensities[a]	Assignment and Remarks
Azides $-N=\overset{+}{N}=N^-$	2170–2080 (s)	NNN antisym stretch
	1345–1175 (s)	NNN sym stretch
Azo —N=N—	1580–1570 (vs)	Nonconjugated compounds
	1420–1410 (vs)	Conjugated to aromatic ring
	1060–1030 (vs)	—C—N— stretch in aromatic azo compounds
Benzenes		
Monosubst	630–610 (s)	CH out-of-plane deformation
1,2-Disubst and 1,2,4-trisubst	750–700 (s)	CH out-of-plane deformation
1,3-Disubst	750–700 (s)	CH out-of-plane deformation
	480–450 (m)	Out-of-plane ring deformation
1,2,3-Trisubst	655–645 (s)	CH out-of-plane deformation
1,3,5-Trisubst	570–550 (s)	CH out-of-plane deformation
Bromo C—Br	650–490 (vs)	C—Br stretch
	310–270 (s)	C—C—Br bend
***tert*-Butyl**	1250–1200 (m-s)	CH_3 deformation (two bands)
	940–920 (s)	CH_3 rocking
Carbonyl C=O	1870–1650 (w–s)	C=O stretch; weaker than in IR
Carboxylic acids	1680–1640 (s)	C=O sym stretch of dimer
Chloro C—Cl	850–650 (s)	C—Cl stretch
	340–290 (s)	C—C—Cl bend
Cyanamides	1150–1140 (vs)	—N=C=N— sym stretch
Cyclobutanes	1000–960 (vs)	Ring breathing
	700–680 (s)	Ring deformation
	180–150 (s)	Ring puckering
Cyclohexanes	1460–1440 (s)	CH_2 scissoring
	825–815 (s)	Ring vibration (boat)
	810–795 (s)	Ring vibration (chair)
Cyclopentanes	1450–1430 (s)	CH_2 scissoring
	900–880 (s)	Ring breathing
Cyclopropanes	1210–1180 (s)	Ring breathing
	830–810 (s)	Ring deformation
Disulfides C—S—S—C	550–430 (vs)	S—S stretch
Epoxides	1280–1260 (s)	Sym ring stretch
Esters R'COOR	1100–1025 (s)	C—O—C sym stretch
Esters		
Aliphatic saturated	1140–1110 (m)	C—O—C stretch
Aliphatic unsaturated	1275–1200 (m)	C—O—C antisym stretch
—C=C—O—C	1075–1020 (s)	C—O—C sym stretch
Aromatic	1310–1210 (m) }	C—O—C stretches
	1050–1010 (m) }	
Isocyanates —N=C=O	1440–1400	N=C=O sym stretch
Isopropyl $(CH_3)_2CH$—	1180–1160 (m)	CH_3 rocking
	835–795 (ms)	C—C stretching
Ketenes C=C=O	2060–2040 (vs)	C=C=O stretch
Ketones R(R')C=O	1725–1705 (m)	C=O stretch in saturated compounds
	1700–1650 (m)	C=O stretch in aromatic compounds
	1750–1705 (m)	C=O stretch in alicyclic ketones

[a]*Note*: s = strong; m = medium; w = weak; v = very; br = broad.

(continued)

TABLE 12-2 Characteristic Raman Frequencies of Some Functional Groups (*continued*)

Group or Class	Frequency Ranges (cm^{-1}) and Intensities[a]	Assignment and Remarks
Lactones (R ring: CH_2–C(=O)–O–CH_2)	1850–1730 (s)	C=O stretch
Mercaptans C—SH	850–820 (vs)	S—H in-plane deformation
	700–600 (vs)	C—S stretch; weak in IR
Methyl —CH_3	2980–2800 (vs)	CH stretch
	1470–1460 (s)	CH_3 deformation
Methylene =CH_2 or —CH_2—	3090–3070 (s)	=CH_2 antisym stretch
	3020–2980 (s)	=CH_2 sym stretch
	2940–2920 (s)	—CH_2— antisym stretch
	2860–2850 (s)	—CH_2— sym stretch
	1350–1150 (m–s)	CH_2 wag and twist; weak or absent in IR
Nitriles —C—C≡N	2260–2240 (s)	C≡N stretch, nonconjugated
	2230–2220 (s)	C≡N stretch, conjugated
	1080–1025 (s–vs)	C—C—C stretch
	840–800 (s–vs)	C—C—CN sym stretch
	380–280 (s–vs)	C—C≡N bend
Nitrites —ONO	1660–1620 (s)	N=O stretch in alkyl nitrites
Nitro —NO_2	1570–1550 (w)	NO_2 antisym stretch
	1380–1360 (s)	NO_2 sym stretch
	920–830 (s)	C—N stretch
	650–520 (m)	NO_2 bend
Organosilicon compounds	1300–1200 (s)	Si—C stretch
Oximes	1680–1620 (vs)	C=N stretch; may be absent in IR
Peroxides —C—O—O—C—	900–850 (variable)	O—O stretch; weak in IR
Phosphines	2350–2240 (m)	P—H stretch
Pyridines	1620–1560 (m)	Ring stretching
	1020–980 (vs)	Ring breathing
Pyrroles	3450–3350 (s)	NH stretch
	1420–1360 (vs)	Ring stretching
Sulfides C—S	705–570 (s)	C—S stretch
Sulfonamides —SO_2NH_2	1155–1135 (vs)	SO_2 stretch
Sulfones —SO_2—	1360–1290 (m)	SO_2 antisym stretch
	1170–1120 (s)	SO_2 sym stretch
	610–545 (s)	SO_2 scissoring
Sulfoxides >SO	1050–1010 (s)	S=O stretch
Sulfonyl chlorides R—OSOCl	1230–1200 (m)	S=O stretch
Thiocyanates —S—C≡N	650–600 (s)	S—CN stretch
Thiols RSH	2590–2560 (vs)	S—H stretch
	700–550 (vs)	C—S stretch
	340–320 (vs)	C—S—H out-of-plane bend
Thiophenes	740–680 (vs)	C—S—C stretch
	570–430 (s)	Ring deformation
Xanthates —O—C(=S)—S—	670–620 (vs)	C=S stretch; not seen in IR
	480–450 (vs)	C—S stretch

[a]*Note*: s = strong; m = medium; w = weak; v = very; br = broad.

12-5 PRELIMINARY ANALYSIS

12-5a Introduction

Table 12-3 gives a list of common absorption regions for important functional groups. The preliminary analysis of a spectrum should determine the presence or absence of bands in these 13 regions.

Certain types of compounds give strong, broad absorptions, which are very prominent in the IR spectrum. The hydrogen-bonded OH stretching bands of alcohols, phenols, and carboxylic acids are easily recognized at the high-frequency end of the spectrum. The stretching of the NH_3^+ group in amino acids gives a very broad, unsymmetrical band, which extends over several hundred wavenumbers. Broad bands associated with bending of NH_2 or NH groups of primary or secondary amines are found at the low-frequency end of the spectrum. Amides also give a broad band in this region. Examples of these characteristic absorptions are illustrated in Figure 12-7.

In the next three subsections, suggestions are given for the preliminary analysis of the IR spectra of hydrocarbons or the hydrocarbon parts of molecules; of compounds containing carbon, hydrogen, and oxygen; and finally of compounds also containing nitrogen. After this preliminary study of the spectrum, the analyst should have some idea of the functionalities under investigation. The spectrum is then examined in more detail. In later sections, some of the detailed structural information that can be obtained from such an examination is discussed.

TABLE 12-3 Regions of the Infrared Spectrum for Preliminary Analysis

Region	Group	Possible Compounds Signified
3700–3100	—OH	Alcohols, aldehydes, carboxylic acids
	—NH	Amides, amines
	≡C—H	Alkynes
3100–3000	=CH or —CH_2—	Aromatic compounds, alkenes, small rings
3000–2800	—CH, —CH_2—, —CH_3	Aliphatic groups
2800–2600	—CHO	Aldehydes (Fermi doublet)
2700–2400	—SH	Mercaptans and thiols
	—PH	Phosphines
2400–2000	C≡N	Nitriles
	$—N=\overset{+}{N}=\overset{-}{N}$	Azides
	—C≡C—	Alkynes[a]
1870–1650	C=O	Acid halides, aldehydes, amides, amino acids, anhydrides, carboxylic acids, esters, ketones, lactams, lactones, quinones
1650–1550	C=C, C=N, NH	Unsaturated aliphatics,[a] aromatics, unsaturated heterocycles, amides, amines, amino acids
1550–1300	NO_2	Nitro compounds
	CH_3 and CH_2	Alkanes, alkenes, arenes, etc.
1300–1000	C—O—C and C—OH	Ethers, alcohols, sugars
	S=O, P=O, C—F	Sulfur, phosphorus, and fluorine compounds
1100–800	Si—O and P—O	Organosilicon and -phosphorus compounds
1000–650	=C—H	Alkenes and aromatic compounds
	—NH	Aliphatic amines
800–400	C—halogen	Halogen compounds
	Aromatic rings	Aromatic compounds

[a]Band may be absent owing to symmetry (see Section 12-2a).

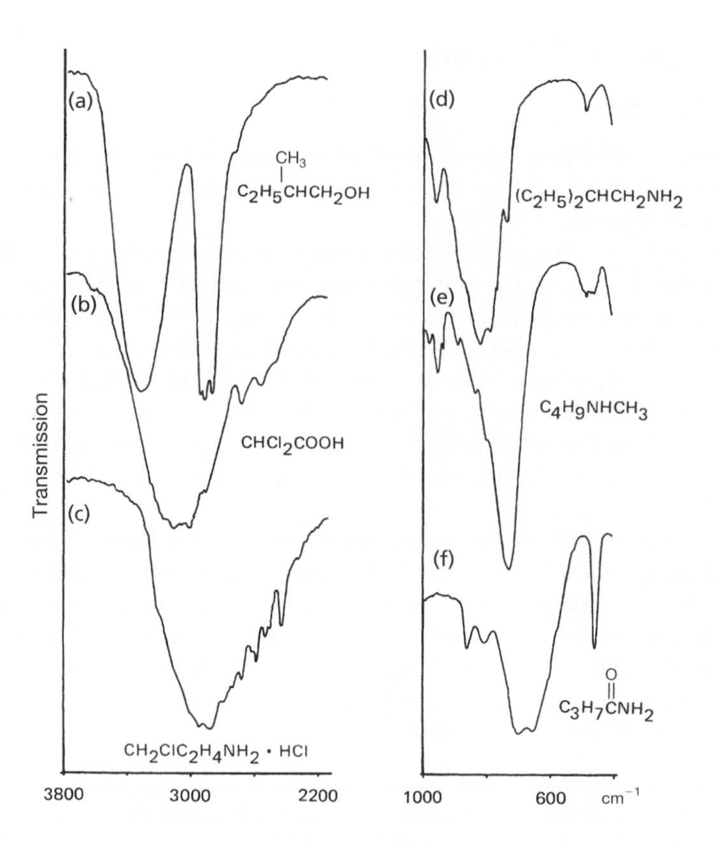

FIGURE 12-7 Some characteristic broad IR bands.

12-5b Hydrocarbons or Hydrocarbon Groups

The nature of a hydrocarbon or the hydrocarbon part of a molecule can be identified by first looking in the region between 3100 and 2800 cm^{-1}. If there is no absorption between 3000 and 3100 cm^{-1}, the compound contains no aromatic or unsaturated aliphatic $=CH$ groups. Cyclopropanes, which absorb above 3000 cm^{-1}, are an exception. If the absorption is entirely above 3000 cm^{-1}, the compound is probably aromatic or contains only $=CH$ or $=CH_2$ groups. Absorptions both above and below 3000 cm^{-1} indicate the presence of both saturated and unsaturated or cyclic hydrocarbon moieties (see Figure 12-8 for an example).

The region between 1000 and 650 cm^{-1} should be examined to confirm these conclusions. Strong bands in this region suggest alkenes or aromatic structures. A small, sharp band near 725 cm^{-1} is indicative of a linear chain containing four or more CH_2 groups. The presence of CH_2 groups also is indicated by a strong band near 1440 cm^{-1} in the IR and Raman spectra.

12-5c Compounds Containing Oxygen

If it is known from elemental analysis or mass spectrometry that there is an oxygen atom present in the molecule, one should look in three regions for bands from the oxygen-containing functional group. A strong, broad band between 3500 and 3200 cm^{-1} is likely due to the hydrogen-bonded OH stretching mode of an alcohol or a phenol (see Section 12-12d). Be aware that water also absorbs in this region. When hydrogen bonding is absent, the OH stretching band is sharp and at higher frequencies (3650–3600 cm^{-1}). Carboxylic acids give very broad OH stretching bands between 3200 and 2700 cm^{-1} (Section 12-12d). Several sharp peaks may be seen on this broad band near 3000 cm^{-1} from CH stretching vibrations.

One should look next for a very strong band between 1850 and 1650 cm^{-1} from the $C{=}O$ stretching of a carbonyl group (Section 12-7). The third region is between 1300 and 1000 cm^{-1}, where bands from $C{-}OH$ stretching of alcohols and carboxylic acids and from $C{-}O{-}C$ stretching of ethers and esters are observed (Sections 12-12b and 12-12e). Some examples of $C{-}O$ stretching bands are shown in Figure 12-9. The presence or

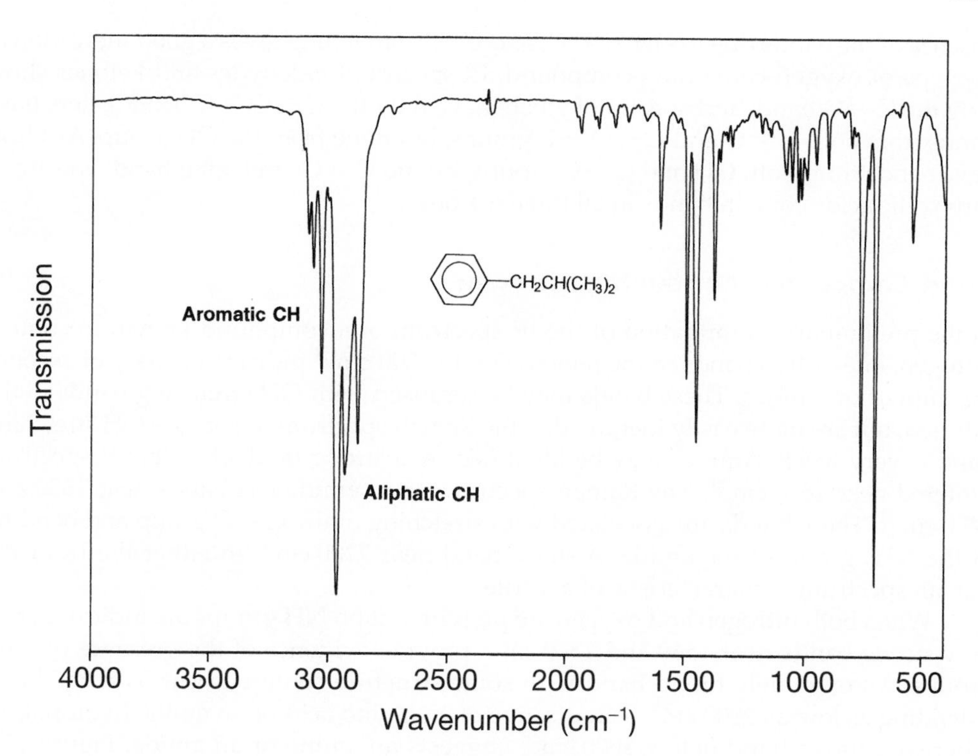

FIGURE 12-8 The IR spectrum of *sec*-butylbenzene.

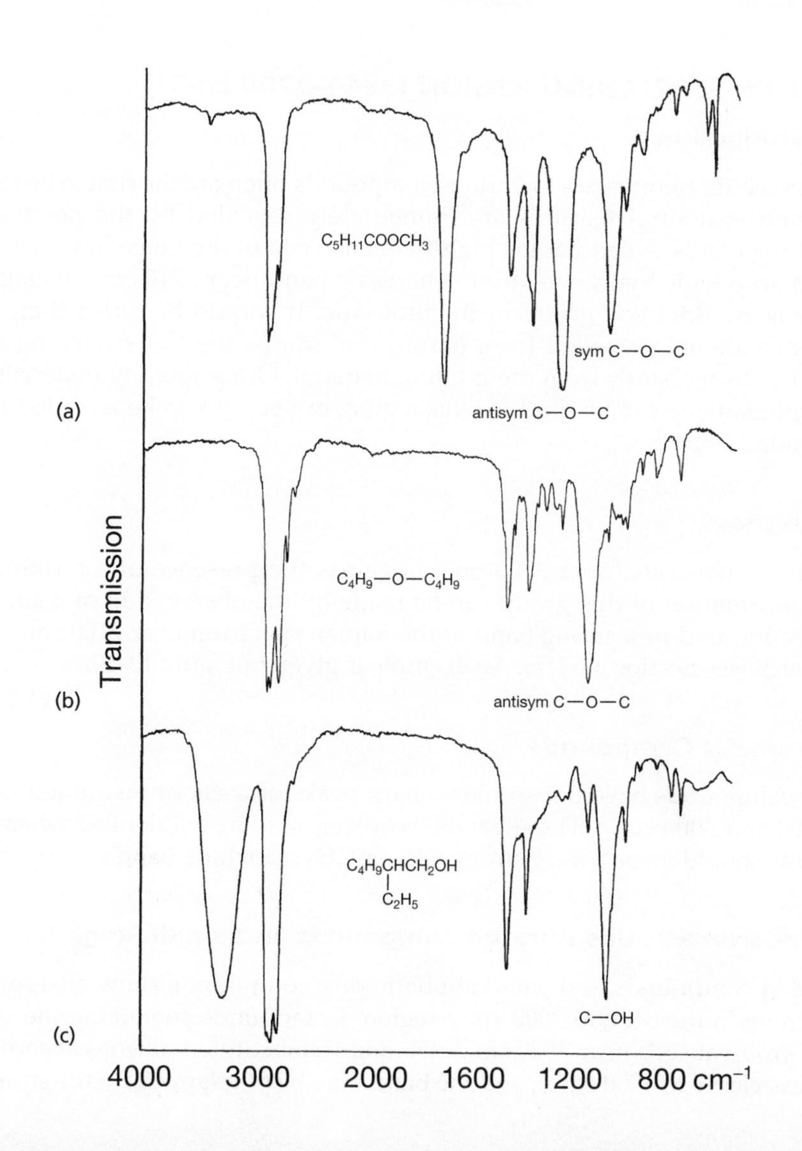

FIGURE 12-9 The IR spectra of (a) an ester, (b) an ether, and (c) an alcohol, showing the C—O stretching bands between 1300 and 1000 cm^{-1}.

absence of the bands due to OH, C=O, and C—O stretching gives a good indication of the type of oxygen-containing compound. IR spectra of aldehydes and ketones show only the C=O band, and spectra of ethers have only the C—O—C band. Esters have bands from both C=O and C—O—C groups, but none from the OH group. Alcohols have bands from both OH and C—O groups, but no C=O stretching band. Spectra of carboxylic acids contain bands in all three regions.

12-5d Compounds Containing Nitrogen

In the preliminary examination of the IR spectrum of a compound known to contain nitrogen, one or two bands in the region 3500 to 3300 cm^{-1} indicate primary or secondary amines or amides. These bands may be confused with OH stretching bands in the infrared, but are more easily identified in the Raman spectrum, where the OH stretching band is very weak. Amides may be identified by a strong doublet in the IR spectrum centered near 1640 cm^{-1}. The Raman spectrum of an amide has bands near 1650 and 1400 cm^{-1}. These bands are associated with stretching of the C=O group and bending of the NH_2 group of the amide. A sharp band near 2200 cm^{-1} in either the IR or the Raman spectrum is characteristic of a nitrile.

When both nitrogen and oxygen are present, but no NH groups are indicated, two very strong bands near 1560 and 1370 cm^{-1} provide evidence of the presence of nitro groups. An extremely broad band with some structure centered near 3000 cm^{-1} and extending as low as 2200 cm^{-1} is indicative of an amino acid or an amine hydrohalide, whereas a broad band below 1000 cm^{-1} suggests an amine or an amide. Figure 12-7 provides examples of these broad bands.

12-6 THE CH STRETCHING REGION (3340–2700 cm^{-1})

12-6a Introduction

The CH stretching frequencies in various compounds often are the first to be examined, since certain structural features are immediately revealed by the position of the CH stretching bands. A band at the high-frequency end of the range indicates the presence of an acetylenic hydrogen atom, whereas a band near 2710 cm^{-1} usually means that there is an aldehyde group in the molecule. It should be noted that, when IR spectra of solids are recorded from paraffin oil mulls, the CH stretching region is obscured by strong bands from the mulling material. Other mulling materials such as hexachlorobutadiene can be used for this region, or spectra can be recorded from KBr discs (Section 11-6b).

12-6b Alkynes

A sharp band observed near 3300 cm^{-1} suggests the presence of a terminal ≡CH group. Confirmation of this group can be made by the observation of a small, sharp peak in the infrared or a strong band in the Raman spectrum near 2100 cm^{-1} from the C≡C stretch (see Section 12-11c). An example is given in Figure 12-10.

12-6c Aromatic Compounds

Aromatic compounds have one or more sharp peaks of weak or medium intensity between 3100 and 3000 cm^{-1}. These bands, however, may be overlooked when they appear only as shoulders on a very strong CH_3 or CH_2 stretching band.

12-6d Nonaromatic Unsaturated Compounds and Small Rings

Unsaturated compounds and small aliphatic ring compounds show absorption from CH stretching in the 3100 to 3000 cm^{-1} region. Compounds containing the vinylidine (=CH_2) group absorb near 3080 cm^{-1}. Di- and trisubstituted alkenes absorb at lower frequencies, closer to 3000 cm^{-1}, and the band may be overlapped by the stronger CH_3

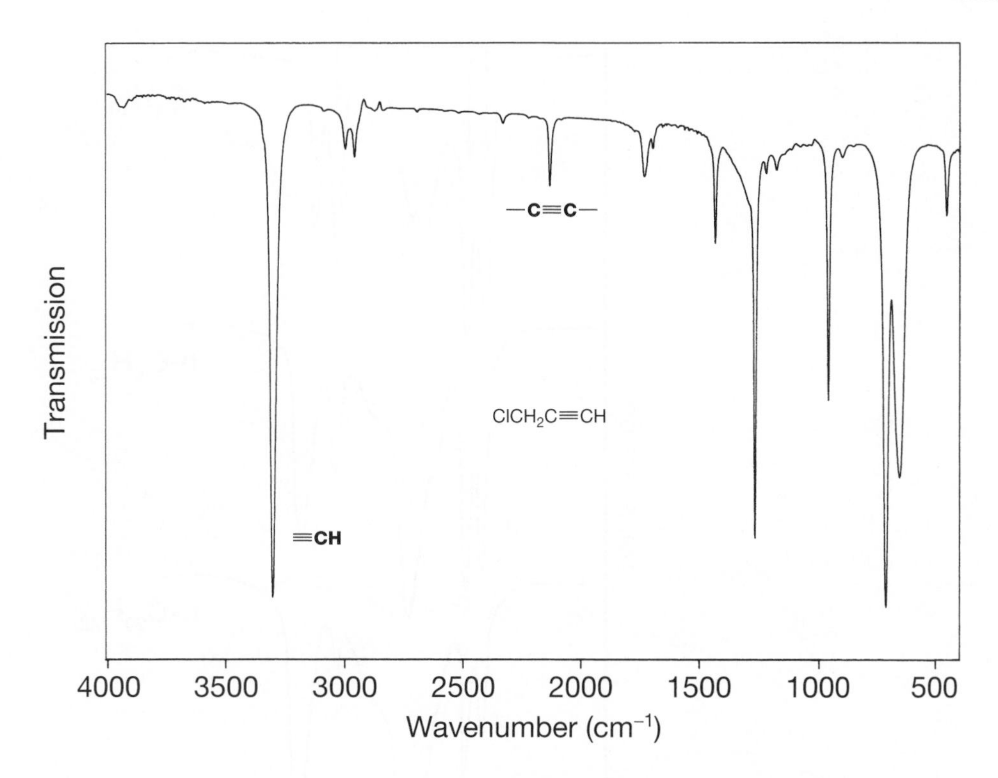

FIGURE 12-10 The IR spectrum of 3-chloropropyne.

or CH_2 absorption (Section 12-6e). Cyclopropane derivatives have a band between 3100 and 3070 cm^{-1}, and epoxides absorb between 3060 and 3040 cm^{-1}.

12-6e Saturated Hydrocarbon Groups

Saturated compounds can have methyl, methylene, or methine groups, each of which has characteristic CH stretching frequencies. Methyl (CH_3) groups absorb near 2960 and 2870 cm^{-1}, and the methylene (CH_2) bands are at 2930 and 2850 cm^{-1}. In many cases, only one band in the 2870 to 2850 cm^{-1} region can be resolved when both CH_3 and CH_2 groups are present in the molecule. Figure 12-11 shows the CH stretching region of three unbranched, saturated molecules. As the carbon chain becomes longer, the CH_2 bands increase in intensity relative to the CH_3 absorptions. The doublet at 2960 cm^{-1} is due to the antisymmetric CH_3 stretching mode, which would be degenerate in a free CH_3 group or in a molecule in which the threefold symmetry of the group was maintained, for example, in CH_3Cl. The degeneracy is removed in the saturated molecules, and consequently two individual antisymmetric CH_3 stretching bands are observed. The methine (CH) stretch can be observed (as a weak peak near 2885 cm^{-1}) only when CH_3 and CH_2 groups are absent. The methoxy group (CH_3O—) has a characteristic sharp band of medium intensity near 2830 cm^{-1}, separate from other CH stretching bands. Saturated cyclic hydrocarbon groups also have characteristic CH stretching bands in both IR and Raman spectra. The frequencies range from 3100 cm^{-1} in cyclopropanes down to 2900 cm^{-1} in cyclohexanes and larger rings.

12-6f Aldehydes

The CH stretching mode of the aldehyde group appears as a Fermi doublet (Section 12-2c) near 2820 and 2710 cm^{-1}. The 2710 cm^{-1} absorption is very useful and characteristic of aldehydes, but the higher frequency component often appears as a shoulder on the 2850 cm^{-1} CH_2 stretching band. In the spectrum of an aromatic aldehyde, there is no overlap with other CH stretching bands and the doublet is clearly resolved. In ortho-substituted aromatic aldehydes, the frequencies of the doublet are about 40 cm^{-1} higher than usual with the high-frequency component stronger and broader (Figure 12-12).

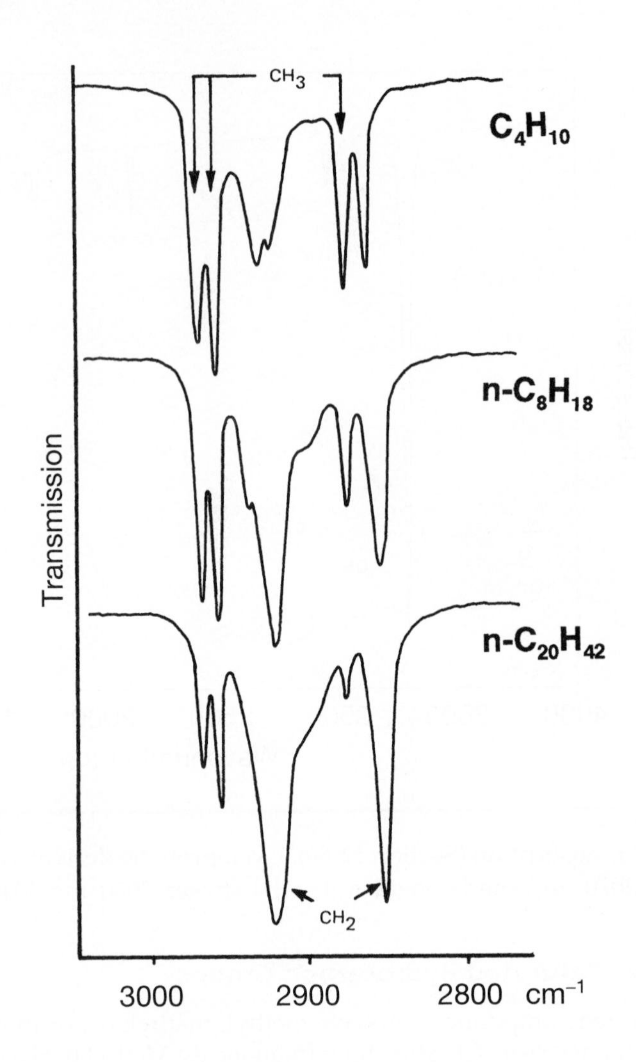

FIGURE 12-11 The C—H stretching region of n-C_4H_{10}, n-C_8H_{18}, and n-$C_{20}H_{42}$.

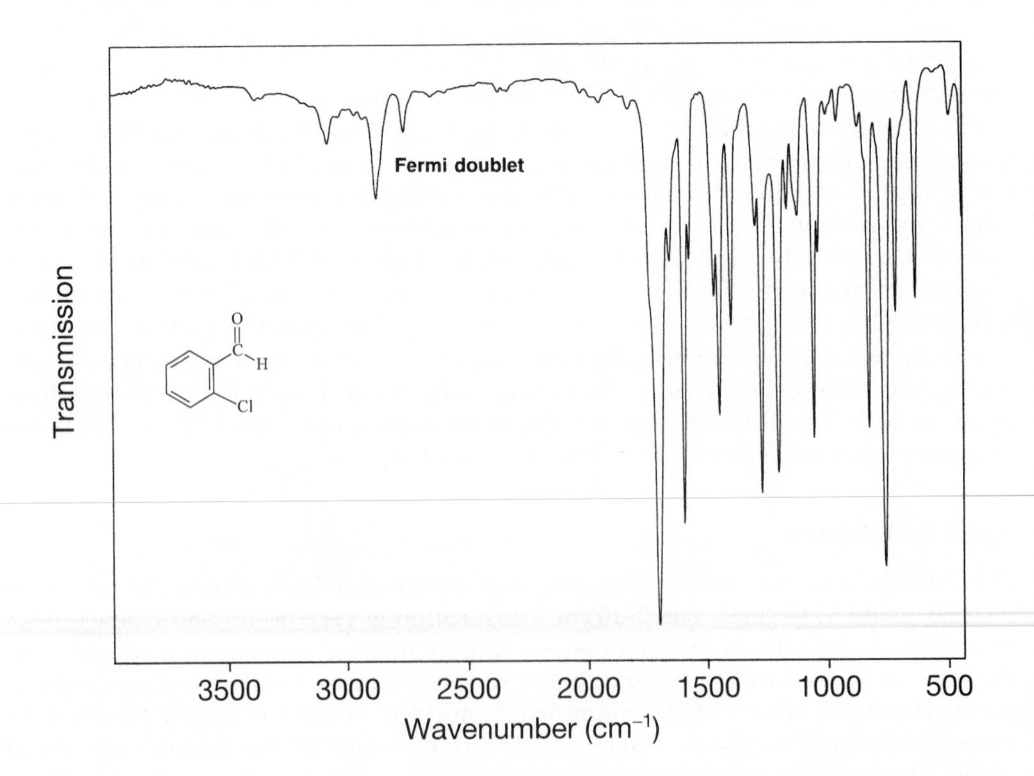

FIGURE 12-12 The IR spectrum of an aromatic aldehyde, 2-chlorobenzaldehyde, showing the Fermi doublet at 2750 and 2870 cm^{-1}.

12-7 THE CARBONYL STRETCHING REGION (1850–1650 cm^{-1})

12-7a Introduction

The carbonyl stretching region is one of the most important regions of the spectrum for structural analysis. If we were to include metal carbonyls and salts of carboxylic acids, the region of the C=O stretching mode would extend from 2200 down to 1350 cm^{-1}. Most organic compounds containing the C=O group, however, show very strong IR absorption in the range of 1850 to 1650 cm^{-1}. The actual position of the peak (or peaks) within this range is characteristic of the type of compound. At the upper end of the range are found bands from anhydrides (two bands) and four-membered cyclic esters (β lactones), while acyclic amides and substituted ureas absorb at the lower end of the range. Table 12-4

TABLE 12-4 Carbonyl Stretching Frequencies for Compounds Having One Carbonyl Group

Range (cm^{-1}) and Intensity[a]	Classes of Compounds	Remarks
1840–1820 (vs)	β lactones	4-Membered ring
1810–1790 (vs)	Acid chlorides	Saturated aliphatic compounds
1800–1750 (vs)	Aromatic and unsaturated esters	C=C stretch higher than normal (1700–1650 cm^{-1})
1800–1740 (s)	Carboxylic acid monomer	Only observed in dil soln
1790–1740 (vs)	γ lactones	5-Membered ring
1790–1760 (vs)	Aromatic or unsaturated acid chlorides	Second weaker combination band near 1740 cm^{-1} (Fermi resonance)
1780–1700 (s)	Lactams	Position depends on ring size
1770–1745 (vs)	α-Halo esters	Higher frequency due to electronegative halogen
1750–1740 (vs)	Cyclopentanones	Unconjugated structure
1750–1730 (vs)	Esters and δ lactones	Aliphatic compounds
1750–1700 (s)	Urethanes	R—O—(C=O)—NHR compounds
1745–1730 (vs)	α-Halo ketones	Noncyclic compounds
1740–1720 (vs)	Aldehydes	Aliphatic compounds
1740–1720 (vs)	α-Halo carboxylic acids	20 cm^{-1} higher frequency if halogen is fluorine
1730–1705 (vs)	Aryl and α,β-unsaturated aliphatic esters	Conjugated carbonyl group
1730–1700 (vs)	Ketones	Aliphatic and large ring alicyclic
1720–1680 (vs)	Aromatic aldehydes	Also α,β-unsaturated aliphatic aldehydes
1720–1680 (vs)	Carboxylic acid dimer	Broader band
1710–1640 (vs)	Thiol esters	Lower than normal esters
1700–1680 (vs)	Aromatic ketones	Position affected by substituents on ring
1700–1680 (vs)	Aromatic carboxylic acids	Dimer band
1700–1670 (s)	Primary and secondary amides	In dil soln
1700–1650 (vs)	Conjugated ketones	Check C=C stretch region
1690–1660 (vs)	Quinones	Position affected by substituents on ring
1680–1630 (vs)	Amides (solid state)	Note second peak due to NH def near 1625 cm^{-1}
1670–1660 (s)	Diaryl ketones	Position affected by substituents on ring
1670–1640 (s)	Ureas	Second peak due to NH def near 1590 cm^{-1}
1670–1630 (vs)	Ortho-OH or -NH_2 aromatic ketones	Frequency lowered by chelation with ortho group

[a] *Note*: s = strong; m = medium; w = weak; v = very.

summarizes the ranges in which compounds with a single carbonyl group absorb. In compounds with two interacting carbonyl groups, out-of-phase and in-phase (antisymmetric and symmetric) C=O stretching vibrations can occur (Section 12-2b). Usually, the out-of-phase mode is found at higher frequencies than the in-phase mode.

12-7b Compounds Containing a Single C=O Group

The type of carbonyl functional group usually cannot be identified from the C=O stretching band alone, because several such groups may absorb within a given frequency range. An initial separation into possible compounds, however, can be achieved with the information in Table 12-4. For example, a single carbonyl peak in the region 1750 to 1700 cm^{-1} could indicate an ester, an aldehyde, a ketone (including cyclic ketones), a large ring lactone, a urethane derivative, an α-halo ketone, or an α-halo carboxylic acid. Other bands then must be examined to specify the functionality more closely. The presence of the halogen could be checked from the elemental analysis, an aldehyde also would have a peak near 2700 cm^{-1} (Section 12-6f), and esters and lactones give a strong band near 1200 cm^{-1}, which is often quite broad in esters (Section 12-12e). Urethanes would have an NH stretching band near 3300 cm^{-1} if hydrogen bonding is present, or near 3400 cm^{-1} if it is absent.

12-8 AROMATIC COMPOUNDS

12-8a General

Bands characteristic of aromatic compounds may be found in five regions of the IR spectrum: 3100 to 3000 cm^{-1} (CH stretching), 2000 to 1700 cm^{-1} (overtones and combinations), 1650 to 1430 cm^{-1} (C=C stretching), 1275 to 1000 cm^{-1} (in-plane CH deformation), and 900 to 690 cm^{-1} (out-of-plane CH deformation). Examples of IR spectra of two aromatic compounds may be seen in Figures 12-8 and 12-12. In these spectra, there are bands in all five regions. The intensities of the CH stretching bands range from medium, as in Figure 12-8, to weak. Occasionally, these bands are seen only as shoulders on a strong aliphatic CH stretching band. Usually, there are sharp bands near 1600, 1500, and 1430 cm^{-1} in benzene derivatives. The 1600 cm^{-1} absorption can be hidden by a strong C=O stretching band in some carbonyl compounds, or an NH deformation band in amines. There also are several sharp bands between 1275 and 1000 cm^{-1} in aromatic compounds. In the 900 to 690 cm^{-1} region, one or two strong bands are observed. These bands, together with the pattern of weak bands between 2000 and 1700 cm^{-1}, give an indication of the type of substitution on the benzene ring.

12-8b Substituted Aromatic Compounds

The monosubstitution pattern is easily recognized by two very strong bands near 750 and 700 cm^{-1} in the IR spectrum from out-of-plane CH bending. These are very prominent in the spectrum of Figure 12-8, which now is recognized as a monosubstituted benzene derivative. Another example may be seen in Figure 11-6 for polystyrene. In addition, monosubstitution is indicated by four weak but clear absorptions between 2000 and 1700 cm^{-1}. Again, Figure 12-8 illustrates this pattern. The patterns of bands in these two regions also can give an indication of the positions of substitution in di- and trisubstituted benzene rings. These regions are less useful for highly substituted derivatives. The characteristic bands of mono-, di-, and trisubstituted aromatic compounds are summarized in Table 12-5. Examples of the three disubstitution patterns are shown in Figure 12-13.

12-9 COMPOUNDS CONTAINING METHYL GROUPS

12-9a General

In Section 12-2a, the vibrations of a methyl group were discussed in terms of the symmetry of the group. These vibrations (stretching, bending [deformation], rocking, and torsion) give rise to IR absorption and Raman scattering in four different regions of the

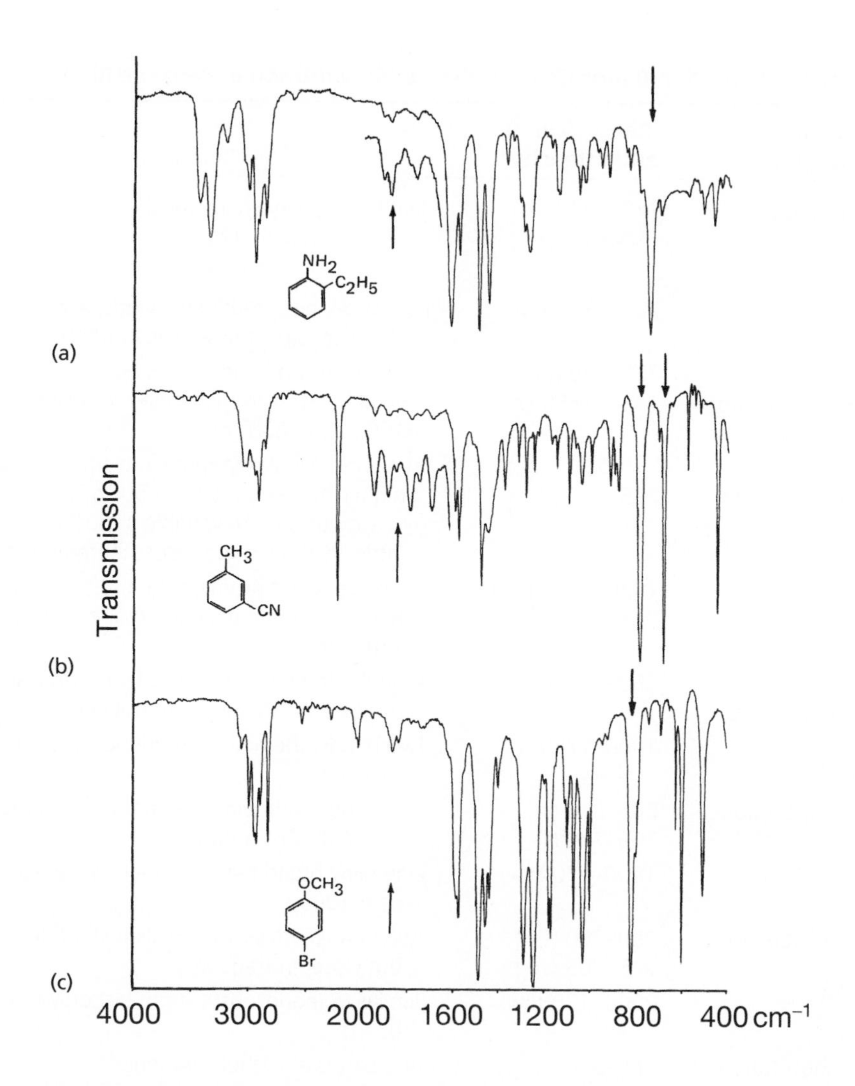

FIGURE 12-13 IR spectra showing characteristic substitution patterns in (a) ortho-, (b) meta-, and (c) para-disubstituted benzenes.

spectrum. The frequencies of CH stretching vibrations of methyl groups have been discussed in Section 12-6e. Antisymmetric deformation of the HCH angles of a CH_3 group gives rise to very strong IR absorption and Raman scattering in the 1470 to 1440 cm^{-1} region. Bending of methylene (CH_2) groups also gives rise to a band in the same region. The symmetric CH_3 deformation gives a strong, sharp IR band between 1380 and 1360 cm^{-1}. This band appears as a doublet when more than one CH_3 group is attached to the same carbon atom and gives a good indication of the presence of isopropyl or *tert*-butyl groups (Section 12-9b). When the methyl group is attached to an atom other than carbon, there is a significant shift in the symmetric CH_3 deformation frequency: OCH_3 (1460–1430 cm^{-1}), NCH_3 (1440–1410 cm^{-1}), SCH_3 (1330–1290 cm^{-1}), PCH_3 (1310–1280 cm^{-1}), and $SiCH_3$ (1280–1250 cm^{-1}).

The CH_3 rocking vibrations are usually coupled with skeletal modes and may be found anywhere between 1240 and 800 cm^{-1}. Medium to strong bands in both IR and Raman spectra may be observed, but these are of little use for structure determination. The methyl torsion vibration has a frequency between 250 and 100 cm^{-1}, but often is not observed in either IR or Raman spectra. In cases for which torsional frequencies can be observed or estimated, information on rotational isomerism, conformation, and barriers to internal rotation can be obtained.

12-9b Isopropyl and *tert*-Butyl Groups

Isopropyl and *tert*-butyl groups give characteristic doublets in the symmetric CH_3 deformation region of the IR spectrum. The isopropyl group gives a strong doublet at 1385/1370 cm^{-1}, whereas the *tert*-butyl group gives a strong band at 1370 cm^{-1} with a weaker peak at 1395 cm^{-1}. Examples of these doublets are seen in Figure 12-14.

TABLE 12-5 Absorption Bands Characteristic of Substitution in Benzene Rings

Type of Substitution	Range (cm^{-1}) and Intensity[a]	Remarks
Monosubstitution	770–730 (vs) 710–690 (s)	Two bands; very characteristic of monosubstitution
	2000–1700 (w)	Four weak but prominent bands; also very characteristic of monosubstitution
1,2-disubstitution (ortho)	770–730 (vs)	A single strong band
	1950–1650 (vw)	Several bands, the two most prominent near 1900 and 1800 cm^{-1}
1,3-disubstitution (meta)	810–750 (vs) 725–680 (s)	Two bands similar to monosubstitution, but the higher-frequency band is 50 cm^{-1} higher, and the lower is more variable in position and intensity than in monosubstituted benzenes
	1930–1740 (vw)	Three weak but prominent bands; the lowest-frequency band may be broader with some structure
1,4-disubstitution (para)	860–800 (vs)	A single strong band, similar to 1,2-disubstitution, but 50 cm^{-1} higher in frequency
	1900–1750 (w)	Two bands; the higher-frequency one is usually stronger
1,3,5-trisubstitution	865–810 (s) 765–730 (s)	Two bands with wider separation than mono- or 1,4-disubstitution
	1800–1700 (w)	One fairly broad band with a much weaker one near 1900 cm^{-1}
1,2,3-trisubstitution	780–760 (s) 745–705 (s)	Two bands, similar to 1,3-disubstitution, but closer in frequency
	2000–1700 (w)	Similar to monosubstitution, but only three bands
1,2,4-trisubstitution	885–870 (s) 825–805 (s)	Two bands at higher frequencies than mono-, 1,3-di-, and the other trisubstituted compounds
	1900–1700 (w)	Two prominent bands near 1880 and 1740 cm^{-1} with a much weaker one between

[a]*Note*: s = strong; m = medium; w = weak; v = very.

12-10 COMPOUNDS CONTAINING METHYLENE GROUPS

12-10a Introduction

There are two kinds of methylene groups: the $—CH_2—$ group in a saturated chain and the terminal $=CH_2$ group in vinyl, allyl, or vinylidene compounds. Diagrams of stretching, bending, wagging, twisting, and rocking motions of a CH_2 group are given in Figure 12-2. The CH stretching vibrations are discussed in Section 12-6e. Bending, wagging, and rocking modes of CH_2 groups also give rise to important group frequencies.

12-10b CH_2 Bending (Scissoring)

The bending (or scissoring) motion of saturated $—CH_2—$ groups gives a band of medium to strong intensity between 1480 and 1440 cm^{-1}. When the CH_2 group is adjacent to a carbonyl or nitro group, the frequency is lowered to 1430 to 1420 cm^{-1}. A vinyl $=CH_2$ group gives a band of medium intensity between 1420 and 1410 cm^{-1}. This band sometimes is assigned as an in-plane deformation, since the two hydrogen atoms are in the same plane as the C=C group.

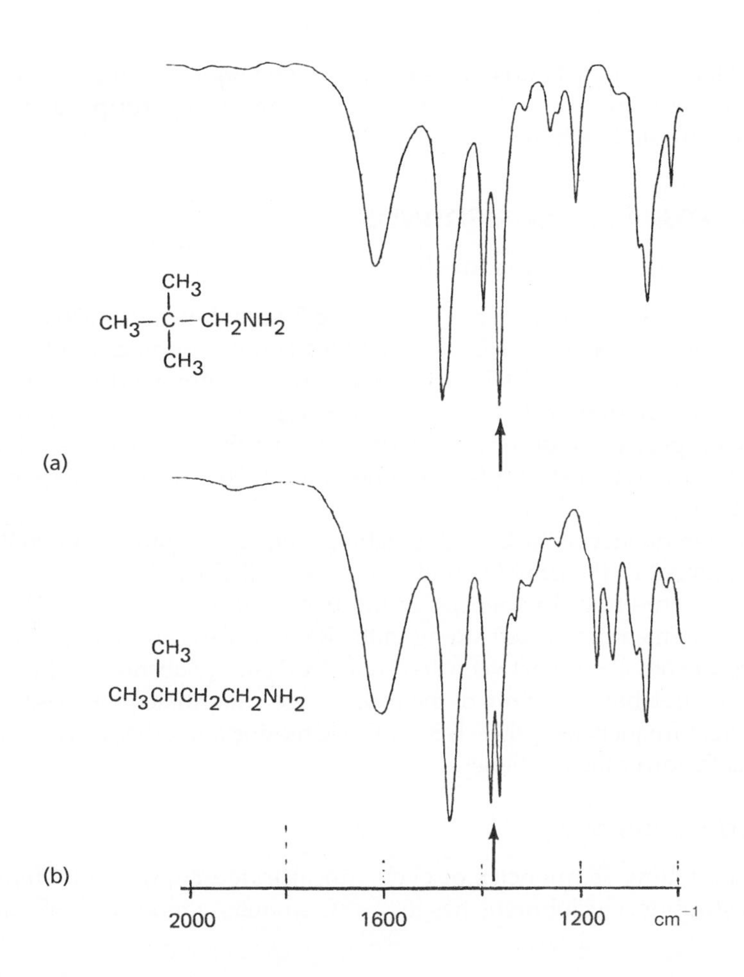

FIGURE 12-14 Examples of the doublets observed in the symmetric CH_3 deformation region for (a) a *tert*-butyl group and (b) an isopropyl group. (Reproduced with permission of Aldrich Chemical Co., Inc., from C. J. Pouchert [ed.], *The Aldrich Library of FT–IR Spectra.*)

12-10c CH_2 Wagging and Twisting

The $—CH_2—$ wagging and twisting frequencies in saturated groups are observed between 1350 and 1150 cm^{-1}. The IR bands are weak unless an electronegative atom such as a halogen or sulfur is attached to the same carbon atom. The CH_2 twisting modes occur at the lower end of the frequency range and give very weak IR absorption.

12-10d CH_2 Rocking

A small band is observed near 725 cm^{-1} in the IR spectrum when there are four or more methylene groups in a chain. The intensity increases with the increasing chain length. This band is assigned to the rocking of the CH_2 groups in the chain. Many compounds, however, have bands in this region, so the CH_2 rocking band is useful only for aliphatic molecules. The band is always present in spectra recorded from Nujol mulls.

12-10e CH_2 Wagging in Vinyl and Vinylidene Compounds

The $=CH_2$ wagging modes in vinyl and vinylidene compounds are found at much lower frequencies than in saturated groups. In vinyl compounds, a strong band is observed in the infrared between 910 and 900 cm^{-1}. For vinylidene compounds the frequency range is 10 cm^{-1} lower. The overtone of the CH_2 wag often can be seen clearly as a band of medium intensity near 1820 cm^{-1} for vinyl and 1780 cm^{-1} for vinylidene compounds. These frequencies are raised above the normal range by halogens or other functional groups on the carbon atom. More is said about vinyl and vinylidene groups in Section 12-11d.

12-10f Relative Numbers of CH_2 and CH_3 Groups

One further useful observation can be made concerning the H—C—H deformation bands in saturated parts of a molecule. When there are more $—CH_2—$ groups than CH_3 groups present, the 1480–1440 cm^{-1} band is stronger than the 1380–1360 cm^{-1} band (symmetric CH_3 deformation). The relative intensities of these two bands, coupled with the 725 cm^{-1} band and the bands in the CH stretching region (Figure 12-11), can give information on the relative numbers of $—CH_2—$ and CH_3 groups, as well as on the length of the saturated carbon chain.

12-11 UNSATURATED COMPOUNDS

12-11a The C=C Stretching Mode

Stretching of a C=C bond gives rise to a strong Raman line near 1650 cm^{-1}. The corresponding IR band often is weak and sometimes is not observed at all in symmetrical molecules (Figures 12-1a and 12-15a). Weak absorption from overtones or combinations may occur in this region and can be mistaken for a C=C stretching mode. In such cases, a Raman spectrum will confirm the presence or absence of a C=C bond, because this group always gives a strong Raman line in the 1690–1560 cm^{-1} region (see Figures 12-1b and 12-15b).

The exact frequency of the C=C stretching mode gives some additional information on the environment of the double bond. Tri- or tetraalkyl-substituted groups and trans-disubstituted alkenes have frequencies in the 1690 to 1660 cm^{-1} range. These bands are weak or absent in the infrared but strong in the Raman. Vinyl and vinylidene compounds as well as cis alkenes absorb between 1660 and 1630 cm^{-1}. Substitution by halogens may shift the C=C stretching band out of the usual range. Fluorinated alkenes have very high C=C stretching frequencies (1800–1730 cm^{-1}). Chlorine and other heavy substituents, in contrast, usually lower the frequency.

12-11b Cyclic Compounds

The C=C stretching frequencies of cyclic unsaturated compounds depend on ring size and substitution. Cyclobutene has its C=C stretching mode at 1565 cm^{-1}, whereas

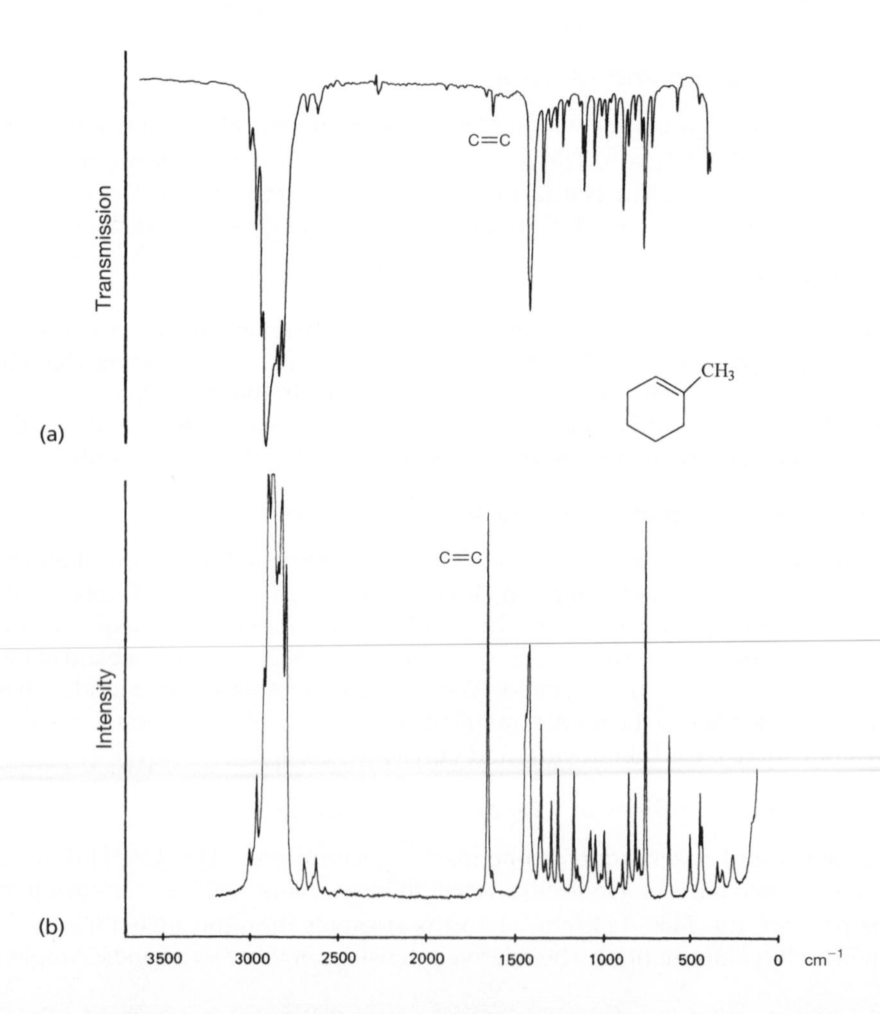

FIGURE 12-15 (a) IR and (b) Raman spectra of 1-methylcyclohexene.

cyclopentene, cyclohexene, and cycloheptene have C=C stretching frequencies of 1610, 1645, and 1655 cm^{-1}, respectively. These frequencies increase with substitution. An example can be seen in Figure 12-15a, in which the C=C stretching band of 1-methylcyclohexene is found at 1670 cm^{-1}. The band is weak in the IR spectrum, but strong in the Raman.

12-11c The C≡C Stretching Mode

A small, sharp peak due to C≡C stretching is observed in the IR spectra of terminal alkynes near 2100 cm^{-1} (Figure 12-10). In substituted alkynes the band is shifted by 100 to 150 cm^{-1} to higher frequencies. If the substitution is symmetric, no band is observed in the infrared because there is no change in dipole moment during the C≡C stretching vibration. Even when substitution is unsymmetrical, the band may be very weak in the infrared and could be missed. In such cases, a Raman spectrum is very valuable, since the C≡C stretching mode always gives a strong line.

12-11d CH= and CH_2= Bending Modes

The CH= and CH_2= wagging or out-of-plane bending modes are very important for structure identification in unsaturated compounds. They occur in the region between 1000 and 650 cm^{-1}. The trans CH bending of a vinyl group gives rise to a strong IR band between 1000 and 980 cm^{-1}, whereas a trans-disubstituted alkene is characterized by a very strong band between 980 and 950 cm^{-1}. Electronegative groups tend to lower this frequency. The cis-disubstituted alkenes give a medium to strong, but less reliable, band between 750 and 650 cm^{-1}. IR spectra of trans- and cis-disubstituted alkenes are shown in Figure 12-16.

The CH_2 out-of-plane wagging vibration of vinyl and vinylidene compounds gives a strong band between 910 and 890 cm^{-1}. This band, coupled with the trans CH bending mode, gives a very characteristic doublet (1000 and 900 cm^{-1}) that distinguishes

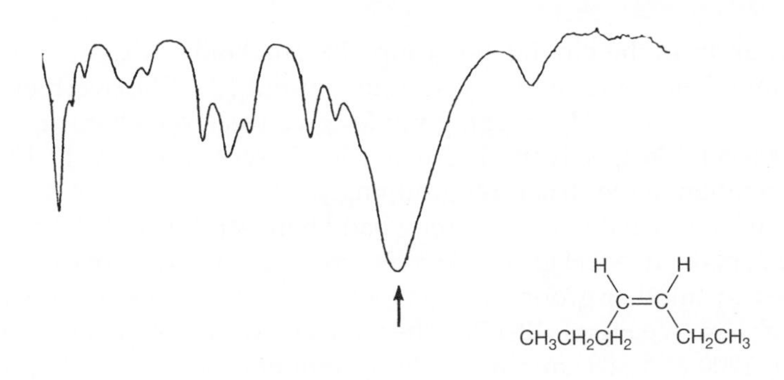

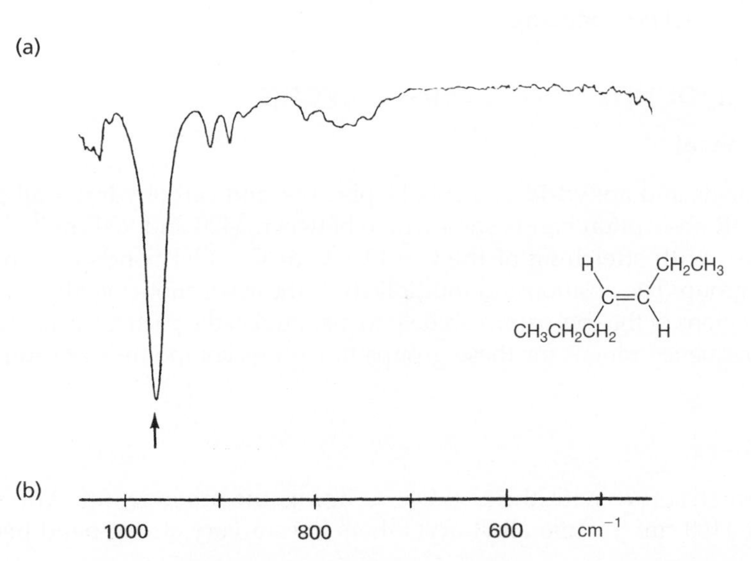

FIGURE 12-16 Portions of IR spectra of (a) cis- and (b) trans-disubstituted alkenes showing the out-of-plane bending modes. (Reproduced with permission of Aldrich Chemical Co., Inc., from C. J. Pouchert (ed.), *The Aldrich Library of FT–IR Spectra.*)

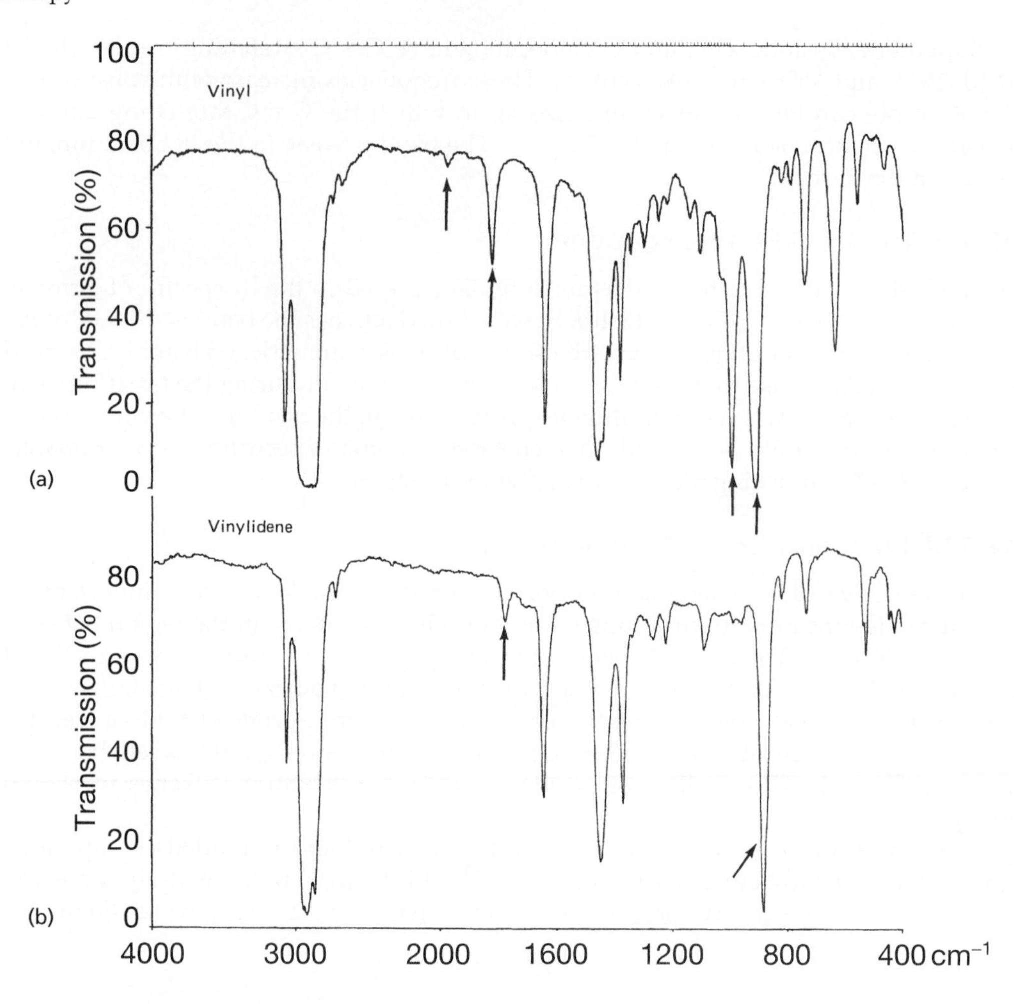

FIGURE 12-17 IR spectra of (a) 3,4-dimethyl-1-hexene, a compound containing a vinyl group, and (b) 2,3-dimethyl-1-pentene, a compound containing a vinylidene group.

the vinyl group from the vinylidene group (900 cm^{-1} only). Examples of IR spectra of vinyl and vinylidene compounds are seen in Figure 12-17. The overtones of the out-of-plane CH bend and the CH_2 wagging modes give weak but characteristic bands near 1950 and 1800 cm^{-1}, respectively. These are clearly seen in Figure 12-17 and provide a useful confirmation of the structural grouping.

Cyclic alkenes usually have a strong band between 750 and 650 cm^{-1} from out-of-plane bending of the two CH groups in a cis arrangement. When one of these hydrogens is substituted by another group such as methyl, however, the band between 750 and 650 cm^{-1} is absent (see Figure 12-15a). There are always several bands of medium intensity between 1200 and 800 cm^{-1} in the IR spectra of cycloalkyl and cycloalkenyl compounds from $—CH_2—$ rocking.

12-12 COMPOUNDS CONTAINING OXYGEN

12-12a General

Carboxylic acids and anhydrides, alcohols, phenols, and carbohydrates all give strong, often broad IR absorption bands somewhere between 1400 and 900 cm^{-1}. These bands are associated with stretching of the C—O—C or C—OH bonds or bending of the C—O—H group. The position and multiplicity of the absorption, together with evidence from other regions of the spectrum, can help to distinguish the particular functional group. The usual frequency ranges for these groups in various compounds are summarized in Table 12-6.

12-12b Ethers

The simplest structure with the C—O—C link is the ether group. Aliphatic ethers absorb near 1100 cm^{-1}, while alkyl aryl ethers have a very strong band between 1280

TABLE 12-6 C—O—C and C—O—H Group Vibrations

Range (cm^{-1}) and Intensity[a]	Group or Class	Assignment and Remarks
1440–1400 (m)	Aliphatic carboxylic acids	C—O—H deformation; may be obscured by CH_3 and CH_2 deformation bands
1430–1280 (m)	Alcohols	C—O—H deformation; broad band
1390–1310 (m–s)	Phenols	C—O—H deformation
1340–1160 (vs)	Phenols	C—O stretch; broad band with structure
1310–1250 (vs)	Aromatic esters	C—O—C antisym stretch
1300–1200 (s)	Aromatic carboxylic acids	C—O stretch
1300–1100 (vs)	Aliphatic esters	C—O—C antisym stretch
1280–1220 (vs)	Alkyl aryl ethers	Aryl C—O stretch; a second band near 1030 cm^{-1}
1270–1200 (s)	Vinyl ethers	C—O—C stretch; a second band near 1050 cm^{-1}
1265–1245 (vs)	Acetate esters	C—O—C antisym stretch
1230–1000 (s)	Alcohols	C—O stretch; see below for more specific frequency ranges
1200–1180 (vs)	Formate and propionate esters	C—O—C stretch
1180–1150 (m)	Alkyl-substituted phenols	C—O stretch
1160–1000 (s)	Aliphatic esters	C—O—C sym stretch
1150–1050 (vs)	Aliphatic ethers	C—O—C stretch; usually centered near 1100 cm^{-1}
1150–1130 (s)	Tertiary alcohols	C—O stretch; lowered by chain branching or adjacent unsaturated groups
1110–1090 (s)	Secondary alcohols	C—O stretch; lowered 10–20 cm^{-1} by chain branching
1060–1040 (s–vs)	Primary alcohols	C—O stretch; often fairly broad
1060–1020 (s)	Saturated cyclic alcohols	C—O stretch; not cyclopropanol or cyclobutanol
1050–1000 (s)	Alkyl aryl ethers	Alkyl C—O stretch
960–900 (m–s)	Carboxylic acids	C—O—H deformation of dimer

[a]*Note*: s = strong; m = medium; w = weak; v = very.

and 1220 cm^{-1} and another strong band between 1050 and 1000 cm^{-1}. In vinyl ethers the C—O—C stretching mode is found near 1200 cm^{-1}. Vinyl ethers can be further distinguished by a very strong C=C stretching band and by the out-of-plane CH bending and CH_2 wagging bands, which are observed near 960 and 820 cm^{-1}, respectively. These frequencies are below the usual ranges discussed in Section 12-11d. Examples of these three types of ether are compared in Figure 12-18. Cyclic, saturated ethers such as tetrahydrofuran have a strong antisymmetric C—O—C stretching band in the range 1250 to 1150 cm^{-1}, whereas unsaturated, cyclic ethers have their C—O—C stretching modes at lower frequencies.

12-12c Alcohols and Phenols

Alcohols and phenols in the pure liquid or solid state have broad bands from hydrogen-bonded OH stretching. For alcohols, this band is centered near 3300 cm^{-1}, whereas in phenols the absorption maximum is 50 to 100 cm^{-1} lower. Phenols absorb near 1350 cm^{-1} from the OH deformation and give a second broader, stronger band from C—OH stretching near 1200 cm^{-1}. This second band always has fine structure from underlying aromatic CH in-plane deformation vibrations. IR spectra of an alcohol and a phenol are compared in Figure 12-19.

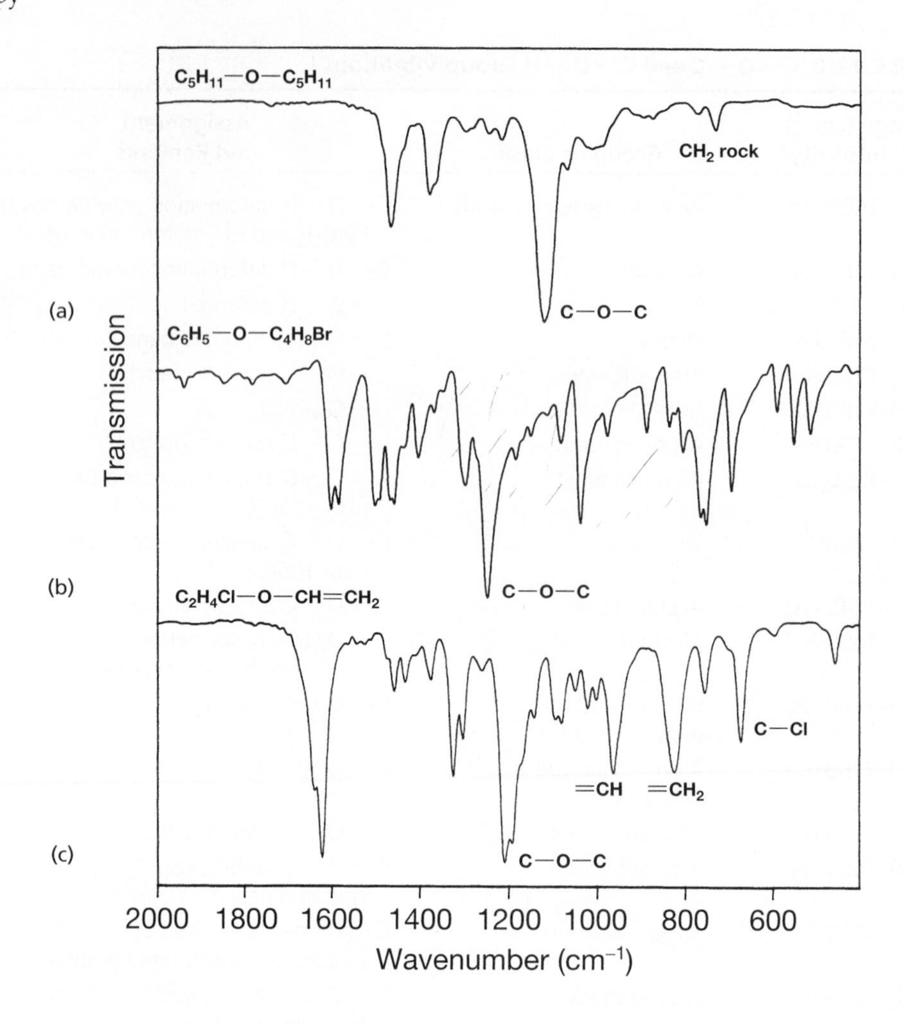

FIGURE 12-18 The IR spectra of three types of ethers: (a) a simple aliphatic ether, (b) an alkyl aryl ether, and (c) a vinyl ether.

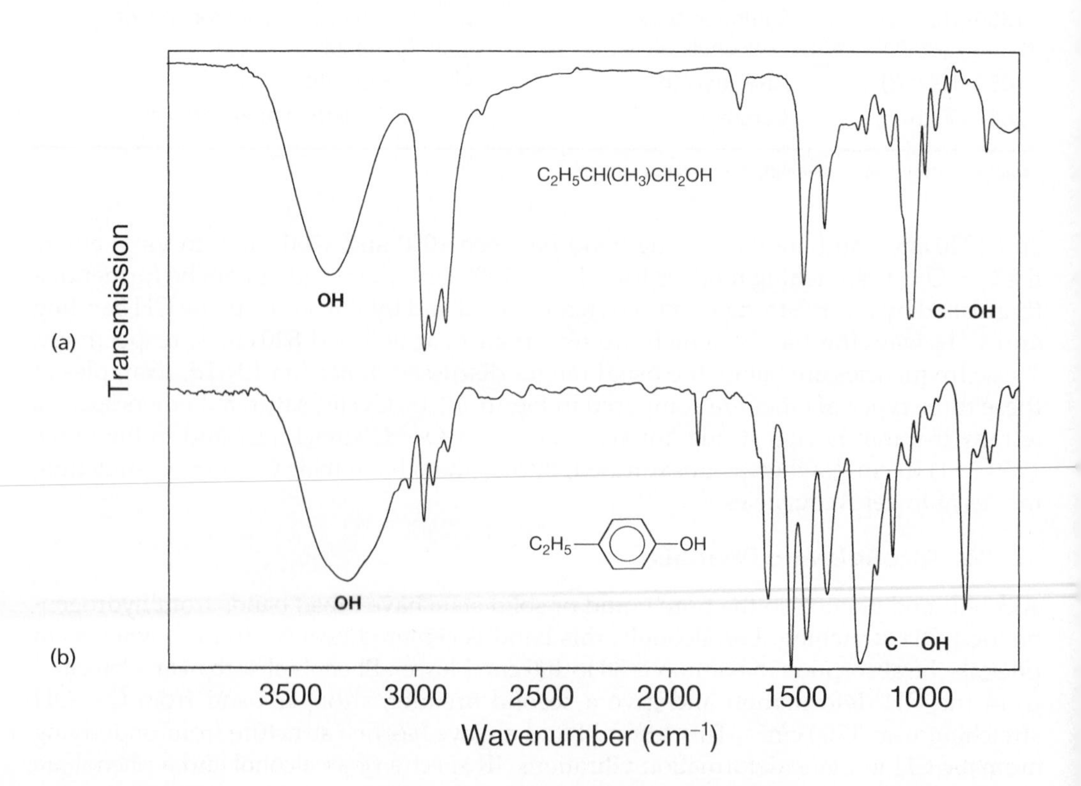

FIGURE 12-19 IR spectra of (a) an alcohol, 2-methyl-1-pentanol, and (b) a phenol, 4-ethylphenol.

In simple alcohols, the frequency of the C—OH stretch is raised by substitution on the C—OH carbon atom (Table 12-6). Sugars and carbohydrates give very broad absorption bands centered near 3300 cm^{-1} (OH stretching), 1400 cm^{-1} (OH deformation), and 1000 cm^{-1} (C—OH stretching).

12-12d Carboxylic Acids and Anhydrides

Carboxylic acids usually exist as dimers except in dilute solution. The carbonyl stretching band of the dimer is found near 1700 cm^{-1}, whereas in the monomer spectrum the band is located at higher frequencies (1800–1740 cm^{-1}). In addition to the very broad OH stretching band mentioned in Section 12-5a, the following three vibrations are associated with the C—OH group in carboxylic acids: a band of medium intensity near 1430 cm^{-1}, a stronger band near 1240 cm^{-1}, and another band of medium intensity near 930 cm^{-1}. The presence of an anhydride is detected by the characteristic absorption in the C=O stretching region, which consists of a strong, sharp doublet with one band at unusually high frequency (1840–1800 cm^{-1}) and a second band about 60 cm^{-1} lower (1780–1740 cm^{-1}). The C—O—C stretch gives rise to a broad band near 1150 cm^{-1} in open-chain anhydrides and at higher frequencies in cyclic structures.

12-12e Esters

The antisymmetric C—O—C stretching mode in esters gives rise to a very strong and often quite broad band near 1200 cm^{-1}. The actual frequency of the maximum of this band can vary from 1290 cm^{-1} in benzoates down to 1100 cm^{-1} in aliphatic esters. There may be structure on this band due to CH deformation vibrations that absorb in the same region. The band may be even stronger than the C=O stretching band near 1750 cm^{-1}. The symmetric C—O—C stretch also gives a strong band at lower frequencies between 1160 and 1000 cm^{-1} in aliphatic esters.

12-13 COMPOUNDS CONTAINING NITROGEN

12-13a General

The presence of primary or secondary amines and amides can be detected by absorption from stretching of NH_2 or NH groups between 3350 and 3200 cm^{-1}. Tertiary amines and amides, in contrast, are more difficult to identify, because they have no N—H groups. Nitriles and nitro compounds also give characteristic IR absorption bands near 2250 and 1530 cm^{-1}, respectively. Isocyanates and carbodiimides have very strong IR bands near 2260 and 2140 cm^{-1}, respectively, where very few absorptions from other groupings occur. Oximes, imines, and azo compounds give weak IR bands in the 1700 to 1600 cm^{-1} region from stretching vibrations of the —C=N— or —N=N— group.

12-13b Amino Acids, Amines, and Amine Hydrohalides

Three classes of nitrogen-containing compounds (amino acids, amines, and amine hydrohalides) give rise to very characteristic broad absorption bands. Some of the most striking of these are found in the IR spectra of amino acids, which contain an extremely broad band centered near 3000 cm^{-1}, and often extending as low as 2200 cm^{-1}, with some structure (Figure 12-20). Amine hydrohalides (ammonium halides) give a similar, very broad band, which has structure on the low-frequency side. The center of the band tends to be lower than in amino acids, especially in the case of tertiary amine hydrohalides, in which the band center may be as low as 2500 cm^{-1}. In fact, this band gives a very useful indication of the presence of a tertiary amine hydrohalide (Figure 12-21).

Both amino acids and primary amine hydrohalides have a weak but characteristic band between 2200 and 2000 cm^{-1} (Figure 12-21), which is believed to be a combination of the $—NH_3^+$ deformation near 1600 cm^{-1} and the $—NH_3^+$ torsion near 500 cm^{-1}. Primary amines have a fairly broad band in their IR spectra centered near 830 cm^{-1}, whereas the frequency for secondary amines is about 100 cm^{-1} lower (see Figures 12-7d and 12-7e). This band is not present in the spectra of tertiary amines or amine hydrohalides.

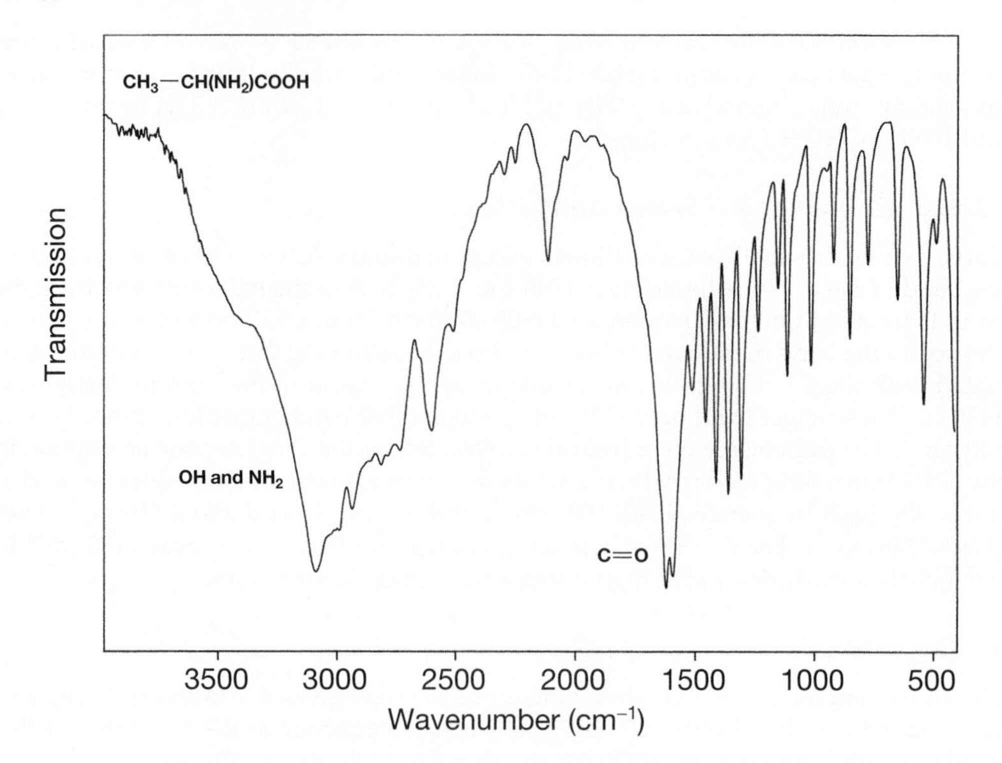

FIGURE 12-20 The IR spectrum of the amino acid L-alanine.

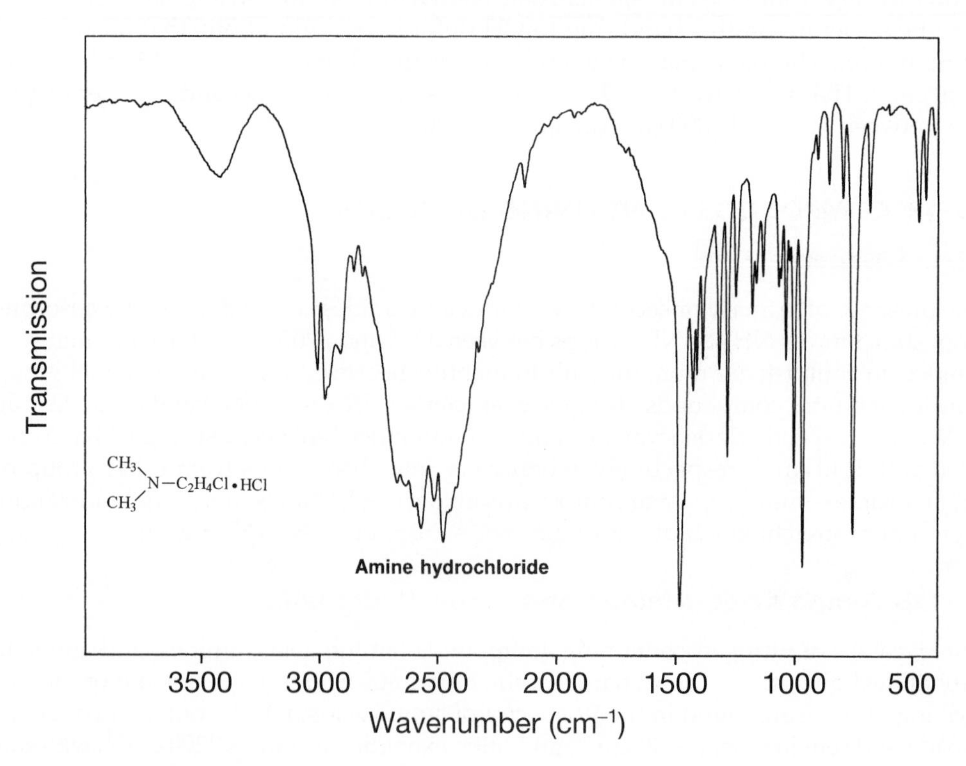

FIGURE 12-21 The IR spectrum of a tertiary amine hydrohalide, (2-chloroethyl)dimethylamine hydrochloride.

12-13c Anilines

In anilines, the characteristic broad band shown by aliphatic amines in the 830–730 cm^{-1} region is not present, and so the out-of-plane CH deformations of the benzene ring can be observed. These bands permit the ring substitution pattern to be determined. Of course, when an aliphatic amine is joined to a benzene ring through a carbon chain, both the characteristic amine band and the CH deformation pattern are present.

Figure 12-22 illustrates the IR spectrum of an aniline derivative, and Figure 12-23 the spectrum of an aliphatic amine joined to a benzene ring. The presence of the benzene ring is identified in both compounds by CH stretching bands between 3100 and

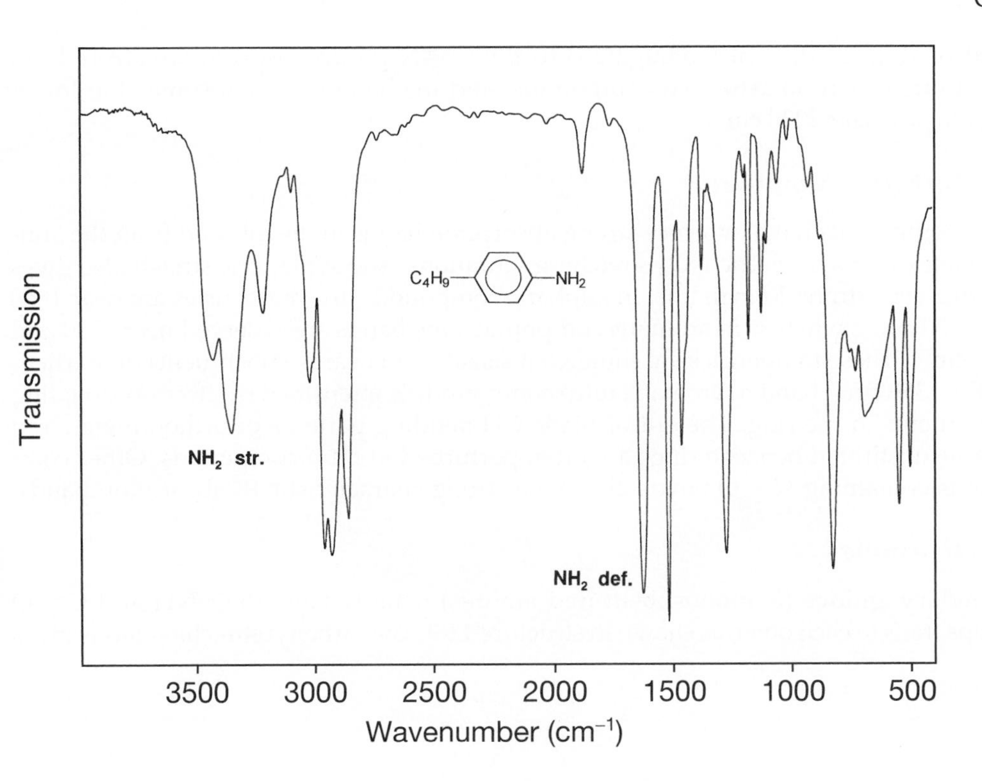

FIGURE 12-22 The IR spectrum of an aniline derivative, 4-butylaniline.

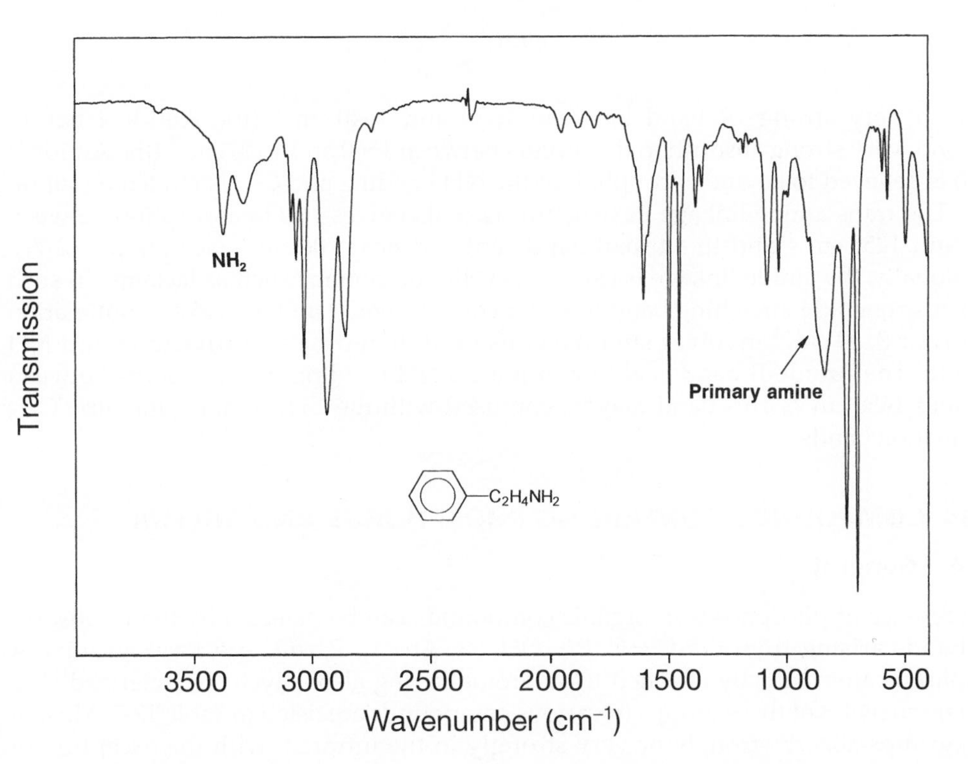

FIGURE 12-23 The IR spectrum of an aliphatic amine joined to a benzene ring, phenethylamine.

3000 cm^{-1}, out-of-plane CH bending bands between 850 and 700 cm^{-1}, and bands diagnostic of the substitution patterns between 2000 and 1700 cm^{-1}. In the out-of-plane bending region, the single band at 825 cm^{-1} in Figure 12-22 indicates the presence of a para-disubstituted benzene ring, whereas the doublet at 740 and 700 cm^{-1} in Figure 12-23 indicates monosubstitution (Section 12-8).

12-13d Nitriles

Saturated nitriles absorb weakly in the infrared near 2250 cm^{-1}, although the band is strong in the Raman spectrum. Unsaturated or aromatic nitriles for which the double bond or ring is adjacent (conjugated) to the C≡N group absorb more strongly in the infrared than do saturated compounds, and the band occurs at somewhat lower frequencies, near 2230 cm^{-1}.

12-13e Nitro Compounds

Nitro compounds have two very strong absorption bands in the infrared from the antisymmetric and symmetric NO_2 stretching vibrations. The symmetric stretch also gives rise to a very strong Raman line. In aliphatic compounds, the frequencies are near 1550 and 1380 cm^{-1}, whereas in aromatic compounds, the bands are observed near 1520 and 1350 cm^{-1}. These frequencies are somewhat sensitive to nearby substituents. In particular, the 1350 cm^{-1} band in aromatic nitro compounds is intensified by electron-donating substituents in the ring. The out-of-plane CH bending patterns of ortho-, meta-, and para-disubstituted benzene rings are often perturbed in nitro compounds. Other compounds containing N—O bonds also have strong characteristic IR absorption bands.

12-13f Amides

Secondary amides (*N*-monosubstituted amides) usually have their NH and C=O groups trans to each other, as shown in structure **12-9**. The carbonyl stretching mode gives

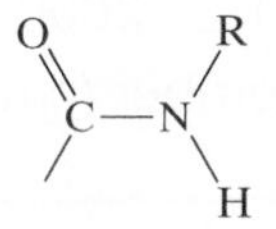

12-9

rise to a very strong IR band between 1680 and 1640 cm^{-1} (the Amide I band). A second, very strong absorption that occurs between 1560 and 1530 cm^{-1} (the Amide II band) is believed to be due to coupling of the NH bending and C—N stretching vibrations. The trans amide linkage in structure **12-9** also gives rise to absorption between 1300 and 1250 cm^{-1} and to a broad band centered near 700 cm^{-1} (see Figure 12-7f). Occasionally, the amide linkage is cis, as in cyclic compounds such as lactams. In such cases, a strong NH stretching band is seen near 3200 cm^{-1}, and a weaker combination band near 3100 cm^{-1} involves simultaneous excitation of C=O stretching and NH bending. The Amide II band is absent, but a cis NH bending mode absorbs between 1500 and 1450 cm^{-1}. This band may be confused with the CH_2 or antisymmetric CH_3 deformation bands.

12-14 COMPOUNDS CONTAINING PHOSPHORUS AND SULFUR

12-14a General

The presence of phosphorus in organic compounds can be detected by the IR absorption bands arising from the P—H, P—OH, P—O—C, P=O, and P=S groups. A phosphorus atom directly attached to an aromatic ring also is well characterized. The usual frequencies of these groups in various compounds are listed in Table 12-7. Most of these groups absorb strongly or very strongly in the infrared, with the exception of P=S. The Raman spectrum is valuable for detecting this group, which has a frequency between 700 and 600 cm^{-1}. There is no characteristic P—C group frequency in aliphatic compounds.

The SO_2 and SO groups give rise to very strong IR bands between 1400 and 1000 cm^{-1}. Other bonds involving sulfur, such as C—S, S—S, and S—H, give very weak IR absorption, and a Raman spectrum is needed to identify these groups. Characteristic frequencies of some sulfur-containing groups also are listed in Table 12-7. The C=S group has been omitted from the table because the C=S stretching vibration is invariably coupled with vibrations of other groups in the molecule. Frequencies in the 1400 to 850 cm^{-1} range have been assigned to this group, with thioamides at the low-frequency end of the range. IR bands involving C=S groups are usually weak.

TABLE 12-7 Characteristic IR Frequencies of Groups Containing Phosphorus or Sulfur

Group and Class	Range (cm^{-1}) and Intensity[a]	Assignment and Remarks
The P—H Group		
Phosphorus acids and esters	2425–2325 (m)	P—H stretch
Phosphines	2320–2270 (m)	P—H stretch; sharp band
	1090–1080 (m)	PH_2 deformation
	990–910 (m–s)	P—H wag
The P—OH Group		
Phosphoric or phosphorus acids, esters, and salts	2700–2100 (w)	OH stretch; one or two broad bands
	1040–920 (s)	P—OH stretch
The P—O—C Group		
Aliphatic compounds	1050–950 (vs)	Antisym P—O—C stretch
	830–750 (s)	Sym P—O—C stretch (methoxy and ethoxy phosphorus compounds only)
Aromatic compounds	1250–1160 (vs)	Aromatic C—O stretch
	1050–870 (vs)	P—O stretch
The P—C Group		
Aromatic compounds	1450–1430 (s)	P joined directly to a ring; sharp band
Quaternary aromatic	1110–1090 (s)	P^+ joined directly to a ring; sharp band
The P=O Group		
Aliphatic compounds RO(P=O)—	1260–1240 (s)	Strong, sharp band
Aromatic compounds ArO(P=O)—	1350–1300 (s)	Lower frequency (1250–1180 cm^{-1}) when OH group is attached to the P atom
Phosphine oxides	1200–1140 (s)	P=O stretch
The S—H Group		
Thiols (mercaptans)	2580–2500 (w)	S—H stretch; strong in Raman
The C—S Group	720–600 (w)	C—S stretch; strong in Raman
The S—S Group		
Disulfides	550–450 (vw or absent)	S—S stretch; strong in Raman
The S=O Group		
Sulfoxides	1060–1020 (vs)	S=O stretch
Dialkyl sulfites	1220–1190 (vs)	S=O stretch
The SO_2 Group		
Sulfones, sulfonamides, sulfonic acids, sulfonates, and sulfonyl chlorides	1390–1290 (vs) 1190–1120 (vs)	SO_2 antisym stretch SO_2 sym stretch
The S—O—C Group		
Dialkyl sulfites	1050–850 (vs)	S—O—C stretching (two bands)
Sulfates	1050–770 (vs)	Two or more bands

[a]*Note*: s = strong; m = medium; w = weak; v = very.

12-14b Phosphorus Acids and Esters

Phosphorus acids have P—OH groups that give one or two broad bands of medium intensity between 2700 and 2100 cm^{-1}. Esters and acid salts that have P—OH groups also absorb in this region. The presence of a PH group is indicated by a small, sharp

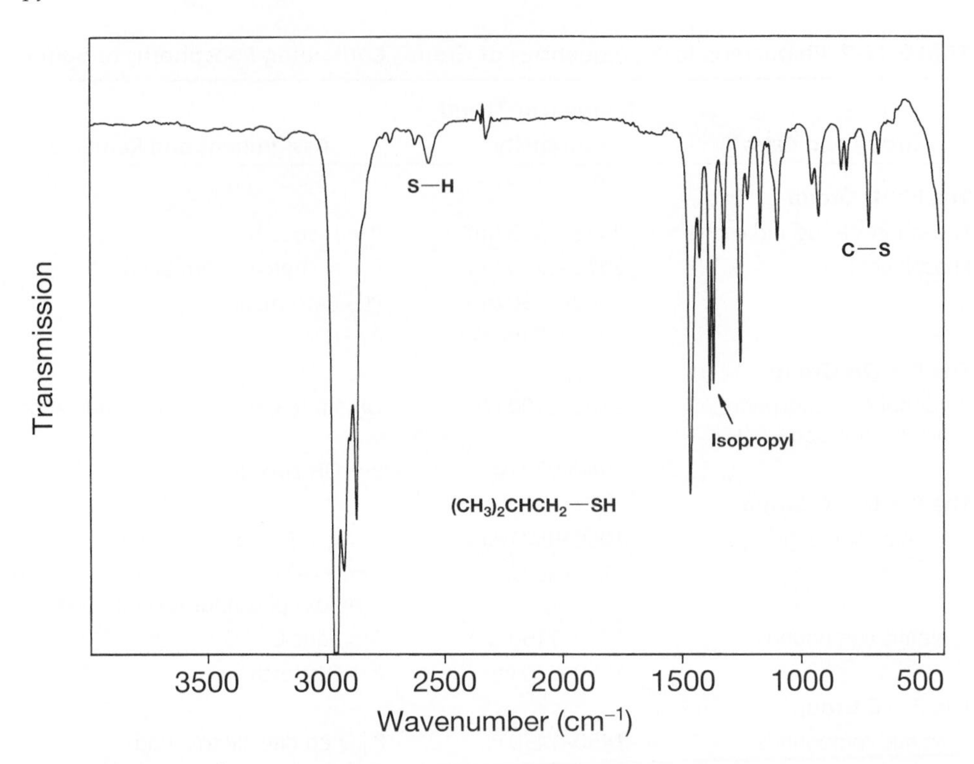

FIGURE 12-24 The IR spectrum of an aliphatic thiol.

band near 2400 cm^{-1}. In ethoxy and methoxy phosphorus compounds, as well as in other aliphatic compounds with a P—O—C linkage, a very strong and quite broad IR band is observed between 1050 and 950 cm^{-1}. The presence of a P=O bond is indicated by a strong band close to 1250 cm^{-1}.

12-14c Compounds Containing C—S, S—S, and S—H Groups

Raman spectra of compounds containing C—S and S—S bonds have very strong lines from these groups, between 700 and 600 and near 500 cm^{-1}, respectively. These group frequencies, especially S—S, are either absent or appear only very weakly in the infrared. The S—H stretching band near 2500 cm^{-1} is normally quite weak in the infrared but shows a high intensity in the Raman spectrum. The spectrum of a simple aliphatic thiol is shown in Figure 12-24, in which the S—H stretch can be seen at 2530 cm^{-1} and the C—S stretch near 700 cm^{-1}. The compound contains an isopropyl group, and the symmetric deformations of the two methyl groups give rise to a strong characteristic doublet centered at 1375 cm^{-1} (Section 12-9b).

12-14d Compounds Containing S=O Groups

The stretching vibration of the S=O group in sulfoxides gives rise to strong, broad IR absorption near 1050 cm^{-1}. In the spectra of alkyl sulfites, the S=O stretch is observed near 1200 cm^{-1}. Sulfones contain the SO_2 group, which gives rise to two very strong bands from antisymmetric (1369–1290 cm^{-1}) and symmetric (1170–1120 cm^{-1}) stretching modes.

The frequencies of the SO_2 stretching vibrations of sulfonyl halides are about 50 cm^{-1} higher than those of sulfones, because of the electronegative halogen atom. In the IR spectra of sulfonic acids, a very broad absorption is observed between 3000 and 2000 cm^{-1} from OH stretching. The antisymmetric and symmetric SO_2 stretching modes, respectively, give two strong bands near 1350 cm^{-1} and between 1200 and 1100 cm^{-1} in both alkyl and aryl sulfonic acids and sulfonates.

12-15 HETEROCYCLIC COMPOUNDS

12-15a General

Heterocyclic compounds containing nitrogen, oxygen, or sulfur may exhibit three kinds of group frequencies: those involving CH or NH vibrations, those involving motion of the ring, and those from the group frequencies of substituents on the ring. The

TABLE 12-8 Characteristic IR Frequencies for Some Heterocyclic Compounds

Classes of Compounds	Range (cm^{-1}) and Intensity[a]	Assignment and Remarks
Azoles (imidazoles, isoxazoles, oxazoles, pyrazoles, triazoles, tetrazoles)	3300–2500 (s, br)	H-bonded NH stretch; resembles carboxylic acids
	1650–1380 (m–s)	Three ring-stretching bands
	1040–980 (s)	Ring breathing
Carbazoles	3490–3470 (vs)	NH stretch (dil soln, nonpolar solvents)
1-4-Dioxanes	1460–1440 (vs)	CH_2 deformation
	1400–1150 (s)	CH_2 twist and wag
	1130–1000 (m)	Ring mode; strong in Raman
	850–830 (w)	Very strong in Raman
Furans	3140–3120 (m)	CH stretch; higher than most aromatics
	1600–1400 (m–s)	Ring stretching (three bands)
	770–720 (vs)	Band weakens as number of substituents increases
Indoles	3470–3450 (vs)	NH stretch
	1600–1500 (m–s)	Two bands
	900–600 (vs)	Substitution patterns due to both 5- and 6-membered rings
Pyridines (general)	3080–3020 (w–m)	CH stretch; several bands
	2080–1670 (w)	Combination bands
	1615–1565 (s)	Two bands due to C=C and C=N stretch in ring
	1030–990 (s)	Ring-breathing
2-Substituted	780–740 (s)	CH out-of-plane deformation
	630–605 (m–s)	In-plane ring deformation
	420–400 (s)	Out-of-plane ring deformation
3-Substituted	820–770 (s)	CH out-of-plane deformation
	730–690 (s)	Ring deformation
	635–610 (m–s)	In-plane ring deformation
	420–380 (s)	Out-of-plane ring deformation
4-Substituted	850–790 (s)	CH out-of-plane deformation
Disubstituted	830–810 (s)	Two bands due to CH out-of-plane deformations
	740–720 (s)	
Trisubstituted	730–720 (s)	CH out-of-plane deformation
Pyrimidines	1590–1370 (m–s)	Ring stretching; four bands
	685–660 (m–vs)	Ring deformation
Pyrroles	3480–3430 (vs)	NH stretch; often a sharp band
	3130–3120 (w)	CH stretch; higher than normal
	1560–1390 (variable)	Ring stretch; usually three bands
	770–720 (s, br)	CH out-of-plane deformation
Thiophenes	1590–1350 (m–vs)	Several bands due to ring stretching modes
	810–680 (vs)	CH out-of-plane deformation; lower than in pyrroles and furans
Triazines	1560–1520 (vs)	Two bands due to ring stretching modes
	1420–1400 (s)	
	820–740 (s)	Out-of-plane ring deformation

[a]*Note*: s = strong; m = medium; w = weak; v = very.

identification of a heterocyclic compound from its IR and Raman spectra is a difficult task. Characteristic frequencies for some heterocyclic compounds are collected in Table 12-8, and the characterization of a few types of compounds is discussed in this section. Heterocyclic nitrogen compounds usually have strong Raman spectra, with very strong lines arising from —C=N— groups when present.

12-15b Aromatic Heterocycles

Hydrogen atoms attached to carbon atoms in an aromatic heterocyclic ring such as pyridine give rise to CH stretching modes in the usual 3100 to 3000 cm^{-1} region, or a little higher in furans, pyrroles, and some other compounds. Characteristic ring stretching modes, similar to those of benzene derivatives, are observed between 1600 and 1000 cm^{-1}. The out-of-plane CH deformation vibrations give rise to strong IR bands in the 1000 to 650 cm^{-1} region. In some cases, these patterns are characteristic of the type of substitution in the heterocyclic ring, for example, with furans, indoles, pyridines, pyrimidines, and quinolines. The in-plane CH bending modes also give several bands in the 1300 to 1000 cm^{-1} region for aromatic heterocyclic compounds. CH vibrations in benzene derivatives and analogous modes in related heterocyclic compounds can be correlated and may be useful in structure determination.

Overtone and combination bands are observed between 2000 and 1750 cm^{-1} in the IR spectra. These bands are similar to those observed for benzene derivatives (see Table 12-5) and are characteristic of the position of substitution. In aromatic heterocyclic compounds involving nitrogen, the coupled C=C and C=N stretching modes give rise to several characteristic vibrations. These are similar in frequency to their counterparts in the corresponding nonheterocyclic compounds and give rise to very strong Raman lines. Ring-stretching modes are found in the 1600 to 1300 cm^{-1} region. Other skeletal ring modes include ring-breathing modes near 1000 cm^{-1}, in-plane ring deformation between 700 and 600 cm^{-1}, and out-of-plane ring deformation modes, which may be observed between 700 and 300 cm^{-1}.

Nonaromatic heterocyclic compounds usually have one or more CH_2 groups present. The stretching and deformation (scissoring) modes give rise to bands in the usual regions (Section 12-10). The wagging, twisting, and rocking modes, however, often interact with skeletal ring modes and may be observed over a wide range of frequencies.

12-15c Pyrimidines and Purines

Pyrimidines (**12-10**) and purines (**12-11**) absorb strongly in the infrared between 1640 and 1520 cm^{-1} due to C=C and C=N stretching of the ring. A band near 1630 cm^{-1} is

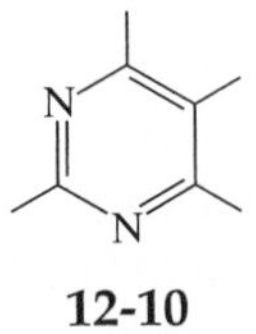

12-10 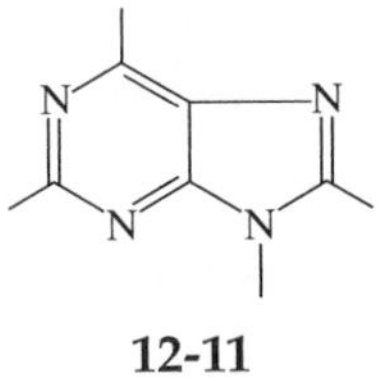 **12-11**

attributed to C=N stretching and a second band between 1580 and 1520 cm^{-1} is assigned to a C=C stretch. The C=N stretch usually gives rise to a very strong Raman line in these and related heterocyclic compounds. Pyrimidines and purines usually have absorption bands between 700 and 600 cm^{-1} from CH out-of-plane bending. Nitrogen heterocycles can form *N*-oxides, which have a characteristic very strong IR band near 1280 cm^{-1}.

12-15d Five-Membered Ring Compounds

Pyrroles, furans, and thiophenes generally have a band in their IR spectra from C=C stretching near 1580 cm^{-1}. A strong band also is observed between 800 and 700 cm^{-1} from an out-of-plane deformation vibration of the CH=CH group, similar to that of cis-disubstituted alkenes. In the spectra of pyrroles, a strong, broad band is observed between 3400 and 3000 cm^{-1} from the H-bonded N—H stretching mode. Furans have medium to strong IR bands in the ranges 1610 to 1560 cm^{-1}, 1520 to 1470 cm^{-1}, and 1400 to 1390 cm^{-1} from ring stretching vibrations. All furans have a strong absorption near 595 cm^{-1}, which is attributed to a ring deformation mode.

Thiophenes absorb in the infrared between 3100 and 3000 cm^{-1} (CH stretching), 1550 and 1200 cm^{-1} (ring stretching), and 750 and 650 cm^{-1} (out-of-plane C—H bending). IR spectra of thiophenes generally have a band between 530 and 450 cm^{-1} from an out-of-plane ring deformation.

12-15e NH Stretching Bands

Spectra of heterocyclic nitrogen compounds may contain bands from a secondary or tertiary amine group. Pyrroles, indoles, and carbazoles in nonpolar solvents have their NH stretching vibrations between 3500 and 3450 cm^{-1}, and the band is very strong in the infrared. In saturated heterocyclics, such as pyrrolidines and piperidines, the band is at lower frequencies. Azoles have a very broad, hydrogen-bonded NH stretching band between 3300 and 2500 cm^{-1}. This band might be confused with the broad OH stretching band of carboxylic acids.

12-16 COMPOUNDS CONTAINING HALOGENS

12-16a General

A halogen atom adjacent to a functional group often causes a significant shift in the group frequency. Fluorine is particularly important in this regard, and special care must be exercised in drawing conclusions from IR and Raman spectra when this element is present. Carbon–fluorine stretching bands are very strong in the infrared, usually between 1350 and 1100 cm^{-1}, but they are often weak in the Raman. Other functional groups that absorb in this region of the spectrum may be hidden by the CF stretching band. These groups often can be detected in the Raman spectrum. There are many known cases of symmetric C—F stretching modes at frequencies much lower than the usual 1350 to 1100 cm^{-1}.

12-16b CH_2X Groups

The CH_2 wagging mode in compounds with a —CH_2X group gives rise to a strong band whose frequency depends on X. When X is Cl, the range is 1300 to 1250 cm^{-1}. For Br, the band is near 1230 cm^{-1}, and for I, a still lower frequency near 1170 cm^{-1} is observed.

12-16c Perhaloalkyl Groups

In perhaloalkyl groups, the presence of more than one halogen atom on a single carbon atom shifts the C—X stretching frequency to the high wavenumber end of the range. The antisymmetric stretching frequency of the —CCl_3 group gives rise to a band in the IR spectrum in the 830 to 700 cm^{-1} range.

12-16d Aromatic Halogen Compounds

Halogen atoms attached to aromatic rings are involved in certain vibrations that are sensitive to the mass of the halogen atom. One of the benzene ring vibrations that involve motion of the substituent atom gives rise to bands between 1250 and 1100 cm^{-1} when the substituent is fluorine, between 1100 and 1040 cm^{-1} for chlorine, and between 1070 and 1020 cm^{-1} for bromine.

12-17 BORON, SILICON, TIN, LEAD, AND MERCURY COMPOUNDS

Boron–carbon and silicon–carbon stretching modes are not usually identifiable, since they are coupled with other skeletal modes. The C—B—C antisymmetric stretching mode in phenylboron compounds, however, gives a strong IR band between 1280 and 1250 cm^{-1}, and a silicon atom attached to an aromatic ring gives two very strong bands near 1430 and 1110 cm^{-1}. Important values are collected in Table 12-9.

TABLE 12-9 Some IR Group Frequencies in Boron and Silicon Compounds

	Group	Range (cm^{-1}) and Intensity[a]	Assignment and Remarks
Boron	—BOH	3300–3200 (s)	Broad band due to H-bonded OH stretch
	—BH and —BH_2	2650–2350 (s)	Doublet for —BH_2 stretch
		1200–1150 (m–s)	—BH_2 deformation of B—H bend
		980–920 (m)	—BH_2 wag
		ca. 1430 (m–s)	Benzene ring vibration
	B—N	1460–1330 (vs)	B—N stretch; borazines and aminoboranes
	B—O	1380–1310 (vvs)	B—O stretch; boronates, boronic acids
	C—B—C	1280–1250 (vs)	C—B—C antisym stretch
Silicon	—SiOH	3700–3200 (s)	OH stretch, similar to alcohols
		900–820 (s)	Si—O stretch
	—SiH, —SiH_2, and —SiH_3	2150–2100 (m)	Si—H stretch
		950–800 (s)	Si—H deformation and wag
	Si—Ar	ca. 1430 (m–s)	Ring mode
		ca. 1100 (vs)	Ring mode
	Si—O—C (aliphatic)	1100–1050 (vvs)	Si—O—C antisym stretch
	Si—O—Ar	970–920 (vs)	Si—O stretch
	Si—O—Si	1100–1000 (s)	Si—O—Si antisym stretch

[a]*Note*: s = strong; m = medium; w = weak; v = very.

The B—O and B—N bonds in organoboron compounds give very strong IR bands between 1430 and 1330 cm^{-1}. The Si—O—C vibration gives a very strong IR absorption, which is often quite broad in the 1100 to 1050 cm^{-1} range. The B—CH_3 and Si—CH_3 symmetric CH_3 deformation modes occur at 1330 to 1280 cm^{-1} and 1280 to 1250 cm^{-1}, respectively. The CH_3 deformations in metal–CH_3 groups give rise to bands between 1210 and 1180 cm^{-1} in organomercury and organotin compounds and between 1170 and 1150 cm^{-1} in organolead compounds.

The mercury–carbon bond in aliphatic organomercury compounds is characterized by a very strong Raman band between 550 and 500 cm^{-1}. For aromatic mercury compounds, a band between 250 and 200 cm^{-1} is assigned to the phenyl–Hg stretch. These very strong bands may be seen in Raman spectra of dilute aqueous solutions.

12-18 ISOTOPICALLY LABELED COMPOUNDS

12-18a The Effect of ^2H and ^{13}C Isotopic Substitution on Stretching Modes

In Section 12-1 the vibrational frequency of a diatomic group was given by the equation $v(cm^{-1}) = 130.3\sqrt{k/\mu}$, in which k is the force constant (N m^{-1}) and μ is the reduced mass $[m_1 m_2/(m_1 + m_2)]$. Isotopic substitution does not change the force constant, so the frequency of the isotopically substituted group is given by eq. 12-2.

$$\nu(\text{isotopic}) = \nu(\text{normal})\sqrt{\mu(\text{normal})/\mu(\text{isotopic})} \qquad \textbf{(12-2)}$$

For example, consider the CH group, for which $\mu(\text{normal}) = 0.923$ amu. The reduced mass for the CD group is $\mu(\text{isotopic}) = 1.714$ amu. For a CH stretch observed at 3000 cm^{-1}, the analogous CD stretch is expected at $3000 \times (0.923/1.714)^{1/2} = 2200$ cm^{-1}.

In general, bands from CD stretching modes are observed in the range 2300 to 2100 cm^{-1}. Similar calculations may be made for stretching modes of other X—D groups. Much smaller changes in the Raman and IR spectra (band shifts) are predicted for ^{13}C isotopic substitution.

12-18b The Effect of Deuterium Substitution on Bending Modes

All types of C—H bending vibrations are displaced by a factor of approximately $0.7\sqrt{1/2} = 0.495$ to lower frequencies when hydrogen atoms are replaced by deuterium. The disappearance of a band from a C—H bending mode is usually of more analytical value than the appearance of the corresponding C—D band. The new CD group vibration is likely to couple with C—C skeletal stretching vibrations and often cannot be identified. The most important CH group vibrations are the methylene bending (scissoring) vibration found between 1470 and 1400 cm^{-1} and the symmetric methyl deformation mode observed between 1385 and 1360 cm^{-1}.

For example, selective deuteration at the α positions of cyclopentanone can be observed in the IR spectrum. In the spectrum of undeuterated cyclopentanone, bands from CH bending are observed at 1455 and 1406 cm^{-1}. In the spectrum of cyclopentanone-2,2,5,5-d_4, only the 1455 cm^{-1} band remains in this region. This observation shows that the 1406 cm^{-1} band is from the active α-methylenes, whereas the 1455 cm^{-1} band is from the β-methylene groups.

The IR spectrum of 3-pentanone [$CH_3CH_2(CO)CH_2CH_3$] contains bands at 1461, 1414, 1379, and 1365 cm^{-1}. Selective deuteration of this molecule is demonstrated by the disappearance of bands in the IR spectrum at 1461 and 1379 cm^{-1} when the methyl groups are deuterated [$CD_3CH_2(CO)CH_2CD_3$] and the disappearance of bands at 1414 and 1365 cm^{-1} when the CH_2 groups are deuterated [$CH_3CD_2(CO)CD_2CH_3$]. The bands at 1461 and 1379 cm^{-1} are from the antisymmetric and symmetric methyl deformations, the band at 1414 cm^{-1} is attributed to the CH_2 scissors vibration, and the 1365 cm^{-1} band is from a methylene vibration coupled with the C—CO stretch.

12-19 USING THE LITERATURE FOR VIBRATIONAL SPECTRA

Positive identification of a compound is obtained when its IR or Raman spectrum exactly matches the spectrum of a known compound. To establish such identities, collections of spectra or references to spectra in the literature are needed. Published spectra, either in the literature or in collections, are currently available for several hundred thousand compounds. In this section, sources of collections of spectra, sources of literature, and references to spectra are listed. In addition to the very extensive *Sadtler Standard Infrared Spectra* collection and other large collections of IR spectra, there are several smaller collections of IR, FT–IR, and Raman spectra.

1. *Sadtler Standard Infrared Spectra*, Sadtler Research Laboratories, Inc., 3314 Spring Garden Street, Philadelphia, PA 19104.

   The main collection consists of prism or grating spectra in loose-leaf volumes, containing 1000 spectra per volume. The format of the earlier spectra is linear in wavelength (2–15 μm). The index to this collection consists of the following four sections: Chemical Classes, Alphabetical, Molecular Formula, and Numerical.
2. *The Aldrich Library of FT–IR Spectra*, C. J. Pouchert (ed.), Aldrich Chemical Co., Inc., P.O. Box 355, Milwaukee, WI 53201.

   This compilation contains FT–IR spectra with alphabetic and molecular formula indices.
3. *Selected Infrared Spectral Data*, American Petroleum Institute (API), Research Project 44, Department of Chemistry, Texas A&M University, College Station, TX 77843.

   This large collection of spectra is continually updated. The presentation is usually linear in wavelength for the older entries in the collection, but linear in wavenumber for the more recent spectra. This is the most extensive collection of spectra of high-purity petroleum hydrocarbons; also included are nitrogen and sulfur compounds found in petroleum.

4. *Coblentz Society Spectra*, P.O. Box 9952, Kirkwood, MO 63122.
   In the book, more than 10,000 spectra are given in volumes of 1000 spectra each, in notebook format. In the computer database, more than 9500 spectra have been presented. http://www.acdlabs.com/products/spec_lab/exp_spectra/spec_libraries/coblentz.html
5. *The Coblentz Society Desk Book of Infrared Spectra*, 2nd ed. C. D. Craver (ed.), The Coblentz Society Inc., P. O. Box 9952, Kirkwood, MO 63122 (1982).
   Eight hundred and seventy-two grating spectra are grouped by chemical classes, with text.
6. *Sadtler Standard Raman Spectra*, Sadtler Research Laboratories, Inc., 3314 Spring Garden Street, Philadelphia, PA 19104.
   Both parallel- and perpendicular-polarized Raman spectra are presented together with the corresponding IR spectrum.
7. *Selected Raman Spectra Data*, American Petroleum Institute (API), Research Project 44, Department of Chemistry, Texas A&M University, College Station, TX 77843.
   This compilation is produced in the same format as the API IR spectra (above reference 3). There are 500 Raman spectra obtained using mercury vapor lamp excitation and 200 laser-excited spectra.
8. *Thermodynamic Research Center Data Project*, Chemistry Department, Texas A&M University, College Station, TX 77843.
   The emphasis is on spectra of petrochemicals and other major industrial chemicals.
9. *Characteristic Raman Frequencies of Organic Compounds*, F. E. Dollish, W. G. Fateley, and F. F. Bentley, New York: Wiley–Interscience, 1973.
   This work includes 108 representative Raman spectra.
10. *Introductory Raman Spectroscopy*, J. R. Ferraro and K. Nakamoto, San Diego: Academic Press, 1994.
    This book contains Raman spectra of 30 solvents.
11. *Analytical Chemistry*, **19**, 700–765 (1947) and **22**, 1074–1114 (1950).
    These two articles contain Raman spectra with tables of frequencies and relative intensities of 291 hydrocarbons and oxygenated compounds. The spectra were obtained using mercury vapor lamp excitation.
12. *Ramanspektren*, K. F. W. Kohlrausch, London: Heyden and Sons Ltd., 1972.
    This work contains data on Raman spectra obtained using mercury vapor lamp excitation.
13. ACD/NIST IR Database. http://www.acdlabs.com/products/spec_lab/exp_spectra/spec_libraries/nist_ir.html
    This database contains 5200 gas-phase spectra.
14. Fiveash Data Management (FDM) Inc. IR Spectral Databases. http://www.acdlabs.com/products/spec_lab/exp_spectra/spec_libraries/fdm_ir.html
    More than 6000 spectra are presented.
15. S. T. Japan ATR/IR and FT-IR Databases. http://www.acdlabs.com/products/spec_lab/exp_spectra/spec_libraries/st_japan.html
    More than 80,000 spectra are divided into 60 separate databases.

To find a spectrum in a collection or in the original print literature, it is necessary to have a reference. The following list of indices is useful in this regard. The ultimate source is the *Chemical Abstracts Index*. The abstract gives the reference to the paper in which the spectrum was published. This procedure can be tedious, leading to many papers in which the complete spectrum may not be included.

1. *American Society for Testing and Materials (ASTM)*, distributed by Sadtler Research Laboratories, Inc., 3314 Spring Garden Street, Philadelphia, PA 19104.
   This source contains comprehensive indices for the IR spectra in the general collections listed above, plus IR spectra abstracted from technical journals through 1972. There is a molecular formula list and a serial number list, each with names and references to published IR spectra. There also is an alphabetical list of compound names, formulas, and references.

2. *Atlas of Spectral Data and Physical Constants for Organic Compounds*, 2nd ed., J. G. Grasselli and W. M. Ritchey (eds.), CRC Press, Inc., 2000 Corporated Blvd. NW, Boca Raton, FL 33431, 1975.

   This index contains coded IR spectra for 22,000 compounds. It lists strong bands in the infrared and includes Raman, UV, NMR, and mass spectral data when available.
3. H. M. Hershenson, *Infrared Absorption Spectra* (Indices for 1945–1957 and 1958–1962), New York: Academic Press, 1959 and 1964.

   There are 36,000 references to IR absorption spectra. The indices are alphabetic, and references are made to 66 journals and one collection of spectra.

Problems

12-1 How many normal modes of vibration does a molecule with 12 atoms have? Why does the IR spectrum of 1,3,5-trifluorobenzene (with 12 atoms) contain only 10 bands due to fundamentals? How could frequencies be obtained for the other 10 fundamentals? Why are there only 20 fundamentals for this molecule?

12-2 Carbon dioxide has three fundamental vibrations: 2350 cm^{-1} (antisymmetric C═O stretch), 1335 cm^{-1} (symmetric C═O stretch), and 667 cm^{-1} (degenerate O═C═O bending). Suggest an explanation for the observation that the Raman spectrum of CO_2 gas contains two bands of equal intensity near 1385 and 1285 cm^{-1}.

12-3 List the effects of hydrogen bonding on the IR spectra of compounds such as phenols, carboxylic acids, and amides.

12-4 The normal C═O stretching frequency in organic compounds occurs at approximately 1730 cm^{-1}. Explain why the frequencies of the C═O stretching bands in the IR spectra of acid chlorides and amides are observed at 1800 and 1650 cm^{-1}, respectively.

12-5 Write out the structures of alanine, ethyl carbamate, and 1-nitropropane, all with formula $C_3H_7O_2N$. Assign some features of the spectra given below to functional groups in these molecules.

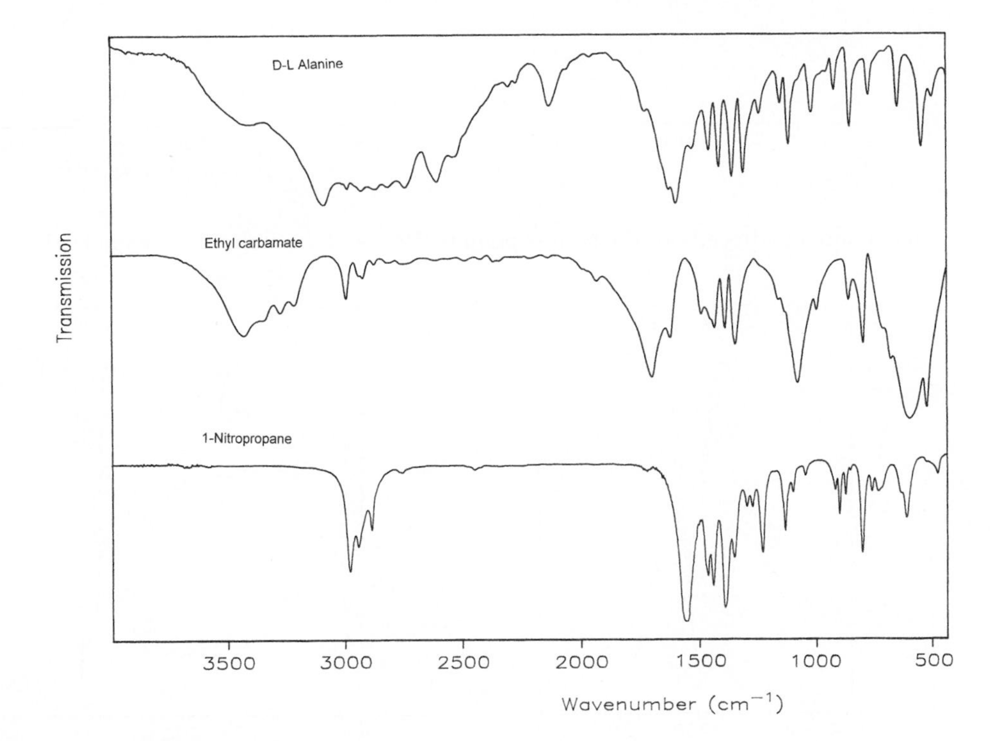

12-6 The compounds represented in the following spectra contain only carbon and hydrogen. Identify the type of compound and suggest possible structures.

(a) A volatile liquid hydrocarbon. The molecular weight determined by mass spectrometry is 84.

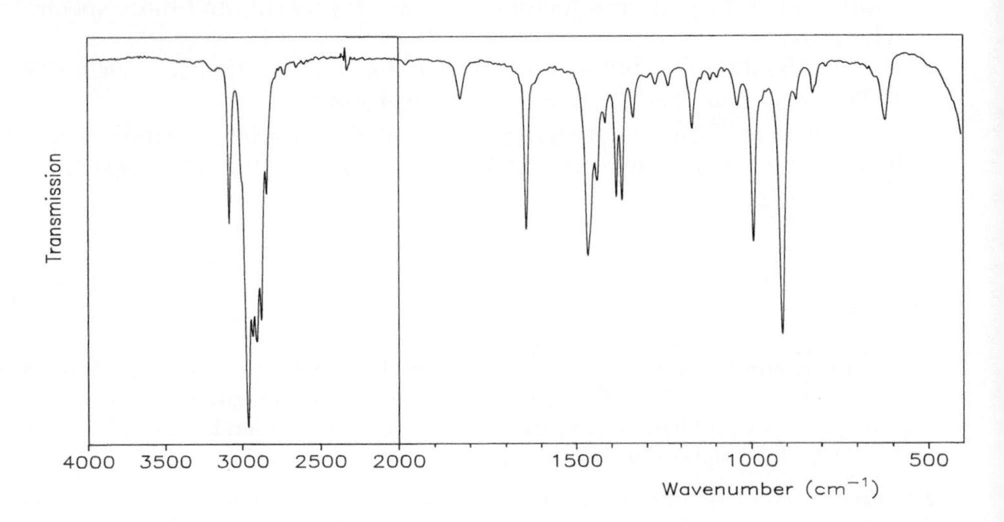

(b) A liquid hydrocarbon with the molecular formula C_7H_{14}.

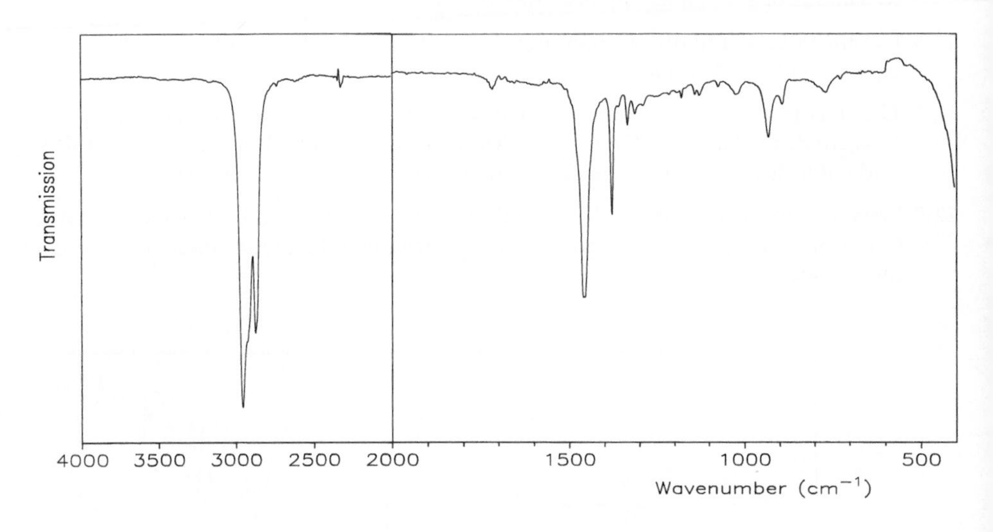

(c) A liquid hydrocarbon. The boiling point is 159°C and the molecular weight is 120.

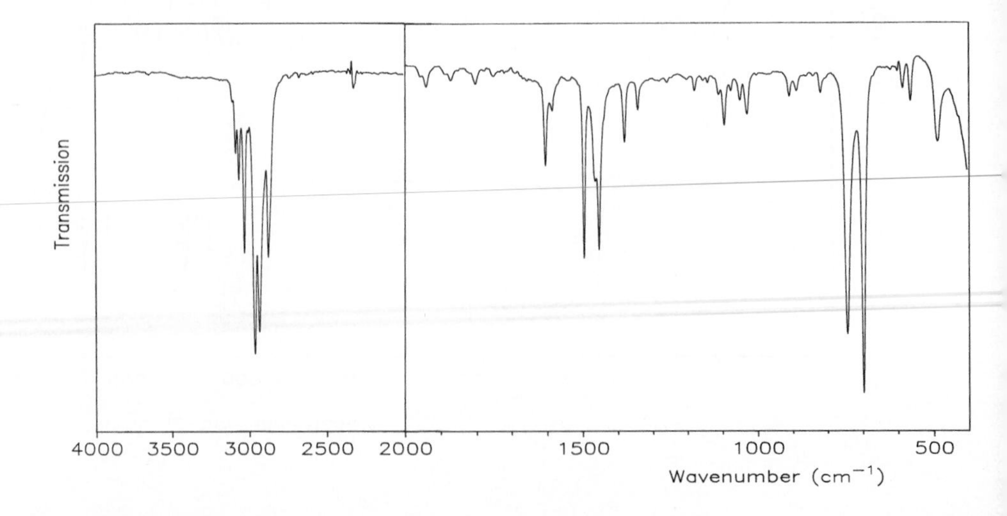

(d) A liquid hydrocarbon with a molecular weight of 104.

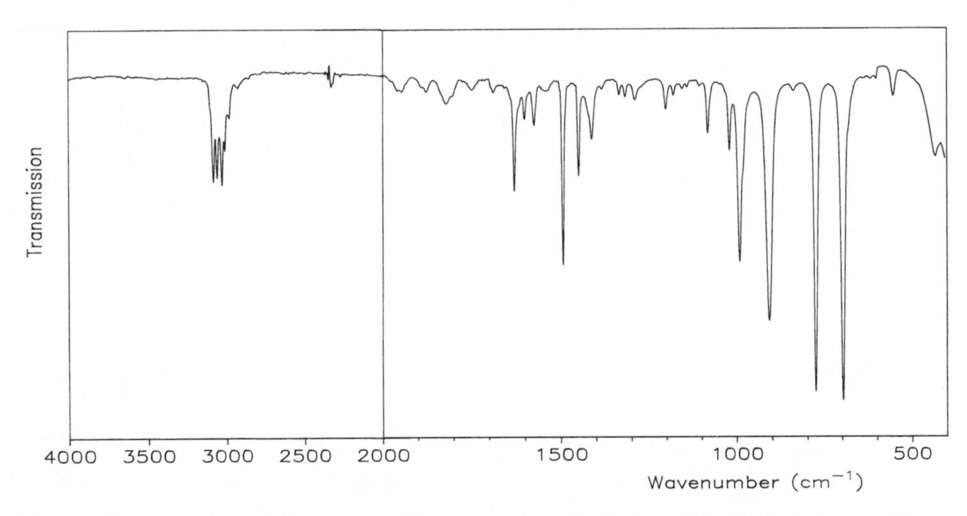

12-7 The following spectra are of compounds containing only C, H, and O. Each compound contains only one kind of functional group involving oxygen. Consult Sections 12-7 and 12-12, as well as the sections on carbon–hydrogen vibrations to deduce the structures.

(a) A liquid with a high boiling point (206°C) and a molecular weight of 138.

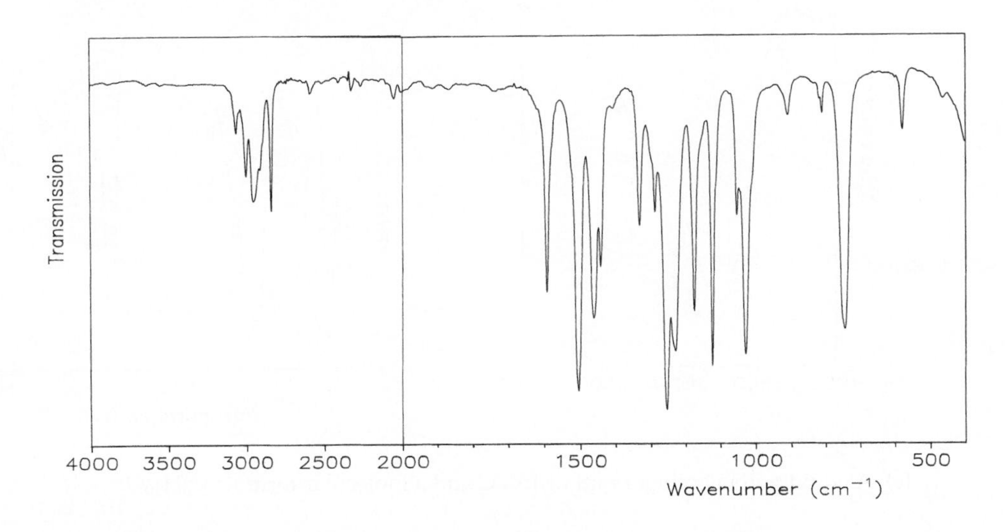

(b) A liquid compound with a molecular weight of 100.

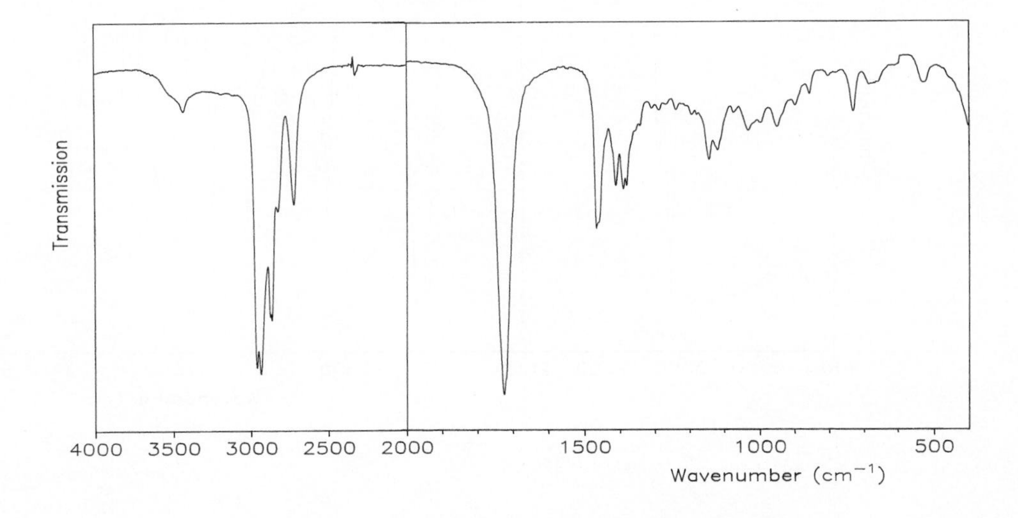

(c) A liquid compound with a molecular weight of 74.

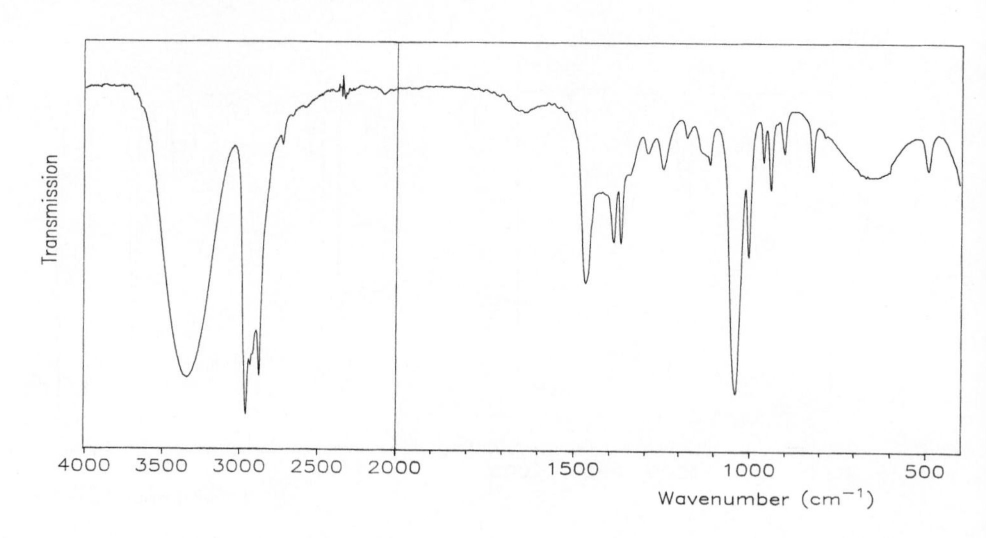

(d) A solid with a low melting point (43°C) and a molecular weight of 122.

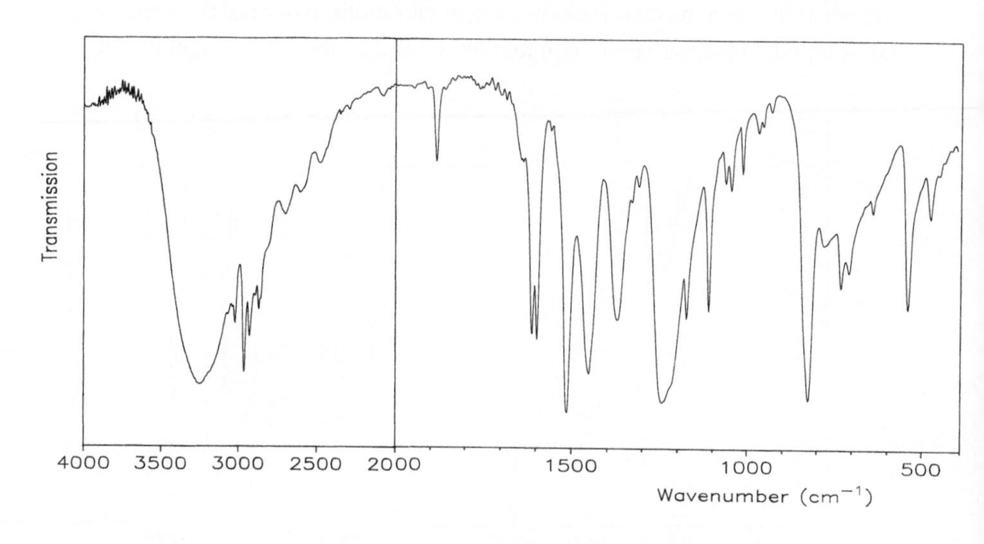

(e) A liquid with a boiling point of 155°C and a molecular formula $C_6H_{10}O$.

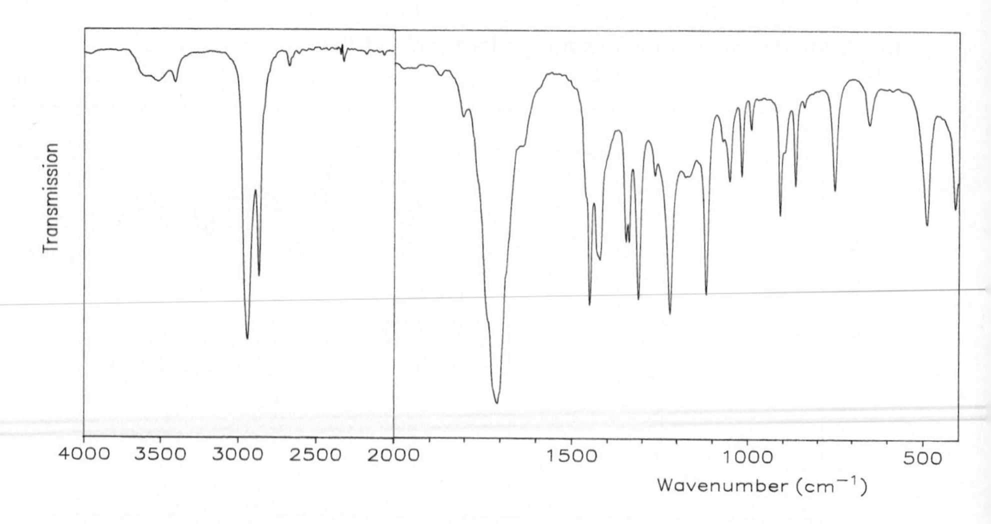

(f) A solid with a low melting point (50–52°C) and molecular weight of 164.

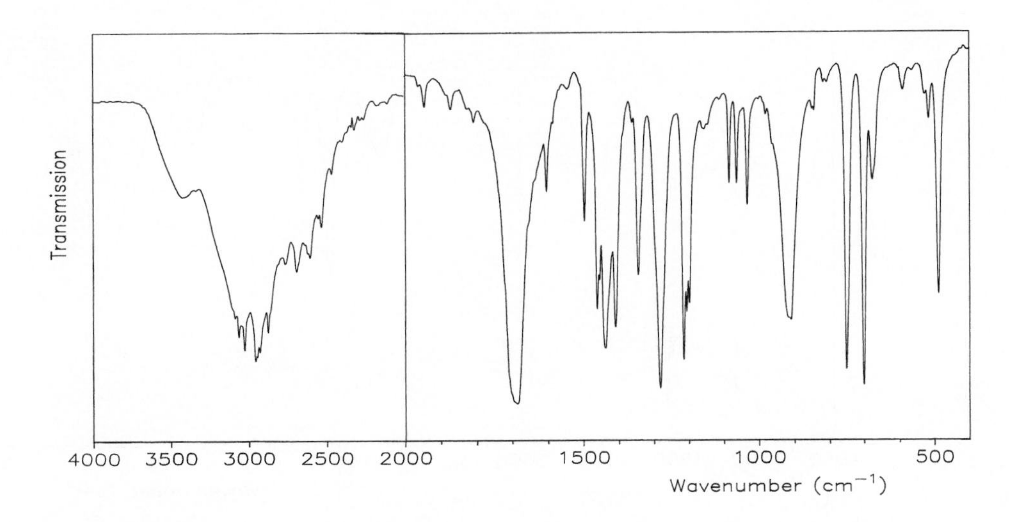

(g) An aromatic compound containing only C, H, and O.

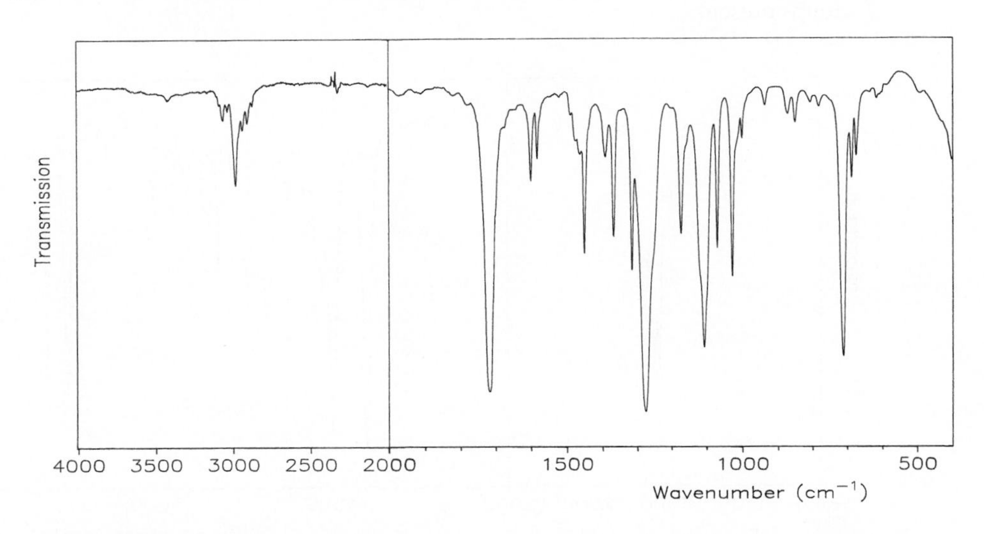

12-8 Compounds containing only C, H, and N yielded the following IR spectra. In each compound, there is only one kind of nitrogen-containing functional group. Consult Section 12-13, as well as the sections on carbon–hydrogen vibrations, to deduce the structures.

(a) A compound with a molecular weight of 103.

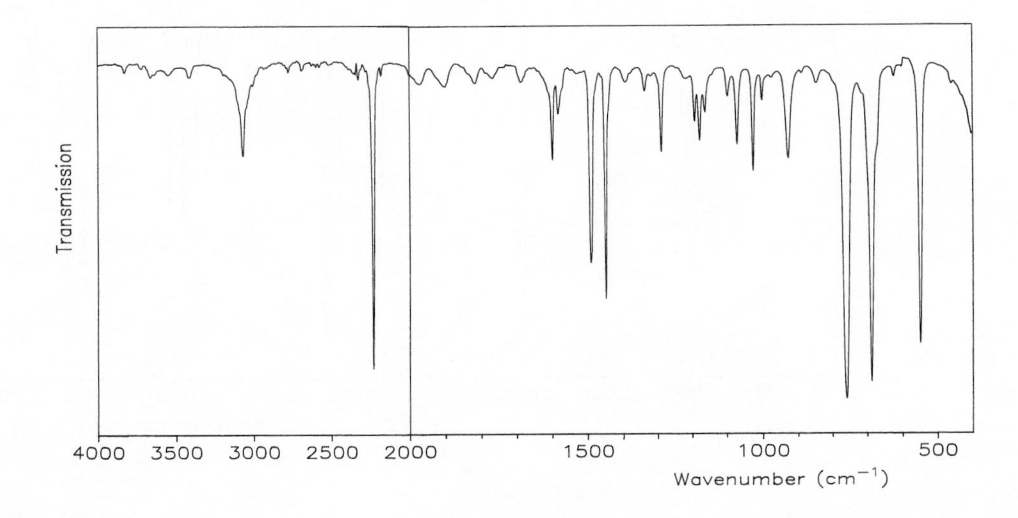

(b) A liquid compound with a molecular weight of 101.

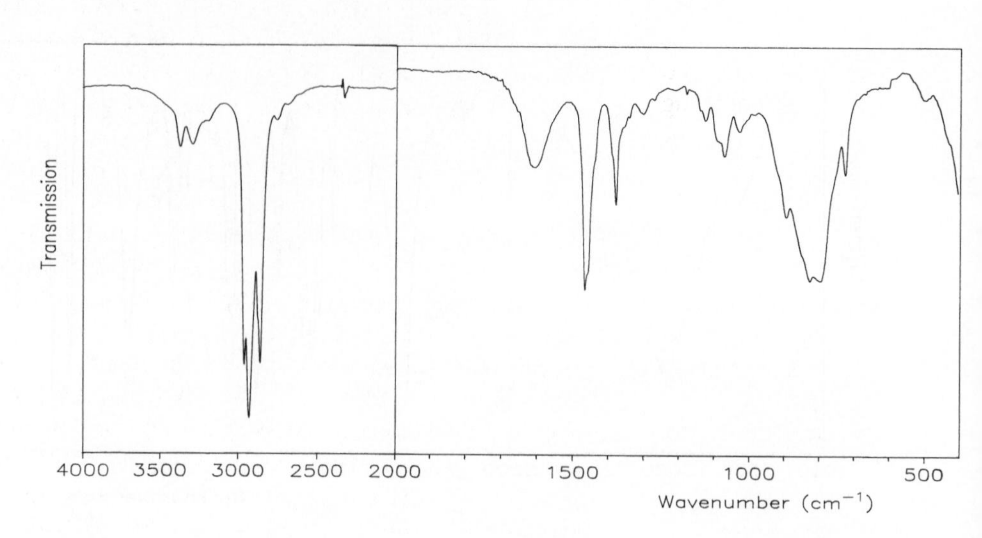

(c) A compound with a molecular formula $C_{10}H_{15}N$. Try first to deduce the functional groups present.

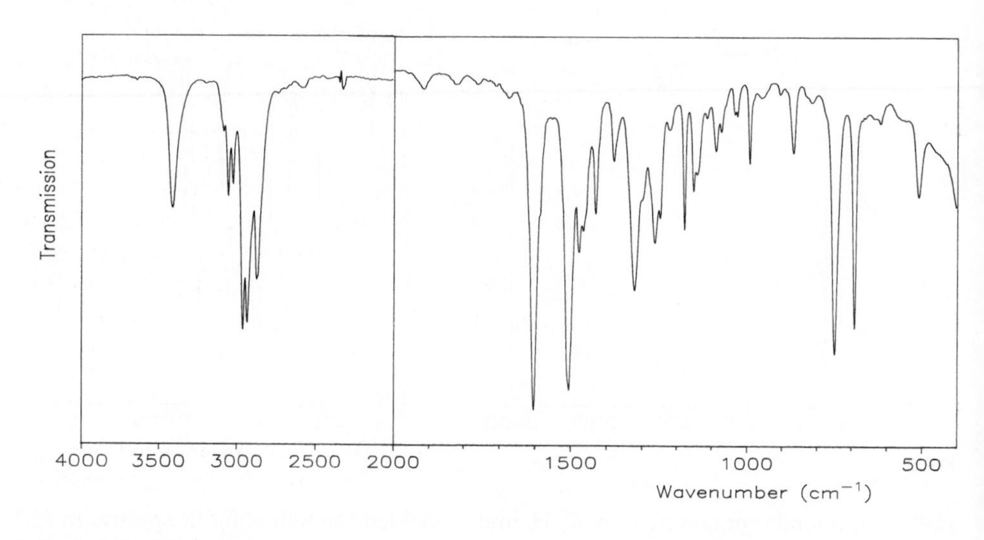

(d) A compound with a molecular weight of 79.

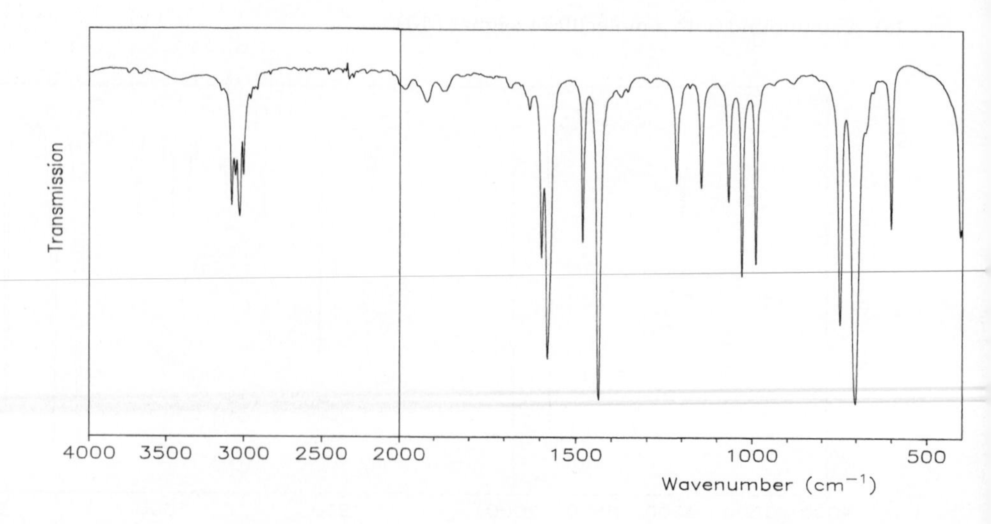

12-9 The spectra that follow are of compounds containing both oxygen and nitrogen in addition to carbon and hydrogen. Suggest a structure for each compound.

(a) A KBr pellet of a compound with the molecular formula $C_3H_7O_3N$.

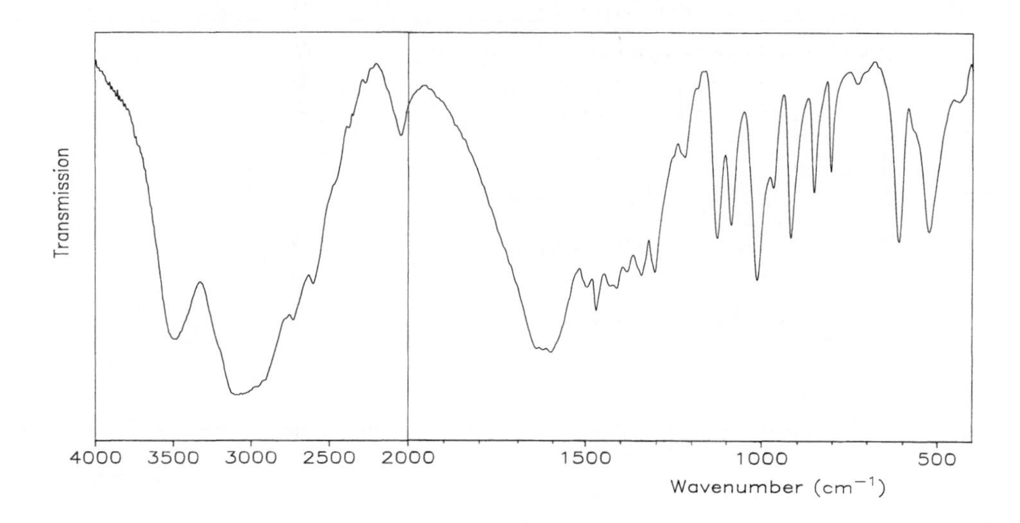

(b) A thin film of a compound with a molecular weight of 153.

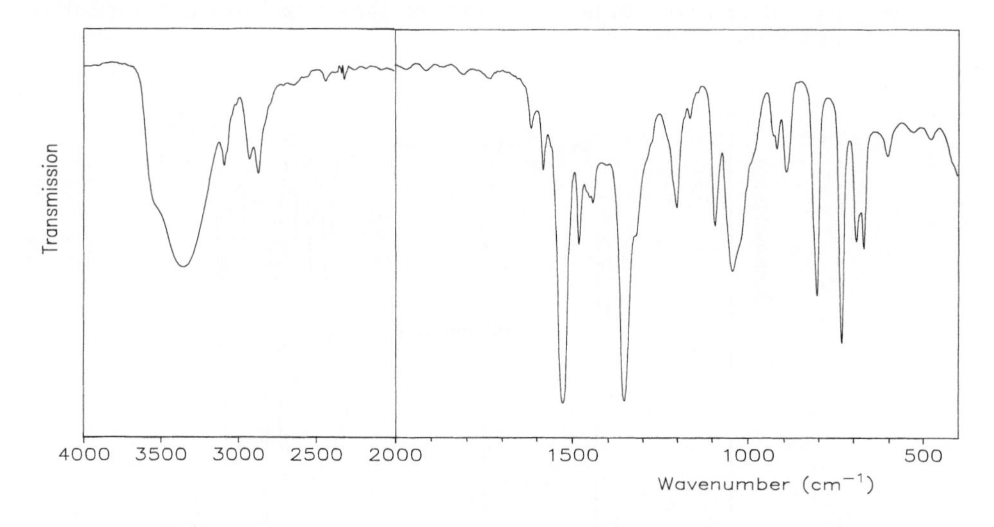

(c) A Nujol mull of a compound with a molecular weight of 87.

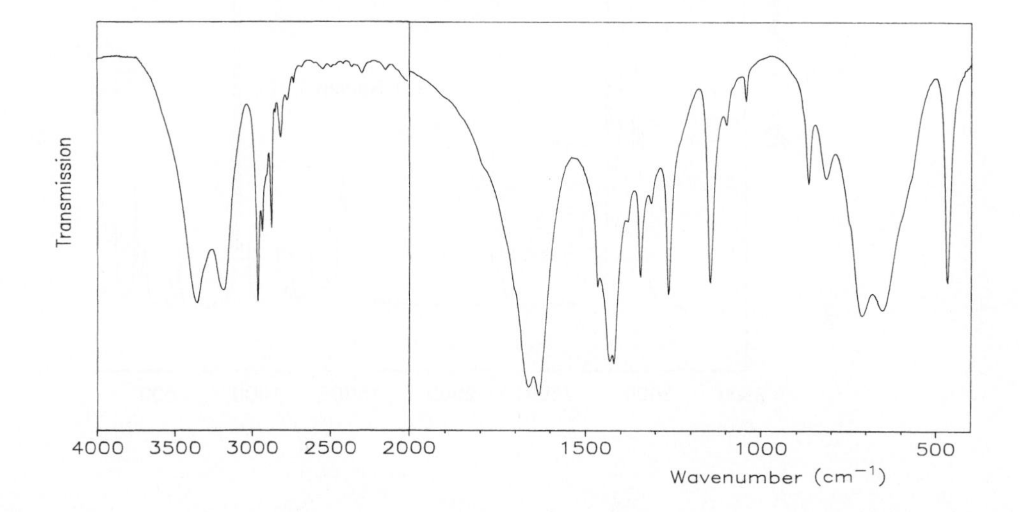

12-10 The spectrum below is of a compound that contains sulfur and oxygen in addition to carbon and hydrogen. The compound is a liquid (with a boiling point between 158 and 160°C) and a molecular weight of 138. Suggest possible structures.

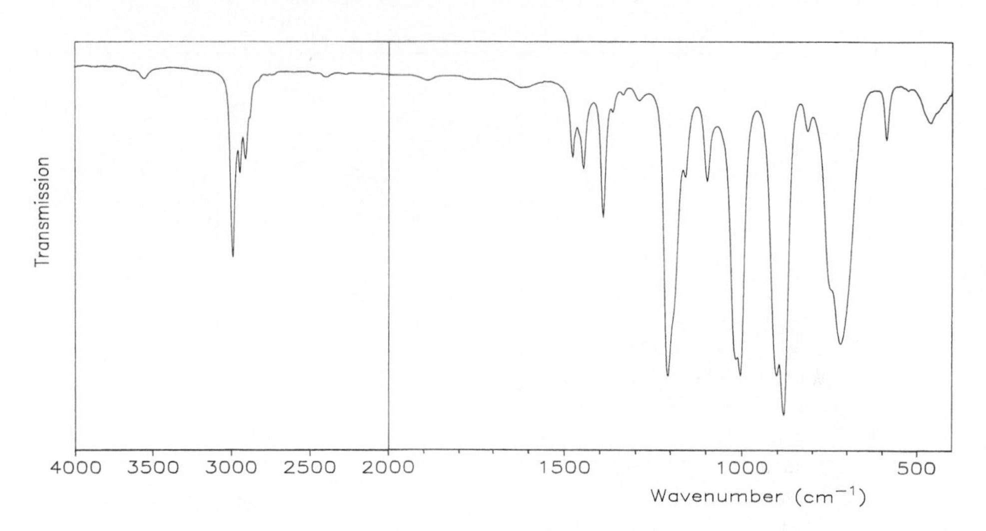

12-11 This problem illustrates that both IR (a) and Raman (b) spectra may be needed to deduce a structure. The compound has a molecular weight of 71 and contains oxygen, nitrogen, carbon, and hydrogen. Identify the compound from the strong Raman line near 2200 cm^{-1} and the bands in the IR spectrum.

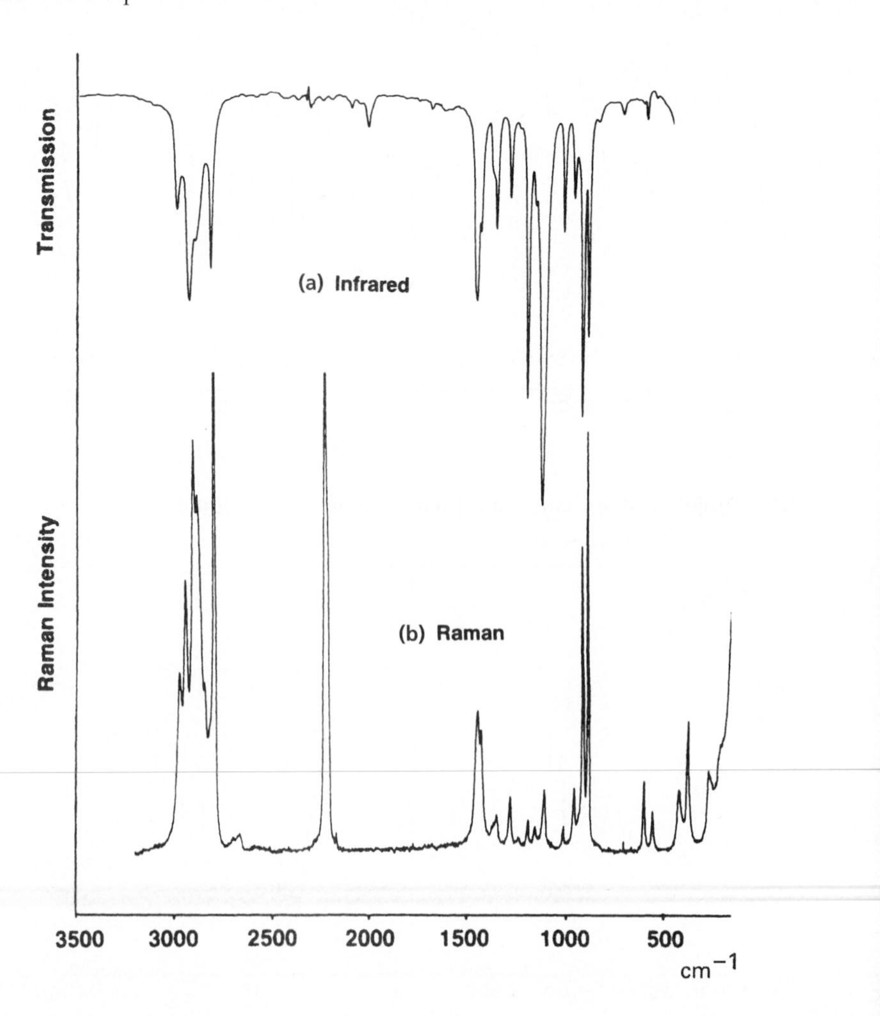

Bibliography

Infrared and Raman Group Frequencies

12.1 G. Socrates, *Infrared and Raman Characteristic Group Frequencies,* 3rd ed., Chichester, UK: John Wiley & Sons Ltd., 2001.

12.2 L. J. Bellamy, *The Infrared Spectra of Complex Molecules,* 3rd ed., London: Chapman & Hall, 1975.

12.3 N. P. G. Roeges, *A Guide to the Complete Interpretation of Infrared Spectra of Organic Structures,* Chichester, UK: John Wiley & Sons Ltd., 1994.

12.4 L. J. Bellamy, *Advances in Infrared Group Frequencies,* 2nd ed., London: Chapman & Hall, 1980.

12.5 H. A. Szymanski and R. F. Erickson, *Infrared Band Handbook* (2 vols.), New York: IFI/Plenum Press, 1970.

12.6 D. Dolphin and A. Wick, *Tabulation of Infrared Spectra Data,* New York: John Wiley & Sons Ltd., 1977.

12.7 B. H. Stuart, *Infrared Spectroscopy: Fundamentals and Applications,* New York: John Wiley & Sons, 2004.

12.8 F. E. Dollish, W. G. Fateley, and F. F. Bentley, *Characteristic Raman Frequencies of Organic Compounds,* New York: Wiley–Interscience, 1974.

12.9 M. C. Tobin, *Laser Raman Spectroscopy,* New York: Krieger Publishing Co., 1981.

12.10 P. Hendra, C. Jones, and G. Warnes, *Fourier Transform Raman Spectroscopy,* Chichester, UK: Ellis Horwood, 1991.

12.11 J. R. Ferraro and K. Nakamoto, *Introductory Raman Spectroscopy,* 2nd ed., New York: Academic Press, 2002.

12.12 E. Smith and G. Dent, *Modern Raman Spectroscopy: A Modern Approach,* Chichester, UK: John Wiley & Sons, 2005.

12.13 N. B. Colthup, L. H. Daly, and S. E. Wiberly, *Introduction to Infrared and Raman Spectroscopy,* 3rd ed., San Diego: Academic Press, 1990.

Symmetry Effects

12.14 F. A. Cotton, *Chemical Applications of Group Theory,* 3rd ed., New York: Wiley–Interscience, 1990.

Hydrogen Bonding

12.15 G. C. Pimentel and A. L. McClellan, *The Hydrogen Bond,* San Francisco: WH Freeman and Co., 1960.

Interpretation of Spectra

12.16 D. Lin-Vien, N. B. Colthup, W. G. Fateley, and J. G. Grasselli, *The Handbook of Infrared and Raman Characteristic Frequencies of Organic Molecules,* San Diego: Academic Press, 1991.

12.17 R. R. Hill and D. A. E. Rendell, *The Interpretation of Infrared Spectra: A Programmed Introduction,* London: Heyden & Son Ltd., 1975.

12.18 K. Nakamoto, *Infrared Spectra of Inorganic and Coordination Compounds,* 4th ed., New York: John Wiley & Sons, Inc., 1986.

12.19 L. C. Thomas, *Interpretation of Infrared Spectra of Organophosphorus Compounds,* London: Heyden, 1975.

Sources of Interpreted Spectra and Problems

12.20 K. Nakanishi and P. H. Solomon, *Infrared Absorption Spectroscopy,* 2nd ed., San Francisco: Holden Day, 1977. (Contains 100 problems with detailed solutions.)

12.21 H. A. Szymanski, *Interpreted Infrared Spectra,* 3 vols., New York: Plenum Press, 1964, 1966, 1967.

12.22 T. Cairns, *Spectroscopic Problems in Organic Chemistry,* London: Heyden & Son Ltd., 1964.

12.23 A. J. Baker, T. Cairns, G. Eglinton, and F. J. Preston, *More Spectroscopic Problems in Organic Chemistry,* London: Heyden & Son Ltd., 1967.

12.24 D. Steele, *The Interpretation of Vibrational Spectra,* London: Chapman and Hall, 1971. (Contains 26 infrared [and other] spectra of organic molecules with interpretation.)

12.25 R. K. Smalley and B. J. Wakefield, "Infrared Spectroscopic Problems and Answers," in *An Introduction to Spectroscopic Methods for the Identification of Organic Compounds,* vol. 1, *Nuclear Magnetic Resonance and Infrared Spectroscopy,* F. Scheinmann (ed.), Oxford, UK: Pergamon Press, 1970. (The chapter contains 14 problems followed by detailed answers.)

12.26 R. Davis and C. H. J. Wells, *Spectral Problems in Organic Chemistry,* New York: Chapman & Hall, 1984. (Contains 56 problems based on IR, ^{1}H and ^{13}C NMR, and mass spectra, together with either analytical data or a molecular formula. Solutions are given in the form of references to the catalog of The Aldrich Chemical Co., Inc., and other commonly available listings of organic compounds.)

PART

IV

Electronic Absorption Spectroscopy

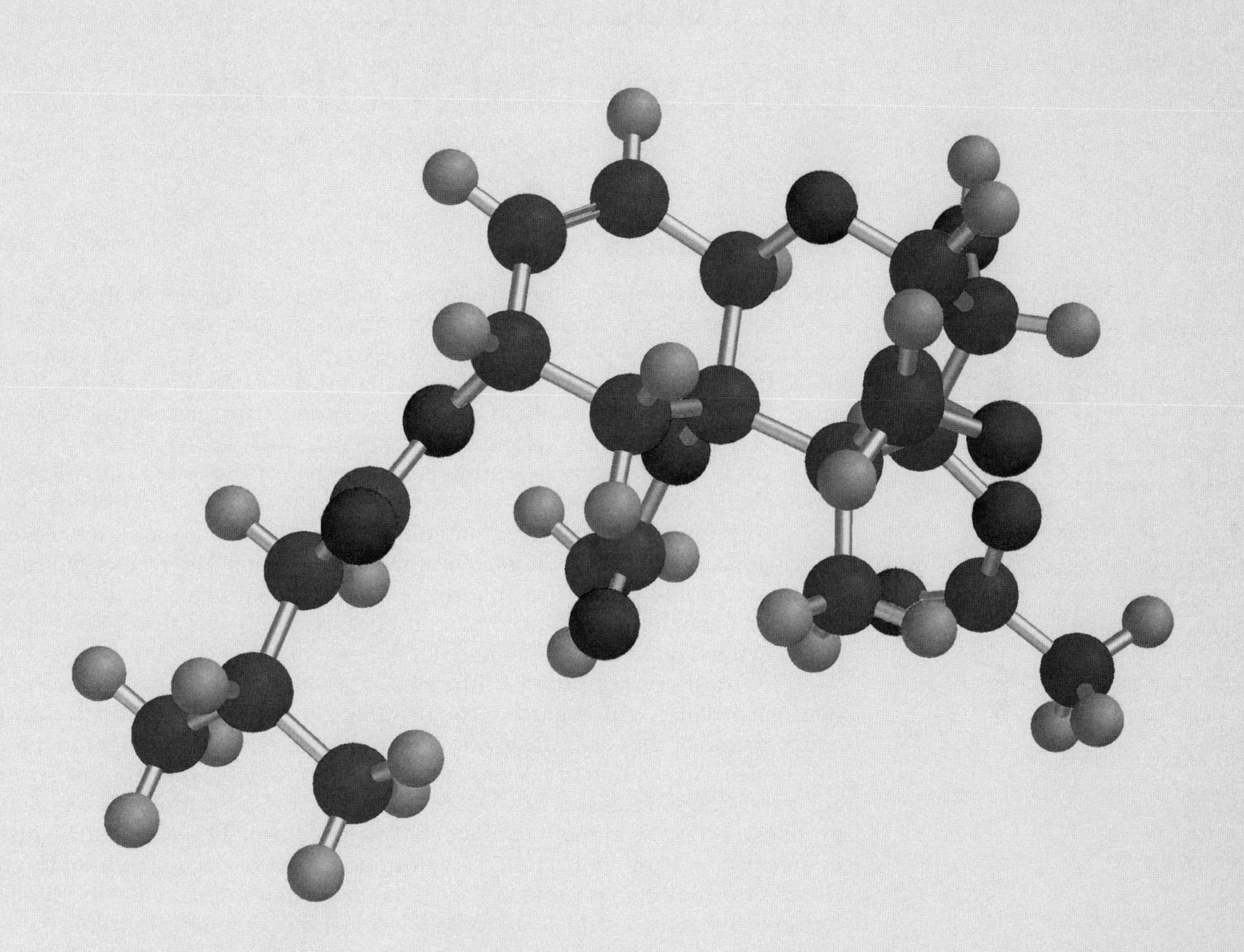

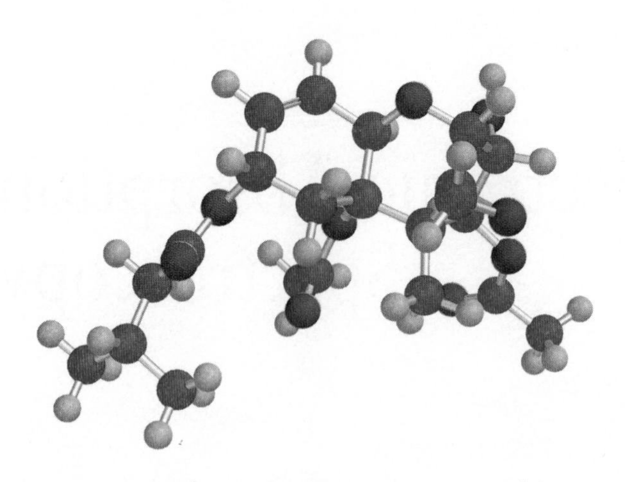

CHAPTER

13

Introduction and Experimental Methods

13-1 INTRODUCTION

Spectroscopy of the ultraviolet (UV) and visible (vis) regions of the electromagnetic spectrum was among the earliest instrumental techniques used in organic structure determination and was predated only by refractive index and optical rotation measurements. UV–vis spectroscopy is used to detect the presence of certain functional groups called *chromophores*, especially in conjugated systems such as aromatics, dienes, polyenes, and α,β-unsaturated ketones.

UV–vis spectroscopy is a prominent member of the broad classification of electronic spectroscopies. It involves measurement of the energy and intensity of ordinary light absorption for any organic compound. Although metal complexes also can absorb UV–vis light, these compounds are not included herein. The results typically are displayed as a plot of absorbance intensity (A) on the vertical axis and the wavelength of light (λ, in nanometers) on the horizontal axis, giving UV–vis absorption curves such as that shown in Figure 13-1 for 3-methylcyclohexanone.

Electronic spectra provide information about the electronic properties of molecules, particularly with regard to conjugation. All organic compounds absorb light in the UV region of the electromagnetic spectrum, and some absorb light in the visible region as well. Absorption of UV or visible light occurs only when the energy of incident radiation is the same as that of an electronic transition in the molecule studied. Such absorption of energy is termed *electronic excitation* and usually is associated with moving a single electron from an occupied to an unoccupied molecular orbital (Figure 13-2), whereby the molecule is promoted from the molecular ground state to a higher energy, electronically excited state. In a given molecule many different electronic transitions are possible. Those important in organic chemistry often involve promoting an electron

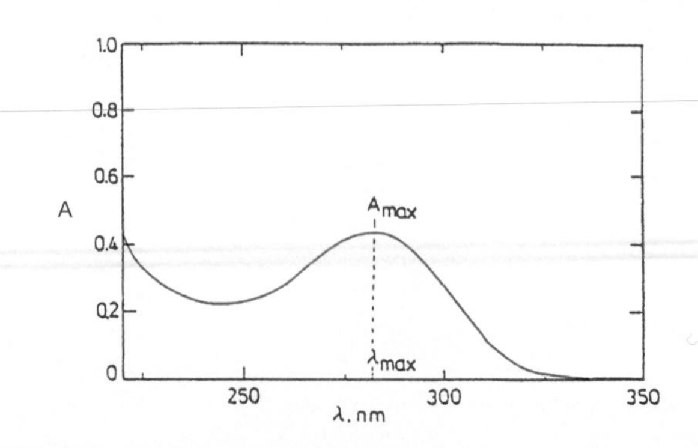

FIGURE 13-1 UV–vis spectrum of 3-methylcyclohexanone, 0.0245 M in methanol, run in a 1.0 cm cuvette. (Spectrum obtained by P. M. Sabido.)

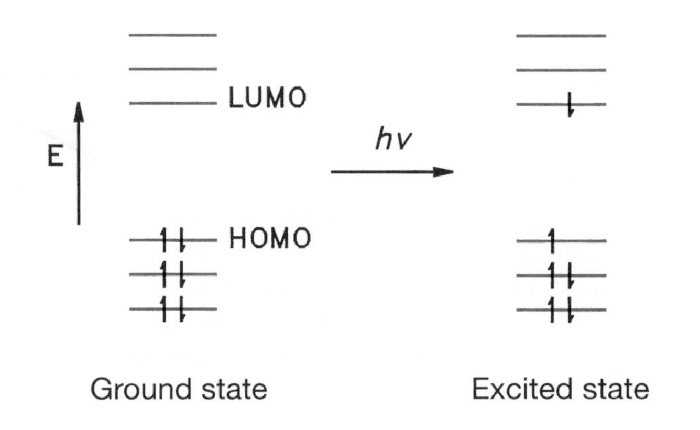

FIGURE 13-2 Idealized representation on a potential energy scale of occupied and unoccupied molecular orbitals in the electronic ground state (left) and electronic configuration of an excited state arising by promotion of an electron from the highest occupied molecular orbital to the lowest unoccupied molecular orbital (right). The electrons and their relative spin orientations are represented by small arrows.

from the **H**ighest **O**ccupied bonding or nonbonding **M**olecular **O**rbital (HOMO) to the **L**owest **U**noccupied **M**olecular **O**rbital (LUMO).

Although electronic transitions arise between ground and excited states of the *entire* molecule, in UV–vis absorption most of the action can usually be assigned to parts of the molecule (*chromophores*) in which valence electrons are found, such as the nonbonding (n) or π electrons. One speaks of an electronic transition in a chromophore, which in organic molecules is typically a functional group such as a carbonyl group, a carbon–carbon double bond, or an aromatic ring. Representative chromophores are shown in Table 13-1, along with their λ_{max}, the wavelength at which light absorption reaches a maximum, and ϵ_{max}, the molar absorptivity constant at which light absorbance reaches a maximum (Figure 13-1). Despite the implication of the term, chromophores ("bearing color") may absorb light in the UV as well as the visible part of the spectrum, and not necessarily convey color (Greek *chroma*).

13-2 MEASUREMENT OF ULTRAVIOLET–VISIBLE LIGHT ABSORPTION

The UV–vis spectrum represents absorption of light as a plot (see Figure 13-1) of energy (usually reported in organic chemistry as wavelength λ, from $E = hc/\lambda$) vs. the intensity

TABLE 13-1 Electronic Absorption Data for Isolated Chromophores

Chromophore	Example	Solvent	λ_{max} (nm)[a]	ϵ (liter mol^{-1} cm^{-1})
C=C	1-Hexene	Heptane	180	12,500
—C≡C—	1-Butyne	Vapor	172	4,500
(benzene ring)	Benzene	Water	254	205
			203.5	7,400
	Toluene	Water	261	225
			206.5	7,000
C=O	Acetaldehyde	Vapor	298	12
			182	10,000
	Acetone	Cyclohexane	275	22
			190	1,000
	Camphor	Hexane	295	14
—COOH	Acetic acid	Ethanol	204	41
—COCl	Acetyl chloride	Heptane	240	34
—COOR	Ethyl acetate	Water	204	60
—$CONH_2$	Acetamide	Methanol	205	160
—NO_2	Nitromethane	Hexane	279	16
			202	4,400
>C=N—	$C_2H_5CH{=}NC_4H_9$	Isooctane	238	200

Adapted from J. B. Lambert, H. F. Shurvell, L. Verbit, R. G. Cooks, and G. H. Stout, *Organic Structural Analysis*, New York: Macmillan Publishing, 1976.

[a]Chromophores often have more than one absorption band.

of absorption (as absorbance, A, or as molar extinction coefficient, ϵ, which is a rough measure of the transition probability). The wavelength at maximum absorbance for each electronic transition is termed λ_{max}.

13-2a Wavelength and λ_{max}

Electromagnetic radiation (see Section 1-2) may be described by the wavelength λ, by the frequency ν (s^{-1}), or by the wavenumber $\bar{\nu}$ (cm^{-1}) ($\lambda\nu = c$, the velocity of light; $\bar{\nu} = 1/\lambda$). Commonly used wavelength units in UV and visible light regions are nanometers (nm) and Ångstroms (Å). According to Planck's equation ($\Delta E = h\nu$), frequency is directly proportional to energy. Table 13-2 lists the units commonly used for λ and ν, and Table 13-3 gives some useful conversion factors. Eq. 13-1,

$$\Delta E = \frac{hc}{\lambda} = \frac{28{,}600}{\lambda}\ \text{kcal mol}^{-1} = \frac{120{,}000}{\lambda}\ \text{kJ mol}^{-1}\ (\text{for } \lambda \text{ in nm}) \tag{13-1}$$

is convenient for calculation of energies in the familiar units of kcal mol^{-1}. Hence light of 300 nm wavelength corresponds to an energy of 95.4 kcal mol^{-1} or 399 kJ mol^{-1}, depending on the units of h.

Wavenumbers are directly proportional to energy, so that a given range of reciprocal centimeters (cm^{-1}) represents the same energy anywhere in the electromagnetic spectrum. For example, a shift of $\bar{\nu}$ of 700 cm^{-1} anywhere in the spectrum corresponds to 2.00 kcal mol^{-1}. On the other hand, wavelength is inversely proportional to energy, and thus the relationship is not linear. For example, an energy change of 2.00 kcal mol^{-1} at 200 nm corresponds to a shift of 2.7 nm, but the same energy change at 800 nm corresponds to a shift of approximately 44 nm.

The UV region is at the lower end of the visible spectrum, below 400 nm. The UV spectrum may be divided into two parts: the near UV, 190 to 400 nm (53,000–25,000 cm^{-1}), and the far or vacuum UV, below 190 nm (>53,000 cm^{-1}). This division is appropriate because atmospheric oxygen begins to absorb around 190 nm and obscures absorption by organic molecules. Oxygen must be removed from the spectrophotometer, either by using a vacuum instrument or by vigorous purging with nitrogen.

13-2b The Beer–Lambert–Bouguer Law and ϵ_{max}

The laws of Beer, Lambert, and Bouguer (or, more simply, Beer's law) state that at a given wavelength the proportion of light absorbed by a transparent medium is

TABLE 13-2 Definitions of Terms and Equations

Quantity	Unit of Measure	Dimensions
Wavelength (λ)	Nanometer (nm)	1 nm = 10^{-9} m
	Ångstrom (Å)	1 Å = 10^{-10} m
Frequency (ν)	Hertz (Hz) or s^{-1}	Cycles per second (s^{-1})
Energy	Depends on the units of h	6.626×10^{-27} erg s
		6.626×10^{-34} J s
		9.534×10^{-14} kcal–s mol^{-1}
		1.583×10^{-34} cal–s $molec^{-1}$

TABLE 13-3 Useful Conversion Factors

cm^{-1}	Hz	kcal mol^{-1}	kJ mol^{-1}
1	3.00×10^{10}	2.86×10^{-3}	1.20×10^{-2}
3.33×10^{-11}	1	9.53×10^{-14}	3.99×10^{-13}
3.50×10^{2}	1.05×10^{13}	1	4.18

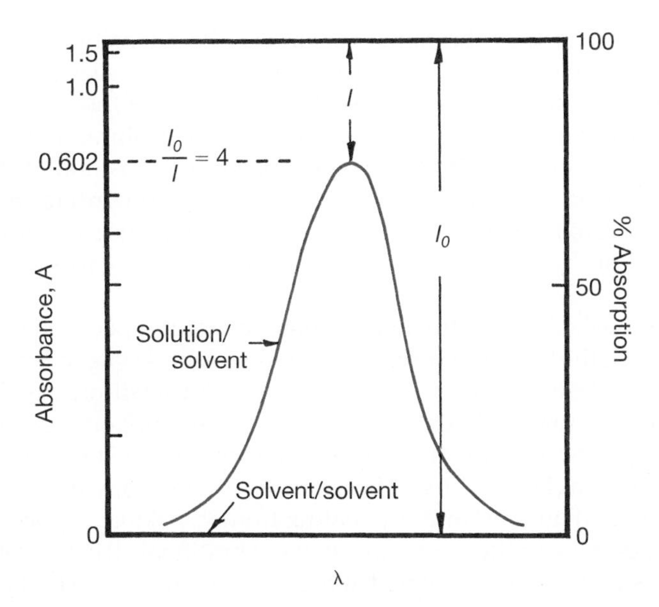

FIGURE 13-3 Measurement of solute absorbance A by a double-beam spectrophotometer.

independent of the intensity of the incident light and is proportional to the number of absorbing molecules through which the light passes. These laws are expressed by eq. 13-2,

$$I = I_0 e^{-kl} \quad \text{or} \quad I = I_0 10^{-\epsilon cl} \quad \text{or} \quad \log(I_0/I) = \epsilon cl \tag{13-2}$$

in which I_0 is the intensity of incident light, I the intensity of transmitted light, k the absorption coefficient, and l the pathlength (cm). The absorption coefficient k is related to the more common molar extinction coefficient (ϵ), or molar absorptivity, by the equation $k \approx 2.303\ \epsilon c$, in which c is the concentration in mol liter^{-1}. Since $\log(I_0/I)$ is defined as the absorbance (A) and A is the quantity actually measured, eq. 13-2 is rewritten as eq. 13-3,

$$A = \epsilon lc \quad \text{or} \quad A = l\sum_i \epsilon_i c_i \tag{13-3}$$

for i absorbing species. The units of ϵ (cm^2 mol^{-1} or liter mol^{-1} cm^{-1}) usually are omitted.

In practice, the quantities actually measured are the relative intensities of the light beams transmitted by a reference cell containing pure solvent and by an identical cell containing a solution of the analyte. When the respective intensities are taken as I_0 and I, the resulting absorption is that of the dissolved solute only (Figure 13-3). One also can see from Figure 13-3 that ϵ is different at different λ, and one refers to ϵ_{max} at λ_{max} for a given absorption.

13-2c Shape of Absorption Curves: The Franck–Condon Principle

Absorption of UV–vis light is typically recorded as broad absorption maxima (Figure 13-1) and not as single, sharp lines representing the absorption in an extremely narrow energy range. The absorption curves are broadened because the electronic levels have vibrational levels superimposed on them.

For simplicity, let us look at the ground and excited electronic states of a diatomic molecule. In the more common case, the bond strength in the excited electronic state is less than that in the ground state, and the equilibrium internuclear distance is longer than in the ground state. A typical potential energy diagram is shown in Figure 13-4.

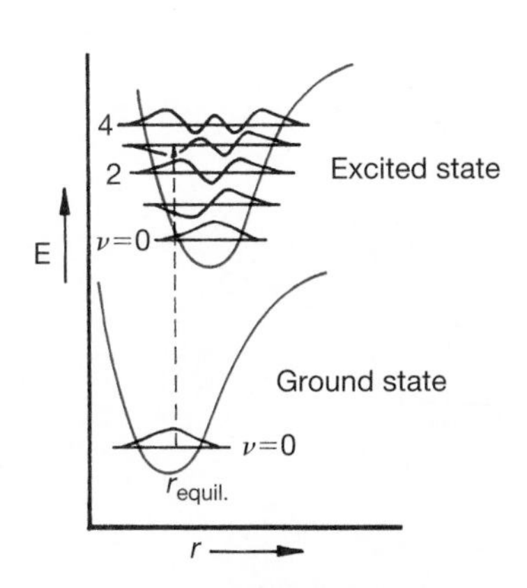

FIGURE 13-4 Potential energy diagram for a diatomic molecule illustrating Franck–Condon excitation. The equilibrium separation is longer in the excited than in the ground state.

Most molecules exist mainly in their ground vibrational state at room temperature. Excitation can occur to any of the excited state vibrational levels, so that the absorption due to the electronic transition consists of a large number of lines. In practice, the lines overlap and a continuous, broad band is observed. Hence, the shape of an absorption band is determined by the spacing of the vibrational levels and by the distribution of the total band intensity over the vibrational subbands. The intensity distribution is determined by the *Franck–Condon principle,* which states that *nuclear motion is negligible during the time required for an electronic excitation.* Another statement of the Franck–Condon principle based on classical mechanics is that the most probable vibrational component of an electronic transition is one that involves no change in the position of the nuclei, a so-called *vertical transition.* As represented by the vertical arrow in Figure 13-4, the most probable transition is to the excited $\nu = 3$ vibrational state. This state has a maximum at the same internuclear distance (r) as that corresponding to the starting point of the transition.

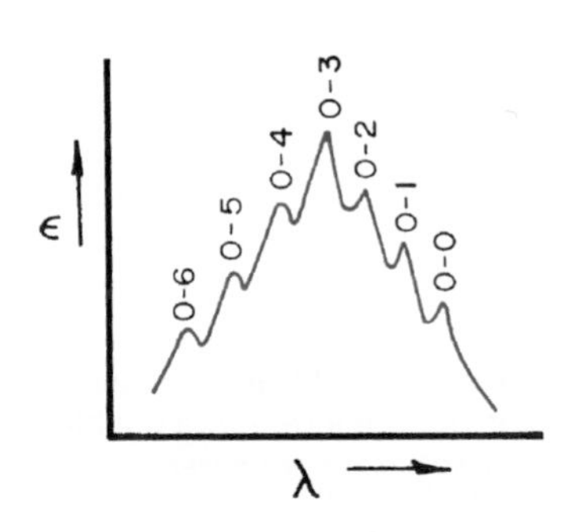

FIGURE 13-5 Intensity distribution among vibronic bands as determined by the Franck–Condon principle.

Figure 13-5 shows the vibrational-electronic (vibronic) spectrum corresponding to Figure 13-4, in which the 0–3 band (from $\nu = 0$ in the ground state to $\nu = 3$ in the excited state) is the most intense spike. Note that the other transitions, including the 0–0 band, have significant probabilities. This situation occurs because, even in the ground electronic state (zeroth vibrational level), the internuclear distance is described by a probability distribution (Figure 13-4). Therefore, transitions may originate over a range of r values, and more than one band originating from $\nu = 0$ may be observed.

Sometimes, on raising the temperature, the vibrational structure of a band is lost. More vibrational states can be populated at higher temperature, so that a larger number of possible vibrational transitions can occur on electronic excitation. Featureless or broad bands also are observed at ambient temperatures, usually in solution spectra, as in Figure 13-1, because solute–solvent vibrational interactions become important.

13-2d Solvent Effects and λ_{max} Shifts

Moving an electron from the ground state to an excited state configuration typically leads to an excited state that is more polar than the ground state and more sensitive to solvation effects. The least polar solvents (e.g., hydrocarbons) have the least effect on UV–vis spectra and are least effective in inhibiting vibrational fine structure (Figure 13-5). Polar solvents, in contrast, interact more strongly with solutes. They tend to smooth out vibrational structure on the absorption bands and to cause spectral shifts.

In polar solute molecules, the excited state usually receives predominant contributions from highly polar structures, such as $R_2C^+—O^-$ in the $\pi \rightarrow \pi^*$ state of ketones, and is more polar than the ground state. In accord with simple electrostatic theory, such polar excited states are expected to be stabilized by polar solvents, thereby facilitating electronic excitation and resulting in a red (bathochromic) shift of the absorption band. Only in the unlikely event that the ground state is more polar than the excited state could one expect to find a blue (hypsochromic) shift with increasing solvent polarity.

Consider the case of the α,β-unsaturated ketone 4-methyl-3-penten-2-one (mesityl oxide). Its UV–vis spectrum (Figure 13-4b) shows two readily accessible UV transitions: (1) a weak absorption near 320 nm ($\epsilon \sim 50$) associated with the promotion of a nonbonding (n) oxygen electron to an antibonding π^* orbital ($n \rightarrow \pi^*$ transition), and (2) a strong absorption near 230 nm ($\epsilon \sim 12{,}000$) associated with the promotion of a bonding π electron to an antibonding π^* orbital ($\pi \rightarrow \pi^*$ transition).

For $\pi \rightarrow \pi^*$ excited states (Section 13-4b), dipole–dipole interactions and hydrogen bonding with solvent molecules tend to lower the energy of the excited state more than the ground state, with the result that the λ_{max} *increases* (red shift) about 10 nm in going from a fairly noninteractive solvent like heptane to methanol (Figure 13-6). For $n \rightarrow \pi^*$ excited states (Section 13-4b), both the ground and the excited states are lowered in energy by dipole–dipole and hydrogen-bonding interaction with solvent. In hydrogen-bonding solvents, the ground state n electrons coordinate with the solvent more strongly than excited state n electrons, with the result that the λ_{max} *decreases* about 15 nm in going from heptane to methanol solvent (Figure 13-7). Solvent effects on $n \rightarrow \pi^*$ and $\pi \rightarrow \pi^*$ transitions are summarized for the example of mesityl oxide in Table 13-4.

Although solvent effects have been utilized historically to characterize absorbances as being, for example, $n \rightarrow \pi^*$ or $\pi \rightarrow \pi^*$, current computational methods often are more convenient and more useful. Time-Dependent Density Functional Theory (TDDFT) provides energy differences between ground and excited states and the

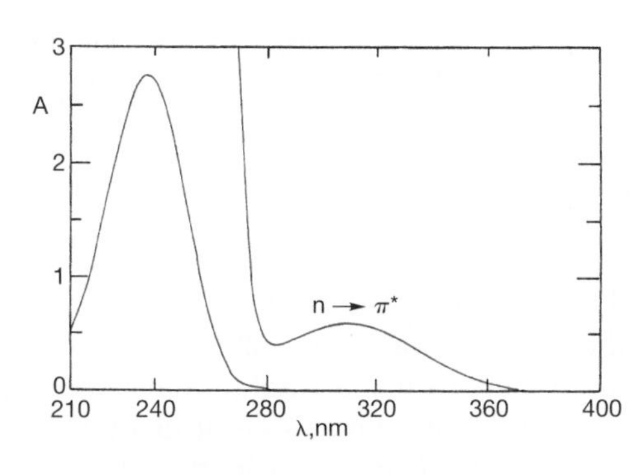

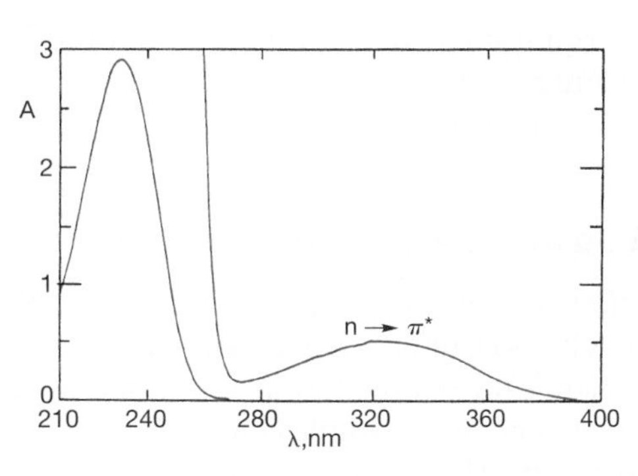

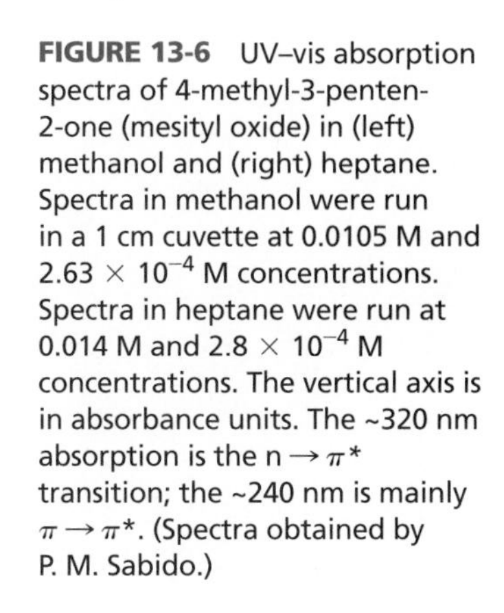

FIGURE 13-6 UV–vis absorption spectra of 4-methyl-3-penten-2-one (mesityl oxide) in (left) methanol and (right) heptane. Spectra in methanol were run in a 1 cm cuvette at 0.0105 M and 2.63×10^{-4} M concentrations. Spectra in heptane were run at 0.014 M and 2.8×10^{-4} M concentrations. The vertical axis is in absorbance units. The ~320 nm absorption is the $n \rightarrow \pi^*$ transition; the ~240 nm is mainly $\pi \rightarrow \pi^*$. (Spectra obtained by P. M. Sabido.)

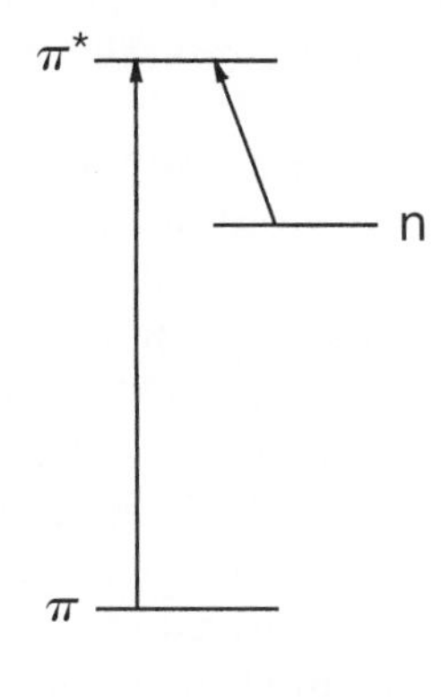

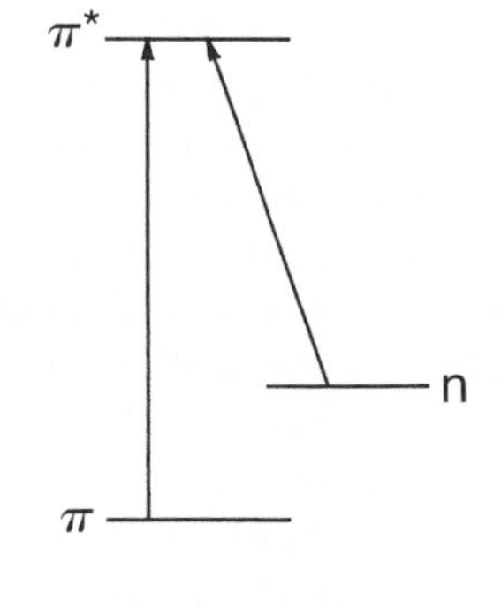

FIGURE 13-7 Influence of solvent on the orbitals involved in $\pi \rightarrow \pi^*$ and $n \rightarrow \pi^*$ electronic transitions.

intensities of the transitions, from which a simulated spectrum is obtained. Although excitations obtained by TDDFT usually are complex, they can be recast in the simple nomenclature described herein and identified, for example, as $\pi \rightarrow \pi^*$. Semiempirical methods such as AM1 have been used in a similar fashion.

13-3 QUANTITATIVE MEASUREMENTS

As explained in Section 13-2b, the proportion of light absorbed by a transparent medium is independent of the incident-light intensity but proportional to the number of

TABLE 13-4 Influence of Solvent on the UV λ_{max} and ϵ_{max} of the $n \rightarrow \pi^*$ and $\pi \rightarrow \pi^*$ Excitations of 4-Methyl-3-penten-2-one (Mesityl Oxide)

$(CH_3)_2C{=}CH{-}C({=}O){-}CH_3$

	$\pi \rightarrow \pi^*$ Transition		$n \rightarrow \pi^*$ Transition	
Solvent	λ_{max} (nm)	ϵ_{max} (liter mol^{-1} cm^{-1})	λ_{max} (nm)	ϵ_{max} (liter mol^{-1} cm^{-1})
Hexane	229.5	12,600	327	97.5
Diethyl ether	230	12,600	326	96
Ethanol	237	12,600	325	78
Methanol	238	10,700	312	74
Water	244.5	10,000	305	60

From H. H. Jaffé and M. Orchin, *Theory and Applications of Ultraviolet Spectroscopy*, New York: John Wiley & Sons, Inc., 1962.

absorbing molecules through which the light passes. When the absorptivity constants are large, such as $\epsilon \sim 10{,}000$, UV–vis spectroscopy becomes ideal for determining concentration using Beer's law (eq. 13-3).

13-3a Difference Spectroscopy

Difference spectroscopy is a sensitive method for detecting and recording changes in a UV–vis spectrum associated with a change in the chromophore, as in solvent perturbation or chemical reaction. In this technique, when a double-beam spectrophotometer is used, the reference and sample compartments each contain identical materials at identical concentrations. The net UV–vis spectrum is a flat line. Next, the material only in the sample compartment is allowed to change (by changing the solvent or as a reaction occurs over time). As it does, a difference spectrum emerges. Alternatively, in a single-beam instrument, the UV–vis spectrum of the reference is recorded, stored electronically, and subtracted from the UV–vis spectrum of the sample as it undergoes change.

For example, the bile pigment bilirubin has an intense UV–vis absorption near 450 nm (Figure 13-8). Bilirubin is known to undergo photoisomerization to give a mixture of two photobilirubins that have UV–vis spectra only slightly different from that of bilirubin. After 50 seconds of photoirradiation, a slight drop and slight broadening of the UV–vis spectrum is observed (dotted lines in Figure 13-8). Spectral changes associated with the photoisomerization, however, may be observed much more easily by sensitive absorbance difference spectroscopy. The solid lines in Figure 13-8 from such an experiment show the emergence of negative peaks near 450 nm and a positive peak near 500 nm. The most rapid spectral changes occur early, and the absorbance difference lessens over time, indicating that photoequilibrium is being approached.

13-3b Deviations from Beer's Law

Beer's law, which states that the absorbance is proportional to the number of absorbing molecules, is valid for a large number of compounds over a considerable range of concentrations. Since the molar absorption coefficient ϵ depends on wavelength, however, Beer's law can be true strictly only for pure monochromatic light. True deviations may be expected when the concentration of absorbing molecules is so high that they interact with each other. Effects such as association and dissociation are the most common causes of deviations from the Beer–Lambert–Bouguer law. For example, compounds that tend to form dimers, such as aqueous dye solutions, seldom follow Beer's law over any extended concentration range. A test of Beer's law can be made by dilution of the sample solution to a different volume, which should then show the correct absorbance corresponding to the dilution. For example, porphyrins are known to aggregate in aqueous solutions. As illustrated in Figure 13-9, coproferriheme solutions in pH 6.97 aqueous phosphate buffer and at constant ionic strength deviate from Beer's law. Plots (ϵ vs. λ) of spectra over a wide range of coproferriheme concentrations, 3.18×10^{-4} to 3.97×10^{-8} M, should be identical and thus coincident if Beer's law were obeyed. They clearly are not, because the solute is forming soluble aggregates.

13-3c Isosbestic Points

Isosbestic points are shared crossover points at which ϵ remains invariant in a series of overlapping UV–vis curves. Note in Figure 13-9 that there are two wavelengths at which the ϵ values are the same for all spectra. These isosbestic points correspond to wavelengths at which the ϵ values of two absorbing species are identical. The presence of a third absorbing species is highly improbable, because it too would be required to have identical ϵ values at the isosbestic points. Isosbestic points therefore are usually a diagnostic for the presence of only two absorbing species—in the example shown, most likely a monomer and dimer. Isosbestic points also are present in Figures 13-8 and 14-4.

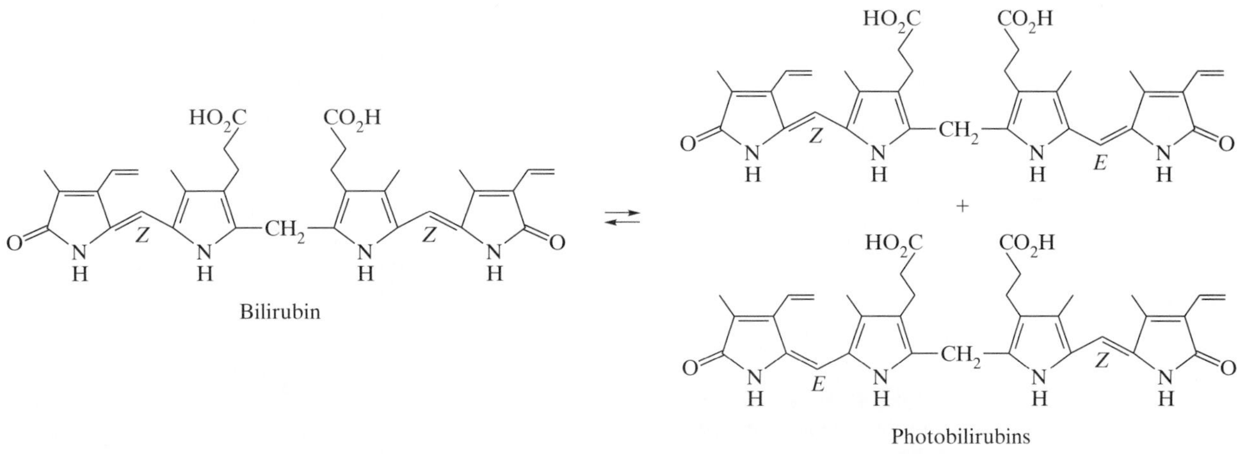

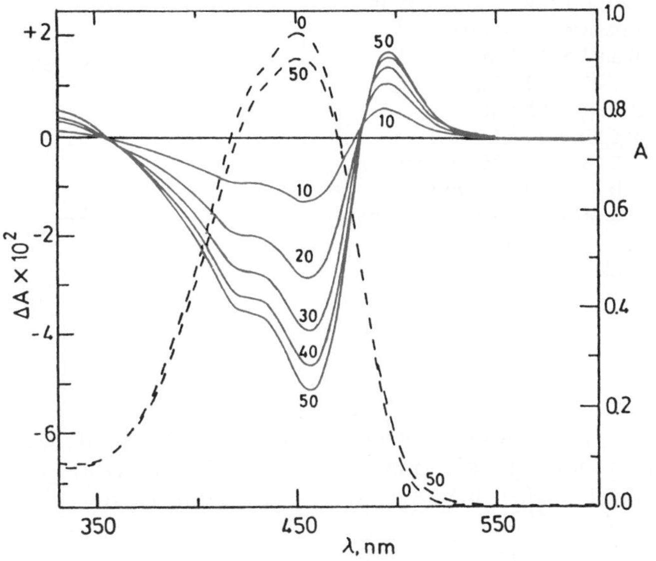

FIGURE 13-8 Absorbance difference spectra (——) obtained from irradiation of a 1.5×10^{-5} M solution of bilirubin in $CHCl_3$/1% ethanol/10% triethylamine at 450 nm (10 nm bandpass). The cumulative irradiation time (s) is indicated on each scan. Absorption curves (– – –) of the sample solution before and after irradiation are superimposed on the absorbance difference spectra. (From the data of D. A. Lightner, T. A. Wooldridge, and A. F. McDonagh, *Biochem. Biophys. Res. Comm.*, **86**, 235 [1979].)

13-4 ELECTRONIC TRANSITIONS

UV–vis spectroscopy measures the probability and energy of exciting a molecule from its ground electronic state to an electronically excited state (promoting an electron from an occupied to an unoccupied molecular orbital). This process is illustrated in Figure 13-10a for the carbonyl chromophore of an aldehyde or ketone, in which several possible excited states are included. An electronically excited state may decay unimolecularly back to the ground state *photophysically* via emitting energy of *fluorescence* or *phosphorescence* (Section 13-4a). Alternatively, it might decay *photochemically* to a different ground state (and hence a different structure). Thus one can measure both absorption and emission from molecules. In the present treatment, we shall be concerned mainly with absorption processes.

A UV–vis absorption spectrum of a given compound may exhibit many absorption bands, with the λ_{max} of each band corresponding roughly to the energy associated with the formation of a particular excited state. This situation is illustrated in Figure 13-1 for a single band and in Figure 13-6 for two bands, in which λ_{max} corresponds to the excitation energy and ϵ_{max} to the intensity of the electronic transition—a measure of the probability of promoting an electron with the requisite excitation energy.

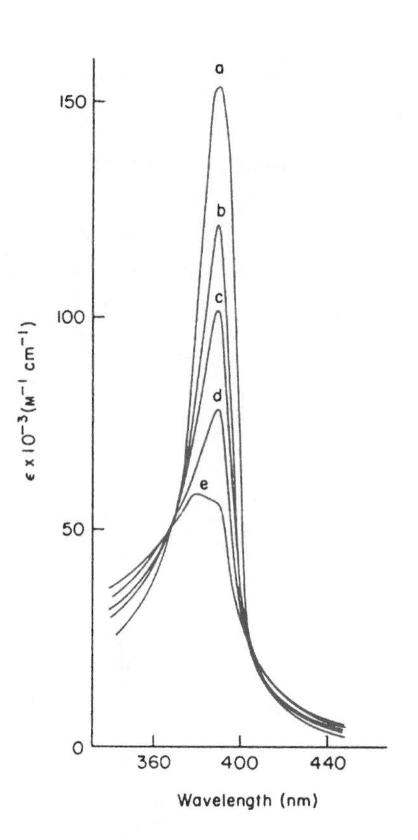

FIGURE 13-9 Spectra of coproferriheme in phosphate buffer (pH 6.97) showing variation due to aggregation. Concentrations were (a) 3.97×10^{-8} M; (b) 6.36×10^{-6} M; (c) 2.78×10^{-5} M; (d) 9.93×10^{-5} M; (e) 3.18×10^{-4} M. (Reproduced with permission from S. B. Brown (ed.), *Introduction to Spectroscopy for Biochemists,* New York: Academic Press, 1980.)

As indicated earlier, electronic excitations are typically assigned to chromophores in a molecule. Although the entire molecule is in the excited energy state, the excited state energy is localized mainly within the chromophore for simple transitions.

13-4a Singlet and Triplet States

Most molecules have ground electronic states in which all electron spins are paired with another electron with the opposite spin, referred to in Figure 13-10 as the S_0 state. Most excited electronic states also have all electron spins paired, even though there may be two orbitals that each possess only one electron (the S_1 and S_2 states in Figure 13-10b). These are known as *singlet states,* as each exists in only one form with no net spin angular momentum.

For molecules having an even number of electrons, regardless of whether or not the ground state is a singlet, there are excited states in which paired electrons have their spins parallel, giving the molecule a net spin angular momentum. Angular components along a given direction, then, can have values of +1, 0, or −1 times the angular momentum. As there are three values, such an electronic configuration is known as a *triplet state* (T_1 in Figure 13-10). Selection rules permit $S \rightarrow S$ (e.g., $S_0 \rightarrow S_1$ in Figure 13-10a) and $T \rightarrow T$ processes, but not $S \rightarrow T$ or $T \rightarrow S$. Because ground states are usually singlets, most excitations are to singlet excited states. Triplet excited states are formed usually by intersystem crossing from an excited singlet state, rather than by direct excitation from the S_0 ground state. The intersystem crossing $S_1 \rightarrow T_1$ in Figure 13-10a is extremely slow in comparison with other, singlet–singlet, processes. Direct excitations from a singlet ground state to a triplet excited state (such as $S_0 \rightarrow T_1$ in Figure 13-10a) typically have very small to vanishing ϵ_{max} values. Emission from an excited singlet state to the ground state (Figure 13-10a, $S_1 \rightarrow S_0$, fluorescence) is a very rapid process, and emission from an excited triplet state to the ground state ($T_1 \rightarrow S_0$, phosphorescence) is slow.

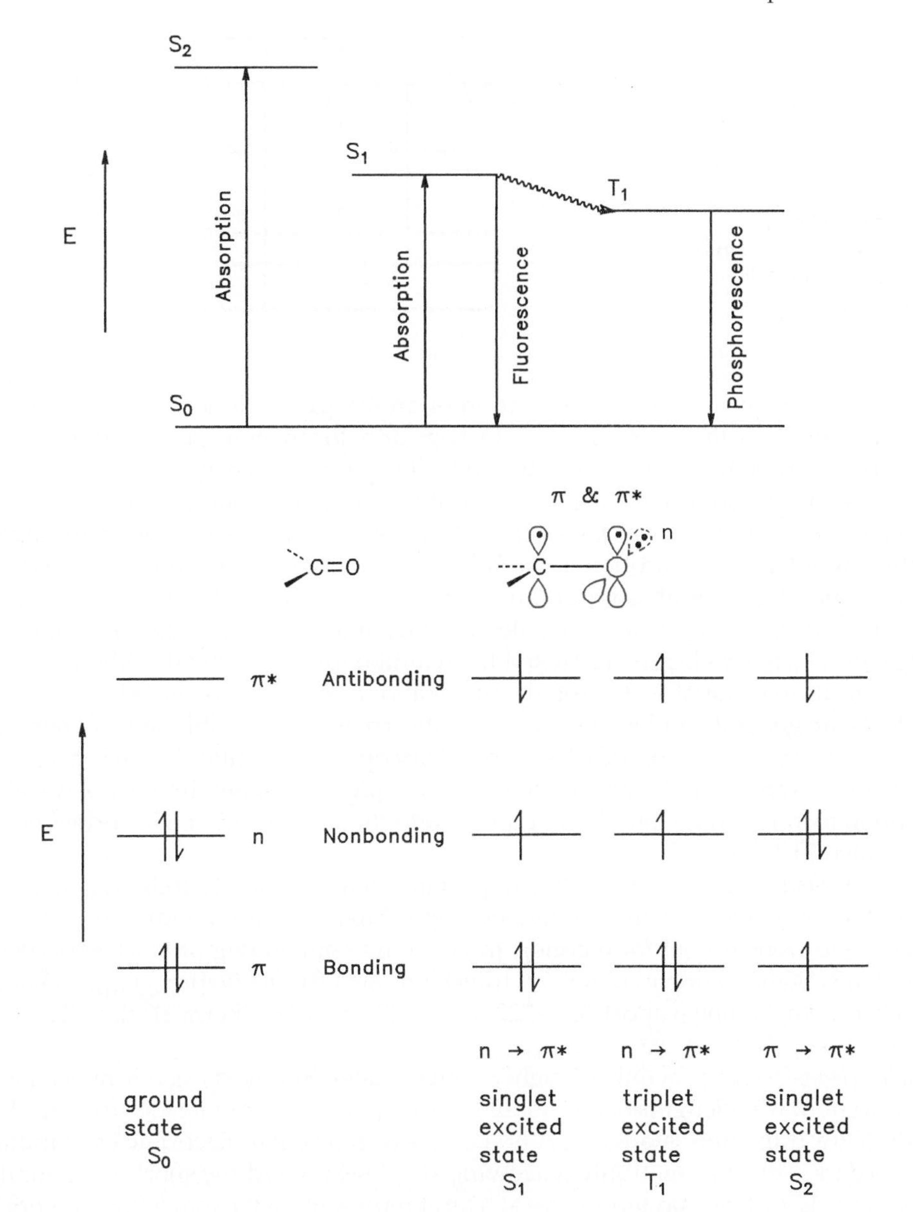

FIGURE 13-10 (a) Energy diagram for electronic excitation and decay processes in the (ketone) carbonyl chromophore. (b) Diagram of selected electronic molecular orbital energies for an isolated carbonyl chromophore, showing the ground state and excited state configurations. Singlet states (S) all have electron spins paired. Triplet states (T) have two spins parallel. Note that the n orbital containing two electrons is orthogonal to the π and π^* orbitals. Subscript 0 refers to ground state, 1 to the first excited state, and 2 to the second excited state.

13-4b Classification of Electronic Transitions

The wavelength of an electronic transition depends on the energy difference between the ground state and the excited state. It is a useful approximation to consider the wavelength of an electronic transition to be determined by the energy difference between the molecular orbital originally occupied by the electron and the higher orbital to which it is excited. Saturated hydrocarbons contain only strongly bound σ electrons. Their excitation to antibonding σ^* orbitals ($\sigma \rightarrow \sigma^*$) or to molecular Rydberg orbitals (involving higher valence shell orbitals, 3s, 3p, 4s, etc.) requires relatively large energies, corresponding to absorption in the far-UV region. Such transitions are rarely of importance in structural elucidation. On the borderline is cyclopropane, which has λ_{max} at 190 nm. Contrast this cycloalkane to propane, which has its λ_{max} at about 135 nm.

Electronic transitions commonly observed in the readily accessible UV (above ~190 nm) and the visible regions have been grouped into the following main classes (Figure 13-11).

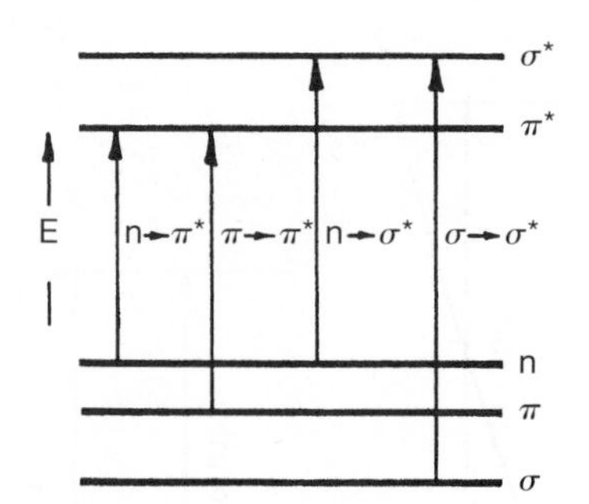

FIGURE 13-11 Relative electronic orbital energies and selected transitions in order of increasing energy.

1. **1.** n → π* *transitions* involve the excitation of an electron in a nonbonding atomic orbital, such as unshared electrons on O, N, S, or halogen atoms, to an antibonding π* orbital associated with an unsaturated center in the molecule. The transitions occur with compounds that possess double or triple bonds containing heteroatoms, for example, C=O, C=N, C≡N, C=S, or N=O. A common example is the low-intensity absorption in the 285 to 320 nm region of saturated aldehydes and ketones, higher with unsaturation (e.g., Figures 13-1 and 13-6).
2. **2.** π → π* *transitions* occur in molecules with double or triple bonds or aromatic rings, in which a π electron is excited to an antibonding π* orbital. Although ethylene itself does not absorb strongly above about 185 nm, conjugated π electron systems are generally of lower energy and absorb in the accessible spectral region. An important application of UV–vis spectroscopy is to define the presence, nature, and extent of conjugation. Increasing conjugation generally moves the absorption to longer wavelengths and finally into the visible region. This principle is illustrated in Table 13-5.
3. **3.** n → σ* *transitions*, which are of less importance than the first two classes, involve excitation of an electron from a nonbonding orbital to an antibonding σ* orbital. Since n electrons do not form bonds, there are no antibonding orbitals associated with them. Some examples of n → σ* transitions are CH_3OH (vapor), λ_{max}, 183 nm, ϵ 150; trimethylamine (vapor), λ_{max} 227 nm, ϵ 900; and CH_3I (hexane), λ_{max} 258 nm, ϵ 380.
4. **4.** *Rydberg transitions* are mainly to higher excited states. For most organic molecules, they occur at wavelengths below about 200 nm and are not of interest structurally. A Rydberg transition is often part of a series of molecular electronic excitations that occurs with systematically narrowing spacings toward the short wavelength side of the UV range and terminates at a limit representing the ionization potential of the molecule.

TABLE 13-5 Effect of Extended Conjugation in Alkenes on the Position of Maximum π → π* Absorption

n in $H(CH{=}CH)_nH$	λ_{max} (nm)	ϵ_{max} (liter $mol^{-1}cm^{-1}$)	Color
1	162	10,000	Colorless
2	217	21,000	Colorless
3	258	35,000	Colorless
4	296	52,000	Colorless
5	335	118,000	Pale yellow
8	415	210,000	Orange
11	470	185,000	Red
15	547[a]	150,000	Violet

From J. B. Lambert, H. F. Shurvell, L. Verbit, R. G. Cooks, and G. H. Stout, *Organic Structural Analysis*, New York: Macmillan Publishing, 1976.

[a]Not a maximum.

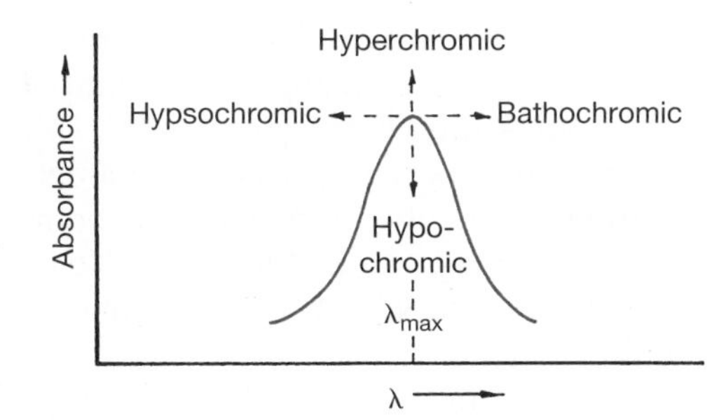

FIGURE 13-12 Terminology of shifts in the position of an absorption band.

In addition to the term *chromophore*, several related terms are used commonly in the present context. The term *auxochrome* (color enhancer) is used for substituents containing unshared electrons (OH, NH, SH, halogens, etc.). When attached to π electron chromophores, auxochromes generally move the absorption maximum to longer wavelengths (lower energies). Such a movement is described as a *bathochromic* or *red shift*. The term *hypsochromic* denotes a shift to shorter wavelength (*blue shift*). Increased conjugation usually results in increased intensity, termed *hyperchromism*. A decrease in intensity of an absorption band is termed *hypochromism*. These terms are summarized in Figure 13-12.

13-4c Allowed and Forbidden Transitions

Electronic transitions may be classed as intense or weak according to their magnitude, measured roughly by ϵ_{max}. These correspond respectively to *allowed* or to *forbidden* transitions. Allowed transitions are those for which (1) there is no change in the orientation of electron spin, (2) the change in angular momentum is 0 or ± 1, and (3) the product of the electric dipole vector and the group theoretical representations (reference 13.1) of the two states is totally symmetric.

The first or *spin selection rule* may be stated as follows: transitions between states of different spin multiplicities (such as S and T) are invariably forbidden, since electrons cannot undergo spin inversion during the change of electronic state. The second or *angular momentum selection rule* usually causes no restrictions, since most states are within one unit of angular momentum of each other. The last rule is the *symmetry selection rule*. If the direct product of the group theoretical representations (reference 13.1) to which the initial and final state functions belong is different from all the representations to which the coordinate axes belong, the transition moment of that transition is zero. A good example of this rule is the $n \rightarrow \pi^*$ transition of saturated alkyl ketones, for which a carbonyl n electron is promoted to an *orthogonal* π^* orbital (90° movement of charge) (structure in Figure 13-10). Such a transition is said to be symmetry forbidden. For most organic molecules, such forbidden transitions are usually observable but of weak intensity. The residual intensity arises because the intensity of the electronic absorption band really depends on the average of the electronic transition moments over all the nuclear orientations of the vibrating molecule and this average is not necessarily zero. When the symmetry of a molecule is periodically changed by some vibration that is not totally symmetric, the symmetry of the electronic wave functions also is changed periodically, since the electrons adapt instantaneously to the motion of the nuclei. Hence a symmetry forbidden transition may become partially allowed. The intensity of a transition that is symmetry forbidden but has become vibrationally allowed is much less than that of an ordinarily allowed transition. Such vibrational contributions are temperature dependent.

13-5 EXPERIMENTAL ASPECTS

13-5a Solvents

Most measurements are carried out on fairly dilute solutions (10^{-2} to 10^{-6} M) of the sample in an appropriate solvent. Such a solvent should not interact with the solute and should not absorb in the spectral region of interest. Older UV–vis spectrophotometers may be double-beam instruments. Most newer UV–vis instruments are computer controlled and single beam. In such instruments, measurements are taken of a solution of the desired compound, followed by rescanning the spectrum with all parameters held the same and using pure solvent in the sample cell to obtain the baseline. Some useful solvents and their short wavelength cutoff limits are given in Table 13-6. Note the significant advantage to be gained by the use of short pathlength cells (1 mm or less), for which there is less solvent in the pathlength.

It is of the utmost importance to use solvents or other reagents with appropriate purity. Commercially available solvents with such designations as "spectral grade" and "for spectroscopy" are not necessarily pure but have been specially prepared to ensure the absence of impurities absorbing in the UV–vis region. Nonabsorbing contaminants may well be present.

Commonly used polar solvents are 95% ethanol, water, and methanol. Aliphatic hydrocarbons (such as hexane, heptane, and cyclohexane) are examples of nonpolar

TABLE 13-6 Short Wavelength Cutoff Limits of Various Solvents

Solvent	Cutoff Point, λ (nm)[a]		Boiling Point (°C)
	10 mm Cell	0.1 mm Cell	
Acetonitrile	190	180	81.6
2,2,2-Trifluorethanol	190	170	79
Pentane	190	170	36.1
2-Methylbutane	192	170	28
Hexane	195	173	68.8
Heptane	197	173	98.4
2,2,4-Trimethylpentane (isooctane)	197	180	99.2
Cyclopentane	198	173	49.3
Ethanol (95%)	204	187	78.1
Water	205	172	100.0
Cyclohexane	205	180	80.8
2-Propanol	205	187	82.4
Methanol	205	186	64.7
Methylcyclohexane	209	180	100.8
Dibutyl ether	210	195	142
EPA[b]	212	190	—
Diethyl ether	215	197	34.6
1,4-Dioxane	215	205	101.4
Bis(2-methoxyethyl) ether (glyme)	220	199	162
Dichloromethane	232	220	41.6
Chloroform	245	235	62
Carbon tetrachloride	265	255	76.9
N,N-Dimethylformamide	270	258	153
Benzene	280	265	80.1
Toluene	285	268	110.8
Tetrachloroethylene	290	278	121.2
Pyridine	305	292	116
Acetone	330	325	56
Nitromethane	380	360	101.2
Carbon disulfide	380	360	46.5

From J. B. Lambert, H. F. Shurvell, L. Verbit, R. G. Cooks, and G. H. Stout, *Organic Structural Analysis*, New York: Macmillan Publishing, 1976.

[a]The cutoff point is taken as the wavelength at which the absorbance in the indicated cell is about 1.

[b]A 5/5/2 mixture by volume of ethyl ether, isopentane, and ethanol.

solvents that have good spectral transparency (low λ cutoff) (Table 13-6) and have boiling points high enough so that solvent evaporation does not become a problem. They must be rigorously purified, however, because these hydrocarbons may contain alkenic impurities or traces of aromatic compounds. It has been observed that fluoroalkanes have enhanced transparency relative to alkanes, and a similar finding has been made for fluorinated alcohols such as 2,2,2-trifluoroethanol. Organic cyanides such as acetonitrile and propionitrile are polar, nonhydroxylic solvents with excellent spectral transparency. Another widely used such solvent is 1,4-dioxane, which is transparent to about 215 nm.

Several mixed solvents have found use in spectroscopic studies, particularly at very low temperatures, usually down to liquid nitrogen temperatures, about −190°C. These solvent systems do not crystallize when cooled but instead become viscous and glassy. Low temperature mixed solvents include (1) EPA, a 5/5/2 mixture by volume of diethyl ether, isopentane, and ethanol; (2) methanol and glycerol, 9/1 v/v; (3) tetrahydrofuran and diglyme, 4/1 v/v; and (4) methylcyclohexane and isopentane, 1/3 v/v. The degree of contraction of these solvents over the range 25° to −190°C is 29.4% for methylcyclohexane/isopentane (1/3), 29% for EPA, 24.4% for methanol/glycerol (9/1), and 26.1% for tetrahydrofuran/diglyme (4/1).

13-5b Cells (Cuvettes) and Sample Preparation

Cells for use in the visible region are usually made of Pyrex or similar glass that is transparent to about 380 nm. In the UV region, quartz cells are necessary, and those made of high-purity fused silica (Ultrasil, Spectrosil, Supersil) are recommended. Rectangular cuvettes are routinely used in absorption work. The preferred pathlength is 10 mm, since l in eq. 13-3 then is equal to unity. Short-pathlength cells are essential when solvent absorption must be minimized. Cells in the range 0.01 to 2 mm are used for work in the far-UV region.

For most purposes, cells may be cleaned by rinsing *at least 10 times* with the solvent to be used in the measurement. A final rinse with methanol or acetone is suitable if the cells are to be air dried. In general, oven drying is recommended, particularly for short-pathlength cells. In this case two final rinses should be with distilled water so that no trace of a flammable organic liquid remains. Although a drying oven set at a high temperature does not damage the cells (remember that they have been fused during manufacture at temperatures above 1000°C), removing a fragile quartz cell from a hot oven can be troublesome. A fast and convenient drying method is to place the cell in a vacuum desiccator, heated to about 50°C if possible. In this manner, cells may be dried in less than 5 minutes.

Above all, cell windows should never be touched with the fingers. It is good practice when cleaning a cell to rinse the outside a few times. Any material that still remains on the outside optical faces should be wiped off with a lintless wiper, such as Kimwipes, *soaked in solvent*. Never use a dry cloth or tissue.

Quantitative analytical techniques are applied to sample preparation. Volumetric glassware and cells must be clean and dry. Solid samples should be dried to constant weight in a desiccator to remove adhering water or solvent. A typical 1 to 10 mm cell holds anywhere from 0.2 to 3 mL of solution, so an appropriate amount of stock sample solution should be prepared. Since the measurement is nondestructive, the sample may be recovered by evaporation of the solvent. If the amount of sample available permits, 10 to 25 mL of solution is a convenient size to prepare. The amount of sample required for this volume is sufficient to minimize errors associated with weighing small quantities. The analytical balance used must be capable of weighing directly to at least 0.1 mg. Because most compounds being measured probably will not have been run previously by the operator, an initial sample concentration should be approximately 0.05% (about 0.5 mg mL^{-1}) for small molecules and about 0.005% for polypeptides and large macromolecules. A peak absorbance in the range of 0.7 to 1.2 absorbance units is desirable for most instruments, since it gives a good pen deflection and the electronics are usually most sensitive in this range.

13-5c Possible Sources of Error

Errors may arise from the nature of the sample being examined, from the instrument, and, last but not least, from the operator. The simplest error may be one in which Beer's law (eq. 13-3) is not obeyed (Section 13-3b).

The problem of stray light occurs in most double monochromator instruments at high sample absorbances. Stray light has a wavelength different from that desired. It can arise from scattering of the light beam from any of the surfaces that it encounters: lenses, prisms, slit edges, and so forth, as well as dirty cell windows. Turbid solutions scatter light, and the scattering becomes more important at shorter wavelengths. Sometimes dust particles are responsible for the scattering. Such solutions should be passed through a fine filter, such as Millipore, or centrifuged.

Oxygen has a series of absorption bands that begin at about 195 nm and extend to shorter wavelength. Hence, to work in this spectral region, air must be excluded from the instrument. Otherwise much of the light is lost in both the sample and the reference beam because of oxygen absorption. Oxygen is most easily removed by flushing the entire optical path of the instrument with pure, dry nitrogen. A liquid nitrogen cylinder having a gaseous take-off valve provides a convenient source of high-purity nitrogen. Optimum flow rates can be determined by observing the disappearance of the oxygen absorption spectrum.

Problems

13-1 Calculate the molar extinction coefficient of 3-methylcyclohexanone at 280 nm and 320 nm from the data of Figure 13-1.

13-2 A compound $C_5H_8O_2$ has a UV λ_{max} at 270 nm ($\epsilon = 32$) in methanol; in hexane, the band shifts to 290 nm ($\epsilon_{max} = 40$). What functional groups could be present? Draw a possible structure.

13-3 Calculate the absorbance (A) of a 0.005 M solution of 3-methylcyclohexanone in isooctane in a 10 cm pathlength quartz cuvette at $\lambda = 280$ nm.

13-4 A compound, 0.0002 M in methanol, shows $\lambda_{max} = 235$ nm with an absorbance (A) of 1.05 when measured in a 0.5 cm pathlength quartz cuvette. Calculate its ϵ_{max} at 235 nm.

13-5 The molar absorptivity of benzoic acid (mol. wt. = 122.1) in ethanol at 273 nm is about 2000. If an absorbance not exceeding 1.35 is desired, what is the maximum allowable concentration in g L^{-1} that can be used in a 2.00 cm cell?

13-6 A 250 mg sample containing a colored component X is dissolved and diluted to 250 mL. The absorbance of an aliquot of this solution, measured at 500 nm in a 1.00 cm cell, is 0.900. Pure X (10.0 mg) is dissolved in 1 L of the same solvent. The absorbance measured in a 0.100 cm cell at the same wavelength is 0.300. What is the percentage of X in the first sample?

13-7 Colorless substances X and Y form the colored compound XY on mixing: $X + Y \rightleftharpoons XY$. When 2.00×10^{-3} mol of X is mixed with a large excess of Y and diluted to 1 L, the solution has an absorbance that is twice as great as when 2.00×10^{-3} mol of X is mixed with 2.00×10^{-3} mol of Y and treated similarly. What is the equilibrium constant for the formation of XY?

13-8 The following absorbances were measured for three solutions containing A and B separately and in a mixture, all in the same cell. Calculate the concentrations of A and B in the mixture.

		Absorbance	
		475 nm	670 nm
0.001 M	*A*	0.90	0.20
0.01 M	*B*	0.15	0.65
Mixture		1.65	1.65

Bibliography

13.1 H. H. Jaffé and M. Orchin, *Theory and Applications of Ultraviolet Spectroscopy*, New York: John Wiley & Sons, Inc., 1962.

13.2 R. L. Pecsok, L. D. Shields, T. Cairns, and I. B. McWilliam, *Modern Methods of Chemical Analysis*, 2nd ed., New York: John Wiley & Sons, Inc., 1976.

13.3 H. H. Perkampus, *UV-vis Atlas of Organic Compounds*, 2nd ed., Weinheim, Germany: Wiley-VCH, 1992.

13.4 Solvent properties: O. Korver and J. Bosma, *Anal. Chem.*, **43**, 1119 (1971).

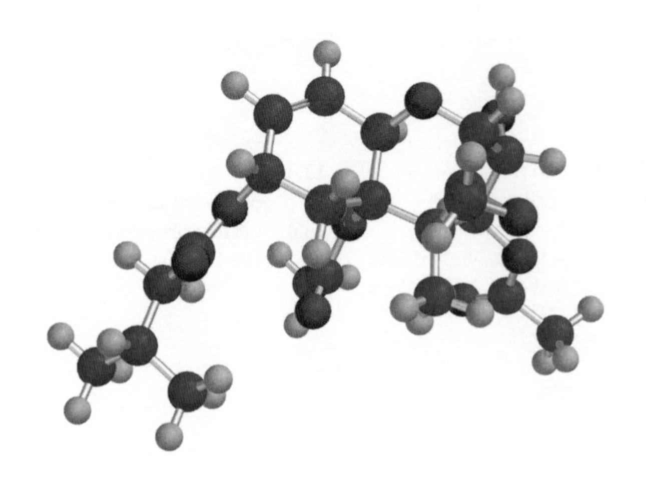

CHAPTER 14

Structural Analysis

14-1 ISOLATED CHROMOPHORES

14-1a The Carbonyl Group: Ketone and Aldehyde Absorption

The longest wavelength transition in aliphatic aldehydes and ketones, the n $\rightarrow \pi^*$ band, is probably the best studied of any electronic transition (reference 14.1 and Figures 13-10 and 14-1). It is a weak ($\epsilon \sim$ 10–20) and rather broad band, occurring in the neighborhood of 270 to 300 nm. As noted in Section 13-2d, its position is quite solvent sensitive. The n $\rightarrow \pi^*$ transition involves the promotion of an electron from a nonbonding orbital on oxygen to the antibonding π^* orbital associated with the entire carbonyl group. The transition is symmetry forbidden (Section 13-4c)—hence the low intensity (Figure 13-1).

A second carbonyl band, the $\pi \rightarrow \pi^*$ transition, occurs near 190 nm in ketones. This allowed transition is considerably more intense than the n $\rightarrow \pi^*$ transition. This wavelength region is near the practical wavelength cutoff of most UV instruments, so that often only the beginning of the band is observed as so-called *end absorption*. Transitions at wavelengths shorter than about 190 nm are most likely due to excitations from the carbonyl σ bond ($\sigma_{CO} \rightarrow \pi^*$) and from Rydberg transitions.

The symmetries of the n, π, and π^* orbitals involved in the transitions are important. The bonding π orbital and the antibonding π^* orbital lie in the same plane, whereas the nonbonding n orbital is in an orthogonal plane. Hence promotion of an electron from the nonbonding orbital is not possible without a significant change in the geometry of the molecule. The weak n $\rightarrow \pi^*$ intensity is due to nonsymmetrical vibrations that slightly deform the molecule and lower its symmetry, allowing the transition to acquire a finite probability.

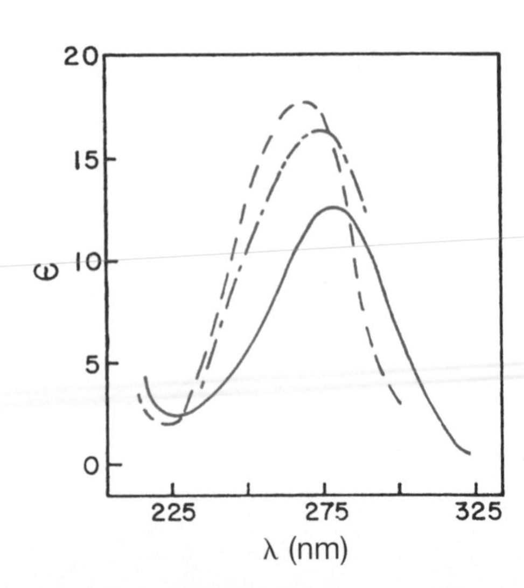

FIGURE 14-1 Solvent effects on the n $\rightarrow \pi^*$ transition of acetone in hexane (—), in 95% ethanol (- . -), and in water (- - -).

Cyclic ketones absorb at longer wavelength than the corresponding open-chain analogues. In addition, there is a variation in the position of the absorption band with ring size in nonpolar solvents, as illustrated in Table 14-1.

The n $\rightarrow \pi^*$ transition of ketones and aldehydes is sensitive to the presence of nonalkyl substituents, which, when located α to the carbonyl group, affect the position (λ_{max}) and intensity (ϵ_{max}) of the absorption band. For example, the presence of an α bromine in the cyclohexanone series causes a bathochromic shift of λ_{max} of about 23 nm when the bromine is axial, but only a 5 nm shift when it is equatorial. Equatorially substituted 2-chloro-4-*tert*-butylcyclohexanone has its n $\rightarrow \pi^*$ maximum at a slightly shorter wavelength than that of the parent ketone, whereas the axial chlorine isomer has a more intense absorption band at a considerably longer wavelength (Table 14-1).

In general, n $\rightarrow \pi^*$ transitions are easily recognizable by their low intensities, by the spectral shifts caused by substitution, and by the sensitivity of the position of the band to solvent effects. Shifts in the position of absorption bands on going from the vapor phase to solution or from one solvent to another are caused mainly by differences in the solvation energies of the solute in the ground and excited electronic states. The effect of solvent on the position of the n $\rightarrow \pi^*$ absorption has served as an important diagnostic tool (Section 13-2d). The fundamental role of the unshared electron pair in this transition can be demonstrated by the disappearance of the n $\rightarrow \pi^*$ band in acid solution, in which the unshared pair is protonated.

Changing from a nonpolar to a polar solvent results in a significant hypsochromic shift in the position of the n $\rightarrow \pi^*$ transition. Hydroxylic solvents of comparable dielectric constant cause a larger blue shift than do nonhydroxylic, polar solvents such as acetonitrile. The larger shifts occasioned by hydroxylic solvents are attributable in part to greater hydrogen bonding to the carbonyl oxygen lone pairs than to the π^* electrons, thus lowering the energy of the ground state relative to that of the excited state. An example of solvent effects on the n $\rightarrow \pi^*$ band of acetone is shown in Figure 14-1, and of an α,β-unsaturated ketone in Table 13-4.

TABLE 14-1 Absorption Data for Aliphatic Aldehydes and Ketones

		n $\rightarrow \pi^*$ Transition		$\pi \rightarrow \pi^*$ Transition	
Compound	**Solvent**	λ_{max} (nm)	ϵ_{max} (liter mol^{-1} cm^{-1})	λ_{max} (nm)	ϵ_{max} (liter mol^{-1} cm^{-1})
Formaldehyde	Vapor	304	18	175	18,000
	Isopentane	310	5		
Acetaldehyde	Vapor	289		182	10,000
Acetone	Vapor	274	13.6	195	9,000
	Cyclohexane	275	22	190	1,000
Butanone	Isooctane	278	17		
2-Pentanone	Hexane	278	15		
4-Methyl-2-pentanone	Isooctane	283	20		
Cyclobutanone	Isooctane	281	20		
Cyclopentanone	Isooctane	300	18		
Cyclohexanone	Isooctane	291	15		
Cycloheptanone	Isooctane	292	17		
Cyclooctanone	Isooctane	291	15		
Cyclononanone	Isooctane	293	17		
Cyclodecanone	Isooctane	288	15		
2-Chloro-4-*tert*-butyl-cyclohexanone					
Equatorial Cl	Isooctane	286	17		
Axial Cl	Isooctane	306	49		

From J. B. Lambert, H. F. Shurvell, L. Verbit, R. G. Cooks, and G. H. Stout, *Organic Structural Analysis*, New York: Macmillan Publishing, 1976.

TABLE 14-2 Effect of Heteroatom Substituents on the Carbonyl n $\rightarrow \pi^*$ Transition

X in $CH_3C(=O)—X$	Solvent	λ_{max} (nm)	ϵ_{max} (liter mol^{-1} cm^{-1})
—H	Vapor	290	10
$—CH_3$	Hexane	279	15
	95% ethanol	272.5	19
—OH	95% ethanol	204	41
—SH	Cyclohexane	219	2200
$—OCH_3$	Isooctane	210	57
$—OC_2H_5$	95% ethanol	208	58
	Isooctane	211	58
$—O—C(=O)—CH_3$	Isooctane	225	47
—Cl	Heptane	240	40
—Br	Heptane	250	90
$—NH_2$	Methanol	205	160

From J. B. Lambert, H. F. Shurvell, L. Verbit, R. G. Cooks, and G. H. Stout, *Organic Structural Analysis*, New York: Macmillan Publishing, 1976.

14-1b The Carbonyl Group: Acid, Ester, and Amide Absorption

Introduction of a heteroatom at the carbon atom of the carbonyl chromophore causes a large blue shift of the n $\rightarrow \pi^*$ band, although the intensity remains about the same as for aldehydes and ketones. The effect of such substituents on the n $\rightarrow \pi^*$ transition is illustrated in Table 14-2.

The heteroatom attached to the carbonyl group in carboxylic acids, esters, acid chlorides, amides, and so on, can donate electron density by conjugation to the carbonyl function. The energy of the antibonding π^* orbital is raised by interaction with the unshared pair of the substituent, whereas the p orbital occupied by the n electrons in the ground state is not affected. The n $\rightarrow \pi^*$ transition energy thus is raised, and the absorption shifts to shorter wavelength.

Two bands have been identified in the UV spectra of aliphatic carboxylic acids and esters. The first is the n $\rightarrow \pi^*$ transition in the vicinity of 210 nm (ϵ = 40–60), and the second is the $\pi \rightarrow \pi^*$ transition at about 165 nm (ϵ = 2500–4000).

Carboxylic acids are extensively dimerized in the liquid state, in nonpolar solvents, and even to some extent in the vapor phase (eq. 14-1).

$$2\,RCOOH \rightleftharpoons R—C\begin{matrix} O\cdots H—O \\ O—H\cdots O \end{matrix}C—R \qquad (14\text{-}1)$$

Hence, depending on solvent and concentration, one may be dealing with pure dimer, pure monomer, or a mixture of the two. In the case of carboxylic acids, it is recommended that spectra be determined at more than one concentration. Hydroxylic solvents such as alcohols shift the above equilibrium toward monomer. Conversion of the acid to an ester eliminates the possibility of hydrogen bonding according to eq. 14-1. Essentially identical spectra from a carboxylic acid and, for example, its methyl or ethyl ester are indicative that dimerization is absent or is not spectroscopically important.

UV spectra of amides have received considerable attention because of the importance of the amide chromophore in polypeptides and proteins. At least five transitions of the amide group have been identified (reference 14.3). The two lowest-energy transitions are a weak n $\rightarrow \pi^*$ band ($\epsilon \sim 100$) near 220 nm in nonpolar solvents and a relatively strong $\pi \rightarrow \pi^*$ transition in the 173 to 200 nm region ($\epsilon \sim 8000$). Both of these transitions

TABLE 14-3 Absorption Data for Unconjugated Alkenes

Compound	Solvent	λ_{max} (nm)	ϵ_{max} (liter mol^{-1} cm^{-1})
Ethylene	Vapor	162	10,000
cis-2-Butene	Vapor	174	—
trans-2-Butene	Vapor	178	13,000
1-Hexene	Vapor	177	12,000
	Hexane	179	—
cis-2-Octene[a]	Vapor	183	13,000
trans-2-Octene[a]	Vapor	179	15,000
Allyl alcohol	Hexane	189	7,600
Cyclohexene	Vapor	176	8,000
	Cyclohexane	183.5	6,800
Cholest-4-ene	Cyclohexane	193	10,000
1,5-Hexadiene	Vapor	178	26,000

From A. E. Hansen and T. D. Bouman, *Adv. Chem. Phys.*, **44**, 545 (1980).
[a]Data from J. R. Platt, H. B. Klevens, and W. C. Price, *J. Chem. Phys.*, **17**, 466 (1949).

are strongly perturbed by the nitrogen 2p orbital of the conjugated π system, which extends over the N, C, and O atoms. The n $\rightarrow$ π^* amide transition exhibits the usual solvent effect, being blue shifted on going from nonpolar to hydroxylic solvents.

The question of whether aliphatic amides protonate on oxygen or nitrogen was examined by UV spectroscopy. It was reasoned (reference 14.4) that if protonation occurs on nitrogen, the nitrogen atom would be removed from the amide system and the spectral properties of the simple carbonyl group should result, for instance, in a large red shift of the n $\rightarrow$ π^* band and a large blue shift of the $\pi \rightarrow \pi^*$ band. Studies of the acidity dependence of the $\pi \rightarrow \pi^*$ absorption of *N,N*-dimethylacetamide near 195 nm showed no significant shifts, indicating that the oxygen-protonated amide is the dominant species in dilute acid solutions.

14-1c Unconjugated Alkenes

Most unconjugated alkenes absorb in the far-UV region, near 200 nm. Substitution of the double bond influences λ_{max} somewhat, with more highly substituted double bonds having λ_{max} bathochromically shifted relative to the less highly substituted (Table 14-3). Ethylene itself absorbs well outside the generally accessible UV region, with a broad absorption maximum at about 162 nm. The rather intense absorption ($\epsilon \sim 10{,}000$) is attributed to a $\pi \rightarrow \pi^*$ absorption maximum. Geometrical isomers of disubstituted alkenes often can be differentiated, although not necessarily systematically. *trans*-2-Butene has a longer wavelength $\pi \rightarrow \pi^*$ absorption than does *cis*-2-butene, but *trans*-2-octene has a shorter wavelength $\pi \rightarrow \pi^*$ transition than do *cis*-2-octene compounds (Table 14-3). When two alkenic chromophores in the same molecule are insulated from each other by saturated carbons, their spectrum approximates the sum of the two chromophores, such as 1,5-hexadiene in Table 14-3.

The UV spectra of unconjugated alkenes have been shown to exhibit more than just the high-intensity $\pi \rightarrow \pi^*$ transition in the accessible spectral region. Other bands have been observed in the vapor-phase spectra of alkyl-substituted alkenes: higher-energy Rydberg transitions and a weaker band somewhat above 200 nm, which has been assigned to a $\pi \rightarrow \sigma^*$ transition (references 14.5, 14.6, and 14.7).

14-2 CONJUGATED CHROMOPHORES

UV–vis spectroscopy has long been used as a structural diagnostic to detect conjugation when two or more chromophores are present, such as dienes, polyenes, α,β-unsaturated carbonyl compounds, and aromatics.

TABLE 14-4 Absorption Data for Alkenes

Compound	Solvent	λ_{max} (nm)	ϵ_{max} (liter mol^{-1} cm^{-1})
1-Hexene	Vapor	177	12,000
	Hexane	179	—
1,3-Hexadiene	Hexane	224	26,400
1,3,5-Hexatriene	Isooctane	268	43,000
1,3,5,7-Octatetraene	Cyclohexane	304	—
1,3,5,7,9-Decapentaene	Isooctane	334	121,000
1,3,5,7,9,11-Dodecahexaene	Isooctane	364	138,000

From A. J. Merer and R. S. Mulliken, *Chem. Rev.*, **63**, 639 (1969).

14-2a Dienes and Polyenes

Conjugation of π electron systems of double bonds results in dramatic bathochromic shifts and increased intensities. Spectral data illustrating some of the above effects are given in Table 14-4. The bathochromic shifts resulting from increased double bond conjugation may be illustrated in terms of Hückel molecular orbital theory (more sophisticated theories lead to the same results). The molecular orbitals and their relative energies for ethylene, 1,3-butadiene, and two higher conjugated homologues are depicted in Figure 14-2. The energies of the highest occupied molecular orbitals (HOMO) increase, while those of the lowest unfilled molecular orbitals (LUMO) decrease with increasing conjugation. The observed transition involves promotion of an electron from the HOMO to the LUMO. Figure 14-2 shows that as the conjugated π system increases in length, the energy required for the transition becomes less; that is, a bathochromic shift results.

The $\pi \rightarrow \pi^*$ absorption band of the diene chromophore is shifted in a systematic way by substitution, as illustrated in Table 14-5. This pattern was recognized more than 70 years ago by Woodward and by the Fiesers and led to the formulation of the Woodward–Fieser rules (Section 14-10a), which have enjoyed wide use in structure determination.

14-2b α, β,-Unsaturated Carbonyl Groups

Many organic molecules of interest contain both carbonyl and alkene chromophores. If the groups are separated by two or more σ bonds, there is generally (but with important exceptions; see Section 14-8) little electronic interaction. Thus the effect of the two chromophores on the observed spectrum is essentially additive. Compounds in which the double bond is conjugated with a carbonyl group exhibit spectra in which both the alkenic $\pi \rightarrow \pi^*$ and the carbonyl n $\rightarrow \pi^*$ absorption maxima of the isolated chromophores undergo bathochromic shifts of 15 to 45 nm, although each band is not necessarily displaced by an equal amount. Photoionization data indicate that the n orbital

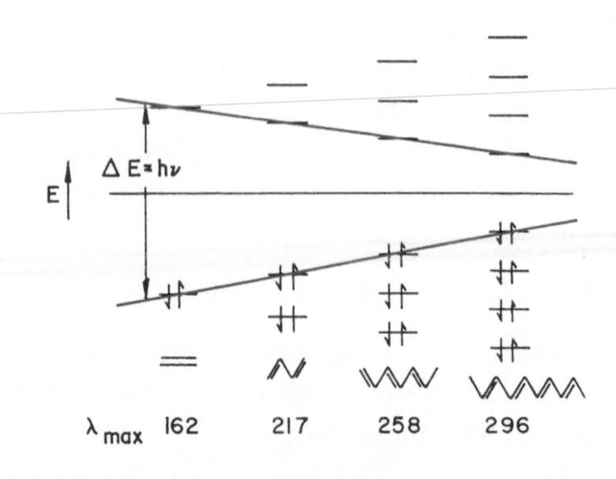

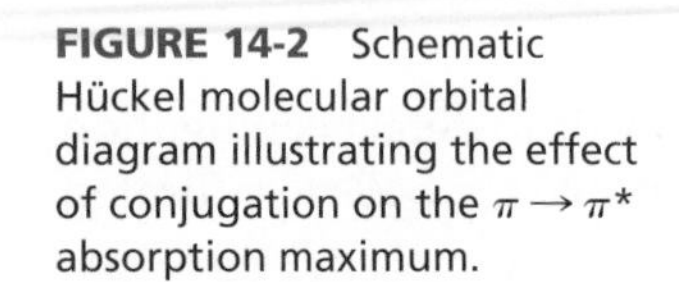
FIGURE 14-2 Schematic Hückel molecular orbital diagram illustrating the effect of conjugation on the $\pi \rightarrow \pi^*$ absorption maximum.

TABLE 14-5 Absorption Data for Substituted Dienes in Ethanol

Compound	λ_{max} (nm)	ϵ_{max} (liter mol^{-1} cm^{-1})
$CH_2{=}CH{-}CH{=}CH_2$	217.5	22,400
$CH_3CH{=}CH{-}CH{=}CH_2$	223	25,000
$CH_2{=}C(CH_3){-}CH{=}CH_2$	222.5	22,800
$CH_3CH{=}CH{-}CH{=}CHCH_3$	226	23,800
$CH_2{=}C(CH_3){-}C(CH_3){=}CH_2$	226.5	20,300
$CH_3CH{=}C(CH_3){-}C(CH_3){=}CH_2$	231.5	19,200

From W. F. Forbes, R. Shilton, and A. Balasubramanian, *J. Org. Chem.*, **29**, 3527 (1964).

energy is relatively constant, so that the red shift is most probably caused by a lowering of the energy of the π^* orbital.

Crotonaldehyde ($CH_3CH{=}CH{-}CH{=}O$) in ethanol solution has an intense band at 220 nm ($\epsilon = 15{,}000$) and a weak band at 322 nm ($\epsilon = 28$). The low intensity and the hypsochromic shift in hydroxylic solvents relative to hydrocarbon solvents suggest that the 322 band is the $n \rightarrow \pi^*$ transition. The bathochromic shift relative to a saturated carbonyl group indicates that the excited π^* orbital extends over all the atoms of the conjugated carbonyl group. The phenomenon is general and may be seen in α,β-unsaturated ketones and aldehydes, in conjugated acids, and in conjugated esters (Table 14-6).

The solvent effect on the $\pi \rightarrow \pi^*$ transition is opposite to that on the $n \rightarrow \pi^*$ peak. The $\pi \rightarrow \pi^*$ absorption shifts to longer wavelength with increasing solvent polarity. The effect of solvent on the $n \rightarrow \pi^*$ and $\pi \rightarrow \pi^*$ transitions of mesityl oxide, $(CH_3)_2C{=}CH{-}(CO){-}CH_3$, is given in Table 13-4.

As the number of double bonds conjugated with the carbonyl group increases, the $\pi \rightarrow \pi^*$ transition shifts to longer wavelength and its intensity increases, causing the much weaker $n \rightarrow \pi^*$ absorption to appear as a shoulder or to become completely obscured by the more intense, overlapping $\pi \rightarrow \pi^*$ band.

TABLE 14-6 UV–Vis Absorption Bands of α,β-Unsaturated Carbonyl Compounds in Ethanol

	$\pi \rightarrow \pi^*$		$n \rightarrow \pi^*$	
Compound	λ_{max} (nm)	ϵ_{max} (liter mol^{-1} cm^{-1})	λ_{max} (nm)	ϵ_{max} (liter mol^{-1} cm^{-1})
$CH_2{=}CH{-}CHO$	207	11,200	322	28
	203[a]	12,000[a]	345[a]	20[a]
$CH_3CH{=}CH{-}CHO$	217	17,900	327	
$CH_2{=}C(CH_3)CHO$	216	11,000		
$CH_3CH{=}C(CH_3)CHO$	226	16,100		
$CH_2{=}CHCOCH_3$	215	3,600		
	203[a]	9,600[a]	331[a]	25[a]
$CH_3CH{=}CHCOCH_3$	221	12,300		
$CH_2{=}C(CH_3)COCH_3$	214	7,550		
$CH_3CH{=}C(CH_3)COCH_3$	223	13,600		
$CH_2{=}CH{-}CO_2H$	200	10,000		
$CH_3CH{=}CHCO_2H$	206	14,000		
$CH_2{=}C(CH_3)CO_2H$	210			
$CH_3CH{=}CHCO_2Et$	210	12,600		
$CH_3CH{=}C(CH_3)CO_2H$	213	12,500		
$CH_3CH{=}C(CH_3)CONH_2$	214	12,100		

[a]In cyclohexane.

Alkyl substitution shifts the $\pi \rightarrow \pi^*$ and $n \rightarrow \pi^*$ maxima in opposite directions, the $\pi \rightarrow \pi^*$ being displaced to longer wavelength. Such effects of substitution on the position of the $\pi \rightarrow \pi^*$ transition, as with dienes, can be predicted through the use of empirical rules first formulated by Woodward and modified by the Fiesers. These rules, which have played an important role in assigning the structures of steroids and other natural products, are discussed in Section 14-10.

Table 14-6 illustrates the effects of conjugation for aldehydes, ketones, acids, esters, and amides. The data indicate a considerable similarity in the locations of the $\pi \rightarrow \pi^*$ transitions and a similar sensitivity toward substitution of the carbon–carbon double bond. In general, the $\pi \rightarrow \pi^*$ transition lies between 200 and 220 nm for α,β-unsaturated carbonyls, and alkyl-substituted systems have their $\pi \rightarrow \pi^*$ transition shifted toward the higher end of the range. Extended conjugation causes bathochromic shifts of the $\pi \rightarrow \pi^*$ band.

14-3 AROMATIC COMPOUNDS

The benzene ring ranks with the carbonyl group and alkenes among the most widely studied chromophores. The spectrum of benzene above 180 nm consists of three well-defined absorption bands due to $\pi \rightarrow \pi^*$ transitions (Figure 14-3). An intense, structureless band occurs at about 185 nm, with a somewhat weaker band (λ_{max} ~200 nm) of poorly resolved vibrational structure overlapping the 185 nm absorption. The longest wavelength transition is a low-intensity system centered near 255 nm, with a characteristic vibrational structure. Data on the benzene absorption bands and some of the various nomenclature systems used to describe them are given in Table 14-7.

The benzene absorptions at 254 and 204, termed ^1L_b and ^1L_a in the Platt notation (Table 14-7), are both forbidden. The superscript 1 indicates that the transition is to a singlet excited state. The ^1L_a band is able to borrow intensity from the allowed ^1B transition, which overlaps it at shorter wavelength. The different transition probabilities relate to configuration interaction because of the degeneracy of the highest occupied and lowest vacant orbitals in benzene. Benzene belongs to the D_{6h} point group, and the intensity of the symmetry forbidden 254 nm ^1L_b transition also should be zero. Vibrational distortions from hexagonal symmetry, however, result in a small net transition dipole moment and the observed low intensity. The ^1L_b, or benzenoid, absorption band usually is easy to identify. It has about the same intensity in benzene and its simple derivatives, $\epsilon = 250\text{–}300$. It also has similar well-defined vibrational structure with up to six vibrational bands, as in benzene itself. The vibrational structure is less evident in polar solvents and more sharply defined in vapor spectra or in nonpolar solvents.

Alkyl substitution shifts the benzene absorption to longer wavelengths and tends to reduce the amount of vibrational structure. Increases in band intensities are observed commonly. In general, substitution can perturb the benzene ring by both polar and resonance effects. A methyl substituent causes the largest hyperchromic wavelength shift and the greatest change in vibrational intensities. The effect decreases as the methyl hydrogens are replaced by alkyl groups. This result often is cited as evidence for the importance of C—H hyperconjugation ($\sigma \rightarrow \pi$ electron interaction) (reference 14.1).

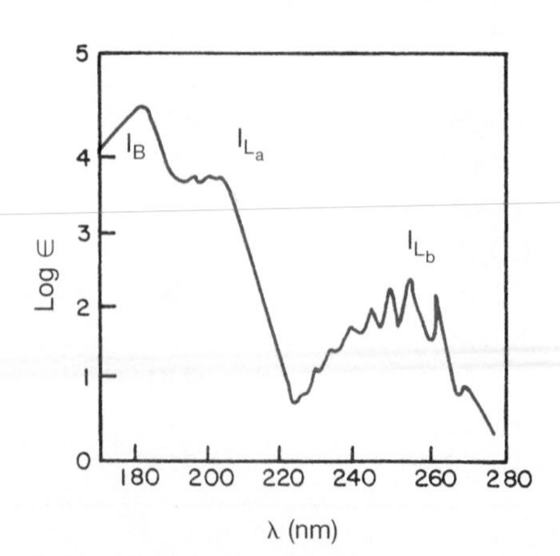

FIGURE 14-3 The UV spectrum of benzene in hexane.

TABLE 14-7 Notation Systems for Benzene Absorption Bands

λ_{max} (ϵ_{max}) in Hexane			Origin of the Spectral Notation[a]
184 nm (68,000)	**204 nm (8800)**	**254 nm (250)**	
^1B	^1L_a	^1L_b	a
β	*p* (para)	α	b
$^1E_{2u}$	$^1B_{1u}$	$^1B_{2u}$	c
Second primary	Primary	Secondary	d
	K (conjugation)	*B* (benzenoid)	e

[a]*Note:* a = Platt free-electron method notation: J. R. Platt, *J. Chem. Phys.*, **17**, 484 (1949); b = empirical notation of Clar based on behavior of bands with temperature: E. P. Clar, *Aromatische Kohlenwasserstoffe*, Berlin: Springer Verlag, 1952; c = molecular orbital approach based on the group theoretical notation of the transitions; d = empirical notation: L. Doub and J. M. Vandenbelt, *J. Am. Chem. Soc.*, **69**, 2714 (1947), **71**, 2414 (1949); e = early empirical notation: A. Burawoy, *J. Chem. Soc.*, 1177 (1939); 20 (1941).

The absorption data for the xylenes given in Table 14-8 indicate that bathochromic shifts caused by alkyl disubstitution are usually in the order para > meta > ortho. Alkylbenzenes, like alkyl-substituted alkenes, normally do not undergo any significant spectral changes when the solvent is varied. On the other hand, polar substituents such as $—NH_2$, —OH, $—OCH_3$, —CHO, —COOH, and $—NO_2$ cause marked spectral changes. With these groups, the intensity of the ^1L_b band is enhanced. Much of the fine structure is lost in polar solvents, although it may be observed to some extent in nonpolar solvents. In addition, the ^1L_a band is shifted bathochromically; for example, in aniline, thiophenol, and benzoic acid, it occurs in the 230 nm region (Table 14-8).

Carbonyl substituents possess nonbonding and π electrons. Their π system can conjugate with the π system of an aromatic ring. Since the energy of the π^* state is lowered by delocalization over the entire conjugated system, both the $\pi \rightarrow \pi^*$ and the $n \rightarrow \pi^*$ absorptions occur at longer wavelength than in the corresponding unconjugated chromophoric substituent. For example, acetophenone, $C_6H_5—(CO)—CH_3$, exhibits an $n \rightarrow \pi^*$ absorption at 320 nm and the ^1L_b aromatic transition is at 276 nm. Both bands are shifted bathochromically and are considerably increased in intensity, partly as the result of conjugation of the benzene π system with the π system of the carbonyl group. A similar effect on the $\pi \rightarrow \pi^*$ benzene transition is seen in styrene, in which a C=C bond is conjugated with the benzene ring.

Substitution of benzene by auxochromes (nonchromophoric groups, usually containing unshared electrons such as —Cl, $—NR_2$, —OH), chromophores, or fused rings has varying effects on the absorption spectrum. Because of their importance, we consider these effects in some detail.

The spectral changes found when phenol is converted to the phenoxide (phenylate) anion and when the anilinium cation is converted to aniline are of considerable interest and practical importance. In the case of aniline, the 280 nm band is most probably due to the benzenoid ^1L_b transition, red shifted and enhanced by electron donation from the amino group to the ring. Resonance structures involving intramolecular charge transfer (eq. 14-2) make a substantial contribution to the ground electronic state, but their

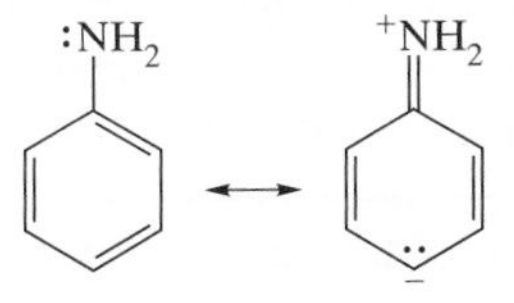

$:NH_2$ ⟷ $^+NH_2$ **(14-2)**

predominant contributions are to the excited state. In general, substituted benzenes for which this type of resonance form can be written have bathochromically shifted spectra relative to benzene and exhibit hyperchromism.

TABLE 14-8 Absorption Data for Benzene and Derivatives[a]

Compound	Solvent	λ_{max} (nm)	ϵ_{max}	λ_{max} (nm)	ϵ_{max}	λ_{max} (nm)	ϵ_{max}	λ_{max} (nm)	ϵ_{max}
Benzene	Hexane	184	68,000	204	8,800	254	250		
	Water	180	55,000	203.5	7,400	254	204		
Toluene	Hexane	189	55,000	208	7,900	262	260		
	Water			206	7,000	261	225		
Ethylbenzene	Ethanol[b]			208	7,800	260	220		
tert-Butylbenzene	Ethanol			207.5	7,800	257	170		
o-Xylene	25% methanol			210	8,300	262	300		
m-Xylene	25% methanol			212	7,300	264	300		
p-Xylene	Ethanol			216	7,600	274	620		
1,3,5-Trimethylbenzene	Ethanol			215	7,500	265	220		
Fluorobenzene	Ethanol			204	6,200	254	900		
Chlorobenzene	Ethanol			210	7,500	257	170		
Bromobenzene	Ethanol			210	7,500	257	170		
Iodobenzene	Ethanol			226	13,000	256	800		
	Hexane			207	7,000	258	610	285(sh)	180
Phenol	Water			211	6,200	270	1,450		
Phenoxide ion	aq NaOH			236	9,400	287	2,600		
Aniline	Water			230	8,600	280	1,400		
	Methanol			230	7,000	280	1,300		
Anilinium ion	aq acid			203	7,500	254	160		
N,N-Dimethylaniline	Ethanol			251	14,000	299	2,100		
Thiophenol	Hexane			236	10,000	269	700		
Anisole	Water			217	6,400	269	1,500		
Benzonitrile	Water			224	13,000	271	1,000		
Benzoic acid	Water			230	10,000	270	800		
	Ethanol			226	9,800	272	850		
Nitrobenzene	Hexane			252	10,000	280(sh)	1,000	330(sh)	140
Benzaldehyde	Hexane			242	14,000	280	1,400	328	55
	Ethanol			240	16,000	280	1,700	328	20
Acetophenone	Hexane			238	13,000	276	800	320	40
	Ethanol			243	13,000	279	1,200	315	55
Styrene	Hexane			248	15,000	282	740		
	Ethanol			248	14,000	282	760		
Cinnamic acid									
cis-	Hexane	200	31,000	215	17,000	280	25,000		
trans-	Hexane	204	36,000	215	35,000	283	56,000		
	Ethanol			215	19,000	268	20,000		
Stilbene									
cis-	Ethanol			225	24,000	274	10,000		
trans-	Heptane	202	24,000	228	16,000	294	28,000		
Phenylacetylene	Hexane	202	44,000	248	17,000	hidden			
2,2′-Dimethylbiphenyl	Hexane	198	43,000	228(sh)	6,000	264	800		
Diphenylmethane	Ethanol			220	10,000	262	500		

From J. B. Lambert, H. F. Shurvell, L. Verbit, R. G. Cooks, and G. H. Stout, *Organic Structural Analysis*, New York: Macmillan Publishing, 1976.

[a]If vibrational structure is present, λ_{max} refers to the subband of highest intensity.

[b]"Ethanol" should be taken to mean 95% ethanol.

Conversion of aniline to the anilinium cation involves attachment of a proton to the nonbonding electron pair, thereby removing it from conjugation with the π electrons of the ring (eq. 14-3).

$$C_6H_5\ddot{N}H_2 \xrightarrow{H_3O^+} C_6H_5\overset{+}{N}H_3 \qquad \textbf{(14-3)}$$

The absorption characteristics of this ion closely resemble those of benzene. The blue shift observed in the conversion of aniline to the anilinium ion is typical of the spectral changes due to protonation of basic groups and can serve as a useful tool in structure elucidation.

Conversion of phenol to the phenoxide anion makes an additional pair of nonbonding electrons available to the conjugated system, and both the wavelengths and the intensities of the absorption bands are increased (Table 14-8). Analogous to the information obtainable in the aniline–anilinium ion conversion, the presence or absence of a phenolic group may be determined by comparing the UV spectra of the compound in a neutral and in an alkaline (pH 13) solution.

The aniline–anilinium or phenol–phenoxide interconversion as a function of pH can demonstrate the presence of the two species in equilibrium by the appearance of an isosbestic point in the UV spectrum (Section 13-3c). If two substances, each of which obeys Beer's law, are in equilibrium, the spectra of all equilibrium mixtures at a constant total concentration intersect at a fixed wavelength, the isosbestic point, at which the absorbances of the two species are equal. An example is shown in Figure 14-4 for 4-methoxy-2-nitrophenol (eq. 14-4).

$$\text{4-}OCH_3\text{-2-}NO_2\text{-}C_6H_3OH \xrightarrow{-H^+} \text{4-}OCH_3\text{-2-}NO_2\text{-}C_6H_3O^- \qquad \textbf{(14-4)}$$

In the UV spectra of the *cis*- and *trans*-cinnamic acids ($C_6H_5CH{=}CHCOOH$), the band at about 280 nm represents the ^1L_b benzenoid absorption displaced to longer wavelength and intensified by conjugation with the double bond and the carbonyl group of the carboxylic acid. The molar absorption coefficient of the trans isomer is more than twice that for the cis and is thought to be related to the longer chromophoric length in

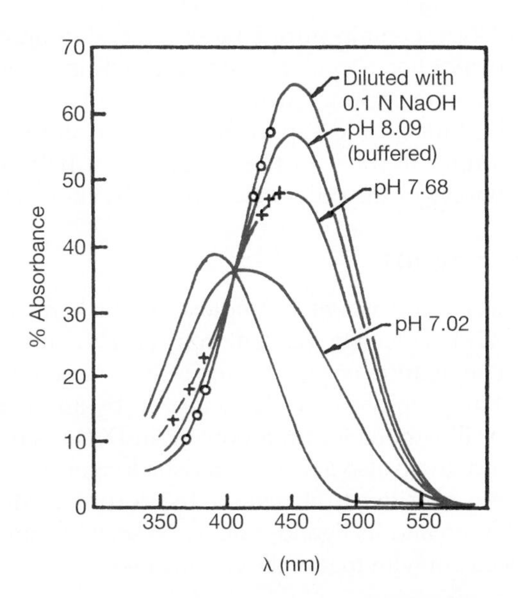

FIGURE 14-4 The spectra of 4-methoxy-2-nitrophenol as a function of pH. The phenol has the shorter-wavelength absorption maximum, the phenoxide the longer. All curves meet at a common point, the isosbestic point, at which the absorbances of the phenol and the phenoxide are equal. (Adapted with permission from H. H. Jaffé and M. Orchin, *Theory and Applications of Ultraviolet Spectroscopy,* New York: John Wiley & Sons, Inc., 1962, p. 562.)

the trans compound. A similar relation is found in the *cis-* and *trans*-stilbenes ($C_6H_5CH{=}CHC_6H_5$). A generally applicable rule for many cis–trans isomer pairs is that the lowest energy $\pi \rightarrow \pi^*$ transition for the trans isomer occurs at longer wavelength and is more intense than that for the cis isomer.

The spectral data for biphenyl ($C_6H_5{-}C_6H_5$) illustrate the effects of conjugation of adjacent benzene chromophores. The spectrum above 185 nm consists of two broad and intense bands at 202 nm (ϵ 44,000) and 248 nm (ϵ 17,000). The strong intensity of the 248 nm band indicates that it corresponds to the 205 nm ^1L_a band of benzene, shifted by conjugation between the two rings. The ^1L_b absorption is concealed beneath the broad envelope of the intense ^1L_a band. The deviation from coplanarity in biphenyl of about 23° still permits considerable conjugation. Ortho substituents, however, increase the deviation of the rings from coplanarity with concomitant loss of conjugation. This effect can be seen in the data for 2,2′-dimethylbiphenyl, which more closely resembles the sum of two independent alkylbenzene systems.

The presence of a saturated methylene group between two chromophores in a molecule may result in complete loss of conjugation. This situation is well illustrated in the data for diphenylmethane ($C_6H_5{-}CH_2{-}C_6H_5$), for which the ^1L_b band at 262 nm has an ϵ of 500, almost exactly the sum of two isolated benzene rings.

14-4 IMPORTANT NATURALLY OCCURRING CHROMOPHORES

14-4a Amino Acids, Peptides, and Proteins

The carboxylate ion chromophore of amino acids exhibits a weak $n \rightarrow \pi^*$ UV–vis absorption near 205 nm and a more intense absorption in the far–UV region near 175 to 190 nm. When aromatic, sulfhydryl, sulfide, or disulfide groups are present, the UV–vis absorption of amino acids is dominated by side chain absorption from these chromophores. For example, in cysteine [$HSCH_2CH(NH_3^+)CO_2^-$] a new, longer-wavelength absorption due to —S— appears at 235 nm (ϵ_{max} 3200); in phenylalanine [$C_6H_5CH_2CH(NH_3^+)CO_2^-$] a benzenoid transition appears near 257 nm (ϵ_{max} 220); and in tryptophan [indole-$CH_2CH(NH_3^+)CO_2^-$] bands are present at 280 nm (ϵ_{max} 5050) and 219 nm (ϵ_{max} 34,000). Tyrosine, which has a phenol residue [4-$HOC_6H_4{-}CH_2CH(NH_3^+)CO_2^-$], exhibits pH-sensitive bands at 274 nm (ϵ 1440) pH 6 and 290 nm (ϵ 2500).

Peptides and proteins may be considered as polymers of amino acids, with the repeating group being the amide or peptide bond, —CO—NH—. Thus, to a first approximation, the UV–vis spectral characteristics of the side chains of amino acids and proteins are the same. The chromophores of proteins that have significant absorption in the UV are given in Table 14-9.

14-4b Nucleic Acids and Polynucleotides

Nucleic acids, the building blocks of DNA, contain purine (adenine and guanine) and pyrimidine (uracil, thymine, and cytosine) heterocycles, which are rich in π electrons and exhibit $\pi \rightarrow \pi^*$ bands in the 240–275 nm region. Nucleosides (with a sugar unit as well as the heterocycle) and nucleotides (nucleoside 5′-phosphates) have almost identical UV–vis spectra, which are very similar to those of their component heterocycles. They also are the component chromophores of the polynucleotides DNA and RNA.

14-4c Porphyrins and Metalloporphyrins

Porphyrins are common natural products found as metal complexes in the heme (14-1) of blood and in the chlorophyll-*a* (**14-2**, p. 426) of green leaves. Chlorophyll-*a* is a dihydroporphyrin (chlorin), but the macrocyclic conjugation of porphyrins is still maintained. The UV–vis spectra of porphyrins and metalloporphyrin are characterized by an intense absorption near 400 nm (the Soret band), as illustrated for protoporphyrin-IX (**14-3**, p. 426) in Figure 14-5. In addition to the Soret band, there also are characteristic longer-wavelength bands that are weak. These bands are very sensitive to changes in the porphyrin structure, especially to the presence of the metal atom and its ligand, as may be seen by comparing the protoporphyrin-IX **14-3** UV–vis spectrum with that of the chlorin **14-2**.

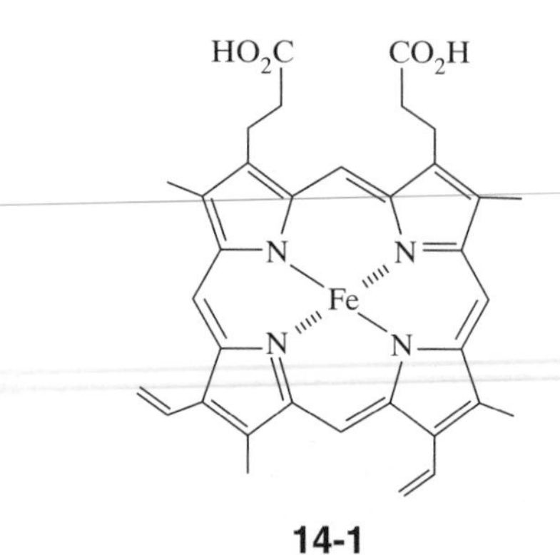

14-1

TABLE 14-9 Important Chromophores of Proteins

Residues	Chromophore	Location (nm)	log ϵ_{max}	Assignment
Peptide bond	CONH	162	3.8	$\pi^{+} \rightarrow \pi^{-}$
		188	3.9	$\pi^{0} \rightarrow \pi^{-}$
		225	2.6	$n \rightarrow \pi^{-}$
Aspartic, glutamic	COOH	175	3.4	$n \rightarrow \pi^{*}$
		205	1.6	$n \rightarrow \pi^{*}$(?)
Aspartate, glutamate	COO^{-}	200	2	$n \rightarrow \pi^{*}$
Lysine, arginine	N—H	173	3.4	$\sigma \rightarrow \sigma^{*}$
		213	2.8	$n \rightarrow \sigma^{*}$
Phenylalanine	Phenyl	188	4.8	
		206	3.9	$\pi \rightarrow \pi^{*}$
		261	2.35	
Tyrosine	Phenolic	193	4.7	
		222	3.9	$\pi \rightarrow \pi^{*}$
		270	3.16	
Tyrosine (ionized)	Phenolate ion	200?	5	
		235	3.97	$\pi \rightarrow \pi^{*}$
		287	3.41	
Tryptophan	Indole	195	4.3	
		220	4.53	$\pi \rightarrow \pi^{*}$
		280	3.7	
		286	3.3	
Histidine	Imidazole	211	3.78	$\pi \rightarrow \pi^{*}$(?)
CysSH	S—H	195	3.3	$n \rightarrow \sigma^{*}$
CysS—	S—	235	3.5	$n \rightarrow \sigma^{*}$
Cystine	—S—S—	210	3	$n \rightarrow \sigma^{*}$
		250	2.5	

Modified from J. Donovan, *Physical Principles and Techniques of Protein Chemistry,* Part A, S. Leach (ed.), New York: Academic Press, 1960.

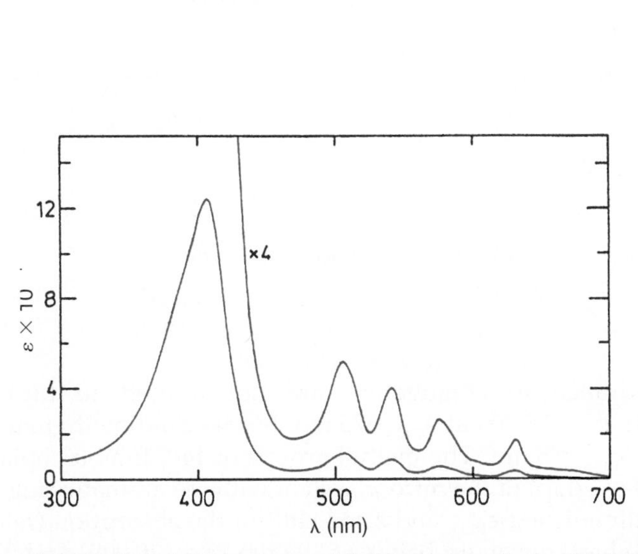

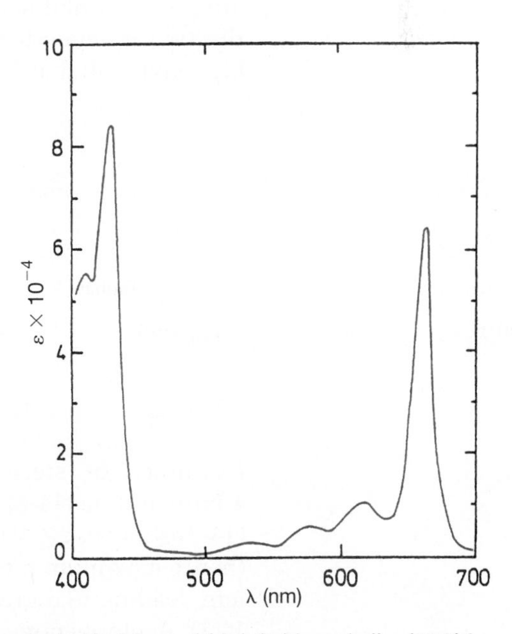

FIGURE 14-5 UV–vis spectra of (left) protoporphyrin-IX (**14-3**) in dimethylsulfoxide, and (right) chlorophyll-*a* (**14-2**) in chloroform. (Right spectrum recorded by P. M. Sabido; left spectrum from data of A. S. Holt, *Chemistry and Physiology of Plant Pigments,* vol. 14, T. W. Goodwin ed., New York: Academic Press, 1976.)

14-2 **14-3**

14-5 STERIC EFFECTS

Stilbene can exist as two stereoisomers: trans (**14-4**) and cis (**14-5**). Only the trans isomer can adopt a conformation easily with maximum coplanarity between the aromatic ring π systems and the central C=C double bond. Consequently, one expects a greater λ_{max} and ϵ_{max} for the long-wavelength electronic transition of the trans isomer.

14-4 **14-5**

14-4 λ_{max} (nm)	14-4 ϵ_{max}	14-5 λ_{max} (nm)	14-5 ϵ_{max}
296	29,000	280	10,500
228	16,500	224	24,000

As explained earlier (Section 14-3), the coupling of two benzene chromophores in biphenyl lowers the energy of the allowed benzene ^1L_a transition from 204 to 248 nm. When coplanarity and hence maximum π overlap of the two biphenyl rings, however, are severely inhibited sterically, the biphenyl derivative behaves more like the sum of the two independent aromatic chromophores. Compare the data for benzene and biphenyl with that for *m*-xylene and 2,2′,6,6′-tetramethylbiphenyl.

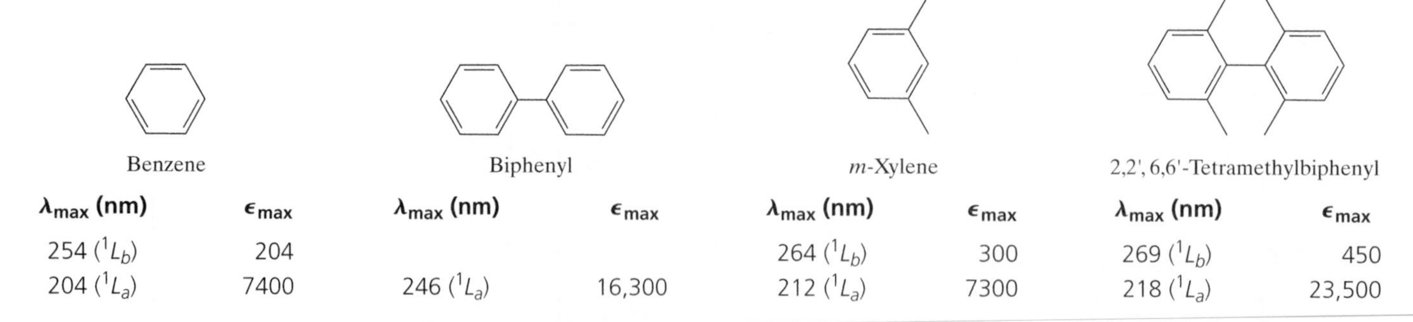

Benzene λ_{max} (nm)	ϵ_{max}	Biphenyl λ_{max} (nm)	ϵ_{max}	*m*-Xylene λ_{max} (nm)	ϵ_{max}	2,2′,6,6′-Tetramethylbiphenyl λ_{max} (nm)	ϵ_{max}
254 (^1L_b)	204			264 (^1L_b)	300	269 (^1L_b)	450
204 (^1L_a)	7400	246 (^1L_a)	16,300	212 (^1L_a)	7300	218 (^1L_a)	23,500

Examples of steric hindrance to conjugation are not limited to biphenyls. 4-Nitroaniline (**14-6**) shows ϵ_{max} 16,000 at λ_{max} 375 nm, but its 2,6-dimethyl analogue (**14-7**) shows ϵ_{max} 4800 at λ_{max} 385 nm. The methyl groups of **14-7** inhibit coplanarity (hence maximum p orbital overlap) of the nitro π system with the aromatic ring π system, leading to a greatly diminished ϵ_{max} and a red shift of the absorption (reference 14.1). Acetophenone (**14-8**) has benzenoid bands at 199 nm (ϵ_{max} 20,000) and 278 nm (ϵ_{max} 1000), an n→π^* conjugated carbonyl absorption at 320 nm (ϵ_{max} 45), and a 243 nm absorption (ϵ_{max} 12,600, ethanol) that has been ascribed (reference 14.8) to an electron transfer (ET) band. A methyl group at the ortho or meta position causes a 3 nm bathochromic shift of the ET band, and a para methyl causes a 10 nm bathochromic shift (ϵ_{max}^{252} 15,100). Yet, in 2,4,6-trimethylacetophenone (**14-9**), the ET band is not shifted (λ_{max} = 242 nm) and the ϵ_{max} is reduced (to 3600). Two ortho methyls effectively inhibit coplanarity of the carbonyl and aromatic π systems.

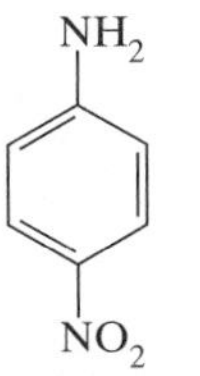

14-6 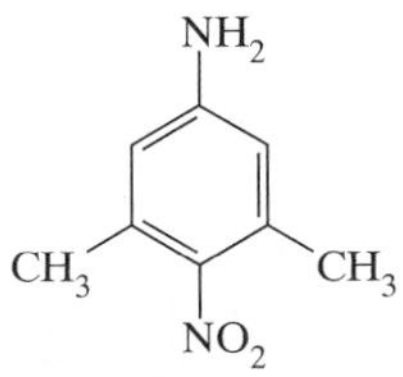 **14-7** 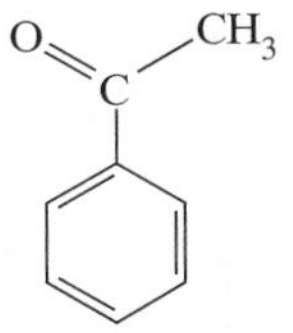 **14-8** 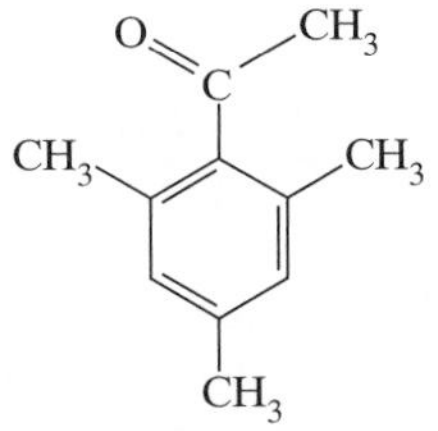 **14-9**

14-6 SOLVENT EFFECTS AND DYNAMIC EQUILIBRIA

Cyclohexane-1,3-dione in cyclohexane solvent exhibits weak absorption near 295 nm (ϵ_{max} ~50), but in ethanol the UV–vis spectrum changes: λ_{max} 255 nm (ϵ_{max} 12,500). In alkaline ethanol the spectrum changes again: λ_{max} 280 nm (ϵ_{max} 20,000). The diketone undergoes the equilibrium shown in eq. 14-5.

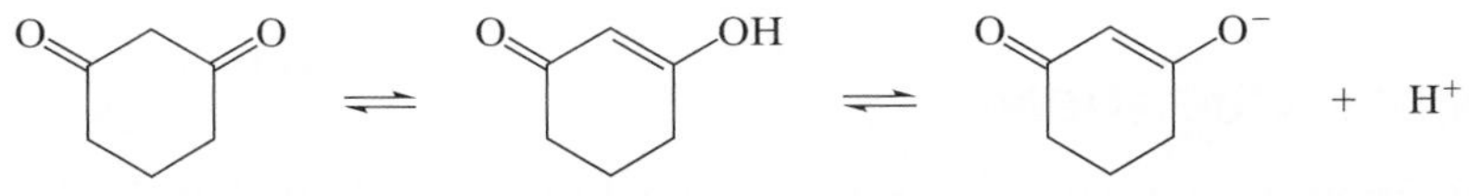

$$\rightleftharpoons \qquad \rightleftharpoons \qquad + \ H^+ \qquad (14\text{-}5)$$

In the hydrocarbon solvent, the equilibrium lies largely in favor of the diketo tautomer, which exhibits the weak λ_{max} = 295 n $\rightarrow \pi^*$ absorption. In the polar, protic solvent, ethanol, the β-hydroxyenone tautomer is favored, and this form shows its λ_{max} at 255 nm for the $\pi \rightarrow \pi^*$ absorption. Deprotonation of this enol under alkaline conditions gives the enolate anion, which absorbs more intensely and at longer wavelengths than the enol form.

When a ketone is dissolved in methanol and a drop of hydrochloric acid is added, an equilibrium is established in which the ketone is converted to its dimethyl ketal, as shown in eq. 14-6.

$$RR'C{=}O \xrightleftharpoons{CH_3OH/H^+} R{-}C(OCH_3)_2{-}R' + H_2O \qquad (14\text{-}6)$$

An analogous reaction can be written for conversion of an aldehyde to an acetal. Although the ketal can be isolated only after removal of the acid catalyst, its formation in solution is monitored readily by UV spectroscopy, since ketal formation causes the carbonyl group and its associated n $\rightarrow \pi^*$ transition to disappear.

14-7 HYDROGEN BONDING STUDIES

Hydrogen bonding between molecules is an important factor that may result in relatively large spectral shifts, which may be used to deduce information about the strength of hydrogen bonds. Experimentally, it is found that absorption bands are blue shifted when the chromophore under investigation functions as a hydrogen bond *acceptor* or are red shifted when it serves as a *donor*. For example, benzthiazoline-2-thione (**14-10**) can function as a hydrogen bond acceptor at the thione group or as a hydrogen bond donor at the NH group. The compound exhibits blue shifts in hydroxylic solvents (—O—H···S=C<) but undergoes bathochromic shifts in the presence of acceptor molecules such as acetone, with which it functions as a hydrogen bond donor (>N—H···O=C<). In indifferent solvents such as CCl_4, there is extensive self-association, the thione acts as both donor and acceptor, and no shifts are observed in the

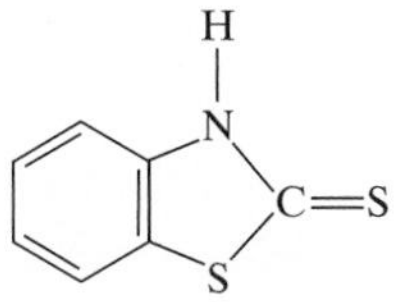

14-10

absorption spectra. The *N*-methyl derivative of benzthiazoline-2-thione, which cannot function as a hydrogen bond donor, exhibits only a blue shift in donor solvents.

The absorption spectra of compounds engaged in *intramolecular* hydrogen bonding are generally solvent insensitive when studied in donor or acceptor solvents. Compare, for example, the behavior of 4-nitrophenol (**14-11**) with that of 2,4- or 2,6-dinitrophenol (**14-12** and **14-13**) in cyclohexane solution. A substantial bathochromic

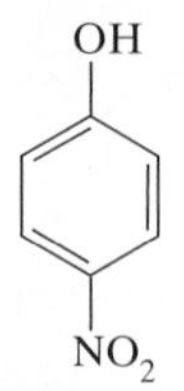

OH NO$_2$ **14-11** | NO$_2$ **14-12** | O_2N **14-13**

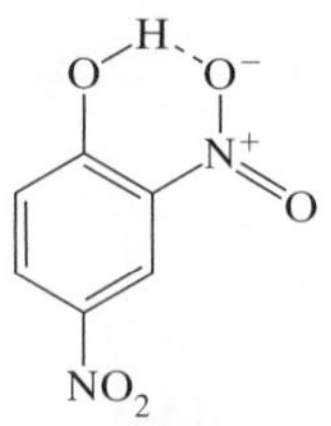

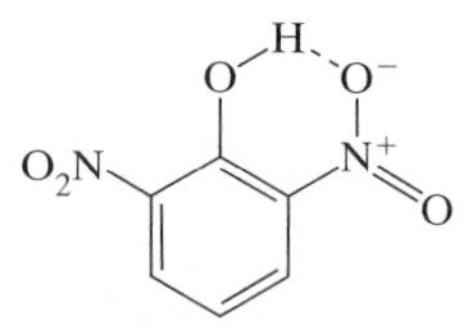

shift is observed when 4-nitrophenol is placed in the presence of a proton acceptor (from 286 to 297 nm with dioxane and from 286 to 307 nm with triethylamine). On the other hand, no spectral shifts are found for 2,4- and 2,6-dinitrophenols when these proton acceptors are added. A vast literature exists on spectral aspects of hydrogen bonding, and several monographs on the topic have been published (reference 14.9).

14-8 HOMOCONJUGATION

Homoconjugation, by which a saturated center intervenes between double bonds, can have a profound effect on the UV–vis spectrum. In 1-norbornenone **(14-14)**, for example, the n → π^* transition is bathochromically shifted and intensified, compared with that in norbornanone **(14-15)**. In addition, a new, moderately intense UV–vis

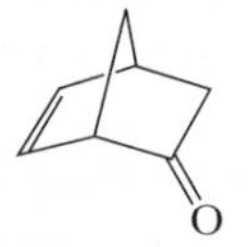

	14-14		14-15
λ_{max} (nm)	304	215	295
ϵ_{max}	290	2,800	23

band appears near 215 nm. The shifts are due to coupling or interaction of locally excited states from the component chromophores. The phenomenon is general, but the magnitudes of the wavelength shifts and of the intensity (ϵ) changes depend very much on the relative orientation of the two chromophores, as demonstrated by the following examples.

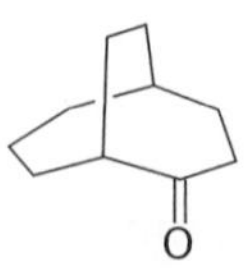

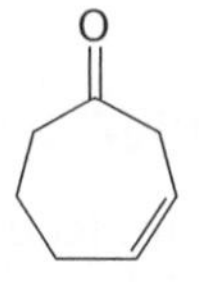

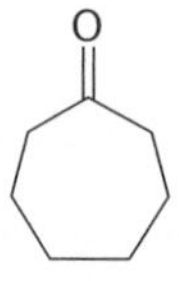

λ_{max} (nm)	307	300	295	274
ϵ_{max}	110	22	450	20

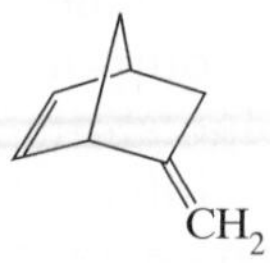

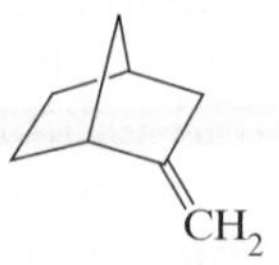

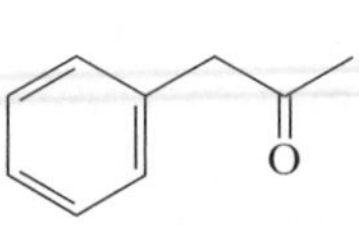

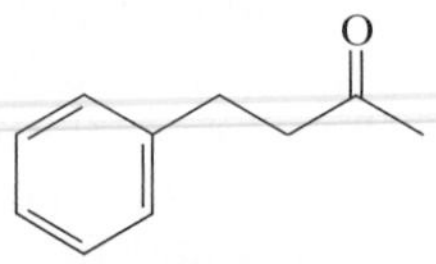

	CH_2	CH_2		
λ_{max} (nm)	207	195	289	281
ϵ_{max}	10,300	10,900	108	32

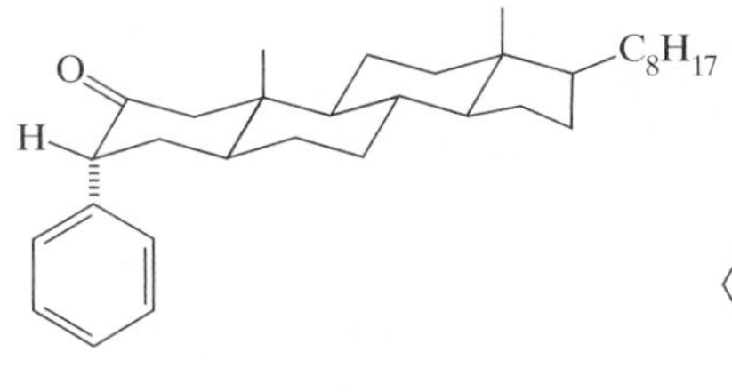

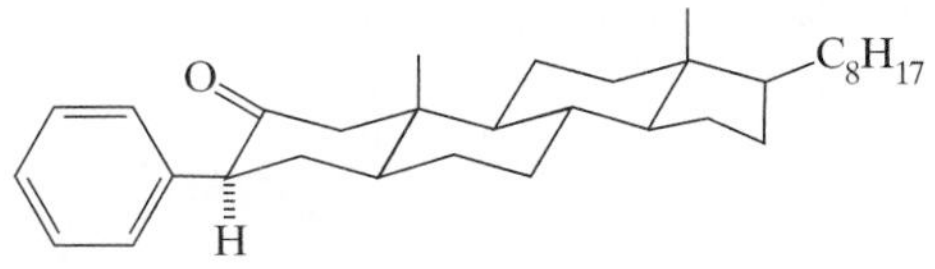

λ_{max} (nm)	299	282
ϵ_{max}	220	60

Even more remote homoconjugation has been detected in the δ,ϵ-unsaturated ketone *trans*-5-cyclodecenone. The transannular interaction between the two chromophores produces a blue-shifted n $\rightarrow \pi^*$ band and a new (charge transfer) band (Section 14-9) near 215 nm. Here again, the alignment of the two chromophores is important, as the transannular interaction appears to be absent in the cis isomer.

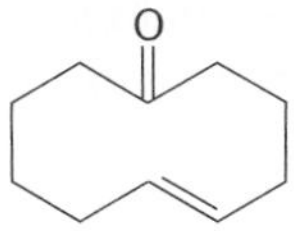

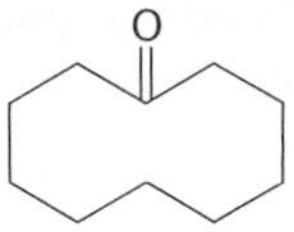

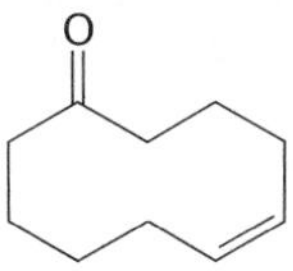

λ_{max} (nm)	279	215	188	290	288
ϵ_{max}	18	2300	8700	15	15

14-9 CHARGE TRANSFER BANDS

Dissolving a sample in a solvent or mixing two compounds in an indifferent solvent for UV measurements may lead to the formation of a new band due to formation of a charge transfer (CT) complex (reference 14.10). For example, iodine in hexane has a violet color, but in benzene it is brown. Tetracyanoethylene (TCNE) and aniline each form colorless solutions in chloroform, but, when mixed, the solution becomes deep blue. Iodine and benzene form a CT complex, as do TCNE and aniline. The observed CT bands arise when a donor molecule, such as iodine or aniline, with a filled orbital, forms a complex with an acceptor molecule, such as benzene or TCNE, which possesses an unoccupied orbital of appropriate symmetry at a slightly higher energy. The CT band is not present in either the isolated donor or the acceptor molecule but is found, usually as a new absorption band, in the CT complex. CT bands usually are broad and quite intense, an important asset since often the equilibrium constants for formation of the complexes are small.

Almost any type of orbital (such as n, σ, π) can function as a donor or an acceptor, but the most common examples involve π orbitals. Some π donors, in increasing order of donor ability (basicity), are benzene < mesitylene < naphthalene < anthracene < *N,N,N′,N′*-tetramethyl-*p*-phenylenediamine. Some examples of π acceptors, also called π acids, in increasing order of acceptor ability (acidity), are *p*-benzoquinone < 1,3,5-trinitrobenzene < chloranil < tetracyanoethylene. The donor–acceptor complex is analogous to the combination of a Lewis acid with a Lewis base. In contrast to the latter complex, however, CT complexes typically are characterized by very weak interactions.

UV spectroscopy has been used in an investigation of the detrimental effects of chlorinated hydrocarbons on living organisms. It is known that these substances attack the central nervous system and block the transport of ions across nerve membranes. The absorption spectrum of the insecticide DDT (**14-16**) has a band at 240 nm, which in the presence of cockroach nerve axons shifts to 245 nm, with a shoulder appearing at 270 nm. DDT may function as a CT acceptor with the nerve as the donor. The resultant CT complex is responsible for deactivation of the nerve function.

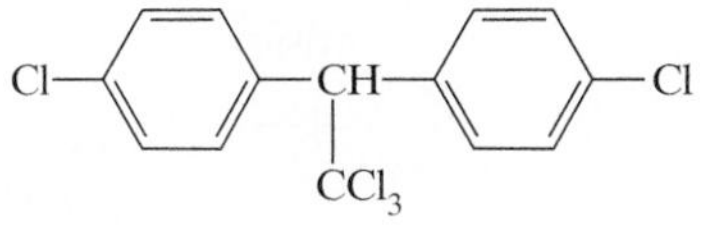

14-16

The strongly electron-deficient compound TCNE forms intermolecular CT complexes with a variety of electron-donor compounds. This marriage between electron donor and acceptor may lead to a new absorption in the UV–vis, typically at longer wavelength than found in the separate components. Such long-wavelength CT bands have been seen in the colored complexes between colorless components: TCNE and *p*-xylene (460 nm), and TCNE and paracyclophane (521 nm).

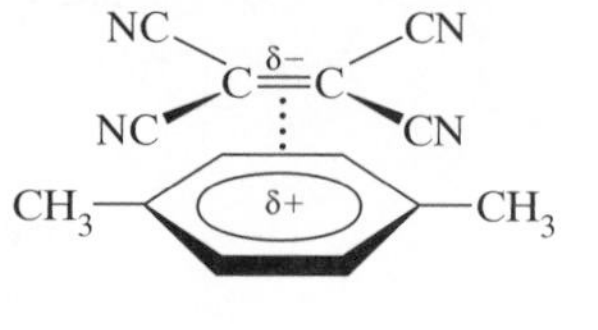

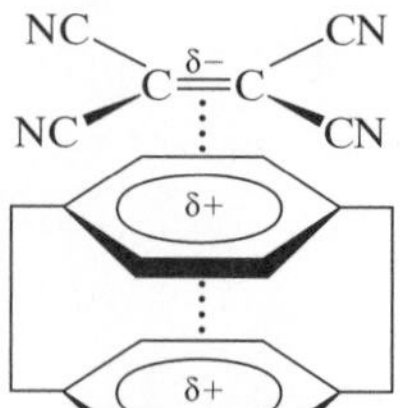

$\lambda_{max}^{CH_2Cl_2}$ (nm)	460	521

Intramolecular CT is thought to be responsible for a weak band corresponding to the inflection near 220 nm in 7-norbornenone (**14-17**) in heptane. In 2,2,3,3-tetrafluoropropanol, the n $\rightarrow \pi^*$ band shifts to 226 nm (ϵ_{max} 49) and the weak band is resolved ($\epsilon_{max}^{235} = 58$) (reference 14.11). No such band is found in the saturated analogue, 7-norbornanone (**14-18**), whose n $\rightarrow \pi^*$ transition lies some 20 nm red shifted relative to 7-norbornenone.

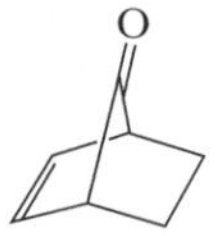

14-17

ϵ_{271}^{max} 35
ϵ_{222}^{infl} 370

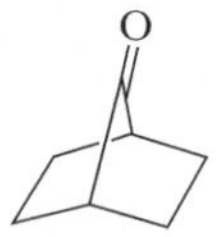

14-18

ϵ_{293}^{max} 18

14-10 THE WOODWARD–FIESER RULES

14-10a Conjugated Dienes and Polyenes

Extensive studies of the UV spectra of alkenes, particularly those of terpenes and steroids, led Woodward and the Fiesers in the early 1940s to formulate empirical rules for the prediction of the wavelength of the $\pi \rightarrow \pi^*$ absorption maxima of various dienes and polyenes. These rules have proved to be quite useful for solving organic structure problems, especially among natural products. The rules assign a base-level absorption maximum for the parent chromophore, and then add specified increments to this value for various substituents attached to the parent π electron system. The values used in the Woodward–Fieser rules for diene absorption are given in Table 14-10.

The following applications of the rules to dienes and trienes clearly reveal the importance of the large contributions from a homoannular diene and from extended conjugation.

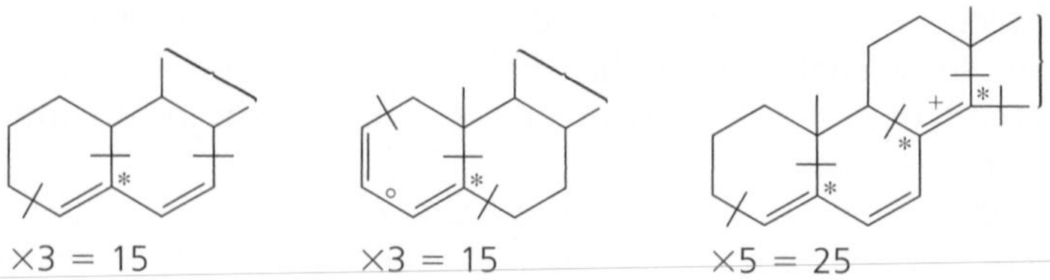

Symbol				
┼	alkyl substituent (5)	×3 = 15	×3 = 15	×5 = 25
*	exocyclic (5)	×1 = 5	×1 = 5	×3 = 15
+	extra conjugation (30)	×0	×0	×1 = 30
o	homoannular (39)	×0	×1 = 39	×0
	Parent	214	214	214
	Predicted λ_{max} (nm)	234	273	284
	Observed[a] λ_{max} (nm)	235	275	283

Data from A. I. Scott, *Interpretation of the Ultraviolet Spectra of Natural Products*, New York: Pergamon Press, 1964.
[a]Steroids.

TABLE 14-10 Woodward–Fieser Rules for the Calculation of Absorption Maximum of Dienes and Polyenes (good to about ±3 nm)

	λ (nm)
Parent chromophore	214
For each alkyl substituent at any position add	5
For each exocyclic double bond add	5
*exocyclic to ring B only	
For each additional conjugated double bond (one end only) add	30
For each homoannular diene (rather than acyclic or heteroannular) add	39
°homoannular (same ring)	

Note: In cases for which both types of diene systems are present, the one with the longer wavelength is designated as the parent system. Do *not* count the additional double bond as a substituent, since this effect is included.

For each polar group	
—O—acyl	0
—OR	6
—SR	30
—Cl, —Br	5
$—NR_2$	60
Solvent correction	0

From R. B. Woodward, *J. Am. Chem. Soc.*, **63**, 1123 (1941); **64**, 72, 76 (1942); L. F. Fieser and M. Fieser, *Natural Products Related to Phenanthrene,* New York: Reinhold, 1949.

Further applications are illustrated by the following examples.

1. Naturally occurring abietic acid and levopimaric acid are both known to have two carbon–carbon double bonds. Their UV–vis spectra clearly indicate that each is a conjugated diene, one with λ_{max} 237.5 (ϵ 10,000) and the other with λ_{max} 272.5 (ϵ 7000). They may be distinguished by the following Woodward–Fieser calculations.

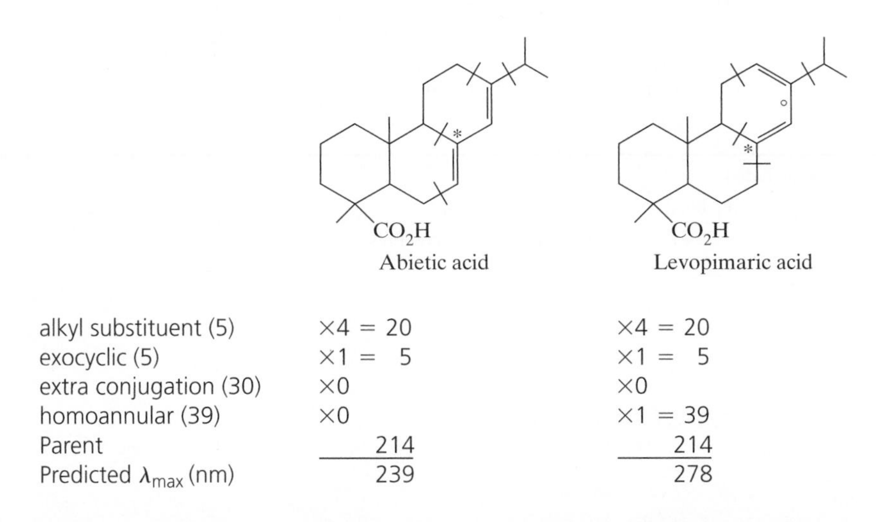

Abietic acid   Levopimaric acid

	Abietic acid	Levopimaric acid
alkyl substituent (5)	×4 = 20	×4 = 20
exocyclic (5)	×1 = 5	×1 = 5
extra conjugation (30)	×0	×0
homoannular (39)	×0	×1 = 39
Parent	214	214
Predicted λ_{max} (nm)	239	278

2. Isomeric trienes may be distinguished by application of the Woodward–Fieser rules. $\Delta^{2,4,6}$-Cholestatriene, $\Delta^{3,5,7}$-cholestatriene, and $\Delta^{5,7,9(11)}$-androstan-3β,17β-diol have UV–vis λ_{max}^{obs} at 324, 306, and 315 nm, respectively. The calculated λ_{max} are obtained as follows:

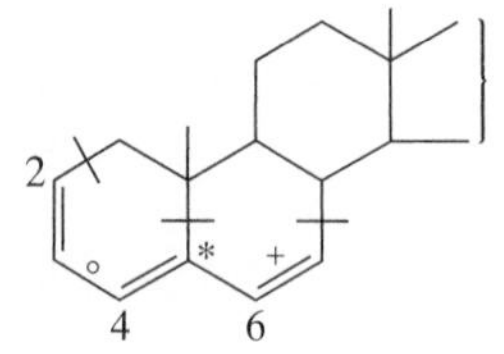

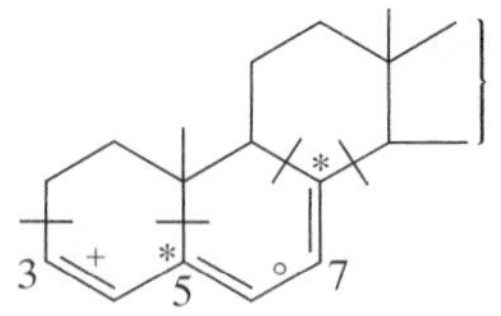

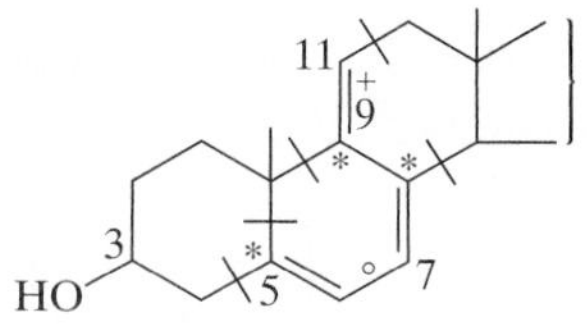

alkyl substituent (5)	×3 = 15	×4 = 20	×5 = 25
exocyclic (5)	×1 = 5	×2 = 10	×3 = 15
extra conjugation (30)	×1 = 30	×1 = 30	×1 = 30
homoannular (39)	×1 = 39	×1 = 39	×1 = 39
Parent	214	214	214
Predicted λ_{max} (nm)	303	313	323
Observed λ_{max} (nm)	306	315	324

Data from A. I. Scott, *Interpretation of the Ultraviolet Spectra of Natural Products*, New York: Pergamon Press, 1964.

14-10b Conjugated Ketones, Aldehydes, Acids, and Esters

As with dienes and polyenes, extensive studies of terpene and steroid enones led Woodward and the Fiesers to formulate empirical rules (Table 14-11) for predicting the wavelength (λ_{max}) of maximum UV–vis absorption for the $\pi \rightarrow \pi^*$ transition. To apply these rules, the base absorption of the parent unit is supplemented according to substituent position and type, extended conjugation, etc. The rules predict λ_{max} for ethanol solutions, but adjustments in λ_{max} can be made for other solvents, as noted.

Examples

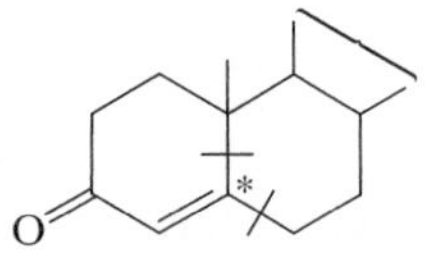

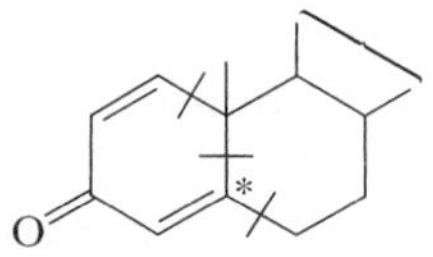

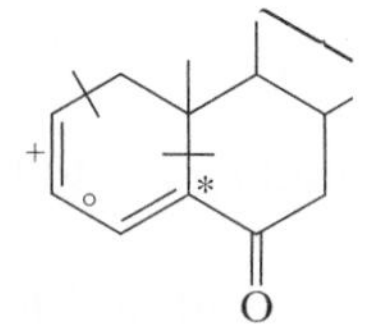

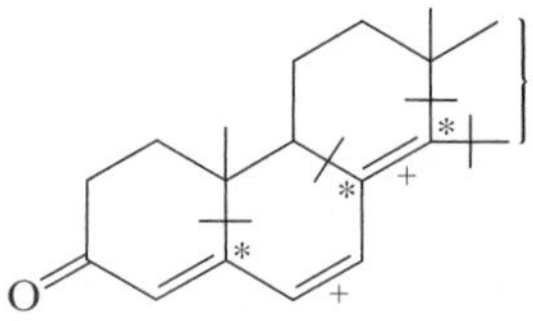

α alkyl (10)	×0	×0	×1 = 10	×0
β alkyl (12)	×2[a] = 24	×2[b] = 24	×0[a]	×1[a] = 12
γ, δ, etc., alkyl (18)	×0	×0	×1 = 18	×3 = 54
exocyclic C=C (5)	×1 = 5	×1 = 5	×1 = 5	×3 = 15
extra conjugation (30)	×0	×0[b]	×1 = 30	×2 = 60
homoannular (39)	×0	×0	×1 = 39	×0
Parent	215	215	215	215
Predicted λ_{max} (nm)	244	244	317	356
Observed[c] λ_{max} (nm)	241	244	314	348

[a]Do not count the double bond as a substituent; this effect is included.

[b]For cross-conjugated systems, count only the more substituted double bond.

[c]Data from A. I. Scott, *Interpretation of the Ultraviolet Spectra of Natural Products*, New York: Pergamon Press, 1964.

TABLE 14-11 Rules for the Calculation of the Position of $\pi \rightarrow \pi^*$ Absorption Maximum of Unsaturated Carbonyl Compounds

β—C(β)=C(α)—C(R)=O and —C(δ)=C(γ)—C(β)=C(α)—C(R)=O		λ (nm)
Parent α,β-unsaturated carbonyl compound		
acyclic, six-membered, or larger ring ketone, R = alkyl		215
five-membered ring ketone		202
aldehyde, R = H		207
acid or ester, R = OH or OR		193
For each alkyl substituent add		
α		10
β		12
If there are other double bonds, for each γ, δ, etc. alkyl substituent, add		18
For each exocyclic carbon–carbon double bond add		5
For each extra conjugation add (do not count the double bond as a substituent, as this effect is included)		30
For each homoannular diene add		39
For each polar group add		
—OH	α	35
	β	30
	δ	50
—O—Ac	α, β, or δ	6
—OR	α	35
	β	30
	γ	17
	δ	31
—SR	β	85
—Cl	α	15
	β	12
—Br	α	25
	β	30
—NR_2	β	95
Solvent correction		
Ethanol, methanol		0
Chloroform		1
Dioxane		5
Diethyl ether		7
Hexane, cyclohexane		11
Water		−8

From R. B. Woodward, *J. Am. Chem. Soc.*, **63**, 1123 (1941); **64**, 72, 76 (1942); L. F. Fieser and M. Fieser, *Natural Products Related to Phenanthrene*, New York: Reinhold, 1949; A. I. Scott, *Interpretation of the UV Spectra of Natural Products*, New York: Pergamon Press, 1964.

The rules do very well in predicting the values of the simple enone, the homoannular dienone, and the cross-conjugated dienone on page 432, but less well for the trienone. Further examples of applications are illustrated as follows.

1. Can the regioisomeric ketones $\Delta^{3,5}$-cholestan-7-one and $\Delta^{3,5}$-cholestan-2-one be distinguished by UV–vis spectroscopy?

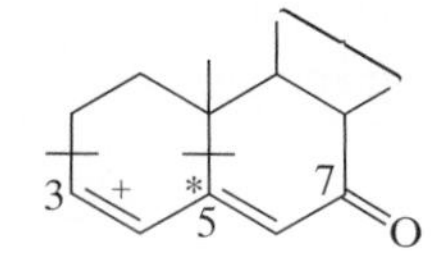

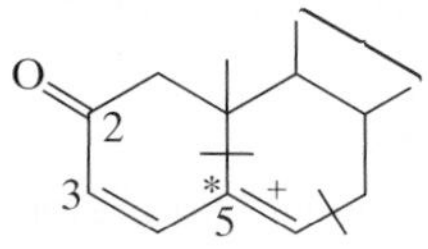

	$\Delta^{3,5}$-cholestan-7-one	$\Delta^{3,5}$-cholestan-2-one
α alkyl (10)	×0	×0
β alkyl (12)	×1 = 12	×0
γ, δ, etc., alkyl (18)	×1 = 18	×2 = 36
exocyclic C=C (5)	×1 = 5	×1 = 5
extra conjugation (30)	×1 = 30	×1 = 30
Parent	215	215
Predicted λ_{max} (nm)	280	286
Observed λ_{max} (nm)	277	290

Data from A. I. Scott, ***Interpretation** of the Ultraviolet Spectra of Natural Products*, New York: Pergamon Press, 1964.

2. The trienone steroids shown below have UV–vis λ_{max} at 348 and 388 nm. Use the Woodward–Fieser rules to distinguish them.

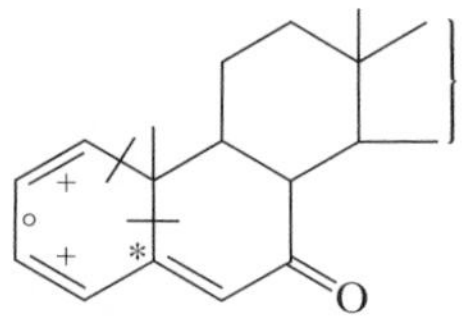

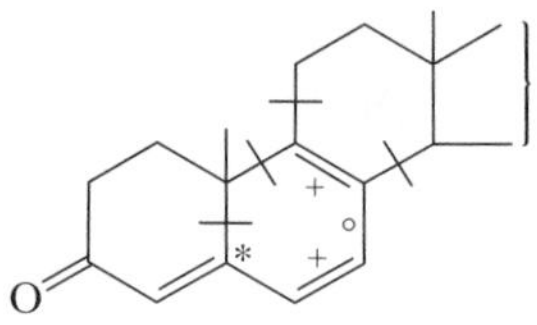

α alkyl (10)	×0	×0
β alkyl (12)	×1 = 12	×1 = 12
γ, δ, etc., alkyl (18)	×1 = 18	×3 = 54
exocyclic C=C (5)	×1 = 5	×1 = 5
extra conjugation (30)	×2 = 60	×2 = 60
homoannular diene (39)	×1 = 39	×1 = 39
Parent	215	215
Predicted λ_{max} (nm)	349	385

14-11 WORKED PROBLEMS AND EXAMPLES

PROBLEM 14-1

Explain the following data for azobenzene ($C_6H_5—N{=}N—C_6H_5$) in isooctane.

trans (anti)	λ_{max} = 318 nm	ϵ_{max} = 22,600
cis (syn)	λ_{max} = 282 nm	ϵ_{max} = 5,200

Answer. As in stilbenes (Section 14-5), steric hindrance in the syn or cis isomer of azobenzene inhibits mutual coplanarity of the two phenyl rings. The connecting —N=N— linkage thereby limits conjugation and raises the excitation energy (decreasing λ_{max}). The reduced ϵ_{max} in the cis isomer follows from a shorter distance between the ends of the conjugated system. As a rough rule of thumb, the greater the distance between the ends of a conjugated chromophore, the greater is ϵ_{max}.

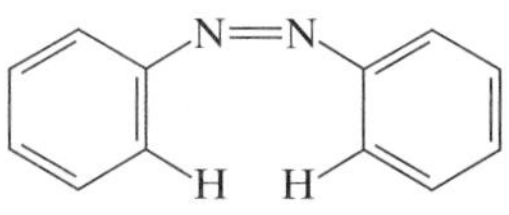

N=N
H H

PROBLEM 14-2

Explain the following UV–vis data.

	CH_3, CH_3, *t*-Bu diene	CH_3, CH_3, CH_3 diene
λ_{max}^{EtOH}	225	231
ϵ_{max}^{EtOH}	6400	10,000

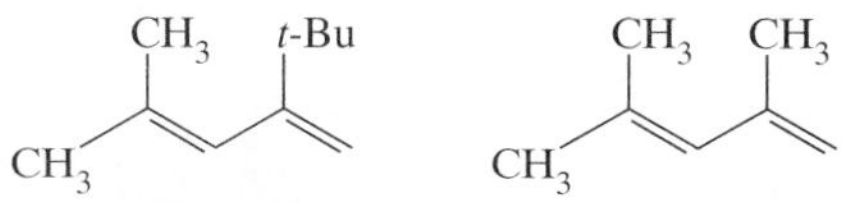

Answer. A nonbonded steric interaction between CH_3 and *tert*-butyl is stronger than a nonbonded CH_3/CH_3 interaction in destabilizing the s-trans conformation shown. In the *tert*-butyl analogue, conjugation is inhibited by rotation about the sp^2-sp^2 C—C bond, leading to a twisted diene.

PROBLEM 14-3

Explain the following UV–vis data for $(CH_2)_n$ cyclic dienes

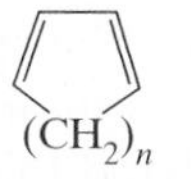

n	λ_{max} (nm)	ϵ_{max}
2	258	10,000
3	248	7,500
4	228	5,600
6	<215	end absorp.

Answer. With increasing ring size, the diene deviates increasingly from coplanarity. As the diene twists out of coplanarity, the $\pi \rightarrow \pi^*$ transition energy increases (decreasing λ_{max}) and the probability of the excitation (ϵ) decreases.

PROBLEM 14-4

How would you distinguish the following pairs of compounds using UV–vis spectroscopy?

(a) $CH_3CH{=}CH{-}CH{=}CH{-}CH_3$ and $CH_3CH{=}CH{-}CH_2{-}CH{=}CH_2$

(b) 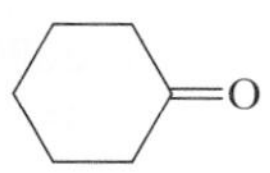 cyclohexanone (=O) and δ-valerolactone (=O, O)

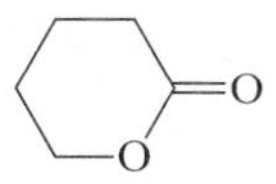

(c) 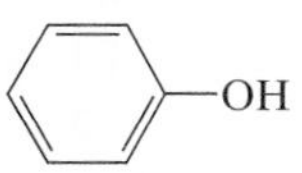 phenol (—OH) and benzaldehyde (—CHO)

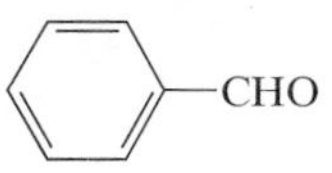

Answer. In (a), the upper diene is conjugated and therefore should absorb longer-wavelength light than does the lower diene, which is not conjugated (Table 14-4). In (b), the n $\rightarrow \pi^*$ absorption of the ketone appears near 290 nm (Table 14-1), but in the ester (lactone) it appears near 210 nm (Table 14-2). In (c), phenol has a moderately strong, long-wavelength absorption near 270 nm, which shifts to near 290 nm and intensifies upon basification. Benzaldehyde has a moderately strong absorption near 280 nm that does not shift appreciably upon basification.

PROBLEM 14-5

UV–vis spectroscopy was an important link in the chain of reasoning used in the structure elucidation of mangostin, the yellow coloring matter of the East Indian mangosteen tree. The UV–vis spectrum of mangostin (Table 14-12) was found to be similar to those of known polyhydroxyl derivatives of xanthone (**14-19**). The IR spectrum con-

14-19

firmed the presence of a highly conjugated carbonyl group. The molecular formula of mangostin indicated six oxygen atoms. The xanthone nucleus accounts for two of them; chemical tests and degradation work showed the remaining oxygens to come from three hydroxyls and one methoxy group. Information on the location of the oxygen functions was provided by UV spectroscopy. Comparison of the spectra of xanthones having only one peri hydroxyl group (that is, ortho to the carbonyl group) with those of xanthones having two peri hydroxyl groups (Table 14-12) indicates that introduction of a second peri hydroxyl results in bathochromic shifts and decreased intensities of the two longest-wavelength bands, while the short wavelength band undergoes a hypsochromic shift. The spectra of mangostin and derivatives do not show these spectral changes. Thus mangostin was considered to possess a single peri hydroxyl group. The side chains of mangostin absorb two moles of hydrogen upon catalytic hydrogenation, indicating the presence of two double bonds (it was possible to rule out a triple bond). The double bonds cannot be conjugated with the xanthone ring or with each other, since the UV spectrum of tetrahydrodimethylmangostin was found to be essentially identical

TABLE 14-12 Absorption Data of Mangostin and Some Polyhydroxyxanthones in 95% Ethanol

Compound	λ_{max} (nm) (log ϵ)			
One peri hydroxyl group				
Mangostin	243(4.5)	259(4.4)	318(4.4)	351(3.9)
1,6-Dihydroxyxanthone	247(4.3)	263(4.0)	305(4.1)	355(3.8)
1,3,6-Trihydroxyxanthone	237(4.6)	251(4.4)	313(4.4)	337(4.1)
1,3,7-Trihydroxy-6-methoxyxanthone	239(4.3)	256(4.5)	310(4.2)	362(4.0)
Two peri hydroxyl groups				
1,8-Dihydroxyxanthone	229(3.8)	252(3.9)	334(3.3)	380(2.9)
HO O OH / CH$_3$ / OH	231(4.0)	263(4.3)	343(3.8)	400(3.2)

From P. Yates and G. H. Stout, *J. Am. Chem. Soc.*, **80**, 1691 (1958); P. Yates and A. Ault, *Tetrahedron*, **23**, 3307 (1967).

to that of the unhydrogenated analogue. Hence it was concluded that the double bonds were isolated. Determination of the number of side chains showed there to be one double bond in each of the two side chains. The overall results of the chemical and spectroscopic work led to structure **14-20** for mangostin.

HO O OCH$_3$ HO O OH

14-20

PROBLEM 14-6

Calculate λ_{max} for the following dienes.

alkyl substituent (5)	0 × 5 = 0	2 × 5 = 10	3 × 5 = 15
Parent	214	214	214
Answer	214	224	229
Observed λ_{max}^{EtOH} (nm)	217	226	231
ϵ_{max}^{EtOH}	22,400	23,800	10,000

PROBLEM 14-7

The compounds below have UV–vis absorption λ_{max} at 234, 244, and 273 nm. Which compound has which λ_{max}?

alkyl substituent (5)	3 × 5 = 15	4 × 5 = 20	4 × 5 = 20
exocyclic (5)	1 × 5 = 5	2 × 5 = 10	0 × 5 = 0
homoannular (39)	0	0	39
Parent	214	214	214
Answer	234	244	273

PROBLEM 14-8

The ketones below have λ_{max} at 236, 244, and 256 nm. Which ketone has which absorption?

O O O

α-substituent (10)	1 × 10 = 10	1 × 10 = 10	0 × 10 = 0
β-substituent (12)	1 × 12 = 12	2 × 12 = 24	2 × 12 = 24
exocyclic (5)	0	2 × 5 = 10	1 × 5 = 5
Parent	215	215	215
Answer	237	259	244
Observed λ_{max}	236	256	244

PROBLEM 14-9

Two isomeric enamines are formed from 10-methyl-$\Delta^{1(9)}$-octalin-2-one. Can they be distinguished by UV–vis spectroscopy?

alkyl substituent (5)	×3 = 15	×3 = 15
exocyclic (5)	×1 = 5	×1 = 5
extra conjugation (30)	×0	×0
homoannular (39)	×1 = 39	×0
polar group (60)	×1 = 60	×1 = 60
Parent	214	214
Answer	333	294

PROBLEM 14-10

Conversion of cholesta-4,6-dien-3-dione (right) to its enol acetate gave a product with $\lambda_{max}^{EtOH} = 302$ nm, $\epsilon_{max}^{EtOH} = 12{,}600$. Which enol acetate was formed?

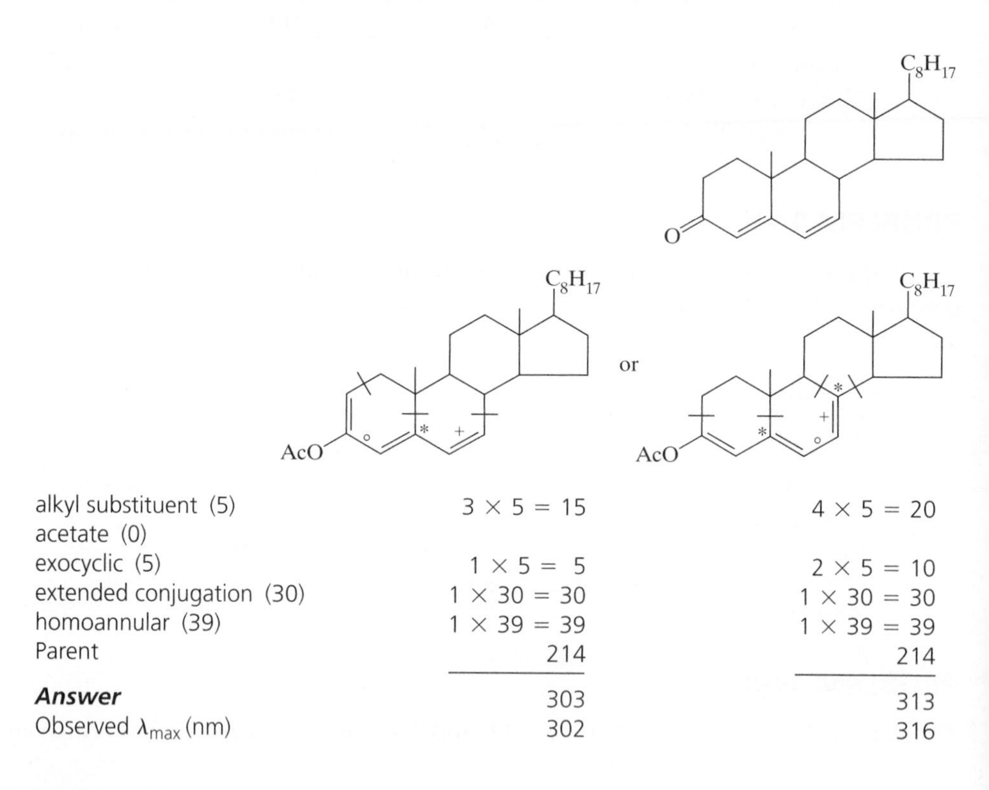

alkyl substituent (5)	3 × 5 = 15	4 × 5 = 20
acetate (0)		
exocyclic (5)	1 × 5 = 5	2 × 5 = 10
extended conjugation (30)	1 × 30 = 30	1 × 30 = 30
homoannular (39)	1 × 39 = 39	1 × 39 = 39
Parent	214	214
Answer	303	313
Observed λ_{max} (nm)	302	316

PROBLEM 14-11

Determine the position of the tautomeric equilibrium between hydroxypyridine and pyridinone from the observed UV–vis spectrum (λ_{max} 224, ϵ_{max} 7200 and λ_{max} 293, ϵ_{max} 5900).

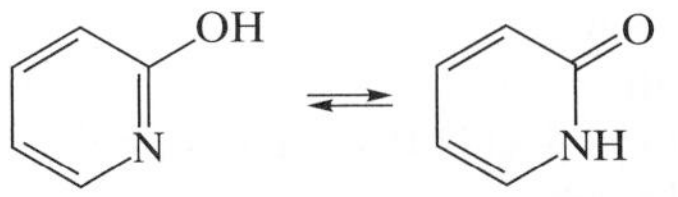

Data for:

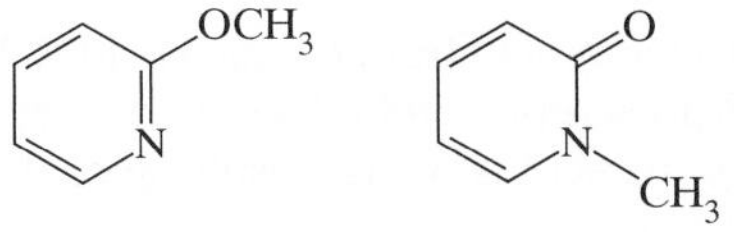

λ_{max}	269	<205	297	226
ϵ_{max}	3200	>5300	5900	6100

Answer. The equilibrium lies largely to the pyridinone side.

PROBLEM 14-12

Explain the following observations.

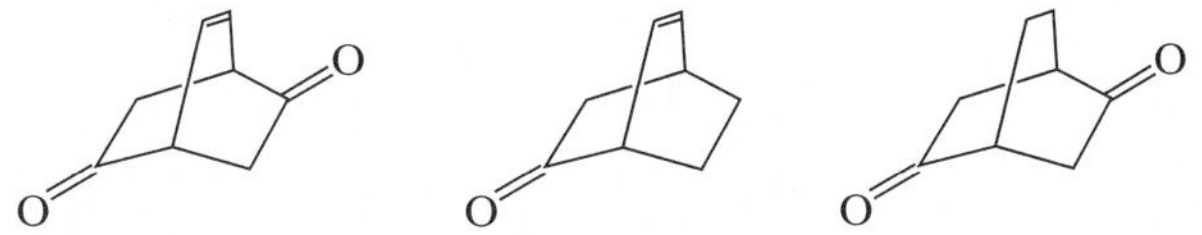

$n \rightarrow \pi^*$	ϵ_{307}^{max} 267	ϵ_{296}^{max} 122	ϵ_{296}^{max} 32
	ϵ_{296}^{max} 267		
$\pi \rightarrow \pi^*$	ϵ_{223}^{max} 2290		ϵ_{203}^{max} 3100

Answer. The enedione exhibits a more intense ϵ_{max} with a longer wavelength λ_{max} than in either the dione or the enone, suggesting that the C=C acts to link the two C=O groups electronically. The ϵ_{max} value of the enedione is slightly more than twice that of the enone, and λ_{max} is shifted to longer wavelength, suggesting that the former is something more than two independent enones. The band near 223 nm in the enedione suggests charge transfer.

Problems

14-1 Using the data in Table 14-1, calculate the electronic transition energies (in kcal mol^{-1}) for the $n \rightarrow \pi^*$ and $\pi \rightarrow \pi^*$ excitations of acetaldehyde.

14-2 At pH 13 the absorbance of a certain phenol is 1.5 at 430 nm and 0.0 at 290 nm. At pH 4, the values for a solution of the same concentration are 0.0 and 0.5, respectively. At pH 8, the values are 0.6 and 0.3, respectively.

(a) Explain the spectral changes.
(b) Calculate the pK_a of the phenol.
(c) When the concentration used is 10.8 mg of phenol in 20 mL of solvent, with ϵ = 300 liter mol^{-1} cm^{-1} at 290 nm and a cell length of 1 cm, calculate the molecular weight of the phenol.
(d) Draw possible structure(s) for the substance that absorbs light at 290 nm.

14-3 Estimate the K_a of a weak acid from the data below. Samples (1 g) are dissolved in equal quantities of the various buffers, and all solutions are measured under the same conditions. The anion of the acid is the only substance that absorbs at the wavelength used.

pH	4	5	6	7	8	9	10	11
A	0.00	0.00	0.06	0.39	0.95	1.13	1.18	1.18

14-4 At pH 13, the absorbance of a particular phenol solution is 1.5 at 400 nm and 0.0 at 270 nm. At pH 4, the values for a solution of the same concentrations are 0.0 and 1.0 at these two wavelengths, respectively. At pH 9, the values are 0.9 and 0.4, respectively.

(a) Explain the spectral change.
(b) Calculate the pK_a of the phenol.
(c) When the concentration used is 18.8 mg of phenol in 20 ml of solvent, for an $\epsilon_{270} = 100$ liter mol^{-1} cm^{-1} and a cell length of 1 cm, calculate the molecular weight of the phenol.
(d) Draw a possible structure.

14-5 (a) Estimate the K_a of a weak acid from the data below. All of the various buffered solutions are one millimolar in sample, and all solutions were measured under the same conditions. The anion of the acid is the only substance that absorbs at the wavelength used.

pH:	4	5	6	7	8	9	10	11
A:	0.00	0.00	0.10	0.75	1.00	1.25	1.50	1.50

(b) What is the value of the molar extinction coefficient for the anion at this wavelength if a 1 cm cell is used?

14-6 Explain the following results.

λ_{max} (nm)	328	313	299
ϵ_{max}	56,200	30,600	29,500

14-7 Account for the following observations.

(a)

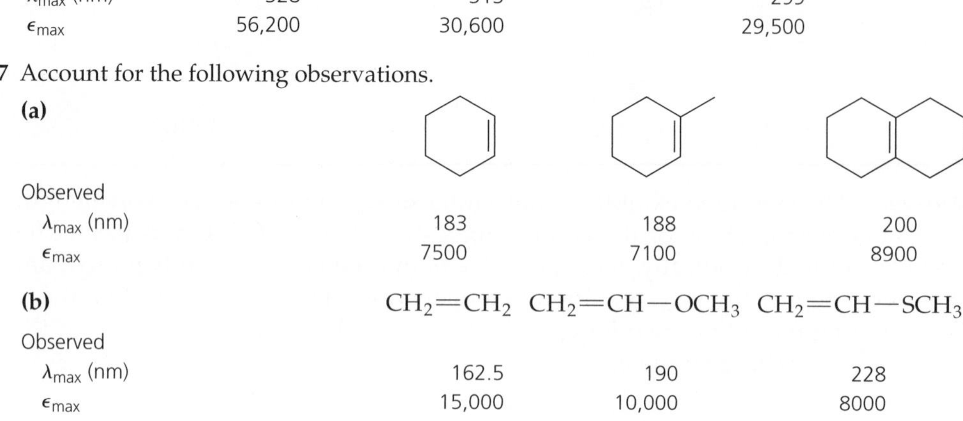

Observed			
λ_{max} (nm)	183	188	200
ϵ_{max}	7500	7100	8900

(b)

	$CH_2{=}CH_2$	$CH_2{=}CH{-}OCH_3$	$CH_2{=}CH{-}SCH_3$
Observed			
λ_{max} (nm)	162.5	190	228
ϵ_{max}	15,000	10,000	8000

(c)

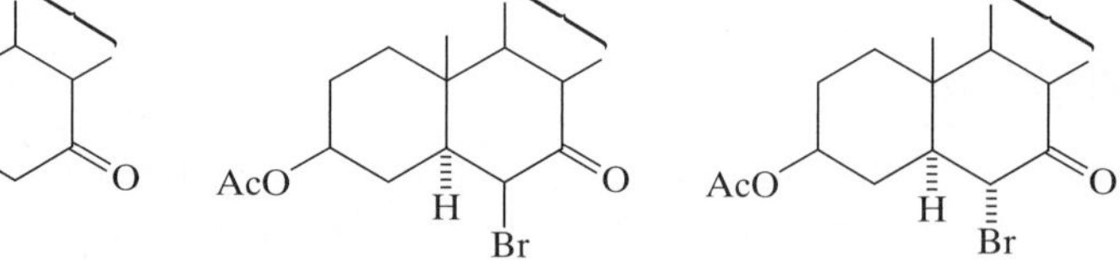

Observed			
λ_{max} (nm)	287	313	282
ϵ_{max}	40	158	40

(d)

Observed					
λ_{max} (nm)	466	380	298	337	299
ϵ_{max}	31	11	29	34	34

(e)

Observed				
λ_{max} (nm)	232	232	245	243
ϵ_{max}	12,500	12,000	6500	1400

14-8 Calculate the λ_{max} for each of the following compounds.

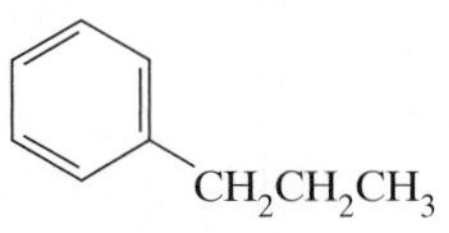

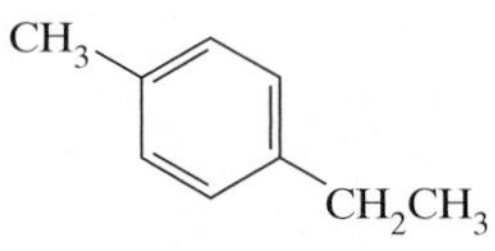

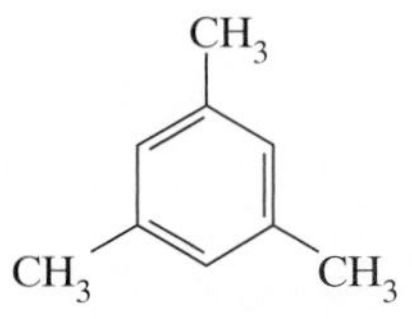

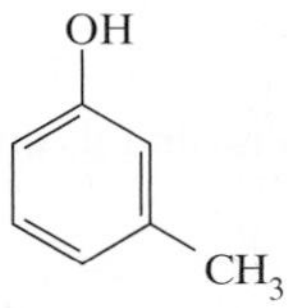

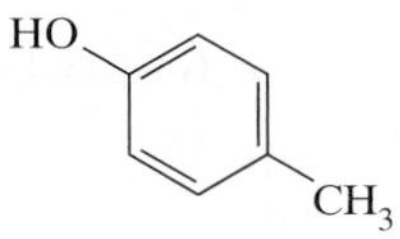

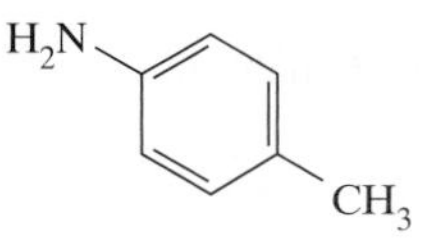

$CH_2CH_2CH_3$ CH_3 CH_2CH_3 CH_3 CH_3 CH_3

OH CH_3 HO CH_3 H_2N CH_3

14-9 Calculate λ_{max} for the following steroids.[a]

(a)

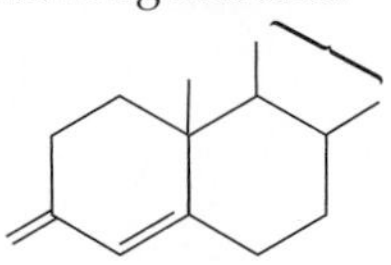

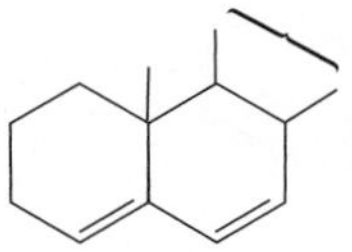

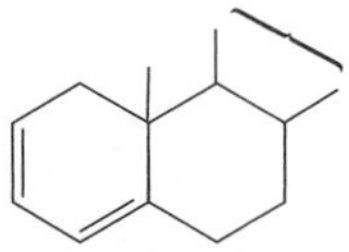

Observed			
λ_{max} (nm)	239	235	275
ϵ_{max}	17,300	19,000	10,000

(b)

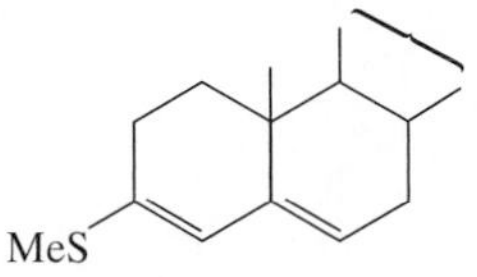

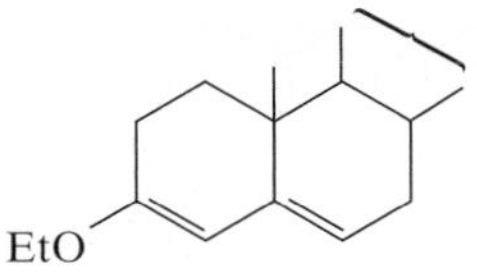

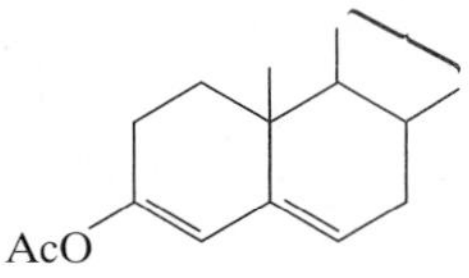

MeS EtO AcO

Observed			
λ_{max} (nm)	268	241	235
ϵ_{max}	22,600	22,600	19,000

(c)

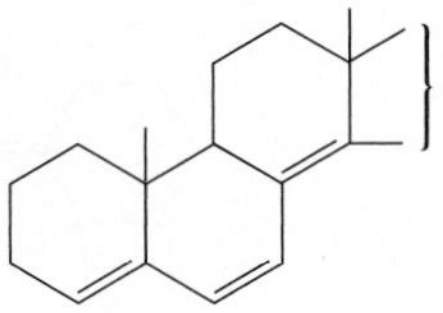

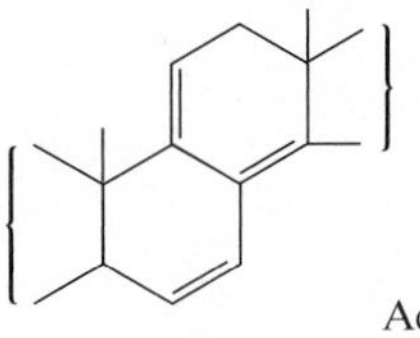

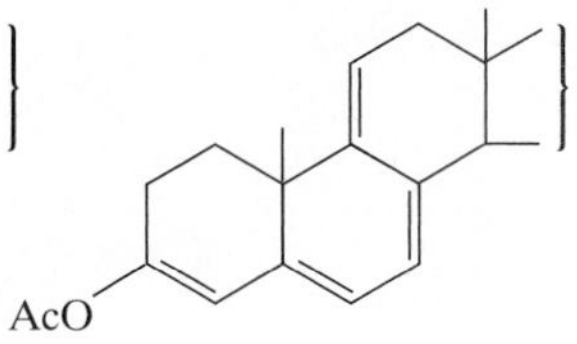

AcO

Observed			
λ_{max} (nm)	283	285	355
ϵ_{max}	33,000	9,100	19,700

(d)

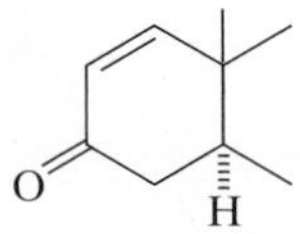

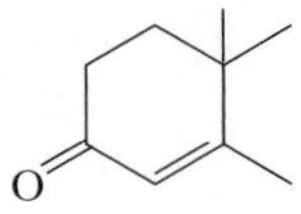

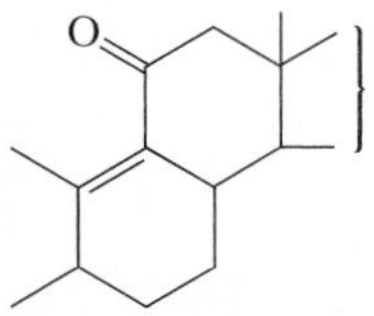

O H O O

Observed			
λ_{max} (nm)	230	241	254
ϵ_{max}	10,000	16,600	9100

(e)

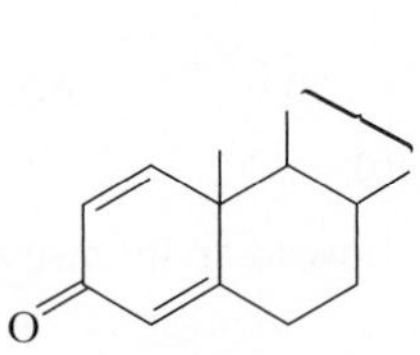

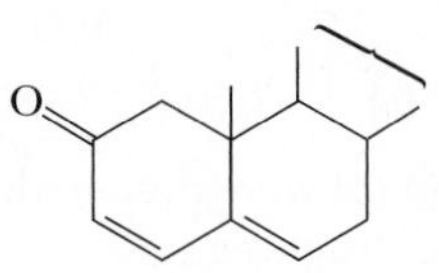

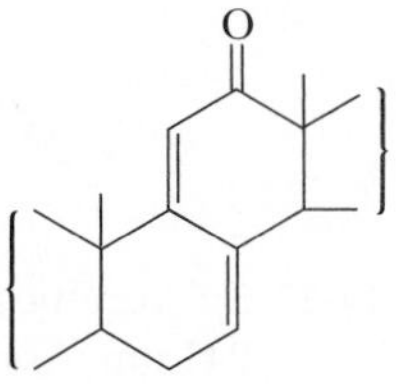

O O O

Observed			
λ_{max} (nm)	244	290	292
ϵ_{max}	15,000	12,600	13,000

[a]Observed data from A. I. Scott, *Interpretation of the Ultraviolet Spectra of Natural Products,* New York: Pergamon Press, 1964.

(f)

Observed			
λ_{max} (nm)	348	348	327
ϵ_{max}	11,000	26,500	

14-10 Calculate the $\pi \rightarrow \pi^*$ transition λ_{max} for cholesta-4,6-diene-3-one **(a)** and its enol acetate **(b).**

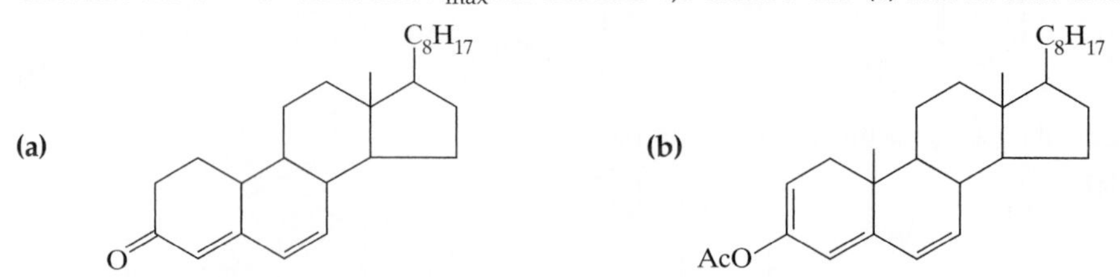

14-11 What UV–vis λ_{max} would you predict for the $\pi \rightarrow \pi^*$ transitions of the following compounds?

14-12 Calculate the approximate λ_{max} for the $\pi \rightarrow \pi^*$ transition of each of the following compounds.

14-13 Calculate the approximate λ_{max} for the $\pi \rightarrow \pi^*$ transitions of each of the following compounds.

14-14 An enol acetate of cholest-4-ene-3-one **(a)** is prepared and has $\lambda_{max} = 238$ nm with $\log \epsilon_{max} = 4.2$. Is the enol acetate **b** or **c**?

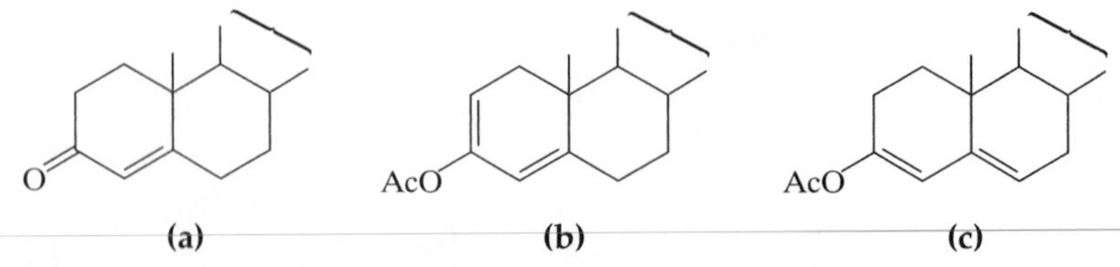

14-15 Spiroenones **(a)** and **(b)** were prepared. One showed an intense λ_{max} at 247 nm, the other at 241 nm. Assign the structures.

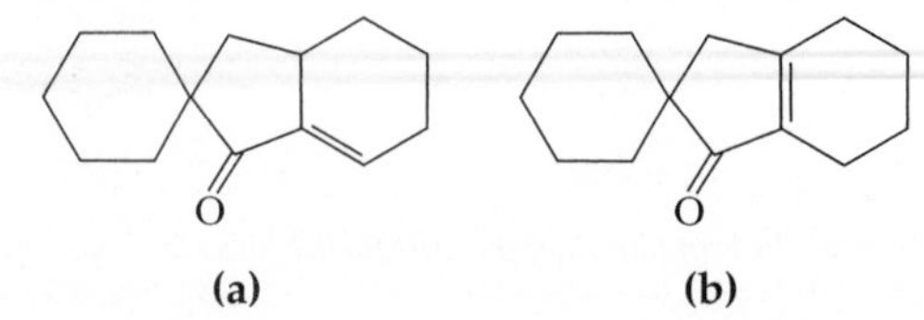

14-16 Predict and explain whether UV–vis spectroscopy can be used for distinguishing members of the following isomeric pairs.

(a) and

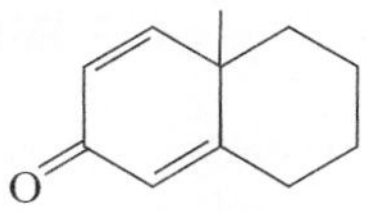

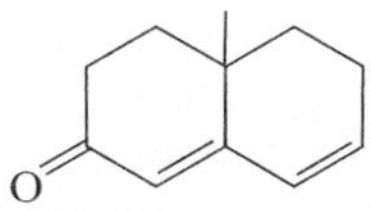

(b) 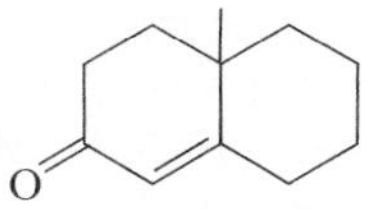 and

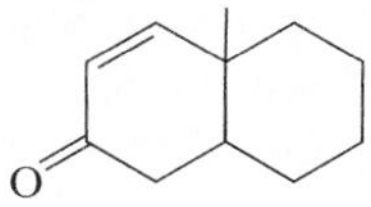

(c) CH_2OH / CH_3 and OH / CH_2CH_3

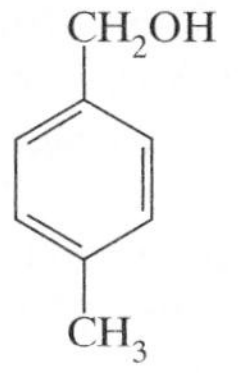

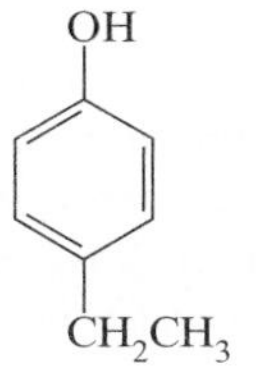

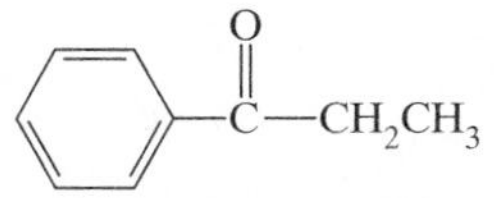

(d) C_6H_5—C(=O)—CH_2CH_3 and C_6H_5—CH_2C(=O)—CH_3

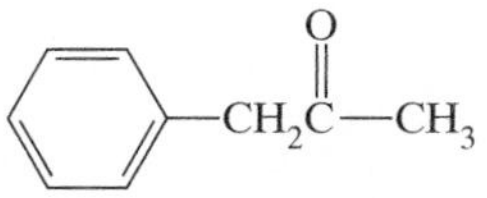

(e) $CH_3CH_2COOCH_3$ and $CH_3COOCH_2CH_3$

(f) $CH_3—CH=CH—CH_2—CH=CH—CH_3$ and $CH_3CH_2—CH=CH—CH=CH—CH_3$

14-17 A compound (A), $C_{11}H_{16}$, has λ_{max} 288 nm. On treatment with Pd/C (which dehydrogenates cyclic compounds completely to aromatic compounds without rearrangement), α-methylnaphthalene (below) is produced. What is the structure of A?

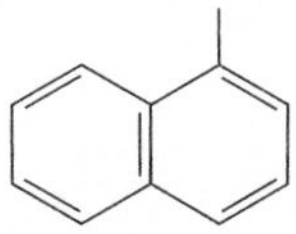

14-18 An unknown monocyclic hydrocarbon A, C_8H_{14}, has a λ_{max} at 234 nm and could be selectively ozonized to yield B, $C_7H_{12}O$, which has λ_{max} at exactly 239 nm. Reduction of B with $LiAlH_4$ and careful elimination of water (dehydration) gave C, C_7H_{12}, which has a λ_{max} at 267 nm. Give structures for A, B, and C.

14-19 Explain the following observations.

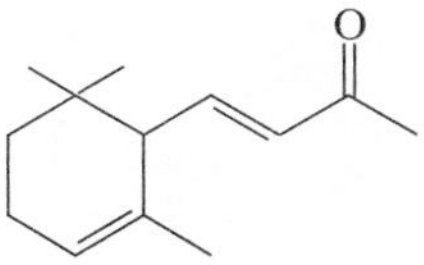

	α-Ionone	β-Ionone	Ψ-Ionone
Observed			
λ_{max} (nm)	228	281	291
ϵ_{max}	14,300	9500	21,800

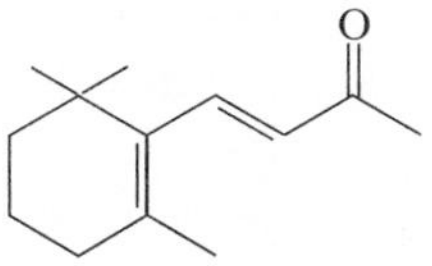

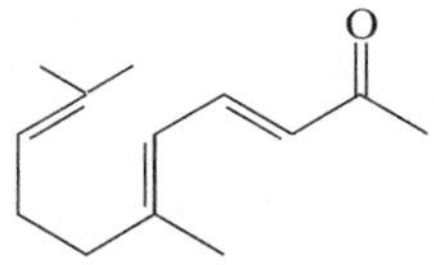

14-20 An unsaturated ketone, $C_9H_{12}O$, was thought to have the structure given below. Its UV spectrum showed λ_{max} 300 nm with $A = 1.34$.

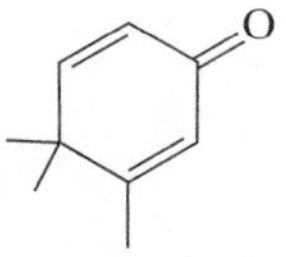

(a) Is the suggested structure correct? Apply the Woodward-Fieser rules.
(b) Write a structure that satisfies the λ_{max} and has the same carbon skeleton.
(c) Determine the molecular extinction coefficient ϵ for the recorded spectrum determined in a 1 cm cell with 4 mg of compound in 200 ml of solvent.

Bibliography

14.1 H. H. Jaffé and M. Orchin, *Theory and Applications of Ultraviolet Spectroscopy*, New York: John Wiley & Sons, Inc., 1962.

14.2 J. B. Lambert, H. F. Shurvell, L. Verbit, R. G. Cooks, and G. H. Stout, *Organic Structural Analysis*, New York: Macmillan Publishing, 1976.

14.3 M. B. Robin, F. A. Bovey, and H. Basch, in *The Chemistry of Amides*, J. Zabichy (ed.), New York: Wiley–Interscience, 1970.

14.4 H. Benderly and K. Rosenheck, *Chem. Commun.*, 179 (1972).

14.5 A. J. Merer and R. S. Mulliken, *Chem. Rev.*, **63**, 639 (1969).

14.6 A. E. Hansen and T. D. Bouman, *Adv. Chem. Phys.*, **44**, 545 (1980).

14.7 T. D. Bouman, A. E. Hansen, B. Voigt, and S. Ruttrup, *Int. J. Quantum Chem.*, **23**, 595 (1983).

14.8 J. N. Murrell, *J. Chem. Soc.*, 3779 (1965).

14.9 S. N. Vinogradov and R. H. Linnell, *Hydrogen Bonding*, New York: Van Nostrand Reinhold, 1971.

14.10 See the following examples: R. Foster (ed.), *Molecular Complexes*, London: Elek Science, 1973; N. Mataga and T. Kubota, *Molecular Interactions and Electronic Spectra*, New York: Marcel Dekker, 1970; R. S. Mulliken and W. Person, *Molecular Complexes*, New York: John Wiley & Sons, Inc., 1969; R. Foster, *Organic Charge Transfer Complexes*, New York: Academic Press, 1969.

14.11 D. A. Lightner, J. K. Gawroński, A. E. Hansen, and T. D. Bouman, *J. Am. Chem. Soc.*, **103**, 4291 (1981).

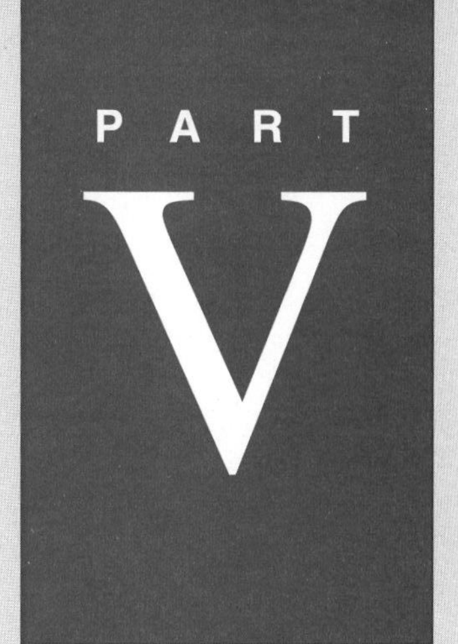

Integrated Problems

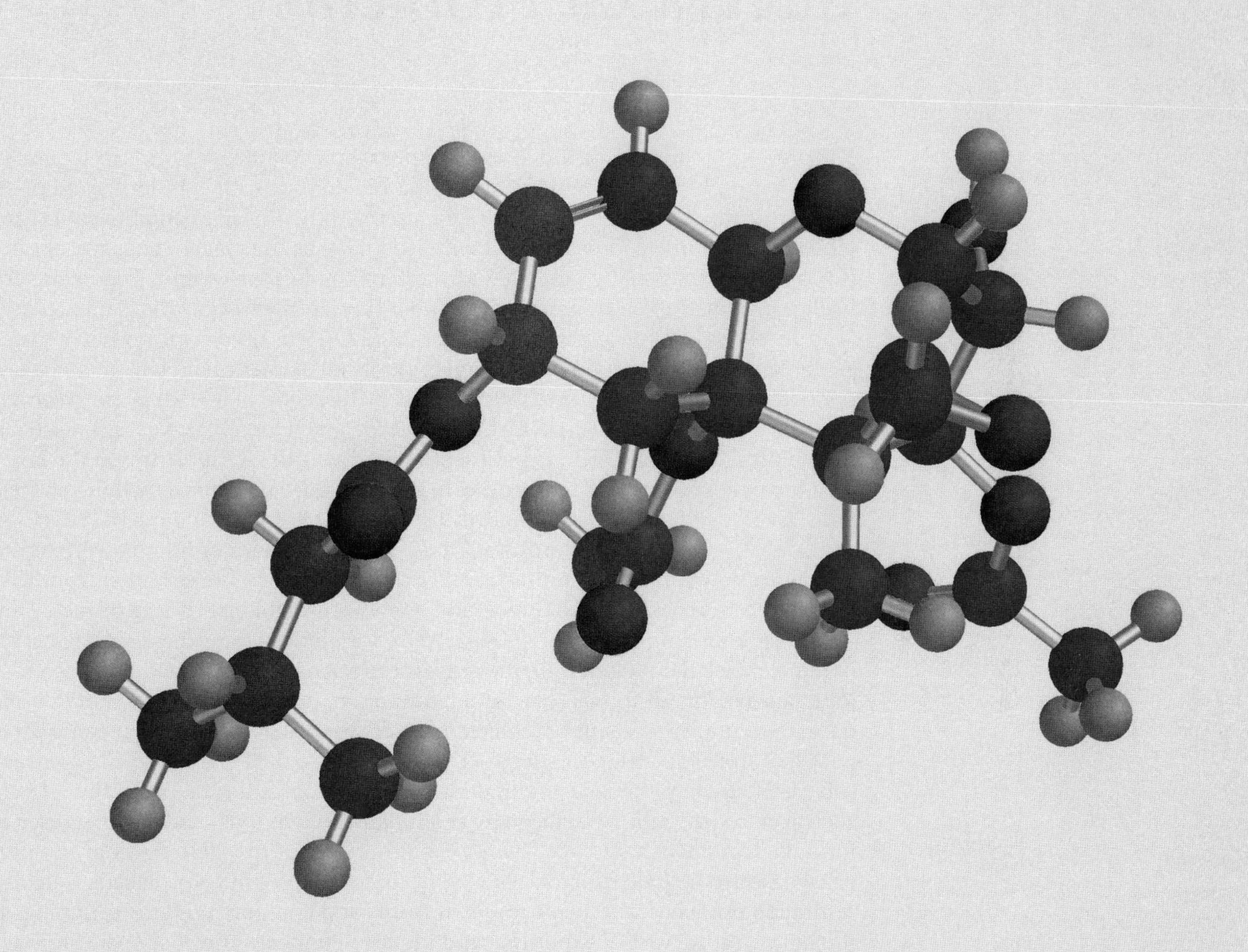

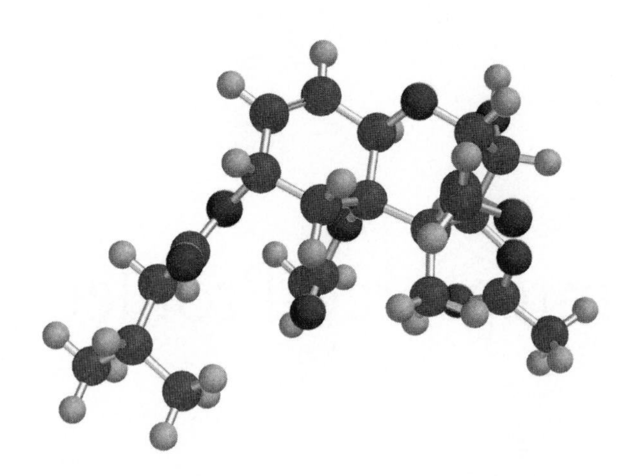

Integrated Problems

The preceding chapters included spectral problems that focused on each separate area. The following problems use all four spectroscopic methods for the derivation of structures. They are arranged loosely in order of increasing difficulty. Elemental formulas are not provided. Both chemical ionization (CI with isobutane as the reagent gas) and electron impact (EI) mass spectra usually are given for nominal molecular weight. The EI spectra provide further information on fragmentation. Other types of ionization were utilized occasionally. Matrixassisted laser desorption ionization (MALDI) was used for nonvolatile materials such as polymers, and desorption ionization (DI) for salts, other nonvolatile materials, and compounds that are thermally unstable. Proton NMR spectra at 300 MHz are given for all exercises but one. Standard ^{13}C spectra at 75 MHz are given for all cases except when DEPT spectra are provided. The DEPT experiment includes a full ^{13}C spectrum at the bottom and a methine-only spectrum in the middle. In the top DEPT spectrum, methine and methyl carbons give positive peaks and methylene carbons give negative peaks. HETCOR spectra provide ^{13}C–^{1}H correlation in some cases. Infrared spectra of liquids were recorded from neat films between KBr windows, and spectra of solids were obtained from KBr discs. A Raman spectrum is given in one case, and ultraviolet–visible spectra are provided for all cases.

The following procedure is suggested for approaching these problems. Obtain the nominal molecular weight from the mass spectra. Analyze fragmentation processes and isotopic clusters for additional structural information. Examine the 1D ^{1}H and ^{13}C NMR spectra to determine the relative numbers of hydrogens and carbons, and compare this result with that from MS. Suggest substructures from these data. When DEPT spectra are given, sort out methyl, methylene, and methinyl groups. Consider the possibility of heteroatoms such as nitrogen, sulfur, or silicon by examination of the mass spectra. Suggest a molecular formula. Use the infrared spectrum to determine what functional groups are present, and use the electronic spectrum to decide what chromophores may be present. Calculate the unsaturation number from the proposed formula and compare with the groupings identified thus far. Use the COSY experiment to identify connectivities between protons, and the HETCOR experiment to identify connectivities between protons and carbons.

Finally, bring the various proposed structural units together in possible structures and reexamine all the data to determine whether they are consistent with the structures. If you reach an impasse after due effort, we provide tables of molecular formulas and of functional groups as hints at the end of the problems. Refer to them only as a last resort. The full structures are given in the *Solutions Manual*. The solution to the first problem is presented here in detail.

Use the spectra as further exercises on the principles developed in the text. Try to identify every peak in the ^{1}H and ^{13}C NMR spectra, and justify chemical shifts and coupling constants. Use HETCOR to assign every hydrogen to an attached carbon, and identify all the COSY cross peaks. Find infrared absorptions for all the functional groups and hydrocarbon components in the molecules. Rationalize the location of the wavelength maxima and the intensities in the electronic spectra. Explain the differences among the various mass spectra (EI, CI, and DI) and identify important fragmentation pathways.

PROBLEM 15-1

Mass spectrum (CI)

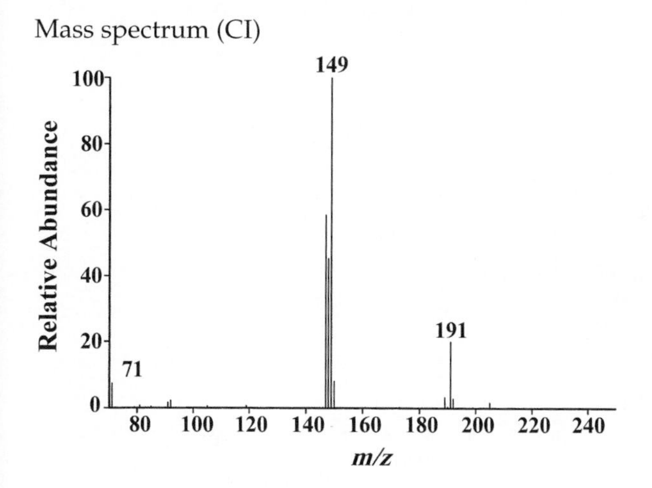

Mass spectrum (EI)

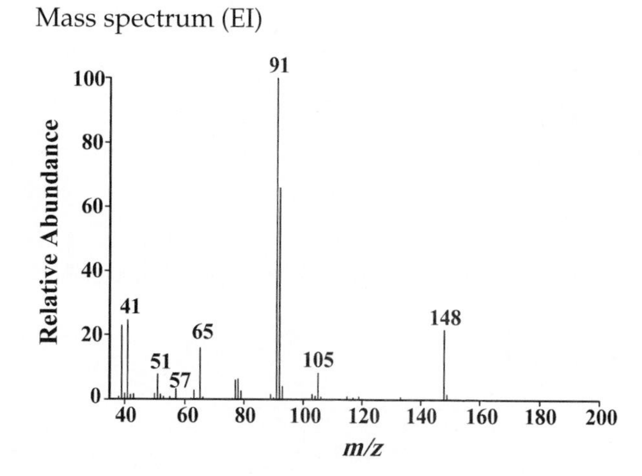

Proton NMR spectrum ($CDCl_3$)

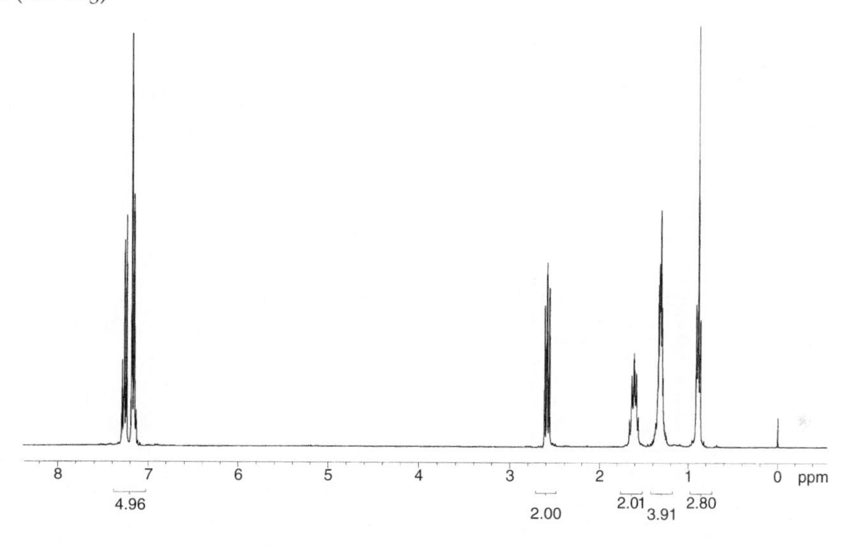

DEPT spectra

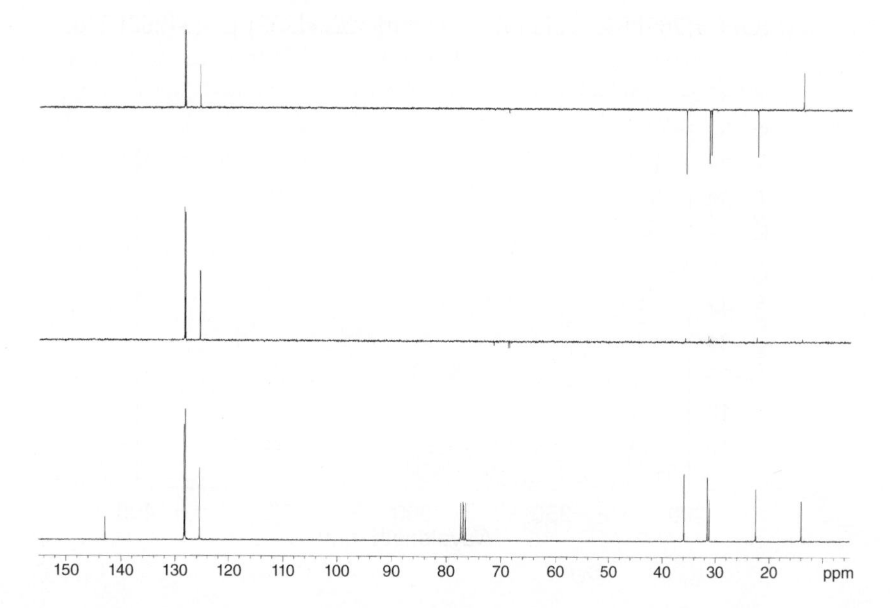

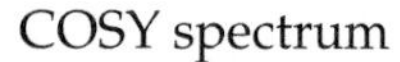

COSY spectrum

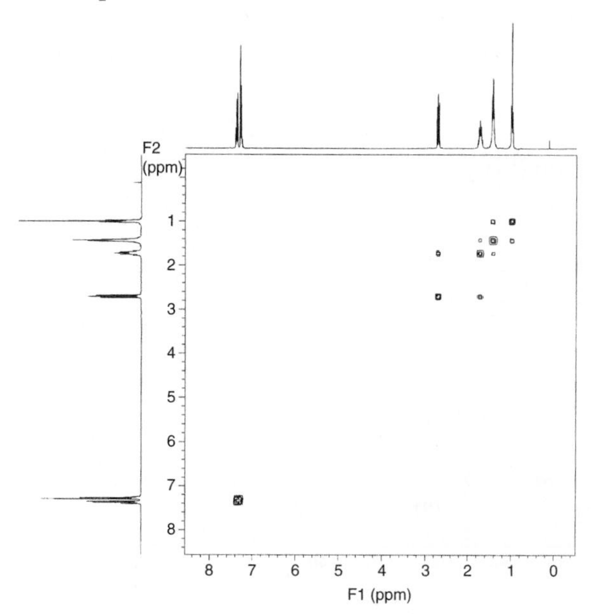

HETCOR spectrum

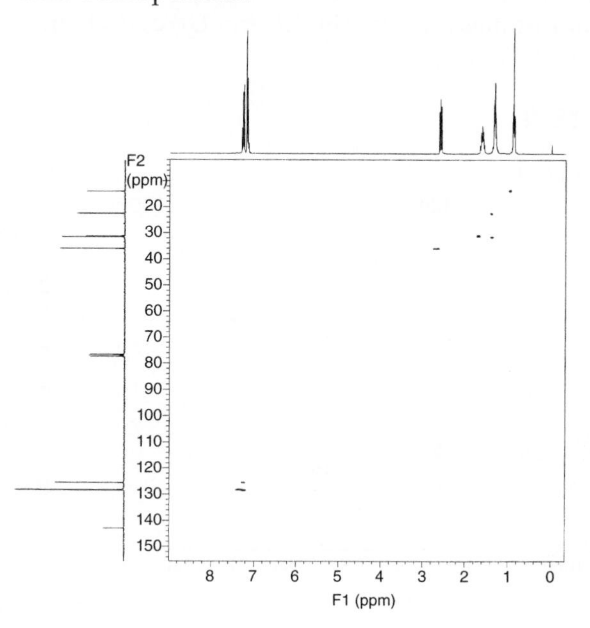

Infrared spectrum (neat)

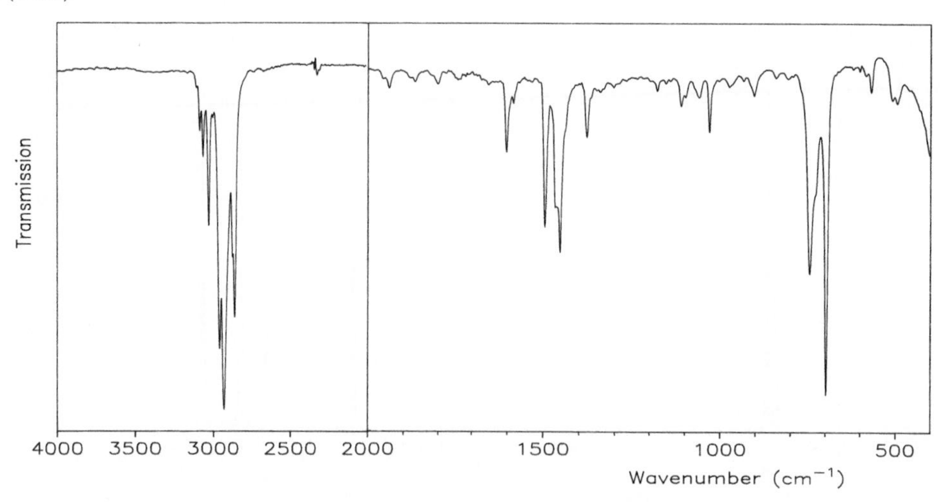

Ultraviolet–visible spectrum (EtOH, ϵ[209] 8400, ϵ[243] 76, ϵ[248] 122, ϵ[253] 174, ϵ[262] 216, ϵ[268] 171)

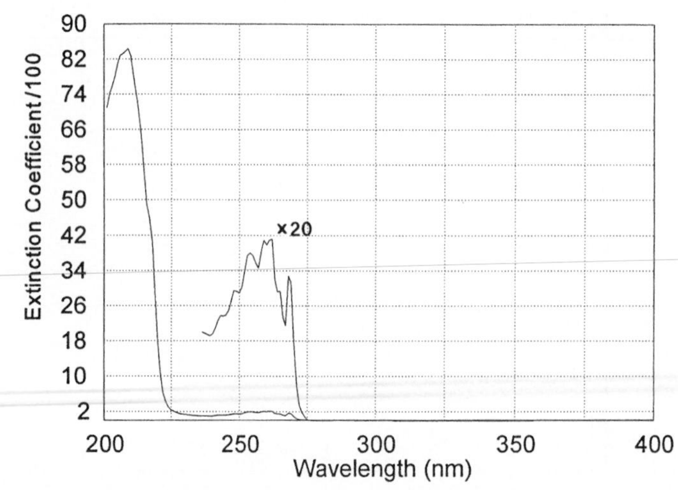

SOLUTION TO PROBLEM 15-1

Mass Spectra

The isobutane CI spectrum shows the presence of ions at *m/z* 191 and at 147 through 149. The molecular ion seems to be 190. Extensive fragmentation in isobutane CI, however, is unusual and needs to be examined carefully: the ion at 191 might be an ion-molecule adduct. In this case, the molecular weight would be 148 and the adduct ion would correspond to the addition of $C_3H_7^+$. Isobutane does form adduct ions, normally in low abundance, by addition of $C_3H_7^+$ (43) or, more usually, $C_4H_9^+$ (57). There is no evidence in the EI mass spectrum for a molecular ion other than 148.

The fragmentation behavior is consistent with a monosubstituted benzene, the abundant benzyl/tropyl ion at 91 dominating the spectrum. The expected further fragmentation to give 65 (loss of acetylene) and competitive fragmentation to give the less favored phenyl cation at 77 also are observed, as are further fragments from both these ions at 39 and 51, respectively. Loss from the molecular ion of a fragment of 43 Da gives the ion at 105, formally the phenonium ion. All these fragmentations, summarized below, are consistent with the *n*-amylbenzene structure as determined by NMR (below). It is worthy of note that the amyl cation is present at low abundance in the CI spectrum (71). β-Cleavage is expected and observed in the EI spectrum, which therefore shows a low-abundance butyl ion at m/z 57.

$C_6H_5CH_2$—C_4H_9 $]^{+\cdot}$ → $C_7H_7^+$ (m/z 91) $\xrightarrow{-C_2H_2}$ $C_5H_5^+$ (m/z 65) $\xrightarrow{-C_2H_2}$ $C_3H_3^+$ (m/z 39)

$C_6H_5CH_2CH_2CH(H)C_2H_5$ $]^{+\cdot}$ → m/z 92 $]^{+\cdot}$ $\xrightarrow{-H^\cdot}$ m/z 91

$C_6H_5CH_2CH_2C_3H_7$ $]^{+\cdot}$ $\xrightarrow{-C_3H_7}$ m/z 105

NMR Spectra

The aromatic resonances in the ^1H spectrum occur at a position unperturbed by conjugating substituents. The integration suggests a monosubstituted benzene and the resonance position a saturated hydrocarbon substituent. The aromatic resonance is too closely coupled for an interpretation by inspection. The resonance at δ 2.6 is from the protons on the carbon attached (α) to the benzene ring. The splitting pattern indicates that the β group is CH_2. The resonance at δ 0.9 is from the terminal methyl group, adjacent to a CH_2 according to its splitting. The resonance at δ 1.6 probably is from the β CH_2, and its quintet pattern indicates that both the α and the γ groups are CH_2. The γ and δ patterns coincide at δ 1.3. Thus the entire structure may be deduced from the ^1H spectrum to be $C_6H_5CH_2CH_2CH_2CH_2CH_3$.

The ^{13}C spectrum contains four aromatic and five aliphatic resonances. The ipso carbon, with reduced intensity, is at δ 143, moved to higher frequency (lower field) by the α and β effects of the side chain. It is absent in the DEPT spectra, which do not detect quaternary carbons. The remaining aromatic carbons appear in the CH-only DEPT spectrum in the middle. The upper DEPT spectrum indicates the presence of one CH_3 and four CH_2 groups, probably progressively at higher frequency with closeness to the ring, with some ambiguity for the γ and δ carbons.

The COSY spectrum shows no connectivities involving the aromatic protons. As two CH_2 groups coincide, only three aliphatic connectivities appear: α-CH_2 to β-CH_2, β-CH_2 to γ,δ-CH_2, and γ,δ-CH_2 to ϵ-CH_3. The HETCOR spectrum shows the obvious connectivities between the aromatic protons and carbons and between the methyl protons and carbon. The α-CH_2 group is seen to give the highest-frequency proton and carbon resonances. The overlap of the γ and δ protons is resolved into the two carbon resonances, and the β protons appear to correlate with the carbon resonance at the third from highest frequency.

Infrared Spectrum

The sharp peaks observed above and below 3000 cm^{-1} indicate the presence of both aromatic (or unsaturated hydrocarbon) and aliphatic hydrocarbon groups. No other functional groups are indicated in the spectrum. The presence of a benzene ring is confirmed by the sharp peaks near 1600 and 1500 cm^{-1} and by the two strong bands near 740 and 695 cm^{-1}, which suggest monosubstitution. This conclusion is supported by the pattern of weak bands between 2000 and 1700 cm^{-1}.

The single sharp peak at 1375 cm^{-1} can be attributed to the symmetric stretching of an isolated methyl group. The absence of absorptions characteristic of isopropyl or *tert*-butyl groups indicates that there is no branching in the aliphatic carbon chain. This conclusion is supported by the absorption in the 725 cm^{-1} region, which in this spectrum appears as a shoulder on the 740 cm^{-1} band.

The material is a monosubstituted benzene ring, whose substituent is a straight-chain, saturated hydrocarbon.

Ultraviolet–Visible Spectrum

The observation of UV bands above 200 nm indicates the presence of a chromophore. The weak ($\epsilon \sim 200$), structured band near 260 nm (^1L_b) and a more intense ($\epsilon \sim 7500$) band near 210 nm (^1L_a) are characteristic of the benzene chromophore. The absorption is not strongly perturbed by groups containing lone pairs or π electrons, such as in phenol or styrene, for which the long-wavelength band would be shifted ($\lambda \sim 270$) and intensified ($\epsilon \sim 1500$). Consequently, the material contains an aromatic ring without conjugating substituents.

PROBLEM 15-2

Mass spectrum (CI)

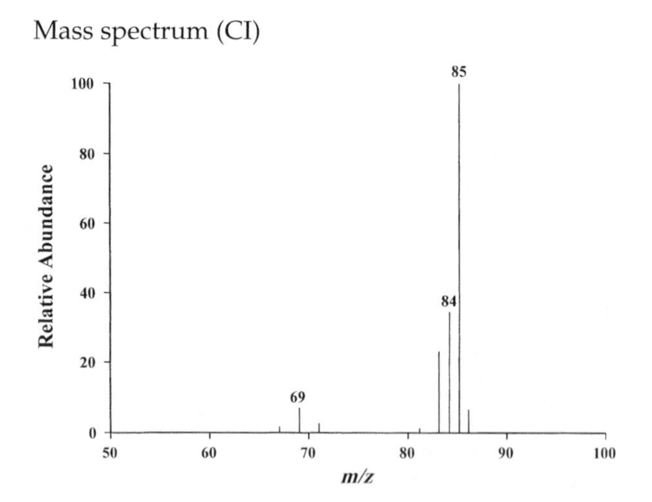

Mass spectrum (EI)

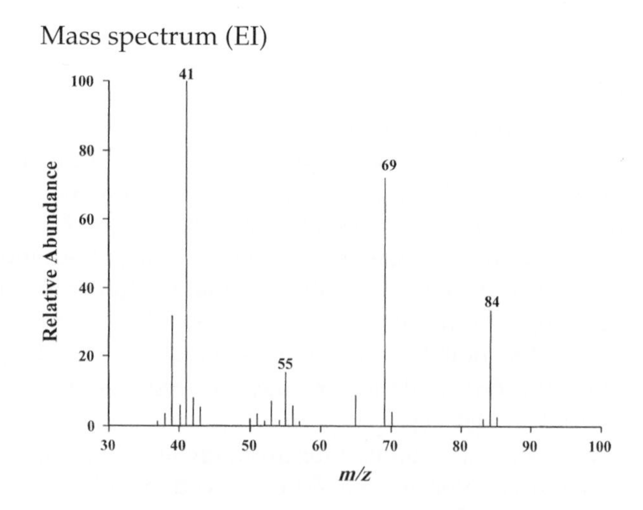

Proton NMR spectrum ($CDCl_3$)

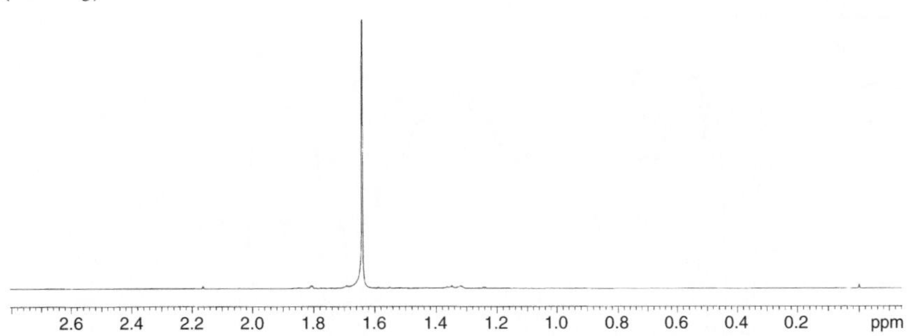

Carbon-13 NMR spectrum

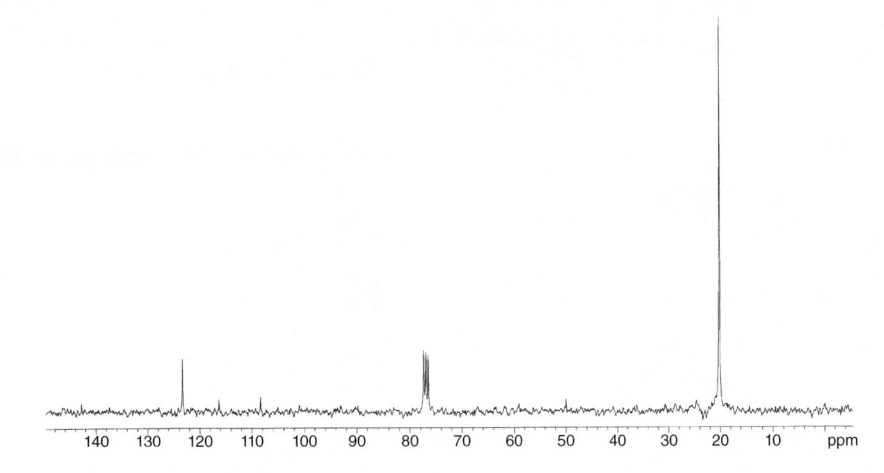

HETCOR spectrum

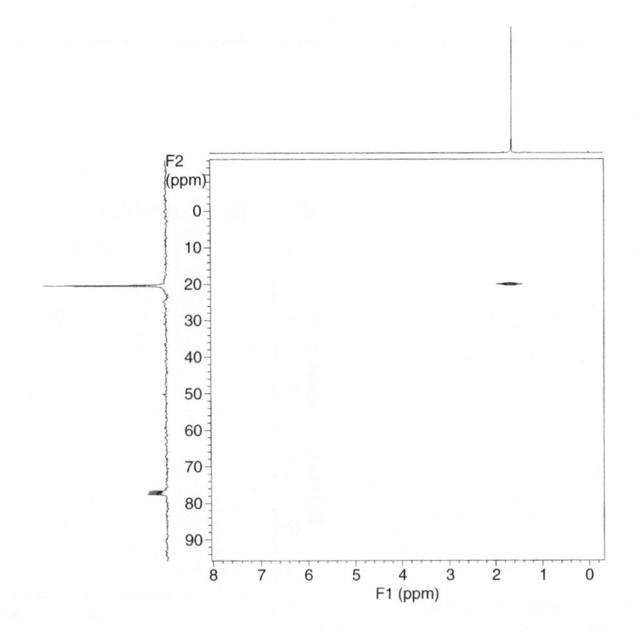

Infrared spectrum (neat)

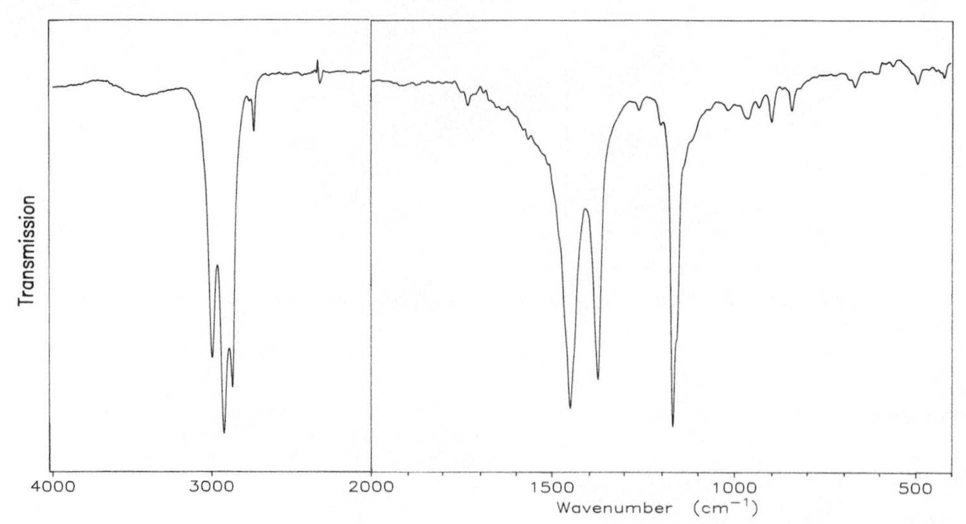

Raman spectrum (neat)

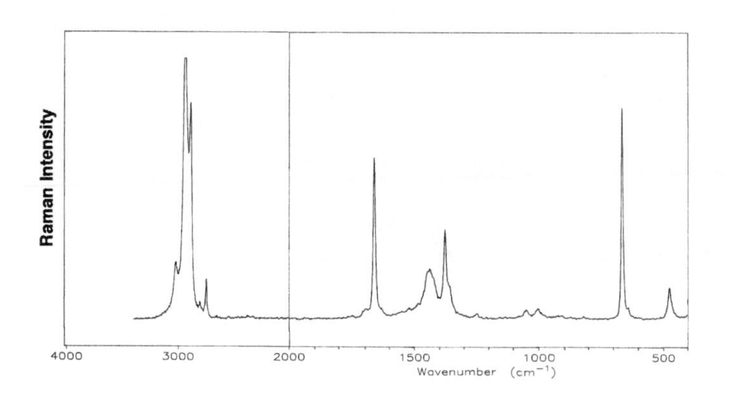

Ultraviolet–visible spectrum (EtOH)

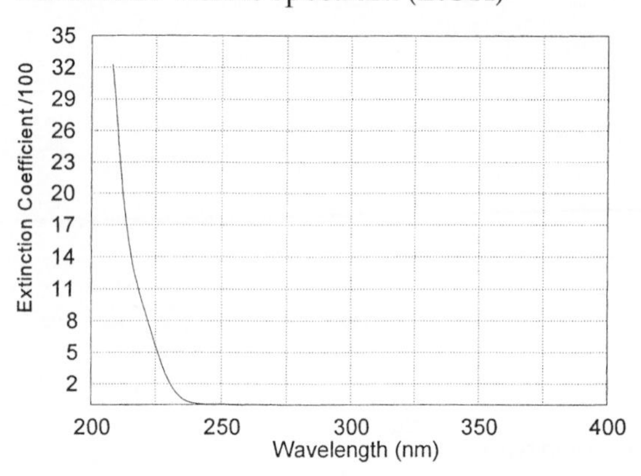

PROBLEM 15-3

Mass spectrum (CI)

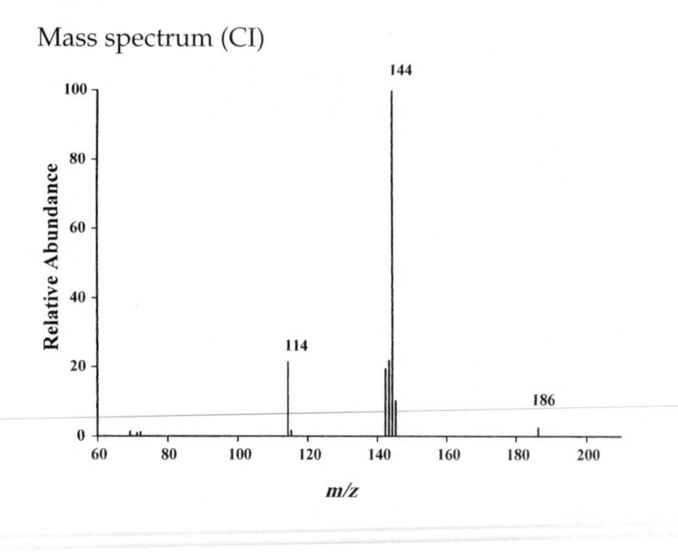

Mass spectrum (EI)

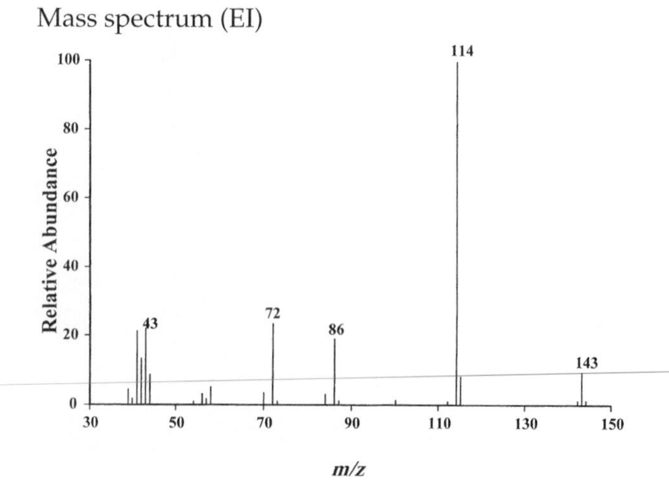

Proton NMR spectrum ($CDCl_3$)

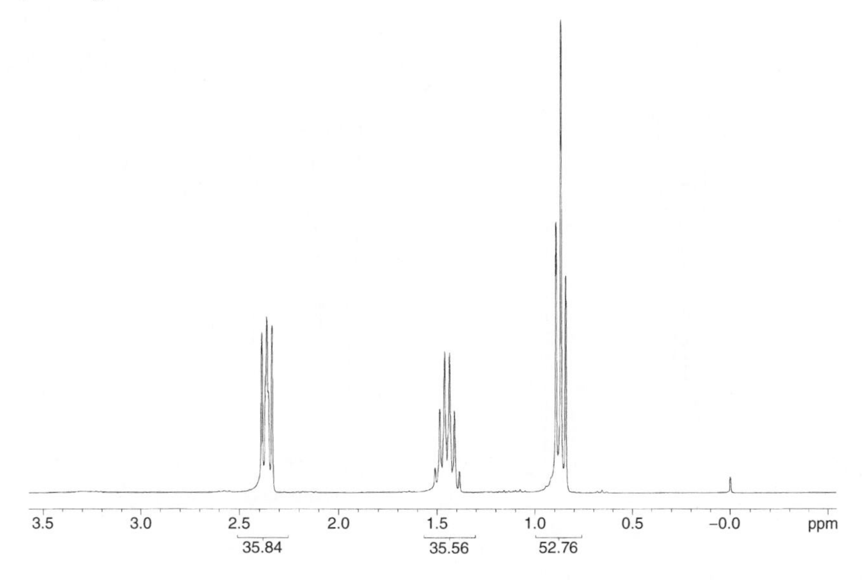

Carbon-13 NMR spectrum

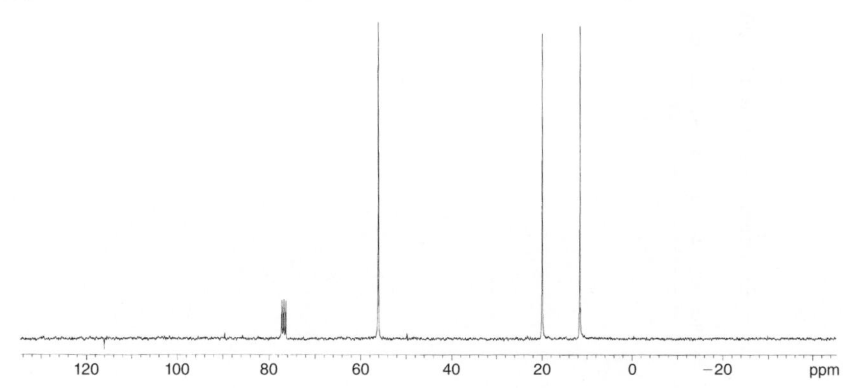

COSY spectrum

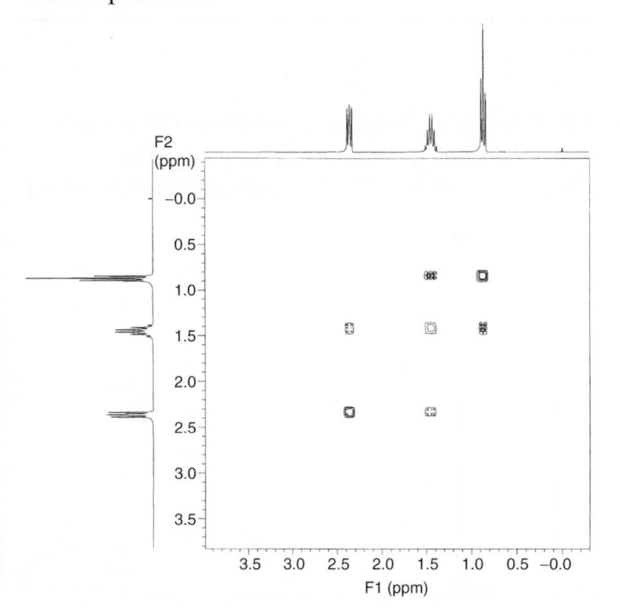

HETCOR spectrum

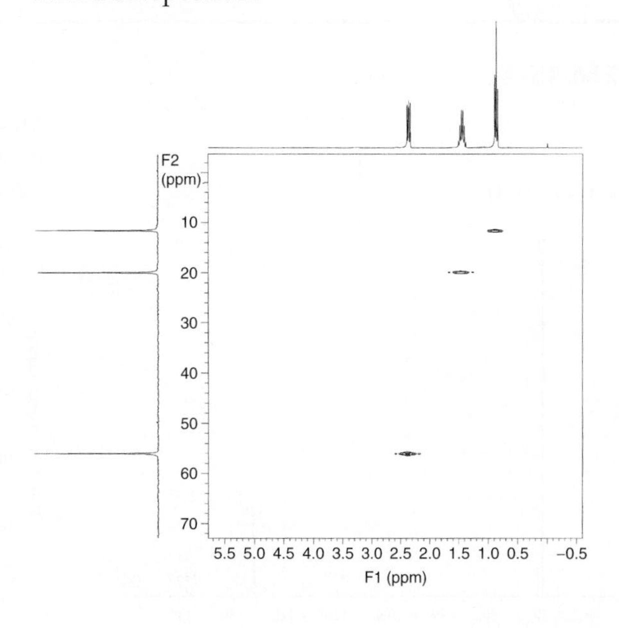

Infrared spectrum (neat)

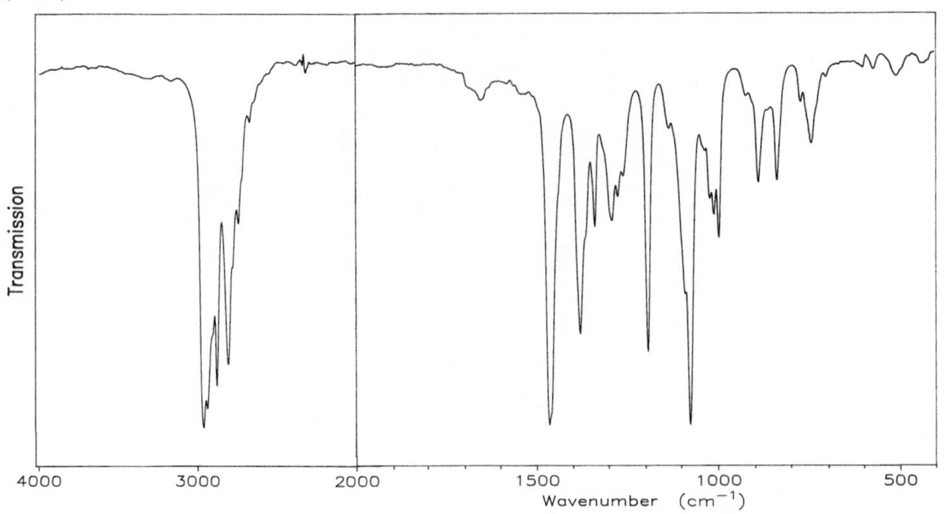

Ultraviolet–visible spectrum (EtOH)

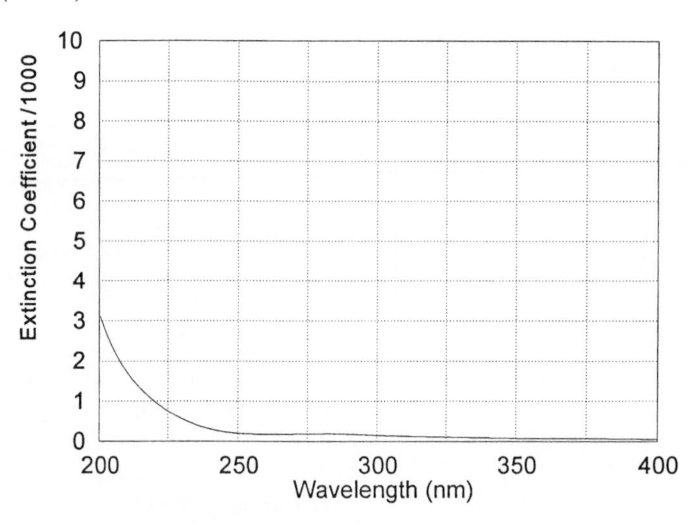

PROBLEM 15-4

Mass spectrum (CI)

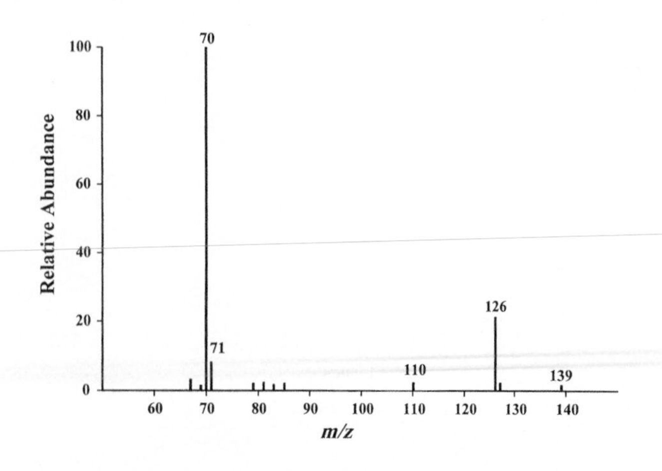

Mass spectrum (EI) [*Note: The molecular ion is absent. Because EI is a hard method of ionization, loss of multiple hydrogen atoms sometimes occurs, that is, H loss followed by one or more H_2 losses.*]

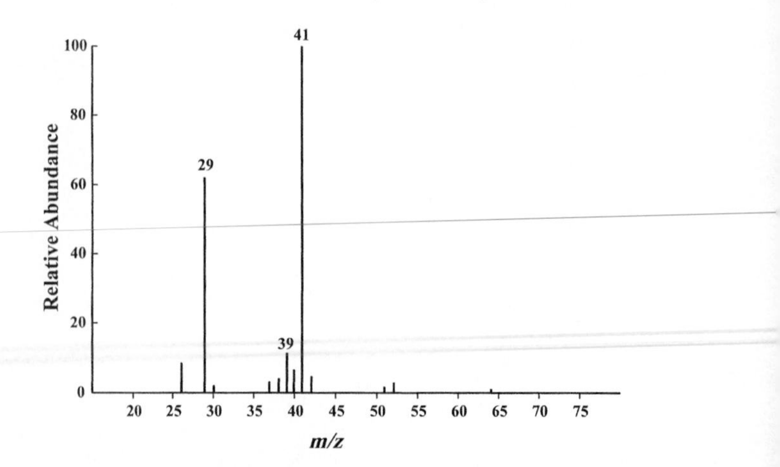

Proton NMR spectrum ($CDCl_3$)

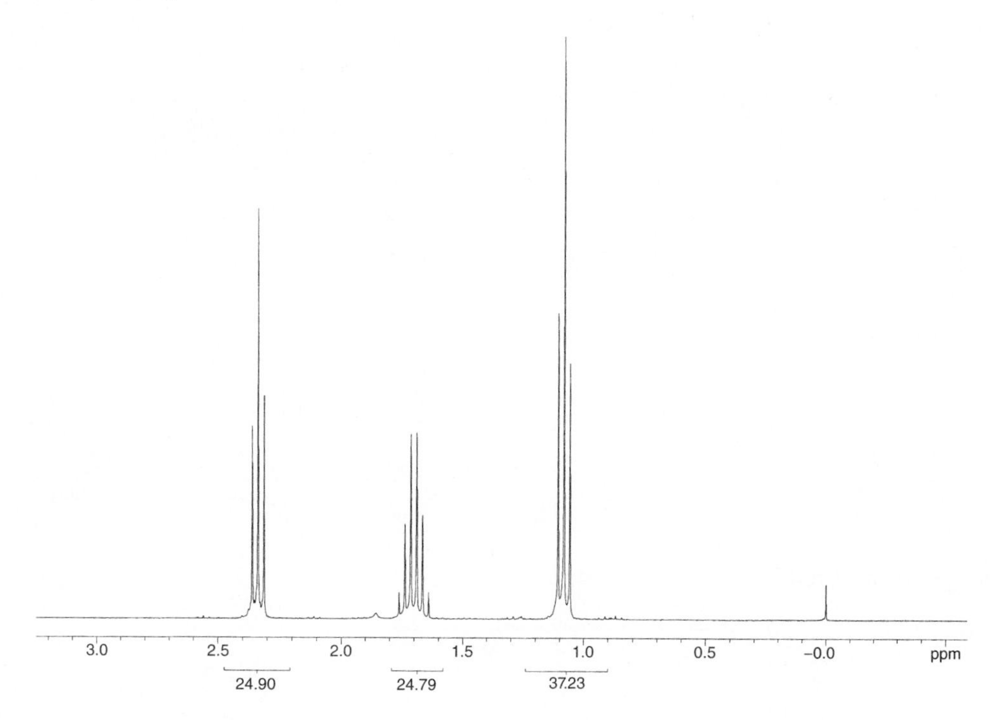

Carbon-13 NMR spectrum

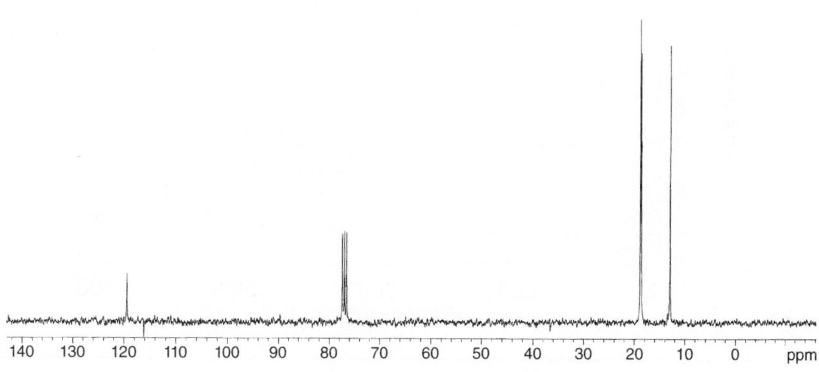

COSY spectrum

HETCOR spectrum

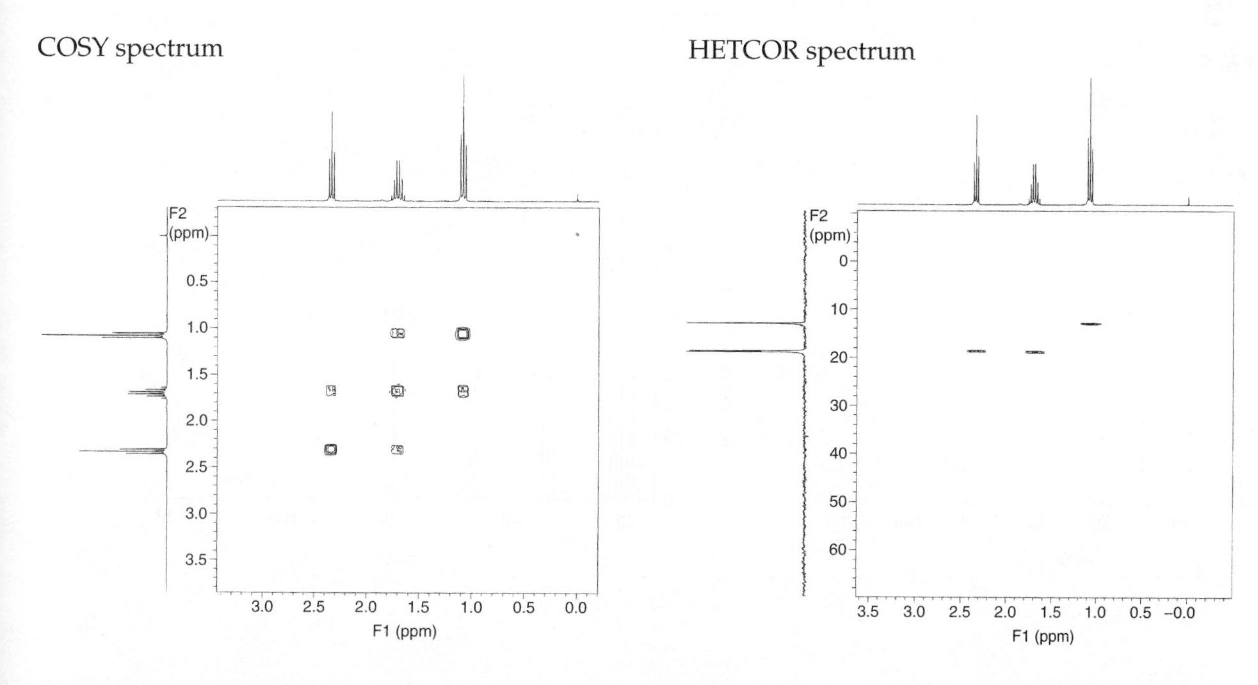

Infrared spectrum (neat)

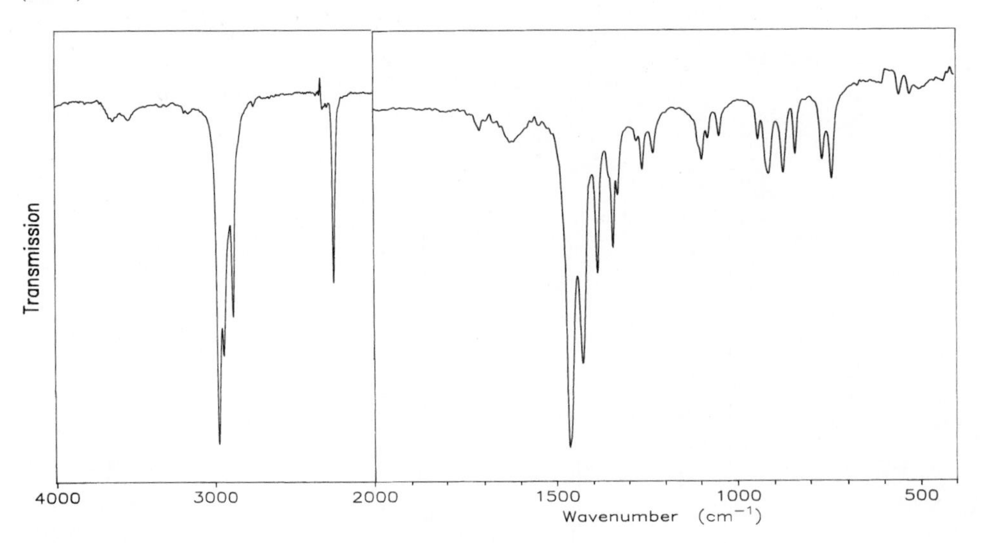

Ultraviolet–visible spectrum (EtOH)

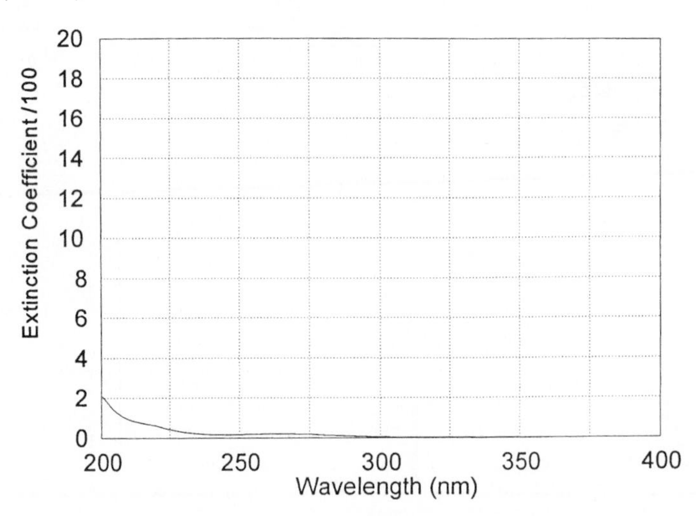

PROBLEM 15-5

Mass spectrum (CI)

Mass spectrum (EI)

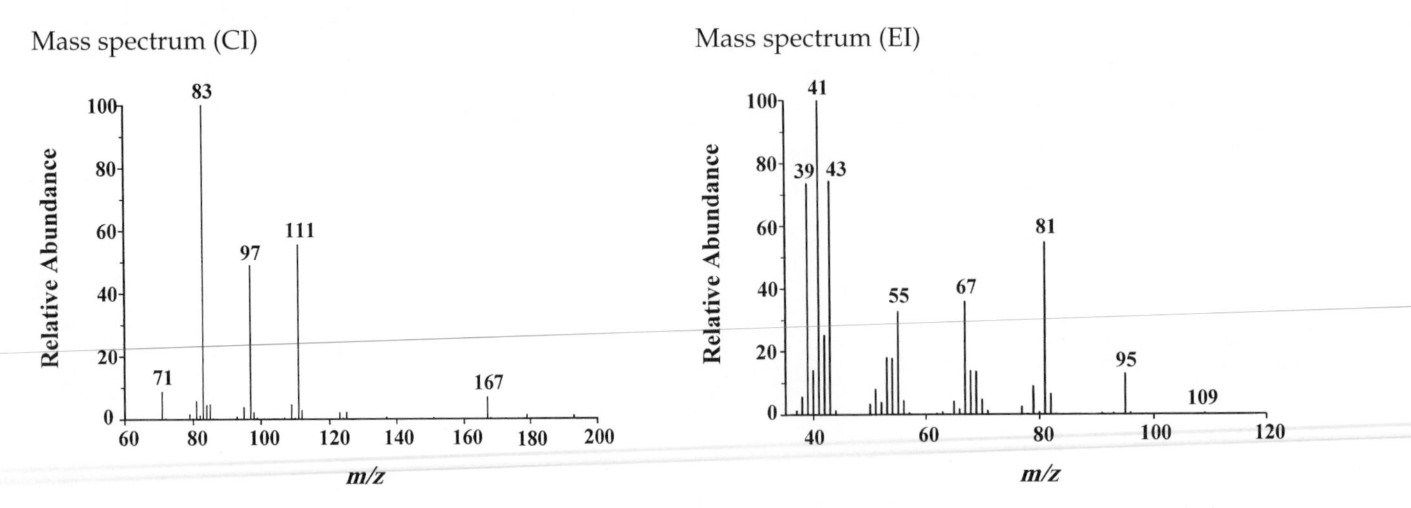

Proton NMR spectrum ($CDCl_3$)

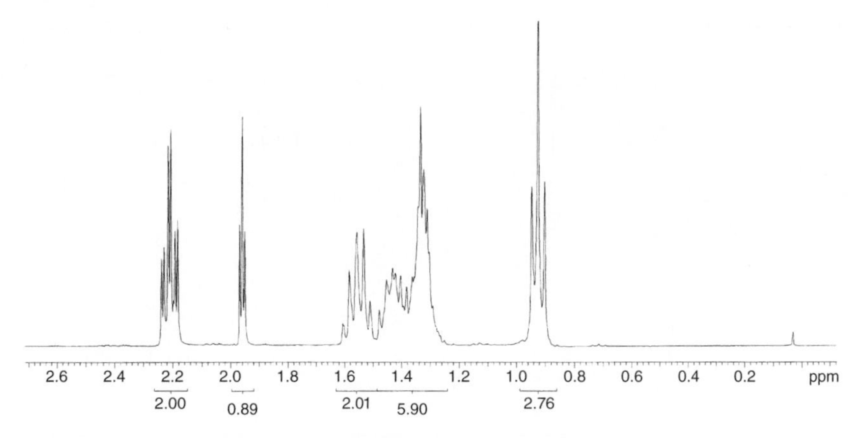

Carbon-13 NMR spectrum

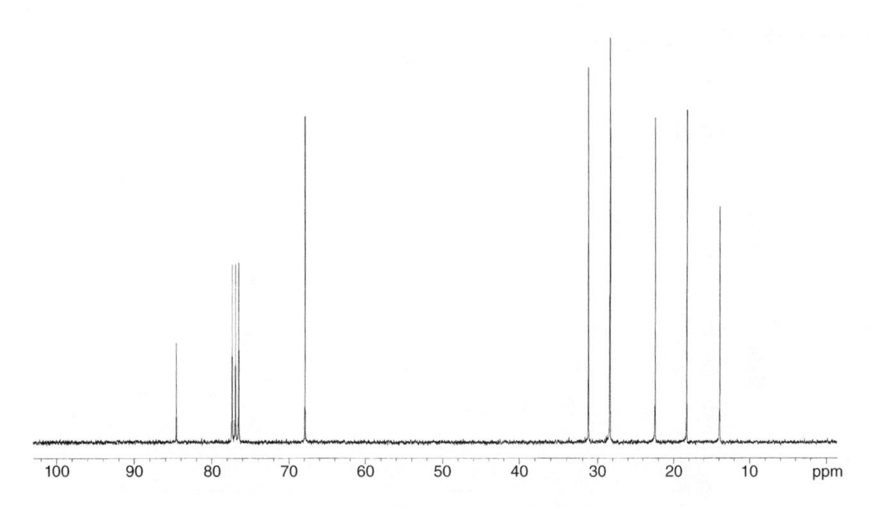

COSY spectrum

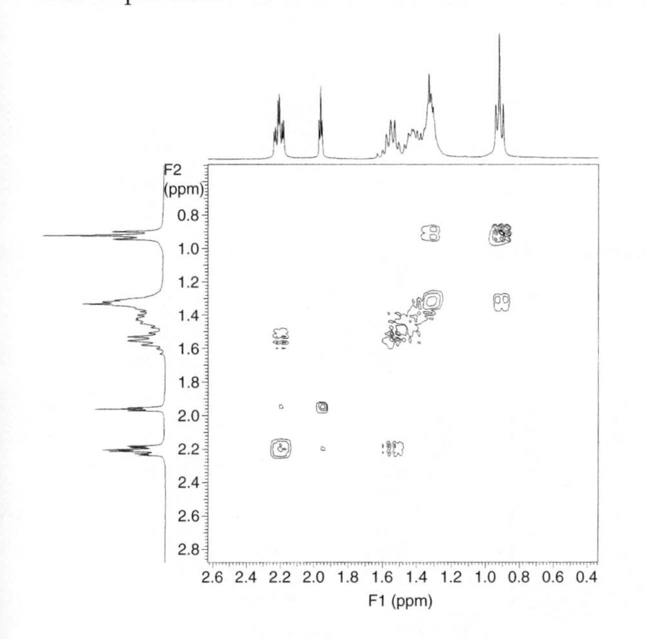

HETCOR spectrum

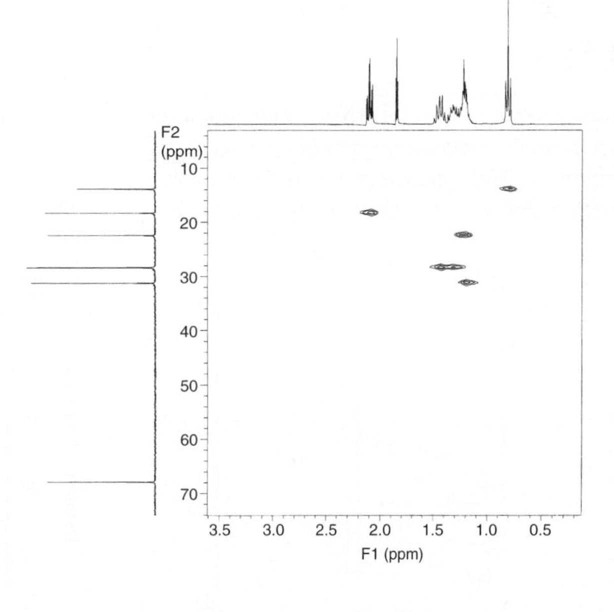

Infrared spectrum (neat)

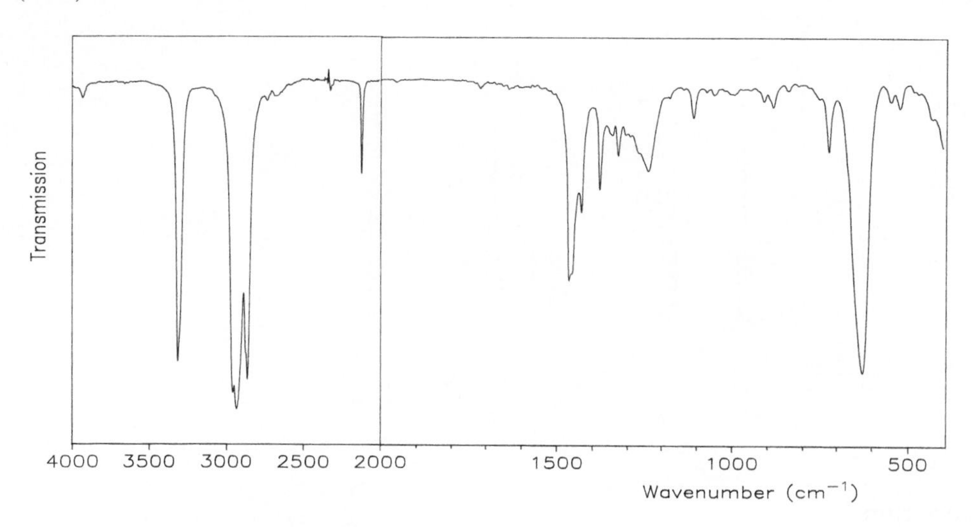

Ultraviolet–visible spectrum (EtOH, ϵ[224] 83)

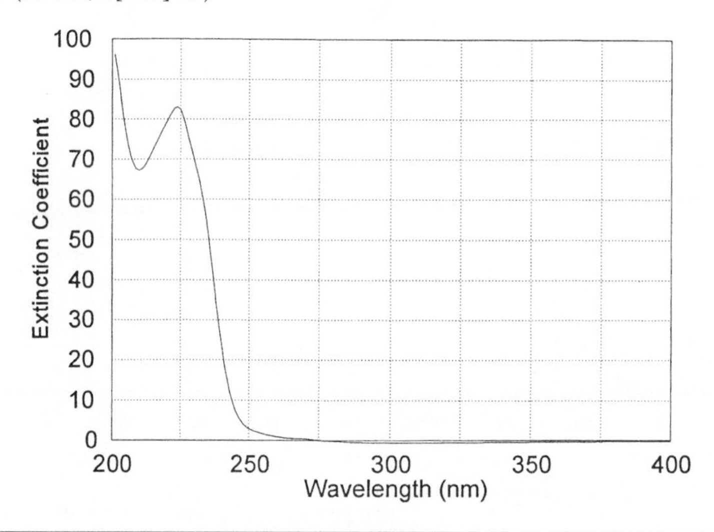

PROBLEM 15-6

Mass spectrum (negative ion CI) [*Note: This unknown contains halogen. Such compounds often lose halogen and give no molecular ion. Halogenated compounds are often best studied with negative rather than positive ionization.*]

Mass spectrum (EI)

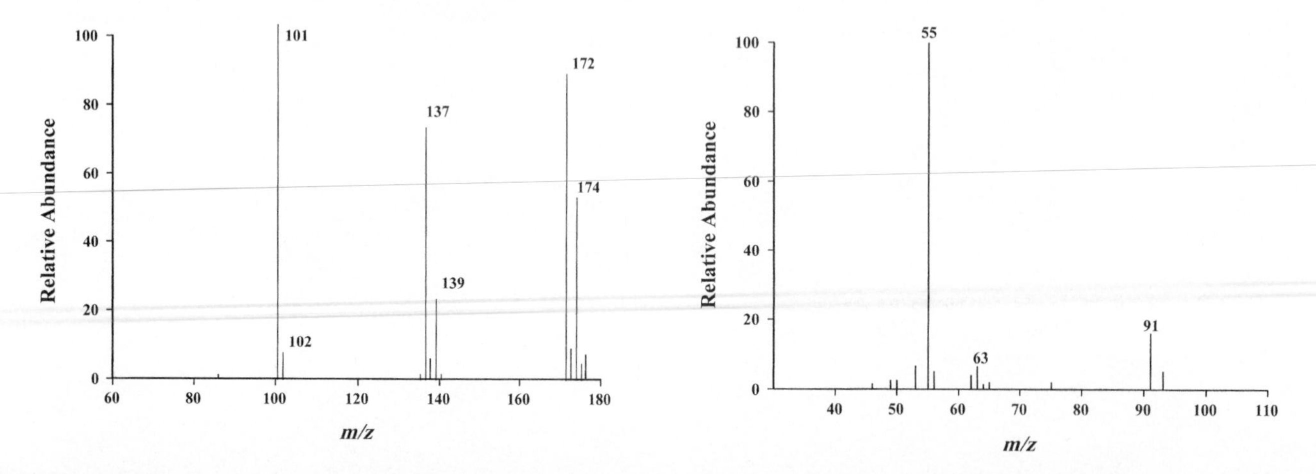

Proton NMR spectrum ($CDCl_3$)

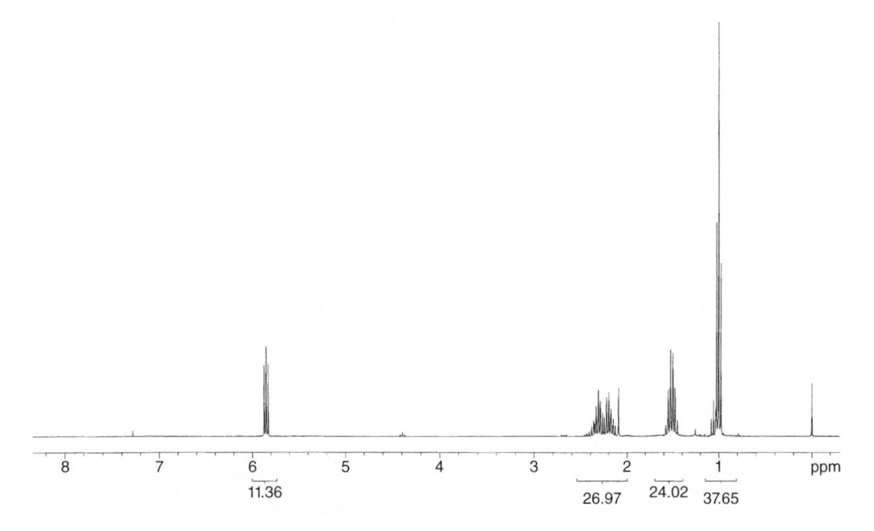

Carbon-13 NMR spectrum

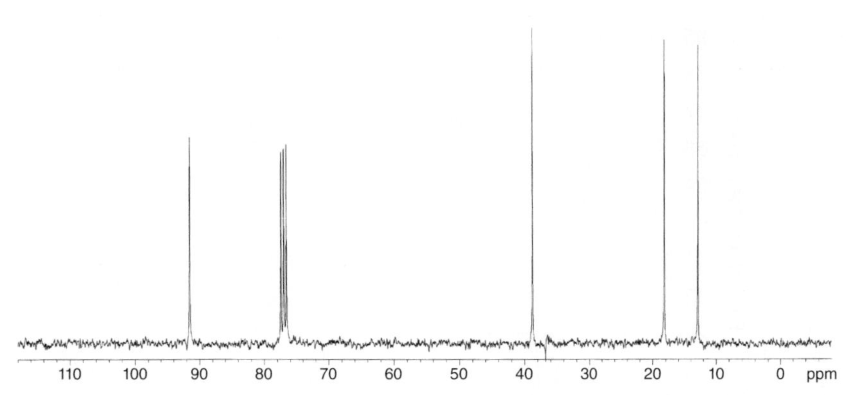

COSY spectrum

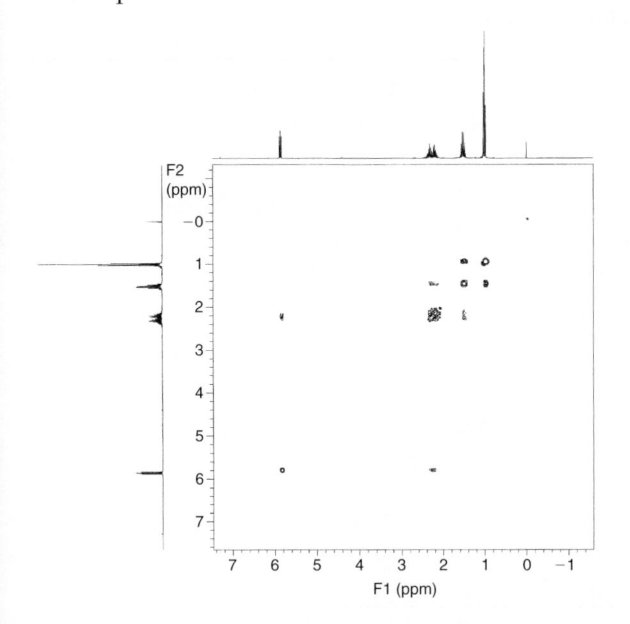

HETCOR spectrum

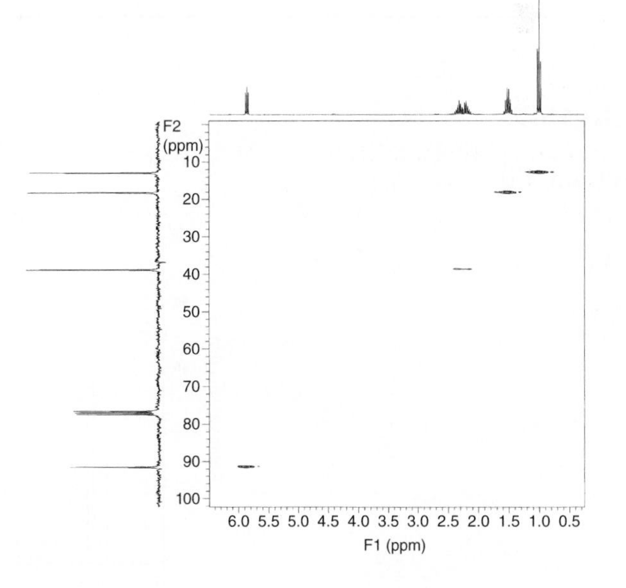

Infrared spectrum (neat)

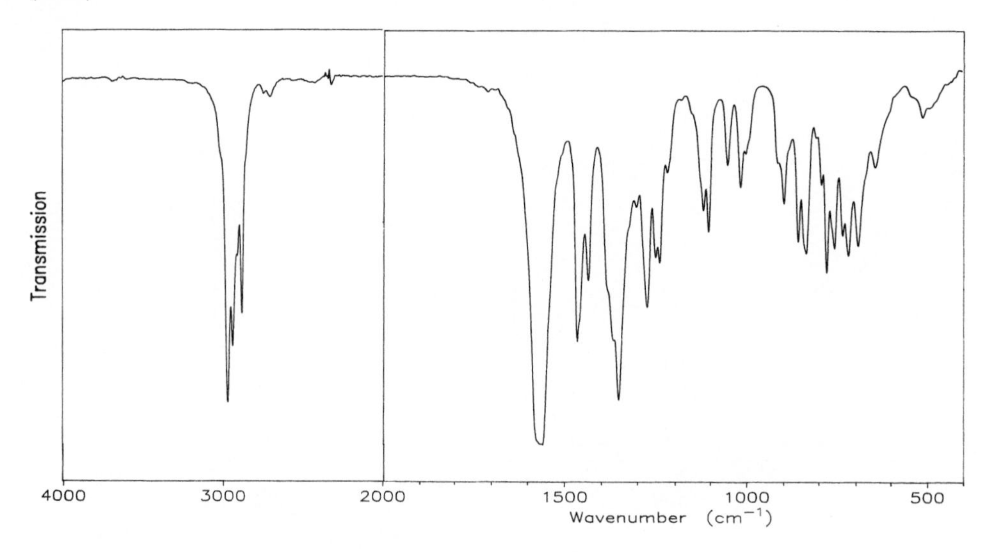

Ultraviolet–visible spectrum (EtOH, ignore the shoulder at 280 nm)

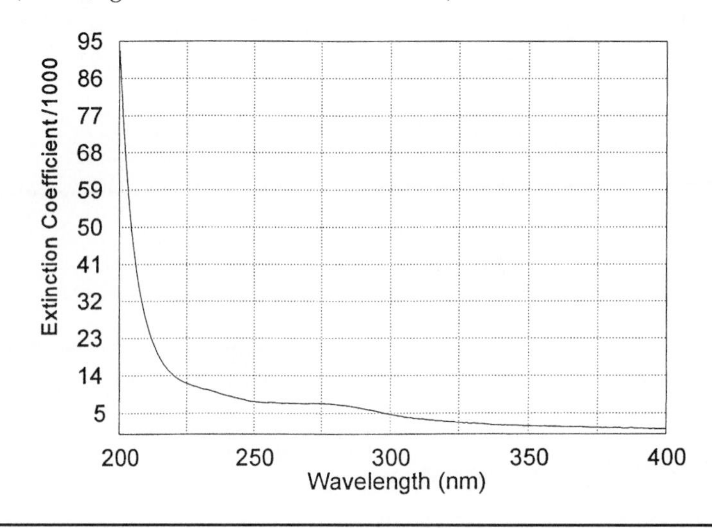

PROBLEM 15-7

Mass spectrum (EI)

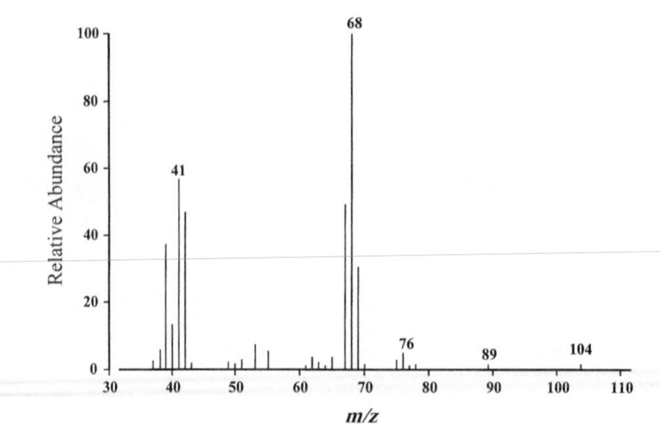

Proton NMR spectrum ($CDCl_3$)

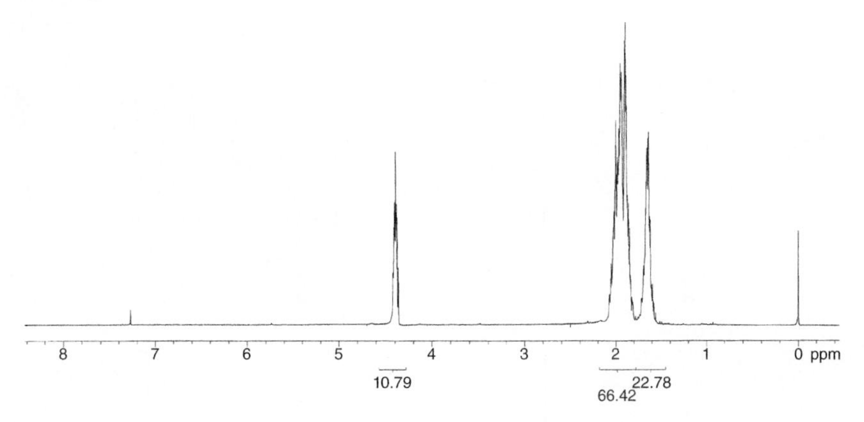

Carbon-13 spectrum

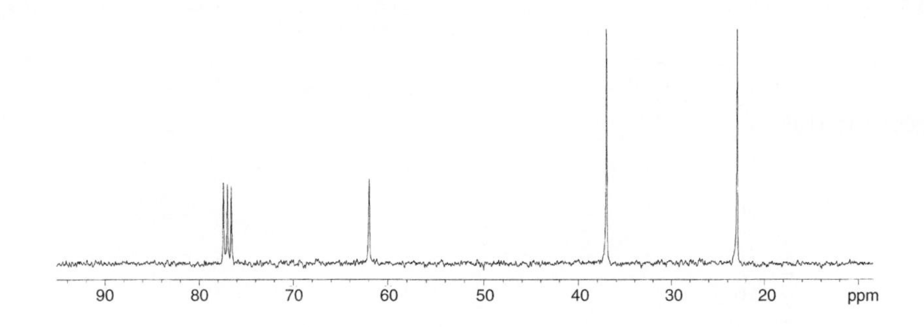

COSY spectrum

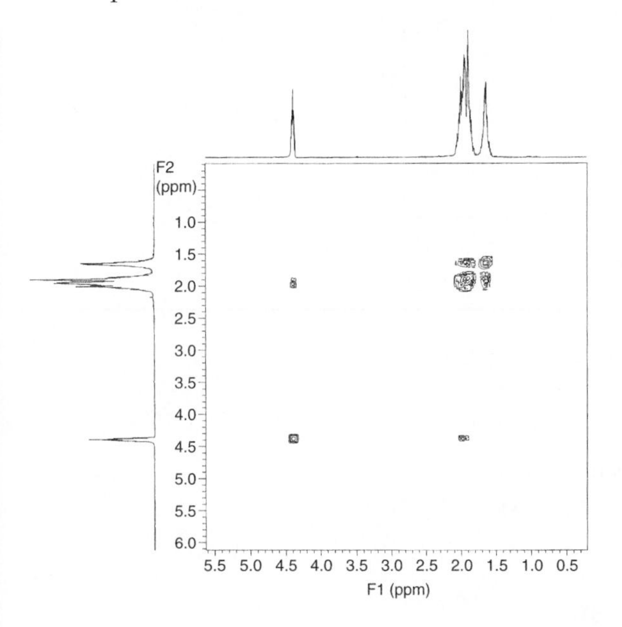

HETCOR spectrum

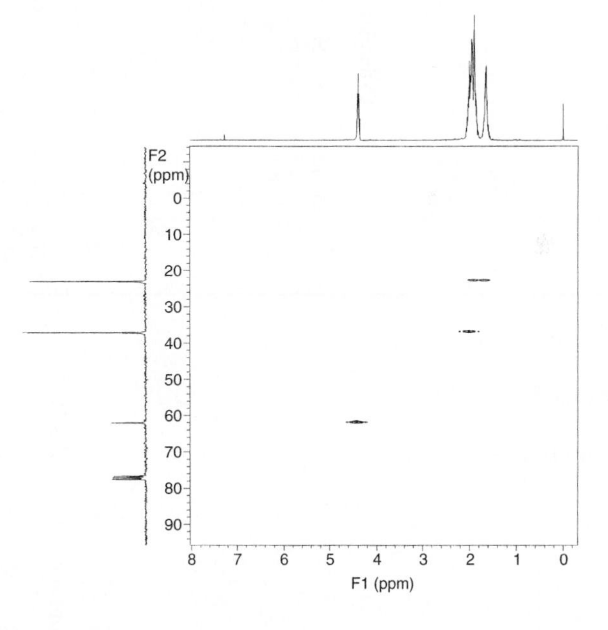

Infrared spectrum (neat)

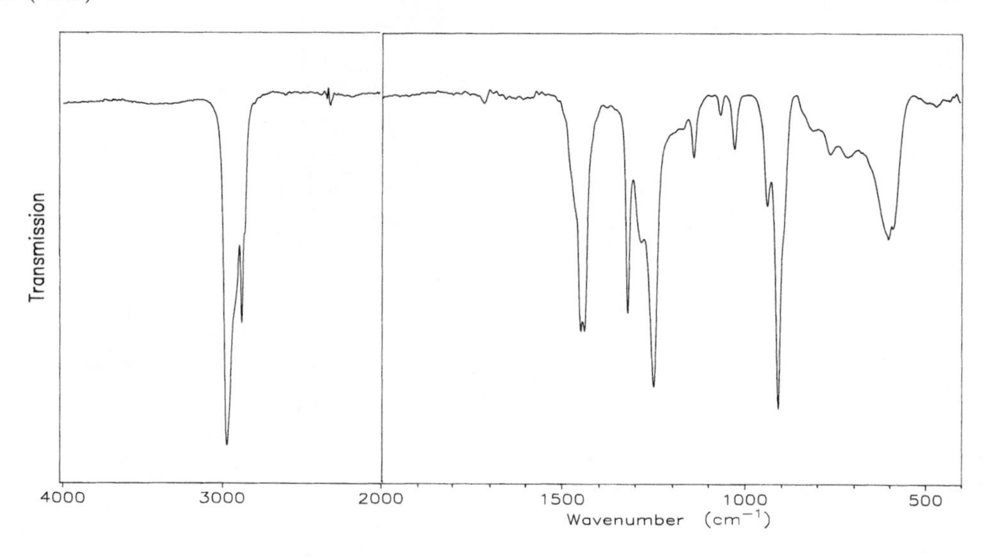

Ultraviolet–visible spectrum (EtOH)

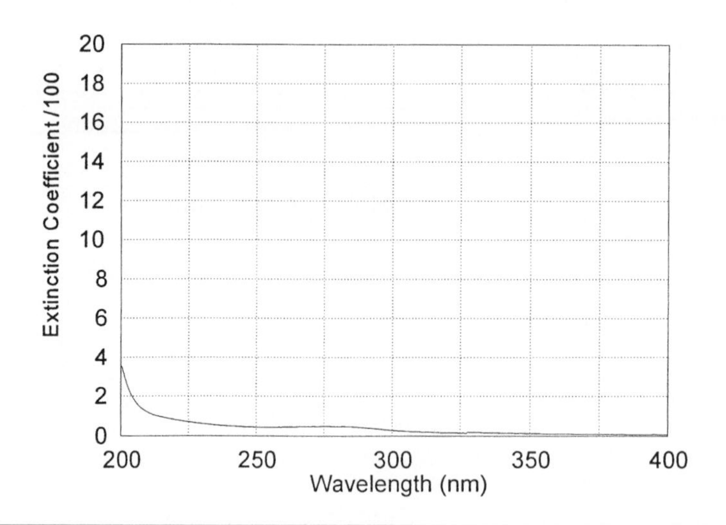

PROBLEM 15-8

Mass spectrum (CI)

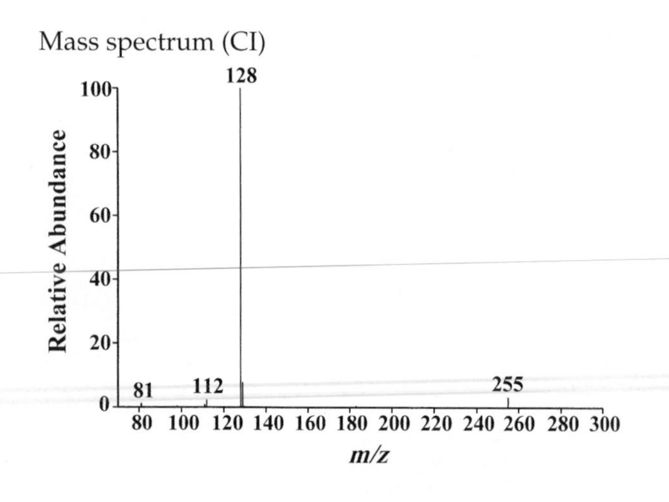

Mass spectrum (EI)

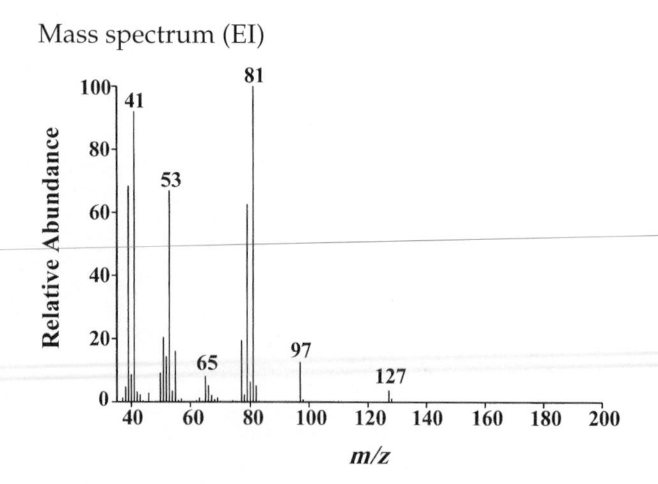

Proton NMR spectrum with expansion ($CDCl_3$)

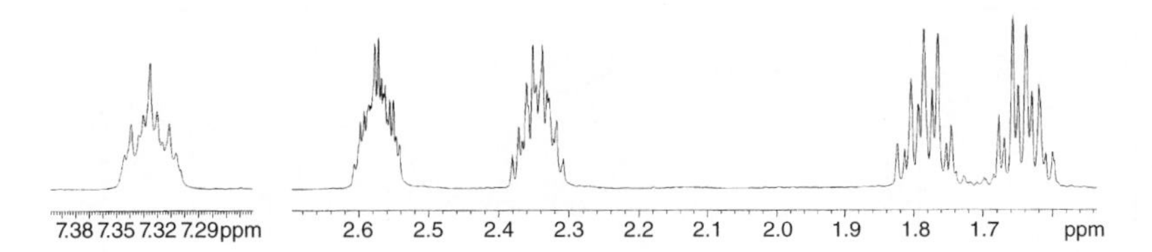

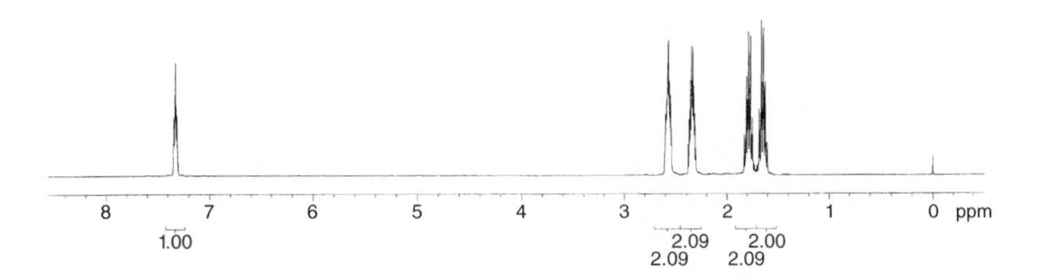

DEPT spectra

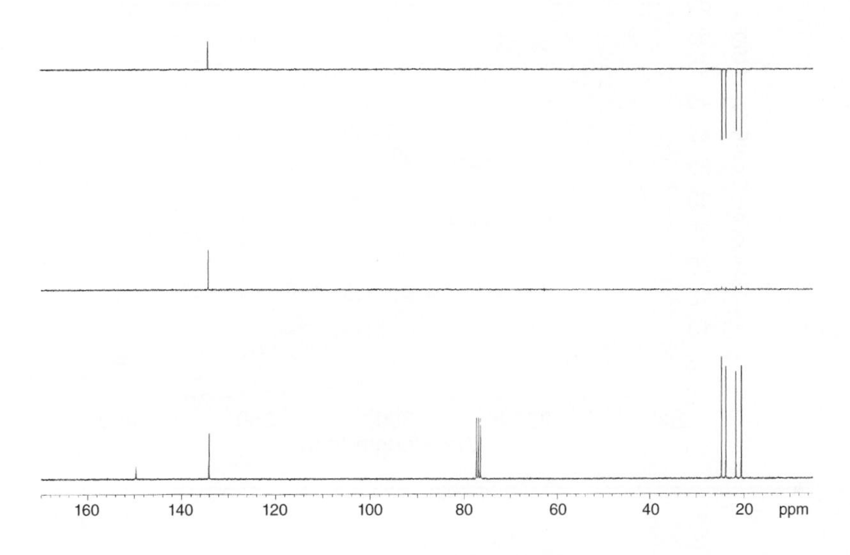

COSY spectrum

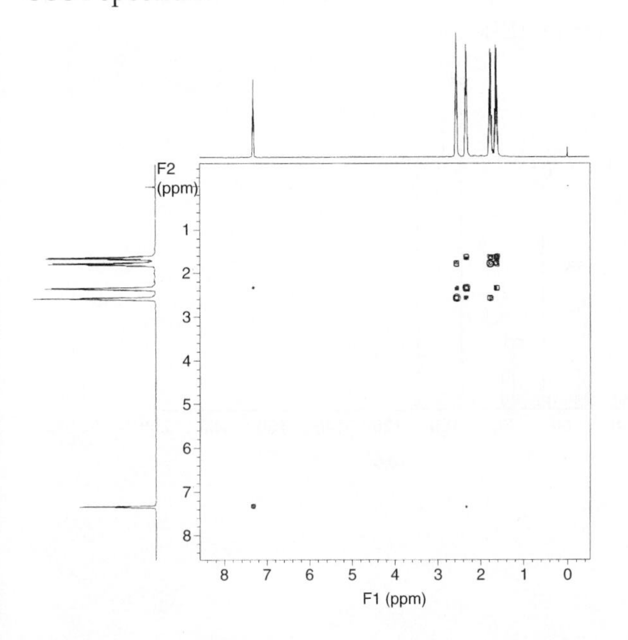

HETCOR spectrum

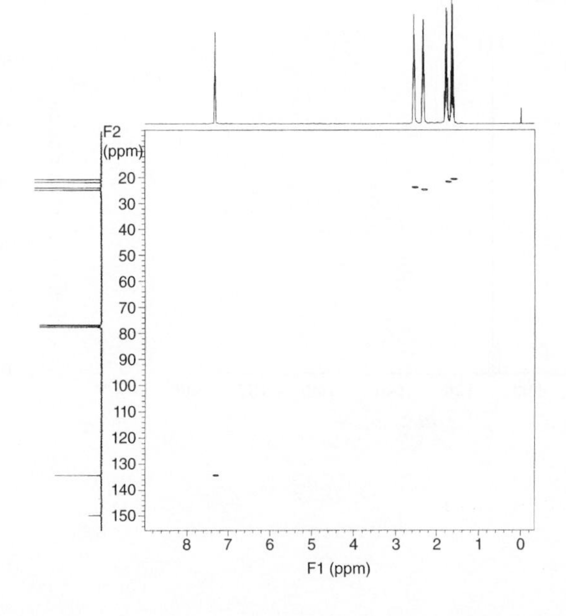

Infrared spectrum (neat)

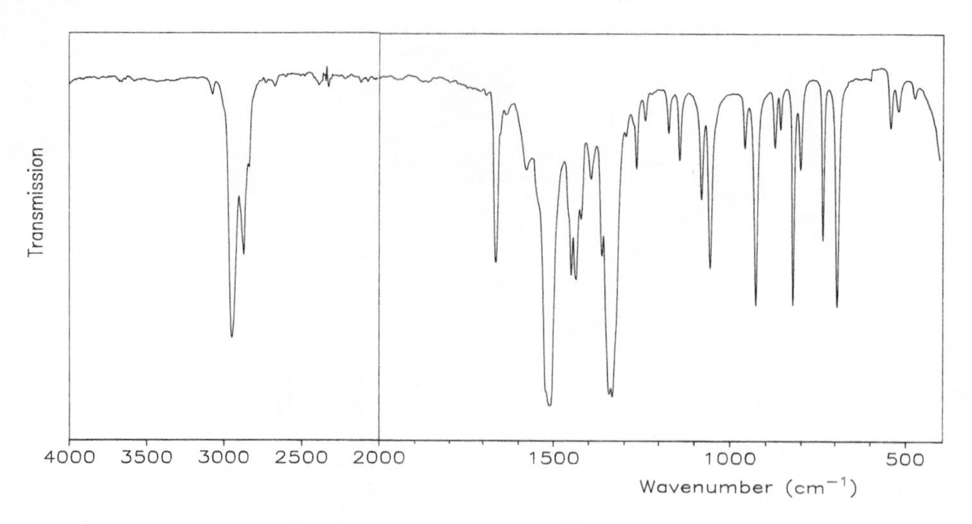

Ultraviolet–visible spectrum (EtOH, ϵ[225] 4300, ϵ[255] 5600, ϵ[322] 73)

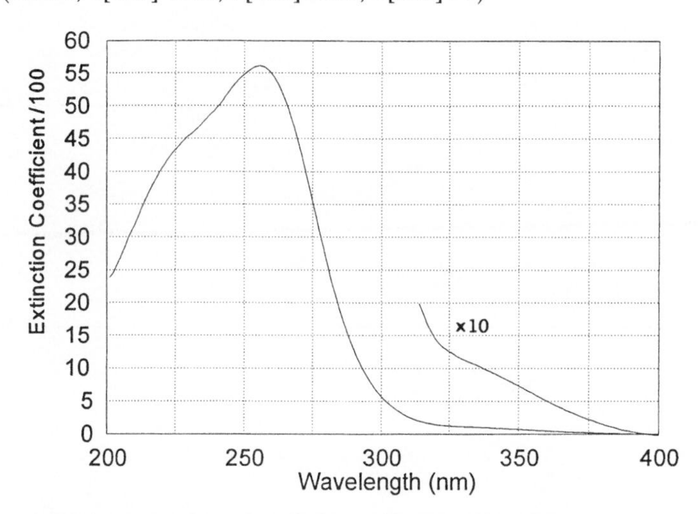

PROBLEM 15-9

Mass spectrum (CI)

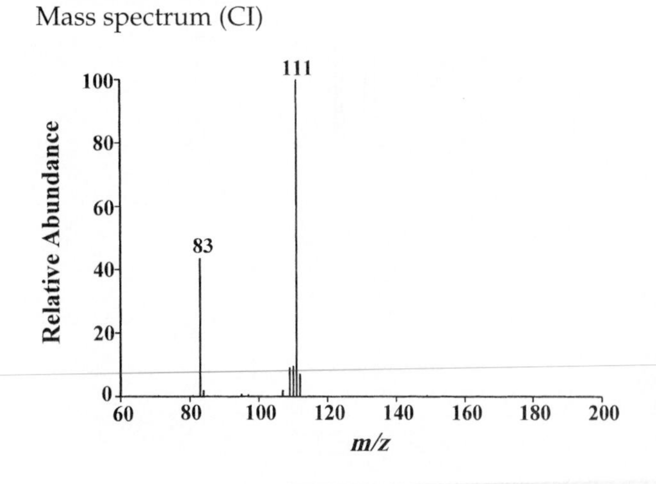

Mass spectrum (EI)

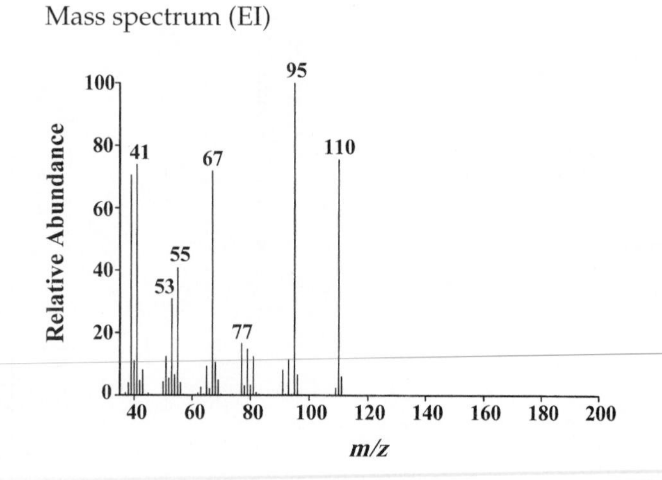

Proton NMR spectrum ($CDCl_3$)

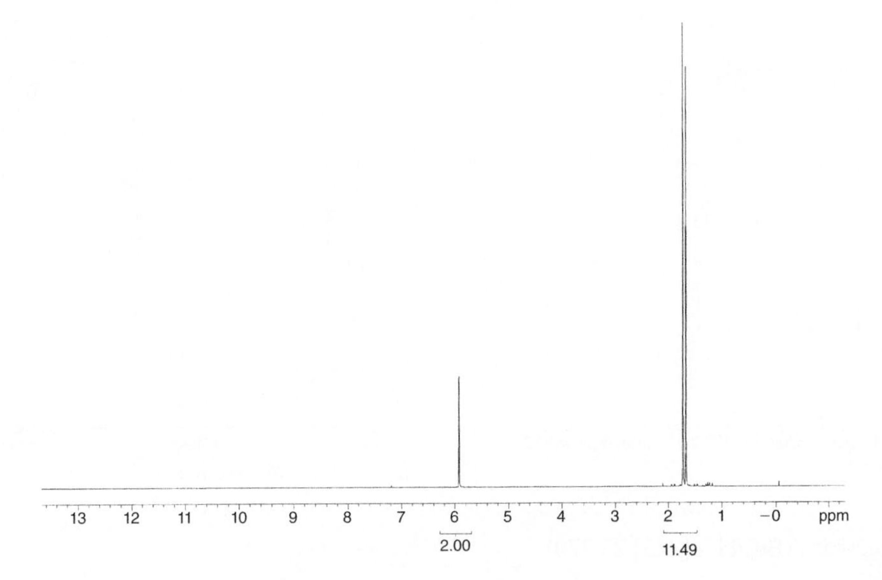

Carbon-13 NMR spectrum

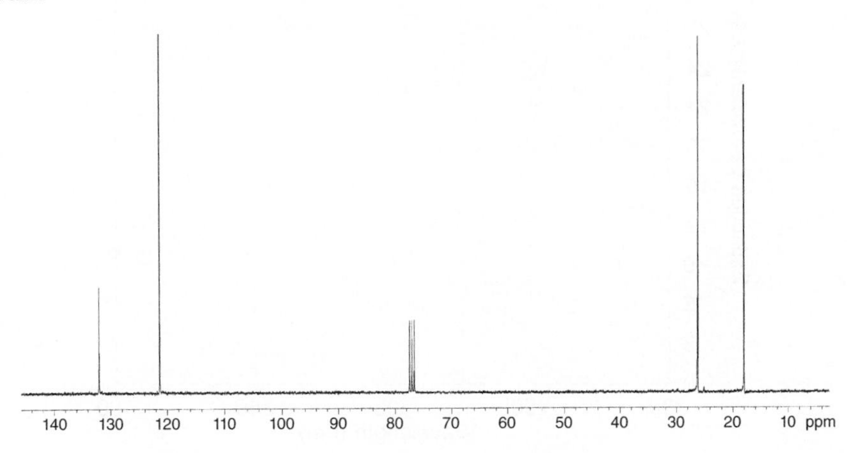

COSY spectrum

HETCOR spectrum

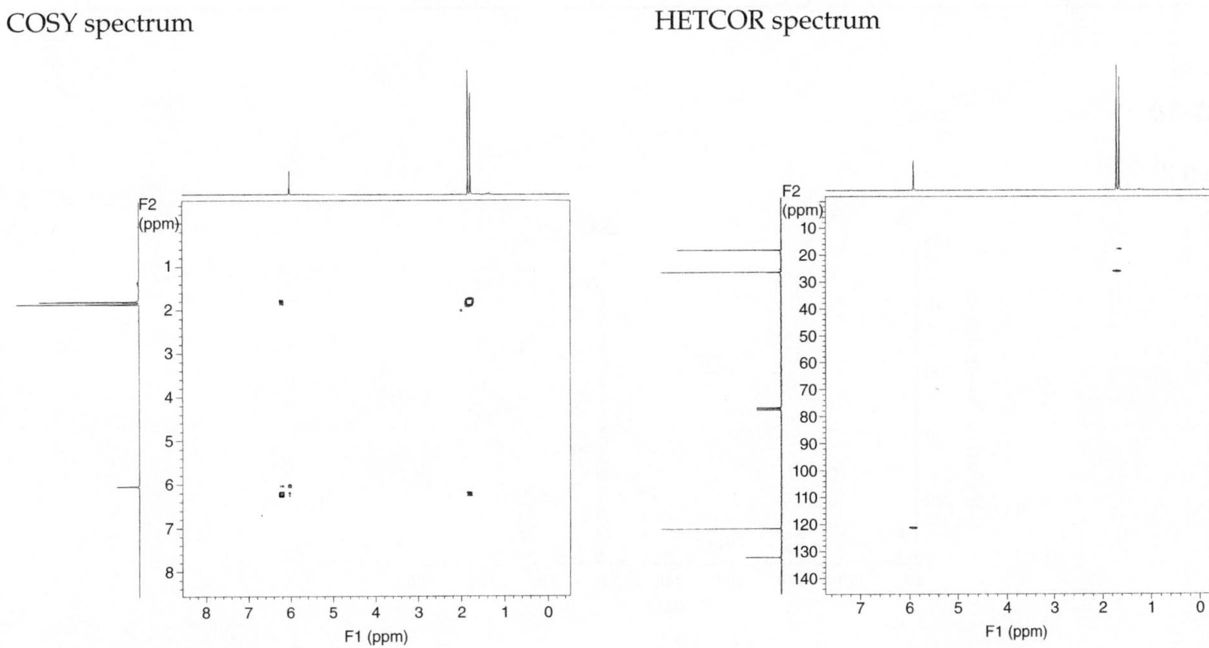

Infrared spectrum (neat)

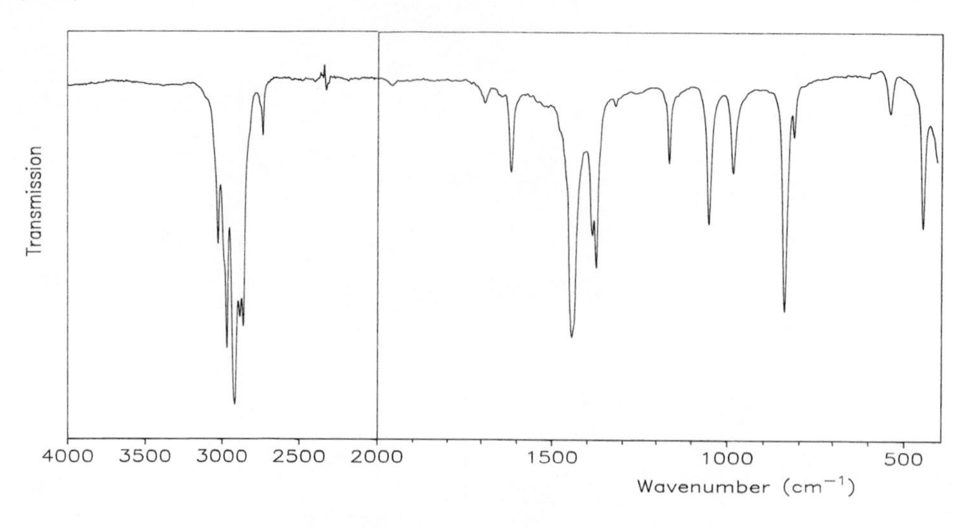

Ultraviolet–visible spectrum (EtOH, ϵ[243] 25,370)

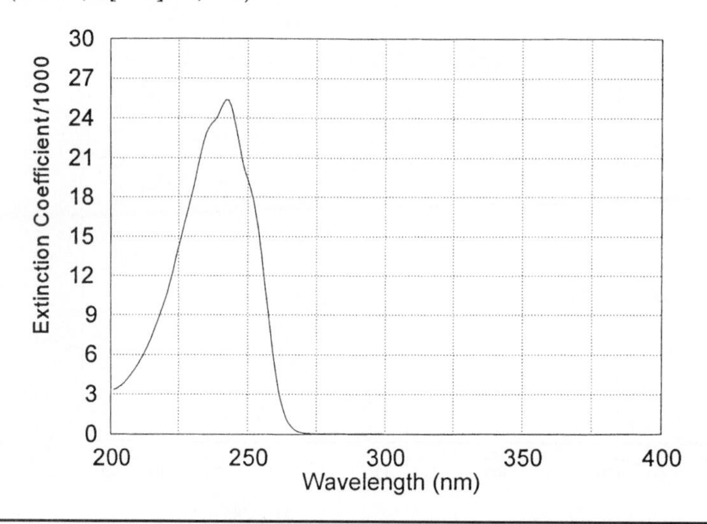

PROBLEM 15-10

Mass spectrum (CI)

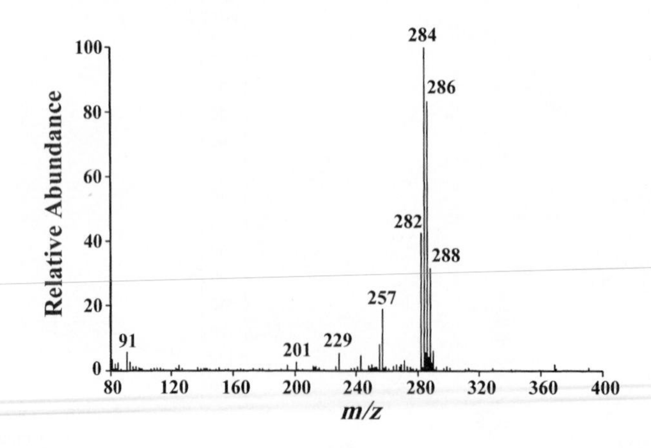

Mass spectrum (EI)

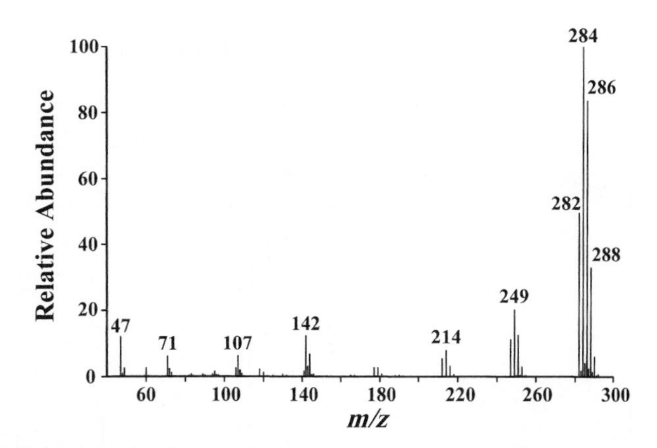

Carbon-13 NMR spectrum (tetrahydrofuran/acetone-d_6)

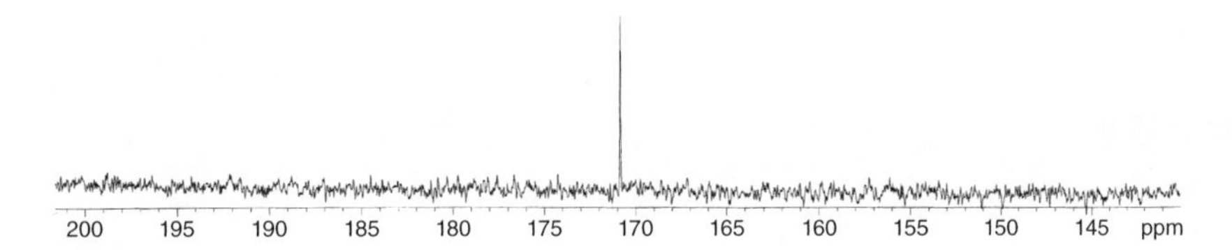

Infrared spectrum (KBr disc)

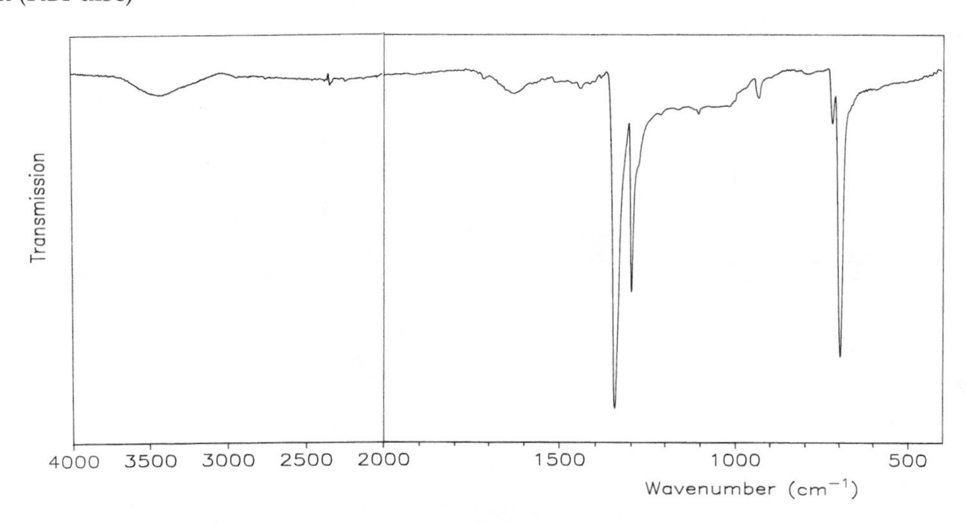

Ultraviolet–visible spectrum (EtOH ϵ[217] 75,000, ϵ[232] 15,000, ϵ[240] 7000, ϵ[251] 3200, ϵ[280] 530, ϵ[291] 261, ϵ[301] 211)

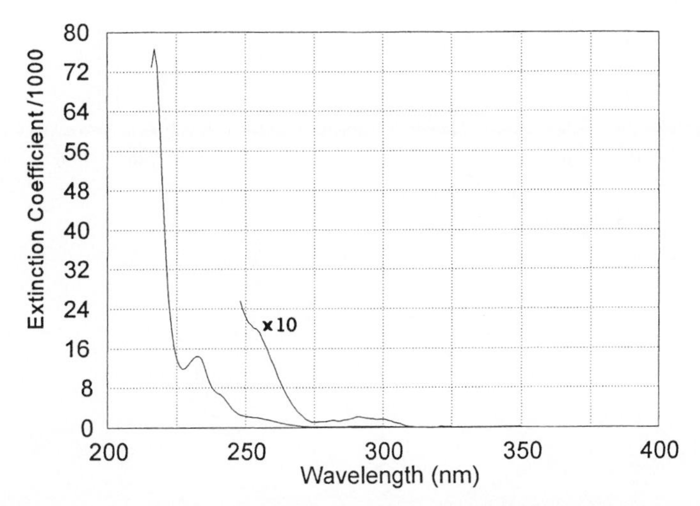

PROBLEM 15-11

Mass spectrum (CI)

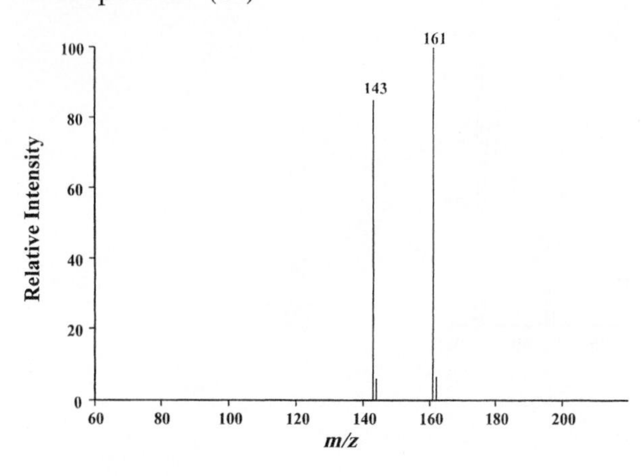

Mass spectrum (EI)

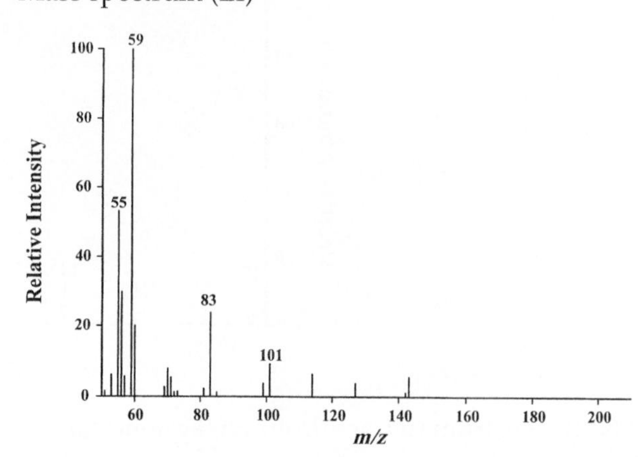

Proton NMR spectrum (D_2O)

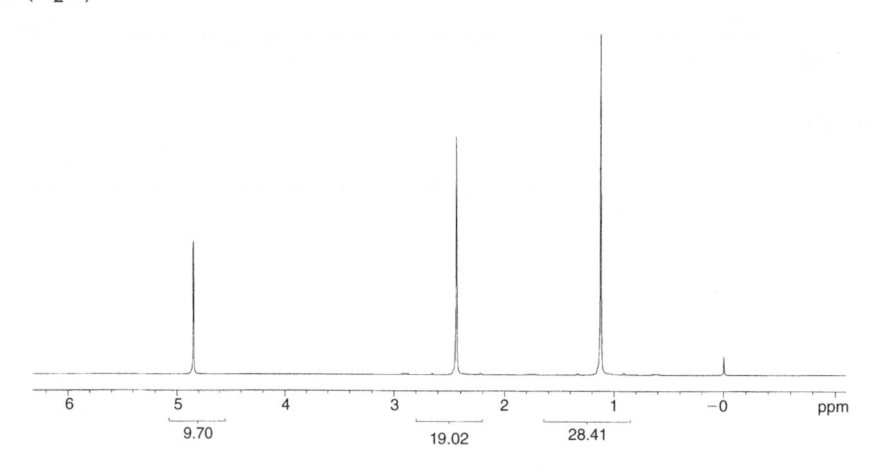

Carbon-13 NMR spectrum

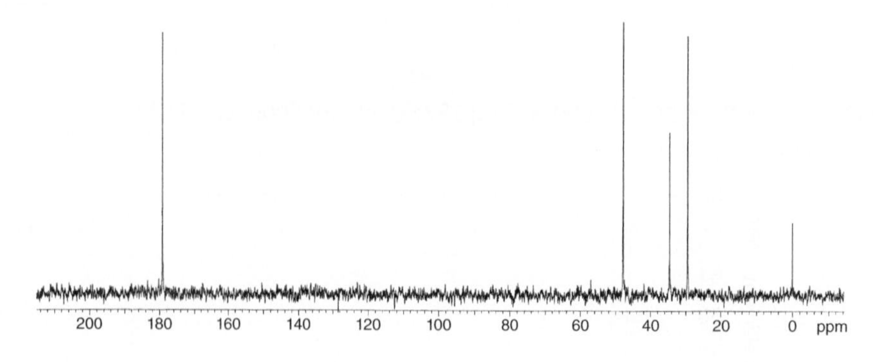

HETCOR spectrum

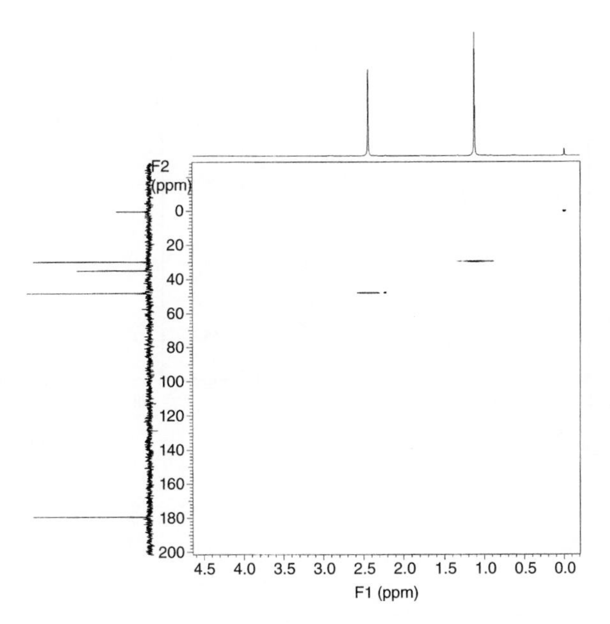

Infrared spectrum (KBr disc)

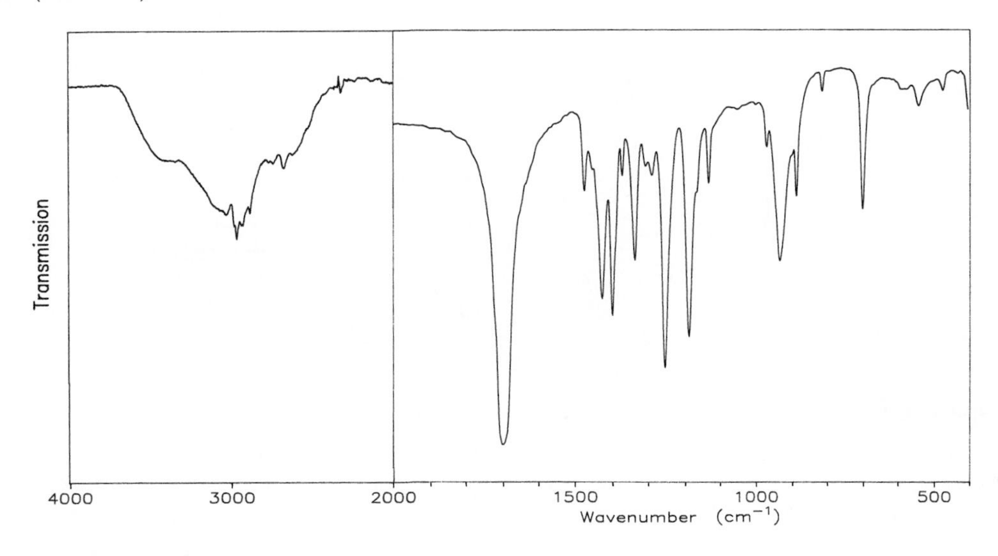

Ultraviolet–visible spectrum (EtOH, ϵ[212] 109)

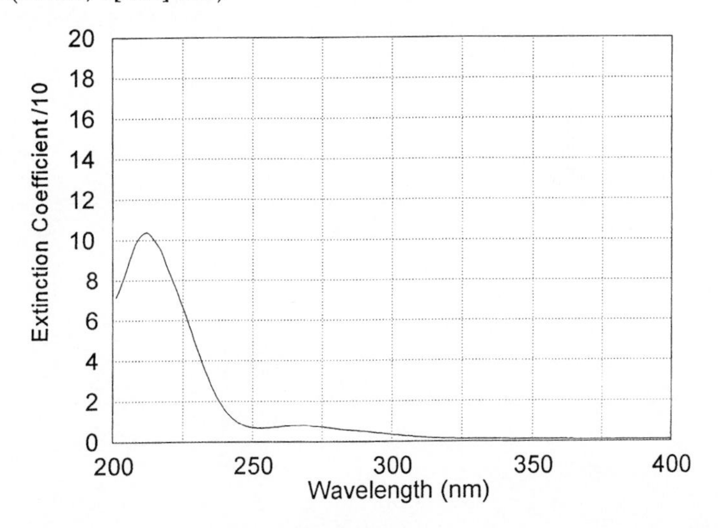

PROBLEM 15-12

Mass spectrum (CI)

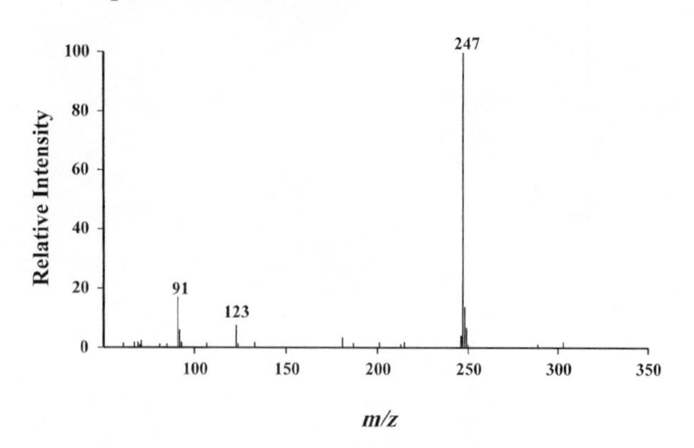

Mass spectrum (EI)

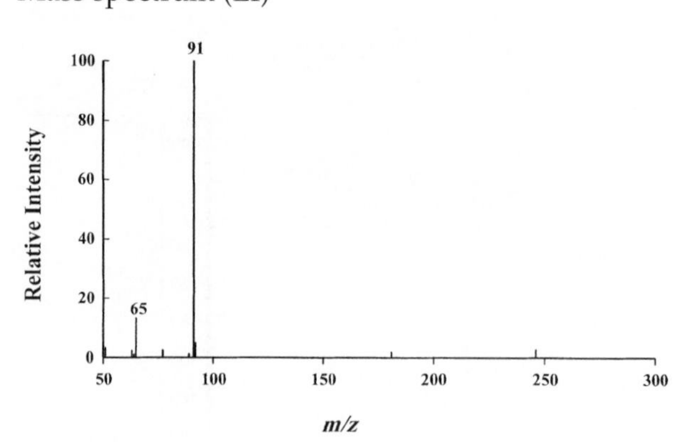

Proton NMR spectrum ($CDCl_3$)

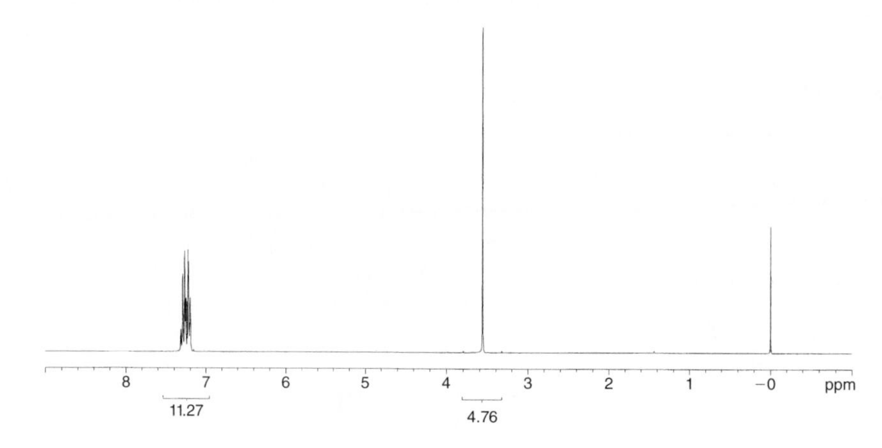

Carbon-13 NMR spectrum

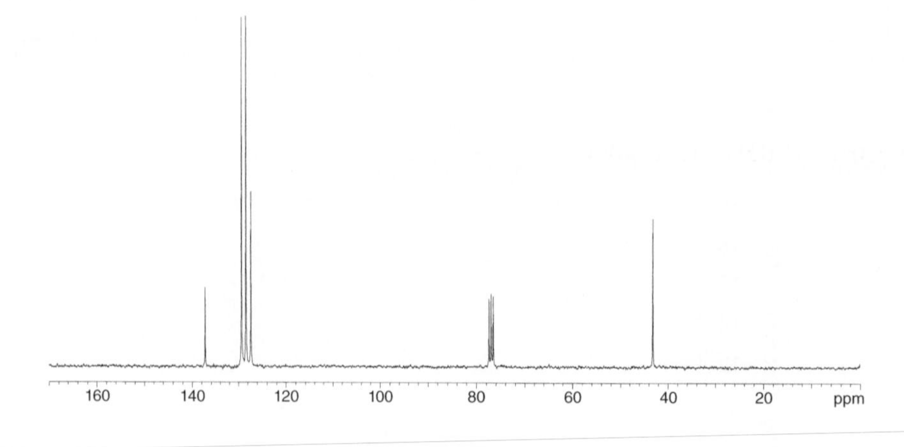

COSY spectrum

HETCOR spectrum

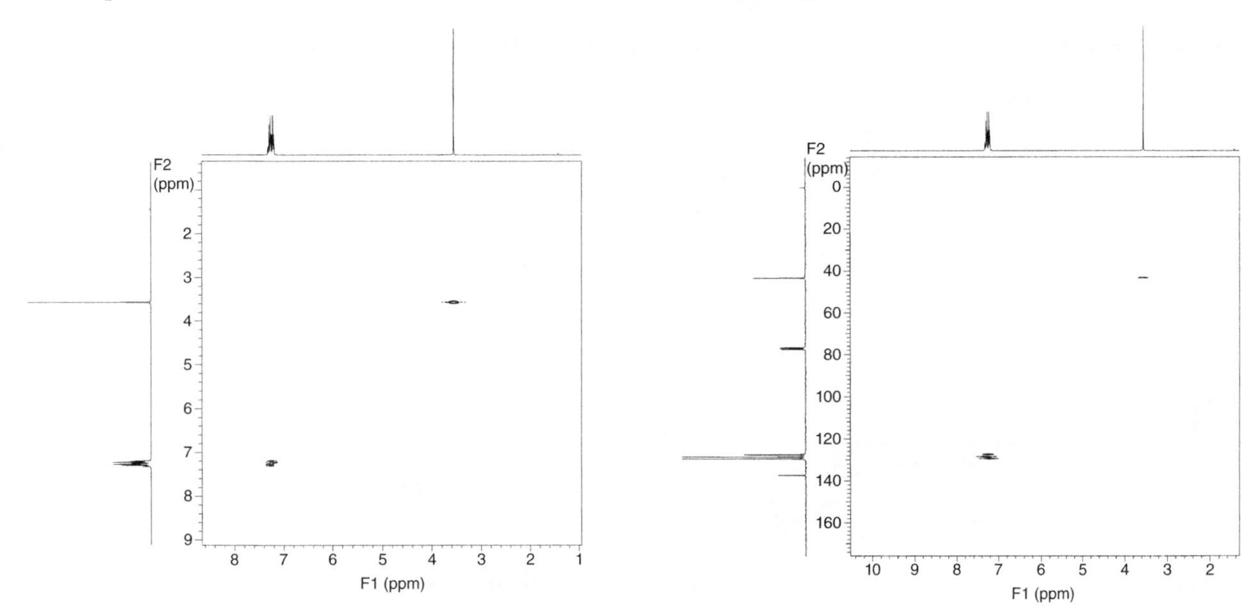

Infrared spectrum (KBr disc)

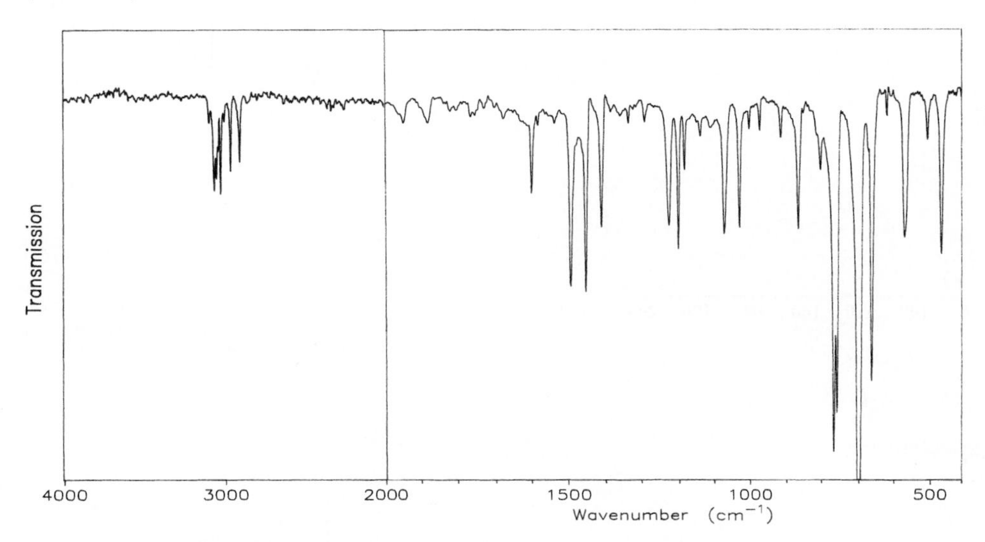

Ultraviolet–visible spectrum (EtOH, ϵ[220] 16,299, ϵ[261] 1652, ϵ[267] 1162)

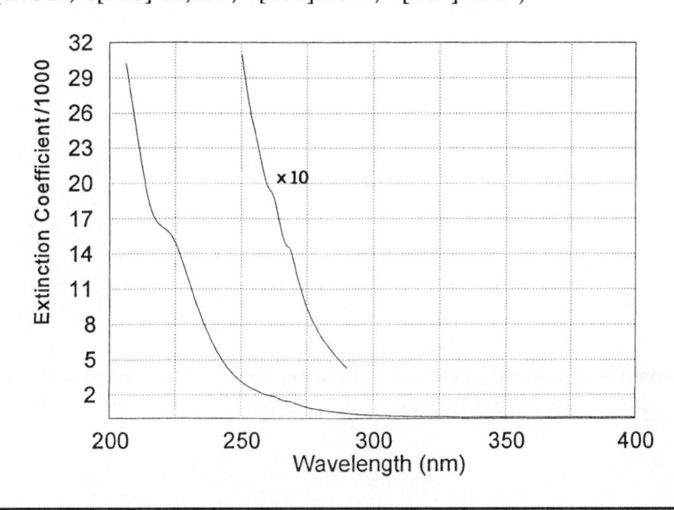

PROBLEM 15-13

Mass spectrum (desorption EI)

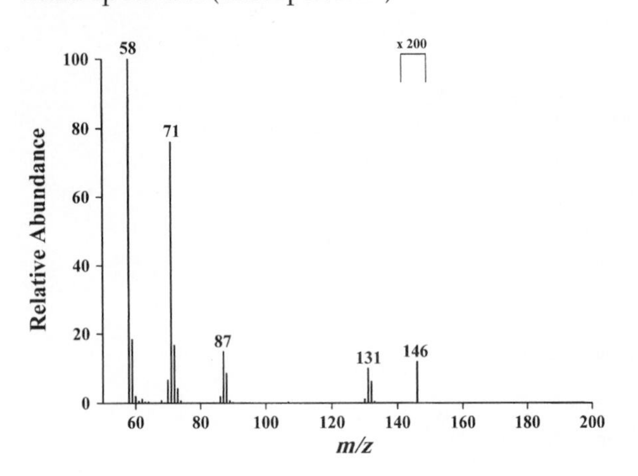

Mass spectrum (CI)

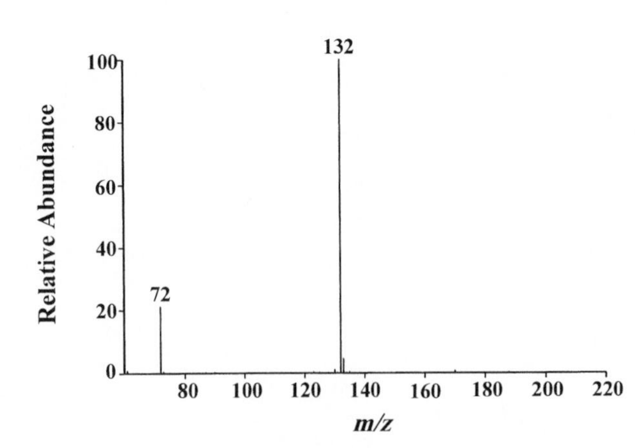

Mass spectrum (EI)

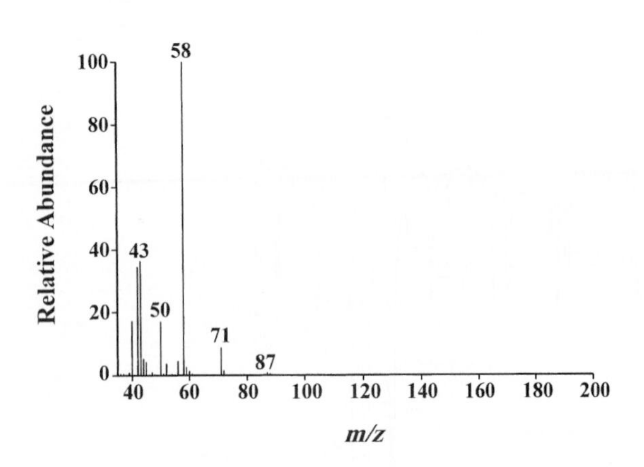

Proton NMR spectrum (D_2O)

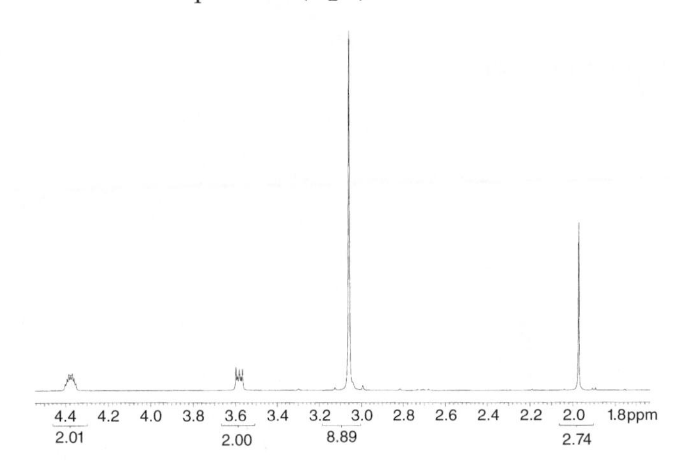

Carbon-13 NMR spectrum

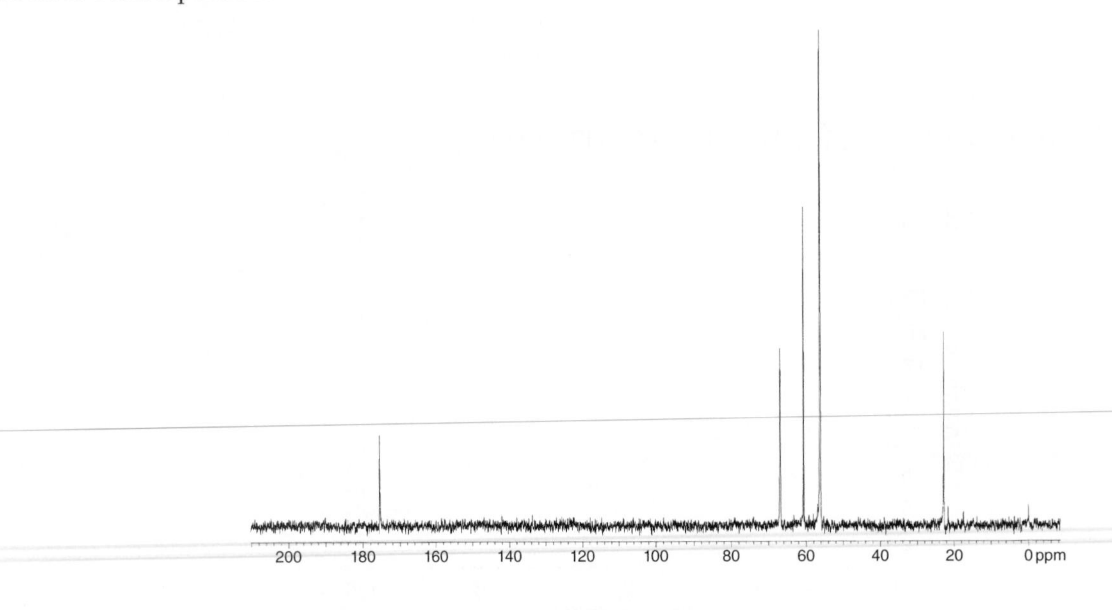

COSY spectrum

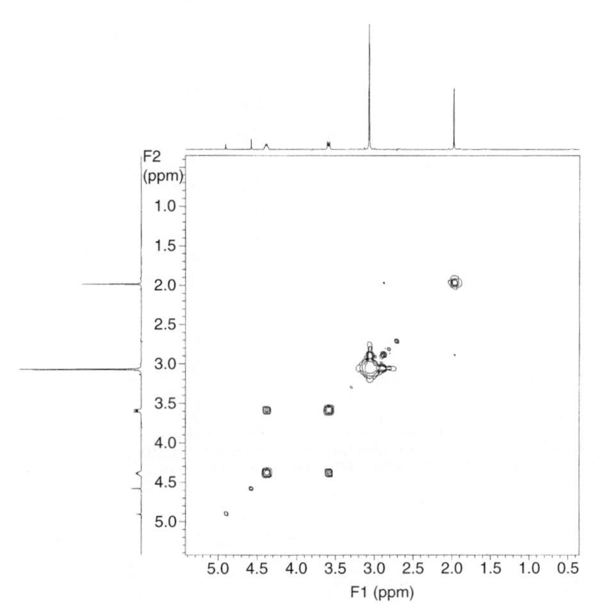

HETCOR spectrum

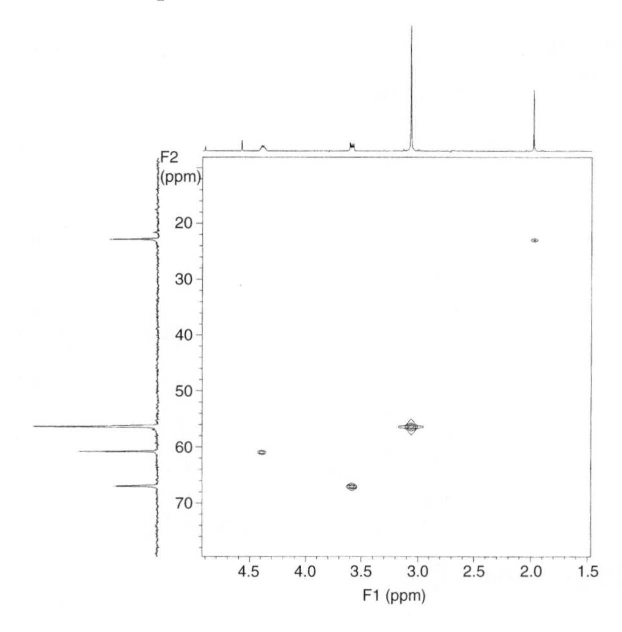

Infrared spectrum (KBr disc)

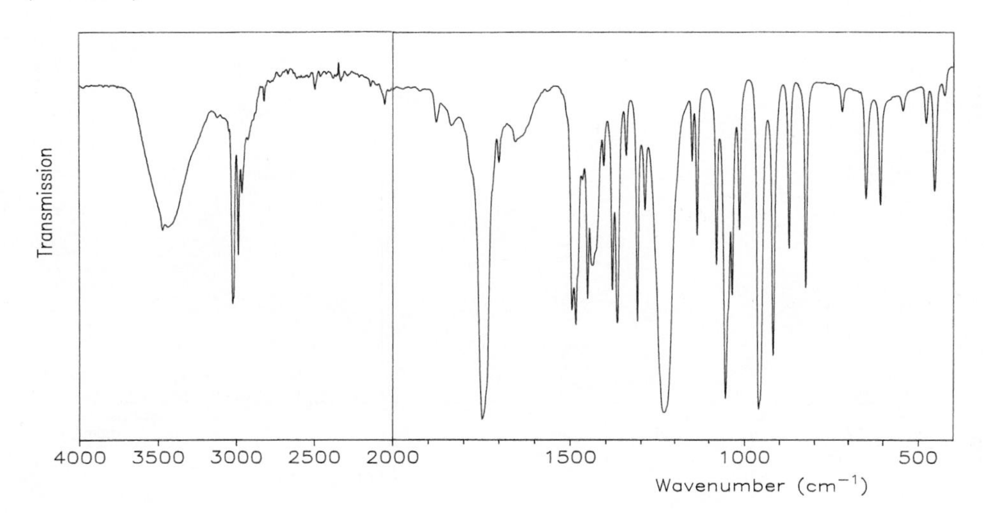

Ultraviolet–visible spectrum (EtOH, ϵ[210] 82)

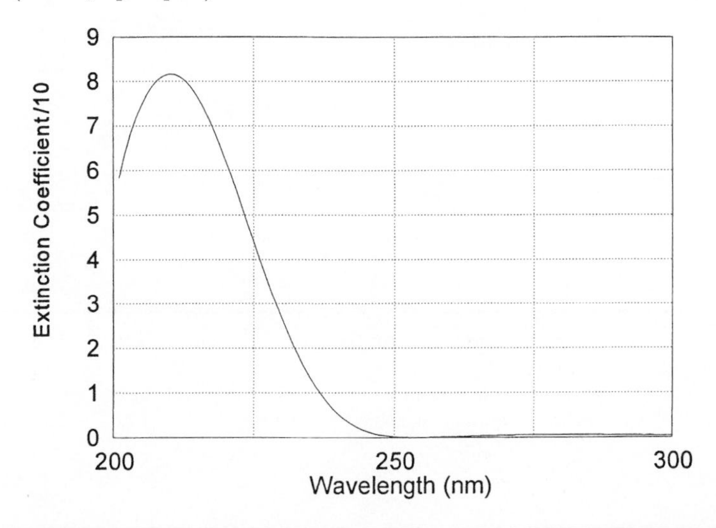

PROBLEM 15-14

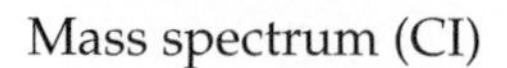

Mass spectrum (CI)

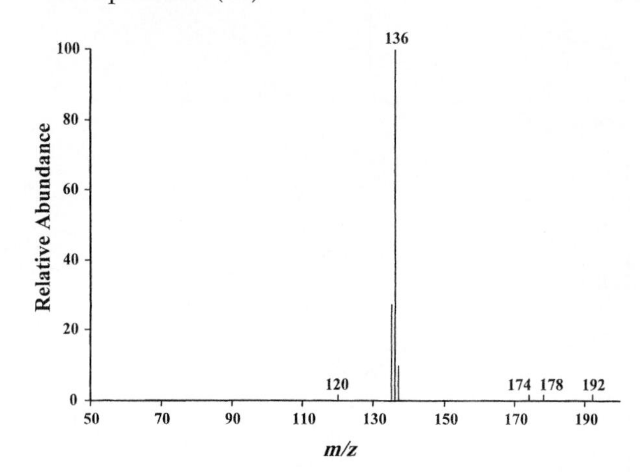

Mass spectrum (EI)

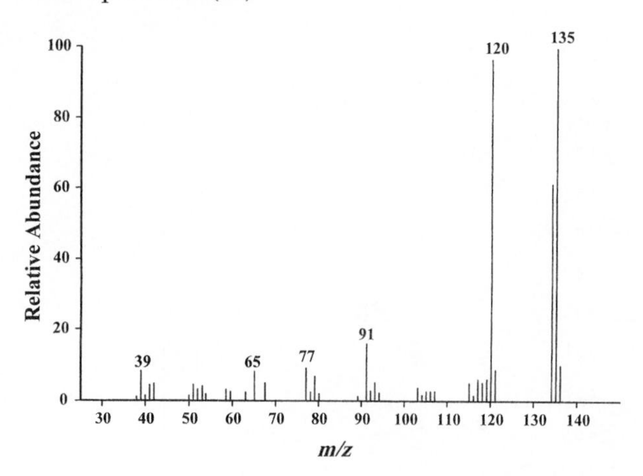

Proton NMR spectrum ($CDCl_3$)

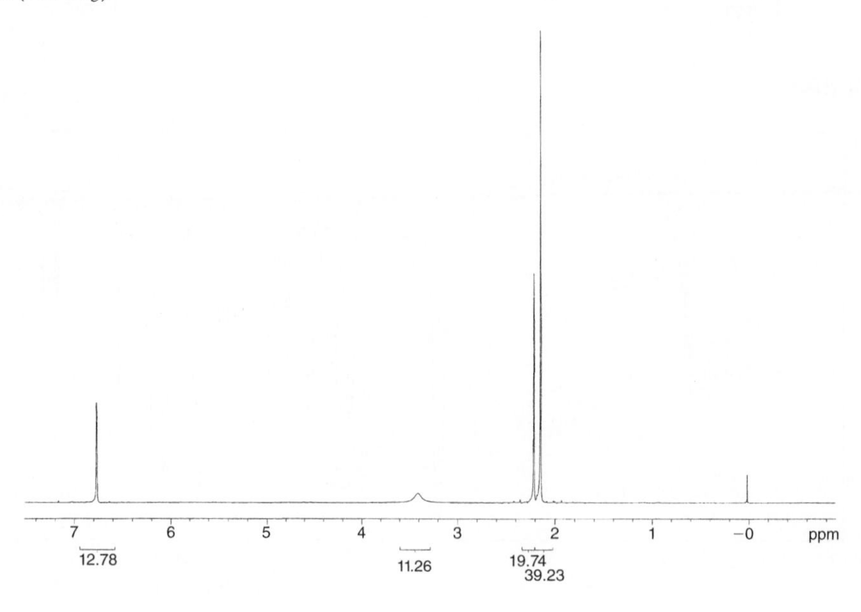

Carbon-13 NMR spectrum

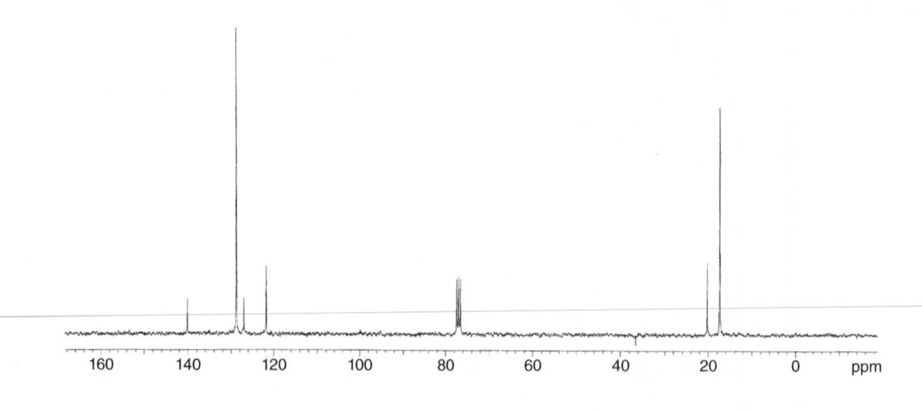

COSY spectrum

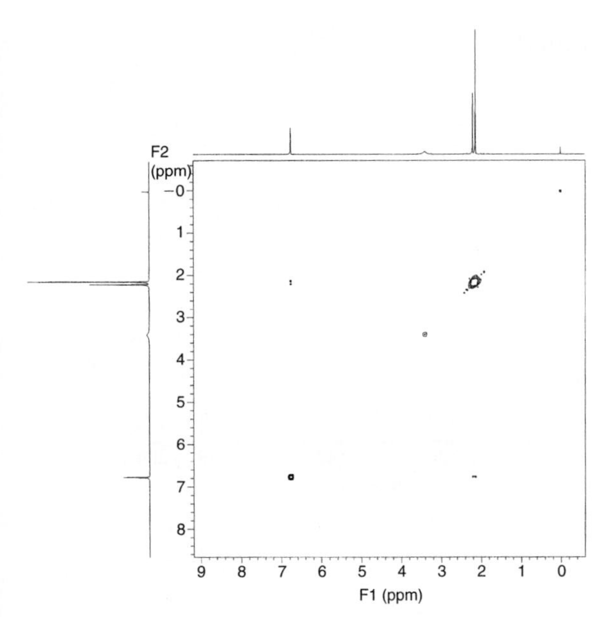

HETCOR spectrum

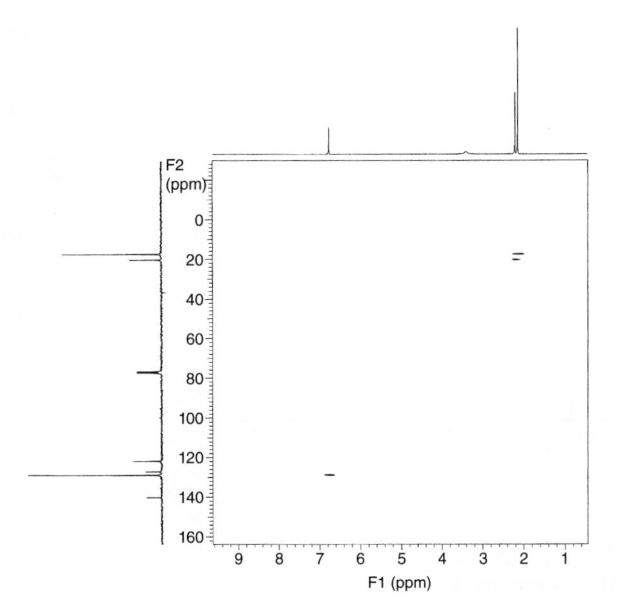

Infrared spectrum (neat)

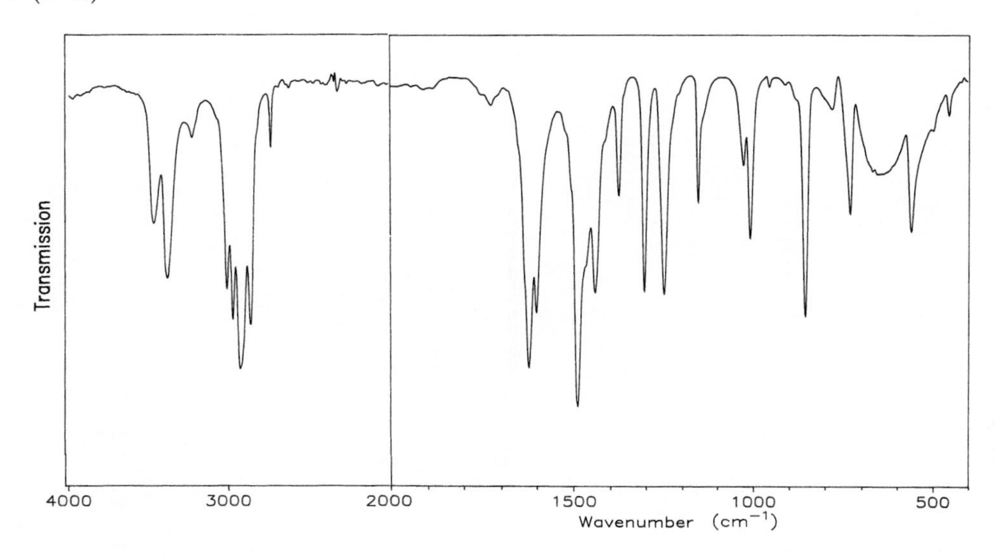

Ultraviolet–visible spectrum (EtOH, ϵ[236] 8123, ϵ[289] 2140)

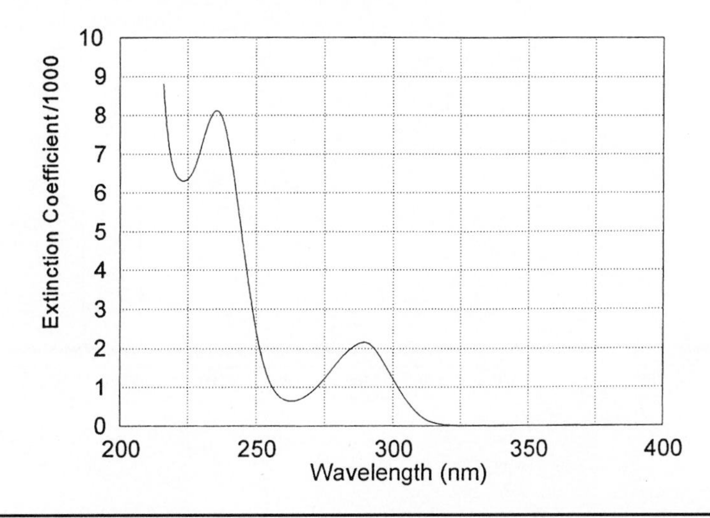

PROBLEM 15-15

Mass spectrum (CI)

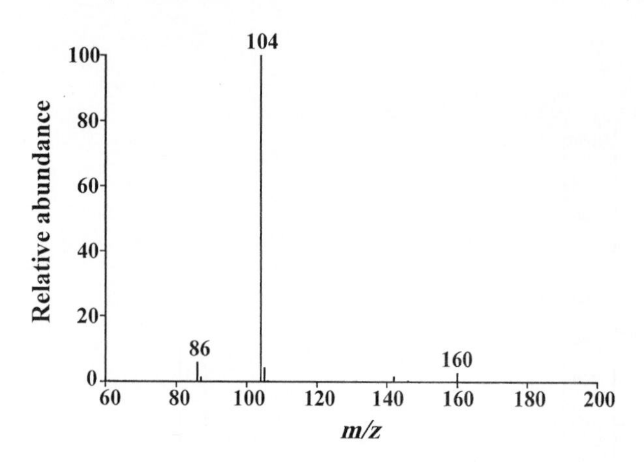

Mass spectrum (EI)

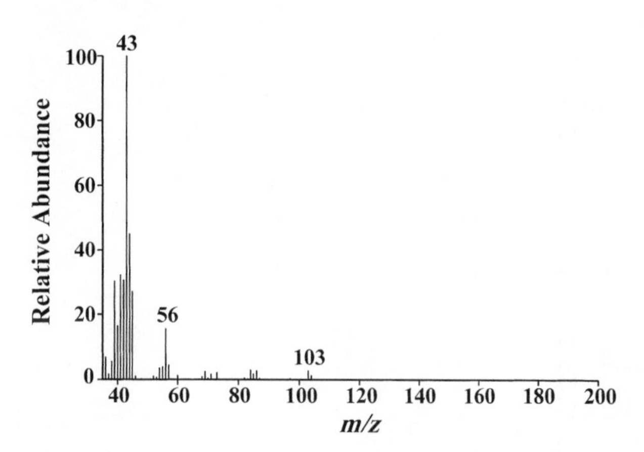

Proton NMR spectrum (D_2O)

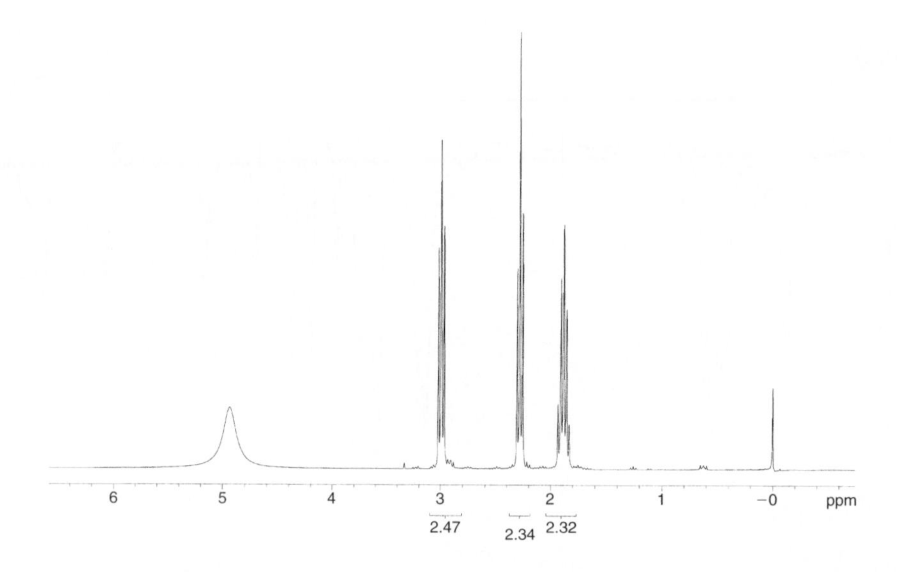

Carbon-13 NMR spectrum

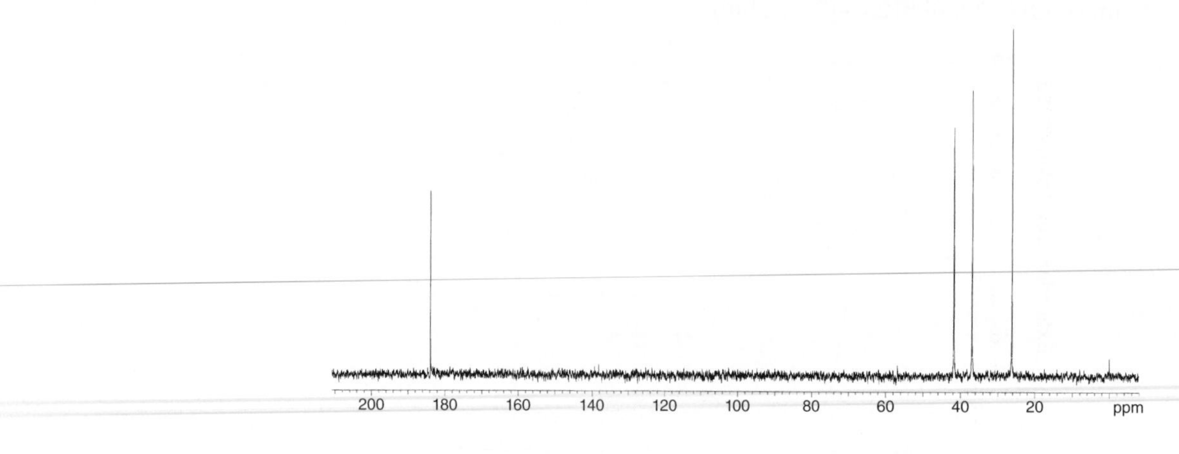

COSY spectrum

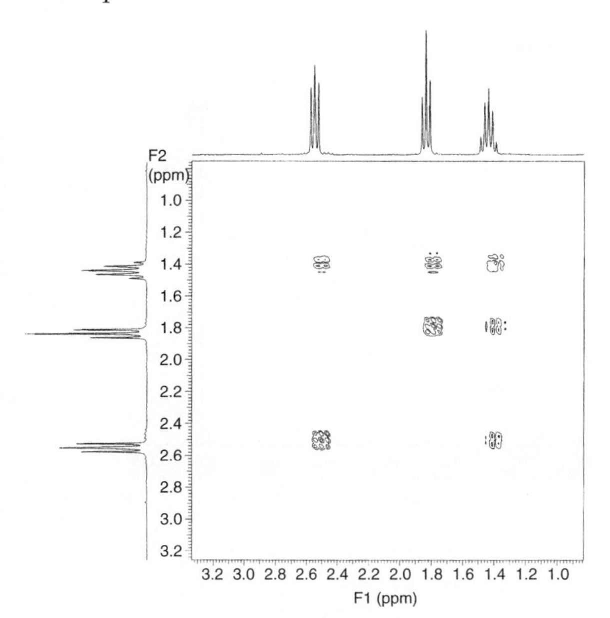

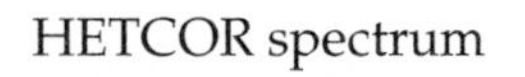

HETCOR spectrum

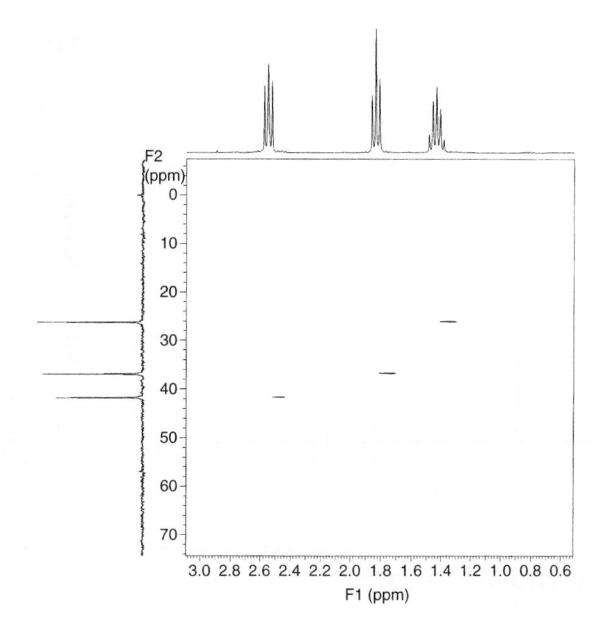

Infrared spectrum (KBr disc)

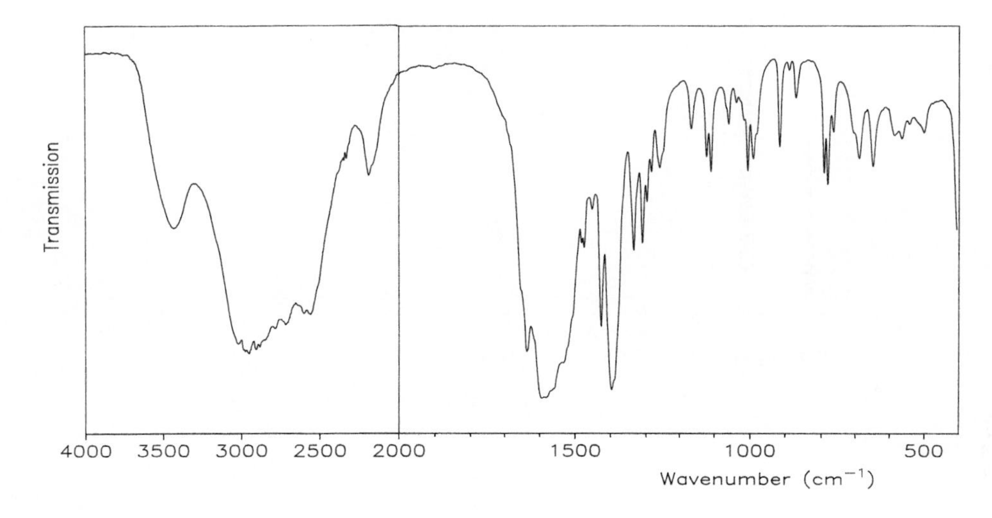

Ultraviolet–visible spectrum (EtOH)

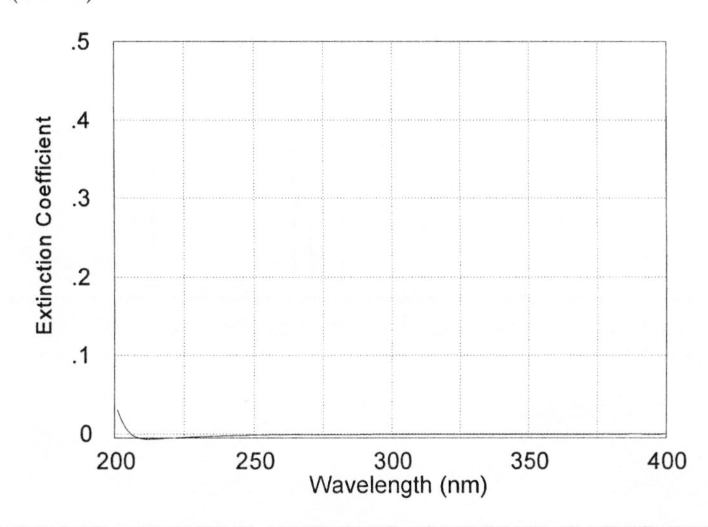

PROBLEM 15-16

Mass spectrum (CI)

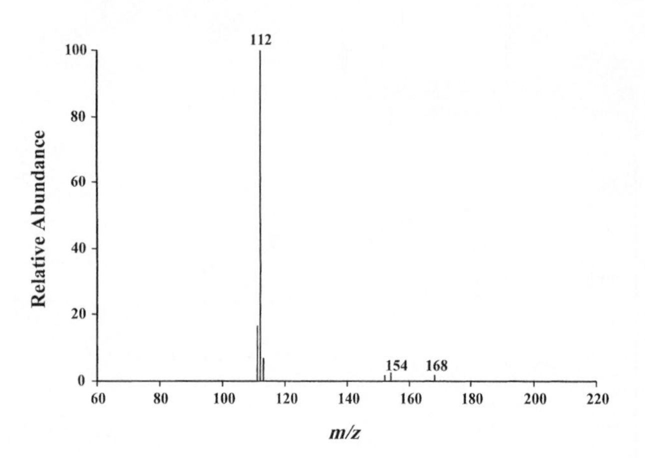

Mass spectrum (EI)

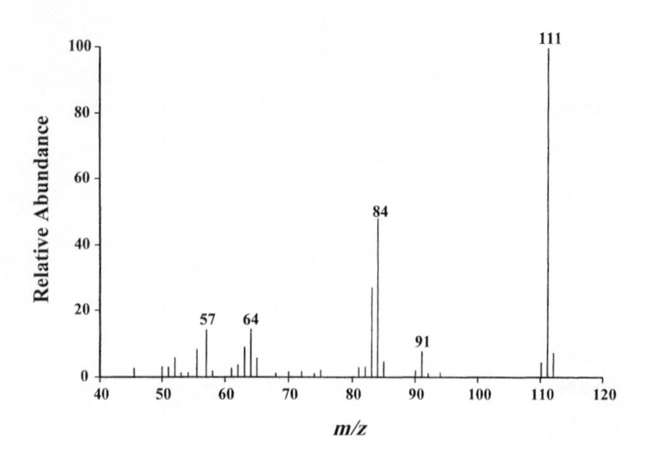

Proton NMR spectrum ($CDCl_3$)

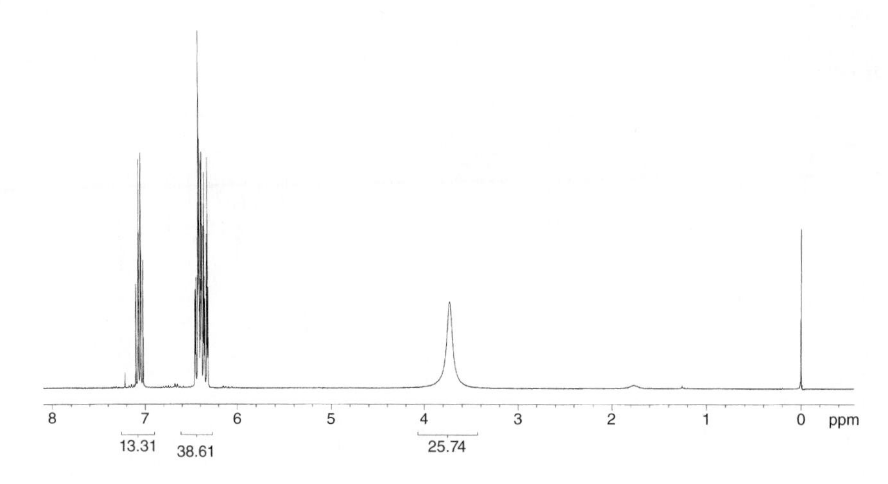

Carbon-13 NMR spectrum

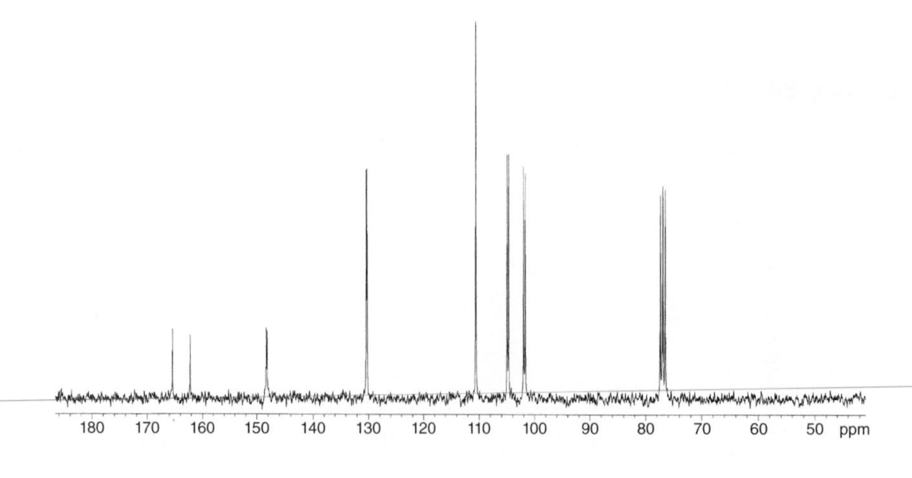

COSY spectrum

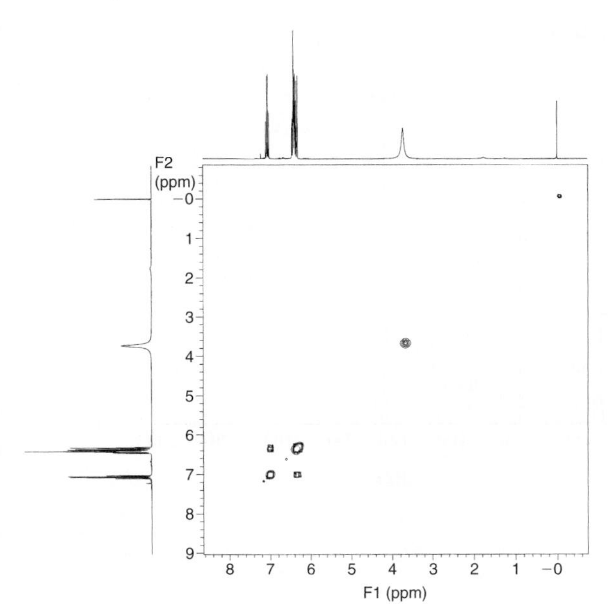

HETCOR spectrum

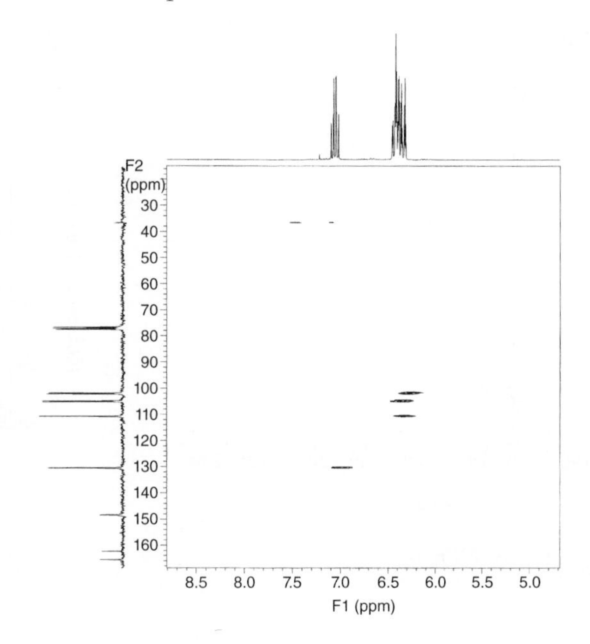

Infrared spectrum (neat)

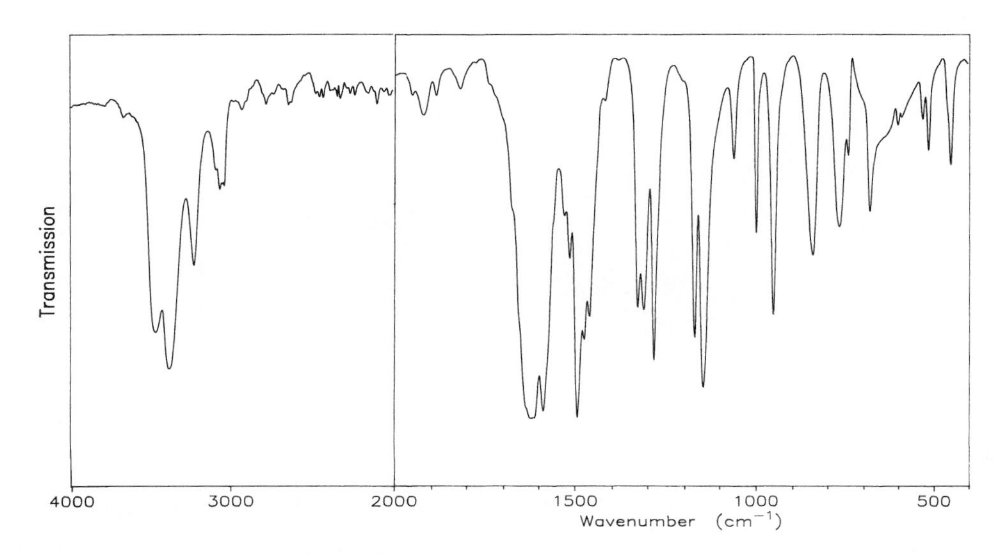

Ultraviolet–visible spectrum (EtOH, ϵ[280] 1610)

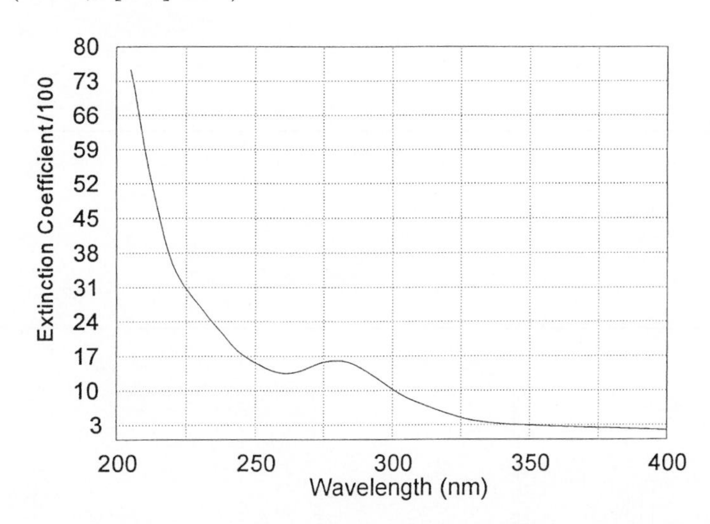

PROBLEM 15-17

Mass spectrum (CI)

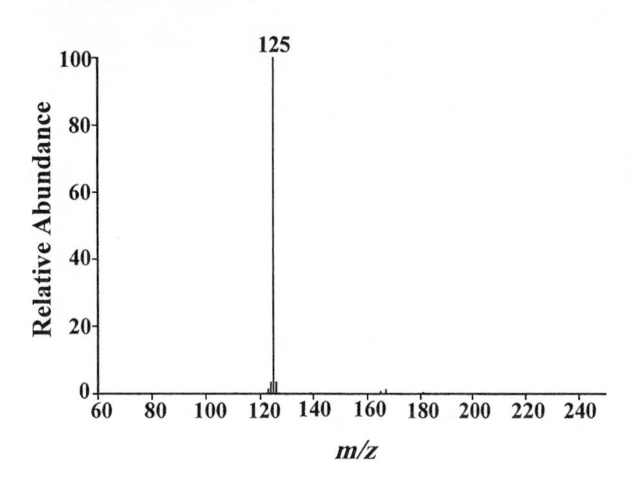

Mass spectrum (EI)

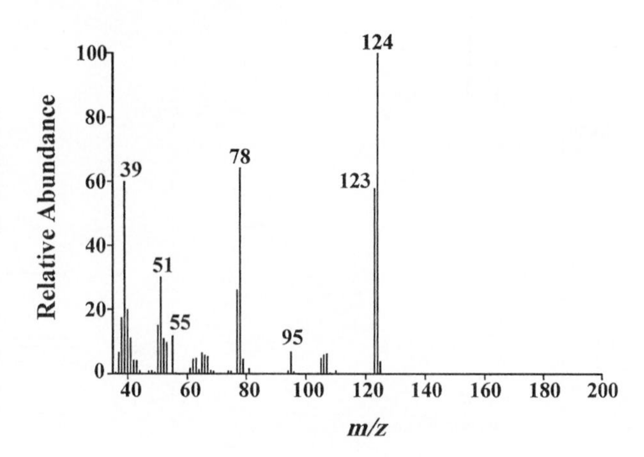

Proton NMR spectrum ($CDCl_3$)

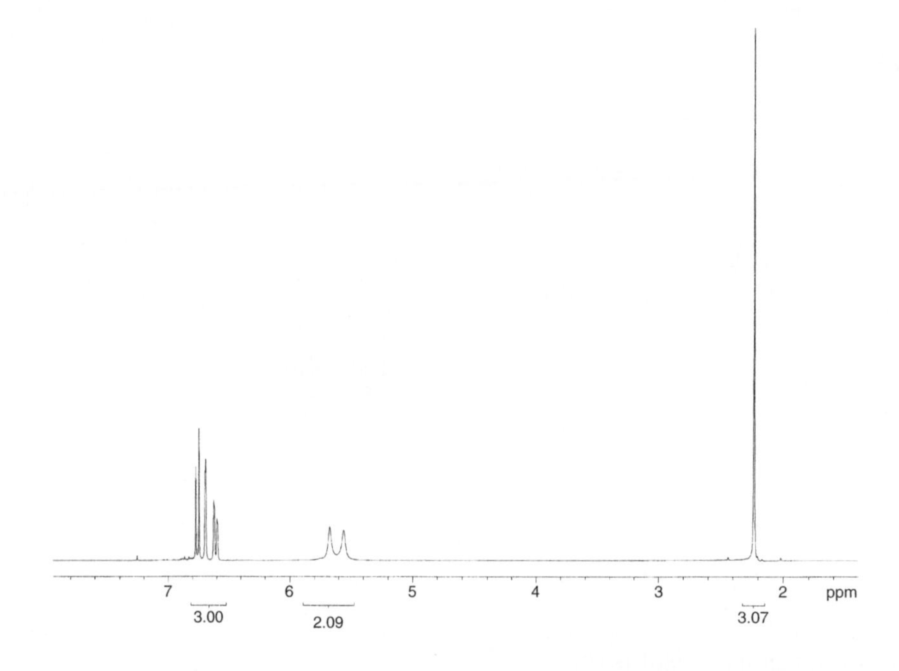

DEPT spectra

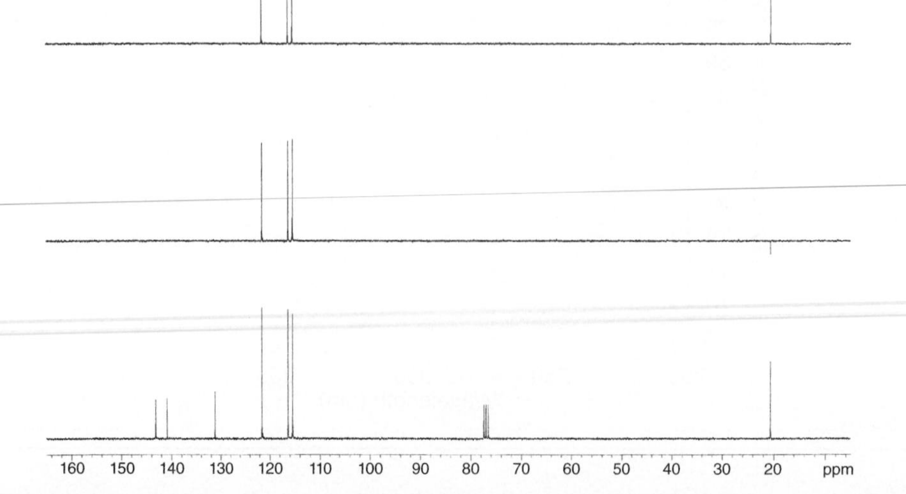

COSY spectrum

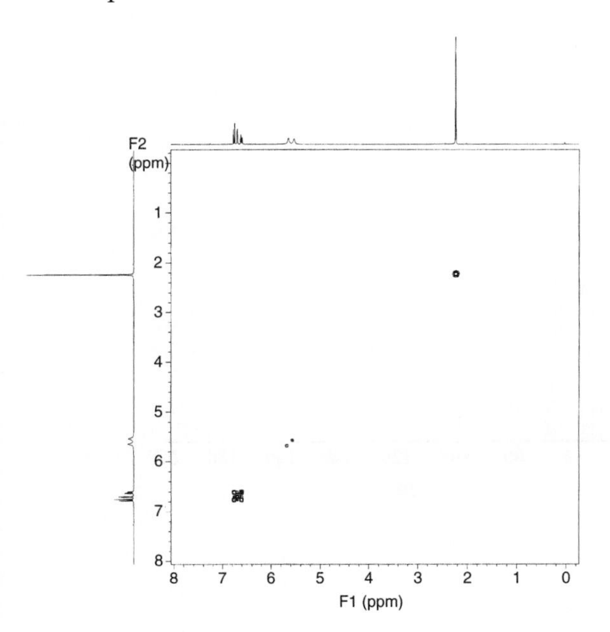

HETCOR spectrum

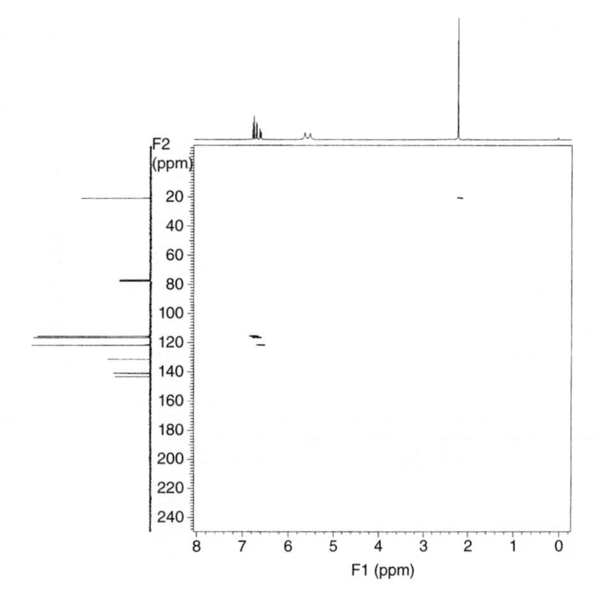

Infrared spectrum (KBr disc)

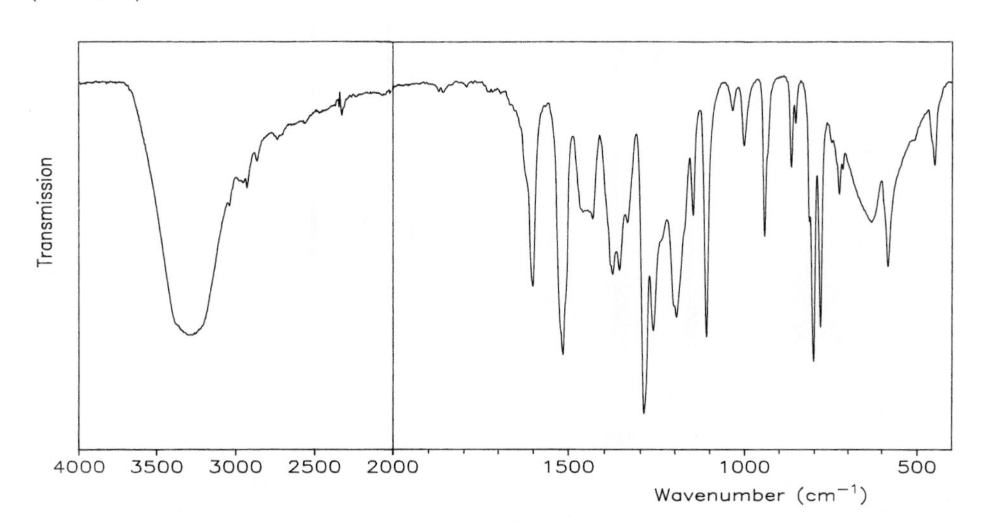

Ultraviolet–visible spectrum (EtOH, ϵ[203] 27,760, ϵ[218] 5336, ϵ[282] 2538)

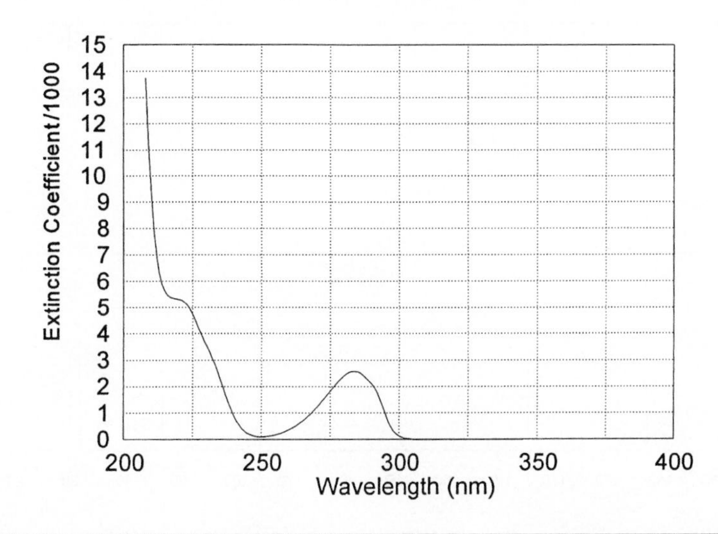

PROBLEM 15-18

Mass spectrum (CI)

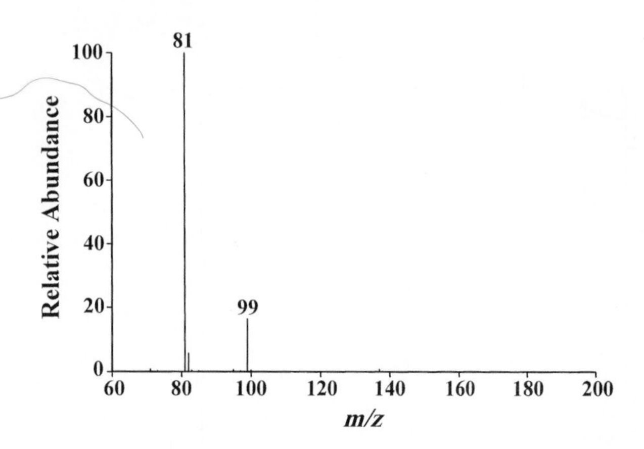

Mass spectrum (EI)

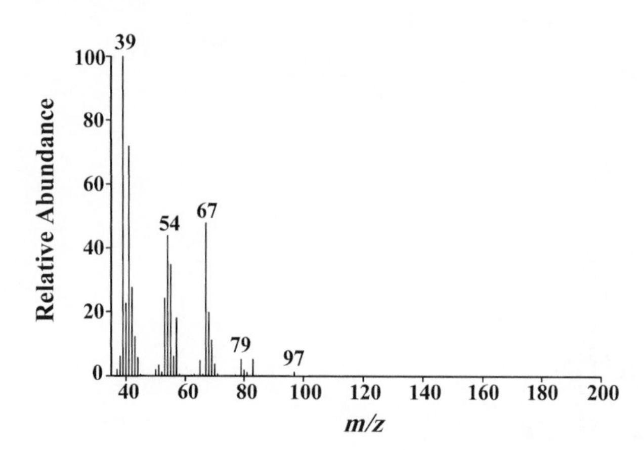

Proton NMR spectrum ($CDCl_3$)

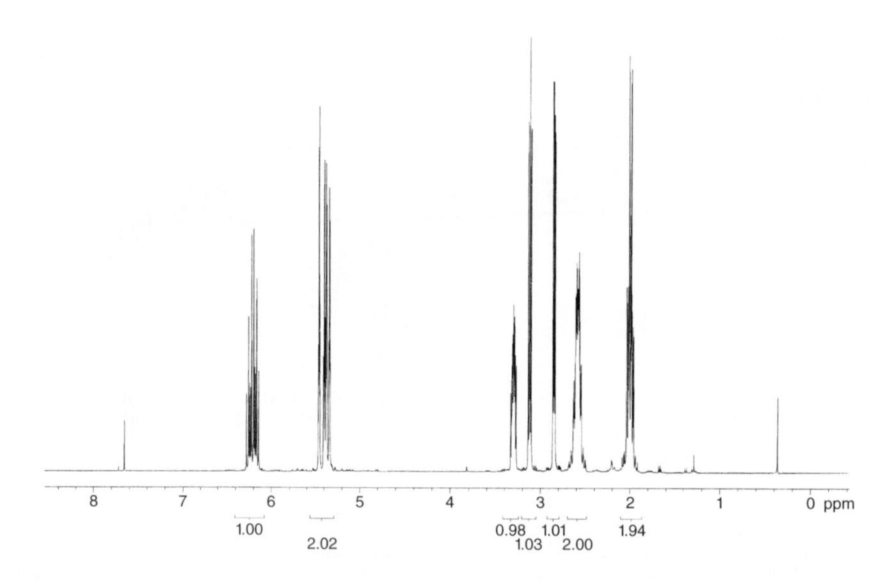

DEPT spectra

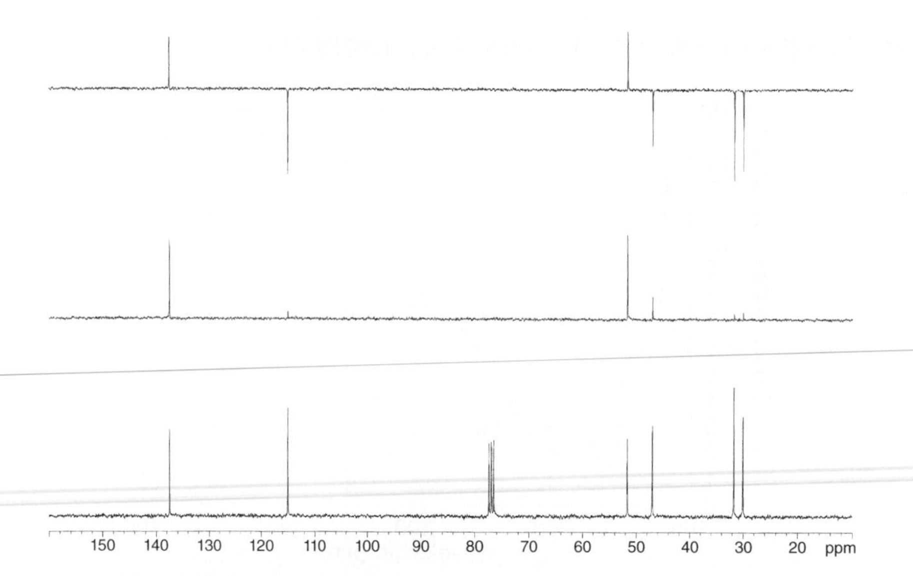

COSY spectrum

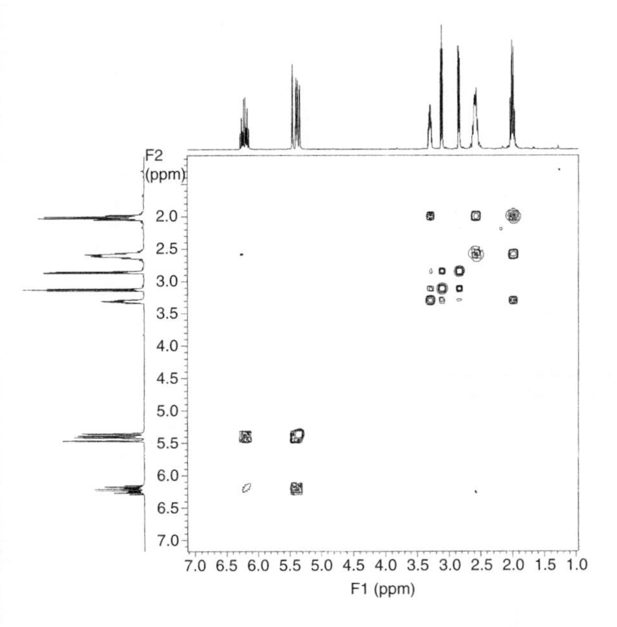

HETCOR spectrum

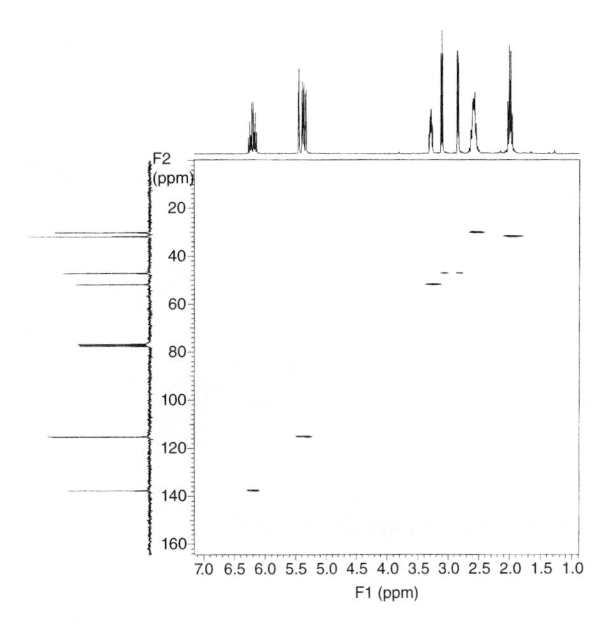

Infrared spectrum (neat)

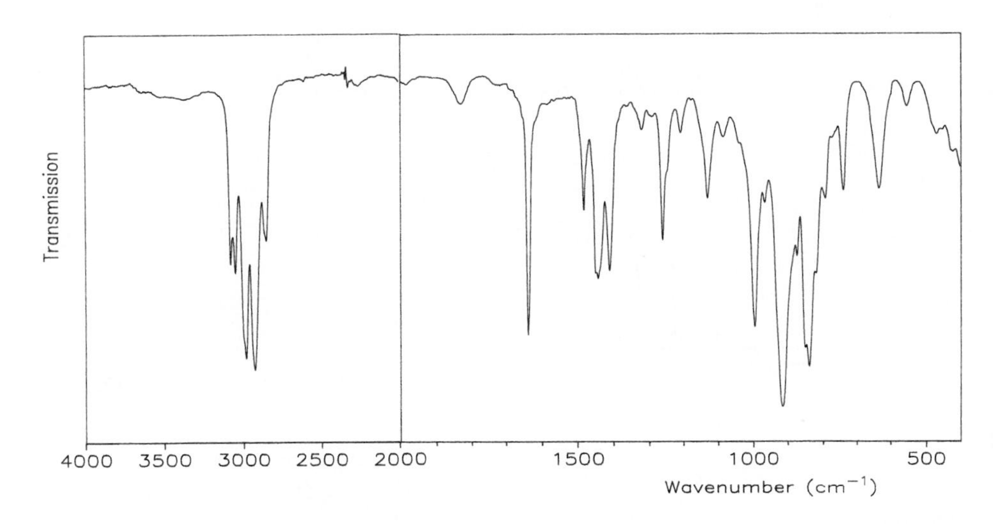

Ultraviolet–visible spectrum (EtOH, $\epsilon[202]$ 49, $\epsilon[211]$ 13, $\epsilon[271]$ 14)

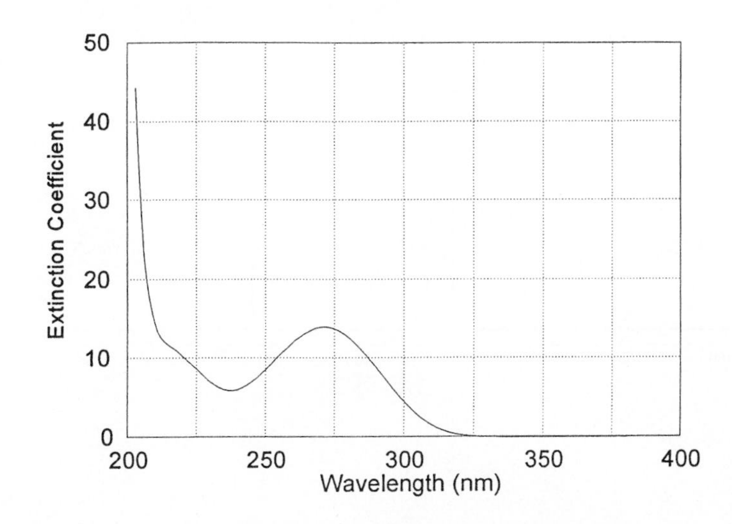

PROBLEM 15-19

Mass spectrum (CI)

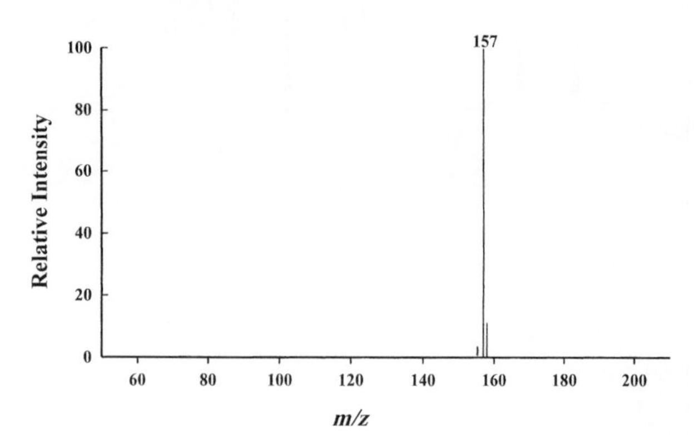

Mass spectrum (EI)

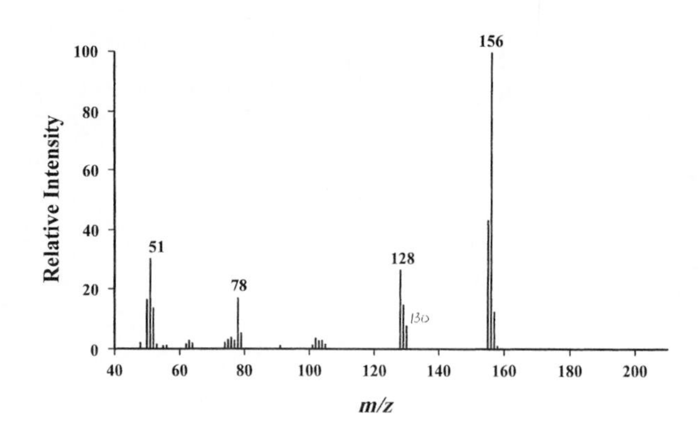

Proton NMR spectrum with expansion ($CDCl_3$)

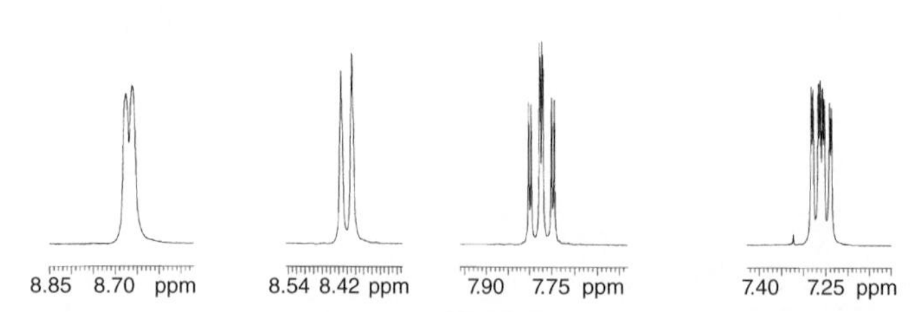

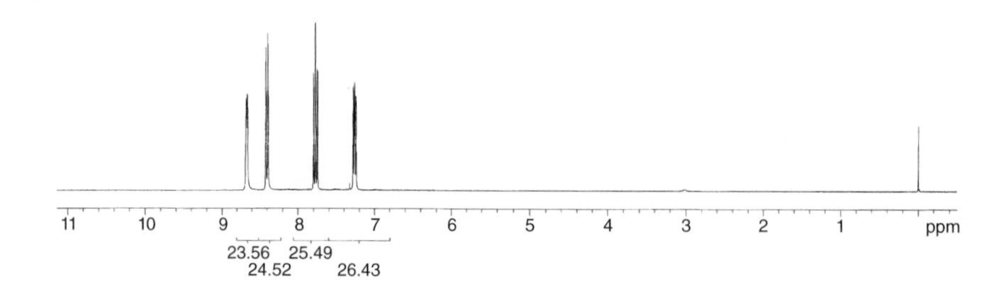

Carbon-13 NMR spectrum

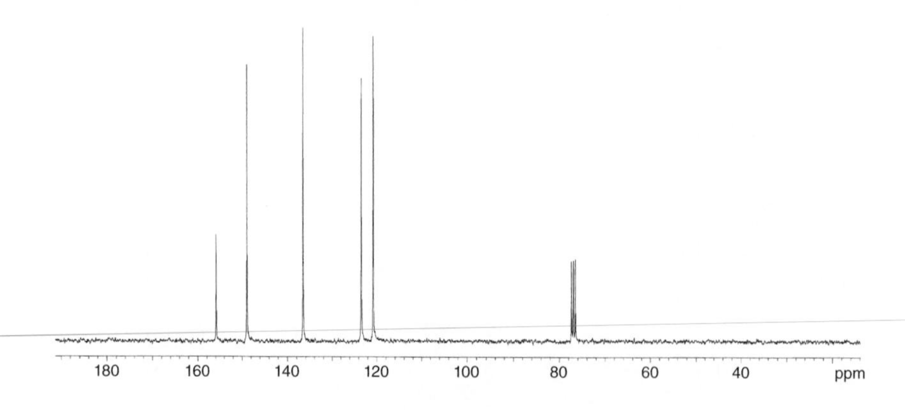

COSY spectrum

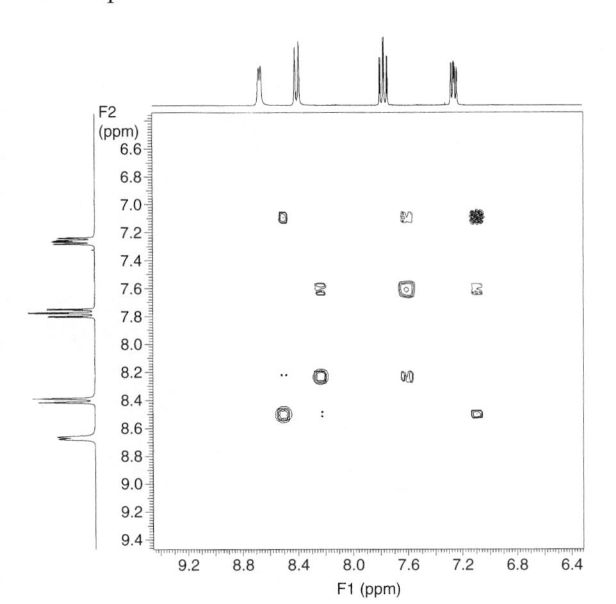

HECTOR spectrum

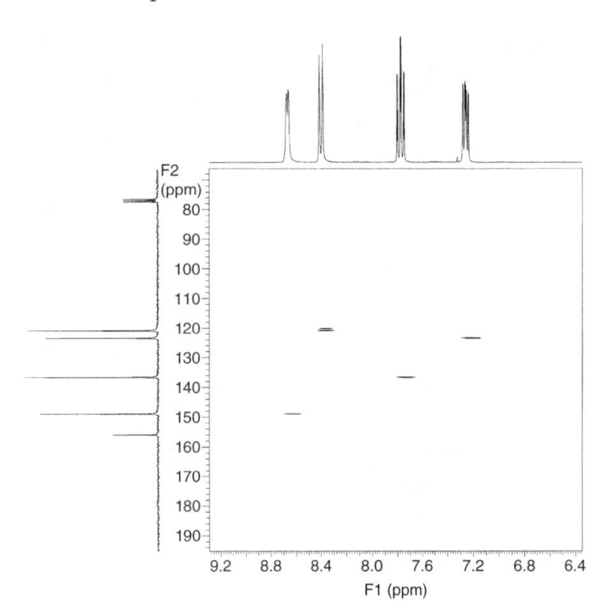

Infrared spectrum (KBr disc)

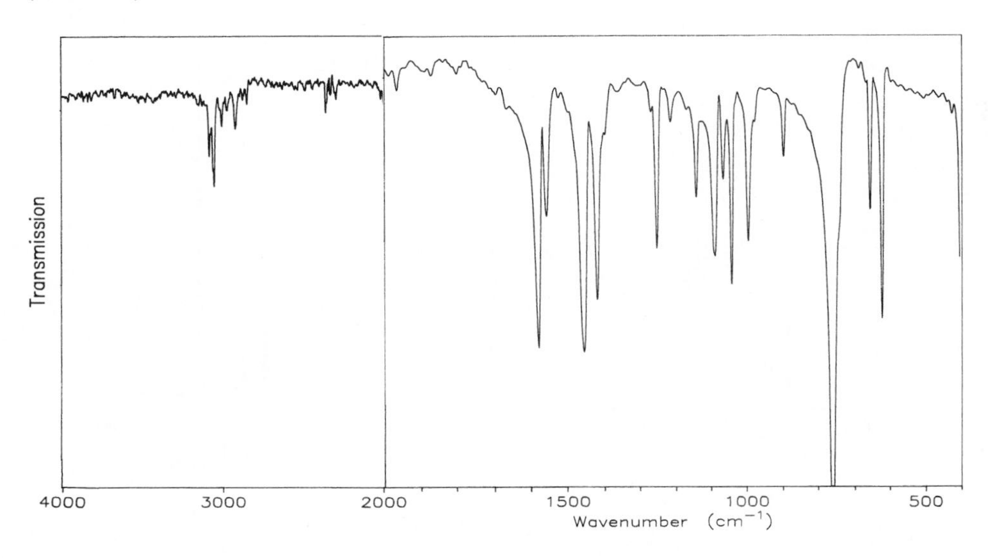

Ultraviolet–visible spectrum (EtOH, ϵ[236] 10,245, ϵ[242] 8595, ϵ[282] 13,419)

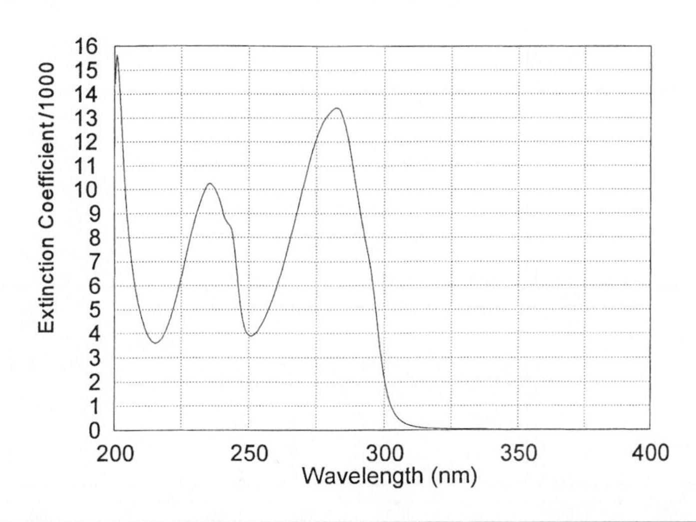

PROBLEM 15-20

Mass spectrum (CI)

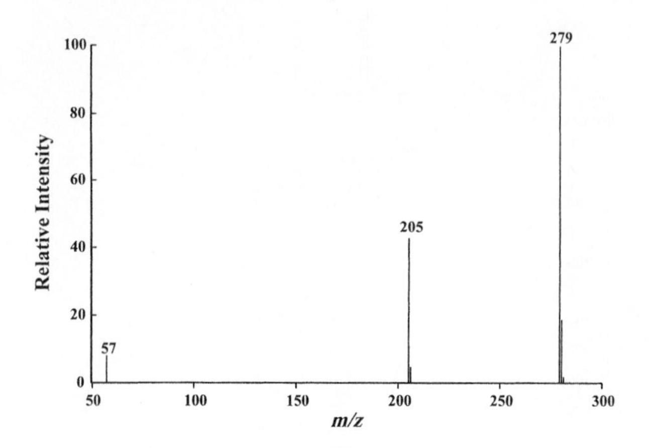

Mass spectrum (EI)

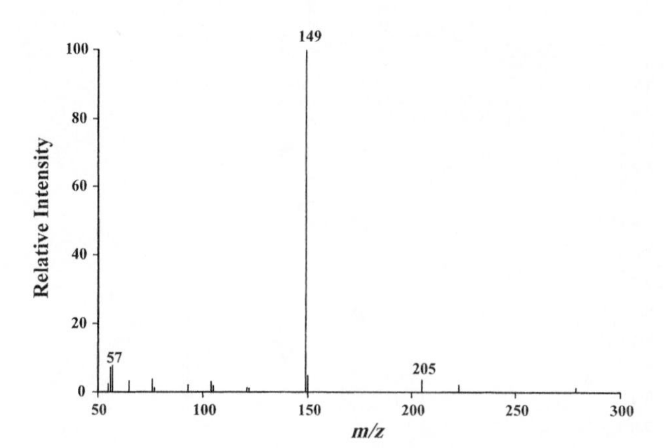

Proton NMR spectrum ($CDCl_3$)

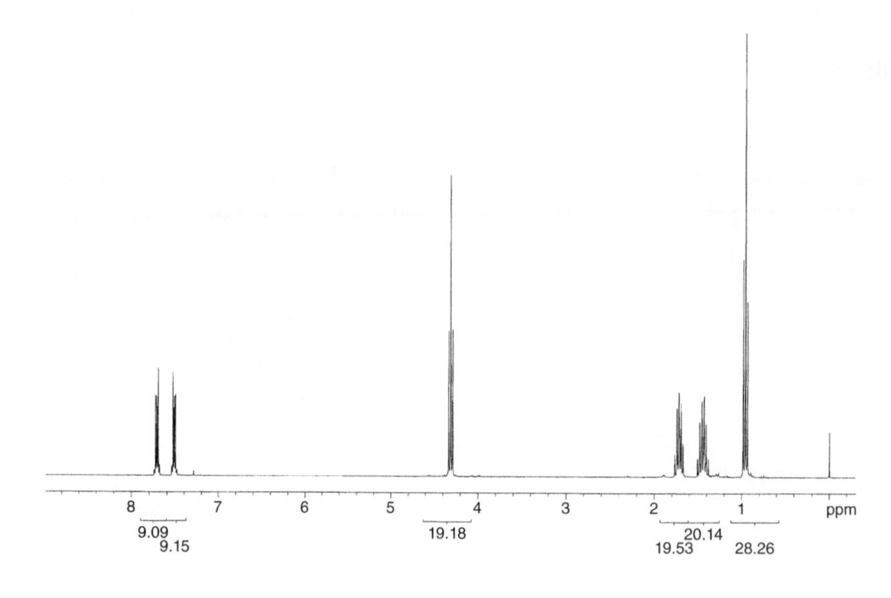

DEPT spectra

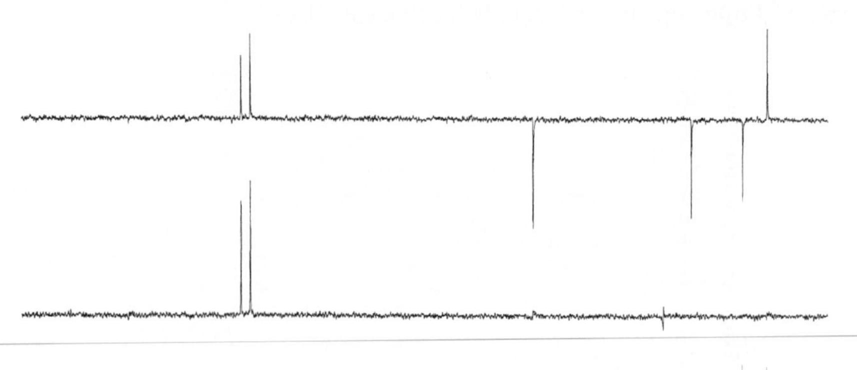

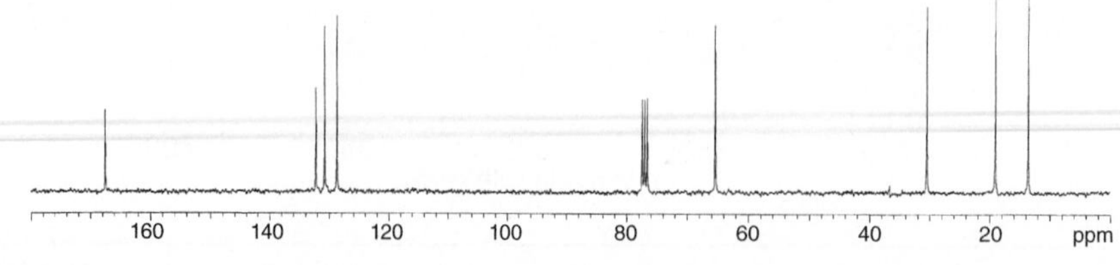

COSY spectrum

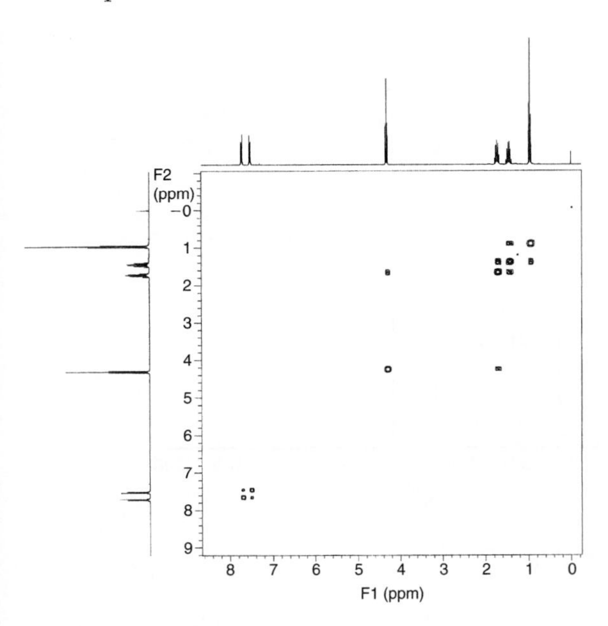

HETCOR spectrum

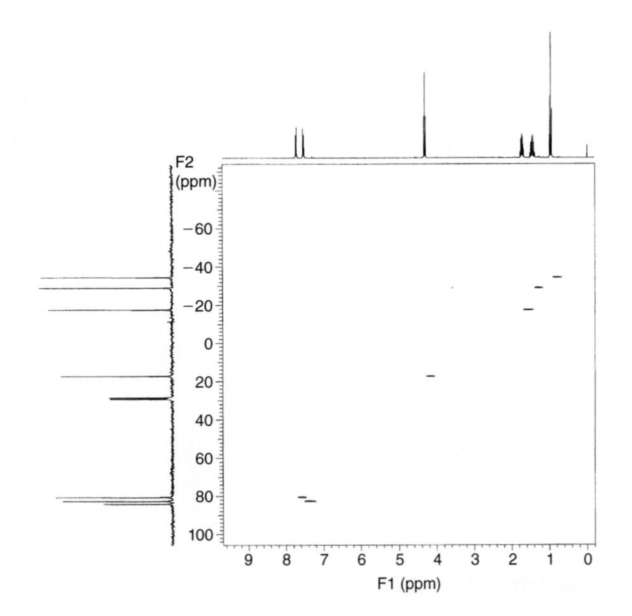

Infrared spectrum (neat)

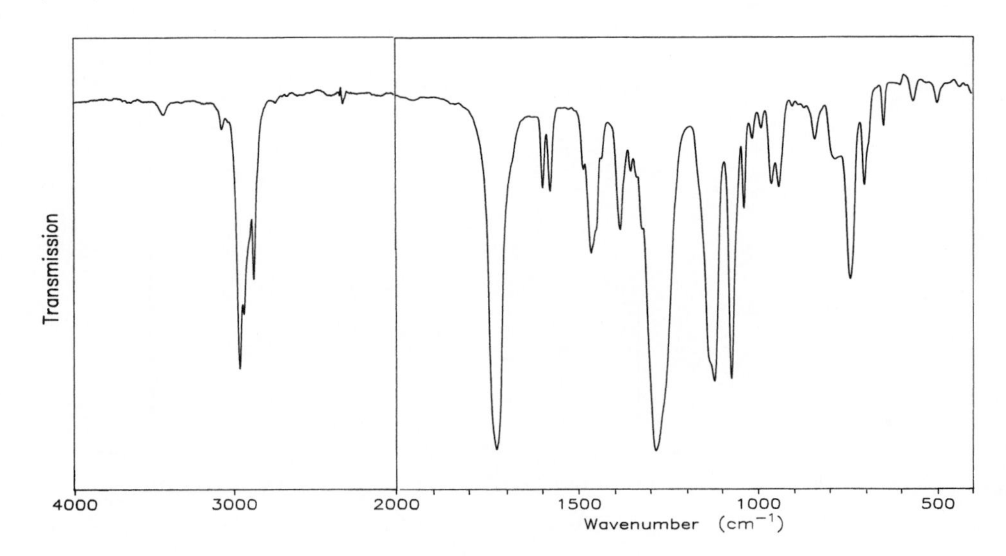

Ultraviolet–visible spectrum (EtOH, ϵ[225] 7313, ϵ[275] 1141, ϵ[280] 1037)

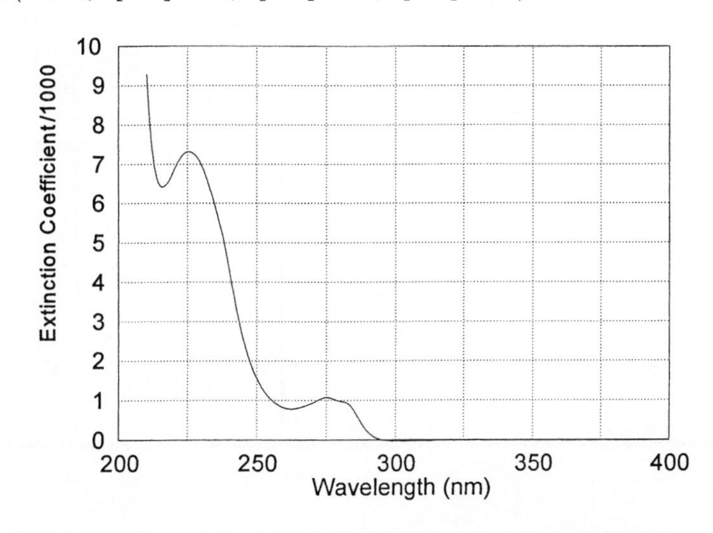

PROBLEM 15-21

Mass spectrum (CI)

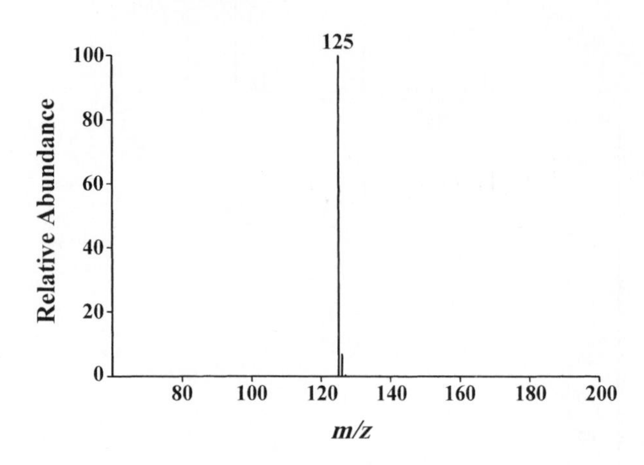

Mass spectrum (EI)

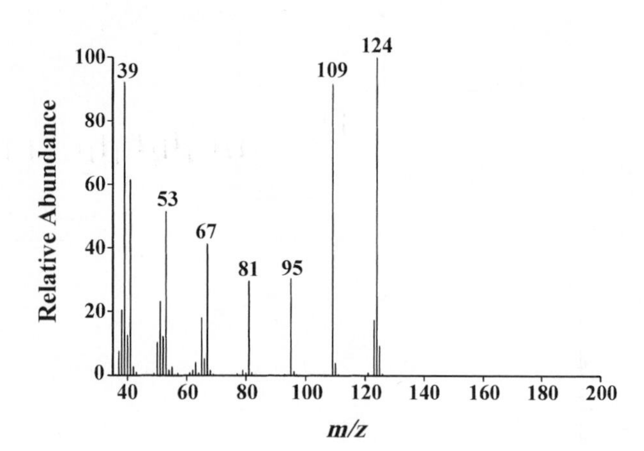

Proton NMR spectrum ($CDCl_3$)

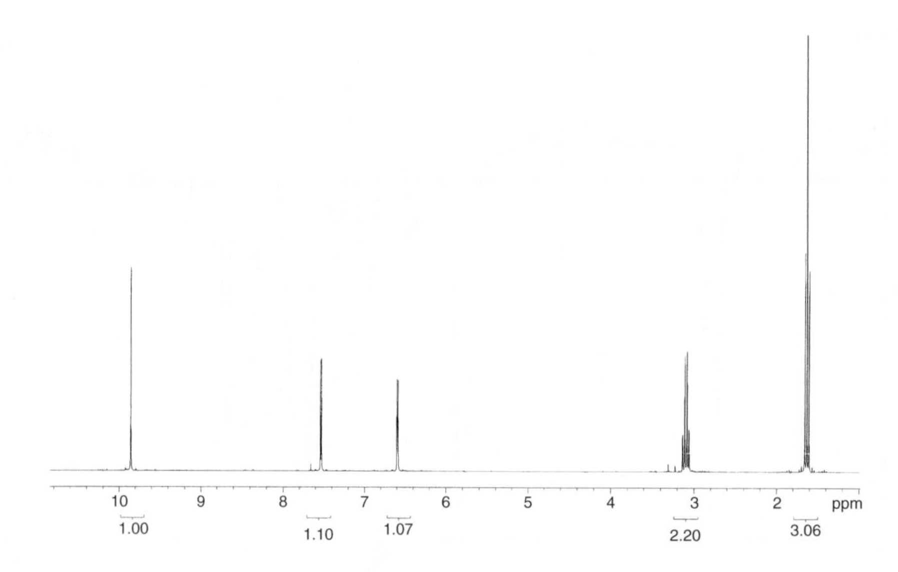

Carbon-13 spectrum

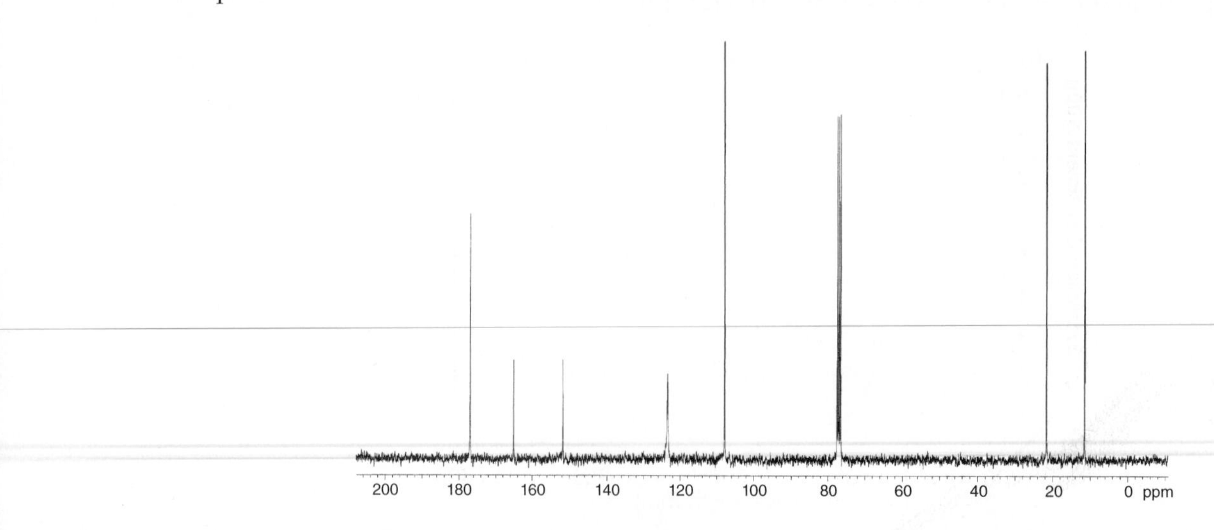

COSY spectrum

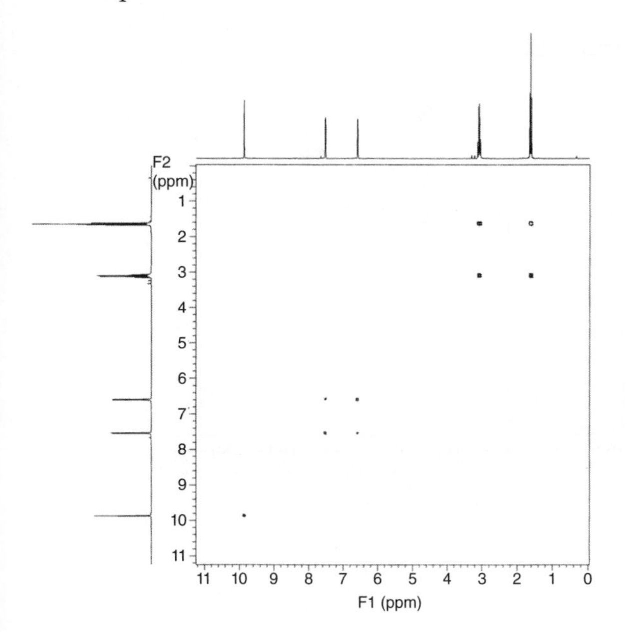

HETCOR spectrum

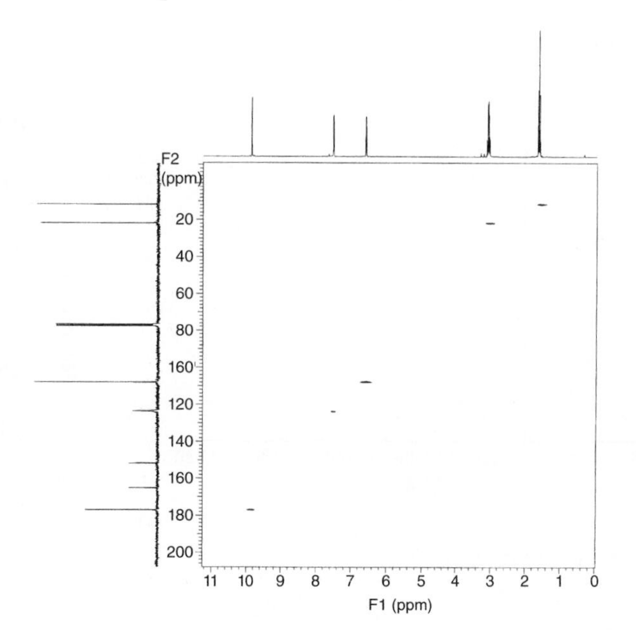

Infrared spectrum (neat)

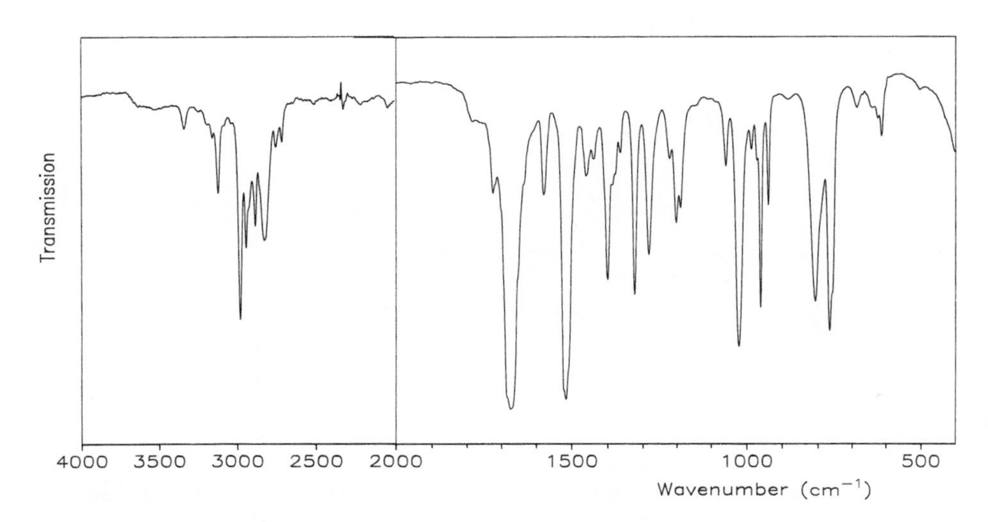

Ultraviolet–visible spectrum (EtOH, ϵ[225] 2525, ϵ[285] 16,421, ϵ[373] 39)

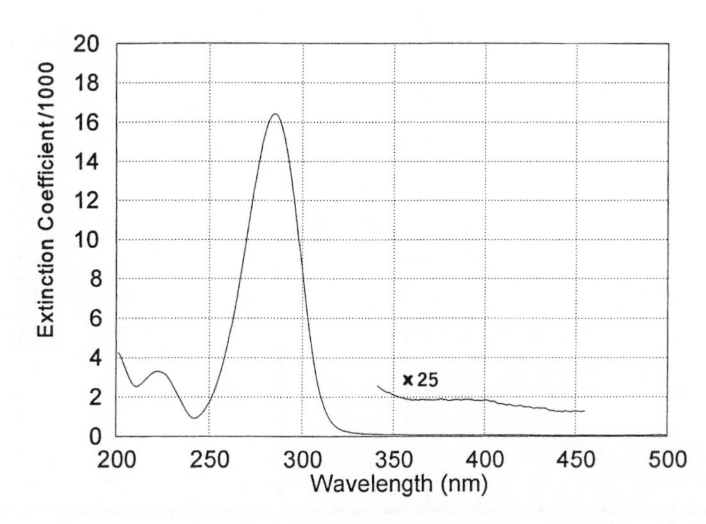

PROBLEM 15-22

Mass spectrum (CI)

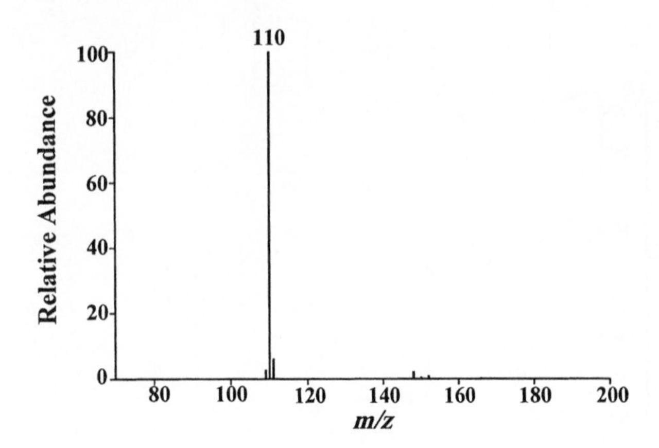

Mass spectrum (EI)

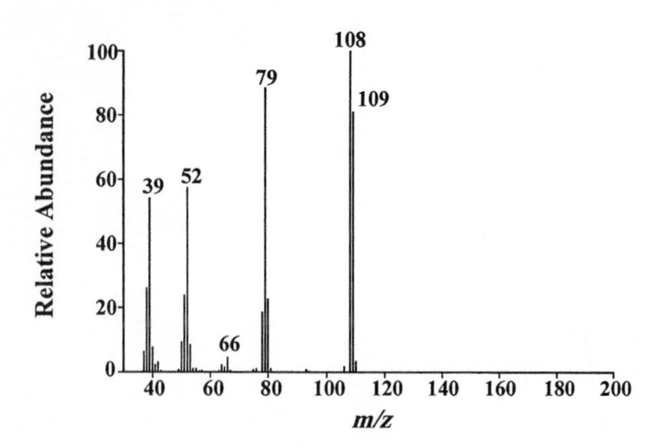

Proton NMR spectrum ($CDCl_3$)

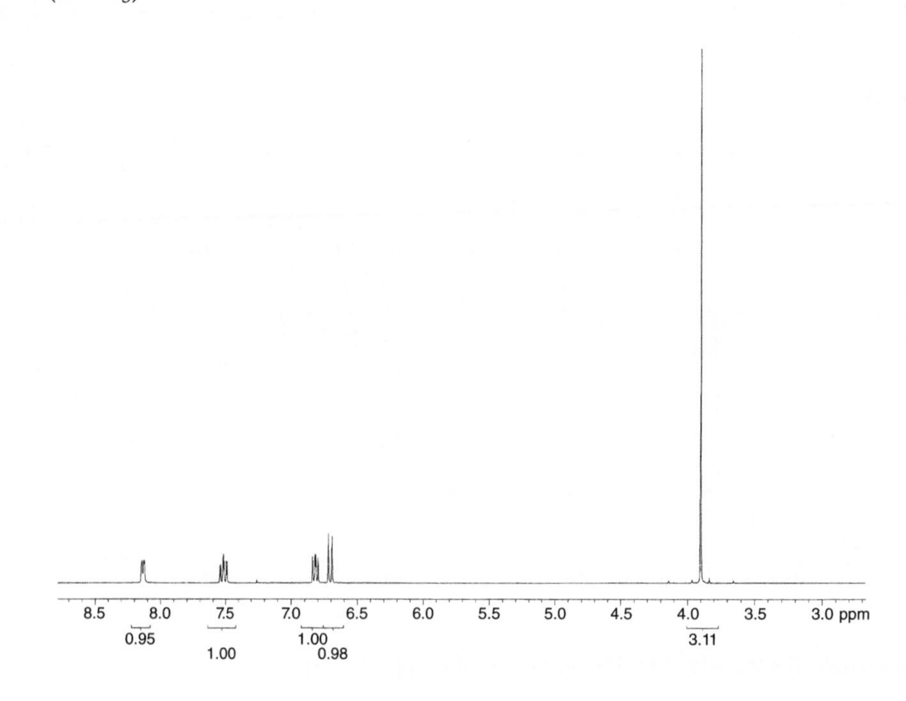

Carbon-13 NMR spectrum

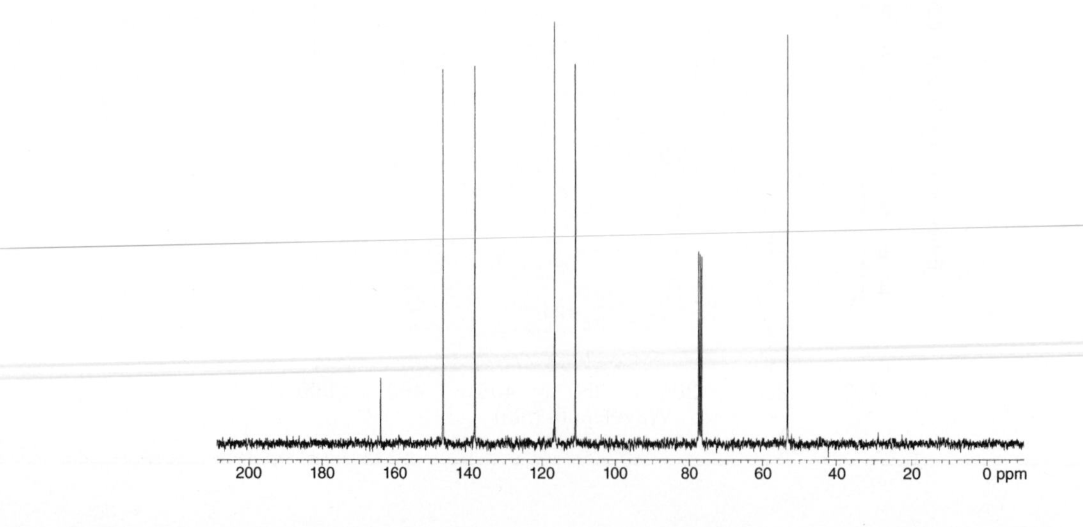

COSY spectrum

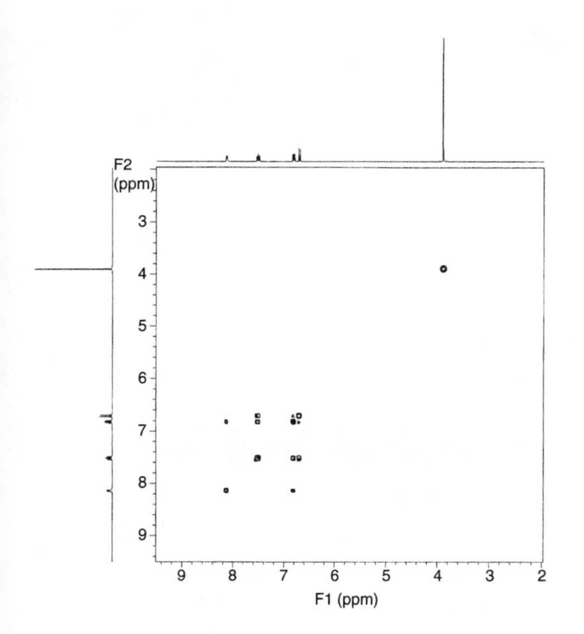

HETCOR spectrum

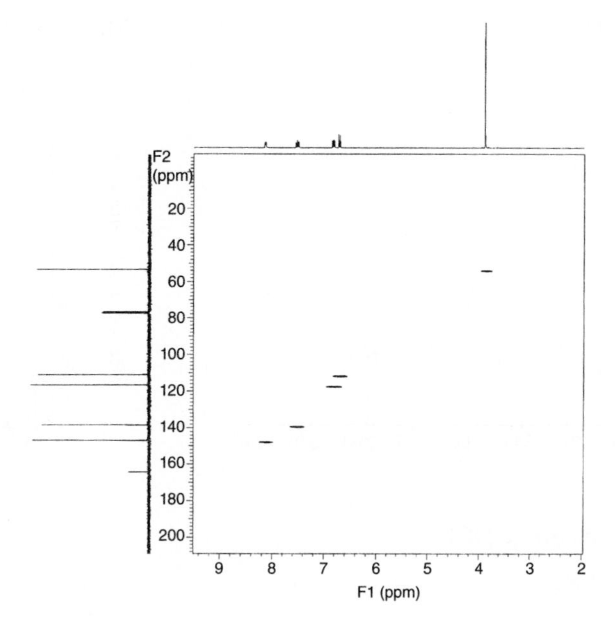

Infrared spectrum (neat)

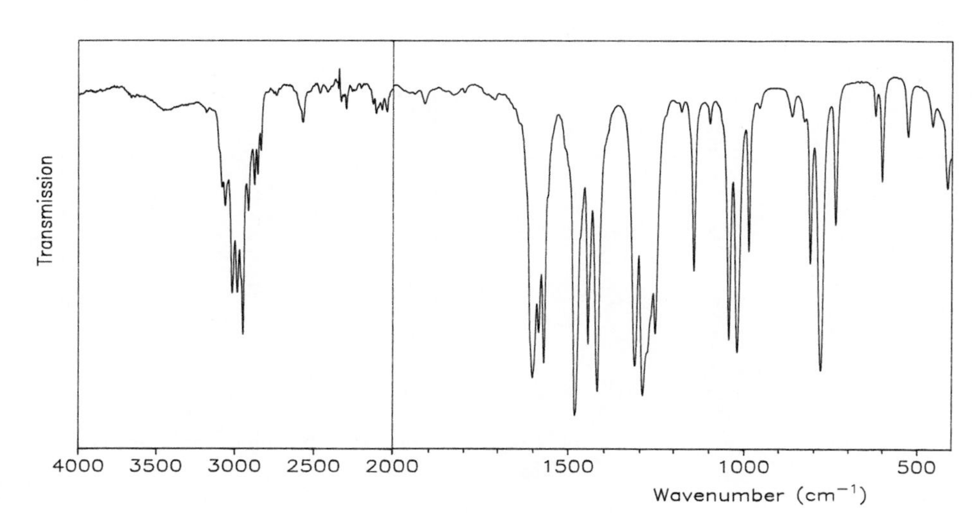

Ultraviolet–visible spectrum (EtOH, ϵ[214] 7600, ϵ[272] 4000)

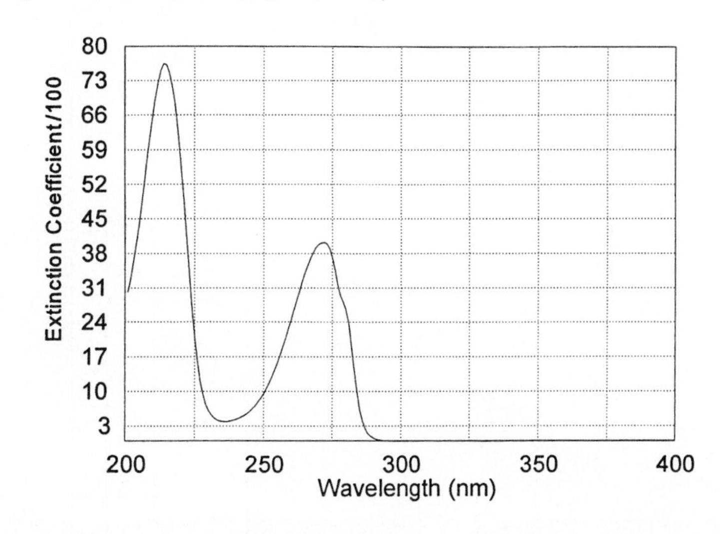

PROBLEM 15-23

Mass spectrum (CI)

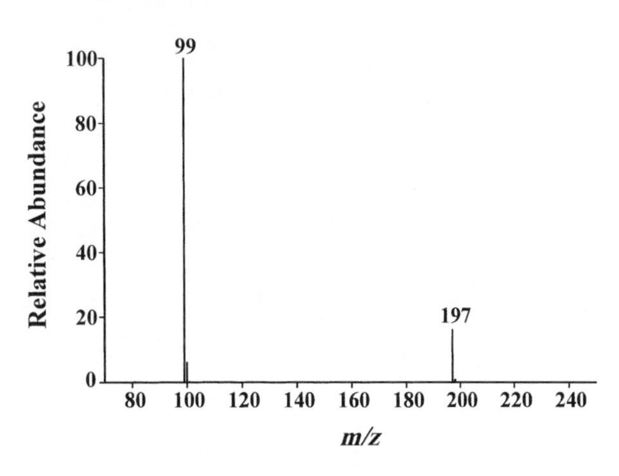

Mass spectrum (EI)

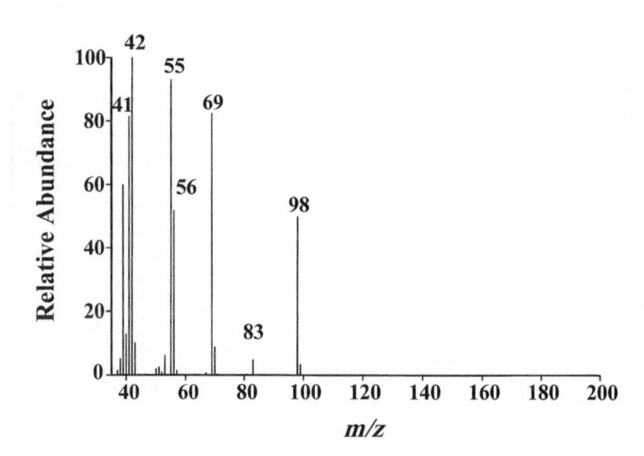

Proton NMR spectrum ($CDCl_3$)

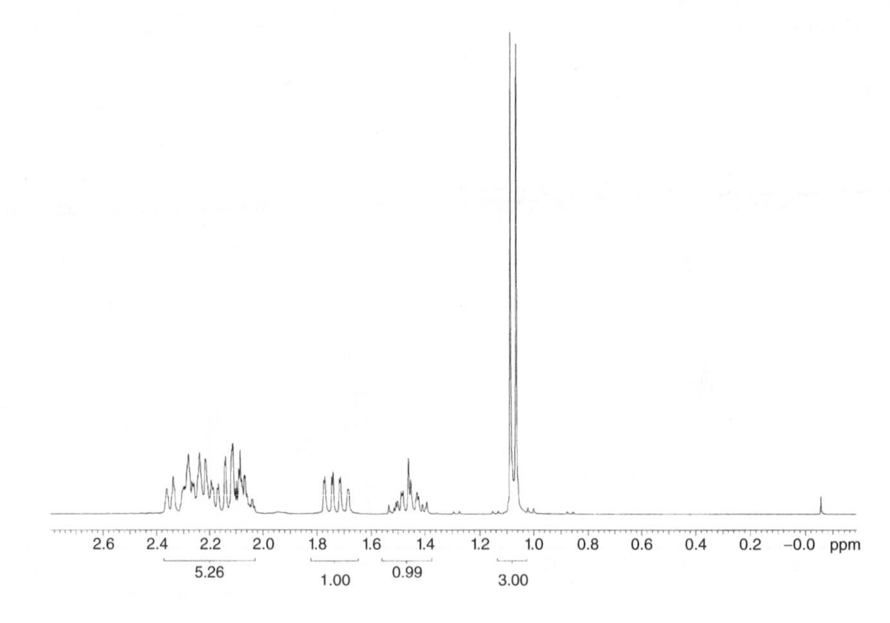

Carbon-13 NMR spectrum

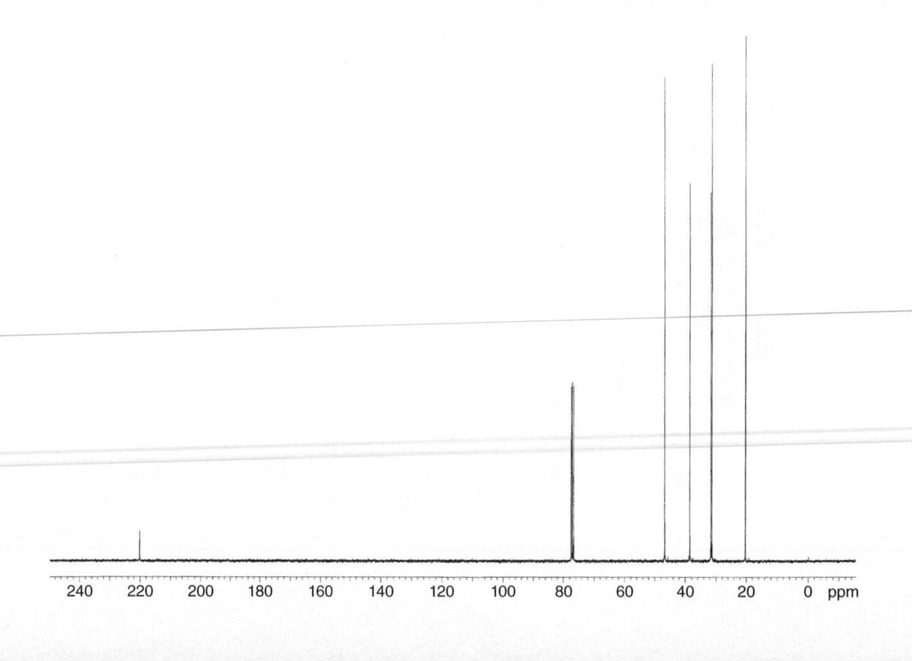

COSY spectrum

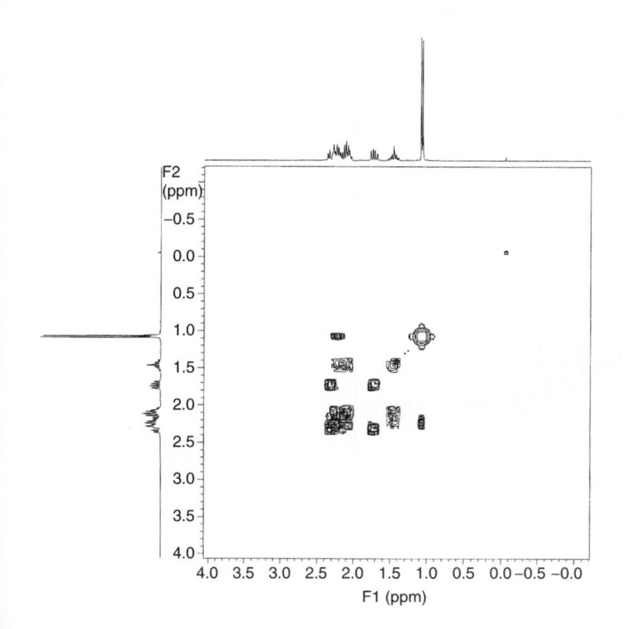

HETCOR spectrum

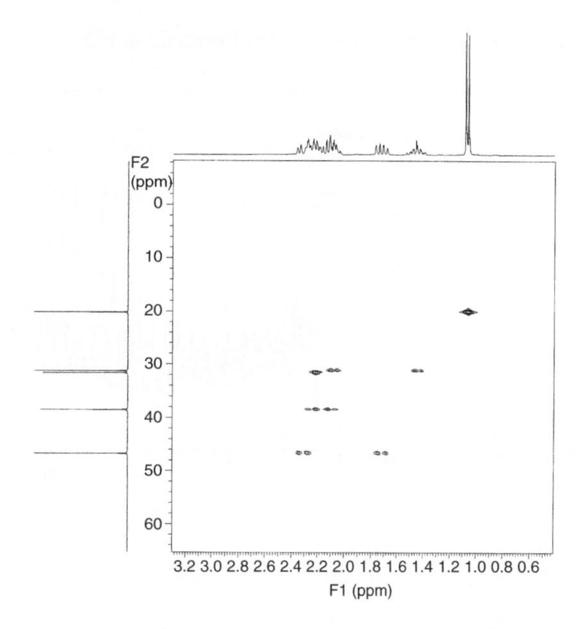

Infrared spectrum (neat)

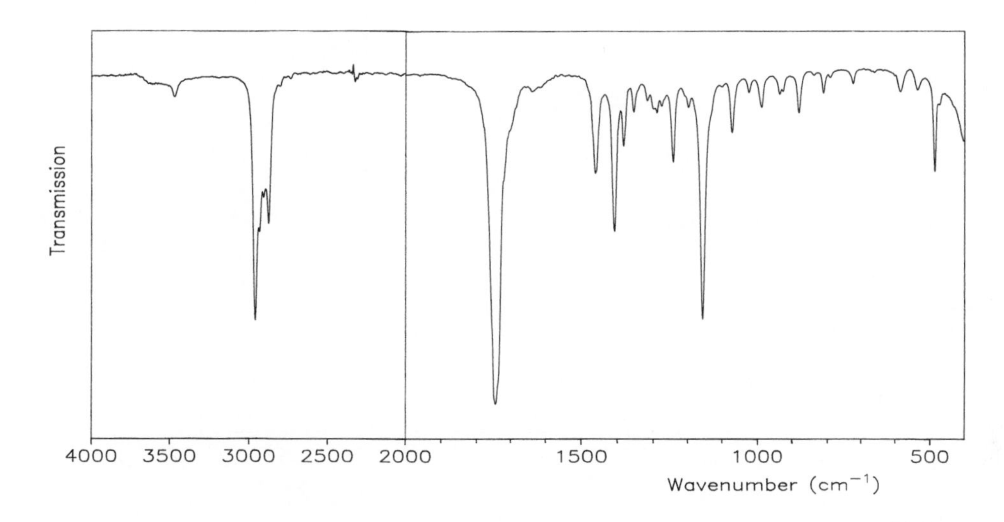

Ultraviolet–visible spectrum (EtOH, ϵ[210] 25, ϵ[288] 22)

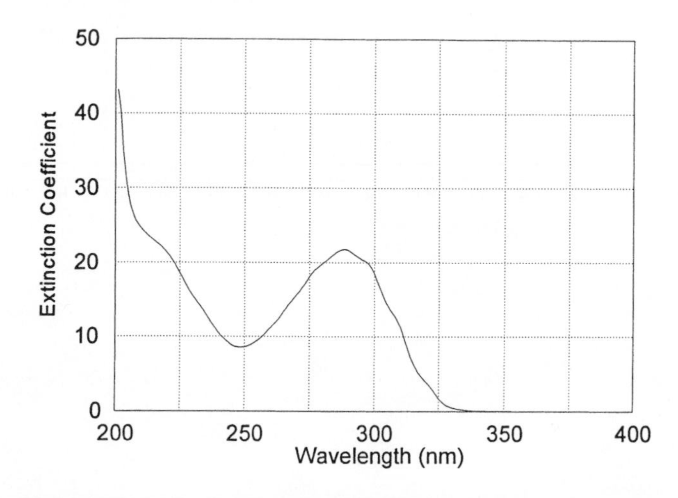

PROBLEM 15-24

Mass spectrum (MALDI in 2,5-dihydroxybenzoic acid)

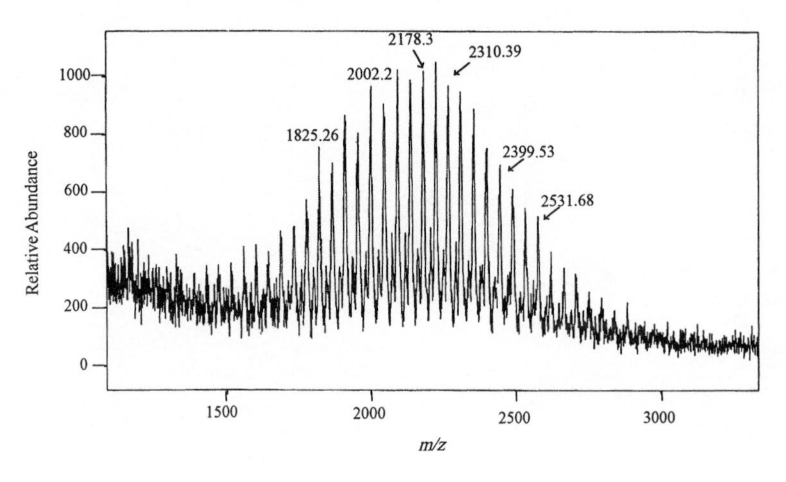

Proton NMR spectrum ($CDCl_3$)

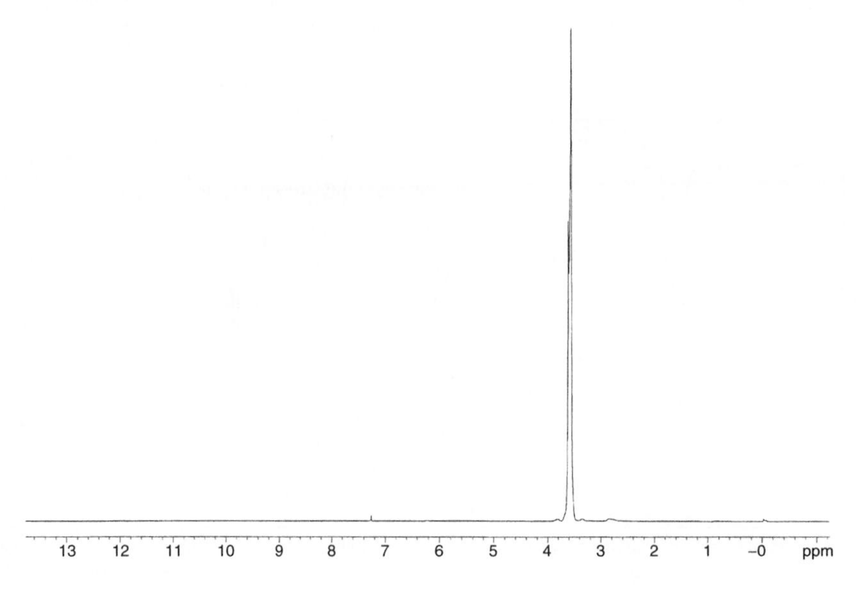

Carbon-13 NMR spectrum

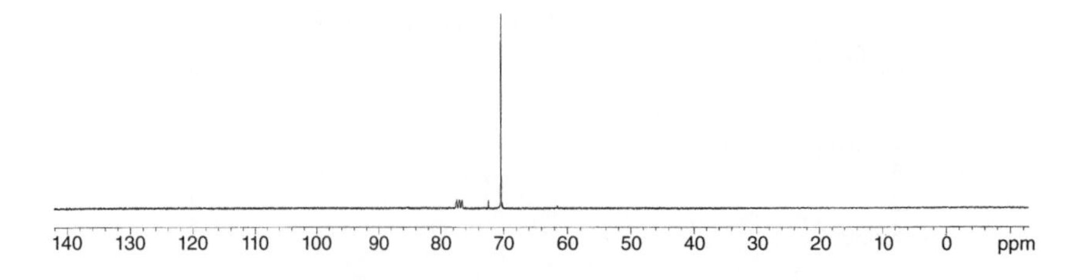

Infrared spectrum (KBr disc)

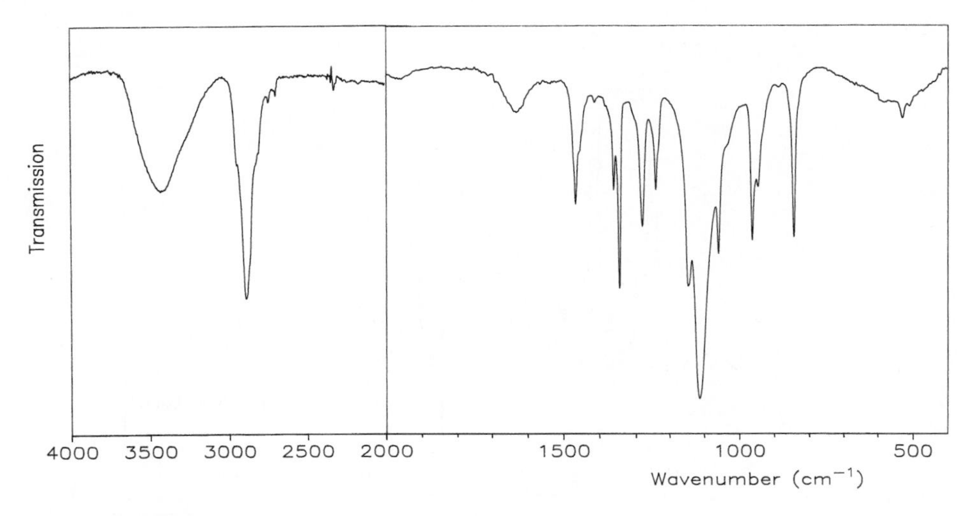

Ultraviolet–visible spectrum (EtOH)

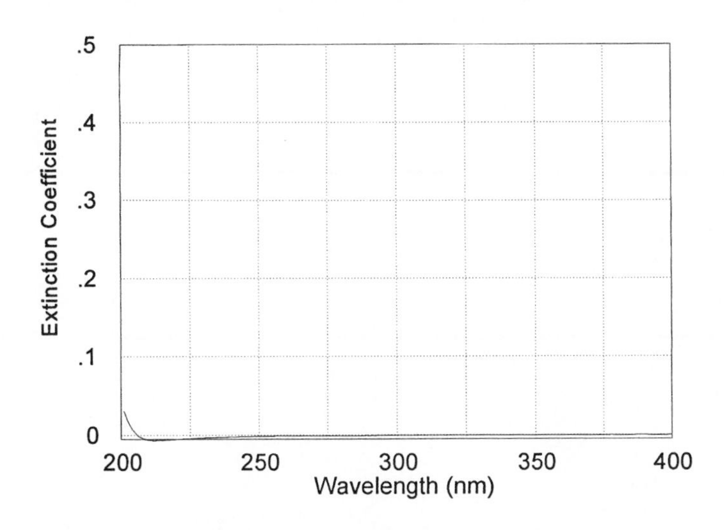

PROBLEM 15-25

Mass spectrum (CI)

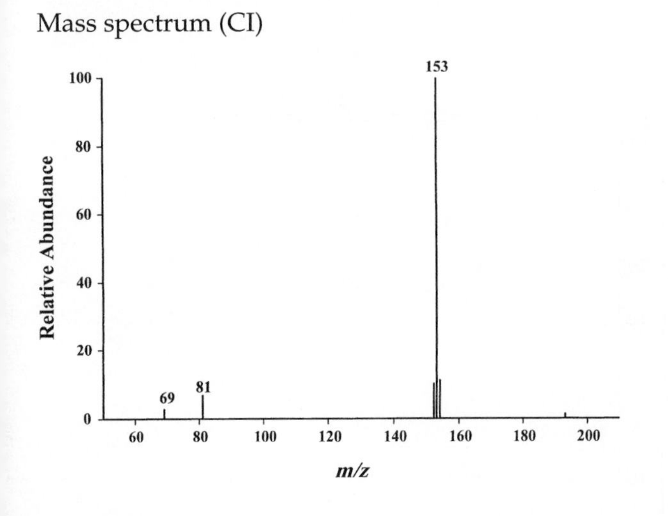

Mass spectrum (EI)

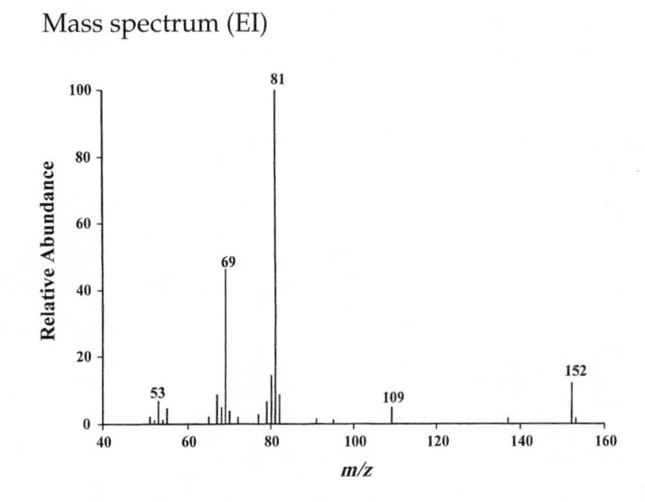

Proton NMR spectrum with expansion ($CDCl_3$)

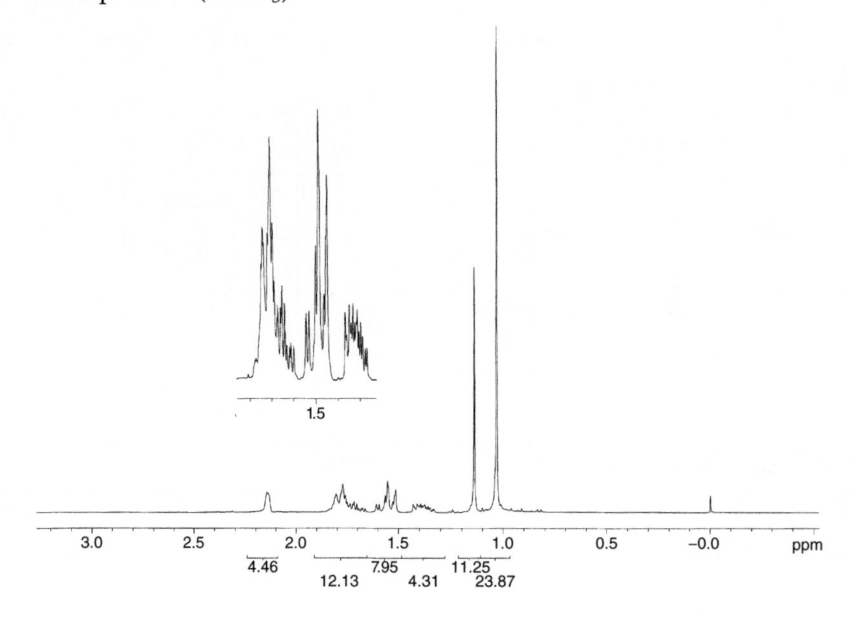

DEPT spectra

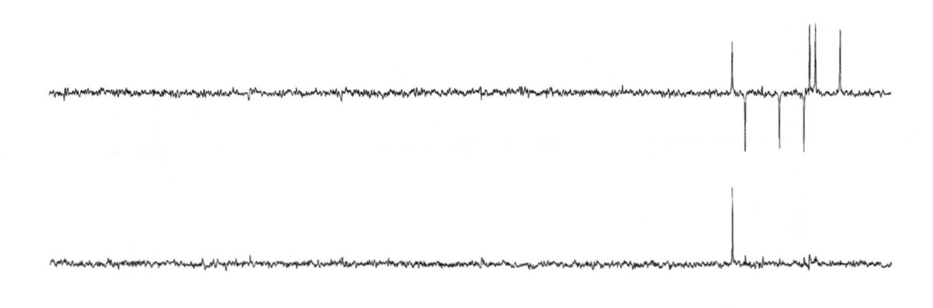

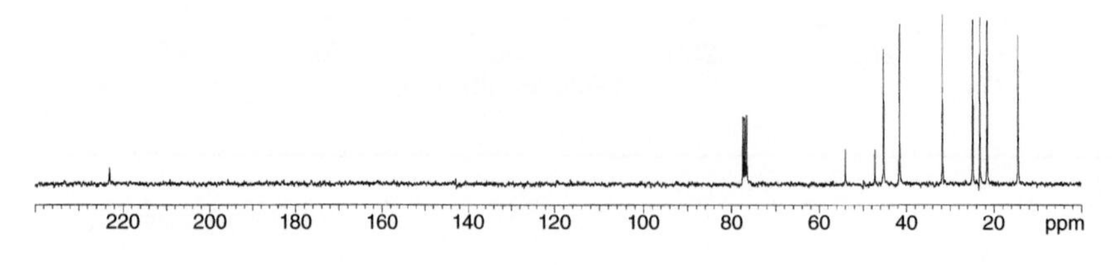

COSY spectrum

HETCOR spectrum

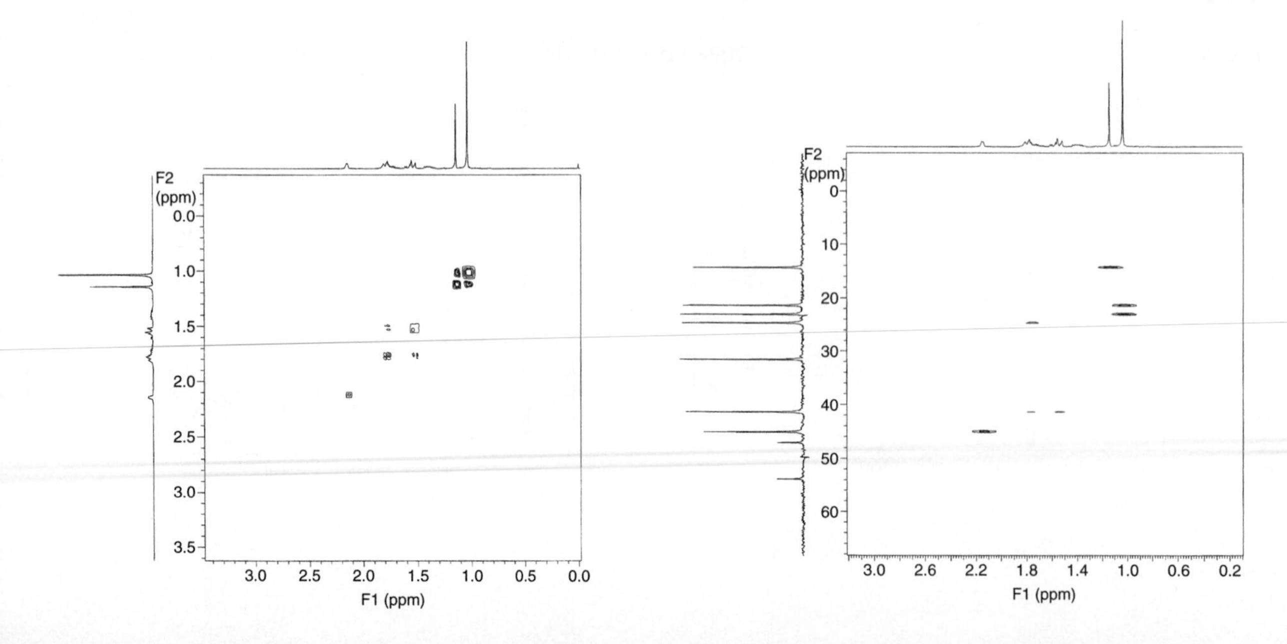

Infrared spectrum (neat)

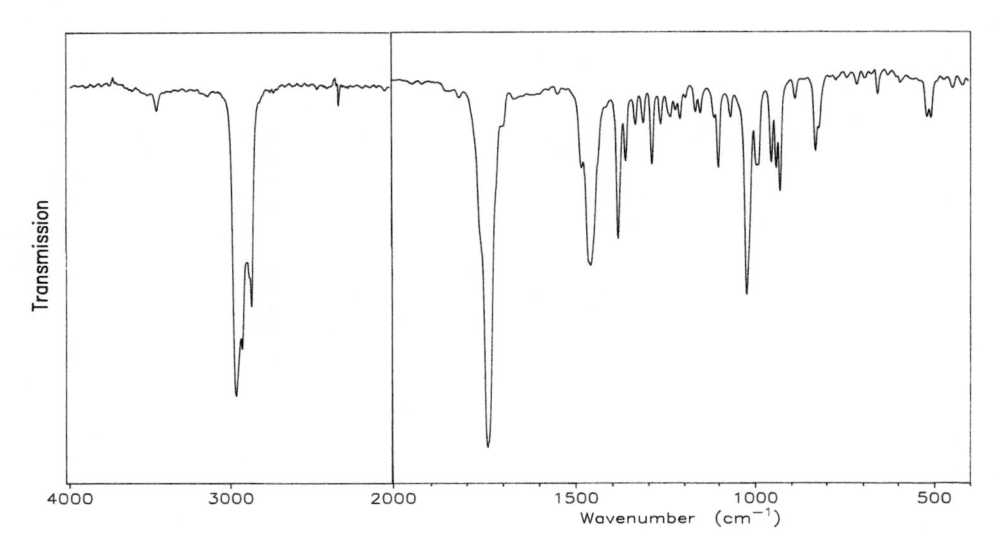

Ultraviolet–visible spectrum (EtOH, ϵ[289] 18)

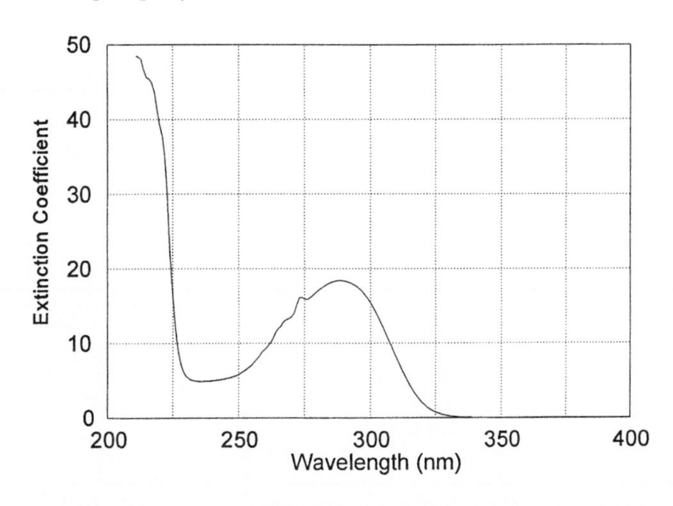

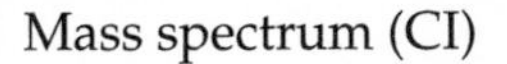

PROBLEM 15-26

Mass spectrum (CI)

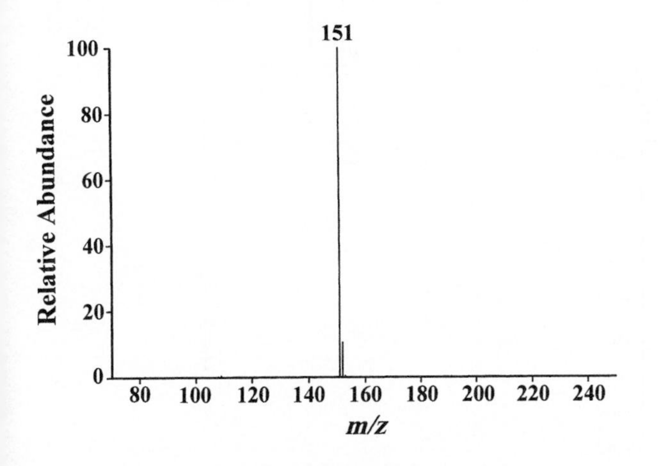

Mass spectrum (EI)

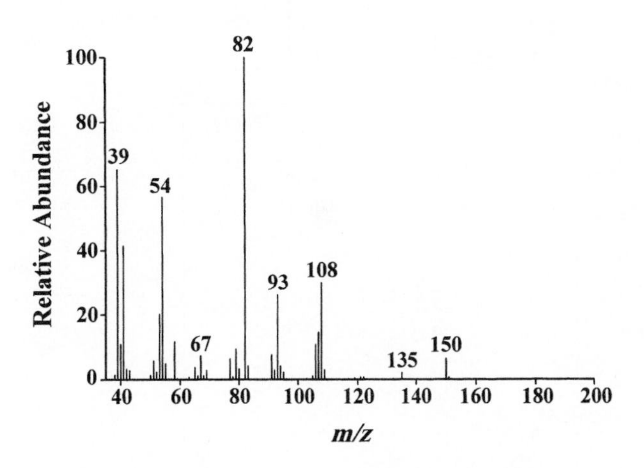

Proton NMR spectrum ($CDCl_3$)

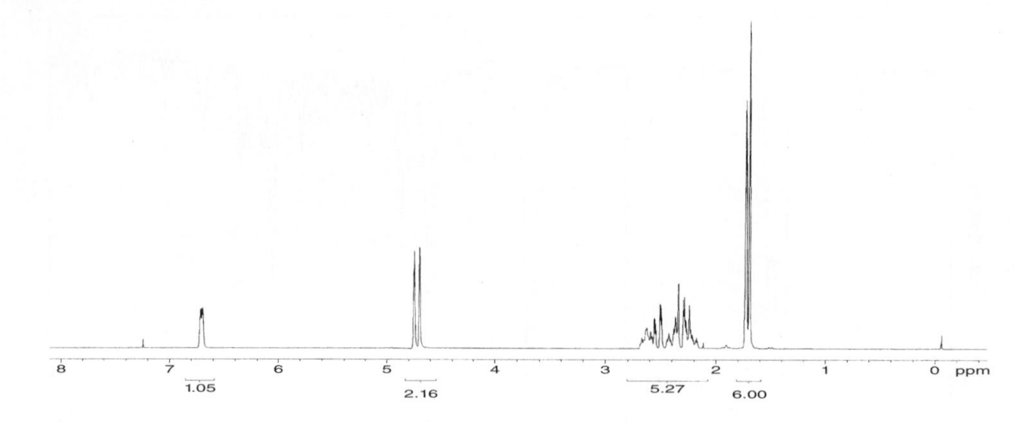

DEPT spectra

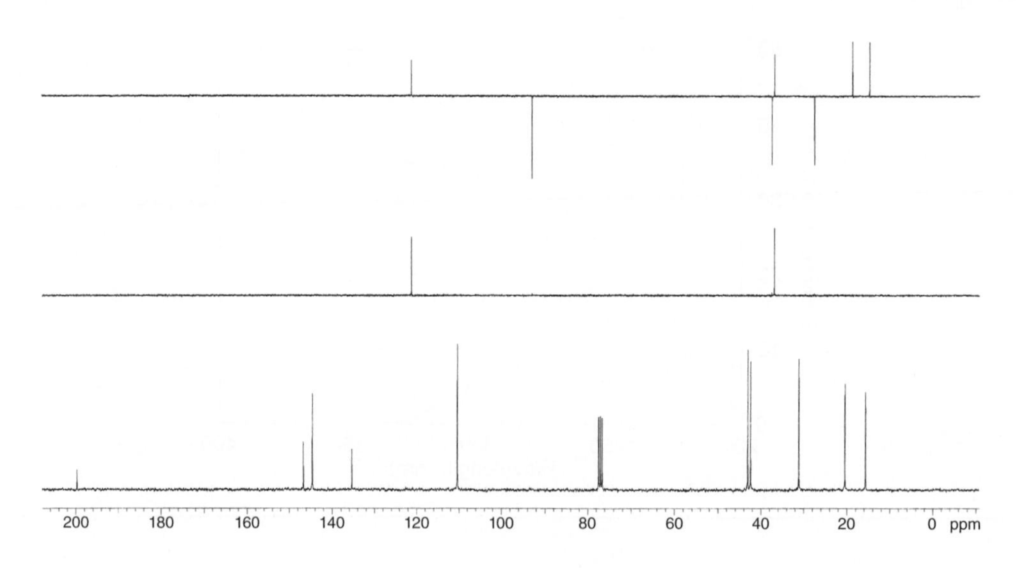

COSY spectrum

HETCOR spectrum

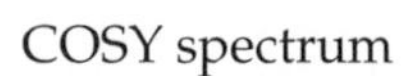

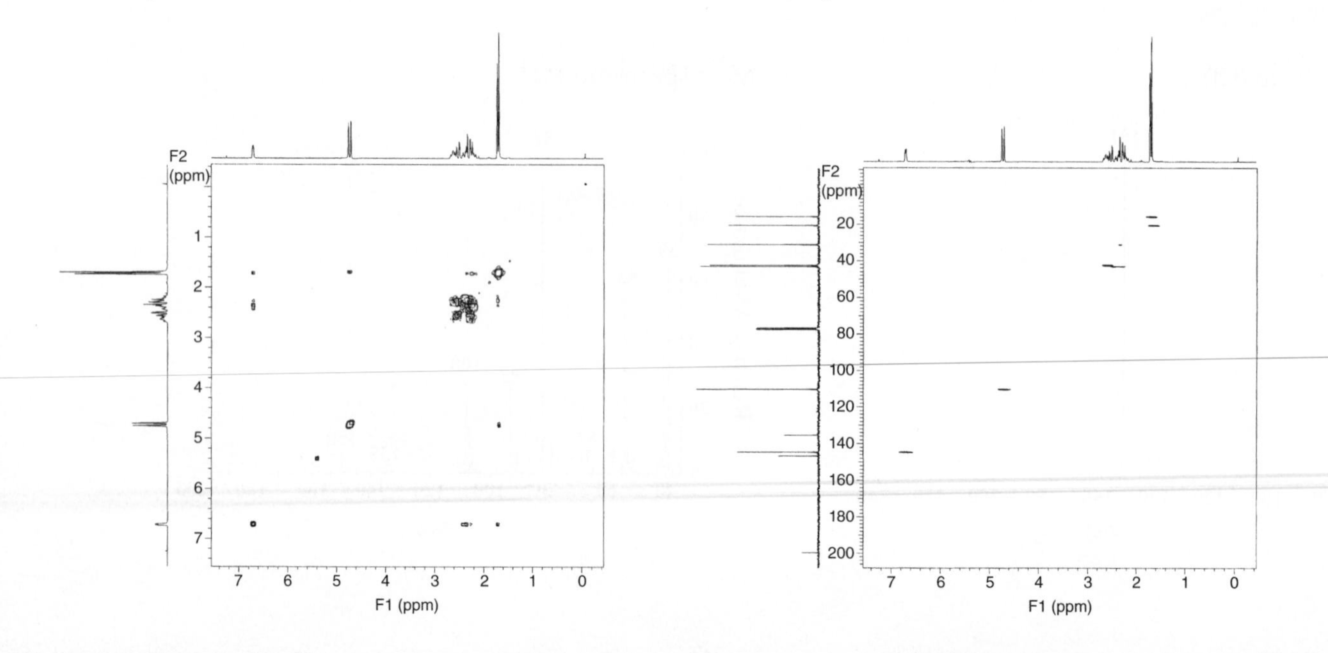

Infrared spectrum (neat)

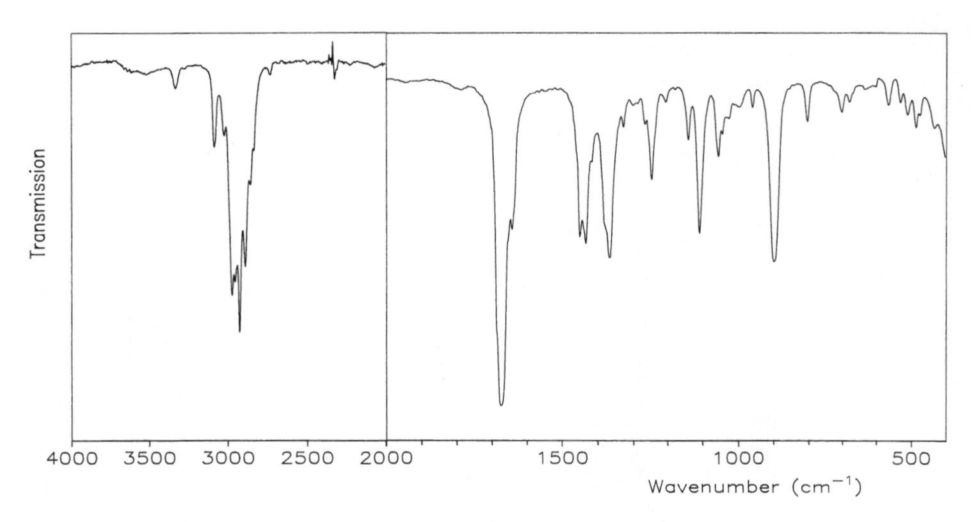

Ultraviolet–visible spectrum (EtOH, ϵ[231] 43, ϵ[235] 9144)

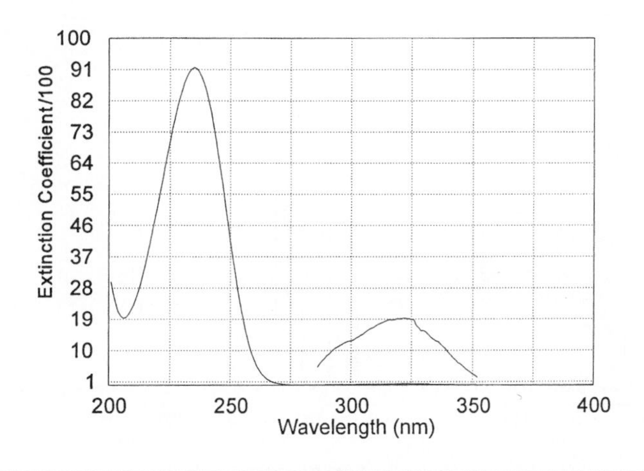

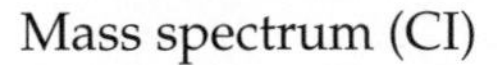

PROBLEM 15-27

Mass spectrum (CI)

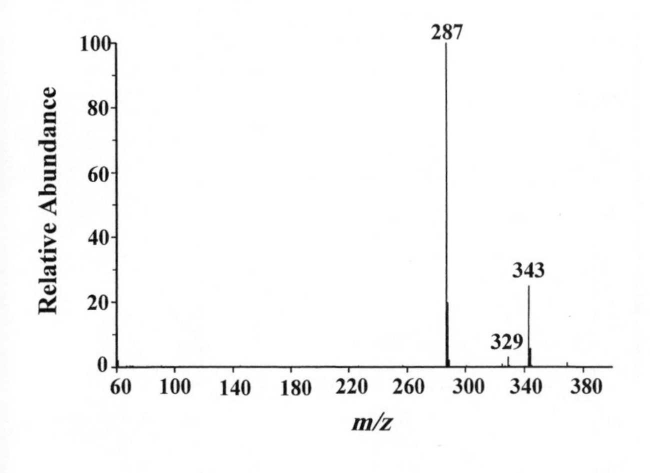

Mass spectrum (EI)

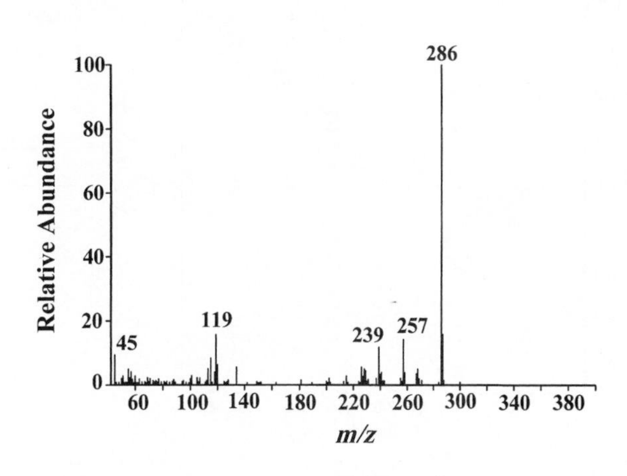

Proton NMR spectrum ($CDCl_3$)

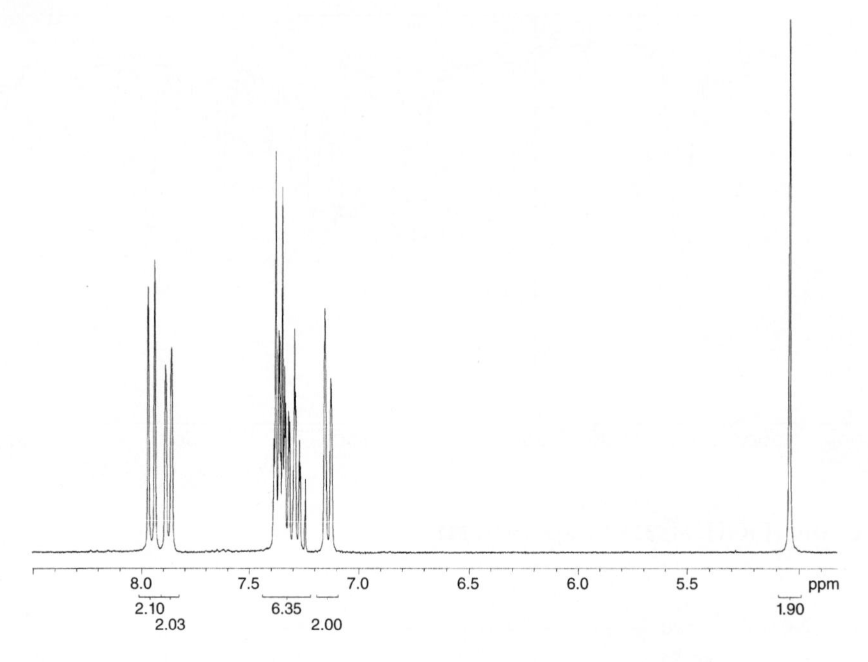

Carbon-13 NMR spectrum

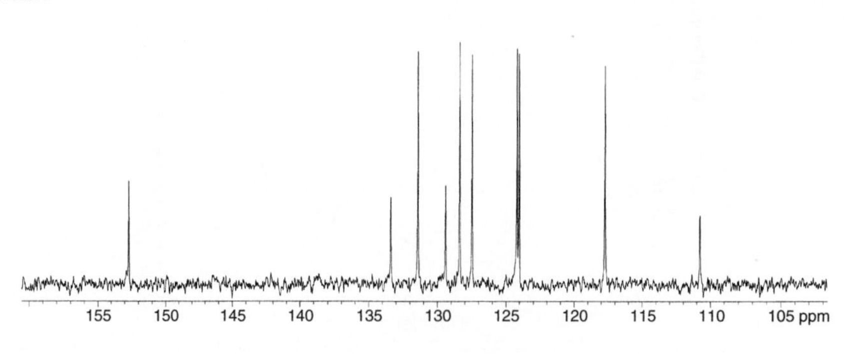

COSY spectrum

HETCOR spectrum

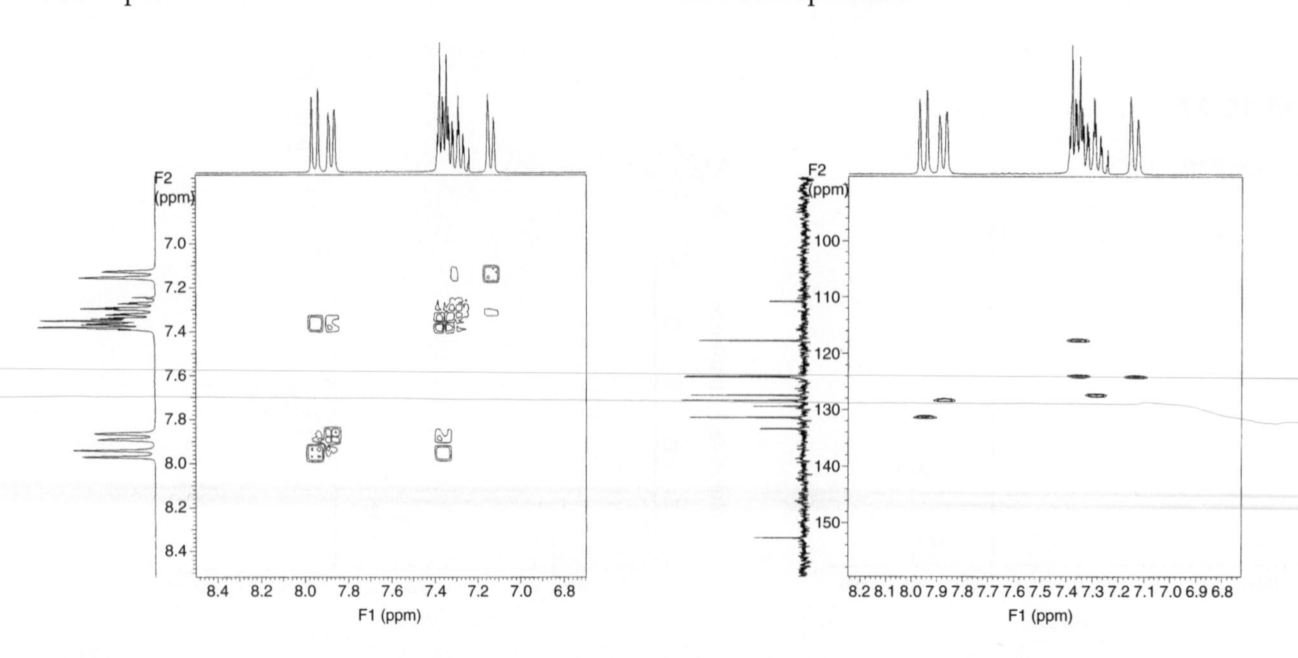

Infrared spectrum (KBr disc)

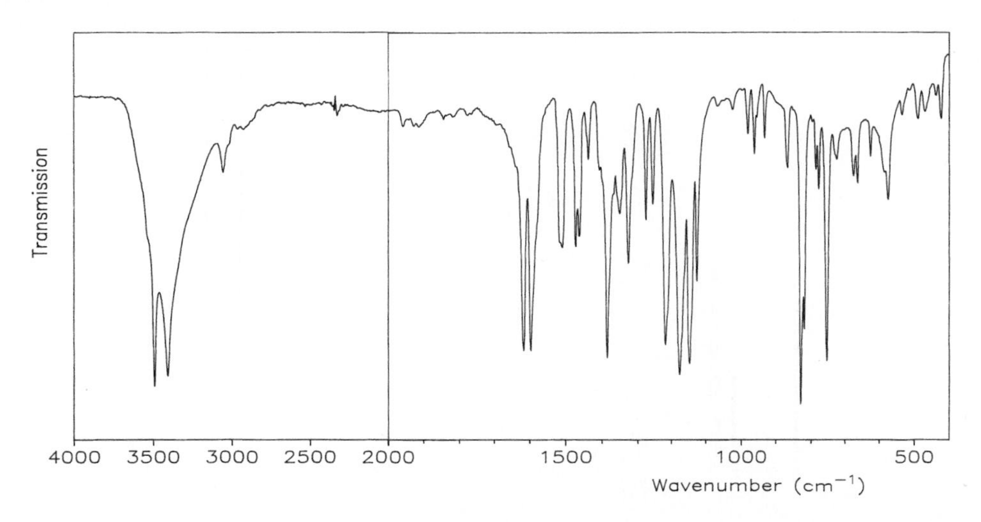

Ultraviolet–visible spectrum (EtOH, ϵ[228] 107,190, ϵ[268] 8079, ϵ[278] 9390, ϵ[289] 8262, ϵ[325] 5662, ϵ[336] 7344)

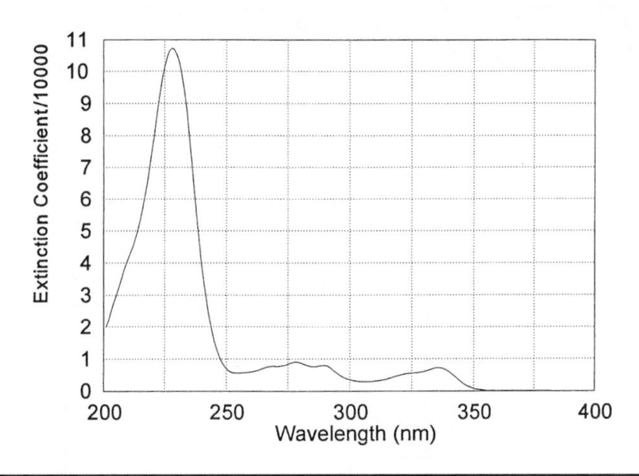

PROBLEM 15-28

Mass spectrum (CI)

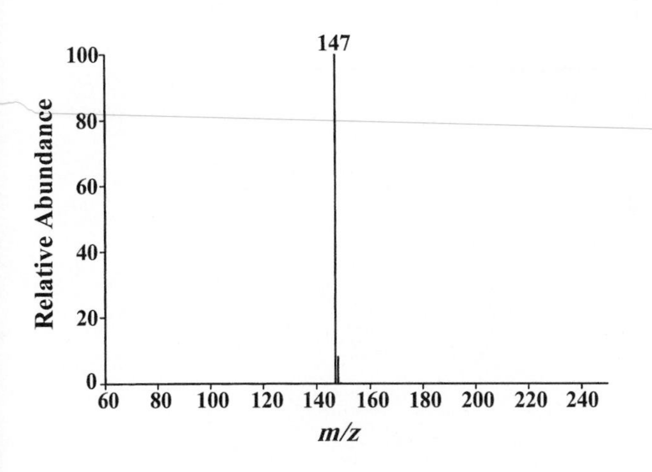

Mass spectrum (EI)

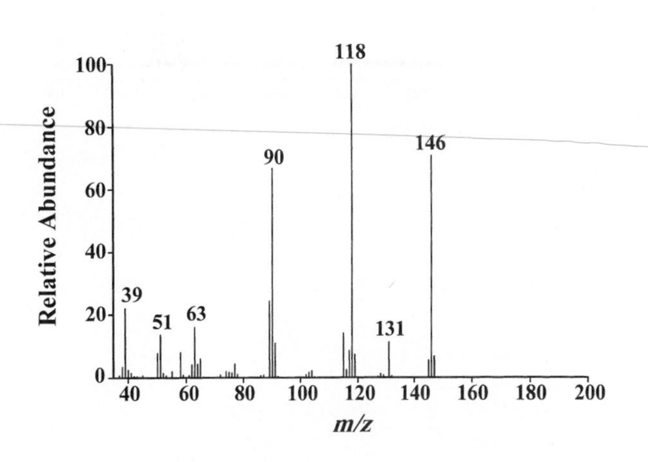

Proton NMR spectrum ($CDCl_3$)

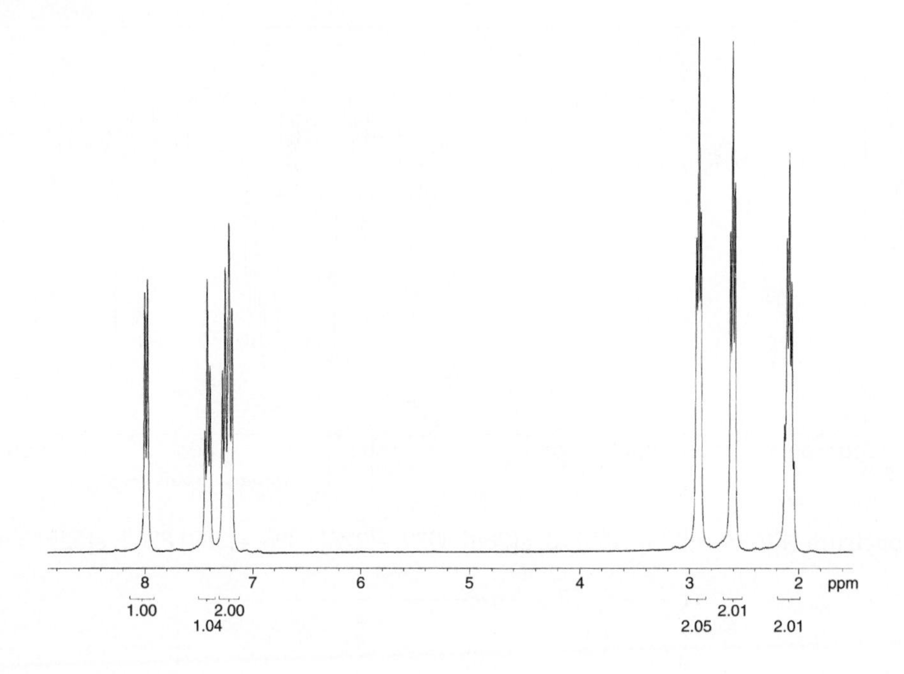

Carbon-13 NMR spectrum

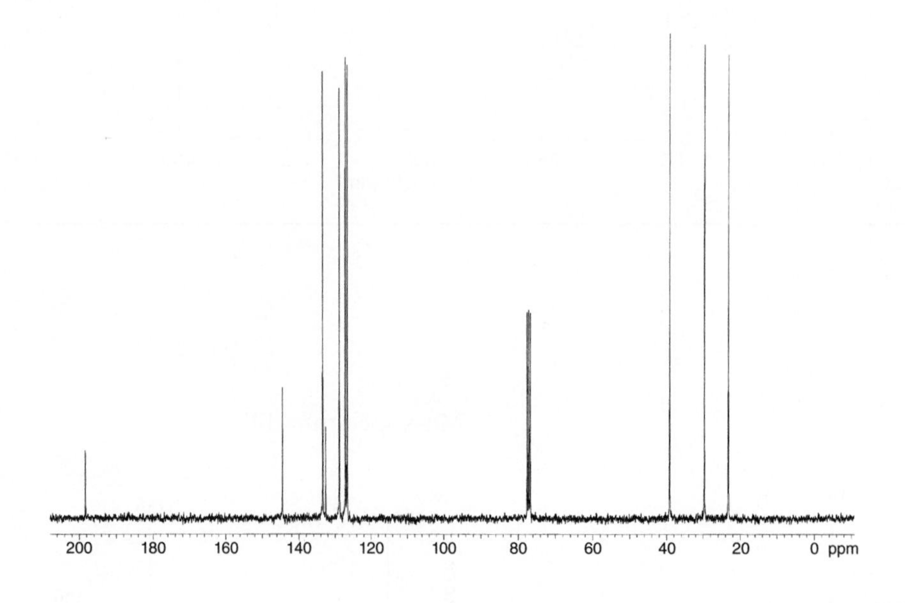

COSY spectrum

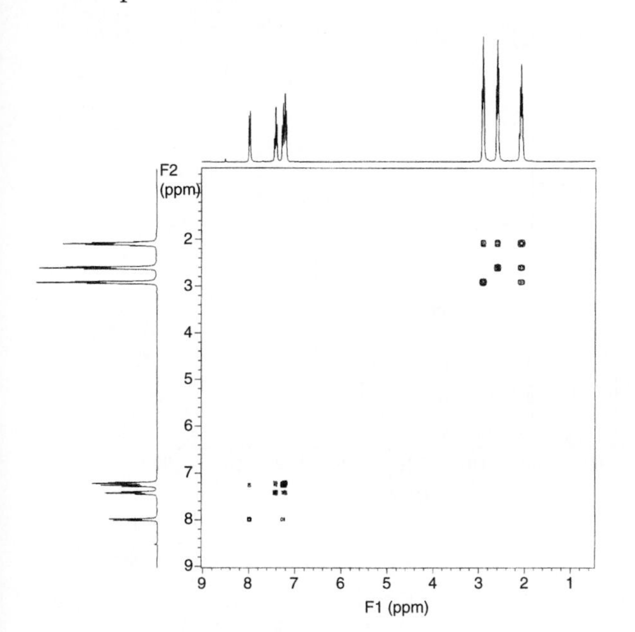

HETCOR spectrum

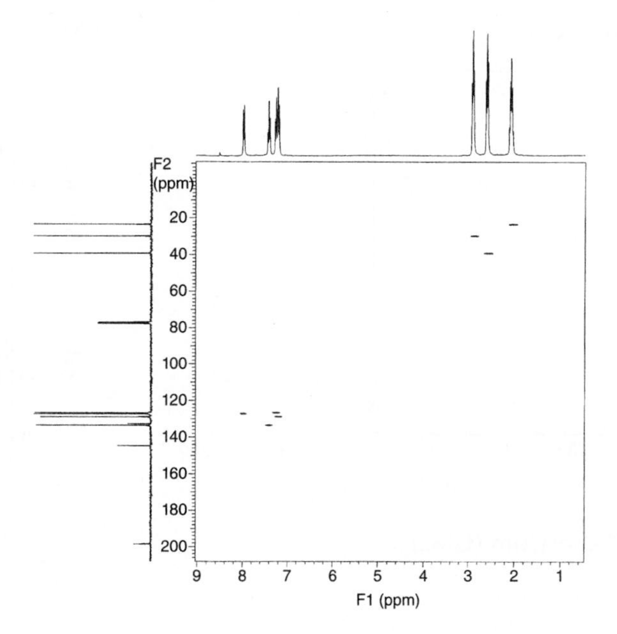

Infrared spectrum (neat)

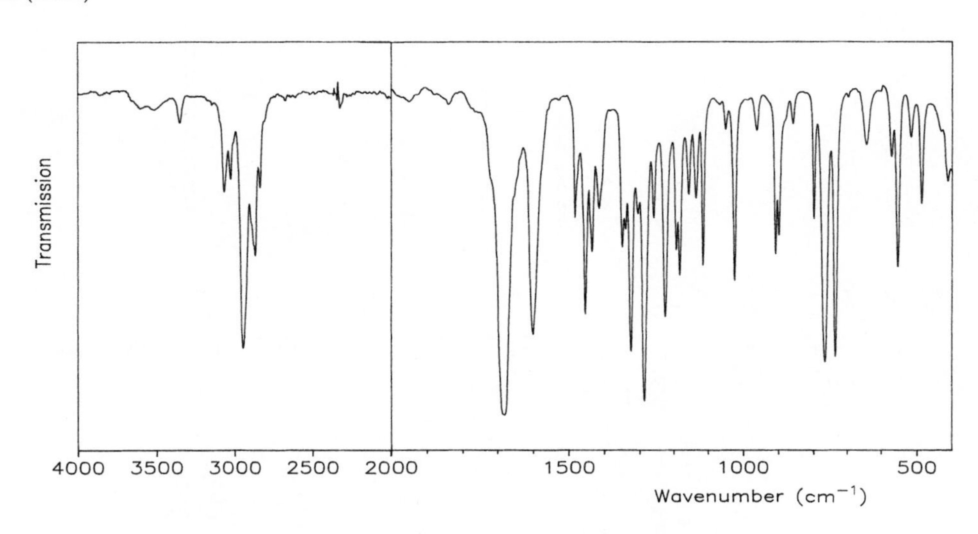

Ultraviolet–visible spectrum (EtOH, ϵ[206] 24,500, ϵ[248] 12,200, ϵ[291] 2200)

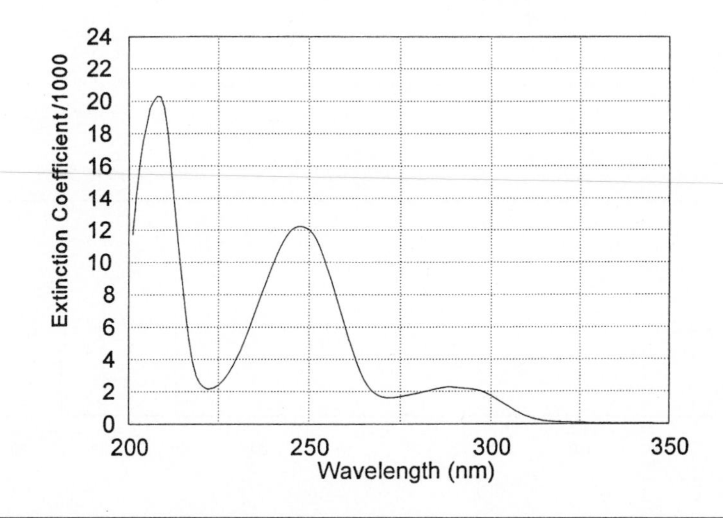

PROBLEM 15-29

Mass spectrum (CI)

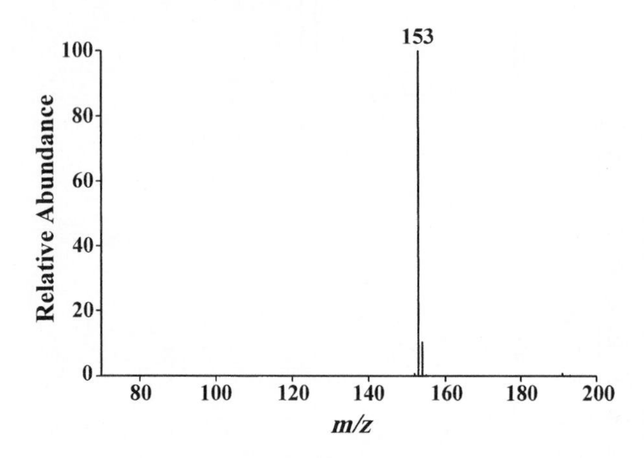

Mass spectrum (EI)

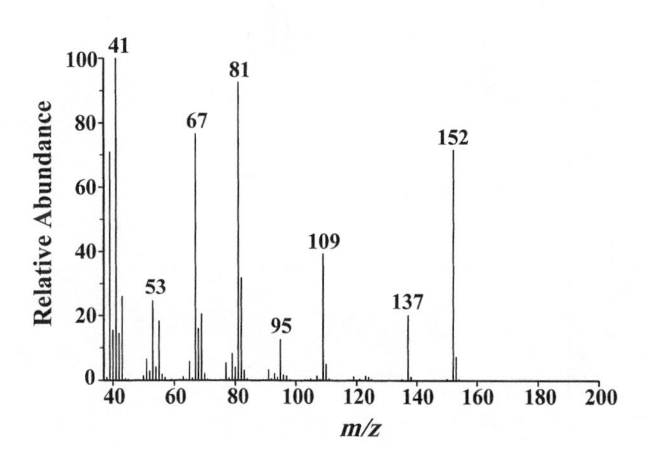

Proton NMR spectrum ($CDCl_3$)

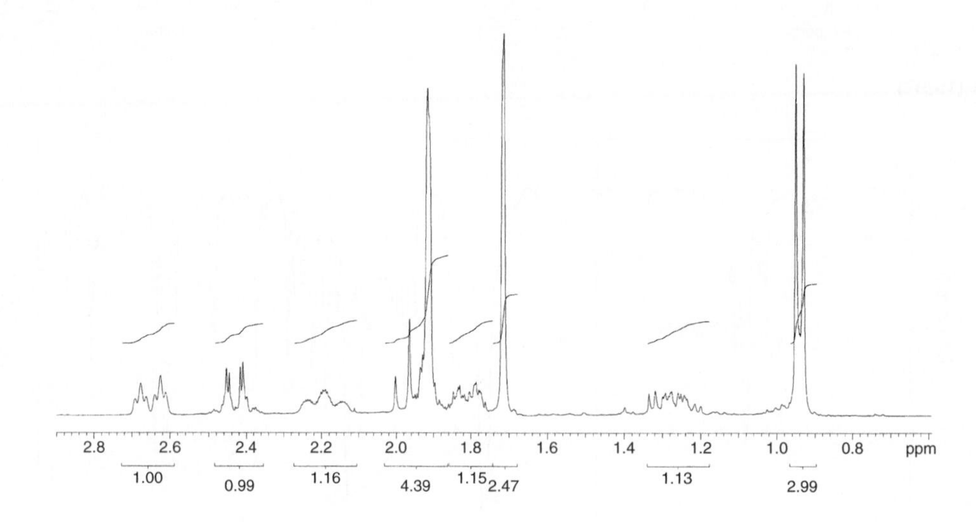

DEPT spectra

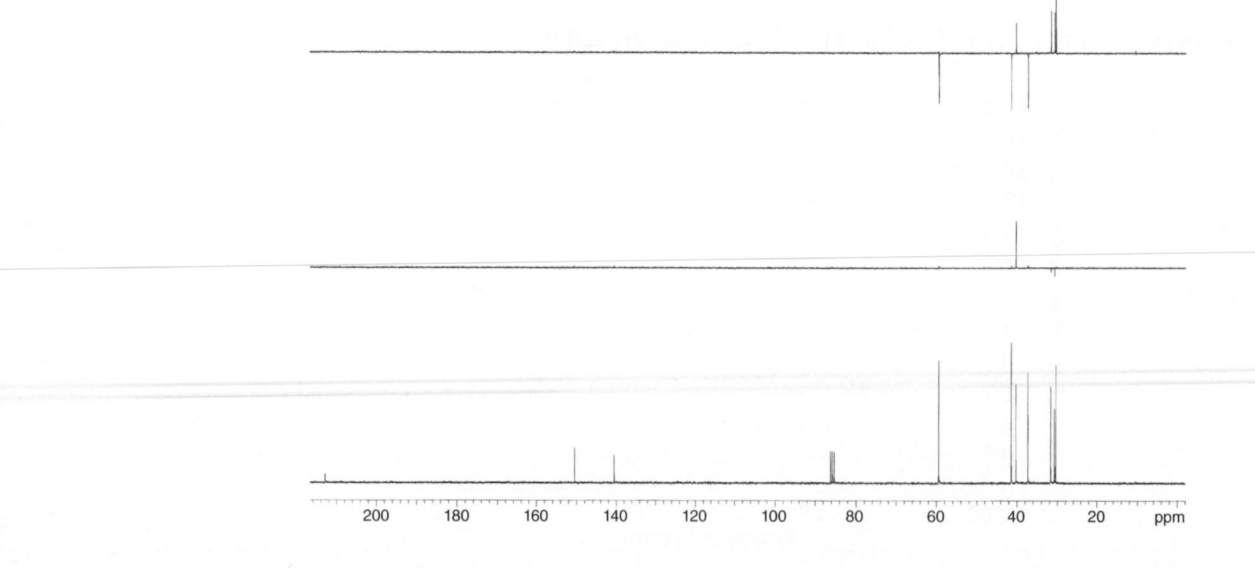

COSY spectrum

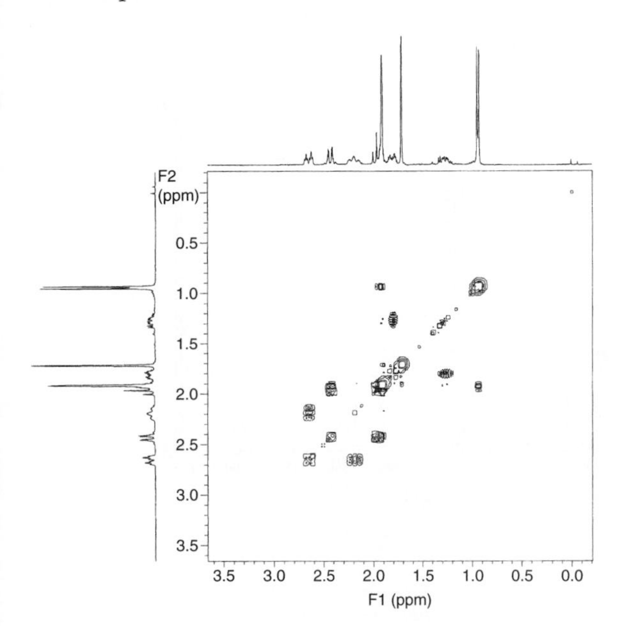

HETCOR spectrum

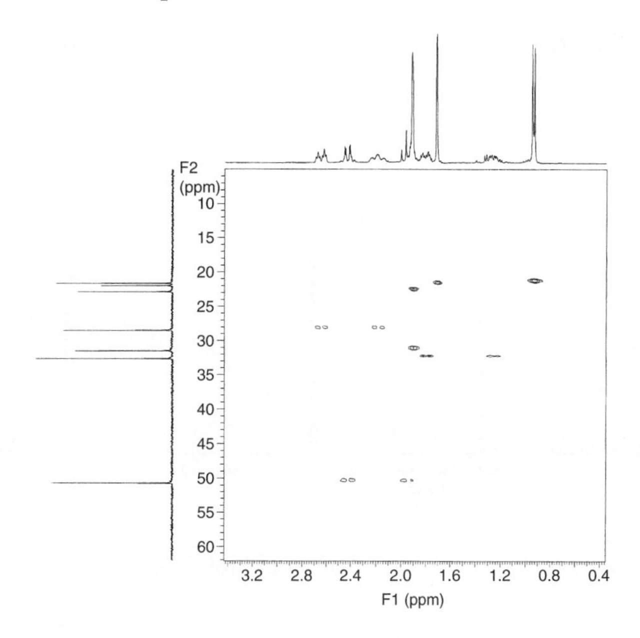

Infrared spectrum (neat)

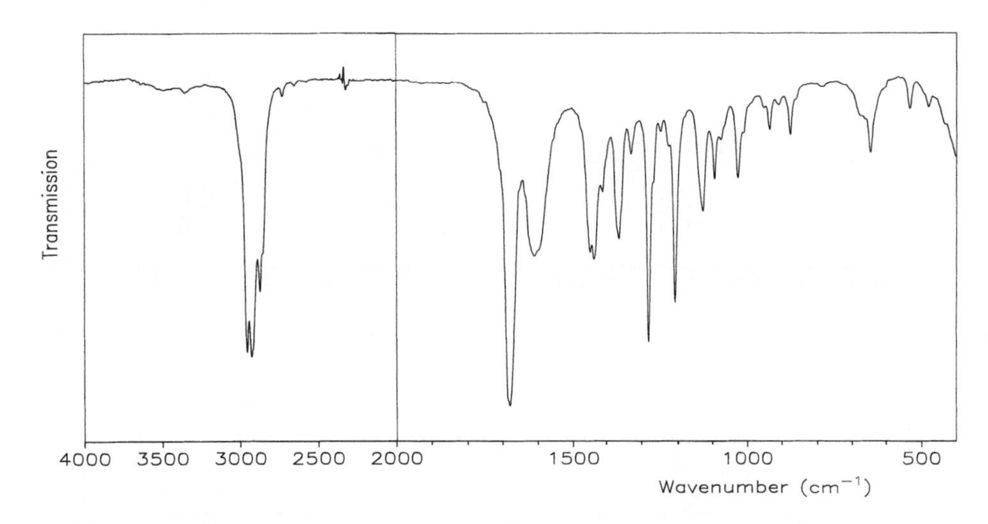

Ultraviolet–visible spectrum (EtOH, ϵ[252] 7040, ϵ[311] 72)

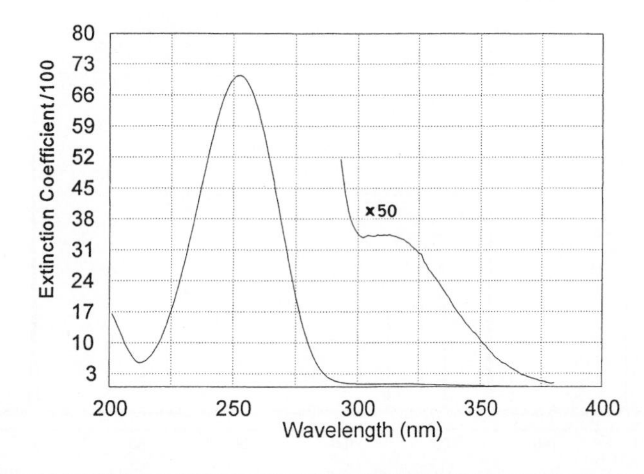

PROBLEM 15-30

Mass spectrum (CI)

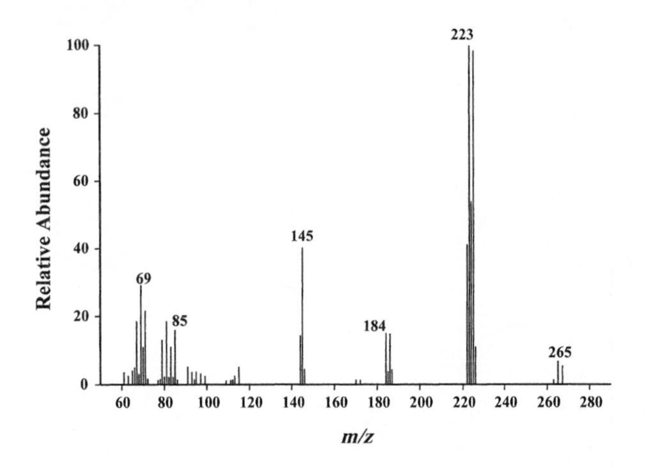

Mass spectrum (EI)

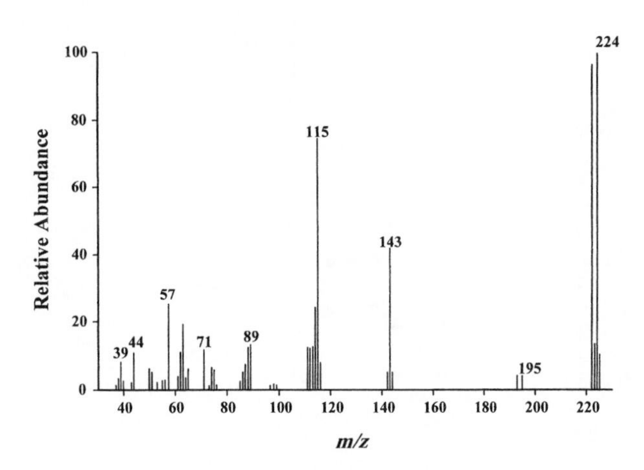

Proton NMR spectrum with expansion ($CD_3(SO)CD_3$)

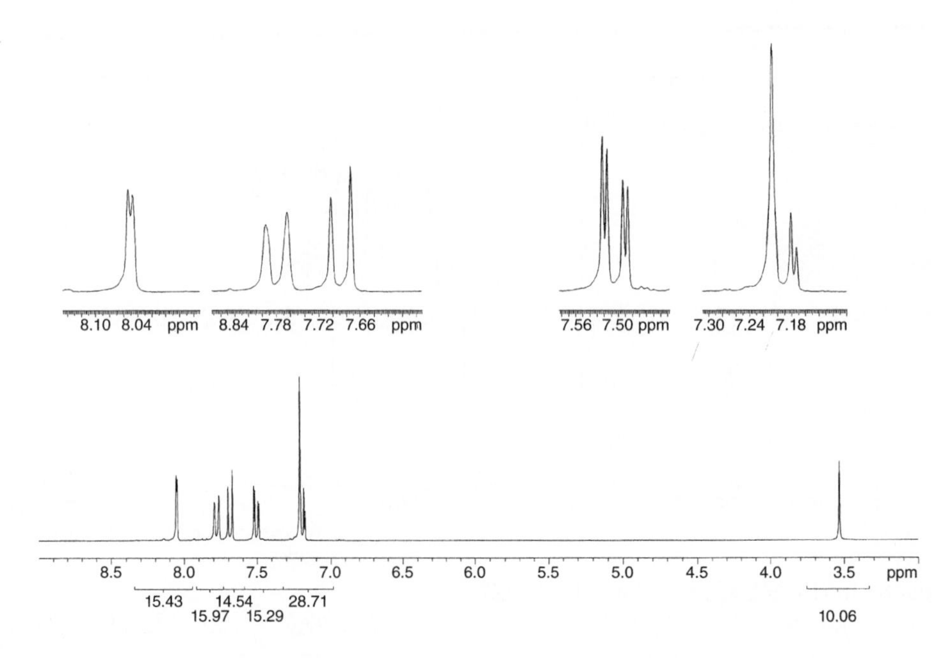

Carbon-13 NMR spectrum

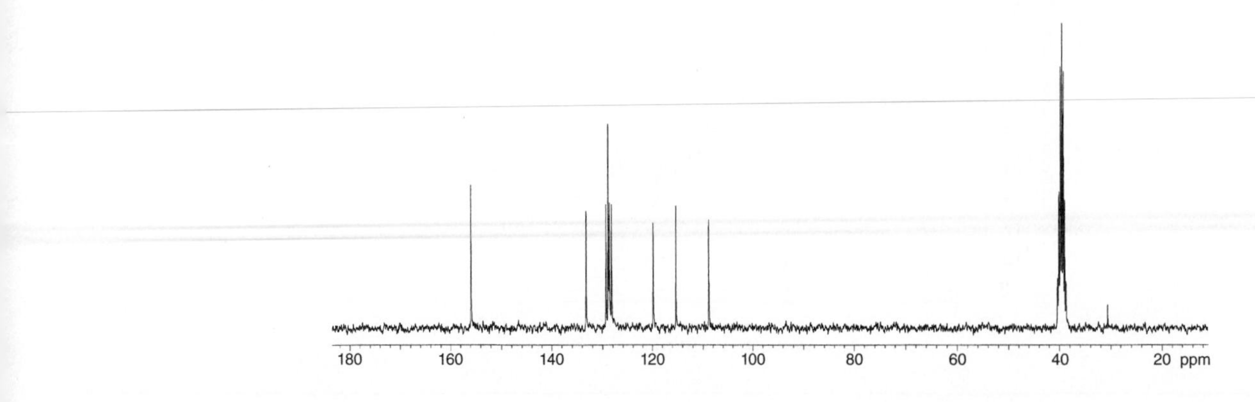

COSY spectrum

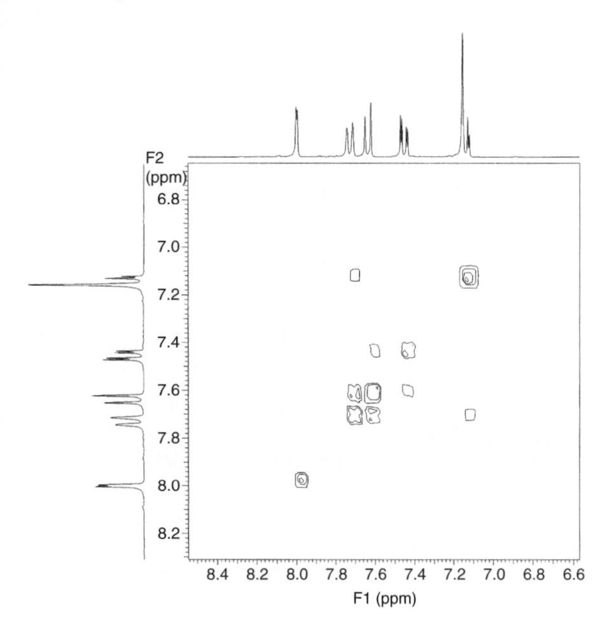

HETCOR spectrum

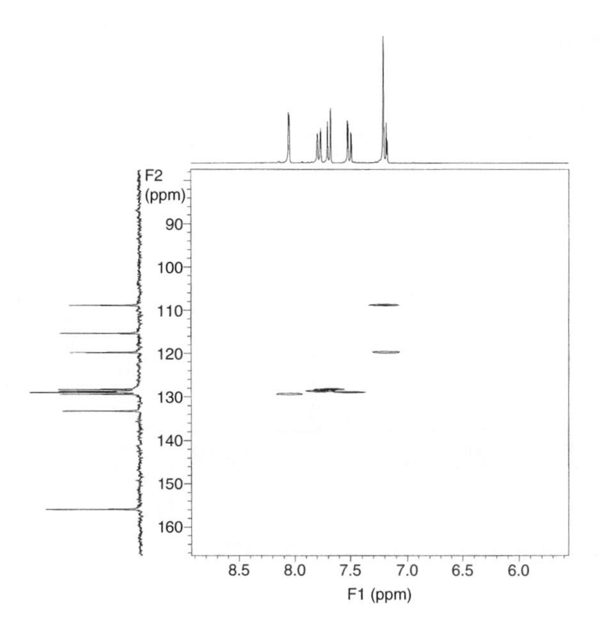

Infrared spectrum (KBr disc)

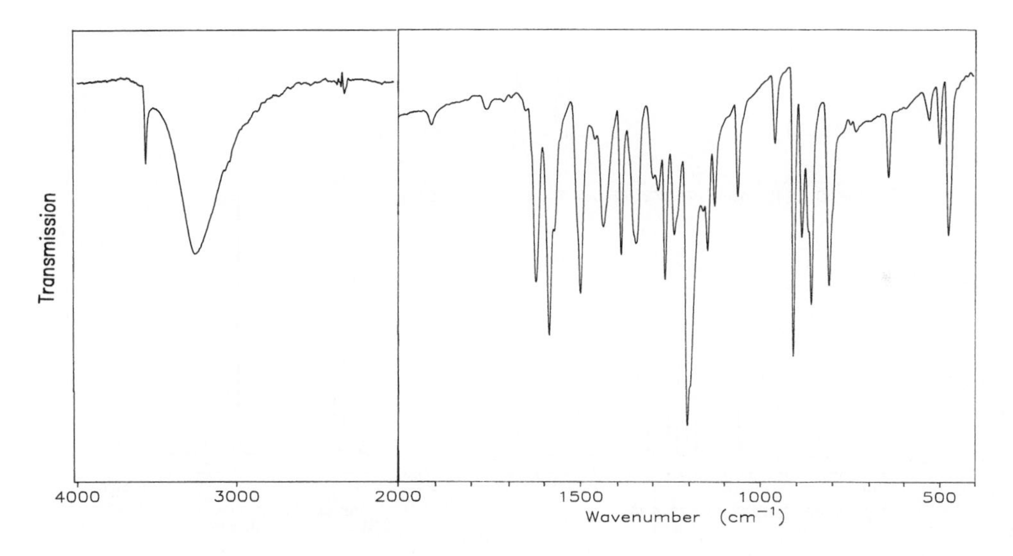

Ultraviolet–visible spectrum (EtOH, ϵ[232] 80,627, ϵ[276] 10,880, ϵ[286] 7669, ϵ[328] 1872, ϵ[340] 2056)

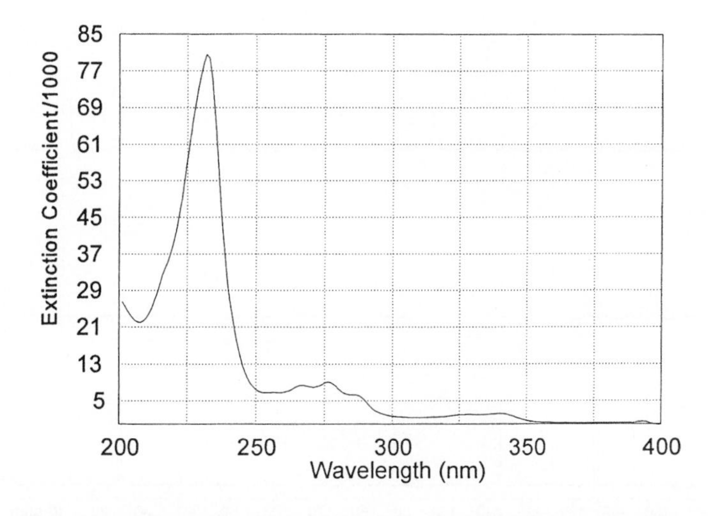

PROBLEM 15-31

Mass spectrum (CI)

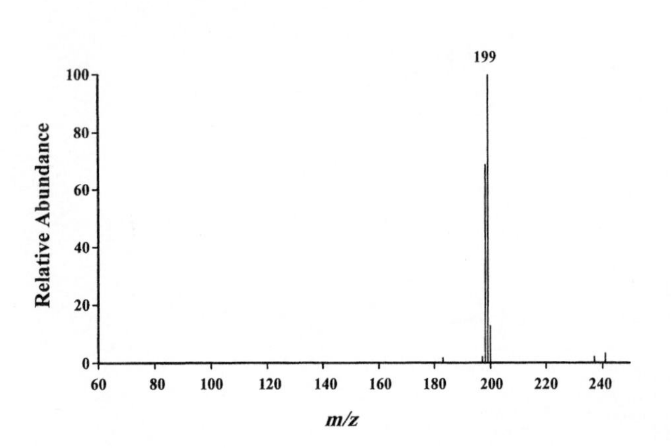

Mass spectrum (EI)

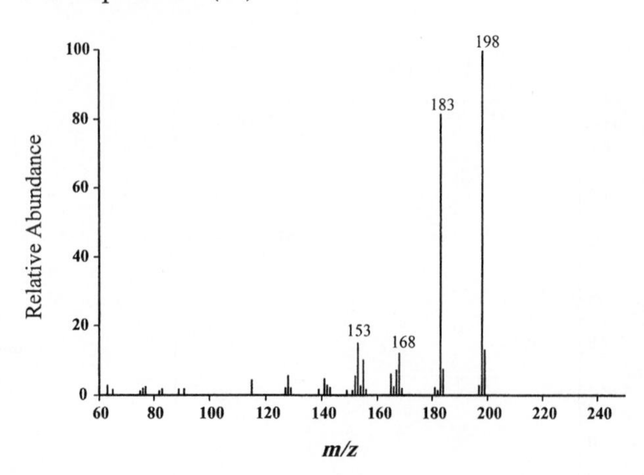

Proton NMR spectrum with expansion ($CDCl_3$)

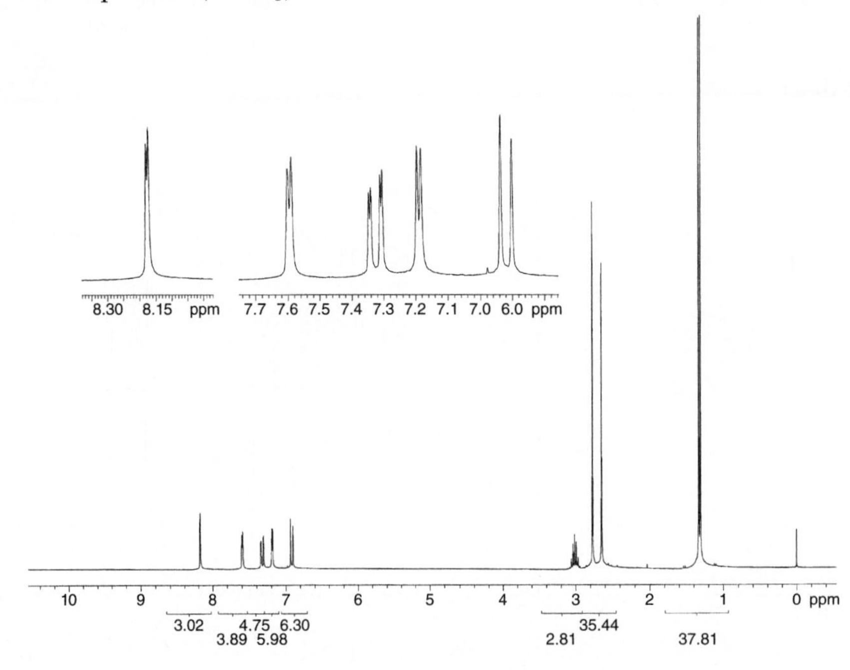

DEPT spectra

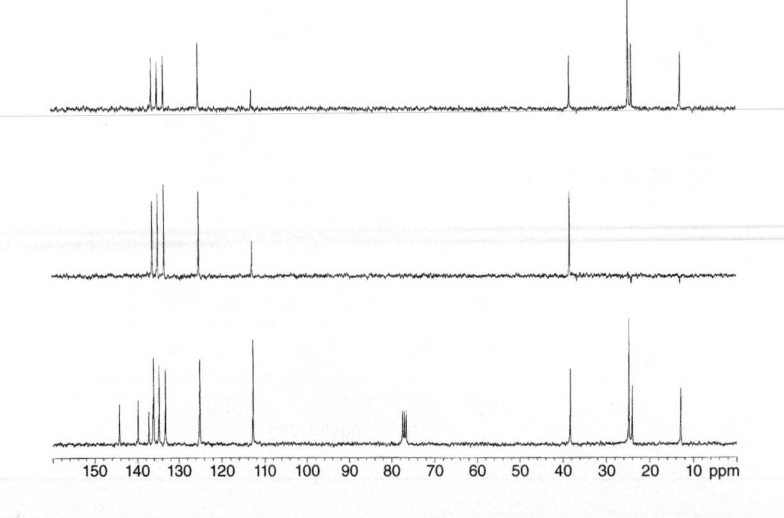

COSY spectrum

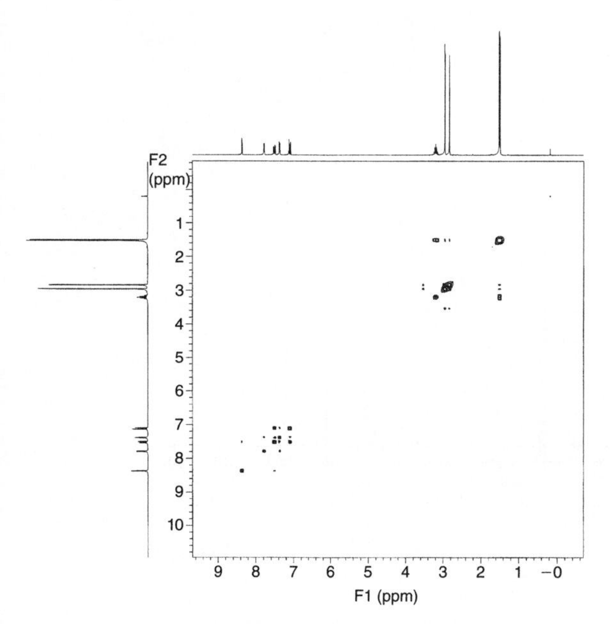

HETCOR spectrum

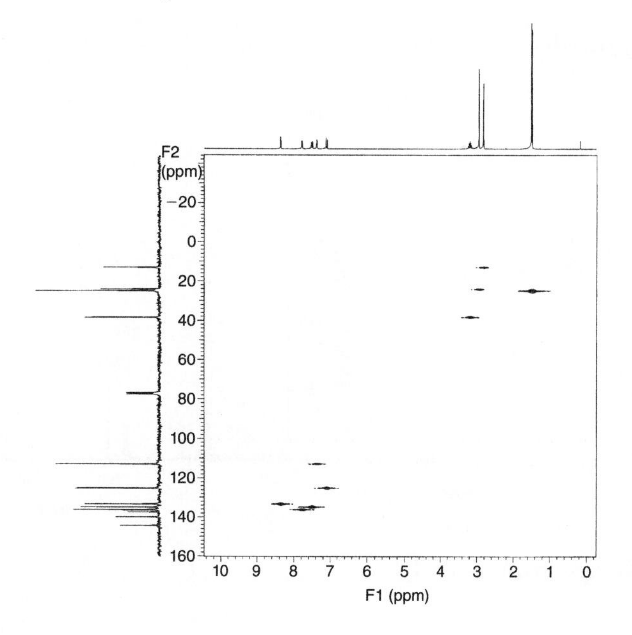

Infrared spectrum (KBr disc)

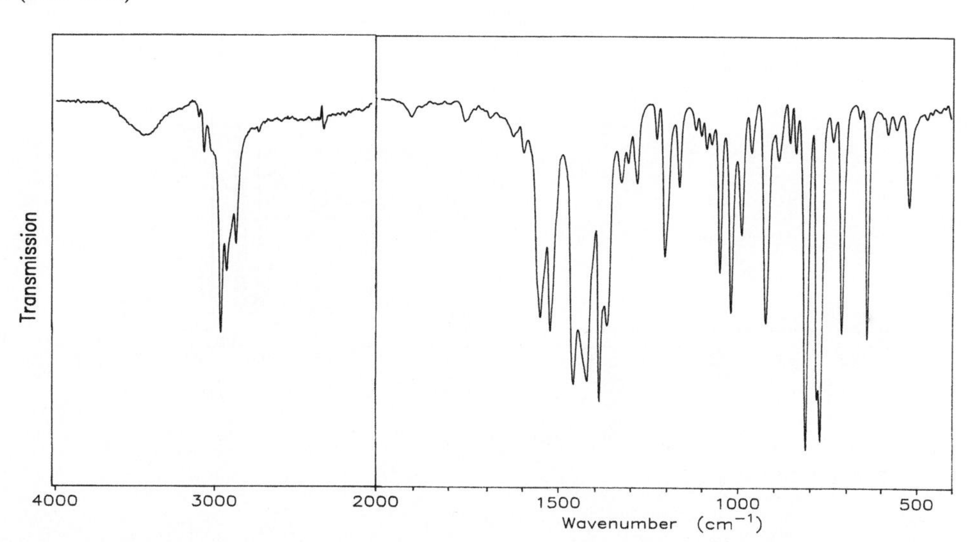

Ultraviolet–visible spectrum (EtOH, ϵ[214] 13,069, ϵ[244] 27,837, ϵ[284] 48,370, ϵ[289] 47,400, ϵ[304] 11,848, ϵ[349] 4664, ϵ[366] 3212, ϵ[602] 455, ϵ[649] 378, ϵ[720] 133)

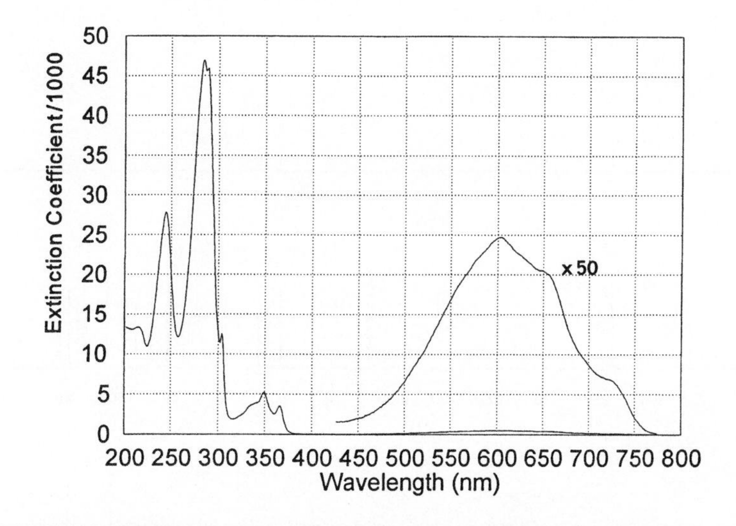

PROBLEM 15-32

Mass spectrum (desorption EI)

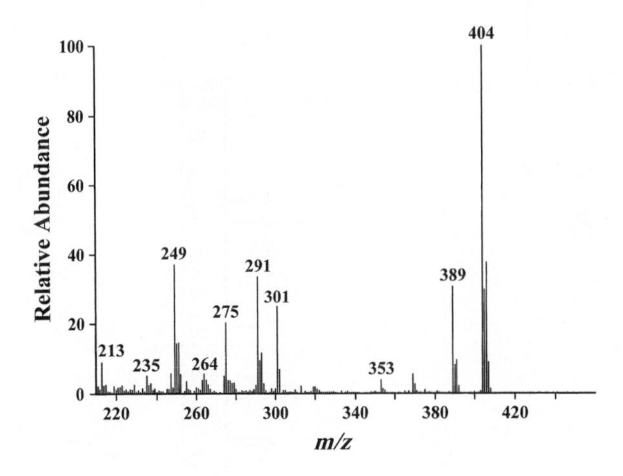

Proton NMR spectrum ($CDCl_3$)

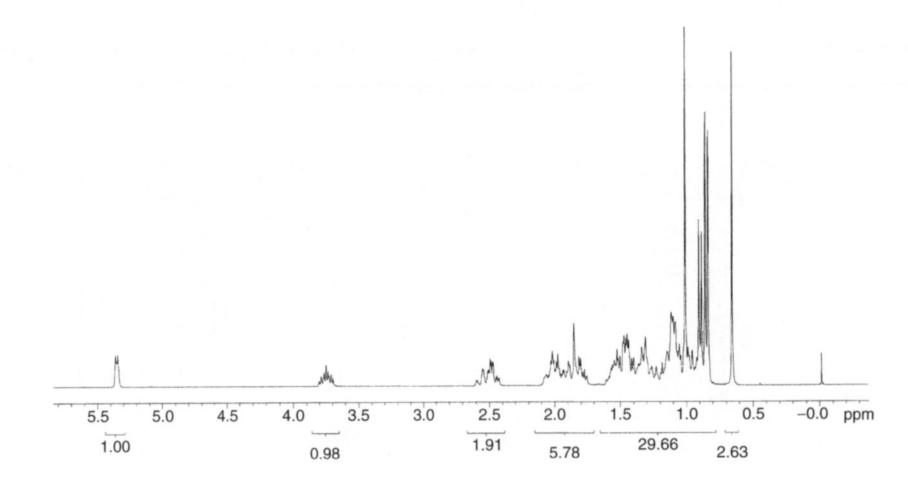

DEPT spectra

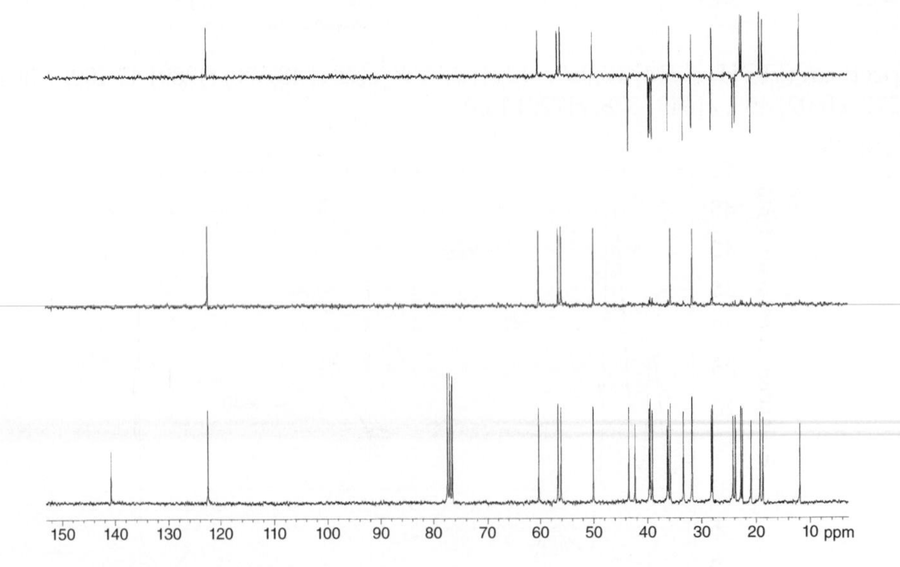

COSY spectrum

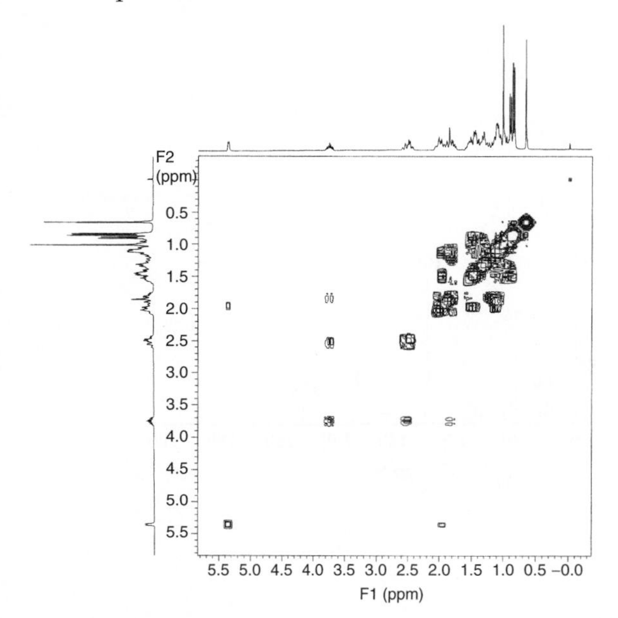

COSY spectrum with expansion

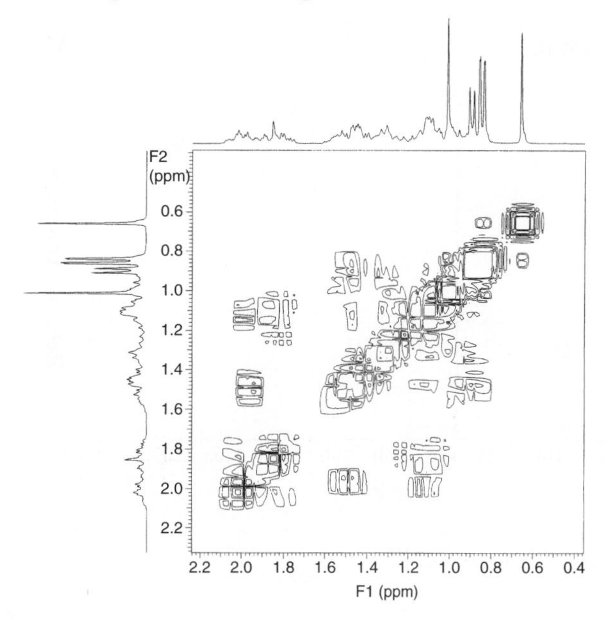

HETCOR spectrum

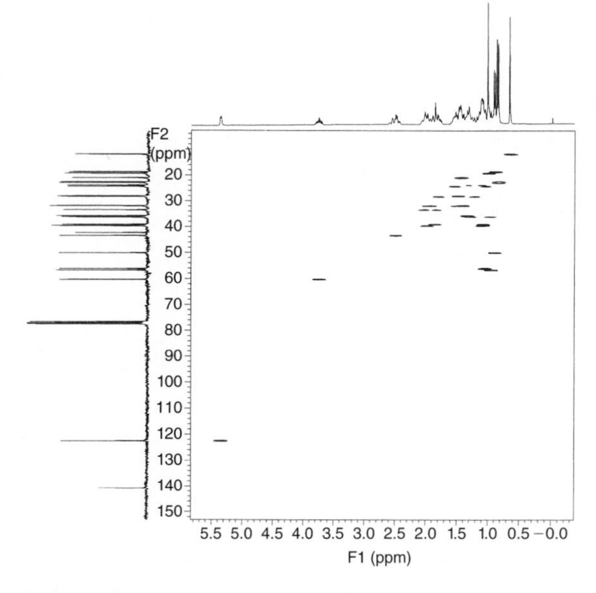

Ultraviolet–visible spectrum (EtOH, ϵ[203] 5500)

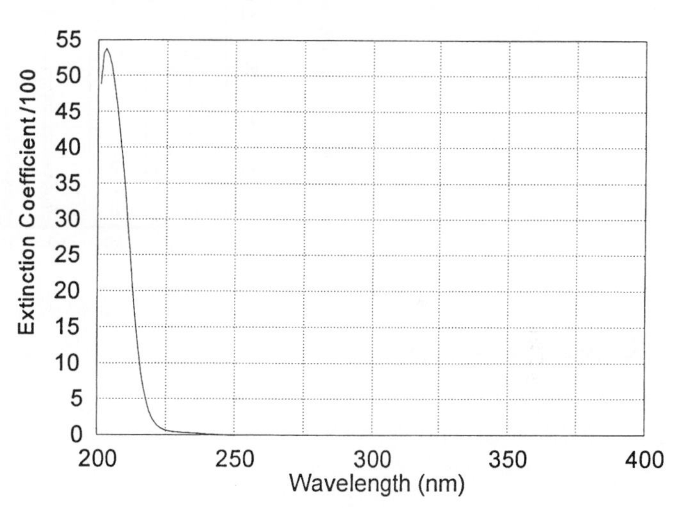

Infrared spectrum (KBr disc)

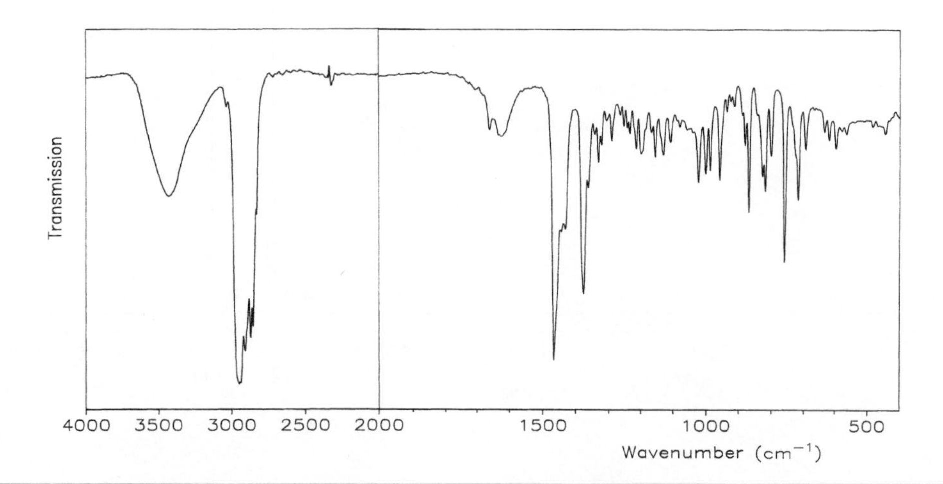

PROBLEM 15-33

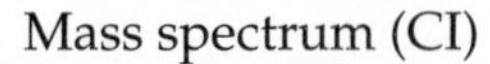
Mass spectrum (CI)

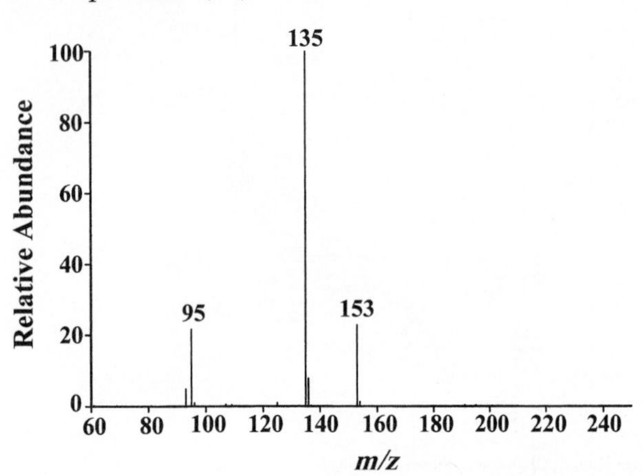

Mass spectrum (EI)

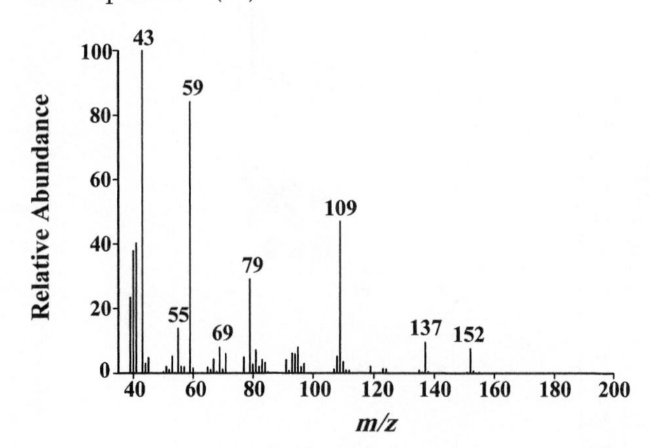

Mass spectrum (desorption EI)

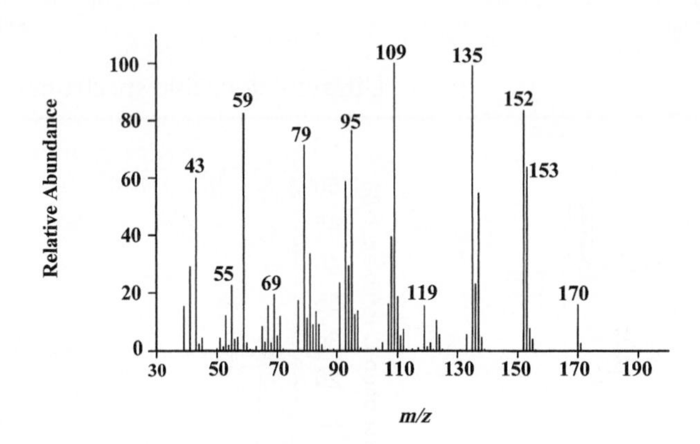

Proton NMR spectrum ($CDCl_3$)

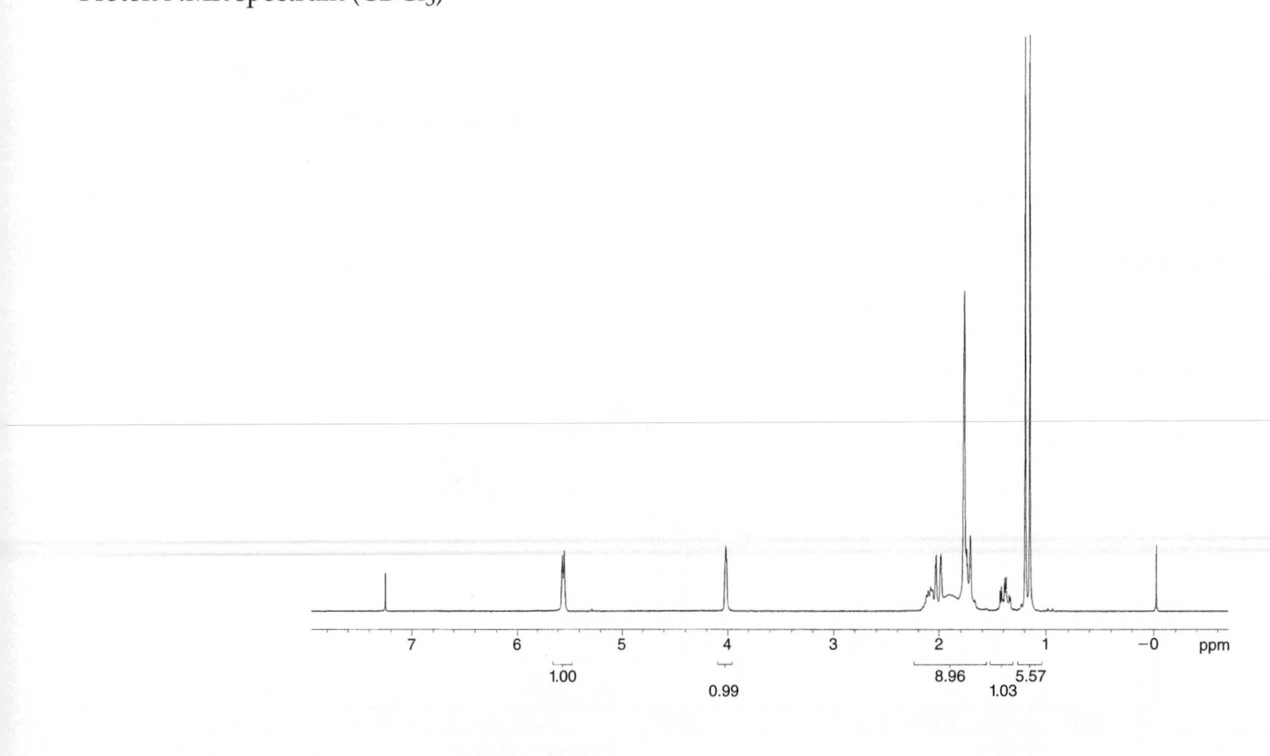

Carbon-13 NMR spectrum

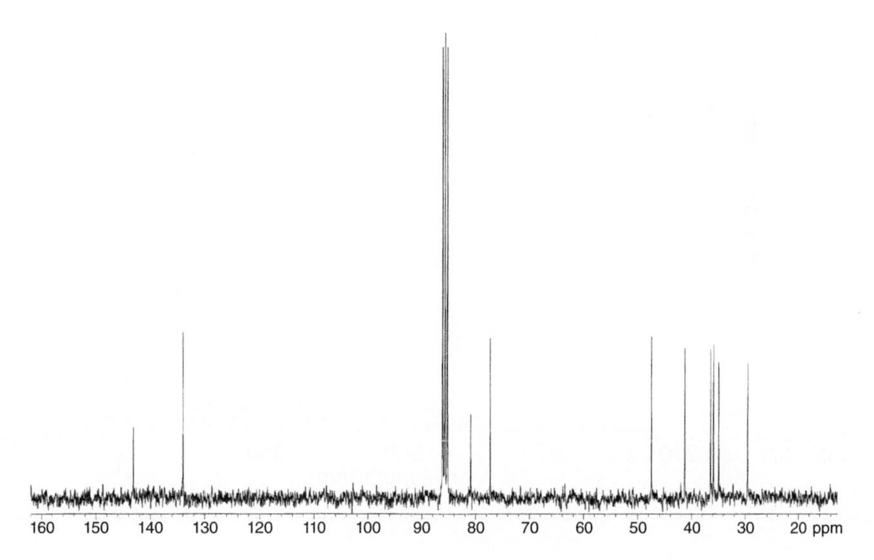

COSY spectrum

HETCOR spectrum

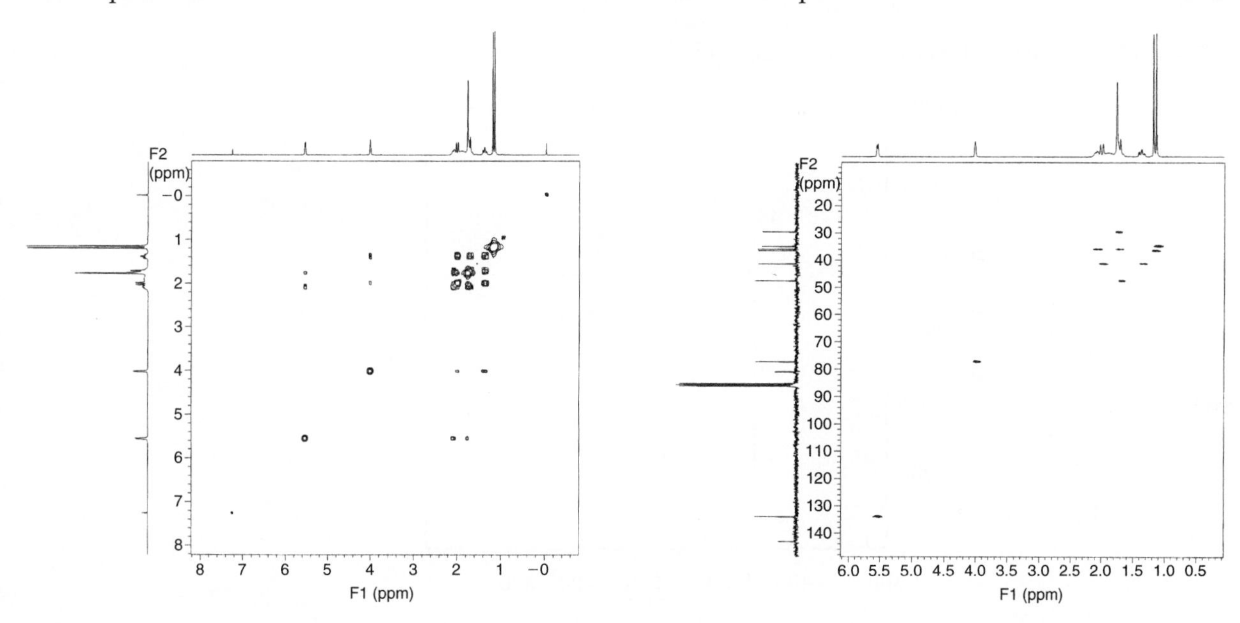

Infrared spectrum (KBr disc)

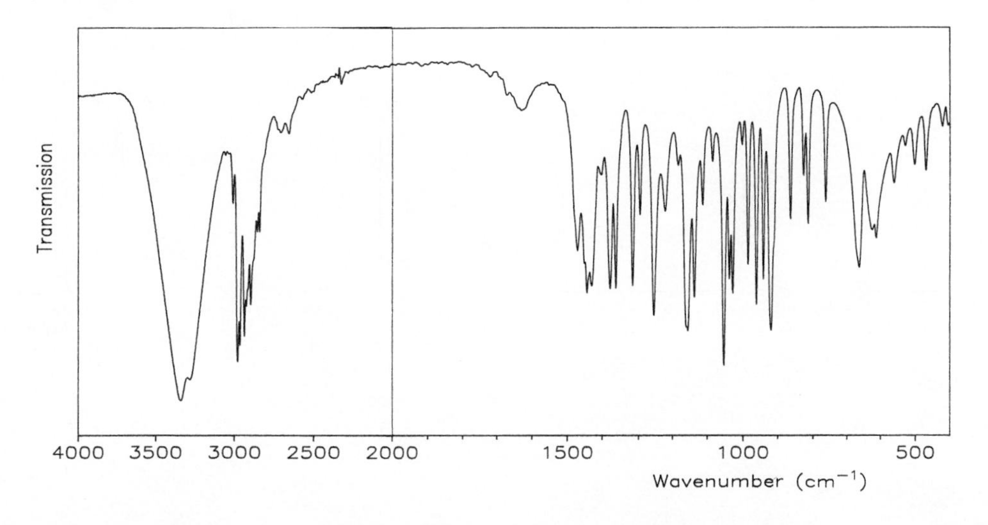

Ultraviolet–visible spectrum (EtOH)

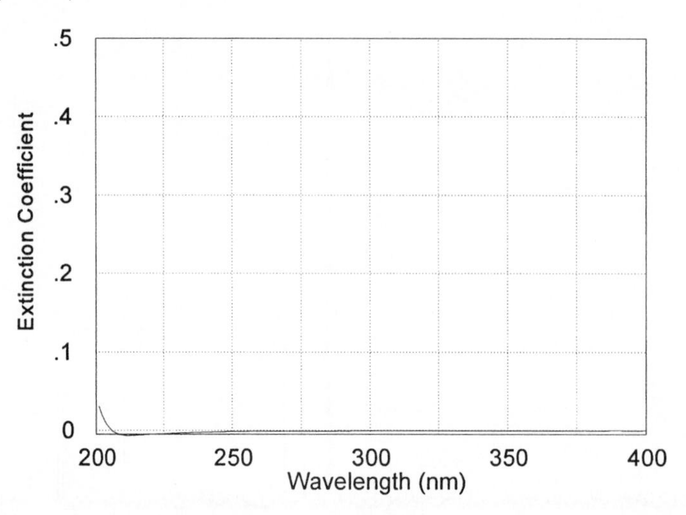

PROBLEM 15-34

Mass spectrum (MALDI in α-cyano-4-hydroxysinapinic acid)

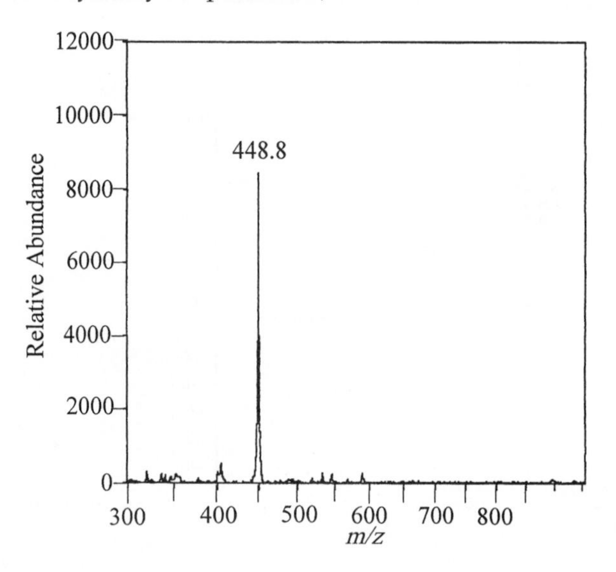

Proton NMR spectrum ($CF_3CO_2D/CD_3(CO)CD_3$)

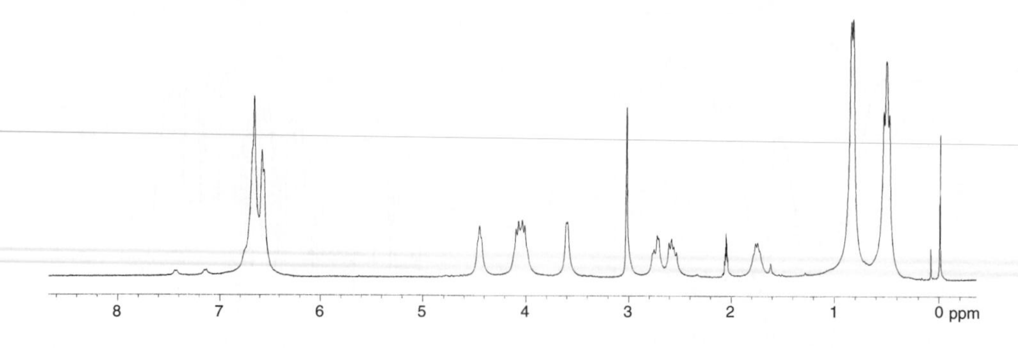

DEPT spectra

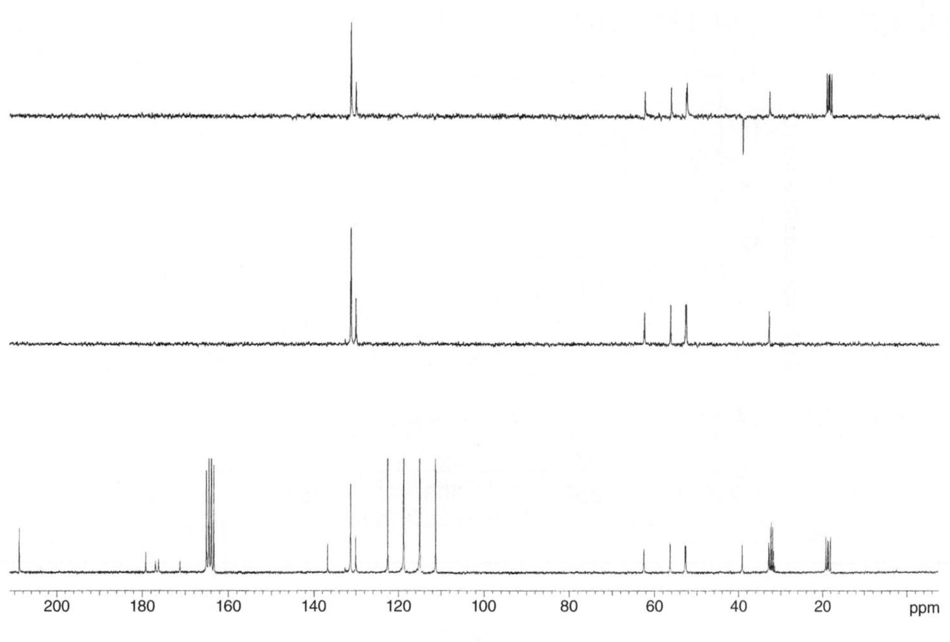

COSY spectrum

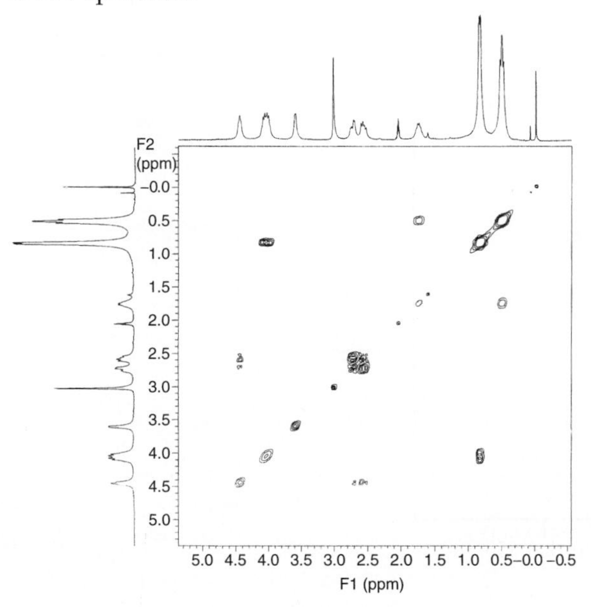

HETCOR spectrum

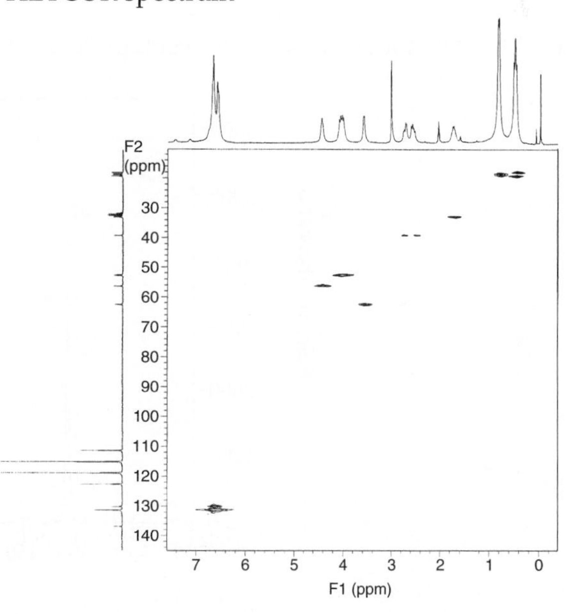

Infrared spectrum (KBr disc)

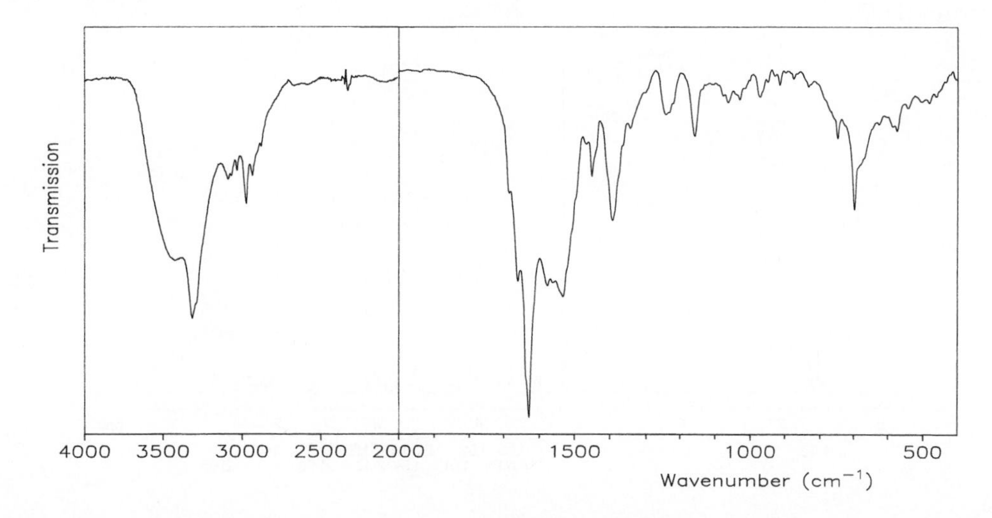

Ultraviolet–visible spectrum (H_2O, [pH 7.4 phosphate buffer], ϵ[247] 114, ϵ[252] 142, ϵ[258] 166, ϵ[264] 129, ϵ[267] 92, ϵ[301] 28)

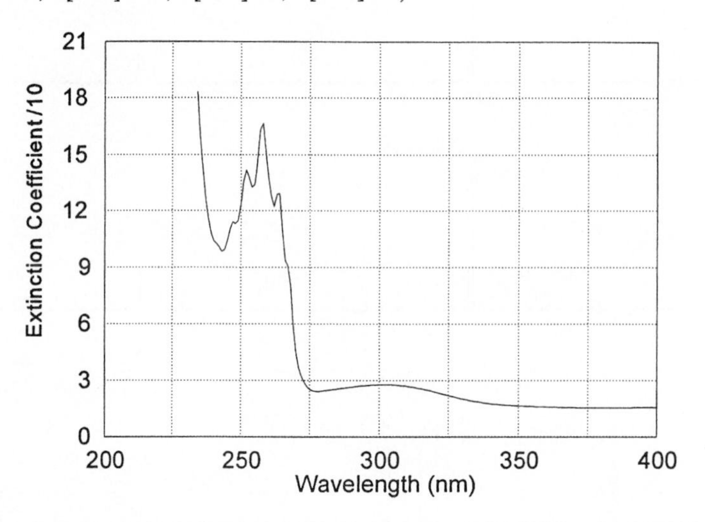

PROBLEM 15-35

Mass spectrum (MALDI in α-cyano-4-hydroxysinapinic acid)

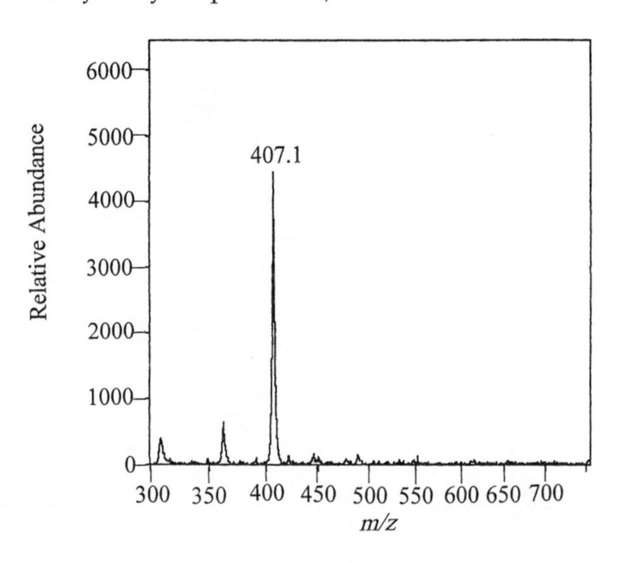

Proton NMR spectrum (D_2O)

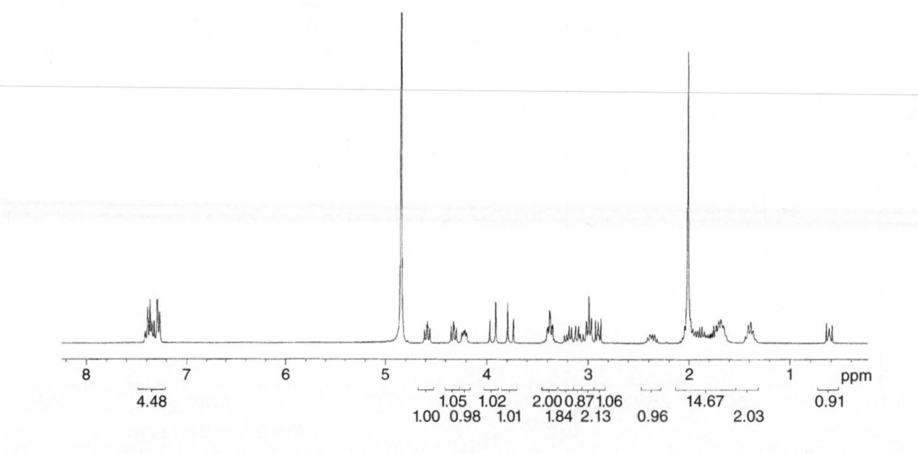

DEPT spectra

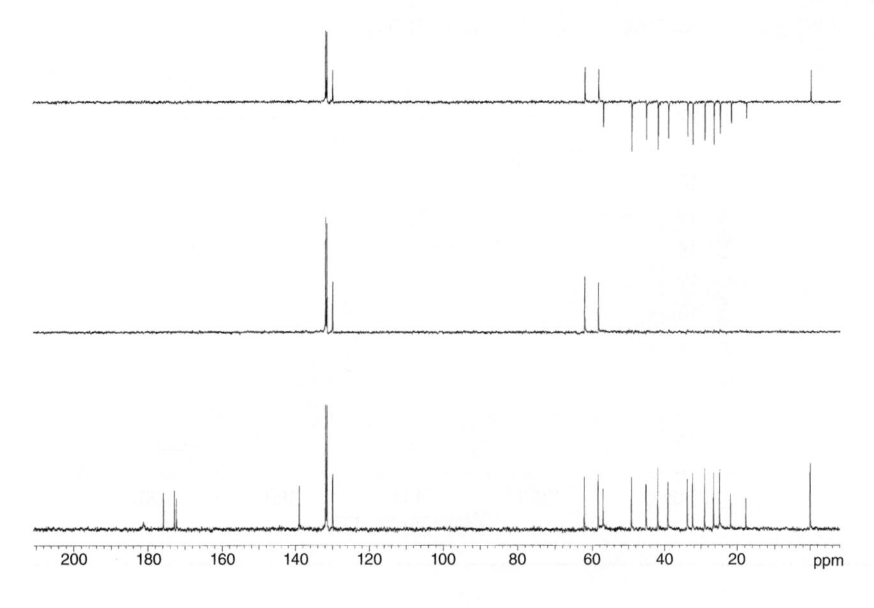

COSY spectrum

HETCOR spectrum

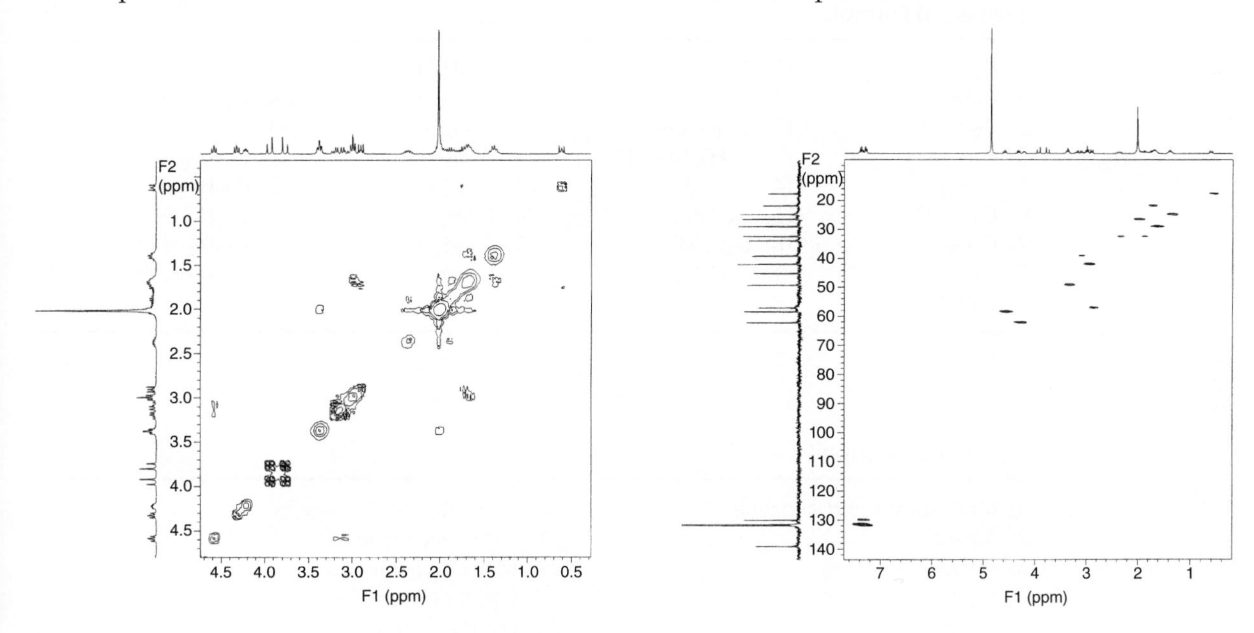

Infrared spectrum (KBr disc)

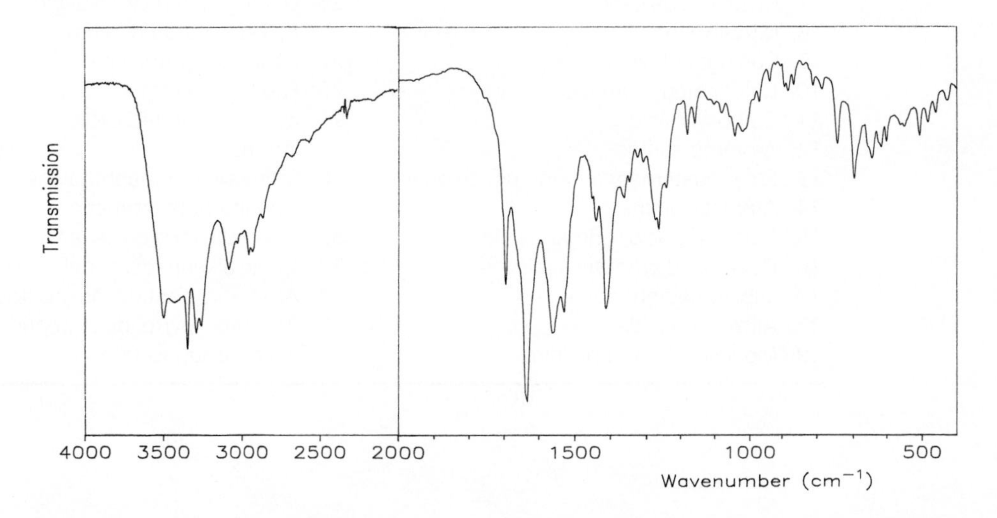

Ultraviolet–visible spectrum (H_2O, [pH 7.4 phosphate buffer], ϵ[247] 143, ϵ[252] 175, ϵ[258] 206, ϵ[264] 164, ϵ[267] 120, ϵ[312] 94)

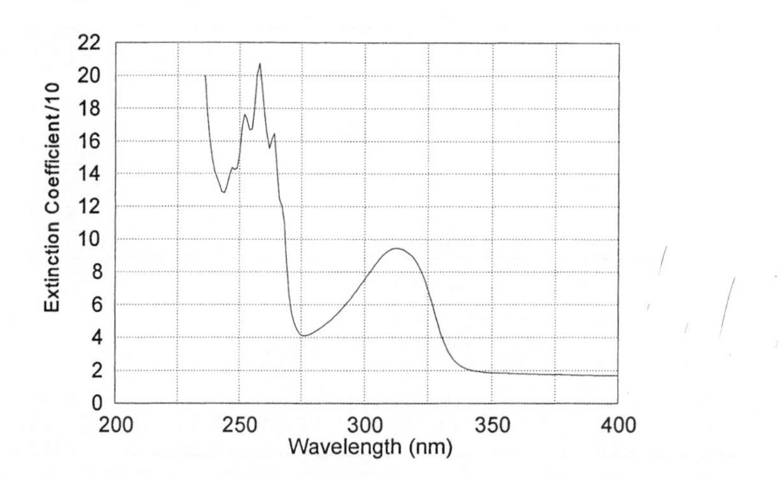

Elemental Formulas			
1. $C_{11}H_{16}$	10. C_6Cl_6	19. $C_{10}H_8N_2$	28. $C_{10}C_{10}O$
2. C_6H_{12}	11. $C_7H_{12}O_4$	20. $C_{16}H_{22}O_4$	29. $C_{10}H_{16}O$
3. $C_9H_{21}N$	12. $C_{14}H_{14}S_2$	21. $C_7H_8O_2$	30. $C_{10}H_7OBr$
4. C_4H_7N	13. $C_7H_{16}NO_2Cl$	22. C_6H_7NO	31. $C_{15}H_{18}$
5. C_8H_{14}	14. $C_9H_{13}N$	23. $C_6H_{10}O$	32. $C_{27}H_{45}Cl$
6. $C_4H_8NO_2Cl$	15. $C_4H_9NO_2$	24. $(C_2H_4O_2)_n$	33. $C_{10}H_{18}O_2$
7. C_5H_9Cl	16. C_6H_6NF	25. $C_{10}H_{16}O$	34. $C_{20}H_{30}N_4O_5$
8. $C_6H_9NO_2$	17. $C_7H_8O_2$	26. $C_{10}H_{14}O$	35. $C_{22}H_{33}N_5O_5 \cdot C_2H_4O_2$
9. C_8H_{14}	18. $C_6H_{10}O$	27. $C_{20}H_{14}O_2$	

Functional Groups	
1. Monosubstituted benzene	20. Aromatic ester
2. Alkene	21. Furan, aldehyde
3. Amine	22. Methoxypyridine
4. Nitrile	23. Cyclic ketone
5. Alkyne	24. Polymeric ether
6. Chloronitroalkane	25. Bicyclic ketone
7. Chlorocycloalkane	26. Unconjugated cyclic diene, α,β-unsaturated ketone
8. Nitroalkene	27. Hydroxynaphthalene
9. Conjugated diene	28. Fused aryl ketone
10. Chlorinated aromatic	29. α,β-Unsaturated cyclic ketone
11. Carboxylic acid	30. Bromohydroxynaphthalene
12. Aromatic sulfide	31. Nonalternate aromatic
13. Ester, quaternary ammonium chloride	32. Polycyclic chloroalkene
14. Aromatic amine	33. Cyclic alkene, dialcohol
15. Carboxylic acid, primary amine	34. Aromatic, peptide, amino acid
16. Fluorinated aromatic amine	35. Aromatic, pyrrolidine, peptide, amino acid, acetate
17. Substituted phenol	
18. Alkene, epoxide	
19. Monosubstituted pyridine	

INDEX

A

B

D

E

J

K

L

M

N

O

P

Q

R

S

T

Ionization Methods

Type	Examples	Analyte State	Typical Ion Types	Fragmentation
Electron ionization	EI	Gas	$M^{+}\cdot$	Extensive
Chemical ionization	CI	Gas	$(M+H)^{+}$, $(M-H)^{-}$	Limited
Desorption	MALDI, FAB, LSIMS	Solid	$(M+H)^{+}$, $(M-H)^{-}$, $(M+Cat)^{+}$	Limited
Spray	ESI, APCI	Solution	$(M+nH)^{n+}$, $(M-nH)^{n-}$, $(M+nCat)^{n+}$	Little

Note: Cat^{+} = permanently charged cation such Na^{+} or R_4N^{+}, MALDI = matrix-assisted laser desorption ionization, FAB = fast-atom bombardment, LSIMS = liquid secondary ion mass spectrometry, ESI = electrospray ionization, APCI = atmospheric pressure chemical ionization.

Exact Masses and Abundances for Isotopes of Common Elements

Element	Isotope	Mass	% Abundance
Hydrogen	^{1}H	1.007 825	99.9885
	^{2}H	2.014 101	0.0115
Lithium	^{6}Li	6.015 122	7.59
	^{7}Li	7.016 004	92.41
Boron	^{10}B	10.012 937	19.9
	^{11}B	11.009 305	80.1
Carbon	^{12}C	12.	98.93
	^{13}C	13.003 354	1.07
Nitrogen	^{14}N	14.003 074	99.632
	^{15}N	15.000 108	0.368
Oxygen	^{16}O	15.994 914	99.757
	^{17}O	16.999 131	0.038
	^{18}O	17.999 160	0.205
Fluorine	^{19}F	18.998 403	100
Sodium	^{23}Na	22.989 769	100
Aluminum	^{27}Al	26.981 538	100
Silicon	^{28}Si	27.976 926	92.2297
	^{29}Si	28.976 494	4.6832
	^{30}Si	29.973 770	3.0872
Phosphorus	^{31}P	30.973 761	100
Sulfur	^{32}S	31.972 070	94.93
	^{33}S	32.971 458	0.76
	^{34}S	33.967 866	4.29
Chlorine	^{35}Cl	34.968 852	75.78
	^{37}Cl	36.965 902	24.22
Bromine	^{79}Br	78.918 338	50.69
	^{81}Br	80.916 291	49.31
Iodine	^{127}I	126.904 468	100

Data from physics.nist.gov.